Low-Noise Microwave Transistors & Amplifiers

OTHER IEEE PRESS BOOKS

Digital MOS Integrated Circuits, *Edited by M. I. Elmasry*
Geometric Theory of Diffraction, *Edited by Robert C. Hansen*
Modern Active Filter Design, *Edited by R. Schaumann, M. A. Soderstrand, and K. R. Laker*
Adjustable Speed AC Drive Systems, *Edited by B. K. Bose*
Optical Fiber Technology, II, *Edited by C. K. Kao*
Protective Relaying for Power Systems, *Edited by S. H. Horowitz*
Analog MOS Integrated Circuits, *Edited by P. R. Gray, D. A. Hodges, and R. W. Brodersen*
Interence Analysis of Communication Systems, *Edited by P. Stavroulakis*
Integrated Injection Logic, *Edited by J. E. Smith*
Sensory Aids for the Hearing Impaired, *Edited by H. Levitt, J. M. Pickett, and R. A. Houde*
Data Conversion Integrated Circuits, *Edited by Daniel J. Dooley*
Semiconductor Injection Lasers, *Edited by J. K. Butler*
Satellite Communications, *Edited by H. L. Van Trees*
Frequency-Response Methods in Control Systems, *Edited by A.G.J. MacFarlane*
Programs for Digital Signal Processing, *Edited by the Digital Signal Processing Committee*
Automatic Speech & Speaker Recognition, *Edited by N. R. Dixon and T. B. Martin*
Speech Analysis, *Edited by R. W. Schafer and J. D. Markel*
The Engineer in Transition to Management, *I. Gray*
Multidimensional Systems: Theory & Applications, *Edited by N. K. Bose*
Analog Integrated Circuits, *Edited by A. B. Grebene*
Integrated-Circuit Operational Amplifiers, *Edited by R. G. Meyer*
Modern Spectrum Analysis, *Edited by D. G. Childers*
Digital Image Processing for Remote Sensing, *Edited by R. Bernstein*
Reflector Antennas, *Edited by A. W. Love*
Phase-Locked Loops & Their Application, *Edited by W. C. Lindsey and M. K. Simon*
Digital Signal Computers and Processors, *Edited by A. C. Salazar*
Systems Engineering: Methodology and Applications, *Edited by A. P. Sage*
Modern Crystal and Mechanical Filters, *Edited by D. F. Sheahan and R. A. Johnson*
Electrical Noise: Fundamentals and Sources, *Edited by M. S. Gupta*
Computer Methods in Image Analysis, *Edited by J. K. Aggarwal, R. O. Duda, and A. Rosenfeld*
Microprocessors: Fundamentals and Applications, *Edited by W. C. Lin*
Machine Recognition of Patterns, *Edited by A. K. Agrawala*
Turning Points in American Electrical History, *Edited by J. E. Brittain*
Charge-Coupled Devices: Technology and Applications, *Edited by R. Melen and D. Buss*
Spread Spectrum Techniques, *Edited by R. C. Dixon*
Electronic Switching: Central Office Systems of the World, *Edited by A. E. Joel, Jr.*
Electromagnetic Horn Antennas, *Edited by A. W. Love*
Waveform Quantization and Coding, *Edited by N. S. Jayant*
Communication Satellite Systems: An Overview of the Technology, *Edited by R. G. Gould and Y. F. Lum*
Literature Survey of Communication Satellite Systems and Technology, *Edited by J. H. W. Unger*
Solar Cells, *Edited by C. E. Backus*
Computer Networking, *Edited by R. P. Blanc and I. W. Cotton*
Communications Channels: Characterization and Behavior, *Edited by B. Goldberg*
Large-Scale Networks: Theory and Design, *Edited by F. T. Boesch*
Optical Fiber Technology, *Edited by D. Gloge*
Selected Papers in Digital Signal Processing, II, *Edited by the Digital Signal Processing Committee*
A Guide for Better Technical Presentations, *Edited by R. M. Woelfle*
Career Management: A Guide to Combating Obsolescence, *Edited by H. G. Kaufman*
Energy and Man: Technical and Social Aspects of Energy, *Edited by M. G. Morgan*
Magnetic Bubble Technology: Integrated-Circuit Magnetics for Digital Storage and Processing, *Edited by H. Chang*
Frequency Synthesis: Techniques and Applications, *Edited by J. Gorski-Popiel*
Literature in Digital Processing: Author and Permuted Title Index (Revised and Expanded Edition), *Edited by H. D. Helms, J. F. Kaiser, and L. R. Rabiner*
Data Communications via Fading Channels, *Edited by K. Brayer*
Nonlinear Networks: Theory and Analysis, *Edited by A. N. Willson, Jr.*
Computer Communications, *Edited by P. E. Green, Jr. and R. W. Lucky*
Stability of Large Electric Power Systems, *Edited by R. T. Byerly and E. W. Kimbark*
Automatic Test Equipment: Hardware, Software, and Management, *Edited by F. Liguori*
Key Papers in the Development of Coding Theory, *Edited by E. R. Berkekamp*

Low-Noise Microwave Transistors & Amplifiers

Edited by
Hatsuaki Fukui
Member, Technical Staff
Bell Laboratories

A volume in the IEEE PRESS Selected Reprint Series, prepared under the sponsorship of the IEEE Microwave Theory and Techniques Society.

The Institute of Electrical and Electronics Engineers, Inc. New York

IEEE PRESS

1981 Editorial Board

S. B. Weinstein, *Chairman*

J. M. Aein
R. W. Brodersen
J. K. Butler
R. E. Crochiere
P. H. Enslow, Jr.
E. W. Herold

A. W. Love
A. G. J. MacFarlane
M. G. Morgan
B. R. Myers
Norman Peach
P. M. Russo
Herbert Sherman

W. R. Crone, *Managing Editor*

Isabel Narea, *Production Manager*

Copyright © 1981 by
THE INSTITUTE OF ELECTRICAL AND ELECTRONICS ENGINEERS, INC.
345 East 47th Street, New York, NY 10017
All rights reserved.

PRINTED IN THE UNITED STATES OF AMERICA

Sole Worldwide Distributor (Exclusive of IEEE):

JOHN WILEY & SONS, INC.
605 Third Ave.
New York, NY 10016

Wiley Order Numbers: Clothbound: 0-471-86589-3
Paperbound: 0-471-86588-5

IEEE Order Numbers: Clothbound: PC0148-7
Paperbound: PP0149-5

Library of Congress Catalog Card Number 81-6994

Contents

Preface ... viii
Introduction ... 1

Part I: Noise Characterization and Measurements ... 5

Noise Figures of Radio Receivers, *H. T. Friis (Proceedings of the IRE,* July 1944) ... 6
Theory of Noisy Fourpoles, *H. Rothe and W. Dahlke (Proceedings of the IRE,* June 1956) ... 10
Available Power Gain, Noise Figure, and Noise Measure of Two-Ports and Their Graphical Representations, *H. Fukui (IEEE Transactions on Circuit Theory,* June 1966) ... 18
The Effect of Feedback on Noise Figure, *S. Iversen (Proceedings of the IEEE,* March 1975) ... 24
Wave Representation of Amplifier Noise, *P. Penfield, Jr. (IRE Transactions on Circuit Theory,* March 1962) ... 26
A Wave Approach to the Noise Properties of Linear Microwave Devices, *R. P. Meys (IEEE Transactions on Microwave Theory and Techniques,* January 1978) ... 28
IRE Standards on Methods of Measuring Noise in Linear Twoports, 1959 *(Proceedings of the IRE,* January 1960) ... 32
Representation of Noise in Linear Twoports, *IRE Subcommittee 7.9 on Noise (Proceedings of the IRE,* January 1960) ... 41
The Determination of Device Noise Parameters, *R. Q. Lane (Proceedings of the IEEE,* August 1969) ... 47
A Microwave Noise and Gain Parameter Test Set, *R. Q. Lane (IEEE International Solid-State Circuits Conference,* 1978) ... 49
On the Determination of Device Noise and Gain Parameters, *M. Sannino (Proceedings of the IEEE,* September 1979) ... 52
An Improved Computational Method for Noise Parameter Measurement, *M. Mitama and H. Katoh (IEEE Transactions on Microwave Theory and Techniques,* June 1979) ... 54
Accurate and Automatic Noise Figure Measurements with Standard Equipment, *N. Kuhn (IEEE MTT-S International Microwave Symposium Digest,* 1980) ... 58
Measurement of Losses in Noise-Matching Networks, *E. W. Strid (IEEE Transactions on Microwave Theory and Techniques,* March 1981) ... 61

Part II: Noise Properties of Bipolar Transistors ... 67

Microwave Transistors: Theory and Design, *H. F. Cooke (Proceedings of the IEEE,* August 1971) ... 68
The Noise Performance of Microwave Transistors, *H. Fukui (IEEE Transactions on Electron Devices,* March 1966) ... 87
Noise Measure of Microwave Transistors, *H. Fukui (Proceedings of the IEEE,* September 1966) ... 100
A Simplified Approach to Noise in Microwave Transistors, *S. D. Malaviya and A. van der Ziel (Solid-State Electronics,* December 1970) ... 102
A More Accurate Expression for the Noise Figure of Transistors, *A. van der Ziel, J. A. Cruz-Emeric, R. D. Livingstone, J. C. Malpass, and D. A. McNamara (Solid-State Electronics,* February 1976) ... 110
Limitations of Nielsen's and Related Noise Equations Applied to Microwave Bipolar Transistors, and a new Expression for the Frequency and Current Dependent Noise Figure, *R. J. Hawkins (Solid-State Electronics,* March 1977) ... 113
Low-Noise Microwave Bipolar Transistor with Sub-Half-Micrometer Emitter Width, *T-H. Hsu and C. P. Snapp (IEEE Transactions on Electron Devices,* June 1978) ... 119

Part III: Noise Properties of Field-Effect Transistors ... 127

Microwave Field-Effect Transistors—1976, *C. A. Liechti (IEEE Transactions on Microwave Theory and Techniques,* June 1976) ... 128
Noise Characteristics of Gallium Arsenide Field-Effect Transistors, *H. Statz, H. A. Haus, and R. A. Pucel (IEEE Transactions on Electron Devices,* September 1974) ... 150
Noise Performance of Gallium Arsenide Field-Effect Transistors, *R. A. Pucel, D. J. Massé, and C. F. Krumm (IEEE Journal of Solid-State Circuits,* April 1976) ... 164
Design of Microwave GaAs MESFET's for Broad-Band Low-Noise Amplifiers, *H. Fukui (IEEE Transactions on Microwave Theory and Techniques,* July 1979) Addendum, *H. Fukui (IEEE Transactions on Microwave Theory and Techniques,* October 1981) ... 176
Optimal Noise Figure of Microwave GaAs MESFET's, *H. Fukui (IEEE Transactions on Electron Devices,* July 1979) ... 184
Optimization of Low-Noise GaAs MESFET's, *H. Fukui, J. V. DiLorenzo, B. S. Hewitt, J. R. Velebir, Jr., H. M. Cox, L. C. Luther, and J. A. Seman (IEEE Transactions on Electron Devices,* June 1980) ... 190
Super Low-Noise GaAs MESFET's with a Deep-Recess Structure, *K. Ohata, H. Itoh, F. Hasegawa, and Y. Fujiki (IEEE Transactions on Electron Devices,* June 1980) ... 194

Highly Reliable GaAs MESFET's with a Statistic Mean NF_{min} of 0.89 dB and a Standard Deviation of 0.07 dB and 4 GHz, *T. Suzuki, A. Nara, M. Nakatani, and T. Ishii (IEEE Transactions on Microwave Theory and Techniques,* December 1979) 200

Performance of GaAs MESFET's at Low Temperatures, *C. A. Liechti and R. B. Larrick (IEEE Transactions on Microwave Theory and Techniques,* June 1976) 205

GaAs Dual-Gate MESFET's, *T. Furutsuka, M. Ogawa, and N. Kawamura (IEEE Transactions on Electron Devices,* June 1978) 211

Failure Mechanisms and Reliability of Low-Noise GaAs FETs, *J. C. Irvin and A. Loya (The Bell System Technical Journal,* October 1978) 218

Determination of the Basic Device Parameters of a GaAs MESFET, *H. Fukui (The Bell System Technical Journal,* March 1979) 230

Experimental Measurement of Microstrip Transistor-Package Parasitic Reactances, *R. J. Akello, B. Easter, I. M. Stephenson (IEEE Transactions on Microwave Theory and Techniques,* May 1977) 244

Part IV: Low-Noise Amplifier Design 251

Description of the Noise Performance of Amplifiers and Receiving Systems, *IRE Subcommittee 7.9 on Noise (Proceedings of the IEEE,* March 1963) 252

Stability and Power Gain of Tuned Transistor Amplifiers, *A. P. Stern (Proceedings of the IRE,* March 1957) 259

Stability and Power-Gain Invariants of Linear Twoports, *J. M. Rollett (IRE Transactions on Circuit Theory,* March 1962) 268

Stability Considerations of Low-Noise Transistor Amplifiers with Simultaneous Noise and Power Match, *L. Besser (IEEE MTT-S International Microwave Symposium,* 1975) 272

Power Waves and the Scattering Matrix, *K. Kurokawa (IEEE Transactions on Microwave Theory and Techniques,* March 1965) 275

Scattering Parameter Approach to the Design of Narrow-Band Amplifiers Employing Conditionally Stable Active Elements, *C. S. Gledhill and M. F. Abulela (IEEE Transactions on Microwave Theory and Techniques,* January 1974) 284

Low-Noise Design of Microwave Transistor Amplifiers, *R. S. Tucker (IEEE Transactions on Microwave Theory and Techniques,* August 1975) 290

Feedback Effects on the Noise Perfomance of MESFETs, *G. D. Vendelin (IEEE MTT-S International Microwave Symposium,* 1975) 294

An Introduction to Simulation and Optimization, *J. W. Bandler (IEEE MTT-S International Microwave Symposium,* 1976) 297

Optimum Gain-Bandwidth Limitations of Transistor Amplifiers as Reactively Constrained Active Two-Port Networks, *W. H. Ku and W. C. Petersen (IEEE Transactions on Circuits and Systems,* June 1975) 300

Microwave Octave-Band GaAs FET Amplifiers, *W. H. Ku, M. E. Mokari-Bolhassan, W. C. Petersen, A. F. Podell, and B. R. Kendall (IEEE MTT-S International Microwave Symposium,* 1975) 311

Part V: Practical Amplifier Techniques 315

A Wide-Band Low Noise *L*-Band Balanced Transistor Amplifier, *R. S. Engelbrecht and K. Kurokawa (Proceedings of the IEEE,* March 1965) 316

An Integrated 4-GHz Balanced Transistor Amplifier, *T. E. Saunders and P. D. Stark (IEEE Journal of Solid-State Circuits,* March 1967) 327

Applications of Integrated Circuit Technology to Microwave Frequencies, *H. Sobol (Proceedings of the IEEE,* August 1971) 334

Design and Performance of Microwave Amplifiers with GaAs Schottky-Gate Field-Effect Transistors, *C. A. Liechti and R. L. Tillman (IEEE Transactions on Microwave Theory and Techniques,* May 1974) 346

Performance of Dual-Gate GaAs MESFET's and Gain-Controlled Low-Noise Amplifiers and High-Speed Modulators, *C. A. Liechti (IEEE Transactions on Microwave Theory and Techniques,* June 1975) 354

Low-Noise Receiver Design Trends Using State-of-the-Art Building Blocks, *H. C. Okean and A. J. Kelly (IEEE Transactions on Microwave Theory and Techniques,* April 1977) 363

Low-Noise and Linear FET Amplifiers for Satellite Communications, *D. A. Cowan, P. Mercer, and A. B. Bell (IEEE Transactions on Microwave Theory and Techniques,* December 1977) 377

A Low-Noise Gallium Arsenide Field Effect Transistor Amplifier for 4 GHz Radio, *R. H. Knerr and C. B. Swan (The Bell System Technical Journal,* March 1978) 383

The Matched Feedback Amplifier: Ultrawide-Band Microwave Amplification with GaAs MESFET's, *K. B. Niclas, W. T. Wilser, R. B. Gold, and W. R. Hitchens (IEEE Transactions on Microwave Theory and Techniques,* April 1980) 389

Cryogenically Cooled GaAs FET Amplifier with a Noise Temperature Under 70 K at 5.0 GHz, *J. Pierro (IEEE Transactions on Microwave Theory and Techniques,* December 1976) 399

Low-Noise Cooled GASFET Amplifiers, *S. Weinreb (IEEE Transactions on Microwave Theory and Techniques,* October 1980) 403

GaAs IC Direct-Coupled Amplifiers, *D. Hornbuckle (IEEE MTT-S International Microwave Symposium Digest,* 1980) 417

20-GHz Band Monolithic GaAs FET Low-Noise Amplifier, *A. Higashisaka and T. Mizuta (IEEE Transactions on Microwave Theory and Techniques,* January 1981) 420

Advanced Microwave Circuits, *E. F. Belohoubek (IEEE Spectrum,* February 1981) 426

Design Considerations for Monolithic Microwave Circuits, *R. A. Pucel (IEEE Transactions on Microwave Theory and Techniques,* June 1981) 430

Author Index 453

Subject Index 455

Editor's Biography 461

Preface

This is a collection of papers on low-noise microwave transistors, low-noise microwave transistor amplifiers, and relevant subjects. This volume is intended to serve both device and circuit engineers as well as graduate students in the field of microwave electronics. It aims to provide a basic understanding of the noise characterization, the representation and measurement of active linear twoports, the noise performance of microwave transistors, and the design and practical techniques of low-noise transistor amplifiers in the microwave region.

The microwave region, particularly the 0.5–20 GHz band, is most desirable for long-distance communication and highly sensitive radar systems because in sky noise the "low-noise window" is provided in this frequency range. Thus the use of a low-noise device is essential to take full advantage of this gift from nature. Microwave transistors have marked a milestone as being a most effective low-noise device for this purpose.

The editor is grateful to Members of the IEEE Microwave Theory and Techniques Society's Administrative and Publication Committees, in particular to Drs. R. H. Knerr and F. J. Rosenbaum, for their consideration and support. He also appreciates Members of the IEEE Press Editorial Board for their approval. Finally he is greatly indebted to Mr. W. R. Crone and other members of the IEEE PRESS staff for their tireless efforts and expertise.

Introduction

Overall View of Low-Noise Amplification at Microwave Frequencies

During the past decade remarkable progress has been made in microwave transistors for both low-noise and high-power applications. The great challenge has been to replace life-limited, bulky vacuum tubes with life-unlimited, compact solid-state devices in the microwave region with ever greater degrees of advanced performance, simplified operation, sound reliability, reduced maintenance, and improved cost-effectiveness for the overall system. In the early stages of this effort, various kinds of two-terminal devices were used in reflection-type amplifiers by virtue of their negative resistance property produced under certain operational conditions. They included varactor and tunnel diodes for low-noise amplification, and IMPATT and Gunn diodes for high-power applications. Some of them are still in use. However, no separation is available between the input and output within a two-terminal device itself. For stable operation, therefore, the separation must be provided by the external circuitry using a unilateralizing component, such as a circulator. Moreover, the two-terminal device inherently lacks the signal controlling electrode. This is the reason that it is used only in the reflection-type amplifier, instead of the transmission-type. Consequently, adjustments of dc bias and device-circuit coupling conditions become crucial to provide suitable performance characteristics as a microwave amplifier. In the case of a parametric amplifier a pumping source is additionally necessary for a varactor diode. On the contrary, a transistor amplifier is free from these operational restrictions, at least in principle. Its input is separated from the output so that they are independently adjusted for their optimum conditions. This is very important not only for the initial optimization of an amplifier but also for long-term stable operation with minimum maintenance effort.

Since the microwave performance of the transistor has significantly been advanced, this device has enjoyed the prime position among solid-state devices in the microwave region. This trend has recently been accelerated in accordance with continuous breakthroughs in the noise performance and CW power handling capability of GaAs MESFETs (MEtal Semiconductor Field-Effect Transistors). For example, a CW output power of 10 watts has been achieved at 10 GHz with a single-chip GaAs MESFET at a reasonable gain. There has also been notable progress in the noise performance, such as a noise figure of less than 0.5 dB at 4 GHz, 1 dB at 10 GHz, or 2 dB at 20 GHz, each at room temperature. Such a low-noise figure is due to the inherent advantage of the FET as a majority-carrier device in contrast with the BPT (BiPolar Transistor) as a minority-carrier device.

The overall performance of a signal transmission system is determined by both signal power level at the transmitter and weak signal reception capability at the receiving end. State-of-the-art reception in the microwave region is represented by the 4K maser followed by the 20K parametric amplifier, both of which are complicated and expensive. A rapid improvement in the noise performance of the GaAs MESFET has made the gap much narrower. The aforementioned values of the noise figure are now approaching those of room-temperature parametric amplifiers. By cryogenically cooling GaAs MESFET amplifiers, their noise figures are becoming competitive with those of cryogenically cooled parametric amplifiers. The GaAs MESFET amplifier is much superior to the parametric amplifier in original cost, space, weight, and maintenance. The noise performance of GaAs MESFET amplifiers is satisfactory for most communication systems. Together with the development of MOMIC (MOnolithic Microwave Integrated Circuit) technology the GaAs MESFET will play an ever increasing role even in the consumer market, such as for the direct reception of TV signals from a broadcasting satellite. Furthermore, the state-of-the-art output power and noise figures previously mentioned indicate by no means the ultimate capabilities of the microwave transistor. Changes in the structure and material, as well as novel uses of the device, will bring further improvements in the microwave performance and thus extend the useful frequency range.

As the microwave performance of the GaAs MESFET has been improved, the usefulness of the Si BPT has appeared to be fading away. However, the noise performance of state-of-the-art Si BPTs is almost as good as GaAs MESFETs at relatively low microwave frequencies; therefore, the Si BPT will play a role as an inexpensive three-terminal amplifying device for low-cost microwave amplifiers.

Classification of Noise Sources and Brief History of Noise Reseach

Noise arising in an electronic component or a device is historically classified as "thermal noise", "shot noise", or "flicker noise". Thermal noise occurs in any conductor due to the random motion of electrons caused by thermal agitation. Shot noise and flicker noise, which are also called "excess noise" as a generic name, are present only when current flows through a conductor by an external means.

After the discovery of the Brownian motion in 1827 [1], a number of investigators accounted for this phenomenon. In the early twentieth century, A. Einstein [2] showed the Brownian motion to be entirely fundamental. He exhibited a clear analysis of the problem as arising from continuous and random bombardments of molecules. He also mentioned that his method was applicable to electrical circuits. In 1918, W. Schottky [3] pointed out that the lowest achievable noise level is limited by the Brownian motion of electrons and other charged particles in an electrical con- Boltzmann's constant k and absolute temperature T. On the other hand, shot noise is specified by the electronic the other hand, shot noise is specified by the electronic charge q and average current I, which is known as Schottky's theorem. Therefore it was expected that the fundamental physical constants, such as k, q and Avogadro's number, were determined from noise voltage measurement. This was then experimentally confirmed by many workers [4] in good agreement with other measurement methods.

Shot noise originates in the discrete nature of electron flow. In a vacuum tube electrons are emitted from the cathode in a series of independent and random events, resulting in shot noise in the plate current. The magnitude of shot noise has a uniform frequency characteristic until a certain high frequency at which the electron transit-time effect becomes appreciable. A random process with constant spectrum density is called "white noise". Thermal noise arising in a resistor is also white noise.

In 1925 an unexpectedly large value of the electronic charge was obtained from shot noise measurements conducted at low frequencies by J. B. Johnson [5]. This observation led to the discovery of flicker noise in conjunction with an analytical treatment shown by Schottky in 1926 [6]. Flicker noise is influenced by surface state and bulk imperfection and cannot be described by simple physical explanations as applied for thermal noise and shot noise. Since flicker noise has a $1/f^n$ frequency spectrum with n close to unity, this noise is often called "1/f noise".

In 1928 Johnson experimentally verified Schottky's theory that thermal noise voltage depends only on resistance, temperature and bandwidth and is independent of material and shape [7]. Because of this discovery thermal noise is also called "Johnson noise". At the same time, H. Nyquist derived the famous formula for thermal noise voltage based on the thermodynamics of a telephone line [8]. He showed that the available noise power is always kTB, where B is the equivalent noise bandwidth. It is convenient to remember that kTB takes a numerical value of 4×10^{-21} watts per cycle bandwidth at 290 K. Nyquist's theorem covers almost all one needs to know about thermal noise. Noise figure was later defined in terms of thermal noise [9].

For semiconductor devices the name "generation-recombination noise (GR noise)" is often used [10]. This noise is caused by the random generation and recombination of hole-electron pairs, the random generation of carriers from traps, or the random recombination of carriers with empty traps. In bulk material, such as in a nearly intrinsic semiconductor, the nomenclature of GR noise is more appropriate than shot noise because spontaneous fluctuations in the carrier density exist even if no electric field is applied. In junction devices, on the other hand, bias voltage is needed across a potential barrier to inject minority carriers or to change the injection level. The GR noise in this case therefore shows much closer resemblance to shot noise than in the bulk material case.

Since diffusion is a random process, fluctuations in the diffusion rate produce localized fluctuations in the carrier density. Such spontaneous fluctuations are called "diffusion noise". Diffusion noise is a more general process than thermal noise and it reduces to thermal noise if Einstein's relation holds as in bulk material. Einstein's relation is valid if charge carriers have a Maxwellian distribution. In junction devices, fluctuations in the diffusion rate give a major contribution to shot noise.

"Modulation noise" refers to noise not directly caused by fluctuations in the transition or diffusion rate. Instead this is due to fluctuations in the carrier density or current flow produced by some modulation mechanisms.

Microwave Transistors and Their Noise Properties

The transistor is traditionally classified into two groups, namely, bipolar and unipolar [11]. The bipolar transistor (BPT) is based on the phenomenon of minority-carrier injection into a region where majority carriers are present. The injected carriers move primarily through diffusion process. In the unipolar group, practically all of the current flowing through a device is due to majority carriers. The field-effect transistor (FET) [12], [13] belongs to this category, in which the conductivity of a conducting channel is modulated by a transverse electric field under the controlling electrode. In the static induction transistor (SIT) [14], a third transistor, carriers are injected into a high-resistivity region from the source due to static induction from the drain. The injection level is modulated by a change in the potential barrier height in front of the source. This change is provided much more effectively by variation in the gate potential than that of the drain. In this device the source-gate channel resistance is deliberately minimized in order to eliminate the negative feedback effect on the channel current. This effort results in no saturation of the drain current. The SIT is primarily unipolar but can also be regarded as a bipolar device with an infinitesimally thin base region.

The major noise sources in small-signal microwave transistors can be arranged as follows [10]: i) shot noise in BPTs at low injection, noise due to generation and recombination in the junction space-charge region, and shot noise in the leakage current of FETs; ii) thermal noise and induced gate noise in FETs, GR noise in FETs and BPTs at low temperatures, noise due to recombination centers in the space-charge regions of FETs; and iii) flicker noise usually neglected at microwave frequencies.

In BPTs at low injection the passage of carriers across potential barriers can be considered as a series of independent, random events. As a consequence full shot noise should be associated with each current flowing in the device. At microwave frequencies this noise may be modified by transit-time or diffusion-time effect. This is called the "corpuscular approach". Another approach considers this noise as being caused by the random processes of diffusion and/or recombination. The noise sources to be introduced are then diffusion noise and GR noise sources. This is called the "collective approach". This approach can be used even at high-injection levels. Since only low-level injection is used for low-noise amplification, the two approaches are equivalent.

In FETs the noise voltage distribution occurs along the channel due to distributed thermal noise sources. At microwave frequencies this noise voltage gives rise to a noise current flowing to the gate by capacitive coupling. This induced gate noise is partially correlated with the drain noise.

A simple shot noise picture can give a reasonable description of most noise phenomena in BPTs, whereas a simple thermal noise model can explain most noise behaviors of FETs. This is the key distinction between the noise properties of BPTs and FETs. The noise sources in SITs have not yet been reported. Based on the operational principle it could be conjectured that the SIT would have a reduced thermal noise and an increased shot noise in comparison with the FET.

Various noise sources in a device are transformed as equivalent noise sources into an equivalent circuit of this device. Then its noise performance is characterized in common terms to electrical engineers, such as noise figure.

Noisy Linear Twoports

From a circuit-theory point of view, an active electronic device or an amplifier can be treated as a noisy linear twoport (or a noisy linear four-terminal network) for small-signal applications. Gain and noise are the most important factors in order to describe the microwave performance of such a twoport. The gain and noise behaviors are then conventionally represented in terms of the available power gain and noise figure, respectively. Both available gain and noise figure depend on the bias condition for the device of interest, as well as on the driving source impedance. In the microwave region the source impedance is usually the characteristic impedance of the input transmission line, which is normally $Z_o = 50 + j0$ ohms. Therefore a matching network (or an impedance transformer) is needed between the line and device input in order to obtain high gain or low noise.

The matching condition necessary for the minimum noise figure is, however, not the same as for the maximum available gain in most cases. Furthermore, when an amplifier is designed in a multi-stage configuration, the overall noise figure is determined by noise contributions from all stages. The most significant contribution, of course, arises in the first stage and some come from the second stage, especially if the gain of the first stage is low. The overall noise figure of this amplifier becomes close to $M + 1$, where M is the noise measure of the first stage.

The noise figure and available gain are expressed in terms of four noise parameters and four gain parameters, respectively. The noise measure can then be expressed as a function of the noise and gain parameters. Variations of the noise figure, available gain, and noise measure with respect to the source impedance are conveniently mapped on the Smith chart on which a constant value of either noise figure, available gain, or noise measure is shown as a circle. This type of representation has long been adopted in the industry to describe the noise and gain performance of a microwave transistor as the function of source impedance or source reflection coefficient.

The optimum source impedance for the minimum noise measure comes between those for the minimum noise figure and for the maximum available gain on the source-impedance Smith chart. In the case of practical microwave transistors, however, the location of the optimum source impedance for the minimum noise measure is usually found to be very close to that for the minimum noise figure. Therefore, the noise matching is much more important than the gain matching in the input circuit of the first stage for low-noise microwave transistor amplifiers.

As mentioned earlier, the noise performance of a linear twoport is completely described in terms of four independent noise parameters. It is clearly shown how these parameters are determined in the IRE standards on methods of measuring noise in linear twoports, 1959. However, this process was extremely tedious two decades ago. Introduction of statistical treatments to the data analysis and application of automated test set-ups have greatly eased the situation. In recent years the noise performance of a microwave transistor has been remarkably improved. Today it is not uncommon to measure low noise figures, such as less than 1 dB. Then the treatment of losses in the input circuit has to be seriously reconsidered in order to obtain an accurate value of the noise figure of a device. Recently researchers and engineers have increasingly paid attention to this problem. Their works will appear in future issues of IEEE publications.

Low-Noise Microwave Transistor Amplifiers

Low-noise microwave transistor amplifiers are primarily used in low-noise receiving systems for radioastronomy, radar, satellite communications, terrestrial communications, and allied purposes. The sensitivity of these receivers is often defined in terms of "noise temperature" [9] which is equivalent to noise figure. In the design of practical amplifiers, stability consideration is essential in addition to those for gain and noise. Along with recent developments in the field of measurement instrumentation, the scattering matrix representation has become popular to characterize the microwave performance. Computer aided techniques are now widely used in the design of microwave circuits, both passive and active networks. The user should be aware of the limitations of each technique for effective application.

A microwave transistor amplifier generally consists of single-ended and/or balanced stages, each of which has competitive merits. Dual-gate MESFET amplifiers extensively widen the linearity range, having an easy gain control. Feedback technique is implemented in the microwave region to extend the useful bandwidth to over an octave. Cooling the GaAs MESFET significantly improves its noise behavior because the major noise sources are thermal noise. It has been demonstrated that the noise performance of a GaAs MESFET amplifier cryogenically cooled to 20K is almost comparable to that of a parametric amplifier operating at the same temperature. Recently considerable efforts have been placed for realization of monolithic microwave amplifiers on GaAs chips. This is the dawn of a new era. Vital groundwork is being laid down for innovative applications in consumer microwave electronics. This, along with more traditional approaches to the conventional communication systems, will make eye-opening progress in leaps and bounds.

Editorial Comments

This collection is divided into five major parts containing a total of 60 papers. Naturally some papers cover more than one subject area so these papers are placed in an expedient manner to be most beneficial to the reader. In each Part, at least one review or tutorial paper is included which contains a comprehensive bibliography. Additionally every paper has its own references. Therefore an overall bibliography is not provided in this volume.

Excellent review papers on the fundamental subjects in the field of electrical noise are compiled in another IEEE PRESS reprint book [15]. The subjects include history of noise research, statistical treatments of noise problems, basic mechanisms of noise in semiconductors and solid-state devices, and standard noise sources.

References

[1] R. Brown in Philosophical Magazine in 1828, 1829 and 1830.
[2] A. Einstein, *Investigations on Theory of the Brownian Movement*, edited with notes by R. Fürth, translated by A. D. Cowper, and published by Dover Publications, Inc., in 1956.
[3] W. Schottky, "Über Spontane Stromschwankungen in verschiedenen Elektrizitätleitern," Ann. d. Phys., vol. 57, pp. 541-567, 1918.
[4] For example, C. A. Hartmann, "Über die Bestimmung des elektrischen Elementarquantums aus dem Schroteffekt," Ann. d. Phys., vol. 65, pp. 51-78, 1921.
[5] J. B. Johnson, "The Schottky effect in low frequency circuit," Phys. Rev., vol. 26, pp. 71-85, 1925.
[6] W. Schottky, "Small shot and flicker effects," Phys. Rev., vol. 28, pp. 74-103, 1926.
[7] J. B. Johnson, "Thermal agitation of electric charge in conductors," Phys. Rev., vol. 32, pp. 97-109, 1928.
[8] H. Nyquist, "Thermal agitation of electricity in conductors," Phys. Rev., vol. 32, pp. 110-113, 1928.
[9] W. W. Mumford and E. H. Scheibe, *Noise Performance Factors in Communication Systems*, 1968, Horizon House-Microwave, Inc., Dedham, MA.
[10] A. van der Ziel, *Noise: Sources, Characterization, Measurement*, 1970, Prentice-Hall, Inc., Englewood Cliffs, NJ.
[11] W. Shockley, "Transistor electronics: imperfections, unipolar and analog transistors," Proc. IRE, vo. 40, pp. 1289-1313, November 1952.
[12] W. Shockley, "A unipolar 'field-effect' transistor," ibid, pp. 1365-1376.
[13] G. C. Dacey and I. M. Ross, "The field-effect transistor," Bell Syst. Tech. J., vol. 34, pp. 1149-1189, November 1955.
[14] J. Nishizawa, T. Terasaki, and J. Shibata, "Field-effect transistor versus analog transistor (static induction transistor)," IEEE Trans. Electron Devices, vol. ED-22, pp. 185-197, August 1975.
[15] IEEE PRESS, *Electrical Noise: Fundamentals and Sources*, edited by M. S. Gupta, 1977, The Institute of Electrical and Electronics Engineers, Inc., New York, NY.

Part I
Noise Characterization and Measurements

Noise Figures of Radio Receivers*

H. T. FRIIS†, FELLOW, I.R.E.

Summary—A rigorous definition of the noise figure of radio receivers is given in this paper. The definition is not limited to high-gain receivers, but can be applied to four-terminal networks in general. An analysis is made of the relationship between the noise figure of the receiver as a whole and the noise figures of its components. Mismatch relations between the components of the receiver and methods of measurements of noise figures are discussed briefly.

INTRODUCTION

THE importance of noise originating in a radio receiver has increased as shorter and shorter wavelengths have come into practical usage. Many papers on the subject, notably those by Llewellyn[1] and Jansky,[2] have appeared since the writer, in 1928, showed experimentally[3] that thermal-agitation noise (Johnson noise) determined the absolute sensitivity of short-wave radio receivers. Early in 1942 North[4] suggested the adoption of a standard for the absolute sensitivity of radio receivers which differed by a factor of 2 from the standard used by us at that time. We adopted his standard, since ours was somewhat limited in that it was based on matched impedances in the input circuit of the receiver.

In this paper a more rigorous definition of the standard of absolute sensitivity, the so-called noise figure, of a radio receiver is suggested. The definition is not limited to high-gain receivers, but can be applied to four-terminal networks in general. It also makes it possible by a simple analysis to give the relationship between the noise figure of the receiver as a whole and the noise figures of its components. In the case of a double-detection receiver these components may be a high-frequency amplifier, a frequency converter, and an intermediate-frequency amplifier. The paper also gives a brief description of methods of measurements of noise figures.

The four-terminal network whose noise figure is to be defined is shown schematically in Fig. 1. A signal generator is connected to its input terminals and an output circuit is also indicated. The input and output impedances of the network may have reactive components and they may be matched or mismatched to the generator and the output circuit, respectively. The four-terminal network may be, for instance, an amplifier, a converter, an attenuator, or a simple transformer. The presence of the signal generator is required for the definition that follows, but the attenuator in the signal generator and the output circuit toward the right are shown only to illustrate measurements of noise figures and gains.

The noise figure will be defined in terms of available signal power, available noise power, gain, and effective bandwidth. The definitions of these terms will be given and discussed next.

AVAILABLE SIGNAL POWER

A generator with an internal impedance R_0 ohms and electromotive force E volts delivers $E^2R_1/(R_0+R_1)^2$ watts into a resistance R_1 ohms. This power is maximum and equal to $E^2/4R_0$ when the output circuit is matched to the generator impedance, that is when $R_1=R_0$. $E^2/4R_0$ is hereafter called the available power of the generator, and it is, by definition, independent of the impedance of the circuit to which it is connected. The output power is smaller than the available power when R_1 is unlike R_0, since there is a mismatch loss. In amplifier input circuits a mismatch condition may be beneficial[5] due to the fact that it may decrease the output noise more than the output signal. It is the presence of such mismatch conditions in amplifier input circuits that makes it desirable to use the term available power in this paper. The symbol S_g will be used for the available signal power at the output terminals of the signal generator shown in Fig. 1. S_g is here equal to V^2/RA watts where V is the voltage across the input terminals

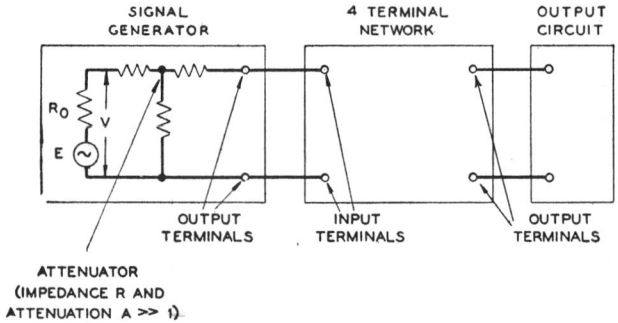

Fig. 1

of the attenuator, R the characteristic impedance of the attenuator, and A the nominal attenuation (A is assumed large).

The output terminals of any network may be considered as being the output terminals of a signal generator. The symbol S will be used for the available signal power at the output terminals of the four-terminal network shown in Fig. 1.

---

* Decimal classification: R261.2. Original manuscript received by the Institute, September 7, 1943.
† Bell Telephone Laboratories, Inc., New York, N. Y.
[1] F. B. Llewellyn, "A study of noise in vacuum tubes and attached circuits," PROC. I.R.E., vol. 18, pp. 243–266; February, 1930.
[2] K. G. Jansky, "Minimum noise levels obtained on short-wave radio receiving systems," PROC. I.R.E., vol. 25, pp. 1517–1531; December, 1937.
[3] Unpublished report.
[4] D. O. North, "The absolute sensitivity of radio receivers," *RCA Rev.*, vol. 6, pp. 332–344; January, 1942. The reader is referred to this paper for reference to other papers on this subject.

[5] That such an improvement might be possible was first discussed in detail by F. B. Llewellyn in his paper "A rapid method of estimating the signal-to-noise ratio of a high gain receiver." PROC. I.R.E., vol. 19, pp. 416–421; March, 1931.

Gain

The gain of the network is defined as the ratio of the available signal power at the output terminals of the network to the available signal power at the output terminals of the signal generator. Hence

$$G = S/S_g. \qquad (1)$$

This is an unusual definition of gain since the gain of an amplifier is generally defined as the ratio of its output and input powers. This new definition is introduced here for the same reason that made it desirable to use the term available power. Note that while the gain is independent of the impedance which the output circuit presents to the network, it does depend on the impedance of the signal generator.

The four-terminal network has generally some kind of band-pass characteristic. The gain G is defined as that at the mid-band frequency.

Available Noise Power

As in the case of signal power, the available noise power between two terminals is defined as the noise power which would be absorbed by a matched output circuit.

The symbol N will be used for the available noise power at the output terminals of the network. This power is due to all the noise sources in the network itself and the Johnson-noise sources in the signal generator, but noise sources in the output circuit toward the right in Fig. 1 are not included.

The Johnson-noise power available from a resistance will be discussed now. Any resistance, R ohms, acts as a Johnson-noise generator with a mean-square electromotive force equal to $4KTRdf$. K is Boltzmann's constant $= 1.38 \times 10^{-23}$, T is the absolute temperature of the resistance, and df is the bandwidth. The available Johnson-noise power is then

$$4KTRdf/4R = KTdf \text{ watts} \qquad (2)$$

and this is the noise power available over the band df at the output terminals of the signal generator in Fig. 1. It is, in fact, the available noise power between any two terminals of a passive network when all its parts have the same temperature T.

Effective Bandwidth

The contribution to the available output noise by the Johnson-noise sources in the signal generator is readily calculated for an ideal or square-top band-pass characteristic and it is $GKTB$ where B is the bandwidth in cycles per second. In practice, however, the band is not flat; i.e., the gain over the band is not constant but varies with the frequency. In this case, the total contribution is $\int G_f KT df$ where G_f is the gain at the frequency f. The effective bandwidth B of the network is defined as the bandwidth of an ideal band-pass network with gain G that gives this contribution to the noise output. Therefore,

$$GKTB = \int G_f KT df$$

or

$$B = \frac{1}{G} \int G_f df. \qquad (3)$$

Noise Figures

The noise figure of the network in Fig. 1 will now be defined in terms of S_g, S, KTB, and N.

It is important to have the highest possible signal-to-noise ratio at the output terminals of the network. The maximum value of this ratio would be as high as the available-signal-to-available-noise ratio at the signal-generator terminals if there were absolutely no noise sources present in the network. Simple lossless transformers or filters are examples of networks with no noise sources. In general, however, noise sources are present and these noise sources reduce the available signal-to-noise ratio at the output terminals of the network. The noise figure F of the network[6] is defined as the ratio of the available signal-to-noise ratio at the signal-generator terminals to the available signal-to-noise ratio at its output terminals.[7] Thus

$$F = (S_g/KTB)/(S/N) = (S_g/KTB)(N/S) \qquad (4)$$

and since

$$G = S/S_g$$
$$F = (1/G)(N/KTB). \qquad (5)$$

Solving for N gives the following expression for the available noise output:

$$N = FGKTB \text{ watts}. \qquad (6)$$

This noise output includes the contribution made by the Johnson-noise source in the signal generator. This contribution is $GKTB$. The available output noise due only to noise sources in the network is, therefore,

$$(F - 1)GKTB \text{ watts}. \qquad (7)$$

All the terms in (4), (5), (6), and (7) have been defined, but a value for the temperature T of the generator terminal impedance must still be chosen before the noise figure is definite. It is suggested that the noise figure be defined for a temperature of 290 degrees Kelvin (63 degrees Fahrenheit). Then

$$KT = 1.38 \times 10^{-23} \times 290$$
$$= 4 \times 10^{-21} \text{ watts per cycle bandwidth.}$$

The relationship between the noise figure and the degree of mismatch that exists between the network and its output and input circuits is important. Definition (4) shows clearly that the output circuit and hence its coupling with the network has no effect on the value of the noise figure. However, it also shows that the noise figure does depend on the degree of mismatch between the generator and the network since both S and N will vary with the magnitude of this mismatch.

---

[6] We have, until now, used the symbol $\overline{NF}$ for the noise figure, but we shall use, hereafter, the symbol F suggested by Dr. S. Roberts.

[7] Because of this definition, the noise figure has also been called "excess noise ratio."

Measurement of the Noise Figure

Although a detailed description of noise-figure-measuring equipment will not be given in this paper, it is believed worth while to outline a method of such measurements.

It is not difficult to measure the noise figure F when the gain of the network is so large that a noise-power reading can be obtained by means of a thermocouple connected between the output terminals of the network (Fig. 1). The measurement procedure is then simply to adjust the attenuation A of the signal attenuator until the output reading is double that due to noise only which is obtained with the signal generator turned off. S is then equal to N and definition (4) gives

$$F = S_g/KTB = V^2/4RAB \times 10^{-21}. \qquad (8)$$

The effective band B is calculated from a gain-versus-frequency curve. The voltage V across the input terminals of the signal attenuator may be measured by means of thermocouples, tube voltmeters, thermistors, etc., and by cross-checking such different equipment the value of V can be obtained with adequate accuracy even in the centimeter-wavelength range. It is more difficult to obtain an accurate value of the attenuation A because of its large magnitude. For a short-wave receiver for which F may be as small as 3, formula (8) gives $A = 5.2 \times 10^{13}$ for $R = 80$ ohms, $V = 1$ volt, and $B = 20{,}000$ cycles. Only very careful work will give satisfactory data with such magnitudes of attenuation. Very thorough shielding of the signal generator is one important requirement.

The noise figure of a network made up of nondissipative elements is unity since it contains no noise sources (expression (7) is equal to zero). The losses in simple transformers and filters are generally sufficiently low to come under this classification. The noise figure of an attenuator at 63 degrees Fahrenheit temperature is by (5) equal to its attenuation when it is matched to the signal generator since under these conditions $N = KTB$ and attenuation $= 1/G$. A network made up of a transmitting and a receiving antenna is equivalent to an attenuator with an attenuation A equal to the ratio of transmitted to received power. Assuming no static or star noise and no circuit losses in the receiving antenna, its noise figure is, by (5), $F = A(N/KTB)$. If T_r is the absolute temperature of the radiation resistance of the receiving antenna, $N = KT_rB$. Hence $F = A(T_r/T) = A(T_r/290)$. The value of T_r is not definitely known, but $T_r = T$ is believed to be a good approximation for antennas whose radiation strikes the earth.[8]

Noise Figures for Two Networks in Cascade

If the gain of the network shown in Fig. 1 is not large, an amplifier following the network is required to obtain a noise-output reading. For this case a noise-figure analysis of two networks in cascade is required. Also

[8] For further information on this subject the reader is referred to a paper by R. E. Burgess, "Noise in receiving aerial systems," *Proc. Phys. Soc.*, vol. 53, pp. 293–304; May, 1941.

from a design point of view it is important to know the relationship between the noise figure of a whole receiver and the noise figures of its components since it indicates the component on which efforts for improvement are worth while.

The two networks are shown schematically in Fig. 2. We are also considering here the general case where the two networks, the generator, and the output circuit may be either matched or mismatched. The definitions given for a single network will now be applied to the network ab made up of the two networks a and b in cascade and to the individual networks a and b.

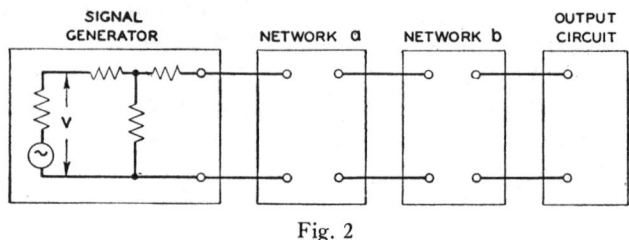

Fig. 2

Expression (6) gives for the available output noise at the output terminals of network b

$$N_{ab} = F_{ab}G_{ab}KTB_{ab} \text{ watts}. \qquad (9)$$

To simplify the analysis, it will be assumed that the two networks have the same ideal or square-top bandpass characteristics ($B_a = B_b = B$). The equivalent band B_{ab} is then equal to B. The total gain G_{ab} is by the definition of gain (1) equal to G_aG_b. Then

$$N_{ab} = F_{ab}G_aG_bKTB \text{ watts}. \qquad (10)$$

A new expression for this noise power may be derived by applying expression (6) to network a and expression (7) to network b. Applying (6) to network a, the available noise power at its output terminals is

$$N_a = F_aG_aKTB \text{ watts}. \qquad (11)$$

Multiplying this power by G_b gives then the following expression for the available noise power at the output terminals of network b due to the noise sources in network a and the Johnson-noise sources in the signal generator

$$F_aG_aG_bKTB \text{ watts}. \qquad (12)$$

Expression (7) applied to network b gives the following expression for the available noise power at the output terminals of network b due to noise sources in network b only,

$$(F_b - 1)G_bKTB \text{ watts}. \qquad (13)$$

The total available noise power N_{ab} at the output terminals of network b is the sum of the noise powers given by (12) and (13), hence,

$$N_{ab} = F_aG_aG_bKTB + (F_b - 1)G_bKTB$$
$$= \left(F_a + \frac{F_b - 1}{G_a}\right)G_aG_bKTB. \qquad (14)$$

Comparing this expression with (10) gives the following simple relationship between the noise figures of the two networks

$$F_{ab} = (F_a + F_b) - 1/G_a. \qquad (15)$$

This relationship is valid for any distribution of noise power throughout the bands of the two networks. The assumptions made in regard to the band-pass characteristics could be made less severe for uniform noise-power distribution. Although they do not seriously limit the usefulness of relationship (15) for practical cases, it is recommended that both the effect of nonuniform noise distribution and the effect of unequal and nonideal bands be studied to make sure whether it is necessary to use correction factors for the different terms of (15).

The rather complicated matter of the effects of nonuniform noise distribution and nonideal band characteristics may be clarified somewhat by pointing out that the relationship (15) may always be applied to an element of band df at a frequency f in the band of the networks. It usually will be found that the noise figure corresponding to an element of band varies somewhat across the band of an actual network.

The noise figures F_a and F_b in (15) will be discussed next. Network b has no effect on the noise figure F_a of network a. This follows from the discussion of a single network. That discussion also pointed out that network a does affect the noise figure F_b of network b. Therefore, if F_b is measured separately by a signal generator, as shown in Fig. 1, then this signal generator must have a terminal impedance which is identical to the output impedance between the output terminals of network a.

Mismatch Relations for Two Networks

The reader is referred to a paper by Burgess[8] and a more recent paper by Herold[9] for detailed discussions of the advantage of mismatch relations, and only a brief discussion will be given here.

The over-all noise figure for two networks has a minimum value when the degree of mismatch between them is made identical to the mismatch which gives the lowest noise figure for the second network when it is connected directly to a signal generator. Offhand, this is not evident, but an analysis of the matching condition between a signal generator and a network shows that the optimum matching condition is independent of any noise sources in the signal generator. For the lowest over-all noise figure, the optimum matching condition between the signal generator and network a does, on the other hand, depend on both networks. When network a is a low-gain converter and network b an intermediate-frequency amplifier, the highest possible gain G_a in the first network, which is obtained when it is matched to the signal generator, will in general give the lowest over-all noise figure.

Noise Figures for Several Networks in Cascade

The analysis for two networks may be easily extended to more than two networks. For example, if three networks are considered, (15) gives

[9] E. W. Herold, "An analysis of the signal-to-noise ratio of ultra high-frequency receivers," *RCA Rev.*, vol. 6, pp. 302–332, January, 1942.

$$F_{abc} = F_{ab} + (F_c - 1)/G_{ab} = F_a \\ + (F_b - 1)/G_a + (F_c - 1)/G_a G_b. \quad (16)$$

In most receivers the gains of the amplification stages are such that the noise figures of only the first two stages must be considered.

Measurement of the Noise Figures of Two Networks

It may be desirable to determine F_a by indirect measurements, particularly if G_a is low. This may be done as follows. The noise figures F_{ab} and F_b may be measured by the method described for a single network. The gain G_a may be obtained, by means of signal generators, as the increase in available signal power required to give a certain signal output reading when network a is left out. The noise figure F_a may then be calculated by means of (15).

A second method of noise-figure measurement includes the measurement of the ratio of the output noise of network b with network a in normal operation to that with network a passive. This ratio is called the Y figure and is especially useful in converter measurements. Network a is said to be passive if its available output noise is only KTB watts. In measuring Y it is usually most convenient to replace the output impedance of network a with an equal passive impedance at room temperature.

An expression for F_a in terms of Y, F_b, and G_a will be developed next. By definition,

$$Y = N_{ab}/N_b. \quad (17)$$

Formula (5) gives

$$F_b = (1/G_b)(N_b/KTB_b) \quad (18)$$

and

$$F_{ab} = (1/G_a G_b)(N_{ab}/KTB_{ab}). \quad (19)$$

The above three formulas give

$$F_{ab} = (F_b Y/G_a)(B_b/B_{ab}). \quad (20)$$

It will also here be assumed that the two networks have the same ideal or square-top band-pass characteristic. Then

$$F_{ab} = F_b Y/G_a. \quad (21)$$

Formulas (15) and (21) give

$$F_a = [F_b(Y - 1) + 1]/G_a. \quad (22)$$

Formula (22) is often simpler to use for the determination of F_a than relationship (15) because it is easier to measure Y than F_{ab}. Note also that F_{ab} may be determined experimentally from the convenient relation given by (21).

Conclusion

All signal and noise powers are in watts. It is confusing to use the decibel scale when noise powers are added, and it has not, therefore, been used in this paper.

In concluding, it is hoped that the definitions and symbols suggested will come into general usage. It should be pointed out that the paper is the result of a great many discussions during the past two years with scientific workers both inside and outside the Bell Telephone Laboratories.

Theory of Noisy Fourpoles*

H. ROTHE†, SENIOR MEMBER, IRE, AND W. DAHLKE†

Summary—The well-known theory of fourpoles only comprises passive fourpoles and active fourpoles with internal sources of sinusoidal currents or voltages of defined frequencies. This theory is now completed for fourpoles with internal noise sources. Simple equivalent circuits are derived for such networks. They consist of the original but noise-free fourpole cascaded with a preceding noise fourpole in which all noise-sources are concentrated. The latter contains the equivalent noise conductance G_n, the equivalent noise resistance R_n and the complex correlation admittance Y_{cor}. With these quantities the noise behavior of any desired fourpole can be described sufficiently. In particular it is possible to calculate the noise figure F and its dependence on the matching conditions to the signal source of a single fourpole or a group of cascaded fourpoles. The methods of experimental determination of the elements of the noise fourpoles are discussed. The same theory is also useful for mixer-circuits as well as for traveling-wave tubes and transistors, as application results are given for grid controlled electron tubes.

ance matrix is more convenient as equivalent circuit. But if there are inner noise sources inside the fourpole, these well-known fourpole equations are no more sufficient. They must rather be completed by two noise currents i_1 and i_2 respectively, by two noise voltages u_1 and u_2 to the form

$$
\begin{aligned}
I_1 &= Y_{11}U_1 + Y_{12}U_2 + i_1 \\
U_1 &= Z_{11}I_1 + Z_{12}I_2 + u_1 \\
&\text{or} \\
I_2 &= Y_{21}U_1 + Y_{22}U_2 + i_2 \\
U_2 &= Z_{21}I_1 + Z_{22}I_2 + u_2.
\end{aligned}
\tag{1}
$$

Here i_1 and i_2 (u_1 and u_2) represent the short circuit

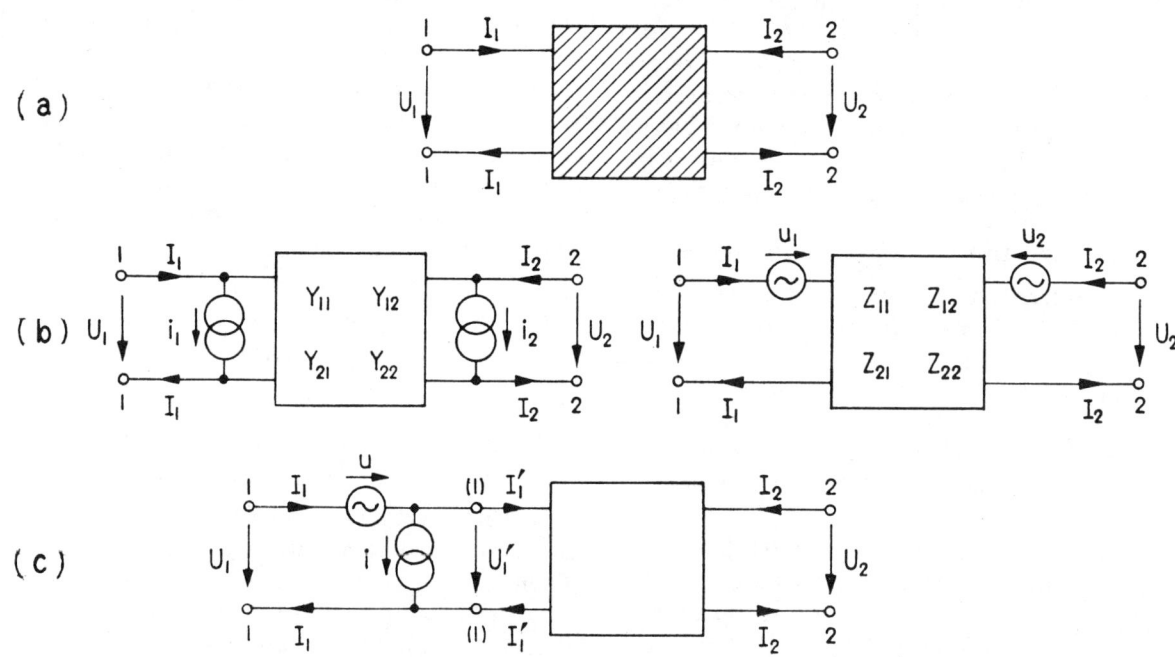

Fig. 1—(a) Fourpole with internal noise sources; (b) equivalent circuit with the outside noise current sources i_1 and i_2 respectively u_1 and u_2; (c) equivalent circuit with noise voltage source u and noise current source i at the input.

FOURPOLE EQUATIONS OF NOISY FOURPOLES

IN FIG. 1(a) is shown the principal scheme of a fourpole with internal noise sources. The electrical behavior of this fourpole will be described by two linear equations between the input voltage and current U_1 and I_1 and the output voltage and current U_2 and I_2. The special form of these equations depends on the network itself, if the Π-admittance matrix or the T-resist-

* Original manuscript received by the IRE August 15, 1955; revised manuscript received November, 1955. Previously published in the "Archiv der elektrischen Übertragung," vol. 9, pp. 117-121, 1955. In extract discussed by the IRE Fall Meeting, Syracuse, October, 1954, and in the Symposium on Fluctuations in Microwave Tubes, New York, November, 1954.
† Telefunken G.m.b.H., Ulm/Donau, Germany.

noise current (open circuit noise voltage) at the input respectively at the output for $U_1 = U_2 = 0$ ($I_1 = I_2 = 0$) caused only by the internal noise sources. Between both noise sources normally a correlation has to be assumed.

The system of Fig. 1 can be represented by the equivalent circuits of Fig. 1(b). In these circuits the noisy four-pole of Fig. 1(a) is replaced by a noise-free but otherwise unchanged fourpole together with the noise current sources i_1 and i_2 (noise voltage sources u_1 and u_2) with an inner infinite (zero) impedance.

In order to characterize the noise qualities of a fourpole it is more convenient to use only noise sources preceding the noise-free fourpole. This is possible by using the chain matrix

$$I_1 = AU_2 + BI_2 + i,$$
$$U_1 = CU_2 + DI_2 + u. \qquad (2)$$

All internal noise sources then will be represented at the input side by a noise current source i and a noise voltage source u, as shown in the equivalent circuit of Fig. 1(c). It consists of the noise-free fourpole between the points $(1)(1)$ and $2\ 2$ and a preceding noise fourpole between the points $1\ 1$ and $(1)(1)$.

$$u = -i_2/Y_{21}$$
$$i = i_1 + uY_{11} = i_1 - i_2(Y_{11}/Y_{21})$$
or
$$u = u_1 + iZ_{11} = u_1 - u_2(Z_{11}/Z_{21})$$
$$i = -u_2/Z_{21}. \qquad (4)$$

As equivalent of a noise current source i_2 (noise voltage source u_2) at the output, therefore, a voltage source u (current source i) and an additional current source $-i_2 Y_{11}/Y_{21}$ (voltage source $-u_2 Z_{11}/Z_{21}$) are necessary at the input.

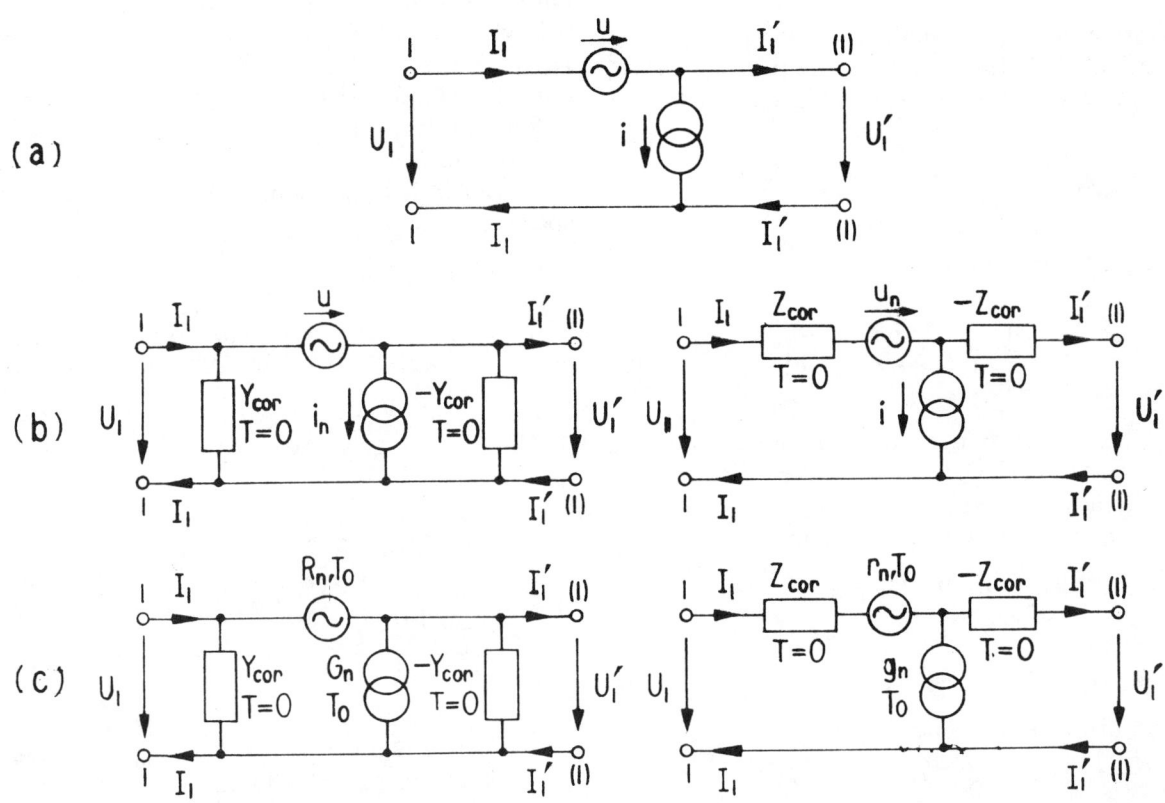

Fig. 2—(a) Noise fourpole with correlated noise sources u and i; (b) noise fourpole with correlation admittance Y_{cor} (correlation impedance Z_{cor}) and uncorrelated noise voltage source u, noise current source i_n (u_n and i); (c) noise fourpole with correlation admittance Y_{cor} (correlation impedance Z_{cor}) and uncorrelated noise sources R_n and G_n (r_n and g_n).

With the current i_1' and the voltage U_1' at the input terminals $(1)(1)$ of the noise-free fourpole Eq. (2) changes to

$$I_1 = I_1' + i,$$
$$U_1 = U_1' + u. \qquad (2a)$$

By introducing the noise sources i and u in Eq. (1) we get

$$I_1 = Y_{11}(U_1 - u) + Y_{12}U_2 + i$$
$$I_2 = Y_{21}(U_1 - u) + Y_{22}U_2$$
or
$$U_1 = Z_{11}(I_1 - i) + Z_{12}I_2 + u$$
$$U_2 = Z_{21}(I_1 - i) + Z_{22}I_2. \qquad (3)$$

A comparison with Eq. (1) gives the transforming formulas for both of the new noise sources

Noise Fourpole and Characteristic Noise Terms

Normally a correlation exists between the two noise sources u and i of the above defined noise fourpole shown in Fig. 2(a). But the noise current i (noise voltage u) can be divided into one part i_n (u_n) not correlated to $u(i)$ and a second part fully correlated to u (i). This second part must be proportional to u (i). As factor of proportionality having the dimension of an admittance (impedance) we introduce the complex correlation admittance $Y_{cor} = G_{cor} + jB_{cor}$ (correlation impedance $Z_{cor} = R_{cor} + jX_{cor}$). We therefore may write

$$i = i_n + uY_{cor}, \quad \text{or} \quad u = u_n + iZ_{cor}. \qquad (5)$$

These new terms Y_{cor} and Z_{cor} corresponding to the well-known correlation coefficient

$$\gamma = \frac{\overline{iu^*}}{\sqrt{\overline{|i|^2}\,\overline{|u|^2}}}$$

are defined by (4, 5, 11)

$$Y_{\text{cor}} = \gamma\sqrt{\frac{\overline{|i|^2}}{\overline{|u|^2}}} = \frac{\overline{iu^*}}{\overline{|u|^2}}$$

$$\text{or}\quad Z_{\text{cor}} = \gamma\sqrt{\frac{\overline{|u|^2}}{\overline{|i|^2}}} = \frac{\overline{i^*u}}{\overline{|i|^2}}. \tag{6}$$

The correlation between i_1 and u (u_1 and i) in (4) which is not identical with the correlation between i and u, can be expressed by another correlation coefficient

$$\alpha = \frac{\overline{i_1 u^*}}{\sqrt{\overline{|i_1|^2}\,\overline{|u|^2}}}\quad\text{resp.}\quad\frac{\overline{u_1 i^*}}{\sqrt{\overline{|u_1|^2}\,\overline{|i|^2}}}.$$

With it we get the relations

$$Y_{\text{cor}} = Y_{11} - \alpha\sqrt{\frac{\overline{|i_1|^2}}{\overline{|u|^2}}}\ \text{or}\ Z_{\text{cor}} = Z_{11} - \alpha\sqrt{\frac{\overline{|u_1|^2}}{\overline{|i|^2}}}. \tag{6a}$$

Therefore, the correlation between i and u and also Y_{cor} (Z_{cor}) can be zero even if $\alpha\neq 0$ and a finite correlation exists between i_1 and i_2 (u_1 and u_2). For i_1 (u_1) uncorrelated to i_2 (u_2) is $\alpha = 0$ and we obtain

$$i_n = i_1\quad\text{or}\quad u_n = u_1 \tag{5a}$$

and

$$Y_{\text{cor}} = Y_{11}\quad\text{or}\quad Z_{\text{cor}} = Z_{11}. \tag{6b}$$

Introducing (5) into (2a) we get

$$\begin{array}{ll}I_1 = I_1' + i_n + uY_{\text{cor}} & I_1 = I_1' + i\\ & \text{or}\\ U_1 = U_1' + u & U_1 = U^1 + u_n + iZ_{\text{cor}}\end{array} \tag{7}$$

and further

$$\begin{array}{l}I_1 = I_1' + i_n + U_1 Y_{\text{cor}} - U_1' Y_{\text{cor}}\\ U_1 = U_1' + u\\ \text{or}\ I_1 = I_1' + i\\ U_1 = U_1' + u_n + I_1 Z_{\text{cor}} - I_1' Z_{\text{cor}}.\end{array} \tag{8}$$

Eq. (8) for the noise fourpole can be realized by the equivalent networks of Fig. 2(b). They only consist of the uncorrelated noise sources u and i_n (u_n and i) and the correlation admittance $+Y_{\text{cor}}$ (correlation impedance $+Z_{\text{cor}}$) at the input side and the correlation admittance $-Y_{\text{cor}}$ (correlation impedance $-Z_{\text{cor}}$) at the output side. Both admittances (impedances) are noise-free and have, therefore, the noise temperature $T = 0$. Using the well-known Nyquist formulas

$$\begin{array}{ll}\overline{|u|^2} = 4kT_0\Delta f\,R_n & \overline{|u_n|^2} = 4kT_0\Delta f\,r_n\\ & \text{or} \\ \overline{|i_n|^2} = 4kT_0\Delta f\,G_n & \overline{|i|^2} = 4kT_0\Delta f\,g_n,\end{array} \tag{9}$$

we express the noise currents and voltages by the characteristic noise terms:

equivalent noise resistance R_n or r_n,
equivalent noise conductance G_n or g_n.

So we get the equivalent networks of Fig. 2(c) for the noise fourpoles. They describe completely the noise behavior of the whole network by the three terms R_n, G_n, and Y_{cor} (r_n, g_n, and Z_{cor}). As Y_{cor} (Z_{cor}) is complex four real characteristic noise terms are needed in fact.

Between the characteristic noise terms of the Π-matrix and those of the T-matrix the following transformation rules exist similar as they exist for fourpole coefficients

$$g_n = G_n + R_n|Y_{\text{cor}}|^2 \qquad R_n = r_n + g_n|Z_{\text{cor}}|^2$$

$$r_n = \frac{G_n}{|Y_{\text{cor}}|^2 + (G_n/R_n)}\quad\text{or}\quad G_n = \frac{r_n}{|Z_{\text{cor}}|^2 + (r_n/g_n)}$$

$$Z_{\text{cor}} = \frac{Y_{\text{cor}}^*}{|Y_{\text{cor}}|^2 + (G_n/R_n)}\qquad Y_{\text{cor}} = \frac{Z_{\text{cor}}^*}{|Z_{\text{cor}}|^2 + (r_n/g_n)}.$$

(a)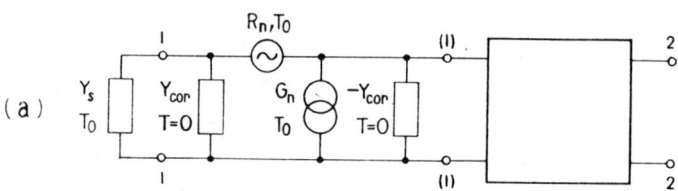

(b)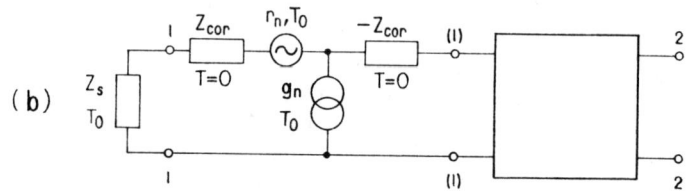

Fig. 3—Equivalent circuit of Fig. 2(c) together with signal source.

The Total Noise Conductance G_{tot}

In the operation of a fourpole a signal source with the inner admittance $Y_s = G_s + jB_s$ (inner impedance $Z = R + jX$) is connected to the input terminals $1\ 1$, as Fig. 3 shows. In the case of the Π-circuit in Fig. 3(a) the inner conductance G_s of the signal source delivers a noise current inflow i_s in the terminals $1\ 1$ that is uncorrelated to all other noise sources. Therefore, the sum of noise power at the output of the fourpole comes from the signal source as well as from the fourpole. But the whole noise power can be assumed as engendered by a single total equivalent noise current i_{tot} flowing into the input terminals. As all noise sources of the equivalent network of Fig. 3(a) are located at the left side of the terminals $(1)(1)$, the short circuit noise current between $(1)(1)$ must be identical with this total equivalent noise current i_{tot}. It is easy calculated to

$$i_{tot} = i_s + i_n + u(Y_s + Y_{cor}). \quad (11)$$

In (11) each component of i_{tot} is uncorrelated to the other ones. Therefore, the mean square value $\overline{|i_{tot}|^2}$ is equal to the sum of the mean square values of each part

$$\overline{|i_{tot}|^2} = \overline{|i_s|^2} + \overline{|i_n|^2} + \overline{|u|^2}|Y_s + Y_{cor}|^2. \quad (12)$$

Introducing the total noise conductance G_{tot} by the Nyquist formula

$$\overline{|i_{tot}|^2} = 4kT_0\Delta f G_{tot} \quad (13)$$

and using (9) we obtain

$$G_{tot} = G_s + G_n + R_n|Y_s + Y_{cor}|^2$$
$$= G_s + G_n + R_n[(G_s + G_{cor})^2 + (B_s + B_{cor})^2]. \quad (14)$$

and in the case that $G_s \to 0$

$$G_{tot}{}^0 = G_N + R_N[G_{cor}{}^2 + (B_s + B_{cor})^2]. \quad (14a)$$

This expression for the total noise conductance shows that a complete characterization of the noise quality needs again the four values R_n, G_n, G_{cor}, and B_{cor} besides the admittance Y_s. The total noise conductance $G_{tot}{}^0$ determines the noise behavior of the network in a very simple way. It solely consists of admittances, respectively, conductances and resistances which are independent of the bandwidth of the network. We notice that G_{tot} does not depend on the loading admittance at the output of the fourpole. But it depends on account of the term $R_n|Y_s + Y_{cor}|^2$ upon the real part G_s as well as on the imaginary part jB_s of the source admittance.

The function $G_{tot} = f(B_s)$ is a quadratic parabola, symmetrical to the ordinate $B_s = -B_{cor}$ as shown by Fig. 4. The second differential quotient of the parabola is $2R_n$. The minimum of G_{tot} at the vertex of the parabola is equal to

$$G_{tot\,min} = G_s + G_n + R_n(G_s + G_{cor})^2 \quad (15)$$

respectively for $G_s \to 0$

$$G_{tot}{}^0{}_{min} = G_n + R_N G_{cor}{}^2. \quad (15a)$$

Analog relations are valid for the dual T-network, as shown in Fig. 3(b). All noise sources can be replaced by the total equivalent noise resistance

$$R_{tot} = R_s + r_n + g_n|Z_s + Z_{cor}|^2. \quad (16)$$

Experimental Determination of the Characteristic Noise Values

The experimental methods to determine the noise current sources f.e. using a noise diode are well known. If G_{tot} is measured as function of the source susceptance

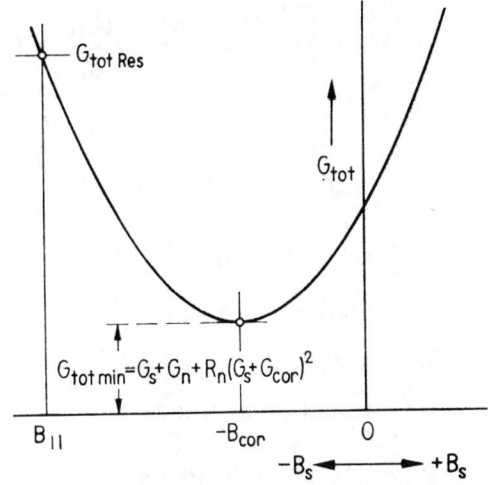

Fig. 4—Total noise admittance G_{tot} as function of the signal source susceptance B_s.

B_s we find the value of B_{cor} by the tuning condition for $G_{tot\,min}$. If $G_{tot\,min} - G_s$ is plotted as function of G_s we find corresponding to (14) a quadratic parabola with the second differential quotient equal to $2R_n$ as shown in Fig. 5. In the vertex of the parabola we have the ordinate value $G_{tot\,min} - G_s = G_n$ and the abscissa value $G_s = -G_{cor}$. By measuring $G_{tot} = f(B_s)$ and $G_{tot\,min} = f(G_s)$ we therefore find the four values R_n, G_n, G_{cor} and B_{cor}. Because usually no negative values of G_s are available the vertex of the parabola must be found by extrapolation if $G_{cor} > 0$. But on principle positive as well as negative values of G_{cor} and B_{cor} are possible and also measured. The methods to determine the characteristic noise values r_n, g_n, R_{cor} and X_{cor} for the T-circuit are the same on principle.

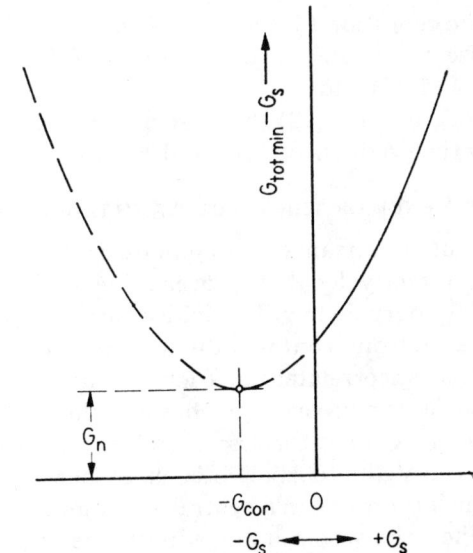

Fig. 5—$G_{tot\,min} - G_s$ as function of the signal source conductance G_s.

Calculation of the Noise Figure

By introducing the total noise conductance G_{tot} (14) into the well-known definition of the excess noise figure [3, 10, 11]

$$F_z = F - 1 = \frac{\overline{|i_{tot}|^2} - \overline{|i_s|^2}}{\overline{|i_s|^2}} = \frac{G_{tot}}{G_s} - 1 \qquad (17)$$

we obtain

$$F_z = \frac{1}{G_s}(G_n + R_n|Y_s + Y_{cor}|^2). \qquad (18)$$

This function of G_s has its lowest value

$$F_{z\,opt} = 2R_n(G_{cor} + G_{s\,opt}) = 2(R_n G_{cor} + \sqrt{R_n G_{tot}^0})$$
$$= 2[R_n G_{cor} + \sqrt{R_n G_n + (R_n G_{cor})^2 + R_n^2(B_s + B_{cor})^2}] \qquad (19)$$

for the optimal internal conductance

$$G_{s\,opt} = \sqrt{\frac{G_n}{R_n} + |jB_s + Y_{cor}|^2} = \sqrt{\frac{G_{tot}^0}{R_n}} \qquad (20)$$

of the signal source. This condition is called *noise matching*. As shown by (17) and (18) the values $F_{z\,opt}$ and $G_{s\,opt}$ depend on the tuning of the signal source. If we choose

$$B_s + B_{cor} = 0 \qquad (21)$$

which condition is independent of the fourpole's input susceptance B_{11} the excess noise figure becomes its absolute minimum

$$F_{z\,min} = 2R_n(G_{cor} + G_{s\,min}) = 2(R_n G_{cor} \sqrt{R_n G_{tot}^0{}_{min}})$$
$$= 2[R_n G_{cor} + \sqrt{R_n G_n + (R_n G_{cor})^2}] \qquad (22)$$

for the corresponding signal source admittance

$$G_{s\,min} = \sqrt{\frac{G_n}{R_n} + G_{cor}^2} = \frac{G_{tot}^0{}_{min}}{R_n} \qquad (23)$$

We call the condition (21) *noise tuning*. To get the minimum noise figure $F_{z\,min}$ the conditions (20) for *noise matching* and (21) for *noise tuning* therefore must be fulfilled together. By (22) the noise figure is solely represented by the products of $R_n G_n$ and $R_n G_{cor}$.

Influence of the Input Admittance Y_{11}

The input admittance Y_{11} composed of several admittances directly located between the terminals *1 1* of the noisy fourpole may be divided into two principal parts. The first one contains the admittances with the noise power uncorrelated to each of the other noise sources, while the second one contains these other admittances being more or less correlated to the inner noise sources of the fourpole. Let us consider f.e. an hf amplifier using a triode in a neutralized common cathode circuit. Then the whole input admittance Y_{11} consists of the admittance $Y_c = G_c + jB_c$ of the resonance circuit between grid and cathode including the cold input admittance of the tube and delivering uncorrelated noise power and further of the electronic input admittance $Y_{el} = G_{el} + j\omega\Delta C_g$ being closely related to the noise of the electron flow inside of the tube.

In the preceding paragraphs both of these principal parts of noise sources were concentrated into a single noise fourpole to get simple expressions for the noise figure. But this procedure has the decisive disadvantage of preventing a separate discussion of the mentioned two parts of noise, so that G_n and Y_{cor} depend on the noise of the input circuit as well as of the inner noise sources f.e. the electron flow, while R_n is only influenced by the latter one.

To get a complete and separate information on the influence of both these parts of noise sources we therefore propose to transfer the admittance Y_c with uncorrelated noise really located inside of the noisy fourpole to the outside of its terminals *1 1* that is parallel to the admittance Y_s of the signal source. To do this outgrouping of circuit noise (see Table I) it is only necessary to introduce into all equations of the sections "The Total Noise Conductance G_{tot}" and "Experimental Determination of the Characteristic Noise Values".

TABLE I

instead of	the new terms
i_s	$i_y = i_s + i_c$
$\overline{\|i_s\|^2}$	$\overline{\|i_y\|^2} = \overline{\|i_s\|^2} + \overline{\|i_c\|^2} = 4kT_0\Delta f(G_s + G_c)$
$Y_s = G_s = jB_s$	$Y = G + jB = Y_s + Y_c = (G_s + G_c) + i(B_s + B_c)$
G_s	$G = G_s + G_c$
$G_s \to 0$	$G_s + G_c \to 0$
B_s	$B = B_s + B_c$

It is easy to prove that G_{tot} resp. G_{tot}^0 is unchanged by this transformation while on the other hand the characteristics G_n and Y_{cor} are changed in quantity and physical interpretation [2]. In the above discussed example of an hf amplifier the new terms G_n and Y_{cor} now represent the noise behavior of the electron flow only, while the influence of the resonance circuit including the cold input admittance of the tube is represented by the admittance Y_c.

The Noise Figure with Separated Y_c

To get the influence of Y_c on the noise figure we have to use in (17) the expression of G_{tot} obtained by introducing the new terms given by (24) into (14). So we find

$$F_z = \frac{1}{G_s}(G_c + G_n + R_n|Y_s + Y_c + Y_{cor}|^2) \qquad (18a)$$

and therefore for *noise matching*

$$F_{z\,opt} = 2R_n(G_c + G_{cor} + G_{s\,opt})$$
$$= 2[R_n(G_c + G_{cor}) + \sqrt{R_n(G_c + G_n) + R_n^2(G_c + G_{cor})^2 + R_n^2(B_s + B_c + B_{cor})^2}] \qquad (19a)$$

and in the case that $G_c \to 0$

$$F_{z\,opt} = 2(R_n G_{cor} + \sqrt{R_N G_{tot}^0})$$
$$= 2[R_n G_{cor} + \sqrt{R_n[G_n + R_n G_{cor}^2 + R_n(B_s + B_c + B_{cor})^2]}], \quad (19b)$$

with the optimal source conductance

$$G_{s\,opt} = \sqrt{\frac{G_c + G_n}{R_n} + |jB_s + Y_c + Y_{cor}|^2} \quad (20a)$$

$$G_{s\,opt} = \sqrt{\frac{G_{tot}^0}{R_n}} \quad (20b)$$

For *noise tuning* the tuning condition

$$B_s + B_c + B_{cor} = 0 \quad (21a)$$

is valid. We obtain

$$F_{z\,min} = 2R_n(G_c + G_{cor} + G_{s\,min})$$
$$= 2[R_n(G_c + G_{cor})$$
$$+ \sqrt{R_n(G_c + G_n) + R_n^2(G_c + G_{cor})^2}] \quad (22a)$$

respectively for $G_c \to 0$

$$F_{z\,min} = 2(R_n C_{cor} + \sqrt{R_n G_{tot}^0{}_{min}})$$
$$= 2[R_n G_{cor} + \sqrt{R_n(G_n + R_n G_{cor}^2)}] \quad (22b)$$

with

$$G_{s\,min} = \sqrt{\frac{G_c + G_n}{R_n} + (G_c + G_{cor})^2}. \quad (23a)$$

with

$$G_{s\,min} = \frac{\sqrt{G_{tot}^0{}_{min}}}{R_n}. \quad (23b)$$

If $G_c \ll G_n, G_{cor}$ (22a) and (23a) are identical with (22) and (23). If further $G_{cor} = 0$ we get

$$F_{z\,min} \to 2\sqrt{R_n G_n}, \quad (24)$$
$$G_{s\,min} \to \sqrt{G_n/R_n}. \quad (25)$$

Both expressions only imply the noise terms R_n and G_n. They are valid for triodes in uhf region.

For triodes at low frequencies $G_c \gg G_n, G_{cor}$ (22a) and (23a) simplify to

$$F_{z\,min} \to 2[R_n G_c + \sqrt{R_n G_c + (R_n G_c)^2}], \quad (26)$$
$$G_{s\,min} \to \sqrt{G_c^2 + (G_c/R_n)} \quad (27)$$

depending on G_c and R_n only.

To study the excess noise figure as function of the noise matching of the signal source it is useful to transform (18a) into the form of a circle. For that purpose we introduce the expressions $F_{z\,min}$ and $G_{s\,min}$ by (22a) and (23a) and receive

$$F_z = F_{z\,min} + R_n G_{s\,min}\left(m + \frac{1}{m} - 2\right) \quad (28)$$

with

$$m + \frac{1}{m} = \frac{G_{s\,min}}{G_s}\left[1 + \left(\frac{G_s}{G_{s\,min}}\right)^2 + \left(\frac{B_s + B_c + B_{cor}}{G_{s\,min}}\right)^2\right]. \quad (29)$$

The coefficient m only depends on the quotients $G_s/G_{s\,min}$, respectively on

$$\frac{B_s + B_c + B_{cor}}{G_{s\,min}}$$

and represents the standing wave ratio U_{max}/U_{min} of a transmission line with a wave-resistance $Z = 1/G_{s\,min}$ being connected to the signal source with the inner admittance $G_s + j(B_s + B_c + B_{cor})$.

In the complex plane with $G_s/G_{s\,min}$ as abscissa and $(B_s + B_c + B_{cor})G_{s\,min}$ as ordinate the curves of constant m and therefore also constant noise figure F_z are circles as shown in the well-known matching diagram of Fig. 6. For $G_s = G_{s\,min}$ and $B_s + B_c + B_{cor} = 0$ we get $m = 1$ and therefore $F_z = F_{z\,min}$. The center point of minimum noise figure was already introduced as fulfilling the conditions of noise matching and noise tuning. The points of every circle with extreme values of $B_s + B_c + B_{cor}$ are fulfilling the condition of noise matching (20a). The locations of this condition are given by the dashed hyperbola

$$\left(\frac{G_{s\,opt}}{G_{s\,min}}\right)^2 - \left(\frac{B_s + B_c + B_{cor}}{G_{s\,min}}\right)^2 = 1. \quad (30)$$

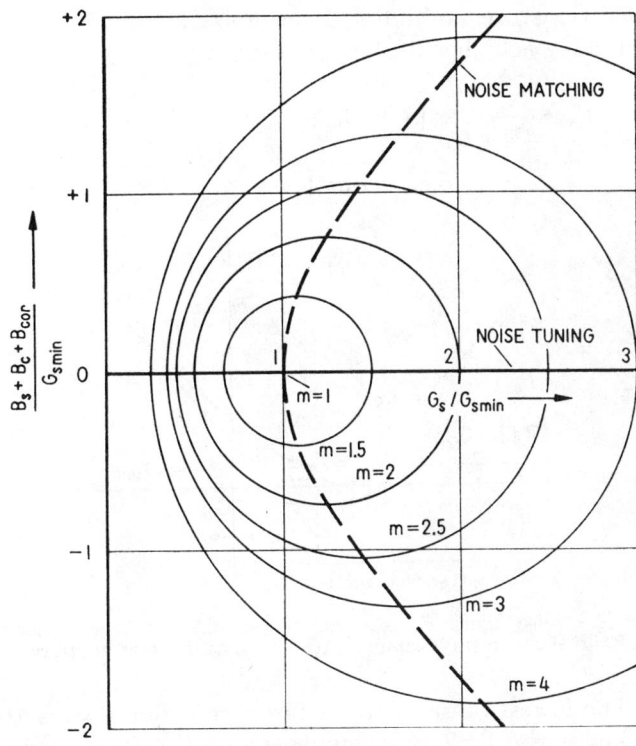

Fig. 6—Noise matching diagram between the signal source and the noise fourpole. The circles are curves for constant noise figure.

Of special interest is the influence of G_{cor} upon the magnitude of $F_{z\,opt}$ and $F_{z\,min}$ along this hyperbola for noise matching. In Fig. 7 $F_{z\,opt}$ corresponding to (19a) is given as function of $B = B_s + B_c$ for a negative value of G_{cor}. This curve is again a quadratic hyperbola symmetrical to $B = -B_{cor}$ and a slope of the asymptotes equal to $\pm R_n$. In the vertex the curve is crossing the center point of Fig. 6 with $F_{z\, opt} = F_{z\, min}$. The point of intersection of the asymptotes has a negative value of $F_{z\, opt} = -2R_n(G_c + G_{cor})$ in our example. The distance

$$2\sqrt{R_n(G_c + G_{cor}) + R_n^2(G_c + G_{cor})^2}$$

between this point and the vertex of the hyperbola is always positive and greater than $2R_n(G_c + G_{cor})$, so that $F_{z\, min}$ remains positive. Positive values of G_{cor} shift the hyperbola to higher values of $F_{z\, opt}$, negative G_{cor} to lower values. For $G_c + G_{cor} \to -\infty$ the vertex and, therefore, $F_{z\, min}$ is going to zero.

We have to notice that the condition for *noise tuning* given by (21a) is independent of the impedance $Y_{11} = G_{11} + jB_{11}$ of the fourpole. For *power matching* the conditions

$$Y_s^* = Y_c + Y_{11} \qquad (31)$$

or

$$B_s + B_c + B_{cor} = 0$$
$$G_s = G_c + G_{11} \qquad (31a)$$

are valid. Therefore, power matching is only identical with noise matching and noise tuning if $G_c + G_{11} = G_{s\, min}$ and $B_{cor} = B_{11}$. In this case the diagram of Fig. 6 is identical with the diagram for power matching. If these conditions are not fulfilled the noise figure for power matching is higher than $F_{z\, min}$ and given by

$$F_z = \frac{1}{G_c + G_{11}}\left[G_c + G_n + R_n(2G_c + G_{11} + G_{cor})^2 + R_n(B_{cor} - B_{11})^2\right]. \qquad (32)$$

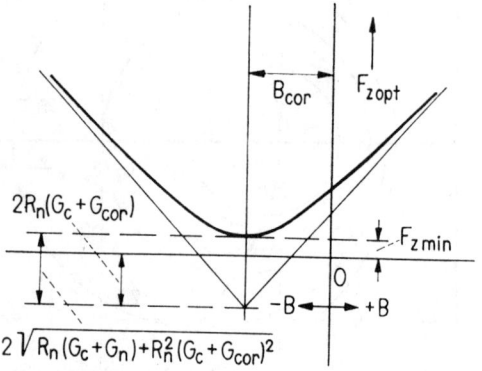

Fig. 7—Noise figure $F_{z\, opt}$ as function of the source susceptance $B = B_s + B_c$. In this example is $(G_c + G_{cor})$ assumed as negative.

The lowest noise figure in this case is not always attained with $G_c = 0$ but sometimes with $G_c > 0$ [5]. Using internal feedback inside of the noisy fourpole the conditions for noise matching, noise tuning and power matching can often be combined. Then the absolute minimum value $F_{z\, min}$ of the excess noise figure occurs together with the reflection-free connection to the signal source.

Chain Connection of Noisy Fourpoles

The noise figure for chain connection of two noisy fourpoles can be calculated by the same principal method [7]. The result

$$F_z = F_z^I + \frac{F_z^{II}}{V_L} \qquad (33)$$

is the same as already given by Friis [3]. F_z^I and F_z^{II} are the noise figures of the two fourpoles alone and V_L the available power gain of the first fourpole. The total noise figure of n equal fourpoles chained together each of them with the noise figure F_z^I is

$$F_z^n = F_z^I \frac{1 - (1/V_L)^n}{1 - (1/V_L)}. \qquad (34)$$

For $n \to \infty$ then results

$$F_z^\infty = F_z^I \frac{V_L}{V_L - 1}. \qquad (35)$$

Fig. 8 shows F_z^n/F_z^I as function of V_L for different numbers of n. Eqs. (33) to (35) show very clearly that the noise figure alone is insufficient for full determination of the quality of a noisy fourpole. But the term F_z^∞ given by (35) seems especially adequate as figure of merit.

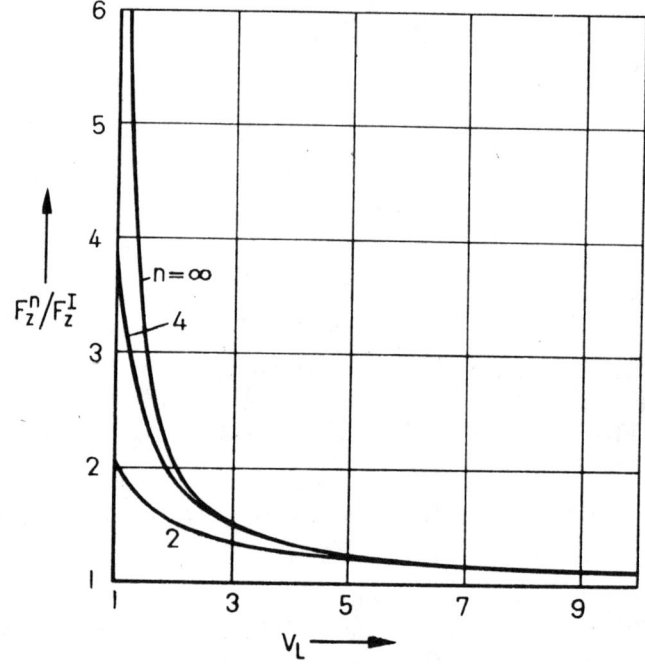

Fig. 8—Relative noise figure of a n-cascaded amplifier as function of the available gain V_L.

Mixer Circuits

The noise properties of mixer stages can also be described by the noise fourpole. Then the short circuit noise currents i_1 and i_2 in Eq. (1) depend on the period of the oscillator frequency ω_0. As i_1 concerns to the input side with the high frequency ω_h and i_2 to the intermediate frequency ω_i in most mixer currents i_1 is practically uncorrelated to i_2. Then we obtain

$$G_n = G_1 \tag{36}$$

with G_1 given by the Nyquist equation

$$\overline{|i_1|^2} = 4kT_0 \Delta f G_1$$

and

$$Y_{cor} = Y_{11} \tag{37}$$

where Y_{11} is the mean value of the input admittance for the high frequency ω_h middled over the period of the oscillator frequency ω_o.

APPLICATION OF THE METHOD

The application of the above considerations on electron tubes was already shown in earlier papers [5, 6, 8]. For triodes at higher frequencies with $C_{ga} = C_{ka} = 0$ respectively in neutralized circuits the short circuit noise current i_i is identical with the induced grid noise current i_g at the input side and i_2 with the space charge suppressed shot noise i_a at the output side. R_n is identical with the well-known equivalent noise resistance R_{eq} and independent of frequency in first approximation. The correlation admittance Y_{cor} as measure of the correlation between the input noise sources i and u is found to be zero in first mostly sufficient approximation while the equivalent noise conductance $G_n > 0$ up to high frequencies in sufficient approximation is proportional to ω^2 (ω = angular frequency) [5, 6]. Only for full correlation between i_g and i_a, G_n would be zero. The two parameters R_n and G_n alone fully prescribe the noise behavior of the electron stream in neutralized triodes.[1]

The minimum noise figure is then given by (25) if G_s is negligibly small. We call this lowest possible value the "electronic noise figure" of the triode. As R_n is independent of frequency while G_n is proportional to the square of the frequency it is possible to calculate this electronic noise figure with help of the low frequency value of R_n and the magnitude of G_n measured f.e. in the 100 mc band. So calculated values agree very well with measured values up to very high frequencies [6].

A feedback over C_{ga} decreases the magnitude of R_n but does not change G_n and gives a finite value of Y_{cor} with a negative conductance G_{cor}. Therefore, the noise figure is lowered by this feedback [2, 8]. In screen grid tubes R_n is again identical with R_{eq}. On behalf of the additional partition noise the admittance G_n is larger than in the comparable triode system. G_n is proportional to ω^n where the exponent n starting with the value 2 for low frequencies increases with frequency. Y_{cor} is no more zero but gets a positive conductance G_{cor} proportional to ω^2 and also a positive susceptance B_{cor}.

This method is also very useful in case of transistors [8, 9] and traveling-wave tubes [1]. It does not only prescribe the noise behavior of the amplifying elements but also gives the possibility of conclusions concerning the location and properties of the noise sources inside of their equivalent networks.

BIBLIOGRAPHY

[1] Bauer, H., and Rothe, H., "Der äquivalente Rauschvierpo als Weilenvierpol." *Archiv der elektrischen Übertragung*, Vol. 10 (1956) in the press.
[2] Dahlke, W., "Transformationsregeln für rauschende Vierpole." *Archiv der elektrischen Übertragung*, Vol. 9 (September, 1955), pp. 391–401.
[3] Friis, H. T., "Noise Figures of Radio Receivers." PROCEEDINGS OF THE IRE, Vol. 32 (July, 1944), pp. 419–422.
[4] Montgomery, H. C., "Transistor Noise in Circuit Applications." PROCEEDINGS OF THE IRE, Vol. 40 (November, 1952), pp. 1461–1471.
[5] Rothe, H., "Die Grenzempfindlichkeit von Verstärkerröhren. Teil III: Äquivalenter Rauschleitwert und Geräuschzahl." *Archiv der elektrischen Übertragung*, Vol. 8 (May, 1954), pp. 201–212.
[6] Rothe, H., "Röhren für Ein- und Ausgangsstufen im 4000-MHz-Gebiet." *Fernmeldetechnische Zeitschrift*, Vol. 7 (October, 1954), pp. 532–539.
[7] Rothe, H., and Dahlke, W., "Theorie rauschender Vierpole." *Archiv der elektrischen Übertragung*, Vol. 9 (March, 1955), pp. 117–121.
[8] Rothe, H., "Die Theorie rauschender Vierpole und ihre Anwendung." *Nachrichtentechnische Fortschritte*, Heft 2 (1955), pp. 24–36.
[9] Schubert, J., "Rauscheigenschaften der Transistoren." Lecture held in the Symposium "Rauschen" ("Noise") of the "Nachrichtentechnische Gesellschaft" in Munich, Germany, (April, 1955).
[10] Standards on Electron Devices: "Methods of Measuring Noise." PROCEEDINGS OF THE IRE, Vol. 41 (July, 1953), pp. 890–896.
[11] van der Ziel, A., "Noise." New York, Prentice-Hall, Inc., 1954

---

[1] This is valid if the circuit admittance $Y_e = G_e + jB_e$ including the "cold" input admittance of the tube are considered as grouped outside of the fourpole parallel to the signal admittance as described in the section "Influence of the Input Admittance Y_{11}." If Y_e remains located inside of the fourpole the noise current i_1 is including the noise current inflow belonging to G_e. Then we get $G_n' = G_n + G_e$ and $Y_{cor}' = Y_{cor} + Y_e$ while $R_n' = R_n$ is unchanged [2]. G_n' and Y_{cor}', therefore, represent no more alone the noise behavior of the electron flow but also the quality of the input circuit.

Available Power Gain, Noise Figure, and Noise Measure of Two-Ports and Their Graphical Representations

H. FUKUI

Abstract—The expressions for available power gain and noise measure of linear two-ports are introduced in terms of the two-port parameters and the gain and the noise parameters, respectively. Their graphical representations on the source admittance plane with rectangular coordinates are also shown. Furthermore, it is shown that the behavior of available power gain, noise figure, and noise measure can be represented on the Smith-chart or the complex reflection coefficient plane of the source admittance. It is more convenient to investigate the gain and noise performance of amplifiers over a wide range of source admittance in this representation than with rectangular coordinates. As an example of the graphical representation the gain and noise performance of a microwave transistor is illustrated on the Smith-chart.

I. Introduction

IT IS WELL KNOWN that the noise figure of linear twoports is a function of the source admittance [1]. The noise figure of a linear two-port driven by a signal source with an admittance, $Y_s = G_s + jB_s$, can be expressed as

$$F = F_{min} + \frac{R_{ef}}{G_s}\{(G_s - G_{of})^2 + (B_s - B_{of})^2\} \quad (1)$$

where F_{min} is the minimum noise figure, R_{ef} is a parameter with units of resistance, and G_{of} and B_{of} are the particular source conductance and susceptance, respectively, which produce F_{min}. Thus, the noise figure characteristics of linear two-ports can be completely described by the four noise parameters, F_{min}, R_{ef}, G_{of}, and B_{of}. Furthermore, a graphical representation of the noise figure on a rectangular coordinate system of source admittance has been given by Rothe and Dahlke [2].

In recent years, the noise measure has been proposed by Haus and Adler [3] as a more significant parameter than the noise figure when considering cascaded two-ports. The noise measure M is defined as [4]

$$M = \frac{F - 1}{1 - \frac{1}{G_a}} \quad (2)$$

where G_a is the available power gain of the two-port.

Since G_a and F are functions of source admittance Y_s, M will also be a function of Y_s. However, this situation has not previously been made clear. This paper will consider the problem and show that constant-M loci can be mapped in the source admittance plane. Another

Manuscript received April 19, 1965; revised November 3, 1965.
The author is with Bell Telephone Laboratories, Inc., Murray Hill, N. J.

purpose of this paper is to demonstrate how the constant G_a, F, and M loci appear on the Smith-chart representation.

II. Available Power Gain

A. The Available Power Gain of Two-ports

The available power gain G_a of the two-port shown in Fig. 1 is defined as the ratio of the available power at

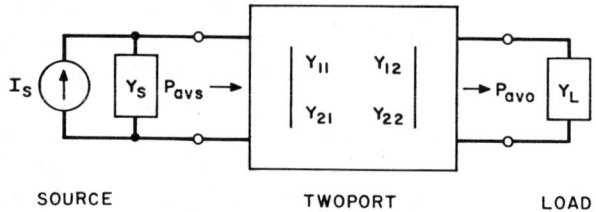

Fig. 1. Available power gain of a twoport.

the output P_{avo} to the available power from the source P_{avs}, i.e.,

$$G_a = \frac{P_{avo}}{P_{avs}}. \quad (3)$$

G_a is a function of the two-port parameters, such as y-parameters, and the source admittance Y_s.

Using the same procedure given by Linvill and Gibbons [5], G_a is expressed as follows.

$$G_a = \frac{|y_{21}|^2 G_s}{g_{22}|y_{11} + Y_s|^2 - \operatorname{Re}[y_{12}y_{21}(y_{11} + Y_s)^*]}. \quad (4)$$

B. Alternate Expression for Available Power Gain

From the expression for G_a its maximum value, $G_{a\max}$, is derived for the particular source admittance $Y_{og} = G_{og} + jB_{og}$ as follows.

$$G_{a\max} = \left|\frac{y_{21}}{y_{12}}\right| \frac{1}{k + \sqrt{k^2 - 1}} \quad (5)$$

$$G_{og} = \frac{|y_{12}y_{21}|}{2g_{22}}\sqrt{k^2 - 1} \quad (6)$$

$$B_{og} = -b_{11} + \frac{\operatorname{Im}(y_{12}y_{21})}{2g_{22}} \quad (7)$$

where

$$k = \frac{2g_{11}g_{22} - \operatorname{Re}(y_{12}y_{21})}{|y_{12}y_{21}|}. \quad (8)$$

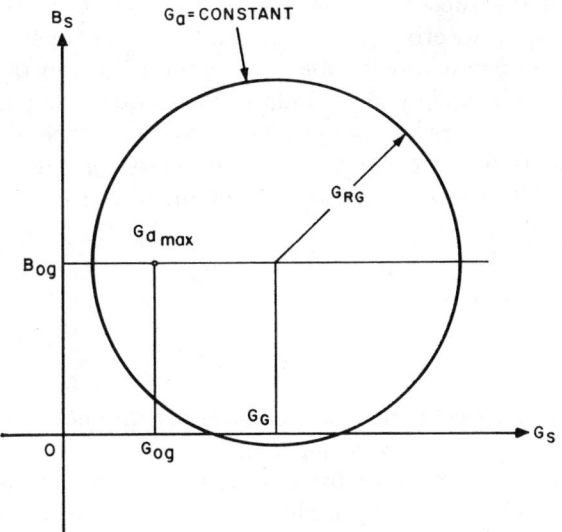

Fig. 2. Constant available power gain locus on rectangular source admittance plane.

Now (4) is rewritten using a set of gain parameters, $G_{a\max}$, G_{og}, B_{og}, and a parameter R_{eg}:

$$\frac{1}{G_a} = \frac{1}{G_{a\max}} + \frac{R_{eg}}{G_s}[(G_s - G_{og})^2 + (B_s - B_{og})^2] \quad (9)$$

where

$$R_{eg} = \frac{g_{22}}{|y_{21}|^2}. \quad (10)$$

It should be noticed that (9) is quite similar to the expression for noise figure given in (1). This suggests that a graphical representation similar to that for the noise figure can be expected for the available power gain.

C. A Graphical Representation of Available Power Gain on the Rectangular Source Admittance Plane

Rearranging (9), the following equation can be derived:

$$(G_s - G_G)^2 + (B_s - B_G)^2 = G_{RG}^2 \quad (11)$$

where

$$G_G = G_{og} + \frac{1}{2R_{eg}}\left(\frac{1}{G_a} - \frac{1}{G_{a\max}}\right) \quad (12)$$

$$B_G = B_{og} \quad (13)$$

$$G_{RG} = \left[\frac{G_{og}}{R_{eg}}\left(\frac{1}{G_a} - \frac{1}{G_{a\max}}\right) + \frac{1}{4R_{eg}^2}\left(\frac{1}{G_a} - \frac{1}{G_{a\max}}\right)^2\right]^{1/2}. \quad (14)$$

Equation (11) represents a family of circles on the rectangular coordinates with G_s as the abscissa and B_s as the ordinate, as shown in Fig. 2. Each circle represents a constant G_a locus and has its center at the point (G_G, B_G) and a radius of G_{RG}. G_G and G_{RG} depend upon the value of G_a, but B_G is independent of G_a. Therefore, the locus of the center points for constant G_a becomes a straight line parallel to and at a distance of B_{og} from the abscissa.

III. Noise Measure

A. An Expression for the Noise Measure in Terms of Two-port Parameters and Source Admittance

Substituting (1) and (9) into (2), the noise measure can be expressed in terms of the gain parameters, noise parameters, and source admittance as follows.

$$M = \frac{F_{\min} - 1 + \frac{R_{ef}}{G_s}[(G_s - G_{of})^2 + (B_s - B_{of})^2]}{1 - \frac{1}{G_{a\max}} - \frac{R_{eg}}{G_s}[(G_s - G_{og})^2 + (B_s - B_{og})^2]}. \quad (15)$$

Rearranging (15), the following equation can be derived,

$$(G_s - G_M)^2 + (B_s - B_M)^2 = G_{RM}^2 \quad (16)$$

where

$$G_M = \frac{M\left(2R_{eg}G_{og} - \frac{1}{G_{a\max}} + 1\right) + 2R_{ef}G_{of} - F_{\min} + 1}{2(MR_{eg} + R_{ef})} \quad (17)$$

$$B_M = \frac{MR_{eg}B_{og} + B_{of}R_{ef}}{MR_{eg} + R_{ef}} \quad (18)$$

$$G_{RM} = \frac{1}{2(MR_{eg} + R_{ef})}\Bigg[\left\{M\left(\frac{1}{G_{a\max}} - 1\right) + F_{\min} - 1\right\}^2$$
$$+ 4(MR_{eg}G_{og} + R_{ef}G_{of})\left\{M\left(\frac{1}{G_{a\max}} - 1\right) + F_{\min} - 1\right\}$$
$$- 4MR_{eg}R_{ef}(|Y_{og}|^2 + |Y_{of}|^2 - 2G_{og}G_{of} - 2B_{og}B_{of})\Bigg]^{1/2} \quad (19)$$

For the condition

$$G_{RM} = 0, \quad (20)$$

M takes on its minimum value $M_{\min}$, i.e.,

$$M_{\min} = \frac{M_2}{M_1}\left[1 + \left(1 - \frac{M_1 M_3}{M_2^2}\right)^{1/2}\right] \quad (21)$$

where

$$M_1 = \left(1 - \frac{1}{G_{a\max}}\right)^2 + 4\left(1 - \frac{1}{G_{a\max}}\right)R_{eg}G_{og} \quad (22)$$

$$M_2 = \left(1 - \frac{1}{G_{a\max}} + 2R_{eg}G_{og}\right)(F_{\min} - 1 - 2R_{ef}G_{of})$$
$$+ 2R_{eg}R_{ef}(|Y_{og}|^2 + |Y_{of}|^2 - 2B_{og}B_{of}) \quad (23)$$

$$M_3 = (F_{\min} - 1)^2 - 4(F_{\min} - 1)R_{ef}G_{of}. \quad (24)$$

The source conductance and susceptance which produce the minimum noise measure are

$$G_{om} = \frac{M_{\min}\left(2R_{eg}G_{og} - \frac{1}{G_{a\max}} + 1\right) + 2R_{ef}G_{fo} - F_{\min} + 1}{2(M_{\min}R_{eg} + R_{ef})} \quad (25)$$

$$B_{om} = \frac{M_{\min}R_{eg}B_{og} + R_{ef}B_{of}}{M_{\min}R_{eg} + R_{ef}}. \quad (26)$$

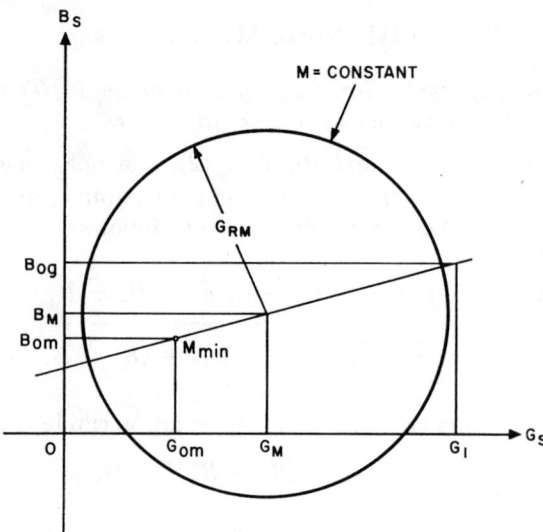

Fig. 3. Constant noise measure locus on rectangular source admittance plane.

B. A Graphical Representation of Noise Measure on the Rectangular Source Admittance Plane

Equation (16) represents a family of circles on the rectangular coordinates with G_s as the abscissa and B_s as the ordinate as shown in Fig. 3. Each circle gives a constant M locus and has its center at the point (G_M, B_M) and a radius of G_{RM}. Since G_M, B_M, and G_{RM} are functions of M, the locus of the center points (G_M, B_M) is not parallel to the abscissa. However, this locus is still given by a straight line through points (G_{om}, B_{om}) and (G_1, B_{og}), corresponding to unity G_a.

IV. BILINEAR TRANSFORMATIONS OF THE GRAPHICAL REPRESENTATIONS

Equations (11) and (16) and a corresponding equation for the noise figure can be rewritten in a form more convenient for representing the behavior of G_a, F, and M on the complex reflection coefficient plane [6], defined as

$$\rho = u + jv = |\rho|\exp(-j\varphi) = \frac{1-y_s}{1+y_s} \quad (27)$$

where

$$y_s = g_s + jb_s = \frac{1}{Y_0}(G_s + jB_s) \quad (28)$$

and Y_0 is the (real) characteristic admittance of the input transmission line. A merit of this plane is that a half-plane with either positive or negative conductance can be represented entirely within a unit circle.

Since ρ is a bilinear function of the complex number y_s, any circle in the Y_s-plane can be transformed into a circle in the ρ-plane. Thus, (11) and (16) and the corresponding equation for the noise figure can be expressed in the form

$$(u - u_j)^2 + (v - v_j)^2 = r_j^2 \quad (29)$$

where the subscript j represents g, f, and m for G_a, F, and M, respectively. (u_j, v_j) and r_j, respectively, give the coordinate points for the center of a constant G_a (or F, M) circle and its radius. The center point is also given by the polar coordinates, $|\rho|$ and φ, in the ρ-plane. Furthermore, the Smith-chart coordinates, g_s and b_s, are related to the rectangular coordinates in the manner [7],

$$g_s = \frac{1-u^2-v^2}{(1+u)^2+v^2} \quad (30)$$

$$b_s = \frac{-2v}{(1+u)^2+v^2}. \quad (31)$$

Therefore, the center point can also be specified in terms of g_s and b_s in the same ρ-plane.

Since φ is constant for all constant G_a and F loci as seen later, the polar coordinates is most convenient for mapping G_a and F in the ρ-plane. On the other hand, φ depends upon M so that a preferable coordinate is no longer offered for drawing constant M loci.

In the polar ρ-plane the center point and the radius of a constant G_a circle are given as follows:

$$|\rho_g| = \frac{[(1-g_{og}^2-b_{og}^2)^2+4b_{og}^2]^{1/2}}{(1+g_{og})^2+b_{og}^2+2\delta_g} \quad (32)$$

$$\varphi_g = \tan^{-1}\left[\frac{2b_{og}}{1-g_{og}^2-b_{og}^2}\right] \quad (33)$$

$$r_g = \frac{2g_{RG}}{(1+g_{og})^2+b_{og}^2+2\delta_g} \quad (34)$$

where

$$\delta_g = \frac{1}{2R_{eg}Y_0}\left(\frac{1}{G_a}-\frac{1}{G_{a\max}}\right) \quad (35)$$

and g_{og}, b_{og}, and g_{RG} are normalized values of G_{og}, B_{og}, and G_{RG} with respect to Y_0. The radial coordinate $|\rho_{og}|$ for $G_{a\max}$ is then given by taking δ_g equal to zero. Figure 4 illustrates the geometrical relations of the above parameters in the ρ-plane.

For the noise figure, completely identical forms to the above are available simply by replacing the subscripts. In this case δ_f should be

$$\delta_f = \frac{F-F_{\min}}{2R_{ef}Y_0}. \quad (36)$$

If the Smith-chart coordinates are employed for the mapping of the noise measure, the center point and the radius of a constant M circle are given by

$$g_m = \frac{(1+g_M)[(1+g_M)g_M+b_M^2-g_{RM}^2]-b_M^2}{(1+g_M)^2+b_M^2} \quad (37)$$

$$b_m = \frac{b_M[(1+g_M)^2+b_M^2-g_{RM}^2]}{(1+g_M)^2+b_M^2} \quad (38)$$

$$r_m = \frac{g_{RM}}{(1+g_M)^2+b_M^2-g_{RM}^2} \quad (39)$$

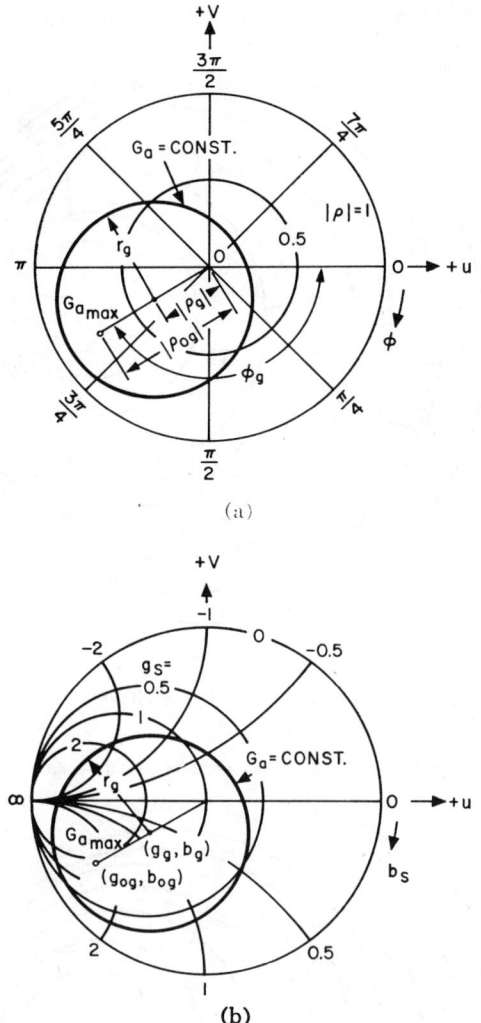

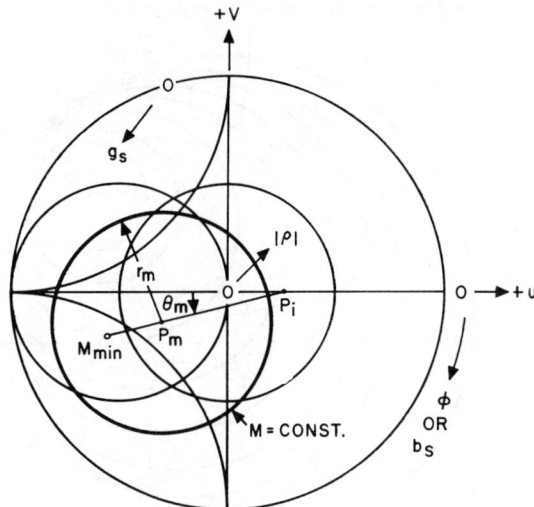

M_{min}: $(|\rho_{om}|, \phi_{om})$ OR (g_{om}, b_{om})
P_i: $(|u_i|, 0)$ OR $(g_i, 0)$
P_m: $(|\rho_m|, \phi_m)$ OR (g_m, b_m)

Fig. 5. Constant noise measure locus on the Smith-chart.

$$g_i = \frac{(1 + g_{om})g_{om} - (\cot\theta_m - b_{om})b_{om}}{1 + g_{om} + b_{om}\cot\theta_m}. \quad (41)$$

θ_m can be determined only by the gain and the noise parameters and is independent of M as seen in (40). For the mapping of constant M circles, the center points can also be given by this line and either g_m or b_m, instead of g_m and b_m. In practice, this line is simply drawn by connecting two representative points, (g_{om}, b_{om}) and (g_1, b_{og}), corresponding to unity G_a. Figure 5 shows the mapping of constant M locus on the Smith-chart where the polar reflection coefficient coordinates also overlap.

V. Exchangeable Power Gain, Extended Noise Figure, and Extended Noise Measure

The noise figure is normally defined in terms of available power gain as given in (1). However, the available power concept leads to difficulties when the input and output admittances of two-ports have negative real parts. Such cases can, in general, be handled in a similar manner by using the exchangeable power and gain concepts in place of available power and gain. This leads to an extended definition of the noise figure [8]. Using the exchangeable power gain G_e and the extended noise figure F_e, an extended noise measure M_e can be defined as [3]

$$M_e = \frac{F_e - 1}{1 - \dfrac{1}{G_e}}. \quad (42)$$

In the above cases, the graphical representations of G_e, F_e, and M_e on the source admittance planes are also available in similar ways. In some cases, minor modifications will be required, for example, use of a negative conductance Smith-chart [9], an expanded Smith-chart outside the unit circle to include both positive and negative conductances, and so on.

Fig. 4. (a) Constant available power gain locus on complex reflection coefficient plane. (b) Constant available power gain locus on the Smith-chart.

where again small letters indicate normalized values of capital letters with respect to Y_0. The coordinates for M_{min} are, of course, designated by g_{om} and b_{om}.

Due to the nature of the bilinear transformation, the locus connecting the centers of the constant-M circles is still a straight line on the ρ-plane. Its angle θ_m and intersecting point g_i with the abscissa are given by

$$\tan\theta_m = \left[\left\{F_{min} - 1 - R_{ef}Y_0\left(\frac{|Y_{of}|^2}{Y_0^2} + \frac{2G_{of}}{Y_0} + 1\right)\right\}R_{eg}B_{og}\right.$$

$$\left. - \left\{\frac{1}{G_{a_{max}}} - 1 - R_{eg}Y_0\left(\frac{|Y_{og}|^2}{Y_0^2} + \frac{2G_{og}}{Y_0} + 1\right)\right\}R_{ef}B_{of}\right]$$

$$\cdot \left[(F_{min} - 1)\left(\frac{|Y_{og}|^2}{Y_0^2} - 1\right)\frac{R_{eg}Y_0}{2} - \left(\frac{1}{G_{a_{max}}} - 1\right)\right.$$

$$\cdot \left(\frac{|Y_{of}|^2}{Y_0^2} - 1\right)\frac{R_{ef}Y_0}{2} + \left\{|Y_{of}|^2\left(\frac{G_{og}}{Y_0} + 1\right)\right.$$

$$\left.\left. - |Y_{og}|^2\left(\frac{G_{of}}{Y_0} + 1\right) + Y_0(G_{of} - G_{og})\right\}R_{eg}R_{ef}\right]^{-1} \quad (40)$$

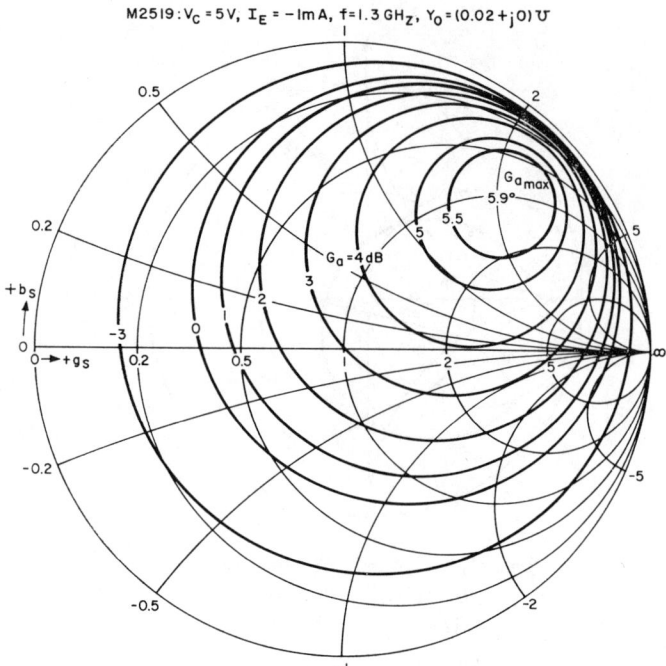

Fig. 6. Available power gain chart.

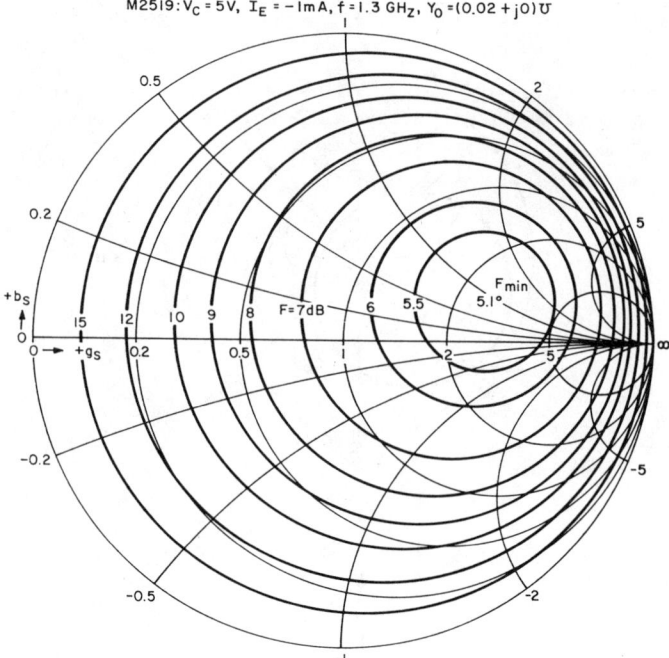

Fig. 7. Noise figure chart.

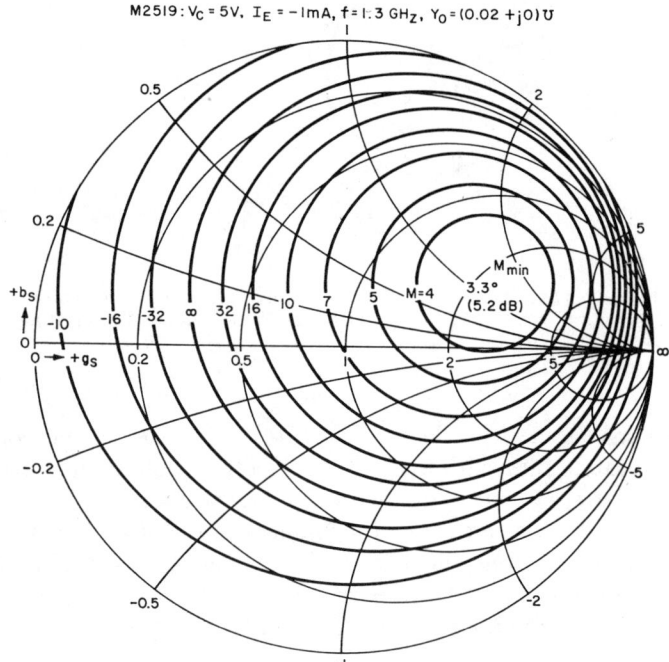

Fig. 8. Noise measure chart.

VI. Examples of Graphical Representations

A. Device Characteristics

A silicon n-p-n double-diffused planar microwave transistor M2519 was used in actual representations of available power gain, noise figure, and noise measure of its equivalent two-port. This transistor was characterized at an emitter current of -1 mA, collector voltage of 5 volts, and frequency of 1.3 GHz. The measurements of the gain and the noise parameters were made using the method recommended by the IRE Standard Committee on Electron Tubes [1], [10]. The gain parameters can also be determined by a rather simple method in which $G_{a_{max}}$, G_{og}, and B_{og} are directly obtained. Using the relation of (12), R_{eg} can then be determined by the aid of another measurement of the available power gain with a given source admittance, for example, $Y_s = (0.02 + j0)$ mho. Another way for determining the gain parameters is to use the measured y-parameters (or any suitable two-port parameters) as seen in (5) through (8) and (10).

Predetermined parameter values are as follows:

$G_{a_{max}}$ = 3.93 (5.9 dB)
R_{eg} = 2.54 ohms
G_{og} = 18.4 millimhos
B_{og} = 44.2 millimhos
F_{min} = 3.25 (5.1 dB)
R_{ef} = 15.6 ohms
G_{of} = 53 millimhos
B_{of} = 20 millimhos

From the above parameters, the following parameters were obtained:

M_{min} = 3.32 (5.2 dB)
G_{om} = 47 millimhos
B_{om} = 30 millimhos.

B. Graphical Representations on the Smith-Chart

The available power gain, noise figure, and noise measure for the same transistor are shown on a Smith-chart representation of source admittance, as shown in Figs. 6, 7, and 8, respectively. From these figures, it can be understood easily and precisely what source admittance is able to give a good performance for the amplifier and how the situation becomes worse as the source admittance differs from the optimum value.

As expected, the locus for infinite M coincides with

that for unity G_a and the coordinate point for M_{min} is located between those for $G_{a_{max}}$ and F_{min}.

VII. Conclusions

In this paper it has been shown that the noise measure of linear two-ports can be expressed in terms of the gain parameters, the noise parameters, and the source admittance. Constant M loci have been represented in the source admittance plane as well as constant G_a and constant F loci. Transformations of these loci into the Smith-chart have been made, since this is a very convenient way of representing a wide range of immittances. Such a mapping is useful when the gain and noise performance of amplifiers is investigated, especially in the microwave region. By looking at the noise measure chart, for example, it is easy to deduce what source admittance gives the best overall noise performance and how it deteriorates as the admittance differs from its optimum value. These effects were demonstrated using a microwave transistor as an example.

References

[1] "IRE Standards on Electron Tubes: Methods of Testing," 1962, 62 IRE 7 S1, pt. 9: "Noise in linear twoports."
[2] H. Rothe and W. Dahlke, "Theory of noisy fourpoles," *Proc. IRE*, vol. 44, pp. 811–818, June 1956.
[3] H. A. Haus and R. B. Adler, *Circuit Theory of Linear Noisy Networks*. New York: Wiley, 1959.
[4] ——, "Invariants of linear noisy networks," *1956 IRE Conv. Rec.*, pt. 2, pp. 53–67, 1956.
[5] J. G. Linvill and J. F. Gibbons, *Transistors and Active Circuits*, New York: McGraw-Hill, 1961. The authors gave a wrong expression for available power gain in their book, probably due to typographical errors.
[6] J. C. Slater, "Microwave Electronics," *Rev. Mod. Phys.*, vol. 18, pp. 441–512, October 1946.
[7] P. H. Smith, "Transmission-line calculator," *Electronics*, vol. 12, pp. 29–31, January 1939; and "An improved transmission-line calculator," *Electronics*, vol. 17, p. 130, January 1944.
[8] H. A. Haus and R. B. Adler, "An extension of the noise figure definition," *Proc. IRE (Correspondence)*, vol. 45, pp. 690–691, May 1957.
[9] H. Fukui, "The characteristics of Esaki diodes at microwave frequencies," *1961 ISSCC Digest of Tech. Papers*, pp. 16–17, 1961; and *J. IECE (Japan)*, vol. 43, pp. 1351–1356, November 1960 (in Japanese).
[10] ——, "The noise performance of microwave transistors," *IEEE Trans. on Electron Devices*, vol. ED-13, pp. 329–341, March 1966.

The Effect of Feedback on Noise Figure

SVEIN IVERSEN

Abstract — Exact formulas for the effect of series and shunt feedback on noise figure are derived. The resulting noise figure is expressed by the feedback impedance, the two-port parameters, and the noise parameters of the device without feedback. The formulas are useful when considering the effect of biasing network and package elements on noise figure.

I. Introduction

The effect of feedback on noise figure has been considered in [1]–[4]. But exact formulas for the noise figure of two-ports with series and shunt feedback have not been published. Bellomo [2] has analyzed the problem with a feedback resistor, but he has neglected the correlation between noise sources. The purpose of this letter is to present exact formulas for the noise figure of a two-port with a feedback impedance in series and shunt connection.

The resulting noise figure is expressed by the noise figure F, the noise sources $\overline{v_n^2}$ and $\overline{i_n^2}$ (Fig. 1) (or equivalently $R_{N_v} = \overline{v_n^2}/4kt\Delta f$ and $R_{N_i} = 4kt\Delta f/\overline{i_n^2}$) and the correlation admittance $Y_\gamma = G_\gamma + jB_\gamma = \overline{v_n^* i_n}/\overline{v_n^2}$ of the two port without feedback [5]. (The asterisk denotes the complex conjugate). The noise figure is then given by

$$F = 1 + \frac{1}{R_{N_i} G_S} + \frac{R_{N_v}}{G_S}[|Y_S|^2 + 2\operatorname{Re}(Y_\gamma^* Y_S)] \qquad (1)$$

where $Y_S = G_S + jB_S$ is the source admittance.

II. Series Feedback

We start by finding the resulting noise figure of 2 three-poles in series connection (Fig. 2) expressed by the noise parameters F, R_{N_v}, R_{N_i}, and Y_γ of each of the 2 three-poles. As an intermediate noise equivalent the one indicated in Fig. 2 is convenient during the computation. The equations relating this equivalent to that in Fig. 1 are easily derived [4], [6].

$$v_1 = v_n - z_{11} i_n, \qquad v_2 = -z_{21} i_n \qquad (2)$$

or

$$v_n = v_1 - \frac{z_{11}}{z_{21}} v_2, \qquad i_n = -\frac{v_2}{z_{21}}. \qquad (3)$$

We immediately find

$$Z' = Z + \hat{Z} \qquad (4)$$

$$v_1' = v_1 + \hat{v}_1$$

$$v_2' = v_2 + \hat{v}_2. \qquad (5)$$

Inserting (2) into (5) and the result into (3) and using (4), we obtain

$$v_n' = v_n + i_n A + \hat{v}_n + \hat{i}_n \hat{A}, \qquad i_n' = i_n B + \hat{i}_n \hat{B} \qquad (6)$$

where

$$A = \frac{\hat{z}_{11} z_{21} - z_{11} \hat{z}_{21}}{z_{21} + \hat{z}_{21}} \qquad B = \frac{z_{21}}{z_{21} + \hat{z}_{21}}$$

$$\hat{A} = \frac{z_{11} \hat{z}_{21} - \hat{z}_{11} z_{21}}{z_{21} + \hat{z}_{21}} \qquad \hat{B} = \frac{\hat{z}_{21}}{z_{21} + \hat{z}_{21}}. \qquad (7)$$

Noise sources in different two-ports are assumed uncorrelated. From (6) we then get

$$R'_{N_v} = R_{N_v}[1 + 2\operatorname{Re}(Y_\gamma A)] + \frac{|A|^2}{R_{N_i}}$$

$$+ \hat{R}_{N_v}[1 + 2\operatorname{Re}(\hat{Y}_\gamma \hat{A})] + \frac{|\hat{A}|^2}{\hat{R}_{N_i}} \qquad (8)$$

$$\frac{1}{R'_{N_i}} = \frac{|B|^2}{R_{N_i}} + \frac{|\hat{B}|^2}{\hat{R}_{N_i}} \qquad (9)$$

$$Y'_\gamma = \frac{1}{R'_{N_v}}\left(R_{N_v} B Y_\gamma + \frac{A^* B}{R_{N_i}} + \hat{R}_{N_v} \hat{B} \hat{Y}_\gamma + \frac{\hat{A}^* \hat{B}}{\hat{R}_{N_i}}\right) \qquad (10)$$

(Re indicates the real part of the complex number). Using (1) and rearranging, we get

$$F' = F + \hat{F} - 1 + \frac{1}{R_{N_i} G_S}(|C|^2 + 2\operatorname{Re} C) + \frac{1}{\hat{R}_{N_i} G_S}(|\hat{C}|^2 + 2\operatorname{Re}\hat{C})$$

$$+ \frac{R_{N_v}}{G_S} 2\operatorname{Re}(Y_\gamma C Y_S^*) + \frac{\hat{R}_{N_v}}{G_S} 2\operatorname{Re}(\hat{Y}_\gamma \hat{C} Y_S^*) \qquad (11)$$

where

$$C = A Y_S - \hat{B} = \frac{Y_S(\hat{z}_{11} z_{21} - z_{11}\hat{z}_{21}) - \hat{z}_{21}}{z_{21} + \hat{z}_{21}}$$

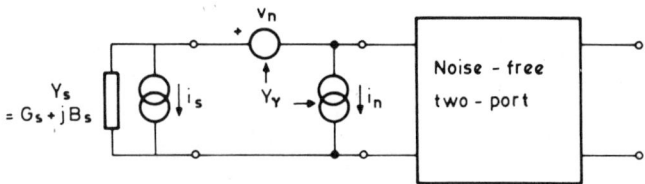

Fig. 1. Noise equivalent of noisy two-port with source admittance Y_S.

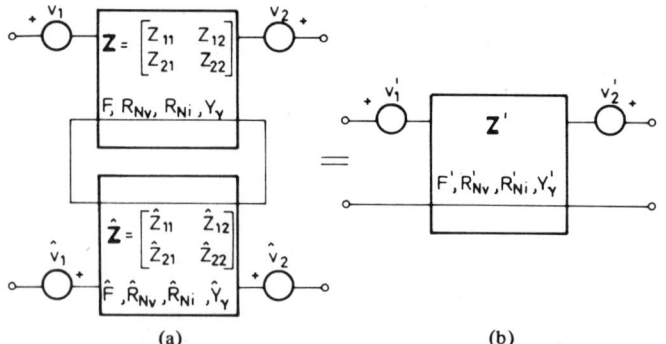

Fig. 2. (a) 2 three-poles in series connection with equivalent noise voltage generators. (b) Resulting noise equivalent.

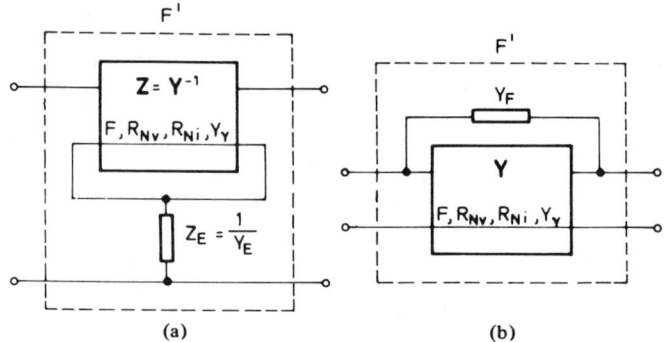

Fig. 3. (a) Three-pole with series feedback. (b) Three-pole with shunt feedback.

$$\hat{C} = \hat{A} Y_S - B = \frac{Y_S(z_{11}\hat{z}_{21} - \hat{z}_{11} z_{21}) - z_{21}}{z_{21} + \hat{z}_{21}}. \qquad (12)$$

If one of the three-poles is a shunt impedance $Z_E = 1/Y_E (Y_E = G_E + jB_E)$ as in Fig. 3(a) (series feedback), we have

$$C = C_E = \frac{Y_S Z_E(z_{21} - z_{11}) - Z_E}{z_{21} + Z_E} = \frac{Y_S(y_{22} + y_{21}) + \Delta y}{Y_E y_{21} - \Delta y} \qquad (13)$$

Manuscript received July 12, 1974.
The author is with the Teleteknisk Laboratorium, the University of Trondheim, 7034 Trondheim-NTH, Norway.

$$\hat{C} = \hat{C}_E = -C_E - 1$$

where $\Delta y = y_{11}y_{22} - y_{21}y_{12}$. We easily find

$$\hat{R}_{N_v} = 0, \quad \hat{R}_{N_i} = \frac{1}{G_E}, \quad \hat{Y}_\gamma = 0, \quad \hat{F} = 1 + \frac{G_E}{G_S}. \quad (14)$$

Putting (13) and (14) into (11), we obtain

$$F' = F + \frac{1}{R_{N_i}G_S}(|C_E|^2 + 2\,\mathrm{Re}\,C_E) + \frac{G_E}{G_S}|C_E|^2 + \frac{R_{N_v}}{G_S} 2\,\mathrm{Re}\,(Y_\gamma C_E Y_S^*). \quad (15)$$

III. Shunt Feedback

By a similar procedure using y parameters and noise-current generators [6], we can find the noise parameters of a parallel connection of 2 three-poles. The results are

$$R'_{N_v} = R_{N_v} D + \hat{R}_{N_v}\hat{D} \quad (16)$$

$$R'_{N_i} = \frac{1}{R_{N_i}} + \frac{1}{\hat{R}_{N_i}} + R_{N_v}[E + 2\,\mathrm{Re}\,(Y_\gamma E^*)] + \hat{R}_{N_v}[\hat{E} + 2\,\mathrm{Re}\,(\hat{Y}_\gamma \hat{E}^*)] \quad (17)$$

$$Y'_\gamma = \frac{R_{N_v}}{R'_{N_v}}(Y_\gamma D^* + D^* E + \hat{Y}_\gamma \hat{D}^* + \hat{D}^* \hat{E}) \quad (18)$$

where

$$D = \frac{y_{21}}{y_{21} + \hat{y}_{21}} \quad E = \frac{\hat{y}_{11}y_{21} - y_{11}\hat{y}_{21}}{y_{21} + \hat{y}_{21}}$$

$$\hat{D} = \frac{\hat{y}_{21}}{y_{21} + \hat{y}_{21}} \quad \hat{E} = \frac{y_{11}\hat{y}_{21} - \hat{y}_{11}y_{21}}{y_{21} + \hat{y}_{21}} \quad (19)$$

The resulting noise figure is

$$F' = F + \hat{F} - 1 + \frac{R_{N_v}}{G_S}[|K|^2 + 2\,\mathrm{Re}\,\{K(Y_S^* + Y_\gamma^*)\}] + \frac{\hat{R}_{N_v}}{G_S}[|\hat{K}|^2 + 2\,\mathrm{Re}\,\{\hat{K}(Y_S^* + \hat{Y}_\gamma^*)\}] \quad (20)$$

where

$$K = E - \hat{D}Y_S = \frac{\hat{y}_{11}y_{21} - \hat{y}_{21}(y_{11} + Y_S)}{y_{21} + \hat{y}_{21}}$$

$$\hat{K} = \hat{E} - DY_S = \frac{y_{11}\hat{y}_{21} - y_{21}(\hat{y}_{11} + Y_S)}{y_{21} + \hat{y}_{21}}. \quad (21)$$

If one of the three-poles is a series admittance $Y_F = G_F + jB_F$ as in Fig. 3(b) (shunt feedback), this formula reduces to

$$F' = F + \frac{R_{N_v}}{G_S}[|K_F|^2 + 2\,\mathrm{Re}\,\{K_F(Y_S^* + Y_\gamma^*)\}] + \frac{G_F}{|Y_F|^2 G_S}|K_F|^2 \quad (22)$$

where

$$K_F = \frac{Y_F(y_{11} + y_{21} + Y_S)}{y_{21} - Y_F}. \quad (23)$$

IV. Conclusions

The formulas presented here may seem rather complicated, but in practical problems they can reduce drastically due to appropriate approximations. Often the three-pole parameters and the source and feedback impedances have values such that $C_E \approx Y_S/Y_E$ and $K_F \approx Y_F$. Equations (15) and (22) then become identically the same as the formulas for the noise figure when an admittance Y_E is connected in series with the input of the three-pole, respectively an admittance Y_F is shunted across the input.

The formulas are believed to be useful when considering the effect of biasing network, contact resistances, lead inductances [7], and stray capacitances on the noise figure of an active device.

References

[1] H. A. Haus and R. B. Adler, *Circuit Theory of Linear Noisy Networks*. New York: Wiley, 1959.
[2] A. F. Bellomo, "Gain and noise considerations in RF feedback amplifier," *IEEE J. Solid-State Circuits*, vol. SC-3, pp. 290–294, Sept. 1968.
[3] R. E. Bogner, "Feedback and noise performance," *Electron. Eng.*, vol. 37, pp. 115–117, Feb. 1965.
[4] H. Bittel and L. Storm, *Rauschen*. Berlin, Germany: Springer, 1971, pp. 261–268.
[5] H. A. Haus et al., "Representation of noise in linear two ports," *Proc. IRE*, vol. 48, pp. 69–74, Jan. 1960.
[6] H. Rothe and W. Dahlke, "Theory of noisy four poles," *Proc. IRE*, vol. 44, pp. 811–818, June 1956.
[7] A. Anastassiou and M. J. O. Strutt, "Effect of source lead inductance on the noise figure of a GaAs FET," *Proc. IEEE* (Lett.), vol. 62, pp. 406–408, Mar. 1974.

Wave Representation of Amplifier Noise*

This letter will call attention to a wave representation of noise in a linear twoport which has been given by Bauer and Rothe,[1] but which apparently has not been described in English. The representation has the advantages that both equivalent noise generators are at the input to the amplifier (or linear twoport), and that they are *uncorrelated*.

Previous Representations

Several schemes have been used to represent noise at a given frequency f in a linear twoport.[2] By Thévenin's theorem, the noise generated internally by the twoport can be represented by two noise voltage sources in series with the input and the output. These two generators are in general correlated, and four real numbers are necessary to specify the noise properties of the twoport; these may be the mean-square values of the two voltages and their complex correlation coefficient.

Unfortunately, in this representation there is one equivalent generator at each port. For many purposes (particularly calculating noise figures of amplifiers) it is advantageous to have both generators at one port, the input. Rothe and Dahlke[3] have shown how to do this with the equivalent circuit of Fig. 1. The two generators are correlated; one can define the four real numbers associated with this representation as the mean-square values of e_n and i_n, and their complex correlation coefficient ρ or,

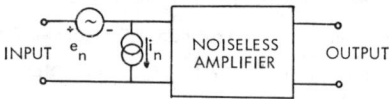

Fig. 1—The Rothe-Dahlke noise model for a noisy twoport has two correlated generators at the input. The advantage is that the noiseless part can be disregarded in calculating the noise figure of the amplifier.

alternatively, the quantities

$$R_n = \frac{\overline{|e_n|^2}}{4kT_0 \, \Delta f} \qquad (1)$$

$$G_n = \frac{\overline{|i_n|^2}}{4kT_0 \, \Delta f} \qquad (2)$$

and

$$\rho = \frac{\overline{i_n e_n^*}}{\sqrt{\overline{|i_n|^2}\,\overline{|e_n|^2}}}, \qquad (3)$$

---

* Received by the PGCT, October 30, 1961. This work was supported in part by the U. S. Army Signal Corps, the Air Force Office of Scientific Research, and the Office of Naval Research; and in part by a Ford Foundation Fellowship.
[1] H. Bauer and H. Rothe, "Der äquivalente Rauschvierpol als Wellenvierpol," *Arch. elekt. Übertragung*, vol. 10, pp. 241–252; June, 1956. (In German).
[2] IRE Subcommittee 7.9 on Noise, H. A. Haus, Chairman, "Representation of Noise in Linear Twoports," *Proc. IRE*, vol. 48, pp. 69–74; January, 1960.
[3] H. Rothe and W. Dahlke, "Theory of noisy fourpoles," *Proc. IRE*, vol. 44, pp. 811–818; June, 1956.

---

where k is Boltzmann's constant, T_0 is standard temperature of 290°K, Δf is a range of frequencies centered about f, the horizontal line indicates an average, and the asterisk indicates the complex conjugate. The correlation coefficient ρ is less than or equal to one in magnitude.

The advantage of this representation is the ease of calculating the amplifier noise figure. The noiseless amplifier in Fig. 1 does not affect the noise figure because it treats all noises and signals identically. The noise-figure expression, therefore, is independent of the properties of the noiseless amplifier except insofar as they help to determine the quantities R_n, G_n, and ρ. If the source impedance is $Z_s = R_s + jX_s$, the excess noise figure is [2,3]

$$F - 1 = \frac{\overline{|e_n + Z_s i_n|^2}}{4kT_0 \, \Delta f R_s}$$

$$= \frac{R_n + |Z_s|^2 G_n + 2\sqrt{R_n G_n}\,Re(\rho Z_s)}{R_s}. \qquad (4)$$

As Z_s is varied over all values such that R_s is positive, the noise figure goes through the minimum value

$$(F-1)_{\min} = 2\sqrt{R_n G_n}$$
$$\cdot \left(\sqrt{1 - (Im\,\rho)^2} + Re\,\rho\right), \qquad (5)$$

where Im and Re stand for "imaginary part" and "real part," respectively. This minimum occurs for the particular value of source impedance $Z_{s,\text{opt}} = R_{s,\text{opt}} + jX_{s,\text{opt}}$

$$|Z_{s,\text{opt}}|^2 = \frac{R_n}{G_n} \qquad (6)$$

$$X_{s,\text{opt}} = \sqrt{\frac{R_n}{G_n}}\,Im\,\rho. \qquad (7)$$

This fact is evident from an alternative form for (4)

$$F - 1 = (F-1)_{\min}$$
$$+ \frac{G_n}{R_s}|Z_s - Z_{s,\text{opt}}|^2. \qquad (8)$$

It has been suggested[2] that the noise characteristics of a linear twoport, such as an amplifier, be specified by the four numbers $(F-1)_{\min}$, R_n, and $Z_{s,\text{opt}}$ ($Z_{s,\text{opt}}$ is a complex number). Probably such a specification is more convenient than simply R_n, G_n, and ρ, since the quantities of engineering interest are $(F-1)_{\min}$ and the optimum source impedance.

Wave Representation

An alternate representation, suggested by Bauer and Rothe,[1] uses wave, or scattering[4] variables, and retains the advantages of the Rothe-Dahlke representation. It has the further advantage that the two noise sources are *uncorrelated*, so that the expression for noise figure is particularly simple.

We define the noise-wave generators a_n and b_n at the input of the amplifier, in terms of the parameters of the Rothe-Dahlke model:

$$a_n = \frac{e_n + Z_\nu i_n}{2\sqrt{Re\,Z_\nu}} \qquad (9)$$

$$b_n = \frac{e_n - Z_\nu^* i_n}{2\sqrt{Re\,Z_\nu}}, \qquad (10)$$

where Z_ν is a normalization impedance. Thus, if the scattering matrix of the twoport is **S**, we obtain

$$\begin{bmatrix} b_1 - b_n \\ b_2 \end{bmatrix} = \begin{bmatrix} S_{11} & S_{12} \\ S_{21} & S_{22} \end{bmatrix} \times \begin{bmatrix} a_1 + a_n \\ a_2 \end{bmatrix} \qquad (11)$$

An equivalent circuit of this representation is shown in Fig. 2. The wave generators are represented by directional couplers and ordinary sources; this schematic is accurate in the limit of small coupling to the line, and large strengths of the sources. Note that this "transmission line" has a complex "characteristic impedance" Z_ν.

The normalization impedance Z_ν used in (9) and (10) is chosen so as to make a_n and b_n uncorrelated:

$$\overline{a_n b_n^*} = 0. \qquad (12)$$

It happens that this impedance is precisely $Z_{s,\text{opt}}$ given in (6) and (7). This point is of importance: if the optimum source impedance is used as the normalization impedance in defining noise-wave generators from the Rothe-Dahlke noise generators, the resulting wave generators are uncorrelated. The simplicity of the formulas given below is entirely the result of this fact. The wave representation here fails when $R_{s,\text{opt}} = 0$; that is, only when $R_n = 0$, $G_n = 0$, or $\rho = \pm j$. In each of these cases, $(F-1)_{\min}$ equals zero and is achieved with a reactive source. Only one equivalent noise source is necessary if it is properly placed.

---

[4] H. J. Carlin, "The scattering matrix in network theory," *IRE Trans. on Circuit Theory*, vol. CT-3, pp. 88–97; June, 1956.

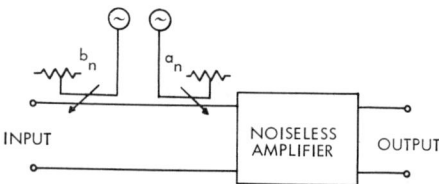

Fig. 2—Proposed wave model of a noisy twoport consists of a noiseless twoport with uncorrelated wave generators at the input. The noiseless twoport can be disregarded in calculating the noise figure of the amplifier.

The strength of a_n defines a temperature T_a

$$T_a = \frac{\overline{|a_n|^2}}{k\,\Delta f}, \qquad (13)$$

and similarly, the strength of b_n defines a temperature T_b

$$T_b = \frac{\overline{|b_n|^2}}{k\,\Delta f}. \quad (14)$$

In terms of the Rothe-Dahlke model, these are

$$T_a = 2T_0\sqrt{R_n G_n}$$
$$\cdot(\sqrt{1-(Im\,\rho)^2} + Re\,\rho) \quad (15)$$
$$T_b = 2T_0\sqrt{R_n G_n}$$
$$\cdot(\sqrt{1-(Im\,\rho)^2} - Re\,\rho), \quad (16)$$

so that T_a is related to the minimum noise figure of (5):

$$T_a = T_0(F-1)_{\min}. \quad (17)$$

On the other hand, R_n and G_n can be derived from the parameters of the wave representation:

$$R_n = \frac{T_a + T_b}{4T_0\,Re\,(1/Z_\nu)} \quad (18)$$

$$G_n = \frac{T_a + T_b}{4T_0\,Re\,(Z_\nu)}. \quad (19)$$

The expression for the noise figure $F-1$ is very simple when written in terms of these wave variables. Consider the amplifier run from a source with impedance $Z_s = R_s + jX_s$. The reflection coefficient of the source Γ_s is

$$\Gamma_s = \frac{Z_s - Z_\nu}{Z_s + Z_\nu^*} \quad (20)$$

when the wave variables are defined with Z_ν as the normalization impedance. The noiseless amplifier shown in Fig. 2 does not affect the noise figure, except insofar as it affects the values of T_a, T_b, or Z_ν. Thus, for calculating the noise figure it may be disregarded, or replaced by a matched load. Because a_n and b_n are uncorrelated, the power output caused by the internal sources is proportional to

$$\overline{|a_n + \Gamma_s b_n|^2} = \overline{|a_n|^2} + |\Gamma_s|^2\,\overline{|b_n|^2}. \quad (21)$$

On the other hand, the power output caused by thermal noise of the source impedance at temperature T_0 is proportional (with the same constant of proportionality) to

$$kT_0\,\Delta f(1 - |\Gamma_s|^2). \quad (22)$$

The ratio of these two is the noise figure, simply

$$F - 1 = \frac{T_a + T_b|\Gamma_s|^2}{T_0(1 - |\Gamma_s|^2)}$$
$$= \frac{T_a}{T_0} + \frac{T_a + T_b}{T_0}\frac{|\Gamma_s|^2}{1 - |\Gamma_s|^2} = \frac{T_a}{T_0}$$
$$+ \frac{T_a + T_b}{T_0}\frac{|Z_s - Z_\nu|^2}{4R_s\,Re\,(Z_\nu)}. \quad (23)$$

The choice of source impedance to minimize the noise figure is, from (23), obviously the value that sets Γ_s to zero, or simply Z_ν. The minimum excess noise figure so achieved is T_a/T_0.

The temperatures T_a and T_b can be easily measured, at least in principle, even at microwave frequencies. To measure T_a connect a zero-temperature (Kelvin) source to the amplifier through an adjustable lossless network (such as an E-H tuner), and adjust this network to obtain the minimum noise power output from the amplifier, say P_1. Then, raise the temperature of this source to some other temperature T_{ref} without changing its impedance or the network that couples it to the amplifier input. Read the resulting power P_2. Then T_a is given by

$$T_a = T_{ref}\frac{P_1}{P_2 - P_1}. \quad (24)$$

Alternatively, T_a can be determined from a standard noise-figure test by adjusting the lossless network to minimize the noise figure.

To measure T_b replace the source by an adjustable reactive load, such as an adjustable length of transmission line or a waveguide plunger. Adjust for a minimum power output P_3; then adjust for a maximum power output P_4. Then

$$T_b = T_{ref}\frac{1}{P_2 - P_1}$$
$$\cdot\left[\frac{4P_3 P_4}{(\sqrt{P_3} + \sqrt{P_4})^2} - P_1\right]. \quad (25)$$

Alternately, the magnitude of the reflection coefficient at the input of the amplifier $|S_{11}|$ with the output load attached, can be determined from a standing-wave measurement on the amplifier as terminated by the coupling network set for finding T_a. Then, T_b can be calculated from

$$T_b = \frac{T_{ref}}{P_2 - P_1}[P_3(1 + |S_{11}|)^2 - P_1]$$
$$= \frac{T_{ref}}{P_2 - P_1}[P_4(1 - |S_{11}|)^2 - P_1]. \quad (26)$$

Other schemes that do not depend on a low-temperature source for measuring these quantities can be devised.

It has been suggested[2] that T_a [or $(F-1)_{\min}$], Z_ν, and one other parameter (let us say R_n) be used to characterize noisy twoports, since these are the parameters of most engineering interest. Perhaps a more meaningful fourth parameter to specify would be T_b, or a quantity related to it, rather than R_n.

The author is indebted to W. E. Dahlke, of Telefunken G. m. b. H., for calling his attention to the paper by Bauer and Rothe.

PAUL PENFIELD JR.
Dept. of Electrical Engineering
and Research Lab. of Electronics,
Mass. Inst. Tech.,
Cambridge, Mass.

A Wave Approach to the Noise Properties of Linear Microwave Devices

R. P. MEYS

Abstract—Noise temperature or noise factor are important parameters for many microwave devices. Their dependence on source characteristics is classically established using low-frequency concepts such as impedance, admittance, voltage, and current sources. This paper presents a derivation of the noise properties of linear two-ports in terms of noise waves, which leads to a convenient measurement method in distributed systems.

I. Introduction

THE classical derivations for noise temperature and noise factor of linear two-ports use the equivalent circuit of Fig. 1(a) and lead to the results [1], [2]

$$T_n = T_{n\,\min} + T_0 \frac{R_n}{G_s} |Y_s - Y_{\text{opt}}|^2 \tag{1}$$

$$F_0 = F_{0\,\min} + \frac{R_n}{G_s} |Y_s - Y_{\text{opt}}|^2. \tag{2}$$

Manuscript received March 22, 1977; revised May 27, 1977.
The author is with the Laboratoire d'Electronique Générale et Radioelectricité, Université Libre de Bruxelles, Brussels, Belgium.

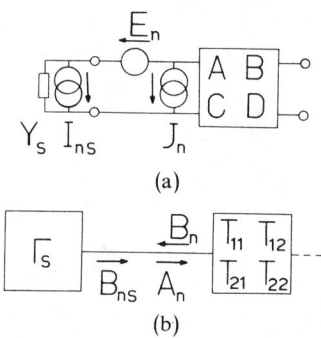

Fig. 1. (a) The classical representation of a linear noisy two-port. (b) The representation of a linear noisy two-port using noise waves.

Sometimes (1) and (2) are written as

$$T_n = T_{n\,\min} + 4T_0 \frac{R_n}{Z_0} \frac{|\Gamma_s - \Gamma_{\text{opt}}|^2}{|1 + \Gamma_{\text{opt}}|^2(1 - |\Gamma_s|^2)} \tag{3}$$

$$F_0 = F_{0\,\min} + 4 \frac{R_n}{Z_0} \frac{|\Gamma_s - \Gamma_{\text{opt}}|^2}{|1 + \Gamma_{\text{opt}}|^2(1 - |\Gamma_s|^2)} \tag{4}$$

Reprinted from *IEEE Trans. Microwave Theory and Tech.*, vol. MTT-26, pp. 34–37, Jan. 1978.

with

$\Gamma_{opt} = (Y_0 - Y_{opt})/(Y_0 + Y_{opt});$
$Z_0 = (1/Y_0)$ = characteristic impedance of the distributed system.

These last expressions are hybrid ones, containing the impedance parameter R_n and the wave parameter Γ_{opt}. The unknown quantities $T_{n\,min}$ ($F_{0\,min}$), R_n, Y_{opt} (Γ_{opt}) are determined either a) by a cut-and-try method using an automatic noise-figure meter, or b) by a set of measurements with various source admittances followed by a computation such as described in [2].

At microwave frequencies, however, a treatment of noise in terms of waves looks attractive. This was already done [3]. The analysis starts with the model of Fig. 1(b) and defines two uncorrelated noise waves A_n, B_n by

$$A_n = -\frac{E_n + Z_v J_n}{2\sqrt{\text{Re}(Z_v)}}$$

$$B_n = \frac{E_n - Z_v^* J_n}{2\sqrt{\text{Re}(Z_v)}}$$

where Z_v is a complex normalization impedance dependent on the device. This choice leads to theoretical simplicity, but does not suggest an easy-to-use measurement method.

II. NOISE REPRESENTATION BY CORRELATED WAVE SOURCES

Let us start with the same model [Fig. 1(b)], but simply define the noise-wave sources with reference to the characteristic impedance Z_0 of the line or waveguide mode used at the input. We now connect a source of reflection factor Γ_s and noise wave B_{ns} to the input. The total noise wave that would be incident on a noiseless matched load substituted for the device input is

$$A_{ns} = A_n + \Gamma_s B_n + B_{ns}$$

the squared modulus of which equals, assuming no correlation between source and two-port noise

$$\overline{|A_{ns}|^2} = \overline{|A_n|^2} + |\Gamma_s|^2 \overline{|B_n|^2} + 2\,\text{Re}(\Gamma_s \overline{A_n^* B_n}) + \overline{|B_{ns}|^2}. \quad (5)$$

If we introduce the following notation:

$$\overline{|A_{ns}|^2} = kT_{ns}\Delta f$$
$$\overline{|A_n|^2} = kT_a \Delta f$$
$$\overline{|B_n|^2} = kT_b \Delta f$$
$$\overline{A_n^* B_n} = kT_c \Delta f \cdot e^{j\phi_c}$$
$$\Gamma_s = |\Gamma_s| e^{j\phi_s}$$

where k is Boltzmann's constant and Δf is the frequency interval of interest, and consider that the source noise wave is given by

$$\overline{|B_{ns}|^2} = (1 - |\Gamma_s|^2) k T_s \Delta f$$

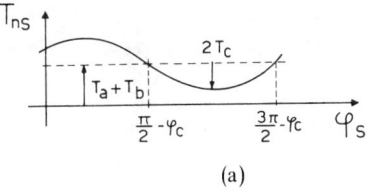
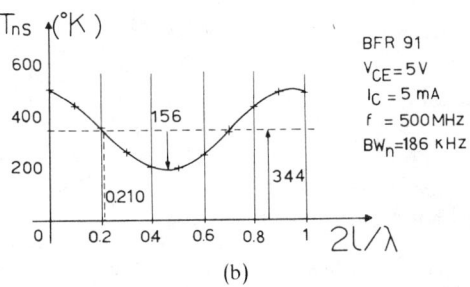

Fig. 2. (a) The total direct noise wave as a function of ϕ_s. (b) An example of experimental results.

(5) becomes

$$T_{ns} = T_a + |\Gamma_s|^2 T_b + 2T_c |\Gamma_s| \cos(\phi_s + \phi_c) + T_s(1 - |\Gamma_s|^2). \quad (6)$$

To entirely define the noise properties of the two-port, we must determine the four quantities which appear in (6): T_a, T_b, T_c, ϕ_c. The following procedure is proposed.

a) Connect an almost lossless source ($|\Gamma_s| \simeq 1$) of variable ϕ_s to the input. Then observe

$$T_{ns} = T_a + T_b + 2T_c \cos(\phi_s + \phi_c)$$

which is represented as a function of ϕ_s in Fig. 2(a). From this graph we deduce T_c, ϕ_c, $T_a + T_b$.

b) Connect a matched load of known temperature T_s to the input, so

$$T_{ns} = T_a + T_s$$

from which we deduce T_a.

III. EXAMPLE OF NOISE-WAVE SOURCES MEASUREMENT

The method has been applied as an example to a BFR91 transistor operating at $V_{CE} = 5\,\text{V}$, $I_c = 5\,\text{mA}$ to determine its noise-wave sources around $f = 500$ MHz. The test setup for the measurement of T_{ns} with $|\Gamma_s| \simeq 1$ is shown in Fig. 3(a). A waveguide below the cutoff attenuator is used in the reversed mode to inject a known sine wave. As the coupling is small and a short is connected at the third port, we realize very closely the condition $|\Gamma_s| = 1$. A variable-length line is placed between the attenuator and the DUT. The receiving system consists of a 500-MHz selective preamplifier, a first mixer and 50-MHz IF amplifier, a step attenuator, a second mixer and 5-MHz IF amplifier with calibrated noise bandwidth, and finally a quadratical diode detector. First, with the IF attenuator on its reference position, the detector output is recorded. Then the IF attenuation is increased by A_{if} and a known sine wave is injected so as to recover the

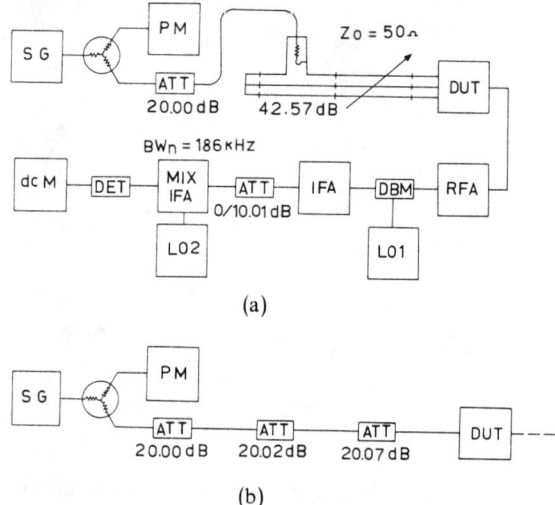

Fig. 3. (a) The measurement setup with $|\Gamma_s| = 1$. (b) The measurement setup with $|\Gamma_s| = 0$.

TABLE I
THE NUMERICAL RESULTS PERTAINING TO THE EXAMPLE

l (mm)	0	30	60	90	120	150	180	210	240	270	300
P_m (dBm)	46.80	47.34	48.28	49.59	50.70	50.80	49.85	48.49	47.78	46.88	46.86
T_{ns} (°K)	499	441	355	263	203	199	247	338	427	490	492

same detector output as before. It is shown in the Appendix that under these conditions

$$T_{ns} = \frac{1}{A_{if}^2 - 1} \frac{1}{A_s^2} \frac{P_m}{kBW_n} \quad (7)$$

A_{if}^2 IF power attenuation;
A_s^2 total power attenuation between power meter and DUT input;
BW_n system-noise bandwidth;
P_m power-meter reading.

The experimental results are shown in Table I. Fig. 2(b) is a plot of T_{ns}, as calculated from (7) with the attenuation and bandwidth values indicated on Fig. 3(a), versus the normalized variable-line length $2l/\lambda$. Conformity of the experimental curve to the theoretical sinusoid is almost perfect. From Fig. 2(b) we find

$$T_a + T_b = 344 \text{ K}$$

$$T_c = 78 \text{ K}$$

$$\frac{3\pi/2 - \phi_c}{2\pi} = 0.703 - 0.210$$

where the 0.703 term in the last equation is the phase of Γ_s for zero extension, which has been measured separately. The minus sign follows from the fact that arg (Γ_s) decreases as the line is extended.

The last step is to measure T_a. This step is, in fact, similar to the conventional measurement of noise temperature with a matched source. It has been performed by modifying the test setup as shown in Fig. 3(b). Using the relation

$$T_a = \frac{1}{A_{if}^2 - 1} \frac{1}{A_s^2} \frac{P_m}{kBW_n} - T_s \quad (8)$$

(which is a special case of (7)), with T_s equal to the ambient temperature (301 K), $P_m = -49.51$ dBm, and with the attenuation values indicated on Fig. 3(a) and (b), we find $T_a = 172$ K.

IV. FROM NOISE WAVES TO NOISE TEMPERATURE AND NOISE FACTOR

The noise temperature T_n is defined as that temperature which must be added to the source temperature to account for the noise introduced by the linear two-port. Using this definition and (6) we obtain

$$T_n = \frac{T_a + |\Gamma_s|^2 T_b + 2T_c |\Gamma_s| \cos(\phi_s + \phi_c)}{1 - |\Gamma_s|^2}. \quad (9)$$

Let us now write (3) under the more compact form

$$T_n = T_{n\min} + T_d \frac{|\Gamma_s - \Gamma_{opt}|^2}{1 - |\Gamma_s|^2}. \quad (10)$$

Identification of (9) and (10) yields the results

$$T_a = T_{n\min} + T_d |\Gamma_{opt}|^2 \quad (11)$$

$$T_b = T_d - T_{n\min} \quad (12)$$

$$T_c = T_d |\Gamma_{opt}| \quad (13)$$

$$\phi_c = \pi - \phi_0 \quad (14)$$

with $\phi_0 = \arg(\Gamma_{opt})$. Inversion of the system of (11)–(14) yields

$$T_d = \tfrac{1}{2}(T_{ab} \pm \sqrt{T_{ab}^2 - 4T_c^2}), \quad (T_{ab} = T_a + T_b) \quad (15)$$

$$T_{n\min} = T_d - T_b \quad (16)$$

$$|\Gamma_{opt}| = \frac{T_c}{T_d} \quad (17)$$

$$\phi_0 = \pi - \phi_c. \quad (18)$$

It is a simple matter to verify that only the + sign in (15) must be retained to satisfy the condition $|\Gamma_{opt}| < 1$. The corresponding equations for noise factor are readily obtained, if we bear in mind that

$$F_0 = 1 + \frac{T_n}{T_0}.$$

When applying (15)–(18) to our previous example, we find successively

$$T_d = 325 \text{ K}$$

$$T_{n\min} = 153 \text{ K}$$

$$|\Gamma_{opt}| = 0.240$$

$$\phi_0 = 87.5°$$

which achieves the noise characterization of the device.

V. Conclusion

A description of the noise behavior of linear microwave devices has been proposed based on the four fundamental wave parameters T_a, T_b, T_c, ϕ_c. As the example has shown, they can be easily measured and lead, through simple formulas, to the traditional quantities T_n and F_0. Sine waves have been used as power reference, but this is in no case a necessity. Use of calibrated noise sources is possible insofar as a sufficient power level is available to inject a noise wave of the same order of magnitude as the device's sources despite the small coupling required by the condition $|\Gamma_s| \simeq 1$. This usually avoids the need for working with a known noise bandwidth.

Appendix

If Γ_i designates the reflection factor at the device's input, the total incident noise wave actually is

$$\frac{1}{1 - \Gamma_s \Gamma_i} A_{ns}$$

giving rise at the output of the measurement setup to a power

$$P_0 = \alpha \frac{1}{|1 - \Gamma_s \Gamma_i|^2} \overline{|A_{ns}|^2}. \tag{19}$$

When a sine wave B_{ss} is injected, the total incident wave becomes

$$\frac{1}{1 - \Gamma_s \Gamma_i} (A_{ns} + B_{ss}).$$

Following the principle of the measurement, this wave will give rise to the same output power as before, after attenuation by A_{if}, so

$$P_0 = \alpha \frac{1}{A_{if}^2} \frac{1}{|1 - \Gamma_s \Gamma_i|^2} (\overline{|A_{ns}|^2} + |B_{ss}|^2). \tag{20}$$

From (19) and (20) it follows that

$$\overline{|A_{ns}|^2} = \frac{1}{A_{if}^2 - 1} |B_{ss}|^2. \tag{21}$$

Substituting

$$|B_{ss}|^2 = \frac{P_m}{A_s^2} \quad \text{and} \quad \overline{|A_{ns}|^2} = kT_{ns} BW_n$$

in (21) leads to (7).

References

[1] H. Rothe and W. Dahlke, "Theory of noisy fourpoles," *Proc. IRE*, vol. 44, pp. 811–818, June 1956.
[2] R. Q. Lane, "The determination of device noise parameters," *Proc. IEEE*, vol. 57, pp. 1461–1462, Aug. 1969.
[3] P. Penfield, "Wave representation of amplifier noise," *IRE Trans. Circuit Theory*, vol. CT-9, pp. 83–84, Mar. 1962.

IRE Standards on Methods of Measuring Noise in Linear Twoports, 1959*

59 IRE 20. S1

COMMITTEE PERSONNEL

Subcommittee on Noise

H. A. Haus, *Chairman* 1956–1959

W. R. Atkinson 1957–1959	W. A. Harris 1956–1959
G. M. Branch 1957–1959	S. W. Harrison 1956–1959
W. B. Davenport, Jr. 1956–1959	W. W. McLeod 1957–1959
W. H. Fonger 1957–1958	E. K. Stodola 1957–1959

T. E. Talpey 1956–1959

Committee on Electron Tubes
1958–1959

G. A. Espersen, *Chairman*
H. J. Reich, *Vice-Chairman*

E. M. Boone	W. J. Kleen	P. A. Redhead
A. W. Coolidge	P. M. Lapostolle	H. Rothe
W. S. Cranmer	V. Learned	W. G. Shepherd
P. A. Fleming	A. S. Luftman	E. S. Stengel
H. B. Frost	R. M. Matheson	R. G. Stoudenheimer
K. Garoff	G. D. O'Neill	W. W. Teich
H. A. Haus	A. T. Potjer	B. H. Vine
T. J. Henry	G. W. Pratt	R. R. Warnecke
E. O. Johnson		S. E. Webber

Standards Committee
1959–1960

R. F. Shea, *Chairman*
J. G. Kreer, Jr., *Vice-Chairman*
C. H. Page, *Vice-Chairman*
L. G. Cumming, *Vice-Chairman*

J. Avins	E. A. Gerber	E. Mittelmann
W. F. Bailey	A. B. Glenn	L. H. Montgomery, Jr.
M. W. Baldwin, Jr.	V. M. Graham	G. A. Morton
J. T. Bangert	R. A. Hackbusch	R. C. Moyer
W. R. Bennett	R. T. Haviland	J. H. Mulligan, Jr.
J. G. Brainerd	A. G. Jensen	A. A. Oliner
D. R. Brown	R. W. Johnston	M. L. Phillips
P. S. Carter	I. Kerney	R. L. Pritchard
A. G. Clavier	A. E. Kerwien	P. A. Redhead
S. Doba, Jr.	G. S. Ley	R. Serrell
P. Elias	Wayne Mason	W. A. Shipman
G. A. Espersen	D. E. Maxwell	H. R. Terhune
R. J. Farber	P. Mertz	E. Weber
D. G. Link	H. I. Metz	R. B. Wilcox
G. L. Fredendall	H. R. Mimno	W. T. Wintringham

Measurements Coordinator

J. G. Kreer, Jr.

---

* Approved by the IRE Standards Committee, June 11, 1959. Reprints of this Standard, 59 IRE 20. S1, together with the tutorial paper, "Representation of Noise in Linear Twoports," which immediately follows the Standard in this issue, may be purchased while available from the Institute of Radio Engineers, 1 East 79 Street, New York, N. Y., at $0.75 per copy. A 20 per cent discount will be allowed for 100 or more copies mailed to one address. Reprinted from *Proc. IRE*, vol. 48, pp. 60-68, Jan. 1960.

1. Introduction

SPURIOUS undesired signals are always present in signaling systems and their components. These spurious signals are usually called noise. Since noise reduces the amount of information that can be transmitted with a specific signal power, quantitative measures of noise are often indispensable to engineering evaluations of signaling systems.

Test measurements for noise might logically be expected to begin with some generally useful quantitative measure of information-handling capacity in the presence of noise, or of the amount of annoyance which it creates. Such tests can be made in specific instances, but numerous difficulties are encountered. Often they are subjective and hard to evaluate accurately. Furthermore, the effects of noise vary enormously, depending on the system in question, and on the levels and characteristics of both signal and noise. It is useful, therefore, to measure noise in the engineering terms commonly used to describe signals.

Measures in terms of power are particularly comprehensive for stationary Gaussian noise, which is completely characterized by its power spectral density. *Thermal noise*[1] and *shot noise* under constant operating conditions are examples of stationary Gaussian noise. Any noise that is steady or stationary in character, and which originates from the linear superposition of a large number of small independent events, is almost certainly Gaussian. Other types of noise exist that may not be Gaussian—for example, "impulse" noise from ignition systems, atmospheric noise, powerline hum, and crosstalk. Complete characterization of these types requires, in addition to spectral density, other information such as the waveform of a typical pulse or the phase of the interference. However, for practical purposes, treatment by the methods below is often adequate; if not, special methods, not described here, will be necessary to deal with these types of non-Gaussian noise.

For simplicity, linearity of system response is assumed in the test methods to be described. Since noise usually enters the system at a point where the signal level is low, linearity will generally exist in the sense that the principle of superposition applies; that is, signals and noise produce currents and voltages that are simply additive, without the complicated intermodulation effects that occur in nonlinear systems. When the noise arises from independent uncorrelated sources, the spectral densities of these independent noises are also additive in a linear system. Heterodyne systems, incorporating frequency shifters, are linear in this sense when the signals are small.

2. Noise Factor

The system whose noise performance is under consideration here is essentially the linear two-port transducer. Signals enter at the input *port*, are processed internally, and leave the transducer at its output port. Any noise input accompanying the signal input is processed in an identical manner. Noise sources internal to the transducer contribute an additional noise at the output port that is independent of the signal or noise input. The noise performance of the transducer is commonly rated by comparing the noise-power outputs of the actual transducer and of its noise-free equivalent.[2] One such measure of performance is the *noise factor* (*noise figure*). The noise factor, at a specified input frequency, is defined as the ratio of 1) the total noise power per unit bandwidth at a corresponding output frequency available at the output port when the noise temperature of the input termination is *standard* (290°K) to 2) that portion of 1) engendered at the input frequency by the input termination. The standard noise temperature 290°K approximates the actual noise temperature of most input terminations. An alternative but related measure of performance useful for very low-noise transducers designed to operate from input terminations with noise temperatures substantially below 290°K is the "effective input noise temperature." (See definition in Section 5.)

2.1 Variation of Noise Factor with Source Admittance

As defined, the noise factor depends upon the internal structure of the transducer and upon its input termination, but not upon its output termination. Thus, the noise performance of a transducer is meaningfully characterized by its noise factor only if the input termination is specified. The noise factor F of any linear transducer, at a given transducer operating point and input frequency, varies with the admittance Y_s of its input termination (called "source admittance" hereafter) in the following manner:[3]

$$F = F_0 + \frac{R_n}{G_s} | Y_s - Y_0 |^2 \qquad (1)$$

where G_s is the real part of Y_s, and the parameters F_0, Y_0, and R_n characterize the noise properties of the transducer and are independent of its input termination. Thus, the noise performance of a transducer can be meaningfully characterized for all input terminations through specification of the parameters F_0, $Y_0 = G_0 + jB_0$ and R_n.

The "optimum noise factor" F_0, at the given transducer operating point and frequency, is the lowest noise factor that can be obtained through adjustment of the source admittance Y_s. The "optimum source admittance" $Y_0 = G_0 + jB_0$ is that particular value of source admittance Y_s for which this optimum noise factor F_0 is realized. These interpretations of F_0 and Y_0 are evident from (1). The parameter R_n is positive and has the dimensions of a resistance. This parameter appears as the

---

[1] IRE standard definitions of italicized terms may be found in "IRE Standards on Electron Tubes: Definitions of Terms, 1957," 57 IRE 7. S2, Proc. IRE, vol. 45, pp. 983–1010; July, 1957.

[2] The noise-free equivalent of a transducer is a fictitious transducer with identical port-to-port transfer properties but with no internal noise sources.

[3] See tutorial paper, Haus, *et al.*, "Representation of noise in linear twoports," this issue, p. 69.

coefficient of the $|Y_s - Y_0|^2$ term in the general expression for F and, therefore, characterizes the rapidity with which F increases above F_0 as Y_s departs from Y_0.

The necessity of using more than one parameter to specify the noise properties of linear transducers was not at first widely recognized. In the low-frequency triode, for example, a single noise generator characterized by an equivalent noise resistance describes its noise properties with sufficient precision over the range of circuit parameters in which the tube is generally employed. The necessity for additional parameters did not become acute until transistors and high-frequency electron tubes were developed.

The parameters F_0, Y_0 and R_n can be calculated if the noise theory of the transducer is known or, alternatively, can be determined empirically from noise measurements. The empirical method is discussed in Section 4. Until this section is reached, it will be assumed 1) that the input termination of the transducer is specified and 2) that the noise factor under discussion is the noise factor appropriate to this particular input termination. Used in this context, the noise factor meaningfully characterizes the noise performance of the transducer.

2.2 Average Noise Factor

In any communication system the signal is distributed over some finite bandwidth over which both signal and noise time averages may vary with frequency. Theoretically, in treating the interfering effect of the noise, it would be necessary to consider the frequency distributions (spectra) of both noise and signal; but in practice many cases are sufficiently well approximated by considering only the total powers of signal and noise. When only the total noise power in the band need be considered, the noise performance of the transducer is again rated by comparing its actual noise output for some standard noise input to that noise output which would have been obtained had the transducer been noiseless. One such measure of performance is the average noise factor. The average noise factor is defined as the ratio of 1) the total noise power delivered into the output termination by the transducer when the noise temperature of the input termination is standard (290°K) at all frequencies to 2) that portion of 1) engendered by the input termination. For heterodyne systems, 2) includes only that portion of the noise from the input termination which appears in the output via the principal-frequency transformation of the system, and does not include spurious contributions such as those from an image-frequency transformation.

The quantitative relation between the average noise factor $\bar{F}$ and the noise factor $F(f)$ is

$$\bar{F} = \frac{\int F(f)G(f)df}{\int G(f)df} \tag{2}$$

where f is the input frequency and $G(f)$ is the *transducer gain*. The average noise factor is the weighted average of the noise factor over the band in question, the weighting factor being the transducer gain. To emphasize that the noise factor, as opposed to the average noise factor, is a point function of frequency, the term *spot noise factor* may be used. Either average or spot noise factor is a numeric, designating a power ratio, which may be expressed in decibels by multiplying its common logarithm by 10.

The average noise factor also depends upon the internal structure of the transducer and upon the admittance of its input termination, but not upon its output termination, except insofar as the output power mismatch varies with frequency, and thus modifies the frequency dependence of transducer gain.

The noise-factor concept is useful over the entire frequency range of engineering interest. Since it compares the actual noise with the fundamental limit set by thermal agitation, the noise factor gives a broad and direct evaluation of the degree to which a system approaches the noise-free ideal.

3. Measurement of Average Noise Factor

In order to measure the average noise factor, it is necessary to obtain a measure of the noise power that is actually delivered to the output termination. This measure is divided by a similar measure of the output noise that would have been obtained if the transducer were noiseless and merely transmitted the *thermal noise* of the input termination at standard temperature. The following general methods of measurement are suitable for performing the evaluation. In each method a measure of the output noise is taken directly, but the methods differ in the ways of determining the reference noise output of the ideal noiseless transducer.

a. CW-Signal-Generator Method (See Section 3.1): In this method a power meter at the system output and a calibrated signal generator at the system input are used to determine the frequency dependence of transducer gain. From this dependence a "noise bandwidth" is determined. By use of this noise bandwidth, that portion of the output noise resulting from the input-termination noise can be determined. Dividing total output noise by this reference noise gives the average noise factor.

b. Dispersed-Signal Source Method (See Section 3.2): In this method a signal generator having its available power dispersed uniformly over the pass band of the system, and calibrated in terms of available power per unit bandwidth,[4] is used to determine that portion of the

[4] The National Bureau of Standards intends to initiate a calibration service for X-band microwave noise sources on January 1, 1960, and to extend this to other common microwave bands as rapidly as possible. Calibration will be in terms of the noise power delivered by the terminals of the customer's source to a matched waveguide of standard dimensions for the appropriate microwave frequency band. This power will be expressed in db above kT_0B, with an accuracy of 0.1 db. Service will be available at three frequencies within the band. Costs are not yet determined, but past experience indicates that cost of calibration is comparable with the cost of the instrument to be calibrated.

output-noise power which results from the input-termination noise. Suitable dispersed-signal generators are thermionic noise diodes, gas discharge tubes, resistors of known temperature, or an oscillator whose frequency is swept through the band at a uniform rate.

c. Comparison Method (See Section 3.3): This method consists of direct comparison between the network being tested and a secondary standard in the form of a network of the same type for which the average noise factor has been determined.

Of these three methods the first, involving the direct measurement of noise bandwidth, is complicated. The noise-source dispersed-signal method is simpler, and therefore preferred in both laboratory and production situations. The gas-discharge-generator dispersed-signal method is especially useful for high frequencies where the noise output of a diode may be affected by transit-time or lead effects. The swept-oscillator dispersed-signal method may have some application as a supplementary method. Comparison methods are primarily of use in production testing.

The general arrangement of the apparatus for any of these methods is shown in Fig. 1. Certain requirements must be met by each of the three components of Fig. 1.

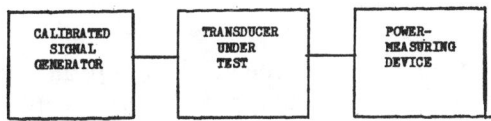

Fig. 1—*Average-noise-factor* test arrangement.

1) Since the calibrated signal generator is the input termination of the transducer, the behavior of the generator output admittance with frequency, over the pass band of the system, must duplicate that of the input termination with which the measured average noise factor is intended to be associated. The signal generator may deliver either power at a single frequency or power distributed over a frequency spectrum. In the latter case, its available power must be uniform over the pass band of the system. In either case, its available power must be accurately known. The calibration of the generator should take account of any elements that have been added to the basic generator to realize desired output admittance characteristics.

2) The transducer under test must be linear in the sense that its available-output-power change for a given available-input-power change is independent of the initial power level for all values used in testing. It may include linear elements or linear frequency shifters, but, particularly, simple envelope detectors must be excluded unless a noise source with a uniform power spectrum is being used as a signal generator. For example, in testing a conventional heterodyne receiver the power-measuring device must be connected ahead of the second detector unless the latter is itself a suitable power meter meeting the requirements described later.

3) The power-measuring device must indicate quantitatively the relative values of a), the noise power at the output of the system with no signal input and b), the total power at the output when an input signal is applied. The measuring device may be either a true power-measuring device, such as a bolometer or thermocouple, or some other type of instrument that has been calibrated to read power for the particular wave forms used in testing.

3.1 CW-Signal-Generator Method

The measurement of *average noise factor* with a CW sinewave signal generator will now be described. The *average noise factor* to be determined may be written

$$\overline{F} = \frac{N_0}{kT_0 \int G(f)df} \quad (3)$$

where

$G(f)$ = transducer gain at frequency f,
N_0 = noise-power output in watts,
k = Boltzmann's constant, 1.38×10^{-23} joules per degree K, and
T_0 = standard noise temperature, 290°K.

It is convenient to introduce quantities G_0 and B such that

$$G_0 B \equiv \int G(f)df \quad (4)$$

where

G_0 = transducer gain at some convenient reference frequency f_0.
B = noise bandwidth of the system in cycles per second (cps), and
$G(f)$ = transducer gain at frequency f.

In general, B is a function of f_0.

With the signal generator connected to the transducer, but with the generator output at zero except for its thermal noise at standard temperature, let the transducer-power output be P_1. With the available signal power of the generator set at P_s at the reference frequency f_0, let the transducer-power output be P_2. Then the transducer gain at f_0 is $G_0 = (P_2 - P_1)/P_s$. Also the noise-power output will correspond to the original value P_1, so that (3) becomes

$$\overline{F} = \frac{1}{\frac{P_2}{P_1} - 1} \frac{P_s}{kT_0 B}. \quad (5)$$

To measure $\overline{F}$ accurately it is desirable to choose P_s so that P_2 is several times greater than P_1. However, for convenience and to avoid saturation, it is common to choose P_s so that $P_2 = 2P_1$. Eq. (5) then becomes

$\bar{F} = P_s/kT_0B$, and P_s may be considered to be a power equivalent to the noise output as referred to the input circuit. It will be noted that, since only the ratio P_2/P_1 enters $\bar{F}$, an absolute calibration of the power-output measuring meter is not necessary, although power ratios must be determined accurately. Alternatively, a calibrated attenuator may be employed between the output of the transducer under test and the power output indicator.

3.1.1 Determination of Noise Bandwidth: The noise bandwidth may be determined from a plot, on linear scales, of the curve of relative transducer gain versus frequency, as shown in Fig. 2. The noise bandwidth is

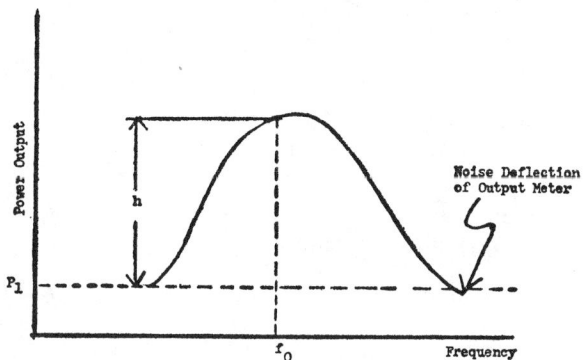

Fig. 2—Transducer relative power output as a function of the frequency of a sine-wave input from a signal generator with fixed *available power* output.

the width, along the frequency axis, of a rectangular response curve with an area M and with a height h at f_0 which are the same as those of the actual response curve. The frequency f_0 is usually, but not always, the frequency of maximum response. The area M and height h are those above the noise contribution to the power output, since only the signal output is of interest here. If the area is M, if the height at f_0 is h, and if the frequency scale is D cps per unit of length, the noise bandwidth B referred to f_0, is MD/h. Since the average noise factor refers only to the principal response of the system, spurious or image responses should be excluded from the response curve.

If the characteristics of the network are such that signals large compared with the noise can be handled without saturation, the effect of noise can be ignored in determining B. That is, the P_1 level in Fig. 2 may become negligibly low. It may be possible to achieve this same effect by reducing the gain of the network to considerably less than its normal value, but it should be ascertained that this does not cause undesirable saturation effects and that bandwidth is not altered by reducing the gain, as may happen if regenerative effects exist.

3.2 Dispersed-Signal-Source Method

Determinations of *average noise* factor may conveniently be made with a signal generator whose available power is distributed uniformly over the response band of the system. This method may eliminate the need for a direct determination of the noise bandwidth, as will be shown below. It is necessary that the available power of the generator in watts per cps of bandwidth be accurately calibrated. If the system is connected as in Fig. 1 with the generator output zero except for the thermal noise at standard temperature, a reading P_1 will be obtained on the power-output meter. (If desired, an equivalent passive termination may be substituted for the generator to obtain P_1.) If the generator is now made to have an available power density of p watts per cps in addition to the initial thermal noise, the power output will be increased by pBG_0 and the new output-meter reading will be P_2 or, in terms of the same symbols as used previously, $G_0 = (P_2 - P_1)/pB$. Also, as before, $N_0 = P_1$, or by substitution in (3),

$$\bar{F} = \frac{1}{\frac{P_2}{P_1} - 1} \frac{p}{kT_0}. \qquad (6)$$

If the temperature T of the internal impedance of the generator (with the generator output zero except for the thermal noise) is not equal to the standard temperature T_0, (6) does not give the correct noise figure expression and a correction has to be introduced:

$$\bar{F} = \left(1 - \frac{T}{T_0}\right) + \frac{1}{\frac{P_2}{P_1} - 1} \frac{p}{kT_0}. \qquad (6a)$$

It should be noted that this measurement may involve spurious or image responses, which will ordinarily give an average noise factor, that is deceptively good, *i.e.*, too low, unless appropriate correction is made. In other words, if the dispersed signal covers pass-band responses not ordinarily used, such as images, which have appreciable gains compared to that of the pass band ordinarily used, the test signal will produce an effect in the output circuit greater than would be produced if the dispersed-signal source were limited so as to include only the response ordinarily used. Hence, if important spurious responses exist, a bandwidth measurement must be made and the *average noise factor* initially determined must be increased in the ratio of 1) the noise bandwidth including all spurious responses, to 2) the noise bandwidth when only the desired response is considered.

3.2.1 Noise-Diode Generators: A temperature-limited diode can be used as a noise generator when connected in a circuit such as that shown in Fig. 3(a). This circuit is typical, but many variations are possible. It is assumed here that the frequency-independent resistive

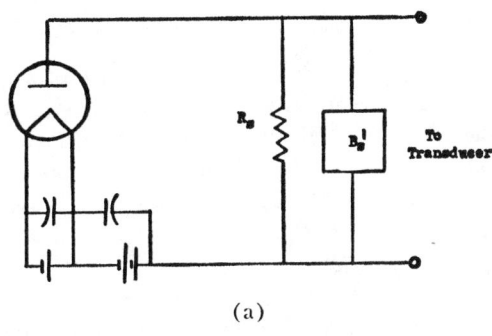

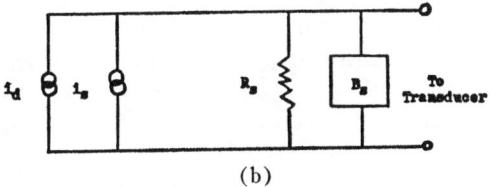

Fig. 3—(a) Typical diode noise-generator circuit; (b) equivalent circuit.

load R_s simulates the real part of the desired transducer source admittance Y_s at every frequency in the pass band of the system. The susceptance B_s' plus any stray susceptance associated with the diode and wiring must simulate the susceptive part B_s of Y_s. The equivalent circuit of this array is shown in Fig. 3(b). The *noise-current generator* i_s represents *thermal noise* in Y_s. For each small element of bandwidth Δf, the mean-square value of i_s for R_s at *standard noise temperature* is

$$\overline{i_s^2} = 4kT_0\Delta f/R_s. \tag{7}$$

The *noise current generator* i_d represents full shot noise in the diode current I_d. If the cathode temperature is adjusted to give a temperature-limited direct plate current of I_d amperes, then, for each small element of bandwidth Δf, the mean-square value of i_d is

$$\overline{i_d^2} = 2eI_d\Delta f, \tag{8}$$

where e is the electronic charge, 1.60×10^{-19} coulomb.

This noise-diode generator circuit has, over and above its thermal noise, a constant available noise power per unit bandwidth of p watts per cps, where

$$p = \frac{eI_dR_s}{2}. \tag{9}$$

Inserting (9) into (6), one obtains for the average noise factor the expression

$$\overline{F} = \frac{1}{\left(\dfrac{P_2}{P_1} - 1\right)} \frac{eI_dR_s}{2kT_0}. \tag{10}$$

Since the numerical value of $kT_0/e = 0.0250$ volt,

$$\overline{F} = 20I_dR_s \frac{1}{\left(\dfrac{P_2}{P_1} - 1\right)} \tag{11}$$

where I_d is in amperes, R_s is in ohms, and P_2/P_1 is the ratio of noise-power outputs.

As before, it is desirable that P_2 be considerably larger than P_1 (preferably several times larger) so that the difference between P_2/P_1 and unity can be determined accurately. For convenience, P_2 is often made equal to $2P_1$. Sometimes a smaller value of P_2 must be used because of limitations imposed on the diode plate current I_d. In such cases additional care must be taken to insure that P_1 and P_2 are stable, repeatable readings to assure reasonable accuracy in the result.

In the foregoing discussions no account has been taken of electron transit time in the diode. If the transit time is an appreciable fraction of a cycle, the noise output of the diode will be lowered.[5] At low frequencies the noise output of diodes may increase above the value (8) because of flicker noise.

3.3 Comparison Methods of Noise Measurement

The methods just described may not be convenient or necessary for production testing. In such cases, noise factors can be checked approximately by carefully chosen comparisons with the performance of a master standard unit of known noise factor. A measurement of signal-to-noise ratio after detection, with a fixed modulated-signal input, may be used for checking individual units, provided that the bandwidths of the networks both preceding and following the detector, and also the detector characteristics, are maintained within close limits.

3.4 Precautions

Care should be taken to insure that the apparatus attached to the network to be measured does not materially affect the bandwidth of the system. For example, if regeneration is introduced that greatly reduces the bandwidth, the noise factor may be markedly altered. In any event, undesired feedback usually makes the measurement much more difficult.

Careful shielding and filtering of input and output elements is essential, particularly when the difference in power level between output and input is large. If the measuring-signal generator has a temperature different from standard temperature, an appropriate correction must be applied.

---

[5] For transit-time correction, see D. B. Fraser, "Noise spectrum of temperature-limited diodes," *Wireless Engr.*, vol. 26, pp. 129–131; April, 1949.

4. Measurement of Spot-Noise Parameters

The noise factor discussed in Section 3 is the weighted average noise factor over the entire pass band of the network. The *spot noise factor*, which is the noise factor at a particular frequency, can be determined by including a very narrow band filter of the desired frequency between the network and the power-measuring device. When the spot frequency is near the center of the band, the factor so obtained may not be greatly different from the average.

4.1 Noise Factor of Transducers in Cascade

Frequently, several networks are connected in cascade, and it is desirable to know how the noise factor of each affects the noise factor of the over-all system. This is necessary both in evaluating the effect of improvement in any part of the system and in measuring the noise factor of a single unit in a system.

For a number of networks in cascade, as shown in Fig. 4, the system spot noise factor is given in terms of

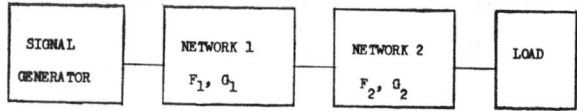

Fig. 4—Networks in cascade.

the component spot noise factors by

$$F = F_1 + (F_2 - 1)/G_1 + (F_3 - 1)/G_1G_2 + \cdots \quad (12)$$

where G_1, G_2, and so forth, are the available power gains of the component networks.[6]

The spot noise factor F_i and the available gain G_i of the ith network are those obtained with a source impedance equal to the impedance presented by the output of the preceding part of the cascade. An analogous expression for average noise factor can be derived, but is more involved. When the frequency for spot noise factor is chosen near the center of the noise-transmission characteristic, (12) for spot noise factor is often a satisfactory approximation for average noise factor.

These formulas also apply to networks that attenuate rather than amplify, in which case the corresponding gains are less than unity. It should be mentioned that the spot noise factor of a passive attenuating network at standard temperature is the factor by which the available power is attenuated in passing through it. It should also be noted that, when each of the various networks has substantial gain and the noise factor of the later stages is not excessive, then the over-all noise factor is largely determined by the noise factor of the first stage as is obvious from (12).

---

[6] If a negative output resistance exists somewhere in the cascade, the noise factor of the over-all system can be calculated as described by H. A. Haus and R. B. Adler, "An extension of the noise figure definitions," Proc. IRE, vol. 45, pp. 690–691; May, 1957.

4.2 The Noise Parameters F_0, G_0, B_0, and R_n

As stated in Section 2, the *noise factor* F of any linear transducer at a given input frequency varies with the admittance $Y_s = G_s + jB_s$ of its input termination in the manner shown in (1), which can be expanded as follows

$$F = F_0 + \frac{R_n}{G_s}[(G_s - G_0)^2 + (B_s - B_0)^2] \quad (13)$$

where G_0 and B_0 are the conductive and susceptive parts respectively of the optimum source admittance Y_0 cited in Section 2.

The four parameters F_0, G_0, B_0, and R_n characterize the noise properties of the transducer and are independent of the source admittance Y_s. These noise parameters can be determined empirically by fitting this four-parameter expression to observed values of F as a function of Y_s.

A suitable program for determining F_0, G_0, B_0, and R_n from noise-factor measurements is the following:

1) With the source conductance G_s maintained constant, measure several values of the *noise factor* F for different values of the source susceptance B_s. Plot the curve F vs B_s and determine the optimum source susceptance B_0.

2) With the source susceptance B_s maintained at its optimum value B_0, measure several values of the *noise factor* F for different values of the source conductance G_s. Plot the curve F vs G_s and determine the optimum source conductance G_0.

3) Using the data already obtained, plot F vs x, where x is the quantity $|Y_s - Y_0|^2/G_s$. These data should lie on a straight line of the form $F = F_0 + R_n x$. The slope of this line is the resistance parameter R_n and its F intercept at $x = 0$ is the optimum noise factor F_0.

If a direct-reading noise-factor meter is available, the noise-factor minima in steps 1) and 2) can be observed directly on this meter, yielding values for B_0 and G_0. Additional data of F vs x for step 3) can then be obtained by using other convenient values of Y_s.

If the points in step 3) do not line on a straight line, then

a) the estimates of B_0 and G_0 in steps 1 and 2 may be inaccurate,[7]
b) the individual noise-factor measurements may be in error, or
c) the transducer under test may be nonlinear.

4.2.1 Equipment: The measuring equipment for determining the noise factor is discussed in Section 3. For the measurement of the noise parameters, it is necessary

---

[7] This can occur if the minima in the F-vs-B_s and -G_s curves are very shallow. In this case, slightly adjusted values of B_0 and G_0 may improve the linearity of the F-vs-x curve.

to provide, in addition, a means for adjusting the source admittance. The type of equipment needed depends on the frequency of the measurement.

An arrangement suitable for use at frequencies below one hundred or two hundred megacycles is shown in Fig. 5. The susceptance is adjustable by means of a calibrated capacitance or inductance and the conductance is varied by using different resistances or a variable resistance. At the higher frequencies of this range, care must be taken that the susceptance B_s is maintained at its optimum value B_0 when the conductance G_s is varied in the measurements specified in step 2). A temperature-limited diode or a resistor at known temperature can be used as a noise source. Gas discharge tubes combined with calibrated attenuators have also been used in the upper part of this frequency range.

Noise generators suitable for use at higher frequencies are, by nature of their construction, devices with fixed output conductance. A calibrated source with variable internal admittance can be made by combining such a generator with a network that acts like a variable-ratio transformer. An arrangement used for measurements on amplifiers at frequencies above 500 mc is shown in Fig. 6.

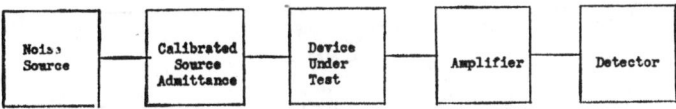

Fig. 5—Test arrangement for determination of noise parameters.

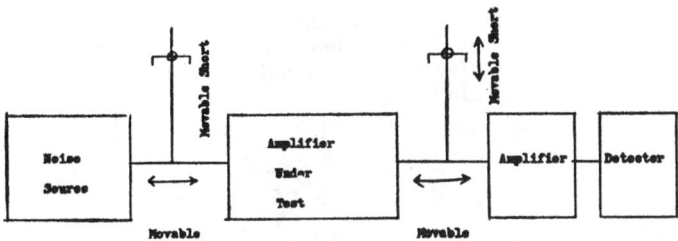

Fig. 6—Test arrangement used for measurements on amplifiers at frequencies above 500 mc.

The transformer consists of a section of transmission line fitted with an adjustable stub that can be moved along the length of the line. Two identical units are used. One unit couples the amplifier to the noise generator or signal generator. The second unit couples the output of the amplifier to the succeeding stage and is used to facilitate the measurement by maximizing the output. The correction due to the noise of the second stage is correspondingly reduced.

4.2.2 Sample Determination of the Noise Parameters at Low Frequencies: The determination of the noise parameters of a germanium alloy junction *p-n-p* transistor, operated in the common-emitter connection at a frequency of 900 cps, is presented in this section.

Fig. 7 shows observed F-vs-B_s data for a constant source conductance of 1.00 millimho. The optimum source susceptance B_0 is zero. Fig. 8 shows observed F-vs-G_s data for $B_s = B_0 = 0$. The optimum source conductance G_0 is 0.5 millimho. Fig. 9 shows the F-vs-x plot obtained from the data used previously in Figs. 7 and 8. The intercept and slope of the line plotted by use of these data yield $F_0 = 1.55$ and $R_n = 540$ ohms. The curves superposed on the experimental data in Figs. 7

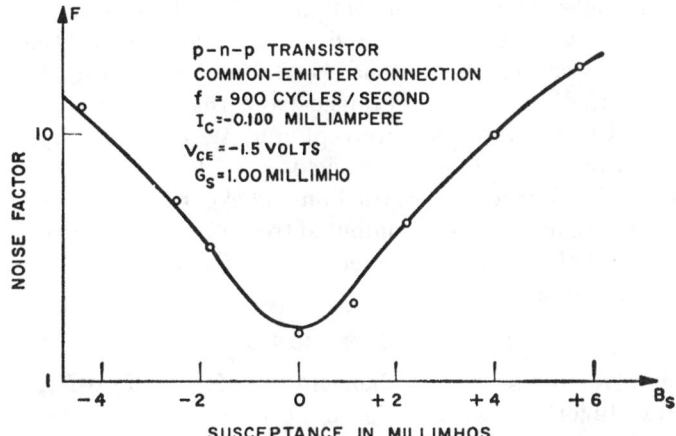

Fig. 7—Noise factor vs source susceptance.

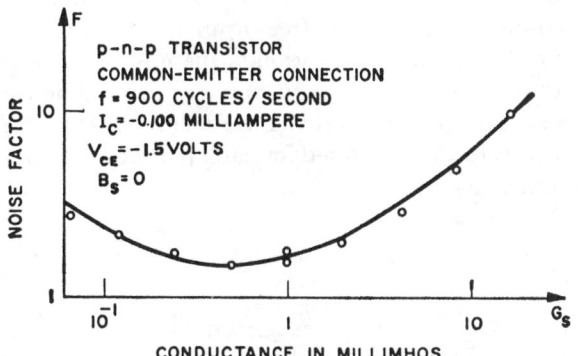

Fig. 8—Noise factor vs source conductance.

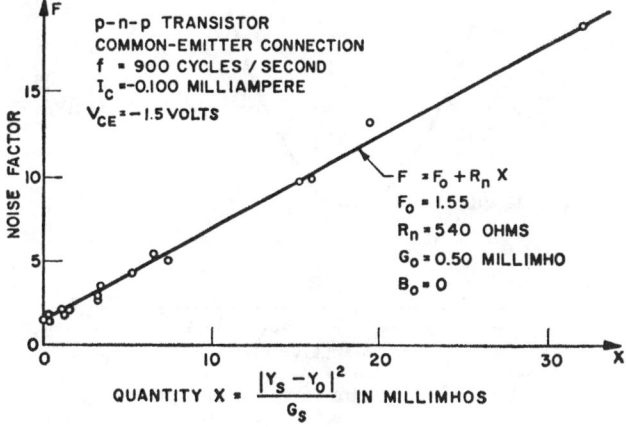

Fig. 9—Noise factor vs quantity x.

and 8 were computed from (13) for the values of the noise parameters F_0, G_0, B_0, and R_n determined above.

4.2.3 Sample Determination of the Noise Parameters at High Frequencies: The determination of the noise parameter for a disk-seal triode, made at a frequency of 870 mc, is described in this section. The arrangement of equipment shown in Fig. 6 was used. A series of readings with input-stub position and length chosen to give $G_s = 38$ millimhos for various values of B_0 provided the data shown in Fig. 10. The optimum source susceptance B_0 is found to be -54.0 millimhos. Fig. 11 shows observed F-vs-G_s data for $B_s = B_0 = -54.0$ millimhos. The optimum source conductance is 32.0 millimhos. Fig. 12 shows the F-vs-x plot obtained from the data used in Figs. 10 and 11. The intercept and slope of the line drawn by use of these data yield $F_0 = 9.9$ and $R_n = 117$ ohms. The curves superposed on the experimental data in Figs. 10 and 11 were computed from (13) by using the values of the noise parameters F_0, G_0, B_0, and R_n determined above.

5. Definition

Effective Input Noise Temperature (of a Two-Port Transducer). The input-termination *noise temperature* which, when the input termination is connected to a noise-free equivalent of the transducer, would result in the same output noise power as that of the actual transducer connected to a noise-free input termination.

Note 1: For heterodyne systems there is, in principle, more than one output frequency corresponding to a single input frequency, and vice versa; an *Effective Input Noise Temperature* is defined for each pair of corresponding frequencies.

Note 2: The *Effective Input Noise Temperature* depends upon the impedance of the input termination.

Note 3: The *Effective Input Noise Temperature*, T_e, in degrees Kelvin is related to the *Noise Factor* F by the equation

$$T_e = 290(F - 1).$$

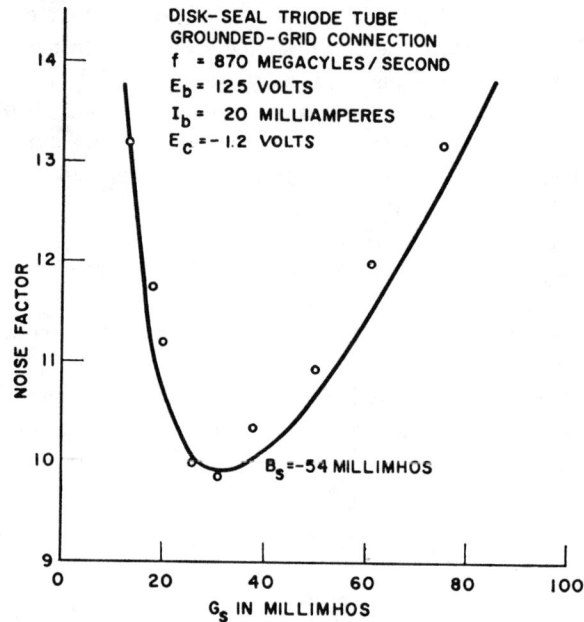

Fig. 11—Noise factor vs source conductance.

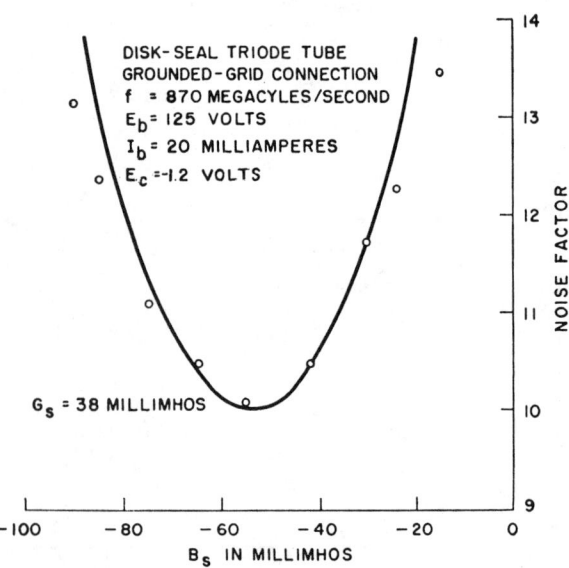

Fig. 10—Noise factor vs source susceptance.

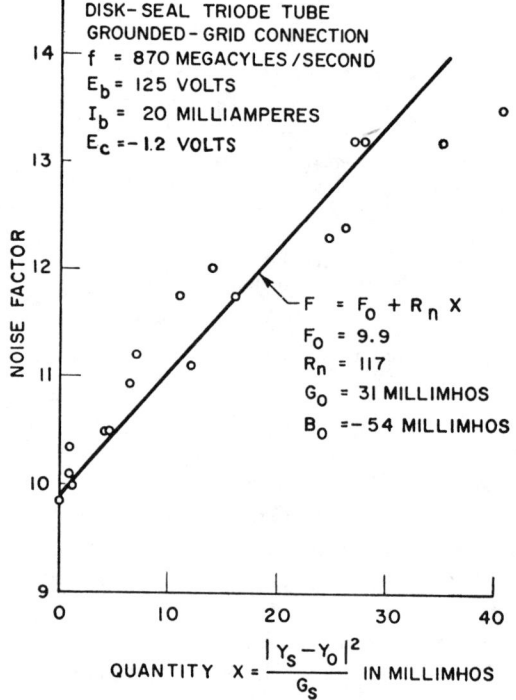

Fig. 12—Noise factor vs quantity x.

Representation of Noise in Linear Twoports*

IRE Subcommittee 7.9 on Noise
H. A. Haus, *Chairman*

W. R. Atkinson	W. H. Fonger	W. W. McLeod
G. M. Branch	W. A. Harris	E. K. Stodola
W. B. Davenport, Jr.	S. W. Harrison	T. E. Talpey

The compilation of standard methods of test often requires theoretical concepts that are not widely known nor readily available in the literature. While theoretical expositions are not properly part of a Standard they are necessary for its understanding and it seems desirable to make them easily accessible by simultaneous publication with the Standard. In such cases a technical committee report provides a convenient means of fulfilling this function.

The Standards Committee has decided to publish this technical committee report immediately following "Standards on Methods of Measuring Noise in Linear Twoports." A copy of this paper will be attached to all reprints of the Standards.

Summary—This is a tutorial paper, written by the Subcommittee on Noise, IRE 7.9, to provide the theoretical background for some of the IRE Standards on Methods of Measuring Noise in Electron Tubes. The general-circuit-parameter representation of a linear twoport with internal sources and the Fourier representations of stationary noise sources are reviewed. The relationship between spectral densities and mean-square fluctuations is given and the noise factor of the linear twoport is expressed in terms of the mean-square fluctuations of the source current and the internal noise sources. The noise current is then split into two components, one perfectly correlated and one uncorrelated with the noise voltage. Expressed in terms of the noise voltage and these components of the noise current, the noise factor is then shown to be a function of four parameters which are independent of the circuit external to the twoport.

I. Introduction

ONE of the basic problems in communication engineering is the distortion of weak signals by the ever-present thermal noise and by the noise of the devices used to process such signals. The transducers performing signal processing such as amplification, frequency mixing, frequency shifting, etc., may usually be classified as twoports. Since weak signals have amplitudes small compared with, say, the grid bias voltage of a vacuum tube, or emitter bias current of a transistor, etc., the amplitudes of the excitations at the ports can be linearly related. Consequently, the description and measurement of noise that is presented in this paper, and which forms the basis for the "Methods of Test" described in this same issue,[1] can be restricted to linear noisy twoports. Even if inherently nonlinear characteristics are involved, as in mixing, etc., linear relations still exist among the signal input and output amplitudes, although these may not be associated with the same frequency. Image-frequency components may be eliminated by proper filtering, but a generalization of our results to cases in which image frequencies are present is not difficult.

The effect of the noise originating in a twoport when the signal passes through the twoport is characterized at any particular frequency by the (*spot*) *noise factor* (*noise figure*). Since this has meaning only if the source impedance used in obtaining the noise factor is also specified, the noise contribution of a twoport is often indicated by a minimum noise factor and the source impedance with which this minimum noise factor is achieved. However, the extent to which the noise factor depends upon the input source impedance, upon the amount of feedback, etc., can be indicated only by a more detailed representation of the linear twoport.[2,3]

As a basis for understanding the representation of networks containing (statistical) noise sources, we shall consider first the analysis of networks containing Fourier-transformable signal sources. We shall then describe noise in two ways: a) as a limit of Fourier-integral transforms, and b) as a limit of Fourier-series transforms. The reasons for the wider use of the latter description will be presented. With this background we shall be ready to show that the four noise parameters used in the standard "Methods of Test" completely characterize a noisy linear twoport.

II. Representations of Linear Twoports

The excitation at either port of a linear twoport can be completely described by a time-dependent voltage $v(t)$ and the time-dependent current $i(t)$. (For waveguide twoports, where no voltage or current can be identified uniquely, an equivalent voltage and current can always

---

* Original manuscript received by the IRE, September 15, 1958; revised manuscript received March 6, 1959.
[1] "IRE Standards on Methods of Measuring Noise in Linear Twoports, 1959," this issue, p. 60.

[2] H. Rothe and W. Dahlke, "Theory of noisy fourpoles," Proc. IRE, vol. 44, pp. 811–818; June, 1956. Also, "Theorie rauschender Vierpole," *Arch. elekt. Übertragung*, vol. 9, pp. 117–121; March, 1955.
[3] A. G. T. Becking, H. Groendijk, and K. S. Knol, "The noise factor of four-terminal networks," *Philips Res. Repts.*, vol. 10, pp. 349–357; October, 1955.

be used.) Let it be assumed that the voltage and current functions can be transformed from the time domain to the frequency domain, and that V and I stand for the Fourier transforms when the function is aperiodic and for the Fourier amplitudes when the function is periodic. The linearity of the twoport *without internal sources* then allows an impedance representation:

$$V_1 = Z_{11}I_1 + Z_{12}I_2$$
$$V_2 = Z_{21}I_1 + Z_{22}I_2. \quad (1)$$

The subscripts 1 and 2 refer to the input and output ports, respectively, and the coefficients Z_{jk} are, in general functions of frequency. The currents are defined to be positive if the flow is into the network as shown in Fig. 1.

If the twoport contains internal sources, then (1) and the equivalent circuit must be modified. By a generalization of Thévenin's theorem, the twoport may be separated into a source-free network and two voltage generators, one in series with the input port and one in series with the output port. If the time-dependent functions $e_1(t)$ and $e_2(t)$ describing these equivalent generators can be transformed to functions of frequency E_1 and E_2, respectively, the impedance representation of the linear twoport *with internal sources* becomes

$$V_1 = Z_{11}I_1 + Z_{12}I_2 + E_1$$
$$V_2 = Z_{21}I_1 + Z_{22}I_2 + E_2 \quad (2)$$

and the equivalent network is that shown in Fig. 2.

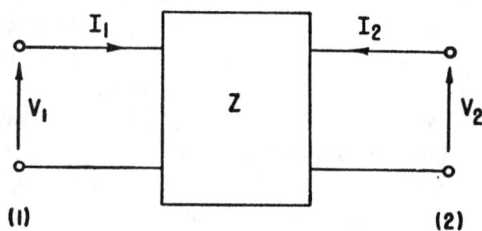

Fig. 1—Sign convention for impedance representation of linear twoport.

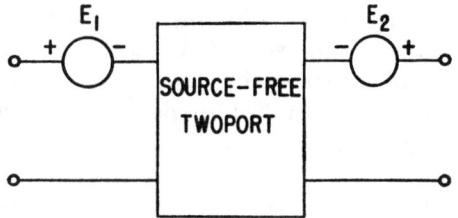

Fig. 2—Separation of twoport with internal sources into a source-free twoport and external voltage generators.

For the purpose of the analysis to follow, it should be emphasized again that such equations in general characterize the behavior of the twoport as a function of frequency. For practical purposes, it should be pointed out that in most cases the impedance parameters of a linear twoport can be measured at a particular frequency by applying sinusoidal voltages that produce outputs large compared with those caused by the internal sources.

The impedance representation of a linear twoport with internal sources has been reviewed because of its familiarity. However, it is well known that other representations, each leading to a different separation of the internal sources from the twoport, are possible. A particularly convenient one for the study of noise is the general-circuit-parameter representation[4]

$$V_1 = AV_2 + BI_2 + E$$
$$I_1 = CV_2 + DI_2 + I \quad (3)$$

where E and I are again functions of frequency which are the Fourier transforms of the time-dependent functions $e(t)$ and $i(t)$ describing the internal sources.

As shown in Fig. 3, the internal sources are now represented by a source of voltage acting in series with the input voltage and a source of current flowing in parallel with the input current. It will be seen that this particular representation of the internal sources leads to four noise parameters that can easily be derived from single-frequency measurements of the twoport noise factor as a function of input mismatch. It has the further advantage that such properties of the twoport as gain and input conductance do not enter into the noise factor expression in terms of these four parameters.

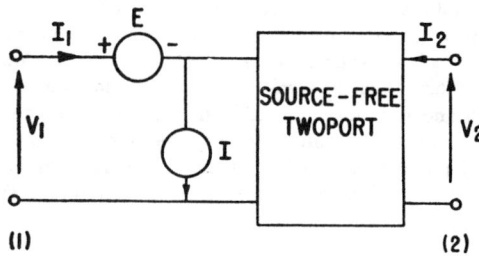

Fig. 3—Separation of twoport with internal sources into a source-free twoport and external input current and voltage sources.

III. Representations of Stationary Noise Sources

When a linear twoport contains stationary internal noise sources, the frequency functions E and I in (3) cannot be found by conventional Fourier methods from the time functions $e(t)$ and $i(t)$, since these are now random functions extending over all time and have infinite energy content. Two alternatives are possible. One may consider the substitute functions

$$e(t, T) = e(t) \quad \text{for } -\frac{T}{2} < t < \frac{T}{2}$$
$$= 0 \quad \text{for } \frac{T}{2} < |t|$$
$$i(t, T) = i(t) \quad \text{for } -\frac{T}{2} < t < \frac{T}{2}$$
$$= 0 \quad \text{for } \frac{T}{2} < |t| \quad (4)$$

[4] E. Guillemin, "Communication Networks," John Wiley and Sons, New York, N. Y., vol. 2, p. 138; 1935.

where T is some long but finite time interval, and use the Fourier transforms

$$E(\omega, T) = \frac{1}{2\pi}\int_{-T/2}^{T/2} e(t, T)e^{-j\omega t}dt$$

$$I(\omega, T) = \frac{1}{2\pi}\int_{-T/2}^{T/2} i(t, T)e^{-j\omega t}dt. \quad (5)$$

Alternatively, one may construct periodic functions,

$$e(t, T) = e(t) \quad \text{for } -\frac{T}{2} < t < \frac{T}{2}$$

$$e(t + nT, T) = e(t, T) \quad \text{with } n \text{ an integer,}$$

$$i(t, T) = i(t) \quad \text{for } -\frac{T}{2} < t < \frac{T}{2}$$

$$i(t + nT, T) = i(t, T) \quad \text{with } n \text{ an integer,} \quad (6)$$

and expand these functions into Fourier series with amplitudes

$$E_m(\omega, T) = \frac{1}{T}\int_{-T/2}^{T/2} e(t, T)e^{-j\omega t}dt$$

$$I_m(\omega, T) = \frac{1}{T}\int_{-T/2}^{T/2} i(t, T)e^{-j\omega t}dt \quad (7)$$

where $\omega = m2\pi/T$ with m an integer, and T is again finite. In either case, the substitute functions can be made to approach the actual functions as closely as desired by making the interval T larger and larger.

In either the Fourier-integral or the Fourier-series approach, we consider a set of substitute functions obtained in principle from a series of measurements a) on an ensemble of systems with identical statistical properties or b) on one and the same system at successive time intervals sufficiently separated so that no statistical correlation exists.[5]

Since noise is a statistical process, we are in general interested in statistical averages[6] rather than the exact details of any particular noise function. These averages are important since they relate to physically measurable stationary quantities. For example, in the Fourier-integral approach the spectral density of the noise-voltage excitation is defined by

$$W_e(\omega) = \lim_{T \to \infty} \frac{\overline{|E(\omega, T)|^2}}{2\pi T} \quad (8)$$

where the bar indicates an arithmetic average of the Fourier transforms of an ensemble of functions described by (4). This spectral density is proportional to the noise power (in a narrow frequency band[7] around a given frequency) in a resistor across which the fluctuating voltage $e(t)$ appears.

Unfortunately, the notation developed in the literature for dealing with spectral densities is unwieldy, since the quantity with which the density is associated is relegated to a subscript. This is one of the reasons why researchers on noise have tended to use the historically older Fourier-series approach. The use of spectral densities is preferred in rigorous mathematical treatments of noise and in questions involving definitions of noise processes, since this assures that all points on the frequency axis within a narrow range are equivalent. For practical purposes, however, the noise amplitudes that are associated with discrete frequencies in the Fourier-series method can be as closely distributed as desired by choosing the time interval T sufficiently large.

IV. Relationship of Spectral Densities and Fourier Amplitudes to Mean-Square Fluctuations

If there is an open-circuit noise voltage $v(t)$ across a terminal pair, the mean-square value $\overline{v^2(t)}$ is related to the spectral density $W_v(\omega)$ and to the Fourier amplitudes $V_m(\omega)$ as follows:

$$\overline{v^2(t)} = \lim_{T \to \infty} \frac{1}{T}\int_{-T/2}^{T/2} v^2(t)dt$$

$$= \int_{-\infty}^{+\infty} W_v(\omega)d\omega = \sum_{m=-\infty}^{\infty} \overline{|V_m(\omega)|^2} \quad (9)$$

where the amplitudes V_m are now those obtained as $T \to \infty$.

Suppose that the stationary time function $v(t)$ is passed through an ideal band-pass filter with the narrow bandwidth $\Delta f = \Delta\omega/2\pi$ centered at a frequency $f^0 = \omega^0/2\pi$. The mean-square value of the voltage appearing at the filter output, usually denoted by the symbol $\overline{v^2}$ and called the mean-square fluctuation of $v(t)$ within the frequency interval Δf" is

$$\overline{v^2} = 4\pi\Delta f W_v(\omega_0) = 2\overline{|V_m|^2}. \quad (10)$$

This equation relates the mean-square fluctuations, the spectral density, and the Fourier amplitude. If the filter is narrow-band but not ideal, the quantity Δf is the noise bandwidth.[1] The factor 2 in (10) arises because $W_v(\omega)$ and V_m have been defined on the negative as well as the positive frequency axis.

Eq. (10) shows that spectral densities and mean-square fluctuations are equivalent. Although mathematical limits are involved in the definitions, these

---

[5] The equivalence of the ensemble and time average is known as the ergodic hypothesis. An interesting discussion of the implications of this hypothesis can be found in Born, "Natural Philosophy of Cause and Chance," Oxford University Press, New York, N. Y., p. 62; 1948. See also W. B. Davenport, Jr. and W. L. Root, "An Introduction to the Theory of Random Signals and Noise," McGraw-Hill Book Company, Inc., New York, N. Y., p. 66 ff.; 1958.

[6] W. R. Bennett, "Methods of solving noise problems," PROC. IRE, vol. 44, pp. 609–637; May, 1956.

[7] A frequency interval Δf is called narrow if the physical quantities under consideration are independent of frequency throughout the interval, and if $\Delta f \ll f_0$, where f_0 is the center frequency.

quantities can be measured to any desired degree of accuracy. For example, if one measures the power flowing into a termination connected to the terminals with which $v(t)$ is associated, one measures essentially $W_v(\omega_0)$, provided that the following requirements are met:

1) The termination is known and has a high impedance so that the voltage across the terminals remains essentially the open-circuit voltage $v(t)$, or the internal impedance associated with the two terminals is also known so that any change in voltage can be computed. The noise voltage contributed by the termination must either be negligible, or its statistical properties must be known so that its effect can be taken into account.

2) The termination is fed through a filter with a pass band sufficiently narrow so that the spectral density and internal impedance are essentially constant throughout the band; or the termination is itself a resonant circuit with a high Q corresponding to a sufficiently narrow bandwidth.

3) The power measurement is made over a period of time that is long compared to the reciprocal bandwidth, $1/\Delta f$, of the filter. In this way the power-measuring device takes an average over many long time intervals that is equivalent to an ensemble average.

In the description of noise transformations by linear twoports that follows, the mean-square fluctuations will be used. Mean-square current fluctuations can be related to the Fourier amplitudes $I_m(\omega)$ by a procedure similar to the one just described. Since fluctuations of the cross products of statistical functions will also be used it should be noted that

$$\lim_{T \to \infty} \frac{1}{T} \int_{-T/2}^{T/2} v(t,T) i(t,T) dt = \sum_{m=-\infty}^{\infty} \overline{V_m I_m^*} \quad (11)$$

where $i(t)$ is a noise current and V_m and I_m are again the amplitudes obtained as $T \to \infty$.

We may interpret the real part of the complex quantity $2\overline{V_m I_m^*}$ as the contribution to the average of vi from the frequency increment $\Delta f = 1/T$ at the angular frequency $\omega = m\Delta\omega$. However, the imaginary part of $\overline{V_m I_m^*}$ also contains phase information that has to be used in noise computations. We shall, therefore, use the expression

$$\overline{vi^*} = 2\overline{V_m I_m^*} \quad \text{for } m > 0, \omega > 0 \quad (12)$$

as the complex cross-product fluctuations of $v(t)$ and $i(t)$ in the frequency increment Δf. They are related to the cross-spectral density by

$$\overline{vi^*} = 4\pi \Delta f W_{iv}(\omega) \quad (13)$$

where

$$W_{iv}(\omega) = \lim_{T \to \infty} \frac{\overline{I^*(\omega,T)V(\omega,T)}}{2\pi T}. \quad (14)$$

There are several ways of specifying the fluctuations (or the spectral densities) that characterize the internal noise sources. A mean-square voltage fluctuation can be given directly in units of volt^2 second. It is often convenient, however, to express this quantity in resistance units by using the Nyquist formula, which gives the mean-square fluctuation of the open-circuit noise voltage of a resistor R at temperature T as

$$\overline{e^2} = 4kTR\Delta f$$

where k is Boltzmann's constant. For any mean-square voltage fluctuation $\overline{e^2}$ within the frequency interval Δf, one defines the equivalent noise resistance R_n as

$$R_n = \frac{\overline{e^2}}{4kT_0\Delta f} \quad (15)$$

where T_0 is the standard temperature, 290°K. The use of R_n has the advantage that a direct comparison can be made between the noise due to internal sources and the noise of resistances generally present in the circuit. Note that R_n is not the resistance of a physical resistor in the network in which e is a physical noise voltage and therefore does not appear as a resistance in the equivalent circuit of the network.

In a similar manner, a mean-square current fluctuation can be represented in terms of an equivalent noise conductance G_n which is defined by

$$G_n = \frac{\overline{i^2}}{4kT_0\Delta f}. \quad (16)$$

V. Noise Transformations by Linear Twoports

The statistical averages discussed in Sections III and IV will now be used to describe the internal noise sources of a linear twoport. We may consider functions of the type given by (6) and represent the noise sources E and I in (3) by the Fourier amplitudes $E_m(\omega, T)$ and $I_m(\omega, T)$. Thus, if the circuit shown in Fig. 3 represents a separation of noise sources from a linear twoport, the noise-free circuit is preceded by a noise network. Since a noise-free network connected to a terminal pair does not change the signal-to-noise ratio (noise factor evaluated at that terminal pair) the noise factor of the over-all network is equal to that of the noise network.

To derive the noise factor, let us connect the noise network to a statistical source comprising an internal admittance Y_s and a current source again represented by a Fourier amplitude $I_s(\omega, T)$. The network to be used for the noise-factor computation is then as shown in Fig. 4. By definition,[8] *the spot noise factor (figure) of a network at a specified frequency is given by the ratio*

[8] "IRE Standards on Electron Tubes: Definitions of Terms, 1957," Proc. IRE, vol. 45, pp. 983–1010; July, 1957.

of 1) the total output noise power per unit bandwidth available at the output port to 2) that portion of 1) engendered by the input termination at the standard temperature T_0.

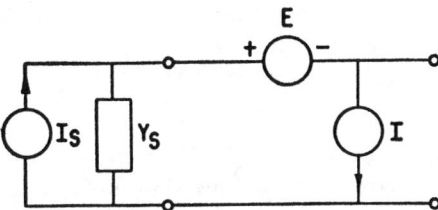

Fig. 4—Truncated network for noise-factor computation.

Now the total short-circuit noise current at the output of the network shown in Fig. 4 can be represented in terms of Fourier amplitudes by

$$I_s(\omega, T) + I_m(\omega, T) + Y_s E_m(\omega, T).$$

Let us assume that the internal noise of the twoport and the noise from the source are uncorrelated. If we then square the total short-circuit noise-current Fourier amplitudes, take ensemble averages and use equations similar to (10) and (12) to introduce mean-square fluctuations, we obtain a mean-square current fluctuation

$$\overline{i_s^2} + \overline{|i + Y_s e|^2} = \overline{i_s^2} + \overline{i^2} + |Y_s|^2 \overline{e^2} + Y_s^* \overline{ie^*} + Y_s \overline{i^* e} \qquad (17)$$

to which the total output noise power is proportional. It should be noted here that when e and i are complex, the symbols $\overline{e^2}$ and $\overline{i^2}$ denote, by convention, $\overline{|e|^2}$ and $\overline{|i|^2}$. Since the noise power due to the source alone is proportional to $\overline{i_s^2}$, the noise factor becomes

$$F = 1 + \frac{\overline{|i + Y_s e|^2}}{\overline{i_s^2}}. \qquad (18)$$

In the denominator, the mean-square source noise current is related to the source conductance G_s by the Nyquist formula

$$\overline{i_s^2} = 4kT_0 G_s \Delta f. \qquad (19)$$

In the numerator of (18), the four real variables involved in $\overline{i^2}$, $\overline{e^2}$ and $\overline{ie^*}$, where e^* is the complex conjugate of e, describe the internal noise sources. If the Fourier transforms, (5), are used to characterize the noise, the self-spectral densities of noise current and voltage and the cross-spectral density of these quantities would have to be specified.

Before proceeding, let us review in general terms the methods employed and the conclusions reached. The representation of the internal noise sources of a noisy linear twoport by a voltage generator and a current generator lumps the effect of all internal noise sources into the two generators. A complete specification of these generators is thus equivalent to a complete description of the internal sources, as far as their contribution to terminal voltages and currents is concerned. Since the sources under consideration are noise sources, their description is confined to the methods applicable to noise. The extent and detail of the description depends on the amount of detail envisioned in the analysis. Since, in the case of the noise factor, only the mean-square fluctuations of output currents are sought, the specification of self and cross-product fluctuations of the generator voltages and currents is adequate.

The expression for the noise factor given in (18) can be simplified if the noise current is split into two components, one perfectly correlated and one uncorrelated with the noise voltage. The uncorrelated noise current, designated by i_u, is defined at each frequency by the relations

$$\overline{e i_u^*} = 0 \qquad (20)$$

$$\overline{(i - i_u) i_u^*} = 0. \qquad (21)$$

The correlated noise current, $i - i_u$, can be written as $Y_\gamma e$, where the complex constant $Y_\gamma = G_\gamma + jB_\gamma$ has the dimensions of an admittance and is called the correlation admittance. The cross-product fluctuation $\overline{ei^*}$ may then be written

$$\overline{ei^*} = \overline{e(i - i_u)^*} = Y_\gamma^* \overline{e^2}. \qquad (22)$$

The noise-voltage fluctuation can be expressed in terms of an equivalent noise resistance R_n as

$$\overline{e^2} = 4kT_0 R_n \Delta f \qquad (23)$$

and the uncorrelated noise-current fluctuation in terms of an equivalent noise conductance G_u as

$$\overline{i_u^2} = 4kT_0 G_u \Delta f. \qquad (24)$$

The fluctuations of the total noise current are then

$$\overline{i^2} = \overline{|i - i_u|^2} + \overline{i_u^2}$$
$$= 4kT_0 [|Y_\gamma|^2 R_n + G_u] \Delta f. \qquad (25)$$

From (18)–(24), the formula for the noise factor becomes

$$F = 1 + \frac{1}{4kT_0 G_s \Delta f} [\overline{i_u^2} + |Y_s + Y_\gamma|^2 \overline{e^2}] \qquad (26)$$

$$= 1 + \frac{G_u}{G_s} + \frac{R_n}{G_s} [(G_s + G_\gamma)^2 + (B_s + B_\gamma)^2]. \qquad (27)$$

Thus, the noise factor is a function of the four parameters G_u, R_n, G_γ and B_γ. These depend, in general, upon the operating point and operating frequency of the twoport, but not upon the external circuitry. In a vacuum tube triode the correlation susceptance B_γ is negligibly small at frequencies such that transit times are small compared to the period. Also, G_u and G_γ are vanishingly small when there is little grid loading. Thus, tube noise at low frequencies is adequately characterized by the single nonzero constant R_n. Tube noise at high frequencies and transistor noise at all frequencies have no such simple representation.

Since the noise factor is an explicit function of the source conductance and susceptance, it can readily be shown that the noise factor has an optimum (minimum) value at some optimum source admittance $Y_o = G_o + jB_o$ where

$$G_o = \left[\frac{G_u + R_n G_\gamma^2}{R_n}\right]^{1/2} \quad (28)$$

$$B_o = -B_\gamma \quad (29)$$

and the value of this minimum noise factor is

$$F_o = 1 + 2R_n(G_\gamma + G_o). \quad (30)$$

In terms of G_o, B_o and F_o, the noise factor for any arbitrary source impedance then becomes

$$F = F_o + \frac{R_n}{G_s}[(G_s - G_o)^2 + (B_s - B_o)^2]. \quad (31)$$

Eq. (31) shows that the four real parameters F_o, G_o, B_o and R_n give the noise factor of a twoport for every input termination of the twoport.[9] As shown in the methods of test that are published in this issue, a measurement of the minimum noise factor and of the source admittance Y_o with which F_o is achieved gives the first three parameters. The parameter R_n can be computed from an additional measurement of the noise factor for a source admittance Y_s other than Y_o.

From the given values, F_o, G_o, B_o and R_n, one can compute, if desired, the noise fluctuations $\overline{i^2}$, $\overline{e^2}$, and $\overline{ei^*}$ (or the corresponding spectral densities). To do this, one uses (28) to (30) to find Y_γ and G_u. From these one evaluates $\overline{e^2}$, $\overline{i^2}$, and $\overline{ei^*}$ from (23), (25), and (22). The fluctuations of any terminal voltage or current of the twoport produced with a given source and load can then be evaluated from the known coefficients, (A, B, C, D) of (3), and the known noise fluctuations. Thus, the noise in a twoport is completely characterized (with regard to the fluctuations or the spectral densities at the terminals) by the noise fluctuations $\overline{e^2}$, $\overline{i^2}$ and $\overline{ei^*}$, or alternately by the four noise parameters F_o, G_o, B_o and R_n.

VI. Conclusion

The preceding discussion showed that, with limited objectives, the noise in a linear twoport can be characterized adequately by a limited number of parameters. Thus, if one seeks only information concerning the mean-square fluctuations or the spectral densities of currents or voltages into or across the ports of a linear twoport at a particular frequency and for arbitrary circuit connections of the twoport, it is sufficient to specify two mean-square fluctuations and one product fluctuation or two spectral densities and one cross-spectral density. This involves the specification of four real numbers. If a band of frequencies is being considered, these quantities have to be given as functions of frequency unless they are approximately constant over the band.

Different separations of internal noise sources will lead, in general, to different frequency dependences of the resulting fluctuations or spectral densities. Thus, a particular separation of the noise sources may be preferable to another separation if it is found that fluctuations (spectral densities) are less frequency-dependent in the band. In particular, if available information about the physics of the noise in a particular device suggests introducing, *inside the twoport*, appropriate noise generators of fluctuations (spectral densities) with no frequency dependence, or with a simple frequency dependence, it may be more advantageous to specify the device in terms of these physically suggestive generators.

However, usually one resorts to the characterizations of noise presented in this paper, since they have the advantage that they do not require any knowledge of the physics of the internal noise. Furthermore, noise-factor measurements performed on the twoport yield (more or less) directly the noise parameters of the general circuit representation, namely the fluctuations (spectral densities) of the voltage and current generators attached to the input of the noise-free equivalent of the twoport. This fact recommends the use of the noise parameters natural to the general-circuit-parameter representation.

---

[9] Reasoning similar to that leading to (31) can be carried out in a dual representation, where impedances and admittances are interchanged. An equation similar to (31) results, again involving four noise parameters, some of which are different from the ones used here.

The Determination of Device Noise Parameters

Abstract—A novel noise measurement technique is outlined which results in data that directly give the noise parameters of the test device when processed by a simple computer program.

The usual method[1] of obtaining device noise parameters ($F_{\min}$, R_n, G_{opt}, B_{opt}) entails an experimental search for the "minimum" noise figure and the source admittance that results in this "minimum."

There are two reasons why this search technique is unsatisfactory in practice. First, it is tedious, and it is inaccurate in determining G_{opt}, B_{opt} since the partial derivative of noise factor, with respect to source admittance, is zero at the noise factor minimum. Second, the presence of transformation-dependent loss in the input matching network biases the obtained value of Y'_{opt}, usually in the direction of $G'_{opt} > G_{opt}$ and $B'_{opt} < B_{opt}$, since this minimizes the noise factor insertion loss product.

The methods that follow avoid these practical difficulties and by use of a simple computer program can simplify noise characterization. The noise behavior of a two-port may be expressed as

$$F = F_{\min} + \frac{R_n}{G_s}|Y_s - Y_{opt}|^2 \qquad (1)$$

where $Y_{opt} = G_{opt} + jB_{opt}$ is the particular source admittance realizing $F_{\min}$, and $Y_s = G_s + jB_s$ is the source admittance.

In principle, four (nonsingular) measurements of noise factor from different source admittances will determine the four real numbers ($F_{\min}$, R_n, G_{opt}, B_{opt}), the noise parameters. However, since there are bound to be some experimental errors both in the measurement of the noise factor ($\sim \pm 10$ percent) and in the measurement of the admittance at which it occurs (± 10 percent), a few (three) redundant measurements will allow a statistical smoothing of the experimentally determined surface.

In order to use a readily available IBM subroutine[2] for simultaneous equation solution, (1) may be cast in a form[3] that is linear with respect to four new parameters A, B, C, and D:

$$F = A + BG_s + \frac{C + BB_s^2 + DB_s}{G_s}. \qquad (2)$$

Equation (2) represents a hyperplane having the four new parameters A, B, C, and D, where

$$F_{\min} = A + \sqrt{4BC - D^2} \qquad (3)$$

$$R_n = B \qquad (4)$$

$$G_{opt} = \frac{\sqrt{4BC - D^2}}{2B} \qquad (5)$$

$$B_{opt} = \frac{-D}{2B}. \qquad (6)$$

We have usually a priori knowledge that both $F_{\min}$ and G_{opt} are positive thus allowing proper selection of the roots in (3) and (5).

A least-squares fit of the seven observed noise factors to the plane of (2) is sought; therefore, the following error criterion is established:

$$\varepsilon \equiv \frac{1}{2}\sum_{i=1}^{7} W_i\left[A + B\left(G_i + \frac{B_i^2}{G_i}\right) + \frac{C}{G_i} + \frac{DB_i}{G_i} - F_i\right]^2 \qquad (7)$$

where W_i is a weighting factor to be used if certain data are known to be less reliable than the average. Then,

$$\frac{\partial \varepsilon}{\partial A} = \sum_{i=1}^{7} W_i P = 0 \qquad (8)$$

$$\frac{\partial \varepsilon}{\partial B} = \sum_{i=1}^{7} W_i\left(G_i + \frac{B_i^2}{G_i}\right)P = 0 \qquad (9)$$

$$\frac{\partial \varepsilon}{\partial C} = \sum_{i=1}^{7} W_i \frac{P}{G_i} = 0 \qquad (10)$$

$$\frac{\partial \varepsilon}{\partial D} = \sum_{i=1}^{7} W_i \frac{B_i}{G_i}P = 0 \qquad (11)$$

where

$$P = \left[A + B\left(G_i + \frac{B_i^2}{G_i}\right) + \frac{C}{G_i} + \frac{DB_i}{G_i} - F_i\right].$$

Equations (8) through (11) are solved by an IBM 1130 computer yielding the noise parameters via (3) through (6).

A set of gain parameters[4] may be obtained by substitution of a signal

TABLE I

Unit No.	V_c(V)	I_c(mA)	G_{opt}(mmhos)	B_{opt}(mmhos)	$F_{\min}$(dB)	R_n(ohms)
TL01 JA155A/1 TO-18 pkg. 500 MHz CE Fixed Bias					
7	5.0	3.0	5.69	−5.32	2.77	61.96
8	5.0	3.0	4.31	−5.07	2.24	53.86
9	5.0	3.0	5.28	−4.32	2.76	66.62
10	5.0	3.0	4.84	−5.15	2.23	51.71
11	5.0	3.0	4.10	−5.27	2.26	59.69
19	5.0	3.0	5.20	−4.94	2.15	50.60
TL01 JA155A/1 TO-18 pkg. 1016 MHz CE Fixed Bias					
19	5.0	3.0	13.20	−6.18	3.11	54.19
21	5.0	3.0	11.29	−6.77	3.02	55.65
23	5.0	3.0	11.59	−6.12	3.22	53.18
24	5.0	3.0	10.49	−7.05	3.04	54.71
25	5.0	3.0	7.26	−7.12	2.85	53.58

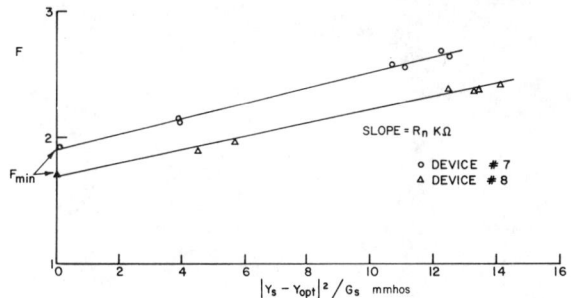

Fig. 1. Experimental data scatter.

generator for the noise source (or a large increase in excess noise power), viz.,

$$\frac{1}{G_{av}} = \frac{1}{G_{MA}} + \frac{R_{eq}}{G_s}|Y_s - Y_{opt,g}|^2 \qquad (12)$$

where $R_{eq} = g_{22}/|Y_{21}|^2$, G_{av} = available gain, and G_{MA} = maximum available gain.

Since the loss of the input network may be obtained for each source admittance, this loss in decibels is subtracted from the measured noise or 1/gain factor in decibels prior to processing, thus, removing the aforementioned bias error.

Manuscript received May 6, 1969; revised June 5, 1969. This work was supported in part by Wright-Patterson AFB under Contract F33615-68-C-1612.

[1] "IRE standards on methods of measuring noise in linear twoports, 1959," *Proc. IRE*, vol. 48, pp. 60–68, January 1960.

[2] Subroutine Sim Q.

[3] H. Fukui, "The noise performance of microwave transistors," *IEEE Trans. Electron Devices*, vol. ED-13, pp. 329–341, March 1966.

[4] ——, "Available power gain, noise figure, and noise measure of two-ports and their graphical representations," *IEEE Trans. Circuit Theory*, vol. CT-13, pp. 137–142, June 1966.

Experimental Results

A developmental microwave transistor type was characterized at 0.5 GHz and 1 GHz using the program described; the results appear in Table I.

Fig. 1 displays the scatter in the experimental data about the computed best straight line.

Shortcoming of the Method as Presently Used

This program was originally used for a UHF MOS structure with somewhat disappointing results due to an apparently high sensitivity to measurement error for devices with large values of R_n (>150 Ω). The author invites suggestions for a more sophisticated fitting routine that may result in lower error sensitivity.

Acknowledgment

The author wishes to thank R. Rohrer for suggesting the least-squares fitting program, M. Purnaiya for supplying the experimental data, A. Keet for writing the FORTRAN program, and J. Gibbons of Stanford University for general encouragement.

Richard Q. Lane
Res. and Develop. Lab.
Fairchild Semiconductor
Palo Alto, Calif. 94304

A Microwave Noise and Gain Parameter Test Set

Richard Q. Lane

California Eastern Laboratory, Inc.

Burlingame, CA

THE DESIGN of low noise amplifier chains requires that the dependence on source reflection coefficient of both the noise figure and the available gain of each stage be known. This source dependence for noise figure takes the form of the parabolic surface of Figure 1 and is described by equation (1).

It has been shown that reciprocal available gain has a similar form; viz equation (2). The minimum noise figure of a chain of like devices is given by the noise measure; equation (3). Loci of constant noise measure map as circles on the source reflection coefficient plane. With the advent of the automatic network analyzer, the gain parameters $1/G_{ma}$, $Y_{S\ opt}$, R_{eq}, have been readily available; derived from the device scattering parameters.

Measurement of the noise parameters has always been more delicate and entailed a search for the minimum noise figure and the source reflection coefficient which engendered it, followed by one or more noise figure (and source reflection coefficient) measurements away from the minimum to find the noise resistance R_n^2.

The noise figure measured however, is the system noise figure F_m, not the device noise figure F_1:

$$F_m = L_1 \frac{(F_1 + F_2 - 1 + \ldots)}{G_{al}}.$$

Furthermore, the input matching loss L_1 depends upon the source reflection coefficient sustained and therefore varies during the search.

A manual method has been described[3] which by using premeasured source reflection coefficients and losses followed by off line data reduction, avoids the foregoing difficulties. The recent advent of a noise meter that outputs both noise figure and gain[4] and a powerful computing controller[5] makes the earlier method noted[3] quite attractive.

The relationship of equation (1) may be restated with reference to four new parameters, A, B, C, D as in equation (4)[6].

In principle, four measurements of source reflection coefficient and noise figure would suffice to determine the four unknown parameters, and in a similar fashion the gain parameters. Equation (4) becomes overdetermined if more than four measurements are taken, but by minimizing the square of the error as expressed in equation (9), many measurements may be used to find those parameters which give the best least squares fit to the surface of Figure 1.

Figure 3 shows two varactor controlled microwave tuners and their equivalent circuit.

Voltage control of a pair of GaAs parametric amplifier varactors, ($f_c \sim 750 GHz$), is used to map a portion of the source reflection coefficient plane as shown in Figure 4.

The diodes, mounted with their bypass capacitors in modified connectors* are only capable of about 0.1pF capacitance change, but the effect of this change is increased by placing each diode as the termination of a short length of transmission line. The tuning screws compensate for the line junction shunt discontinuities.

The tuner varactors are controlled by a pair of D/A converters**; Figure 5. The resulting system noise figure and gain is read from the noise meter by the calculator in BCD. This sequence is repeated for up to ten values of source reflection coefficient and the true device noise figure and gain calculated each time by removing the known tuner loss and second stage noise figure contribution. The noise parameters which cause the partial derivative of equation (9) to be zero are derived, and in a similar fashion the gain parameters are found.

Strictly, for bilateral devices the output of the test device should be rematched for each source reflection coefficient. Most interest is in operation close to the noise figure minimum, therefore the output is arranged to maximize the available gain in this vicinity and not changed as the source is switched. The error accrued by so doing is small in practically useful devices[7]:

$$F_1 = \frac{1}{L_1}\left[F_m - \frac{L_c F_2 - 1}{G_{tl}} - \left(\frac{1^{error}}{G_{tl}} - \frac{1}{G_{al}}\right)\right]$$

It would be feasible to employ a second varactor tuner which stepped over several output transformations for each source setting and so remove the above small error.

The test set uses double side band mixing with a 10MHz first IF in order to reduce second stage contribution.

A cassette tape is prepared which contains the source reflection coefficient, losses and D/A voltage pairs required. The main program requests the second stage noise figure, then uses the taped values to drive the varactors and correct the data. The results tabled in Figure 6 took about six seconds per device.

---

*APC-7 HPIB

[1] Fukui, H., "Available Power Gain, Noise Figure and Noise Measure of Two-ports and Their Graphical Representations", *IEEE-CT*, Vol. CT-13, p. 137-142; June, 1966.

[2] "IRE Standards on Methods of Measuring Noise in Linear Twoports", *Proc. IRE*, p. 61-68; Jan., 1960.

[3] Lane, R.Q., "The Determination of Device Noise Parameters", *Proc. IEEE*, Vol. 57, No. 8, p. 1461-1462; Aug., 1969.

[4] Ailtech System Noise Monitor 7380.

[5] Hewlett Packard 9825A Calculator.

[6] Fukui, H., "The Noise Performance of Microwave Transistors", *IEEE-ED*, Vol. ED-13, No. 3, p. 329-341; March, 1966.

[7] Lane, R.Q., "Microwave Noise Figure Measurements", *CEL Technical Note*.

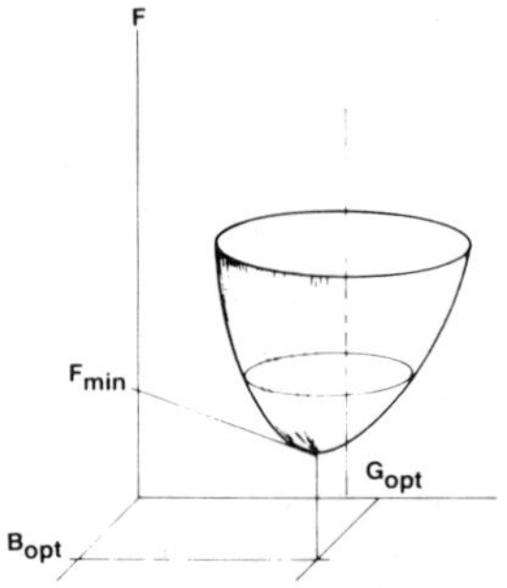

$$F = A + BG_s + (C + BB_s + DB_s)/G_s \qquad (4)$$

$$F_{min} = A + \sqrt{4BC - D^2} \qquad (5)$$

$$R_n = B \qquad (6)$$

$$G_{opt} = \sqrt{4BC - D^2}/2B \qquad (7)$$

$$B_{opt} = -D/2B \qquad (8)$$

$$\epsilon = \sum_{}^{n} W_i [A + B(G_i + B_i^2/G_i) + C/G_i + DB_i/G_i - F_i]^2 \qquad (9)$$

$$F = F_{min} + R_n |Y_s - Y_{opt}|^2/G_s \qquad [1]$$

$$1/G_a = 1/G_{ma} + R_{eq}|Y_s - Y_{opt}|^2/G_s \qquad [2]$$

$$M + 1 = (FG_a - 1)/(G_a - 1) \qquad [3]$$

FIGURE 1—Noise surface and equations 1 through 3.

FIGURE 2—Transformed noise equations 4 through 9.

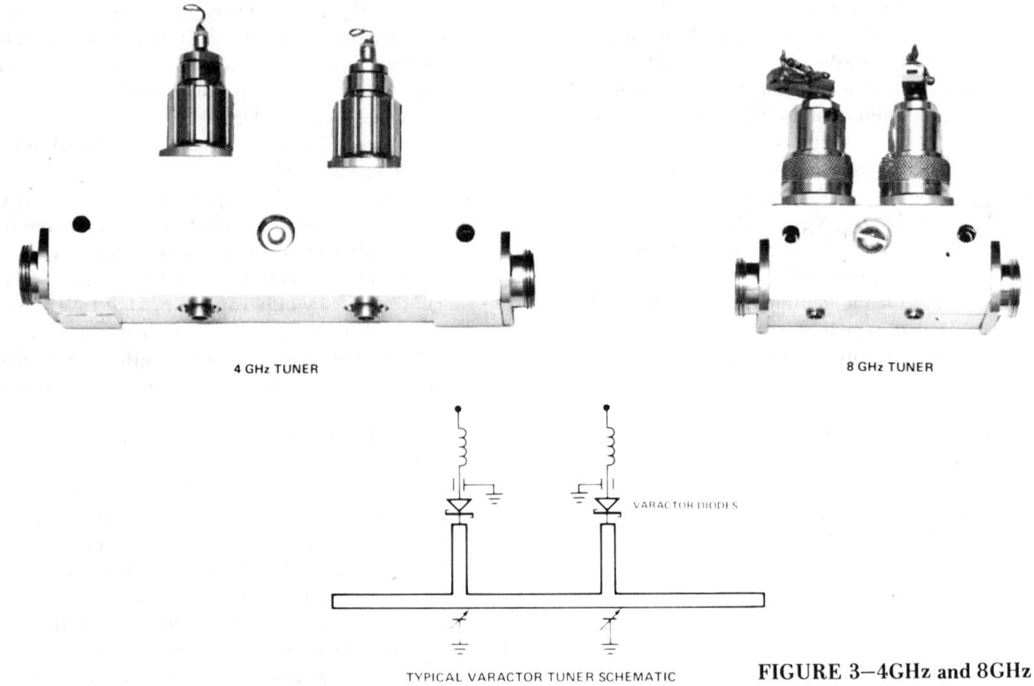

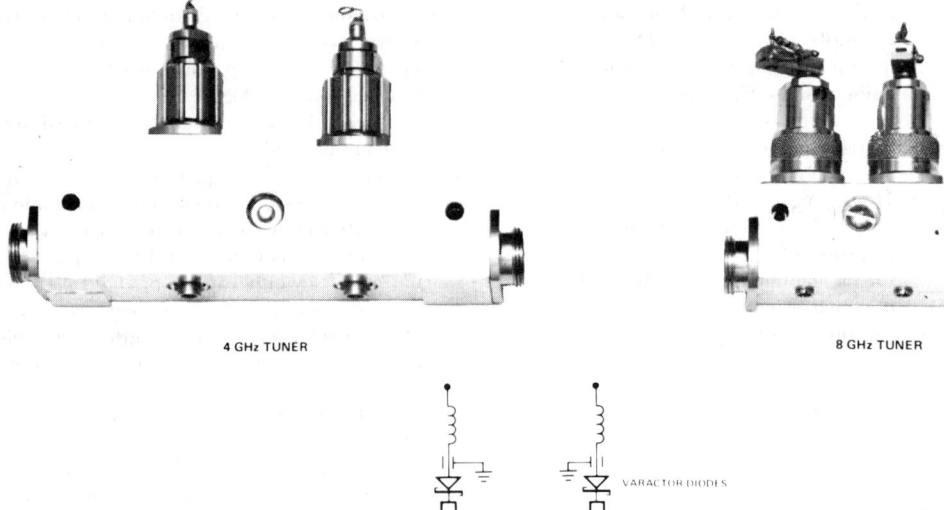

FIGURE 3—4GHz and 8GHz varactor tuners.

FIGURE 4—(*below*) Regions mappable by the tuners.

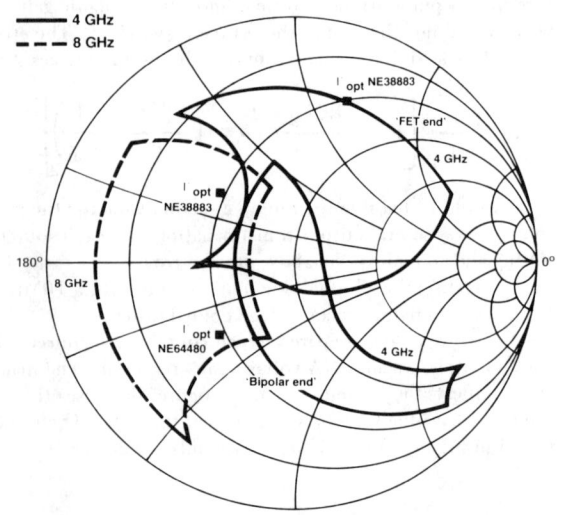

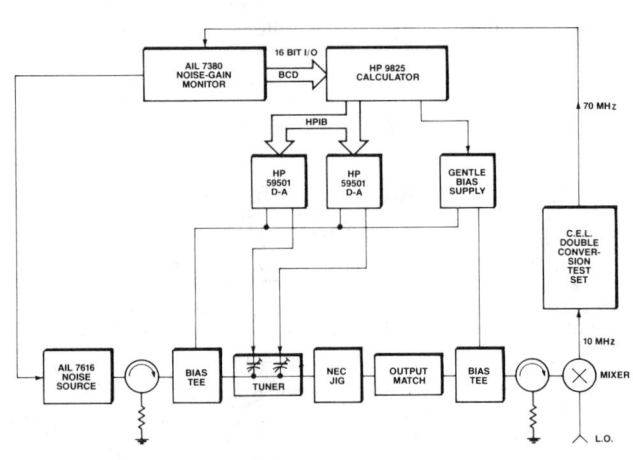

FIGURE 5—Schematic of complete test set.

```
Freq 4.0 GHz              Freq 4.0 GHz               Freq 4.0 GHz
   NE64480                   NE24483                    NE38883
   5.0 mA                    10.0 mA                    10.0 mA
   7.0 volts                 3.0 volts                  3.0 volts
NF2= 4.3 dB               NF2= 4.3 dB                NF2= 4.3 dB
Track 1                   Track 1                    Track 1
File 1                    File 1                     File 1
S/N                       S/N                        S/N
  1                         9                          7
NFmin dB         2.99     NFmin dB          1.19     NFmin dB          0.99
Ref Coef         0.31     Ref Coef          0.69     Ref Coef          0.64
at deg         231.98     at deg           70.24     at deg           65.12
Rn/50            0.45     Rn/50             0.41     Rn/50             0.33
Gma dB           9.07     Gma dB           16.46     Gma dB           17.80
Ref Coef         0.54     Ref Coef          0.53     Ref Coef          0.52
at deg         240.58     at deg           93.30     at deg           96.22
Req/50           0.06     Req/50            0.08     Req/50            0.05


                          Freq 8.0 GHz               Freq 8.0 GHz
                             NE24483                    NE38883
                             10.0 mA                    10.0 mA
                             3.0 volts                  3.0 volts
                          NF2= 6.3 dB                NF2= 6.3 dB
                          Track 1                    Track 1
                          File 1                     File 1
                          S/N                        S/N
                            9                          7
                          NFmin dB          2.81     NFmin dB          2.03
                          Ref Coef          0.62     Ref Coef          0.76
                          at deg          144.48     at deg          136.67
                          Rn/50             0.26     Rn/50             0.16
                          Gma dB            6.29     Gma dB            8.08
                          Ref Coef          0.56     Ref Coef          0.58
                          at deg          155.97     at deg          155.46
                          Req/50            0.10     Req/50            0.07
```

FIGURE 6—Test set output.

On the Determination of Device Noise and Gain Parameters

MARIO SANNINO

Abstract—The least-squares fitting of measured noise figures and gains versus input termination admittance is an established procedure to determine linear two-port noise and gain parameters. Unfortunately, the method is liable to the serious inconvenience of yielding often erroneous results or even results without physical meaning. Some criteria are suggested which allow the carrying out of measurements in such a manner as to safely avoid the above drawbacks.

It is known that the noise figure of a linear two port can be related to the input termination (source) admittance $Y_s = G_s + jB_s$ through the relationship

$$F(Y_s) = F_0 + R_n \frac{|Y_s - Y_0|^2}{G_s} \tag{1}$$

where the minimum noise figure F_0, the optimum source admittance $Y_0 = G_0 + jB_0$, and the noise resistance R_n are the two-port noise parameters.

In 1969, Lane presented a computer-aided analytical method to determine the above parameters [1], which offers many advantages from both the accuracy and the experimental convenience viewpoints as compared with the graphic procedures suggested earlier by the IRE Subcommittee on Noise [2].

Lane's method is an application of the least-squares technique which reduces the derivation of the noise parameters to the solution of a four linear equation system obtained as fit of some (redundant) measured noise figures versus source admittances chosen without stringent rules. The procedure is also used to derive gain parameters, and recently, it has been shown in describing a fully automated computer-controlled noise and gain measuring system [3].

Unfortunately, the method, in spite of its advantages, has not always been received with favor by experimenters, because it is liable to the serious inconvenience of yielding often erroneous results or, in some cases, even results without physical meaning. On the other hand, this was noted by the author himself, who consequently concluded his paper by inviting suggestions for a more sophisticated data processing technique with lower error sensitivity.

Manuscript received March 12, 1979; revised April 4, 1979. This work was supported by the Italian Research Council (C.N.R.).
The author is with the Istituto di Elettrotecnica ed Elettronica, Università di Palermo, viale delle Scienze, Palermo, Italy 90128.

Actually, under some experimental conditions, i.e., depending on the admittances chosen to perform measurements, the ill-conditioning of the coefficient matrix of the equation system to solve may occur, with a consequent large increase in the error sensitivity. In this case, the only way to obtain correct results is to repeat all the measurements for different values of the source admittance.

Allowing for the above considerations, the present note may prove useful in that:

1) it shows that the above mentioned ill-conditioning situation occurs when the chosen admittances do lie in the neighborhoods of some *singular loci* on the G_s, B_s plane;
2) it proposes some measurement criteria to follow in order to surely avoid such drawbacks, together with suggestions for practical application of these criteria.

To prove this, let us consider the symmetric coefficient-matrix of the four linear-equations system which is to be solved in accordance with the least-squares method as applied in [1]. It is given by

$$\begin{bmatrix} n & & & (a_{jk} = a_{kj}) \\ \sum_{i=1}^{n}\left(G_{si} + \frac{B_{si}^2}{G_{si}}\right) & \sum_{i=1}^{n}\left(G_{si} + \frac{B_{si}^2}{G_{si}}\right)^2 & & \\ \sum_{i=1}^{n}\frac{1}{G_{si}} & \sum_{i=1}^{n}\left(1 + \frac{B_{si}^2}{G_{si}^2}\right) & \sum_{i=1}^{n}\frac{1}{G_{si}^2} & \\ \sum_{i=1}^{n}\frac{B_{si}}{G_{si}} & \sum_{i=1}^{n} B_{si} + \frac{B_{si}^3}{G_{si}^2} & \sum_{i=1}^{n}\frac{B_{si}}{G_{si}^2} & \sum_{i=1}^{n}\left(\frac{B_{si}}{G_{si}}\right)^2 \end{bmatrix} \tag{2}$$

where n is the number of the sets of experimental data F_i, G_{si}, B_{si} processed (with $i = 1 \div n$).

By inspection of this matrix, some values of the input termination admittance Y_s can be recognized which can cause the matrix to become singular, in which case, solutions do not exist because the surface (1) is not completely defined. In general, singularity of the matrix occurs for every locus of Y_s which represents a projection on the plane G_s, B_s (or on the Smith chart) of a curve which can be thought of as an intersection of the paraboloid $F(Y_s)$ given by (1) with other ones of the same family.

Some of these values are the ones for which a column (row) is proportional to another one or can be obtained as a linear combination of other ones. Therefore, loci of Y_s which cause singularity of the matrix

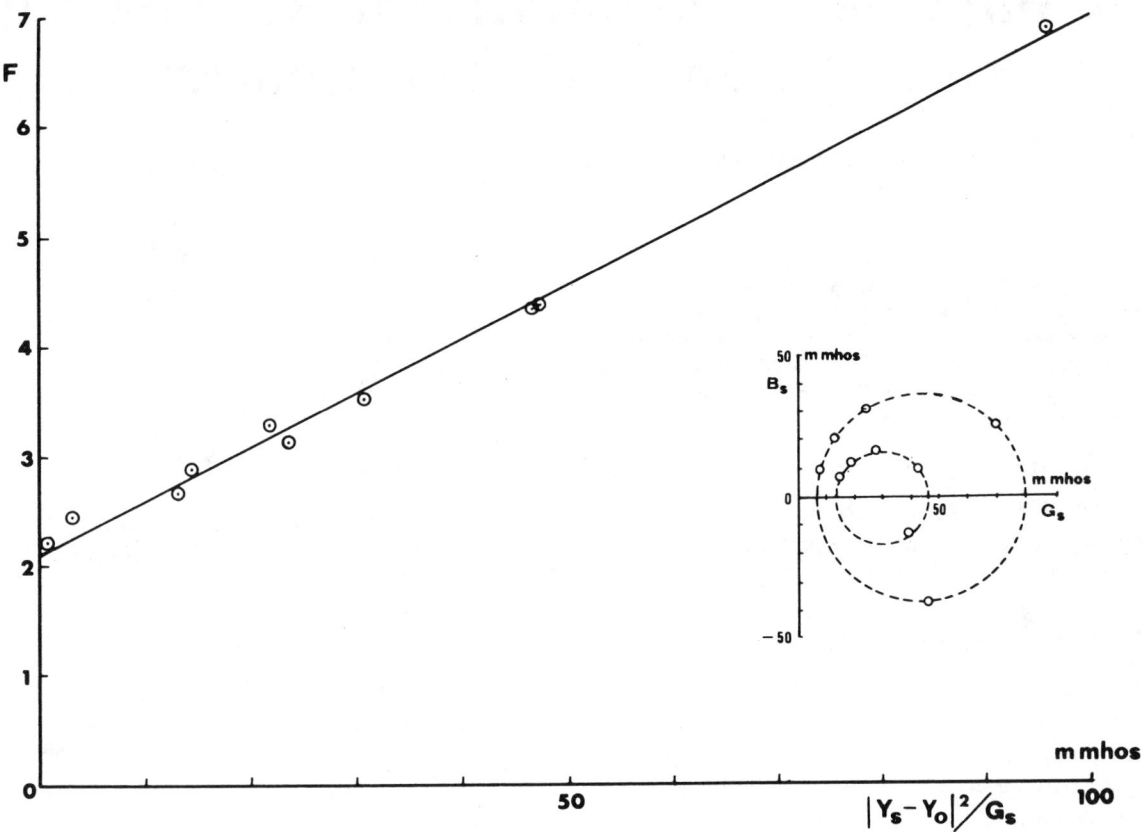

Fig. 1. Scatter of the experimental data of Table I with respect to the computed best straight line for a transistor AT-4642 (Avantek; I_c = 5 mA, V_{ce} = 10 V, f = 3.5 GHz). The noise parameters are F_0 = 3.26 dB, G_0 = 35.3 mmhos, B_0 = 9.18 mmhos and R_n = 48.7 Ω.

(2) are

$$G_S = \text{const}; B_S = \text{const}; |Y_S|^2 = \text{const};$$
$$|Y_S|^2/G_S = \text{const}; |Y_S|^2/B_S = \text{const} \qquad (3)$$

$$\cdot |Y_S|^2/G_S = c + c'/G_S \quad (c \text{ and } c' \text{ are constants}) \qquad (4)$$

which consequently do not define completely the surface given by (1). Note that particular curves belonging to family (4) are the ones for which $c' = -1$, i.e.

$$|Y_S|^2/G_S = c - 1/G_S \qquad (5)$$

which represent circles on the G_S, B_S plane and centered circles on the Smith chart.

Obviously, the values of Y_S chosen to perform measurements in practice never happen to belong exactly to a singular curve, but often lie in the neighborhood of such a locus, giving rise to ill-conditioning of the matrix (2) and, consequently, to unacceptable errors in the parameter determination. This has been proven by several experimental and computer-simulated tests, which have also shown that, under the above conditions, results can vary very strongly versus the number (say $n = 4, \ldots, 10$) of the experimental data processed.

A way to avoid such inconveniences is to choose the admittances Y_S in the neighborhoods of two (or more) different singular loci, belonging either to the same family or to different families, because this corresponds to completely defining the surface (1).

A suggestion which may be followed in practice is the use of a (coaxial or waveguide) slide-screw tuner as noise source admittance transformer, which allows the realization of admittances defined by (5) for two different values of the constant c simply by sliding the carriage for two different penetrations of the tuning slug. Obviously, admittances corresponding to low noise figures, i.e., near the predictable optimum value Y_0, are to be selected to improve accuracy.

To prove the effectiveness of the procedure suggested, experimental results regarding a low noise transistor (Avantek, AT-4642; I_c = 5 mA, V_{ce} = 10V; frequency 3.5 GHz) are reported. The data processed,

TABLE I

G_S	B_S	F	G_S	B_S	F
mmhos	mmhos	dB	mmhos	mmhos	dB
38.6	-13.7	4.25	48.2	-36.9	6.44
43.2	8.78	3.50	71.6	25.7	5.27
28.6	16.8	3.95	24.0	30.6	4.91
19.0	13.7	4.64	13.4	21.2	6.39
14.2	8.37	5.45	7.87	9.99	8.38

corresponding to two different circles on the G_S, B_S plane, are reported in two columns in Table I. In Fig. 1 they are displayed together with the computed best straight line. The parameters have been also computed by processing a number of data less than 10, provided that, as above suggested, some of them belong to one locus of Table 1 and some to the other one. This has shown that only slight variations of noise parameters occur versus redundancy if the number of data processed is sufficiently large (say $n \geqslant 7$).

The measurements have been performed through a measuring setup equipped with variable attenuator, test receiver, and gas-discharge noise source. The data have been processed with an HP 9830 A desk computer using the well-known IBM subroutine SIM Q translated into BASIC language.

In conclusion, it is noteworthy that the loci defined by the conditions G_S = const and B_S = const, on which the IRE graphical fitting procedure is based, belong to the families of singular loci discussed above. Therefore, the procedure proposed here covers also one of the IRE methods as a particular case, with the important difference that the measurement conditions as required by the IRE suggestions are to be *exactly* fulfilled, while the results of the procedure proposed here are largely independent of the accuracy with which the above conditions are realized. This may be useful when (coaxial or strip-line) stub tuners are used as admittance transformer networks [3].

An Improved Computational Method for Noise Parameter Measurement

MASATAKA MITAMA, MEMBER, IEEE, AND HIDEHIKO KATOH

Abstract—Conventional methods for noise parameter measurement for linear noisy two-ports have been improved by introducing a computational method for evaluating measured admittance errors. Derivation and comparison with a conventional method are given. Noise parameters of a packaged 0.5-μm gate-length GaAs MESFET (NE38806) were successfully measured using the proposed technique.

I. Introduction

AS IS WELL KNOWN, the noise behavior of a linear noisy two-port network can be characterized by the four noise parameters, F_0, G_0, B_0, and R_n, as

$$F = F_0 + \frac{R_n}{G_s}\{(G_s - G_0)^2 + (B_s - B_0)^2\} \quad (1)$$

where

- F noise figure,
- Y_s $G_s + jB_s$ = source admittance,
- F_0 minimum noise figure,
- Y_0 $G_0 + jB_0$ = optimum source admittance that gives minimum noise figure,
- R_n equivalent noise resistance.

The usual method [1] of obtaining noise parameters requires a special source admittance setting procedure, which is tedious.

An alternate method [2] consists of performing noise figure measurements for more than four arbitrary source admittances with a least squares method used for data processing. In this method, an estimated error (residue) ϵ'_i is defined for the ith measured data set (G_{si}, B_{si}, F_i) ($i = 1, 2, \cdots, N$) as

$$\epsilon'_i = \left| F_0 + \frac{R_n}{G_{si}}\{(G_{si} - G_0)^2 + (B_{si} - B_0)^2\} - F_i \right| \quad (2)$$

which is shown by the dotted line ϵ'_i in Fig. 1. The weighted square sum of the estimated error is minimized. However, of the three measurement values G_{si}, B_{si}, and F_i, (2) considers the measured error in F only. That is, no consideration is made for measured G_s and B_s errors. Thus measured F errors at Y_s values away from Y_0 tend to be overvaluated, resulting in unsatisfactory computed noise parameter values, especially at higher microwave frequencies. In particular, according to Lane [2], computed results are highly sensitive to measurement errors in case of a large R_n.

Manuscript received September 7, 1978; revised January 2, 1979.
The authors are with the Central Research Laboratories, Nippon Electric Company, Ltd., 4-Miyazaki, Takatsu-ku, Kawasaki 213, Japan.

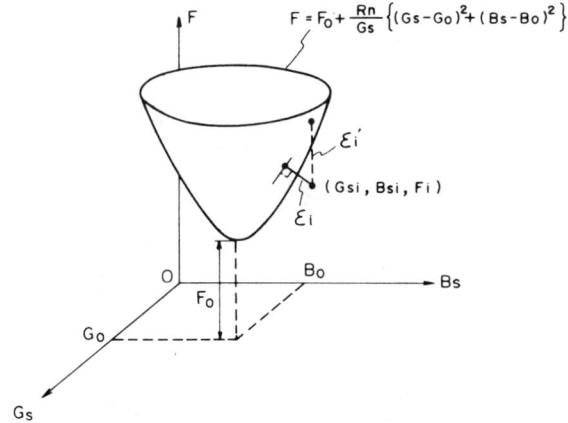

Fig. 1. Estimated error ϵ_i for the present least squares method. The dashed line designated by ϵ'_i represents the corresponding error used in the conventional method [2].

Assigning weighting values according to each measurement accuracy has been suggested [3], but is deemed impractical.

In the present paper, a computational method is introduced for evaluating measured source admittance errors, as well as the measured F error.

Another reason for unsatisfactory results with the conventional methods is the random differences between measured Y_s values and actual Y_s values imposed onto the two-port network. In the present method, a low-loss microstrip tuner is used, which helps to decrease the errors introduced by these differences.

II. Least Squares Method Considering Y_s Errors

Let $G_s = x$, $B_s = y$, $F = z$, $F_0 = a_1$, $R_n = a_2$, $G_0 = a_3$, and $B_0 = a_4$ for simplicity of notation. Let the estimated values for the measured set (X_i, Y_i, Z_i) ($i = 1, 2, \cdots, N$) in the x, y, and z coordinates be $(\hat{x}_i, \hat{y}_i, \hat{z}_i)$ and the estimated values for parameters a_1, a_2, a_3, and a_4 be $\hat{a}_1$, $\hat{a}_2$, $\hat{a}_3$, and $\hat{a}_4$. Furthermore, define the estimated errors (residues) as

$$V_{xi} = \hat{x}_i - X_i \quad V_{yi} = \hat{y}_i - Y_i \quad V_{zi} = \hat{z}_i - Z_i \quad (3)$$

$$V_{a1} = \hat{a}_1 - a_1^0 \quad V_{a2} = \hat{a}_2 - a_2^0 \quad V_{a3} = \hat{a}_3 - a_3^0 \quad V_{a4} = \hat{a}_4 - a_4^0 \quad (4)$$

where a_1^0, a_2^0, a_3^0, and a_4^0 are zeroth-order approximation values. These estimated values must satisfy the functional relation of

$$G(\hat{x}_i, \hat{y}_i, \hat{z}_i; \hat{a}_1, \hat{a}_2, \hat{a}_3, \hat{a}_4) = 0, \quad i = 1, 2, \cdots, N \quad (5)$$

where

$$G(x,y,z;a_1,a_2,a_3,a_4)$$
$$= -z + a_1 + \frac{a_2}{x}\{(x-a_3)^2 + (y-a_4)^2\}. \quad (6)$$

Equation (6) is derived by transposing the left-hand side value of (1) to the right-hand side values.

The term "estimated values" is defined such that $\hat{x}_i, \cdots, \hat{a}_4$ minimize

$$S = \sum_i \left(w_{xi}V_{xi}^2 + w_{yi}V_{yi}^2 + w_{zi}V_{zi}^2\right)_{\min} \quad (7)$$

where w_{xi}, w_{yi}, and w_{zi} are weights to be determined according to measurement accuracies, and the subscript min denotes the minimum value within the parentheses.

When $w_{xi} = w_{yi} = w_{zi} = 1$, the quantity

$$\epsilon_i = \left\{\left(w_{xi}V_{xi}^2 + w_{yi}V_{yi}^2 + w_{zi}V_{zi}^2\right)_{\min}\right\}^{1/2} \quad (8)$$

represents the length of a segment of a line normal to the quasi-elliptic paraboloid represented by (5) projected from the measured point (X_i, Y_i, Z_i), as shown by the solid line ϵ_i in Fig. 1.

Note that the minimum value within the parentheses in (7) occurs when the line drawn from the measured point is perpendicular to the quasi-elliptic paraboloid. With (8), the evaluation of the measured Y_s error as well as that of the measured F error becomes possible, as contrasted to the conventional method [2].

Assuming that estimated errors are small, (5) can be expanded in a Taylor series to a first-order approximation as

$$G_0^i + G_x^i V_{xi} + G_y^i V_{yi} + G_z^i V_{zi}$$
$$+ G_{a1}^i V_{a1} + G_{a2}^i V_{a2} + G_{a3}^i V_{a3} + G_{a4}^i V_{a4} = 0 \quad (9)$$

where

$$G_0^i = G(X_i, Y_i, Z_i; a_1^0, a_2^0, a_3^0, a_4^0) \quad (10)$$

$$G_x^i = \left.\frac{\partial G}{\partial x}\right|_{x=X_i, y=Y_i, z=Z_i; a_1=a_1^0, a_2=a_2^0, a_3=a_3^0, a_4=a_4^0} \quad (11)$$

and $G_y^i, \cdots, G_{a4}^i$ are similar to the above. Using (9), (8) can be rewritten as

$$\epsilon_i = \sqrt{w_i} |d_i| \quad (12)$$

where

$$d_i = -\left(G_{a1}^i V_{a1} + G_{a2}^i V_{a2} + G_{a3}^i V_{a3} + G_{a4}^i V_{a4} + G_0^i\right) \quad (13)$$

$$1/w_i = \left(G_x^i\right)^2/w_x + \left(G_y^i\right)^2/w_y + \left(G_z^i\right)^2/w_z. \quad (14)$$

The derivation of (12) is shown in the Appendix.

Then, by the least squares method, a set of linear equations is obtained as

$$\frac{\partial S}{\partial V_{a1}} = 2\sum_i \left[w_i d_i G_{a1}^i\right] = 0$$

$$\vdots$$

$$\frac{\partial S}{\partial V_{a4}} = 2\sum_i \left[w_i d_i G_{a4}^i\right] = 0. \quad (15)$$

Fig. 2. Microstrip tuner.

In order to solve (15), the initial values a_1^0, a_2^0, a_3^0, and a_4^0 were obtained by using the conventional method described in [2]. Then (15) was solved by an iteration method using (4) to obtain the V_{a1}, V_{a2}, V_{a3}, and V_{a4} that make S minimum. In the present calculation, w_{xi}, w_{yi}, and w_{zi} were set to unity.

III. Measurement

A. Microstrip Tuner

Usually, commercial stub tuners are widely used to provide Y_s values necessary for noise parameter measurements. However, good reproducibility is difficult to obtain with these stub tuners. In addition, dissipation loss must be calibrated for each Y_s value, which is tedious. To avoid these requirements, an integrated microstrip tuner, shown in Fig. 2, has been used. To a 50-Ω main line, eight open shunt stubs (lands) were constructed on a 0.8-mm-thick teflon-glass fabric board (Di Clad 522). Each stub consisted of ten 0.8-mm $\times$ 1-mm lands with 0.2-mm separation. By connecting appropriate lands with an adhesive conductive sheet, necessary Y_s values can be obtained on an arbitrary basis. Y_s values were recorded with an $X-Y$ recorder prior to the noise figure measurement. Fig. 3 shows an example. Circles in the figure indicate the source admittance points realized. Each solid line shows the locus of discrete change of the source admittance when one particular stub length is adjusted by connecting the lands one by one with an adhesive conductive sheet. The reproducibility using this arrangement was superior to that using a stub tuner. The dissipation loss, calculated by using generalized scattering matrices [4], was 0.02 dB for the particular Y_0 circuit configuration; hence the circuit loss was neglected throughout.

B. Available Gain

In order to minimize the Y_s value inaccuracy caused by connector disconnection, a two-port network available gain was calculated using both the premeasured Y_s value and the two-port network S parameters. Since circuit losses will result in an actual available gain smaller than that computed, receiving system noise contribution will not be overevaluated. The output matching circuit con-

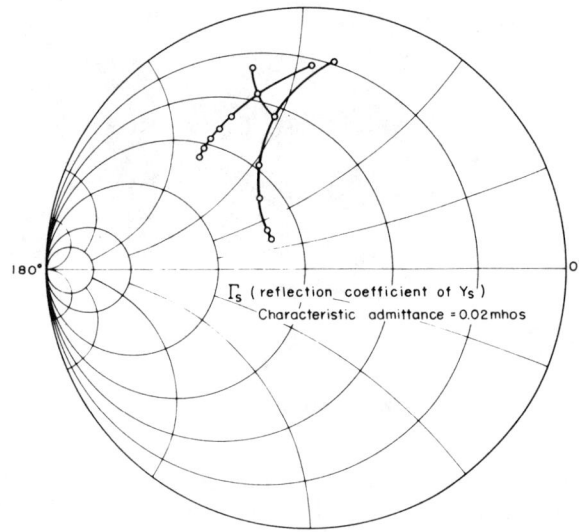

Fig. 3. An example of source admittance ($f = 6$ GHz).

sisted of two quarter-wave transformer sections, common throughout the measurement, with less than 2.5 output VSWR. In a unilateral two-port network case, as is the case of an FET, the input impedance of the two-port network can be considered as being independent of the load impedance. Output circuit loss was neglected throughout.

C. NF Measurement

Noise figure (NF) measurements were made at 6, 7, and 8 GHz with an HP 342A automatic NF meter for a packaged 0.5-μm gate-length GaAs MESFET (NE 38806). To reduce the receiving system noise contribution, a low-noise preamplifier (NF < 3 dB, Gain > 17 dB) was placed directly after the FET output matching circuit. Although the measured NF was the double-sideband (signal and image) value, it can be considered as a single-sideband value, since signal and image separation is only 60 MHz.

IV. Calculation Results

A. Comparison with Conventional Method

In Fig. 4, the computed noise parameter standard deviation as a function of the number of measurement points N is shown both for the present and for the conventional method [2]. For each N, ten different trains of data sets were randomly selected from a total of fourteen measured points. Then, noise parameters and standard deviations were calculated for each N case. The figure shows that a smaller noise parameter deviation is obtainable with the present method for any given number of measurement points. Fig. 5 shows computed noise parameter sensitivities to an individual data set (G_{si}, B_{si}, F_i) accuracy. Noise parameters are calculated by extracting each data set one by one sequentially from fourteen data sets. It is clear that the individually measured data is less sensitive to the computed parameters with the present method. The small deviation in

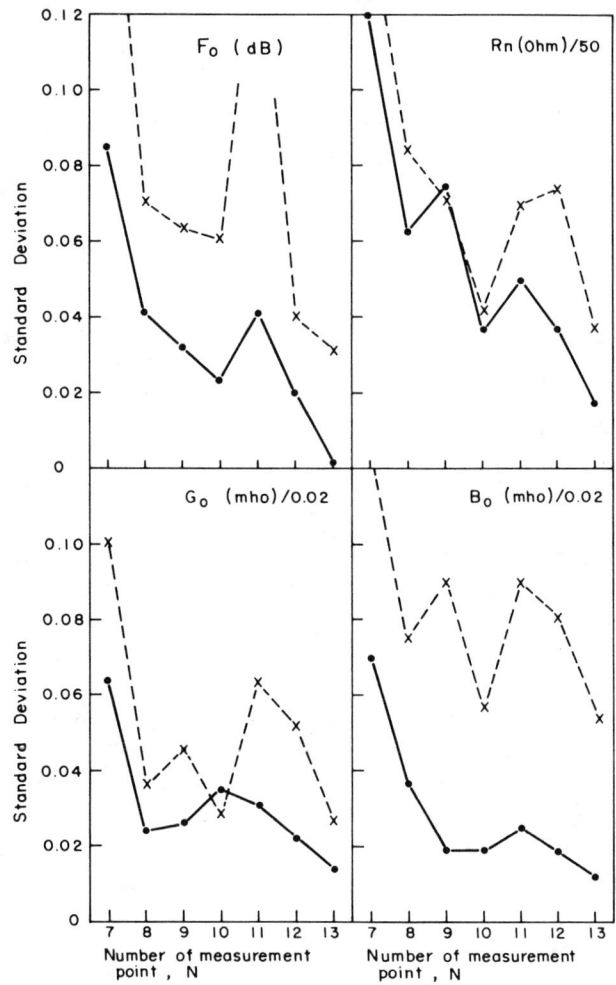

Fig. 4. Computed noise parameter standard deviation comparison with conventional method ($f = 6$ GHz) [2]. The solid line curve corresponds to the present method, while the dashed line curve corresponds to the conventional method.

TABLE I
Noise Parameters of a 0.5-μm Gate-Length GaAs MESFET
(NE 38806 or 2SK124). $V_p = -2.21$ V, $I_{DSS} = 51$ mA

FET	I_{DS} (mA)	V_{DS} (V)	f (GHz)	F_0 (dB)	$|\Gamma_0|$	ϕ_0 (deg.)	$R_n/50$ (ohm)
NE 38806	10	3	6	1.88	0.660	114.3	0.53
"	"	"	7	1.93	0.642	140.2	0.30
"	"	"	8	2.10	0.592	164.0	0.12

computed noise parameters and the insensitivity to individual data with the present method should be attributable to the new estimated error evaluation method described in Section II.

B. Measured Noise Parameters for NE 38806

Noise parameters for a packaged 0.5-μm gate-length GaAs MESFET (NE 38806) were measured with the present method. Results are listed in Table I. Measurement points for 6, 7, and 8 GHz were 14, 13, and 15, respectively. Because of rather small R_n values, 26.5–6 Ω, constant noise figure circles [5] indicated a mild dependence on Γ_s (reflection coefficient of Y_s) planes.

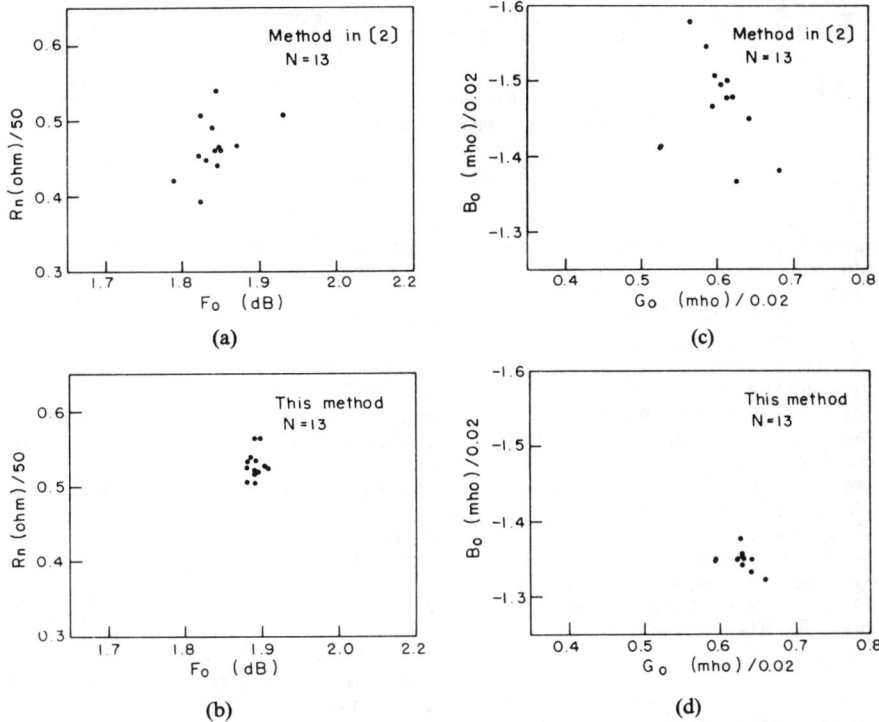

Fig. 5. Comparison of the sensitivity in computed noise parameters to individual data ($f=6$ GHz). (a) and (c) are the conventional method [2]; (b) and (d) are the present method.

V. Conclusion

Conventional methods [1], [2] for noise parameter measurement have been improved by introducing a new computational method for evaluating measured source admittance errors as well as the measured noise figure error and by utilizing an integrated microstrip tuner. It was demonstrated that a smaller deviation in the computed noise parameters and insensitivity to individual data were achieved with the present method when compared with the conventional method [2]. Noise parameters of a packaged 0.5-μm gate-length GaAs MESFET (NE 38806) have been successfully measured with the new method.

Appendix
Derivation of (12)

Consider the terms within the parentheses of (8),

$$s_{hi} = w_{xi} V_{xi}^2 + w_{yi} V_{yi}^2 + w_{zi} V_{zi}^2. \quad (A.1)$$

In the following, the superscripts and subscripts i's are dropped throughout for simplicity. Using (9) and (13), (A.1) can be rewritten as

$$s_h = w_x V_x^2 + w_y V_y^2 + \frac{w_z}{G_z^2}(d - G_x V_x - G_y V_y)^2. \quad (A.2)$$

The minimum of s_h occurs when $\partial s_h/\partial V_x = 0$ and $\partial s_h/\partial V_y = 0$, simultaneously. Thus we obtain a set of linear equations as

$$\left(w_x + w_z \frac{G_x^2}{G_z^2}\right) V_x + \left(w_z \frac{G_x G_y}{G_z^2}\right) V_y = \left(w_z \frac{G_x}{G_z^2}\right) d \quad (A.3)$$

$$\left(w_z \frac{G_x G_y}{G_z^2}\right) V_x + \left(w_y + w_z \frac{G_y^2}{G_z^2}\right) V_y = \left(w_z \frac{G_y}{G_z^2}\right) d. \quad (A.4)$$

Solving (A.3) and (A.4), we obtain

$$V_x = w\left(\frac{G_x}{w_x}\right) d \quad V_y = w\left(\frac{G_y}{w_y}\right) d \quad \text{and} \quad V_z = w\left(\frac{G_z}{w_z}\right) d. \quad (A.5)$$

Then, from (A.1) and (A.5), we have

$$\epsilon = \sqrt{(s_h)_{\min}} = \sqrt{w}\, |d|.$$

Acknowledgment

The authors would like to thank Dr. K. Ayaki and the members of the Microwave Circuit Group, Central Research Laboratories, Nippon Electric Company, Ltd., for their encouragement and guidance.

References

[1] IRE Subcommittee on Noise, "IRE standards on methods of measuring noise in linear twoports, 1959," *Proc. IRE*, vol. 48, pp. 60–68, Jan. 1960.
[2] R. Q. Lane, "The determination of device noise parameters," *Proc. IEEE*, vol. 57, pp. 1461–1462, Aug. 1969.
[3] M. S. Gupta, "Determination of the noise parameters of a linear 2-port," *Electron. Lett.*, vol. 6, pp. 543–544, Aug. 20, 1970.
[4] G. E. Bodway, "Two port power flow analysis using generalized scattering parameters," *Microwave J.*, vol. 10, pp. 61–69, May 1967.
[5] H. Fukui, "Available power gain, noise figure, and noise measure of two-ports and their graphical representations," *IEEE Trans. Circuit Theory*, vol. CT-13, pp. 137–142, June 1966.

ACCURATE AND AUTOMATIC NOISE FIGURE
MEASUREMENTS WITH STANDARD EQUIPMENT

Nick Kuhn
Hewlett-Packard Co. Inc.
Stanford Park Division
1501 Page Mill Road
Palo Alto, CA. 94304

Noise figure measures the amount of noise a receiver or a component adds to a system. Minimizing such noise, because it obscures low-level signals, is an important concern in most of today's microwave systems. Noise figure measurements are necessary at the device level, the component level and the system level. They are necessary during design, manufacture and maintenance. Yet noise measurements have been fraught with non-repeatability. If, for example, a vendor and his customer are to independently measure noise figure so they agree within 0.2 dB, they each must usually use a specially calibrated noise source and tediously and manually remove several insidious effects that must otherwise be accepted as measurement errors. Even then, they often go through a time-consuming and expensive period of exchanging components and measurement equipment to achieve repeatability from system to system.

This paper describes a noise figure measurement system built of ordinary commercial components that routinely makes measurements from one system to another within 0.1 dB. Figure 1 shows the noise figure of a broadband receiver measured three times with three noise sources from different production runs and three sets of the noise power ratio measuring instruments. Because the noise sources were manufactured several months apart, Figure 1 also demonstrates the time stability of the noise source output. The absolute accuracy of the measurement from all known sources of error, although dependent on the SWR of the unit under test, is about ±0.22 dB, a factor of two better than traditional systems assembled from standard instruments

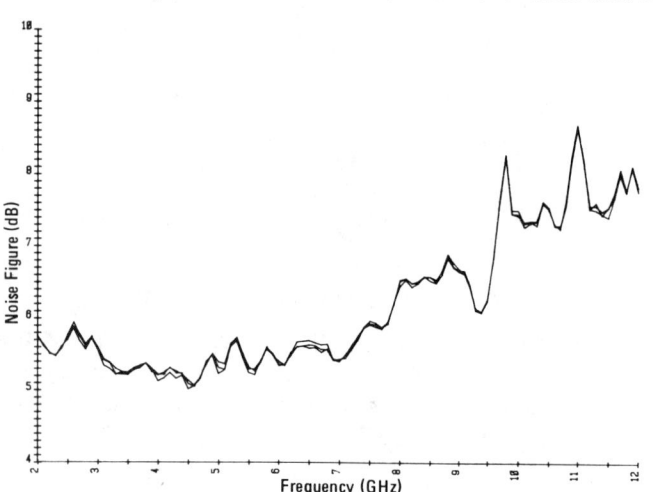

Figure 1. Measurement repeatablity as shown by three measurements with three sets of measurement equipment.

This system uses a desktop computer to process data and account for many small effects that, in manual noise measurement systems, are very bothersome to correct, and are often accepted as measurement errors. Such effects include ambient temperature variation and the variation with frequency of the excess noise ratio of the noise source. Correction for these effects is now so routine and simple, that noise figure measurements will likely become an often used tool to replace old, seriously questioned, often postponed routines. These routines were usually dreaded, especially by the occasional user who felt he had to re-educate himself at every new encounter.

Another very significant advantage of this system is that it can measure the gain of the unit under test and display its noise figure without the noise contributions of the more permanent parts of the system. Another way to describe this is that the data is corrected for the second stage noise contributions. Thus when characterizing a microwave preamplifier, for example, the noise contributions of the mixer and IF amplifier are removed from the final result.

Noise Measurement

Figure 2 is a block diagram for the general purpose microwave noise figure measurement system. The unit under test (UUT) can be all of or any part of the large block shown as the UUT/receiver. The remainder of that block is then considered as part of the measurement system whose noise contributions can be removed before final data presentation. When testing a microwave amplifier, for example, the mixer, local oscillator and IF amplifier are part of the system. When testing a receiver, however, the mixer, local oscillator, and IF amplifier are part of the unit under test.

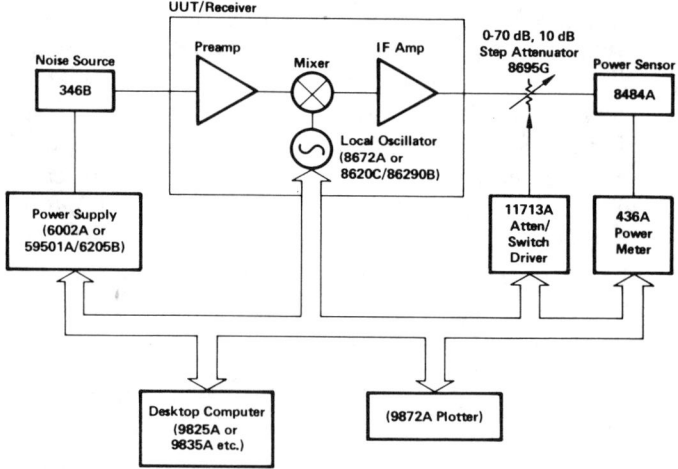

Figure 2. Block diagram of noise measurement system.

Most of the equipment is of a general purpose nature and likely to be available in many organizations This allows the system to be assembled for trial use at minimum expense. The noise source, power sensor, and power meter are the principal accuracy-determining elements of the system and should not be exchanged for alternatives. Otherwise, however, there is great flexibility in selecting equipment.

This system, like most noise figure meters, works by comparing the noise power output from the receiver for two different levels of input noise to the receiver. The ratio of the noise power outputs, corresponding to a noise source being first at temperature

Reprinted from *IEEE MTT-S Int. Microwave Symp. Digest*, 1980, pp. 425-427.

T_H (turned ON) and then at T_C (turned OFF), yields the effective input noise temperature of the UUT/receiver according to

$$T_e = \frac{T_H - T_C Y}{Y-1} \quad (1)$$

The Y factor is the ratio of the two powers, that is, $Y = N_H/N_C$. For solid state noise sources T_C is the ambient temperature and T_H is related to the excess noise ratio (ENR) of the source by

$$ENR = 10 \log \frac{T_H - T_o}{T_o} \quad (2)$$

where T_o is the reference temperature of 290 Kelvins. The noise figure in dB is related to the effective input noise temperature by

$$F(dB) = 10 \log(\frac{T_e}{T_o} + 1) \quad (3)$$

Equations (1) and (3) hold for the noise measurement system by itself to yield the second stage noise contribution, T_{e2} and F_2. They also hold when the unit-under-test (UUT) is attached to the input to yield the total noise characteristics T_{e12} and F_{12}. The effective input noise temperature of the UUT alone, T_{e1}, is given by

$$T_{e1} = T_{e12} - \frac{T_{e2}}{G_1} \quad (4)$$

where G_1 is the available gain of the UUT. But, as will soon be shown, the same measurements of N_H and N_C for the measurement system alone (N_{H2} and N_{C2}) and for the UUT attached to the input (N_{H12} and N_{C12}) can also yield the gain G_1. Then equations (4) and (3) can be solved for the effective input noise temperature and noise figure of the UUT alone.

The difference between the hot and cold noise power outputs leads to gain G_1. For the measurement system by itself

$$N_{H2} - N_{C2} = kBG_2(T_H - T_C) \quad (5)$$

where G_2 is the gain of the measurement system. The noise added by the system, being the same when the noise source is hot and cold, disappears from the difference in equation (5). When the UUT is connected to the input of the system

$$N_{H12} - N_{C12} = kBG_1 G_2 (T_H - T_C) \quad (6)$$

The ratio of equation (6) to equation (5) is

$$\frac{N_{H12} - N_{C12}}{N_{H2} - N_{C2}} = G_1 \quad (7)$$

The above equations show that accurately known noise output from the noise source and accurate power measurement equipment combined with the above equations can yield the noise properties of a unit-under-test. Although the equations can be solved by hand, this is likely to be done at only one or two frequencies. The desktop computer of this system solves those equations, and also synchronizes the operation of the noise source, gathers the power measurement data, tunes the frequency of the local oscillator, interpolates the proper ENR of the noise source from stored calibration data and outputs the results to a printer and/or plotter in scarcely any more time than it takes to make the power measurements alone.

Equipment Alternatives

An important aspect of this system is the great flexibility in selecting equipment. The IF frequency can be anywhere in the 10 MHz to 18 GHz range of the power sensor. The IF gain·bandwidth product should be large enough for the power meter to read above -49 dBm where it reads most accurately and quickly. This means a minimum gain of about 50 dB for a 10 MHz bandwidth or 60 dB for a 1 MHz bandwidth.

The attenuator at the output of the IF is used to keep the power meter readings from going above -20 dBm, the maximum for the power sensor. The maximum allowed value of attenuation during measurement should be about 30 dB. If a larger value is needed, the IF input power is above +10 dBm and is likely to be approaching saturation and decreased accuracy.

Signal generators and sweep oscillators usually make good general purpose local oscillators for use with double balanced mixers. The important quality of a local oscillator is that the noise level at the IF frequency away from the local oscillator frequency be low. Yig tuned and cavity tuned oscillators tend to have low enough noise. Signal generators are sometimes needed for good frequency resolution or for power level control. An example of such a case is characterizing the noise figure and conversion loss of a mixer vs. local oscillator drive level. For many receiver applications, such as some TV and EW applications, an electronically tuned local oscillator can be tuned through its frequency range by the desktop computer with an appropriate D to A converter.

Accuracy

The high inherent accuracy of the system arises from the noise source, which has low SWR and a stable, calibrated excess noise ratio, and from the power meter and its power sensor. The system measures the output noise power ratio for the hot and cold noise source to an accuracy of ±0.04 dB. Traditional noise figure meters can measure the noise power output ratio to an accuracy of ±0.15 dB. Re-reflections of noise power between the UUT and noise source cause the measurement uncertainties shown in Figure 3. A typical value, for a noise source SWR of 1.1 and a UUT

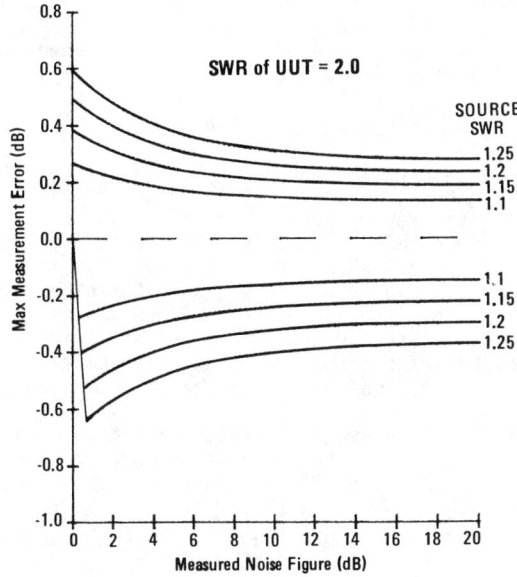

Figure 3. Mismatch uncertainty limits in noise figure measurement due to re-reflection between the noise source and unit under test.

SWR of 2.0, yields an uncertainty of ±0.15 dB. Doubling the noise source reflection coefficient, as must be done with many models of noise sources, approximately doubles the uncertainty. ENR calibration uncertainty arises from many different sources that tend to combine like random variables. The RSS measurement uncertainty due to ENR, including an allowance for interpolation between calibration frequencies, is about ±0.15 dB. If an allowance for receiver non-linearities of ±0.05 dB is included, the overall root-sum-of squares measurement uncertainty is ±0.22 dB.

Operation

The special function keys of the desktop computer, pictured in Figure 4, are the operating controls of the measurement system. They allow selection of measurement frequencies, the form of presenting output data, initiating a new calibration, and certain manual operations for checking operating power levels, integrity of connections, etc. during startup.

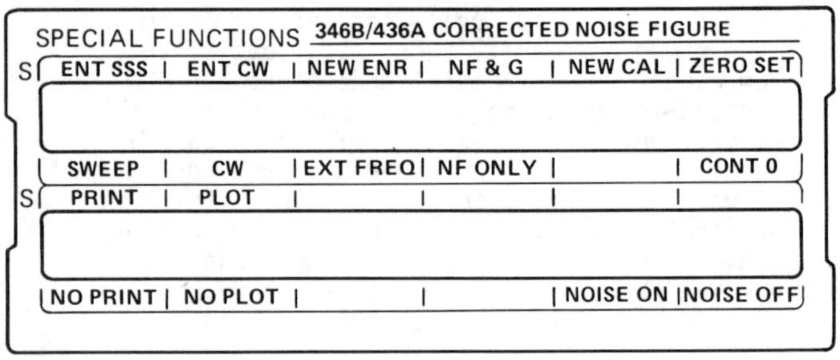

Figure 4. Special function keys of the desktop computer for operating the system.

Many users of this system will find it best to modify the program. Consider, for example, a production test application where only a few measurement situations are encountered. The tasks performed by the special function keys may be changed. Several of the keys, for example, might each direct the program to a specific CW measurement frequency for making tuning adjustments while observing the calculator LED output of noise figure and gain. Then another key might be programmed to make a final broadband sweep at pre-programmed frequencies while both plotting and printing the measurement results at each frequency. The operator would not need to be asked to input the several frequencies and plotting limits.

Measurement Results

Although Figure 1 showed the repeatability of the measurement equipment, some of the traditional non-repeatability when measuring amplifiers is associated with non-repeatable interactions between the mixer and the UUT. Figure 5 shows the corrected noise figure and gain of an X-band amplifier as measured by this system for two different mixers. There was a 3 dB attenuator included on the RF port of each mixer to minimize the interaction. But the noise contribution of the attenuator was easily corrected by the computer.

Similar measurements are also possible on mixers. Figure 6 shows the corrected noise figure and conversion loss of a doubly balanced mixer. Here the broadband property of the 10 MHz to 18 GHz noise source is used to measure the IF noise contribution.

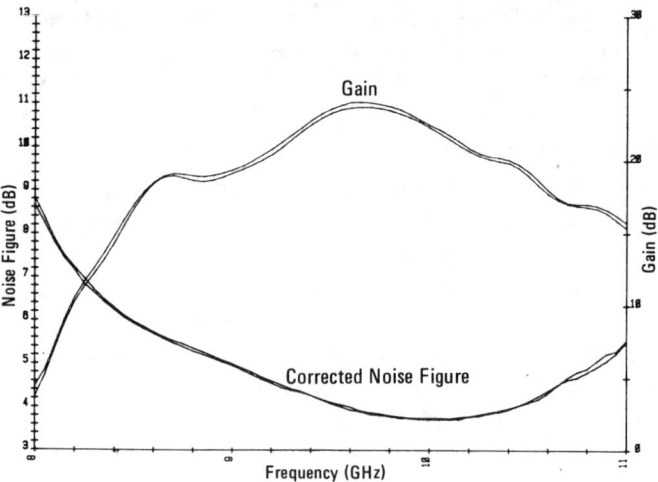

Figure 5. Corrected noise figure and gain of an X-band amplifier measured with two different mixers on the receiver.

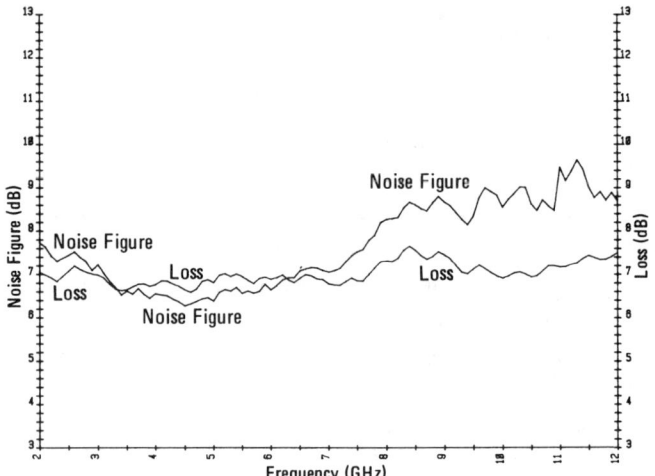

Figure 6. Corrected single sideband noise figure and conversion loss of a double balanced mixer as measured by the system from 2 to 12 GHz with 100 MHz frequency spacing.

Although noise figure and conversion loss are usually assumed to be identical because of traditional difficulty in making such measurements, Figure 6 shows that they are quite different. The relative position of the two curves is a function of local oscillator power. The time for measuring and plotting the results of Figure 6 is about 70 seconds.

This noise figure measurement is not only applicable to production testing. Its high accuracy, arising mainly from a low SWR, well-calibrated noise source and a very accurate power ratio detector, makes it a factor of two more accurate than traditional systems assembled from standard instruments. This system ought to be considered in any critical noise measurement application, especially where repeatability is necessary.

Reference

1. Mumford, W.W. and Scheibe, E.H., "Noise Performance Factors in Communication Systems," Horizon House, 1968

Measurement of Losses in Noise-Matching Networks

ERIC W. STRID

Abstract — The noise contribution of an input-matching network to a low-noise amplifier is equal to the inverse of the network's available gain. The available gain of various networks at 4 GHz was computed from high-accuracy S-parameter measurements. The available gain of a typical tuner was experimentally found to be a strong function of its tuning, which shows that "back-to-back" measurements of two tuners to obtain the loss of each tuner can be inaccurate. Measurement of the available gain of an amplifier's input-matching circuit is shown to give quick insight into its minimum noise contribution before the actual amplifier stage is built.

I. Introduction

SOME APPLICATIONS of microwave amplifiers require squeezing out every bit of low-noise performance from an active device. One such application is a satellite earth station low-noise amplifier, whose contribution to system operating noise temperature is relatively large due to the low equivalent temperature of the antenna [1]. Development of such amplifiers requires selection of the best active devices, circuit materials, and techniques.

The noise of active two-ports is usually characterized with an impedance-substitution setup such as that shown in Fig. 1 [2]. Small losses in the components between the noise generator and the device under test (DUT) contribute directly to the total noise figure, and the accuracy to which this loss is known contributes directly to the accuracy of the noise-figure measurement. In current practice, the losses of typical input circuits used in microwave-transistor noise-test setups are in the range of 0.1–1 dB. This range is normally much greater than the uncertainty in the calibration of the noise source. Because this uncertainty leads to inaccurate noise data for the device, especially F_{min} data,

Manuscript received June 3, 1980; revised October 28, 1980.

The author was with Farinon Transmission Systems, San Carlos, CA. He is now with Applied Research Group, Tektronix Laboratories, Tektronix, Inc., Beaverton, OR 97077.

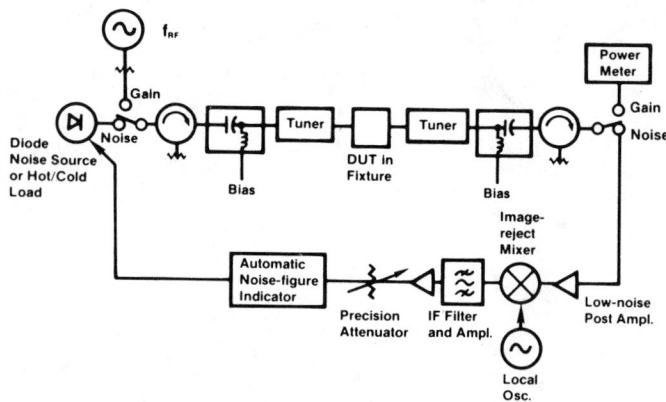

Fig. 1. Microwave transistor noise characterization system [2].

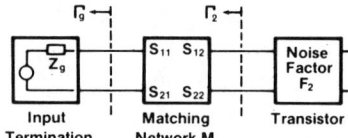

Fig. 2. Block diagram of a low-noise input match.

an amplifier stage designed with the rough data must sometimes be built before the full capabilities of the device are known. Such cut-and-try techniques are very inefficient.

Although the input noise-matching networks of low-noise amplifiers are selected on the basis of their relative noise-figure performances, the absolute losses of these networks are usually not known. Transistor chips used in such an amplifier normally cannot be removed for testing or for performance comparisons in another circuit. To compare circuits, several amplifiers must be made with chips from the same lot in order to obtain a statistical sample of circuit performance. This procedure is very tedious and inefficient, especially when the best devices available are used.

In the following section, the appropriate loss of a noise-matching network is discussed. Various methods of measuring these losses are thus examined, including the popular "back-to-back" method. The assumptions inherent in the back-to-back method are shown to be poor, but the direct measurement of the network S-parameters with sufficient accuracy is shown to lead to useful loss data. The loss of a typical slide-screw tuner is shown to be a strong function of the magnitude and angle settings, and the losses of input circuits for a 4-GHz low-noise amplifier are shown.

II. Loss in the Matching Network

Fig. 2 shows the block diagram of the circuit elements for a low-noise input stage. Network M in Fig. 2 refers to the following components of Fig. 1: switch, isolator, bias network, input tuner, and any adaptors and/or cables between the noise generator and the DUT. For an amplifier stage, network M consists of all the components that are used to effect a noise match between the input and the first-stage transistor and to provide any necessary bias. In Fig. 2, S_{ij} refers to the S-parameters of network M, Γ_g is the reflection coefficient of the noise generator (or input termination), and Γ_2 is the reflection coefficient "looking into" network M from the active device. The characteristic impedances of the networks are assumed to be equal, so no Z_0 normalization factors are included in the following equations.

The noise factor F (noise figure = noise factor expressed in decibels) of a cascade of noisy two-ports [3] is given by

$$F = F_1 + \frac{F_2 - 1}{G_1} + \frac{F_3 - 1}{G_1 G_2} + \cdots \quad (1)$$

where F_n and G_n are the noise factor and the available gain, respectively, of the nth two-port. The available gain of a two-port is the ratio of its available output power to the power available from its input source. The available gain α of a passive two-port is less than 1, and its noise factor is the factor by which available power is attenuated in passing through it, i.e., $1/\alpha$ [4]. Thus, the noise factor of the cascade of a passive two-port and a two-port with noise factor F_2 is simply F_2/α.

For network M in Fig. 2, the available gain is [5]

$$\alpha = \frac{|S_{21}|^2 (1 - |\Gamma_g|^2)}{(1 - |\Gamma_2|^2)|1 - S_{11}\Gamma_g|^2} \quad (2)$$

where

$$\Gamma_2 = S_{22} + \frac{S_{21} S_{12} \Gamma_g}{1 - S_{11} \Gamma_g}. \quad (3)$$

Thus, the noise contribution of a passive network depends on its source immittance but is independent of its load immittance. It should be noted that this is different than a two-port's power gain or transducer gain.

The noise factor of the two-port F_2 (in this case, a transistor) has the familiar dependence [4] on the source reflection coefficient Γ_2

$$F_2 = F_{\min} + \frac{r_n}{\mathrm{Re}(Y_2)} |Y_2 - Y_{\min}|^2 \quad (4)$$

$$= F_{\min} + 4 r_n \frac{|\Gamma_2 - \Gamma_{\min}|^2}{(1 - |\Gamma_2|^2)|1 + \Gamma_{\min}|^2} \quad (5)$$

where $F_{\min}$ = transistor's optimum noise factor, R_n = transistor's noise resistance, r_n = normalized noise resistance = R_n/Z_o, $Y_{\min}$ = transistor's optimum source admittance, corresponding to $\Gamma_{\min}$, and Y_2 = source admittance, corresponding to Γ_2.

For calculation of the cascaded noise figure given in (1), the available gain of the first stage G_1 must be known and is also dependent on its source reflection coefficient Γ_2. In practice, the gain measurement is simplified by measuring both the gain and the noise figure with an isolator after the stage(s) under test, as in Fig. 1. This isolator effectively makes the (measured) power gain equal to the (desired) available power gain, regardless of how the DUT's input and output tuners are adjusted.

For the case of an ideal generator impedance $\Gamma_g = 0$, the available gain simplifies to

$$\alpha_0 = \frac{|S_{21}|^2}{1-|S_{22}|^2} \quad (6)$$

where the subscript 0 denotes $\Gamma_g = 0$.

The assumption of ideal generator impedance leads to some error, even with relatively small $|\Gamma_g|$. Equations (2) and (3) can be rearranged to give

$$\alpha = \frac{|S_{21}|^2(1-|\Gamma_g|^2)}{(1-|S_{22}|^2)+|\Gamma_g|^2(|S_{11}|^2-|D|^2)-2\operatorname{Re}(\Gamma_g M)} \quad (7)$$

where

$$D = S_{11}S_{22} - S_{21}S_{12}$$

and

$$M = S_{11} - DS_{22}^*.$$

For $|\Gamma_g| \ll 1$, the first order effect is from the denominator

$$\alpha \simeq \frac{\alpha_0}{1 - 2\frac{\operatorname{Re}(\Gamma_g M)}{1-|S_{22}|^2}} \quad (8)$$

and α ranges from $\alpha_{\min}$ to $\alpha_{\max}$, depending on the relative phase of Γ_g and M

$$\alpha_{\min.\max} \simeq \frac{\alpha_0}{1 \pm 2|\Gamma_g||S_{11} + \alpha_0 S_{22}^* \frac{S_{12}}{S_{21}^*}|} \quad (9)$$

Here, α would seem to have a very large Γ_g dependence, but as M approaches a lossless network, [6] the factor multiplying $|\Gamma_g|$ approaches 0 and α_0, $\alpha_{\min}$, and $\alpha_{\max}$ all approach unity. For a typical case of the tuner discussed in the following, if $|\Gamma_g| = 0.05$ and $\alpha_0 = -0.55$ dB, then $\alpha_{\min} = -0.58$ dB and $\alpha_{\max} = -0.52$ dB.

The available and the delivered powers have also been discussed for the cases of different ambient temperatures, [3], [7] for multiports, [8] and for n-cascaded two-ports [9]. Other sources of noise-measurement error, including noise-source calibration and mismatch, Y-factor accuracy, second-stage noise contribution, ambient-temperature corrections, etc., are also very important to the noise measurement, but are beyond the scope of this discussion.

III. Tuner Loss Experiments

Probably the most popular method of measuring what is loosely called "tuner loss" is to lock the tuner at its setting, then to cascade it with another tuner of the same type that is tuned to conjugate match the first tuner, and finally, to measure the substitution loss [6] of the cascade in a 50-Ω (or Z_0) system. The second tuner matches the S_{22} of the first tuner back to Z_0, and the assumption is that half of the measured loss was caused by each tuner, since they are similarly constructed.

Two tuners shown thus connected in Fig. 3 have S-parameters S_{ija} and S_{ijb} with port 2 of tuner a connected to

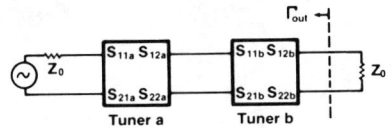

Fig. 3. "Back-to-back" measurement of tuners.

port 1 of tuner b. The transmission of the cascade is

$$S_{21tot} = \frac{S_{21a}S_{21b}}{1-S_{22a}S_{11b}}. \quad (10)$$

We seek the necessary conditions on $[S_a]$ and $[S_b]$ for the available gain of the first tuner with ideal source impedance to be equal to half of this attenuation

$$\frac{|S_{21a}|^2}{1-|S_{22a}|^2} = |S_{21tot}|. \quad (11)$$

An illustrative sufficient condition is

$$S_{11b} = S_{22a}^* \quad (12a)$$
$$|S_{21a}| = |S_{21b}|. \quad (12b)$$

That is, if tuner b is adjusted so that its S_{11} is the conjugate of tuner a's S_{22} and simultaneously has the same $|S_{21}|$, then the attenuation of the cascade is indeed twice the available gain (in decibels) of tuner a at that setting. However, tuning tuner b either for $\Gamma_{out} = 0$ or for maximum power delivered to the load does not, in general, exactly satisfy either (12a) or (12b). Other complications with this method include inaccuracies due to nonideal source and load impedances [6]. The error in α from nonideal source impedance is not measurable using this method. The main problem with this technique, however, is that a tuner can have a value of α_0 for one tuning that is rather different from the value for the conjugate tuning.

Another method used for double-slug tuners is to conjugate tune one slug to tune out the other, then measure the attenuation of the tuner alone. This does not correctly account for losses in the line and connectors of the tuner beyond the two slugs.

The available gain of a matching network can be calculated from (2) or from (6), if the S-parameters are measured with high accuracy. The available gain of the matching network with ideal generator impedance α_0, requires only accurate $|S_{21}|$ and $|S_{22}|$. The accuracy in α_0 varies directly with that of $|S_{21}|^2$ and is very sensitive for large $|S_{22}|$. Fig. 4 shows the error in α_0 versus $|S_{22}|$ for various absolute errors in $|S_{22}|$ when there is no $|S_{21}|$ error. For typical GaAsFET's in the 2–12-GHz range, $|S_{22}|$ of a low-noise input-matching network ranges as high as 0.8 to effect minimum noise match.

S-parameter measurements with a magnitude accuracy of < 0.01 are within the capability of computer-controlled, network-analyzer systems if multiple measurements are averaged [10], [11]. The system (ANA) used in the experiments is basically an HP 8409 system but is driven by PDP-11/70 minicomputer. A CW or narrow-band swept RF source was used to avoid "harmonic skipping" prob-

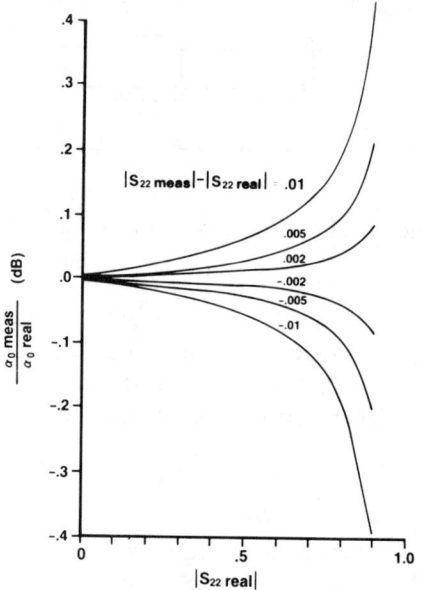

Fig. 4. Available gain error versus $|S_{22}|$ for various $|S_{22}|$ errors, when the available gain with an ideal source impedance is calculated from measured S-parameters. Subscript meas refers to the measured value, and subscript real refers to the actual value.

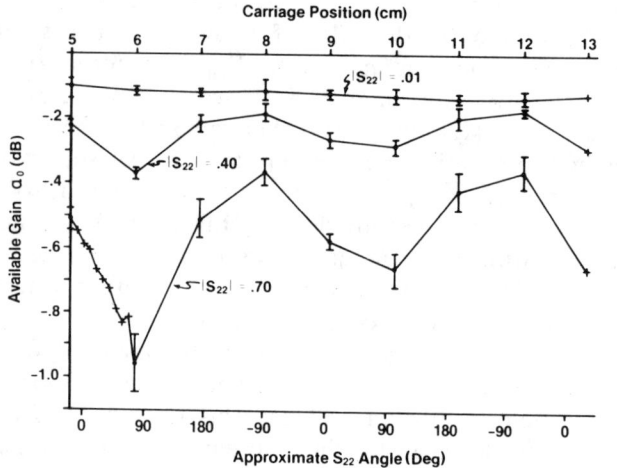

Fig. 5. Available gain with an ideal source impedance of a coaxial slide-screw tuner, measured on ANA at 4.0 GHz. + denotes three measurements averaged. ● denotes at least six measurements averaged.

this figure show the spread of α_0 from individual measurements (S_{21}, S_{22} pairs). As expected from Fig. 4, the spread is larger for larger $|S_{22}|$. This tuner is equipped with vernier scales for resetting a tuning condition. At 4 GHz, backlash was undetectable from carriage movement and was just detectable from stub micrometer positioning. Thus, settings were made by approaching a setting from one direction, as is done with a precision IF attenuator. As with most tuners, the loss is higher for higher $|S_{22}|$. Without appropriately correcting for the higher loss with higher $|S_{22}|$, the $\Gamma_{\min}$ apparent from tuning for minimum total noise figure has too small a magnitude. (GaAs MESFET's present more measurement difficulty than silicon bipolar transistors because a MESFET's $|\Gamma_{\min}|$ is almost always much larger.)

The available gain was also a function of the S_{22} angle. After averaging numerous measurements, a half-wavelength (3.75-cm) periodicity in the available gain became apparent. This periodicity might be caused by radiation from the tuning element. It is this dependence on the S_{22} angle that makes back-to-back measurements of this tuner inaccurate. For example, if this tuner is tuned for $|S_{22}|=0.70$, where $\alpha_0 = -0.9$ dB, an electrically identical tuner tuned for the same $|S_{22}|$ but conjugate angle will have $\alpha_0 = -0.4$ dB. This phenomenon was confirmed in back-to-back measurements. The attenuation of the tuner pair was a strong function of S_{22} angle, and there was no way of telling which tuner had what loss. If there were no S_{22} angle dependence and if the different tuners had similar enough $|S_{21}|$ versus $|S_{22}|$ characteristics, then the back-to-back measurement with very good source and load matches would yield accurate available-gain data.

IV. NOISE-MATCHING CIRCUIT MEASUREMENTS

Available-gain measurements were done on two microstrip matching networks for the input stage of a 3.7–4.2-GHz low-noise amplifier (LNA). Calibration standards that take into account the connectors on the microstrip circuits were used, and care was taken to shield the circuits in order to avoid radiation losses. Both circuits presented $|S_{22}|=0.75$ to a GaAsFET, but the minimum total noise figure (usually at midband) averaged 0.2 dB lower for circuit B than for circuit A in eight amplifiers built with each circuit from the same lot of FET's. Fig. 6 shows the available gain of circuits A and B as measured on the ANA. The difference in the available gain of circuits A and B agrees very well with the difference in noise performance. This available-gain measurement ascertained that circuit A was simply more lossy than circuit B, and that the difference in noise performance was not due to the circuit not presenting $\Gamma_{\min}$ to the FET or causing some type of spurious oscillation. Circuit A was built on alumina, while circuit B was built on teflon-fiberglass and had a different circuit topology. Circuit B was used for the input curcuit of an economical satellite earth station LNA having 120 K maximum effective input noise temperature over the 3.7- to 4.2-GHz band.

lems, and full error modeling [12] was necessary to eliminate the effects of the source and load impedances. Repeatability was greatly improved by simply checking the calibration standards often between measurements. When the measured magnitude of any of the calibration standards had drifted more than 0.005, the system was recalibrated. The rms difference between the measured S_{21} and S_{12} of the matching circuits was typically 0.004 in magnitude and 0.6° in phase, which compares well with the expected reciprocity. For calculations of α, the average of the magnitudes of the measured S_{21} and S_{12} was used for $|S_{21}|$ in (6).

The available gain of a 2–18-GHz slide-screw tuner, Maury Microwave type 2640D, was measured on the ANA for various angle and magnitude settings at 4 GHz. Fig. 5 shows the results of these measurements. The error bars in

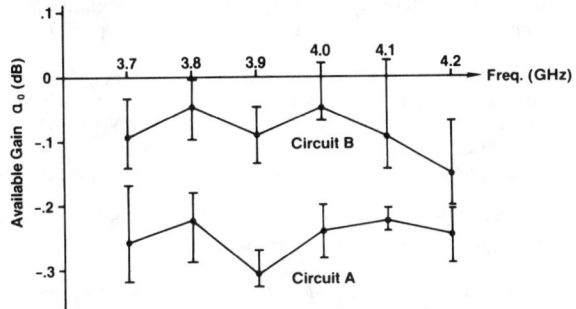

Fig. 6. Available gain with an ideal source impedance of two input matching circuits. Three measurements of circuit A were averaged, and seven measurements for three different tuning conditions of circuit B were averaged. Error bars show the spread of individual measurements.

Finally, the optimum noise figure of an FET (a selected NEC NE388) at 4 GHz was measured with two different networks. First, the FET was installed in circuit *B* and was biased and tuned for optimum noise figure at 4.0 GHz. The total noise figure of this amplifier at 4 GHz was 1.25 dB at the waveguide input. Subtraction of noise contributions from the input isolator, input matching network, and later stages implied a noise figure of 1.0 dB for the first-stage transistor. The FET was then removed and measured in the test setup of Fig. 1 using the tuner shown in Fig. 5 and the same input isolator and post-amplifier. When the input tuner and FET bias were adjusted for minimum total noise figure, the transistor's noise figure was 1.3 dB. The problem here is that the minimum total noise figure indicated does not correspond to the minimum first-stage noise figure for the following reasons. Since the later stages contribute to the total noise, the input tuning and FET biasing tend to optimize the gain of the first stage by compromising its noise figure. As noted above, the decrease in the tuner's available gain with increasing $|S_{22}|$ also tends to make the input tuning misleading.

Instead of randomly searching for the real minimum, the normal procedure is to measure the first-stage noise figure for about seven or more source admittances and then least-squares fit the data to (4) to find F_{min}, Γ_{min}, and r_n [13]. This procedure was performed with the tuner's corresponding measured available gain subtracted from the measured total noise figure for each tuning case. The F_{min} from the fitting program was 1.1 dB, and the calculated $|\Gamma_{min}|$ was considerably larger than for the minimum total noise-figure tuning. Two other factors that probably contribute to the remaining 0.1-dB discrepancy are 1) the loss of an APC-7 coax-to-microstrip transition between the input tuner and the FET measurement plane, and 2) the transistor's bias may not have been exactly equal in both cases.

V. Conclusions

The noise-factor contribution of a passive two-port is equal to the inverse of its available gain and is given by (2). For low-loss passive circuits, the effect of non-Z_0 source impedance on the available gain is smaller with lower loss in the circuit. The back-to-back attenuation measurement of two tuners to determine the available gain of one of them assumes that the attenuation of the tuner pair is twice the available gain of either tuner. In practice, this can be a poor assumption.

Averaged measurements of a passive two-port on a computer-corrected network analyzer were successfully used to determine the available gain at 4 GHz of a slide-screw tuner and two matching circuits for a low-noise amplifier. Much higher accuracy is necessary as $|S_{22}|$ approaches unity, but data for $|S_{22}|=0.75$ agreed well with the noise performance of amplifiers using these networks. Measurement of the available gain of a matching network gives immediate insight into its best noise performance, whereas testing the circuit with its intended active device does not conclude whether extra noise is from circuit losses, Γ_{min} mismatch, spurious oscillations, nonoptimal biasing of the active device, or the device itself if it is not changeable.

The tuner problem is sometimes circumvented by selecting tuners with losses low enough to be ignored. Such a selection is not easy, depending on the accuracy desired, and usually results in limited tuning range and ease [14], [15]. The best double-slug coaxial tuners are less lossy than the slide-screw tuner measured here, but a slide-screw tuner has the advantage of almost independent magnitude and angle control. A major motivation for studying lossy noise-matching networks, however, is to show that the extra losses and source mismatch effects can be removed from the noise readings in a manner analogous to the removal of cable and adapter reflections by a computer-corrected network analyzer. Several decibels of loss can easily be removed as long as it is repeatable, as is done on a corrected network analyzer.

An automatic device-noise-parameter measurement system [16] can be assembled that electrically tunes the input, takes noise figure measurements, and quickly fits the data to find the noise parameters F_{min}, r_n, and Γ_{min}. A major problem with this approach is that the loss of an electrically tuned input tuner is much higher than that of mechanical tuners. This loss difference makes available-gain characterization of the tuner critically necessary. If the tuner has repeatable S-parameters, its available gain can be measured for each setting and then used to correct each noise figure measurement, as was done here for the slide-screw tuner example.

For low-noise amplifier development, what is needed is not a noise meter that measures the noise figure from a source impedance of approximately $50+j0$ Ω. Instead, the noise system should measure $F(\Gamma)$ for several Γ values and read out F_{min}, Γ_{min}, r_n, and $F(\Gamma)$ for any given Γ automatically over a frequency band. Such a system is necessary to collect sufficient noise data on active devices to define useful design windows. Measurement of an input-matching circuit cascaded with an active device on such a system would directly reveal the noise contribution of the matching circuit, how much figure improvement is possible from matching Γ_{min}, and what vector direction to tune to decrease $|\Gamma_{min}|$ at the input.

ACKNOWLEDGMENT

The author wishes to thank C. Hsieh, R. Q. Lane, P. Estabrook, and L. Lockwood for many helpful discussions.

REFERENCES

[1] H. C. Okean and P. P. Lombardo, "Noise performance of M/W and MM-wave receivers," *Microwave J.*, vol. 16, no. 1, Jan. 1973.
[2] R. A. Pucel, D. J. Masse, and C. F. Krumm, "Noise performance of gallium–arsenide FET's," *IEEE J. Solid-State Circuits*, vol. SC-11, April 1976.
[3] W. W. Mumford and E. H. Scheibe, *Noise Performance Factors in Communications Systems*. Dedham, MA: Horizon House, 1968.
[4] "IRE standards on methods of measuring noise in linear twoports, 1959," *Proc. IRE*, vol. 48, pp. 60–68, Jan. 1960.
[5] C. K. S. Miller, W. C. Daywitt, and M. G. Arthur, "Noise standards, measurements, and receiver noise definitions," *Proc. IEEE*, vol. 55, June 1967.
[6] D. M. Kerns and R. W. Beatty, *Basic Theory of Waveguide Junctions and Introductory Microwave Network Analysis*. New York: Pergamon, 1967.
[7] T. Y. Otoshi, "The effect of mismatched components on microwave noise-temperature calibrations," *IEEE Trans. Microwave Theory Tech.*, vol. MTT-16, pp. 675–686, Sept. 1968.
[8] D. F. Wait, "Thermal noise from a passive linear multiport," *IEEE Trans. Microwave Theory Tech.*, vol. MTT-16, pp. 687–691, Sept. 1968.
[9] T. Mukaihata, "Applications and analysis of noise generation in N-cascaded mismatched two-port networks," *IEEE Trans. Microwave Theory Tech.*, vol. MTT-16, pp. 699–708, Sept. 1968.
[10] B. P. Hand, "Developing accuracy specifications for automatic network analyzer systems," *Hewlett-Packard J.*, pp. 16–19, Feb. 1970.
[11] E. F. Da Silva and M. K. McPhun, "Repeatability of computer-corrected network analyzer measurements of reflection coefficients," *Electron. Lett.*, vol. 14, no. 25, pp. 832–834, Dec. 1978.
[12] J. Fitzpatrick, "Error models for systems measurements," *Microwave J.*, pp. 63–66, May 1978.
[13] R. Q. Lane, "The determination of device noise parameters," *Proc. IEEE*, (Letters), vol. 57, pp. 1461–1462, Aug. 1969.
[14] G. Caruso and M. Sannino, "Computer-aided determination of microwave two-port noise parameters," *IEEE Trans. Microwave Theory Tech.*, vol. MTT-26, pp. 639–642, Sept. 1978.
[15] M. Mitama and H. Katoh, "An improved computational method for noise parameter measurement," *IEEE Trans. Microwave Theory Tech.*, vol. MTT-27, pp. 612–615, June 1979.
[16] R. Q. Lane, "A microwave noise and gain parameter test set," *ISSCC Dig. Tech. Pap.*, pp. 172–173, Feb. 1978.

Part II
Noise Properties of Bipolar Transistors

Microwave Transistors: Theory and Design

HARRY F. COOKE, MEMBER, IEEE

Abstract—Microwave transistors are useful as small-signal amplifiers to 6 GHz and power amplifiers to 4 GHz. Nearly all microwave transistors are of the silicon planar type. Power transistors use three types of geometries—interdigitated, overlay, and mesh—while small-signal transistors use interdigitated only. The general theory of the frequency response of transistors is reviewed, including active and inactive elements. A condensed description of the design and processing steps for a silicon microwave transistor is given. A final section deals with the types of high-frequency measurements used in the design and analysis of transistors.

I. INTRODUCTION

MICROWAVE transistors have come to be regarded as a pinnacle in the transistor art. Not only are they difficult to build, but most of the more desirable parameters have been optimized. Because of the very fine geometry and the shallow diffusions which are used, yields have remained relatively modest compared to low-frequency types. This is in spite of the fact that a 1.5-in-diam silicon slice has a potential of almost 8000 transistors on 0.015-in centers.

The maximum frequency of oscillation now exceeds 15 GHz for the typical small-signal transistor. Noise figures less than 1 dB are now attained at 500 MHz, and less than 6 dB at 6 GHz. Power output is also impressive—a peak of 100 W at 1 GHz and 5 W CW at 4 GHz. Fig. 1 gives a plot of noise figure and power output versus frequency as of November 1970.

If oscillation or useful gain above 1 GHz is taken as the criterion for microwave operation, the germanium mesas and microalloy transistors of the 1958–1959 era were the first microwave transistors. The advantages of germanium over silicon as a semiconductor material, however, remained largely theoretical, and by 1963 silicon microwave transistors were beginning to become competitive. From the point of view of the user, the relative ruggedness of the silicon transistor was decisive and the phaseout of germanium was begun.

Manuscript received January 2, 1971; revised March 15, 1971.
The author is with Avantek, Incorporated, Santa Clara, Calif. 95051.

Reprinted from *Proc. IEEE*, vol. 59, pp. 1163–1181, Aug. 1971.

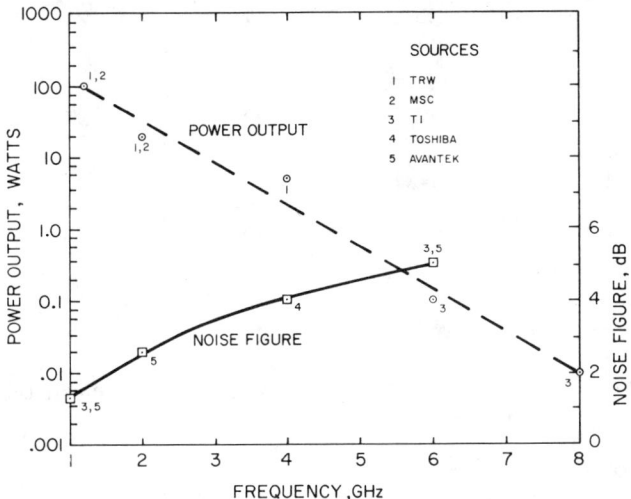

Fig. 1. Noise figure and power output of microwave transistors, November 1970.

Today a few germanium, L-band, planar transistors are still in use, but their numbers are decreasing rapidly. One very important reason for the demise of the high-frequency germanium transistor is that germanium does not have a natural passive oxide. For this reason, germanium planar transistors use silicon oxide as the passivating agent.

This paper will attempt to give a broad picture of the theory, construction, and testing of microwave transistors. The details of some areas which are largely descriptive (such as power transistor design) will be discussed only from the basic theoretical point of view. Where processes or theory are common to low-frequency transistors, only a brief mention will be given.

The Introduction will be completed with a brief description of the general structure of microwave transistors.

All microwave transistors are now planar in form and almost all are silicon n-p-n. The geometry falls into three general types (Fig. 2): 1) interdigitated, for small signal and power; 2) overlay, for power only; and 3) mesh (also called "matrix" and "emitter grid"), for power only. Several analyses [1] have been made comparing these three structures on an idealized basis; the conclusions reached tend to become less clear when practical devices are considered. It can be shown, however, that where one structure displays a definite superiority in some respect, it is achieved at the expense of a degradation of one or more other parameters. This is particularly true with respect to capacitance, where reduced C_{TE} is obtained at the expense of high MOS capacitance, etc.

It appears now that the overlay and mesh structures are beginning to dominate the VHF–UHF power device market while the interdigitated transistor is still the best structure for S- and C-band small-signal applications.

Most power transistors now include some form of integral emitter resistors to aid in equalizing the current distribution over the larger area which a power device must cover. The overlay transistor uses an integral diffused resistor as part of the emitter itself, while the interdigitated generally use thin-film resistors as a part of the contact system. Mesh devices [1], [2] have been constructed using both techniques. Some earlier devices used separate diffused resistors mounted on the package.

The use of silicon nitride as an additional passivating agent is fast becoming universal. It has a particular value in microwave transistors where the junctions are extremely shallow and thus highly susceptible to contamination.

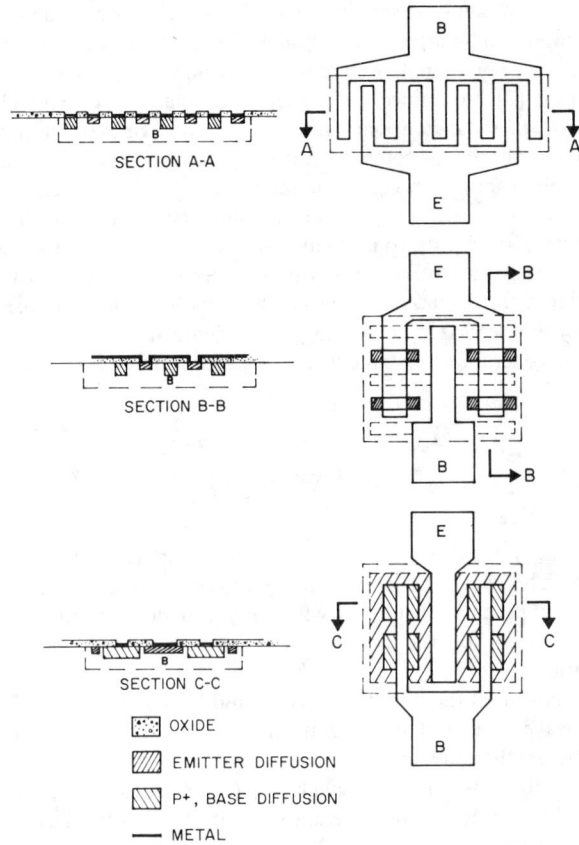

Fig. 2. Three general types of microwave transistor geometry.

Silicon nitride is a superior passivating agent (compared to silicon dioxide), although it cannot be placed directly over a bipolar transistor junction. A thin layer of SiO_2 is usually placed between it and the silicon. Since it is not etched at the same rate as SiO_2 by hydrofluoric acid etches, it has some advantages as a secondary mask. Junctions passivated with it have a lower failure rate and can operate at a higher temperature.

The current carrying capability of a transistor is proportional to the emitter periphery, base resistance is inversely proportional to the emitter width, and f_T (at lower currents) is inversely proportional to the emitter area. Thus a good microwave transistor has a very narrow emitter (e.g., ≈ 1.0 μm, for state of the art). To keep the length to a practical size, the device is broken up into emitter "sites." These sites may be grouped together (overlay) or simply connected in parallel (interdigitated). Some of the more sophisticated mesh and interdigitated types for high power also group the emitters into secondary sites.

II. Transistor Theory

A. Performance

1) Gain: In this section the theory of transistor action, insofar as the high-frequency characteristics are concerned, will be reviewed. Appropriate equations will be used where required, but proofs and developments will be shortened or presented by reference only.

A good microwave transistor must, first of all, be a good dc transistor. Most of the requirements for good dc operation can be satisfied independently of the high-frequency requirements and where exceptions occur, such as the effect of a drift field on the dc current gain, these will be mentioned.

Before discussing gain, it should be pointed out that there are several definitions of gain which are useful for transistor characterization. Maximum available gain (MAG), G_{max}, is obtained when both input and output are simultaneously conjugately matched. G_{max} exists only when the device is unconditionally stable. Unilateral gain, or U [3], is obtained when a device is unilateralized with a lossless network and matched at both ports. U is the same common base, common collector or common emitter. In the microwave region, U is 1–3 dB higher than G_{max}, common emitter, for a typical small-signal transistor. Both G_{max} and U are often defined in terms of f_{max}, and both should drop to unity at about that frequency. Since common-emitter microwave transistors may have power gain with no impedance transformation, they can have useful gain when inserted directly into a 50-Ω system. This gain is identical to $|S_{21}|^2$.

The general form for the gain as a function of frequency is

$$G(f) \approx \frac{G_0}{\left[1 + G_0^2\left(\frac{f}{f_{max}}\right)^4\right]^{1/2}}. \tag{1}$$

When $G_0^2(f/f_{max})^4 \gg 1$, $G \approx (f_{max}/f)^2$ and gain falls at 6 dB per octave. The gain function is actually discontinuous, and there are regions of potential instability where gain is undefined. Fig. 3 shows these regions for the common-base and common-emitter configurations.

Equation (1) shows that f_{max} is a suitable frequency for defining microwave gain and it is useful, then, to examine f_{max} and the parameters that govern it.

If feedback is considered negligible (or neutralized by a lossless network), a common-emitter transistor can be considered as having the following simplified parameters:

$$\text{Re}(Y_{in}) \approx \frac{1}{r_b'}$$

$$\text{Re}(Y_{out}) \approx \omega_T C_C$$

current gain

$$|h_{fe}| = \frac{\omega_T}{\omega}.$$

The equivalent circuit of Fig. 4 depicts how the parameters fit into a simplified high-frequency equivalent circuit.

The gain of this two-port at ω is

$$G \approx \frac{\omega_T}{4\omega^2 r_b' C_C}. \tag{2}$$

At f_{max} the gain G will have dropped to one, and

$$f_{max} \approx \sqrt{\frac{f_T}{8\pi r_b' C_C}}. \tag{3}$$

A more practical expression is

$$f_{max_{GHz}} \approx 6.3 \sqrt{\frac{f_T \text{ in GHz}}{(r_b' C_C) \text{ in ps}}}. \tag{3a}$$

This is an approximate expression based on a simplified model; a later section on modeling will present a more accurate model. Note the three critical parameters, f_T, r_b', and C_C.

2) Noise Figure: In addition to having gain, small-signal transistors are usually required to have a low enough noise figure to meet certain systems requirements. The noise figure is simply a measure of the degradation in the signal-to-noise ratio that a signal undergoes when passing through a signal processing element, such as a transistor amplifier.

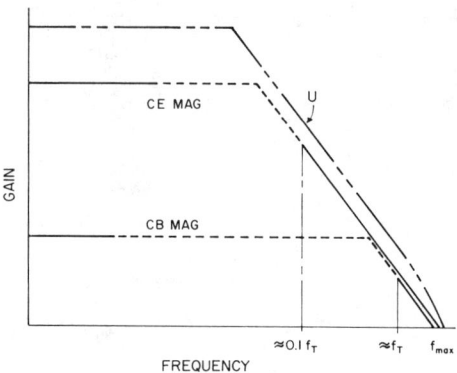

Fig. 3. Maximum available gain and U versus frequency. Broken line represents regions of potential instability.

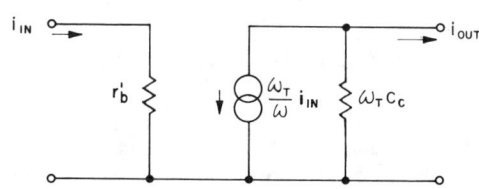

Fig. 4. Simplified equivalent circuit of a transistor for gain calculations. $\text{Re}(Y_{in}) \cong 1/r_b'$ and $\text{Re}(Y_{out}) \cong \omega_T C_C$.

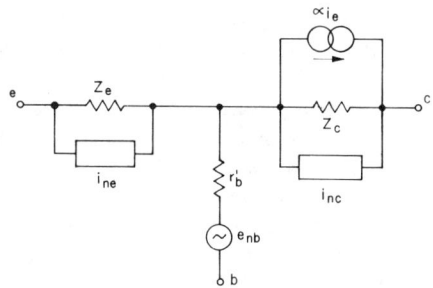

Fig. 5. Common-base transistor noise equivalent circuit.

$$\text{Noise figure } F = \frac{\text{signal-to-noise ratio at input}}{\text{signal-to-noise ratio at output}}.$$

It will be convenient here, as in the previous section on gain, to present a simplified model in order to develop an expression for noise figure. A number of papers have been written on the noise figure of transistors but Nielsen's [4] was the first to present it in terms of generally known transistor parameters. Cooke [5] manipulated Nielsen's equations to give a version which uses more accessible parameters. Fukui's paper [6] presents an exhaustive investigation into transistor noise figure including parasitics.

Fig. 5 is a simplified noise model of a common-base amplifier. The common-emitter amplifier can be treated in the same manner, but the result is the same, and the calculations are more tedious. Fig. 5 shows that in the microwave region the transistor has three principal sources of noise.

a) Shot noise in the emitter:

$$\overline{i_{ne}^2} = 2qI_E \Delta f = 2kTg_e \Delta f. \tag{4}$$

b) *Shot noise in the collector:*

$$\overline{i_{nc}^2} = 2qI_c\Delta f. \tag{5}$$

c) *Thermal noise in the base resistance:*

$$\overline{e_{nb}^2} = 4kTr_b'\Delta f. \tag{6}$$

Terms a) and b) are correlated by alpha.

Using the Cooke method [5], the noise figure can be shown to be as follows:

$$F = 1 + \frac{r_b'}{R_g} + \frac{r_e}{2R_g} + \frac{(R_g + r_b' + r_e)^2}{2\alpha_0 r_e R_g}\left[\left(\frac{f}{KF_T}\right)^2 + \frac{1}{h_{FE}} + \frac{I_{co}}{I_E}\right]. \tag{7}$$

This has the general form

$$F = F_p\left(1 + \left(\frac{f}{f_c}\right)^2\right) \tag{8}$$

where f_c and F_p are as defined in Fig. 6.

The factor K that appears in (7) needs some explanation. Actually, noise figure depends upon f_α and not f_T. The reason for this is that noise voltages and currents appear simply as magnitudes in noise calculations. The frequency f_T determines gain and differs from f_α primarily because of the relatively large phase shift in h_{fe} compared to α. By definition, then, the factor K is

$$K \doteq \frac{f_\alpha}{f_T}. \tag{9}$$

The value of K depends upon the magnitude of the drift field in the base of the transistor, and is explained in detail in the section on characteristic frequencies. For most silicon transistors, $K = 1.2$.

Equation (7) does not include the effect of $1/f$ noise nor does it include the effect of parasitics, such as feedback capacity. Feedback capacity can reduce noise figure (as in the common-emitter connection) but gain is also reduced and the noise measure M^1 remains the same.

The importance of the dc common-emitter gain (h_{FE}) in noise figure (7) is often overlooked. For example, to obtain a noise figure of 1 dB or less at a frequency where

$$\left(\frac{f}{KF_T}\right)^2 \ll \frac{1}{h_{FE}}$$

h_{FE} must be greater than 100 for a typical transistor operated from a realistic source resistance. For modern planar devices, the shot noise from I_{co} is very small, and this term is often left out of (7).

B. Frequency Variation of Current Gain

1) Characteristic Frequencies: In the previous section it was pointed out that performance (i.e., gain and noise figure) both vary in the microwave region. This variation is due to the frequency dependence of a number of elements in the transistor. These in turn can best be characterized in terms of a number of characteristic or "cutoff" frequencies.

The gain of a common-emitter bipolar transistor is constant up

---

[1] Noise measure is defined as

$$M = G(F - 1)/(G - 1). \tag{10}$$

M was brought into use as a noise parameter since noise figure alone is a poor figure of merit in low-gain systems [7]. A somewhat more useful version of (10) is the noise figure of an infinite cascade of identical stages designated as F_∞,

$$F_\infty = M + 1 = F + (F - 1)/(G - 1). \tag{11}$$

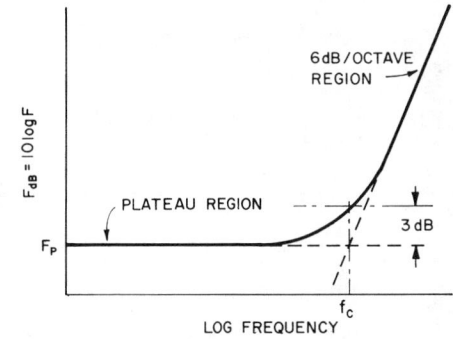

Fig. 6. Noise figure versus frequency.

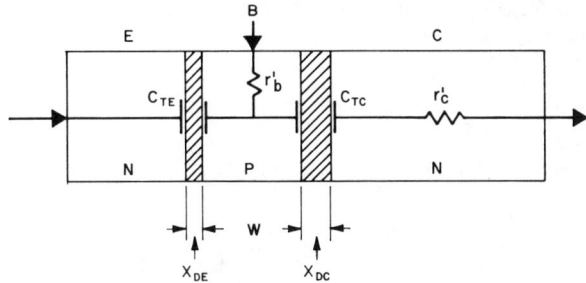

Fig. 7. One-dimensional view of a junction transistor.

to a "lower critical frequency" where it becomes conditionally stable. As frequency is further increased, it will become stable again at approximately $0.1 f_T$, and from there on the gain will continue to fall at approximately 6 dB per octave. For the gain to fall at exactly 6 dB per octave, the falloff in current gain would have to be the only frequency-dependent element in the transistor. This is not necessarily true. Near the limit of a transistor's usefulness where the gain is very low, gain may fall at less than 6 dB per octave due to capacitive feedthrough, or more due to other frequency-dependent elements.

The microwave range is in the 6-dB-per-octave region, and it is this frequency dependence that will now be examined. If a signal is put into the base or the emitter and taken out of the collector, it will encounter four successive principal regions of delay or attenuation. Fig. 7 shows these regions in a simplified cross-sectional view of a transistor. Since all silicon microwave transistors are presently n-p-n, the transistor is shown as such. Following the transit of carriers from the emitter we have, successively, the following.

1) The emitter–base transition capacity C_{TE} shunts the active emitter region.

2) Carriers must next cross the base region W through a combination of diffusion and electric field. During this transition the identity of a group of carriers will be lost and the signal attenuated.

3) Carriers next cross the collector depletion layer, X_{dc} wide, under the influence of an electric field (no diffusion).

4) If there is any appreciable resistance between the collector depletion-layer and the external collector terminal, a final RC delay is encountered.

Each of these elements can be regarded as a time delay.

$$\tau_{ec} = \tau_e + \tau_b + \tau_d + \tau_c. \tag{12}$$

τ_{ec} total transit time emitter to collector
τ_e emitter–base junction capacity charging time
τ_b base transit time
τ_d collector depletion-layer transit time
τ_c collector-capacitance resistance charging time.

The time constants can be more conveniently expressed in terms

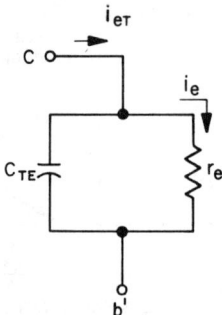

Fig. 8. Equivalent circuit of emitter–base junction.

of characteristic frequencies, as will be seen from the discussion that follows. Therefore, let

$$\tau = \frac{1}{\omega}. \tag{13}$$

Anticipating the following sections slightly we then obtain

$$\frac{1}{\omega_T} = \frac{1}{\omega_e} + \frac{1}{\omega_b} + \frac{1}{\omega_d} + \frac{1}{\omega_c}. \tag{14}$$

The frequency $\omega_T = 2\pi f_T$ is the most important parameter in a high-frequency transistor since it dominates both the gain and the noise performance (refer to Section I). Each of these characteristic frequencies will now be discussed.

a) The emitter–base junction cutoff frequency: When a transistor is operated in the normal mode, the emitter–base junction is in forward bias. The junction has a finite capacity associated with it as well as a resistance as shown in Fig. 8.

If a terminal emitter current i_{et} flows into the emitter terminal, it will divide between the capacitance C_{TE} and the resistance r_e. However, only the current that flows through r_e (sometimes called the space-charge resistance) gets injected into the base and is amplified. Thus the current in C_{TE} is a true parasitic. This is a simple RC-type cutoff so that

$$i_e = i_{et} \frac{1}{1 + j\omega r_e C_{TE}}. \tag{15}$$

The emitter current can now be normalized to the terminal current and called the high-frequency emitter injection efficiency γ_{hf}. Assuming that $\gamma_0 = 1$, at a cutoff frequency we shall define as ω_e,

$$\frac{|\gamma_{hf}|}{\gamma_0} = \frac{\sqrt{2}}{2}. \tag{16}$$

Thus from (15), (16)

$$\frac{1}{\omega_e} \cong r_e C_{TE}. \tag{17}$$

Equation (17) is valid for large devices but not for small-signal types intended for the gigahertz range. The collector capacitance C_{TC} must also be charged as well as some of the other miscellaneous capacitances such as those in the package. Therefore, (17) should be modified to give

$$\frac{1}{\omega_e} = r_e(C_{TE} + C_{TC} + C_x). \tag{17a}$$

b) Base transit characteristic frequency: The time for carriers to cross the base region is obtained by solving the transport equations [8], [9]. The solution is generally given in terms of β^*, called the base-transport factor. It is the ratio of current at the collector edge of the base to that at the emitter edge. β^* is complex since the carriers undergo a phase shift as well as a reduction in amplitude as they cross the base region.

If the carriers are first assumed to cross the base by diffusion alone, then

$$\beta^* = \text{sech}\left[\frac{W}{L}(1 + j\omega\tau_B)^{\frac{1}{2}}\right] \tag{18}$$

where τ_B = lifetime for minority carriers in the base. At low frequencies, $\omega\tau_B \ll 1$ and

$$\beta^* = \beta_0^* \cong \text{sech}\left(\frac{W}{L}\right). \tag{19}$$

β_0^* can be written as a truncated power series

$$\beta_0^* \approx 1 - \frac{1}{2}\left(\frac{W}{L}\right)^2 + \cdots \tag{20}$$

but $(W/L)^2$ must be much less than 1 for the transistor to be useful. Therefore, $\beta_0^* \approx 1$.

We now wish to obtain a suitable cutoff frequency. The frequency at which $|\beta^*|$ is 3 dB below its low-frequency value (which is ≈ 1) is designated as ω_β. Thus the magnitude of (18) must be obtained and set equal to $\sqrt{2}/2$ and solved for ω.

Before doing this is will be useful to make a slight transformation of the imaginary term in (18). The transit time across the base, t_B, can be written in terms of base width and diffusion constant:

$$t_B = \frac{W^2}{D}. \tag{21}$$

However, D is by definition

$$D = \frac{L^2}{\tau_B}. \tag{22}$$

Thus in (18) the imaginary (or frequency-dependent) term is $j\omega t_B$, where

$$t_B = \frac{W^2}{D} \tag{23}$$

or

$$t_B = \frac{W^2}{L^2}\tau_B. \tag{24}$$

Equation (18) then becomes

$$\beta^* = \text{sech}\left[\left(\frac{W}{L}\right)^2 + j\omega t_B\right]^{\frac{1}{2}}. \tag{25}$$

Equation (25) is then evaluated using a hyperbolic secant series to obtain $\omega_\beta t_B$ when $|\beta^*| = \sqrt{2}/2$. Using Pritchard's result [8], we obtain

$$\omega_\beta t_B = 2.43. \tag{26}$$

Substituting (24) into (26) gives

$$\omega_\beta\left(\frac{W^2}{D}\right) = 2.43 \tag{27}$$

or

$$\frac{1}{\omega_\beta} = \frac{W^2}{2.43D}. \tag{28}$$

The frequency ω_β is the frequency where the magnitude of the base transport factor is down 3 dB from its low-frequency value. If $\gamma_{hf} = 1$, it is also approximately equal to ω_α, the frequency where the common-base current gain is down 3 dB from its low-frequency value. ω_β is identical to $\omega_{\alpha i}$, the intrinsic (base only) cutoff frequency. A number of expressions for the frequency response of a transistor are written in terms of a characteristic frequency designated by ω_0. It is by definition

$$\frac{1}{\omega_0} \doteq \frac{W^2}{2D}. \tag{29}$$

Then

$$\omega_\beta = 1.2\,\omega_0. \tag{30}$$

There has been some confusion resulting from the similarity between (28) and (29), particularly since the solution of (25) using a hyperbolic cosine can lead to a factor of 2 instead of 2.43.

Equation (28) describes the cutoff frequency of the base transport factor for a transistor where the minority carriers move through the base by diffusion only. Carriers also can be caused to drift in the presence of an electric field, and even more important by a combination of field and diffusion [11]. The electric field in conventional planar transistors[2] is obtained through a grading of the impurities in the base [12], [13], i.e.,

$$\varepsilon = -\frac{kT}{q} \cdot \frac{1}{N(X)} \cdot \frac{dN(X)}{dX}. \tag{31}$$

An exponential distribution is a good approximation for $N(X)$ for diffused bases. The magnitude of the field obtained is then defined by a field factor η

$$\eta \doteq \ln \frac{N_{BE}}{N_{BC}}. \tag{32}$$

η can also be defined in terms of the electric field and the base width normalized to kT/q:

$$\eta = \frac{\varepsilon W}{\dfrac{kT}{q}}. \tag{33}$$

η can assume values of $\eta = 0$ (i.e., no field) to $\eta \cong 9$. The maximum value is set by other considerations in transistor design [13]. The effect of the electric field in reducing the transit time and raising the cutoff frequencies is described in a classic paper by Te Winkel [15]. This paper was written before the parameter f_T came into common usage, but it is quite apparent that f_T, as a new parameter, simplifies calculations a great deal.

The frequency[3] ω_{Ti} is greater than ω_0 when a drift field is present, and can be defined in terms of η and ω_0.

$$\omega_{Ti} = \omega_0 \left[\frac{\tfrac{1}{2}\eta^2}{\eta - 1 + e^{-\eta}} \right] \tag{35}$$

which is useful for all but small values of η.

---

[2] This is not the only means of obtaining an aiding field in the base of a transistor. Long reported using a separate base electrode which was biased to produce an aiding field in the base of a transverse transistor [30].

[3] The frequency ω_{Ti} refers to the internal base characteristics, while ω_T is used to describe the terminal properties and may include the effect of other frequency-sensitive elements. If $\tau_{ec} \cong \tau_e + \tau_b$, then ω_T is related to ω_{Ti} by

$$\frac{1}{\omega_T} \cong r_e C_{TE} + \frac{1}{\omega_{Ti}}. \tag{34}$$

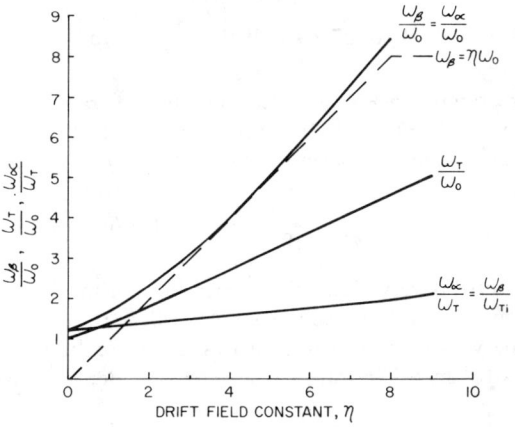

Fig. 9. Effect of drift field on characteristic frequencies.

The frequency ω_{Ti} has other interesting properties; at $\omega = \omega_{Ti}$, the magnitude of the internal common-emitter current gain = 1. The presence of an aiding field affects the various cutoff frequencies as shown in Fig. 9. Note that ω_α and ω_β are more affected by the drift field than is ω_{Ti}. The frequency ω'_{Ti} is identical to ω_1 of Te Winkel [15]. Today ω_1 is usually used to designate the frequency where the *terminal* value of h_{fe} has dropped to one. It is usually higher than Te Winkel's because of capacitive feedthrough.

Since the presence of the drift field increases ω_α more than ω_{Ti}, the ratio of $\omega_\alpha/\omega_{Ti}$ also changes with η, i.e.,

$$\omega_\alpha = (1.21 + 0.09\eta)\omega_{Ti}. \tag{36}$$

Fig. 9 also shows the ratio of $\omega_\alpha/\omega_{Ti}$ as a function of η.

The frequency ω_α was at one time regarded as the principal high-frequency parameter. Since common-base operation was the only mode possible, this made some sense. Today, however, common-base operation is seldom used and ω_T is the dominant parameter. In fact, ω_α is only relevant as it applies to noise figure (refer to Section II-A).

A number of papers define τ_b (12) as $1/\omega_0$ and $\omega_b = \omega_0$. It is actually more rigorous to use ω_β instead of ω_0, since this is the actual cutoff frequency rather than a mathematically convenient time constant. Thus for a transistor with no drift field

$$\tau_b = \frac{1}{\omega_b} = \frac{W^2}{2.43D} \tag{37}$$

and for a transistor with a field [15]

$$\tau_b = \frac{1}{\omega_{Ti}} = \frac{W^2}{2D(0.8 + 0.46\eta)}. \tag{37a}$$

c) Collector depletion-layer transit time and characteristic frequency: The collector junction is operated in reverse bias when the transistor is normally used as an amplifier. Power devices can be driven into saturation (collector junction in forward bias) but the effect is unfortunate rather than planned. Because the junction is in reverse bias, there is a well-defined depletion layer, X_d wide. Although the junction is formed by diffusion in most microwave transistors, it has been found that the step-junction equations still apply and are simpler to use than those for graded junctions; therefore, in general,

$$X_d \cong \left[\frac{2\varepsilon\varepsilon_0(V + \phi)}{qN} \right]^{\frac{1}{2}} \tag{38}$$

or for silicon,

$$X_d|_{Si} \cong 3.64 \times 10^3 \left[\frac{V+\phi}{N}\right]^{\frac{1}{4}} \text{ cm.} \tag{39}$$

For example, a typical silicon microwave transistor built on 5-$\Omega \cdot$ cm (i.e., $N=1.0 \times 10^{15}$) epitaxial material, and operated at 10 V, will have a depletion-layer width of about 3.6 μm. The field is then

$$\varepsilon = \frac{10.7}{3.6 \times 10^{-6}} = 3 \times 10^6 \text{ V/m.}$$

In spite of this very high field, carriers do not move through the depletion layer at the speed of light. At fields in the order of 1×10^6 V/m, the velocity in silicon saturates due to scattering effects at $V_{SL} \cong 8 \times 10^4$ m/s. The transit time across the depletion layer then becomes

$$\tau_m \approx \frac{X_d}{V_{SL}}. \tag{40}$$

The velocity V_{SL} is much greater than the equivalent velocity of carriers crossing the base region, and the time delay is insignificant in low-frequency devices. However, in microwave transistors, $X_d \gg W$; as a result, for state-of-the-art devices,

$$\tau_m \cong \tau_b.$$

The expression given in (40) is not correct if the sinusoidal response of the depletion layer is required. To obtain this, Pritchard [8] defines a new parameter β_m, the depletion-layer transport factor, by

$$\beta_m = \frac{\text{current leaving the depletion layer}}{\text{current entering the depletion layer}}. \tag{41}$$

With the assumption that the collector multiplication is zero, the sinusoidal response is [8]

$$\beta_m \cong \left[1 - j\omega \frac{\tau_m}{2}\right]. \tag{42}$$

Since the depletion-layer transit time is a simple delay, it affects the transistor parameters by changing the transfer admittance Y_{21} to $\beta_m Y_{21}$. If the Y matrix is then evaluated for h_{21e}, it will be found that the denominator contains the term $\tau_m/2$ in addition to $1/\omega_\beta$ and the emitter-base charging time. We may then define ω_d as

$$\frac{1}{\omega_d} = \frac{\tau_m}{2} = \frac{X_d}{2V_{SL}} \tag{43}$$

also

$$X_d = K\left(\frac{V}{N}\right)^{\frac{1}{4}}$$

so that

$$\frac{1}{\omega_d} = \frac{K}{2V_{SL}}\left(\frac{V}{N}\right)^{\frac{1}{4}}. \tag{43a}$$

Early obtained the same results as (43) from a similar calculation [16].

d) Collector RC cutoff frequency: When transistors were first made, the substrate material was of a single resistivity. Thus to obtain reasonable breakdown voltages it was necessary that the entire substrate be in the order of 1–10 $\Omega \cdot$ cm. As a result, there was a large high-resistivity region in the collector not covered by the collector depletion layer. The high-frequency resistance of this region is designated as r'_c. If a current is to flow out of the collector terminal, the collector capacitance must be charged through r'_c. This leads to a simple RC-type cutoff; thus

$$\frac{I_{c\text{ out}}}{I_{c\text{ in}}} = \frac{1}{1+j\omega r'_c C_C} = \frac{1}{1+j\frac{\omega}{\omega_c}} \tag{44}$$

where

$$\omega_c \doteq \frac{1}{r'_c C_C} \tag{45}$$

then

$$\tau_c = \frac{1}{\omega_c} = r'_c C_C. \tag{46}$$

Present-day microwave transistors are fabricated using a thin high-resistivity epitaxial layer on a lower resistivity substrate. In the structure, τ_c is usually negligible compared to the other three time constants in the transistor.

e) Summary: With the results obtained in the foregoing sections, we may now rewrite (14), using (17a), (37), (43), and (46). Common-emitter operation is also assumed.

For $\eta=0$ (no field),

$$\frac{1}{\omega_T} = r_e(C_{TE} + C_C + C_X) + \frac{W^2}{2.43D} + \frac{X_d}{2V_{SL}} + r'_c C_C. \tag{47}$$

Or if a drift field is present and $\eta > 2.5$,

$$\frac{1}{\omega_T} = r_e(C_{TE} + C_C + C_X) + \frac{W^2}{2D(0.8 + 0.46\eta)} + \frac{X_d}{2V_{SL}} + r'_c C_C. \tag{48}$$

Table I summarizes the various characteristic frequencies discussed and how they are interrelated.

2) Behavior of Current Gain: The previous section develops certain characteristic frequencies that are related to the frequency limitations on current flow in the transistor. To complete the picture on active parameters, these frequencies will now be used to set up expressions for the current gain of the transistor.

The common-emitter current gain h_{fe} is the most important current gain parameter. It is intuitively easier, however, to first relate α to the characteristic frequencies already discussed. From α the common-emitter current gain can be derived. The parameter α is simply the ratio of the current leaving the collector to that entering the emitter:

$$\alpha \cong \gamma \beta^* \beta_m. \tag{49}$$

The emitter efficiency, like the base transport factor, has a low-frequency limiting value and a frequency-dependent term as well.

$$\gamma_0 \cong \frac{1}{1 + \frac{R_{EE}}{R_{BB}}}. \tag{50}$$

At high frequencies, γ_0 is reduced by the emitter transition capacitance.

$$\gamma_{hf} \cong \frac{\gamma_0}{1 + j\omega r_e C_{TE}}. \tag{51}$$

Equation (51) is actually a restatement of (15) in a slightly different form.

The base transport factor β^* (sometimes called the internal α) has already been discussed from the point of view of establishing its critical frequencies. The exact expression for α or β^* is long (see

TABLE I

Equation (1):	$\omega_0 \doteq \dfrac{2D}{W^2}$	a characteristic frequency appearing in the solution of the transport equations
Equation (2a):	$\omega_\beta = K\omega_0$	frequency where β^* is $\sqrt{2}/2$ of the low-frequency value
Equation (2b):	$\omega_\beta = 1.2\,\omega_0 = \dfrac{2.43 D}{W^2}$	diffusion only, no drift field
Equation (2c):	$\omega_\beta = \omega_0 \dfrac{\tfrac{1}{2}\eta^2(1.21+0.09\eta)}{\eta-1+e^{-\eta}}$	general case where drift field is present $1.2\omega_0 \le \omega_\beta \le 8.1\omega_0$
Equation (3):	$\omega_\alpha = \omega_\beta$	frequency where $|\alpha|$ is $\sqrt{2}/2$ of the low-frequency value, assuming $\gamma = 1$
Equation (4a):	$\omega_\alpha = 1.2\,\omega_0$	no drift field
Equation (4b):	$\omega_\alpha = \omega_0 \dfrac{\tfrac{1}{2}\eta^2(1.21+0.09\eta)}{\eta-1+e^{-\eta}}$	general case; see (2b) with drift field
Equation (5a):	$\omega_{Ti} = \alpha_0\omega_0 \cong \omega_0$	frequency where $|h_{fe}|$ is 1.0; diffusion only, no drift field
Equation (5b):	$\omega_{Ti} = \dfrac{\tfrac{1}{2}\eta^2}{\eta-1+e^{-\eta}}\omega_0$	general case with drift field
Equation (6a):	$\omega_\beta = \omega_\alpha = (1.21+0.09\eta)\omega_{Ti}$	general case, including drift field
Equation (6b):	$\omega_\beta = \omega_\alpha = 1.21\,\omega_{Ti}$	diffusion only, no drift field
Approximation 1:	$\omega_\alpha \cong \omega_0$	only valid when $\eta = 0$
Approximation 2:	$\omega_\alpha \cong \eta\omega_0$	only valid when $\eta > 2$
Approximation 3:	$\omega_{Ti} = \omega_0(0.8+0.46\eta)$ $= \dfrac{2D}{W^2}(0.8+0.46\eta)$	only valid when $\eta > 2.5$

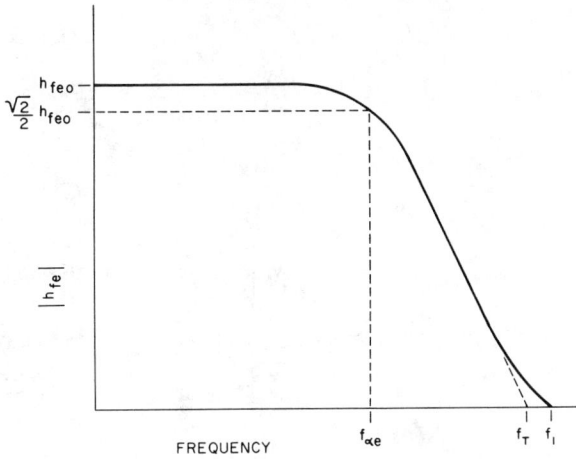

Fig. 10. Plot of $|h_{fe}|$ versus frequency.

[15]) and too complex to give any insight into its behavior. A simpler approximation is given as

$$\beta^* = \alpha_i = \dfrac{\alpha_0 \exp\left[-j(0.22+0.098\eta)\dfrac{\omega}{\omega_\alpha}\right]}{1+j\dfrac{\omega}{\omega_\alpha}}. \quad (52)$$

Since ω_α and ω_β are essentially identical, ω_β could also be used in (52). The term in the parenthesis on the right side of the equation is the so-called "excess phase" term and accounts for the additional phase shift caused by the field, if present. If no drift field is present, $\eta = 0$ and

$$\alpha_i|_{\eta=0} = \dfrac{\alpha_0 \exp\left[-j0.22\dfrac{\omega}{\omega_\alpha}\right]}{1+j\dfrac{\omega}{\omega_\alpha}}. \quad (53)$$

The common-emitter current gain[4] h_{fe} is

$$h_{fe} = \dfrac{\alpha}{1-\alpha}. \quad (54)$$

Also, from [15],

$$h_{fe} = \dfrac{\tfrac{1}{2}\eta^2}{\eta-1+e^{-\eta}} h_{fe0} \quad (55)$$

---

[4] The use of β for h_{fe} is avoided because β was used for the base transport efficiency.

which also can be written

$$h_{fe} \cong \dfrac{h_{fe0}}{1+jh_{fe0}\left(\dfrac{\omega}{\omega_{Ti}}\right)}. \quad (56)$$

Equation (55) shows how the drift field influences h_{fe}, while (56) includes the effect by using ω_{Ti} as a frequency parameter. The magnitude of (56) is also useful.

$$|h_{fe}| \cong \dfrac{h_{fe0}}{\left[1+(h_{fe0})^2\left(\dfrac{\omega}{\omega_{Ti}}\right)^2\right]^{\tfrac{1}{2}}}. \quad (57)$$

Fig. 10 is a plot of (57). If (57) is solved for the frequency where $|h_{fe}| = (h_{fe0})\sqrt{2}/2$ (which is called $\omega_{\alpha e}$), we obtain

$$\omega_{\alpha e} = \dfrac{\omega_T}{h_{fe0}}. \quad (58)$$

It is interesting to note at this point that the presence of a drift field multiplies both ω_{Ti} and h_{fe0} by the same factor, thus the 3-dB cut-off frequency $\omega_{\alpha e}$ is independent of the drift field [15].

Equations (56) and (57) can also be used with ω_T as well as with ω_{Tf}. If (57) is set equal to one and solved for ω, it will be found that ω equals ω_{Ti}. The *terminal* value of h_{fe} is altered by the effect of collector capacity and other parasitics as well, and will usually drop to one at a somewhat higher frequency. This frequency has now been designated as ω_1, the frequency where the terminal magnitude of h_{fe} actually becomes one. The same standard [17] also sets f_T as the frequency where the extrapolated value of $|h_{fe}| = 1$, with the requirement that the extrapolation is made from the region where $|h_{fe}|$ is falling at 6 dB per octave.

C. Parasitics

1) Base Resistance: The active or intrinsic transistor is confined to a small region under the emitter and extending across the base to the collector. It is customary to designate the intrinsic base as b' to differentiate from the external base b. The intrinsic base is a physically inaccessible region and cannot be contacted directly. As a result, there is a volume of resistive base material between the intrinsic base and the so-called base contact in a planar transistor (Fig. 11).

The current that flows in the base consists of 1) the α, or active current, which flows by diffusion, field, or a combination of the two; and 2) the $(1-\alpha)$ or "defect" current. The latter is a drift current and therefore attenuated by resistive losses.

Base resistance causes several undesirable effects:

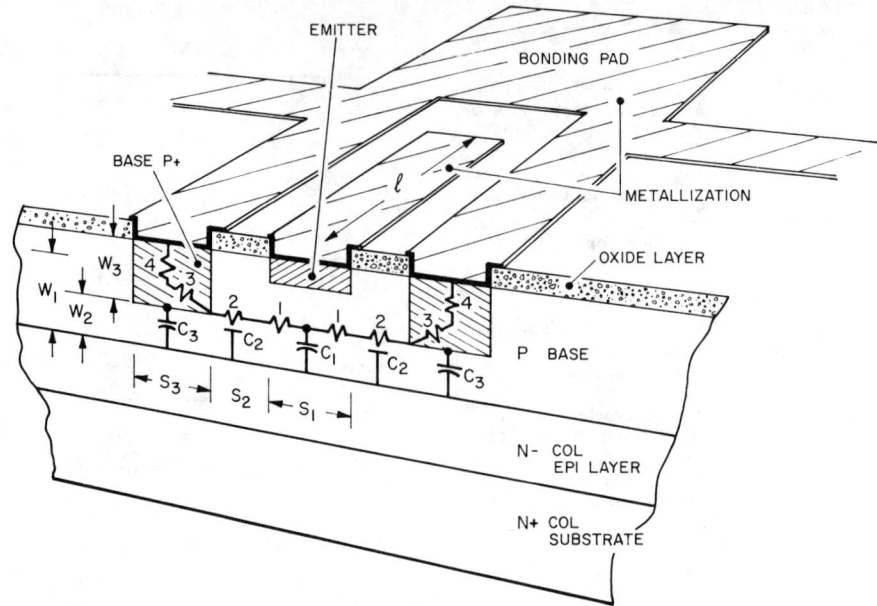

Fig. 11. Cross section of a planar transistor showing r_b' and capacitance. 1, 2, and 3 are parts of r_b', and 4 is R_c (contact resistance).

1) attenuation or loss signal in the base lead;
2) de-biasing of the region under the emitter (for example, the majority, or $(1-\alpha)$ current in an n-p-n transistor, biases the edges of the emitter positively with respect to the center of the emitter, causing the minority carriers (electrons) to concentrate near the edge; this also can be regarded as a form of feedback);
3) increased noise figure;
4) miscellaneous low-frequency effects, such as increase in saturated collector resistance, etc.

Because base resistance is a distributed and not a lumped resistance, it has been called the base "spreading" resistance. In a planar transistor, base resistance is really part of an RC transmission line. Fig. 11 shows a cross section of a modern microwave transistor, indicating that r_b' may be broken down into the following four regions:

1) base resistance under the emitter;
2) base resistance between the emitter and the p^+ region;
3) base resistance in the p^+ region itself;
4) contact resistance between the metal and the p^+ region.

Contact resistance is not properly a part of r_b', but from a circuit viewpoint, the effects on gain and noise are similar. Reference [18] shows how complex an equivalent circuit for the base region can be.

In early low-frequency devices, the capacitive reactances shunting r_b' were relatively high and r_b' was regarded as a frequency-independent element. This is not so in microwave transistors and certain other special types[5] where r_b' decreases with frequency due to capacitive shunting.

The calculation of base resistance from geometry is fairly straightforward if the resistivity and mobility are assumed constant throughout the base region under the emitter. When the impurity distribution is known, such as complementary error function, exponential, linear grade, or hyperbolic, the base sheet resistance can be calculated, and hence the base resistance [12]:

$$\frac{1}{R_{BB}} = q\mu \int_0^W N(X)\,dX \quad (59)$$

and

$$r_b' = \frac{R_{BB}}{W}. \quad (60)$$

The calculation of r_b' for more complex elements is given by Phillips [9]. A simple outline for estimating the low-frequency value of r_b' in a planar transistor is given here. (Refer to Fig. 11.)

r_{b1}' under the emitter: $\dfrac{\rho_1 S_1}{12Wl} = \dfrac{R_{BB1} S_1}{12l}$,

r_{b2}' between the emitter and p^+: $\dfrac{\rho_2 S_2}{2W_2 l} = \dfrac{R_{BB2} S_2}{2l}$,

r_{b3}' in the p^+: $\dfrac{\rho_3 S_3}{12W_3 l} = \dfrac{R_{BB3} S_3}{12l}$.

Total $r_b' = \dfrac{1}{\text{number of emitters}} [r_{b1}' + r_{b2}' + r_{b3}' + R_c]$.

S_1, S_2, and S_3 are widths of the emitter, emitter-to-p^+ space, and the p^+ opening, respectively, as shown in Fig. 11.

Ordinarily mobility [μ in (59)] is considered constant in making an estimate of r_b'. Irwin has published graphs of the resistivity of diffused layers where mobility is changing with depth [19]. His curves cover coerror functions and hyperbolic distributions in silicon.

Some recent data seem to indicate that very shallow junctions do not follow accurately in any of the impurity distributions usually described in the literature. The estimate of r_b' obtained as shown in Fig. 11 is still of considerable value in the design of a microwave transistor.

2) Capacitance: As would be expected, the internal capacitances of a transistor play an important role in the frequency response of the device as an amplifier. Extrinsic capacitances, such as those in the bonding pads and package, also affect performance. Those in the device itself will be considered first.

The active portion of a transistor has two types of capacitance, transition and diffusion, both of which are associated with the two junctions.

a) Emitter–base junction capacitance:

1) Transition capacitance C_{TE} arises from the fact that the

---

[5] The 2N929-930 type has an extremely lightly doped base region to enhance h_{fe}. Base resistance is 750–1000 Ω at dc and about 200 Ω at 200 MHz.

emitter has a finite area and a depletion layer. Since the silicon transistor junction is formed by diffusion and graded to some degree, it shows a 1/3-power voltage dependence when measured under reverse bias conditions. However, the emitter–base junction is normally operated in forward bias, and under this condition the step- (or 1/2-power) junction equations apply. The step-junction equation is simpler to handle:

$$C_{TE} \cong A_E \left[\frac{\varepsilon\varepsilon_0 q N_{BE}}{2(V+\phi)}\right]^{\frac{1}{2}} \quad (61)$$

or for silicon

$$C_{TE} \cong A_E \times 4.12 \times 10^3 \left(\frac{N_{BE}}{V+\phi}\right)^{\frac{1}{2}}. \quad (62)$$

For a typical silicon microwave transistor the contact potential ϕ is approximately 0.7 V. The applied voltage V is positive when the device is in forward bias. In other words, forward bias reduces the total voltage and increases C_{TE}.

If, for design purposes, it is desired that C_{TE} be reduced, only the emitter area A_E can be changed without significantly affecting the other parameters. The relative dielectric constant ε is set by the material used and N_{BE} is usually determined by other considerations, such as r_b' and h_{fe}. Occasionally N_{BE} can be reduced to reduce C_{TE} provided the emitter width is also narrowed to keep r_b' from increasing.

Since C_{TE} is shunted by r_e, the space-charge resistance of the emitter, and C_{DE}, the diffusion capacitance, it cannot be measured directly under forward bias. It is usually inferred from a plot of $1/I_E$ versus $1/f_T$ (refer to measurements).

2) Regarding the diffusion capacitance C_{DE}, when an ac voltage is applied to the forward biased emitter–base junction, the injected charge in the base is modulated. This change in charge with voltage defines a capacitance C_{DE}:

$$C_{DE} \doteq \frac{\Delta q_B}{\Delta V_{BE}}. \quad (63)$$

This equation can be manipulated to give C_{DE} in terms of base width and the diffusion constant:

$$C_{DE} = \frac{1}{r_e}\left(\frac{W^2}{2D_B}\right). \quad (64)$$

The expression in the parenthesis has the dimensions of $1/\omega$ and is characteristic base frequency, as defined in the section on characteristic frequencies. Therefore,

$$C_{DE} = \frac{1}{r_e \omega_0} = \frac{qI_E}{kT} \cdot \frac{1}{\omega_0} \quad (65)$$

which is correct for a transistor where the carrier is moved by diffusion only. In the case where there is a drift field, ω_{Ti} must be substituted for ω_0:

$$C_{DE} = \frac{1}{r_e \omega_{Ti}}. \quad (65a)$$

Since $\omega_{Ti} > \omega_0$, it can be said that drift field reduces C_{DE} in the same proportion that it increases ω_{Ti} over ω_0. At medium and higher currents, C_{DE} tends to mask out C_{TE} because C_{DE} varies directly with emitter current (64).

b) Collector–base junction capacitance: In this section it will be necessary to refer to the physical cross section of a planar transistor (Fig. 11) in order to differentiate the various capacitances. The base consists of three major regions: one under the emitter, one between the base and p$^+$, and one associated with the p$^+$ itself. The capacitances associated with these regions where they form a junction with a collector are in the form of an RC transmission line. A rigorously accurate representation of this region can result in a very complex equivalent circuit [18]. For the purposes of this paper, two or three regions will be sufficient. The part under the emitter is most important and it is C_1 (Fig. 11) that should be used in (2) in the expression for gain. Often the collector capacitance is arbitrarily divided into an inner and an outer region designated as C_{CI} and C_{CO} where $C_{CI} = C_1$ and $C_{CO} = C_2 + C_3$.

As in the emitter–base junction there is both a transition capacity and a diffusion capacity associated with the junction. For collector diffusion capacity to be present, the base width must modulate with changes in collector voltage. Diffusion capacity is significant in alloy transistors where the collector is metallic, and the depletion layer lies almost entirely in the base region. The situation is quite different in a silicon microwave transistor where $N_{BC} > N_C$. Hence, the depletion layer lies almost entirely in the collector epitaxial region and the base width does not change significantly with collector voltage except near saturation conditions [20], [21].

The total value of C_{TC} for a silicon transistor can be written as

$$C_{TC}\bigg|_{\text{total}} = A_{c_{cm^2}}(4.12 \times 10^3)\left(\frac{N_C}{V+\phi}\right)^{\frac{1}{2}} \quad (66)$$

where N_C is the impurity density in the collector epitaxial region. Epitaxial material is generally specified in terms of resistivity ρ rather than impurity density. The relationship between ρ and N is

$$\rho = \frac{1}{q\mu N}. \quad (67)$$

Mobility μ is fairly constant for values of ρ between 1 and 10 $\Omega \cdot$ cm. Then for n-type silicon,

$$\rho \cong \frac{5.1 \times 10^{15}}{N} \quad (68)$$

which can be substituted in (66) to give

$$C_{TC} \text{ total} = \frac{A_{cm^2} \times 2.94 \times 10^{11}}{[\rho(V+\phi)]^{\frac{1}{2}}}. \quad (66a)$$

An estimate of the fractional values for the various portions of C_{TC} (i.e., C_1, C_2, and C_3 in Fig. 11) is made by taking proportions by area. The value of C_{CI} is 20 to 25 percent of the C_{TC} total as shown in the figure. In the past it was a practice to measure the collector base time constant $r_b C_{CI}$ and r_b'; and from these two measurements a value of C_{CI} could be inferred. For the older mesa transistor this was reasonably accurate, but now the collector base time constant itself is complex (see measurements and modeling) and this method is no longer valid. Also, the fact that at low frequencies Im (Y_{22E}) $\cong j\omega C_{C1}$ and Im (h_{22e}) $\cong j\omega(C_{C0}+C_{C1})$ can be used to separate C_C into its components.

Fig. 11 also shows the bonding pad used to connect the base of the transistor to the outside world. It forms a capacitance with the collector region which lies underneath it. The dielectric in this case is silicon dioxide, silicon nitride, or a combination of the two. Also the oxide may be a combination of grown and deposited material. The dielectric constant will vary then with the type of dielectric and how it was treated after it was formed. The thickness of this layer is usually not very great, in the order of 2000 to 10 000 Å. It is also the practice to minimize the pad area in order to keep down the pad capacitance. A typical value for pad capacitance is in the order of 0.05–0.1 pF/mil^2.

The pad and package capacitances lie outside of any significant loss resistance and therefore affect bandwidth but not the intrinsic device gain. "Extrinsic" capacitances such as those between collector and base do reduce common-emitter gain, but theoretically can be neutralized. However, wide-band neutralization at microwave frequencies is very difficult, and good transistor design will be such as to minimize bonding pad and other parasitic capacitance.

D. Transistor Modeling

This section will briefly describe equivalent circuits or models as applied to microwave transistors. In general, models tend to fall into three categories: device, measurement, or circuit oriented. The parameters derived from these models are often similar, and certain figures of merit such as the maximum stable gain have exact counterparts in the different systems.

Fig. 12(a) is an example of a device-oriented model because most of its elements have direct physical counterparts in the real transistor. The frequency dependence of this model, except for the effect of C_{TE}, is implied in the expression that must be used for the α current generator. While not shown in this figure, the addition of direct capacitances between all three terminals and inductances would improve the accuracy somewhat.

The T equivalent model is particularly useful because it can be used for noise calculations by simply adding the appropriate noise voltage or current generators (refer to Fig. 5).

Another approach to this type of model is the hybrid π shown in Fig. 12(b). The frequency dependence of the output current generator is eliminated by adding the diffusion capacity to the emitter-base emittance, where

$$C_{DE} = \frac{1}{\omega_{Ti} r_e}.$$

Note that this is a common-emitter circuit and consequently r_e is divided by $(1 - \alpha_0)$. This model readily yields the usual approximations to the h parameters [8], [9]. For example, at low frequencies

$$h_{11e} = r_b' + \frac{r_e}{1 - \alpha_0} \qquad (69)$$

and therefore, if $\alpha_0 \cong 1$,

$$h_{11e} \cong r_b' + h_{fe0} r_e \qquad (70)$$

or at high frequencies, when $C_{DE} \gg C_{TE}$,

$$h_{11e} \cong r_b + \frac{1}{\dfrac{1-\alpha_0}{r_e} + \dfrac{\omega}{\omega_{Ti} r_e}}$$

and if $(1 - \alpha_0) \ll \omega/\omega_{Ti}$,

$$\left. h_{11e} \right|_{hf} \cong r_b' + r_e \left(\frac{\omega_{Ti}}{\omega} \right). \qquad (71)$$

A measurement-oriented model is usually based on considerations of accuracy and ease of measurement at a particular band of frequencies. As the transistor evolved and became applicable to higher and higher frequencies, the system of measurements also changed. Early transistors were usually characterized by the hybrid parameters; later systems used Y parameters, while today scattering parameters are most popular. In the microwave range it is difficult to realize an effective short or open circuit and thus S parameters are based on nominal and such easily achievable terminations as 50 Ω. The present popularity in scattering parameters is also due to the availability of accurate and easy-to-use measuring equipment. Some of the scattering parameters have a direct relationship to performance in a real system. For example, the 50-Ω insertion-power gain is $|S_{21}|^2$. The equivalent circuit [Fig. 12(c)] is very simple.

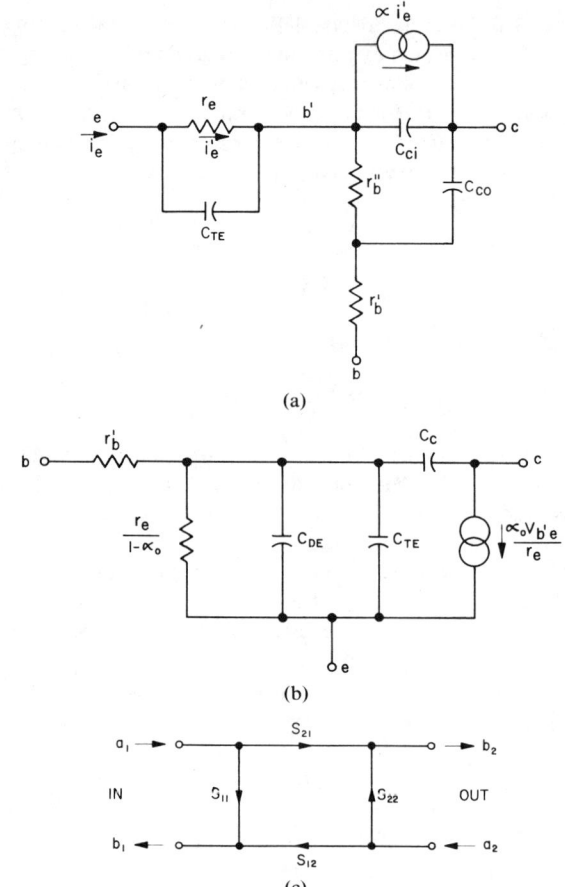

Fig. 12. (a) T equivalent circuit. (b) Hybrid equivalent circuit. (c) Scattering parameter equivalent circuit.

Parameters a_1, b_1, etc., are waves entering or leaving the terminals and have the dimensions of the square root of power. Several excellent references on the scattering matrix are available [22]. Scattering parameters are most useful for circuit design but are very poor as an adjunct to transistor design since they do not relate easily to intrinsic device parameters. The (ABCD) matrix is a model that is adapted to the cascading of stages. This is an example of a circuit-oriented model. The parameters are usually obtained by transformation from another model, such as the scattering parameters.

III. Transistor Design and Fabrication

This section will describe briefly the most important steps in the design-fabrication cycle. In order to keep the size of the section tractable, techniques not specifically characteristic of microwave transistors will be given only brief mention. Other details like cleanup procedures will be left out completely. Fig. 13 is an outline covering the main steps.

A. Selection of Geometry

Within the limitations of a given technology a designer should attempt to make a device that will meet the exact design goals as closely as possible. Over design would result in a device that would be uneconomical and probably unstable at the frequency of application. Thus a design using the maximum possible f_T and the minimal possible geometry should be avoided, except where actually needed to meet performance goals.

For a power transistor, the objective power, efficiency, and operating voltage are used to obtain an average current. From this the emitter periphery can be calculated, using 1.5 mA/mil as a first-order estimate. Typical emitter widths for competitive devices are approximately

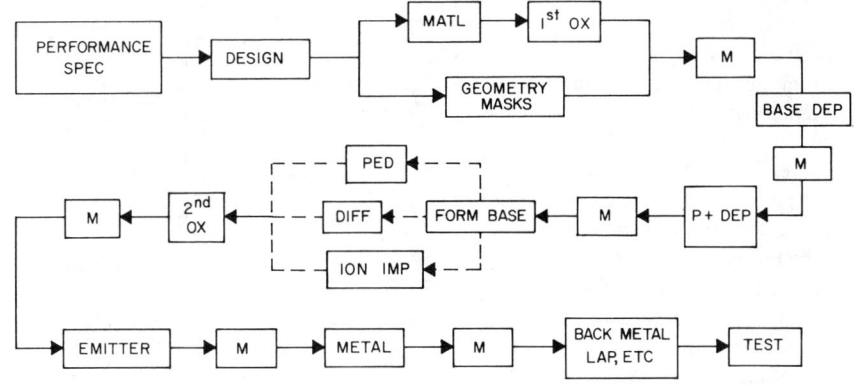

Fig. 13. Process flow for silicon microwave transistors (simplified). *M* is photomasking.

1.0 GHz   0.2 mil
1–2 GHz   0.1 mil
2–4 GHz   0.07 mil
4–6 GHz   0.05 mil.

The length-to-width ratio of an emitter should not exceed about a 20-to-1 ratio to minimize the voltage-drop down the finger. The physical arrangement of the emitters (i.e., device aspect ratio) should be determined from the thermal resistance versus ease of fabrication considerations. Long devices have lower thermal resistance but also lower fabrication yields.

A small-signal transistor design will be approached somewhat differently. Using the noise figure (7) together with an estimate of the f_T, which can be obtained, a value for r_b' is calculated. The base sheet resistance required for estimating f_T is usually already known, and thus r_b' can be used to obtain an emitter width and spacing (refer to Section II-C).

The p$^+$ area (Fig. 11) must be determined next. Its width is based on the contact resistance that is realizable with the metal system to be used. Contact resistance is usually expressed in $\Omega \cdot \text{cm}^2$. Hooper *et al.* [23] published data on contact resistance versus base impurity concentration for molybdenum. Fujinuma [24] reported values for platinum silicide somewhat lower than those for molybdenum.

The width of the p$^+$ region with a metal system must then be selected to give a contact resistance at least as low as r_b' and preferably lower. A wide p$^+$ region lowers contact resistance and makes photomasking easier, but adds to the transistor collector capacity.

Once the configuration of the p$^+$ and emitters has been determined, the basic geometrical design is set. Contact pads should be made 1 mil square or even less, when multiple pads must be used to handle the current. For very small transistors, with active areas less than 1 square mil, the pads should be less than 0.25 mil^2. Pad capacity is also minimized by using a relatively thick oxide.

B. Material

Because of the very shallow junctions used, the breakdown voltage of microwave transistors is almost always determined by the radius of curvature at the edge of the base [25]. This limitation can be circumvented to some degree by including a guard ring as a part of the p$^+$ structure and driving it in much deeper than the active base under the emitter. Mesa etching could theoretically provide a usable increase in breakdown voltage if the edges of the mesa could be passivated effectively. Microwave transistors are further limited in selection of material by the collector capacitance (66) versus depletion-layer transit time (43a) tradeoff. It is usually resolved by taking the overall required terminal f_T and breaking it down into the base and collector depletion-layer transit times. With the collector-depletion-layer transit time and the operating voltage known, the resistivity of the epitaxial layer can be determined. The thickness of the layer must be somewhat greater than the depletion layer or C_c will be higher than calculated. A good design will seek a reasonable proportion between τ_b and τ_d. An obviously poor design might have a low base cutoff frequency and a high depletion-layer cutoff frequency. Such a device would have a relatively high collector capacity and a low gain.

If a transistor is to deliver a substantial amount of power, the resistivity of the epitaxial region must be lower than that for an equivalent small-signal transistor. Power output is inversely proportional to r_c' and to the resistance of the column of material with a cross-sectional area equal to the emitter area and length equal to that of the epitaxial layer thickness,

$$P_{\text{out}} \propto \frac{1}{r_c'} \propto \frac{A_e}{t_{\text{epi}} \rho_{\text{epi}}}. \tag{72}$$

Present-day microwave transistors are usually built on 0.25-to-5-$\Omega \cdot$ cm material. Power devices have a p$^+$ depth that yields a breakdown voltage in the order of 35 to 50 V. Small-signal transistors usually have a lower breakdown voltage (≈ 20 V), usually because of the shallower diffusions used. State-of-the-art high-f_T (>7 GHz) transistors are built on 0.25–1-$\Omega \cdot$ cm epitaxial material in order to keep the collector depletion layer narrow. When this is done, epitaxial layer is operated only partially depleted.

C. Base Formation

Following the first oxide, the base area is opened up and deposited. After depositing the p$^+$ part of the base, the entire base region can be diffused. The primary base will have a higher (up to 1500 Ω/□) sheet resistance in order to minimize emitter dip effect (i.e., enhanced diffusion boron under the emitter) and to keep mobility high.

Alternately, the primary base can be implanted [26], [27], or first deposited, and then the actual diffusion accomplished through the bombardment with protons [28].

D. Emitter

Following the base formation, a second oxide is deposited on the slice. This oxide is generally of the type formed by the thermal decomposition of a gas or liquid. It is a low-temperature process so as to minimize the possibility of driving in the base farther. The emitter is then defined and diffused. Emitter diffusion times tend to be very short for microwave transistors compared to the hours used for low-frequency transistors. The temperatures are also low, generally less than 950°C.

E. Impurity Profiles

It is useful to plot the concentration of impurities in the wafer as a function of the distance from the surface. Since the impurities are introduced from the surface, the concentrations are highest there (with the exception of the implanted devices). Plots typically show the concentration at an emitter site, although plots can also be made starting in the p^+ region. An impurity plot can show the impurity concentration by type or sign or the net concentration, i.e., $|N_D - N_A|$. The net concentration shows which type of impurity predominates, as well as giving some indication of the base resistance.

Fig. 14(a) is a plot of the impurity concentration versus depth for an ideal device, and a diffused base with two types of emitters. To keep the comparison meaningful, the base width W has been kept constant. The "ideal" emitter has a concentration that is constant and close to solid solubility. The ideal base is also rectangular and extends from X_e to X_b below the surface. This base is ideal from the viewpoint of minimizing base resistance only. Some grading would be useful if f_T were the parameter to be optimized.

The boron or base impurities (designated as N_b) extend into the base to where they equal the epitaxial layer impurity level N_C at the base depth X_b. The base width is in the order of 0.10–0.2 μm for state-of-the-art devices.

The emitter impurities are shown for two types of dopants, phosphorous and arsenic. Phosphorous usually produces a higher surface concentration (N at $X=0$), but has a much less abrupt profile compared to arsenic. The arsenic profile is best for at least three reasons.

1) There is less compensation in the base region, and therefore r_b' is lower. Compensation means that where there are equal quantities of donors and acceptors, the impurities are thus neutralized and cannot transport current. The shaded area in Fig. 14 is this region.

2) Emitter push effect is eliminated, and narrower bases (higher f_T) are obtainable.

3) In the region of compensation, the base impurities have a negative grading, i.e., they increase with depth, resulting in a retarding field. It is thought that this retarding field cancels the effect of the aiding field in the remainder of the base resulting in what is essentially a "diffusion" transistor. The only silicon transistor that exhibits a net positive drift field also has both an implanted base and an arsenic emitter [26]. It is uncertain at this time which of these was responsible for the presence of the drift field.

The originally grown junction transistors had the impurities added during the growing of the crystal. When the diffused base or mesa transistor first appeared, the doping was added by diffusion, usually in a quartz tube at an elevated temperature. Now with the use of an accelerator, impurities may be implanted or caused to diffuse into the crystal by radiation.

Implantation [Fig. 14(b)] is unique in that it can place the impurities so that the peak of the distribution (which is Gaussian) is below the surface. Because the distribution can be controlled to some degree by controlling the energy of the ions, a more "rectangular" profile with lower base resistance and less retarding field can be obtained. Ion implantation, however, has some very serious disadvantages.

1) Because the relatively large boron ions damage the crystal during the radiation process, the slice must be annealed at a temperature close to that at which diffusion occurs. Annealing temperatures of 600–700°C are typical.

2) Even after annealing, a relatively large part of the boron impurities are tied up in Frenkel pairs, which are inactive and are not available to carry current.

3) When the emitter is diffused, the emitter impurities should intersect the peak of the base impurities. The emitter diffusion is very

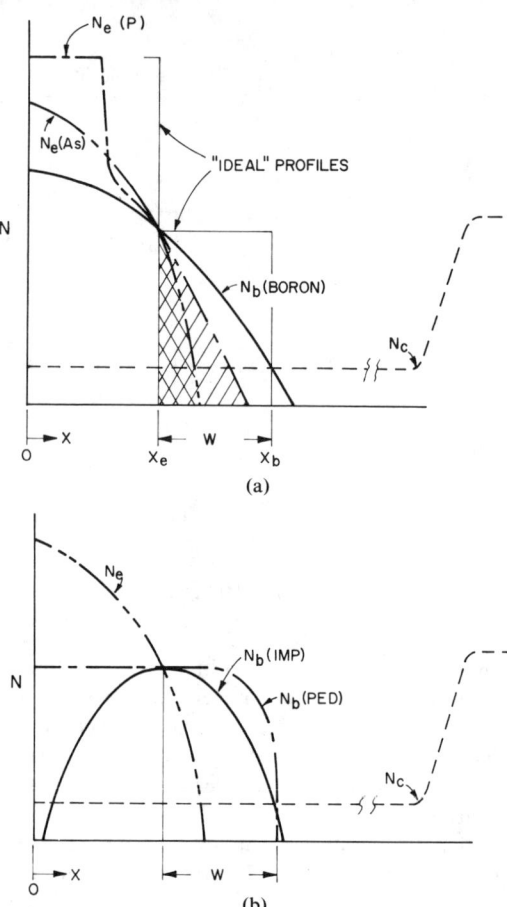

Fig. 14. (a) Comparison of arsenic and boron emitter profiles. (b) Comparison of impurity profiles obtained by ion implantation and proton enhanced diffusion.

critical since diffusions too shallow or too deep can result in no transistor action at all.

Proton enhanced diffusion (PED) appears to offer some advantages over both diffusion and ion implantation. PED uses the relatively small nuclei of the hydrogen atom to generate vacancies in the crystal. The process includes deposition of the impurity (e.g., boron) on the surface of the slice followed by bombardment with protons at a temperature in the 400–700°C range. The boron atoms quickly diffuse into the crystal to a depth accurately determined by the energy of the protons. Fig. 14(b) indicates that the diffusion front could be close to ideal. Compared to the implanted profile, there is more compensation in the emitter, but this does not appear to be a serious drawback. A comparison of the diffused and PED transistor profiles with the same base width showed a reduction in r_b' of about 1.8 to 1 for the PED device.

F. Contacts

The slice is again photomasked and the reverse contact mask is used to open up the base contact areas. A number of contact systems are suitable for microwave transistors. Aluminum in general is not suitable because of the difficulty in obtaining clean shallow junctions. Systems based on the following metals have been used successfully on a variety of microwave devices: contacting metals (Al, Mo, Ni, Cr, Ti, PtSi); barrier metals (W, Mo, Pt); and top metals (Al, Au). When platinum silicide is used, the platinum must be deposited first and the silicide formed either during the deposition process or by the subsequent heat treatment. Silicide will form only in the contact opening over the bare silicon and the remainder of the platinum can be removed chemically. The refractory metal is deposited next and then the gold. An electron gun or sputtering is used in the deposition of the refractory metal.

Sputtering of the metal results in a contact that adheres well, but excessive heat can cause sufficient damage to make the dc current gain of the device quite low. After photographically defining the contacts, the transistor is ready for the final processing.

G. Final Processing

These steps are essentially the same as for lower frequency transistors. Slices are normally processed at about a 0.01-in thickness. During the processing, the back becomes contaminated so that it is later lapped off to a final thickness of approximately 0.005 in. For high-power transistors, the final lapping has an additional benefit in that it reduces the thermal resistance of the die. After lapping, the slice may or may not be gold backed depending upon the die attachment system used. Backside gold is usually doped with phosphorous to insure that the metal semiconductor interface is nonrectifying, i.e., the n-type dopant on the n substrate will not produce an unwanted junction. The slice is then diced in the usual manner and is ready for packaging.

H. Other Variations

Conventional photomasking, which uses 3000–4000-Å light, has a limitation due to interference effects. This limit is in the order of 1.0 μm. An electron beam is now being used to define the geometry of smaller elements (for example, the emitter stripes). The beam is guided electronically, or the slice is moved mechanically, to trace out the desired pattern. Electron beam definition can probably be extended to give lines of less than 0.10 μm. Fig. 13 summarizes the process flow for a silicon planar microwave transistor.

I. Packaging

Transistors have conventionally been supplied in packages to protect the die and to increase the size of the unit so that a typical user can handle it conveniently. At the higher frequencies (i.e., >4 GHz), it is becoming apparent that conventional packaging will not be practical, and the die must be mounted directly in a microwave integrated circuit. For example, a chip mounted in the 0.07-in package will give 1 dB more gain at 4 GHz than it will in a package twice the size, but even then the chip parameters are measurably degraded by the smaller package. The main reason for this degradation is that the sealing system must occupy a physical space that separates the chip from the external circuit. This separation results in an increased lead inductance that reduces gain and bandwidth. The physical configuration of the package influences this loss to some extent and a coaxial package will perform slightly better. However, the coaxial package does not lend itself well to other than coaxial circuitry, while the trend today is more and more to microstrip.

Microwave transistor packages, where they can be used, should give adequate protection from the environment in which they are to be used. As a package is made smaller it becomes increasingly difficult to obtain a reliable hermetic seal. Three sealing technologies are presently in use: glass-to-metal, ceramic-to-metal, and plastic. Of the two hermetic systems, a glass-to-metal seal is the easier to realize, but it is also the more fragile. Ceramic-to-metal seals rely on a sophisticated brazing technique and are somewhat more expensive to build, but are superior mechanically. A package built using the latter technique is shown in Fig. 15(c). Plastic seals are economical but nonhermetic and, in addition, degrade the characteristics of higher frequency transistors. The loss in gain suffered at these higher frequencies in a small-signal transistor seems to be due more to increased capacity than to the losses in the potting dielectric. A new sealing technique, where the seals are sandwiched into the

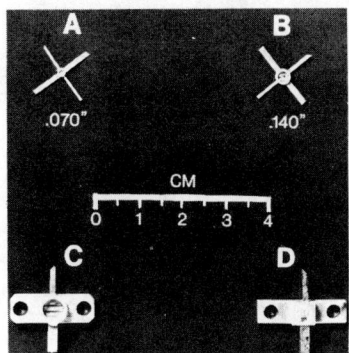

Fig. 15. Microwave transistor packages. *A* and *B* are small signal, and *C* and *D* are power. (Courtesy Ceradyne, Inc.)

TABLE II

Material	Approximate Thermal Conductivity (cal cm^{-1}°C^{-1})
BeO	0.56
Cu	0.97
Al$_2$O$_3$	0.045
Glass	0.002
Si	0.24–0.35
Diamond	1.51

uncured ceramic or brazed into slots in the ceramic, gives promise of a more rugged package for small-signal applications.

If there is sufficient dissipation within the transistor to raise the temperature of the chip over 125°C, the package design must also include means of dissipating the heat. The thermal resistance θ of a package is usually given in °C/W; thus

$$T_j = T_s + P_D \theta. \tag{73}$$

A junction temperature T_j of 125°C is now considered a good number for a high-reliability system. The thermal resistance θ will include the resistance in the chip itself plus that of the package, and any other resistance in the heat path to the final sink which is at T_s.

Most circuit applications require that the collector be isolated from the ground plane, which means that the package must include an insulator. Beryllium oxide (BeO) is the most commonly used material. Table II gives a short list of the thermal characteristics of several materials. It shows why BeO is preferred to Al$_2$O$_3$ and why diamond is regarded as the ultimate in an insulating heat sink. Present-day power devices are thermally limited by the resistance of the chip itself. Most of the dissipation in a transistor takes place in the collector depletion layer. This is less than 1 μ from the upper surface, while the die thickness is in the order of 75–125 μm. It is difficult to thin chips to less than 50 μm, and the handling and assembly problems are multiplied. An inverted mounting technique, such as used for avalanche diodes, will be the ultimate approach to minimizing thermal resistance in transistors. A reliable system for separating the emitter and base contacts is the biggest obstacle yet to be overcome before inverted mounting will become practical. There is a further benefit to this type of mounting if the emitter is made the common element. The advantage is that the common lead inductance will also be reduced with the result that stability and gain are both improved.

IV. Measurements

This section will be concerned primarily with measurements that are used to determine basic parameters and control the fabrication process. Noise figure measurement would usually be considered a

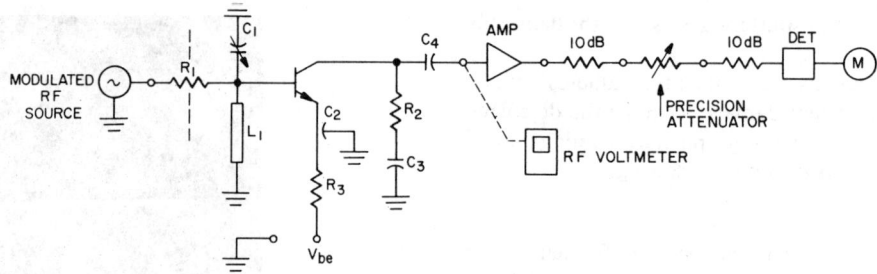

Fig. 16. Test setup for measuring f_T. Here $R_1 \cong 10$ kΩ; $R_3 = V_{be}/I_E$; $R_2 = 1$ Ω; $C_1 = 0.5$–5 pF; C_2, $C_3 = 0.05$ μF; and $C_4 = 0.001$ μF.

performance criterion, except that for microwave transistors it is the only means of obtaining values for certain parameters. The measurements discussed here are an important part of the information "feedback loop" in transistor design. The discussion will include measurement of f_T, f_α, η, r_b', C_c, ρ_{epi}, C_{TE}, and $r_b'C_c$.

A. Measurement of f_T

In the section on current gain it was pointed out that f_T is an extrapolated parameter [17] obtained by measuring $|h_{fe}|$ in the 6-dB-per-octave falloff region, i.e.,

$$f_T = f_{\text{test}}|h_{fe}| \tag{74}$$

where $|h_{fe}|$ is obtained at the test frequency. The first step in the measurement of f_T is to determine a proper test frequency so that [17]

$$2 < |h_{fe}| < 10.$$

If the approximate value of f_T is not known, the test frequency must be obtained by trial and error. The validity of a measurement is checked by changing the frequency by one octave and looking for a 6-dB change in $|h_{fe}|$.

For most microwave transistors, a suitable f_T test frequency is in the range of 400–1000 MHz. If measurements are made only occasionally a transfer bridge (GR 1607A) can be used. However, if f_T must be monitored continuously, two other alternatives are possible. The first is to measure the scattering parameters and then from these calculate $|h_{fe}|$, where

$$h_{fe} = \frac{-2S_{21}}{(1 + S_{11})(1 + S_{22}) + S_{12}S_{21}}. \tag{75}$$

The calculation is not simple and is best done using a computer. It is also possible to measure $|h_{fe}|$ directly using a special text fixture. A suitable setup for 400 MHz is shown in Fig. 16.

C_1 and L_1 are used to tune out the package and fixture input capacity. R_1 is the constant current source. It is made by evaporating tantalum on a 0.140- by 1.00-in ceramic rod and anodizing it to the desired resistance. For 1 percent accuracy,

$$R_1 \geq 100\left[|h_{fe}|\frac{0.026}{I_E} + r_b'\right]. \tag{76}$$

The resistor R_2 is a 1-Ω disk or several chip resistors in parallel. Note that the system shown uses an AM-modulated 400-MHz test signal. This makes possible a very small input test current into the transistor itself. An RF voltmeter can also be used, but sensitivity may be a problem.

B. Measurements of f_α

This parameter is not directly accessible in microwave transistors because the terminal value of α may never drop to $\sqrt{2}/2$ due to parasitics. However, f_α can be obtained from a noise figure plot.

Fig. 6 shows how the corner frequency f_c is obtained. From this and (77), f_α is obtained.

$$f_\alpha = \left[\frac{Kf_c^2}{1 + \dfrac{r_b'}{R_g} + \dfrac{r_e}{2R_g} + \dfrac{K}{h_{FE}}}\right]^{\frac{1}{2}} \tag{77}$$

where

$$K = \frac{(R_g + r_b' + r_e)^2}{2\alpha_0 r_e R_g}. \tag{78}$$

C. Measurement of η

This parameter is the drift field factor and, like f_α, must be inferred from noise measurements. First, f_T and f_α are obtained as outlined in Sections IV-A and B. Referring to (35) we obtain

$$\eta = \left[\frac{f_\alpha}{f_T} - 1.21\right]11.1. \tag{79}$$

D. Measurement of r_b'

The classical method of obtaining base resistance was simply to measure h_{11} at a relatively high current and frequency, i.e.,

$$h_{11e} \cong r_b' + \frac{0.026}{I_E}\frac{\omega_T}{\omega}. \tag{71}$$

If I_E is large,

$$h_{11e} \cong r_b'. \tag{80}$$

This is again a case of the older method no longer being usable. If the second term on the right-hand side of (71) is to be negligible, then the test frequency must be in the gigahertz range. As a result h_{11e} becomes very complex. A more accurate expression is then

$$h_{11e} = r_b' + \omega L_b + \frac{\omega_T}{\omega}\frac{0.026}{I_E} + \omega_T L_e. \tag{81}$$

Thus the measurement of h_{11e} is not a useful approach to obtaining r_b' in a modern microwave transistor.

Referring to the noise figure expression (7), let

$$\frac{I_{c0}}{I_E} \ll \left(\frac{f}{f_\alpha}\right)^2 \ll \frac{1}{h_{FE}}$$

and

$$\alpha_0 \cong 1.$$

Under these assumptions, the plateau frequency noise figure F_p is

$$F_p = 1 + \frac{r_b'}{R_g} + \frac{r_e}{2R_g} + \frac{(R_g + r_e + r_b')^2}{2R_g r_e h_{FE}}. \tag{82}$$

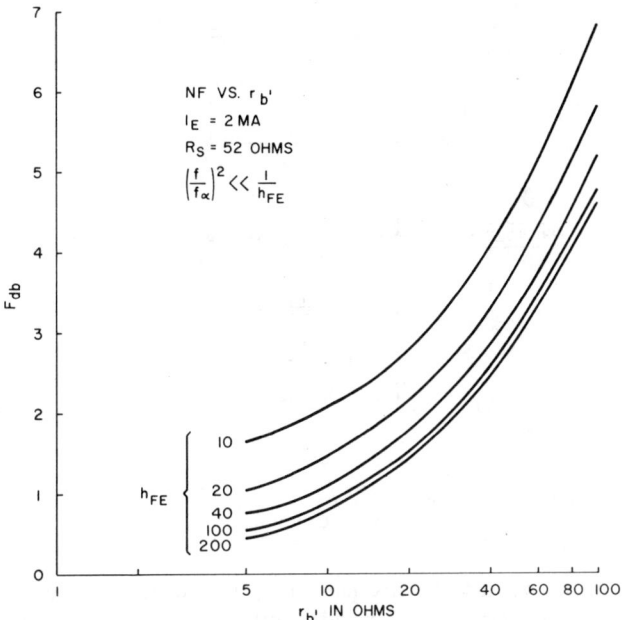

Fig. 17. Graphical solution to r_b' when $R_g = 50\ \Omega$.

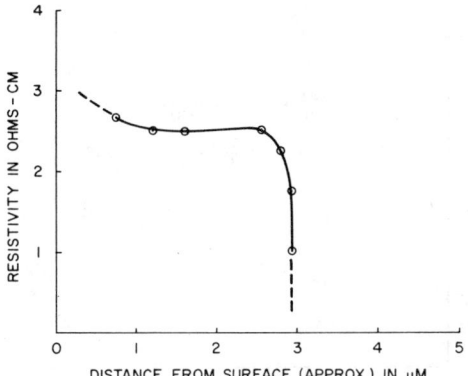

Fig. 18. Plot of ρ versus X_d for an epitaxial film.

A suitable frequency for measuring this noise figure is 1.0 MHz. Since R_g and r_e are known and h_{FE} is easily obtained, (82) can be solved for r_b'. Although the solution is not difficult, it is tedious, and a graphical approach is indicated. Fig. 17 is a graph for one set of test conditions.

Since r_b', or part of it, is frequency dependent, the value obtained by the procedure is the asymptotic low-frequency value. However, the results, when substituted into the noise figure expression (7) and the calculations of Fig. 11, agree very well. This technique is particularly useful in evaluating the contact resistance, since it affects noise figure in the same manner as does r_b'.

E. Measurement of C_{TC}

The total collector capacitance (exclusive of contacts) is obtained most easily by probing the base diffusion before the emitter has been put in. A bridge or direct reading meter can be used for the measurement. Care should be taken to balance the instrument with the probe in position but not quite touching the base area.

The contact capacitance can be obtained by probing the completed chip and subtracting out the device collector capacitance.

Collector capacitance under the emitter (C_{C1} or C_{Ci}) can be obtained for a simple low-frequency transistor by measuring the $r_b'C_c$ time constant and dividing by r_b'. For microwave transistors the time constant is somewhat more complex and should actually be referred to as $h_{12}/j\omega$. Section IV-H gives more detail on h_{12}.

F. Measurement of ρ_{epi}

This is the resistivity of the collector epitaxial region. It is obtained by making a diffusion into the epitaxial layer, either as a special diode or as the base of a transistor. The capacitance is then measured as a function of the applied voltage (i.e., reverse bias) from a fraction of a volt up to breakdown, or to where $\Delta C/\Delta V$ becomes 0. Which occurs first will depend upon ρ, the depth of the diffusion, and the thickness of the epitaxial film. For n-type silicon in the 1–10-$\Omega \cdot$cm range,

$$\rho \cong \frac{1.77(A_{mil^2})^2 \times 10^{-2}}{(V + \phi)C_{pF}^2}\ \Omega \cdot \text{cm}. \tag{83}$$

Note that ϕ, the contact potential or barrier voltage, appears in (83). It can be eliminated from the equation by taking the differential and then evaluating the slope of the capacitance versus voltage curve. The differential method is extremely tedious and requires a a computer for accurate evaluation. On the other hand, if ϕ is known with a reasonable degree of accuracy and the epitaxial film is not so thin that it depletes out at a very small voltage, then a value for ϕ can be assumed without impairing the usefulness of this approximation.

From the same measurement the depletion-layer width X_d, can be obtained.

$$X_d = \frac{A_{mil^2} \times 6.7 \times 10^{-2}}{C_{pF}}\ \mu\text{m}. \tag{84}$$

A typical curve of ρ versus X_d is shown in Fig. 18. This is an actual plot made from a transistor base diffusion. Note the very abrupt transition from n to n^+. This is characteristic of epitaxial material grown by a low-temperature process. It is obvious that this measurement also yields another important parameter, the thickness of the epitaxial region.

G. Measurement of C_{TE}

The emitter transition capacitance can be obtained from the plot of $1/I_E$ versus $1/f_T$. If the depletion-layer transit time and collector time constant are neglected, then

$$\frac{1}{\omega_T} \cong \frac{kT}{qI_E} \cdot (C_{TE} + C_{TC} + C_x) + \frac{1}{\omega_{Ti}}. \tag{85}$$

The slope of (85) is

$$\frac{d\left(\frac{1}{f_T}\right)}{d\left(\frac{1}{I_E}\right)} = \frac{kT}{q}(C_{TE} + C_{TC} + C_x). \tag{86}$$

Therefore, the slope divided by a constant kT/q is $(C_{TE}+C_{TC}+C_x)$. A typical plot is given in Fig. 19. C_x and C_{TC} are determined from the previous section and must be subtracted from (86) to obtain C_{TE} alone. The extrapolation of the linear portion of the curve through 0 (i.e., $I_E = \infty$) yields ω_{Ti}. The value thus obtained for ω_{Ti} will not be the intrinsic value but will also include the collector depletion-layer transit time, i.e.,

$$\frac{1}{\omega_{Ti\ meas}} = \frac{1}{\omega_{Ti}} + \frac{1}{\omega_d}. \tag{87}$$

The correction can easily be made remembering that

$$\frac{1}{\omega_d} = \frac{X_d}{2V_{SL}} \tag{43}$$

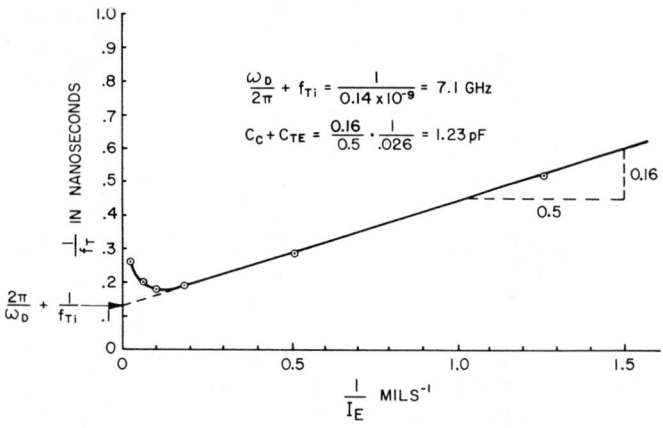

Fig. 19. $1/I_E$ versus $1/f_T$ plot.

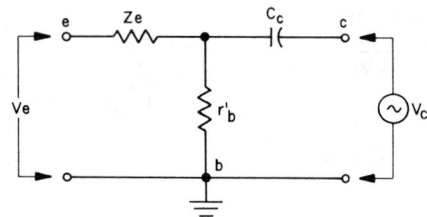

Fig. 20. Equivalent circuit for deriving h_{rb} and $r'_b C_c$.

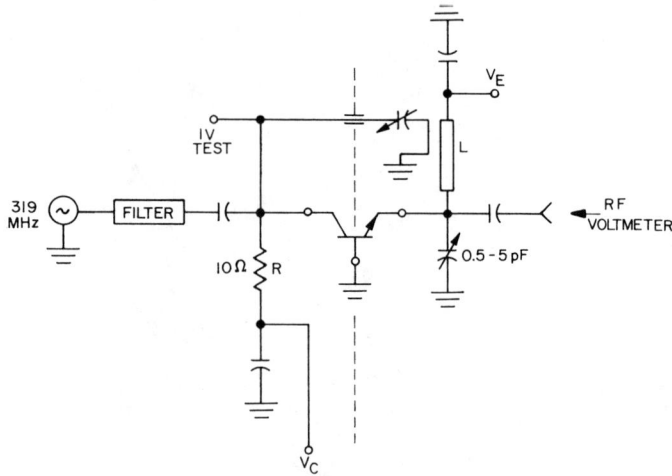

Fig. 21. $r'_b C_c$ test fixture.

and

$$X_d = \left[\frac{2\varepsilon\varepsilon_0(V+\phi)}{qN}\right]^{\frac{1}{2}}. \quad (88)$$

These can be combined with (68) to yield for n-type silicon,

$$\frac{1}{\omega_d} = 3 \times 10^{12}[\rho(V+\phi)]^{\frac{1}{2}}. \quad (89)$$

H. Measurement of $r'_b C_c$

This parameter is related to h_{12b} such that

$$|h_{12b}| = \omega r'_b C_c \quad (90)$$

where h_{12b} is the common base reverse voltage transfer ratio. It is derived from the simplified model given in Fig. 20, where it is obvious that

$$h_{12b} \doteq \frac{V_e}{V_c} = \frac{r'_b}{r'_b + \dfrac{1}{j\omega C_c}}. \quad (91)$$

If ω is selected properly,

$$1/\omega C_c \gg r'_b,$$

and

$$|h_{12b}| \cong \omega r'_b C_c. \quad (92)$$

The model given above is much too simple for a microwave transistor. However, the total $r'_b C_c$ product that will be obtained by this measurement is still of use. Base on Fig. 11 a more exact expression for h_{12} is

$$h_{12} \cong j\omega(r'_{b_1}C_{C_1} + r'_{b_2}C_{C_2} + r'_{b_3}C_{C_3} + R'_c(C_{C\text{total}}) \\ + r'_{b_2}C_{C_1} + r'_{b_3}C_{C_1} + r'_{b_3}C_{C_2}). \quad (93)$$

To dissect (93) into its various components is a tedious job re-quiring many measurements. The contact resistance is best obtained from a test pattern on the chip and the other portions of C_C by scaling from known dimensions.

The measurement of h_{12b} is simple in theory but difficult to implement due to the very high ratio of V_c/V_e for a good microwave transistor. For example, if r'_b equals 20 Ω, C_C equals 0.02 pF, and f equals 320 MHz, then

$$|h_{12b}| \cong 8 \times 10^{-4}.$$

A test signal of 1 V for V_c is near the maximum allowable while still maintaining small-signal conditions; this makes V_e for the above calculation equal to approximately 0.8 mV. Standard RF voltmeters usually have a maximum sensitivity of 1.0 mV full scale, and 0.8 mV is read on the meter scale with the poorest accuracy. This indicates that a higher test frequency is desirable, but with the attendant problems in isolation.

An $r'_b C_c$ test fixture requires some form of neutralization to cancel out the residual capacity of the fixture itself and that of the transistor package. Fig. 21 shows one form of such a test circuit. The circuit also includes a filter to attenuate the harmonics of the test signal. An output tank circuit allows tuning of the probe and other distributed capacity. The test frequency 319 MHz, was selected so that with $V_c = 1$ V, $r'_b C_C$ in picoseconds is $V_e/2$. In order to maintain the high isolation required, this fixture should be built in the form of a box with milled cavities.

V. Conclusion

The theory of microwave transistors and their limitations has been well established, and performance can be predicted with reasonable accuracy. New technologies developed in the past five years have made it possible to approach frequency limits more closely than was first thought possible. The present era is one of new fabrication technologies rather than breakthroughs in theory. Radiation as an adjunct or replacement for diffusion, electron beam masking, etc., is being used to make better transistors, but the materials limitations still exist and are not likely to change. In the next year or so, small-signal transistors will become useful in the X band and power devices in the C band. Power transistors will not replace tubes where the very high peak powers are required. However, in phased array systems where modest average powers are needed, the transistor will eventually predominate. At the lower end of the microwave region near 1 GHz, small-signal transistors will continue to replace paramps as the 1-dB-maximum noise figure limit moves into the 1–2-GHz band.

Nomenclature

α	Common-base current gain.
α_0	Common-base current gain at low frequencies.
α_i	Intrinsic alpha determined by base transport only.
β^*	Base transport factor; the ratio of minority carriers injected from the emitter into the base to the number arriving at the base edge of the collector–base depletion layer.
β_0^*	Low-frequency value of β^*.
β_m	Collector–depletion-layer transport factor; the ratio of carriers leaving the collector–base depletion layer to those entering it from the base.
γ	Emitter injection efficiency; the ratio of the minority carrier current injected into the base from the emitter to the total emitter current.
γ_{hf}	High-frequency value of γ.
γ_0	Low-frequency value of γ.
Δf	Bandwidth (Hz).
ε	Relative dielectric constant.
ε_0	Permittivity of free space.
η	Drift field factor.
θ	Thermal resistance (°C/W).
μ	Mobility.
μm	Distance in microns.
π	The constant, $3.1416\cdots$
ρ	Resistivity ($\Omega \cdot$ cm).
τ	A time constant.
τ_B	Minority carrier lifetime in base region.
τ_b	Base transit time.
τ_c	Collector capacitance–resistance charging time.
τ_d	Collector depletion-layer time constant.
τ_e	Emitter capacitance–resistance changing time.
τ_{ec}	Transit time for carriers from emitter to collector.
τ_m	Collector depletion-layer transit time.
ϕ	Contact potential of a junction.
ω	Radian frequency, $2\pi f$.
$\omega_\alpha, \omega_\beta, \omega_0, \omega_T, \omega_{Ti}$	See f_α, f_β, etc.
A	Area in general.
A_{mil^2}	Area (mil^2).
A_c	Collector area.
A_e	Emitter area.
C_C	Capacitance of collector–base junction.
C_E	Capacitance of emitter–base junction.
C_{DC}	Diffusion capacitance of collector–base junction.
C_{DE}	Diffusion capacitance of emitter–base junction.
C_{TC}	Transition capacitance of collector–base junction.
C_{TE}	Transition capacitance of emitter–base junction.
C_{CI}	Collector capacitance (inner) under emitter.
C_{CO}	Collector capacitance (outer) outside emitter.
C_{C1}, C_{C2}, C_{C3}	Components of collector capacity.
D	Diffusion constant, for minority carriers.
E	Electric field (V/m).
e_{nb}	Thermal noise voltage due to base resistance.
$\overline{e_{nb}^2}$	Mean-square value of base thermal noise voltage.
F	Noise factor (figure).
F_p	Plateau noise figure; noise figure of a transistor in middle-frequency range, where $dF/df = 0$.
f	Frequency.
f_α	Frequency where $|\alpha|$ has dropped to $\sqrt{2}/2$.
$f_{\alpha e}$	Frequency where $|h_{fe}|$ has dropped to $(\sqrt{2}/2)h_{fe0}$.
f_β	Frequency where $|\beta^*|$ has dropped to $(\sqrt{2}/2)\beta_0^*$.
f_c	Upper noise corner frequency, where $F = F_p + 3$ dB.
f_0	A characteristic frequency [see (28)], $= 1/2\pi(W^2/2D)$.
f_T	Frequency where the extrapolated $|h_{fe}|$ goes to unity.
f_{Ti}	Intrinsic value of f_t as determined by base transit time only.
$f_{\max}$	Maximum frequency of oscillation; the highest frequency at which a device will oscillate in a lossless circuit, or the frequency where the unilateral gain is unity.
G	Gain.
G_0	Low-frequency gain.
$G_{\max}$	Maximum available gain (sometimes called MAG); gain of an unconditionally stable device when input and output are simultaneously conjugately matched.
h_{fe}	Common-emitter current gain.
h_{FE}	Dc common-emitter current gain.
h_{fe0}	Low-frequency common-emitter current gain.
I_c	Dc collector current.
I_{co}	Collector cutoff (a leakage) current.
I_E	Dc emitter current.
i_{ne}	Emitter shot noise current.
$\overline{i_{ne}^2}$	Emitter shot noise current, mean-square value.
i_{nc}	Collector shot noise current.
$\overline{i_{nc}^2}$	Collector shot noise current, mean-square value.
i_e	Signal current flowing into active part of emitter admittance.
i_{et}	Total signal current flowing into emitter terminal.
K	An arbitrary constant.
k	Boltzmann's constant, $= 1.38 \times 10^{-23}$ J/°K.
L	Diffusion length; average distance carriers will diffuse before recombining.
L_b	Base lead inductance.
L_e	Emitter lead inductance.
l	Length.
M	Noise measure.
N	Impurity concentration (atoms/cm^3).
N_{BC}	Impurity concentration in base at collector edge.
N_{BE}	Impurity concentration in base at emitter edge.
N_C	Impurity concentration in collector adjacent to base.
P_D	Power dissipated.
q	Electronic charge, $= 1.6 \times 10^{-19}$ C.
R_{BB}	Base sheet resistance.
R_{EE}	Emitter sheet resistance.
R_g	Source resistance, for noise measurement.
r'_c	High-frequency collector series resistance.
r'_b	Base spreading resistance.
$r'_{b1}, r'_{b2}, r'_{b3}$	Components of r'_b.
r_e	Emitter resistance, $= kT/qT_E$.
S	Stripe dimension, in general.
S_1, S_2, S_3	Refer to Fig. 11.
t	Time.
T	Temperature (°K).
T_j	Transistor junction temperature.
T_s	Transistor heat sink temperature.
U	Unilateral gain; the gain of a device with a lossless unilateralizing network and both ports conjugately matched.

V_{SL} Scattering limited velocity of carriers.
W Base width.
W_1
W_2 Widths in transistor geometry (refer to Fig. 11).
W_3
X_d Depletion-layer width.
X_{dc} Collector–base junction depletion-layer width.
X_{de} Emitter–base junction depletion-layer width.
Z_{in} Input impedance of a transistor.
Z_{out} Output impedance of a transistor.
Z_c Collector impedance.
Z_e Emitter impedance.

ACKNOWLEDGMENT

The author wishes to thank R. Webster for many helpful discussions, and G. Pierson and P. C. Lindsey for assistance in editing and correcting the manuscript.

REFERENCES

[1] M. Fukuta, H. Kisaki, and S. Maekawa, "Mesh emitter transistor," *Proc. IEEE* (Lett.), vol. 56, Apr. 1968, pp. 742–743.
[2] J. Andeweg and T. H. J. van den Hurk, "A discussion of the design and properties of a high-power transistor for single sideband applications," *IEEE Trans. Electron Devices*, vol. ED-17, Sept. 1970, pp. 717–724.
[3] S. J. Mason, "Power gain in feedback amplifiers," *IRE Trans. Circuit Theory*, vol. CT-1, June 1954, pp. 20–25.
[4] E. G. Nielsen, "Behavior of noise figure in junction transistors," *Proc. IRE*, vol. 45, July 1957, pp. 957–964.
[5] H. F. Cooke, "Transistor noise figure," *Solid State Design*, vol. 4, Feb. 1963, pp. 137–142.
[6] H. Fukui, "The noise performance of microwave transistors," *IEEE Trans. Electron Devices*, vol. ED-13, Mar. 1966, pp. 329–341.
[7] H. A. Haus and R. B. Adler, "Optimum noise performance of linear amplifiers," *Proc. IRE*, vol. 46, Aug. 1958, pp. 1517–1533.
[8] R. L. Pritchard, *Electrical Characteristics of Transistors.* New York: McGraw-Hill, 1967.
[9] A. B. Phillips, *Transistor Engineering.* New York: McGraw-Hill, 1962.
[10] R. L. Pritchard, "Frequency variations of current-amplification factor for junction transistors," *Proc. IRE*, vol. 40, Nov. 1952, pp. 1476–1481.
[11] W. Shockley, "Transistor electronics: Imperfections, unipolar and analog transistors," *Proc. IRE*, vol. 40, Nov. 1952, pp. 1289–1313.
[12] J. L. Moll and I. M. Ross, "The dependence of transistor parameters on the distribution of base layer resistivity," *Proc. IRE*, vol. 44, Jan. 1956, pp. 72–78.
[13] H. Kroemer, "The drift transistor," *Transistor I*, RCA Labs., Princeton, N. J., 1956, pp. 202–220.
[14] A. H. Marshak, "Optimum doping distribution for minimum base transit time," *IEEE Trans. Electron Devices*, vol. ED-14, Apr. 1967, pp. 190–194.
[15] J. Te Winkel, "Drift transistor," *Electron Radio Eng.*, vol. 36, Aug. 1959, pp. 280–288.
[16] J. M. Early, "PNIP and NPIN junction transistor triodes," *Bell Syst. Tech. J.*, vol. 33, May 1954, pp. 517–533.
[17] R. L. Pritchard, J. B. Angell, R. B. Adler, J. M. Early, and W. M. Webster, "Transistor internal parameters for small-signal representation," *Proc. IRE*, vol. 49, Apr. 1961, pp. 725–738.
[18] H. N. Ghosh, F. H. De La Moneda, and N. R. Dono, "Computer-aided transistor design, characterization, and optimization," *Proc. IEEE*, vol. 55, Nov. 1967, pp. 1897–1912.
[19] J. C. Irwin, "Resistivity of bulk silicon and of diffusion layers in silicon," *Bell Syst. Tech. J.*, vol. 41, Mar. 1962, pp. 387–410.
[20] C. T. Kirk, Jr., "A theory of transistor cutoff frequency (f_T) falloff at high current densities," *IRE Trans. Electron Devices*, vol. ED-9, Mar. 1962, pp. 164–174.
[21] R. J. Whittier and D. A. Tremere, "Current gain and cutoff frequency falloff at high currents," *IEEE Trans. Electron Devices*, vol. ED-16, Jan. 1969, pp. 39–57.
[22] G. E. Bodway, "Circuit design and characterization of transistors by means of three port scattering parameters," *Microwave J.*, vol. 11, May 1968.
[23] R. C. Hooper, J. A. Cunningham, and J. G. Harper, "Electrical contacts to silicon," *Solid-State Electron.* (Notes), vol. 8, Oct. 1965, pp. 831–833.
[24] K. Fujinuma, private communication.
[25] S. M. Sze and G. Gibbons, "Effect of junction curvature on breakdown voltage in semiconductors," *Solid-State Electron.*, vol. 9, Sept. 1966, pp. 831–845.
[26] K. Fujinuma *et al.*, "A low noise microwave transistor made by ion implantation," presented at the Int. Electron Devices Meeting, Washington, D. C., Oct. 1969, Paper 16.7.
[27] J. F. Gibbons, "Ion implantation in semiconductors—Part I: Range distribution theory and experiments," *Proc. IEEE*, vol. 56, Mar. 1968, pp. 295–319.
[28] T. Tsuchimoto and T. Tokuyama, "Enhanced diffusion of boron and phosphorous in silicon during hot substrate ion implantation," presented at the 1st Int. Conf. Implantation, May 4, 1970.
[29] W. E. Beadle, K. E. Daburlos, and W. H. Eckton, Jr., "Design, fabrication, and characterization of a germanium microwave transistor," *IEEE Trans. Electron Devices*, vol. ED-16, Jan. 1969, pp. 125–138.
[30] E. L. Long, "A high-gain 15-W monolithic power amplifier with internal fault protection," presented at the 1970 IEEE Int. Solid-State Circuits Conf., Philadelphia, Pa.

The Noise Performance of Microwave Transistors

H. FUKUI

Abstract—Expressions for the noise parameters of microwave transistors are derived. The theory is based on a small-signal common-emitter equivalent circuit which includes a new basic noise equivalent circuit and the dominating header parasitics. The theory is verified experimentally in the L-band (1 to 2 Gc/s) frequency range using Ge and Si microwave transistors. It is found that the header parasitics have little influence on the minimum noise figure, but do have large effects on the equivalent noise resistance and the optimum source admittance in the frequency region above about one-half of the series-resonant frequency resulting from the parasitics in conjunction with wafer parameters. For a quick evaluation of the noise performance, new approximate expressions are also given for the noise figure and for the optimum current which produces the lowest value.

Principal Symbols

A, B, C, D = Noise parameters
a = Drift potential
B_0 = Optimum source susceptance
B_s = Source susceptance
C_{BE} = Base-emitter header stray capacitance
C_{CB} = Collector-base header stray capacitance
C_{CE} = Collector-emitter header stray capacitance
C_c = $C_{c_1} + C_{c_2}$
C_{c_1} = Inner collector-base capacitance
 = $C_{T_{c_1}} + C_{D_c}$
C_{c_2} = Outer collector-base capacitance
 = $C_{T_{c_2}}$
C_{D_c} = Collector diffusion capacitance
C_{D_e} = Emitter diffusion capacitance
 = $1/\omega_1 r_1$
C_E = $C_{T_e} + C_{BE}$
C_e = $C_{D_e} + C_{T_e}$
$C_{T_{c_1}}$ = Inner collector-base transition region capacitance
$C_{T_{c_2}}$ = Outer collector-base transition region capacitance
C_{T_e} = Emitter-base transition region capacitance
D_0 = Diffusion constant of the minority carrier in the base region
E = Built-in field strength in the base region
e_B = Thermal noise voltage of the base resistance
F = Noise figure
$F_{\min}$ = Minimum noise figure
$(F_{\min})_{HF}$ = Approximate high-frequency minimum noise figure
Δf = Narrow frequency interval
G_0 = Optimum source conductance
G_s = Source conductance
g_e = Real part of y_e

Manuscript received August 11, 1965.
The author is with Bell Telephone Laboratories, Inc., Murray Hill, N. J.

g_{e0}	= Low-frequency values of $g_e = 1/r_{e0}$	α	= Ac intrinsic common-base short-circuited forward current gain
I_B	= Dc base current		
I_C	= Dc collector current	α_0	= Low-frequency value of α
I_E	= Dc emitter current	α_{DC}	= Dc common-base short-circuited forward current gain
i_B	= Base current noise generator		
i_C	= Collector current noise generator	β	= Ac intrinsic common-emitter short-circuited forward current gain
i_b	= Ac intrinsic base current		
i_{br}	= Ac resultant base current flowing through r_{b_1}		= $\alpha/1-\alpha$
		β_0	= Low-frequency value of β
i_0	= Ac short-circuited output current		= $\alpha_0/1-\alpha_0$
i_s	= Thermal noise current generator of the source conductance	β_{DC}	= Dc common-emitter short-circuited forward current gain
K	= Drift factor		
k	= Boltzmann's constant		= $\dfrac{\alpha_{DC}}{1-\alpha_{DC}}$
L_{B_1}	= Inner base lead inductance		
L_{B_2}	= Outer base lead inductance	β_T	= Ac resultant common-emitter short-circuited forward current gain
L_C	= Collector lead inductance		
L_E	= Emitter lead inductance	ϵ	= Base of natural logarithm
M, N	= Multiplying factors	ω	= Angular frequency = $2\pi f$
m	= Excess phase factor	ω_1	= Angular unity gain frequency at which $|\beta_1|$ is unity
N_c	= Impurity density at collector extreme of the base		
			= $2\pi f_1$
N_e	= Impurity density at emitter extreme of the base	ω_T	= Angular total unity gain frequency
q	= Electronic charge		
R_n	= Equivalent noise resistance		= $\left(\dfrac{1}{\omega_1} + C_{Te}r_1 + \dfrac{x_m}{2v_{SC}}\right)^{-1} = 2\pi f_T$
r_b	= $r_{b_1} + r_{b_2}$		
$r_{b\infty}$	= Ultimate value of r_b where I_C is assumed to be infinity	ω_α	= Angular α-cutoff frequency = $2\pi f_\alpha$
r_{b_1}	= Inner base resistance		
r_{b_2}	= Outer base resistance		
r_{e0}	= Differential emitter resistance		
	= $\left(\dfrac{\partial I_E}{\partial V_{EB}}\right)^{-1} = \dfrac{kT}{qI_E}\cdot\dfrac{\alpha_0}{\alpha_{DC}} = \dfrac{kT}{qI_C}\alpha_0$		
r_1	= $kT/qI_C = r_{e0}/\alpha_0$		
s	= Slope of r_b vs. $1/I_C$ line		
T	= Temperature in °K		
v_{SC}	= Scattering limited velocity in the collector-base transition region		
W	= Effective base width between emitter- and collector-base transition regions		
x_m	= Width of collector-base transition region		
Y_{BE}	= $j\omega C_{BE}$		
Y_{be}	= $1/\beta_0 r_1 + j\omega C_e$		
Y_{CE}	= $j\omega C_{CE}$		
Y_s	= $G_s + jB_s$		
y_{be}	= Ac intrinsic common-emitter short-circuited input admittance		
y_{ce}	= Ac intrinsic common-base short-circuited forward transfer admittance		
y_e	= Ac intrinsic common-base short-circuited input admittance		
Z_B	= $j\omega L_{B_2}$		
Z_b	= $r_b + j\omega L_{B_1}$		
Z_C	= $j\omega L_C$		
Z_E	= $j\omega L_E$		

I. Introduction

RECENT TECHNOLOGICAL improvements in the manufacturing process of microwave transistors have resulted in units which are capable of excellent performance in low-noise microwave amplifiers [1].

The theory of shot noise in junction transistors has been verified over a wide range of operating conditions for both germanium and silicon units at frequencies up to VHF [2]–[13]. The agreement between theory and experiment, however, becomes poorer above 1 Gc/s [14]. This deterioration seems to be partly due to the effect of transistor header parasitics. Taking this effect into account, a noise analysis has been initiated by the author [15] and Thommen and Strutt [16], [17], independently. The former has developed his theory based on a new common-emitter noise equivalent circuit, whereas the latter has used the conventional common-base configuration. This paper will describe the above mentioned theory in detail. A comparison of the theory with experimental results will also be given.

II. Theory

In this paper, the noise performance of transistors in the common-emitter configuration will be discussed based on the small-signal theory. For currently available microwave transistors, only the common-emitter configuration gives good stability over a wide range of operating conditions, especially in wide band amplifiers.

A. Equivalent Circuit

Figure 1 shows an equivalent circuit for the microwave transistor wafer in the common-emitter configuration. Here the following relations are used to describe the high-frequency performance of graded-base transistors [18]–[21],

$$\omega_0 = \frac{2D_0}{W^2} \quad (1)$$

$$a = \ln\left(\frac{N_e}{N_c}\right) = \frac{qEW}{kT} \quad (2)$$

$$\omega_1 = \frac{\frac{1}{2}a^2}{a - 1 + \epsilon^{-a}} \omega_0 \quad (3)$$

$$\omega_\alpha = K\omega_1 \quad (4)$$

$$K = 1.21 + 0.09a \quad (5)$$

$$m = 0.22 + 0.1a \quad (6)$$

$$\alpha = \frac{\alpha_0 \epsilon^{-jm(\omega/\omega_\alpha)}}{1 + j\frac{\omega}{\omega_\alpha}} \quad (7)$$

$$\beta = \frac{\alpha}{1-\alpha} \simeq \frac{1}{\frac{1}{\beta_0} + j\frac{\omega}{\omega_1}} \quad (8)$$

$$|\beta|^2 \simeq \left(\frac{\omega_1}{\omega}\right)^2 \quad (9)$$

$$r_{e0} = \left(\frac{\partial I_E}{\partial V_{BE}}\right)^{-1} = \frac{kT}{qI_E} \cdot \frac{\alpha_0}{\alpha_{DC}} = \frac{1}{g_{e0}} \quad (10)$$

$$r_1 = \frac{kT}{qI_C} = \frac{r_{e0}}{\alpha_0} \quad (11)$$

$$C_{D_e} = \frac{1}{\omega_1 r_1} \quad (12)$$

$$\frac{1}{\omega_T} = \frac{1}{\omega_1} + r_1 C_{T_e} + \frac{x_m}{2v_{SC}} \quad (13)$$

$$|\beta_T|^2 \simeq \left(\frac{\omega_T}{\omega}\right)^2. \quad (14)$$

In the above representation, the trapping-recombination effects of carriers in the emitter-base transition region are taken into account [18], [12]. In the following analysis of the noise figure, feedback parameters between the collector and the base, such as C_{c_1} and C_{c_2}, will be neglected for sake of simplicity. The simplification will be evaluated later on.

The common-emitter noise equivalent circuit was first developed by Giacoletto [22]. Here the common-base noise equivalent circuit given by van der Ziel [3], [6], [12] will be transformed into the common-emitter configuration as shown in the Appendix.[1] As a result, three noise generators are introduced into the common-emitter equivalent circuit shown in Fig. 1. These noise generators are due to

[1] Chenette also used a common-emitter noise equivalent circuit from a different viewpoint than the above with no verification [23].

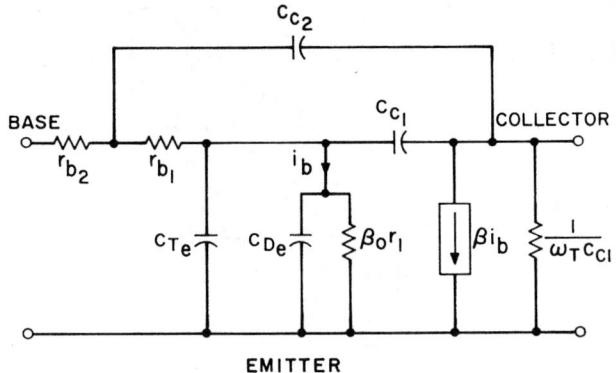

Fig. 1. Equivalent circuit for the microwave transistor wafer in the common-emitter configuration.

fluctuations in the dc base current, fluctuation in the dc collector current, and thermal noise of the base resistance. The mean square values of these generators in a narrow frequency interval are given by

$$\overline{i_B^2} = 2qI_B\Delta f \quad (15)$$

$$\overline{i_C^2} = 2qI_C\Delta f \quad (16)$$

$$\overline{e_B^2} = 4kTr_b\Delta f. \quad (17)$$

These three noise generators are uncorrelated to one another as verified in the Appendix,

$$\overline{e_B^* i_B} = 0, \quad \overline{e_B^* i_C} = 0, \quad \text{and} \quad \overline{i_B^* i_C} = 0. \quad (18)$$

This is a feature of the common-emitter configuration.

Since the high-frequency noise behavior is of interest in this paper, flicker noise is ignored.

In order to obtain the complete noise equivalent circuit for the actual transistor in the microwave region, the effect of header parasitics should be added to the original circuit which is related to the wafer alone. These parasitics are L_{B_1}, L_{B_2}, L_E, L_C, C_{BE}, and C_{CE}. C_{CB} is neglected, as well as C_{c_1} and C_{c_2}, for the sake of mathematical simplicity. The effect of this simplification on the noise performance will be discussed later.

For the evaluation of the noise performance, the signal source driving the transistor should also be taken into consideration because its internal conductance generates thermal noise and its susceptance affects the noise figure through noise tuning [24]. The mean square value of this thermal noise in a narrow frequency interval is given by

$$\overline{i_s^2} = 4kTG_s\Delta f. \quad (19)$$

As a result of the foregoing, a new noise equivalent circuit can be given as shown in Fig. 2.

B. Analysis of Noise Figure

It is well known that the noise figure of a circuit such as shown in Fig. 2 is given by the ratio of the actual total mean squared noise current in the ac short-circuited output $\overline{i_0^2}(\text{Total})$ to that portion which results from the thermal noise (at 290°K) originating in the source conductance $\overline{i_0^2}(\text{Source})$, i.e.,

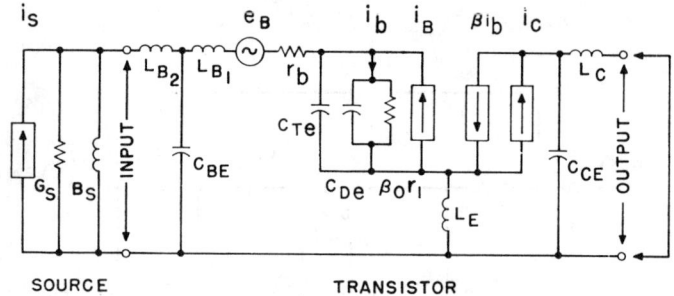

Fig. 2. Noise equivalent circuit for the actual microwave transistor.

$$F = \frac{\overline{i_0^2}(\text{Total})}{\overline{i_0^2}(\text{Source})}. \quad (20)$$

If it is assumed that

$$\frac{x_m}{2v_{SC}} \ll \frac{1}{\omega_1} + r_1 C_{T_e}, \quad (21)$$

βi_b becomes approximately equal to $\beta_T i_{b_T}$ and

$$|\beta_T Y_{be}|^2 = \frac{1}{r_1^2}. \quad (22)$$

Hence, from (15) through (22),

$$F = 1 + \frac{r_b}{G_s} |Y_s + Y_{BE} + Y_s Y_{BE} Z_B|^2$$
$$+ \frac{qI_B}{2kTG_s} |1 + (Y_s + Y_{BE})(Z_b + Z_E)$$
$$+ Y_s Z_B \{1 + Y_{BE}(Z_b + Z_E)\}|^2$$
$$+ \frac{qI_C r_1^2}{2kTG_s} |Y_s + Y_{BE} + Y_{be} + Y_{be}(Y_s + Y_{BE})(Z_b + Z_E)$$
$$+ Y_s Z_B \{Y_{BE} + Y_{be} + Y_{BE} Y_{be}(Z_b + Z_E)\}|^2$$
$$\times \left| 1 + \frac{Z_E}{Z_b + \frac{1}{Y_{be}} + \frac{1 + Y_s Z_B}{Y_s + Y_{BE} + Y_s Y_{BE}(Z_b + Z_E)}} \right|^2. \quad (23)$$

If it is assumed that

$$\left| \frac{Z_E}{Z_b + \frac{1}{Y_{be}} + \frac{1 + Y_s Z_B}{Y_s + Y_{BE} + Y_s Y_{BE}(Z_b + Z_E)}} \right| \ll 1, \quad (24)$$

(23) reduces to

$$F = 1 + \frac{r_b}{G_s} |Y_s + Y_{BE} + Y_s Y_{BE} Z_B|^2$$
$$+ \frac{qI_B}{2kTG_s} |1 + (Y_s + Y_{BE})(Z_b + Z_E)$$
$$+ Y_s Z_B \{1 + Y_{BE}(Z_b + Z_E)\}|^2$$
$$+ \frac{kT}{2qI_C G_s} |Y_s + Y_{BE} + Y_{be} + Y_{be}(Y_s + Y_{BE})(Z_b + Z_E)$$
$$+ Y_s Z_B \{Y_{BE} + Y_{be} + Y_{BE} Y_{be}(Z_b + Z_E)\}|^2. \quad (25)$$

Substituting (11) through (14) into (25) and referring to the list of principal symbols, (25) is of the following form:

$$F = A + BG_s + \frac{C + BB_s^2 + DB_s}{G_s} \quad (26)$$

where

$$A = 1 + \frac{1}{\beta_0} + \frac{qI_C r_b}{kT} \left[\frac{1}{\beta_{DC}} + \frac{1}{\beta_0^2} + \left(\frac{\omega}{\omega_T}\right)^2 \right]. \quad (27)$$

$$B = \left[r_b \left(1 + \frac{1}{\beta_0}\right) + \frac{kT}{2qI_C} \left\{ 1 + \left(\frac{qI_C}{kT} r_b\right)^2 \right. \right.$$
$$\left. \left. \times \left[\frac{1}{\beta_{DC}} + \frac{1}{\beta_0^2} + \left(\frac{\omega}{\omega_T}\right)^2 \left(1 + \frac{\omega^2(L_{B_2} + L_E)^2}{r_b^2}\right) \right] \right\} \right]$$
$$\times (1 - \omega^2 L_{B_1} C_{BE})^2 + \frac{qI_B}{2kT} [\omega(L_{B_1} + L_{B_2} + L_E)$$
$$- \omega^3 L_{B_1}(L_{B_2} + L_E) C_{BE}]^2 + \frac{qI_C}{2kT} \left[\left\{ \frac{1}{\beta_0^2} + \left(\frac{\omega}{\omega_T}\right)^2 \right\} (\omega L_{B_1})^2 \right.$$
$$+ 2 \left\{ \frac{1}{\beta_0^2} + \left(\frac{\omega}{\omega_T}\right)^2 \left(1 + \frac{C_{BE}}{C_e}\right) \right\} \omega^2 L_{B_1}(L_{B_2} + L_E)$$
$$\times (1 - \omega^2 L_{B_1} C_{BE}) - 2 \left(\frac{kT}{qI_C}\right) \left(\frac{\omega}{\omega_T}\right)$$
$$\left. \times \omega(L_{B_1} + L_{B_2} + L_E)(1 - \omega^2 L_{B_1} C_{BE}) \right]. \quad (28)$$

$$C = \frac{qI_C}{2kT} \left[\frac{1}{\beta_{DC}} \{1 - \omega^2 C_{BE}(L_{B_2} + L_E)\} + \frac{1}{\beta_0^2} \right.$$
$$+ \left(\frac{\omega}{\omega_T}\right)^2 \left(1 + \frac{2C_{BE}}{C_e}\right) \right] \{1 - \omega^2 C_{BE}(L_{B_2} + L_E)\}$$
$$+ (\omega C_{BE})^2 \left[r_b \left(1 + \frac{1}{\beta_0}\right) + \frac{kT}{2qI_C} \right.$$
$$\left. \left. \cdot \left\{ 1 + \left(\frac{qI_C r_b}{kT}\right)^2 \left(\frac{1}{\beta_{DC}} + \frac{1}{\beta_0^2} + \left(\frac{\omega}{\omega_T}\right)^2\right) \right\} \right] \right]. \quad (29)$$

$$D = \left(\frac{\omega}{\omega_T}\right) + 2\omega C_{BE} \left[r_b \left(1 + \frac{1}{\beta_0}\right) \right.$$
$$+ \frac{kT}{2qI_C} \left\{ 1 + \left(\frac{qI_C r_b}{kT}\right)^2 \left(\frac{1}{\beta_{DC}} + \frac{1}{\beta_0^2} + \left(\frac{\omega}{\omega_T}\right)^2\right) \right\} \right]$$
$$\cdot (1 - \omega^2 L_{B_1} C_{BE}) - \frac{qI_C}{kT} \left[\left\{ \frac{1}{\beta_{DC}} + \frac{1}{\beta_0^2} + \left(\frac{\omega}{\omega_T}\right)^2 \right\} \right.$$
$$\left. \cdot \{1 - \omega^2 C_{BE}(L_{B_2} + L_E)\} + \left(\frac{\omega}{\omega_T}\right)^2 \left(\frac{2C_{BE}}{C_e}\right) \right]$$
$$\times [\omega(L_{B_1} + L_{B_2} + L_E) - \omega^3 C_{BE} L_{B_1}(L_{B_2} + L_E)]. \quad (30)$$

The coefficients A, B, C, and D form a set of noise parameters [15]. Only the parameter A is independent of the header parasitics. The others are affected by the header.

The noise figure is also given by an alternate expression as follows:

$$F = F_{\min} + \frac{R_n}{G_s} [(G_s - G_0)^2 + (B_s - B_0)^2]. \quad (31)$$

$F_{\min}$ is the minimum noise figure, R_n an equivalent noise resistance, and G_0 and B_0 the particular source conductance and susceptance, respectively, which produce $F_{\min}$. $F_{\min}$, G_0, B_0, and R_n form another set of noise parameters. For a given wafer on different typical high quality microwave headers, $F_{\min}$ of the encapsulated transistor is almost independent of the header.

As is well known, both sets of noise parameters can completely specify the noise performance of linear two-ports. The inter-relationships between the sets are given as follows:

$$F_{\min} = A + \sqrt{4BC - D^2} \qquad (32)$$

$$G_0 = \frac{\sqrt{4BC - D^2}}{2B} \qquad (33)$$

$$B_0 = -\frac{D}{2B} \qquad (34)$$

$$R_n = B \qquad (35)$$

$$A = F_{\min} - 2R_n G_0 \qquad (36)$$

$$C = R_n(G_0^2 + B_0^2) \qquad (37)$$

$$D = -2R_n B_0. \qquad (38)$$

C. No-Parasitic Case

If all header parasitics are neglected, the expressions for the noise figure become (the suffixes NP indicate the case of no parasitics):

$$(F)_{NP} = \left(1 + \frac{1}{\beta_0}\right)\left[1 + r_b G_s\left(1 + \frac{B_s^2}{G_s^2}\right)\right] + \frac{kTG_s}{2qI_c}\left(1 + \frac{B_s^2}{G_s^2}\right)$$

$$+ \frac{qI_c r_b}{2kT}\left(\frac{1}{\beta_{DC}} + \frac{1}{\beta_0^2}\right)\left[2 + \frac{1}{r_b G_s}\right.$$

$$\left. + r_b G_s\left(1 + \frac{B_s^2}{G_s^2}\right)\right] + \left(\frac{B_s}{G_s}\right)\left(\frac{\omega}{\omega_T}\right)$$

$$+ \frac{qI_c r_b}{2kT}\left[2 + \frac{1}{r_b G_s} + r_b G_s\left(1 + \frac{B_s^2}{G_s^2}\right)\right]\left(\frac{\omega}{\omega_T}\right)^2 \qquad (39)$$

$$(A)_{NP} = A \qquad (40)$$

$$(B)_{NP} = r_b\left(1 + \frac{1}{\beta_0}\right)$$

$$+ \frac{kT}{2qI_c}\left[1 + \left(\frac{qI_c r_b}{kT}\right)^2\left\{\frac{1}{\beta_{DC}} + \frac{1}{\beta_0^2} + \left(\frac{\omega}{\omega_T}\right)^2\right\}\right] \qquad (41)$$

$$(C)_{NP} = \frac{qI_c}{2kT}\left[\frac{1}{\beta_{DC}} + \frac{1}{\beta_0^2} + \left(\frac{\omega}{\omega_T}\right)^2\right] \qquad (42)$$

$$(D)_{NP} = \frac{\omega}{\omega_T} \qquad (43)$$

$$(F_{\min})_{NP} = 1 + \frac{1}{\beta_0} + \frac{qI_c r_b}{kT}\left[\frac{1}{\beta_{DC}} + \frac{1}{\beta_0^2} + \left(\frac{\omega}{\omega_T}\right)^2\right]$$

$$+ \left[\frac{1}{\beta_{DC}} + \frac{1}{\beta_0^2} + \frac{2qI_c r_b}{kT}\left(1 + \frac{1}{\beta_0}\right)\left\{\frac{1}{\beta_{DC}} + \frac{1}{\beta_0^2} + \left(\frac{\omega}{\omega_T}\right)^2\right\}\right.$$

$$\left. + \left(\frac{qI_c r_b}{kT}\right)^2\left\{\frac{1}{\beta_{DC}} + \frac{1}{\beta_0^2} + \left(\frac{\omega}{\omega_T}\right)^2\right\}^2\right]^{1/2} \qquad (44)$$

$$(R_n)_{NP} = r_b\left(1 + \frac{1}{\beta_0}\right)$$

$$+ \frac{kT}{2qI_c}\left[1 + \left(\frac{qI_c r_b}{kT}\right)^2\left\{\frac{1}{\beta_{DC}} + \frac{1}{\beta_0^2} + \left(\frac{\omega}{\omega_T}\right)^2\right\}\right] \qquad (45)$$

$$(G_0)_{NP} = \left[\frac{1}{\beta_{DC}} + \frac{1}{\beta_0^2} + \frac{2qI_c r_b}{kT}\left(1 + \frac{1}{\beta_0}\right)\right.$$

$$\cdot \left\{\frac{1}{\beta_{DC}} + \frac{1}{\beta_0^2} + \left(\frac{\omega}{\omega_T}\right)^2\right\} + \left(\frac{qI_c r_b}{kT}\right)^2\left\{\frac{1}{\beta_{DC}} + \frac{1}{\beta_0^2} + \left(\frac{\omega}{\omega_T}\right)^2\right\}^2\right]^{1/2}$$

$$\div \left[\frac{kT}{qI_c} + 2r_b\left(1 + \frac{1}{\beta_0}\right) + \frac{qI_c r_b^2}{kT}\left\{\frac{1}{\beta_{DC}} + \frac{1}{\beta_0^2} + \left(\frac{\omega}{\omega_T}\right)^2\right\}\right]$$

$$(46)$$

$$(B_0)_{NP} = \frac{-\left(\frac{\omega}{\omega_T}\right)}{\frac{kT}{qI_c} + 2r_b\left(1 + \frac{1}{\beta_0}\right) + \frac{qI_c r_b^2}{kT}\left\{\frac{1}{\beta_{DC}} + \frac{1}{\beta_0^2} + \left(\frac{\omega}{\omega_T}\right)^2\right\}} \qquad (47)$$

Furthermore, for the case $B_s = 0$, (39) reduces to $(F_{NP})_0$, i.e.,

$$(F_{NP})_0 = \left(1 + \frac{1}{\beta_0}\right)(1 + r_b G_s) + \frac{kTG_s}{2qI_c}$$

$$+ \frac{qI_c G_s}{2kT}\left(r_b + \frac{1}{G_s}\right)^2\left[\frac{1}{\beta_{DC}} + \frac{1}{\beta_0^2} + \left(\frac{\omega}{\omega_T}\right)^2\right]. \qquad (48)$$

This expression coincides with that given by Chenette [23].

D. Quick Estimate of High-Frequency Noise Figure

With increasing frequency, the contribution of the terms $1/\beta_{DC}$, $1/\beta_0$, and $1/\beta_0^2$ to the minimum noise figure in (44) becomes small. Thus, the following expression can be used for a quick estimate of the high-frequency minimum noise figure:

$$(F_{\min})_{HF} \simeq 1 + h\left(1 + \sqrt{1 + \frac{2}{h}}\right) \qquad (49)$$

where

$$h = \frac{qI_c r_b}{kT}\left(\frac{\omega}{\omega_T}\right)^2 \qquad (50)$$

or

$$h = 0.04 I_c r_b \left(\frac{f}{f_T}\right)^2 \qquad (50a)$$

(I_c in mA, r_b in ohms, $T = 290°$).

$(F_{\min})_{HF}$ increases monotonically with increasing h as shown in Fig. 3. The above estimate is good if the transistor has sufficiently good dc characteristics, i.e., a dc current gain h_{FE} of

$$h_{FE} > 10\left(\frac{\omega_T}{\omega}\right)^2. \qquad (51)$$

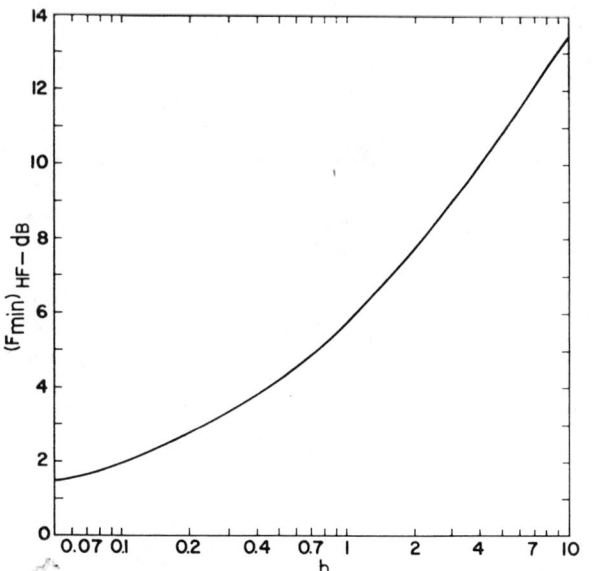

Fig. 3. Approximate high-frequency minimum noise figure vs. a parameter h.

E. Optimum Current for Minimum Noise Figure

The minimum noise figure is a function of collector current. As the collector current decreases, the shot noises caused by fluctuations in the collector and the base currents decrease, but f_T also decreases and the base resistance increases in certain cases [25]. Therefore, $F_{\min}$ becomes minimum at a certain collector current $(I_C)_0$. Because the general solution for $(I_C)_0$ is too complicated, the particular case given in II-D will be considered below.

$(I_C)_0$ will be derived by differentiating (49) with respect to I_C and equating it to zero. That equation becomes equivalent to

$$\frac{dh}{dI_C} = 0 \qquad (52)$$

because

$$\frac{d(F_{\min})_{HF}}{dh} \neq 0.$$

For more realistic cases C_{BE} must be taken into consideration. Under the assumption of (21), the total unity gain frequency is then described approximately by a new expression

$$\frac{1}{\omega_T} = \frac{1}{\omega_1} + \frac{kT}{eI_C} C_E \qquad (53)$$

where

$$C_E = C_{T_e} + C_{BE}. \qquad (54)$$

Since r_b is also generally a function of I_C, using (50) and (53), (52) becomes

$$\left[\frac{1}{\omega_1} + \frac{kTC_E}{q(I_C)_0}\right]\left[1 + \frac{(I_C)_0}{r_b} \cdot \frac{dr_b}{dI_C}\right] - \frac{2kTC_E}{q(I_C)_0} = 0. \qquad (55)$$

Here r_b is still an unknown function of I_C. If it is assumed that, for example [25],

$$r_b = r_{b\infty} + \frac{s}{I_C} \qquad (56)$$

where $r_{b\infty}$ is the imaginary value of r_b at $1/I_C = 0$ and s the slope of r_b vs. $1/I_C$ line, $(I_C)_0$ can be derived from (55) as follows:

$$(I_C)_0 = \frac{kT\omega_1 C_E}{2q}\left[1 + \sqrt{1 + \frac{8qs}{kT\omega_1 C_E r_{b\infty}}}\right] \qquad (57)$$

or

$$(I_C)_0 = 0.08 f_1 C_E [1 + \sqrt{1 + (50s/f_1 C_E r_{b\infty})}]\ \text{mA} \qquad (57a)$$

(f_1 in Gc/s, C_E in pF, $r_{b\infty}$ in ohms, $T = 290°$K).

For constant r_b, i.e., $s = 0$, (57) reduces to

$$(I_C)_0 = \frac{kT\omega_1 C_E}{q} \qquad (58)$$

or

$$(I_C)_0 = 0.16 f_1 C_E\ \text{mA} \qquad (58a)$$

(f_1 in Gc/s, C_E in pF, $T = 290°$K).

Both (57) and (58) should apply for transistors having a sufficiently high dc current gain as given in (51).

III. Comparison with Experiment

A. Samples Used and Their Optimum Currents

In order to verify the theory presented in this paper, measurements have been made on two types of microwave transistors, a M2519 *NPN* silicon double-diffused planar unit and a L2254 *PNP* germanium epitaxial mesa units. These transistors were encapsulated in modified TO-18 packages. The measuring technique used was based on the method recommended by the IRE Standards [26], [27].

From preliminary measurements, the smallest noise figures were found to be obtained at an emitter current of about -1 mA for the M2519 and at about 2 mA for the L2254, as shown in Fig. 4 for the case of conjugate match. These numbers are very close to the predicted values of -1.0 mA for the M2519 and 2.4 mA for the L2254, respectively.[2] The predicted values correspond to collector currents of 0.96 mA and -1.8 mA, respectively, which were determined by (57) using predetermined parameters tabulated in Table I.

B. M2519 Transistor

For the M2519 transistor mentioned in Section III-A, the noise figure was measured as a function of source admittance at an emitter current of -1 mA and frequencies of 1.3 and 1.95 Gc/s. From the measurements with G_s constant and B_s variable and vice versa, B_0 and G_0 were determined. The behavior of noise figure was then plotted as a function of $|Y_s - Y_0|^2/G_s$, and $F_{\min}$ and R_n were determined. The measured noise parameters are tabulated in Table II.

[2] The theoretical expression of (57) has been derived for the minimum noise figure. According to experimental data, however, the optimum current for smallest noise figure is almost independent of source admittance.

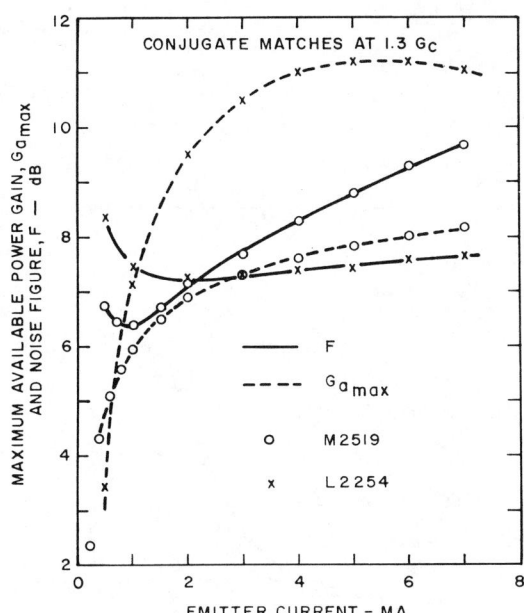

Fig. 4. Gain and noise performance as a function of emitter current.

TABLE I
PREDETERMINED PARAMETERS FOR OPTIMUM CURRENTS

Unit No.	f_1 (Gc/s)	C_E (pF)	$r_{b\infty}$ (ohms)	s (ohms-mA)
M2519-A63P3	1.7	1.9	7.1	2.84
L2254-2032-40	6.3	1.8	—	0

TABLE II
MEASURED NOISE PARAMETERS

Freq. (Gc/s)	$F_{\min}$ (dB)	R_n (ohms)	G_0 (m.mhos)	B_0 (m.mhos)
1.3	5.1	15.6	53	20
1.95	7.0	66.2	20	28

From the characterization of this transistor the following device parameters were obtained at $V_{CB} = 5\text{V}$ and $I_E = -1$ mA:

$\beta_{DC} = 17.0 \qquad \beta_0 = 18.4$
$f_T = 1.02 \text{ Gc/s} \qquad r_b = 9.2 \text{ ohms}$
$L_{B_1} = 1.1 \text{ nH} \qquad L_{B_2} = 1.1 \text{ nH}$
$L_E = 1.1 \text{ nH, and} \qquad C_{BE} = 0.4 \text{ pF.}$

The value of C_{BE} was taken from its statistical average. The header inductance values were first determined by measured y-parameters [25] to be 1.0 nH each, and these numbers were modified later to 1.1 nH to get better agreement between the measured and calculated noise performance. f_T was determined by the intrinsic common-emitter current gain obtained from the y-parameters after modifying them for the header parasitics [25]. The f_T value was also obtained from the overall common-emitter current gain measurement. This f_T of 1.0 Gc/s is almost the same as the former. r_b was determined by the power gain method [25].

Substituting the above numbers into (27) through (35) and (44) through (47), all noise parameters were calculated in the frequency region of 0.1 to 4 Gc/s both with and without parasitics, as shown in Figs. 5 through 9.

This comparison shows that the present expressions for the noise parameters agree well with experimental results. This example also reveals several other important factors:

1) Present header parasitics have little effect on $F_{\min}$ in the low microwave region (up to about 2 Gc/s), but become important at S-band and above.

2) The header parasitics have strong effects on R_n, G_0, and B_0, which show a resonance in their frequency characteristics. This resonant frequency f_r can be determined from the wafer parameters and header parasitics, i.e., by putting $B_0 = 0$ and solving it for f. A rough general estimate of f_r is given by the following expression:

$$f_r \simeq \left[\frac{kT f_T}{2\pi (L_{B_1} + L_{B_2} + L_E) q I_C} \right]^{1/2}. \quad (59)$$

For the modified TO-18 package used for the M2519 and L2254 transistors, (59) reduces approximately to

$$f_r(\text{Gc}) \simeq 1.2 \sqrt{\frac{f_T (\text{Gc})}{I_C (\text{mA})}}. \quad (60)$$

Since f_T is a function of I_C as given in (53), f_r varies only slowly with changing I_C for a given device.

3) In the frequency region below one tenth f_r, the effect of the header parasitics can be neglected, but above about one-half f_r the effect becomes considerable.

4) In the vicinity of f_r, R_n decreases, G_0 increases, and B_0 changes sign from negative to positive, as a result of the series resonance.

5) At frequencies well above f_r, the behavior of R_n, G_0, and B_0 with parasitics is quite different from that without parasitics.

When minimum noise figure operation is required at frequencies above the UHF region, the effect of header parasitics must therefore be taken into account by utilizing the appropriate source admittance, i.e., Y_0.

When only a rough estimate of $F_{\min}$ is required, $(F_{\min})_{HF}$ of (49) may be useful. As shown in Fig. 5, $(F_{\min})_{HF}$ is a good approximation to $(F_{\min})_{NP}$ in the high-frequency region.

C. L2254 Transistors

The noise performance of L2254 transistors was investigated at L-band, both at 1.0 and 1.3 Gc/s. Table III shows the device parameters required for the characterization. Relatively simple methods were employed for determination of the device parameters as compared with the previous example. f_T was deduced from a h_{fe} measurement at 1.0 Gc/s. r_b was determined by the power gain method at 1.3 Gc/s. For the header parasitics, typical values of the modified TO-18 package were used, i.e., $L_{B_1} = L_{B_2} = L_E = 1.0$ nH and $C_{BE} = 0.4$ pF. Using the above-mentioned parameters, the noise performance was calculated at 1.0 and 1.3 Gc/s.

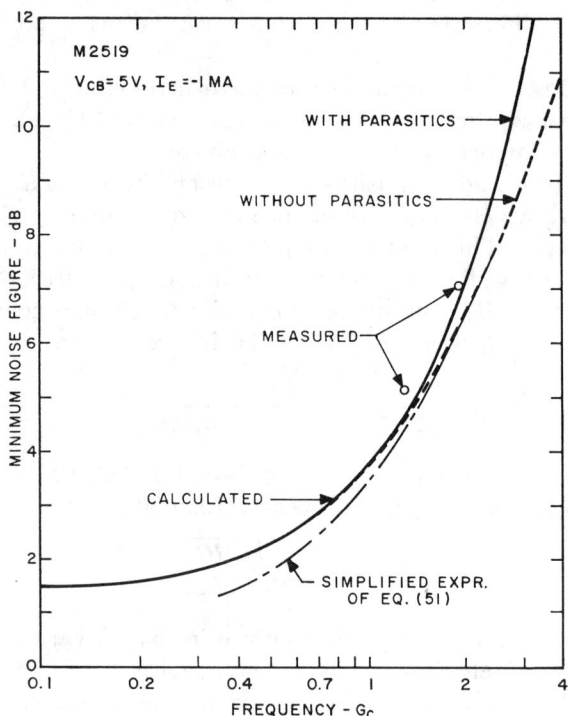

Fig. 5. Minimum noise figure vs. frequency.

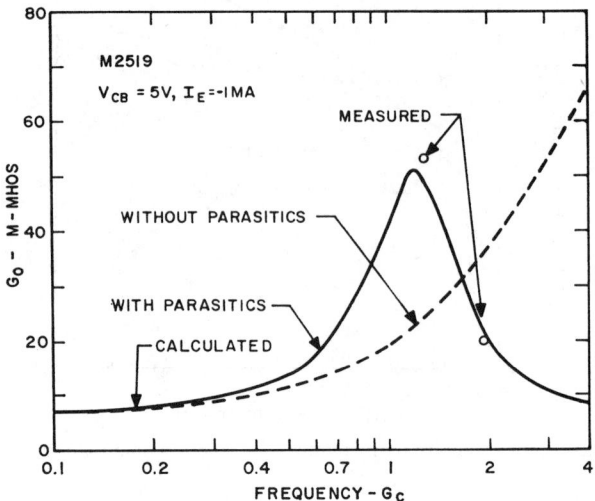

Fig. 7. Optimum source conductance vs. frequency.

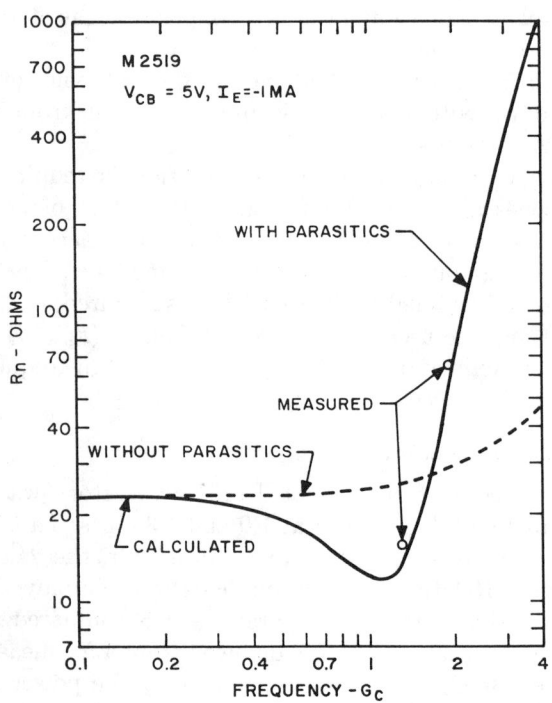

Fig. 6. Noise equivalent resistance vs. frequency.

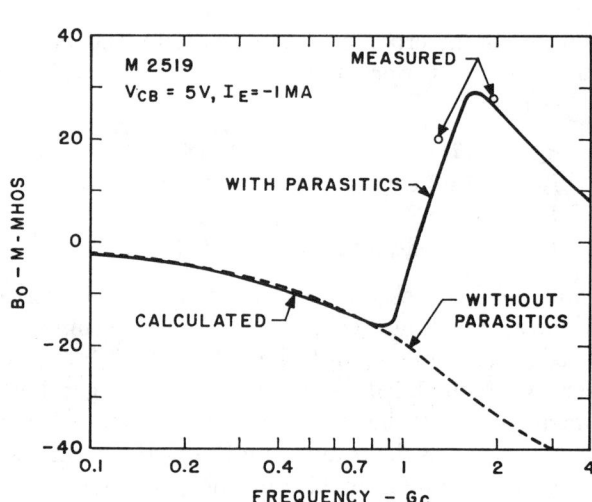

Fig. 8. Optimum source susceptance vs. frequency.

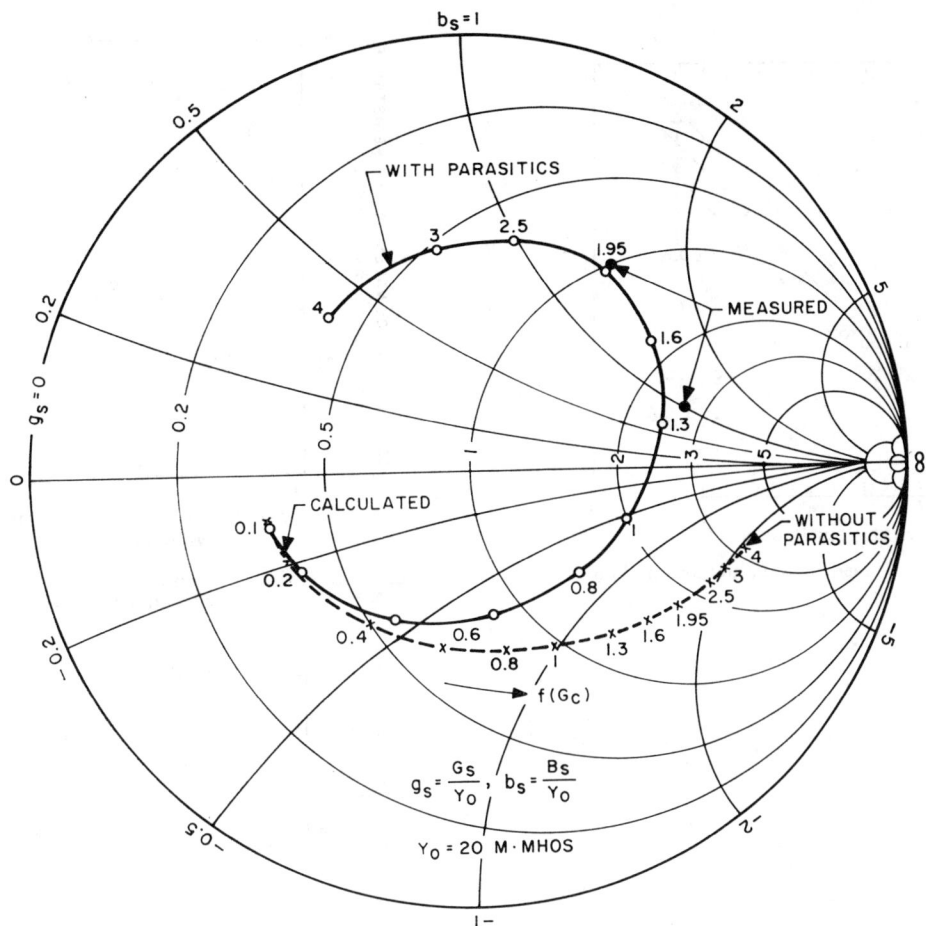

Fig. 9. Optimum source admittance loci on the Smith-chart.

TABLE III
DEVICE PARAMETERS AT $V_{CB} = -5V$ AND $I_E = 2$ mA

Unit No.	β_{DC}	β_0	f_T (Gc/s)	r_b (ohms)	C_c (pF)
A6943	3.9	6.2	2.19	16.3	0.42
A7639	21.2	37.4	2.20	18.9	0.32
A7746	5.7	8.4	1.75	18.6	0.28
A8899	9.3	13.1	2.22	17.8	0.31
B4227	5.7	11.3	2.00	15.6	0.47
B4243	10.8	21.7	2.15	16.2	0.41
B8238	8.1	12.0	1.94	16.1	0.49
B8241	5.7	12.0	1.98	15.8	0.51
B8245	5.9	10.1	2.05	16.7	0.45
B8250	4.0	10.6	1.93	15.6	0.46
B8867	8.1	13.1	1.91	16.7	0.31
B8906	5.9	12.5	1.86	14.8	0.35
B9070	5.9	11.8	1.95	16.2	0.31
B9235	5.1	10.2	1.75	14.9	0.27
B9275	2.8	5.9	1.54	15.3	0.28

The noise figure measurements on these units were made at a fixed 50-ohm source impedance, i.e., $Y_s = G_s = 0.02$ mho. Figure 10 shows good agreement between the calculated and measured noise figures at both 1.0 and 1.3 Gc/s for this condition.

In the following paragraphs, two sets of calculated noise parameters, with and without parasitics (abbreviated WP and NP, respectively), will be compared to each other. First, Fig. 11 shows a comparison between the two 50-ohm noise figures, WP and NP. The difference between them is about 0.1 dB at the lower end and about 0.05 dB at the higher end of the noise figure range. Secondly, the minimum noise figures are compared as shown in Fig. 12, where the difference is only a few hundredths of a decibel. Third is a comparison of R_n as shown in Fig. 13. With parasitics, R_n is about 5.5 ohms lower than without. The reason is that the input series resonance takes place at L-band for these units when operating at $I_E = 2$ mA. For the same reason, $G_{0(WP)}$ is larger than $G_{0(NP)}$ by about 5 millimhos and most of the $B_{0(WP)}$ values change their sign from negative at 1.0 Gc/s to positive at 1.3 Gc/s, whereas the $B_{0(NP)}$ values increase in (negative) magnitude with increasing frequency as shown in Figs. 14 and 15. Again, strong effects of the header parasitics on R_n, G_0, and B_0 are observed with the L2254 transistors.

Next, the current dependence of noise figure with 50-ohm source impedance for a L2254 transistor (No. 2032-40) will be given at 1.3 Gc/s. Predetermined device parameters at $V_{CB} = -5V$ are tabulated in Table IV.

Figure 16 shows the measured noise figure and the calculated noise figures with and without parasitics. The calculated curves for the two cases are almost identical and agree well with the measured values.

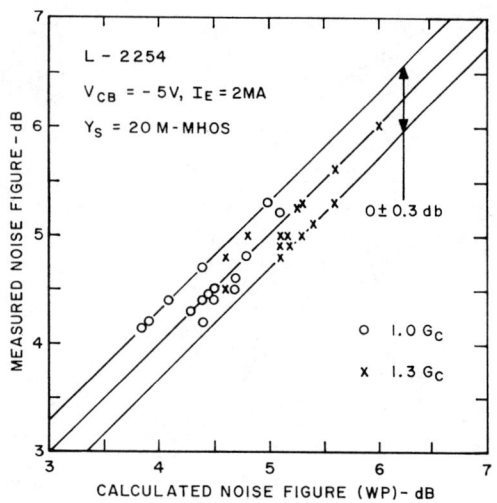

Fig. 10. Comparison between calculated and measured noise figure.

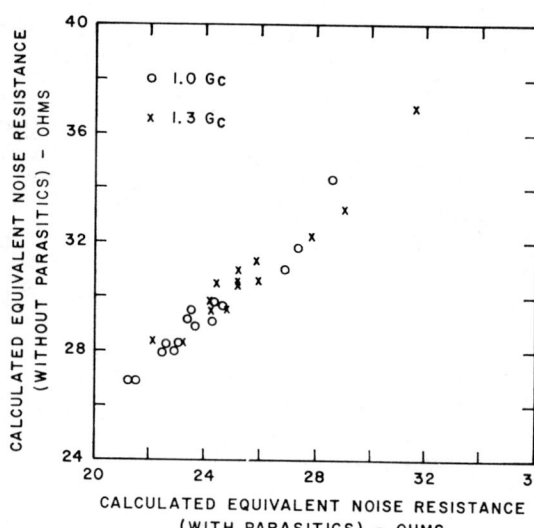

Fig. 13. Comparison of equivalent noise resistances, with and without parasitics.

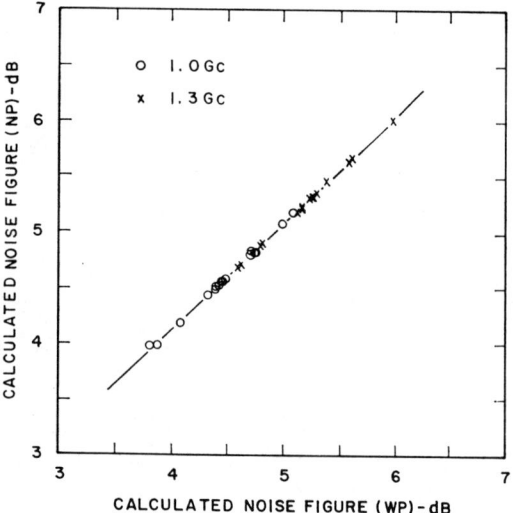

Fig. 11. Comparison of calculated noise figures, with and without parasitics.

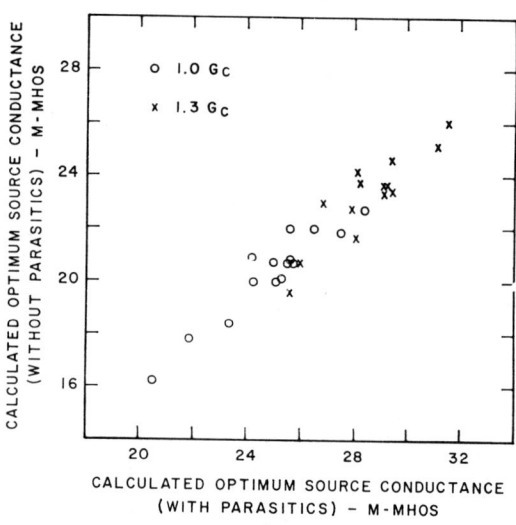

Fig. 14. Comparison of optimum source conductances, with and without parasitics.

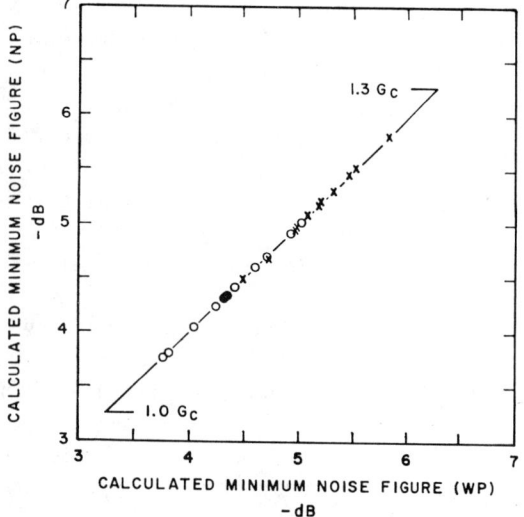

Fig. 12. Comparison of calculated minimum noise figures, with and without parasitics.

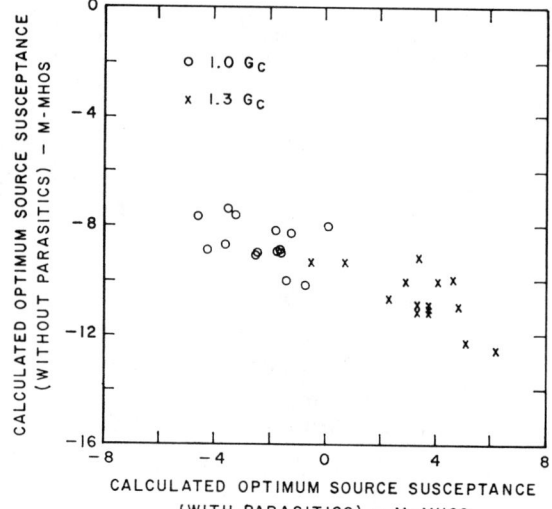

Fig. 15. Comparison of optimum source susceptances, with and without parasitics.

TABLE IV
PREDETERMINED DEVICE PARAMETERS AT $V_{CB} = -5V$

I_E (mA)	β_{DC}	β_0	f_T (Gc/s)	r_b (ohms)	s (ohms-mA)	C_c (pF)
0.5	1.1	2.9	0.76			
0.7	1.4	3.6	1.10			
1	1.8	4.5	1.55			
1.5	2.4	5.8	2.20			
2	2.9	7.0	2.70	28	0	0.20
3	3.7	9.1	3.45			
4	4.5	11.0	3.87			
5	5.2	13.0	4.10			
6	5.8	14.7	4.26			

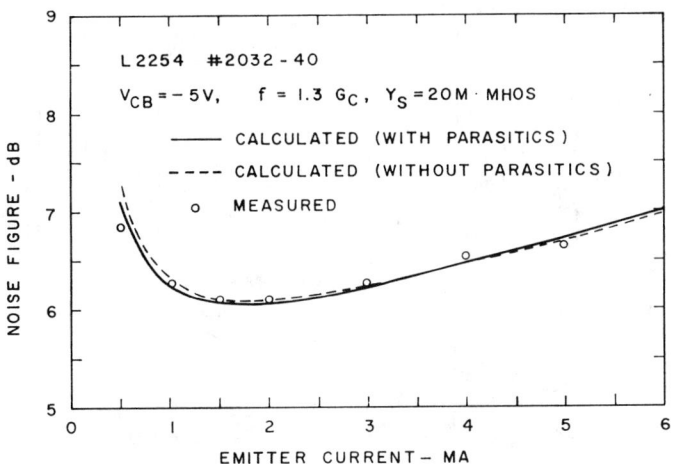

Fig. 16. Calculated and measured current dependence of noise figure.

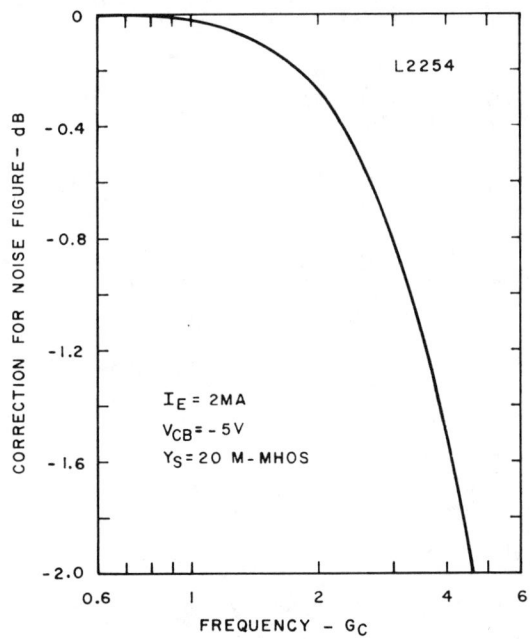

Fig. 17. Calculated correction factor for the noise figure.

D. Feedback Effect Due to Collector-Base Capacitances

In the derivation of the noise figure expressions in Section II, the effects caused by components between collector and base, such as C_c and C_{CB}, have been neglected for the sake of mathematical simplicity. From preliminary work on this problem, it has been found that the above simplification is usually justified at frequencies up to L-band for present well-designed microwave transistors, but that the effect should be taken into consideration at higher frequencies. This effect reduces the noise figure since it provides negative feedback. The noise figure expressions given in Section II are therefore somewhat pessimistic in the central-microwave region. As an example, Fig. 17 shows the calculated correction for the noise figure of a typical L2254 transistor with a 50-ohm source impedance.

E. The Effect of Collector Body Resistance and of Lossy Header Parasitics

Some microwave transistors may have a considerable collector body resistance, due mainly to a high resistivity in their epitaxial layers. Furthermore, all lossy elements associated with the headers have been ignored in the previous analysis. Their effects should be taken into account in a detailed discussion of the noise performance, especially at relatively high frequencies. It is obvious that these effects are deleterious. However, they are probably negligible with current microwave transistors in their useful frequency regions. At higher frequencies they will tend to cancel the reduction of the noise figure due to feedback effect discussed in Section III-D. This may be the case for the M2519 transistors.

IV. CONCLUSIONS

Expressions for the noise parameters of microwave transistors have been derived and analyzed. The theory is based on the small-signal linear operation in the common-emitter configuration and includes a new basic noise equivalent circuit and the dominating header parasitics. The theory was verified experimentally in the L-band range using Ge and Si microwave transistors.

The theoretical and experimental investigation revealed the following important effects caused by header parasitics: 1) The parasitics of good present headers have a negligible effect on the minimum noise figure F_{min} in the low microwave region (up to about 2 Gc/s), but they become important at S-band and above. 2) The header parasitics have strong effects on the equivalent noise resistance R_n, the optimum source conductance G_0, and the optimum source susceptance B_0. In conjunction with wafer parameters, the header parasitics give rise to a series resonance at f_r in the frequency characteristics of R_n, G_0, and B_0. At frequencies below one tenth f_r the header parasitics can be neglected, but above about one-half f_r the effect becomes considerable. In the vicinity of f_r, R_n decreases, G_0 increases, and B_0 changes its sign from negative to positive. On the other hand, for the case of no parasitics R_n, G_0, and B_0 increase monotonically in magnitudes with increasing frequency. At frequencies well above f_r, the values of R_n, G_0, and B_0 with parasitics are quite different from those without parasitics.

Minor parasitic effects due to the collector-base capacitances and lossy parasitic elements were also discussed briefly. As the frequency increases, the former tends to reduce the noise figure, whereas the latter increases the noise figure.

Furthermore, new approximate expressions for the noise figure and the optimum current which produces the lowest value of the high-frequency minimum noise figure were given. These results are useful for a quick evaluation of the noise performance.

For optimum noise performance of microwave transistor amplifiers, not only the noise figure but also the gain of the transistors is important [28]. Thus, the noise measure introduced by Haus and Adler [29] is more meaningful than the noise figure for describing the overall performance of the amplifier, especially when cascaded. For example, the feedback effect due to the collector-base capacitances reduces the noise figure as described above. It also reduces the gain. Therefore, the feedback effect may not improve the overall noise performance of the microwave transistor amplifier. The optimization of the noise measure in microwave transistors is presently being investigated by the author and will be reported later.

Appendix

Derivation of Equations (15), (16), and (18)

Figure 18 shows the noise equivalent circuit for the transistor as given by van der Ziel [3], [6], [12]. It consists of two current generators, i_1 and i_2, connected across the emitter-base and the collector-base junction, respectively. In addition, the basic resistance r_b is assumed to have full thermal noise. Assuming the filling and subsequent emptying of a trap in the emitter-base transition region to be a single random event and neglecting the hole-electron pair generation in the base region, van der Ziel found that

$$\overline{i_1^2} = [2qI_E + 4kT(g_0 - g_{e0})]\Delta f \qquad (61)$$

$$\overline{i_2^2} = 2qI_C\Delta f \qquad (62)$$

$$\overline{i_1^* i_2} = 2kTy_{ce}\Delta f. \qquad (63)$$

Figure 19(a) shows van der Ziel's noise equivalent circuit for the intrinsic transistor rewritten in the common-emitter configuration. Here, y_c in Fig. 18 is assumed to be negligible. Figure 19(b) is the basic portion of the new noise equivalent circuit introduced in this paper. Comparing the two circuits, the following relations can be derived:

$$i_B = i_1 - i_2 \qquad (64)$$

$$i_C = i_2 \qquad (65)$$

$$\overline{i_B^* i_C} = \overline{i_1^* i_2} - \overline{i_2^* i_2}. \qquad (66)$$

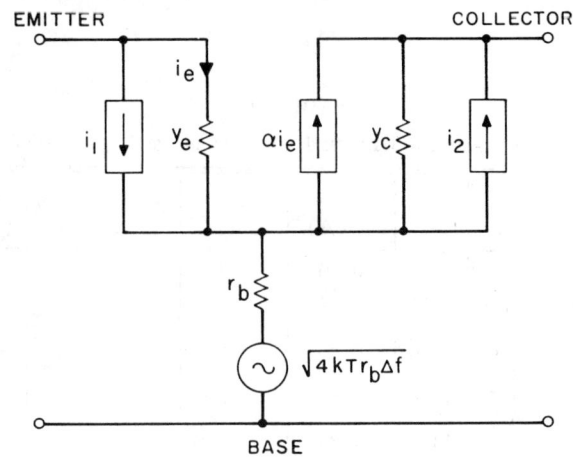

Fig. 18. Van der Ziel's noise equivalent circuit for junction transistors in the common-base configuration.

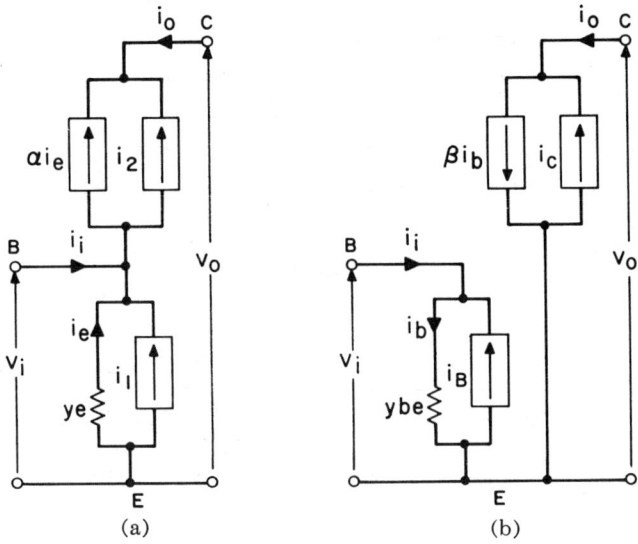

Fig. 19. Comparison of two basic noise equivalent circuits. (a) van der Ziel's circuit. (b) Author's circuit.

Substitution of (61), (62), and (63) into (64), (65), and (66) yields

$$\overline{i_B^2} = 2q(I_E + I_C)\Delta f + 4kT[g_e - g_{e0} - R_e(y_{ce})]\Delta f \qquad (67)$$

$$\overline{i_C^2} = 2qI_C\Delta f \qquad (68)$$

$$\overline{i_B^* i_C} = 2kTy_{ce}\Delta f - 2qI_C\Delta f. \qquad (69)$$

y_e and y_{ce} are given as follows [20]:

$$y_e = g_{e0}\left(1 + jM\frac{\omega}{\omega_1}\right) \qquad (70)$$

and

$$y_{ce} = \frac{\alpha_0 g_{e0}}{1 + jN\frac{\omega}{\omega_1}} \qquad (71)$$

where M and N depend only on a. M ranges from 0.67 for $a = 0$ to about 0.12 for $a = 9$, and N varies from 0.33 for $a = 0$ to 0.87 for $a = 9$. From (70)

$$g_e = g_{e0}. \qquad (72)$$

In the frequency range

$$\omega \ll \frac{\omega_1}{N}. \quad (73)$$

Equation (71) reduces to

$$y_{ce} \simeq \alpha_0 g_{e0}. \quad (74)$$

Applying (10) and (11) in the text to (74) results in

$$y_{ce} \simeq \frac{1}{r_1} = \frac{qI_C}{kT}. \quad (75)$$

Substituting (72) and (75) into (67) and (69) yields

$$\overline{i_B^2} = 2qI_B \Delta f \quad (76)$$

$$\overline{i_B^* i_C} = 0. \quad (77)$$

Since the thermal noise of the base resistance is supposed to be independent on both i_B and i_C, it is usually assumed that

$$\overline{e_B^* i_B} = 0 \quad \text{and} \quad \overline{e_B^* i_C} = 0 \quad (78)$$

where e_B represents the thermal noise voltage.

If the frequency is so high that the assumption of (73) no longer holds, the substitution of (75) into (69) has to be replaced by the original equation (71). Thus, the expression for $\overline{i_B^* i_C}$ is changed as follows:

$$\overline{i_B^* i_C} = -2qI_C \Delta f \frac{\left(N \frac{\omega}{\omega_T}\right)^2 + j\left(N \frac{\omega}{\omega_T}\right)}{1 + \left(N \frac{\omega}{\omega_T}\right)^2}. \quad (79)$$

This cross-correlation term, which is neglected in the text, may tend to partly cancel shot noise components in the output current at the high frequencies mentioned above.

Acknowledgment

The author is indebted to K. D. Bowers and R. S. Engelbrecht for their encouragements during the course of this work. He also wishes to thank Mrs. G. M. Souren and R. D. Tibbetts for making the computer program for Equations (26) through (34), and for carrying out the numerical computation using IBM 7094 computer.

References

[1] R. S. Engelbrecht and K. Kurokawa, "A wide-band low-noise L-band balanced transistor amplifier," *Proc. IEEE*, vol. 53, pp. 237–247, March 1965.
[2] A. van der Ziel, "Noise in junction transistors," *Proc. IRE*, vol. 46, pp. 1019–1038, June 1958.
[3] ——, *Fluctuation Phenomena in Semiconductors*. London: Butterworth, 1959.
[4] W. Guggenbuehl and M. J. O. Strutt, "Theory and experiments of shot noise in semiconductor junction diodes and transistors," *Proc. IRE*, vol. 45, pp. 839–857, June 1957.
[5] E. C. Nielsen, "Behavior of noise figure in junction transistors," *Proc. IRE*, vol. 45, pp. 957–963, July 1957.
[6] G. H. Hanson and A. van der Ziel, "Shot noise in transistors," *Proc. IRE*, vol. 45, pp. 1538–1542, November 1957.
[7] B. Schneider and M. J. O. Strutt, "Theory and experiments on shot noise in silicon P-N junction diodes and transistors," *Proc. IRE*, vol. 47, pp. 546–554, April 1959.
[8] A. van der Ziel, "Shot noise in transistors," *Proc. IRE (Correspondence)*, vol. 48, pp. 114–115, January 1960.
[9] E. R. Chenette, "The influence of inductive source reactance on the noise figure of a junction transistor," *Proc. IRE (Correspondence)*, vol. 47, pp. 448–449, March 1959.
[10] E. R. Chenette, "Frequency dependence of the noise and the current amplification factor of silicon transistors," *Proc. IRE (Correspondence)*, vol. 48, pp. 111–112, January 1960.
[11] B. Schneider and M. J. O. Strutt, "Shot and thermal noise in germanium and silicon transistors at high-level current injections," *Proc. IRE*, vol. 48, pp. 1731–1739, October 1960.
[12] E. R. Chenette and A. van der Ziel, "Accurate noise measurements on transistors," *IRE Trans. on Electron Devices*, vol. ED-9, pp. 123–128, March 1962.
[13] H. Shimomura and Y. Yamakawa, "Noise in transistors at VHF," (in Japanese), *Research on Transistors*, Rept. of the Technical Committee of the Transistor Group of the Institute of Electrical Communication Engineers of Japan, no. 1, pp. 8–13, 1962.
[14] G. E. Hambleton and V. Gelnovatch, "L-band germanium mesa transistors," *Microwave J.*, vol. 8, pp. 42–46 and pp. 67–68, January 1965.
[15] H. Fukui, "Noise performance of microwave transistors," December 1964 (unpublished). This paper is referred to in Thommen [16].
[16] W. F. Thommen, "Beitraege zur Kenntnis des Signal—und Rauscherstazshaltbildes von UHF-Transistoren," Ph.D. dissertation, Eidgenössische Technische Hochschule, Institut für höhere Elektrotechnik, Zurich, 1965.
[17] W. Thommen and M. J. O. Strutt, "Kleinsignal—und Rauschersatzschaltbild von Germanium-UHF-Transistoren bei Kleinen Stromdichten," *A. E. Ü.*, vol. 19, pp. 169–177, April 1965.
[18] C. T. Sah, R. N. Noyce, and W. Shockley, "Carrier generation and recombination in p-n junctions and p-n junction characteristics," *Proc. IRE*, vol. 45, pp. 1228–1243, September 1957.
[19] D. F. Thomas and J. L. Moll, "Junction transistor short-circuit current gain and phase determination," *Proc. IRE*, vol. 46, pp. 1177–1184, June 1958.
[20] J. te Winkel, "Drift transistor," *Electronic and Radio Engr.*, vol. 36, pp. 280–288, August 1959.
[21] A. R. Boothroyd and F. N. Trofimenkoff, "Determination of the physical parameters of transistors of single- and double-diffused structure," *IRE Trans. on Electron Devices*, vol. ED-10, pp. 149–163, May 1963.
[22] L. J. Giacoletto, *Noise Factor of Junction Transistors, Transistors I*. Princeton, N. J.: RCA Labs., pp. 296–308, 1956.
[23] E. R. Chenette, "Low-noise transistor amplifiers," *Solid-State Design*, vol. 5, pp. 27–30, February 1964.
[24] H. Rothe and W. Dahlke, "Theory of noisy fourpoles," *Proc. IRE*, vol. 44, pp. 811–818, June 1956.
[25] H. Fukui, "On the base resistance measurement for microwave transistors," *IEEE Trans. on Electron Devices*, to be published.
[26] IRE Standards on Electron Tubes, pt. 9, Noise in linear twoports, 1962.
[27] H. Fukui, "Noise measurement of transistors at microwave frequencies," May 1964 (unpublished).
[28] H. Fukui, "Available power gain, noise figure, and noise measure of twoports and their graphical representations," *IEEE Trans. on Circuit Theory*, to be published.
[29] H. A. Haus and R. B. Adler, *Circuit Theory of Linear Noisy Networks*. Cambridge, Mass.: M. I. T. Technology Press, 1959, and New York: Wiley, 1959.

Noise Measure of Microwave Transistors

In recent years, considerable improvements in the manufacturing processes of high frequency transistors have made it possible to use them in the microwave region. The gain obtained with these units in single-stage amplifiers, however, is often not sufficient for practical applications. A number of cascaded stages is therefore generally required. The overall noise performance of a multistage amplifier is thus a function of both the gain and the noise figure of the individual stages. In the case where all stages have identical available gains G and identical noise figures F, the overall noise performance improves directly with the noise measure M of the individual stages, which is defined as

$$M = \frac{F-1}{1-\frac{1}{G}} \quad (1)$$

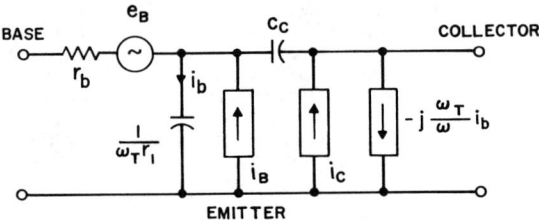

Fig. 1. An equivalent circuit of unencapsulated transistors in their high-frequency region.

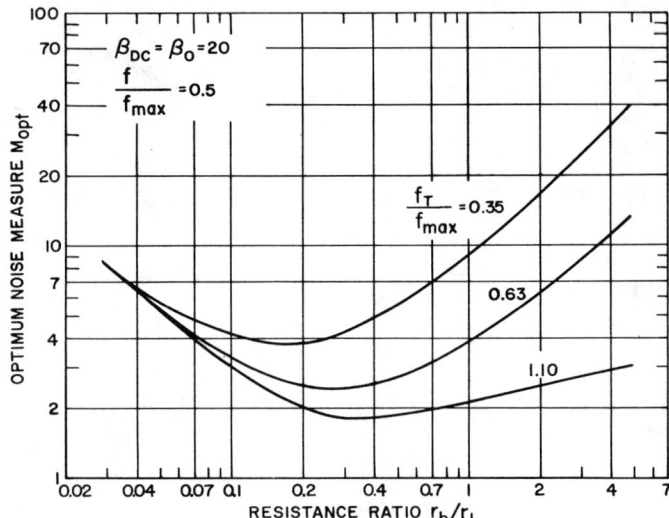

Fig. 3. M_{opt} as a function of resistance ratio r_b/r_1.

by Haus and Adler [1]. Thus, the lowest value of M, M_{opt}, of microwave transistors is a good indication of their inherent capabilities as low-noise amplifiers, and is very useful for their comparison [2]. The purpose of this letter is to give a theoretical expression for M_{opt} in terms of device parameters, comparing it with experiments, and to show some calculated results.

M_{opt} is an invariant for reciprocal lossless transformations [1]; therefore, if it can be assumed that only reactive lossless parasitics are associated with an unencapsulated transistor, M_{opt} must be independent of the header parasitics and can be determined from an equivalent circuit for the unencapsulated transistor. On the other hand, the optimum source admittance which produces M_{opt} would be affected by the parasitics.

Let I_C be the dc collector current, β_{DC} the dc current gain, β_0 the low-frequency current gain, ω_T the angular total unity gain frequency, r_b the total base resistance, C_c the total collector capacitance [3], and let $r_1 = (kT/qI_C)$. Microwave transistors operating in the common-emitter configuration at frequencies where their G values are relatively low but still greater than unity can be represented by an equivalent circuit shown in Fig. 1. It includes three noise generators given by

$$\overline{i_B^2} = 2qI_B\Delta f \quad (2)$$

$$\overline{i_C^2} = 2qI_C\Delta f \quad (3)$$

$$\overline{e_B^2} = 4kTr_b\Delta f \quad (4)$$

where Δf is a narrow frequency interval. These generators are uncorrelated with one another [2]. By calculating both the noise and the gain parameters previously introduced by the author, an approximate expression for M_{opt} can be given in the form [4]

$$M_{opt} = \frac{M_2}{M_1}\left[1 + \sqrt{1 - \frac{M_1 M_3}{M_2^2}}\right] \quad (5)$$

where

$$M_1 \approx \left[1 - \left(\frac{f}{f_{max}}\right)^2\right]^2 \quad (6)$$

$$M_2 \approx \frac{1}{\beta_0} + \frac{1}{\beta_{DC}}\frac{r_b}{r_1} + \frac{1}{4}\left(2 + \frac{r_1}{r_b} + \frac{1}{\omega_T C_c r_1}\right)\left(\frac{f}{f_{max}}\right)^2 \quad (7)$$

$$M_3 \approx -\left[\frac{1}{\beta_{DC}}\left(1 + 2\frac{r_b}{r_1}\right) + \frac{1}{2\omega_T C_c r_1}\left(\frac{f}{f_{max}}\right)^2\right] \quad (8)$$

$$f_{max} = \left[\frac{f_T}{8\pi C_c r_b}\right]^{1/2} = f\sqrt{G_{max}}. \quad (9)$$

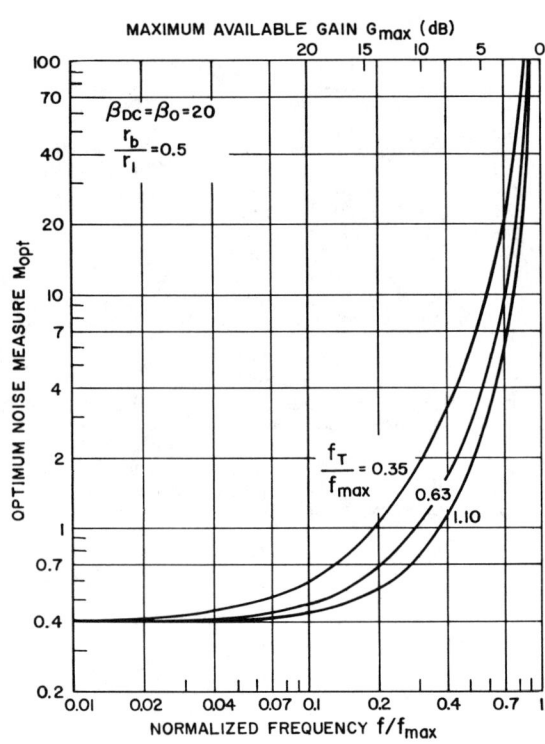

Fig. 2. M_{opt} as a function of frequency.

Reprinted from *Proc. IEEE*, vol. 54, pp. 1204–1205, Sept. 1966.

In order to compare the above expression with experimental results, a planar Si microwave transistor M2519 has been employed. The device parameters were as follows: $\beta_{DC}=17$, $\beta_0=18.4$, $f_T=1.0$ GHz, $C_c=0.54$ pF, and $r_b=9.2$ ohms, all at $V_{CB}=5$ V and $I_C=1$ mA. Substituting these values into (5) yields $M_{opt}=3.56$ at 1.3 GHz, which compares with the experimental value of 3.3 determined from both the noise figure and the available power gain measurements with a number of source admittances [4]. This is a fairly good agreement.

Figures 2 and 3 show calculated M_{opt} as a function of frequency and of resistance ratio r_b/r_1, respectively, for three transistors having average β_{DC} and β_0 and different $f_T/f_{max}(=2\sqrt{\omega_T C_c r_b})$ values. It can be seen in Fig. 2 that M_{opt} increases rather rapidly above about $f_{max}/3$. Since r_1 is inversely proportional to I_C, the curves in Fig. 3 roughly indicate the bias current dependence of M_{opt}. From Figs. 2 and 3, it can be deduced that higher f_T, smaller C_c, and lower r_b can result in lower M_{opt}, provided no restriction is applied to all the individual parameters. On the other hand, a larger $f_T C_c r_b$ product gives better noise performance, provided the gain is unchanged. Furthermore, M_{opt} in the lower frequency region tends to $(F_{min}-1)$, where F_{min} is the minimum noise figure [2].

H. Fukui
Bell Telephone Laboratories, Inc.
Murray Hill, N. J.

References

[1] H. A. Haus and R. B. Adler, *Circuit Theory of Linear Noisy Networks*. Cambridge, Mass.: M.I.T. Technology Press, 1959; and New York: Wiley, 1959.
[2] H. Fukui, "The noise performance of microwave transistors," *IEEE Trans. on Electron Devices*, vol. ED-13, pp. 329–341, March 1966.
[3] H. Fukui, "On the base resistance measurement for microwave transistors," *IEEE Trans. on Electron Devices*, to be published.
[4] H. Fukui, "Available power gain, noise figure, and noise measure of two-ports and their graphical representations," *IEEE Trans. on Circuit Theory*, vol. CT-13, pp. 137–142, June 1966.

A SIMPLIFIED APPROACH TO NOISE IN MICROWAVE TRANSISTORS*

S. D. MALAVIYA† and A. VAN DER ZIEL

Department of Electrical Engineering, University of Minnesota, Minneapolis, Minnesota 55455, U.S.A.

(*Received* 19 *February* 1970; *in revised form* 23 *April* 1970)

Abstract—The theory of noise in transistors is applied to microwave transistors at microwave frequencies and an expression for the noise figure is obtained in terms of easily measurable parameters of the transistor. By making some simplifying assumptions, the expression for the noise figure can be reduced to a simple form containing only one parameter, which is useful for engineering applications. The effects of feedback and parasitics introduced by the encapsulation are taken into account. Friiss' formula is modified in order to take the effects of image frequency response on the measurements into account. A method of measuring gain and noise figure simultaneously is also described. Experimental verification of the theory up to 4 GHz is presented.

Résumé—La théorie du bruit dans les transistors est appliquée aux transistors à micro-ondes à des fréquences de micro-ondes et une expression est obtenue pour la valeur du bruit en fonction de paramètres facilement mesurables du transistor. En faisant quelques suppositions de simplifications, l'expression pour la valeur du bruit peut être réduite à une forme simple ne contenant qu'un paramètre, ce qui est utile pour les applications d'engineering. Les effets de rétroaction et de parasites introduits par encapsulation sont pris en considération. La formule de·Friiss est modifiée afin de prendre en considération les effets de la réponse de fréquence d'image sur les mesures. Une méthode pour mesurer le gain et la valeur du bruit simultanément est également décrite. On présente une vérification expérimentale de la théorie jusqu'à 4 GHz.

Zusammenfassung—Die Rauschtheorie für Transistoren im Mikrowellenbereich wird auf Mikrowellentransistoren angewendet und ein Ausdruck für den Rauschfaktor als Funktion einfach messbarer Transistorparameter erhalten. Mit einigen vereinfachenden Annahmen kann dieser Rauschfaktor auf eine Form gebracht werden, wo er nur noch einen Parameter enthält. Diese ist für den Anwendungsingenieur sehr nützlich. Rückwirkungseinflüsse und parasitäre Schaltelemente, die durch das Gehäuse verursacht werden, sind berücksichtigt. Die Friis'sche Formel wird so modifiziert, dass Einflüsse der Bildfrequenz auf die Messung enthalten sind. Eine Methode zur gleichzeitigen Bestimmung der Verstärkung und des Rauschfaktors wird beschrieben. Die experimentelle Bestätigung für die Theorie bis zu 4 GHz herauf wird aufgezeigt.

1. INTRODUCTION

THE THEORY of noise in transistors at low and high frequencies is well developed and experimental verifications by several workers have shown very good agreements between the theoretical predictions and actual results.[1-11] With the advent of microwave transistors, extensions of the theory to cover the microwave range have been proposed by several workers.[12-18] For the theory of transistor noise see Refs. (19)–(24).

The theoretical expression for noise figure is generally quite complicated because of the effects of feedback and the parasitic elements introduced by the encapsulation. This paper presents an approach whereby a relatively simple expression for the noise figure is obtained which is quite accurate for engineering applications and is expressed in terms of easily measurable parameters of the transistor. The use of the theory in the measurement of base resistance and alpha cut off

---

* Supported by U.S. Army Electronics Command.
† Now at I.B.M. Corporation, Hopewell Junction, New York 12533.

frequency is discussed and the results of a few practical measurements are presented. It is also shown that Friiss' formula has to be modified when image frequency effects are present and the necessary theoretical expressions are derived. Experimental verification of the new formula is also presented. A method whereby the noise figure and gain can be measured simultaneously in some cases is briefly described. It is concluded that the simplified formula is useful and accurate enough for practical purposes.

2. EVALUATION OF THE NOISE FIGURE OF GROUNDED EMITTER TRANSISTOR CIRCUITS

According to VAN DER ZIEL[1] the noise of transistors in common base configuration can be represented by two current generators i_1 and i_2, in parallel with the emitter and collector junctions, respectively, with the polarity as shown in Fig. 1a. If small saturation currents are neglected he finds

$$\overline{i_1^2} = 2qI_E\Delta f + 4kT(g_{eb}-g_{eb0})\Delta f \quad (1)$$

$$\overline{i_2^2} = 2qI_C\Delta f \quad (2)$$

$$\overline{i_1^* i_2} = 2kTY_{ce}\Delta f = 2kT\alpha Y_{eb}\Delta f. \quad (3)$$

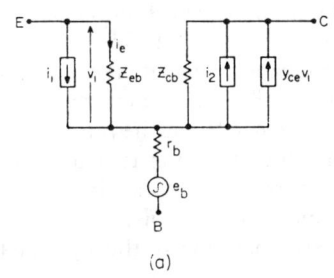

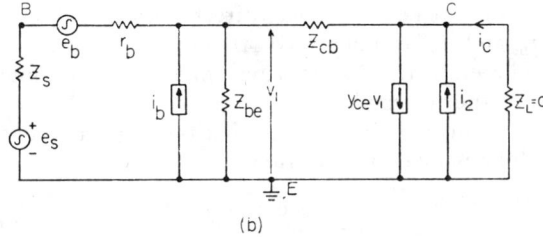

Fig. 1. (a) Equivalent circuit of grounded base transistor.
(b) Equivalent circuit of grounded emitter transistor.

Here Y_{ce} is the (complex) transconductance, $Y_{eb} = g_{eb}+jb_{eb}$ the emitter admittance, g_{eb0} the low frequency value of g_{eb} and α the current amplification factor. The thermal noise of the base resistance r_b is represented by an emf e_b, where

$$\overline{e_b^2} = 4kTr_b\Delta f. \quad (4)$$

Since microwave transistors are generally used in the more stable common emitter configuration, the equivalent circuit is transformed accordingly. The result is shown in Fig. 1b. Here i_2 has the same meaning as before,

$$i_b = i_1 - i_2 \quad (5)$$

and

$$Y_{be} = Y_{eb} - Y_{ce} = Y_{eb}(1-\alpha). \quad (6)$$

In addition, the feedback admittance $Y_{cb} = j\omega C_{cb}$ is connected between collector and base. To complete the circuit used for noise figure calculations, a noise emf $e_s = \sqrt{(4kTR_s\Delta f)}$ in series with an impedance $Z_s = R_s + jX_s$ is added and the output is short-circuited. Calculating the mean square value of short-circuited current i_c and dividing by the contribution of the noise emf e_s, yields for the noise figure F

$$F = 1 + \frac{r_b}{R_s} + \frac{1}{4kTR_s\Delta f} \\ + \overline{\left| i_b(Z_s+r_b) + \frac{i_2(Z_s+r_b+Z_{be}')}{Z_{be}'(Y_{ce}-j\omega C_{cb})} \right|^2} \quad (7)$$

where

$$1/Z_{be}' = Y_{be} + j\omega C_{cb} \quad (7a)$$

Substituting $i_b = i_1 - i_2$ and introducing the quantities

$$e = i_1 Z_{eb}; \quad i = i_2 - \alpha' i_1; \quad \alpha' = \alpha - j\omega C_{cb} Z_{eb} \quad (8)$$

yields after some mathematical manipulations

$$F = 1 + \frac{r_b}{R_s} + \frac{\overline{\left|e' + e'' + \frac{i}{\alpha'}(Z_s + r_b + Z_{eb})\right|^2}}{4kTR_s\Delta f}. \quad (9)$$

We have here split e into a part e' fully correlated with i and a part e'' uncorrelated with i.

We now introduce the noise parameters R_n', Z_{cor}' and g_n' by

$$\overline{e''^2} = 4kTR_n'\Delta f \qquad (10)$$

$$\frac{\overline{\alpha'e'}}{i} = Z_{cor}' = R_{cor}' + jX_{cor}' \qquad (11)$$

$$\frac{\overline{i^2}}{|\alpha'|^2} = 4kTg_n'\Delta f. \qquad (12)$$

This yields

$$F = 1 + \frac{r_b + R_n'}{R_s} + \frac{g_n'}{R_s}|Z_s + r_b + Z_{eb} + Z_{cor}'|^2 \qquad (13)$$

The noise figure can be minimized with respect to the source reactance X_s by making

$$X_s + X_{eb} + X_{cor}' = 0. \qquad (14)$$

The noise figure then becomes

$$F = 1 + \frac{r_b + R_n'}{R_s} + \frac{g_n'}{R_s}(R_s + r_b + R_{eb} + R_{cor}')^2. \qquad (15)$$

Equation (15) has the minimum value

$$F_{min} = 1 + 2g_n'(r_b + R_{eb} + R_{cor}') \\ + 2[g_n'(r_b + R_n') \\ + g_n'^2(r_b + R_{eb} + R_{cor}')^2]^{1/2} \qquad (16)$$

for

$$R_s = \left(\frac{r_b + R_n'}{g_n'} + (r_b + R_{eb} + R_{cor}')^2\right)^{1/2}. \qquad (17)$$

For low frequencies

$$\frac{r_b + R_n'}{g_n'} \gg (r_b + R_{eb} + R_{cor}')^2 \qquad (18)$$

whereas for high frequencies the reverse may be true.

The minimum noise figure is hereby expressed in terms of the parameters R_n', g_n' and R_{cor}', that are evaluated in appendix 1. It is shown there that $R_{eb} + R_{cor}'$ and R_n' are relatively small at high frequencies so that they can be neglected, equation (16) is then simplified as

$$F_{min} = 1 + u + (2u + u^2)^{1/2} \qquad (19)$$

where

$$u = [(1 - \alpha_0 + f^2 f_\alpha^2)\alpha_0](r_b R_{eb0}) \qquad (20)$$

so that F_{min} can be expressed in terms of the single parameter u. This approximation is already mentioned by FUKUI.[16]

3. HEADER PARASITICS AND DEVICE PARAMETERS

The discussion so far relates to the basic transistor. But it is well known that the series inductances and shunt capacitances, generally referred to as header parasitics, which are introduced at the input and output terminals of the transistor by its encapsulation, play an important role in the behavior of the transistor at microwave frequencies. In some cases[26] the circuit is so designed that the input and output parasitics can be represented by short lengths of 50 Ω transmission lines.

While it is possible to measure the circuit elements of these parasitics, we wish to point out here that such measurements are usually not needed in order to interpret the measured values of F_{min}. For it may be assumed for most of the frequency range of a microwave transistor that the input and output parasitics act as lossless transformers; only at the highest frequencies can it happen that they act as attenuators.* It is well known, however, that lossless transformers do not alter the minimum noise figure of the device, even though they change the impedance needed at the input terminals of the transformer for obtaining that minimum in noise figure. Of course, if the parasitics act as attenuators, they will increase the noise figure of the device. We shall see that this is the case at the highest frequencies.

The parasitics make direct measurements of r_b and f_α at microwave frequencies impossible. Therefore these parameters must be determined by other methods.

It is generally assumed that the high frequency alpha is given the relation

$$\alpha = \frac{\alpha_0 \exp(-j\omega\tau)}{1 + jf/f_\alpha} \qquad (21)$$

---
* It may also be shown that the parasitics of the emitter lead and the feedback parasitics hardly affect the value of F_{min}.

where τ is a time constant associated with the base region. Though it is sometimes held that this expression is not valid at high injection and in the presence of drift fields,[27] one may assume that there exists an effective f_α which gives a good approximation over a limited range of frequencies.

Noise figure measurements only require knowledge* of $|\alpha|^2$, so that the value of τ does not affect the noise data. However, if one measures f_α by the input impedance method,[28] the term $\exp(-j\omega\tau)$ has an effect; for that reason the method may not give meaningful values for f_α.

The base resistance can be measured by various methods.[16,28,29] They generally lead to inconsistent results; the power gain method, used at frequencies where the effect of parasitics is still negligible, seems to be the most reliable.

r_b/R_{eb0} and f_α can also be determined from noise figure data. For according to equation (19) the measured value of $F_{\min}$ gives u; if one knows the values of u at two frequencies one can determine r_b/R_{eb0} and f_α. As will be seen in Section 5, good agreement between the theoretical and experimental values of $F_{\min}$ can be obtained over a wide frequency range with the help of this method.

4. EXPERIMENTAL PROCEDURE

Measurements were conducted on microwave transistors up to 4 GHz in order to verify the validity of the theory of noise presented in Section 2. The noise figure was measured with the H.P. Noise Figure Meter type 342 A.

When trying to determine the narrow band noise figure of a microwave amplifier stage with the help of a wideband noise source, one runs into difficulties caused by image response. It is shown in appendix 2 if the receiver has been so adjusted that it has the same power gain for the operating frequency and the image frequency, if F_1 is the narrow band noise figure of the stage, and if F and F_a are the wideband noise figures of amplifier plus receiver and of the receiver alone, respectively, then

$$F = F_1 + \frac{F_a - 1}{(G_1 + G_1')/2} \quad (22)$$

where G_1 and G_1' are available gains of the amplifier stage at the operating and the image

---
* See expressions for g_n' in the appendix.

frequencies, respectively. That is, Friiss' formula holds provided that the power gain is replaced by the average gain $(G_1 + G_1')/2$. Hence by measuring F, F_a, and G_1 and G_1' one can determine F_1.

This procedure was followed in our measurements. To test the validity of equation (22) the tuning stubs of the preamplifier were adjusted to obtain minimum noise figure and its gains G_1 and G_1' were recorded. The output stubs were then readjusted to increase G_1 and decrease G_1' while the input stubs were tuned for minimum noise figure. By repeating the process, several combinations of $G_1 + G_1'$ and the corresponding noise figure F were obtained for the same collector voltage and current. In fact it was possible to go the extremes where either G_1 or G_1' could be reduced to almost zero. Application of equation (22) to these results gave a very consistent value of F_1. It is obvious that ignoring G_1' in such a case would have given widely differing values of F_1.

The measurement of noise in a transistor generally requires that the circuit be adjusted for minimum noise figure and that its gain be recorded without disturbing the circuit. Usually the noise is measured with the calibrating signal turned off to avoid interference and the gain of the system is then reduced to accommodate the relatively large calibrating signal. The output power is measured with and without the transistor in the circuit and their ratio gives the insertion gain of the device. Unless proper precautions are taken, the circuit conditions can easily be disturbed during these changes and the measured gain would then be different from the gain which existed at the time of the noise measurement. The situation becomes more critical at higher frequencies. Also, the change over is time consuming and it is therefore desirable to use a set up where both the noise and the gain are measured simultaneously. Such a system is shown in Fig. 2.

Referring to the figure, signal generator A is set to a frequency of, say, 1.06 GHz. The local oscillator B is then set at 1.00 GHz. The mixer output is fed to two I.F. amplifiers, one is tuned to 30 MHz and is connected to the noise figure meter while the other is tuned to 60 HMz and is connected to a power meter. With this arrangement, the noise meter responds to noise at 1.00 ± 0.03 GHz whereas the power meter responds to the 1.06 GHz

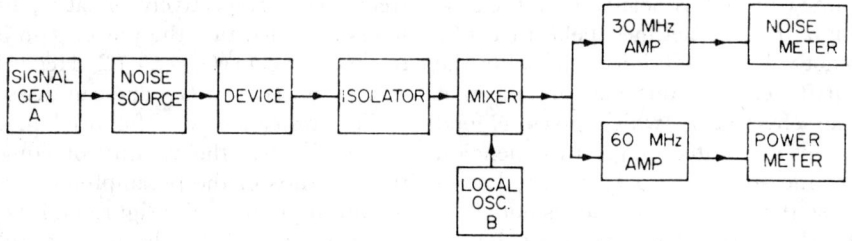

FIG. 2. Circuit for the simultaneous measurement of gain and noise figure.

input signal. Harmonics of the 60 MHz signal cannot interfere with the noise measurement.

The system therefore measures noise and gain simultaneously. The method fails if the gain at 1.06 GHz is significantly different from the gain of the system at 0.97 and 1.03 GHz. This could happen if the circuit Q is high.

5. EXPERIMENTAL RESULTS

In order to compare the experimental values of the noise figure with the theoretical ones, one must know the device parameters $\alpha_0, r_b/R_{eb0}$ and f_α. α_0 was determined accurately by simple measurements at 1 KHz. The base resistance was measured by the input impedance method[28] over a frequency range of 0.5–100 MHz and by the power gain method[29] at 400 MHz; the latter method is more reliable. The measurements did not agree well, as expected. The alpha cut-off frequency f_α was also measured by the input impedance method.

When the values of r_b/R_{eb0} and f_α, thus obtained were substituted in the expression for F_{min}, discrepancies between theory and experiment were noted. But when the values of r_b/R_{eb0} and f_α were deduced from the measurement of F_{min},

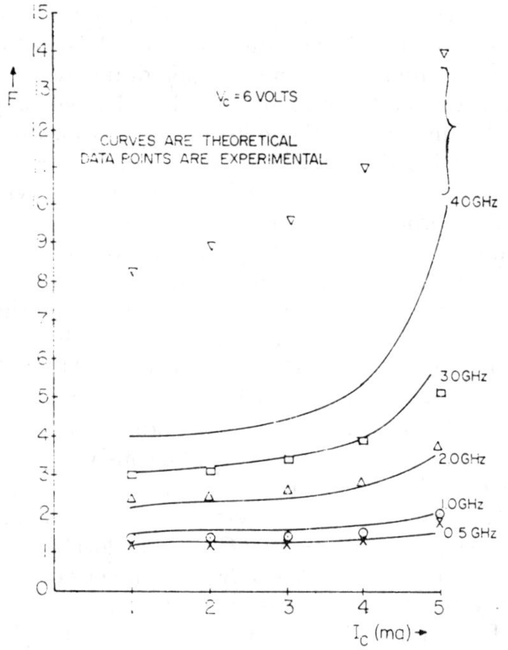

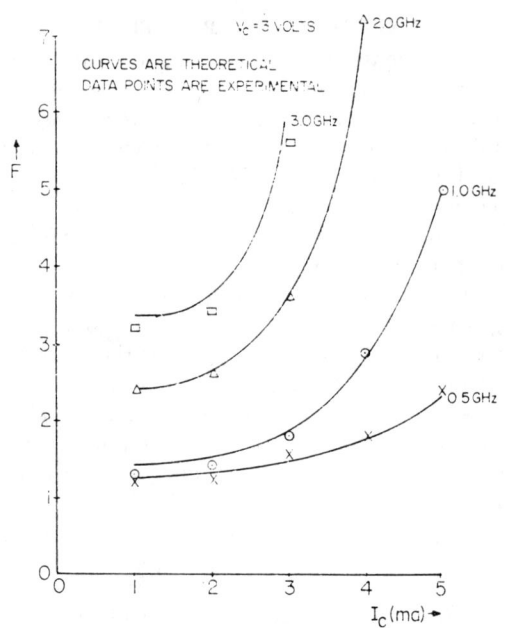

FIG. 3. F_{min} of transistor No. 8 as a function of the collector current at $V_c = 6$ V, at 0.5 GHz, 1.0 GHz, 2.0 GHz, 3.0 GHz and 4.0 GHz; fully drawn curves theory, triangles etc., experiment.

FIG. 4. F_{min} of transistor No. 8 as a function of the collector current at $V_c = 3$ V; at 0.5 GHz, 1.0 GHz, 2.0 GHz and 3.0 GHz.

excellent agreement between theory and experiment could be obtained from 0.5 to 3.0 GHz, whereas a significant discrepancy was noted at 4 GHz (compare Fig. 3 and 4). The noise measurements thus give the most reliable values of the two parameters in question. The good agreement up to 3.0 GHz indicates that the parasitics do not affect $F_{\min}$ up to that frequency.

The discrepancy at 4 GHz is attributed to the fact that at the highest frequencies the header parasitics cease to operate as lossless transducers, but act as attenuators instead. The power gain of the transistor fell rapidly above 3 GHz and was far below unity at 4 GHz, so that the useful frequency range of the transistor did not extend to 4 GHz.

6. CONCLUSIONS

The simplified theoretical expression for noise figure in a microwave transistor presented here is suitable for engineering applications because it involves the knowledge of only a few small-signal parameters of the transistor and the header parasitics need not be evaluated for the purpose. The theory has been verified to be valid over the useful operating range of the transistor and suitable for determining its base resistance and alpha cut off frequencies—the two most important parameters of the microwave transistor. It has also been shown that when Friiss' formula is used in the presence of image frequency response, the average gain at the signal and image frequencies should be used in place of the gain at the signal frequency alone.

Acknowledgment—The experimental microwave transistors were supplied by Texas Instruments Inc., and the authors wish to express their thanks for this essential donation.

REFERENCES

1. A. VAN DER ZIEL and A. G. TH. BECKING, *Proc. IRE* **46**, 589 (1958).
2. A. VAN DER ZIEL, *Proc. Inst. Radio Engrs.* **46**, 1019 (1958).
3. B. SCHNEIDER and M. J. O. STRUTT, *Proc. Inst. Radio Engrs.* **47**, 546 (1959).
4. A. VAN DER ZIEL and E. R. CHENETTE, *Trans. Inst Radio Engrs.* **ED-9**, 123 (1962).
5. J. M. STEWART, *Proc. IEE London* **106B**, 1056 (1959).
6. A. BAELDE, Ph.D. Thesis, Technical University of Delft, The Netherlands (1964).
7. E. R. CHENETTE, *Proc. Inst. Radio Engrs.* **48**, 111 (1960).
8. G. H. HANSON and A. VAN DER ZIEL, *Proc. Inst. Radio Engrs.* **45**, 1538 (1957).
9. W. GUGGENBUEHL and M. J. O. STRUTT, *Proc. Inst. Radio Engrs.* **45**, 839 (1957).
10. E. G. NIELSEN, *Proc. Inst. Radio Engrs.* **45**, 957 (1957).
11. A. VAN DER ZIEL, *Proc. Inst. Radio Engrs.* **48**, 114 (1960).
12. G. J. POLICKY and H. F. COOKE, Northeast Electronics Research Engineering Meeting, Boston (1965).
13. H. F. COOKE, Symposium on Fluctuations in Solids, University of Minnesota, June (1966).
14. H. F. COOKE, *Solid St. Design* **4**, (2), (1963).
15. H. FUKUI, *Trans. IEEE* **ED-13**, 329 (1966).
16. H. FUKUI, *Trans. IEEE* **CT-13**, 137 (1966).
17. H. FUKUI, *Proc. IEEE* **54**, 1204 (1966).
18. W. THOMMEN and M. J. O. STRUTT, *Trans. IEEE* **ED-12**, 499 (1965).
19. A. VAN DER ZIEL, *Trans. Inst. Radio Engrs.* **ED-8**, 525 (1961).
20. A. VAN DER ZIEL, *Proc. Inst. Radio Engrs.* **43**, 1639 (1955).
21. A. VAN DER ZIEL, *Proc. Inst. Radio Engrs.* **45**, 1011 (1957).
22. A. VAN DER ZIEL, *Fluctuation Phenomena in Semiconductors*. Butterworth, London (1959).
23. A. UHLIR, *Proc. Inst. Radio Engrs.* **44**, 557 (1956); Erratum **44**, 1591 (1956).
24. K. S. CHAMPLIN, *Trans. Inst. Radio Engrs.* **ED-7**, 29 (1960).
25. D. C. AGOURIDIS, Ph.D. Thesis, University of Minnesota (1965).
26. T. I.-Line Package, Texas Instruments.
27. J. LINDMAYER and C. WRIGLEY, *Solid-St. Electron.* **2**, 247 (1961).
28. S. D. MALAVIYA, Ph.D. Thesis, University of Minnesota (1969).
29. J. LINDMAYER, *Solid-St. Electron.* **5**, 171 (1962).

APPENDIX 1: EVALUATION OF THE TRANSISTOR PARAMETERS

1. *Evaluation of* $|\alpha'|^2$

According to the definition (8)

$$\alpha' = \alpha - j\omega C_{cb}(R_{eb} - jX_{eb}). \qquad (A.1)$$

Usually $\omega C_{cb} R_{eb}$ and $\omega C_{cb} X_{eb}$ are small for all frequencies of practical interest, so that in good approximation

$$\alpha' \simeq \alpha. \qquad (A.2)$$

But the high-frequency alpha of the transistor can be expressed as

$$\alpha = \frac{\alpha_0 e^{-j\omega\tau}}{1 + jf/f_\alpha}. \qquad (A.3)$$

Hence

$$|\alpha'|^2 \simeq \frac{\alpha_0^2}{1 + f^2/f_\alpha^2}. \qquad (A.4)$$

2. Evaluation of g_n'

We have from the definitions (8) and (12)

$$4kTg_n'\Delta f = \frac{\overline{i_2^2} + |\alpha'|^2\overline{i_1^2} - 2\text{Re}(\overline{\alpha'^* i_1^* i_2})}{|\alpha'|^2}$$

so that

$$g_n' = \frac{\alpha_0}{2|\alpha'|^2}\left\{g_{eb0} + g_{eb} - \frac{1}{2}g_{eb0} - \frac{\text{Re}[(\alpha^* + j\omega C_{cb}Z^*_{eb})\alpha Y_{eb}]}{|\alpha'|^2}\right\}. \quad (A.5)$$

Usually $\omega C_{cb}Z_{eb}^*$ is a small quantity that can be neglected, whereas $|\alpha'|^2$ can be approximated by equation (A.4); this yields

$$g_n' = \frac{1}{2}\frac{g_{eb0}}{\alpha_0}(1 - \alpha_0 + f^2/f_\alpha^2). \quad (A.6)$$

3. Evaluation of Z_{cor}'

The definition (11) can be rewritten as follows

$$Z_{cor}' = \frac{\overline{\alpha'e'i^*}}{\overline{ii^*}} = \frac{\overline{\alpha'e'i^*}}{\overline{ii^*}} = \frac{\overline{\alpha'\alpha'^*e'i^*}}{\overline{\alpha'^*ii^*}}$$

$$= \frac{1}{4kTg_n\Delta f}\frac{\overline{ei^*}}{\alpha'^*} = \frac{Z_{eb}}{4kTg_n\Delta f}\frac{\overline{i_1 i^*}}{\alpha'^*} \quad (A.7)$$

because $\overline{e''i} = 0$.

$$\overline{i_1 i^*} = \overline{i_1(i_2^* - \alpha'^* i_1^*)} = \overline{i_1 i_2^*} - \alpha'^*\overline{i_1^2} \quad (A.8)$$

and

$$\overline{i_1^2} = 2kT\Delta f(2g_{eb} - g_{eb0}) \simeq 2kTg_{be0}\Delta f \quad (A.8a)$$

whereas since $\alpha \simeq \alpha'$

$$\overline{i_1 i_2^*} = 2kT\alpha^* Y_{eb}\Delta f \simeq 2kT\alpha'^* Y_{eb}\Delta f \quad (A.9)$$

so that, if we put $Y_{eb} = g_{eb0} + j\omega C_{eb}$

$$Z_{cor}' = -\frac{j\omega Z_{eb}C_{cb}}{2g_n'} = R_{cor}' + j\omega C_{cor}'. \quad (A.10)$$

Substituting for Z_{cb} yields

$$R_{cor}' = -\frac{\omega^2 C_{eb}^2}{g_{eb0}^2 + \omega^2 C_{eb}^2}\cdot\frac{1}{2g_n'}. \quad (A.10a)$$

We see that this is negative and increases with increasing frequency. At high frequencies it may approach $-R_{eb0}$. Consequently at high frequencies $R_{eb} + R_{cor}'$ can be neglected.

4. Evaluation of R_n

From the definition

$$4kTR_n'\Delta f = \overline{e''^2} = \overline{e^2} - \overline{e'^2} \quad (A.11)$$

and, according to the definition (11)

$$\overline{e'^2} = \frac{\overline{i^2}}{|\alpha|^2}|Z_{cor}'|^2 = 4kTg_n'|Z_{cor}'|^2\Delta f \quad (A.12)$$

and

$$\overline{e^2} = \overline{i_1^2}|Z_{eb}|^2.$$

We thus have

$$R_n' = |Z_{eb}|^2(g_{eb} - \frac{1}{2}g_{eb0}) - g_n'|Z_{cor}|^2. \quad (A.13)$$

Using the approximations

$$|Z_{eb}| \simeq R_{eb0}; \qquad g_{eb} \simeq g_{eb0} = 1/R_{be0}$$

this may be written

$$R_n' \simeq \frac{1}{2}R_{eb0} - g_n'|Z_{cor}'|^2. \quad (A.13a)$$

R_n' is thus equal to $\tfrac{1}{2}R_{eb0}$ at low frequencies, and will decrease at higher frequencies, though it can never become negative. It may therefore be neglected at higher frequencies.

APPENDIX II: EFFECT OF IMAGE RESPONSE

In microwave noise measurements a complication arises because the stage under test, the preamplifier, if any, and the mixer not only respond to the signal frequency $f_i = f_p - f_0$ but also to the image frequency $f_i' = f_p + f_0$. We shall now investigate how this alters the measurements.

Let the amplifier (preamplifier+mixer+i.f. amplifier) have available power gains for G_a and G_a' at the signal and the image frequency, respectively. The thermal noise of the source at the input of the preamplifier thus gives a contribution $(G_a + G_a')kT\Delta f$ to the available output noise power of the amplifier. If the noise figure F_a of the amplifier is defined as

$$F_a = \frac{\text{total output noise power}}{\text{output noise power due to source}} \quad (A.14)$$

then this is indeed the noise figure measured for a wideband noise source. The available output noise power of amplifier+source is therefore $F_a(G_a + G_a')kT\Delta f$, of which the part $(F_a - 1)(G_a + G_a')kT\Delta f$ comes from the amplifier itself.

Let now the stage under test be added, and let it have available gains G_1 and G_1' for the signal and the image frequencies, respectively. If F is the noise figure of the total system, as measured with a wideband noise source, then the total output noise power of the system is

$$P_{\text{tot}} = F(G_1 G_a + G_1' G_a')kT\Delta f. \quad (A.15)$$

What we want to know is the 'narrow band' noise figure of the stage under test, defined as the noise figure measured with a narrow band preamplifier that responds to only *one* of the two frequencies f_i and f_i'. Let F_1 and F_1' be the narrow band noise figures for the stage for the frequencies f_i and f_i', respectively. We then have also

$$P_{\text{tot}} = (F_1 G_1 G_a + F_1' G_1' G_a') kT\Delta f$$
$$+ (F_a - 1)(G_a + G_a') kT\Delta f. \quad (A.16)$$

Equating the two expressions yields

$$F = \frac{F_1 G_1 G_a + F_1' G_1' G_a'}{G_1 G_a + G_1' G_a'} + \frac{(F_a - 1)(G_a + G_a')}{G_1 G_a + G_1' G_a'}. \quad (A.17)$$

We can now adjust the preamplifier plus mixer in such a way that $G_a = G_a'$. If this has been done, equation (A.17) reduces to

$$F = \frac{F_1 G_1 + F_1' G_1'}{G_1 + G_1'} + \frac{2(F_a - 1)}{G_1 + G_1'}. \quad (A.17a)$$

If we further assume that $F_1 = F_1'$, we obtain

$$F = F_1 + \frac{F_a - 1}{(G_1 + G_1')/2} \quad (A.18)$$

so that F_1 can be determined if F, F_a, G_1 and G_1' have been measured. Furthermore we see that Friiss' formula holds in this case, provided that the available gain of the stage is replaced by the *average* available gain: $(G_1 + G_1')/2$. This agrees with experiment (Section 4).

A MORE ACCURATE EXPRESSION FOR THE NOISE FIGURE OF TRANSISTORS†

A. VAN DER ZIEL‡
Electrical Engineering Department, University of Minnesota, Minneapolis, MN 55455, U.S.A.

and

J. A. CRUZ-EMERIC, R. D. LIVINGSTONE, J. C. MALPASS and D. A. MCNAMARA
Electrical Engineering Department, University of Florida, Gainesville, FL 32611, U.S.A.

(*Received 9 June 1975; in revised form 7 July 1975*)

Abstract—A more accurate schematic for the calculation of the minimum noise figure of transistors is presented. It involves two functions $G_1(f)$ and $G_2(f)$ of frequency that can be tabulated. An earlier expression for the minimum noise figure, that involves only a single parameter x is shown to be reasonably accurate, provided that the proper expression for x is used.

Noise in the common base transistor configuration is usually represented by an input emf e in series with the emitter junction and an output current generator i in parallel with the collector junction (between collector and internal base). Noise in the common emitter transistor configuration is usually represented by an input current generator i_b in parallel with the base-emitter junction and an output current generator i_c connected between collector and emitter. It is not difficult to express e, i, i_b and i_c in terms of the basic shot noise sources of the transistor; it is also a simple matter to convert the one equivalent circuit into the other. In addition, the base resistance r_b is assumed to have thermal noise in both cases. These representations are reasonably accurate up to moderate injection levels.

It is easily shown[1] that, for a given signal source impedance $Z_s = R_s + jX_s$, both circuits have the same noise figure. Historically, however, one has made different approximations for the common base[1, 2] and the common emitter[3, 4] configurations, and as a consequence one has obtained somewhat different expressions for the minimum noise figure. It is the aim of this article to investigate the magnitude of the errors made in these approximations and to develop a simple expression for the minimum noise figure that is quite accurate at elevated frequencies. First of all, however, we must discuss these approximations.

In the common base configuration one usually assumes[1, 2] that e and i are uncorrelated, that $\overline{e^2} = 2kTR_{e0}\Delta f$ at all frequencies of practical interest, where $R_{e0} = kT/eI_E$ and I_E is the emitter current. In addition one assumes, when the emitter junction impedance Z_e is written as $Z_e = R_e + jX_e$, that $R_e = R_{e0}$ for all frequencies of practical interest. If it is assumed that the d.c. current amplification factor α_f is reasonably close to unity, after some manipulations one then obtains for the minimum noise figure,

$$F_{\min} = 1 + \left(1 - \alpha_f + \frac{f^2}{f_\alpha^2}\right)\left(1 + \frac{r_b}{R_{e0}}\right) + \left[\left(1 - \alpha_f + \frac{f^2}{f_\alpha^2}\right)\left(1 + \frac{2r_b}{R_{e0}}\right)\right.$$
$$\left. + \left(1 - \alpha_f + \frac{f^2}{f_\alpha^2}\right)^2\left(1 + \frac{r_b}{R_{e0}}\right)^2\right]^{1/2} \quad (1)$$

Here $\alpha = \alpha_0/(1 + jf/f_\alpha)$ is the high frequency current amplification factor and f_α is the alpha cut-off frequency. We shall call this the *common base approximation*.

In the common emitter configuration[3, 4] one usually assumes $X_s = 0$, that i_b and i_c are uncorrelated, that $\overline{i_b^2} = 2eI_B\Delta f$ at all frequencies of practical interest, where I_B is the base current, and that the input impedance can be represented by a fixed resistance R_{beo} and a fixed capacitance C_{beo} in parallel. If one converts to the same notation as in (1), one finds after some manipulations

$$F_{\min} = 1 + \left(1 - \alpha_f + \frac{f^2}{f_\alpha^2}\right)\frac{r_b}{R_{e0}} + \left[\left(1 - \alpha_f + \frac{f^2}{f_\alpha^2}\right)\left(1 + \frac{2r_b}{R_{e0}}\right)\right.$$
$$\left. + \left(1 - \alpha_f + \frac{f^2}{f_\alpha^2}\right)^2\left(\frac{r_b}{R_{e0}}\right)^2\right]^{1/2} \quad (2)$$

so that the only difference is that $(1 + r_b/R_{e0})$ has been replaced by r_b/R_{e0}. Next it is observed that usually $2r_b/R_{e0} \gg 1$, so that little error is made by writing

$$F_{\min} = 1 + x + (2x + x^2)^{1/2},$$

where

$$x = \left(1 - \alpha_f + \frac{f^2}{f_\alpha^2}\right)\frac{r_b}{R_{e0}} \quad (2a)$$

The advantage of this expression is that $F_{\min}$ now depends on a *single* parameter x, so that comparison between theory and experiment is simplified. We shall call this the *common emitter approximation*.

To obtain a more accurate expression for $F_{\min}$ one can introduce the following parameters: First one splits e into a part e' that is fully correlated with i and a part e'' that is uncorrelated with i. One then introduces an equivalent

---
†Partly supported by N.S.F. contract with the University of Florida.

‡This work was performed in the Electrical Engineering Department of the University of Florida, while on leave of absence from the University of Minnesota.

noise resistance R_{ne}, an equivalent noise conductance g_{ne}, and a correlation impedance $Z_{cor} = R_{cor} + jX_{cor}$ by the definitions[1]

$$\overline{e^{n2}} = rkTR_{ne}\Delta f; \quad \frac{\overline{i^2}}{|\alpha|^2} = 4kTg_{ne}\Delta f; \quad Z_{cor} = \frac{e'}{i} = \frac{\overline{ei^*}}{\overline{i^2}}. \quad (3)$$

If one now adjusts X_s and R_s such that F is minimized, one obtains

$$F_{min} = 1 + 2g_{ne}(r_b + R_e + R_{cor}) + 2[g_{ne}(r_b + R_{ne}) + g_{ne}^2(r_b + R_e + R_{cor})^2]^{1/2}. \quad (4)$$

Up to here no new approximations have been introduced. It should be noted that the discussion does not use parallel RC combinations for Z_e and Z_{cor}, because they would complicate the calculations, and would not be sufficiently accurate. Moreover, this fits better with eqn (11).

In good approximation[1] one has for α_f close to unity

$$g_{ne} = \frac{1 - \alpha_f + f^2/f_\alpha^2}{2R_{e0}} \quad (5)$$

Introducing the parameters

$$G_1(f) = \frac{R_e + R_{cor}}{R_{e0}}; \quad G_2(f) = \frac{2R_{ne}}{R_{e0}} \quad (6)$$

the expression for F_{min} may be written

$$F_{min} = 1 + \left(1 - \alpha_f + \frac{f^2}{f_\alpha^2}\right)\left[G_1(f) + \frac{r_b}{R_{e0}}\right] + \left[\left(1 - \alpha_f + \frac{f^2}{f_\alpha^2}\right) \times \left[G_2(f) + \frac{2r_b}{R_{e0}}\right] + \left(1 - \alpha_f + \frac{f^2}{f_\alpha^2}\right)^2\left[G_1(f) + \frac{r_b}{R_{e0}}\right]^2\right]^{1/2} \quad (7)$$

This equation is valid up to moderate injection levels. We see that (7) reduces to (1) if $G_1(f) = G_2(f) = 1$, to (2) if $G_1(f) = 1$, $G_2(f) = 0$, and to (2a) if $G_1(f) = G_2(f) = 0$.

We must now evaluate $G_1(f)$ and $G_2(f)$. We observe that [1, 5]

$$\overline{ei^*} = \alpha^*2kT\Delta f\left(-1 + \frac{Z_e}{R_{e0}}\right); \quad \overline{i^2} = |\alpha|^2 4kTg_{ne}\Delta f \quad (8)$$

$$R_{cor} = \frac{R_{e0}(-1 + R_e/R_{e0})}{1 - \alpha_f + f^2/f_\alpha^2}, \quad \text{or} \quad G_1(f) = \frac{R_e}{R_{e0}} + \frac{-1 + R_e/R_{e0}}{1 - \alpha_f + f^2/f_\alpha^2}. \quad (9)$$

Since [5]

$$R_{ne} = R_e - \frac{|Z_e|^2}{2R_{e0}} - \frac{R_{e0}/2}{1 - \alpha_f + f^2/f_\alpha^2}\left(1 + \frac{|Z_e|^2}{R_{e0}^2} - 2\frac{R_e}{R_{e0}}\right) \quad (10)$$

$$G_2(f) = \frac{2R_e}{R_{e0}} - \frac{|Z_e|^2}{R_{e0}^2} - \frac{1 + |Z_e|^2/R_{e0}^2 - 2R_e/R_{e0}}{1 - \alpha_f + f^2/f_\alpha^2}. \quad (10a)$$

Finally $|Z_e|^2/R_{e0}^2$ and R_e/R_{e0} follow from [6]

$$\frac{Z_e}{R_{e0}} = \frac{\tanh[(1+j)(f/f_\alpha)^{1/2}]}{(1+j)(f/f_\alpha)^{1/2}}. \quad (11)$$

Table 1 shows $G_1(f)$ and $G_2(f)$ as functions of f/f_α for $1 - \alpha_f = 0.0100$. We see that $G_1(f) = G_2(f) = 1.00$ at low frequencies, so that the common base approximation is valid in that case. For $f/f_\alpha > 0$, $G_1(f)$ and $G_2(f)$ lie between 0 and 1, and neither approximation is very

Table 1. $1 - \alpha_f = 0.0100$

f/f_α	$G_1(f)$	$G_2(f)$
0.00	1.000	1.000
0.05	0.892	0.910
0.10	0.730	0.775
0.15	0.624	0.687
0.20	0.563	0.636
0.25	0.526	0.605
0.30	0.501	0.585
0.40	0.468	0.557
0.50	0.445	0.537
0.60	0.425	0.521
0.70	0.407	0.505
0.80	0.390	0.491
0.90	0.374	0.478
1.00	0.359	0.465
1.10	0.345	0.454
1.20	0.332	0.443
1.30	0.321	0.433
1.40	0.310	0.424

accurate. In particular we have for $f/f_\alpha = 1$ and $r_b/R_{e0} = 1$ that $F_{min} = 5\cdot69$ from eqn (1), $F_{min} = 3\cdot75$ from eqn (2a) and $F_{min} = 4\cdot46$ from Table 1. It is therefore recommended that Table 1 be used when good accuracy is required.

We finally observe that the first half of eqn (2a) is quite accurate if a slightly different expression is used for x. Since little error is made if $G_2(f)$ is replaced by $2G_1(f)$, eqn (7) can be written

$$F_{min} = 1 + x + (2x + x^2)^{1/2} \quad \text{with } x$$
$$= \left(1 - \alpha_f + \frac{f^2}{f_\alpha^2}\right)\left[G_1(f) + \frac{r_b}{R_{e0}}\right]. \quad (12)$$

Substituting again $1 - \alpha_f = 0\cdot0100$, $f/f_\alpha = 1$ and $r_b/R_{e0} = 1$ yields $F_{min} = 4\cdot52$ in good agreement with the result from Table 1. This approximation can thus be used if one wants adequate accuracy and an expression for F_{min} that contains only a *single* parameter x.

At high injection levels R_{cor} is positive, and at relatively low frequencies R_{cor}/R_{e0} has a significant value[7]. This behavior was recently explained by Min and van Vliet[8].

REFERENCES

1. Compare e.g. A. van der Ziel, *Noise: Sources, Characterization, Measurement*. Prentice Hall, Englewood Cliffs, N.J. (1970).
2. E. G. Nielsen, *Proc. IRE* **45**, 957 (1957).
3. H. Fukui, *IEEE Trans.* **ED-13**, 329 (1966), *IEEE Trans.* **CT-13**, 137 (1966).
4. S. D. Malaviya and A. van der Ziel, *Solid-St. Electronics* **13**, 1511 (1970), (see especially Appendix 1).
5. A. van der Ziel, *Fluctuation Phenomena in Semiconductors*. Butterworth, London (1959).
6. Compare e.g., A. van der Ziel, *Solid State Physical Electronics*. 3rd Edn. Prentice Hall, Englewood Cliffs, N.J. (1975).
7. H. A. Tong and A. van der Ziel, *IEE Trans.* **ED-15**, 307 (1968).
8. H. S. Min and K. M. van Vliet, *Solid-St. Electronics* **17**, 285 (1974).

LIMITATIONS OF NIELSEN'S AND RELATED NOISE EQUATIONS APPLIED TO MICROWAVE BIPOLAR TRANSISTORS, AND A NEW EXPRESSION FOR THE FREQUENCY AND CURRENT DEPENDENT NOISE FIGURE

R. J. HAWKINS

British Telecom Research Laboratories, Martlesham Heath, Ipswich, Suffolk, IP5 7RE, England

(*Received* 18 *March* 1976; *in revised form* 7 *July* 1976)

Abstract—In many low power microwave bipolar transistors the high frequency performance is largely controlled by the current-dependent time constant of the emitter base junction. This invalidates the use of Nielsen's expression which was derived for the noise figure of transistors controlled by the base transit time, which is substantially independent of emitter current. Values of noise figure obtained from this equation by substituting a current dependent cut-off frequency f_α, derived from the transition frequency f_T, are considerably higher than the measured values. The more recent approximation due to Malaviya and van der Ziel gives better agreement, but is still not entirely consistent with the measurements. A new expression is developed to take into account the emitter time constant, and this gives good agreement with measurements of minimum noise figure up to 3 GHz, and over a wide range of emitter current.

NOTATION

α current transport factor of the base alone
α_0 low frequency value of α
C_{Te} depletion capacitance of the emitter–base diode
C_{De} diffusion capacitance
e_s, e_b thermal noise voltage of the source and base resistors
e_e equivalent noise voltage in the emitter
f_α common base cut off frequency of the complete transistor
f_b cut off frequency of the base
f_e cut off frequency of the emitter circuit
f'_e cut off frequency of the input circuit
f_M characteristic frequency in Malaviya and van der Ziel's expression
f_N characteristic frequency in Nielsen's expression
f_T common emitter gain-bandwidth product
F noise figure
F_{min} minimum noise figure
F_x noise figure minimised with respect to source reactance
i_{cp} collector partition noise current generator
I_e emitter bias current
k Boltzmann's constant
q electronic charge
r_e emitter–base diode slope resistance
R_b base resistance
R_{opt} source resistance for minimum noise figure
R_s source resistance
T absolute temperature
X_{opt} source reactance for minimum noise figure
X_s source reactance.

1. INTRODUCTION

In recent years several circuit models[1–3] have been developed for calculations of noise figure in microwave bipolar transistors, taking into account the various parasitic reactances due to the transistor chip and the package in which it is mounted. However there still seems to be considerable interest[4, 5] in simple models of the type considered by Nielsen[6], Hibberd[7] and Malaviya and van der Ziel[8, 9].

These models make use of van der Ziel's analysis[10] of noise in the active base region, with the addition of the base series resistance R_b and its associated noise generator.

Nielsen's expression for noise figure as a function of source resistance R_s is

$$F = 1 + \frac{R_b}{R_s} + \frac{r_e}{2R_s} + \left(1 - \alpha_0 + \frac{f^2}{f_b^2}\right)\frac{(R_s + R_b + r_e)^2}{2\alpha_0 R_s r_e}. \quad (1)$$

Here r_e is the slope resistance of the emitter base diode, α_0 is the low frequency value of the common base current gain α, and f_b is its 3 dB cut-off frequency assuming a single pole function of the form

$$\alpha = \frac{\alpha_0}{1 + jf/f_b}. \quad (2)$$

By differentiating eqn (1) with respect to R_s the minimum noise figure F_{min} and the corresponding source resistance R_{opt} can be found.

Hibberd[7], Fukui[1] and Malaviya and van der Ziel[8] analysed circuits similar to Nielsen's, but took into account the frequency dependence of the emitter impedance, which he neglected. For frequencies approaching f_b, say

$$f^2 > 10(1 - \alpha_0)f_b^2,$$

where the noise figure has begun to increase significantly with frequency, these authors have given the following simplified formula:

$$F_{min} = 1 + u + \sqrt{(2u + u^2)}. \quad (3)$$

In Hibberd's expression the parameter u is given by

$$u = \frac{R_b}{r_e} \cdot \frac{f^2}{f_b^2}. \quad (4)$$

However, Fukui replaced f_b by f_T, the frequency at which the common emitter current gain has magnitude unity. f_T is usually rather lower than f_b, because it contains the effects of phase-shifts in the transistor[4] which affect neither the magnitude of α nor, therefore, its cut-off frequency f_b. Thus Fukui's expression would predict a more rapid increase with frequency. However, it is clear from his derivation that he expected these phase-shifts to be negligible, so that f_T and f_b would be approximately equal.

Malaviya and van der Ziel retained some terms of order $(1 - \alpha_0)$ and suggested that a closer approximation would be

$$u = \frac{R_b}{\alpha_0 r_e}\left(1 - \alpha_0 + \frac{f^2}{f_b^2}\right). \quad (5)$$

Clearly these simple equations can be applied without detailed knowledge of the full equivalent of the transistor and its package. They should give reliable values for F_{min} because, with one important exception which is the subject of this paper, the parasitic reactances omitted from the simple circuit model have little effect on minimum noise figure. (They do, however, affect the optimum impedance at which the minimum occurs, and the noise figure at a fixed source impedance).

The exception is the capacitance C_{Te} of the emitter-base junction space-charge region. In many present day transistors under typical operating conditions, for example an emitter current of 10 mA or less for a low power microwave device, the time constant $C_{Te}r_e$ of the junction is greater than the base transit time. Thus f_α increases with current until the onset of the Kirk effect. Surprisingly it is seldom noted in the literature that the simple expressions referred to earlier were derived for transistors limited entirely by the base transit time, which is substantially independent of emitter current. It cannot be assumed that they are valid when the emitter time constant has a significant effect on the high frequency performance. The effect of additional emitter capacitance can be included where a frequency dependent emitter impedance has been considered, bearing in mind that in van der Ziel's formulation of the noise sources[10] the intrinsic emitter impedance is derived from an analysis of the active base alone. This does not apply to Nielsen's treatment where the emitter impedance is assumed to be entirely resistive. In spite of this, Nielsen's equation is widely used and it is shown here that the effect of the emitter time constant can be introduced by the addition of a simple term. The new expression shows that the noise figure for a transistor controlled by emitter and base time constants should increase less rapidly with frequency than for a device limited by one time constant alone. Conversely, if Nielsen's expression is fitted to experimental results, artificially high values of the characteristic frequency must be assumed.

2. NOISE FIGURE OF THE EQUIVALENT CIRCUIT

2.1 Circuit analysis

Figure 1 shows the simple T-equivalent circuit which will be analysed in the Appendix. It differs from Nielsen's circuit only in that we include the emitter junction

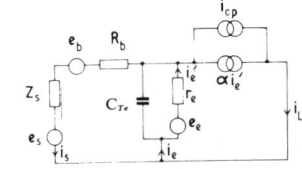

Fig. 1. Nielsen's T-equivalent circuit with the addition of the emitter capacitance C_{Te}.

capacitance C_{Te}, and consider the more general case of a complex source impedance $Z_s \equiv R_s + jX_s$.

In the circuit, e_s and e_b represent the thermal noise in the source and base resistances, and e_e and i_{cp} are the equivalent emitter generator and the collector "partition noise" generator which Nielsen simplified from van der Ziel's analysis[10].

Thus

$$\overline{e_s^2} = 4kTR_s \qquad \overline{e_b^2} = 4kTR_b$$
$$\overline{i_{cp}^2} = 2kT(\alpha_0 - |\alpha|^2)/r_e \qquad \overline{e_e^2} = 2kTr_e, \quad (6)$$

taking $r_e = kT/qI_e$ where T is the absolute temperature.

The advantage of this configuration is that it eliminates the correlation between the emitter and collector noise current generators except at frequencies close to the cut-off of the base transport factor. The greater part of the collector noise current is wholly correlated with the emitter noise current, because the normal transistor action transfers a fraction α to the collector. The generator i_{cp} provides the remainder, which is uncorrelated with e_e.

The frequency dependence of the intrinsic emitter impedance and the small, imaginary, correlation term are ignored in this treatment, as in Nielsen's analysis, though perhaps with greater justification; since the performance of the transistor is limited by the collector transit time and the emitter transition capacitance typical operating frequencies are well below the cut-off of the base. In addition, Abraham[11] has shown that the diffusion capacitance C_{De}, which represents the frequency dependence of the intrinsic emitter impedance, is much lower for a graded base than for a uniform base with the same cut-off frequency.

With these assumptions, the following equation for the noise figure is derived in the Appendix:

$$F = 1 + \frac{R_b}{R_s} + \frac{r_e}{2R_s} + \left(\frac{\alpha_0}{|\alpha|^2} - 1\right)\frac{(R_s + R_b + r_e)^2 + X_s^2}{2r_e R_s}$$
$$+ \frac{\alpha_0}{|\alpha|^2} \cdot \frac{r_e}{2R_s}[\omega^2 C_{Te}^2 X_s^2 - 2\omega C_{Te} X_s + \omega^2 C_{Te}^2 (R_s + R_b)^2]. \quad (7)$$

2.2 Real source impedance—Nielsen's equation

In the case of a real source impedance, e.g. 50 Ω, $X_s = 0$ and eqn (7) becomes

$$F = 1 + \frac{R_b}{R_s} + \frac{r_e}{2R_s} + \left(\frac{\alpha}{|\alpha|^2} - 1\right)\frac{(R_s + R_b + r_e)^2}{2r_e R_s}$$
$$+ \frac{\alpha_0}{|\alpha|^2}\omega^2 C_{Te}^2 r_e^2 \frac{(R_s + R_b)^2}{2r_e R_s}. \quad (8)$$

114

Substituting for α from eqn (2), where f_b is the cut-off frequency of the base alone, and introducing an emitter cut-off frequency $f_e = 1/2\pi C_{Te}r_e$,

$$F = 1 + \frac{R_b}{R_s} + \frac{r_e}{2R_s} + \left\{1 - \alpha_0 + \frac{f^2}{f_b^2}\right\}\frac{(R_s + R_b + r_e)^2}{2\alpha_0 r_e R_s}$$
$$+ \left\{1 + \frac{f^2}{f_b^2}\right\}\frac{f^2}{f_e^2} \cdot \frac{(R_s + R_b)^2}{2\alpha_0 r_e R_s}. \quad (9)$$

The first four terms are readily identified as Nielsen's expression (eqn 1), which corresponds to the case of a very short emitter time constant, so that the transistor is base limited and $f_e \gg f_b$ in eqn (9). In contrast, if the transistor is entirely emitter limited, $f_b \gg f_e$ and eqn (9) becomes

$$F = 1 + \frac{R_b}{R_s} + \frac{r_e}{2R_s} + (1 - \alpha_0)\frac{(R_s + R_b + r_e)^2}{2\alpha_0 r_e R_s} + \frac{f^2}{f_e^2} \cdot \frac{(R_s + R_b)^2}{2\alpha_0 r_e R_s}. \quad (10)$$

If we introduce a modified cut-off frequency

$$f'_e = f_e \frac{R_s + R_b + r_e}{R_s + R_b}, \quad (11)$$

eqn (10) can be rewritten in the same form as Nielsen's expression:

$$F = 1 + \frac{R_b}{R_s} + \frac{r_e}{2R_s} + \left(1 - \alpha_0 + \frac{f^2}{f_e'^2}\right)\frac{(R_s + R_b + r_e)^2}{2\alpha_0 r_e R_s}.$$

Since $f_e = 1/2\pi C_{Te}r_e$, substitution in eqn (11) shows that

$$f'_e = \frac{R_s + R_b + r_e}{2\pi C_{Te}r_e(R_s + R_b)},$$

which is the characteristic frequency of C_{Te}, r_e and $(R_s + R_b)$ in parallel, and is always greater than f_e. Thus the noise figure at a fixed source resistance for an emitter limited transistor should rise less rapidly with frequency than for a base limited device with the same value of the cut-off frequencies (f_e and f_b respectively).

Equation (9) shows that the effect of the emitter time constant is to add a single term to Nielsen's expression. However, if we rewrite the equation in terms of f'_e (eqn 11) a particularly simple form emerges:

$$F = 1 + \frac{R_b}{R_s} + \frac{r_e}{2R_s} + \left\{\left(1 + \frac{f^2}{f_b^2}\right)\left(1 + \frac{f^2}{f_e'^2}\right) - \alpha_0\right\}\frac{(R_s + R_b + r_e)^2}{2\alpha_0 r_e R_s}. \quad (12)$$

It should be noted that eqn (12) does not give the same result as calculating a composite value for the cut-off frequency f_α and substituting it in Nielsen's equation. An overall time constant can be obtained by summing individual time constants [4], so that the overall common base cut-off frequency is given by

$$\frac{1}{f_\alpha} = \frac{1}{f_e} + \frac{1}{f_b}.$$

The frequency dependent factor in Nielsen's equation would then be

$$1 + \frac{f^2}{f_\alpha^2} - \alpha_0 = 1 - \alpha_0 + \frac{f^2}{f_e^2} + \frac{f^2}{f_b^2} + \frac{2f^2}{f_e f_b}, \quad (13)$$

whereas the corresponding term in eqn (12) is

$$\left(1 + \frac{f^2}{f_b^2}\right)\left(1 + \frac{f^2}{f_e'^2}\right) - \alpha_0 = 1 - \alpha_0 + \frac{f^2}{f_e'^2} + \frac{f^2}{f_b^2} + \frac{f^4}{f_e'^2 f_b^2}. \quad (14)$$

Since f is in practice less than f_e and f_b, and $f_e < f'_e$, eqn (14) is always less than eqn (13). The difference is greatest when f_b and f'_e are equal, and since $f'_e \simeq f_e$ for emitter currents of a few mA or more and values of $(R_b + R_s)$ of the order of 50 Ω, this means that

$$f_b \simeq f_e = 2f_\alpha.$$

For example, suppose $f = f_\alpha/2 = f_b/4$. Then the factor in Nielsen's expression, eqn (13) gives $1 - \alpha_0 + 0.25$, while the full expression (eqn 14) for equal emitter and base time constants gives $1 - \alpha_0 + 0.129$. Thus the noise figure of a transistor limited equally by the base and emitter should increase only about half as rapidly with frequency as for one limited entirely by the base (or emitter), for the same value of f_α.

2.3 Minimum noise figure

The minimum noise figure $F_{\min}$ and the corresponding optimum source impedance $Z_{opt} = R_{opt} + jX_{opt}$ are found by differentiating eqn (7) with respect to X_s and then R_s. This equation is of the form

$$F = A + BX_s + CX_s^2,$$

where, by introducing the factor

$$a \equiv \left(1 - \frac{|\alpha|^2}{\alpha_0} + \omega^2 C_{Te}^2 r_e^2\right)\frac{\alpha_0}{|\alpha|^2}, \quad (15)$$

the coefficients can be written

$$A = a\frac{(R_s + R_b)^2}{2r_e R_s} + \frac{\alpha_0}{|\alpha|^2}\left(1 + \frac{R_b}{R_s} + \frac{r_e}{2R_s}\right)$$

$$B = -\frac{\alpha_0}{|\alpha|^2} \cdot \frac{\omega C_{Te} r_e}{R_s}$$

$$C = \frac{a}{2r_e R_s}.$$

Differentiating with respect to X_s and setting dF/dX_s to zero for the optimum source reactance

$$\left.\frac{dF}{dX_s}\right|_{X_{opt}} = 0 = B + 2CX_{opt}.$$

Thus

$$X_{opt} = \frac{-B}{2C} = \frac{\alpha_0}{|\alpha|^2}\frac{\omega C_{Te} r_e^2}{a}, \quad (16)$$

and the corresponding noise figure is

$$F_x = A - CX_{opt}^2 = a\frac{(R_s + R_b)^2}{2r_e R_s} + \frac{\alpha_0}{|\alpha|^2}\left\{1 + \frac{R_b}{R_s} + \frac{r_e}{2R_s}\right\}$$
$$- a\frac{X_{opt}^2}{2r_e R_s}. \quad (17)$$

This must be optimised with respect to source resistance to give F_{min}. Separating powers of R_s, eqn (17) can be written in the form

$$F_x = A + \frac{B}{R_s} + CR_s,$$

where now

$$A = a\frac{R_b}{r_e} + \frac{\alpha_0}{|\alpha|^2}$$

$$B = a\frac{R_b^2 - X_{opt}^2}{2r_e} + \frac{\alpha_0}{|\alpha|^2}\left(R_b + \frac{r_e}{2}\right)$$

$$C = \frac{a}{2r_e}.$$

Differentiating with respect to R_s, at the minimum:

$$\left.\frac{dF}{dR_s}\right|_{R_{opt}} = 0 = \frac{-B}{R_{opt}^2} + C.$$

Thus

$$R_{opt}^2 = \frac{B}{C} = R_b^2 - X_{opt}^2 + \frac{\alpha_0}{|\alpha|^2} \cdot \frac{r_e(2R_b + r_e)}{a} \quad (18)$$

$$F_{min} = A + 2CR_{opt} = a\frac{R_b + R_{opt}}{r_e} + \frac{\alpha_0}{|\alpha|^2}. \quad (19)$$

The factor a is a simple symmetrical function of f_e and f_b, as found by substituting for $|\alpha|$ and $C_{Te}r_e$ in eqn (15)

$$a = \left\{1 + \frac{f^2}{f_b^2} - \alpha_0 + \left(1 + \frac{f^2}{f_b^2}\right)\frac{f^2}{f_e^2}\right\}\frac{1}{\alpha_0}$$

$$= \left\{\left(1 + \frac{f^2}{f_b^2}\right)\left(1 + \frac{f^2}{f_e^2}\right) - \alpha_0\right\}\frac{1}{\alpha_0}. \quad (20)$$

This is similar to the factor which appeared in the noise figure for a fixed source resistance (eqn 12), with the difference that it contains the characteristic frequency f_e of the emitter junction alone, rather than that of the whole input loop, f'_e.

For the base limited case considered by Nielsen, C_{Te} and X_{opt} are zero and eqns (20), (18) and (19) become:

$$a = \left(1 + \frac{f^2}{f_b^2} - \alpha_0\right)\frac{1}{\alpha_0}$$

$$R_{opt}^2 = R_b^2 + \frac{1 + f^2/f_b^2}{1 + f^2/f_b^2 - \alpha_0} r_e(2R_b + r_e) \quad (21)$$

$$= (R_b + r_e)^2 + \frac{\alpha_0 r_e(2R_b + r_e)}{1 + f^2/f_b^2 - \alpha_0} \quad (22)$$

$$F_{min} = \left(1 + \frac{f^2}{f_b^2} - \alpha_0\right)\frac{R_b + R_{opt}}{\alpha_0 r_e} + \left(1 + \frac{f^2}{f_b^2}\right)\frac{1}{\alpha_0}. \quad (23)$$

Equations (22) and (23) can be shown to be the same as those given by Nielsen. In the emitter limited case, $\alpha = \alpha_0$ and eqns (20), (16), (18) and (19) become

$$a = \left\{1 + \frac{f^2}{f_e^2} - \alpha_0\right\}\frac{1}{\alpha_0}$$

$$X_{opt} = \frac{f}{f_e} \cdot \frac{r_e}{\alpha_0 a}$$

$$R_{opt}^2 = R_b^2 + \frac{r_e(2R_b + r_e)}{\alpha_0 a} - \frac{f^2}{f_e^2} \cdot \frac{r_e^2}{\alpha_0^2 a^2} \quad (24)$$

$$F_{min} = \left\{1 + f^2/f_e^2 - \alpha_0\right\}\frac{R_b + R_{opt}}{\alpha_0 r_e} + \frac{1}{\alpha_0}. \quad (25)$$

Comparison of eqns (23) and (25) shows that minimum noise figure increases more rapidly with frequency for the base limited device, for the same values of R_{opt} and cut-off frequency f_e and f_b. In addition, eqns (21) and (24) show that R_{opt} is larger in the base limited case, and this increases the difference between the values of F_{min} calculated from the two equations.

3. COMPARISON WITH EXPERIMENT

In this section we compare calculations using the new expression (eqn 19) with noise measurements of an Avantek AT4641 transistor at 1, 2 and 3 GHz. The experimental points shown in Fig. 2 are measurements of minimum noise figure, using a slide-screw tuner to transform the characteristic impedance of 50 Ω to the optimum impedance.

The curves in Fig. 2 have been calculated from eqns (18)–(20). To do this, it is necessary to know the emitter capacitance C_{Te} (to give f_e and X_{opt}), the base cut-off frequency f_b, and the current gain α_0 and the base resistance R_b at each value of emitter current I_e.

3.1 Derivation of component values

$R_b(I_e)$ was derived from measurements of 50 Ω noise at 200 MHz, which is well below the common-base cut-off frequency. Between 3 and 20 mA, R_b lies in the range 10.5–11.5 Ω, and β_0 in the range 50–58. The measurements

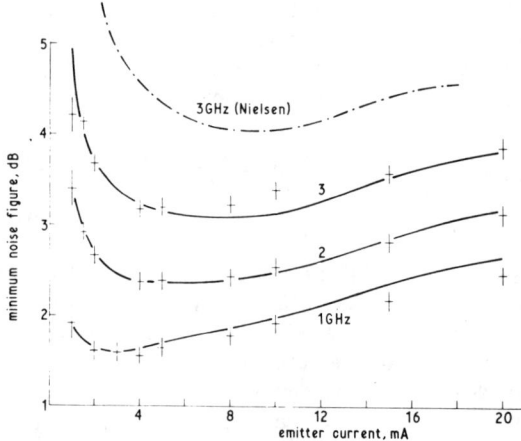

Fig. 2. Measurements of minimum noise figure (+) compared with calculations using the new expression (———) and Nielsen's expression (—·—).

indicate that the resistance falls at lower currents, to about 8.5 Ω at 1 mA, where $\beta_0 = 46$. However, the uncertainty in these measurements is greater at the low currents; typically the error due to ±0.1 dB in noise figure is ±1.5 Ω, and the measurements are consistent with a constant base resistance of about 10.5 Ω over the range of current investigated.

Measurements of other microwave transistors have shown that this is typical of devices with low base resistance (10–15 Ω) and moderate gain ($\beta_0 \simeq 50$). Transistors with higher base resistance (30–40 Ω) and higher gain ($\beta_0 \simeq 100$) at low currents show a marked fall in resistance and gain as the current increases.

Emitter junction capacitance C_{Te} can be obtained from a reciprocal plot of the transition frequency f_T against I_e [4] as shown in Fig. 3. Values of f_T were obtained by extrapolating the 6 dB/octave slope of the common emitter current gain, derived from measurements of S-parameters at 1 GHz. The points give a good straight line with a slope which corresponds to an emitter capacitance of 2.65 pF, and the intercept represents a total fixed time constant of 18–19 ps. As well as the base transit time this comprises any fixed time constants, in the emitter or the collector, and delay times due to the excess phase in the base and the collector depletion layer transit.

These time delays should not affect the magnitude of α or the noise figure, which in the van der Ziel model depends on $|\alpha|^2$ (eqn 3). Hence a reciprocal plot of f_α against I_e should be parallel to the f_T line, but with a smaller intercept which corresponds to the fixed time constants. It is difficult to measure f_α directly for microwave devices because of the effect of the transistor package, and so the intercept $1/2\pi f_b$ has been estimated by comparing the calculated value of $F_{\min}$ with the

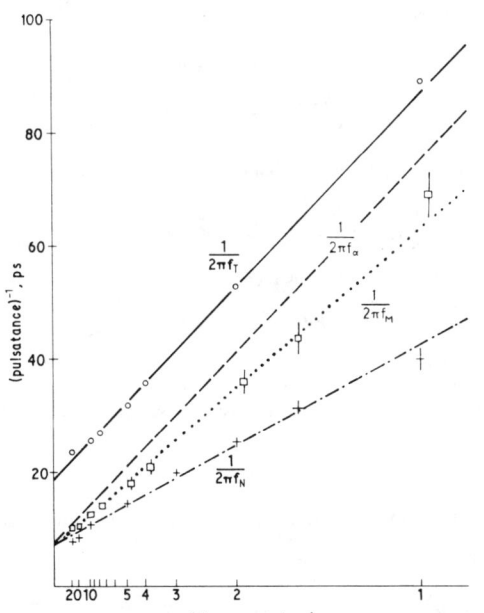

Fig. 3. Reciprocal plots of characteristic frequency against emitter current; (———)f_T measured at 1 GHz, (— —)f_α from new expression for best agreement with noise measurements, (—·—)f_N from Nielsen's eqn for best agreement with noise measurements, (········)f_M from Malaviya and van der Ziel.

measurement at 10 mA and 2 GHz. The best fit was obtained with a fixed time constant of 7 ps, which corresponds to $f_b \simeq 23$ GHz. The corresponding reciprocal plot of $1/2\pi f_\alpha = (1/f_b + 1/f_e)/2\pi$ is shown by the broken line in Fig. 3.

These values of C_{Te}, f_b and $R_b(I_e)$ were used to calculate minimum noise figure for comparison with the experimental results as shown by the curves in Fig. 2. These agree well, except for 15 and 20 mA at 1 GHz. However, this discrepancy is within the error due to the measurement of R_b. A better overall fit is obtained by taking $R_b = 11$ Ω and a fixed time constant of 7.5 ps. Thus of the 18–19 ps intercept of the $1/2\pi f_T$ line, 10.5–12 ps is a pure time delay which does not affect the magnitude of f_α.

We justify these values in the following way. The measured collector base capacitance with 10 V bias is 0.22 pF. Allowing 0.02 pF for the transistor package (measured on an empty package of the same type) and 0.05 pF for the bonding pad (estimated from the pad dimensions), the remaining 0.15 pF capacitance from a measured base area of 2200 μm^2 implies a depletion layer thickness of 1.5 μm. This gives a collector transit time of 18.6 ps, assuming a scattering limited velocity of 80 $\mu m/ns$. According to Early[12], the principal effect of this should be to introduce a time delay of 9.3 ps, and this contributes to the $1/2\pi f_T$ intercept.

Cooke[4] gives two approximate formulae for the total time delay and the 3 dB time constant of the base, as a function of the field factor η due to the doping gradient. For the two cases of base widths of 0.15 and 0.2 μm and a field factor of 4, the 3 dB time constants are 3.5 and 6.2 ps respectively, and the total delays are 5.5 and 9.7 ps, giving excess phase delays of 2.0 and 3.5 ps. Adding these delays to the estimated collector time delay of 9.3 ps gives total delays of 11.3 and 12.8 ps, compared to the measured values of 10.5–12 ps.

The 3 dB time constant estimated for the base is somewhat lower than the value of 7–7.5 ps deduced from the intercept of $1/2\pi f_\alpha$, and this discrepancy may be due to additional fixed time constants in the circuit. De Man et al.[13], and Henderson and Scarbrough[14] have suggested an additional time constant due to the heavy impurity doping in the emitter. A small contribution arises from the charging of the collector-base depletion capacitance of 0.2 pF through the collector series resistance, which was found to be 1.5–2 Ω[15]; it is believed that the thin epitaxial layer is fully depleted at the higher emitter currents with 10 V collector-emitter bias. However this only yields 0.3–0.4 ps, and a larger contribution of about 0.7 ps to the measured intercept arises from the increase in junction temperature at higher values of current, which partly offsets the fall in emitter slope resistance.

3.2 Comparison with earlier equations

Because of the similarity between Nielsen's equation and the new expression, the former can be fitted to experimental results for a transistor controlled by both base and emitter time constants. However it follows from Sections 2.2 and 2.3 that an artificially high value of characteristic frequency (denoted here by f_N) must be substituted in order to achieve this. Values of f_N have been obtained by fitting Nielsen's expression for minimum noise figure to the measurements shown in Fig. 2, and a reciprocal plot against current is shown in Fig. 3.

Within the probable error the points lie on a straight line with a similar intercept to the f_α line but of about half the slope. Thus one might expect f_N to approach f at high currents, and Nielsen's equation and the new expression would give similar results. However at all currents within the range of practical interest, f_α is determined by both base and emitter time constants, and an artificially high value of f_N has to be substituted in Nielsen's equation to agree with the measurements. Substituting $f_N = f_\alpha$ exaggerates the noise figure at 3 GHz by more than 2 dB at 1 mA to 0.9 dB at 20 mA, as shown by the chain-dotted line in Fig. 3.

In a similar way, the expression given by Malaviya and van der Ziel, eqns (2) and (5), has been fitted to the measurements. The required values of characteristic frequency (f_b in eqn (5)) are denoted here by f_M and are shown in Fig. 3. The slope of the dotted line drawn through the points is about 20% less than that of the lines $(1/2\pi f_\alpha)$ and $(1/2f_T)$, showing that their expression, though closer than Nielsen's, is still not entirely consistent with a characteristic frequency derived from f_T.

These results cast some doubt on derivations of f_α from noise measurements, and may explain why very high values have apparently been obtained in this way[16].

CONCLUSIONS

The new expression for noise figure takes into account both fixed (base) and current-dependent (emitter) time constants, yet is little more complicated than the expression derived by Nielsen for wholly base limited transistors. It gives good agreement with measurements of minimum noise figure over a wide range of emitter current, with reasonable values of common base cut-off frequency, consistent with the observed variation of f_T with current. In contrast, earlier expressions required artificially high values of cut-off frequency to fit experimental results, especially at low values of emitter current. The new expression should give a much closer estimate of high frequency noise figure from lower frequency measurements.

Acknowledgements—The author wishes to thank Mr. J. C. Henderson and Mr. T. J. Burslem for their helpful comments, and the Director of Research of the Post Office for permission to make use of the information contained in this paper.

REFERENCES

1. H. Fukui, *Trans IEEE* **ED-13**, 329 (1966).
2. K. Hartmann and M. J. O. Strutt, *Trans IEEE* **ED-20**, 874 (1973).
3. R. J. Hawkins, *4th Int. Conf. Physical Aspects of Noise in Solid State Devices*, Noordwijkerhout, The Netherlands (Sept. 1975).
4. H. F. Cooke, *Proc. IEEE* **59**, 1163 (1971).
5. E. Hughes, *Colloque Int. du CNRS sur le Bruit de Fond des Composants Actifs Semiconducteurs*, Toulouse (Sept. 1971).
6. E. G. Nielsen, *Proc. IRE* **45**, 957 (1957).
7. F. Hibberd, *Electron. Engng* **32**, 163 (1960).
8. S. D. Malaviya and A. van der Ziel, *Solid-St. Electron.* **13**, 1511 (1970).
9. A. van der Ziel, J. A. Cruz-Emeric, R. D. Livingstone, J. C. Malpass and D. A. McNamara, *Solid-St. Electron.* **19**, 149 (1976).
10. A. van der Ziel, *Proc. IRE* **43**, 1639 (1955).
11. R. P. Abraham, *Trans. IRE* **ED-7**, 56 (1960).
12. J. M. Early, *BSTJ* **XXXIII**, 517 (1954).
13. H. DeMan, R. Mertens and R. van Overstraeten, *Electron. Lett.* **9**, 174 (1973).
14. J. C. Henderson and R. J. D. Scarbrough, *ESSDERC*, Munich (1973).
15. B. Kulke and S. L. Miller, *Proc. IRE* **45**, 90 (1957).
16. J. A. Archer, *Solid-St. Electron.* **17**, 387 (1974).

APPENDIX

Considering the loop containing Z_s, R_b and r_e in Fig. 1:

$$i_s(Z_s + R_b) + i'_e r_e = e_s + e_b + e_e. \qquad (26)$$

This will be rewritten in terms of the load current i_L in order to give an expression for the noise output power. At the collector terminal

$$i_L = \alpha i'_e + i_{cp},$$

and in the emitter circuit

$$i_e = i'_e(1 + j\omega C_{Te} r_e) - j\omega C_{Te} e_e. \qquad (27)$$

For zero current into the ground (emitter) node,

$$i_s = i_e - i_L,$$

and using eqns (27)

$$i_s = i'_e(1 + j\omega C_{Te} r_e) - j\omega C_{Te} e_e - i_L$$
$$= \frac{i_L - i_{cp}}{\alpha}(1 + j\omega C_{Te} r_e) - j\omega C_{Te} e_e - i_L.$$

Substituting for i_s and i'_e in eqn (26) and rearranging:

$$\frac{i_L}{\alpha}\{(1 - \alpha + j\omega C_{Te} r_e)(Z_s + R_b) + r_e\} = e_s + e_b$$
$$+ e_e\{1 + j\omega C_{Te}(Z_s + R_b)\}$$
$$+ \frac{i_{cp}}{\alpha}\{(1 + j\omega C_{Te} r_e)(Z_s + R_b) + r_e\}. \qquad (28)$$

We use the definition of noise figure as the ratio of the output noise power to that from a noiseless but otherwise identical device, i.e.

$$F = \overline{i_L^2}/\overline{i_{Lo}^2},$$

where i_{LO} is the value of i_L due to the source generator e_s alone. Thus from eqns (2) and (28)

$$F = 1 + \frac{R_b}{R_s} + \frac{r_e}{2R_s}\{(1 + \omega C_{Te} X_s)^2 + \omega^2 C_{Te}^2(R_s + R_b)^2\} + \left\{\frac{\alpha}{|\alpha|^2} - 1\right\}$$
$$\frac{[(R_s + R_b + r_e(1 - \omega C_{Te} X_s)]^2 + [X_s + \omega C_{Te} r_e(R_s + R_b)]^2}{2r_e R_s}.$$

Combining the frequency-dependent parts of the third term with the fourth yields eqn (7).

Low-Noise Microwave Bipolar Transistor with Sub-Half-Micrometer Emitter Width

TZU-HWA HSU, MEMBER, IEEE, AND CRAIG P. SNAPP, MEMBER, IEEE

Abstract—This paper presents details of the fabrication process and performance of an n-p-n silicon microwave bipolar transistor with emitter opening widths as small as 0.3 μm. The fabrication process involves local oxidation, ion implantation, and lateral etching techniques for emitter definition. Noise figure as low as 1.0 dB at 1.5 GHz, 2.0 dB at 4 GHz, and 3.3 dB at 6 GHz were achieved. Measured noise figures and S-parameters are shown to be in approximate agreement with modeled performance based on device structure and process parameters. Prospects for further reductions in bipolar transistor noise figures are discussed.

I. Introduction

MANY UHF and microwave receivers have requirements for small-signal amplification with as low a noise figure as technologically possible. The interdigitated emitter n-p-n bipolar transistor is one of the principal solid-state devices for achieving very-low-noise amplification, particularly for frequencies less than 4 GHz.

The requirements for achieving low noise and high gain from a bipolar transistor at microwave frequencies are well known [1]. Gain can be effectively increased by reducing parasitic bonding pad capacitance and increasing the emitter periphery to base area ratio by reducing the emitter stripe periodicity to the limits defined by the existing technology. Noise figure can be reduced by optimizing the emitter and base impurity profiles to minimize the emitter-base capacitance charging time τ_e and the netural base delay τ_b and by minimizing both the active and inactive base resistances. The active base resistance under the emitter can be reduced by reducing the width of the emitter stripe while the external inactive base resistance can be reduced by having the p^{++} base insert as close to the emitter as possible. Very narrow emitters are also advantageous for reducing noise figure because they result in small values of the current density dependent time constant τ_e at low values of current and corresponding low values of injected shot noise.

Past development efforts have successfully resulted in small-dimension bipolar transistors with excellent noise figure performance in the 1- to 6-GHz frequency range. Noise figures as low as 2.3 dB at 4 GHz and 3.9 dB at 6 GHz were achieved by Archer [2] using local oxidation and controlled lateral base diffusion to produce a transistor structure with an effective emitter width less than the 1.0-μm emitter opening dimension.

Manuscript received November 17, 1977; revised February 7, 1978.
The authors are with the Microwave Semiconductor Division, Hewlett-Packard Company, Palo Alto, CA 94304.

Electron-beam lithography has been used to fabricate transistors with 0.5-μm emitter openings which exhibited minimum noise figures of 4.1 dB at 6 GHz [3]. Minimum noise figures of 1.8 dB at 1.7 GHz have been obtained from 1.0-μm emitter opening transistors fabricated by self-aligned local oxidation techniques combined with electron-beam lithography [4]. With proper control and self-aligning techniques, conventional contact optical lithography can readily produce 0.7-μm emitter openings and typical transistor noise figure performance of 1.45 dB at 1.5 GHz and 2.7 dB at 4 GHz [5].

This paper presents details of the structure and performance of an n-p-n microwave transistor (HP-505) with emitter opening widths of 0.3 to 0.5 μm fabricated by a process which does not require either submicrometer optical or electron-beam lithography. Device characterization data are presented which show noise figure performance as low as 1.0 dB at 1.5 GHz, 2.0 dB at 4 GHz, and 3.3 dB at 6 GHz. Measured noise figures and S-parameters are shown to be in good agreement with modeled performance based on a regional analysis of device structure and process parameters. A computer program (SUPREM) [6], used to model and simulate the process, results in calculated junction depths and sheet resistances of each region together with doping profiles. The advantages of the fabrication process will be discussed along with prospects for even further reductions in bipolar transistor noise figures at microwave frequencies.

II. Fabrication Technology

The ability to achieve the lateral geometries, vertical impurity profiles, and low parasitic capacitances necessary for good low-noise microwave performance is highly dependent upon the details of interacting fabrication steps. The process developed for the HP-505 involves local oxidation, ion implantation, and a technique similar to that described by Kamioka *et al.* [7] to laterally etch an SiO$_2$ layer sandwiched between two nitride layers to define sub-half-micrometer emitters. The key steps in the fabrication sequence are illustrated by the cross sections shown in Fig. 1 and are discussed below in detail.

A. Base Mesa Formation

Silicon nitride and low-temperature CVD SiO$_2$ layers are first deposited on an n-type epitaxial layer grown on an n$^+$ ⟨111⟩ substrate. Using conventional photolithographic and etching techniques, the base pattern is defined and a silicon mesa formed. The top Silox is etched away, and a 9000-Å

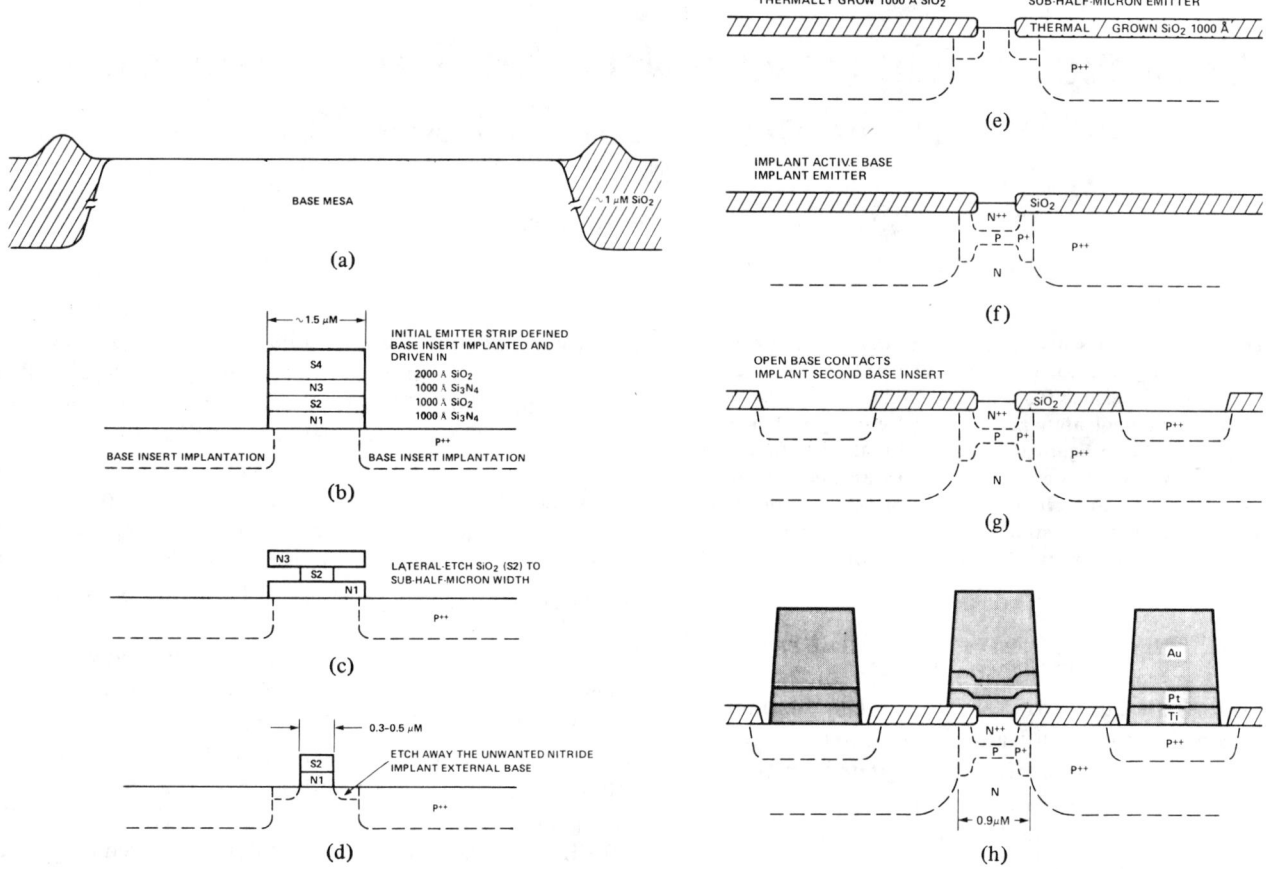

Fig. 1. Cross sections showing the fabrication sequence.

field SiO_2 is thermally grown. The nitride cap protects the base area from oxidation. After removing the nitride cap, the field oxide is coplanar with the surface of the base region except for a small "bird's head" at the edge of the base.

B. Initial Emitter Definition and Base-Insert Implantation

Four alternating nitride and SiO_2 layers are deposited and etched to form the initial 1.5-μm wide emitter defining stripes. A high dose boron p^{++} base-insert implantation is performed, followed by annealing and drive-in.

C. Lateral Etching

The SiO_2 layer ($S2$) sandwiched between the two nitride layers is laterally etched to the desired emitter width. Using a buffered HF etch, the side-etching rate is approximately 1000 Å/min. One point worth noting is that the width of the Silox stripe can be visually inspected and measured under an optical microscope, as illustrated in Fig. 2, thus making the etching fairly controllable, even for $S2$ widths less than 0.5 μm. An SEM photograph of the stripe sandwich at this point is shown in Fig. 3(a).

D. External Base Implantation

The top nitride layer is etched away and an external boron base implantation is performed, followed by annealing and drive-in.

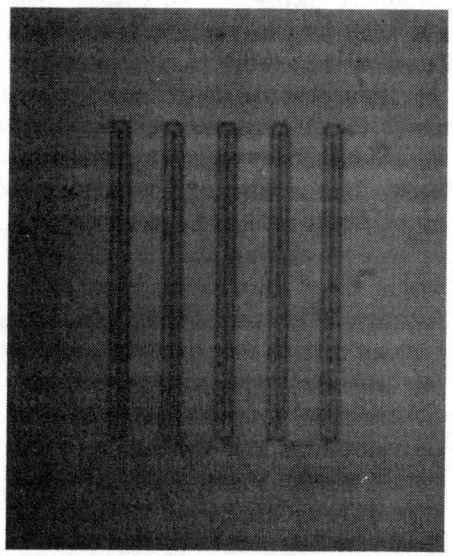

Fig. 2. During lateral etch, the sandwiched narrow SiO_2 strip ($S2$) is clearly visible through the wider top nitride stripe.

E. Final Emitter Definition

The SiO_2 layer is etched away to expose the nitride stripe $N1$. An SEM photograph of the nitride stripe is shown in Fig. 3(b) prior to the growth of 1000 Å of SiO_2. Fig. 3(c) shows the nitride stripe after oxide growth. The nitride stripes

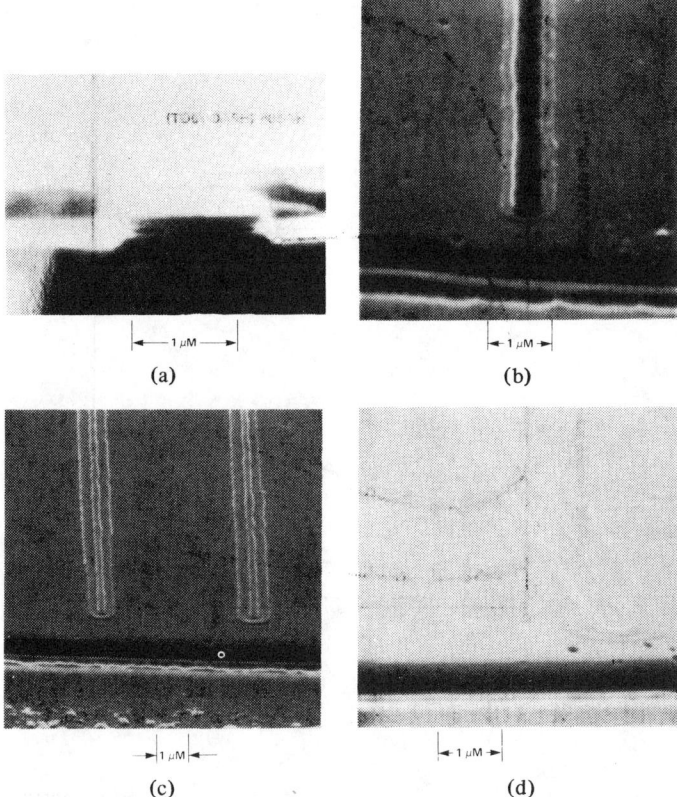

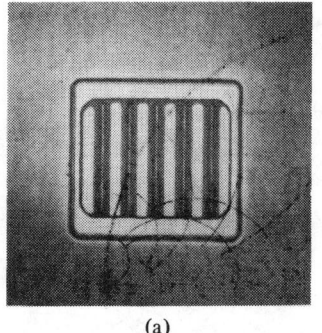
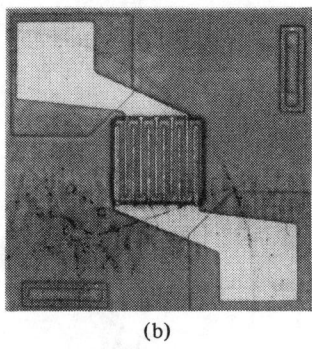

Fig. 4. (a) Emitter and base contact windows prior to metallization. (b) Final metallized transistor. Periodicity is 5 μm.

Fig. 3. SEM photographs at different process steps. (a) Lateral etch of SiO$_2$ (S2) results in SiO$_2$ strip width slightly less than 0.5-μm (step c). (b) Nitride stripe (N1) prior to thermal oxidation (step e). (c) Nitride stripes after thermal oxidation. (d) Final emitter opening after etching off nitride stripe (step f). Width of opening is approximately 0.35 μm.

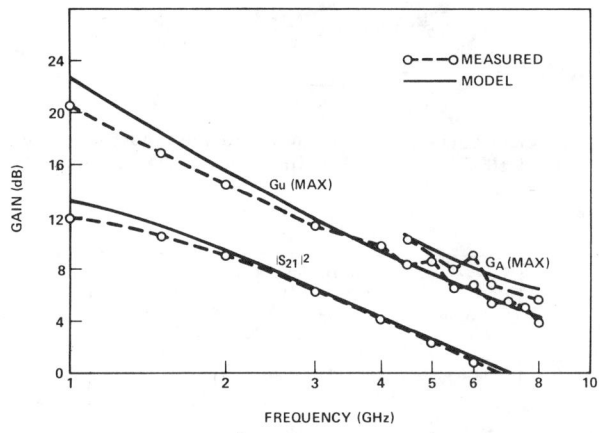

Fig. 5. Measured and modeled G_A(MAX), G_U(MAX), and $|S_{21}|^2$ gains as functions of frequency of typical HP-505 mounted in HPac-70GT package. Bias: V_{CE} = 10 V, I_C = 2 mA.

are removed chemically to open up the final sub-half-micrometer oxide-walled emitter openings as shown in Fig. 3(d).

F. Active Base and Emitter Implantation

The active base implantation is the same as that of the external base. After annealing and drive-in, the n^{++} arsenic emitter is implanted and annealed.

G. Base Contact Opening and Implantation

The base contact pattern is aligned and the thermal oxide etched off to expose the base contact area. To compensate for the depletion of boron surface concentration during the thermal oxidation in step (e), another p^{++} boron implantation is performed, followed by annealing and drive-in. A photograph of the emitter and base contact windows prior to metallization is shown in Fig. 4(a).

H. Metallization

The titanium–platinum–gold system is used for metallization. A top layer of titanium is used as a sputter-etch mask. The four metal layers are sequentially sputter deposited. The metal pattern is defined using standard photolithography, and the top titanium is then etched. After the photoresist is stripped off, the metals are sputter etched down to the bottom titanium layer, followed by a dilute HF etch to remove excess top and bottom titanium. A photograph of the metallized transistor is shown in Fig. 4(b). After metallization, the wafers are glassivated, back-lapped to 4 mils, and the backside metallized to complete the fabrication.

III. Performance

Finished chips were mounted into low parasitic 70-mil diameter ceramic packages (HPac-70GT) [8] and characterized using an automatic network analyzer. The measured $|S_{21}|^2$ gain, maximum available gain G_A (MAX), and maximum unilateral gain G_u (MAX) are shown in Fig. 5 as functions of frequency for a typical HP-505 biased near minimum noise conditions. Measured values of the S_{11} and S_{22} parameters are shown on the Smith chart in Fig. 6.

The frequency dependence of minimum noise figure is of particular interest since the device was designed specifically for low-noise amplification. The minimum-tuned noise figure[1] and associated gain are shown in Fig. 7 for typical units as well as for the best HP-505 that was characterized. For comparison,

---
[1] The noise figure was measured using an AilTech Type 75 automatic noise-figure meter in a system which was calibrated using an AilTech 7009 hot-cold noise source and Type 136 test receiver. Low-loss slide-screw tuners were used to match the transistor input and output.

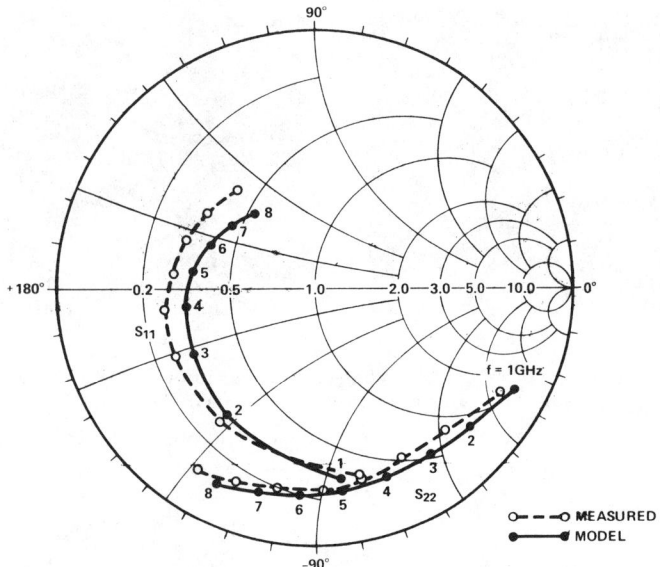

Fig. 6. Frequency dependence of measured and modeled S_{11} and S_{22} for typical HP-505 mounted in HPac-70GT package. Bias: V_{CE} = 10 V, I_C = 2 mA.

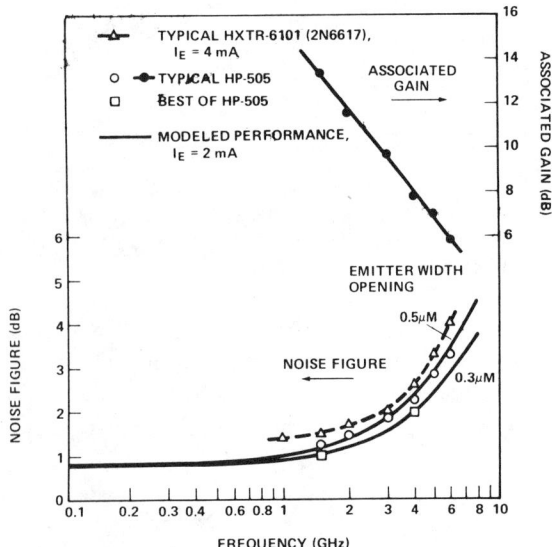

Fig. 7. Measured and calculated noise figure of HP-505 as a function of frequency compared to typical HXTR-6101 (2N6617). Measured associated gain is shown for typical HP-505. All units were mounted in the HPac-70GT package.

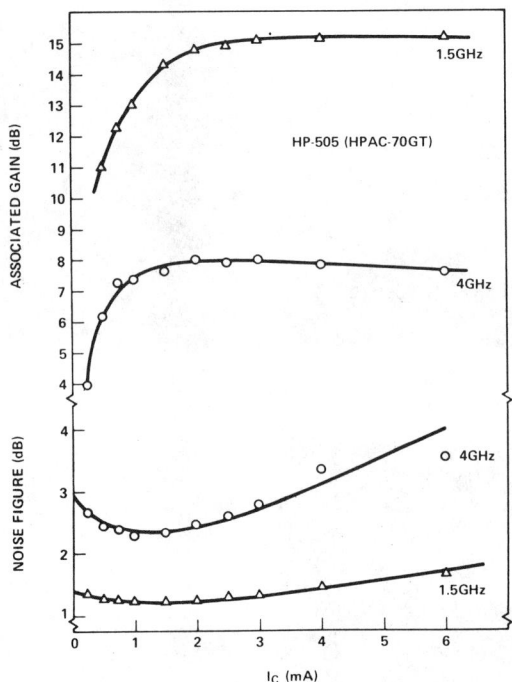

Fig. 8. The minimum noise figure and associated gain of a typical HP-505 at 1.5 GHz and 4 GHz as a function of collector current.

typical data are also shown for the commercial 0.7-μm emitter width HXTR-6101 (2N6617) made by a more conventional self-aligning process [5].

An interesting feature of the HP-505 is its low noise figure at relatively low-bias currents. Fig. 8 shows the measured noise figures and associated gains at 1.5 GHz and 4 GHz as a function of bias current I_c. We see that the noise-figure minimum occurs at an I_c of about 1.5 mA and does not increase significantly until I_c is less than 0.25 mA. This characteristic is particularly attractive for low-noise applications where low power consumption is desired. Optimum noise performance for the HP-505 occurs at very low currents because of the extremely small active-emitter area ($\sim$ 50 μm^2 for five 0.4-μm wide emitters). A bias current of 1.5 mA corresponds to an active-emitter current density of about 3000 A/cm^2.

IV. Discussion

A. S-Parameter Modeling

The familiar lumped T-equivalent circuit based on the transistor physical structure has been found to be effective for modeling the small-signal S-parameters of larger geometry microwave transistors [5], [9]. We have found that this approach is also useful for modeling transistors with sub-half-micrometer emitter stripe widths. The equivalent circuit used for the HP-505 is shown in Fig. 9(a). The passive elements in this circuit can be obtained by a regional analysis of contact and spreading resistances and geometric junction and oxide capacitances [1]. The necessary transistor design and process information is given in Table I.

The current-dependent current source in Fig. 9(a) is driven by the current through the dynamic resistance r_e of the forward-biased emitter-base junction. The common-base current gain α can be approximated by the single-pole expression

$$\alpha = \frac{\alpha_0}{1 + jf/f_b} \exp[-j 2\pi f \tau] \quad (1)$$

where f_b is the base cutoff frequency and τ is assumed to be equal to the collector depletion region delay time τ_d. τ_d is approximately 10 ps for a collector drift region width of 1.7 μm, assuming a saturated drift velocity of 8.5 $\times$ 10^6 cm/s.

The values of the critical emitter-base charging time τ_e and the neutral base delay τ_b can be obtained empirically from measurements of the common-emitter gain-bandwidth product f_T as a function of emitter current I_e. A plot of the total

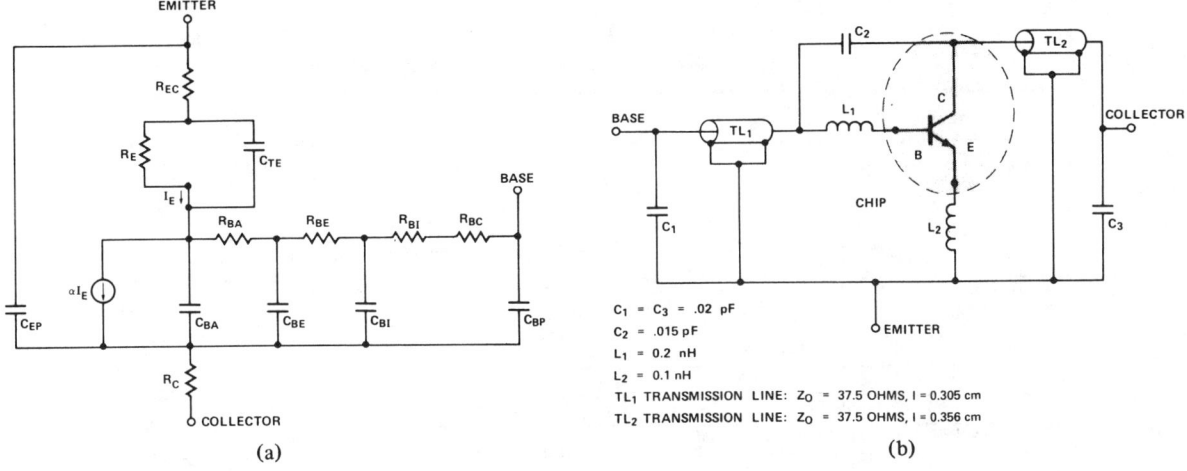

Fig. 9. Equivalent circuits used to model the device S-parameters. (a) Chip. (b) HPac-70GT package.

TABLE I
DESIGN AND PROCESS PARAMETERS

PARAMETER	VALUE	HOW OBTAINED
NUMBER OF EMITTER FINGERS	5	GEOMETRY
EMITTER OPENING WIDTH	0.4 μm	GEOMETRY
EMITTER LENGTH	20 μm	GEOMETRY
p++ BASE INSERT WIDTH	4.1 μm	GEOMETRY
EMITTER PERIODICITY	5 μm	GEOMETRY
ACTIVE BASE SHEET RESISTANCE	25,000 ohms/$\square$	ADJUSTED FOR NF FIT
INACTIVE BASE SHEET RESISTANCE	2,700 ohms/$\square$	CALCULATED (SUPREM)
COLLECTOR DRIFT REGION WIDTH W_d	1.7 μm	FROM C-V MEASUREMENT
h_{FE} AT V_{CE} = 10 V, I_e = 2 mA	100	MEASURED
I_{CBO} AT V_{CE} = 10 V	< 1nA	MEASURED

delay, $\tau_T = 1/(2\pi f_T)$, against reciprocal emitter current is shown in Fig. 10 for a typical HP-505. The infinite current intercept indicates a total current independent time constant τ_0 of 17.5 ps. τ_0 equals $\tau_b + \tau_d + \tau_c$, where τ_c is the charging time of the base-collector capacitance through the collector resistance.

All of the time constants plus the parasitic capacitances and resistances needed to model S-parameters and noise figure versus frequency are listed in Table II for the case of a 0.4-μm emitter opening width. The value for the active base resistances R_{BA} was calculated assuming the effective emitter width was 0.1 μm wider than the emitter opening width due to lateral arsenic diffusion. An active base sheet resistance of 25 000 $\Omega/\square$ was required to give good agreement between measured and predicted noise figure.

The equivalent circuit used for the HPac-70GT package is shown in Fig. 9(b). Combining the chip and package equivalent circuits allows the calculation of all S-parameters and the associated gains. The modeled values of G_A(MAX), G_u(MAX), and $|S_{21}|^2$ gains as functions of frequencies are shown in Fig. 5. The modeled values of S_{11} and S_{22} are shown in Fig. 6.

Good agreement between measured and modeled parameters is obtained over the frequency range of interest.

B. Noise-Figure Modeling

The noise figure of the HP-505 was calculated using the approach taken by Hawkins [10] for the general case in which the emitter-base charging time τ_e is comparable to the neutral base delay τ_b. Using the notation of Table II, Hawkins' equation [10, eq. (19)] for minimum noise figure is

$$F_{\min} = a\,\frac{R_B + R_{OPT}}{r_e} + \left(1 + \frac{f^2}{f_b^2}\right)\frac{1}{\alpha_0}. \qquad (2)$$

The optimum source resistance is

$$R_{OPT} = \left\{R_B^2 - X_{OPT}^2 + \left(1 + \frac{f^2}{f_b^2}\right)\frac{r_e(2R_B + r_e)}{\alpha_0 a}\right\}^{1/2} \qquad (3)$$

and optimum source reactance is

TABLE II
HP-505 Equivalent Circuit Parameters

PARAMETER	SYMBOL	VALUE	HOW OBTAINED
ACTIVE BASE RESISTANCE	R_{BA}	12.5 OHMS	CALCULATED
EXTERNAL BASE RESISTANCE	R_{BE}	2.7 OHMS	CALCULATED
BASE INSERT RESISTANCE	R_{BI}	0.2 OHMS	CALCULATED
BASE CONTACT RESISTANCE	R_{BC}	< 0.2 OHMS	CALCULATED
EMITTER CONTACT RESISTANCE	R_{EC}	< 1 OHM	MEASURED
COLLECTOR RESISTANCE	R_C	5 OHMS	MEASURED AT DC
EMITTER-BASE DYNAMIC RESISTANCE	r_e	13 OHMS	kT/qI_e
ACTIVE BASE C-B JUNCTION CAPACITANCE	C_{BA}	.0040 pF	CALCULATED
EXTERNAL BASE C-B JUNCTION CAPACITANCE	C_{BE}	.0032 pF	CALCULATED
BASE INSERT C-B JUNCTION CAPACITANCE	C_{BI}	.050 pF	CALCULATED
BASE BONDING PAD CAPACITANCE	C_{BP}	.05 pF	CALCULATED
EMITTER BONDING PAD CAPACITANCE	C_{EP}	.05 pF	CALCULATED
EMITTER-BASE JUNCTION CAPACITANCE	C_{Te}	0.7 pF	τ_e/r_e
EMITTER-BASE JUNCTION CHARGING TIME	τ_e	9 psec	$\tau_T - (\tau_b + \tau_c + \tau_d)$
NEUTRAL BASE DELAY TIME	τ_b	7 psec	FROM τ_T VERSUS $1/I_e$
CHARGING TIME OF C_{BC} THROUGH R_C	τ_c	0.5 psec	$R_C C_{BC}$
COLLECTOR DEPLETION REGION DELAY TIME	τ_d	10 psec	$W_d/2v_s$ ($v_s = 8.5 \times 10^6$ cm/sec)
LOW FREQUENCY COMMON-BASE GAIN	α_o	.990	$h_{FE}/(1 + h_{FE})$
TOTAL BASE RESISTANCE	R_B	15.6	$R_{BA} + R_{BE} + R_{BI} + R_{BC}$
TOTAL BASE-COLLECTOR CAPACITANCE	C_{BC}	.107 pF	$C_{BA} + C_{BE} + C_{BI} + C_{BP}$
TOTAL DELAY TIME	τ_T	26.5 psec	$(\tau_e + \tau_b + \tau_c + \tau_d) = 1/2\pi f_T$
EMITTER-BASE JUNCTION CUT-OFF FREQUENCY	f_e	17.7 GHz	$1/2\pi\tau_e$
BASE CUT-OFF FREQUENCY	f_b	22.7 GHz	$1/2\pi\tau_b$
COMMON-EMITTER GAIN-BANDWIDTH PRODUCT	f_T	6 GHz	MEASURED

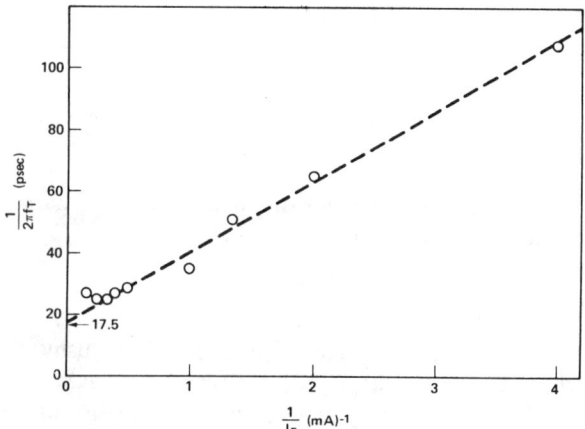

Fig. 10. Plot of $1/(2\pi f_T)$ versus the reciprocal of I_e for typical HP-505.

$$X_{OPT} = \left(1 + \frac{f^2}{f_b^2}\right) \frac{2\pi f C_{Te} r_e^2}{\alpha_0 a} \quad (4)$$

where

$$a = \left\{\left(1 + \frac{f^2}{f_b^2}\right)\left(1 + \frac{f^2}{f_e^2}\right) - \alpha_0\right\} \frac{1}{\alpha_0}.$$

It should be noted that the expressions for optimum source resistance and reactance were derived neglecting the parasitic bonding pad capacitances and collector-base junction capacitances in the equivalent circuit shown in Fig. 9(a). This simplification has little effect on the minimum noise figure but will affect the optimum chip source impedance to some extent.

The calculated minimum noise figures as a function of frequency are shown in Fig. 7 for HP-505 transistors with assumed emitter opening widths of 0.3 and 0.5 μm. Good agreement between experimental and calculated noise figure is obtained over a broad frequency range if, as mentioned above, an active base sheet resistance of 25 000 Ω/□ is assumed.

It is also possible to obtain equally good agreement between experimental and calculated noise figure using the general expression derived by van der Ziel et al. [11] under the assumption that $\tau_e \ll \tau_b$. It is necessary, however, to assume a rather large value for f_α (15 GHz). In addition to eliminating the need to know f_α, Hawkins' formulation also has been found to be more successful in modeling the current dependence of the minimum noise figure.

C. Process Simulation

In order to better understand the dependence of the critical impurity profiles on the fabrication process, we have used the Stanford University Process Engineering Models (SUPREM) program [6] for first-order process simulation. When the processing steps such as oxidation, implantation, and drive-in, and operation conditions such as temperature and time are sequentially specified, the program will output, at the end of each step, the one-dimensional profiles of all the dopants present in the silicon and SiO_2. The junction depths and

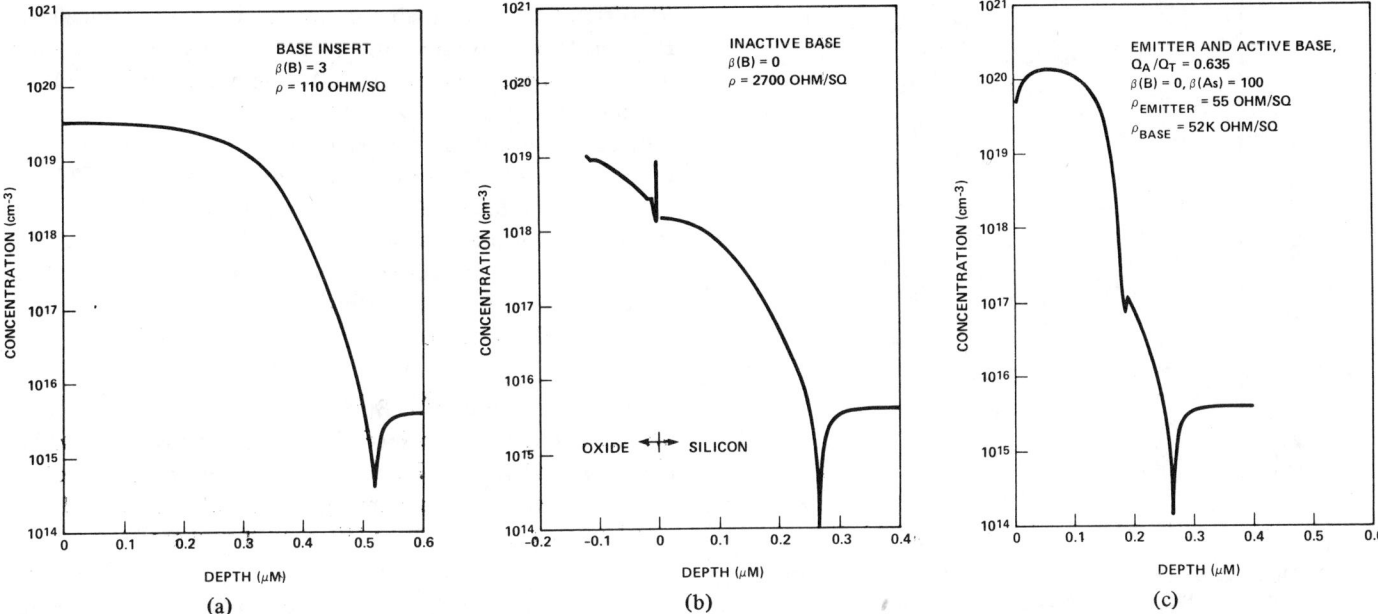

Fig. 11. Doping profiles from SUPREM simulation. (a) Base insert. (b) Inactive base. (c) Active base and emitter.

sheet resistances are also calculated. The use of SUPREM for simulation of the impurity profiles of microwave transistors is highly attractive compared to difficult and costly experimental techniques.

In order to obtain a reasonable first-order simulation, it is necessary, however, to account for the concentration dependence of impurity diffusion and the fractional electrical activity of implanted arsenic. The concentration dependence of the diffusion constant is simulated by

$$D = D_i \left(1 + \beta \frac{C}{n_i}\right) \bigg/ (1 + \beta) \qquad (5)$$

where D_i is the intrinsic diffusivity, C is the impurity concentration, β is a measure of the relative impurity diffusion effectiveness of charged vacancies as compared to neutral ones, and n_i is the intrinsic carrier concentration. From Fair and Tsai [12], it is calculated that at the end of the process cycle the electrically active arsenic dose Q_A is approximately 0.64 the total implanted dose Q_T. Thus the value of Q_A, rather than Q_T, is used in the simulation.

The final simulated impurity profiles at the end of the process cycle for the base insert, the inactive base, and the active emitter-base are shown in Fig. 11. In spite of uncertainties involving the concentration dependence of diffusion constants and impurity electrical activity, it is felt that the simulated profiles are accurate enough for the purposes of this paper. The only major discrepancy is between the simulated active base sheet resistance of 52 000 $\Omega/\square$ and the 25 000-$\Omega/\square$ value required to accurately model noise figure and S-parameters. This discrepancy may be the result of a dependence of effective active sheet resistance on emitter width which arises in the limit of very narrow emitters due to geometrical effects [9], a current dependence of active base resistance or inaccuracies in simulating the emitter diffusion.

D. Prospects for Further Noise-Figure Reduction

From the above discussion on noise-figure modeling, it is clear that there are two directions to go which might result in further reductions in noise figure. One option would be to modify the emitter and base impurity profiles in order to reduce τ_e and τ_b. This would cause the most dramatic reduction in noise figure at the higher frequencies. Any change in the impurity profiles to reduce τ_e and τ_b would, however, be likely to increase R_{BA} and thus increase the plateau noise figure even if the dc current gain h_{FE} were kept constant. Clearly, any attempt to improve noise performance by experimenting with the emitter and base profiles would be complicated by the interdependence of τ_e, τ_b, R_{BA}, and h_{FE}. Nevertheless, this may be a fruitful area for future investigation.

A more straightforward, but technology-limited, approach to further noise-figure reduction is to reduce the active base resistance R_{BA} and external base resistance R_{BE} by further reducing the critical lateral dimensions. R_{BE} could be reduced by decreasing the separation between the p^{++} insert and the emitter. No dramatic improvement would result for the HP-505, however, since, as outlined in Table II, only about 20 percent of the total base resistance resides in the external inactive base. A more effective approach would be to further reduce the active base resistance R_{BA} by going to effective emitter widths less than 0.3 μm. A corresponding reduction in the optimum bias current would, of course, have to occur. The effectiveness of going to even smaller emitter widths is suggested by Fig. 12, which shows predicted noise figures at 0.1, 1, 2, and 4 GHz as a function of emitter opening width for an HP-505 with a constant p^{++} base insert width of 4 μm. In all cases, the effective emitter width was assumed to be 0.1 μm wider than the opening due to lateral arsenic diffusion. Even lower noise figures than those shown in Fig. 12 should be achievable by reducing the inactive external base resistance

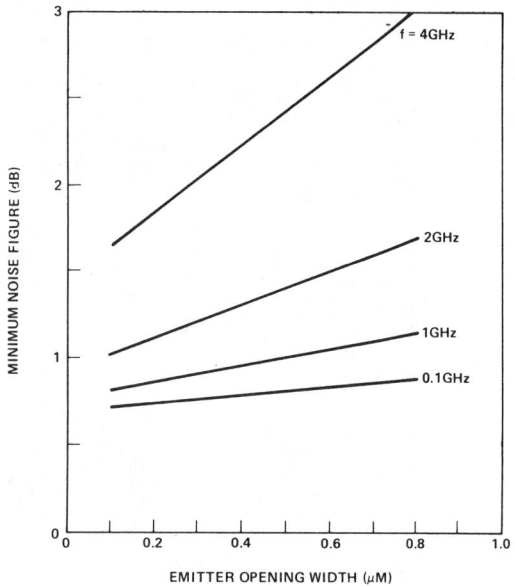

Fig. 12. Modeled minimum noise figure at 0.1, 1, 2, and 4 GHz versus emitter opening width for HP-505 chip fabricated by lateral etch process.

and the amount of lateral arsenic diffusion from the emitter window.

IV. Summary and Conclusions

A process has been investigated which uses ion implantation, local oxidation, and lateral etching in the fabrication of microwave transistors with sub-half-micrometer emitters. From the fabrication sequence that has been illustrated, we can see that the process allows the formation of 0.3-μm emitter openings using noncritical (1.5-μm) resolution masks. This eliminates the necessity of using very sophisticated and expensive electron-beam lithography. In addition to self-aligning the emitter and base inserts, the process enables the spacing between base insert and emitter to be very small (~ 0.3 μm). Also the formation of emitters by local oxidation results in a high emitter-base breakdown voltage ($BV_{EB0} \sim 4$ V) without compromising noise performance.

Typical experimental noise figures are 1.1 dB at 1.5 GHz and 2.3 dB at 4 GHz. The best units, which probably had emitter opening widths of $\simeq 0.3$ μm, exhibited 1.0-dB noise figures with 13-dB associated gain at 1.5 GHz, and 2.0-dB noise figures with 8-dB associated gain at 4 GHz. The low-current bias required for minimum noise figure is an attractive feature for applications requiring micropower dissipation.

It was found that a noise-figure model based on empirical time constants and a regional analysis of transistor structure is still effective for predicting the noise performance of these bipolar transistors with their extremely small critical dimensions. First-order approximations of the critical emitter and base impurity profiles were obtained by use of a process simulation computer program.

Finally, although the reported device performance advances the state of the art for low-noise bipolar transistors [2], it is clear that no theoretical performance barrier has yet been reached. Projections were made which indicate that noise figures < 1.7 dB at 4 GHz and < 0.8 dB at 1 GHz should be achievable with emitter opening widths of 0.1 μm.

Acknowledgment

We are very grateful to B. Feaver and S. Richards for wafer fabrication and to L. Nevin for many discussions on noise-figure and S-parameter modeling. R. Stewart developed the equivalent circuit used for the HPac-70GT package.

References

[1] H. F. Cooke, "Microwave transistors: Theory and design," *Proc. IEEE*, vol. 59, pp. 1163–1181, Aug. 1971.
[2] J. A. Archer, "Low-noise implanted-base microwave transistor," *Solid-State Electron.*, vol. 17, pp. 387–393, 1974.
[3] J. M. Pankyatz and H. T. Yuan, "A high-gain low-noise transistor fabricated with electron beam lithography," in *IEEE Int. Electron Devices Meet. Dig.*, pp. 44–46, Dec. 1973.
[4] T. N. Tsai and L. D. Yau, "Low noise microwave bipolar transistors fabricated by electron and photon lithography," in *IEEE Int. Electron Devices Meet. Dig.*, pp. 202–204, Dec. 1976.
[5] C. P. Snapp, T. H. Hsu, and R. W. Wong, "Process technology and modeling of a low-noise silicon bipolar transistor with sub-micron emitter widths," in *IEEE Int. Microwave Symp. Dig.*, pp. 104–106, June 1976.
[6] D. A. Antoniadis, S. E. Hansen, R. W. Dutton, and A. G. Gonzalez, "SUPREM 1—A program for IC process modeling and simulation," Stanford Univ., Tech. Rep. 5019-1, May 1977.
[7] H. Kamioka, H. Takeda, and M. Takagi, "A new submicron emitter formation with reduced base resistance for ultra high speed devices," *IEEE Int. Electron Devices Meet. Dig.*, pp. 279–282, Dec. 1974.
[8] R. W. Wong and R. D. Stewart, U.S. Patent 3 828 228, Aug. 6, 1974.
[9] R. L. Kronquist, J. Y. Forrier, J. P. Pestil, and M. E. Brilman, "Determination of a microwave transistor model based on an experimental study of its internal structure," *Solid-State Electron.*, vol. 18, pp. 949–963, 1975.
[10] R. J. Hawkins, "Limitations of Nielsen's and related noise equations applied to microwave bipolar transistors, and a new expression for the frequency and current dependent noise figure," *Solid-State Electron.*, vol. 20, pp. 191–196, 1977.
[11] A. Van der Ziel, J. A. Cruz-Emeric, R. D. Livingstone, J. C. Malpass, and D. A. McNamara, "A more accurate expression for the noise figure of transistors," *Solid-State Electron.*, vol. 19, pp. 149–151, 1976.
[12] R. B. Fair and J. C. C. Tsai, "The diffusion of ion-implanted arsenic in silicon," *J. Electrochem. Soc.*, vol. 22, no. 12, pp. 1689–1696, Dec. 1975.

Part III
Noise Properties of Field-Effect Transistors

Microwave Field-Effect Transistors—1976

CHARLES A. LIECHTI, SENIOR MEMBER, IEEE

Abstract—A review of recent and current work on microwave FET's and amplifiers is presented, and an extensive bibliography of recent articles is appended (250 references). First, the various FET structures (MESFET's, JFET's, and IGFET's) and their performances are reviewed. Second, the principle of operation is outlined for Si- and GaAs-MESFET's; the basic device physics, equivalent circuit, high-frequency limitations, and noise behavior are treated. Third, the design principles and performance of microwave MESFET amplifiers are summarized.

I. INTRODUCTION

IT TOOK seventeen years from the initial introduction of the bipolar transistor in 1948 until microwave transistors with practical gain and noise figure became available. In 1965 germanium transistors invaded L band with noise figures under 6 dB. A milestone was reached in 1968 when the AT&T System adopted the balanced transistor amplifier, developed by Engelbrecht and Kurokawa [A1], for use in its S- and C-band microwave communication links. Since 1968 significant progress has been made in obtaining low-noise performance and high-power capability from bipolar transistors for frequencies reaching up to X band. At 8 GHz a silicon n-p-n transistor with 3.9-dB noise figure and 3.8-dB associated gain [A2], and a power transistor with 1-W CW output power[1] and 6-dB power gain [A3] have been reported. At 2 GHz a single (silicon) transistor chip delivers up to 30-W CW output power with 7-dB power gain and 32-percent power-added efficiency [A7]. Bipolar transistors fabricated in GaAs are still in an early stage of development [A8]–[A10].

By 1971, however, breakthroughs had been made in the development of field-effect transistors. Today, GaAs *m*etal *s*emiconductor *f*ield-*e*ffect *t*ransistors (MESFET's) have higher gain, higher power-amplification efficiency, and lower noise figure than bipolar transistors above 4 GHz. More significant is the fact that FET's promise a great deal of potential for further advances. Substantial progress can be expected in the near future because of the following.

1) A large variety of FET structures (MESFET, JFET, IGFET) are suitable for microwave amplification and power generation, and some promising candidates are only in an early stage of development.

2) A variety of semiconductor materials (GaAs, InP, $In_xGa_{1-x}As$, $InAs_xP_{1-x}$) with majority-carrier transport properties[2] superior to silicon are competing for application in FET's [F5].

3) Further miniaturization to submicron dimensions can be realized in most FET structures.

4) Monolithic integration of circuits on semi-insulating substrates enables device isolation with low parasitic capacitances, low-loss interconnections, and high packing density.

This paper reviews recent and current developments in high-frequency FET's and FET amplifiers.[3] In Section II the various microwave FET structures are discussed, and their performance characteristics are summarized. Section III gives an introduction to the device physics, the small-signal characteristics, and the noise behavior of MESFET's. In Section IV the design principles and the performance of microwave amplifiers are reviewed. An extensive bibliography of recent articles is appended. Because of limited space, topics such as fabrication, packaging, reliability, oscillators, mixers, modulators, and digital applications are not treated here.

II. MICROWAVE FET STRUCTURES AND THEIR PERFORMANCE

Today MESFET's, p-n junction FET's (JFET's), and insulated-gate FET's (IGFET's) are used at microwave frequencies [B2]. The cross sections of the various FET structures are illustrated in Fig. 1. Their performance characteristics are tabulated in Table I. Fig. 1 serves as a guide for the following discussion; i.e., first the MESFET's, then the JFET's, and finally the IGFET's are described.

A. MESFET

In 1969 Middelhoek realized a silicon MESFET with 1-μm gate length by projection masking [X1], [X2]. This FET had a 12-GHz maximum frequency of oscillation [C1] which was considerably higher than for previously known FET's and comparable to f_{max} of the best bipolar transistors at that time. The next significant step was the fabrication of 1-μm MESFET's on GaAs. As a result, FET's with f_{max} of 50 GHz and useful gain up to 18 GHz became available in 1971 [C2]–[C6]. This substantial improvement in device performance is due to the following reasons. 1) In GaAs the conduction electrons have a six times larger mobility and a two times larger peak drift velocity than in silicon [F1]. Therefore, parasitic resistances are smaller, the transconductance is larger, and the transit time of electrons in the high-field region is shorter. 2) The active layer is grown on a semi-insulating GaAs substrate with resistivity larger than $10^7 \Omega \cdot cm$. The large parasitic capacitance of the gate bonding pad can thus be eliminated by positioning the pad on the substrate.

In 1972 it became apparent that GaAs-MESFET's are capable of very low-noise amplification [C4]–[C6]. Liechti

---

Manuscript received October 1, 1975; revised January 15, 1976.
The author is with the Solid State Laboratory, Hewlett-Packard Company, Palo Alto, CA 94304.
[1] Combined power from two chips.
[2] Minority-carrier lifetime and mobility are of no concern in unipolar transistors. Therefore, materials with optimized transport properties for electrons but poor performance for holes can be chosen.

[3] The reader is also referred to a review paper on the same topic by Turner [B1].

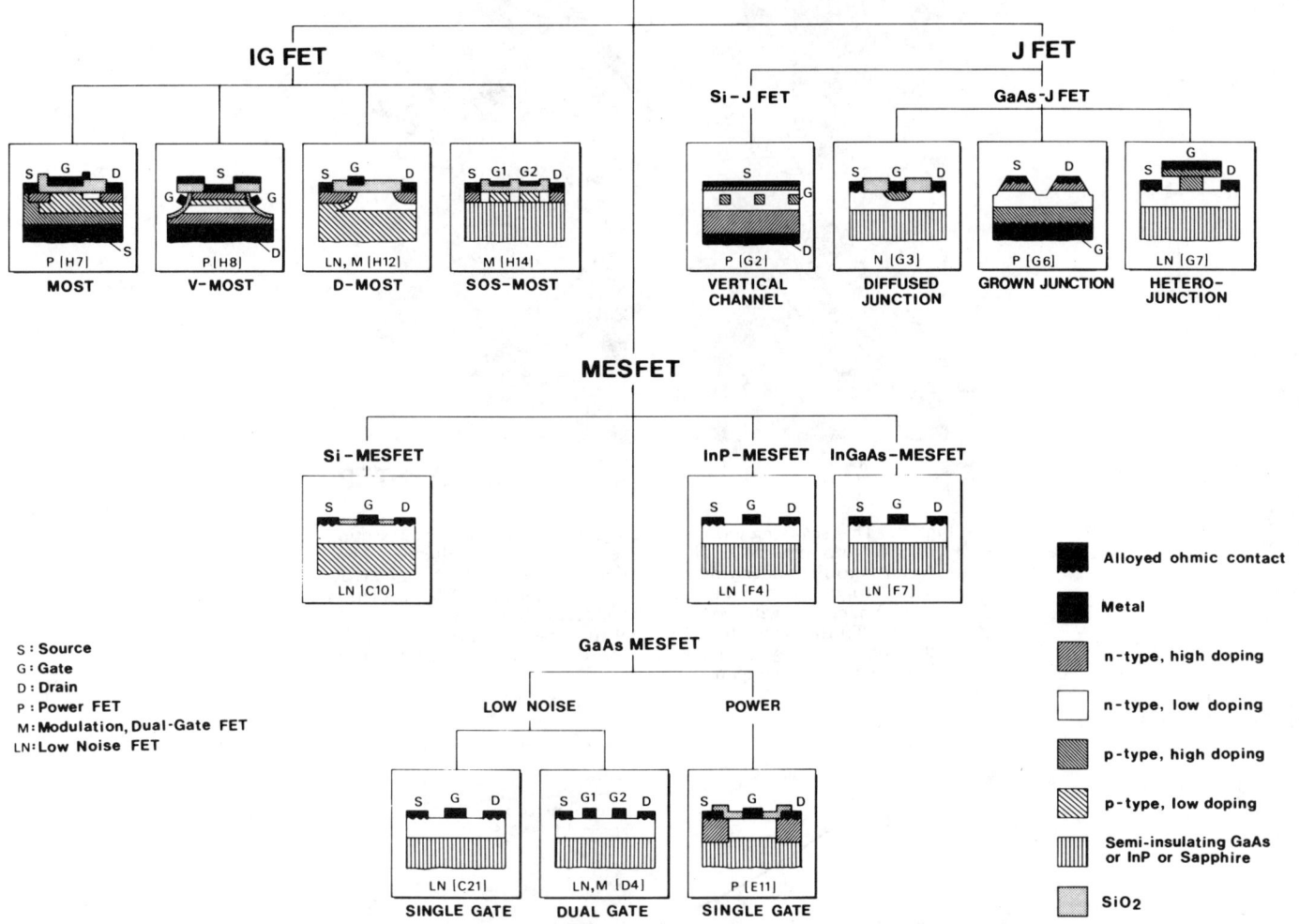

Fig. 1. This "family tree" of microwave FET's shows the cross sections of the various FET structures. Their performance characteristics are listed in Table I.

TABLE I
PERFORMANCE DATA OF THE MICROWAVE FET STRUCTURE SHOWN IN FIG. 1

Type*	Semi-conductor	Single/Dual Gate	Type	Channel Length (μm)	Channel Width (mm)	Application*	Frequency (GHz)	Output Power CW (W)	Assoc. Power Gain (dB)	Small Signal Gain (dB)	Power** Added Effic. (%)	Noise Figure (dB)	Assoc. Gain (dB)	Casc. Noise Figure† (dB)	Max. Avail. Gain (dB)	Reference
MESFET																
Silicon	Si	SG	n	0.5		LN	10					5.8			5.9	[C10]
LN SG GaAs	GaAs	SG	n	0.5		LN	10					2.7	10.5	3.1	13	[C21]
LN SG GaAs	GaAs	SG	n	1		LN	10					3.2	8.0	3.6	10	[D4]
LN DG GaAs	GaAs	DG	n	1		LN,M	10					4.0	12	4.2	18	[D4]
P SG GaAs	GaAs	SG	n	1.5	5.2	P	8	2.2	3.2	4.2	22					[E11]
P SG GaAs	GaAs	SG	n	1.2	0.6	P	22	0.14	4.8	5.6	9					[E17]
LN SG InP	InP	SG	n	1		LN	10					4.7	6.6	5.4	7.8	[F4]
LN SG InGaAs	InGaAs	SG	n	1		LN	7					5.7	5.0	7.0	18	[F7]
JFET																
Silicon	Si	SG	n	1	1.8	P	2.7	0.2	6.0		19					[G2]
Diff. Junction	GaAs	SG	n	2		LN	4					2.5	10	2.7	10	[G3]
Grown Junction	GaAs	SG	n	1.5	6.1	P	6	1.0	6.0	7.0	26					[G6]
Heterojunction	GaAs	SG	n	2		LN										[G7]
IGFET																
MOST	Si	SG	n	5	20	P	0.7	16	6	10	26					[H7]
V-MOST	Si	SG	n	1	18	P	2	4.0	5	6	32					[H9]
SG D-MOST	Si	SG	n	1		LN	1					3.0	9.0	3.3	–	[H13]
DG D-MOST	Si	DG	n	1		LN,M	1					4.5	14	4.6	15	[H13]
DG SOS-MOST	Si	DG	n	4		M	0.5								25	[H14]

* LN - low noise amplification; P - power amplification; M - modulation, switching or amplification with controlled gain; SG - single-gate FET; DG - dual-gate FET.

** The power added efficiency is defined as the RF output power minus the RF input power divided by the dissipated dc power.

† The cascaded noise figure is defined by Eq. (15).

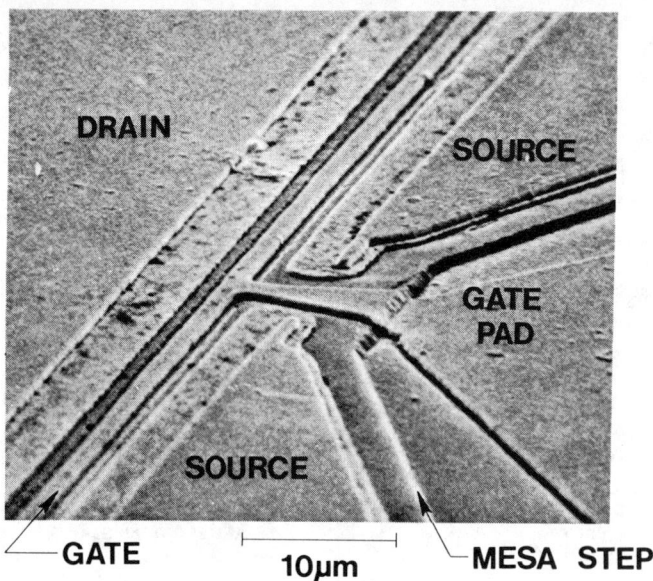

Fig. 2. This scanning electron micrograph shows the center section of a low-noise MESFET. The source and drain are alloyed ohmic contacts to the underlying conductive layer. The gate is the narrow metal stripe forming a Schottky contact. The width of the depletion layer under the gate controls the current flowing from drain to source. To the right, the gate metal runs over a mesa step (edge of the conductive layer) and widens on the semi-insulating substrate into a large bonding pad.

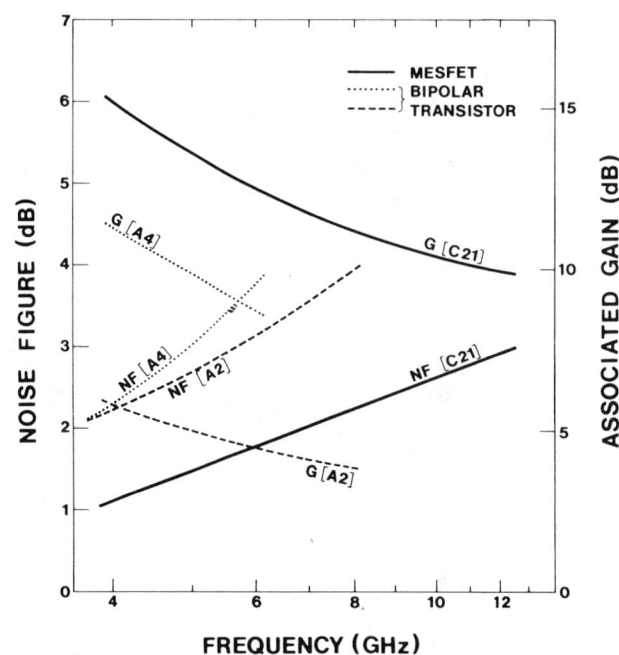

Fig. 3. Lowest reported noise figures and associated gains of microwave transistors are plotted versus frequency. The GaAs-MESFET reported by Ogawa et al. [C21] has 0.5-μm gate length. The dashed line represents the Si bipolar transistor with the lowest published noise figure [A2]. The bipolar transistor with the dotted line [A4] has a considerably higher gain.

et al. [C5] reported a noise figure of 3.5 dB with 6.6-dB associated gain at 10 GHz. Baechtold [M7], [M8] proposed a noise model that agrees well with measurements taking carrier velocity saturation and intervalley scattering into account. In the past few years, small-signal GaAs-MESFET's have been fabricated and characterized by various laboratories [C7]–[C28]. A scanning-electron micrograph of a typical low-noise MESFET is shown in Fig. 2. Lowest reported noise figures and associated gains are plotted in Fig. 3 versus frequency. Best noise performance is achieved with 1) a high-purity buffer layer between the substrate and the active layer [C26]; 2) a high doping level in the active n-layer (2.5×10^{17} cm^{-3} [C21]); 3) smallest possible source and gate-metal resistance [C21], [C22]; and 4) short gate length (0.5 μm [C10], [C21]).

A MESFET structure with two gates is shown schematically

in Fig. 1. This FET has a higher gain and a lower feedback capacitance than the single-gate counterpart [C21], [D1]–[D7]. In addition, the gain can be controlled over a wide range (44 dB [D4]) by varying the dc bias of the second gate. This feature can be used for automatic gain control in amplifiers [D6]. The gain modulation response is very fast. Pulse-amplitude modulation of an RF carrier with less than 100-ps fall and rise times has been demonstrated [D3], [D4].

The GaAs-MESFET is not limited to small-signal low-noise applications. The first power MESFET's appeared in 1973 and were of planar construction as shown in Fig. 4. Fukuta et al. [E1] designed a MESFET with 20 gates, each 1 μm long and 400 μm wide, operated in parallel and interconnected with a second metallization layer. At 2 GHz this MESFET exhibited 1.6-W output power[4] with 5-dB power gain and 21-percent power-added efficiency. At the same time, Napoli et al. [E2] presented a power transistor with self-aligned[5] gates [Fig. 4(c)]. Multiple gate, source, and drain pads had to be interconnected with bonding wires. The planar power FET has been further developed by various laboratories [E3]–[E19]. The techniques yielding high-power capability per unit gate width are:

1) the use of a high-resistivity epitaxial buffer layer to isolate the active layer from the bulk-grown substrate [E16];
2) the addition of inlaid n$^+$-regions under the source and drain electrodes, shown in Fig. 4(b), to increase the drain-source breakdown voltage and decrease the parasitic contact resistances [E11];
3) the design of short gate branches to prevent current crowding and to lower the gate-metal resistance [E11];
4) the flip-chip mounting of the transistor to decrease the thermal resistance and the source-to-ground lead inductance [E7], [E15].

A photomicrograph of a power MESFET from Fujitsu Laboratories [E9] with 104 gate branches, each 1.5 μm long and 50 μm wide, is shown in Fig. 5. Recently reported performance data of power MESFET's are summarized in Table II. The output powers from single chips, operated CW Class A, range from 4 W at 4 GHz to 0.14 W at 22 GHz and power-added efficiencies vary between 44 and 9 percent. Highest efficiency has been obtained by Huang et al. [E18] in Class B operation; 68 percent at 4 GHz and 41 percent at 8 GHz have been measured. These are the highest reported efficiencies of all microwave solid-state devices in this frequency range. The performance figure of merit M_F proposed by Drukier et al. [E15] is also listed in Table II. M_F measures the added RF power per unit gate width times the square of the operating frequency. The MESFET

---
[4] Measured at 1-dB gain compression.
[5] The location of the gates between the source and drain edges is defined by a fabrication method which does not require a critical realignment step.

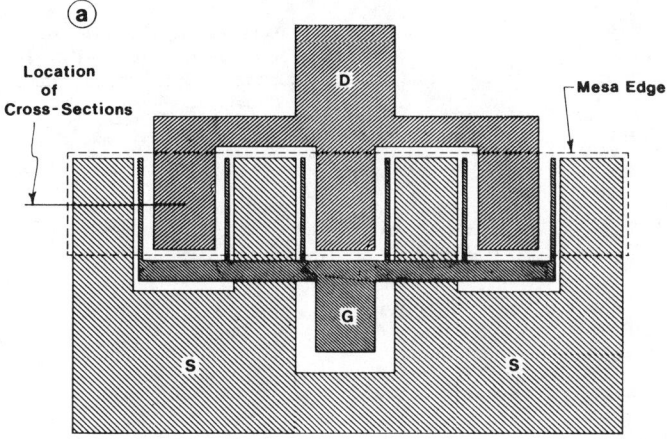

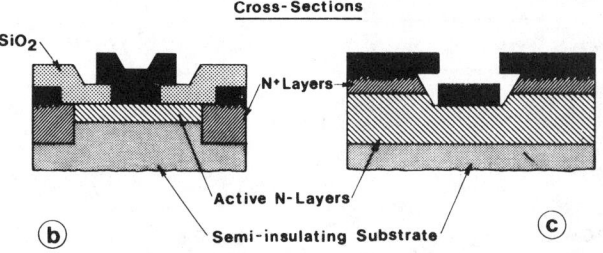

Fig. 4. (a) Illustrates schematically the metallization layout of the planar power MESFET shown in Fig. 5. The gate branches are interconnected with a metal line that crosses over the source. Multigate MESFET's with cross sections (b) [E11] and (c) [E1], [E12]–[E18] have been realized.

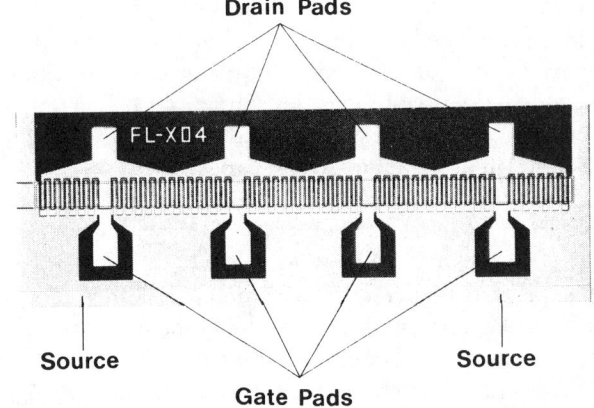

Fig. 5. This photomicrograph shows a power MESFET with 104 gate branches, each 1.5 μm long and 50 μm wide, connected in parallel. The chip delivers 0.9-W output power with 6-dB power gain and 30-percent power-added efficiency at 6 GHz [E9]. (Courtesy of M. Fukuta, Fujitsu Laboratories.)

with the largest M_F delivers 140-mW output power with 4.8-dB power gain and 9-percent power-added efficiency at 22 GHz [E17].

A new device with better heat-sink properties has been proposed by Blocker et al. [E5] for power MESFET's.[6] The metallization layout and the cross section of this MESFET are

---
[6] A similar structure has been realized by Vergnolle et al. [G6] in the form of a GaAs p-n junction FET (see below).

TABLE II
PERFORMANCE OF EXPERIMENTAL MICROWAVE POWER TRANSISTORS
CW DATA OF SINGLE-CHIP GaAs-MESFET's,[1] GaAs-JFET's,[2] AND Si BIPOLAR TRANSISTORS[3]

Frequency (GHz)	Transistor Type	Output Power (W)	Power Gain[4] (dB)	Small-Signal Gain[5] (dB)	Power Added Efficiency[6] (%)	Figure of Merit[7] (WGHz²/mm)	Gate Length[8] (µm)	Total Gate Width[9] (mm)	Number of Cells on Chip	Operating Conditions (Class A,B,C)	Company [Reference]
2.0	Bipolar	30	7.0	-	32	2.2	1.5	43	10	C	MSC [A7]
4.0	Bipolar	8.0	7.0	-	25	4.3	1.0	24	8	C	TRW [A6]
	MESFET	4.0	6.0	7.0	44	9.2	1.5	5.2	2	A	Fujitsu [E11]
6.0	Bipolar	1.5	4.4	-	23	11	1.5	3.1	4	C	Hewlett-Packard [A5]
	JFET	1.0	6.0	7.0	26	4.4	1.5	6.1	2	A	Thomson-CSF [G6]
	MESFET	2.7	5.0	6.0	31	13	1.5	5.2	2	A	Fujitsu [E11]
8.0	Bipolar	0.5	6.0	-	22	22	0.5	1.1	2	B	Texas Instruments [A3]
	MESFET	0.6	6.0	7.5	34	21	1.5	1.4	3	A	Plessey [E19]
	MESFET	2.2	3.2	4.2	22	14	1.5	5.2	2	A	Fujitsu [E11]
15.0	MESFET	0.45	5.2	6.7	13	59	1.5	1.2	2	A	RCA [E17]
22.0	MESFET	0.14	4.8	5.6	9	76	1.5	0.6	1	A	RCA [E17]

[1] Operated with common source.
[2] Operated with common gate.
[3] Operated with common base.
[4] Associated with the stated output power.
[5] Only applicable for Class A operation.
[6] Defined as RF output power minus RF input power divided by the dissipated dc power.
[7] Defined as added RF power × (frequency)² divided by the total gate width or emitter periphery.
[8] Or emitter finger width.
[9] Or total emitter periphery.

shown in Fig. 6. Interdigitated source and drain fingers are located on the top side of the chip, and the gate with plated heat sink is located on the bottom side. The channel is confined to an area within the constricted cross section in the n-layer. This structure has the following advantages: 1) it places the active region in intimate thermal contact with the heat sink; 2) it enables an interdigitated structure without overcrossing, since the ohmic contacts and the gate are located on different sides of the chip; 3) it reduces the parasitic source-to-gate resistance; and 4) it makes a self-aligned process possible. Disadvantages are the higher gate-to-drain and gate-to-source capacitances.

Besides Si and GaAs, InP has been investigated for application in MESFET's. InP has a 50-percent higher maximum drift velocity than GaAs[7] [F2], [F3]. Therefore, one expects a higher current-gain bandwidth f_T for InP-MESFET's. Barrera and Archer [F4] have measured an f_T that is 1.6 times larger than in analogous GaAs devices. However, the maximum frequency of oscillation, f_{max}, is 20 percent lower. Degenerate feedback resulting from a large gate-to-drain capacitance and a small output resistance degrades the gain performance. Noise figures of InP-MESFET's are slightly higher.

The ideal semiconductor for FET's has simultaneously a large mobility, large maximum drift velocity, and large avalanche breakdown field. Consequently, a small electron effective mass, a large intervalley separation, and a large energy gap are required. The first and last requirements are conflicting, since a large energy gap implies a large effective mass; but it is possible to design a better compromise than is found in the binary compounds GaAs and InP.

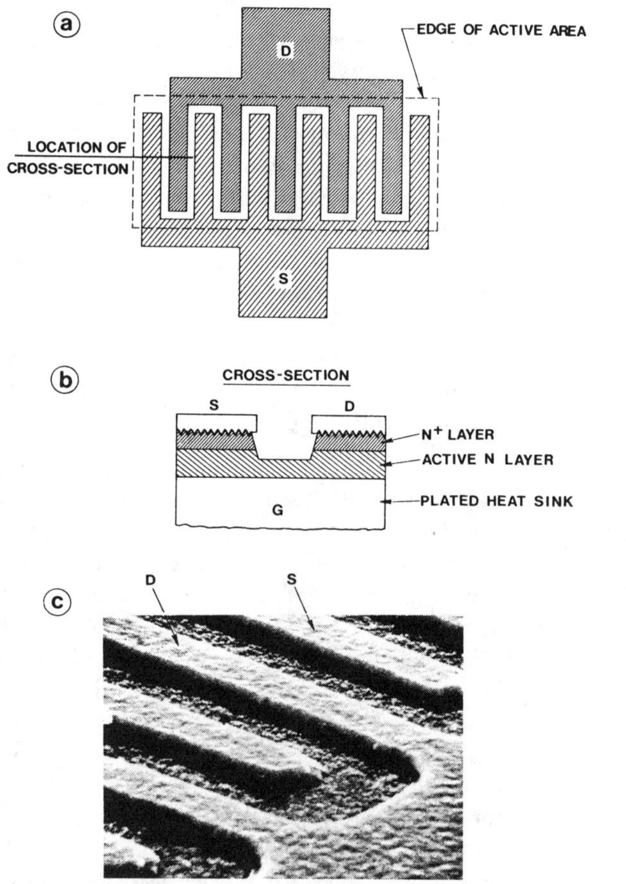

Fig. 6. Power MESFET structure with plated heat sink on the gate as proposed in [E5]. (a) Shows the metallization on the top side of the chip with interdigital source and drain fingers. Outside the active area, the GaAs is converted into semi-insulating material by proton bombardment. (b) Shows the cross section of the device with the source and drain on top and the gate with plated heat sink on the bottom of the chip. (c) Shows the top view of the FET chip [G6]. (SEM courtesy of C. Vergnolle, Thomson-CSF.)

[7] The low field mobility of InP is 25 percent lower, however.

Two systems which merit consideration are the mixed crystals InAs–InP and InAs–GaAs [F5], [F6]. Decker et al. [F7] have grown thin films of $In_{0.04}Ga_{0.96}As$ directly on GaAs. MESFET's fabricated on this material are very similar to their GaAs counterpart with the exception of a larger output resistance. The step in the bandgap at the InGaAs–GaAs interface constrains the electrons to the InGaAs layer and prevents penetration of hot electrons into the substrate at the high-field channel region.

B. JFET

A great deal of effort has been spent on developing p-n junction FET's for microwave applications. In the past, JFET's have not moved to higher frequencies as rapidly as MESFET's primarily because of the difficulty in realizing closely defined p-conducting regions by diffusion or implantation.[8] In more recent developments, such as the power JFET with ion-milled channel [G6] or the heterojunction JFET [G7], these limitations do not apply and rapid advances can be expected.

In 1972 Teszner [G1] described an Si-JFET with a vertical channel for power amplification. This multielement transistor has buried p^+-gate fingers, and the ohmic contacts for source and drain are located on opposite sides of the chip (Fig. 1). The structure was later realized by high-energy ion implantation of the gate [G2]. In this vertical geometry, no implantation-induced defects are generated in the channel. Consequently, low annealing temperatures were used and patterns with 1-μm-long channels were realized. Test devices, with 1.8-mm total gate width, delivered 200-mW output power with 6-dB power gain at 2.7 GHz.

In GaAs, Zuleeg et al. [G3] have realized a low-noise JFET with a diffused p-n junction of 2-μm length. The transistor exhibits 2.5-dB noise figure and 10-dB associated gain at 4 GHz. This JFET has a high tolerance to fast neutron radiation. A 1-MeV neutron fluence of 5×10^{16} neutrons/cm^2 is required to degrade the transconductance by 10 percent [G4], [G5].

A power GaAs-JFET has been built by Vergnolle et al. [G6] [Fig. 6(c)]. The construction is similar to the one shown in Fig. 6(b), except that the Schottky barrier is replaced by a p-n junction grown on a p^+-substrate. This structure has the same advantages as the MESFET of Fig. 6(a) and (b), except that the thermal resistance is limited by the spreading resistance in the GaAs substrate. This JFET with 1.5-μm gate length and 6-mm total gate width delivers 1-W output power[9] with 6-dB power gain and 26-percent power-added efficiency at 6 GHz.

A low-noise FET with a heterojunction gate is now under development at Matsushita Electric [G7]. The junction consists of a p-type $Ga_{0.5}Al_{0.5}As$ layer grown on top of an n-type GaAs layer. The p-layer can be selectively etched to the shape of a 1-μm gate strip. This feature combined with a self-aligned process enables simple fabrication of GaAs-JFET's.

C. IGFET

Field-effect transistors with insulated gates are of interest for power amplification. They offer the following advantages over MESFET's or JFET's. 1) In the active region of an enhancement-mode MOSFET, the input capacitance and the transconductance are almost independent of gate voltage, and the output capacitance is independent of the drain voltage.[10] This leads to very linear (Class A) power amplification with low amplitude and phase distortion. 2) The active gate-voltage range can be larger because n-channel depletion-type IGFET's can be operated from the depletion-mode region $(-V_{GS})$ to the enhancement-mode region $(+V_{GS})$. Unfortunately, practical IGFET's have only been made on Si. Attempts to realize a usable device in GaAs have had only limited success [H1], [H2]. It is difficult to fabricate an insulating film which produces an interface to n-GaAs with a low density of electron states [H2]–[H4]. Recent advances with anodic native oxides of GaAs show promising interface and dielectric properties [H5]. A density of fast interface states of 1 to 2×10^{11} cm^{-2}eV^{-1} has been observed. This is still 10 times higher than the best results in Si [H6], but sufficiently low for MOSFET operation.

Morita et al. [H7] have developed an n-channel depletion-type Si-MOSFET which delivers 16-W output power[11] with 6-dB associated power gain and 26-percent power-added efficiency at 700 MHz. The device has a conventional planar structure with diffused n^+ drain and source regions and a 5-μm-long and 20-mm-wide channel. A very different approach is the vertical-gate MOS transistor (VMOST) designed for power amplification [H8]. This device consists of mesa strips with control gates on both sides of each strip (Fig. 1). The source contact is made to the top surface of the mesa, and the n^+-substrate acts as a common drain. The length of the vertical n-channel is controlled by epitaxial growth and the metal-gate length by angular metal deposition. For a 1-μm-long and 18-mm-wide channel, a CW output power of 4 W[12] with 5-dB power gain has been measured at 2 GHz by Heng [H9]. Also, a double-diffused MOSFET (D-MOST) can be fabricated with channel lengths of less than 1 μm using standard photolithography [H10]. The short channel results from subsequent diffusions of the channel and source impurities under the same oxide layer. This technique gives accurate control of the channel length comparable to the control over the base width in diffused bipolar transistors. The D-MOST has been developed for low-noise UHF amplification by Sigg et al.

---

[8] Lateral diffusion or lateral spreading of the ions during implantation and during the subsequent high-temperature annealing step widen the p-regions with respect to mask dimensions.

[9] Measured at 1-dB gain compression. For a performance comparison with MESFET's, see Table II.

[10] This applies to MOSFET's with n^- drift region.

[11] Measured at 4-dB gain compression.

[12] Measured at 1-dB gain compression.

[H11], [H12]. Single-gate FET's exhibit 3.0-dB noise figure and 9.0-dB associated gain at 1 GHz [H13]. Dual-gate FET's suitable for mixers and for amplifiers with automatic gain control have 4.5-dB noise figure with 14-dB gain [H13]. Silicon-on-sapphire MOS transistors (SOS-MOST) have also been developed for UHF applications with the advantage of very low drain-to-channel and drain-to-source capacitances. A small-signal dual-gate SOS-MOST has been built by Ronen and Strauss [H14] with a 4-μm channel length yielding 25-dB gain at 0.5 GHz.

D. Conclusion

The MESFET's have been the most successful among the microwave FET's in low noise and in power amplification above 2 GHz. The reasons for this are easy realization on GaAs and the fact that the two critical dimensions, the gate length and channel thickness, can be accurately controlled. With advanced fabrication processes, such as E-beam lithography [C3], [C17], ion implantation [C13], [C16], [C27], [W4], [X4], and molecular-beam epitaxy [C28], further improvement in dimensional control is obtained. In comparison with silicon bipolar transistors, MESFET's have higher maximum frequency of oscillation, lower noise figures, and larger associated gain at microwave frequencies (Fig. 3). They also have higher reverse isolation and lower third-order intermodulation distortion. Above 4 GHz, GaAs MESFET's have better efficiency as power amplifiers than bipolar transistors (Table II). In addition, FET's do not exhibit secondary breakdown, are self-ballasting,[13] and have inherently higher input impedance. Also, the gate-to-source and drain-to-source impedances are fairly insensitive to temperature variations [S6]. As majority-carrier devices, FET's are more immune to the effects of neutron and gamma radiation than bipolar transistors [G4], [G5], [Q13]. Currently, much effort is spent in determining the reliability, failure modes, and stability of GaAs MESFET's [I1]–[I9]. Preliminary results indicate a meantime to failure in excess of 10^7 h at 70°C channel temperature as reported by Irie et al. [I3] and Abbott and Turner [I5]. Ch'en et al. [I7], [I8] observed an improvement in the long-term stability of MESFET parameters after passivating the GaAs surface with a thin coating of polycrystalline GaAs.

III. MESFET Principle of Operation

In this section, the physical principles in the operation of a silicon MESFET are explained. Then the differences between silicon and GaAs-MESFET's are outlined and effects occurring in FET's with very short gate length are discussed. Next, the equivalent circuit is presented and high-frequency limitations are described. Finally, the principles of the noise behavior are presented.

---

[13] With rising temperature, the channel and source resistances increase, preventing "thermal runaway."

A. Principles of Silicon MESFET Operation

The current-voltage characteristic of a thin n-type silicon layer in which electrons are carrying the current is plotted in Fig. 7(a). This layer is supported by an insulating silicon substrate. At the surface of the conducting layer, two ohmic contacts are made, called the source and drain. A cross section of this device is shown in Fig. 7(a).[14] If a positive voltage V_{DS} is applied to the drain, electrons will flow from source to drain. Hence the source acts as the origin of carriers and the drain as a sink. For small voltages, the silicon layer behaves like a linear resistor. For larger voltages, the electron drift velocity does not increase at the same rate as the electric field E (Fig. 8). As a result, the current-voltage characteristic falls below the initial resistor line. As V_{DS} is further increased, E reaches a critical field, E_c,[15] for which the electrons reach a maximum velocity, v_s (Fig. 8). At this drain voltage, the current starts to saturate.

In Fig. 7(b) a metal-to-semiconductor contact, called the gate, has been added between source and drain. This contact creates a layer in the semiconductor that is completely depleted of free-carrier electrons. This depletion layer acts like an insulating region and constricts the cross section available for current flow in the n-layer. The width of the depletion region depends on the voltage applied between the semiconductor and the gate. In Fig. 7(b) the gate is shorted to the source and a small drain voltage is applied. Under these conditions, the depletion layer has a finite width and the conductive channel beneath has a smaller cross section d than in Fig. 7(a). Consequently, the resistance between source and drain is larger, as shown in Fig. 7(b). The current I_{DS} flowing from drain to source is given by

$$I_{DS} = wqn(x)v(x)d(x) \tag{1}$$

where w is the gate width (see Fig. 11), q the charge of an electron, n the density of conduction electrons, v their drift velocity, d the conductive layer thickness, and x the co-ordinate in the direction of the electron drift. The electron density n is equal to the constant donor density N_D as long as the field does not exceed the critical value E_c. The voltage along the channel increases from zero at the source to V_{DS} at the drain. Thus the metal-to-semiconductor junction becomes increasingly reverse biased, and the depletion layer becomes wider as we proceed from source to drain. The resulting decrease in conductive cross section d must be compensated by an increase of electric field and electron velocity v to maintain a constant current through the channel. As the drain voltage is increased further, the electrons reach the maximum limiting velocity v_s under the drain end of the gate. This is illustrated in Fig. 7(c). The channel is constricted to the smallest cross section d_0 under the gate

---

[14] For simplicity, the bending of the bands at the free surface of the n-layer and the depleted region at the substrate interface are neglected. Also the electric field is assumed to be uniform in the n-region between the contacts.

[15] In silicon, the value of E_c cannot be accurately defined.

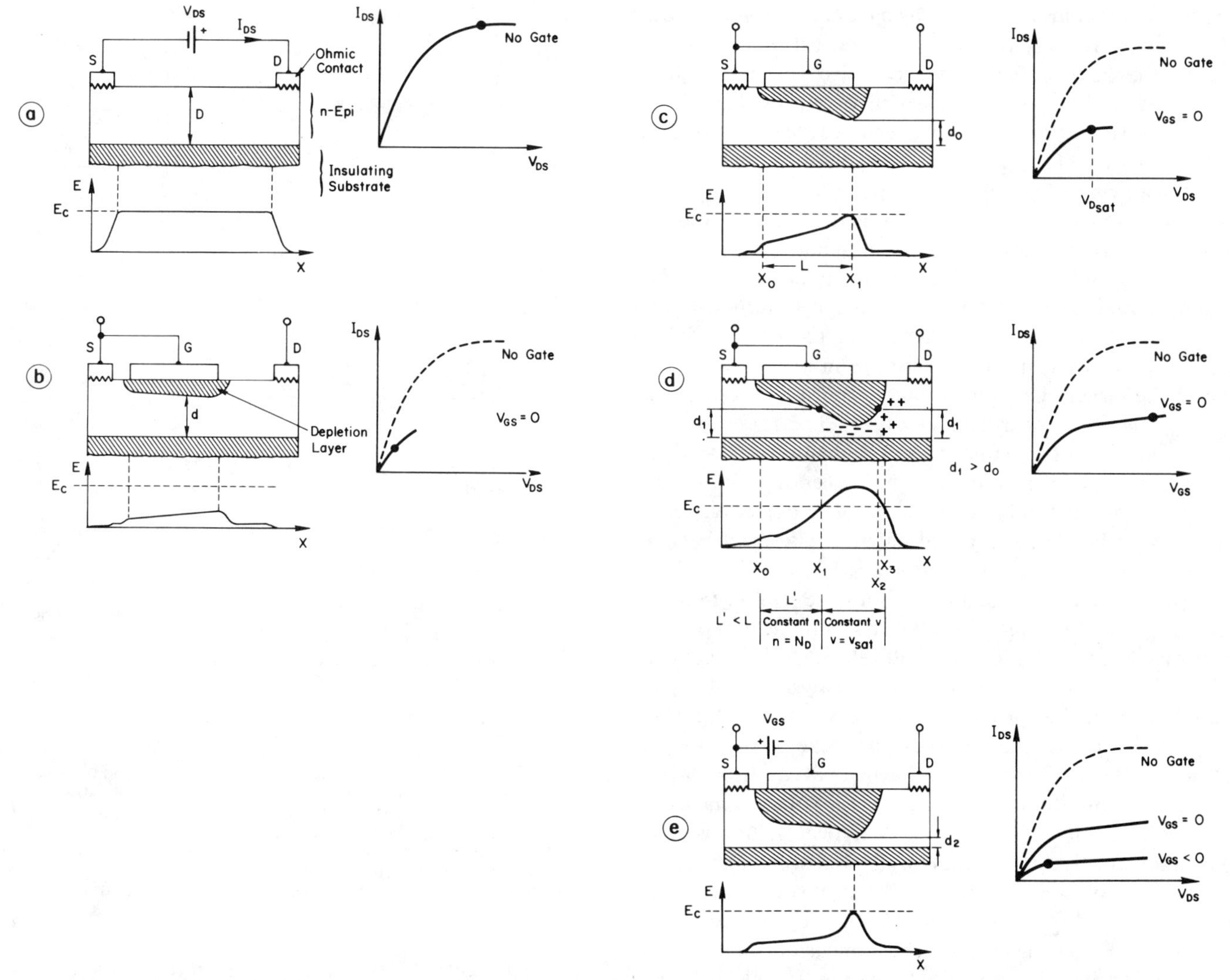

Fig. 7. (a) Shows the I–V characteristic of an n-type silicon layer with two ohmic contacts. The current saturates because the electrons reach a maximum drift velocity at the critical field E_c. In (b)–(d), the current is controlled by the depletion layer under a Schottky gate, shorted to the source. In (c) the current starts to saturate at $V_{D_{sat}}$, and (d) shows the formation of a stationary dipole layer in the channel for $V_{DS} > V_{D_{sat}}$ [J12], [K1]. (e) Illustrates the condition for a negative gate bias. The depletion layer is wider, it constricts the conductive cross section further, and causes the current to saturate at a lower level.

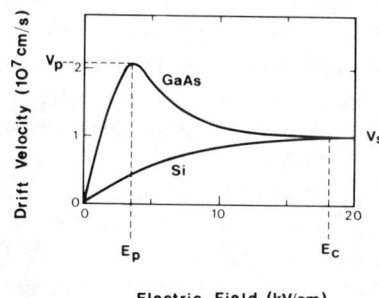

Fig. 8. Equilibrium electron drift velocity versus electric field in GaAs and silicon (after Ruch [K11]).

edge, the electric field reaches the critical value at this point, and the current starts to saturate.

If the drain voltage is increased beyond $V_{D_{sat}}$, the depletion region widens toward the drain. The point x_1, where the electrons reach the limiting velocity, moves slightly toward the source [Fig. 7(d)]. As x_1 moves closer to the source, the voltage at x_1 decreases.[16] Consequently, the conductive cross section d_1 widens and more current is injected into the velocity-limited region. This results in a positive slope of the I_{DS} curve and a finite drain-to-source resistance beyond current saturation [J8], [J12]. The effect is particularly pronounced in microwave MESFET's with short gate lengths.

Proceeding from x_1 toward the drain, the channel potential increases, the depletion layer widens, and the channel cross section d becomes narrower than d_1. Since the electron velocity is saturated, the change in channel width must be compensated for by a change in carrier concentration to maintain constant current. According to (1), an electron accumulation layer forms between x_1 and x_2, where d is smaller than d_1. At x_2 the channel cross section is again d_1 and the negative space charge changes to a positive space charge to preserve constant current. The

[16] The average field between x_0 and x_1 remains nearly unchanged while the distance between x_0 and x_1 decreases.

positive space charge is caused by partial electron depletion. The electron velocity remains saturated between x_2 and x_3 due to the field added by the negative space charge. In short, the drain voltage applied in excess of $V_{D_{sat}}$ forms a dipole layer in a channel that extends beyond the drain end of the gate [J12], [K1].

When a negative voltage is applied to the gate [Fig. 7(e)], the gate-to-channel junction is reverse biased, and the depletion region grows wider. For small values of V_{DS}, the channel will act as a linear resistor, but its resistance will be larger due to a narrower cross section available for current flow. As V_{DS} is increased, the critical field is reached at a lower drain current than in the $V_{GS} = 0$ case, due to the larger channel resistance. For a further increase in V_{DS}, the current remains saturated. In essence, the MESFET consists of a semiconducting channel whose thickness can be varied by widening the depletion region under the metal-to-semiconductor junction. The depletion region widening is the effect of a field or voltage applied between gate and channel of the transistor.

Various analytical solutions for the voltage-current characteristics of short-gate MESFET's with field-dependent electron velocity have been developed. The majority [J2]–[J7] follow a one-dimensional analysis based on the gradual-channel approximation proposed by Shockley [J1]. They compute the drain current at the onset of current saturation [Fig. 7(c)]. Two-dimensional approximations of the field distribution for large drain voltages have also been derived [J8]–[J12]. These analytical solutions make allowance for space charges in the channel and enable calculations of the small-signal drain-to-source resistance in the saturated current region. Much effort has also been concentrated on accurate two-dimensional numerical solutions for Si [K1]–[K7], for GaAs and InP [K8]–[K14].

B. Principles of Gallium Arsenide MESFET Operation

In GaAs, the analysis in the high-field region is considerably more complicated than in Si because 1) the equilibrium electron velocity versus electric field reaches a peak value at about 3 kV/cm, then decreases and levels off at a saturated velocity that is about equal to the limiting velocity in silicon (Fig. 8) [F1], [K11]; 2) for gate lengths shorter than 3 µm, a nonequilibrium velocity-field characteristic has to be considered [K13].

A rigorous treatment of the electron transport in GaAs-MESFET's, based on the equilibrium velocity-field characteristic, has been carried out by Himsworth [K9]. Fig. 9 summarizes the key features of a transistor with 3-µm gate length operated far in the saturated current region. The narrowest channel cross section is located under the drain end of the gate. The drift velocity rises to a peak at x_1, close to the center of the channel, and falls to the low saturated value under the gate edge. To preserve current continuity according to (1), heavy electron accumulation has to form in this region because the channel cross section is narrowing and, in addition, the electrons are moving progressively slower with increasing x. Exactly the opposite occurs between x_2 and x_3. The channel widens and the

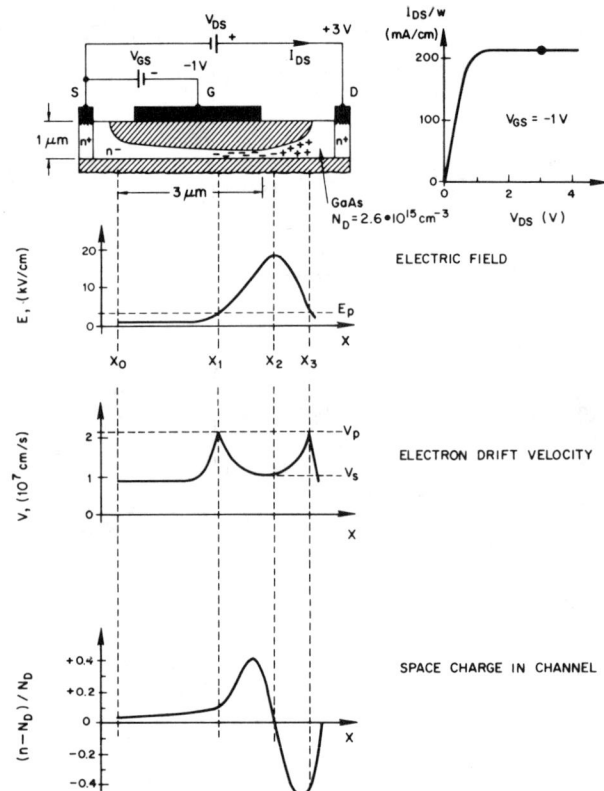

Fig. 9. The channel cross section, electric field, electron drift velocity, and space-charge distribution in the channel are illustrated for a GaAs-MESFET operated in the current-saturated region (data from Himsworth [K9]). Proceeding from x_1 to x_2, the channel cross section becomes narrower and, in addition, the electrons "slow down." To preserve current continuity, a heavy electron accumulation has to form. The opposite occurs between x_2 and x_3.

electrons move faster causing a strong depletion layer.[17] The charges in the accumulation and depletion layers are nearly equal and most of the drain voltage drops in this stationary dipole-layer.

In microwave FET's with very short gate length, the electrons do not reach equilibrium transport conditions in the high-field region of the channel. Nonequilibrium velocity-field characteristics in GaAs have been studied by computer simulations using Monte Carlo methods [K11]–[K14]. In a simplified approach, slow electrons are injected into a constant-field region and their drift velocity is monitored [K11]–[K13]. The situation is schematically illustrated in Fig. 10. As long as E is below the threshold field E_p, the electrons remain in equilibrium conditions. If the electrons enter a high-field region ($E > E_p$), they are accelerated to a higher velocity before relaxing to the equilibrium velocity.[18] This overshoot to more than twice the

---

[17] This region is not fully depleted of free electrons in contrast to the cross-hatched depletion layers.

[18] For $E < E_p$, electrons remain in the "lower valley" where they have a high mobility. For $E \gg E_p$, almost all electrons are transferred to a "satellite valley," a state in which they have a low mobility; i.e., low velocity at a given field. If the field changes suddenly from a value below to above E_p, a time period of approximately 1 ps passes before the carriers are transferred from the lower to the upper valley. During this time, the electrons remain in the high-mobility state in which they can acquire a high velocity in the high field.

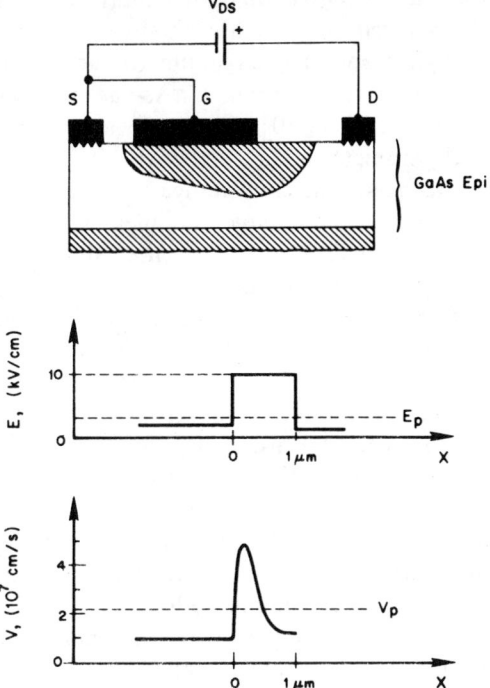

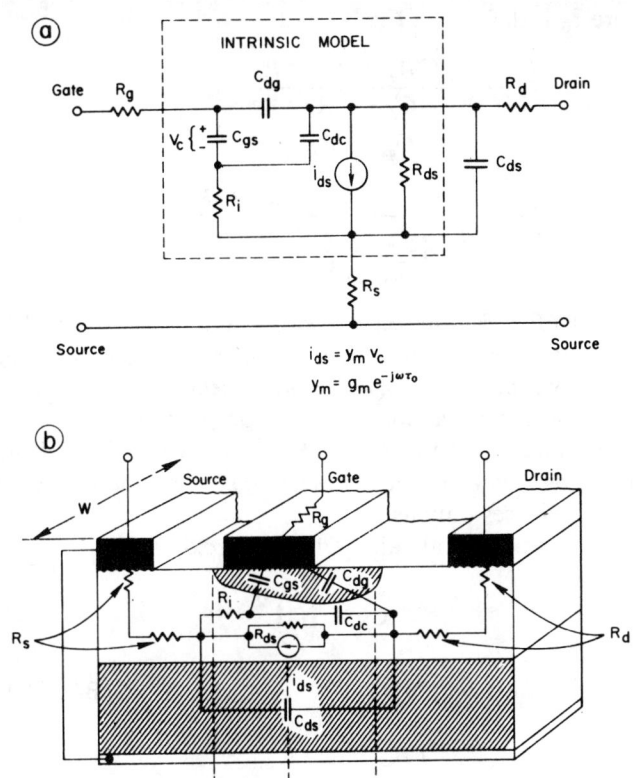

Fig. 10. This figure shows schematically the velocity overshoot of electrons as they enter the high-field region ($E > E_p$) under the gate (data from Ruch [K11]). Under equilibrium conditions, the maximum drift velocity is v_p (Fig. 8).

Fig. 11. (a) Is the equivalent circuit of a MESFET. Typical element values are listed in Table III. (b) Shows the physical origin of the circuit elements.

peak equilibrium velocity v_p, and the relaxation to the equilibrium condition, after traveling over a 0.6-μm path length [K11], is shown in Fig. 10. The effect is only noticeable in MESFET's with less than 3-μm gate length [K12], [K13]. The overshoot shortens the electron transit time through the high-field region and shifts the accumulation layer into the gap between gate and drain [K14].

C. Small-Signal Equivalent Circuit

An RF equivalent circuit of the MESFET should model the channel as a distributed RC network. However, a simple lumped-element circuit is capable of describing the FET's s-parameters accurately up to 12 GHz [C1], [C5], [L1]–[L4]. The equivalent circuit for operation in the saturated current region in common-source configuration is shown in Fig. 11(a). The location of the elements in the FET structure is illustrated in Fig. 11(b). In the intrinsic FET model, the elements ($C_{dg} + C_{gs}$) represent the total gate-to-channel capacitance; C_{dc} models the capacitance of the dipole layer; R_i and R_{ds} show the effects of the channel resistance; and i_{ds} defines the voltage-controlled current source. The transadmittance y_m relates i_{ds} to the voltage across C_{gs}. Up to 12 GHz, y_m is characterized by a frequency-independent magnitude, the transconductance g_m, and by a phase delay τ_0, reflecting the carrier transit time in the channel section where $E > E_p$. The extrinsic (parasitic) elements are: R_s the source resistance, R_d the drain resistance, R_g the gate-metal resistance, and C_{ds} the substrate capacitance. Typical element values for a GaAs-MESFET with 1-μm gate length and 500-μm gate width are listed in Table III.

The analysis of the equivalent circuit yields a critical frequency f_k, above which the MESFET is unconditionally stable. f_k can be approximated by[19]

$$f_k \approx \frac{1}{2\pi(\tau_0 + \tau_1 + \tau_2)} \qquad (2)$$

[19] Equations (2)–(4) were derived as outlined in [B4].

TABLE III
EQUIVALENT-CIRCUIT PARAMETERS OF A LOW-NOISE GaAs-MESFET WITH A 1-μm × 500-μm GATE
(HP EXPERIMENTAL, $N_D = 1 \times 10^{17}$ cm^{-3})

Intrinsic Elements	Extrinsic Elements
g_m = 53 mmho	C_{ds} = 0.12 pF
τ_0 = 5.0 ps	R_g = 2.9 Ω
C_{gs} = 0.62 pF	R_d = 3 Ω
C_{dg} = 0.014 pF	R_s = 2.0 Ω
C_{dc} = 0.02 pF	L_g = 0.05 nH*
R_i = 2.6 Ω	L_d = 0.05 nH*
R_{ds} = 400 Ω	L_s = 0.04 nH*

dc Bias
V_{DS} = 5 V
V_{GS} = 0
I_{DS} = 70 mA

* Contacting inductances of the test fixture in series with R_g, R_d and R_s, respectively.

where τ_0 is defined in Fig. 11 and

$$\tau_1 = \frac{C_{dg}(2R_g + R_i + R_s)}{\frac{C_{dg}}{C_{gs}} + \frac{R_s}{R_{ds}}} \tag{3}$$

$$\tau_2 = \frac{2}{\frac{g_m}{C_{gs}}\left[\frac{C_{dg}}{C_{gs}} + \frac{R_s}{R_{ds}}\right]} \frac{R_g + R_i + R_s}{R_{ds}}. \tag{4}$$

f_k is 6.1 GHz for the MESFET with the parameters listed in Table III. The MESFET with a complex-conjugate-matched input port becomes unstable with decreasing frequency because a larger fraction of the output voltage is fed back to the input over the $C_{dg} - R_{in}$[20] voltage divider. With decreasing frequency, R_{in} rises as $1/\omega^2$ while the reactance of C_{dg} increases only as $1/\omega$.

Mason's unilateral gain [Q1] is approximately

$$G_u \approx \left(\frac{f_u}{f}\right)^2 \tag{5}$$

where f_u is the maximum frequency of oscillation [B4], [C1]

$$f_u \approx \frac{f_T}{2\sqrt{r_1 + f_T\tau_3}} \tag{6}$$

f_T the frequency at unity current gain

$$f_T \approx \frac{1}{2\pi} \frac{g_m}{C_{gs}} \tag{7}$$

r_1 the input-to-output resistance ratio

$$r_1 = \frac{R_g + R_i + R_s}{R_{ds}} \tag{8}$$

and τ_3 the time constant

$$\tau_3 = 2\pi R_g C_{dg}. \tag{9}$$

Equation (5) shows a gain decreasing with 6 dB/octave as the frequency increases. At f_u, unity gain is reached.[21] To maximize f_u, the frequency f_T and the resistance ratio R_{ds}/R_i must be optimized in the intrinsic MESFET. In addition, the extrinsic resistances R_g and R_s and the feedback capacitance C_{dg} have to be minimized.

D. High-Frequency Limitations

The high-frequency limitations of MESFET's are dependent on device geometry and material parameters. In silicon and GaAs, electrons have a higher mobility than holes. Therefore, only n-channel FET's are used in microwave applications (see Table I). Electrons have six times higher low-field mobility[22] and two times higher maximum drift velocity in GaAs as opposed to silicon. The saturated velocities are about equal in both materials. As a consequence, the realized current-gain bandwidths f_T are about two times higher and the maximum frequencies of oscillation f_u three times higher in GaAs- as opposed to Si-MESFET's [C4], [C9], [C10]. In the device geometry, the most critical parameter is the gate length L. Decreasing the gate length decreases the capacitance C_{gs} and increases the transconductance g_m; consequently, there is an improved current-gain bandwidth f_T. For the short-gate-length microwave MESFET's, f_T is proportional to $1/L$ [J5]. High-speed operation is achieved by shrinking the gate length to the minimum size that can be realized with a given technology. Conventional photolithographic contact or projection-masking limits the smallest features to approximately a 1-μm size. An order of magnitude smaller gate length can be realized with X-ray or electron-beam lithography [X3]. A computer study of submicron Si-MESFET's by Reiser and Wolf [K6] reveals that f_T increases while the output resistance R_{ds} decreases with shrinking gate length. The limit for useful gate reduction is reached when the gate length is about equal to the channel thickness D. To keep $L/D > 1$, the channel thickness has to be decreased together with the gate length. This implies a higher doping level. In practical devices, the highest doping level is about 4×10^{17} cm^{-3} because of breakdown phenomena. The conclusion is that the gate length for Si-MESFET's should be larger than 0.1 μm. This geometry limits the current-gain bandwidth to about 70 GHz [K6]. In GaAs, quantitative high-frequency limitations need to be established. In very short gate devices ($L < 0.2$ μm), the field is above the threshold value E_p over the entire gate length [K9] and electrons are expected to remain in their high-mobility state for the entire flight through the channel [K12], [K13].

E. Principles of Noise Behavior

The noise properties of any linear two-port can be represented by a noiseless two-port with noise-current generators connected across the input and output ports [M1]. This is a physically meaningful way of describing the noise behavior of the intrinsic MESFET (Fig. 12).

The noise-current generator at the output represents the short-circuit channel noise generated in the drain-source path. The mean square of i_{nd} can be expressed by [M2]

$$\overline{i_{nd}^2} = 4kT_0\Delta f g_m P \tag{10}$$

with k the Boltzmann constant, T_0 the lattice temperature, Δf the bandwidth, g_m the transconductance, and P a factor depending on the device geometry and the dc bias. For zero drain voltage, i_{nd} characterizes the thermal noise generated by the drain conductance G_{ds}; i.e., $P = G_{ds}/g_m$. For positive drain voltages, the noise generated in the channel is larger than the thermal noise generated by G_{ds} for the following reasons. First, a thermal noise voltage generated locally in the channel modulates the conductive cross section of the channel and results in an amplified noise voltage at the drain [M2]. Second, the electrons are accelerated in the electric field, then scattered in all directions due to interactions with lattice phonons. Their random drift-velocities and the attributed free-carrier temperature

---

[20] R_{in} is the effective resistance between gate and source after the conjugate-impedance-matched generator has been connected; i.e.,

$$R_{in} \approx \frac{1}{2\omega^2 C_{gs}^2(R_g + R_i + R_s)}.$$

[21] f_u is 46 GHz for the MESFET of Table III.
[22] The comparison is made for a doping density of 1×10^{17} cm^{-3} [F1], [F8].

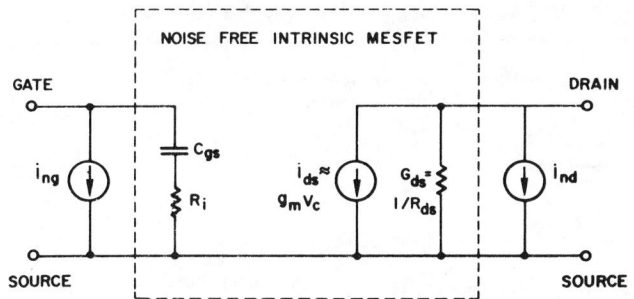

Fig. 12. Equivalent circuit of the (simplified) intrinsic MESFET with noise-current sources at the input and output port.

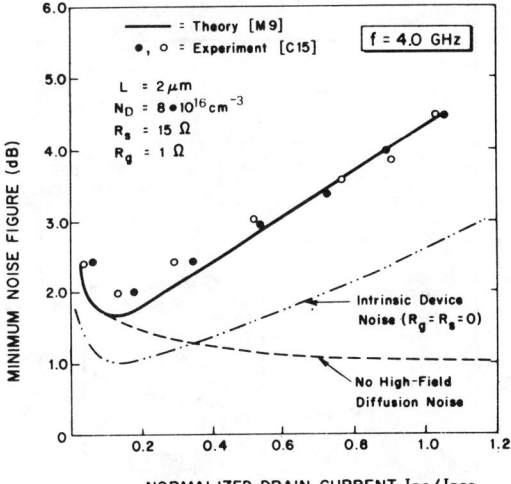

Fig. 13. Theoretical and measured minimum noise figures as a function of normalized drain current for a GaAs-FET with a 2-μm gate. (Courtesy of R. Pucel [C22].)

increase with the applied field to values considerably higher than the lattice temperature (hot-electron noise [M6], [M7]). Third, in GaAs carriers undergo field-dependent transitions from the central valley in the conduction band to satellite valleys and vice versa. A transferred electron experiences an abrupt velocity change. These transitions cause statistical drift-velocity fluctuations and thus generate field-dependent "intervalley-scattering noise" [M8]. Fourth, for large drain voltages, the electrons reach their limiting velocity on the drain side of the channel. In this region, the field has no influence on the carrier drift velocity. Therefore, this channel section cannot be treated as an ohmic conductor. Here, the noise is formulated as high-field diffusion noise[23] [M5], [M9], [M10], and the mean square of the noise current is porportional to the high-field diffusion coefficient in the semiconductor.

A noise voltage, generated locally in the channel, causes a fluctuation in the depletion-layer width. The resulting charge fluctuation in the depletion layer in turn induces a compensating charge variation on the gate electrode. The total induced-gate charge fluctuation is described in Fig. 12 by a noise-current generator i_{ng} at the gate terminal [M3] where

$$\overline{i_{ng}^2} = 4kT_0\Delta f \frac{\omega^2 C_{gs}^2}{g_m} R. \quad (11)$$

C_{gs} is the gate-source capacitance and R a factor depending on the FET geometry and the bias conditions.[24] The two noise currents, i_{nd} and i_{ng}, are caused by the same noise voltages in the channel. Therefore, partial correlation has to be expected.[25] A correlation factor C is defined as [M3]

$$jC = \frac{\overline{i_{ng}^* \cdot i_{nd}}}{\sqrt{\overline{i_{ng}^2} \cdot \overline{i_{nd}^2}}} \quad (12)$$

where j is the imaginary unit and the asterisk defines the complex conjugate. The correlation coefficient is purely imaginary because i_{ng} is caused by capacitive coupling of the gate circuit to the noise sources in the drain circuit. The factors P, R, and C in (10)–(12) have been computed versus normalized gate voltage by Baechtold [M8] for GaAs-MESFET's with various channel length-to-height ratios operated at the onset of current saturation. Statz et al. [M9] have extended the computation of P, R, and C to large drain voltages taking diffusion noise in the velocity-saturated channel region into account.

Using the model of Fig. 12, the minimum noise figure of the intrinsic MESFET can be expressed by [M7], [M10]

$$F_{min} = 1 + 2\sqrt{PR(1-C^2)}\frac{f}{f_T} + 2g_m R_i P\left(1 - C\sqrt{\frac{P}{R}}\right)\left(\frac{f}{f_T}\right)^2. \quad (13)$$

Low-noise MESFET's are normally operated at frequencies below f_T in order to yield sufficient gain. In this case, the linear frequency term in (13) is dominant. Short-gate MESFET's can exhibit very low noise figures for the following reasons. For an optimized drain current ($I_{DS}/I_{DSS} \approx 0.15$), the diffusion-noise contribution is small (Fig. 13), f_T is close to its maximum value [W3], and the correlation coefficient approaches unity ($C \approx 0.9$ [M9]). Substantial noise cancellation occurs at the drain which is expressed by the factor $(1 - C^2)$ in (13). The amplified input-noise current (αi_{ng}) destructively interferes with the correlated i_{nd} if the MESFET's gain and transmission phase are properly adjusted with an optimized input termination

$$\overline{|\alpha i_{ng} + i_{nd}|^2} \ll \overline{|\alpha i_{ng}|^2} + \overline{|i_{nd}|^2}. \quad (14)$$

In a practical MESFET, the parasitic resistances R_g and R_s, shown in Fig. 11, decrease the effectiveness of this noise cancellation [C22], [M9] and, in addition, they generate thermal noise themselves. A comparison between theoretical and experimental noise figures versus drain current is shown in Fig. 13. The increase of F_{min} with drain current is caused by the diffusion noise in the velocity-saturated

---

[23] The diffusion-noise theory is also valid at low fields and is a more general formulation than Johnson's formula for the thermal noise of a conductance [M5].

[24] For zero drain voltage, i_{ng} is the thermal noise of the input conductance g_{11} ($g_{11} \approx \omega^2 C_{gs}^2 R_i$), and R is equal to the product $g_m R_i$.

[25] Complete correlation results for a channel with uniform conductive cross section (e.g., for zero drain voltage).

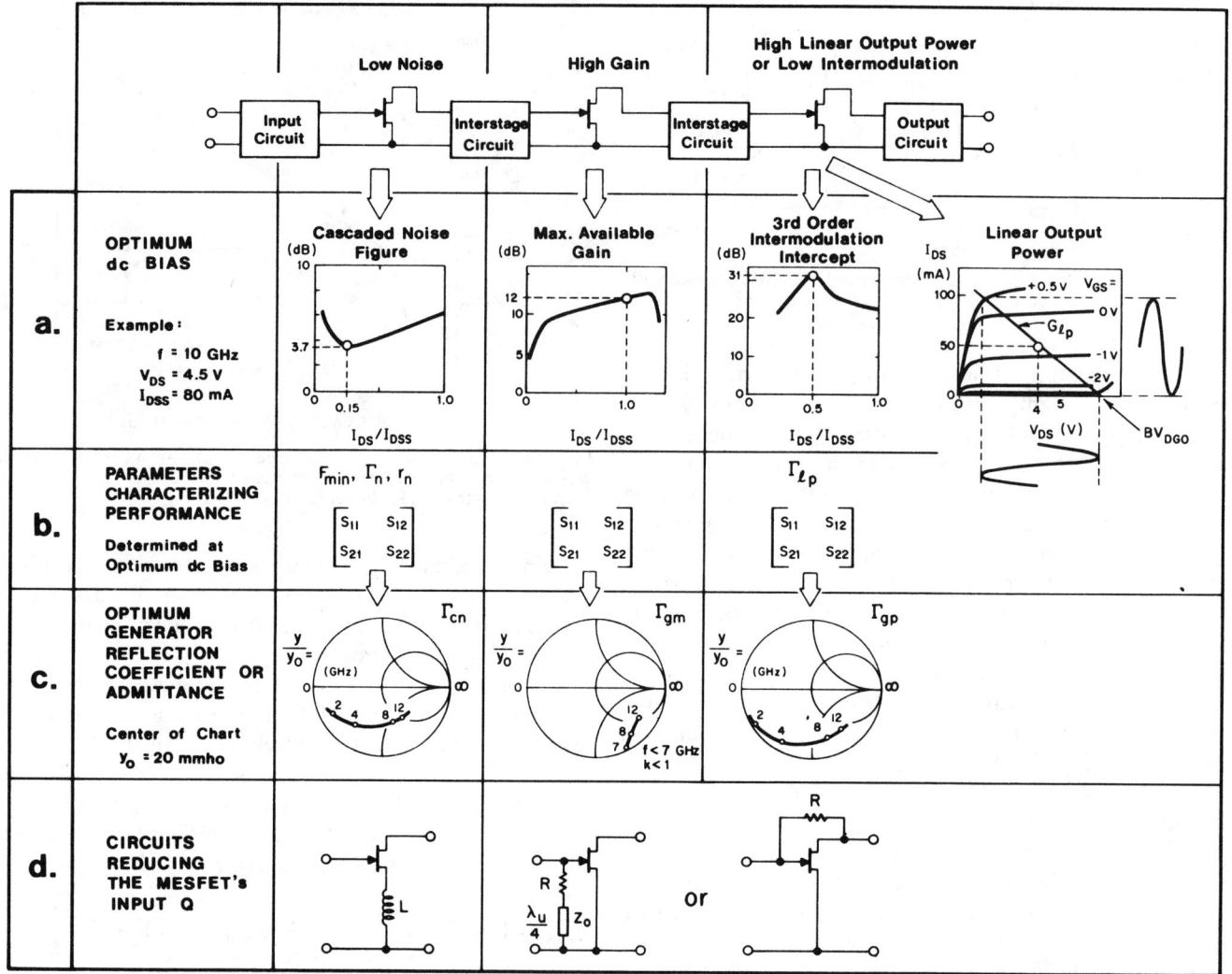

Fig. 14. The figure shows the key parameters in operating a MESFET in either a low-noise front stage, a high-gain stage, or a linear-power stage. (a) Illustrates the FET characteristics leading to the optimum dc bias. (b) Lists the parameters characterizing the performance (e.g., as computer input for CAD). (c) Shows optimum generator reflection coefficient to be synthesized by the circuits. The example is based on an HP MESFET with 1-μm gate length and 500-μm gate width. (d) Shows circuits to reduce the MESFET's input Q or to stabilize the transistor at low frequencies.

region [M9]. The noise-figure rise for small drain currents is caused by the rapid decrease of g_m and consequently of f_T. Also shown is the computed noise figure of the intrinsic MESFET.

IV. MESFET Amplifiers

In this section, basic concepts in the design of low-noise and linear medium-power MESFET amplifiers are reviewed.[26] The optimum dc bias, the parameters for device characterization, and the optimum generator admittance are discussed. Principles for network synthesis are described. Finally, performance characteristics of low-noise MESFET amplifiers are presented. As an example, an amplifier with a low-noise front stage, a high-gain stage, and a power-output stage is considered (Fig. 14). The amplifier shall be designed for low noise figure, flat gain, and high linear output-power capability across a specified band.

A. Optimized Operating Conditions for the MESFET's

The MESFET in the first-amplifier stage has to be operated at the dc bias yielding the lowest cascaded noise figure[27]

$$F_c = \frac{FG - 1}{G - 1} \qquad (15)$$

where F_c is the noise figure of an infinite number of cascaded stages each having a noise figure F and gain G. F_c is nearly independent of the drain voltage as long as $V_{DS} > V_{D_{sat}}$; however, strongly dependent on the drain current [C15], [M7], [M9]. Typically, F_c reaches a minimum at 0.1–$0.2 I_{DSS}$ [C15], [D4], [M9] where I_{DSS} is the

---

[26] The reader is also referred to recent review papers on solid-state microwave amplifiers by Cuccia [O1], Magarshack [O2], and Osbrink et al. [O3].

[27] The term "noise measure," M, proposed by Haus and Adler [N1], is omitted here. Instead, the "cascaded noise figure," F_c is adopted. The relationship between F_c and M is $F_c = 1 + M$.

saturated drain current at zero gate voltage [Figs. 13 and 14(a)]. The drain current is adjusted to this optimum value with the gate bias. In this operating point, the MESFET's noise behavior is characterized by 1) the minimum noise figure $F_{\min}$; 2) the reflection coefficient of the generator Γ_n, which produces $F_{\min}$; and 3) a dimensionless coefficient r_n [N2]–[N9]. For an arbitrary generator reflection coefficient Γ_g, the noise figure is then determined by

$$F = F_{\min} + 4r_n \frac{|\Gamma_g - \Gamma_n|^2}{(1 - |\Gamma_g|^2)|1 + \Gamma_n|^2}. \quad (16)$$

If the s-parameters are known, all parameters of interest can be computed. These are, e.g., the optimum generator reflection coefficient for lowest cascaded noise figure Γ_{cn} [N3], [N9], the optimum load reflection coefficient, the associated transistor gain (17), etc. Γ_{cn} is in general different from the reflection coefficient for maximum gain Γ_{gm} [Fig. 14(c)]. A combination of lossless parallel and series feedback is capable of making Γ_{cn} and Γ_{gm} identical without changing F_c.[28] [N10]–[N14].

The MESFET in the second stage is operated with maximum small-signal gain [Fig. 14(a)]. This condition is obtained at approximately zero gate voltage ($I_{DS} \approx I_{DSS}$), where f_T reaches a maximum, and at a drain voltage that maximizes the output resistance and the resistance ratio $1/r_1$ in (8). The small-signal behavior is fully characterized with the four s-parameters [Q3]. The transducer gain [29] for any generator and load reflection coefficients, Γ_g and Γ_l, respectively, is [Q4]

$$G = \frac{|s_{21}|^2(1 - |\Gamma_g|^2)(1 - |\Gamma_l|^2)}{|(1 - s_{11}\Gamma_g)(1 - s_{22}\Gamma_l) - s_{12}s_{21}\Gamma_g\Gamma_l|^2}. \quad (17)$$

The s-parameters determine also Rollett's [Q2] stability factor k

$$k = \frac{1 + |s_{11}s_{22} - s_{12}s_{21}|^2 - |s_{11}|^2 - |s_{22}|^2}{2|s_{12}s_{21}|}. \quad (18)$$

If k is larger than unity,[30] an optimum combination of Γ_{gm} and Γ_{lm} simultaneously image-matches the two MESFET ports and maximizes the gain [Q4]. If k is smaller than unity, the MESFET is only conditionally stable. In this case, the terminations Γ_g and Γ_l must be carefully chosen to operate the transistor in a stable range [Q4], [Q5] or the resistive stabilization networks described below must be applied. Frequently, $k \gg 1$ and $|s_{12}|$ are small enough for the MESFET to be treated as a unilateral two-port ($s_{12} = 0$; $k = \infty$). In this case, the optimum generator and load terminations are

$$\Gamma_{gm} = s_{11}^* \qquad \Gamma_{lm} = s_{22}^* \quad (19)$$

and the maximum available gain obtained from (17) is

$$G_{\max} = \frac{|s_{21}|^2}{(1 - |s_{11}|^2)(1 - |s_{22}|^2)}. \quad (20)$$

The asterisk in (19) defines the complex conjugate of the s-parameters.

The MESFET in the output stage is intended to operate as a linear (Class A) amplifier. The design objectives can be either lowest intermodulation distortion [P1]–[P3], largest added RF power [Q6], or largest linear output power [T1]. In the last case, the dc bias is graphically determined from the static drain-current versus drain-voltage characteristic. The bias and load conductance line are chosen to maximize the product of linear voltage and current swing [Fig. 14(a)]. The limitations are determined by the maximum dc power dissipation, the drain-to-gate breakdown voltage BV_{DGO}, and the positive gate bias ($V_{GS} \approx 0.5$ V) above which appreciable gate current flows. The optimum load conductance is typically much larger than the MESFET's output conductance. Consequently, the MESFET is not matched to the load and does not deliver maximum gain. However, the large load conductance shunting the MESFET's nonlinear output admittance reduces intermodulation distortion.[31] The optimum load conductance and susceptance determine the optimum reflection coefficient of the load Γ_{lp}. The generator reflection coefficient that provides a complex-conjugate match at the input is then

$$\Gamma_{gp} = \left[s_{11} + \frac{s_{12}s_{21}\Gamma_{lp}}{1 - s_{22}\Gamma_{lp}}\right]^* \quad (21)$$

and the associated gain is determined from (17).

Optimum generator reflection coefficients for low noise, high gain, and high linear-power operation are shown in Fig. 14(c). The plotted data are typical for an unpackaged small-signal GaAs-MESFET with 1-μm gate length and 500-μm gate width. At low frequencies, the FET has a high-Q input admittance and wide-band matching is difficult. Simple circuits that lower the Q value are illustrated in Fig. 14(d). A series-feedback inductance between source and ground increases the input series resistance, decreases the input reactance, and leaves F_c unchanged. A resistor in-series with a short-circuited shunt stub, connected between gate and source, lowers the Q and stabilizes the FET at the low-frequency end of the band [T1]. If the shunt stub is a quarter-wavelength long at the upper band edge, the circuit does not load the FET input and does not decrease the gain at this frequency. Also, a parallel feedback resistance has been proposed for stabilization and input Q lowering [Q7].

B. Amplifier Network Synthesis

The input-matching network transforms the 50-Ω generator impedance to the optimum impedance with

---

[28] The MESFET's noise figure F and gain G change, but the cascaded noise figure F_c remains invariant.
[29] The transducer gain is defined as the power delivered to the load divided by the available power from the generator.
[30] i.e., the MESFET is operated above the critical frequency f_k discussed in Section III.

[31] If low third-order intermodulation is desired [Fig. 14(a)], a drain current is chosen that minimizes the distortion from the nonlinear transconductance and input capacitance [P3].

reflection coefficient Γ_{cn}. The transformed impedance versus frequency must have a negative-reactance slope, $dX/df < 0$. For narrow-band amplifiers, the circuit is derived from a simple graphical design using the Smith chart [L4], [N3], [Q9]–[Q12]. For moderate-bandwidth MESFET amplifiers, impedance-matching bandpass networks with quarter-wave resonators have been applied successfully [Q13]. These networks are derived from low-pass filter prototypes; the circuit topology is well defined, and the element values are optimized by simple computations. The design procedures treat the MESFET as a unilateral device ($s_{12} = 0$). For more accurate optimizations and for large bandwidths, computer-aided design procedures are used in which the MESFET is characterized by all noise- and s-parameters [Q14]–[Q19]. In general, a particular circuit topology is chosen by the designer. The values of the circuit elements are then optimized by the computer. The optimization routine searches in a systematic way for the global minimum of an error function E defined as

$$E = \sum_{f_1}^{f_n} [\text{calculated } H(f) - \text{required } H(f)]^2 \quad (22)$$

where f is the frequency and H is a performance function defined as a weighted sum of gain, noise figure, reflection coefficients, etc. The matching networks are normally built as microstrip circuits. A monolithic amplifier stage, combining lumped matching elements with the MESFET on a single GaAs chip, has also been reported [B1], [R13]. The amplifier modules exhibit very broad-band performance (6–12.4 GHz).

In the interstage networks, the insertion loss versus frequency response has to compensate the MESFET's gain slope [Fig. 15(a)] to achieve a flat amplifier gain. Generally, this is done by matching the output of the preceding FET to the input of the following transistor at the upper band edge and providing an increasing mismatch with decreasing frequency [Fig. 15(b)]. Analytical methods have been developed for the synthesis of reactive networks yielding the desired insertion loss versus frequency characteristic [Q20]–[Q24]. Since the gain slope is provided by reactive mismatch, high standing waves result between the stages at the lower band end. The high voltages generated in the standing waves enhance feedback in the FET's and the large reactance versus frequency slopes of the networks cause high group-delay variations. These problems can be avoided with dissipative coupling networks [Q13], [Q25], [Q26] which provide a lossless impedance match at the highest frequency in the band and introduce increasing resistive loss (attenuation) with decreasing frequency [Fig. 15(c)]. The amount of gain compensation has to be individually chosen for each interstage network to achieve lowest amplifier noise figure across the entire band [Q25] and to prevent premature power saturation in the driver stage [Q26].

MESFET amplifiers are built with balanced and unbalanced circuits. Balanced amplifiers consist of a pair of amplifiers or single-tuned stages whose inputs are connected to the

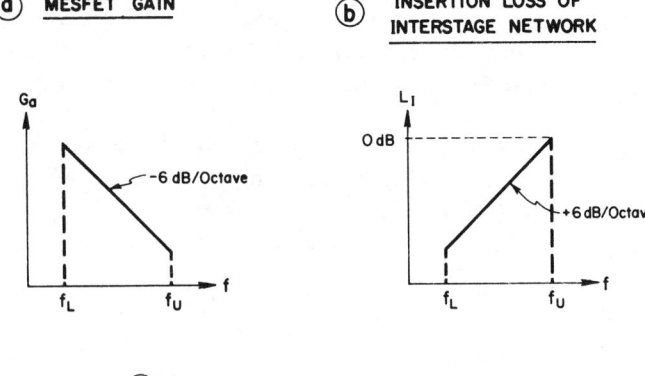

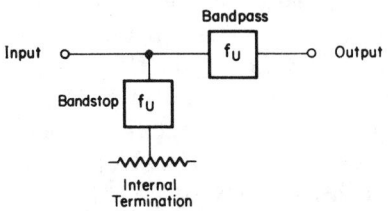

Fig. 15. For a flat amplifier gain, the MESFET's gain versus frequency slope (a) has to be compensated by an inverse slope in the interstage network's insertion loss (b). The insertion-loss characteristic can be obtained with reactive mismatch in an L–C network or with frequency-dependent attenuation in a diplexer resulting in dissipation of the excess power in an internal termination (c).

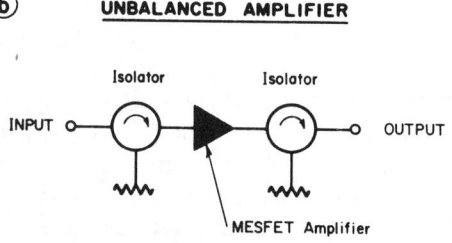

Fig. 16. MESFET amplifiers are built in balanced form with 3-dB hybrid couplers at the input and output of each stage or as an unbalanced chain of stages with isolators on both amplifier ports.

conjugate ports of a 3-dB hybrid coupler and whose outputs are similarly connected to another 3-dB coupler [Fig. 16(a)] [Q27]. The signal applied at the input port of the first coupler splits into two equal parts and is fed to the two amplifiers. The amplified signals from the two amplifier outputs recombine in the second coupler and emerge at one of the coupler's output ports. The advantages of the balanced

amplifier over an unbalanced amplifier are improvement in:
1) input and output impedance matching in an amplifier optimized for noise figure or output power, 2) short- and open-circuit stability, 3) phase linearity, 4) gain compression, 5) intermodulation characteristics, and 6) reduced sensitivity to transistor impedance variations, provided the MESFET's are selected in similar pairs. Unbalanced amplifiers, on the other hand, need only half as many MESFET's, matching networks, and dc power. In general, isolators are required at the input and output of broad-band unbalanced MESFET amplifiers to meet low VSWR specifications and to make the noise figure independent of the source admittance [Fig. 16(b)]. 3-dB hybrid couplers are also used for power combining [Q28] as shown in Fig. 17.

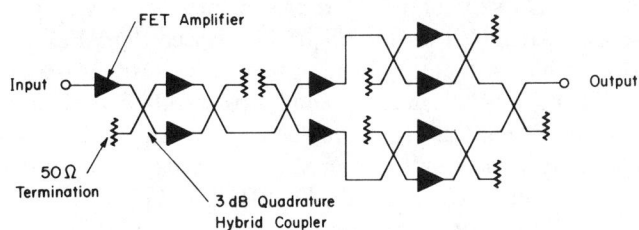

Fig. 17. In transistor power amplifiers, 3-dB hybrid couplers are also used as power combiners (after S. Lazar [Q28]).

C. MESFET Amplifier Performance

Various small-signal FET amplifiers have been described in the literature [C14], [D6], [G3], [P3], [Q7], [Q13], [Q22], [R1]–[R15]. Most are low-noise designs. Noise figures of laboratory prototypes are plotted versus frequency in Fig. 18. The solid lines represent narrow-band amplifiers. Lowest noise figures are 2.2 dB at 4 GHz, 3.6 dB at 8 GHz, and 5.0 dB at 12 GHz. The single data points (circles, squares, etc.) show noise figures of the 1-μm MESFET's used in the first amplifier stages. The data points lie about 1.0 dB below the amplifier noise figures. The noise-figure difference is caused by the insertion loss of the input circuit and by the noise contribution of the following stages. The noise figures of wide-band amplifiers, plotted with dashed lines, are typically 1.0–1.5 dB higher than the narrow-band circuits. At room temperature, thermal-noise sources dominate the noise performance of GaAs MESFET's in microwave amplifiers [M8], [M9]. By cooling the amplifier, a significant noise reduction can be obtained [S1]–[S6]. 1.6-dB noise figure (i.e., 130 K input noise temperature) was measured at 60 K for a 12-GHz amplifier [S5], [S6].

Octave-band MESFET amplifiers covering the 4.0–8.0-GHz frequency range have been built in balanced form [C14], [R5], [R6], [T3]. Typically, a three-stage small-signal amplifier has 24-dB gain, ± 0.7-dB gain variation, 1.8:1 maximum VSWR at the input and output port, +13-dBm output power for 1-dB gain compression, and +20-dBm third-order intermodulation intercept [R16], [R17]. In the 8.0–12.0-GHz band, a three-stage unbalanced amplifier without isolators exhibits 20 ± 1.3-dB gain, 2.5:1 maximum VSWR, +13-dBm output power, and +26-dBm intermodulation intercept [Q13]. Less gain variation, e.g., 28.5 ± 0.5 dB, can be obtained with a balanced design [R15].

A few medium-power amplifiers have been reported [T1]–[T6]. At 6 GHz, a four-stage amplifier with 26-dB gain and 1-W output power,[32] using a single MESFET chip in the output stage, has been built [T2]. More common are balanced output stages which combine the output power of two transistors [T1], [T3], [T6]. Also, wide-band medium-power amplifiers have been designed [T1], [T3],

[32] Measured at 1-dB gain compression.

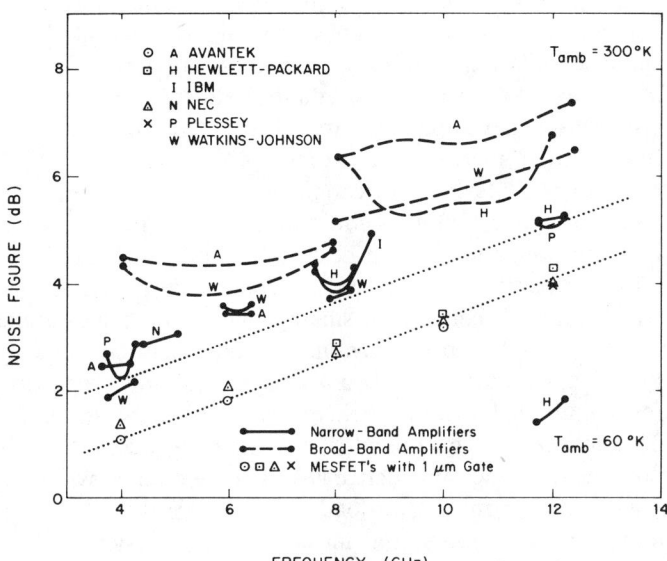

Fig. 18. This noise figure versus frequency graph summarizes performance of experimental MESFET amplifiers. The solid lines represent narrow-band amplifiers, and the single data points show noise figures of the 1-μm MESFET's used in the first stages. The broken lines illustrate the noise performance of wide-band amplifiers.

and one covers the 2–6-GHz band [T1]. In this frequency range, MESFET's with 1-μm gate length have a high-Q input impedance, and matching circuits with resistive components, illustrated in Fig. 14(d), must be used.

V. OTHER APPLICATIONS

So far, comparatively little effort has been spent on the development of GaAs-MESFET oscillators [U1]–[U8]. This application should prove to be of interest since MESFET's combine the advantages of low bias voltage (< 10 V), relatively low noise measure (< 23 dB),[33] and high efficiency (> 15 percent at 10 GHz [U2]). Also GaAs-MESFET mixers [V1]–[V5] are expected to receive more attention in the near future. Good noise performance ($F = 7.4$ dB)[34] and large dynamic range (third-order intermodulation intercept $= +18$ dBm)[34] can be achieved with conversion gain ($G = 6$-dB)[34] [V3]. Another application is the use

[33] This oscillator noise measure is measured at 1-MHz separation from the X-band carrier [U3]. This noise measure is high in comparison to amplifier noise figures because of upconverted 1/f noise. Reduction of traps at the active-layer surface and at the substrate interface is expected to improve the low-frequency noise performance of GaAs-MESFET's in the near future.

[34] Measured at 8 GHz with a balanced MESFET mixer.

of Si-MESFET's [W1], [W2] and GaAs-MESFET's [W3]–[W7] or GaAs-JFET's [W8], [W9] in high-speed digital circuits. Monolithic integrated-logic gates with 100-ps signal-propagation delay (fanout 2) and 4-pJ speed-power product have been built using 1-μm GaAs-MESFET's [W3], [W4]. This is less than half the propagation delay measured on highest speed bipolar logic [W10], [W11]. In addition, the feasibility for medium-scale integration of these GaAs circuits has been demonstrated [W4].

VI. Conclusions and Outlook

GaAs-FET's are capable of low-noise amplification, high-efficiency power amplification and generation, high-speed modulation, and logic. Since these are areas of major microwave-system needs, substantial efforts in device and application developments are anticipated. In the near future, rapid advances are expected in the areas of 1) FET reliability, 2) device fabrication with high yield, uniform and reproducible unit-to-unit parameters by means of ion implantation, and electron-beam lithography, and 3) higher power capability and efficiency due to improved device structures with better heat sinking, thermally stable ohmic contacts, and more burn-out-resistant gates. Looking further ahead, a strong trend toward monolithic integration for digital, analog, and hybrid applications is now apparent. The monolithic approach is attractive because MESFET's fabricated on the same active layer can be used as switches, logic gates with active loads, impedance transformers, amplifiers, oscillators, and mixers; and the devices can be supported, isolated, and interconnected with low parasitic capacitances on the semi-insulating substrate. Monolithic integration is required to handle the complexity of tasks and to serve high-volume low-cast markets. Integrated high-speed logic will be needed in digital communications with gigabit-per-second data rates [W12]–[W14], in multi-phase-shift-keyed modulation and demodulation, time multiplexing, frequency division, counting, frequency synthesis, and waveform synthesis. A need is foreseen for microwave analog and hybrid circuits, on a single chip, such as 1) combinations of a preamplifier, mixer, local oscillator, and IF amplifier, 2) wide-band signal and operational amplifiers, 3) sample and hold circuits, and 4) high-speed D/A and A/D converters. Advances in III–V materials preparation and in submicron device processing will make microwave monolithic circuits with FET's a reality.

Acknowledgment

The author wishes to thank Dr. D. Ch'en, G. Gilbert, Dr. T. Heng, Dr. C. Snapp, Dr. J. Turner, and M. Walker for supplying new, unpublished data, and Dr. R. Archer, Dr. G. Bechtel, Dr. R. Engelmann, Dr. U. Gysel, Dr. R. Lee, Dr. J. Magarshack, Dr. C. Stolte, B. Lizenby, A. Podell, and R. Van Tuyl for many helpful discussions and for their assistance in reviewing the manuscript. The author is also grateful to Mrs. S. Ybarra for drafting the figures and to Ms. E. Miller for typing the manuscript at times of heavy work load.

References

A. Bipolar Transistors

[A1] R. Engelbrecht and K. Kurokawa, "A wide-band low-noise L-band balanced transistor amplifier," *Proc. IEEE*, vol. 53, pp. 237–247, March 1965.

[A2] J. Archer, "Low-noise implanted-base microwave transistors," *Solid-State Electron.*, vol. 17, pp. 387–393, April 1974.

[A3] H. Yuan, J. Kruger, and Y. Wu, "X-band silicon power transistor," in *1975 Int. Microwave Symp., Dig. Tech. Papers*, pp. 73–75.

[A4] C. Snapp, Hewlett-Packard Associates, private communication (HP-515).

[A5] J. Chen and K. Verma, "A 6 GHz silicon bipolar power transistor," in *1974 Int. Electron Devices Meeting, Dig. Tech. Papers*, pp. 299–301.

[A6] J. Steenbergen, A. Harrington, and G. Schreyer, "Broadband power transistor 4.4–5.0 GHz," U.S. Army Electronics Command, Final Technical Report, Contract No. DAAB07-73-C-0283, Fort Monmouth, February 1976.

[A7] G. Gilbert, Microwave Semiconductor Corporation, private communication.

[A8] C. Nuese, J. Gannon, R. Dean, H. Gossenberger, and R. Enstrom, "GaAs vapor-grown bipolar transistors," *Solid-State Electron.*, vol. 15, pp. 81–91, Jan. 1972.

[A9] W. Dumke, J. Woodall, and V. Rideout, "GaAs-GaAlAs heterojunction transistor for high frequency operation," *Solid-State Electron.*, vol. 15, pp. 1339–1343, Dec. 1972.

[A10] M. Konagai and K. Takahashi, "(GaAl)As-GaAs heterojunction transistors with high injection efficiency," *J. Appl. Phys.*, vol. 46, pp. 2120–2124, May 1975.

B. Review Papers on Microwave FET's

[B1] J. Turner, "Microwave FET's and their applications," in *Proc. 1975 Cornell Conference on Active Semiconductor Devices for Microwave and Integrated Optics*, pp. 13–22.

[B2] C. Liechti, "Recent advances in high-frequency field-effect transistors," in *1975 Int. Electron Devices Meeting, Dig. Tech. Papers*, pp. 6–10.

[B3] R. Pringle, "Microwave transistor and monolithic integrated circuit technology," *Microelectron. J.*, vol. 6, pp. 33–41, June 1975.

[B4] S. Ohkawa, K. Suyama, and H. Ishikawa, "Low noise GaAs field-effect transistors," *Fujitsu Sci. Tech. J.*, vol. 11, pp. 151–173, March 1975.

C. Low-Noise Si and GaAs MESFET's (Single-Gate)

[C1] P. Wolf, "Microwave properties of Schottky-barrier field-effect transistors," *IBM J. Res. Develop.*, vol. 14, pp. 125–141, March 1970.

[C2] K. Drangeid, R. Sommerhalder, and W. Walter, "High-speed gallium-arsenide Schottky-barrier field-effect transistors," *Electron. Lett.*, vol. 6, pp. 228–229, April 1970.

[C3] J. Turner, A. Waller, R. Bennett, and D. Parker, "An electron beam fabricated GaAs microwave field-effect transistor," *1970 Symp. GaAs and Related Compounds* (Inst. Phys., Conf. Series No. 9, London, 1971), pp. 234–239.

[C4] W. Baechtold, W. Walter, and P. Wolf, "X and Ku band GaAs MESFET," *Electron. Lett.*, vol. 8, pp. 35–37, Jan. 1972.

[C5] C. Liechti, E. Gowen, and J. Cohen, "GaAs microwave Schottky-gate FET," in *1972 Int. Solid-State Circuits Conf., Dig. Tech. Papers*, pp. 158–159.

[C6] G. Bechtel, W. Hooper, and D. Mock, "X-band GaAs FET," *Microwave J.*, vol. 15, pp. 15–19, Nov. 1972.

[C7] F. Doerbeck, "A planar GaAs Schottky-barrier field-effect transistor with a self-aligned gate," *1970 Symp. GaAs and Related Compounds* (Inst. Phys., Conf. Series No. 9, London, 1971), pp. 251–258.

[C8] M. Driver, H. Kim, and D. Barrett, "Gallium arsenide self-aligned gate field-effect transistors," *Proc. IEEE*, vol. 59, pp. 1244–1245, Aug. 1971.

[C9] W. Baechtold and P. Wolf, "An improved microwave silicon MESFET," *Solid-State Electron.*, vol. 14, pp. 783–790, Sep. 1971.

[C10] W. Baechtold et al., "Si and GaAs 0.5 μm-gate Schottky-barrier field-effect transistors," *Electron. Lett.*, vol. 9, pp. 232–234, May 1973.

[C11] J. Jahncke, "Hoechstfrequenzeigenschaften eines GaAs-MESFET's in Streifenleitungstechnik," *Nachr. Tech. Z.*, vol. 26, pp. 193–199, May 1973.

[C12] R. Hunsperger and N. Hirsch, "GaAs field-effect transistors with ion-implanted channels," *Electron. Lett.*, vol. 9, pp. 577–578, Dec. 1973.

[C13] ——, "Ion-implanted microwave field-effect transistors in GaAs," *Solid-State Electron.*, vol. 18, pp. 349–353, April 1975.

[C14] D. Ch'en and A. Woo, "A practical 4 to 8 GHz GaAs FET amplifier," *Microwave J.*, vol. 17, pp. 26, 72, Feb. 1974.

[C15] G. Brehm and G. Vendelin, "Biasing FET's for optimum performance," *Microwaves*, vol. 13, pp. 38–44, Feb. 1974.
[C16] B. Welch, F. Eisen, and J. Higgins, "Gallium arsenide field-effect transistors by ion implantation," *J. Appl. Phys.*, vol. 45, pp. 3685–3687, Aug. 1974.
[C17] F. Ozdemir *et al.*, "Electron beam fabricated 0.5 μm gate GaAs Schottky-barrier field-effect transistor," *1974 Int. Electron Devices Meeting (late news paper), Suppl. Dig. Tech. Papers*, p. 5.
[C18] E. Kohn, "V-shaped-gate GaAs MESFET for improved high-frequency performance," *Electron. Lett.*, vol. 11, p. 160, April 1975.
[C19] E. Kohn, R. Wueller, R. Stahlmann, and H. Beneking, "High-speed 1 μm GaAs MESFET," *Electron. Lett.*, vol. 11, pp. 171–172, April 1975.
[C20] Data sheets: Fairchild, FMT 940/941, FMT 980/981; Plessey, GAT 3/4; Nippon Electric Company, NE 244/388; Hitachi, HCRL 81-84.
[C21] M. Ogawa, K. Ohata, T. Furutsuka, and N. Kawamura, "Submicron single-gate and dual-gate GaAs MESFET's with improved low noise and high gain performance," *IEEE Trans. Microwave Theory and Techniques*, this issue, pp. 300–305.
[C22] R. Pucel, D. Masse, and C. Krumm, "Noise performance of gallium-arsenide field-effect transistors," in *Proc. 1957 Cornell Conference on Active Semiconductor Devices for Microwaves and Integrated Optics*, pp. 265–276.
[C23] J. Barrera, "The importance of substrate properties on GaAs FET performance," in *Proc. 1975 Cornell Conference on Active Semiconductor Devices for Microwaves and Integrated Optics*, pp. 135–144.
[C24] W. Kellner, H. Kniepkamp, D. Ristow, and H. Boroffka, "Microwave field-effect transistors from sulphur-implanted GaAs," in *1975 Int. Electron Devices Meeting, Dig. Tech. Papers*, pp. 238–242.
[C25] P. Baudet, M. Binet, and D. Boccon-Gibod, "Submicrometer self-aligned GaAs MESFET," *IEEE Trans. Microwave Theory and Techniques*, this issue, pp. 372–376.
[C26] T. Nozaki, M. Ogawa, H. Terao, and H. Watanabe, "Multi-layer epitaxial technology for the Schottky-barrier gate GaAs field-effect transistor," *1974 Conf. GaAs and Related Compounds* (Inst. Phys., Conf. Ser. No. 24, London, 1975), pp. 46–54.
[C27] J. Higgins, B. Welch, F. Eisen, and G. Robinson, "Performance of ion-implanted GaAs FET's," in *1975 Int. Electron Devices Meeting (late news paper), Suppl. Dig. Tech. Papers*, p. 5.
[C28] A. Cho and D. Ch'en, "GaAs MESFET prepared by molecular beam epitaxy (MBE)," *Appl. Phys. Lett.*, vol. 28, pp. 30–31, Jan. 1976.

D. Dual-Gate GaAs MESFET's

[D1] J. Turner, A. Waller, E. Kelley, and D. Parker, "Dual-gate GaAs microwave FET," *Electron. Lett.*, vol. 7, pp. 661–662, Nov. 1971.
[D2] S. Asai, F. Murai, and H. Kodera, "The GaAs dual-gate FET with low noise and wide dynamic range," in *1973 IEEE Int. Electron Devices Conf., Dig. Tech. Papers*, pp. 64–67.
[D3] C. Liechti, "Characteristics of dual-gate GaAs MESFETs," in *Proc. 1974 European Microwave Conf.*, pp. 87–91.
[D4] ——, "Performance of dual-gate GaAs MESFETs as gain-controlled, low-noise amplifiers and high-speed modulators," *IEEE Trans. Microwave Theory and Techniques*, vol. MTT-23, pp. 461–469, June 1975.
[D5] R. Dean and R. Matarese, "Submicrometer self-aligned dual-gate GaAs FET," *IEEE Trans. Electron Devices*, vol. ED-22, pp. 358–360, June 1975.
[D6] M. Maeda and Y. Minai, "Application of dual-gate GaAs FET to microwave variable-gain amplifiers," in *1974 IEEE Int. Microwave Symp., Dig. Tech. Papers*, pp. 351–353.
[D7] S. Asai, F. Murai, and H. Kodera, "GaAs dual-gate Schottky-barrier FET's for microwave frequencies," *IEEE Trans. Electron Devices*, vol. ED-22, pp. 897–904, Oct. 1975.

E. Power GaAs MESFET's

[E1] M. Fukuta, T. Mimura, I. Tujimura, and A. Furumoto, "Mesh source type microwave power FET," in *1973 Int. Solid-State Circuits Conf., Dig. Tech. Papers*, pp. 84–85.
[E2] L. Napoli *et al.*, "High power GaAs FET amplifier—A multi-gate structure," in *1973 Int. Solid-State Circuits Conf., Dig. Tech. Papers*, pp. 82–83.
[E3] M. Driver, M. Geisler, D. Barrett, and H. Kim, "S-band microwave power FET," in *1973 IEEE Int. Electron Devices Meeting, Dig. Tech. Papers*, pp. 393–395.
[E4] H. Yamasaki *et al.*, "S-band microwave GaAs field-effect transistors," in *Proc. 1975 Cornell Conf. on Active Semiconductor Devices for Microwaves and Integrated Optics*, pp. 287–296.
[E5] T. Blocker, H. Macksey, and R. Adams, "X-band RF power performance of GaAs FET's," in *1974 IEEE Int. Electron Devices Meeting, Dig. Tech. Papers*, pp. 288–291.
[E6] H. Macksey and R. Adams, "Fabrication processes for power FET's," in *Proc. 1975 Cornell Conf. on Active Semi-conductor Devices for Microwave and Integrated Optics*, pp. 255–264.
[E7] J. Angus, R. Butlin, D. Parker, R. Bennett, and J. Turner, "The design and evaluation of GaAs power MESFET's," in *Proc. 1975 European Microwave Conf.*, pp. 291–295.
[E8] M. Fukuta, H. Ishikawa, K. Suyama, and M. Maeda, "GaAs 8 GHz-band high power FET," in *1974 IEEE Int. Electron Devices Meeting., Dig. Tech. Papers*, pp. 285–287.
[E9] M. Fukuta, K. Suyama, H. Suzuki, and H. Ishikawa, "GaAs microwave power FET," *IEEE Trans. Electron Devices*, vol. ED-23, pp. 388–394, Apr. 1976.
[E10] M. Fukuta, K. Suyama, H. Suzuki, Y. Nakayama, and H. Ishikawa, "X-band GaAs Schottky-barrier power FET with a high drain-source breakdown voltage," in *1976 Int. Solid-State Circuits Conf., Dig. Tech. Papers*, pp. 166–167.
[E11] ——, "Power GaAs MESFET with a high drain-source breakdown voltage," *IEEE Trans. Microwave Theory and Techniques*, this issue, pp. 312–317.
[E12] L. Napoli and R. DeBrecht, "Performance and limitations of FETs as microwave power amplifiers," in *1973 IEEE Int. Microwave Symp., Dig. Tech. Papers*, pp. 230–232.
[E13] L. Napoli, J. Hughes, W. Reichert, and S. Jolly, "GaAs FET for high power amplifiers at microwave frequencies," *RCA Rev.*, vol. 34, pp. 608–615, Dec. 1973.
[E14] R. Camisa, I. Drukier, H. Huang, J. Goel, and S. Narayan, "GaAs MESFET linear amplifiers," in *1975 IEEE Int. Solid-State Circuits Conf., Dig. Tech. Papers*, pp. 70–71.
[E15] I. Drukier, R. Camisa, S. Jolly, H. Huang, and S. Narayan, "Medium power J-band MESFET's," in *Proc. 1975 Cornell Conf. on Active Semiconductor Devices for Microwaves and Integrated Optics*, pp. 297–304.
[E16] ——, "Medium power GaAs field-effect transistors," *Electron. Lett.*, vol. 11, pp. 104–105, March 1975.
[E17] H. Huang *et al.*, "GaAs MESFET performance," in *1975 Int. Electron Devices Meeting, Dig. Tech. Papers*, pp. 235–237.
[E18] H. Huang, I. Drukier, R. Camisa, S. Narayan, and S. Jolly, "High-efficiency GaAs MESFET amplifiers," *Electron. Lett.*, vol. 11, pp. 508–509, Oct. 1975.
[E19] J. Turner, The Plessey Company, private communication.

F. Electron Transport Properties in GaAs, InP and InGaAs; InP and InGaAs MESFET's

[F1] J. Ruch and W. Fawcett, "Temperature dependence of the transport properties of gallium arsenide determined by a Monte Carlo method," *J. Appl. Phys.*, vol. 41, pp. 3843–3849, Aug. 1970.
[F2] H. Lam and G. Acket, "Comparison of the microwave velocity field characteristics of n-type InP and n-type GaAs," *Electron. Lett.*, vol. 7, pp. 722–723, Dec. 1971.
[F3] W. Fawcett and D. Herbert, "High-field transport in gallium arsenide and indium phosphide," *J. Phys. C: Solid State Phys.*, vol. 7, pp. 1641–1654, May 1974.
[F4] J. Barrera and R. Archer, "InP Schottky-gate field-effect transistors," *IEEE Trans. Electron Devices*, vol. ED-22, pp. 1023–1030, Nov. 1975.
[F5] W. Fawcett, C. Hilsum, and H. Rees, "Optimum semiconductor for microwave devices," *Electron. Lett.*, vol. 5, pp. 313–314, July 1969.
[F6] M. Glicksman, R. Enstrom, S. Mittleman, and J. Appert, "Electron mobility in $In_xGa_{1-x}As$ alloys," *Phys. Rev. B*, vol. 9, pp. 1621–1626, Feb. 1974.
[F7] D. Decker, R. Fairman, and C. Nishimoto, "Microwave InGaAs Schottky-barrier-gate field-effect transistors—Preliminary results," in *Proc. 1975 Cornell Conf. on Active Semiconductor Devices for Microwaves and Integrated Optics*, pp. 305–314.
[F8] S. Sze, *Physics of Semiconductor Devices*. New York: Wiley, 1969, pp. 40, 59.

G. Si and GaAs JFET's

[G1] S. Teszner, "Gridistor development for the microwave power region," *IEEE Trans. Electron Devices*, vol. ED-19, pp. 355–364, March 1972.
[G2] D. Lecrosnier and G. Pelous, "Ion-implanted FET for power applications," *IEEE Trans. Electron Devices*, vol. ED-21, pp. 113–118, Jan. 1974.
[G3] R. Zuleeg, E. Bledl, and A. Behle, "Broadband GaAs field-effect transistor amplifier," Air Force Avionics Lab., Technical Report AFAL-TR-73-109, Wright-Patterson AFB, March 1973.
[G4] A. Behle and R. Zuleeg, "Fast neutron tolerance of GaAs JFET's operating in the hot electron range," *IEEE Trans. Electron Devices*, vol. ED-19, pp. 993–995, Aug. 1972.
[G5] R. Zuleeg and K. Lehovec, "Radiation effect on GaAs interface," Air Force Cambridge Research Labs., Scientific Report AFCRL-TR-74-0495, Sep. 1974.
[G6] C. Vergnolle, R. Funck, and M. Laviron, "An adequate structure for power microwave FETs," in *1975 Int. Solid-State*

Circuits Conf., Dig. Tech. Papers, pp. 66–67.
[G7] S. Umebachi, K. Asahi, M. Inoue, and G. Kano, "A new heterojunction gate GaAs FET," *IEEE Trans. Electron Devices,* vol. ED-22, pp. 613–614, Aug. 1975.

H. Si and GaAs IGFET's

[H1] H. Becke and J. White, "Gallium arsenide insulated-gate field-effect transistors," *1966 Symp. GaAs* (Inst. Phys., Conf. Series No. 3, London, 1967), pp. 219–227.
[H2] T. Miyazaki, N. Nakamura, A. Doi, and T. Tokuyama, "N-channel gallium arsenide MISFET," in *1973 Int. Electron Devices Meeting, Dig. Tech. Papers,* pp. 164–166.
[H3] A. Adams and B. Pruniaux, "GaAs surface film evaluation by ellipsometry and its effect on Schottky barriers," *J. Electrochem. Soc.,* vol. 120, pp. 408–414, March 1973.
[H4] R. Singh and H. Hartnagel, "New method of passivating GaAs with Al_2O_3," in *1974 Int. Electron Devices Meeting, Dig. Tech. Papers,* pp. 576–578.
[H5] H. Hasegawa, K. Forward, and H. Hartnagel, "New anodic native oxide of GaAs with improved dielectric and interface properties," *Appl. Phys. Lett.,* vol. 26, pp. 567–569, May 1975.
[H6] G. Declerck, T. Hattori, G. May, J. Beaudouin, and J. Meindl, "Some effects of trichloroethylene oxidation on the characteristics of MOS devices," *J. Electrochem. Soc.,* vol. 122, pp. 436–439, March 1975.
[H7] Y. Morita, H. Takahashi, H. Matayoshi, and M. Fukuta, "Si UHF MOS high-power FET," *IEEE Trans. Electron Devices,* vol. ED-21, pp. 733–734, Nov. 1974.
[H8] J. G. Oakes, R. A. Wickstrom, D. A. Tremere, and T. M. S. Heng, "A power silicon microwave MOS transistor," *IEEE Trans. Microwave Theory and Techniques,* this issue, pp. 305–311.
[H9] T. Heng, Westinghouse Research Laboratories, private communication.
[H10] T. Cauge, J. Kocsis, H. Sigg, and G. Vendelin, "Double-diffused MOS transistor achieves microwave gain," *Electronics,* vol. 44, pp. 99–104, Feb. 1971.
[H11] H. Sigg, G. Vendelin, T. Cauge, and J. Kocsis, "D-MOS transistor for microwave applications," *IEEE Trans. Electron Devices,* vol. ED-19, pp. 45–53, Jan. 1972.
[H12] H. Sigg, D. Pitzer, and T. Cauge, "D-MOS for UHF linear and nanosecond switching applications," in *1974 IEEE INTERCON, Dig. Tech. Papers,* Session 32/4, pp. 1–7, March 1974.
[H13] H. Sigg, Signetics Corp., private communication (see also Signetics Data Sheet, SD-203/SD-308).
[H14] R. Ronen and L. Strauss, "The silicon-on-sapphire MOS tetrode, some small-signal features LF to UHF," *IEEE Trans. Electron Devices,* vol. ED-21, pp. 100–109, Jan. 1974.

I. GaAs MESFET Reliability

[I1] K. Ohta and M. Ogawa, "Degradation of gold-germanium ohmic contact to n-GaAs," in *1974 IEEE Reliability Physics Symp., Dig. Tech. Papers,* pp. 278–283.
[I2] H. Kohzu, I. Nagasako, M. Ogawa, and N. Kawamura, "Reliability studies of one-micron Schottky-gate GaAs FET," in *1975 Int. Electron Devices Meeting, Dig. Tech. Papers,* pp. 247–250.
[I3] T. Irie, I. Nagasako, H. Kohzu, and K. Sekido, "Reliability study of GaAs MESFET's," *IEEE Trans. Microwave Theory and Techniques,* this issue, pp. 321–328.
[I4] D. Abbott and J. Turner, "Some aspects of GaAs FET reliability," in *1975 Int. Electron Devices Meeting, Dig. Tech. Papers,* pp. 243–246.
[I5] ——, "Some aspects of GaAs MESFET reliability," *IEEE Trans. Microwave Theory and Techniques,* this issue, pp. 317–321.
[I6] S. Bellier, R. Haythornthwaite, J. May, and P. Woods, "Reliability of microwave GaAs field-effect transistors," in *Proc. 1975 Reliability Physics Symp.,* pp. 193–199.
[I7] D. Ch'en, H. Cooke, and J. Wholey, "Long-term stabilization of microwave FET's," *Microwave J.,* vol. 18, pp. 60–61, Nov. 1975.
[I8] ——, "Microwave FET's with improved amplifier stability," in *1976 Int. Solid-State Circuits Conf., Dig. Tech. Papers,* pp. 160–161.
[I9] S. Bearse, "GaAs FET's: Device designers solving reliability problems," *Microwaves,* vol. 15, pp. 32–52, Feb. 1976.

J. JFET and MESFET Theory of Operation (Analytic Solutions)

[J1] W. Shockley, "A unipolar field-effect transistor," *Proc. IRE,* vol. 40, pp. 1365–1376, Nov. 1952.
[J2] J. Turner and B. Wilson, "Implications of carrier velocity saturation in a gallium arsenide field-effect transistor," in *Proc. 1968 Symp. GaAs* (Inst. Phys., Conf. Series No. 7, London, 1969), pp. 195–204.
[J3] K. Drangeid and R. Sommerhalder, "Dynamic performance of Schottky-barrier field-effect transistors," *IBM J. Res. Develop.,* vol. 14, pp. 82–94, March 1970.
[J4] K. Lehovec and R. Zuleeg, "Voltage-current characteristics of GaAs J-FET's in the hot electron range," *Solid-State Electron.,* vol. 13, pp. 1415–1426, Oct. 1970.
[J5] P. Hower and G. Bechtel, "Current saturation and small-signal characteristics of GaAs field-effect transistors," *IEEE Trans. Electron Devices,* vol. ED-20, pp. 213–220, March 1973.
[J6] K. Lehovec and W. Seeley, "On the validity of the gradual channel approximation for junction field-effect transistors with drift velocity saturation," *Solid-State Electron.,* vol. 16, pp. 1047–1054, Sep. 1973.
[J7] R. Fair, "Graphical design and iterative analysis of the dc parameters of GaAs FET's," *IEEE Trans. Electron Devices,* vol. ED-21, pp. 357–362, June 1974.
[J8] A. Grebene and S. Ghandi, "General theory for pinched operation of the junction-gate FET," *Solid-State Electron.,* vol. 12, pp. 573–589, July 1969.
[J9] P. Rossel and J. Cabot, "Output-resistance properties of the GaAs Schottky-gate field-effect transistor in saturation," *Electron. Lett.,* vol. 11, pp. 150–152, April 1975.
[J10] D. Mo and H. Yanai, "Current-voltage characteristics of the junction-gate field-effect transistor with field-dependent mobility," *IEEE Trans. Electron Devices,* vol. ED-17, pp. 577–586, Aug. 1970.
[J11] G. Alley and H. Talley, "A theoretical study of the high frequency performance of a Schottky-barrier field-effect transistor fabricated on a high-resistivity substrate," *IEEE Trans. Microwave Theory and Techniques,* vol. MTT-22, pp. 183–189, March 1974.
[J12] K. Lehovec and R. Miller, "Field distribution in junction field-effect transistors at large drain voltages," *IEEE Trans. Electron Devices,* vol. ED-22, pp. 273–281, May 1975.

K. JFET and MESFET Theory of Operation (Numerical Solutions)

[K1] D. Kennedy and R. O'Brien, "Computer-aided two dimensional analysis of the junction field-effect transistor," *IBM J. Res. Develop.,* vol. 14, pp. 95–116, March 1970.
[K2] ——, "Two-dimensional analysis of J-FET structures containing a low-conductivity substrate," *Electron. Lett.,* vol. 7, pp. 714–716, Dec. 1971.
[K3] C. Kim and E. Yang, "An analysis of current saturation mechanism of junction field-effect transistors," *IEEE Trans. Electron Devices,* vol. ED-17, pp. 120–127, Feb. 1970.
[K4] C. Kim, "Differential drain resistance of field-effect transistors beyond pinchoff: A comparison between theory and experiment," *IEEE Trans. Electron Devices,* vol. ED-17, pp. 1088–1089, Dec. 1970.
[K5] M. Reiser, "Two-dimensional analysis of substrate effects in junction FET's," *Electron. Lett.,* vol. 6, pp. 493–494, Aug. 1970.
[K6] M. Reiser and P. Wolf, "Computer study of submicrometre FET's," *Electron. Lett.,* vol. 8, pp. 254–256, May 1972.
[K7] M. Reiser, "A two-dimensional numerical FET model for dc, ac, and large-signal analysis," *IEEE Trans. Electron Devices,* vol. ED-20, pp. 35–45, Jan. 1973.
[K8] B. Himsworth, "A two-dimensional analysis of indium-phosphide junction field-effect transistors with long and short channels," *Solid-State Electron.,* vol. 16, pp. 931–939, Aug. 1973.
[K9] ——, "A two-dimensional analysis of gallium-arsenide junction field-effect transistors with long and short channels," *Solid-State Electron.,* vol. 15, pp. 1353–1361, Dec. 1972.
[K10] J. Barnes and R. Lomax, "Two-dimensional finite element simulation of semiconductor devices," *Electron. Lett.,* vol. 10, pp. 341–343, Aug. 1974.
[K11] J. Ruch, "Electron dynamics in short channel field-effect transistors," *IEEE Trans. Electron Devices,* vol. ED-19, pp. 652–654, May 1972.
[K12] T. Maloney and J. Frey, "Effects of nonequilibrium velocity-field characteristics on the performance of GaAs and InP field-effect transistors," in *1974 Int. Electron Devices Meeting, Dig. Tech. Papers,* pp. 296–298.
[K13] ——, "Frequency limits of GaAs and InP field-effect transistors," *IEEE Trans. Electron Devices,* vol. ED-22, pp. 357–358, June 1975; also corrections in *IEEE Trans. Electron Devices,* vol. ED-22, p. 620, Aug. 1975.
[K14] R. Hockney, R. Warriner, and M. Reiser, "Two-dimensional particle models in semiconductor device analysis," *Electron. Lett.,* vol. 10, pp. 484–486, Nov. 1974.

L. Equivalent Circuits for MESFET's

[L1] R. Dawson, "Equivalent circuit of the Schottky-barrier field-effect transistor at microwave frequencies," *IEEE Trans. Microwave Theory and Techniques,* vol. MTT-23, pp. 499–501, June 1975.
[L2] G. Vendelin and M. Omori, "Circuit model for the GaAs MESFET valid to 12 GHz," *Electron. Lett.,* vol. 11, pp. 60–61, 1975.
[L3] ——, "Try CAD for accurate GaAs MESFET models," *Microwaves,* vol. 14, pp. 58–70, June 1975.
[L4] G. D. Vendelin, "Feedback effects in the GaAs MESFET

model," *IEEE Trans. Microwave Theory and Techniques*, this issue, pp. 383–385.

M. MESFET Noise Theory

[M1] H. Rothe and W. Dahlke, "Theory of noisy fourpoles," *Proc. IRE*, vol. 44, pp. 811–818, June 1956.
[M2] A. Van der Ziel, "Thermal noise in field-effect transistors," *Proc. IRE*, vol. 50, pp. 1808–1812, Aug. 1962.
[M3] ——, "Gate noise in field-effect transistors at moderately high frequencies," *Proc. IEEE*, vol. 51, pp. 461–467, March 1963.
[M4] A. Van der Ziel and J. Ero, "Small-signal, high-frequency theory of field-effect transistors," *IEEE Trans. Electron Devices*, vol. ED-11, pp. 128–135, April 1964.
[M5] A. Van der Ziel, "Thermal noise in the hot electron regime in FET's," *IEEE Trans. Electron Devices*, vol. ED-18, p. 977, Oct. 1971.
[M6] F. Klaassen, "On the influence of hot carrier effects on the thermal noise of field-effect transistors," *IEEE Trans. Electron Devices*, vol. ED-17, pp. 858–862, Oct. 1970.
[M7] W. Baechtold, "Noise behavior of Schottky barrier gate field-effect transistors at microwave frequencies," *IEEE Trans. Electron Devices*, vol. ED-18, pp. 97–104, Feb. 1971.
[M8] ——, "Noise behavior of GaAs field-effect transistors with short gate lengths," *IEEE Trans. Electron Devices*, vol. ED-19, pp. 674–680, May 1972.
[M9] H. Statz, H. Haus, and R. Pucel, "Noise characteristics of gallium arsenide field-effect transistors," *IEEE Trans. Electron Devices*, vol. ED-21, pp. 549–562, Sep. 1974.
[M10] R. Pucel, H. Haus, and H. Statz, "Signal and noise properties of gallium arsenide microwave field-effect transistors," in *Advances in Electronics and Electron Physics*, vol. 38. New York: Academic Press, 1975, pp. 195–265.

N. Characterization of FET Noise Performance

[N1] H. Haus and R. Adler, *Circuit Theory of Linear Noisy Networks*. New York: Wiley, 1959.
[N2] H. Haus, "IRE standards on methods of measuring noise in linear two ports, 1959," *Proc. IRE*, vol. 48, pp. 60–68, Jan. 1960.
[N3] H. Fukui, "Available power gain, noise figure and noise measure of two-ports and their graphical representations," *IEEE Trans. Circuit Theory*, vol. CT-13, pp. 137–142, June 1966.
[N4] A. Leupp and M. Strutt, "High-frequency FET noise parameters and approximation of the optimum source admittance," *IEEE Trans. Electron Devices*, vol. ED-16, pp. 428–431, May 1969.
[N5] R. Lane, "The determination of device noise parameters," *Proc. IEEE*, vol. 57, pp. 1461–1462, Aug. 1969.
[N6] R. Kaesser, "Noise factor contours for field-effect transistors at moderately high frequencies," *IEEE Trans. Electron Devices*, vol. ED-19, pp. 164–171, Feb. 1972.
[N7] A. Anastassiou and M. Strutt, "Experimental gain and noise parameters of microwave GaAs FET's in the L and S bands," *IEEE Trans. Microwave Theory and Techniques*, vol. MTT-21, pp. 419–422, June 1973.
[N8] ——, "Experimental and computed four scattering and four noise parameters of GaAs FET's up to 4 GHz," *IEEE Trans. Microwave Theory and Techniques*, vol. MTT-22, pp. 138–140, Feb. 1974.
[N9] J. Eisenberg, "Designing amplifiers for optimum noise figure," *Microwaves*, vol. 13, pp. 36–44, April 1974.
[N10] J. Engberg, "Simultaneous input power match and noise optimization using feedback," in *Proc. 1974 European Microwave Conf.*, pp. 385–389.
[N11] G. Vendelin, "Feedback effects on the noise performance of GaAs MESFET's," in *1975 Int. Microwave Symp., Dig. Tech. Papers*, pp. 324–326.
[N12] S. Iversen, "The effect of feedback on noise figure," *Proc. IEEE*, vol. 63, pp. 540–542, March 1975.
[N13] L. Besser, "Stability considerations of low-noise transistor amplifiers with simultaneous noise and power match," in *1975 Int. Microwave Symp., Dig. Tech. Papers*, pp. 327–329.
[N14] A. Anastassiou and M. Strutt, "Effect of source lead inductance on the noise figure of a GaAs FET," *Proc. IEEE*, vol. 62, pp. 406–408, March 1974; also, for corrections, see S. Iversen, *Proc. IEEE*, vol. 63, pp. 983–984, June 1975.

O. Review Papers on Solid-State Microwave Amplifiers

[O1] L. Cuccia, "Status report: Modern low noise amplifiers in communication systems," *Microwave System News (MSN)*, vol. 4, pp. 120–132, Aug/Sep 1974 and pp. 79–90, Oct/Nov 1974.
[O2] J. Magarshack, "Design and applications of solid-state microwave amplifiers," in *Proc. 1975 European Microwave Conf.*, pp. 153–167.
[O3] N. Osbrink *et al.*, "Review of microwave amplifiers," *Microwave J.*, vol. 18, pp. 27–34, Nov. 1975.

P. Signal Distortion in Transistor Amplifiers

[P1] R. Meyer, M. Shensa, and R. Eschenbach, "Cross modulation and intermodulation in amplifiers at high frequencies," *IEEE J. Solid-State Circuits*, vol. SC-7, pp. 16–23, Feb. 1972.
[P2] R. Fair, "Harmonic distortion in the junction field-effect transistor with field-dependent mobility," *IEEE Trans. Electron Devices*, vol. ED-19, pp. 9–13, Jan. 1972.
[P3] C. Liechti and R. Tillman, "Application of GaAs Schottky-gate FET's in microwave amplifiers," in *1973 IEEE Int. Solid-State Circuits Conf., Dig. Tech. Papers*, pp. 74–75.

Q. Amplifier Network Analysis and Synthesis

[Q1] S. Mason, "Power gain in feedback amplifier," *IRE Trans. Circuit Theory*, vol. CT-1, pp. 20–25, June 1954.
[Q2] J. Rollett, "Stability and power-gain invariants of linear two ports," *IRE Trans. Circuit Theory*, vol. CT-9, pp. 29–32, March 1962.
[Q3] K. Kurokawa, "Power waves and the scattering matrix," *IEEE Trans. Microwave Theory and Techniques*, vol. MTT-13, pp. 194–202, March 1965.
[Q4] G. Bodway, "Two port power flow analysis using generalized scattering parameters," *Microwave J.*, vol. 10, pp. 61–69, May 1967.
[Q5] C. Gledhill and M. Abulela, "Scattering parameter approach to the design of narrow-band amplifiers employing conditionally stable active elements," *IEEE Trans. Microwave Theory and Techniques*, vol. MTT-22, pp. 43–48, Jan. 1974.
[Q6] K. Kotzebue, "Microwave-transistor power-amplifier design by large-signal y parameters," *Electron. Lett.*, vol. 11, pp. 240–241, May 1975.
[Q7] L. Besser, "Design considerations of a 3.1–3.5 GHz GaAs FET feedback amplifier," in *1972 IEEE Int. Microwave Symp., Dig. Tech. Papers*, pp. 230–232.
[Q8] W. Leighton, R. Chaffin, and J. Webb, "RF amplifier design with large-signal s-parameters," *IEEE Trans. Microwave Theory Tech.*, vol. MTT-21, pp. 809–814, Dec. 1973.
[Q9] F. Weinert, "Scattering parameters speed design of high-frequency transistor circuits," *Electron.*, vol. 39, pp. 78–88, Sep. 1966.
[Q10] R. Anderson, "S-parameter techniques for faster, more accurate network design," *Hewlett-Packard J.*, vol. 18, pp. 13–24, Feb. 1967.
[Q11] W. Froehner, "Quick amplifier design with scattering parameters," *Electron.*, vol. 40, pp. 100–109, Oct. 1967.
[Q12] R. Tucker, "Low-noise design of microwave transistor amplifiers," *IEEE Trans. Microwave Theory and Techniques*, vol. MTT-23, pp. 697–700, Aug. 1975.
[Q13] C. Liechti and R. Tillman, "Design and performance of microwave amplifiers with GaAs Schottky-gate field-effect transistors," *IEEE Trans. Microwave Theory and Techniques*, vol. MTT-22, pp. 510–517, May 1974.
[Q14] M. Mokari-Bolhassan and T. Trick, "Computer-aided design of distributed-lumped-active networks," *IEEE Trans. Circuit Theory*, vol. CT-18, pp. 187–190, Jan. 1971.
[Q15] J. Bandler, "Computer aided circuit optimization," in *Modern Filter Theory and Design*, G. Temes and S. Mitra, Ed. New York: John Wiley, 1973.
[Q16] J. Bandler, J. Popovic, and V. Jha, "Cascaded network optimization program," *IEEE Trans. Microwave Theory and Techniques*, vol. MTT-22, pp. 300–308, March 1974.
[Q17] E. Sanchez-Sinencio and T. Trick, "CADMIC-computer aided design of microwave integrated circuits," *IEEE Trans. Microwave Theory and Techniques*, vol. MTT-22, pp. 309–316, March 1974.
[Q18] C. Charalambous, "A unified review of optimization," *IEEE Trans. Microwave Theory and Techniques*, vol. MTT-22, pp. 289–300, March 1974.
[Q19] N. Kuhn, "CAD with graphics make circuit design a science," *Microwaves*, vol. 13, pp. 42–50, June 1974.
[Q20] W. Ku, W. Petersen, and A. Podell, "New results on the design of broadband microwave bipolar and FET amplifiers," in *1974 IEEE Int. Symp. Microwave Theory and Techniques, Dig. Tech. Papers*, pp. 357–359.
[Q21] W. Ku *et al.*, "Microwave octave-band GaAs FET amplifiers," in *1975 IEEE Int. Symp. Microwave Theory and Techniques, Dig. Tech. Papers*, pp. 69–72.
[Q22] D. Mellor, "Insertion loss synthesis of matching networks for microwave amplifiers," in *1975 IEEE Int. Solid-State Circuits Conf., Dig. Tech. Papers*, pp. 68–69, 214.
[Q23] D. Mellor and J. Linvill, "A complete computer program for the synthesis of matching networks for microwave amplifiers," in *1975 IEEE Int. Symp. Microwave Theory and Techniques, Dig. Tech. Papers*, pp. 191–193.
[Q24] ——, "Synthesis of interstage networks of prescribed gain

versus frequency slopes," *IEEE Trans. Microwave Theory and Techniques*, vol. MTT-23, pp. 1013–1020, Dec. 1975.
[Q25] N. Marshall, "Optimizing multi-stage amplifiers for low noise," *Microwaves*, vol. 13, pp. 62–64, April 1974.
[Q26] ——, "Optimizing multi-stage amplifiers for linearity," *Microwaves*, vol. 13, pp. 60–64, May 1974.
[Q27] K. Kurokawa, "Design theory of balanced transistor amplifiers," *Bell System Tech. J.*, vol. 44, pp. 1675–1698, Oct. 1965.
[Q28] S. Lazar, "Solid-state power amplifiers for *S*-band phased array radar," *Microwave System News*, vol. 5, pp. 77–84, Feb./Mar. 1975.

R. Performance of Small-Signal MESFET Amplifiers
[R1] P. Clouser and V. Risser, "*C*-band FET amplifiers," in *1970 IEEE Int. Solid-State Circuits Conf., Dig. Tech. Papers*, pp. 52–53.
[R2] W. Baechtold, "*Ku*-band GaAs FET amplifier and oscillator," *Electron. Lett.*, vol. 7, pp. 275–276, May 1971.
[R3] ——, "*X*- and *Ku*-band amplifiers with GaAs Schottky-barrier field-effect transistors," *IEEE J. Solid-State Circuits*, vol. SC-8, pp. 54–58, Feb. 1973.
[R4] S. Arnold, "Single and dual-gate GaAs FET integrated amplifiers in *C*-band," in *1972 IEEE Int. Microwave Symp., Dig. Tech. Papers*, pp. 233–234.
[R5] J. Eisenberg and R. Disman, "Design a 4 to 8 GHz FET amplifier with a 7 dB noise figure," *Microwaves*, vol. 12, pp. 52–56, Feb. 1973.
[R6] C. Ch'en and A. Woo, "A low-noise *C*-band GaAs FET amplifier," 1974 IEEE Int. Microwave Symposium, Session 20, Late News Paper.
[R7] G. Vendelin, J. Archer, and G. Bechtel, "A low-noise integrated *S*-band amplifier," in *1974 IEEE Int. Solid-State Circuits Conf., Dig. Tech. Papers*, pp. 176–177.
[R8] D. James, R. Douville, R. Breithaupt, and A. Van Koughnett, "A 12 GHz field-effect transistor amplifier for communications satellite applications," in *Proc. 1974 European Microwave Conf.*, pp. 97–101.
[R9] N. Slaymaker and J. Turner, "Microwave FET amplifiers with centre frequencies between 1 and 11 GHz," in *Proc. 1973 European Microwave Conf.*, vol. 1, Paper A.5.1.
[R10] N. A. Slaymaker, R. A. Soares, and J. A. Turner, "GaAs MESFET small-signal *X*-band amplifiers," *IEEE Trans. Microwave Theory and Techniques*, this issue, pp. 329–337.
[R11] H. Luxton, "Gallium-arsenide field-effect transistors—Their performance and application up to *X*-band frequencies," in *Proc. 1974 European Microwave Conf.*, pp. 92–96.
[R12] R. Pengelly, "Broadband lumped-element *X*-band GaAs FET amplifiers," *Electron. Lett.*, vol. 11, pp. 58–60, Feb. 1975.
[R13] ——, "Broadband lumped-element *X*-band GaAs FET amplifiers," in *Proc. 1975 European Microwave Conf.*, pp. 301–305.
[R14] R. Soares and J. Turner, "Tunable *X*-band GaAs FET amplifier," *Electron. Lett.*, vol. 11, pp. 474–475, Sep. 1975.
[R15] M. Walker, F. Mauch, and T. Williams, "Cover *X*-band with an FET amplifier," *Microwaves*, vol. 14, pp. 36–45, Oct. 1975.
[R16] D. Ch'en, Avantek, private communication.
[R17] M. Walker, Watkins-Johnson Company, private communication.

S. Performance of MESFET Amplifiers at Low Temperatures
[S1] B. Loriou, M. Bellec, and M. LeRouzic, "Performances à basse température d'un transistor hyperfréquences faible bruit à effet de champ," *Electron. Lett.*, vol. 6, pp. 819–820, Dec. 1970.
[S2] J. Leost and B. Loriou, "Propriétés à basse température des transistors GaAs à effet de champ hyperfréquence," *Hyperfréquences*, vol. 54, pp. 514–522, December 1974.
[S3] J. Jimenez, J. Oliva, and A. Septier, "Very low noise cryogenic MESFET amplifier," in *Proc. 1973 European Microwave Conf.*, vol. 1, Paper A.5.2.
[S4] P. Bura, "Operation of 6 GHz FET amplifier at reduced ambient temperature," *Electron. Lett.*, vol. 10, pp. 181–182, May 1974.
[S5] C. Liechti, R. Larrick, and D. Mellor, "A cooled GaAs MESFET amplifier operating at 12 GHz with 1.6 dB noise figure," in *Proc. 1975 European Microwave Conf.*, pp. 306–309.
[S6] C. A. Liechti and R. B. Larrick, "Performance of GaAs MESFET's at low temperatures," *IEEE Trans. Microwave Theory and Techniques*, this issue, pp. 376–381.

T. Performance of Medium-Power MESFET Amplifiers
[T1] D. Hornbuckle and L. Kuhlman, "Broad-band medium-power amplification in the 2–12.4-GHz range with GaAs MESFET's," *IEEE Trans. Microwave Theory and Techniques*, this issue, pp. 338–342.
[T2] Y. Arai, T. Kouno, T. Horimatsu, and H. Komizo, "A 6-GHz four-stage GaAs MESFET power amplifier," *IEEE Trans. Microwave Theory and Techniques*, this issue, pp. 381–383.
[T3] R. E. Neidert and H. A. Willing, "Wide-band gallium arsenide powder MESFET amplifier," *IEEE Trans. Microwave Theory and Techniques*, this issue, pp. 342–350.
[T4] P. Bura and D. Cowan, "Highly linear medium power 11 GHz FET amplifier," in *1976 Int. Solid-State Circuits Conf., Dig. Tech. Papers*, pp. 158–159.
[T5] F. Sechi, "High-gain 1 W FET amplifier operating in *G* band," in *1976 Int. Solid-State Circuits Conf., Dig. Tech. Papers*, pp. 162–163.
[T6] R. Camisa, J. Goel, and I. Drukier, "GaAs MESFET linear power amplifier stage giving 1 W," *Electron. Lett.*, vol. 11, pp. 572–573, Nov. 1975.

U. MESFET Oscillators
[U1] M. Maeda, S. Takahashi, and H. Kodera, "CW oscillation characteristics of GaAs Schottky-barrier gate field-effect transistors," *Proc. IEEE*, vol. 63, pp. 320–321, Feb. 1975.
[U2] M. Maeda, K. Kimura, and H. Kodera, "Design and performance of *X*-band oscillators with GaAs Schottky-gate field-effect transistors," *IEEE Trans. Microwave Theory and Techniques*, vol. MTT-23, pp. 661–667, Aug. 1975.
[U3] P. Pucel, R. Bera, and D. Masse, "Experiments on integrated gallium-arsenide FET oscillators at *X*-band," *Electron. Lett.*, vol. 11, pp. 219–220, May 1975.
[U4] N. Slaymaker and J. Turner, "Alumina microstrip GaAs FET 11 GHz oscillator," *Electron. Lett.*, vol. 11, pp. 300–301, July 1975.
[U5] M. Omori and C. Nishimoto, "Common-gate FET oscillator," *Electron. Lett.*, vol. 11, pp. 369–371, Aug. 1975.
[U6] D. James, G. Painchaud, E. Minkus, and W. Hoefer, "Stabilized 12 GHz MIC oscillator using GaAs FET's," in *Proc. 1975 European Microwave Conf.*, pp. 296–300.
[U7] H. Abe, Y. Takayama, A. Higashisaka, R. Yamamoto, and M. Takeuchi, "A high-power microwave GaAs FET oscillator," in *1976 Int. Solid-State Circuits Conf., Dig. Tech. Papers*, pp. 164–165.
[U8] R. Pucel, R. Bera, and D. Masse, "An evaluation of GaAs FET oscillators and mixers for integrated front-end applications," in *1975 IEEE Int. Solid-State Circuits Conf., Dig. Tech. Papers*, pp. 62–63.

V. MESFET Mixers
[V1] J. Sitch and P. Robson, "The performance of GaAs field-effect transistors as microwave mixers," *Proc. IEEE*, vol. 61, pp. 399–400, March 1973.
[V2] R. Pucel, D. Masse, and R. Bera, "Integrated GaAs FET mixer performance at *X*-band," *Electron. Lett.*, vol. 11, pp. 199–200, May 1975.
[V3] ——, "Performance of GaAs MESFET mixers at *X*-band," *IEEE Trans. Microwave Theory and Techniques*, this issue, pp. 351–360.
[V4] O. Kurita and K. Morita, "Microwave MESFET mixer," *IEEE Trans. Microwave Theory and Techniques*, this issue, pp. 361–366.
[V5] S. Komaki, O. Kurita, and T. Memita, "GaAs MESFET regenerator for phase-shift keying signals at the carrier frequency," *IEEE Trans. Microwave Theory and Techniques*, this issue, pp. 367–372.

W. High-Speed Logic
[W1] K. Drangeid *et al.*, "A memory-cell array with normally off-type Schottky-barrier FET's," *IEEE J. Solid-State Circuits*, vol. SC-7, pp. 277–282, Aug. 1972.
[W2] O. Cahen, G. Cachier, and J. Puron, "A subnanosecond switching circuit," in *1974 IEEE Int. Solid-State Circuits Conf., Dig. Tech. Papers*, pp. 110–111.
[W3] R. Van Tuyl and C. Liechti, "High speed integrated logic with GaAs MESFET's," *IEEE J. Solid-State Circuits*, vol. SC-9, pp. 269–276, Oct. 1974.
[W4] ——, "High speed GaAs MSI," in *1976 IEEE Int. Solid-State Circuits Conf., Dig. Tech. Papers*.
[W5] E. Kohn, "Normally-off MESFET with fast switching behavior," *Electron. Lett.*, vol. 10, p. 505, Nov. 1974.
[W6] H. Beneking and E. Kohn, "High-speed GaAs MESFET differential amplifier stage with integrated current source," in *1974 Int. Electron Devices Meeting, Dig. Tech. Papers*, pp. 292–295.
[W7] H. Beneking and W. Filensky, "The GaAs MESFET as a pulse regenerator in the gigabit per second range," *IEEE Trans. Microwave Theory and Techniques*, this issue, pp. 385–386.
[W8] J. Notthoff and R. Zuleeg, "High speed, low power GaAs JFET integrated circuits," in *1975 Int. Electron Devices Meeting, Dig. Tech. Papers*, p. 624.

[W9] V. Vodicka and R. Zuleeg, "Ion implanted GaAs enhancement mode JFET's," in *1975 Int. Electron Devices Meeting, Dig. Tech. Papers*, pp. 625–628.

[W10] T. Sudo et al., "A monolithic 8 pJ/2 GHz logic family," *IEEE J. Solid State Circuits*, vol. SC-10, pp. 524–529, Dec. 1975.

[W11] D. DiPietro, "A 5 GHz f_T monolithic IC process for high-speed digital circuits," in *1975 Int. Solid-State Circuits Conf., Dig. Tech. Papers*, pp. 118–119.

[W12] L. Cuccia, J. Spilker, and D. Magill, "Digital communication at gigahertz data rates," Part I, *Microwave J.*, vol. 13, pp. 80–93, Jan. 1970; Part II, *Microwave J.*, vol. 13, pp. 87–92, Feb. 1970; Part III, *Microwave J.*, vol. 13, pp. 75–80, April 1970.

[W13] L. Cuccia, "A technology status report on high-speed MPSK digital modulation systems," in *Proc. 1974 European Microwave Conf.*, p. 505.

[W14] C. Ryan, "Bipolar IC's for microwave signal processing," in *1975 IEEE Int. Microwave Symposium, Dig. Tech. Papers*, pp. 37–39.

X. Fabrication Technologies

[X1] S. Middelhoek, "Projection masking, thin photoresist layers and interface effects," *IBM J. Res. Develop.*, vol. 14, pp. 117–124, March 1970.

[X2] ——, "Metallization processes in fabrication of Schottky-barrier FET's," *IBM J. Res. Develop.*, vol. 14, pp. 148–151, March 1970.

[X3] H. Smith, "Fabrication techniques for surface-acoustic-wave and thin-film optical devices," *Proc. IEEE*, vol. 62, pp. 1361–1387, Oct. 1974.

[X4] C. Stolte, "Device quality n-type layers produced by ion implantation of Te and S into GaAs," in *1975 Int. Electron Devices Meeting, Dig. Tech. Papers*, pp. 585–587.

Noise Characteristics of Gallium Arsenide Field-Effect Transistors

HERMANN STATZ, HERMANN A. HAUS, FELLOW, IEEE, AND ROBERT A. PUCEL, SENIOR MEMBER, IEEE

Abstract—Small signal and noise characteristics for GaAs field-effect transistors are derived with the saturated drift velocity of the carriers underneath the gate taken into account. The noise contributed by the saturated carriers is nonnegligible and in most cases, exceeds the noise generated by the unsaturated region. Parasitic elements contribute importantly by preventing the full cancellation of the correlated noise of the intrinsic transistor and by adding their own Johnson noise. The theory predicts the experimentally observed trend of noise figure dependence on drain current and on source-to-drain voltage. The present theory does not take into account the effects of a possible short negative resistance region underneath the gate.

I. INTRODUCTION

HIGH-FREQUENCY gallium arsenide field-effect transistors have shown astonishingly low noise figures of approximately 3–4 dB at 10 GHz [1], [2]. Even though considerable work in understanding this noise performance has been done, most explanations omit important features of the microwave FET. Van der Ziel, in pioneering work, first analyzed noise in field-effect transistors [3], [4]. Van der Ziel, however, restricted himself to source–drain voltage differences below the pinch-off value. All microwave transistors operate above the so-called pinch-off point.

Van der Ziel omitted, in addition, the nonohmic conductivity of gallium arsenide. The nonohmic behavior is of importance in virtually all GaAs microwave transistors. For example, a voltage drop of 1 V across a typical gate length of 10^{-4} μm corresponds to an average field strength of 10^4 V/cm. The velocity of carriers as a function of field in GaAs reaches a maximum at 3×10^3 V/cm followed by a negative resistance region and then a nearly constant saturated drift velocity. Baechtold [5], [6] corrected for the nonconstant mobility by assuming a constant mobility up to a critical field, and then for fields larger than the critical field, he assumed a constant velocity. This approach neglects the effects of the negative resistance region. Since the negative resistance region occupies in general only a small space under the gate, the approximation may be justified.

Baechtold also allows for a field dependent electron temperature. For gallium arsenide, he assumes

$$\frac{T_e}{T_o} = 1 + \delta \left(\frac{E}{E_{\text{sat}}}\right)^3 \qquad (1)$$

where T_e is the effective noise temperature, T_o is the reference temperature, i.e., 300 K, δ is an empirical constant, E is the electric field, and E_{sat} is the saturation field. Baechtold confirmed (1) by experiments. However, Baechtold neglects the saturated velocity region of the transistor and its contribution to the noise behavior. As stated before, the average field in the transistor under the gate is generally much larger than the saturation field E_{sat}. The region within which the carriers drift at saturated velocity can occupy a major fraction of the total channel length. Therefore, the characteristics of the transistor in the current saturation regime should strongly depend on this satuated velocity region.

This paper includes the effect upon the signal and noise parameters of the saturated drift region. We derive the small signal parameters of the transistor such as transconductance, drain resistance, etc., including the effects of the drift region. The noise analysis in the drift region in some ways follows a suggestion by Shockley et al. [7] according to which the high-field diffusion constant describes the formation of dipole layers which drift through the high-field region to the drain contact. Ruch and Kino [8] have measured the diffusion constant in GaAs at high fields, and Fawcett and Rees [9] have calculated it. These measurements and calculations show a peak in the parallel diffusion constant near the saturation field followed by a rapid drop to rather low values. The measurement of the diffusion constant is complicated by trapping effects. We obtain rather good agreement with experimental noise figure measurements if we use in our expressions values which are consistent with the calculations of Fawcett and Rees [9]. We neglect variations of the diffusion constant with electric field.

It turns out that the noise in the saturated velocity region is rather important and may be larger than the one contributed by the unsaturated region. On the other hand, the correlation between drain noise and gate noise current is rather large, so that the overall noise figure is still attractively low. Parasitic resistances consisting of contact resistance at the source and the series resistance between the source and the region under the gate as well as gate metallization resistance are very important. They prevent the full cancellation of the correlated noise of the intrinsic field-effect transistor and contribute their own Johnson noise. It appears that a reduction of these parasitic resistances would substantially lower the noise figure.

II. THE DC CONDITIONS

In this paper, we consider both the symmetric FET as well as the FET constructed on an insulating substrate.

The latter has a field configuration which approximately corresponds to the field configuration of the symmetric FET on one side of the symmetry plane; the nonconducting substrate surface plays the role of the symmetry plane. Fig. 1 illustrates the situation. Instead of placing the source and drain contacts on the side of the gate, as would normally be done, the contacts are placed below the gate in the positions corresponding to the idealized symmetric FET.

We follow van der Ziel's [3], [4] convention of considering positive charge carriers in order to avoid unnecessary confusion between his and our paper. Nevertheless, all GaAs transistors are made of n-type material, and we, therefore, use in our numerical examples saturated velocities, noise temperatures, mobilities, etc., which are characteristic of electrons. In contrast to van der Ziel's transistor biased below pinchoff, we consider a transistor biased beyond pinchoff. The onset of pinchoff is here defined to occur when the maximum longitudinal electric field first reaches the saturation value E_{sat}. The velocity-E field characteristic of the medium (GaAs) is assumed to follow the idealized plot of Fig. 2, which has already been used by Baechtold [5], [6]. The analysis of the FET is carried out separately in two regions: Region I of ohmic conductivity and Region II of saturated velocity.

We use in our paper some results which have previously been derived by Shockley [10] and van der Ziel [3], [4]. For clarity and continuity of the analysis, however, we repeat these results where needed. It is convenient to introduce the gate potential with respect to the source V_g and the channel potential V_p at the end of Region I, also measured with respect to the source. The effective potential difference W between the gate and the channel is then $W = W_s = V_g + V_{dif}$ at the source side and $W = W_p = V_g + V_{dif} - V_p$ at the pinch-off point. V_{dif} is the built-in diffusion potential due to the doping difference between channel and gate. Most of the expressions to be used in the paper become simplified if we introduce the reduced potentials

$$s = \left(\frac{W_s}{W_{oo}}\right)^{1/2} \qquad (2)$$

$$p = \left(\frac{W_p}{W_{oo}}\right)^{1/2} \qquad (3)$$

where W_{oo} is the value of the gate-to-channel potential required to deplete the channel of carriers, as seen by

$$W_{oo} = \frac{\rho}{2\epsilon} a^2. \qquad (4)$$

In (4), ρ is the space-charge density in the channel region due to the built-in acceptor centers, ϵ is the dielectric constant of the material, and $2a$ is the undepleted channel thickness of the symmetrical transistor. As mentioned, we divide the region under the gate of length L into Region I of length L_1 which is ohmic, and a region of length L_2 which

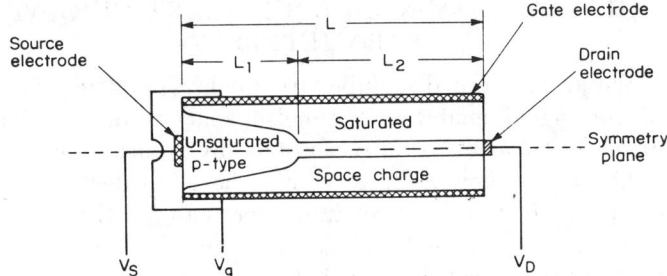

Fig. 1. Idealized geometry of a field-effect transistor.

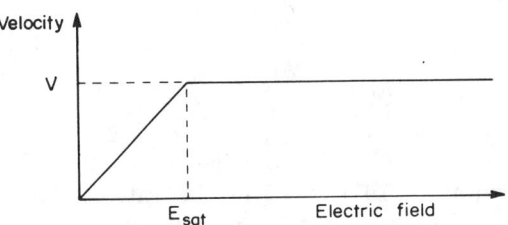

Fig. 2. Simplified velocity versus electric field relationship used in text.

is saturated. In Region I, we have [3]

$$I_d = \frac{lg_o W_{oo}}{L_1} \{p^2 - s^2 - (2/3)(p^3 - s^3)\}. \qquad (5)$$

In (5), I_d is the source-to-drain current, l is the width of the transistor; g_o is defined for the symmetric transistor by

$$g_o = 2\sigma a \qquad (6)$$

where σ is the conductivity of the channel material. For the asymmetric FET, $g_o = \sigma a$.

The undepleted channel width $2b_p$ at the pinch-off point and beyond is given by

$$2b_p = 2a(1 - p). \qquad (7)$$

For applied voltages exceeding the pinch-off value, the longitudinal channel field exceeds the saturation field E_{sat}. By definition, the end of Region I has a longitudinal channel field E_{sat}. The total current that can be carried at this point is equal to

$$I_d = lg_o E_{sat}(1 - p). \qquad (8)$$

By equating (5) and (8), we can determine the position of L_1 as a function of gate and pinch-off potentials, as seen by

$$L_1 = \frac{W_{oo}}{E_{sat}} \frac{p^2 - s^2 - (2/3)(p^3 - s^3)}{1 - p}. \qquad (9)$$

Up to the pinch-off point, ohmic conduction occurs. Beyond the pinch-off point, the carriers drift at constant velocity, occupying the width $2b_p$, as seen in (7). The longitudinal electric field is equal to the saturation field E_{sat} at the pinch-off point. The electric field continues to increase beyond the pinch-off point reaching its maximum value at the drain contact, thus assuring saturated drift throughout Region II.

III. EVALUATION OF EQUIVALENT CIRCUIT PARAMETERS

We are concerned mainly with the low-frequency limit of the signal analysis. The noise analysis takes intermediate frequencies into account by considering the induced gate noise. Fig. 3 shows the equivalent circuit elements of the field-effect transistor. Van der Ziel evaluated the transconductance g_m and the drain conductance for the transistor below pinchoff and found

$$g_m = -\frac{\partial I_d}{\partial V_g} = \frac{g_o l}{L}(d-s) \quad (10)$$

$$g_d = -\frac{\partial I_d}{\partial V_d} = \frac{g_o l}{L}(1-d) \quad (11)$$

where $d = (W_d/W_{oo})^{1/2}$, and W_d is the channel potential at the drain with respect to the gate. V_d is the drain voltage. Here we want to go through an analysis that includes the effect of the saturated velocity pinch-off region. The drain resistance r_d is defined by

$$r_d = -\frac{\partial V_{sd}}{\partial I_d}. \quad (12)$$

In (12), V_{sd} is the total voltage difference between source and drain including the voltage contribution V_p for Region I, plus the voltage between L_1 and L. From the definition of p and s, we have

$$V_p = -W_{oo}(p^2 - s^2). \quad (13)$$

In Region II, we assume the stream of carriers to continue with constant density and uniform cross-section $2b_p$. We have to solve Poisson's equation assuming a uniform acceptor density and including the space charge of the flowing carriers. We have to satisfy the boundary conditions at both ends of Region II. A particular solution of Poisson's equation is obviously that of a constant potential $-W_{oo}(p^2 - s^2)$ everywhere in the carrier stream, and outside of the stream, a potential increasing parabolically toward the gate electrode. To satisfy the boundary condition at the drain end of Section II, we have to add a homogeneous solution of Poisson's equation. Obviously, a series expansion of the form

$$V = \sum_{n=-\infty}^{n=\infty} A_n \cos(2n+1)\frac{\pi y}{2a} \exp(2n+1)\frac{\pi}{2a}x,$$

$$n \text{ integer} \quad (14)$$

is the most general solution [11]. This form satisfies the boundary conditions along the gate. We shall approximate (14) by the lowest space harmonics, with $n = 0, -1$. For reasons of continuity of potential and field ($E_x = E_{sat}$ at L_1) in the middle of the structure ($y = 0$), the homogeneous solution in Region II is of the form

$$V = -\frac{2a}{\pi} E_{sat} \sinh\frac{\pi x}{2a} \cos\frac{\pi y}{2a} \quad (15)$$

where the origin of x is at the pinch-off point. For a drain

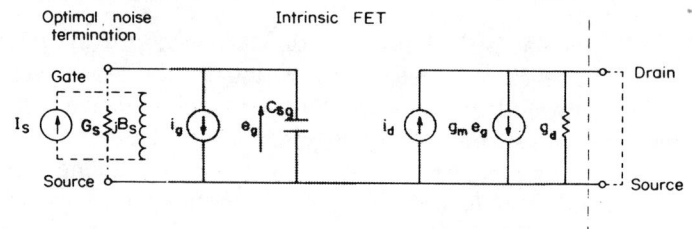

Fig. 3. Equivalent circuit for minimum noise figure calculation.

contact placed at the end of the channel at $y = 0$, the source-to-drain voltage therefore becomes

$$V_{sd} = -W_{oo}(p^2 - s^2) - \frac{2a}{\pi} E_{sat} \sinh\frac{\pi}{2a}L_2. \quad (16)$$

Differentiation of V_{sd} with respect to current I_d gives the drain resistance. We have to keep in mind that p and L_2 depend upon I_d, therefore,

$$r_d = -\frac{dV_{sd}}{dI_d} = 2pW_{oo}\frac{dp}{dI_d} + E_{sat}\left(\cosh\frac{\pi}{2a}L_2\right)\frac{dL_2}{dI_d}.$$

$$(17)$$

The quantity dp/dI_d is evaluated by differentiating (8), seen by

$$\frac{dp}{dI_d} = -\frac{1}{g_o E_{sat} l}. \quad (18)$$

The quantity dL_2/dI_d is also easily obtained, as seen by the equation

$$\frac{dL_2}{dI_d} = \frac{-dL_1}{dI_d} = \frac{-dL_1}{dp}\frac{dp}{dI_d} = \frac{1}{g_o E_{sat} l}\left\{\frac{2pW_{oo}}{E_{sat}} + \frac{L_1}{1-p}\right\}$$

$$(19)$$

where dL_1/dp follows from differentiation of (9). Inserting (18) and (19) into (17) finally gives

$$r_d = \frac{2pW_{oo}}{g_0 E_{sat} l}\left(\cosh\frac{\pi}{2a}L_2 - 1\right) + \frac{L_1}{g_o l(1-p)}\cosh\frac{\pi}{2a}L_2.$$

$$(20)$$

For $L_2 = 0$, (20) reduces to the well-known formula for the drain resistance [3] of a nonsaturated FET, with $p \rightarrow d$, $L_1 \rightarrow L$, and is seen as

$$r_d = \frac{L}{lg_o}\frac{1}{1-d}. \quad (21)$$

We next evaluate the transconductance under velocity saturation conditions. The transconductance g_m is defined as the change in drain current dI_d divided by the change in gate voltage dV_g holding the potential difference between source and drain V_{sd} constant. Using (8) for the drain current in the pinch-off region and remembering the definition of s in (2), we obtain

$$g_m = -\frac{dI_d}{dV_g} = g_o E_{sat} l\frac{dp}{dV_g} = \frac{g_o E_{sat} l}{2sW_{oo}}\frac{dp}{ds}. \quad (22)$$

In our composite transistor consisting of Regions I and II, the value of the potential at the pinch-off point, and thus p, must change in order to accommodate a varying current in the region of saturated velocity. The saturated stream of carriers has to change in width, and according to (8), this leads to a variation of p. Likewise, L_1 and L_2 also change. To evaluate dp/ds, we first use the fact that dV_{sd} is zero. From (16), with $dL_1 = -dL_2$,

$$dV_{sd} = 0 = W_{\infty}(2p\,dp - 2s\,ds) - E_{sat}dL_1 \cosh \frac{\pi}{2a} L_2. \quad (23)$$

To eliminate dL_1 from (23), we need another relationship between dL_1, dp, and ds. We obtain this additional relationship by differentiating (9), seen by

$$\frac{E_{sat}}{W_{oo}} dL_1 = dp \left\{ 2p + \frac{E_{sat}L_1}{W_{oo}} \frac{1}{1-p} \right\} - ds \frac{2s(1-s)}{1-p}. \quad (24)$$

From (23) and (24), one may eliminate dL_1, and an expression for dp/ds is obtained. Inserting this expression for dp/ds into (22), we obtain the desired expression for the transconductance

$$g_m = \frac{lg_o E_{sat}}{W_{oo}} \frac{(1-s)\cosh(\pi/2a)L_2 - (1-p)}{\{2p(1-p) + E_{sat}L_1/W_{oo}\}\cosh(\pi/2a)L_2 - 2p(1-p)}. \quad (25)$$

In the limit of large L_2, (25) reduces to

$$g_m = \frac{lg_o E_{sat}}{W_{oo}} \frac{1-s}{\{2p(1-p) + E_{sat}L_1/W_{oo}\}}. \quad (26)$$

The latter expression is identical, for $p = d$, to the one used in [6]. In the limit of $L_2 = 0$, and thus $L_1 = L$ and $p = d$, we arrive at

$$g_m = \frac{lg_o}{L}(d-s).$$

This is the expression derived by Shockley [10] and van der Ziel [3]. It has already been presented in (10).

IV. THE NOISE EQUIVALENT CIRCUIT AND THE MINIMUM NOISE FIGURE

In this section, we consider the noise of the ideal transistor. Later, we shall include the noise contributions due to parasitic resistances. The equivalent circuit is shown by the solid lines in Fig. 3. The basic transistor equivalent circuit is identical to the one commonly used. In the input circuit, we have the gate–source capacity, C_{sg}. In the output circuit, we have a current generator $g_m e_g$, where e_g is the signal voltage across the gate–source capacitor. Parallel with the output generator is the output conductance g_d of the transistor. As noise sources we use two generators. In the input we use a noise current generator $\overline{i_g^2}$. This type of noise generator has first been introduced and analyzed by van der Ziel [4]. Basically, noise current flows to the gate because of Johnson-type noise voltage fluctuations in the channel region. As Baechtold has shown, this noise is enhanced by hot electrons in the channel. In addition, we introduce the diffusion noise of the pinch-off region and its contributions to the gate current noise. In the output circuit, we use a drain noise voltage generator $\overline{v_d^2}$. This noise voltage results from hot carrier enhanced Johnson noise in Region I, and, as we shall show, from spontaneously generated dipole layers drifting to the drain contact in Region II.

To calculate the noise figure of the circuit in Fig. 3 as a function of the noise generators, we must include an input and output circuit. By dashed lines we show an input admittance circuit $Y_s = G_s + jB_s$. Since the noise figure F will be independent of the particular output circuit, we simply use a short circuit. The quantity $F - 1$ is defined as the mean-square noise current in the output due to the two noise generators, divided by the mean-square noise current in the output circuit produced by the Johnson noise of the input admittance G_s. This is seen as

$$F - 1 = \frac{\overline{|i_g g_m/(Y_s + j\omega C_{gs}) + v_d g_d|^2}}{4kTG_s \Delta f g_m^2 |1/(Y_s + j\omega C_{gs})|^2}. \quad (27)$$

In general, there will be some correlation between i_g and v_d. This is the case because the Johnson noise in the channel contributes both to the drain voltage and to the gate current. We define a correlation coefficient C through the equation

$$\overline{v_d i_g^*} = jC(\overline{v_d^2} \cdot \overline{i_g^2})^{1/2}. \quad (28)$$

Thus $F - 1$ may be written as

$$F - 1 = \frac{1}{4kTG_s \Delta f g_m^2} \{\overline{i_g^2} g_m^2 + g_d^2 \overline{v_d^2} |Y_s + j\omega C_{gs}|^2$$
$$+ jC(\overline{v_d^2})^{1/2}(\overline{i_g^2})^{1/2} g_d g_m (Y_s + j\omega C_{gs})$$
$$- jC(\overline{v_d^2})^{1/2}(\overline{i_g^2})^{1/2} g_d g_m (Y_s^* - j\omega C_{gs})\}. \quad (29)$$

Since we are interested in the minimum possible noise figure, we first find the optimum input susceptance B_s. Taking the derivative of $F - 1$ with respect to B_s, and setting the resulting expression equal to zero, we get

$$B_{s\,opt} = -\omega C_{gs} + \frac{g_m}{g_d} C \frac{(\overline{i_g^2})^{1/2}}{\overline{v_d^2}} \quad (30)$$

and $F - 1$ becomes

$$F - 1 = \frac{1}{4kTG_s \Delta f g_m^2} \{\overline{i_g^2} g_m^2 + g_d^2 \overline{v_d^2} G_s^2 - C^2 g_m^2 \overline{i_g^2}\}. \quad (31)$$

Next, we minimize with respect to G_s and obtain the optimum source conductance

$$G_{s\,opt} = \frac{g_m(\overline{i_g^2})^{1/2}}{g_d(\overline{v_d^2})^{1/2}}(1 - C^2)^{1/2} \quad (32)$$

and minimum excess noise figure

$$(F - 1)_{\min} = \frac{1}{2kTg_m\Delta f} (\overline{i_g^2})^{1/2} (\overline{i_d^2})^{1/2} (1 - C^2)^{1/2}. \quad (33)$$

In (33), we introduced the drain noise current i_d through

$$\overline{i_d^2} = g_d^2 \overline{v_d^2}. \quad (34)$$

In the following sections of the paper, we evaluate the quantities $\overline{i_g^2}$, $\overline{i_d^2}$, and C as a function of the device parameters.

V. OPEN-CIRCUIT VOLTAGE FLUCTUATION DUE TO ENHANCED JOHNSON NOISE

In this section, we calculate voltage fluctuations at the drain due to enhanced Johnson noise in the nonsaturated Region I. In order to save space, we make use of earlier work by van der Ziel [3], [4]. Essentially, we have to add two features to van der Ziel's work. First, the noise temperature in the present treatment is field dependent as shown in (1). The implications of a field dependent noise temperature in Region I have been considered briefly by Klaassen [12] and van der Ziel [13], [14], and in more detail by Baechtold [6]. Second, the noise voltage as calculated at the end of Region I propagates and becomes enhanced in Region II.

In Appendix I, we give the derivation of the noise voltage at the end of Region I, Δv_1, due to the Johnson noise of an infinitesimal section of channel Δx located at a position x where the channel potential with respect to the gate is $W_o(x)$. The result is

$$\overline{\Delta v_1^2} = \frac{4kT_e \Delta f}{I_d} \left[\frac{1 - w(x)}{1 - p}\right]^2 2W_{oo} w(x) \Delta w. \quad (35)$$

We introduced in (35) the reduced potential

$$w(x) = \left(\frac{W_o(x)}{W_{oo}}\right)^{1/2} \quad (36)$$

and the voltage drop due to the current I_d flowing from source to gate in the channel section Δx is equal to $\Delta W_o(x) = 2W_{oo} w(x) \Delta w$.

We now modify (35) by introducing the field dependent noise temperature. The longitudinal electric field E in the channel due to the current I_d may be written as

$$E = \frac{I_d}{lg_o} \frac{1}{1 - w}. \quad (37)$$

If we use (5) for I_d and if we introduce

$$f_1 = p^2 - s^2 - \tfrac{2}{3}(p^3 - s^3) \quad (38)$$

then we may reexpress (1) in the form

$$T_e = T_o \left[1 + \delta \left(\frac{E}{E_{\text{sat}}}\right)^3\right]$$

$$= T_o \left[1 + \delta f_1^3 \left(\frac{W_{oo}}{E_{\text{sat}} L_1}\right)^3 \frac{1}{(1 - w)^3}\right]. \quad (39)$$

Inserting (39) into (35) and using once more (5) for I_d then gives

$$\overline{\Delta v_1^2} = \frac{4kT_o\Delta f}{l(g_o/L_1)} \frac{1}{f_1(1 - p)^2} 2w(1 - w)^2$$

$$\cdot \left[1 + \delta f_1^3 \left(\frac{W_{oo}}{E_{\text{sat}} L_1}\right)^3 \frac{1}{(1 - w)^3}\right] \Delta w. \quad (40)$$

The total voltage fluctuations at the end of Region I follow from a straightforward integration over w between the limits of $w = s$ and $w = p$. The final result may be written

$$\overline{v_1^2} = \frac{4kT_o\Delta f}{l(g_o/L_1)} \frac{P_o + P_\delta}{(1 - p)^2}. \quad (41)$$

The quantities P_o and P_δ are defined as

$$P_o = (f_1)^{-1} [(p^2 - s^2) - \tfrac{4}{3}(p^3 - s^3) + \tfrac{1}{2}(p^4 - s^4)] \quad (42)$$

and

$$P_\delta = 2\delta \left(\frac{W_{oo}}{E_{\text{sat}} L_1}\right)^3 (f_1)^2 \left[(s - p) + \ln \frac{1 - s}{1 - p}\right]. \quad (43)$$

Equation (41) with $P_\delta = 0$ is identical to van der Ziel's equation (13) in [3]. P_δ contains the new noise contributions. Note that our expressions for P_o and P_δ are different from the expressions P_1 and P_2 defined by Baechtold [6].

Next, we evaluate the open-circuit voltage fluctuations at $x = L_2$. Because the current is fixed, the width of the pinch-off channel or p remains constant also (8). Since p is fixed at the pinch-off point, so is the voltage (as seen in 3). In other words, noise fluctuations occurring at a fixed point in Region I have as a consequence that L_1 and L_2 fluctuate. Near the pinch-off point, the electrical field is equal to E_{sat}, thus a noise voltage of amplitude Δv_1 will cause a change of L_1 by ΔL_1 where

$$\Delta L_1 E_{\text{sat}} = \Delta v_1. \quad (44)$$

A variation of (16) gives, with channel width constant and hence, $p = $ constant,

$$\Delta V_{sd} = -E_{\text{sat}} \Delta L_2 \cosh \frac{\pi}{2a} L_2. \quad (45)$$

Inserting (44) into (45) with $\Delta L_1 = -\Delta L_2$ gives

$$\Delta V_{sd} = \Delta v_1 \cosh \frac{\pi}{2a} L_2. \quad (46)$$

The noise voltage at $x = L_1$ given by (41) thus becomes transformed into

$$\overline{v_{d1}^2} = \frac{4kT_o\Delta f}{l(g_o/L_1)} \frac{P_o + P_\delta}{(1 - p)^2} \cosh^2 \frac{\pi}{2a} L_2 \quad (47)$$

when referred to the true drain $x = L_2$. Our notation v_{d1} refers to noise at the true drain caused by fluctuations in Region I. Equation (47) also holds for the asymmetric transistor, provided the value of g_o appropriate for a one-sided device is used. Numerically, (47) gives twice as large a value for the asymmetric transistor.

VI. INDUCED GATE NOISE CURRENTS WITH SHORT-CIRCUITED DRAIN

We have shown in the previous section that there are noise voltage fluctuations along the channel. Since the gate is capacitatively coupled to the channel, there will be induced noise charges on the gate, and likewise, since these charges are time dependent, there will be noise currents flowing into the gate. In the present section, we calculate induced gate currents due to elementary noise fluctuations in Region I only.

Van der Ziel has evaluated these noise currents for a nonsaturated transistor. Because of the additional saturated transistor Region II, the boundary conditions at the end of Region I are different in the present case. Since the drain is assumed short circuited, the noise voltage is zero at the end of Region II, while in van der Ziel's single region transistor, the noise voltage was assumed to be zero at the end of Region I. When we evaluate the induced gate current, we have to add effects due to the "breathing" of the channel in Region II. Finally, we have to allow for a field dependent noise temperature.

Van der Ziel has given a relationship (as derived in Appendix I, (I.6) between the voltage jump ΔW_p at the end of Region I and the noise voltage fluctuation $-\Delta W_x$ at position x in Region I, as seen by the equation

$$\Delta W_p = -\Delta W_x \frac{1 - w(x)}{1 - p}. \quad (48)$$

In (48), $w(x)$ is the reduced channel potential defined in (36). Similarly, p represents the value of w at the end of Region I, i.e., at $x = L_1$. Equation (48) is valid under open circuit conditions. We may obtain the noise voltage at the end of Region II by multiplying ΔW_p in (48) by $\cosh (\pi/2a)L_2$ in exact analogy to (46). Thus

$$\Delta v_d = -\Delta W_x \frac{1 - w(x)}{1 - p} \cosh \frac{\pi}{2a} L_2. \quad (49)$$

Under short-circuit conditions, the voltage change of (49) transforms itself into a drain current fluctuation Δi_d, where

$$\Delta i_d = -\frac{\Delta v_d}{r_d} = \Delta W_x \frac{1}{r_d} \frac{1 - w(x)}{1 - p} \cosh \frac{\pi}{2a} L_2. \quad (50)$$

Equation (50) replaces (11) of [4]. We may obtain (50) from (11) of [4] by multiplying the latter by the factor B, where

$$B = \frac{L_1 \cosh (\pi L_2/2a)}{lg_o(1 - p)r_d}. \quad (51)$$

For the calculation of the induced charge, we need the detailed behavior of the potential disturbance ΔW as a function of x along the channel for an elementary noise fluctuation $-\Delta W_x$ at position x_o. Due to the new boundary conditions, we find now in Region I (see Appendix II)

$$(1 - w)\Delta W = \frac{\Delta i_d}{lg_o} x \text{ for } 0 < x < x_o \quad (52)$$

$$(1 - w)\Delta W = \frac{\Delta i_d}{lg_o} (x - \gamma L_1) \text{ for } x_o < x < L_1 \quad (53)$$

where $x = 0$ now corresponds to the beginning of Region I. Equation (52) is identical to the corresponding equation in [4], while (53) has now a factor of γ in front of L_1, since ΔW no longer needs to be equal to zero at the end of Region I. There is a discontinuity in ΔW at $x = x_o$. After inserting Δi_d from (50) into (52) and (53), we can determine γ from the fact that the discontinuity in ΔW is equal to $-\Delta W_x$. We find with (20)

$$\gamma = 1 + 2p(1 - p) \frac{W_{oo}}{E_{sat}L_1} \left\{ 1 - \frac{1}{\cosh (\pi L_2/2a)} \right\}. \quad (54)$$

For $L_2 = 0$, γ becomes unity, and the original van der Ziel expressions are obtained.

From the potential perturbation ΔW given by (52) and (53), van der Ziel calculated the induced charge ΔQ in Region I. By tracing the additional factor of γ, we obtain the new expression

$$\Delta Q = \frac{2a\rho L_1 l \Delta i_d}{I_d} [-k + \gamma w(x_o)] \quad (55)$$

where

$$k = (f_1)^{-1}[-\tfrac{1}{3}(p^3 - s^3) + \tfrac{1}{6}(p^4 - s^4) + (s^2 - \tfrac{2}{3}s^3)(p - s)] + \gamma p. \quad (56)$$

We next have to add the induced charge in Region II which results from a change Δi_d of the current flowing through the conducting channel. Since the velocity of the carriers in Region II is saturated, the additional current Δi_d requires additional carriers with a charge per unit length of $(\Delta i_d)/u$, where u is the saturated velocity. The induced charge on the gate has the opposite sign and is therefore given by $-(\Delta i_d L_2)/u$. This additional charge can be added to (55) to give a revised expression ΔQ_{rev}, as seen by

$$\Delta Q_{\text{rev}} = \frac{2a\rho L_1 l \Delta i_d}{I_d} [-k' + \gamma w(x_o)] \quad (57)$$

where

$$k' = k + \frac{L_2}{L_1} (1 - p). \quad (58)$$

Equations (57) and (58) become evident when one remembers that $I_d = 2a(1 - p)l\rho u$.

Finally, we can proceed to the revised expression for the square of the induced charge. However, we have to remember that the noise temperature is field dependent, as given in (39). Van der Ziel's expression, (21) in [4], can thus be rewritten and extended by multiplying it by B^2

because of our modification of (51), by replacing his factor $[-p + (W_o(x_o)/W_{oo})^{1/2}]$ by $(-k' + \gamma w)$, and by modifying his temperature T by (39), as seen by the equation

$$\overline{\Delta Q_{\text{rev}}^2} = \frac{64kT_o\Delta f}{g_o/L_1} \frac{L_1^2\epsilon^2}{a^2} lB^2(f_1)^{-3}$$

$$\cdot \left[1 + (f_1)^3 \delta\left(\frac{W_{oo}}{E_{\text{sat}}L_1}\right)^3 \frac{1}{(1-w)^3}\right]$$

$$\cdot (k' - \gamma w)^2(1-w)^2 2w\Delta w. \quad (59)$$

The total mean square of the charge fluctuation may be obtained by integrating (59) over w between the limits of s and p. The total mean square of the gate current simply follows by multiplying the resulting expression by ω^2, seen as

$$\overline{i_{g1}^2} = \frac{64kT_o\Delta f}{g_o/L_1} \frac{L_1^2\epsilon^2\omega^2}{a^2} lB^2(R_o + R_\delta). \quad (60)$$

The notation i_{g1} is chosen to designate the total gate noise current due to all sources in Region I. The expressions R_o and R_δ follow from integration of (59) and are given by

$$R_o = (f_1)^{-3}\{(k')^2(p^2-s^2) - \tfrac{4}{3}k'(k'+\gamma)(p^3-s^3)$$

$$+ \tfrac{1}{2}[(k')^2 + 4k'\gamma + \gamma^2]\cdot(p^4-s^4)$$

$$- \tfrac{4}{5}(k'\gamma + \gamma^2)(p^5-s^5) + (1/3g^2)(p^6-s^6)\} \quad (61)$$

$$R_\delta = \delta\left(\frac{W_{oo}}{E_{\text{sat}}L_1}\right)^3\left[-2(k'-\gamma)^2\left(p - s + \ln\frac{p-1}{s-1}\right)\right.$$

$$\left. + (2k'\gamma - \gamma^2)(p^2-s^2) - \tfrac{2}{3}\gamma^2(p^3-s^3)\right]. \quad (62)$$

For an asymmetric transistor, $\overline{i_{g1}^2}$ in (60) first has to be divided by a factor of 2 since there is only one gate. Furthermore, if we introduce $g_o(\text{asym})$, where $g_o(\text{symm}) = 2g_o(\text{asym})$, then the numerical factor of 64 in (60) has to be changed to 16.

VII. DIFFUSION NOISE IN VELOCITY SATURATED PINCH-OFF REGION

In this section, we shall investigate the noise mechanism in the saturated drift region of the transistor. The noise mechanism can be described in several different ways. For example, the impedance field method of Shockley et al. [7] is suitable to describe the noise in the pinch-off region. We shall look here at the problem in a somewhat different way.

We are dealing here with a nonequilibrium situation, and the normal Johnson noise formulas are not valid. As has been shown by Shockley [7] and van der Ziel et al. [14], [15], the basic noise current density source may be written as

$$\overline{J_n(x)J_n^*(x')} = \frac{4q^2Dn\Delta f}{A}\delta(x-x'). \quad (63)$$

In (63), $J_n(x)$ is the noise current density at x, D is the diffusion constant of the carriers at the high dc electric field existing in the sample, n is the carrier density, A is the cross-sectional area, and $\delta(x-x')$ is the Dirac delta function. By using the Einstein relation $D = (kT/q)\mu$, with μ = mobility, (63) goes into the conventional Johnson noise formula. The Einstein formula, however, is only valid at low fields.

The noise current J_n may be interpreted as a distribution of spatial and time impulses that are mutually uncorrelated. Indeed, (63) can be rewritten as

$$\overline{(AJ_n(x))^2} = \frac{4q^2DnA\Delta f}{\Delta x} \quad (64)$$

for the autocorrelation of the current density at the point x, where we have used the fact that $\delta(x-x') = 1/\Delta x$ for $x = x'$. We now compare (64) with the shot noise expression

$$\overline{i^2} = 2qI_o\Delta f. \quad (65)$$

The shot noise spectrum is the result of current impulses of content q occurring at a rate $r = I_o/q$.

Similarly, (64) may be interpreted as a sequence of current impulses occurring at the rate

$$r = \frac{2DnA}{\Delta x} \quad (66)$$

where each current impulse extends over the distance Δx. Each of the current impulses thus results in a displacement of a charge q over a distance Δx, or, in other words, each current impulse leads to the formation of a dipole layer of strength $(q\Delta x)/A$. Since these current impulses occur in the saturated velocity regime, where we neglect mobility and diffusion effects, the resulting dipole layers are unable to relax, and they drift unchanged to the drain contact. We shall use this interpretation for the analysis of the diffusion noise in the field-effect transistor. The interpretation of high-field noise in FET's as diffusion noise also has been suggested by van der Ziel [14].

VIII. THE DIFFUSION-NOISE INDUCED OPEN-CIRCUIT DRAIN VOLTAGE

We have identified the diffusion noise as spontaneous generation of dipole layers $(q\Delta x)/A$, generated at the rate $r = (2DnA)/\Delta x$, which drift from their generation point $x_0 > 0$ to the drain contact at $x = L_2$. In this section, we take the origin of x at the pinch-off point. Under ac open-circuit drain conditions, fluctuating fields and potentials at the pinch-off point, generated by the drifting dipole layers, must be nullified by drain voltage fluctuations so that a constant dc current is maintained.

We first calculate the potential and field of a dipole layer in the channel. For this purpose, it is advantageous to consider first the potential due to a charge layer of density σ at $x = x_o$ extending from $y = -b$ to $y = +b$. As may be verified, the corresponding potential Φ is given by

$$\Phi = \frac{\sigma}{\epsilon a} \sum_{n=1,3,5}^{\infty} \left(\frac{2a}{n\pi}\right)^2 \sin \frac{n\pi}{2a} b \cos \frac{n\pi}{2a} y \exp \pm \frac{n\pi}{2a} (x - x_o). \quad (67)$$

At the metal gate electrodes, the potential is zero as required by the boundary conditions. The plus sign applies to the region $x < x_o$ and the minus sign to the region $x > x_o$. The potential Ψ due to a dipole layer consisting of a charge density $+\sigma$ at x_o and $-\sigma$ at $x_o + \Delta x_o$ is obtained from the preceding by differentiation, as seen by

$$\Psi = -\Delta x_o \frac{\partial \Phi}{\partial x_o} = \pm \frac{\sigma \Delta x_o}{\epsilon a} \sum_{n=1,3,5}^{\infty} \frac{2a}{n\pi} \sin \frac{n\pi}{2a} b \cos \frac{n\pi}{2a} y$$

$$\cdot \exp \pm \frac{n\pi}{2a} (x - x_o). \quad (68)$$

In (68), the plus sign applies for $x < x_o$ and the minus sign for $x > x_o$. The x component of the electric field due to the dipole layer becomes

$$E_x = -\frac{\sigma \Delta x_o}{\epsilon a} \sum_{n=1,3,5}^{\infty} \sin \frac{n\pi}{2a} b \cos \frac{n\pi}{2a} y \exp \pm \frac{n\pi}{2a} (x - x_o). \quad (69)$$

Next, we match boundary conditions at the pinch-off point. In keeping with the approximations already made in evaluating the small-signal parameters of the transistor, we retain only the fundamental components of potential and field. The potential at the center of the pinch-off point in Region I is $-W_{oo}(p^2 - s^2)$. This potential is consistent with the solution of Poisson's equation in the transverse direction taking into account the space charge of the ionized acceptors. In Region II at $x = 0$, we also have the potential $-W_{oo}(p^2 - s^2)$ due to the particular solution of Poisson's equation taking into account the ionized acceptor and mobile hole space charges. In addition, we have the potential due to charges or voltages on the drain contact. As before, they are of the form $\alpha \cos(\pi y/2a) \exp(\pi x/2a)$ and $\beta \cos(\pi y/2a) \exp(-\pi x/2a)$. Finally, we have the potentials and fields of the dipole layer at $x = x_o$. Thus matching of the potential at the pinch-off point $y \approx 0$, $x = 0$ gives

$$-W_{oo}(p^2 - s^2) = -W_{oo}(p^2 - s^2) + \frac{\sigma \Delta x_o}{\epsilon} \frac{2}{\pi} \sin \frac{\pi b}{2a}$$

$$\cdot \exp \frac{-\pi x_o}{2a} + \alpha + \beta. \quad (70)$$

Similarly, matching the electric field gives

$$E_{\text{sat}} = \frac{-\sigma \Delta x_o}{\epsilon a} \sin \frac{\pi b}{2a} \exp\left(-\frac{\pi x_o}{2a}\right) - \frac{\pi}{2a} \alpha + \frac{\pi}{2a} \beta. \quad (71)$$

As stated, we are calculating open-circuit noise voltages. The quantities α and β adjust themselves to cancel out the effect of the dipole layer. Calling the $\delta \beta$ and $\delta \alpha$ the perturbations of α and β due to the dipole layer, we obtain from (70) and (71)

$$\delta \alpha = -\frac{\sigma \Delta x_o}{\epsilon} \frac{2}{\pi} \sin \frac{\pi b}{2a} \exp -\frac{\pi x_o}{2a}$$

$$\delta \beta = 0. \quad (72)$$

The potential perturbation at the drain contact $x = L_2$ due to the dipole layer is, therefore,

$$\Delta v_{d2} = -\frac{\sigma \Delta x_o}{\epsilon} \frac{2}{\pi} \sin \frac{\pi b}{2a} \exp \frac{\pi (L_2 - x_o)}{2a}. \quad (73)$$

In (73), we have omitted the contribution of the potential from (68) for $x > x_o$. Due to an image dipole layer at $x = L_2 + (L_2 - x_o)$ beyond the drain contact, this portion of the potential approximately gets cancelled out. The notation v_{d2} again signifies that the drain voltage is due to a source in Region II.

Now, the dipole layer drifts with a constant speed u from x_o, starting at its instant of generation t_o. Therefore, (73) gives a time-dependent induced voltage

$$\Delta v_{d2}(x_o, t - t_o) = -\frac{\sigma \Delta x}{\epsilon} \frac{2}{\pi} \sin \frac{\pi b}{2a} \exp \frac{\pi}{2a}$$

$$\cdot [L_2 - x_o - u(t - t_o)] \quad (74)$$

where

$$0 \leq t - t_o \leq (L_2 - x_o)/u.$$

Next, we want to calculate the spectrum due to the drifting dipole layers. A random process of uncorrelated events, occurring at the rate r, each having a response $h(t)$, produces a spectral density

$$\phi(\omega) = \frac{1}{2\pi} r |H(\omega)|^2 \quad (75)$$

where $H(\omega)$ is the Fourier transform of $h(t)$, as seen by

$$H(\omega) = \int_{-\infty}^{+\infty} \exp(-i\omega) h(t) dt. \quad (76)$$

Since we are not interested in frequencies where the transit time through the pinch-off channel becomes comparable to $1/\omega$, we can evaluate (76) in the limit of $\omega \to 0$.

We obtain with (74), using $\sigma = q/A = q/(2bl)$,

$$H(\omega)_{\omega \to 0} = -\frac{1}{\pi} \frac{q \Delta x_o}{\epsilon b l} \sin\left(\frac{\pi b}{2a}\right)$$

$$\cdot \frac{2a}{\pi u}\left[\exp \frac{\pi}{2a}(L_2 - x_o) - 1\right]. \quad (77)$$

Finally, we may obtain the drain voltage fluctuations due to dipole layers starting between x_o and $x_o + \Delta x_o$ by inserting (77) and (66) into (75) as seen by the equation

$$\overline{\Delta v_{d2}^2(x_o)} = \frac{32a^2}{\pi^4 u^2} \frac{q^2 D n \Delta f}{\epsilon^2 bl} \sin^2 \frac{\pi b}{2a}$$

$$\cdot \left[\exp \frac{\pi}{2a}(L_2 - x_o) - 1 \right]^2 \Delta x_o. \quad (78)$$

Equation (78) contains an additional factor of 2, since we do not distinguish between positive and negative frequencies as in (76) but consider the total fluctuations in a bandwidth $\Delta f = \Delta \omega / 2\pi$. The total drain voltage fluctuations due to dipole layers in Region II follows from an integration over x_o from 0 to L_2. If we introduce the dc current I_d, seen by

$$I_d = qun2bl \quad (79)$$

then, we can finally write

$$\overline{v_{d2}^2} = I_d \frac{16}{\pi^5} \sin^2\left(\frac{\pi b}{2a}\right) \frac{qD\Delta f a^3}{\epsilon^2 b^2 l^2 u^3} \left[\exp \frac{\pi}{a} L_2 - 4 \exp \frac{\pi}{2a} L_2 \right.$$

$$\left. + 3 + \frac{\pi L_2}{a} \right]. \quad (80)$$

For the asymmetric transistor with I_d referring to the current flowing in the one-sided transistor, the numerical factor of 16 in (80) has to be replaced by 64.

IX. DIFFUSION NOISE INDUCED DRAIN AND GATE CURRENTS

As outlined in the previous two sections, the diffusion-noise contribution can be attributed to dipole layers drifting through Region II. Since the dipole layers carry no net charge, there is no directly induced charge on the gate. There is, however, an indirect contribution. To be consistent, we have to evaluate the induced gate currents under short-circuit drain conditions. The noise voltage calculated in Section VIII produces under these short-circuit conditions an induced drain current. This drain current causes a breathing of Channels I and II and, therefore, an induced charge on the gate.

Let us call the induced drain current i_{d2}, where

$$i_{d2} = \frac{v_{d2}}{r_d}. \quad (81)$$

We have already calculated, in essence, the induced charge in the gate due to a current change Δi_d. We can adapt (52)–(57) by simply omitting the potential jump at x_o not needed in the present context. As may be seen from (52) and (53), there is no potential jump or discontinuity at $x = x_o$, if we set the parameter γ equal to zero. We can, therefore, take over (57) with the proviso that γ is set equal to zero also within the parameter k', defined in (56) and (58). We thus obtain from (57), for the induced charge Q_2,

$$Q_2 = -\frac{2a\rho L_1 l i_{d2}}{I_d} k'(\gamma = 0). \quad (82)$$

We obtain, finally, the mean-square gate noise current $\overline{i_{g2}^2}$, due to fluctuations in Region II, by taking the mean square of (82), multiplying it by ω^2, and inserting (80) and (81), as seen by

$$\overline{i_{g2}^2} = \frac{1}{r_d^2} \frac{16}{\pi^5} \sin^2\left(\frac{\pi b}{2a}\right) \frac{qI_d D \Delta f a^3 L_1^2 \omega^2}{\epsilon^2 b^2 l^2 u^5} \frac{(k'(\gamma = 0))^2}{(1-p)^2}$$

$$\cdot \left[\exp \frac{\pi}{a} L_2 - 4 \exp \frac{\pi}{2a} L_2 + 3 + \frac{\pi L_2}{a} \right]. \quad (83)$$

For the asymmetric transistor, the numerical value of $\overline{i_{g2}^2}$ is half as large as given by (83). If we simultaneously introduce into (83) the drain resistance and the drain current of the asymmetric transistor, then the numerical factor $16/(\pi^5)$ has to be changed to $64/(\pi^5)$.

X. EVALUATION OF CORRELATION COEFFICIENT

The gate noise currents and the drain noise currents are correlated to some extent, since any elementary noise voltage produced in the channel gives a contribution to both the drain and the gate currents. As before, let us call i_{g1} and i_{d1} the gate and drain currents produced by a fluctuation in the ohmic Region I, and, similarly, let us call i_{g2} and i_{d2} the currents produced by a fluctuation in Region II. Obviously, there is no cross correlation between i_{d1} and i_{g2} or i_{d2} and i_{g1}. The derivation in Section IX shows that there is a very strong correlation between i_{g2} and i_{d2}. Obviously, a drain voltage fluctuation produced by Region II gives rise to a drain current which can be evaluated using Ohm's law and the value of the drain resistance. Similarly, the gate current follows from the breathing of the channel due to drain current. There is a capacitative 90° phase shift between gate and drain currents with full correlation. In other words,

$$\overline{i_{g2}^* i_{d2}} = j[\overline{i_{g2}^2} \cdot \overline{i_{d2}^2}]^{1/2}. \quad (84)$$

Similarly, we can evaluate the correlation between i_{g1} and i_{d1} following closely the treatment by van der Ziel [4]. Because of our boundary conditions at the end of Region I, we have to multiply van der Ziel's equation (20) in [4] by B^2; furthermore, we have to replace his expression $p - u^{1/2}$ by our $k' - \gamma w$ in exact analogy to (57). Finally, we have to insert the field dependent noise temperature of (39) in full analogy with our treatment in Section V. After integration over the reduced channel potential, we obtain the expression

$$\frac{\overline{i_{g1}^* i_{d1}}}{[\overline{i_{g1}^2} \cdot \overline{i_{d1}^2}]^{1/2}} = j \frac{S_o + S_\delta}{(R_o + {}_\delta)^{1/2}(P_o + P_\delta)^{1/2}} \quad (85)$$

where

$$S_o = (f_1)^{-2}\{k'[(p^2 - s^2) - \tfrac{4}{3}(p^3 - s^3) + \tfrac{1}{2}(p^4 - s^4)]$$

$$+ \gamma[-\tfrac{2}{3}(p^3 - s^3) + (p^4 - s^4) - \tfrac{2}{5}(p^5 - s^5)]\} \quad (86)$$

and

$$S_\delta = 2\delta \left(\frac{W_{oo}}{E_{sat}L_1}\right)^3 f_1 \left[(k' - \gamma)\left(s - p + \ln\frac{1-s}{1-p}\right) + \tfrac{1}{2}\gamma(p^2 - s^2)\right]. \quad (87)$$

The overall correlation coefficient defined in (28) can then finally be written as

$$C = \frac{1}{j}\frac{\overline{i_g^* i_d}}{[\overline{i_g^2}\,\overline{i_d^2}]^{1/2}} = \frac{S_o + S_\delta}{(R_o + R_\delta)^{1/2}(P_o + P_\delta)^{1/2}}$$

$$\cdot \left(\frac{\overline{i_{g1}^2}}{\overline{i_g^2}}\right)^{1/2}\left(\frac{\overline{i_{d1}^2}}{\overline{i_d^2}}\right)^{1/2} + \left(\frac{\overline{i_{g2}^2}}{\overline{i_g^2}}\right)^{1/2}\left(\frac{\overline{i_{d2}^2}}{\overline{i_d^2}}\right)^{1/2}. \quad (88)$$

In (88), we understand by i_g and i_d the total gate and drain noise currents produced by fluctuations in Regions I and II.

XI. NOISE FIGURE, EXPERIMENTAL DATA, AND DISCUSSION

A test of our noise theory may be made by applying it to a practical device for which experimental data are available. A proper test of the theory, however, must include the thermal noise contributions of the parasitic elements associated with a practical device. Unfortunately, the experimental noise data on GaAs FET's are often not accompanied with sufficient information about the parasitics, so that we are limited at this time to applying our theory to one specific device design for which the published information is adequate.

The equivalent circuit model for the FET which includes the noisy parasitic elements is shown in Fig. 4. The intrinsic device parameters are included in the shaded box. The capacitances C_{sg} and C_{dg} are, respectively, the gate–source and gate–drain capacitances. The latter is of the order of a few percent of the former and will be neglected in our noise analysis. The parameter R_i is an intrinsic resistance in the undepleted region of the channel under the source side of the gate and represents the ohmic path for the displacement current between the gate and source. Its ohmic losses are included in the gate noise source i_g [16]. The parameter r_d is the drain resistance evaluated in Section III. Shunting it is the drain noise source. The transadmittance y_m of the current generator of the intrinsic transistor shown in Fig. 4 can be approximated by the form $y_m = g_m \exp(-j\omega\tau)$, where $\omega\tau$ is a small phase shift [2], [17]. We shall neglect this phase shift in our analysis.

The extrinsic thermal noise sources of importance are those associated with the gate metallization resistance R_m and the source–gate resistance R_f which includes the source contact resistance and the bulk resistance of the epitaxial layer between the gate and source contacts. Both R_m and R_f show full thermal noise as indicated in Fig. 4.

The gate and drain intrinsic noise sources will be represented in terms of equivalent noise conductances g_{gn} and g_{dn}, respectively, where

$$g_{gn} = \frac{\overline{i_g^2}}{4kT_o\Delta f} \quad (89)$$

$$g_{dn} = \frac{\overline{i_d^2}}{4kT_o\Delta f}. \quad (90)$$

The noise figure can be evaluated by short circuiting the output (drain–ground) terminals and evaluating the noise current contributions of each noise generator, including that of the source e_s. Using the expression for the noise generators in Fig. 4, we obtain the expression for the noise figure, as seen by

$$F = 1 + \frac{1}{R_s}\left[r_n + g_n|Z_s + Z_c|^2\right] \quad (91)$$

where $Z_s = R_s + jX_s$ is the source impedance and r_n, g_n, and Z_c are noise parameters whose expressions will be given later. As a function of the source impedance, F exhibits a minimum when Z_s has its optimum value $Z_{s,\text{opt}}$ given in terms of its components, by

$$R_{s,\text{opt}} = \left((\operatorname{Re} Z_c)^2 + \frac{r_n}{g_n}\right)^{1/2} \quad (92)$$

$$X_{s,\text{opt}} = -\operatorname{Im} Z_c. \quad (93)$$

The minimum value of F can be expressed as

$$F_{\min} = 1 + 2g_n(\operatorname{Re} Z_c + R_{s,\text{opt}}). \quad (94)$$

This expression will serve as the basis of our discussion which follows. The noise quantities g_n, R_n, and Z_c are expressible in terms of the equivalent circuit parameters as

$$g_n = \left|\frac{y_{11}}{y_{21}}g_{dn} - jC(g_{gn})^{1/2}\right|^2 + (1 - C^2)g_{gn} \quad (95)$$

$$r_n = R_m + R_f + \frac{g_{dn}}{|y_{21}|^2}\frac{(1 - C^2)g_{gn}}{g_n} \quad (96)$$

$$Z_c = R_m + R_f + \frac{g_{dn}y_{11}^*/|y_{21}|^2 + jC(g_{gn}g_{dn})^{1/2}/y_{21}}{g_n}. \quad (97)$$

Here

$$y_{11} = \frac{j\omega C_{sg}}{1 + j\omega C_{sg}R_i}, \quad y_{21} \approx \frac{g_m}{1 + j\omega C_{sg}R_i} \quad (98)$$

where y_{11} and y_{21} are, respectively, the short-circuit input and transfer admittance of the intrinsic device. (We have neglected the small effort of the drain–gate capacitance C_{dg} on y_{11} and y_{21}.) Note that for vanishing values of R_i, R_m, and R_f, the expression for $F_{\min}$ and $Z_{s,\text{opt}}$ reduce to those given in Section IV when one sets $Y_{s,\text{opt}} = Z_{s,\text{opt}}^{-1} - 1$.

The parasitic resistances R_m and R_f appear as a series

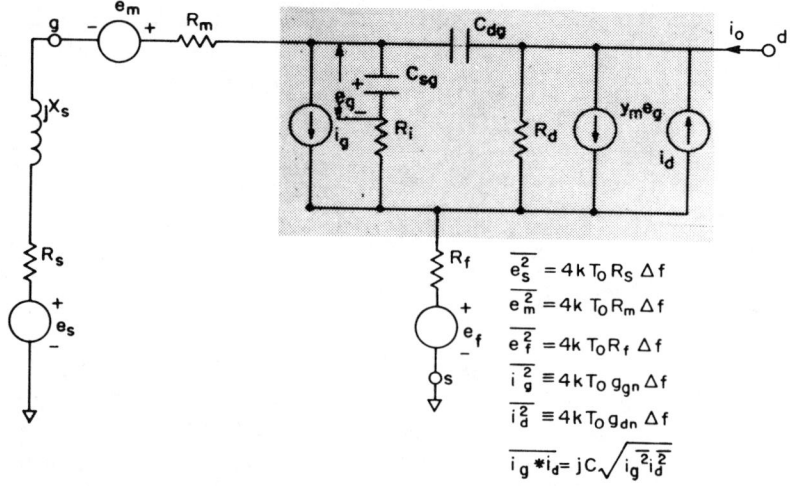

Fig. 4. Equivalent circuit for minimum noise figure calculation with parasitic elements included.

combination in both the equivalent noise resistance r_n and the correlation impedance Z_c, but not in the equivalent noise conductance g_n. One may show that the noise contributions of the parasitics enter in r_n, whereas the ohmic properties of R_m and R_f are relevant in Z_c. Since both Z_c and r_n appear in the expression for $R_{s,\text{opt}}$, the optimum source resistance is a strong function of the parasitic resistances. The ohmic property of the intrinsic resistance R_i also plays a role in $R_{s,\text{opt}}$ via the real part of Z_c.

We have applied these expressions to a practical n-type GaAs FET with a 2-μm gate length whose published design values stated by Brehm and Vendelin are [18], [19]: $a = 2 \times 10^{-5}$ cm, $l = 2.85 \times 10^{-2}$ cm, $L = 2 \times 10^{-4}$ cm, $g_o = \sigma_o a = 1.4 \times 10^{-3}$ mhos (corresponding to a doping level $N_d = 10^{17}$ cm^{-3}), and $W_{oo} = 2.9$ V. The values of the parasitic resistances used are $R_f = 15$ ohms and $R_m = 0.8$ ohms [19]. This design closely parallels geometries studied at other laboratories including our own. At this doping level, the saturation field is approximately 3 kV/cm. Using a value slightly lower, 2.9 kV/cm, and a low field mobility of 4500 cm^2/V·s, the piecewise-linear approximation to the velocity field characteristic in Fig. 2 requires a saturation velocity $u = 1.3 \times 10^7$ cm/s, only slightly higher than the actual value [8].

Evaluation of the noise conductances g_{gn} and g_{dn}, (89) and (90), appearing in the noise figure expression requires a knowledge of the dummy variables (s,p). To obtain solutions for this pair as a function of the external drain-source and gate-source potentials, it was necessary for us to take into account all of the parasitic dc voltage drops in the source–drain region, as well as the self-biasing of the gate due to the voltage drop across R_f produced by the dc drain current. Using our dc and small-signal analysis outlined in Section II and III, and including the parasitic resistances in the source–drain circuit, we have written a computer program which yields values of the variables (s,p) as a function of gate and drain potentials. As a byproduct, I–V characteristics and small-signal parameters were also obtained. For example, at a drain–source bias of 3 V, and a gate-source bias of 0 V, we obtain for the saturation drain current I_{dss}, a value of 28.3 mA and for the transconductance a value of 19 mmhos, which are in very good agreement with Brehm's data [19]. We have also obtained excellent agreement with measurements taken on a variety of device geometries fabricated at our laboratory.

It might be of interest to point out that our computer results show for this device design that for drain-source bias values corresponding to current saturation, i.e., above the "knee" of the I_d–V_d curves, the extent of the saturated velocity region L_2 is of the order of 30 percent of the gate (channel) length l. For shorter gate lengths (of the order of 1 μm), this fraction can reach as high as 90 percent. The extent of this region increases with the source-drain potential but is nearly independent of the source-gate bias. Furthermore, the channel opening under current saturated conditions is almost uniform along the entire length of the gate, that is $s \approx p$, the difference being of the order of 10 percent or less. This means that the longitudinal drop along the channel in Region I is negligible compared to the corresponding drop in Region II. The values (s,p) are strong functions of the gate-source potential, but nearly independent of the source–drain bias. For the device under consideration, typical values of s and p range from 0.65–0.75 at zero gate bias, increasing towards unity at low drain currents.

Knowledge of the (s,p) pairs, obtained from the numerical solutions, also allows us not only to calculate the small-signal parameter g_m, which appears in the noise figure expression, but also the small-signal parameter, C_{sg}, which was obtained by a small-signal analysis of the charge induced on the gate by the field of the depleted impurity charges in the channel and by the Laplace field given by (14). Using the approximation $s = p$, we obtain the simplified expression for $L_2 \approx L_s$

$$C_{sg} \approx \frac{\epsilon l L}{a p} + 1.56 \epsilon l. \qquad (99)$$

The first term is simply that of a "parallel plate" capacitor of thickness ap. The second term is a fringing field correction [20] for either edge of the gate. The fringing accounts for about 10 percent of the total capacitance for the device being considered but can be as high as 25 percent of the total capacitance for a shorter gate.

We have postulated for the gate charging resistance R_i a bias dependence inverse to that of C_{sg}, so that the product $R_i C_{sg}$ is approximately constant. There are good theoretical grounds for this assumption since it can be shown that this time constant is proportional to (though smaller than) the channel transit time $\tau \approx L/u$. We have set this time constant equal to 4.0 ps corresponding to the published data [19].

As a preliminary to our noise figure presentation, we illustrate in Fig. 5 the calculated dependence of the noise conductances g_{gn} and g_{dn} on the gate bias, or more precisely on the normalized drain current I_d/I_{dss} for a particular value of the high field diffusion constant, D. For other values of D our computations show that these conductances are proportional to D except at the low end of the drain current scale. Also shown in Fig. 5 is the magnitude of the correlation coefficient, C. Note that C is nearly constant in value. (For a 1-μm gate, C approaches unity in value over most of the current range.) This fact (which is the result of the strong correlation between i_{g2} and i_{d2} in the expression for C, as seen in (88) and the linear dependence of the noise conductances on D show that the dominant source of noise in the device, at least in the saturated current regime, is the diffusion noise of the saturated velocity region. We have used for the hot electron temperature parameter δ, a value of 1.2 based on Baechtold's experiments, and our choice of E_{sat}.

Fig. 6 illustrates the dependence of F_{min} on the normalized drain current for a particular value of source–drain potential and several values of the high-field diffusion constant. Also shown are the experimental results for the two devices reported by Brehm and Vendelin [19]. Notice that our theoretical results exhibit all of the features of the experimental data, in particular, the rapid rise of F_{min} at high drain currents, a feature not shared by other noise theories. This rapid rise in F_{min} at high drain currents and the strong dependence on the diffusion constant is a reflection of the influence of diffusion noise on the noise conductances, as illustrated in Fig. 5. The fast increase in F_{min} to the left of the minimum is a consequence of the drop-off in transconductance at low drain currents.

We also show in Fig. 6 the value of F_{min} in the absence of parasitics ($R_m = R_f = 0$) for one of the D values. The low level derives from the strong cancellation of noise due to correlation as predicted by (33). The effectiveness of this noise cancellation is diminished drastically when parasitic elements are introduced, hence the large increase in F_{min}. It is obvious then that parasitic resistances should be kept at a minimum, not only because they introduce noise of their own, but also because they indirectly cause an increase in the contributions of the noise sources associated with the intrinsic device.

The minimum noise figure is only a mild function of the source–drain potential in the current saturation region as

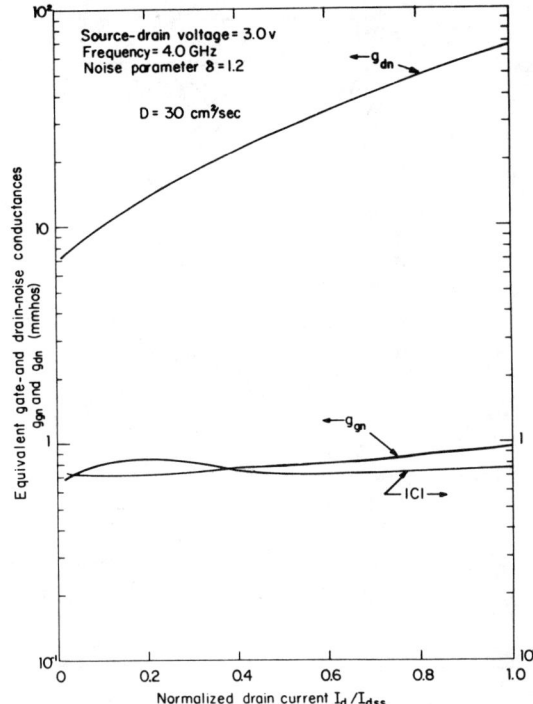

Fig. 5. Equivalent drain noise conductance g_{dn} and equivalent gate noise conductance g_{gn} and correlation coefficient as a function of normalized drain current.

illustrated by the experimental results in Fig. 7 [18], [19]. Our theory also shows this same dependence in good agreement with the data.

The important feature to observe in Figs. 6 and 7 is that to obtain noise figures corresponding to experimental values, one must use a diffusion coefficient in the vicinity of $D = 30$–40 cm^2/s, substantially lower than the low-field value of $D = 110$ cm^2/s. This observation conforms to the fact that theoretically, D shows a sharp drop off at high electric fields to levels in general agreement with the range indicated above [9]. More experimental data are necessary, of course, to permit us to make a more specific choice of D. Note that for $D \approx 35$ cm^2/s our theory provides excellent agreement with experiment.

We have also applied our theory to devices with shorter gate lengths (1 μm) operating at X band. Although our agreement with the experimental results is not quite as good (we feel because of unaccounted circuit losses and some still inaccurate device parameter values), nevertheless diffusion coefficients in the range above are indicated.

Summarizing, the present paper shows that the saturated velocity region in field-effect transistors cannot be neglected. To date, only limited experimental data exist on the noise behavior of GaAs FET's, but it apears that these data are in good agreement with our calculations. While we have not shown plots of $R_{s,opt}$ and $X_{s,opt}$, nevertheless, their calculated values are also in the range of the observations. It will be necessary to compare more extensive experimental measurements with our theory to establish its validity and limitations. Extensive definitive experiments are now in progress at our laboratory to obtain the necessary data.

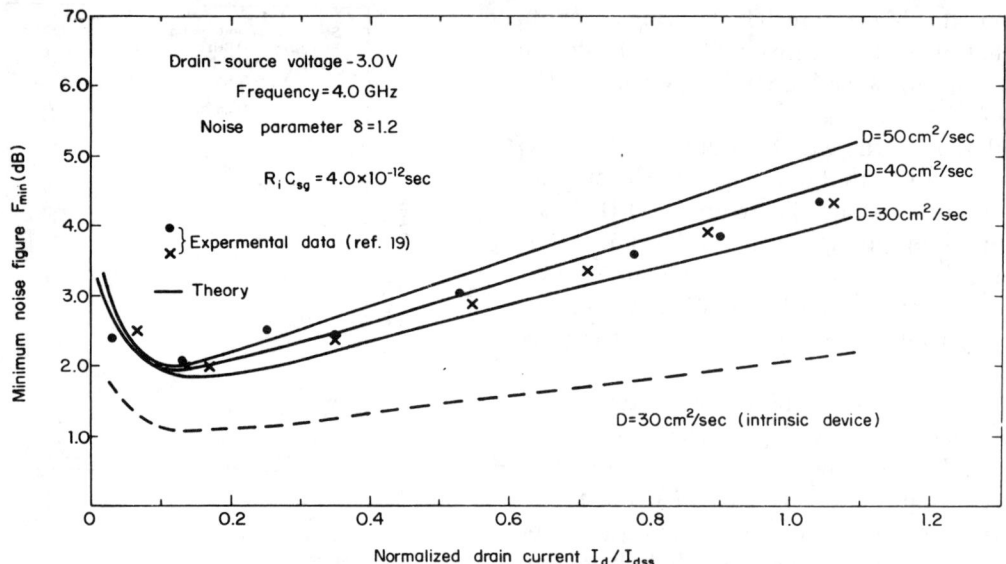

Fig. 6. Minimum noise figure as a function of normalized drain current for various diffusion constants.

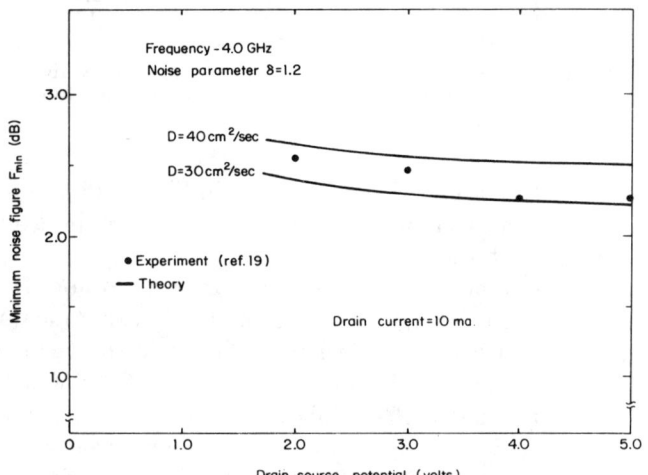

Fig. 7. Minimum noise figure as a function of source–drain potential for a particular diffusion constant.

APPENDIX I

In this Appendix, we derive the noise voltage at L_1 due to Johnson noise within an infinitesimal section of channel. We shall follow van der Ziel's analysis. The width of the ohmic channel, $2b$, is given by

$$2b = 2a[1 - (W/W_{oo})^{1/2}]. \quad (I.1)$$

Taking a perturbation, one obtains

$$\Delta W = -\frac{2}{a}(W_{oo}W)^{1/2}\Delta b. \quad (I.2)$$

Noting that the current is kept constant and the longitudinal field is dW/dx, one has

$$I_d = 2\sigma_o b \frac{dW}{dx} = 2\sigma_o(b_o + \Delta b)\left(\frac{dW_o}{dx} + \frac{d\Delta W}{dx}\right)$$

$$= 2\sigma_o b_o \frac{dW_o}{dx}. \quad (I.3)$$

One obtains to first order in the variation

$$\Delta b \frac{dW_o}{dx} + b_o \frac{d\Delta W}{dx} = 0. \quad (I.4)$$

Substituting (I.2) into (I.4) and introducing the normalized dc voltage seen in (36), one obtains

$$\frac{d\Delta W}{\Delta W} = \frac{dw}{1-w}. \quad (I.5)$$

Integration between the limits x and L_1 gives

$$\Delta W_p = -\frac{1-w(x)}{1-p}\Delta W_x \quad (I.6)$$

where p is the value of w at the end of the ohmic region, $-\Delta W_x$ is the fluctuation in the section Δx, and ΔW_p is the resulting fluctuation at the pinch-off point L_1. (I.6) is equation (10) of [3]. ΔW_x is caused by thermal noise and its mean-square value is (as seen by (11), [3])

$$\overline{\Delta W_x^2} = 4kT\Delta f \frac{\Delta x}{2\sigma_o bl} = \frac{4kT\Delta f}{I_d}\Delta W_o \quad (I.7)$$

where

$$I_d = 2\sigma_o bl \frac{dW_o}{dx} = 4\sigma_o al(1-w)W_{oo}wdw \quad (I.8)$$

and dW_o/dx is the dc electric field at x. Introducing (I.7) into the square of (I.6), one obtains (35) of the text.

APPENDIX II

The induced charge on the gate is obtained in the same way as the one employed by van der Ziel, part of whose derivation is reproduced in the following equations. We supplement van der Ziel's equations to account for the presence of Region II.

The distributed negative charge stored in the depleted region above and below the ohmic channel segment of length dx is

$$Q' \, dx = -2(a-b)\rho_o \, dx = -2a\rho_o\left(1 - \frac{b}{a}\right)dx. \quad \text{(II.1)}$$

This charge is compensated by an equal and opposite charge $Q \, dx = -Q' \, dx$, on the gate contact. Therefore, the fluctuation Δb in b between x and $x + \Delta x$ produces a fluctuating charge

$$d_\Delta Q = -2a\rho_o\left(\frac{\Delta b}{a}\right)dx. \quad \text{(II.2)}$$

The problem is now to compute the perturbation of ohmic channel width $\Delta b/a$ as a function of position. This perturbation exists ahead of $(x > x_o)$ of the perturbing potential ΔW_x at x_o as well as behind it. Ahead of x_o there is the effect of the change in current Δi_d; behind x_o there is the combined influence of Δi_d and ΔW_x. Because the current is given by (I.3), we obtain

$$I_d + \Delta i_d = 2\sigma_o l(b_o + \Delta b)\frac{d}{dx}(W_o + \Delta W). \quad \text{(II.3)}$$

Substituting (I.1), (I.2), and (36) leads to (compare equation (8) of [4])

$$\Delta i_d = 2\sigma_o a l \frac{d}{dx}(1-w)\Delta W. \quad \text{(II.4)}$$

Integrating this equation over x, we obtain (52) of the text after determining the integration constant such that $\Delta W = 0$ for $x = 0$. In (53), the integration constant is chosen for convenience in the form $-\Delta i_d \gamma L_1/(lg_o)$. γ is then determined in the text. The induced charge can now be calculated in Region I by inserting Δb from (I.2) into (II.2). The detailed dependence of ΔW on x is given in (52) and (53). For further details, the reader should consult [4], equations (12)–(17).

ACKNOWLEDGMENT

The authors wish to thank Mrs. Laura Spiniello and Mrs. Janet Newell who performed the extensive and, at times, exasperating computations of the various small-signal parameters and noise functions which were so helpful in our theoretical analysis.

REFERENCES

[1] W. Baechtold, K. Daetwyler, T. Forster, T. O. Mohr, W. Walter, and P. Wolf, "Si and GaAs 0.5 μm gate Schottky barrier field-effect transistors," *Electron. Lett.*, vol. 9, pp. 232–234, May 17, 1973.
[2] C. A. Liechti, E. Gowen, and J. Cohen, "GaAs microwave Schottky gate FET," *ISSCC Digest Tech. Papers*, p. 158, 1972.
[3] A. van der Ziel, "Thermal noise in field-effect transistors," *Proc. IRE*, vol. 50, pp. 1808–1812, 1972.
[4] ——, "Gate noise in field-effect transistors at moderately high frequencies," *Proc. IEEE*, vol. 51, pp. 461–467, Mar. 1963.
[5] W. Baechtold, "Noise behavior of Schottky barrier gate field-effect transistors at microwave frequencies," *IEEE Trans. Electron Devices*, vol. ED-18, pp. 97–104, Feb. 1971.
[6] ——, "Noise behavior of GaAs field-effect transistors with short gate lengths," *IEEE Trans. Electron Devices*, vol. ED-19, pp. 674–680, May 1972.
[7] W. Shockley, J. A. Copeland, and R. P. James, "The impedance field method of noise calculations in active semiconductor devices," in *Quantum Theory of Atoms, Molecules, and the Solid State*, Per-Olov Lowdin, Ed. New York: Academic Press, 1966.
[8] Y. G. Ruch and G. S. Kino, "Transport properties of gallium arsenide," *Phys. Rev.*, vol. 174, pp. 921–931, Oct. 1968.
[9] W. Fawcett and H. D. Rees, "Calculation of the hot electron diffusion rate for GaAs," *Phys. Lett.*, vol. 29 A, pp. 578–579, Aug. 11, 1969.
[10] W. Shockley, "A unipolar field-effect transistor," *Proc. IRE*, vol. 40, pp. 1365–1376, Nov. 1952.
[11] A. B. Grebene and S. K. Ghandhi, "General theory for pinched operation of the junction-gate FET," *Solid-State Electron.*, vol. 12, pp. 573–589, July 1969.
[12] F. M. Klaassen, "Comments on hot carrier noise in field-effect transistors," *IEEE Trans. Electron Devices*, vol. ED-18, pp. 74–75, Jan. 1971.
[13] A. van der Ziel, "Noise resistance of FET's in the hot electron regime," *Solid-State Electron.*, vol. 14, pp. 347–350, Apr. 1971.
[14] ——, "Thermal noise in the hot electron regime in FET's," *IEEE Trans. Electron Devices*, vol. ED-18, p. 977, Oct. 1971.
[15] A. van der Ziel and K. M. van Vliet, "HF thermal noise in space-charge limited solid-state diodes—II," *Solid-State Electron.*, vol. 11, pp. 508–509, 1968.
[16] A. van der Ziel and J. W. Ero, "Small-signal high-frequency theory of field-effect transistors," *IEEE Trans. Electron Devices*, vol. ED-11, pp. 128–135, Apr. 1964.
[17] P. Wolf, "Microwave properties of Schottky-barrier field-effect transistors," *IBM J. Res. Develop.*, vol. 14, pp. 125–141, Mar. 1970.
[18] G. E. Brehm, "Variation of microwave gain and noise figure with bias for GaAs FET's," 1973 Cornell Conf. Microwave Semiconductor Devices, Circuits, and Applications, Cornell University, Ithaca, N. Y.
[19] G. E. Brehm and G. D. Vendelin, "Biasing FET's for optimum performance," *Microwaves*, vol. 13, pp. 38–44, Feb. 1974.
[20] E. Wasserstrom and J. McKenna, "The potential due to a charged metallic strip on a semiconductor surface," *Bell Syst. Tech. J.*, vol. 49, pp. 853–877, May–June 1970.

Noise Performance of Gallium Arsenide Field-Effect Transistors

ROBERT A. PUCEL, SENIOR MEMBER, IEEE, DANIEL J. MASSÉ, MEMBER, IEEE,
AND CHARLES F. KRUMM, MEMBER, IEEE

Abstract — The Schottky-barrier gate gallium arsenide field-effect transistor (GaAs FET) is the first three-terminal, solid-state amplifying device to have demonstrated low-noise performance at X-band and higher. For example, noise figures approaching 3 dB at 10 GHz have been reported, while theory predicts still lower values.

After a brief review of the noise-generating mechanisms intrinsic to the GaAs FET, an enumeration is given of the various parasitic elements associated with the FET which affect the noise performance. These elements include, among others, the gate metallization and source contact resistances, drain-gate feedback capacitance, and source lead inductance. Numerous graphs are presented to illustrate the effects of these elements and the various design parameters on the noise performance.

A comparison is made between the theoretically predicted and the measured noise performance of microwave GaAs FET's.

The best state-of-the-art noise performance as reported by various laboratories is illustrated graphically for single-stage and multistage FET amplifiers.

Finally, some speculation is attempted in regard to the possible reductions in noise figure to be expected from technological and design improvements of GaAs FET's.

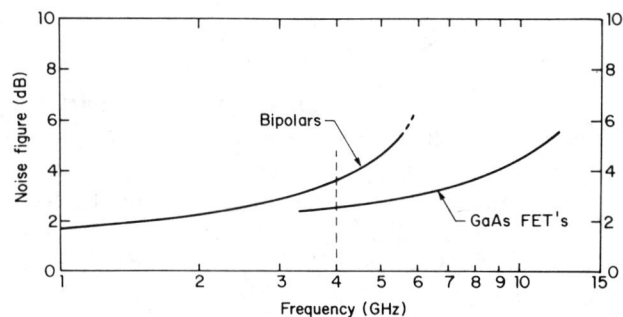

Fig. 1. Noise performance of cascaded (narrow-band) transistor amplifier stages as of July 1975.

I. INTRODUCTION

THE GALLIUM arsenide Schottky-barrier field-effect transistor (FET) is the first three-terminal solid-state device to exhibit linear power amplification at X-band frequencies and higher. Its unique signal-handling capabilities and low-noise properties have been demonstrated by many workers. For example, noise figures approaching 3 dB at 10 GHz have been reported, while theory predicts still lower values.

The GaAs FET is now being used in low-noise amplifiers from low C-band and up. As such it nicely supplements the silicon bipolar transistor which still dominates at frequencies below C-band. However, with the noise reductions now being achieved with buffered-layer FET's, this frequency range will not long remain the sole province of bipolars. Fig. 1 is a comparison of the state-of-the-art performance of low-noise, narrow-band amplifiers using silicon bipolar transistors and GaAs FET's as of July 1975.

Gallium arsenide field-effect transistors also show potential as low-noise microwave mixers and oscillators [1]-[3]. In this paper we shall restrict ourselves to their performance as small-signal amplifiers.

Manuscript received September 15, 1975. This work was based on an oral presentation given at the 5th Biennial Conference on Active Semiconductor Devices for Microwave and Integrated Optics held at Cornell University, Ithaca, NY, August 19-21, 1975.

The authors are with the Research Division, Raytheon Company, Waltham, MA 02154.

As an introduction only a brief review of the present theory of noise of microwave GaAs FET's will be given, since a comprehensive description of the development of this theory has been given [4], [5]. Using this theory we shall assess the relative contributions to the noise performance by sources both intrinsic and extrinsic to the FET. With this as a background, we shall show how these contributions depend on the various material and design parameters at one's disposal. This will allow us to estimate the improvements in noise performance likely to be made in the future with advances in materials and device technology.

Finally, we will compare the theoretical predictions and measured results, and present a summary of the best noise performance obtained with FET devices and multistage amplifiers as of the writing of this paper.

II. SYNOPSIS OF THE NOISE THEORY OF THE GaAs FET

The basic principle of operation of the field-effect transistor was first described by Shockley [6] who assumed a constant mobility throughout the conducting channel region. Van der Ziel, in a series of classic papers, used Shockley's model to derive the small-signal parameters [7] and intrinsic noise properties of the FET [8], [9]. Van der Ziel showed that the intrinsic noise is thermal in origin, and can be represented by two white noise generators, one in the drain circuit, and one in the gate circuit. The gate noise generator, which represents the noise induced on the gate electrode by the passing thermal fluctuations in the drain current, is partially correlated with the drain noise generator.

The constant mobility model of Shockley and van der Ziel, though applicable to long-gate devices, does not apply to microwave devices whose gate lengths are in the micron range. For these devices, when biased in the current saturation

Reprinted from *IEEE J. Solid-State Circuits*, vol. SC-11, pp. 243-255, Apr. 1976.

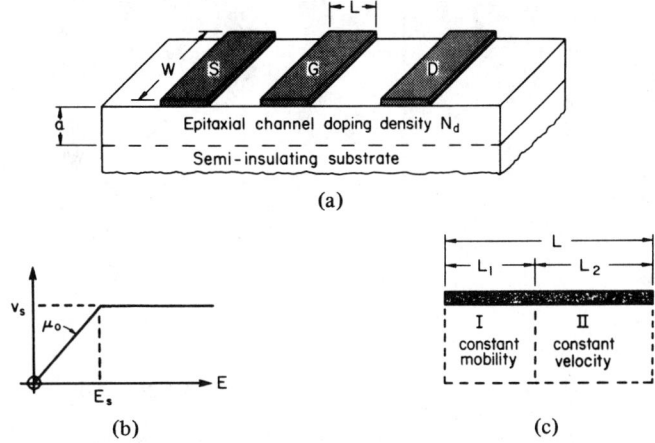

Fig. 2. Perspective sketch and two-section model of FET used in noise analysis. (a) FET model. (b) Assumed velocity-field characteristic. (c) Two-region model of channel.

regime, the average value of the longitudinal dc field in the channel is in the range where the mobility is a decreasing function of field, and indeed, where the carrier velocity is approaching a constant ("saturated") value. Consider, for example, a typical case of a GaAs FET with a 1 μm gate operating with a drain voltage of 3 V. The average longitudinal channel field is 30 kV/cm, approximately ten times the threshold value at which the velocity begins to saturate. Thus, the effects of velocity saturation must be included in any model of a GaAs FET designed for microwave operation.

Velocity saturation within the channel not only modifies the small-signal parameters, but the noise performance as well. Many workers have introduced some aspects of velocity saturation into their FET models, though none of these models include the diffusion noise introduced by electrons experiencing velocity saturation. In the noise and small-signal model developed at the authors' laboratory by Statz et al. and Pucel et al. [4], [5] this high-field diffusion noise is taken into account. It is the dominant intrinsic noise of microwave GaAs FET's.

A brief description of this model will be given now with the help of Fig. 2. Fig. 2(a) is a perspective sketch of a planar FET consisting of a source electrode (S), gate electrode (G), and drain electrode (D), all of width W. The gate length is denoted by L. The conducting n-type epitaxial channel of thickness a, situated on a semi-insulating substrate, is assumed to be uniformly doped at density N_d cm^{-3} with a low-field mobility μ_0. Typical values for these material parameters are $N_d \sim 10^{17}$ cm^{-3}, $a \sim 0.2 - 0.4$ μm, and $\mu_0 \sim 3000$–4500 cm^2/V·s.

Following Turner and Wilson [10], Statz et al. idealized the velocity-field characteristic by a piecewise linear approximation shown in Fig. 2(b). To obtain good agreement with experimental FET data, and reasonable agreement with experimental and theoretical velocity-field data [11], [12], the critical field E_s denoting the onset of velocity saturation was chosen to be 2.9 kV/cm, and the limiting velocity v_s to be 1.3×10^7 cm/s at room temperature.

This piecewise linear approximation to the velocity-field

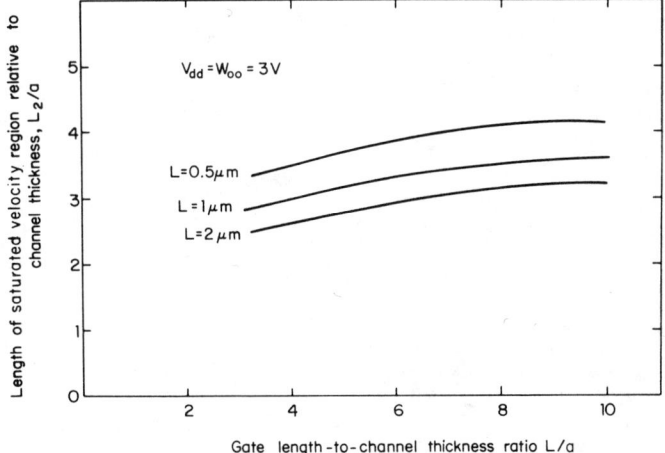

Fig. 3. Length of velocity-saturated zone relative to epi-layer thickness as a function of the gate length and the ratio of the gate length to epi-layer thickness.

characteristic allows one to divide the conducting channel underneath the gate region into two zones as Grebene and Ghandhi [13] have suggested. In this two-zone model, shown in Fig. 2(c), a portion of the channel near the source end is assumed to be in the constant mobility regime, while the remaining portion near the drain end is postulated to be in velocity saturation. The position of the boundary between these two zones, representing the onset of velocity saturation, is a strong function of the source-drain bias, but a weaker function of the gate-source bias. The length of the velocity-saturated zone increases monotonically with source-drain bias.

By a correct application of this model, it can be shown that when the FET is biased in current saturation, that is, above the knee of the drain-voltage-current characteristic, the length of the velocity-saturated zone L_2 is of the order of two to four times the epitaxial layer thickness a [5]. Fig. 3 shows how the length of the velocity-saturated zone, relative to the channel thickness, varies as a function of the geometric ratio L/a for various gate lengths. The drain voltage V_{dd} is assumed to be

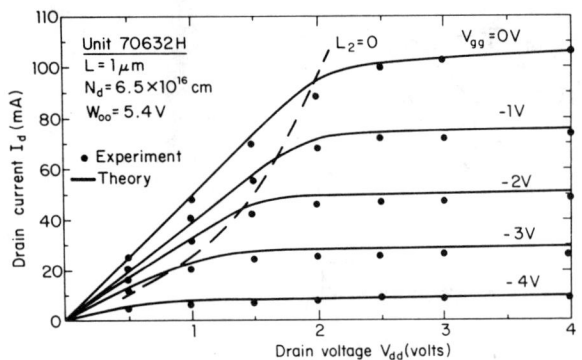

Fig. 4. Comparison between the theoretical and measured drain current-voltage characteristic for an X-band GaAs FET.

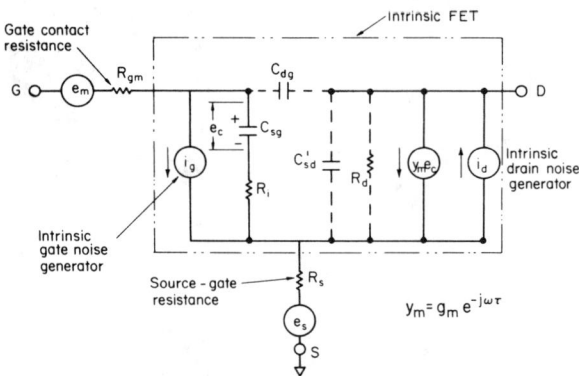

Fig. 6. Noise equivalent circuit of FET showing intrinsic and extrinsic noise sources.

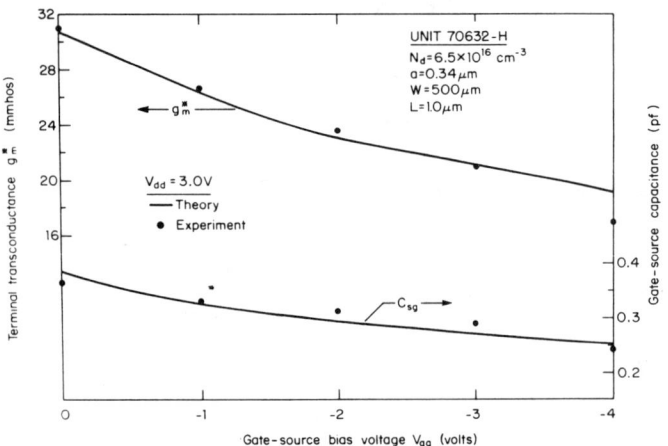

Fig. 5. Comparison between the theoretical and measured values of gate-source capacitance C_{sg} and terminal transconductance g_m^* for an X-band device.

equal to the intrinsic (internal) pinch-off voltage $W_{00} = 3$ V. With contemporary device designs using channel thicknesses of the order of 0.2-0.4 μm, the velocity-saturated zone comprises most of the channel length for gate lengths one micrometer or less. Thus, velocity saturation plays an important role in microwave GaAs FET's.

The piecewise linear approximation, chosen for analytic convenience, is an extreme idealization of the actual $v(E)$ characteristic in that it eliminates any negative resistance regime. In short-channel devices, the assumption of velocity saturation itself may be difficult to justify since the transit time of the electrons in the channel is comparable to the relaxation time of electrons, as Ruch [14] and later Maloney and Frey [15] have pointed out. Despite these recognized limitation, there *does* appear to be an appreciable degree of velocity saturation since this assumption works extremely well for the dc and small-signal characteristics. Fig. 4 demonstrates the agreement between the theoretical model and the measured I-V characteristic. Fig. 5 demonstrates this agreement for the small-signal terminal transconductance g_m^* and source-gate capacitance. The locus $L_2 = 0$ in Fig. 4 denotes the bias conditions for which velocity saturation first begins to manifest itself at the drain end of the gate. To the left of this line, that is, below the "knee" of the I-V characteristic, the channel is entirely in the constant mobility mode of operation. Thus, in the current "saturation" regime, i.e., to the right of the locus, the channel is always in velocity saturation over a portion of its extent. The FET is normally operated in this current-saturated mode.

We shall show later that using the two-zone model for the noise analysis, the agreement between the predicted and measured noise performance of GaAs FET's is equally as good as it is for the dc and small-signal properties, as demonstrated by Figs. 4 and 5.

Statz *et al.* [4] assume that the noise in zone I is thermal, as in the van der Ziel treatment, but enhanced by hot electron effects as postulated by Baechtold [16], [17]. Zone II, however, cannot be treated as an ohmic conductor. Its noise contribution (which is new in FET theory) is dominant in microwave devices and must be represented as a high-field diffusion noise as Shockley *et al.* [18] and van der Ziel [19] have shown.[1] This diffusion noise is proportional to the high-field diffusion coefficient and is linearly dependent on drain current [4], [5]. On the other hand, the thermal noise of region I decreases with increasing drain current. Although the high-field diffusion noise is high, a strong correlation (approaching unity) exists between the drain noise and the induced gate noise. This correlation leads to a high degree of cancellation in the noise output of the GaAs FET.

Fig. 6 is a noise equivalent circuit of the FET, valid for high frequencies. The noise generator i_g represents the induced gate noise of the intrinsic device (shown in dotted lines). Its mean-square value varies as the square of the frequency, i.e., ω^2. The intrinsic drain noise generator i_d has a flat spectrum. The coupling between these noise generators, represented by the correlation coefficient C

$$jC = \frac{\overline{i_g^* i_d}}{\sqrt{|i_g^2||i_d^2|}} \qquad (1)$$

[1] Actually, as van der Ziel [19] has shown, the noise of the constant mobility zone also can be represented as diffusion noise. Since the Einstein relation $D_0 = kT\mu_0/q$ holds in this zone, the diffusion noise expression can be transformed into the more familiar thermal or Johnson noise form. This transformation, of course, is invalid when velocity saturation occurs.

where (*) denotes the complex conjugate, and the overbar (¯) represents a statistical average, approaches unity in magnitude for short-gate devices. (By comparison, in a constant mobility model, $|C| \sim 0.3$–0.4 [9].) In addition to the intrinsic noise sources, the parasitic source-gate resistance R_s and gate metallization resistance R_{gm} introduce thermal noise. This thermal noise is represented, respectively, by the generators labeled e_s and e_m. The resistance R_i represents the resistive charging path for the gate capacitance in the intrinsic FET. The noise associated with R_i is imbedded in the gate noise generator i_g [7].

It is not necessary to include all of the equivalent circuit parameters of the FET since some have a small effect on the noise figure. For example, for simplicity we shall neglect the (small) feedback drain-gate capacitance C_{dg} as well as the source-drain capacitance C_{sd}. The small perturbation of the noise figure produced by these capacitances can be added later if necessary. We may also neglect the small effect of the output drain resistance R_d, and any source lead inductance. We shall show later that inclusion of these parameters, for a well designed device, alters the minimum noise figure by at most a few tenths of a decibel. Thus, as a first approximation C_{dg}, C_{sd}, and R_d^{-1} will be assumed equal to zero. With these approximations, the equivalent circuit used in the noise figure derivation reduces to that shown in Fig. 7. This circuit also includes the signal source impedance Z_g and its associated thermal noise source e_g.

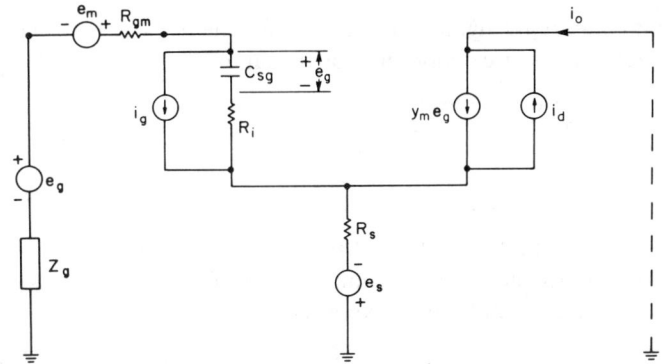

Fig. 7. Simplified equivalent circuit used in noise analysis of FET.

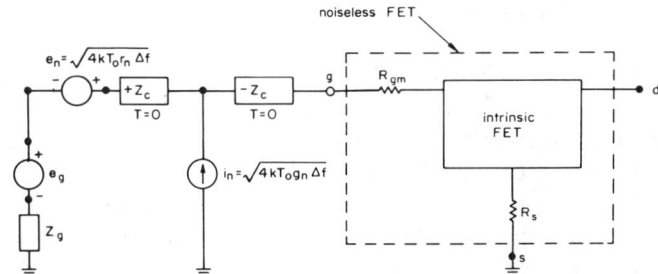

Fig. 8. Representation of noisy FET by a noiseless FET preceded by a noise network.

III. NOISE FIGURE

The configuration shown in Fig. 7 with the source terminal common to input and output is often called the grounded-source or common source connection. Although we shall present the expression for the noise figure for this circuit, our results should apply with negligible error to the common-gate and common-drain configurations [20], [21].

The noise figure F can be expressed as the ratio of the sum of the mean-square noise components in the short-circuited drain-source path produced by all of the noise sources in Fig. 7 to the mean-square thermal noise current component produced by the signal source e_g alone.

By a straightforward (but lengthy) circuit analysis the noise figure can be written in the form

$$F = 1 + \frac{1}{R_g}(r_n + g_n |Z_g + Z_c|^2) \qquad (2)$$

where R_g is the real part of the source impedance (assumed to be at the reference temperature $T_0 = 290$ K). The parameters r_n and g_n are the so-called noise resistance and noise conductance, respectively, and Z_c the correlation impedance [22].

In terms of r_n, g_n, and Z_c all the noise properties of the FET with parasitic resistances are embodied in a very simple noisy network shown in Fig. 8, which precedes the FET (now considered noise-free). Thus, r_n represents a thermal noise voltage generator at the reference temperature; g_n, a shunt thermal noise current generator at the same temperature; and Z_c, an impedance at absolute zero (noiseless). The noise figure of this combined network is the same as that of the original noisy FET, Fig. 7.

The noise functions are given by the simple expressions [5]

$$r_n = (R_s + R_{gm})\frac{T_d}{T_0} + K_r \left(\frac{1 + \omega^2 C_{sg}^2 R_i^2}{g_m}\right) \qquad (3a)$$

$$g_n = K_g \frac{\omega^2 C_{sg}^2}{g_m} \qquad (3b)$$

$$Z_c = R_s + R_{gm} + \frac{K_c}{Y_{11}} \qquad (3c)$$

where T_d is the temperature of the FET. The parameters K_g, K_r, and K_c are numerical noise coefficients which represent the properties of the intrinsic noise generators i_g, i_d and their correlation (1). For an FET not at room temperature, these noise coefficients, as well as the small-signal parameters g_m, R_i, C_{sg}, the parasitic resistances R_s and R_{gm}, and the input impedance Y_{11}^{-1} of the intrinsic device, Fig. 7, given by

$$Y_{11}^{-1} = R_i + \frac{1}{j\omega C_{sg}} \qquad (4)$$

are assumed to be evaluated at the device temperature T_d.

IV. MINIMUM NOISE FIGURE

The first stage of a low-noise amplifier chain is often designed to have a minimum noise figure. The noise figure is optimized by the proper choice of the source impedance $Z_g = R_g + jX_g$. This optimization can be achieved by a suitable lossless matching network between the signal source and the input (gate-source) terminals of the FET. It is easy to show that the minimum noise figure is achieved when the real and imaginary parts of the source impedance are equal to

$$R_g = R_{g0} = \sqrt{R_c^2 + \frac{r_n}{g_n}} \quad (5a)$$

$$X_g = X_{g0} = -X_c \quad (5b)$$

where R_c and X_c are the real and imaginary parts of the correlation impedance. The minimum value of F corresponding to this "noise match" can be expressed as

$$F_{min} = 1 + 2g_n(R_c + R_{g0}). \quad (6)$$

In decibels, F_{min} (dB) = $10 \log_{10} F_{min}$.

When the source is not optimized for best noise performance, the noise figure is given by

$$F = F_{min} + \frac{g_n}{R_g} \{(R_g - R_{g0})^2 + (X_g - X_{g0})^2\}. \quad (7)$$

We shall refer to this equation later when we discuss the experimental procedure for determining F_{min}.

The expression for F_{min} given by (6) can be written to a very good approximation by the simple three-term power series expansion in frequency

$$F_{min} = 1 + 2\left(\frac{\omega C_{sg}}{g_m}\right)\sqrt{K_g[K_r + g_m(R_s + R_{gm})]}$$

$$+ 2\left(\frac{\omega C_{sg}}{g_m}\right)^2 [K_g g_m(R_{gm} + R_s + K_c R_i)] + \cdots \quad (8)$$

valid at room temperature. This simplified form delineates the roles played by the noise sources intrinsic to the device, embodied in the noise coefficients K_g, K_c, and K_r and the noise sources corresponding to the parasitic resistances R_s and R_{gm}.

The frequency dependence of F_{min} is a consequence of the ω^2 dependence of the induced gate noise. Note that F_{min} decreases with increasing gain-bandwidth factor g_m/C_{sg} of the FET. The gain-bandwidth factor is a function of gate bias, eventually decreasing as the gate bias approaches the pinch-off condition. In terms of g_m and C_{sg} individually, note that F_{min} increases with gate capacitance but decreases approximately in proportion to the inverse of the transconductance.

The noise coefficients are frequency-independent numerical factors which are gate-bias dependent, and to a lesser extent, drain-bias dependent. A typical bias dependence of these coefficients is shown in Fig. 9. Note that K_r is an order of magnitude lower than K_g and K_c. The bias dependence is expressed in terms of the drain current I_d normalized to its value I_{dss} at zero gate bias. All three noise coefficients depend in a complicated manner on gate length, channel thickness, and other parameters [5]. Observe that K_g is a strong function of the drain current, increasing at a rate faster than linear with I_d at high currents.

It is evident from (8) that F_{min} can be lowered by minimizing the parasitic resistances R_s and R_{gm}. As we shall see later, it also can be lowered by a proper choice of gate bias, or equivalently, drain current since the noise coefficients as well as the small signal parameters (mainly g_m) are bias-dependent.

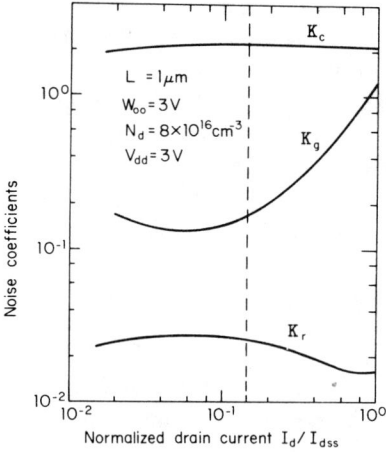

Fig. 9. Drain current dependence of noise coefficients of a GaAs FET for a specific set of design parameters and drain voltage.

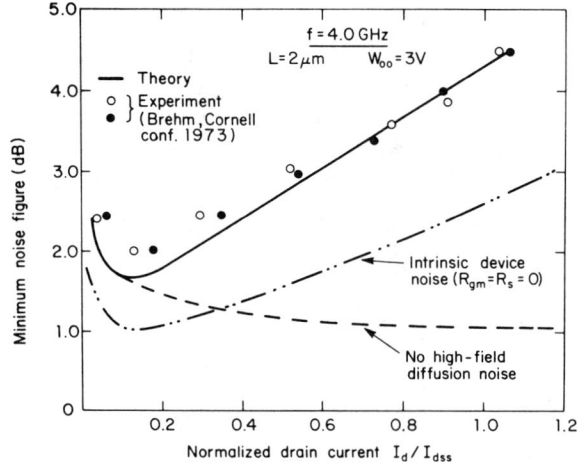

Fig. 10. Theoretical and measured noise figure for a GaAs FET with a 2 μm gate.

V. COMPARISON OF THEORY AND EXPERIMENT

The applications of the two-zone noise model to practical GaAs FET's will be exemplified now.

The solid line in Fig. 10 illustrates the validity of the noise model applied to a device with a 2 μm gate [23]. The nearly linear current dependence of F_{min} in the high current range demonstrates clearly the contribution of the high-field diffusion noise produced in the velocity-saturated zone. This fact is emphasized further by the lowest dashed line which represents F_{min} if this diffusion noise were set equal to zero. The increase in F_{min} at low currents is attributable to the decrease in g_m and the increase in the noise contribution of zone I as pinch-off is approached. The important role played by the parasitic resistances is illustrated by the middle dashed line representing the intrinsic noise obtained by setting R_{gm} and R_s equal to zero. Note that at the minimum the parasitic resistances contribute nearly half of the noise. Fig. 11 illustrates the agreement obtained with the noise data reported for a 1 μm gate device [24]. Again the general features of the bias dependence of F_{min} are reproduced. If we allow for the

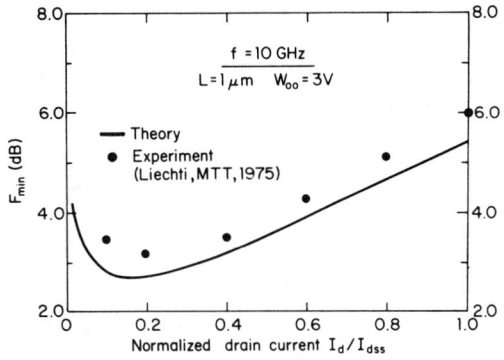

Fig. 11. Theoretical and measured noise figure for a GaAs FET with a 1 μm gate.

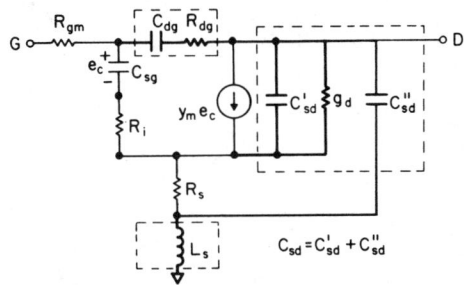

Fig. 12. Inclusion of neglected elements in equivalent circuit for noise figure analysis.

possible measurement uncertainty in the value of the noise figure, which may amount to as much as ±0.4–0.5 dB, the difficulty in accounting for the circuit losses accurately, and the errors introduced by use of the simplified equivalent circuit, the agreement between theory and experiment can be considered satisfactory.

VI. Effects of Neglected Parasitics

We shall assess now the effects of including the equivalent circuit elements neglected in the derivation of the noise figure in Section III. Fig. 12 illustrates these parasitics (delineated by dashed lines). They are, principally, the source-drain capacitance C_{sd}, drain conductance $g_d = R_d^{-1}$, source lead inductance L_s, and the drain-gate feedback admittance Y_{12} represented by a resistance R_{dg} in series with the drain-gate capacitance C_{dg}. This resistance (which is assumed to generate thermal noise) represents the resistive charging path for C_{dg} between the drain and gate terminals as suggested by Vendelin [25] and others. It is possible to include the effects of these parasitics in an exact manner; however, the resultant expression for the noise figure is unwieldy. Fortunately, for the small values of these neglected elements, typical of well-designed FET's, the perturbations to the minimum noise figure are linear functions of the element values, and can be added algebraically to the expression for F_{min}, (6).

Furthermore, since these perturbations are small in comparison to F_{min}, as we shall show, the corrections to F_{min}, expressed in decibel form, are also linear. If we denote the perturbation to F_{min} by ΔF, where the latter represents the inclusion of one of the neglected elements, or any combination of them, then the correction ΔF, expressed in decibels, is given by

$$\Delta F \text{ (dB)} = 10 \log_{10} \frac{F_{min} + \Delta F}{F_{min}} \quad (9a)$$

$$= 4.34 \ln\left(1 + \frac{\Delta F}{F_{min}}\right) \quad (9b)$$

$$\approx 4.34 \frac{\Delta F}{F_{min}} \quad (9c)$$

where the last equation arises from the assumption $|\Delta F| \ll F_{min}$. If this is not true (9a) must be used.

We shall now demonstrate the magnitude and sign of these corrections to F_{min} for the 1 μm gate device discussed earlier. The corrections as evaluated here apply for the bias conditions corresponding to the lowest value of F_{min}, namely 2.75 dB. See Fig. 11.

Unfortunately, we do not have the values of the neglected parasitic elements for the specific 1 μm device discussed earlier. However, since we are discussing small perturbations, we may use the element values obtained by Vendelin [25] for a similar device [26] by a computer optimized fit to the S-parameters. These are $C_{sd} = 0.16$ pF, $R_d = g_d^{-1} \approx 200$ Ω, $C_{dg} \approx 0.014$ pF, $R_{dg} \approx 660$ Ω, and $L_s \approx 26$ pH.

Consider first the inclusion of the drain-gate feedback, illustrated in Fig. 13(a) for a range of feedback admittances Y_{12}. As one might expect, the resistive feedback increases the noise figure. On the other hand, the capacitive component decreases it, in accordance with the findings of others [27]. For the specific values of the feedback elements, Re Y_{12} = 0.38 m℧, Im Y_{12} = 0.66 m℧, the corresponding corrections to the noise figure are $\Delta F = 0.45$ dB and $\Delta F = -0.12$ dB, or a net change of +0.33 dB.

We turn next to the inclusion of the output admittance consisting of g_d and C_{sd} in shunt, shown in dotted lines in Fig. 12. Since there is still some uncertainty amongst workers in the field as to the fraction of the source-drain capacitance which should be terminated at the upper end of R_s, as shown in Fig. 12, and the fraction that should tie to the lower end, we shall only consider the perturbation caused by inclusion of the drain output conductance. This is illustrated in Fig. 13(b). As is evident, the correction is negative. For the stated output resistance $g_d = 5$ m℧, $\Delta F = -0.24$ dB.

The effect of source lead inductance is of second-order importance. For values of this inductance in the range below 200 pH, the noise figure decreases slightly. This range exceeds by almost an order of magnitude the values of parasitic source lead inductance in a well designed device. For the specific value of inductance of the device under consideration, $\Delta F \approx -0.01$ dB.

Thus taken together, all of the neglected parasitics considered increase the noise figure by about 0.1 dB. This is a negligible error. Therefore, for a well designed device, use of the simplified model for noise analysis shown in Fig. 7 is justified.

It must be cautioned that the corrections implied by Fig. 13(a) and (b) apply only to the device considered in the text.

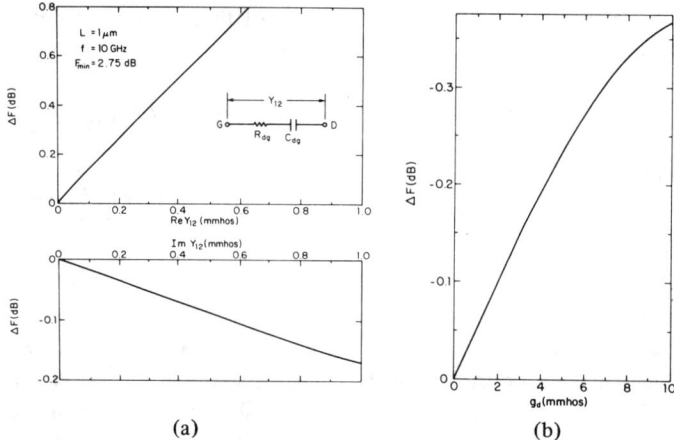

Fig. 13. Corrections to minimum noise figure attributable to neglected equivalent circuit elements. (a) Corrections to F_{min} at 10 GHz attributable to drain-gate feedback. (b) Corrections to F_{min} attributable to drain resistance.

Although corrections for other devices of similar design will be comparable, the quantitative values necessarily will be different.

It should be mentioned in passing that since the noise figure decreases with capacitive feedback, one might use this as a means for improving the noise performance of an FET amplifier by external feedback [27]. We do not believe this to be a satisfactory approach for several reasons. First, increasing feedback in this manner reduces the stability factor of the FET [28]. Second, the available gain decreases. Third, since it is the noise measure, rather than the noise figure that one should minimize, as we shall argue later, no improvement is achieved since noise measure does not change under capacitive feedback, or for that matter, for any lossless feedback scheme, as Haus has shown [29].

VII. Theoretical Dependence of Noise Figure on Device Geometry and Parasitic Resistances

We shall use the results of the noise theory described earlier to show how the noise sources intrinsic and extrinsic to the FET depend on the various material and geometrical parameters at the disposal of the device designer. With this as a background, we estimate the improvements likely to be made in the future with advances in materials and device technology. Specifically, we shall discuss the dependence of the noise performance on gate length, frequency, and extrinsic parasitic resistances.

We will limit ourselves, for convenience, to a specific design based on a channel doping density $N_d = 8 \times 10^{16}$ cm^{-3} and intrinsic pinch-off voltage $W_{00} = qN_d a^2/2\kappa\epsilon_0 = 3$ V typical of contemporary microwave devices where $\kappa = 12.5$ is the dielectric constant of GaAs. However, the general conclusions to be drawn will also apply to other microwave devices with similar, though not identical, design parameters.

Dependence of Minimum Noise Figure on Gate Length

The dependence of F_{min} on gate length is embodied in the source-gate capacitance and transconductance, and in a more complicated manner in the noise coefficients. The theoretical

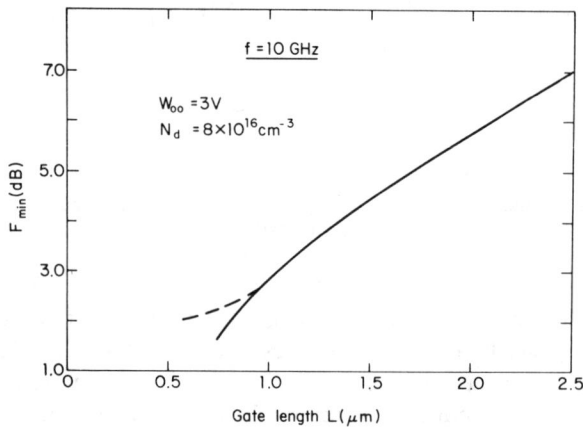

Fig. 14. Theoretical minimum noise figure as a function of gate length.

value of F_{min} at $f = 10$ GHz, as a function of gate length, is illustrated in Fig. 14. Notice the rapid rate of decrease of F_{min} as the gate length approaches 1 μm. Gate length reduction is the single most productive means of improving the noise performance of an FET—up to a point! Although our theoretical curve extends down to $L = 0.5$ μm, we show an additional, arbitrarily drawn dashed line, since we believe the validity of our theory becomes questionable below $L = 1$ μm, for the channel thickness ($a \approx 0.225$ μm) corresponding to the assumed value of N_d and W_{00}.

Below $L = 0.5$ μm, there are other, more fundamental reasons why we believe that the rate of decrease in F_{min} will "flatten out" as implied by the dashed line.

First, as the gate length continues to decrease below a micron, the fringing capacitance of the gate (which does not decrease with gate length) [5] puts a lower asymptote on the gate capacitance; and hence on F_{min} [see (8)]. For example, for $L = 0.5$ μm, this fringing capacitance is over 30 percent of the gate capacitance.

Second, unless the channel thickness is reduced correspondingly, in accordance with the gate length, the electric field in the channel will begin to deviate markedly from a longitudinal field configuration, to one conforming more to a cylindrical

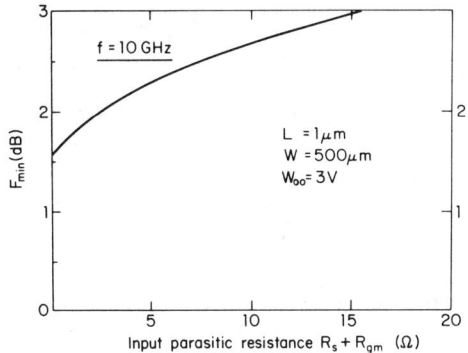

Fig. 15. Theoretical dependence of minimum noise figure on parasitic resistances for a GaAs FET with a 1 μm gate.

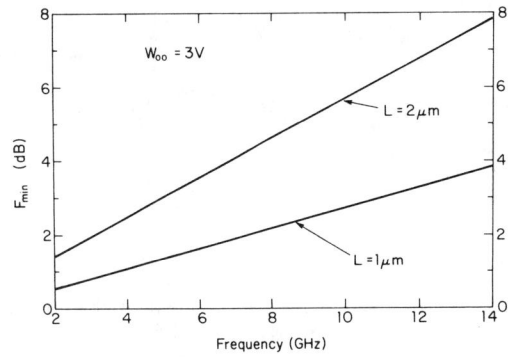

Fig. 16. Predicted frequency dependence of minimum noise figure of a GaAs FET for two values of gate length.

pattern about the gate electrode. The reduced longitudinal component of the electric field leads to a diminished control of the electron flow by the gate potential and to a "softer" drain current-voltage saturation characteristic [30], [32]. Although the noise performance and gain will still improve with decreasing gate length, the rate of improvement should decrease.

If the channel thickness is reduced in proportion to the gate length to reestablish a longitudinal field pattern, this requires use of epitaxial layers 0.1 μm thick, or less. Most of the channel doping profile, in this case, will not be constant, but will be decreasing rapidly toward the substrate. This means that the rate of decrease of transconductance with gate bias will be faster than it would be for an ideal (step) profile. Thus the upturn in F_{min} with decreasing drain current will occur at a higher value of drain current. Hence, the advantages of reducing both gate length and channel thickness, simultaneously, will be partially nullified. Although there appears to be some promise of growing epitaxial layers with a steeper transition zone, eventually one is limited to a lower value of the transition zone fixed by the Debye length [32].

There is another limitation imposed by thinner channel layers, namely, the increase in source gate resistance which must accompany a reduced epitaxial layer.

All of the above considerations must be taken into account in matching the possible advantages of reducing gate lengths much below a micrometer against the additional cost and complexity of producing submicron gate devices with acceptable yield.

Dependence of Minimum Noise Figure on Parasitic Gate and Source Resistance

The theoretical dependence of F_{min} on the parasitic resistances is shown in Fig. 15. Values of $R_s + R_{gm}$ typical of contemporary devices fall in the range from 8-11 Ω for a 500 μm wide gate device. A reduction of the order of perhaps 0.5 dB might be possible with improvements in the design and technology of contacts.

Dependence of Minimum Noise Figure on Frequency

The predicted frequency dependence of F_{min} is illustrated in Fig. 16 for two gate lengths. Note that the curves are nearly linear with frequency. This frequency dependence is exhibited by experimental data also, as we shall demonstrate. The noise degradation with increasing frequency is attributable to the frequency dependence of the induced gate noise.

The theoretical noise figures displayed in the previous graphs were all computed for channel doping profiles exhibiting a slope near the epi-substrate interface—that is, for a nonrectangular profile, representative of present epitaxial layers. Some further reductions in the noise figure can be expected as "steeper" transition zones are achieved by improvements in the technology of epitaxial growth.

VIII. MEASUREMENT OF NOISE FIGURE

Introduction

In this section we shall discuss some of the practical problems in determining the minimum noise figure peculiar to the FET, and the means for overcoming these difficulties.

Earlier we had demonstrated the strong bias dependence of the minimum noise figure and had mentioned that the gain is also bias-dependent. We also pointed out that not only are the lowest noise figure and the maximum available gain achieved at different gate bias values, but that at a given bias condition, the matching conditions for the best noise figure and highest available gain also differ. This is one problem.

Next, the gain associated with the lowest value of the noise figure of present GaAs FET's is not high enough, at least at the upper end of the microwave band, to permit one to neglect in a noise measurement the correction for postamplifier or mixer noise. Thus it is a very tedious procedure to determine the minimum noise figure of an FET by simply varying the tuning adjustments of the input matching circuit because the correction also varies, nor is it a very precise method.

Equation (7) suggests a more direct approach. Note that this equation contains four unknowns, F_{min}, g_n, R_{g0}, and X_{g0}. Thus, at each bias one may, in principle, ascertain F_{min} and the remaining noise parameters by measuring the noise figure and gain for four selected source impedances.[2] However, unless these impedances are chosen judiciously, so that the resultant noise figures do not all cluster near F_{min}, or conversely, be all far removed from F_{min}, large errors can be

---

[2] It is necessary to measure the gain in order to correct for the postamplifier and/or mixer noise.

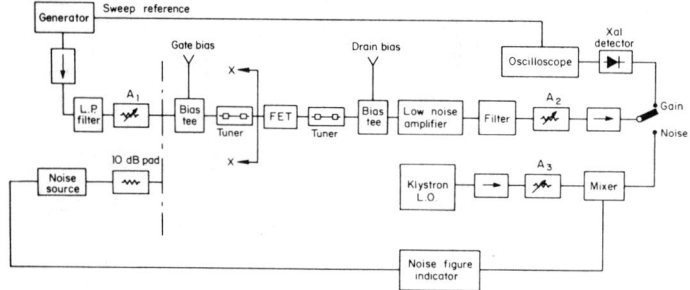
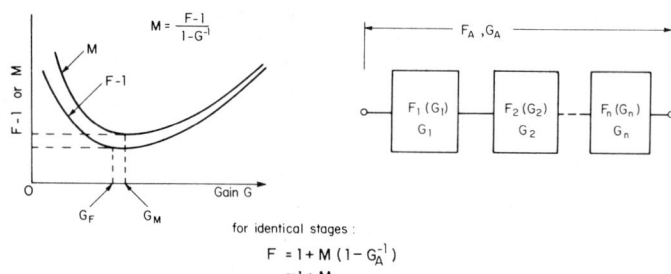

Fig. 17. Experimental set-up for measurement of the microwave gain and noise figure of an FET.

Fig. 18. Relevant to the design of a low-noise multistate, narrow-band FET amplifier.

introduced in the computation of $F_{\min}$. To avoid this pitfall, it is advisable to use more than four measurements.

One such method, widely used [33] is based on seven chosen source impedances and seven noise figure and gain measurements. These seven values, and the corresponding noise figures, are used to obtain the best fit to the four unknown noise parameters (actually four others derived from these) in the minimum mean-square error sense. Naturally, more than seven source impedances may be used, but this lengthens the measurement procedure, and the point of diminishing returns is soon reached.

Noise Measurement Setup

A typical microwave setup to measure the gain and noise figure of FET's is shown in Fig. 17.

The FET is tuned with two coaxial double slug tuners which present very low loss (<0.2 dB) thus reducing the error correction in the noise figure measurement. The low-pass filter at the input eliminates errors in gain measurements due to harmonics; the narrow-band tunable filter (a high-Q cavity) at the output eliminates the image frequencies which would affect the noise measurement.

The gain is measured by a substitution method with a constant level maintained at the output of the crystal detector shown on the oscilloscope. The input level to the FET is adjusted with the attenuator A_1. With the FET and tuners removed, the attenuator A_2 is set at zero and a convenient level set on the oscilloscope. Then the FET is introduced in the circuit and A_2 adjusted to reestablish the original level. The gain is read directly on A_2. With this procedure one must be careful to adjust A_1 such that the saturation level of the low noise amplifier is never approached.

The low-noise amplifier is an essential part of the setup. It facilitates the tuning of the FET for minimum noise and reduces the postamplifier correction. If the noise figure of the postamplifier is F_2 and the noise figure and gain of the FET stage are F_1 and G, respectively, the measured noise figure at the input is given by

$$F = F_1 + \frac{F_2 - 1}{G}. \tag{10}$$

At high microwave frequencies, 10 GHz and higher, the gain G is generally low, less than 6 dB. If F_2 is large, the second term on the right of (10) can be comparable to F_1. In that case, by tuning the FET, one would more likely minimize F by maximizing G rather than minimizing F_1.

The noise figure is measured either with an automatic noise figure indicator such as AILTECH Model 75 or with a receiver and calibrated attenuator, by the so-called Y-factor method. The pad in front of the noise source is necessary only if the source impedance varies with its state (on or off). The attenuation must be taken into account in the noise calculations.

First, the noise figure F_2 of the postamplifier-mixer stages must be determined carefully. Then for a given bias condition of the FET, the output tuner is adjusted for maximum gain and the input tuner for minimum noise. The value of the noise figure measured is recorded together with its associated gain, and F_1 is calculated from (10). It can then be corrected if necessary for input circuit losses. The impedance seen by the FET input is measured (at the plane $X-X$ on Fig. 17) with a network analyzer. This series of measurements is repeated seven times with the input tuner adjusted for slightly different positions each time.

The data are then processed by a computer to obtain $F_{\min}$, g_n, and the optimum source impedance $Z_{g0} = R_{g0} + jX_{g0}$, as described earlier.

This procedure is long and tedious. It can be simplified if many measurements have to be made in the same frequency range. In that case, one can use a set of seven preadjusted tuners which are interchanged for each measurement.

IX. Design Considerations for Cascaded Amplifier

It was mentioned earlier that the lowest noise figure and the highest power gain do not occur at the same bias and tuning conditions. Since the gain at the minimum noise figure condition is not usually high enough to allow one to neglect the noise of the second and succeeding stages, one should not design the first stage of an FET amplifier to have its minimum noise figure if a minimum noise figure for the overall amplifier is to be achieved.

We shall illustrate why this is true with the help of Fig. 18. Shown is a block diagram of a cascaded stage amplifier, assumed to be narrow-band.[3] Since to each value of gain, G, there corresponds a noise figure, F, we have denoted this correspondence as $F(G)$. From the formula for the noise

---

[3]For wide-band amplifiers, other considerations enter besides noise figure.

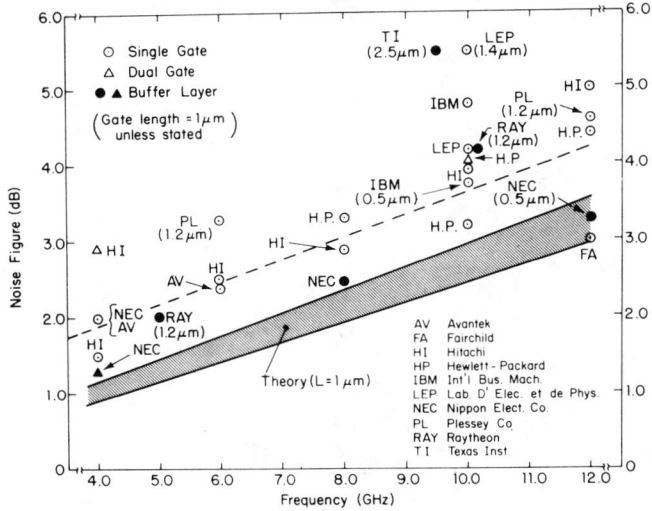

Fig. 19. Noise performance of GaAs FET's obtained at various laboratories (July 1975).

figure of cascaded stages, we find for the overall noise figure of the amplifier

$$F_A = F_1(G_1) + \frac{F_2(G_2) - 1}{G_1} + \frac{F_3(G_3) - 1}{G_1 G_2} +$$

$$\cdots + \frac{F_n(G_n) - 1}{G_1 G_2 G_3 \cdots G_{n-1}} \quad (11)$$

where the subscript denotes the amplifier stage number. For convenience of our discussion we shall assume identical stages $G_1, G_2, G_3, \cdots, G_n = G; F_1, F_2, F_3, \cdots, F_n = F$. Then (11) becomes

$$F_A = 1 + (F - 1)\left(1 + \frac{1}{G} + \frac{1}{G^2} + \cdots \frac{1}{G^{n-1}}\right) \quad (12a)$$

$$= 1 + M\left(1 - \frac{1}{G_A}\right) \quad (12b)$$

where $G_A = G^n$ is the power gain of the amplifier, and $M = M(G)$ is the noise measure of each stage,

$$M(G) = \frac{F(G) - 1}{1 - \frac{1}{G}}. \quad (13)$$

Equation (12b) is equivalent to the statement that the noise measure of n identical cascaded stages is equal to the noise measure of an individual stage [29].

It is evident that in cascade design, where the overall gain G_A is prescribed, one should minimize the noise measure rather than the noise figure of each stage to minimize the overall amplifier noise figure. When the overall gain is high, $F_A = 1 + M(G_A^{1/n})$. The sketches in Fig. 18 show qualitatively how the noise measure and noise figure vary as a function of stage gain. The minimum noise measure usually occurs at a slightly higher current and gain than the minimum noise figure. Also the value of the minimum noise measure exceeds the minimum value of the excess noise figure of a stage, i.e., $M_{min} > F_{min} - 1$. It follows that the lowest possible value of the amplifier noise figure is greater than the minimum noise figure of any individual stage, that is $F_{A, min} > F_{min}$. This, of course, is what one would expect. However, when the gain per stage is of the order of 6 dB or more at the bias condition corresponding to minimum stage noise figure, then the difference between $F_{A, min}$ and F_{min} is small. For example, for the 1 μm gate device described by Fig. 11, the lowest measured value of $F_{min} = 3.2$ dB [24]. The computed value of $M_{min} = 3.6$ dB. Hence for a three-stage amplifier, with ≈ 7.5 dB gain per stage, the amplifier noise figure $F_A = 3.63$ dB, only 0.43 dB greater than the single-stage minimum noise figure.

X. Summary of Noise Performance Obtained at Various Laboratories

We shall present now a compilation of the best noise performance reported by laboratories around the world as of the time of this writing (July 1975). First, the results for single-stage amplifiers (devices) will be given, then cascaded narrow-band amplifiers, and finally, wide-band amplifiers.

Fig. 19 is a graphical presentation of the device performance reported. All devices have 1 μm gates, except where noted. The circles refer to single-gate FET's, the triangles to dual-gate devices. Also shown (by the shaded strip) is the theoretical noise figure for a 1 μm gate for a spread of parasitic resistance values typical of present devices. Note that the buffered-layer device performance is within 0.5 dB of the theoretical.[4] Inclusion of circuit losses and corrections for neglected parasitics will reduce this gap.

Use of buffered layers not only improves the performance of single-gate devices, but also of dual-gate devices as the low noise figure for the NEC device at 4 GHz testifies.

The advantages of buffering are emphasized even more dramatically by the noise performance reported for cascaded narrow-band amplifiers (bandwidth < 20 percent) shown in

---

[4] A buffered-layer device is one that has an epitaxial growth of a high resistivity layer, of the order of 5-10 μm thick, over the substrate prior to channel epitaxial growth.

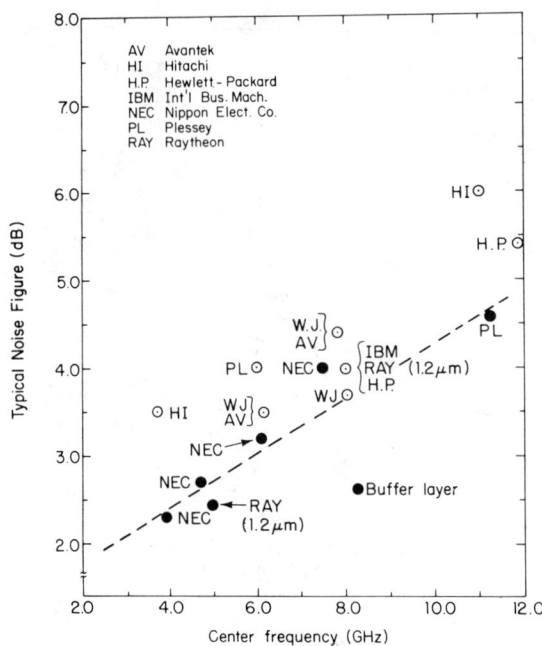

Fig. 20. Noise performance of narrow-band GaAs FET amplifiers as reported by various laboratories (July 1975).

Fig. 20. Notice in particular that the improvement is most pronounced in C-band and in the lower end of X-band.

Buffering appears to improve the noise performance in several ways. First, it covers or "shields" interface traps from the channel. (Present conjecture is that these traps, their nature unknown, may be ionized by the high channel fields and generate a noise spectra extending up to at least the low microwave band.) Second, with a buffer layer, the channel mobility near the substrate side increases substantially above the values with no buffer layer [34]. This not only increases the transconductance of the FET, but also decreases the source-gate parasitic resistance R_s. Reductions in R_s by nearly a factor of two are observed. These latter two improvements also lead to a higher power gain.

It seems reasonable to assume that the noise improvement in C-band and below can be attributed mainly to the reduction in trap noise. On the other hand, in the upper end of C-band and higher, where the trap noise would be expected to have diminished significantly, it is the increase in g_m and the reduction in R_s that is primarily responsible for the improvement in the noise performance. (Note the greater sensitivity of the noise figure to variation in parasitic resistance at the higher end of the frequency band exhibited by the theoretical shaded region in Fig. 19.)

It is interesting to note that the dashed lines through the experimental data in Figs. 19 and 20 both have approximately the same slope, namely 0.3–0.35 dB/GHz, as the theoretical lines in Fig. 19. However, the amplifier noise figures, on the average, exceed the single device values by approximately 0.5–0.6 dB.

Fig. 21 is a sampling of the noise performance reported for wide-band amplifiers. The upper and lower values of F_{min} in each case are not to be construed as the value of F_{min} at the band edges but merely the upper and lower values within the band. Since it is impossible to obtain a good noise match over a wide frequency range, the average noise figures are substantially higher than the narrow-band results.

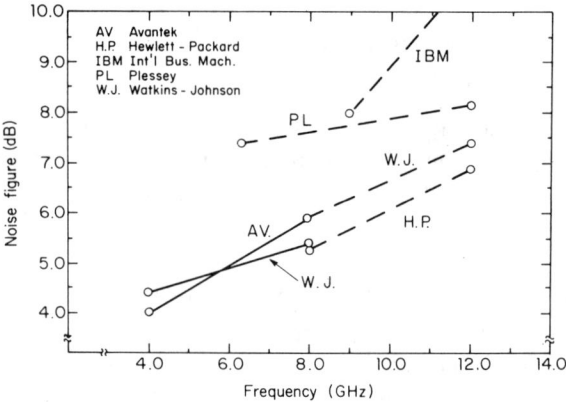

Fig. 21. Noise performance of some wide-band GaAs FET amplifiers (early 1975).

XI. Conclusions

The measured noise figures of GaAs FET's with buffer layers are approaching the theoretically predicted values based on presently realizable channel doping profiles. With some advances in the design and technology of contacts and the achievement of steeper slopes in the doping profile at the substrate-channel interface, still further improvements in the noise performance should be possible.

It is believed that with the present planar device configuration, gate length reductions substantially below a micron will reach a point of diminishing returns. The reasons are 1) fringing gate capacitance, 2) slower rate of increase in transconductance, 3) increased series resistance of the channel layer, and

4) the need for an extremely narrow doping transition zone at the channel-substrate interface.

ACKNOWLEDGMENT

The authors wish to express their appreciation to Dr. J. Thompson and S. R. Steele who supplied the excellent epitaxial material, to J. Curtis who took the measurements, and to R. W. Bierig for his constant encouragement.

The authors also wish to convey a special note of gratitude to the various laboratories and individuals who were willing to share their best noise figure results. Without these, the last three graphs would not have been possible. Sincere apologies are extended to those laboratories which were inadvertently omitted in the survey.

REFERENCES

[1] R. A. Pucel, D. Massé, and R. Bera, "Integrated GaAs FET mixer performance at X-band," *Electron. Lett.*, vol. 11, pp. 199–200, May 1975.
[2] R. A. Pucel, R. Bera, and D. Massé, "Experiments on integrated gallium-arsenide FET oscillators at X-band," *Electron. Lett.*, vol. 11, pp. 219–220, May 1975.
[3] M. Maeda, K. Kimura, and H. Kodera, "Design and performance of X-band oscillators with GaAs Schottky-gate field-effect transistors," *IEEE Trans. Microwave Theory Tech.*, vol. MTT-23, pp. 661–666, Aug. 1975.
[4] H. Statz, H. A. Haus, and R. A. Pucel, "Noise characteristics of gallium arsenide field-effect transistors," *IEEE Trans. Electron Devices*, vol. ED-21, pp. 549–562, Sept. 1974.
[5] R. A. Pucel, H. A. Haus, and H. Statz, "Signal and noise properties of gallium arsenide field-effect transistors," in *Advances in Electronics and Electron Physics*, vol. 38. New York: Academic, 1975.
[6] W. Shockley, "A unipolar 'field-effect' transistor," *Proc. IRE*, vol. 40, pp. 1365–1376, Nov. 1952.
[7] A. van der Ziel and J. W. Ero, "Small signal, high-frequency theory of field-effect transistors," *IEEE Trans. Electron Devices*, vol. ED-11, pp. 128–135, Apr. 1964.
[8] A. van der Ziel, "Thermal noise in field-effect transistors," *Proc. IRE*, vol. 50, pp. 1808–1812, Aug. 1962.
[9] ——, "Gate noise in field-effect transistors at moderately high frequencies," *Proc. IEEE*, vol. 51, pp. 461–467, Mar. 1963.
[10] J. A. Turner and B. L. H. Wilson, "Implications of carrier velocity saturation in a gallium arsenide field effect transistor," in *Proc. 2nd Intl. Symp. Gallium Arsenide*, 1968, pp. 195–204.
[11] J. G. Ruch and G. S. Kino, "Transport properties of gallium arsenide," *Phys. Rev.*, vol. 174, pp. 921–931, Oct. 1968.
[12] J. G. Ruch and W. Fawcett, "Temperature dependence of the transport properties of gallium arsenide determined by a Monte Carlo method," *J. Appl. Phys.*, vol. 41, pp. 3843–3849, Aug. 1970.
[13] A. B. Grebene and S. K. Ghandhi, "General theory for pinched operation of the junction gate FET," *Solid-State Electron.*, vol. 12, pp. 573–589, July 1969.
[14] J. G. Ruch, "Electron dynamics in short channel field-effect transistors," *IEEE Trans. Electron Devices*, vol. ED-19, pp. 652–654, May 1972.
[15] T. J. Maloney and J. Frey, "Effects of nonequilibrium velocity-field characteristics on the performance of GaAs and InP field-effect transistors," in *1974 Dig. Tech. Papers, Int. Electron Devices Meeting*, pp. 296–298.
[16] W. Baechtold, "Noise behavior of Schottky barrier gate field-effect transistors at microwave frequencies," *IEEE Trans. Electron Devices*, vol. ED-18, pp. 97–104, Feb. 1971.
[17] W. Baechtold, "Noise behavior of GaAs field-effect transistors with short gate lengths," *IEEE Trans. Electron Devices*, vol. ED-19, pp. 674–680, May 1972.
[18] W. Shockley, J. A. Copeland, and R. P. James, "The impedance field method of noise calculations in active semiconductor devices," in *Quantum Theory of Atoms, Molecules, and the Solid State*, P.-O. Löwdin, Ed. New York: Academic, 1966.
[19] A. van der Ziel, "Thermal noise in the hot electron regime in FET's," *IEEE Trans. Electron Devices* (Corresp.), vol. ED-18, p. 977, Oct. 1971.
[20] ——, "Equivalence of the noise figures of common source and common gate FET circuits," *Electron. Lett.*, vol. 5, pp. 161–162, Apr. 1969.
[21] R. D. Kässer, "Noise factor contours for field-effect transistors at moderately high frequencies," *IEEE Trans. Electron Devices*, vol. ED-19, pp. 164–171, Feb. 1972.
[22] H. Rothe and W. Dahlke, "Theory of noisy fourpoles," *Proc. IRE*, vol. 44, pp. 811–818, June 1956.
[23] G. E. Brehm, "Variation of microwave gain and noise figure with bias for GaAs FET's," in *Proc. 4th Biennial Cornell Elect. Eng. Conf.*, 1973, pp. 77–85.
[24] C. A. Liechti, "Performance of dual-gate GaAs MESFET's as gain-controlled low-noise amplifiers and high-speed modulators," *IEEE Trans. Microwave Theory and Tech.*, vol. MTT-23, pp. 461–469, June 1975.
[25] G. D. Vendelin, "Circuit model for the GaAs M.E.S.F.E.T. valid to 12 GHz," *Electron. Lett.*, vol. 11, pp. 60–61, Feb. 1975.
[26] C. A. Liechti and R. L. Tillman, "Design and performance of microwave amplifiers with GaAs Schottky-gate field-effect transistors," *IEEE Trans. Microwave Theory Tech.*, vol. MTT-22, pp. 510–517, May 1974.
[27] G. D. Vendelin, "Feedback effects on the noise performance of GaAs FET's," in *1975 Dig. Tech. Papers, IEEE S-MTT Int. Microwave Symp.*, pp. 324–326.
[28] P. Wolf, "Microwave properties of Schottky barrier field effect transistors," *IBM J. Res. Develop.*, vol. 14, pp. 125–141, Mar. 1970.
[29] H. A. Haus and R. B. Adler, *Circuit Theory of Linear Noisy Networks*. New York: Wiley, 1959.
[30] J. R. Hauser, "Characteristics of junction field effect devices with small channel length-to-width ratios," *Solid-State Electron.*, vol. 10, pp. 577–587, June 1967.
[31] T. L. Chiu and H. N. Ghosh, "Characteristics of the junction-gate field effect transistor with short channel length," *Solid-State Electron.*, vol. 14, pp. 1307–1317, Dec. 1971.
[32] C. P. Wu, E. C. Douglas, and C. W. Meuller, "Limitations of the CV technique for ion-implanted profiles," *IEEE Trans. Electron Devices*, vol. ED-22, pp. 319–329, June 1975.
[33] R. Q. Lane, "The determination of device noise parameters," *Proc. IEEE* (Lett.), vol. 57, pp. 1461–1462, Aug. 1969.
[34] T. Nozaki, M. Ogawa, H. Terao, and H. Watanabe, "Multi-layer epitaxial technology for the Schottky barrier GaAs field-effect transistor," in *Proc. 5th Intl. Symp. Gallium Arsenide and Related Compounds*, 1975, pp. 46–54.

Design of Microwave GaAs MESFET's for Broad-Band Low-Noise Amplifiers

HATSUAKI FUKUI, SENIOR MEMBER, IEEE

Abstract—As a basis for designing GaAs MESFET's for broad-band low-noise amplifiers, the fundamental relationships between basic device parameters, and two-port noise parameters are investigated in a semiempirical manner. A set of four noise parameters are shown as simple functions of equivalent circuit elements of a GaAs MESFET. Each element is then expressed in a simple analytical form with the geometrical and material parameters of this device. Thus practical expressions for the four noise parameters are developed in terms of the geometrical and material parameters.

Among the four noise parameters, the minimum noise figure $F_{\min}$, and equivalent noise resistance R_n, are considered crucial for broad-band low-noise amplifiers. A low R_n corresponds to less sensitivity to input mismatch, and can be obtained with a short heavily doped thin active channel. Such a high channel doping-to-thickness (N/a) ratio has a potential of producing high power gain, but is contradictory to obtaining a low $F_{\min}$. Therefore, a compromise in choosing N and a is necessary for best overall amplifier performance. Four numerical examples are given to show optimization processes.

I. Introduction

THE GaAs Schottky-barrier gate field effect transistors (GaAs MESFET's) have demonstrated excellent noise and gain performance at microwave frequencies through K band [1]. The excellent microwave performance of GaAs MESFET's is certainly related to their channel properties. GaAs MESFET's to be used for broad-band low-noise amplifier applications, must have special requirements on their channel properties for optimum performance. The purpose of this paper is to investigate the fundamental relationships between the noise and small-signal properties, and the basic channel parameters of GaAs MESFET's. This information should be useful as a basis for device design.

II. Representation of Noise Properties

A. Noise Parameters

From the circuit point of view, the GaAs MESFET can be treated as a black box of noisy two port. The noise properties of such a black box are then characterized by a set of four noise parameters in the binomial form [2]. A derivation of this form can be written as

$$F = F_{\min} + \frac{R_n}{R_{ss}}\left[\frac{(R_{ss} - R_{op})^2 + (X_{ss} - X_{op})^2}{R_{op}^2 + X_{op}^2}\right] \quad (1)$$

Manuscript received August 14, 1978; revised January 15, 1979.
The author is with Bell Laboratories, Murray Hill, NJ 07974.

where

- F noise figure,
- F_{min} minimum (or optimum) noise figure,
- R_n equivalent noise resistance,
- R_{ss} signal source resistance,
- R_{op} optimum signal source resistance,
- X_{ss} signal source reactance,
- X_{op} optimum signal source reactance.

In this expression, F_{min}, R_n, R_{op}, and X_{op} are the characteristic noise parameters of the device. Since the noise figure F is a function of its driving source impedance, the minimum noise figure F_{min} is achieved only when the driving source impedance is exactly at the optimum signal source impedance.

As has been well known, (1) can be represented on the source impedance Smith chart as a family of circles, each of which corresponds to a constant F value [3]. The spatial distance between two circles is related to R_n. The greater R_n corresponds to the shorter distance. In other words, the noise figure of a device with a small value of R_n is relatively insensitive to the variation in the signal source impedance. Thus small R_n is essential for a device to be used in a broad-band amplifier where a large tolerance is desirable in the input match. Furthermore, as will be seen later, R_n has a close relationship with power gain. Usually, the smaller R_n value corresponds to the higher gain in a given gate structure. In the design of a low-noise MESFET, therefore, obtaining a low R_n should be considered to be just as crucial as achieving a low F_{min}.

It may be noted that an equivalent expression of (1), in terms of the source-reflection coefficient, is found elsewhere [4]. This expression is based on the circuit analysis using the s parameters. The s-parameter representation would be convenient to use in conjunction with the test facilities available for two-port investigation these days. However, the impedance parameter would provide us with direct insight into the operation of a device under consideration. Therefore, the noise parameters in the impedance form are adopted in this paper. If necessary, the noise parameters of this form can be transformed into other forms in a straightforward manner.

B. Noise Equivalent Circuit

It has been well accepted that the noise properties of a GaAs MESFET would be described by an equivalent circuit as shown in Fig. 1 [5], if the effect of reactive parasitic elements on the noise performance could be ignored. The major reactive parasitic elements are lead inductances and header stray capacitances. As the operating frequency increases, the impedance of such external elements may become comparable to those of the corresponding internal elements, and then the reactive parasitic effects are no longer negligible. Such a critical frequency, that the reactive parasitic effect begins to participate in the determination of the noise performance parameters, may vary from one noise parameter to another for a given

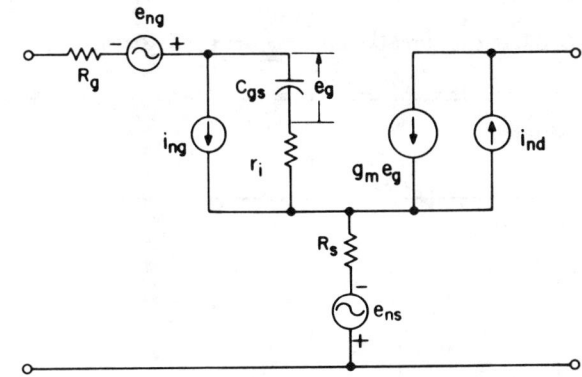

Fig. 1. Noise equivalent circuit of GaAs MESFET's. g_m is the transconductance which is assumed to be constant over the frequency range of interest. Element C_{gs} is the gate-source capacitance, r_i, the associated charging resistance, R_g, the ac gate metallization resistance, and R_s, the total source resistance series. Noise sources i_{ng}, i_{nd}, e_{ng}, and e_{ns} represent the induced gate noise, drain-circuit noise, thermal noise of R_g, and R_s, respectively.

device. For example, the critical frequency for F_{min} may be much higher than those for R_n, R_{op}, and X_{op}, analogous to microwave bipolar transistors [6].

III. Formulation of Noise Parameters in Terms of Device Geometrical and Material Parameters

A. Experimental Procedure and Results

Relationships between the noise parameters and equivalent circuit elements have been given in rigorous but complicated forms [5]. For practical purposes, however, it would be much more convenient if simple analytical forms of such relationships were available with reasonable accuracy.

In order to carry out this search, GaAs MESFET's with nominal gate length of 2 μm were used in the experiments described below. The MESFET's were mounted in the package which had a total input lead inductance of approximately 1 nH, a total output lead inductance of approximately 1 nH, a common-source lead inductance of approximately 0.12 nH, and a header stray capacitance of approximately 0.08 pF at both input and output [7]. As a result of such parasitic element values, these devices nearly satisfied the aforementioned requirement for possible elimination of the reactive parasitic effects on any of the four noise parameters at a test frequency of 1.8 GHz. Therefore, only the equivalent circuit elements shown in Fig. 1 will be referred to in the analyses described later on.

Table I shows the geometrical and material parameters of the six GaAs MESFET's used in the experiments. In Table I, N is the free carrier concentration in the active channel in units of 10^{16} cm^{-3}, L is the gate length in micrometers, a is the active channel thickness in micrometers, Z is the total device width in millimeters, and z is the unit gate width in millimeters. Since z was 0.25 mm for all devices, each device had either two or six paralleled unit

TABLE I
GEOMETRICAL AND MATERIAL PARAMETERS OF SAMPLE GaAs MESFET's

DEVICE	N (10^{16} cm^{-3})	L (μm)	a (μm)	Z (mm)	z (mm)
a : K864-1-02	11.0	1.85	0.174	0.5	0.25
b : K976-1-04	7.5	1.85	0.40	0.5	0.25
c : K949-1-12	5.5	1.85	0.44	0.5	0.25
d : K949-3-02	5.5	2.3	0.35	1.5	0.25
e : C75B-1-05	5.0	2.3	0.46	0.5	0.25
f : C75B-3-04	5.0	2.3	0.38	1.5	0.25

TABLE II
MEASURED VALUES OF NOISE PARAMETERS AND EQUIVALENT CIRCUIT ELEMENTS OF SAMPLE GaAs MESFET's

DEVICE	F_{min} (dB)	R_n (Ω)	R_{op} (Ω)	X_{op} (Ω)	g_m (℧)	C_{gs} (pF)	R_g (Ω)	R_s (Ω)
a	1.80	31	40	85	0.031	1.0	2	8.5
b	1.29	-	45	145	0.021	0.65	2	6
c	1.56	-	45	140	0.019	0.62	3	6
d	1.70	13.3	23	47	0.047	2.0	1	3.5
e	1.73	94	50	125	0.018	0.68	2	8
f	2.04	15.3	35	45	0.044	2.2	4	3.5

gates. These devices were an early version of the low-noise GaAs MESFET reported elsewhere [8]. An epitaxial film, as the active channel, was grown directly on a semi-insulating substrate in the first four devices. In the last two devices, however, there was an additional undoped buffer layer between the substrate and active layer.

Four noise parameters F_{min}, R_n, R_{op}, and X_{op} were measured at 1.8 GHz under optimum gate-bias condition for each device, using the standard technique [9]. Equivalent circuit elements g_m, C_{gs}, R_g, and R_s were evaluated from the s-parameter measurement taken as a function of frequency under the zero gate-bias condition, in a similar way to that described in [10]. All the above parameters were measured at room temperature. The results are shown in Table II.

B. Derivation of Expressions for Noise Parameters in Terms of Equivalent Circuit Elements

It was assumed that the four noise parameters could be expressed in terms of equivalent circuit elements as follows:

$$F_{min} = 1 + k_1 f C_{gs} \sqrt{\frac{R_g + R_s}{g_m}} \quad (2)$$

$$R_n = \frac{k_2}{g_m^2} \quad (3)$$

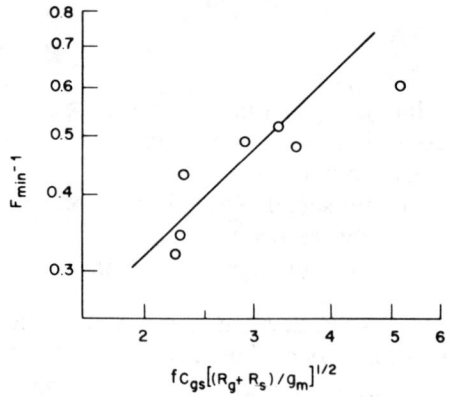

Fig. 2. Correlation between the minimum noise figure F_{min} and equivalent circuit elements, C_{gs}, g_m, R_g, and R_s.

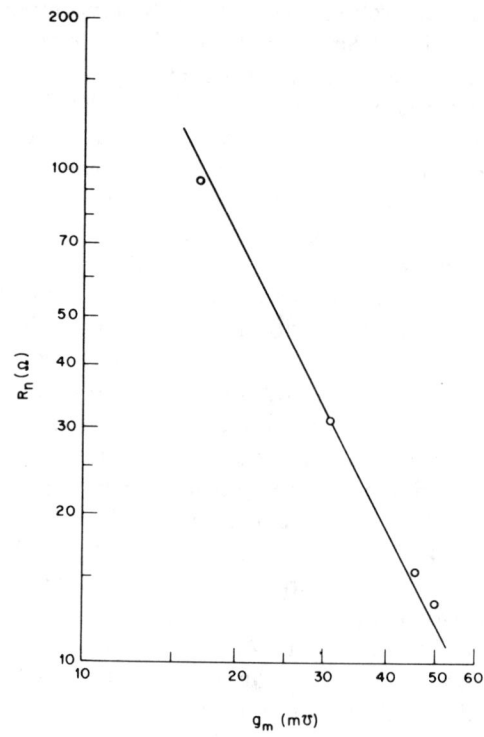

Fig. 3. Correlation between the equivalent noise resistance R_n and transconductance g_m.

$$R_{op} = k_3 \left[\frac{1}{4 g_m} + R_g + R_s \right] \quad (4)$$

$$X_{op} = \frac{k_4}{f C_{gs}} \quad (5)$$

where k_1, k_2, k_3, and k_4 are fitting factors, and f is frequency.

Comparing these expressions with the experimental data shown in Table II would yield determination of the fitting factors. As seen in Figs. 2–5, it has been found that the expressions would well represent the measured values of the noise parameters if the fitting factors were chosen as follows:

$$k_1 = 0.016$$
$$k_2 = 0.030$$

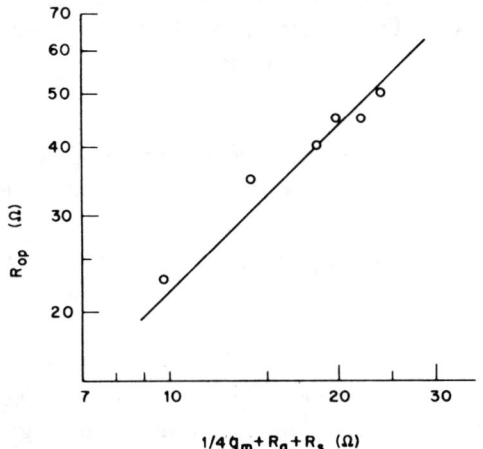

Fig. 4. Correlation between the optimum source resistance R_{op} and equivalent circuit elements g_m, R_g, and R_s.

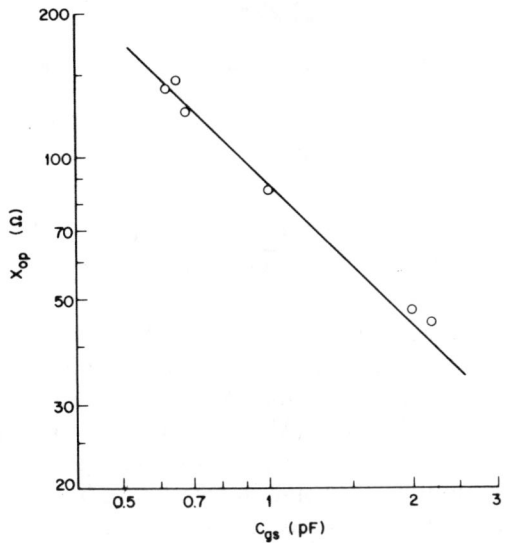

Fig. 5. Correlation between the optimum source reactance X_{op} and gate-source capacitance C_{gs}.

$$k_3 = 2.2$$
$$k_4 = 160$$

provided that R_n, R_{op}, X_{op}, R_g, and R_s are all in ohms, g_m in mhos, C_{gs} in picofarads, and f in gigahertz.

It may be remarked that the equivalent circuit elements as obtained with null gate bias are used in the above expressions for the noise parameters, although the noise parameters are provided with a certain gate bias. In spite of such a difference in the gate-bias conditions, the above relationships were empirically found to be present. If the equivalent circuit elements were measured under a gate-bias condition different from the null gate bias, the fitting factors would have to be modified.

A deviation of X_{op}, with increasing C_{gs}, from the linear relationship as seen in Fig. 5 was probably caused by an increased participation of the input lead inductance in X_{op}.

C. Semiempirical Expressions for Transconductance, Gate-Source Capacitance, and Cutoff Frequency

For design purposes, g_m and C_{gs} must be expressed in terms of the geometrical and material parameters of a device. Approximate expressions were then derived on a semiempirical basis as follows:

$$g_m = k_5 Z \left[\frac{N}{aL}\right]^{1/3} \mho \quad (6)$$

$$C_{gs} = k_6 Z \left[\frac{NL^2}{a}\right]^{1/3} \text{pF} \quad (7)$$

$$f_T = \frac{10^3 g_m}{2\pi C_{gs}} = \frac{9.4}{L} \text{ GHz} \quad (8)$$

in which fitting factors k_5 and k_6 were found to be 0.020 and 0.34, respectively, for the sample devices under the zero gate-bias condition.

Figs. 6 and 7 show comparisons of g_m and C_{gs}, respectively, between the calculated values using L, N, and a given in Table I and the measured values shown in Table II. They are in good agreement in both cases.

D. Simplified Expressions for Parasitic Resistances

Simplified expressions for R_g and R_s of the sample MESFET's would be given by [11]

$$R_g = \frac{17z^2}{hLZ} \, \Omega \quad (9)$$

$$R_s = \frac{1}{Z}\left[\frac{2.1}{a^{0.5}N^{0.66}} + \frac{1.1 L_{sg}}{(a-a_s)N^{0.82}}\right] \Omega \quad (10)$$

where h is the gate metallization height in micrometers, L_{sg} is the distance between the source and gate electrodes in micrometers, and a_s is the depletion layer thickness in micrometers at the surface in the source-gate space.

E. Practical Expressions for Noise Parameters

The substitution of (6)–(10) into (2)–(5) in association with the practical values of the fitting factors yields

$$F_{\min} = 1 + 0.038 f \left[\frac{NL^5}{a}\right]^{1/6}$$
$$\cdot \left[\frac{17z^2}{hL} + \frac{2.1}{a^{0.5}N^{0.66}} + \frac{1.1L_{sg}}{(a-a_s)N^{0.82}}\right]^{1/2} \quad (11)$$

$$R_n = 75 Z^{-2} \left[\frac{aL}{N}\right]^{2/3} \Omega \quad (12)$$

$$R_{op} = 2.2 Z^{-1} \left[12.5\left(\frac{aL}{N}\right)^{1/3}\right.$$
$$\left. + \frac{17z^2}{hL} + \frac{2.1}{a^{0.5}N^{0.66}} + \frac{1.1L_{sg}}{(a-a_s)N^{0.82}}\right] \Omega \quad (13)$$

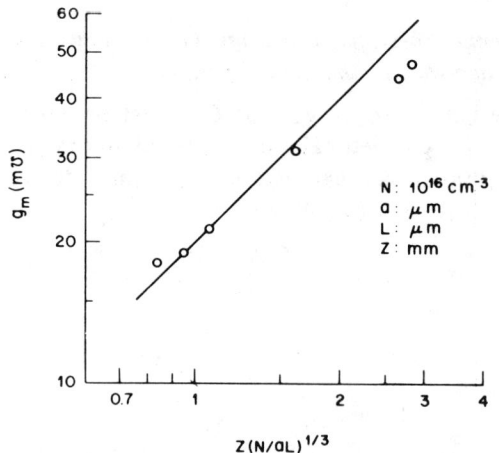

Fig. 6. Transconductance g_m as a function of channel parameters Z, L, a, and N.

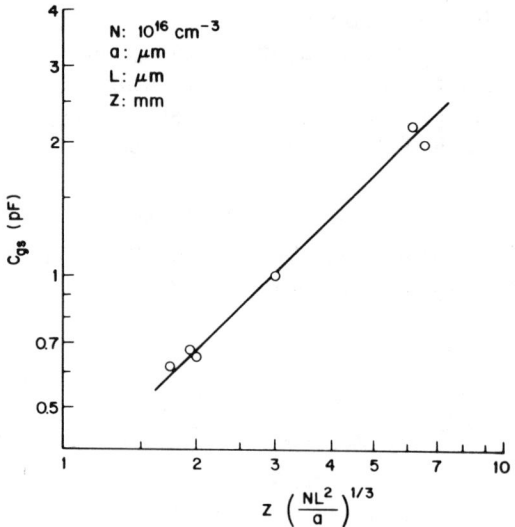

Fig. 7. Gate-source capacitance C_{gs} as a function of channel parameters Z, L, a, and N.

$$X_{op} = \frac{450}{fZ}\left[\frac{a}{NL^2}\right]^{1/3} \Omega. \quad (14)$$

Again, units are gigahertz (GHz) for f, millimeter (mm) for z and Z, micrometer (μm) for a, a_s, h, L, and L_{sg}, and 10^{16} cm^{-3} for N.

Since $F_{\min}$ is dominated by the parasitic resistances outside the gate region, $F_{\min}$ is structure sensitive [12]. Remember that (11) is suitable for the simplest structure of MESFET's. As sophistication increases in the structure, the proper expression for $F_{\min}$ may be obtained after the corresponding modification primarily to (10), and hence to (11). Applications of the gate recess structure [8], [12] and n$^+$-GaAs epitaxial layer [13] are the major examples of structural variations so far reported.

In (12), it is seen that *a small R_n value can be obtained with a short gate device having a heavily doped thin active channel.* The expression for R_n given in (12) may hold, regardless of the structural modification applied to a section of the channel outside the gate region.

IV. Discussions on Minimum Noise Figure, Associated Power Gain, and Input Impedance Match

A. Example 1

In order to see the characteristic variations of $F_{\min}$ and R_n as functions of N and a, the following conditions are assumed:

$$L = L_{sg} = 1.0 \; \mu\text{m}$$
$$R_g = 4.0 \; \Omega$$
$$f = 0.4 f_T = 3.76 \; \text{GHz}.$$

Furthermore, a_s is assumed to be approximately equal to the gate depletion layer thickness under null gate-bias condition a_0 in numerical value. This parameter has been given by [11]

$$a_0 = \left[\frac{0.706 + 0.06 \log N}{7.23 N}\right]^{1/2} \mu\text{m}$$

for aluminum gates, as far as $a > a_0$. Under such conditions, (11) reduces to

$$F_{\min}|_{f=0.4f_T} = 1 + 0.15[N/a]^{1/6}$$
$$\cdot \left[1.82 + \frac{1.9}{a^{0.5}N^{0.66}} + \frac{1}{(a-a_0)N^{0.82}}\right]^{1/2}. \quad (15)$$

The calculated values of $F_{\min}$ by (15) and of R_n by (12) are shown in Fig. 8, both as the contour mapping on the a-N plane. It can be seen that $F_{\min}$ is a weak function of N for a given value of a. Also, $F_{\min}$ takes a low value in the region where the a/N ratio is high. This is the contradictory condition to obtaining the low value of R_n. The high a/N ratio tends to provide not only the critical noise tuning but also the low power gain. Therefore, there is a compromise in choosing a and N for the best overall performance as an amplifying device.

In Fig. 8, there are two other curves, as shown with the dash–dotted line, which indicate practical limits for the selection of a and N. The upper curve corresponds to the a and N values which provide a drain–source breakdown voltage V_B, of 10 V if the gate is biased at the pinchoff voltage, $-V_p$. The lower curve indicates the thickness of the gate depletion layer at zero bias for aluminum gates.

B. Example 2

In order to see a clear distinction between the low a/N device and high a/N device, the following two representatives were assumed under the same conditions, as used in the previous section.

Device A: $N = 12.5 \times 10^{16}$ cm^{-3} and $a = 0.2 \; \mu$m.
Device B: $N = 5.0 \times 10^{16}$ cm^{-3} and $a = 0.5 \; \mu$m.

The noise performance of these devices was calculated using (12)–(15). The results are then plotted on the signal

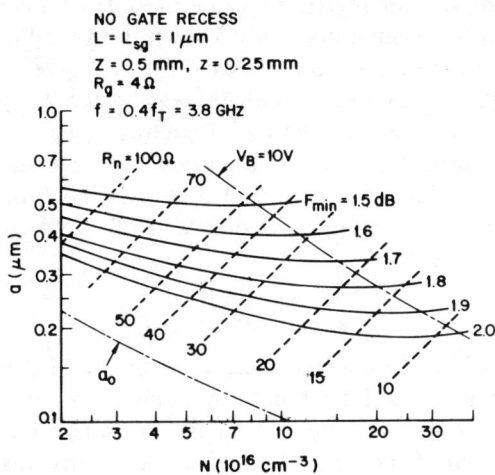

Fig. 8. Contours of equi-F_{min} and equi-R_n on the a-N plane for 1-μm gate devices.

$L = L_{sg} = 1 \mu m$	DEVICE	$N(10^{16} cm^{-3})$	$a(\mu m)$
$Z = 0.5$ mm, $z = 0.25$ mm	---- A	12.5	0.2
$R_g = 4 \Omega$, $f = 3.8$ GHz	——— B	5.0	0.5
$Z_0 = 50 \Omega$			

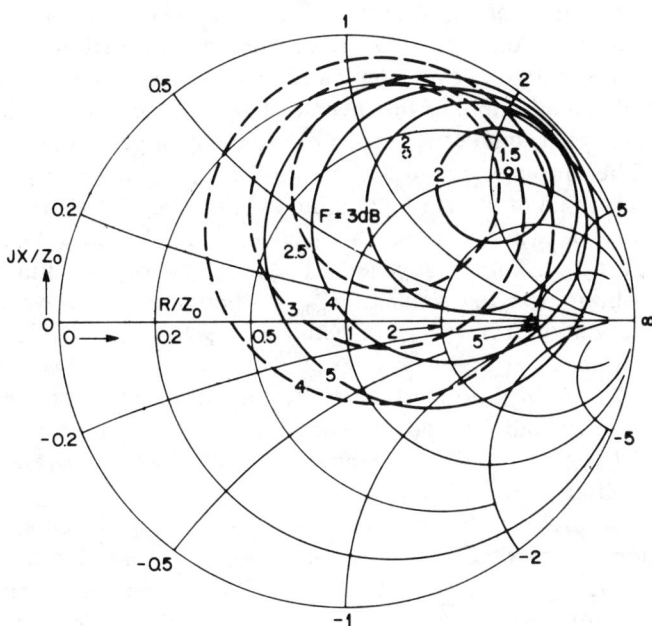

Fig. 9. The noise performance chart for two 1-μm gate devices.

source impedance chart [3], as shown in Fig. 9. It may be concluded that device B may reach the better noise performance in the case of a narrow-band application with the individually designed circuitry. However, device A may be more suitable in such a case that relatively large tolerances for the impedance variation are demanded. In addition, device A usually gives gain higher than that of device B, because of the higher value of g_m with device A.

C. Example 3

As seen in Fig. 9, device B, in spite of the smaller value of F_{min}, would exhibit much worse noise figure than

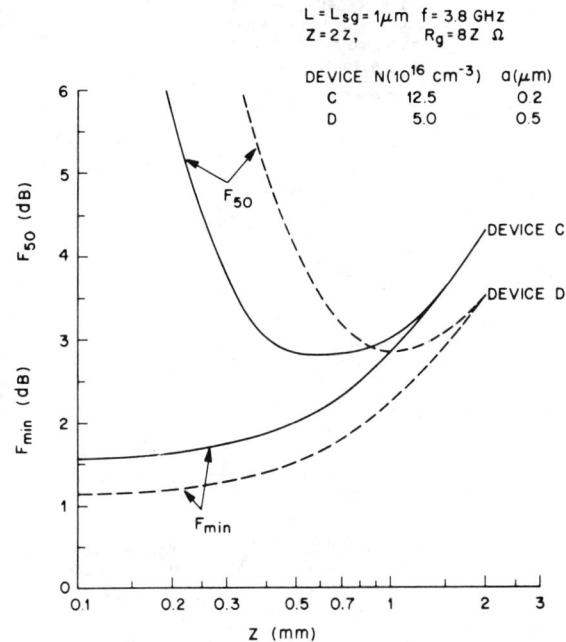

Fig. 10. Noise figures F_{min} and F_{50} as functions of total device width Z for two 1-μm gate devices.

device A when they are simply inserted into a 50-Ω coaxial system. Such a 50-Ω system insertion noise figure can be designated as F_{50}, which is

$$F_{50} = F_{min} + \frac{R_n}{50} \left[\frac{(50 - R_{op})^2 + X_{op}^2}{R_{op}^2 + X_{op}^2} \right]. \quad (16)$$

Next shown are the dependence of F_{min} and F_{50} on Z, provided that $Z = 2z$. Two representative devices C and D were chosen, which were the same as devices A and B, respectively, except for an additional assumption that $R_g = 8Z$, in which R_g is in units of ohms and Z is in millimeters. Parameters F_{min} and F_{50} were then calculated as functions of Z, using (11)–(14) and (16). The results are shown in Fig. 10, in which F_{min} gradually increases with increasing Z for both devices. On the contrary, F_{50} varies as a strong function of Z and takes a minimum for moderate values of Z, as also shown in Fig. 10. For the overall performance as a general purpose low-noise MESFET with two paralleled unit-gates 1 μm long, an optimum value of the total gate width could be evaluated to be 0.4–0.5 mm for device C, and 0.8–1.0 mm for device D.

D. Example 4

In [14], the noise and gain performance as measured at 8 GHz for GaAs MESFET's with 0.5-μm-long gates was described. The device consists of four sections of 70-μm-wide units, i.e., $z = 0.07$ mm and $Z = 0.28$ mm. The height of aluminum as the gate metallization was 0.5 μm. The source-to-gate distance was 0.8 μm. The active layer thickness was claimed to be between 0.1 and 0.2 μm. The gate pinchoff voltage V_p of typical devices was 2.0 V. Since the representative value of N was 25×10^{16} cm^{-3}, the corre-

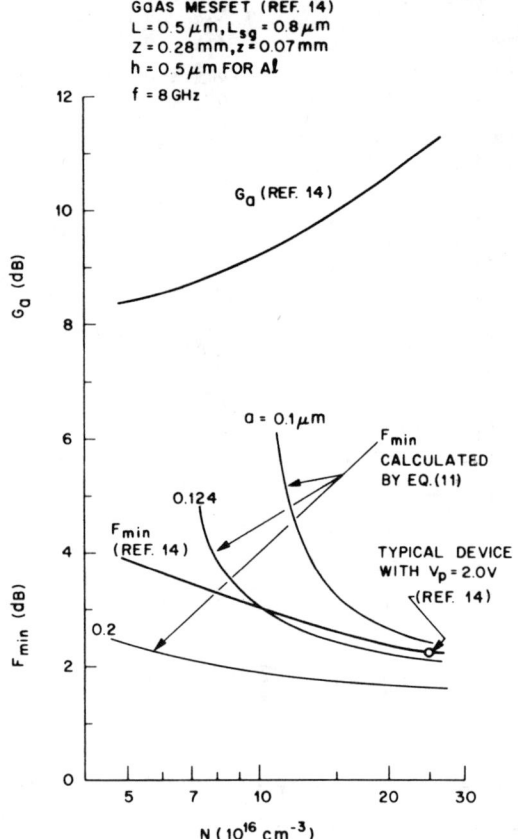

Fig. 11. Calculated values of $F_{\min}$ and measured values of $F_{\min}$ and associated power gain G_a for 0.5-μm gate devices.

sponding value of a was evaluated to be 0.124 μm. Thus $F_{\min}$ at 8 GHz was calculated using (11) as a function of N for the three values of a, i.e., 0.1, 0.124, and 0.2 μm.

As shown in Fig. 11, the measured $F_{\min}$ values happened to be between the two calculated curves of $F_{\min}$ for $a=0.1$ and 0.2 μm. Moreover, the measured $F_{\min}$ for $N \geqslant 10 \times 10^{16}$ cm^{-3} agreed well with the calculated $F_{\min}$ for $a=0.124$ μm. However, the measured $F_{\min}$ for $N < 10 \times 10^{16}$ cm^{-3} appeared to be better than the calculated $F_{\min}$ for $a=0.124$ μm. This discrepancy could have been caused by an increased active layer thickness of the actual devices with $N < 10 \times 10^{16}$ cm^{-3}, in order to maintain the zero gate-bias drain current to be finite in the positive direction. Such an increase in a should result in an improvement of $F_{\min}$ from that predicted for $a=0.124$ μm, as seen in Fig. 11. Therefore, it can be concluded that the experimental data on the noise performance, [14, fig. 2], is well explained by (11) of this paper.

In the same reference figure as mentioned above the associated power gain G_a, was also shown as a function of N. The data are replotted in Fig. 11 in which G_a increases with increasing N and hence N/a, as has been predicted in the previous sections of this paper.

E. Remarks

In general, with increasing N/a, G_a increases and R_n decreases. If R_g is small enough as compared to R_s, $F_{\min}$ is improved with an increased value of N for a given value of a. This has been empirically known in the industry [15]. The reason is that R_s dominates $F_{\min}$ for a given value of L, and that R_s decreases with increasing the Na product. In the simple planar channel structure used here, $F_{\min}$ increases with decreasing a for a given N. However, in a sophisticated channel structure, this may no longer hold partly due to a possible decrease in the effective gate length [11], [13].

V. Conclusions

A set of four noise parameters for GaAs MESFET's of the planar channel structure were semiempirically found in terms of simple functions of its equivalent circuit elements. Each of the equivalent circuit elements was further expressed by the device geometrical and material parameters in a simple analytical form. Combining these two things together yielded the practical expressions for the four noise parameters in terms of the device geometrical and material parameters.

Among the four noise parameters, $F_{\min}$ and R_n were regarded as most crucial for a device to be used in a broad-band low-noise amplifier. Because a device with a small value of R_n behaves rather insensitively to the variation in the signal source impedance, the variation in the noise figure over the band can be expected to be small. In addition, as the inverse of R_n is closely related to power gain, smaller R_n can enhance the gain performance.

The aforementioned expression for R_n indicates that a small value of R_n can be obtained with a short gate device having a heavily doped thin active channel (i.e., a high N/a ratio). This is, in general, the contradictory condition to obtaining a low value of $F_{\min}$. Although $F_{\min}$ is a weak function of N for a given value of a, $F_{\min}$ takes a low value in the region where the N/a ratio is low. However, the low N/a ratio tends to make the noise tuning critical and to degrade the power gain. Therefore, a compromise in choosing a and N is necessary for the best overall amplifier performance.

The proper selection of a and N may vary depending upon the particular purpose of an amplifier. Four examples were given to show the dependence of the noise, gain and input matching properties on a and N in a practical manner. If the gate metallization resistance were designed to be small enough to the total source series resistance, the minimum noise figure, associated power gain, and input matching sensitivity would all be improved with increasing N for a given a. This is a case which has been often observed in the industry.

Acknowledgment

The author is grateful to D. E. Iglesias and W. O. Schlosser for their help in measuring the noise properties and s parameters of the sample devices. He is also thankful to J. V. DiLorenzo, R. H. Knerr, R. Trambarulo, and H. Wang for their careful reading of the original manuscript.

References

[1] H. F. Cooke, "Microwave FET's—A status report," in *IEEE ISSCC Dig. Tech. Papers*, 1978, pp. 116–117.
[2] H. Rothe and W. Dahlke, "Theory of noisy fourpoles," *Proc. IRE*, vol. 44, pp. 811–818, June 1956.
[3] H. Fukui, "Available power gain, noise figure, and noise measure of twoports and their graphical representation," *IEEE Trans. Circuit Theory*, vol. CT-13, pp. 137–142, June 1966.
[4] J. A. Eisenberg, "Systematic design of low-noise, broad band microwave amplifiers using three terminal devices," in *Microwave Semiconductor Devices, Circuits and Applications, Proc. Fourth Cornell Conf.*, 1973, pp. 113–122.
[5] R. A. Pucel, H. A. Haus, and H. Statz, "Signal and noise properties of gallium arsenide microwave field-effect transistors," in *Advances in Electronics and Electron Physics*. New York: Academic, vol. 38, 1975, pp. 195–265.
[6] H. Fukui, "The noise performance of microwave transistors," *IEEE Trans. Electron Devices*, vol. ED-13, pp. 329–341, Mar. 1966.
[7] W. O. Schlosser, private communication.
[8] B. S. Hewitt et al., "Low-noise GaAs M.E.S.F.E.T.S.," *Electron. Lett.*, vol. 12, pp. 309–310, June 10, 1976.
[9] "IRE Standards on Electron Tubes: Methods of Testing," 1962, 62 IRE 7 S1, pt. 9: Noise in linear twoports.
[10] J. Jahncke, "Höchstfrequenzeigenshaften eines GaAs MESFET's in Steifenleitungstechnik," *Nachrichtentech. Z.*, vol. 5, pp. 193–199, May 1973.
[11] H. Fukui, "Determination of the basic device parameters of a GaAs MESFET," *Bell Syst. Tech. J.*, vol. 58, pp. 771–797, Mar. 1979.
[12] B. S. Hewitt et al., "Low-noise GaAs MESFET's: Fabrication and performance," in *Gallium Arsenide and Related Compounds (Edinburg) 1976, Conf. Series No. 33a*, The Inst. Physics, Bristol and London, 1977, pp. 246–254.
[13] H. Fukui, "Optimal noise figure of microwave GaAs MESFET's," *IEEE Trans. Electron Devices*, vol. ED-26, pp. 1032–1037, July 1979.
[14] M. Ogawa, K. Ohata, T. Furutsuka, and N. Kawamura, "Submicron single-gate and dual-gate GaAs MESFET's with improved low noise and high gain performance," *IEEE Trans. Microwave Theory Tech.*, vol. MTT-24, pp. 300–306, June 1976.
[15] H. F. Cooke, "Microwave field effect transistors in 1978," *Microwave J.*, vol. 21, no. 4, pp. 43–48, Apr. 1978.

Addendum to "Design of Microwave GaAs MESFET's for Broad-Band Low-Noise Amplifiers"

HATSUAKI FUKUI

It has been called to the author's attention that (3) in the above paper[1] appears to be inadequate [1], especially for scaling [2]. Considering this situation the expression should read

$$R_n = \frac{k_2}{g_m} \qquad (3)$$

where $k_2 = 0.8$.

This modification leads to rewriting (12) as follows:

$$R_n = \frac{40}{Z}\left[\frac{aL}{N}\right]^{1/3} \Omega. \qquad (12)$$

Consequently, the numerical values for R_n in Fig. 8 should be, in descending order, 46, 39, 33, 29, 25, 21, 18, and 15. Figs. 9 and 10 are also slightly affected by the revised expression. However, the principal statement and conclusions remain unchanged.

The author wishes to thank Dr. R. A. Pucel for his encouragement concerning this amendment.

References

[1] S. Weinreb, "Low-noise cooled GASFET amplifiers," *IEEE Trans. Microwave Theory Tech.*, vol. MTT-28, pp. 1041–1054, Oct. 1980.
[2] A. F. Podell, "A functional GaAs FET noise model," *IEEE Trans. Electron Devices*, vol. ED-28, pp. 511–517, May 1981.

Manuscript received June 3, 1981.
The author is with Bell Laboratories, Murray Hill, NJ 07974.
[1]Hatsuaki Fukui, *IEEE Trans. Microwave Theory Tech.*, vol. MTT-27, pp. 643–650, July 1979.

Optimal Noise Figure of Microwave GaAs MESFET's

HATSUAKI FUKUI, SENIOR MEMBER, IEEE

Abstract—The optimal value of the minimum noise figure F_O of GaAs MESFET's is expressed in terms of either representative equivalent circuit elements or geometrical and material parameters in simple analytical forms. These expressions are derived on a semiempirical basis. The predicted values of F_O for sample GaAs MESFET's using these expressions are in good agreement with the measured values at microwave frequencies.

The expressions are then applied to show design optimization for low-noise devices. This exercise indicates that shortening the gate length and minimizing the parasitic gate and source resistances are essential to lower F_O. Moreover, a simple shortening of the gate length may not bring an improved F_O unless the unit gate width is accordingly narrowed. The maximum value of the unit gate width is defined as the width above which the gate metallization resistance becomes greater than the source series resistance.

Short-gate GaAs MESFET's with optimized designs promise a superior noise performance at microwave frequencies through K band. The predicted values of F_O at 20 GHz, for example, for a half-micrometer gate device and a quarter-micrometer gate device are 3 and 2 dB, respectively. These devices could be fabricated with the current technology.

I. INTRODUCTION

IT IS WELL KNOWN that the noise performance of linear two-ports is characterized by four noise parameters, i.e., the minimum noise figure, equivalent noise resistance, optimum source resistance, and optimum source reactance [1]. An active device as a two-port usually consists of the intrinsic part and surrounding parasitics. Each of the noise parameters of this device can be expressed in terms of the intrinsic device parameters and extrinsic parasitic parameters. The parasitic parameters are then classified into two categories; resistive and reactive. In microwave transistors, the minimum noise figure is relatively insensitive to the reactive parasitics while the other noise parameters are strongly affected by them [2]. As the operating frequencies increase, the three noise parameters become increasingly influenced by the reactive parasitics, such as lead inductances and package capacitances. In contrast, the frequency characteristic of the minimum noise figure is reasonably well described regardless of the reactive parasitics.

The purpose of this paper is to present the practical forms of expressions for the minimum noise figure of GaAs MESFET's for a quick estimate of their noise performance. The study was carried out on a simplified model without reactive parasitics, as supported by experimental evidence. The optimal value of the minimum noise figure will be represented by either equivalent circuit elements or a set of geometrical and material parameters of a device.

Manuscript received May 25, 1978; revised February 5, 1979.
The author is with Bell Laboratories, Murray Hill, NJ 07974.

II. MINIMUM NOISE FIGURE EXPRESSED IN TERMS OF DEVICE EQUIVALENT CIRCUIT ELEMENTS

Essentially, the GaAs MESFET is a low-noise device because only the majority carriers participate in its operation. However, in practical GaAs MESFET's, extrinsic resistances are unavoidably added to the intrinsic part. These parasitic resistances then dominate the noise property of the devices.

The noise study of field-effect transistors was initiated by van der Ziel [3], [4], based on Shockley's gradual-channel approximation [5]. Later the noise behavior of GaAs MESFET's in the common-source configuration has been investigated by a number of workers, as summarized by Liechti in his excellent review article [6]. In this article, an expression for the minimum noise figure of the *intrinsic* GaAs MESFET is given by

$$F_{\min} = 1 + 2\sqrt{PR(1-C^2)}\left[\frac{f}{f_T}\right] + 2g_m R_i P(1 - C\sqrt{P/R})\left[\frac{f}{f_T}\right]^2 \quad (1)$$

where

- $f_T = g_m \times 10^3/2\pi C_{gs}$ cutoff frequency, in GHz;
- f operating frequency, in GHz;
- g_m transconductance, in ℧;
- C_{gs} gate-source capacitance, in pF;
- R_i gate-source internal channel resistance, in Ω;
- P drain noise coefficient;
- R gate noise coefficient;
- C noise correlation coefficient.

For actual GaAs MESFET's, however, this expression is no longer effective. With an exhaustive treatment using an equivalent circuit shown in Fig. 1, Pucel, Haus, and Statz [7] obtained an expression for the minimum noise figure of a real device as follows:

$$F_{\min} = 1 + 2(2\pi f C_{gs}/g_m \times 10^3)\sqrt{K_g[K_r + g_m(R_g + R_s)]} + 2(2\pi f C_{gs}/g_m \times 10^3)^2 [K_g g_m(R_g + R_s + K_c R_i)] + \cdots \quad (2)$$

where

$$K_g = P[(1 - C\sqrt{R/P})^2 + (1 - C^2)R/P]$$

$$K_r = \frac{R(1-C^2)}{(1 - C\sqrt{R/P})^2 + (1 - C^2)R/P}$$

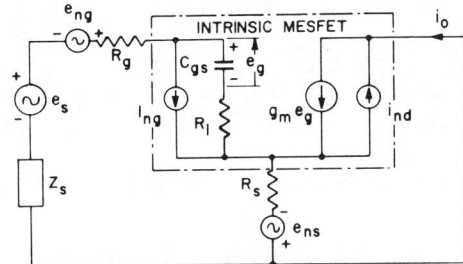

Fig. 1. Equivalent circuit of a GaAs MESFET used for noise analysis by Pucel, Haus, and Statz [7]. Noise sources i_{ng}, i_{nd}, e_{ng}, and e_{ns} represent the induced gate noise, drain circuit noise, thermal noise of the gate metallization resistance R_g, and thermal noise of the source series resistance R_s, respectively. Parameters e_s and Z_s represent the signal source voltage and source impedance. Note that (1) in the text corresponds to the intrinsic MESFET, which is shown within the semibroken lines.

$$K_c = \frac{1 - C\sqrt{R/P}}{(1 - C\sqrt{R/P})^2 + (1 - C^2)R/P}$$

R_g is the ac gate series resistance in ohms and R_s is the source series resistance in ohms.

In (2), R_g and R_s are only parasitic and remain unchanged in the normal operating range of conditions. The other parameters are all intrinsic and vary with the bias conditions in a complicated manner. As a result, the minimum noise figure also changes with the gate and drain bias voltages at a given frequency. The optimal value of the minimum noise figure is usually obtained with a relatively deep gate bias and a relatively low drain bias voltage in the current saturation region.

III. Optimal Noise Figure of the GaAs MESFET

A. Optimal Noise Figure Expressed in Terms of Device Equivalent Circuit Elements

The GaAs MESFET is assumed to operate in the microwave region below the cutoff frequency at room temperature here. The optimal value of the minimum noise figure is designated as F_o. A simple expression for F_o has been empirically found in terms of the equivalent circuit elements which were evaluated at null gate bias and operating drain bias (usually in the vicinity of 5 V). This expression can be written as

$$F_o = 1 + 2\pi K_f f C_{gs} \sqrt{\frac{R_g + R_s}{g_m}} \times 10^{-3} \tag{3}$$

where K_f is a fitting factor of approximately 2.5, representing the quality of channel materials. It is an empirical finding that (3) holds good regardless of gate bias required for F_o. Note that (3) is, in appearance, equivalent to a limiting case of $R = 0$ and/or $C = 1$ in (2), neglecting higher order terms. While the parameter values in (2) vary with bias conditions, those in (3) are fixed as obtained in the aforementioned manner. Therefore, (3) may be regarded as a special case of (2).

In order to examine the validity of (3), the published data [8] on the equivalent circuit elements of a GaAs MESFET with a gate length of 0.9 μm and a pinchoff voltage of ~3.8 V were used as an example. The elements were given as C_{gs} = 0.50 pF, g_m = 0.033 ℧, R_g = 3.0 Ω, and R_s = 5.5 Ω, as measured at null gate bias. Substituting these values into (3) with K_f = 2.5 yields a predicted F_o of 3.54 dB at 10 GHz. This value agrees well with a measured F_o of 3.5 dB as obtained with a gate bias of -1.7 V at such a frequency close to an f_T of 10.5 GHz.

Since $f_T = g_m \times 10^3/2\pi C_{gs}$ by definition, (3) can be rewritten as

$$F_o = 1 + K_f \frac{f}{f_T} \sqrt{g_m(R_g + R_s)}. \tag{4}$$

Furthermore, f_T is related to the gate length L so that

$$F_o = 1 + K_l L f \sqrt{g_m(R_g + R_s)} \tag{5}$$

where K_l is another fitting factor of approximately 0.27, provided that L is in micrometers. Note that g_m is also related to L, as shown in (6) later.

In order to examine the adequacy of (5), five GaAs MESFET's were sampled and used for experiments. These devices were fabricated in a similar manner to those described in [9]. The sample devices had an optimized gate recess structure formed by proper fabrication techniques. As a result, the effective value of gate length was made substantially smaller than the gate metallization length, as will be seen in Table II. It was also found that this effective gate length coincided with the gate length value as obtained by means of the determination method described in [10]. Therefore, L should be regarded as the effective gate length in general. However, this would reduce to the physical length of the gate metal in the case of a plain gate on a planar channel. A further description concerning the effective gate length is given in Appendix II.

The optimal noise figure of the sample devices were measured at 5.92 GHz which was approximately one-third of f_T of these devices. The other parameters, L, g_m, R_g, and R_s, were independently determined by using the technique as described in [10]. The measured values of these parameters are shown in Table I, as well as the predicted value of F_o calculated by (5). The measured and predicted values of F_o are in excellent agreement. In conjunction with the previous examination of (3), (5) can be used in the frequency range up to at least f_T. Equation (5) indicates that *the short gate is essential for low noise, as well as minimizing the parasitic resistances.*

TABLE I
COMPARISON BETWEEN THE PREDICTED AND MEASURED VALUES OF OPTIMAL NOISE FIGURE FOR SAMPLE GaAs MESFET's.

	PARAMETER		DEVICE				
	symbol	units	A	B	C	D	E
Measured	F_o	dB	1.75	1.76	2.22	1.51	1.74
	L	μm	0.46	0.51	0.57	0.51	0.50
	g_m	m℧	54	56	33	48	59
	R_g	Ω	4.5	3.8	13.7	3.8	4.0
	R_s	Ω	3.8	2.9	2.7	1.5	2.3
Predicted	F_o	dB	1.75	1.76	2.21	1.50	1.73

B. Expressions for Transconductance and Parasitic Resistances in Terms of Geometrical and Material Parameters

It has been empirically found that the transconductance can be expressed in terms of the geometrical and material parameters of a device as follows:

$$g_m \approx K_m Z (N/aL)^{1/3} \quad (\mho) \tag{6}$$

where

- Z total device width, in mm;
- a effective thickness of the active channel, in μm;
- N effective free-carrier concentration in the active channel, in 10^{16} cm^{-3};
- K_m fitting factor of 0.023 for low-noise devices.

In [10], an expression for R_g is given in terms of the geometrical and material parameters as

$$R_g \approx 17 z^2 / Z h L_g \quad (\Omega) \tag{7}$$

where

- z unit gate width, in mm;
- h average gate metallization height, in μm;
- L_g average gate metallization length, in μm.

For a recessed-gate device, R_s can be approximated as the sum of three component resistances so that

$$R_s = R_1 + R_2 + R_3 \tag{8}$$

where R_1 represents the source contact resistance, and R_2 and R_3 are partial resistances of the channel between the source and gate electrodes. These component resistances are given by [10]

$$R_1 \approx 2.1/Z a_1^{0.5} N_1^{0.66} \quad (\Omega) \tag{8a}$$

$$R_2 \approx 1.1 L_2 / Z a_2 N_2^{0.82} \quad (\Omega) \tag{8b}$$

$$R_3 \approx 1.1 L_3 / Z a_3 N_3^{0.82} \quad (\Omega) \tag{8c}$$

where

- a_1 effective channel thickness under the source electrode, in μm;
- N_1 effective free-carrier concentration in the channel under the source electrode, in 10^{16} cm^{-3};
- L_2, L_3 effective length of each sectional channel between the source and gate electrodes, in μm;
- a_2, a_3 effective thickness of the sectional channel, in μm;
- N_2, N_3 effective free-carrier concentration of the sectional channel, in 10^{16} cm^{-3}.

In the case of a device with an n$^+$-GaAs layer between the ohmic metal and n-GaAs, for example, L_2 can be approximated as the distance between the source electrode and the edge of the n$^+$ layer and L_3 as the distance of the n layer between the edge of the n$^+$ layer and the effective edge of the gate electrode. Remember that there is always the surface depletion in n-GaAs and n$^+$-GaAs. This has to be taken into account when a_3 and a_2 are evaluated.

For an examination of the adequacy of these expressions, the same five GaAs MESFET's as shown in Table I were used as the samples. While devices C, D, and E were fabricated with n$^+$-GaAs epitaxial layers under the ohmic contacts, devices A and B lacked such layers. For all the devices, N, a, and L were determined by using the technique described in [10]. Parameters L_g and h were evaluated from SEM measurements. For N_1, N_2, N_3, a_1, a_2, a_3, L_2, L_3, Z, and z, either nominal or most appropriate values were applied. Such predetermined values of these parameters are summarized in Table II.

Using the data shown in Table II, g_m, R_g, R_1, R_2, R_3, and R_s were calculated by the above expressions. As shown in Table III, the calculated values of g_m, R_g, and R_s are then compared with the measured values of g_m, R_g, and $(R_s + R_d)/2$ where R_d is the drain series resistance [10]. There is good agreement in all the three cases. It can be seen that the application of an n$^+$ layer has improved the source contact re-

TABLE II
PREDETERMINED VALUES OF GEOMETRICAL AND MATERIAL PARAMETERS FOR SAMPLE GaAs MESFET's

PARAMETER symbol	units	A	B	C	D	E
N	10^{16}cm^{-3}	7.4	8.1	3.5	6.6	9.8
N_1	10^{16}cm^{-3}	7.4	8.1	100	100	100
N_2	10^{16}cm^{-3}	7.4	8.1	100	100	100
N_3	10^{16}cm^{-3}	7.4	8.1	3.5	6.6	9.8
a	μm	0.155	0.138	0.273	0.165	0.140
a_1	μm	0.35	0.45	0.15	0.15	0.15
a_2	μm	0.25	0.35	0.10	0.10	0.10
a_3	μm	0.155	0.138	0.273	0.165	0.140
L	μm	0.46	0.51	0.57	0.51	0.50
L_g	μm	0.8	0.8	0.8	0.8	0.8
L_2	μm	0.75	0.75	0.85	0.75	0.85
L_3	μm	0.4	0.3	0.4	0.3	0.4
h	μm	0.65	0.65	0.2	0.65	0.65
z	mm	0.25	0.25	0.25	0.25	0.25
k		0.04	0.04	0.04	0.04	0.04

TABLE III
COMPARISON BETWEEN THE CALCULATED AND MEASURED VALUES OF TRANSCONDUCTANCE AND PARASITIC RESISTANCES FOR SAMPLE GaAs MESFET's

	PARAMETER symbol	units	A	B	C	D	E
Calculated	g_m	m$\mho$	54	56	33	49	61
Measured	g_m	m$\mho$	54	56	33	48	59
Calculated	R_g	Ω	4.1	4.1	13.3	4.1	4.1
Measured	R_g	Ω	4.5	3.8	13.7	3.8	4.0
Calculated	R_1	Ω	1.89	1.57	0.52	0.52	0.52
	R_2	Ω	1.28	0.85	0.68	0.59	0.68
	R_3	Ω	1.13	0.86	1.15	0.86	1.00
	R_s	Ω	4.3	3.3	2.4	2.0	2.2
Measured	$\frac{R_s + R_d}{2}$	Ω	4.3	3.2	2.7	1.7	2.3

sistance by a factor of more than 3. This has led to an overall improvement of about 40 percent in R_s. Also it is clearly understood that much thinner gate metallization in device C severely degraded F_o. The data shown in Table III suggest that narrowing the gate in the devices with the n⁺ layer would further improve the noise property.

C. Optimal Noise Figure Expressed in Terms of the Geometrical and Material Parameters

Based on a good experimental verification of the analytical expressions for g_m, R_g, and R_s shown in the last section, these expressions can be used to describe the optimal noise figure of a GaAs MESFET in terms of its geometrical and material parameters. Substituting (6)-(8c) into (5) yields the following result:

$$F_o = 1 + kf \left[\frac{NL^5}{a}\right]^{1/6}$$
$$\times \left[\frac{17z^2}{hL_g} + \frac{2.1}{a_1^{0.5}N_1^{0.66}} + \frac{1.1L_2}{a_2N_2^{0.82}} + \frac{1.1L_3}{a_3N_3^{0.82}}\right]^{1/2} \quad (9)$$

where

$$k = K_l\sqrt{K_m} \approx 0.040.$$

It should be noted that F_o is invariant to the total device width but varies with the unit gate width.

As the operating frequency increases, the skin effect on the gate metallization may no longer be ignored and (9) may be rewritten as

$$F_o = 1 + kf \left[\frac{NL^5}{a}\right]^{1/6} \left[\frac{17z^2}{hL_g} + 1.3z^2\left(\frac{f}{hL_g}\right)^{1/2}\right.$$
$$\left. + \frac{2.1}{a_1^{0.5}N_1^{0.66}} + \frac{1.1L_2}{a_2N_2^{0.82}} + \frac{1.1L_3}{a_3N_3^{0.82}}\right]^{1/2}. \quad (10)$$

Again, by using the data given in Table II for the five devices, their F_o values were calculated by both (9) and (10). As seen in Table IV, the predicted values of F_o are in good agreement with the directly measured values at 5.92 GHz for all devices. The data indicate that there was a slight degradation in F_o due to the skin effect.

Expressions (9) and (10) are very useful not only to calculate the optimal noise figure of a GaAs MESFET for a given set of geometrical and material parameters, but also to obtain a guide of optimization to the device design for an improved noise performance.

IV. Examples of Low-Noise Device Design

A. Maximum Unit Gate Width

The maximum value of unit gate width z_m can be defined as the limit above which R_g becomes greater than R_s. Thus equating (7) and (8) yields

$$z_m = 0.25\sqrt{hL_g}\left[\frac{1.9}{a_1^{0.5}N_1^{0.66}} + \frac{L_2}{a_2N_2^{0.82}} + \frac{L_3}{a_3N_3^{0.82}}\right]^{1/2} \text{(mm)}. \quad (11)$$

When $z = z_m$, (10) reduces to

TABLE IV
COMPARISON OF THE PREDICTED VALUE OF OPTIMAL NOISE FIGURE FROM THE GEOMETRICAL AND MATERIAL PARAMETERS, WITH THE DIRECTLY MEASURED VALUE FOR SAMPLE GaAs MESFET's

PARAMETER			DEVICE				
	symbol	units	A	B	C	D	E
Predicted							
by (9)	F_o	dB	1.72	1.80	2.12	1.56	1.70
by (10)	F_o	dB	1.77	1.85	2.18	1.62	1.76
Measured							
directly	F_o	dB	1.75	1.76	2.22	1.51	1.74

TABLE V
DESIGN PARAMETERS OF FIVE REPRESENTATIVE GaAs MESFET's USED FOR CALCULATION OF THE OPTIMAL NOISE FIGURE AS A FUNCTION OF FREQUENCY, AS SHOWN IN FIG. 2

PARAMETER		DEVICE				
symbol	units	a	b	c	d	e
L	μm	0.9	0.9	0.5	0.5	0.25
L_g	μm	0.9	1.2	0.8	0.8	0.4
L_2	μm	1.0	0.75	0.75	0.75	0.4
L_3	μm	0	0.4	0.3	0.3	0.2
h	μm	0.5	1.0	0.65	0.65	0.4
N	10^{16}cm^{-3}	7	4	8	8	18
N_1	10^{16}cm^{-3}	7	200	200	200	200
N_2	10^{16}cm^{-3}	7	200	200	200	200
N_3	10^{16}cm^{-3}	-	4	8	8	18
a	μm	0.3	0.27	0.15	0.15	0.1
a_1	μm	0.3	0.15	0.15	0.15	0.15
a_2	μm	0.17	0.12	0.12	0.12	0.12
a_3	μm	-	0.27	0.15	0.15	0.1
z	mm	0.25	0.25	0.25	0.1	0.065
z_m	mm	0.24	0.23	0.14	0.14	0.065

$$F_o = 1 + 1.5kf\left[\frac{NL^5}{a}\right]^{1/6}$$
$$\times \left[\frac{1.9}{a_1^{0.5}N_1^{0.66}} + \frac{L_2}{a_2N_2^{0.82}} + \frac{L_3}{a_3N_3^{0.82}}\right]^{1/2}\sqrt{1+s} \quad (12)$$

where

$$s = 0.08\sqrt{fhL_g}. \quad (13)$$

The new parameter s represents the perturbation due to the skin effect.

B. Frequency Dependence of Optimal Noise Figure

In the previous section, (3) and (5) were verified to be accurate enough to predict the optimal noise figure of a GaAs MESFET at frequencies up to at least f_T. Thus (10) could be used to calculate the frequency dependence of the optimal noise figure. Five low-noise GaAs MESFET's of various designs, as given in Table V, were then used for this study. The calculated results are shown in Fig. 2.

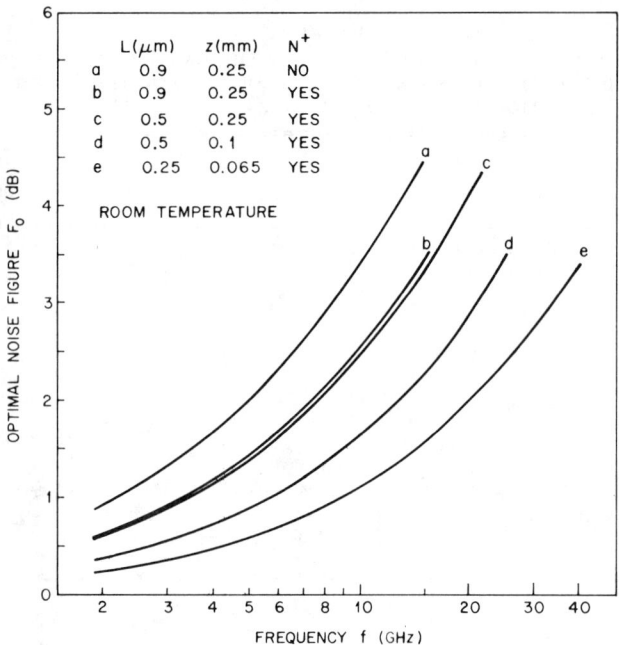

Fig. 2. Calculated optimal noise figure as a function of frequency for GaAs MESFET's with various design parameters, as given in Table V.

In Fig. 2, device a is a simulation of the 0.9-μm gate GaAs MESFET, as referred to in Section III-A. All design parameter values used for this device, except for a_2, were taken from the data given in [8]. For the determination of a_2, the surface depletion in the channel was taken into account. In the calculation of this simulated device, an effective value of Cr-Au gate metallization resistivity of 2.5×10^{-6} $\Omega \cdot$ cm was assumed after a measured R_g of 3 Ω for the corresponding real device. Note again that a calculated F_o of 3.42 dB at 10 GHz agrees well with the F_o value of the real device measured at its nearly cutoff frequency (see Section III-A.).

Devices $b-e$ have n$^+$ layers under the ohmic contacts. An effective value of aluminum gate metallization resistivity of 5.0×10^{-6} $\Omega \cdot$ cm is assumed for these devices, as was used for (7). Device b represents an optimized design of the 0.9-μm gate device. Device c is similar to those used as the samples ($A-E$). In device c, the z value is much greater than z_m specifically required for this device. This shortcoming results in cancellation of the merit of a short gate length of 0.5 μm. Device d is an improved version of device c, simply by narrowing the unit gate width from 0.25 to 0.1 mm. This modification is quite effective in this case. In other words, *a simple shortening of the gate length may not bring the improved optimal noise figure unless the unit gate width is accordingly narrowed below z_m.* The final example is a quarter-micrometer gate device which promises a superior noise performance in microwave frequencies through K band. Although the process control in various stages of fabrication would become extremely critical, device e should not be unrealizable with the current fabrication technology.

V. Conclusions

A simple expression for the optimal value of the minimum noise figure F_o of the GaAs MESFET was derived in terms of its equivalent circuit elements on a semiempirical basis. This expression of F_o was examined by using the known element values of a 0.9-μm gate GaAs MESFET. The calculated value of F_o at 10 GHz agreed well with the reported measured value at this frequency, which was nearly the cutoff frequency of this device. As a next step, the device equivalent circuit elements were shown as functions of the geometrical and material parameters with good experimental verification. The optimal noise figure was thus expressed in analytical forms using the geometrical and material parameters. These expressions were examined with 0.5-μm gate GaAs MESFET's whose geometrical and material parameter values were independently determined. The predicted value of F_o calculated by using these parameter values was in good agreement with the directly measured value at 6 GHz for all sample devices with different design parameters.

The expressions for F_o were then used to show examples of the design optimization for low-noise devices. This exercise indicated that shortening the gate length and minimizing the parasitic gate and source resistances are essential to lower F_o. Moreover, a simple shortening of the gate length would not bring an improved F_o unless the unit gate width was accordingly narrowed. The maximum value of the unit gate width was then defined as the width above which the gate metallization resistance became greater than the source series resistance.

Short-gate GaAs MESFET's with optimized designs have promised a superior noise performance at microwave frequencies through K band. The predicted values of F_o at 20 GHz, for example, for a half-micrometer gate device and a quarter-micrometer gate device were 3 and 2 dB, respectively. These devices could be fabricated with the current technology.

In the Appendixes, a definite advantage of the GaAs MESFET as a low-noise device over the bipolar transistor all through the microwave spectrum is discussed. Also described is the effective gate length which is a key factor to understand the noise behavior of a recessed-gate device properly. Furthermore, the effective gate length provides a crucial guide toward an optimized design of low-noise GaAs MESFET's, minimizing a possible increase in fabrication difficulties for an improved noise performance.

Appendix I

Advantage of the GaAs MESFET Over the Bipolar Transistor as Microwave Low-Noise Devices

It may be worthwhile to mention that there is a characteristic difference in the frequency dependence of the minimum noise figure between the bipolar transistor and the GaAs MESFET. As was described in [2], a quick estimate of the minimum noise figure for the bipolar transistor at room temperature is

$$[F_{\min}]_B \approx 1 + bf^2 \left[1 + \sqrt{1 + \frac{2}{bf^2}}\right] \quad \text{(A1)}$$

where

$$b = 40 I_c r_b / f_T^2 \quad \text{(A2)}$$

r_b is the parasitic base resistance in ohms and I_c is the collector bias current in amperes. In contrast, an equivalent figure for the MESFET is

$$[F_{\min}]_M \approx 1 + mf \quad (A3)$$

where

$$m = \frac{2.5}{f_T} \sqrt{g_m(R_g + R_s)}. \quad (A4)$$

As seen in (A3) and (A1), the minimum noise figure increases with frequency linearly in the MESFET whereas it increases with frequency in a quadratic fashion in the bipolar transistor. In other words, the noise performance of the MESFET tends to degrade with increasing frequency at much smaller rate than that of the bipolar transistor. Since the GaAs MESFET has already demonstrated the better noise performance than the bipolar device even at the low end of microwave frequencies, the former has a definite advantage as the low-noise device over the latter in the entire microwave spectrum.

APPENDIX II
Effective Gate Length

In a recent work [11], the maximally available value of channel current in a GaAs MESFET was studied and its analytical expressions were given in terms of the basic channel parameters, such as the gate length, active channel thickness, and channel free-carrier concentration. This theory was developed on a simple planar channel structure. However, the cross-sectional view of the channel from source to drain is rather complicated in the actual device. The recessed-gate structure is an example of this kind. Sometimes the contour of the gate metal-to-semiconductor contact is not linear but has a curvature. In such a case, there is an effective value of the gate length upon which the above theory still holds [10]. This effective value can be either smaller or greater than the physical value of gate metallization length, depending upon the shape of the contact contour. As mentioned earlier in the text, such an effective value has a good fit to L in (5), although (5) was originally obtained from the planar channel devices. At present, however, no analytical expression is available to show the relationship between the effective gate length and gate metallization length. Nevertheless, the effective gate length can be experimentally determined from dc parameters measured in accordance with the technique described in [10].

One of the most important factors for realizing a low-noise GaAs MESFET is to compromise the two conflicting requirements; shortening the gate length and lowering the gate metallization resistance. This can be done by narrowing the unit gate width through a gate paralleling scheme with much increased fabrication complexity. However, this is also achieved, without such a penalty, by minimizing the ratio of effective gate length to gate metallization length with a correct shape of the gate recess structure formed by proper fabrication techniques.

Acknowledgment

The GaAs MESFET's used as the samples were fabricated in cooperation with J. V. DiLorenzo, H. M. Cox, L. A. D'Asaro, B. S. Hewitt, W. Robertson, J. A. Seman, and J. R. Velebir. The SEM measurements were carried out by L. C. Luther. The noise measuring test equipment used were constructed by D. E. Iglesias and W. O. Schlosser. The author wishes to thank all the members for their help, and J. E. Kunzler and L. J. Varnerin for their continuous interest in the project.

References

[1] IRE Standards on Electron Tubes, pt. 9, Noise in Linear Two-ports, 1962.
[2] H. Fukui, "The noise performance of microwave transistors," *IEEE Trans. Electron Devices*, vol. ED-13, pp. 329–341, Mar. 1966.
[3] A. van der Ziel, "Thermal noise in field-effect transistors," *Proc. IRE*, vol. 50, pp. 1808–1812, Aug. 1962.
[4] —, "Gate noise in field-effect transistors at moderately high frequencies," *Proc. IEEE*, vol. 51, pp. 462–467, Mar. 1963.
[5] W. Shockley, "A unipolar field-effect transistor," *Proc. IRE*, vol. 40, pp. 1365–1376, Nov. 1952.
[6] C. A. Liechti, "Microwave field-effect transistors–1976," *IEEE Trans. Microwave Theory Tech.*, vol. MTT-24, pp. 279–300, June 1976.
[7] R. A. Pucel, H. A. Haus, and H. Statz, "Signal and noise properties of gallium arsenide microwave field-effect transistors," in *Advances in Electronics and Electron Physics*, vol. 38. New York: Academic Press, 1975, pp. 195–265.
[8] C. A. Liechti, E. Gowen, and J. Cohen, "GaAs microwave Schottky-gate FET," in *IEEE ISSCC Dig. Tech. Papers*, pp. 158–159, 1972.
[9] B. S. Hewitt, H. M. Cox, H. Fukui, J. V. DiLorenzo, W. O. Schlosser, and D. E. Iglesias, "Low-noise GaAs MESFETs: Fabrication and performance," in *Gallium Arsenide and Related Compounds (Edinburgh) 1976* (Conf. Ser. 33a, The Institute of Physics, Bristol and London, England, 1977), pp. 246–254.
[10] H. Fukui, "Determination of the basic device parameters of a GaAs MESFET," *Bell Syst. Tech. J.*, vol. 58, pp. 771–797, Mar. 1979.
[11] —, "Channel current limitations in GaAs MESFETs," accepted for publication in *Solid-State Electron*.

Optimization of Low-Noise GaAs MESFET's

HATSUAKI FUKUI, SENIOR MEMBER, IEEE, JAMES V. DILORENZO, BERT S. HEWITT, JAMES R. VELEBIR, JR., HERBERT M. COX, LARS C. LUTHER, AND JOHN A. SEMAN

Abstract—This paper presents a device design which is an effective way of reconciling the two conflicting requirements for low-noise GaAs MESFET's. Decreasing the effective value of gate length can be achieved, without penalty of increased gate metallization resistance, by the virtue of a proper gate-recess structure. This effect can be explained by the "effective gate length" concept. The pertinent fabrication techniques and the optimal noise-figure expression are given for an optimized structure with illustrated examples.

I. Introduction

THE PROGRESS of low-noise GaAs MESFET's has been remarkable in the last few years and still continues to be [1]-[3]. Such an advancement has been achieved by structural changes of the devices in conjunction with improved quality of epitaxial materials [4]. These modifications have brought higher cutoff frequency, lower parasitic resistances, and lower noise coefficient, and have resulted in much improved noise properties of GaAs MESFET's. Shortening the gate length, introducing the gate-recess structure, applying the n^+-GaAs layer, and employing the buffer layer have been the major contributing factors [5]-[7].

There have been conflicting requirements, however. Simply shortening the gate length at a fixed-unit gatewidth would result in an increased gate metallization resistance. This would cancel the merit of the shortened gate length for an improved noise figure. Of course, more unit gates of lesser unit gatewidth could be paralleled but this would add complications to device design. A practical solution for this problem has been found based on a concept called "effective gate length." This concept has been realized with a special form of the gate-recess structure fabricated by appropriate processing steps. This paper will describe such an optimization of low-noise GaAs MESFET's.

II. Minimum Noise Figure

For actual GaAs MESFET's their minimum noise figure is well described by the following expression [4]:

$$F_{\min} = 1 + kfL \sqrt{g_{m0}(R_g + R_s)} \tag{1}$$

where

k noise coefficient of approximately 0.27

Manuscript received November 5, 1979; revised January 25, 1980.
H. Fukui, J. V. DiLorenzo, J. R. Velebir, Jr., H. M. Cox, L. C. Luther, and J. A. Seman are with Bell Laboratories, Murray Hill, NJ 07974.
B. S. Hewitt was with Bell Laboratories, Murray Hill, NJ. He is now with Raytheon Company, Special Microwave Devices Operation, Waltham, MA 02154.

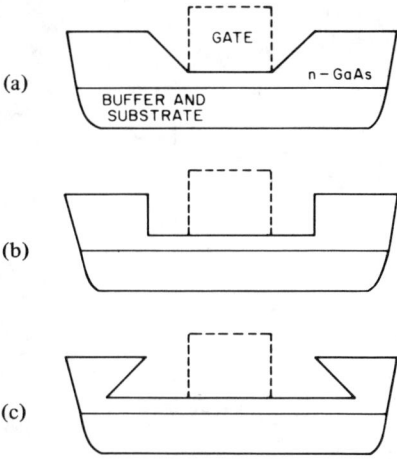

Fig. 1. A schematic view of three representative profiles of the channel cross section of GaAs MESFET's.

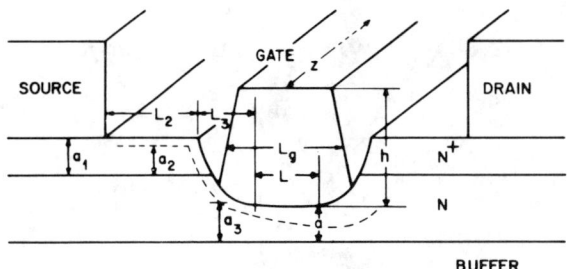

Fig. 2. An optimized low-noise MESFET structure indicating the definition for geometrical and material parameters.

f	frequency, in gigahertz
L	gate length, in micrometers
g_{m0}	transconductance at null gate bias, in mhos
R_g	ac gate metallization resistance, in ohms
R_s	total source series resistance, in ohms.

This expression indicates the dependence of $F_{\min}$ on device parameters. Parameter R_g is further expressed as [8]

$$R_g = \frac{17z^2}{LhZ} \qquad (2)$$

where

h	gate metallization height, in micrometers
z	unit gate finger width, in millimeters
Z	total gatewidth, in millimeters.

Shortening the gate length obviously increases the gate metallization resistance. Therefore, as the gate length is reduced, the unit gatewidth also has to be decreased to keep the gate metallization resistance as low as possible. This has introduced a complication of paralleling an increased number of unit gates, to maintain the device impedance at a favorable level [9]. Instead of this scheme, a mushroom-like cross section of the gate has been reported [10]. However, this approach also has involved sophisticated fabrication processes.

III. Optimum Gate-Recess Structure

On a certain series of low-noise GaAs MESFET's, it was found that some devices did not follow (1). The measured noise figures appeared to be considerably better than the predicted ones. The predictions were based on the measured values of L using a scanning-electron microscope (SEM), and on the ac and dc characterizations of g_{m0}, R_g, and R_s [11], [8]. Then these devices were thoroughly investigated in many respects, including the cross-sectional view of their channels, in comparison with normal devices.

There were three representative configurations in the cross-sectional view of the channel, as illustrated in Fig. 1. The gate recess was formed by the same technique in all the three cases. The entire wafer surface was covered with an AZ-photoresist. After patterning a gate window in the photoresist, the unmasked GaAs was etched away by a solution of H_2O_2 adjusted to a pH of 7.2 with NH_4OH. This etching was continued to remove portion of the epitaxial GaAs until the thickness of the active channel reached a specified value.

The difference in the profiles shown in Fig. 1 was found to be caused by the difference in the gate orientations relative to the crystallographic directions of the substrate. This was due to the faceting nature of the chemical etch used. Its etch rate was selective to the crystallographic orientation of the GaAs. The substrate was typically oriented 6° off the {100} toward the {111}A. The rates for the {111}A, {100}, and {111}B directions were in decreasing order. When the gates were oriented on the wafer in such a manner that the slow etching {111}B orientations developed on the sidewalls, a gate-recess profile schematically shown in Fig. 1(a) was obtained. When the orthogonal direction was chosen for the gate orientation, a gate-recess profile similar to that shown in either Fig. 1(b) or (c) was obtained. It was obvious that type (a) was most advantageous because it would produce the lowest possible R_s.

Moreover, SEM analyses revealed the following important fact. The aforementioned devices which showed superior noise performance had a profile as illustrated in Fig. 2. The gate contact to the channel was not straight as schematically shown in Fig. 1(a) but had a curvature in the cross-sectional view. From this observation, the concept of "effective gate length" was deduced; namely, that the effective gate length was much shorter than the physical length of gate metallization in this case.

It was found later that the effective gate length could be evaluated by a dc determination technique described in [8]. Henceforth, the effective gate length L would be differentiated from the gate metallization length L_g. There is no analytical expression available for L in terms of the geometrical and material parameters of a device at present.

As was defined in [4], the minimum noise figure of a GaAs MESFET operating under the optimum bias conditions for low noise would be designated as the optimal noise figure F_0. Then this parameter has been expressed in terms of the geometrical and material parameters. In accordance with the definitions given to these parameters as illustrated in Fig. 2, F_0 at room temperature is expressed as

$$F_0 = 1 + fK\left[\frac{NL^5}{a}\right]^{1/6}\left[\frac{17z^2}{hL_g}(1+s)\right. $$
$$\left. + \frac{2.1}{a_1^{0.5}N_1^{0.66}} + \frac{1.1L_2}{a_2N_2^{0.82}} + \frac{1.1L_3}{a_3N_3^{0.82}}\right]^{1/2} \qquad (3)$$

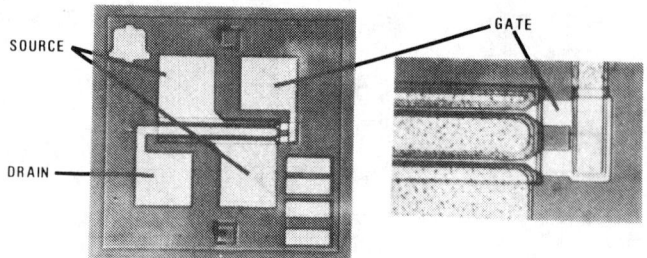

Fig. 3. A layout of the low-noise GaAs MESFET.

where

$$s = 0.08 \sqrt{fhL_g} \qquad (4)$$

and K is a noise coefficient of approximately 0.040. Note that the skin effect on the gate metal and the free-carrier depletion at the GaAs surface are taken into account in (3). The edge of the surface depletion layer is schematically shown by the dashed line in Fig. 2, having a_2 and a_3 as being the effective channel thickness in the sections.

The decrease in F_0 due to the optimum gate-recess structure might have been explained by an effect other than the effective gate length. In this case, either a reduced value of the noise coefficient in (3) or a different form of the expression for F_0 would be required to meet the experimental result.

IV. PRACTICAL EXAMPLES

The pattern layout of a low-noise GaAs MESFET is shown in Fig. 3. The nominal values of source–gate distance, gate metallization length, and gate–drain distance were all 1 μm. There were two 250-μm-wide gate fingers in parallel. The gate metal was aluminum 0.7 μm thick. This device was fabricated in a similar manner to that described in [7]. To avoid development of the so-called "purple plague" phenomenon, the aluminum gate feeders were carefully separated from the final gold metallization by a Ti/Pt multilayer, as shown in Fig. 3. The wafer consisted of an n$^+$-GaAs layer, an n-GaAs layer, and an undoped buffer layer, all consecutively grown on a Cr-doped semi-insulating GaAs substrate by the vapor-phase epitaxy method [6].

Distributions of the measured optimal noise figure F_0 and associated power gain G_a of low-noise GaAs MESFET's from a given slice are shown in Fig. 4. The median values of F_0 and G_a were 1.61 and 8.2 dB, respectively, at a test frequency of 5.9 GHz. The corresponding average values of the participating device parameters were evaluated to be as follows: L = 0.6 μm, L_g = 1.0 μm, g_{mo} = 0.048 mhos, R_g = 3.0 Ω, and R_s = 1.6 Ω. On the other hand, if the conventional channel profile, as shown in Fig. 1(b) at $L = L_g$ = 1.0 μm, were employed, an F_0 of 2.3 dB would be expected with slightly degraded gain at the same frequency.

This example clearly demonstrated the effectiveness of reducing the ratio of L to L_g by the proper technique. In addition, an extended n$^+$-GaAs edge, as close to the gate metal as possible, also contributed to the low F_0 value through a reduced value of R_s in this case. For a further improvement of F_0, narrowing the unit gatewidth should be effective [4]. Halving the unit gatewidth in this case, for example, would

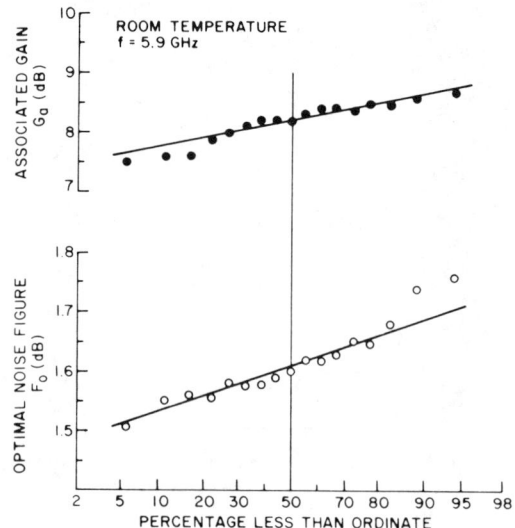

Fig. 4. Distributions of the optimal noise figure and associated power gain of GaAs MESFET's from a slice.

bring an improved value of 1.23 dB for F_0 at 6 GHz without any other modifications. This value is only a couple of tenths of a decibel higher than the state of the art of low-noise GaAs MESFET's at room temperature [3].

V. CONCLUSIONS

A major conflicting requirement in designing a low-noise GaAs MESFET was to shorten the gate length without increasing the gate metallization resistance. This problem has been solved by adoption of a concept called "effective gate length." By choosing the crystallographic orientation of a GaAs epitaxial film properly with respect to the gate axis, the effective value of gate length was made much shorter than the physical length of gate metallization.

The effectiveness of this concept was demonstrated by the superior microwave noise and gain performance of actual devices fabricated along this concept. Although these devices had an average value of gate metallization length of 1 μm, the median value of their measured noise figures at 5.9 GHz was 1.6 dB with a median associated gain of 8.2 dB.

An advanced structure of the low-noise GaAs MESFET was illustrated based on the "effective gate length" concept. Using the geometrical and material parameters of this structure, its optimal noise figure was analytically expressed for ready use in the design work. This expression has suggested further improvement in F_0 of the aforementioned devices. Narrowing the unit gatewidth from 250 to 125 μm, for example, would bring an F_0 of 1.2 dB at 6 GHz even with an average gate metallization length of 1 μm.

REFERENCES

[1] J. V. DiLorenzo, "Progress in the development of low noise and high power GaAs FETs," in *Proc. 6th Cornell Conf. on Active Microwave Semiconductor Devices and Circuits*, pp. 1–28, 1977.

[2] H. F. Cooke, "Microwave field effect transistors in 1978," *Microwave J.*, vol. 21, no. 4, pp. 43–48, Apr. 1978.

[3] C. A. Liechti, "GaAs FET technology: A look into the future," *Microwaves*, vol. 17, pp. 44–49, Oct. 1978.

[4] H. Fukui, "Optimal noise figure of microwave GaAs MESFETs,"

IEEE Trans. Electron Devices, vol. ED-26, pp. 1032–1037, July 1979.

[5] T. Nozaki, M. Ogawa, H. Terao, and H. Watanabe, "Multilayer epitaxial technology for the Schottky-barrier GaAs field-effect transistors," in *Gallium Arsenide and Related Compounds 1974* (Conf. Ser no. 24). London, England: Inst. Phys., 1975, pp. 46–54.

[6] H. M. Cox and J. V. DiLorenzo, "Characteristics of an $AsCl_3$/Ga/H_2 two-bubbler GaAs CVD system for MESFET applications," in *Gallium Arsenide and Related Compounds* (St Louis, MO) 1976 (Conf. Ser. no. 33b). London, England: Inst. Phys., 1977, pp. 11–22.

[7] B. S. Hewitt, H. M. Cox, H. Fukui, J. V. DiLorenzo, W. O. Schlosser, and D. E. Iglesias, "Low-noise GaAs MESFETs: fabrication and performance," in *Gallium Arsenide and Related Compounds* (Edinburgh, Scottland) 1976 (Conf. Ser. no. 33a). London; England: Int. Phys., 1977, pp. 246–254.

[8] H. Fukui, "Determination of the basic device parameters of a GaAs MESFET," *Bell Syst. Tech. J.*, vol. 58, pp. 771–797, Mar. 1979.

[9] J. R. Anderson, M. Omori, and H. F. Cooke, "A unique GaAs MESFET for low noise application," in *1978 IEEE IEDM Tech. Dig.*, pp. 133–135.

[10] S. Takahashi, F. Murai, H. Kurono, M. Hirao, and H. Kodera, "A half-micron gate GaAs FET fabricated by chemical dry etching," in *Proc. 8th Conf. Solid State Devices* (Tokyo, Japan), 1976, pp. 115–118.

[11] H. Fukui, "Design of microwave GaAs MESFETs for broad-band low-noise amplifiers," *IEEE Trans. Microwave Theory Tech.*, vol. MTT-27, pp. 643–650, July 1979.

Super Low-Noise GaAs MESFET's with a Deep-Recess Structure

KEIICHI OHATA, HITOSHI ITOH, FUMIO HASEGAWA, MEMBER, IEEE, AND YOSHINORI FUJIKI

Abstract—Super low-noise GaAs MESFET's for replacement of parametric amplifiers have been successfully developed by adopting a deep-recess structure. The structure of a 0.5-μm gate in a deeply recessed region with a cylindrical edge shape has enabled reduction of the source resistance to a half of that of conventional flat-type MESFET's. The noise figure was improved by more than 0.5 dB by this reduction of the source resistance, and less than 2.0-dB noise figure has been reproducibly obtained at 12 GHz.

The best noise figures were 0.7 dB (14.9-dB gain) at 4 GHz and 1.68 dB (10.7-dB gain) at 12 GHz. The developed MESFET's were applied to two-stage amplifiers of 11.7–12.2-GHz band, and the noise figure obtained was 2.16 dB (T_e: 185 K) at room temperature and 1.94 dB (T_e: 163 K) at 0°C. This performance is good enough to replace some of parametic amplifiers.

Manuscript received November 1, 1979; revised January 28, 1980.
K. Ohata, H. Itoh, and F. Hasegawa are with Central Research Laboratories, Nippon Electric Company, Ltd., Miyazaki, Takatsu-ku, Kawasaki, Japan.
Y. Fujiki is with Microwave and Satellite Communications Division, Nippon Electric Company, Ltd., Ikebe-cho, Midori-ku, Yokohama, Japan.

I. Introduction

ABILITY of low-noise and high-gain performance of GaAs MESFET's and the feasibility of their applications to communications equipment at microwave frequencies up to K-band have already been proved by practical amplifiers for past several years [1]. The noise figure of GaAs MESFET's has been remarkably improved year after year [2]–[5], and low-noise MESFET's are now extensively used for many purposes. However, parametric amplifiers are still used in the front end of the receiver in the satellite communications systems although low-noise GaAs MESFET's are used in the second and the later stages.

There have been great demands to replace the parametric amplifier by the GaAs MESFET amplifier because it gives a lower cost, simpler and compact systems, and easier maintenance. In order to use GaAs MESFET's for the first-stage amplifier of the receiver the noise figure must be significantly improved over the best performance of commercially available GaAs MESFET's. For example, in the 12-GHz satellite communications band, the noise figure of the total system should be less than 2.5 dB at room temperature and less than 2.0 dB when cooling down to about −50°C. Therefore, the noise performance of MESFET's is required to be less than 2.0 dB at 12 GHz which is better than the best value of the commercially available GaAs MESFET's, such as NE388, by more than 0.5 dB.

Although the noise performance of about 2.0 dB at 12 GHz was obtained in a few MESFET's from special lots, the average noise figure was usually 2.5 dB for the conventional flat-type MESFET's. The MESFET of about 2.0-dB noise figure at 12 GHz was also obtained by the introduction of the selectively ion-implanted n⁺ source region [3], however, the fabrication process was too complicated and too difficult for reproducible production.

The purpose of this paper is to demonstrate that such a super low-noise GaAs MESFET can be reproducibly obtained by adopting a deep-recess structure preserving the 0.5-μm gate and a high-quality epitaxial layer. The source resistance was reduced to a half of that of the conventional flat-type FET's by recessing the active layer more than 0.3 μm and by optimizing the shape of the recessed region. A simple analysis of the source resistance of the recessed structure will be also presented.

The developed low-noise GaAs MESFET's are applied to the amplifiers for satellite communications. Some of the preliminary results of such a two-stage amplifier are also presented.

II. Device Design

The source resistance of conventional flat-type FET's was found to be much larger than that expected due to the surface depletion layer. The source resistance of the 0.5-μm-gate low-noise FET (NE 388) is more than 5 Ω whereas the gate resistance is only 1 Ω. The theoretical analysis predicts that the noise figure of less than 2.0 dB at 12 GHz should be reproducibily obtained if the source resistance is reduced to a half of this value, even if the 0.5-μm gate is used.

There are two kinds of methods to reduce the source resistance: preparation of a selective n⁺ contact layer [3] and a recessed structure. The latter has an easier fabrication process, but its effectiveness has not been quantitatively proved. Theoretical and experimental dependence of the source resistance on the recessed depth will be presented in this section.

A. Adopted Recessed Shape

There are several reports on the recessed structures [5]–[8]. Although it is difficult to see the details of their structures, the possible recessed structures can be classified as shown in Fig. 1 (a)–(c). The type of Fig. 1 (a) is the so-called recessed-gate structure, whose recessed region is almost the same as the gate metallization [6]. In this structure, it is quite possible for the gate fringing capacitance to become abnormally large due to contact of the gate metal to the side of the recessed region. The gate breakdown voltage may also be low due to a very short effective gate-drain spacing as suggested by Ogawa et al. [2].

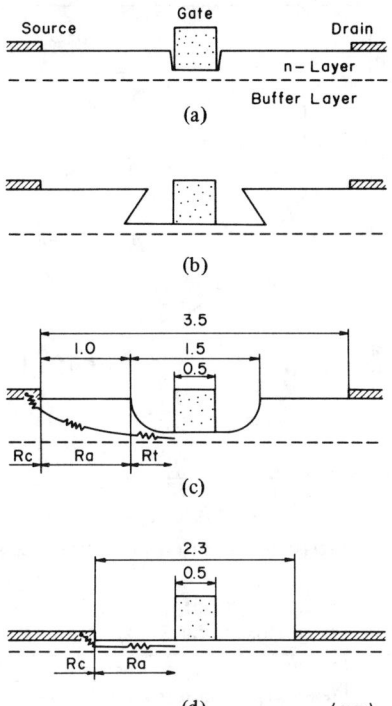

Fig. 1. Cross section of several types of low-noise GaAs MESFET's. (a) and (b) Recessed structure reported so far. (c) Structure developed in this work. (d) Conventional flat-type MESFET.

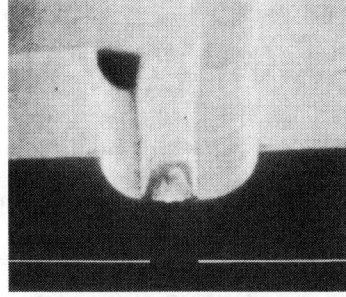

Fig. 2. SEM view of the cross section of the developed MESFET. The marker indicates the 0.5-μm length.

The type of Fig. 1 (b) is the recessed structure which can be made by preferential etching [5], [7]. Inevitably, the region of the thin active layer becomes large, therefore, it is difficult to reduce the source resistance effectively.

The type of Fig. 1 (c) is the one adopted in this work. An SEM photograph of the cross section of the developed FET with this structure is shown in Fig. 2. It has a smooth change of the thickness of the active layer from the gate to the source and drain, still preserving the shortest spacing between the effective source and the gate. It also has almost the same gate fringing capacitance and gate breakdown voltage as those of the conventional flat-type FET's. The structure suggested by Fukui [9] might correspond to this one.

B. Dependence of the Source Resistance on the Recessed Depth

The source resistance of the recessed structure shown in Fig. 1(c) was estimated by dividing the source region into three parts and compared with that of the flat-type FET shown in Fig. 1(d). The resistance of the transition region R_t was cal-

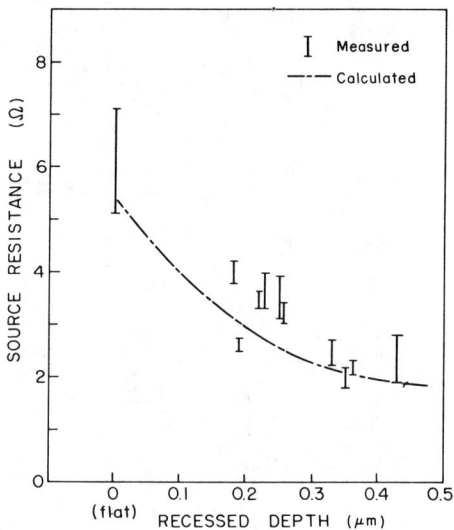

Fig. 3. Source resistance against depth of the recessed region.

TABLE I
THE ESTIMATION OF THE SOURCE RESISTANCE

Type	Contact Resistance R_c (Ω)	Resistance of Active Layer		Total Source Resistance R_s (Ω)
		Uniform Region R_a (Ω)	Transition Region R_t (Ω)	
Recessed	0.45	0.57	1.0	2.0
Recessed (n⁺ Contact Layer)	0.25	0.36	1.0	1.6
Flat	1.0	4.3		5.3

$\rho_c = 1 \times 10^{-6}\, \Omega\text{cm}^2$, $\rho_{cn^+} = 5 \times 10^{-7}\, \Omega\text{cm}^2$ $Z = 280\,\mu m$

culated by integrating the minute resistances in that region. The resistance of other parts were estimated in a similar way as in [9], [10].

The estimated resistances of each part and the total source resistance of the recessed-structure FET with a 0.4-μm-deep recessed region are listed in Table I compared with those of the conventional flat-type FET and the recessed FET with a surface n⁺ contact layer. The calculated dependence of the source resistance on the recessed depth of the recessed FET without the n⁺ contact layer is shown in Fig. 3. The parameters used in the calculation are as follows: $Z = 280$ μm, carrier concentrations of the active epi-layer and the surface contact layer are 2×10^{17} and 2×10^{18} cm^{-3}, respectively, the specific contact resistances for these layers are 1×10^{-6} and 5×10^{-7} $\Omega \cdot$cm^2, respectively. The thickness of the active layer under the gate is 0.1 μm. The thickness of the n⁺ contact layer is 0.2 μm if it is present. The surface depletion layer due to the surface states was assumed to be about 500 Å from the experimental result on the source resistance of the flat-type FET's.

As it can be seen in Fig. 3 and Table I, the source resistance of the recessed-structure FET can be reduced to less than a half of that of the conventional flat-type FET even if there is no surface n⁺ contact layer, if the depth of the recessed region is large enough. This reduction of the source resistance is mainly due to the reduction of the resistance between the source and the gate metals. Introduction of the surface n⁺ contact layer is not very effective compared with the difficulty of production and evaluation of the n⁺-n-n⁻ triple epitaxial layers.

The experimental results on the dependence of the source resistance on the recessed depth are also shown in Fig. 3. The source resistance was evaluated by the same method as the one reported by Wholey and Omori [4]. The device parameters are almost the same as those described above but there is no surface n⁺ contact layer. The source resistance decreases with an increase in the depth of the recessed region, as expected, and reaches about 2 Ω when the recessed depth is more than 0.3 μm, as previously discussed. The scattering is probably due to the scattering of the length of the thin active layer at the bottom of the recessed region, but the scattering is almost the same or lesser than that of the conventional flat-type FET.

III. DEVICE FABRICATION AND CHARACTERIZATION

In order to keep the gate resistance low, the pattern layout of the developed MESFET was made almost the same as that of the previously reported 0.5-μm-gate FET (NE 388) [2]. It has two gate pads on the 280-μm-wide gate. The starting wafer normally has a 0.5-μm-thick active n layer on a high-resistivity buffer layer. The carrier concentration of the active layer is 2.5-3×10^{17} cm^{-3} which is optimum for a 0.5-μm-gate MESFET [2]. The process was performed by the successive steps of mesa isolation, formation of the recessed region, gate lifting off, and formation of the source and drain ohmic contacts. A 0.5-μm-long and 0.5-μm-thick Al gate was formed in the 1.5-μm-wide deeply ($\geqslant 0.3$ μm) recessed region by liftoff technique with a self-alignment method. The recessed width must be controlled well in order to suppress the gate fringing capacitance, to keep the gate breakdown voltage high, and, of course, to decrease the source resistance.

The source and drain electrodes were those of the Ni/AuGe system alloyed at 450°C for 30 s. The source-drain electrode spacing was 3.5 μm which had an appropriate alignment margin of 1 μm against the recessed region. Finally the whole surface was passivated by SiO$_2$.

All FET's were characterized by evaluating the gate resistance R_g, the source resistance R_s, and the dc characteristics. The equivalent gate resistance R_g, was evaluated by measuring the resistance between two gate pads. DC characteristics were checked by a transistor curve tracer. The microwave performance was measured both in a packaged form at 4 and 8 GHz and in a chip form at 8 and 12 GHz. The noise figure was measured by an automatic noise-figure indicator and precisely evaluated by the Y-factor method.

IV. DC CHARACTERISTICS

The developed MESFET exhibited a very high transconductance and a low saturation voltage due to a low source resistance. Fig. 4 shows the correlation between the transconductance and the source resistance. The transconductance of many FET's of many lots both at the zero gate bias and at the noise optimum gate bias (the drain current of 10 mA) are plotted against the source resistance. The same symbol represents the FET's from the same lot. The transconductance is increased by the decrease of source resistance. The transconductance at the noise optimum gate bias is about 40 percent higher than the typical value of the conventional flat-type FET. This rate of the transconductance increase is much higher

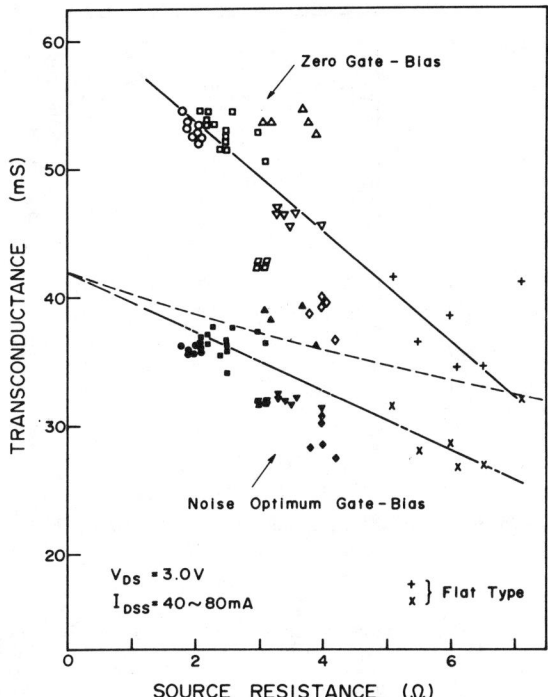

Fig. 4. Correlation between the dc transconductance and the source resistance. The transconductances both at the zero gate bias and the noise optimum gate bias are plotted. The broken line represents the theoretical effect of the source resistance on the transconductance.

TABLE II
TYPICAL DC CHARACTERISTICS OF THE DEVELOPED MESFET COMPARED WITH THE CONVENTIONAL FLAT-TYPE MESFET (NE 388)

Type	Recessed Depth (μm)	R_g (Ω)	R_s (Ω)	g_m (mS)	g_d (mS)	V_P (V)	BV_{GD} (V)	BV_{DS} (V)
New MESFET	0.36	1.2	2.1	37	2.0	-1.5	-12	12
Conventional MESFET	0	1.2	5.5	28	2.0	-2.0	-12	8

conductance g_m, the drain conductance g_d, the pinchoff voltage V_p,[1] the gate-to-drain breakdown voltage BV_{GD}, and the drain-to-source breakdown voltage BV_{DS} of the developed low-noise FET compared with the conventional flat-type FET. The g_m and g_d were measured at the noise optimum gate bias. The BV_{DS} was measured at zero gate bias. The R_s and g_m were significantly improved but the other parameters, such as the gate-to-drain breakdown voltage BV_{GD}, were kept the same. Furthermore, the drain-to-source breakdown voltage BV_{DS} was increased greatly due to the recess structure [11].

For replacement of the parametric amplifier, the first stage GaAs FET amplifier is normally operated in low ambient temperature in order to achieve the best noise figure possible. DC characteristics of the developed low-noise FET were also characterized at low ambient temperatures. Fig. 5 shows the temperature dependence of the parasitic resistances R_s and R_g. The R_g decreased linearly with the decrease of the temperature, as expected. The temperature coefficient was 4.8×10^{-3} K^{-1}. However, the decrease of the source resistance was only about 0.4 Ω for the temperature decrease of 200 K. Therefore, the reduction of the source resistance is more important for the FET's used at low temperatures.

V. MICROWAVE PERFORMANCE

Fig. 6 shows the minimum noise figures NF_{min} and the associated gains G_a of the developed MESFET's compared with the conventional flat-type MESFET's (NE 388). The bars indicate the distribution of the measured values of many FET's from many lots. The minimum noise figure has been improved in the wide frequency range, especially in the higher frequencies. The best noise figure obtained at 12 GHz is 1.68 dB with the associated gain of 10.7 dB. Most of the FET's fabricated from ten wafers exhibit a noise figure of less than 2.0 dB at 12 GHz, as expected. The MESFET's from the best lots exhibited noise figures of 0.7, 1.2, and 1.7 dB at 4, 8, and 12 GHz, respectively.

Fig. 7 shows the drain-current dependence of NF_{min} and G_a at 8 and 12 GHz of the developed MESFET. The low-noise and the high-gain properties are observed in the wide range of the drain current. The small dependence of NF_{min} on the drain current in Fig. 7 is probably due to a small diffusion noise in the developed FET [12]. A high-field dipole layer between the gate and drain as reported by Engelmann and Liechti [13] must have been decreased in the deep and narrow recessed structure.

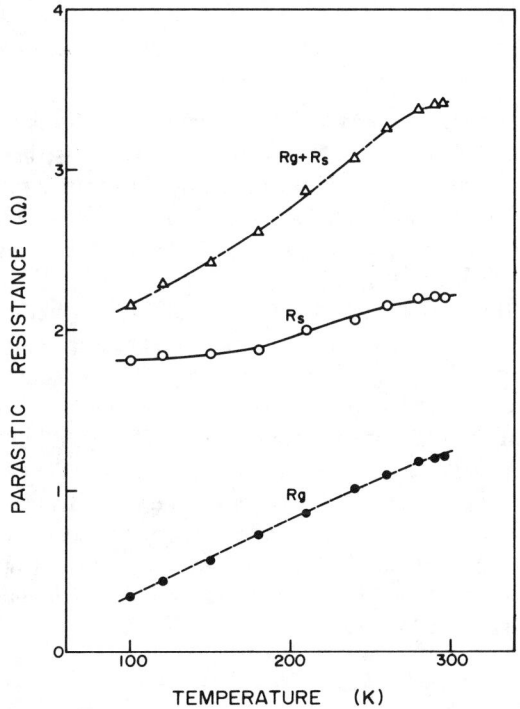

Fig. 5. Temperature dependence of parasitic resistances, the gate resistance: R_g, the source resistance: R_s, the total parasitic resistance: $R_g + R_s$.

than that expected from the equation $g_m = g_{m0}/(1 + R_s g_{m0})$ shown by the broken line in Fig. 4. It might be due to the decrease of effective gate length as suggested by Fukui [9].

Table II is a summary of the typical dc characteristics such as the gate resistance R_g, the source resistance R_s, the trans-

---

[1] The pinchoff voltage of the developed FET was designed to be lower than that of the conventional one in order to give the same drain saturation curent level.

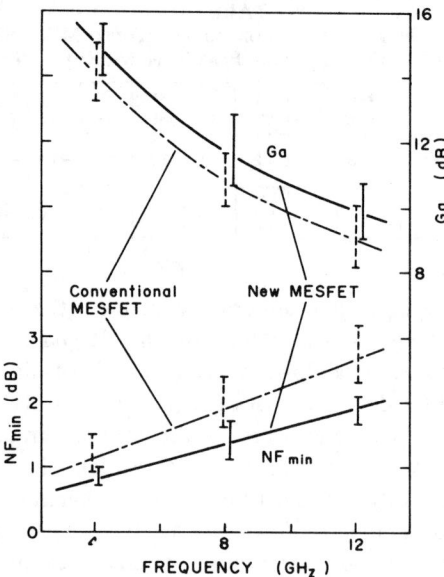

Fig. 6. Minimum noise figures NF_{min} and the associated gains G_a of the developed MESFET's compared with the conventional flat-type MESFET's.

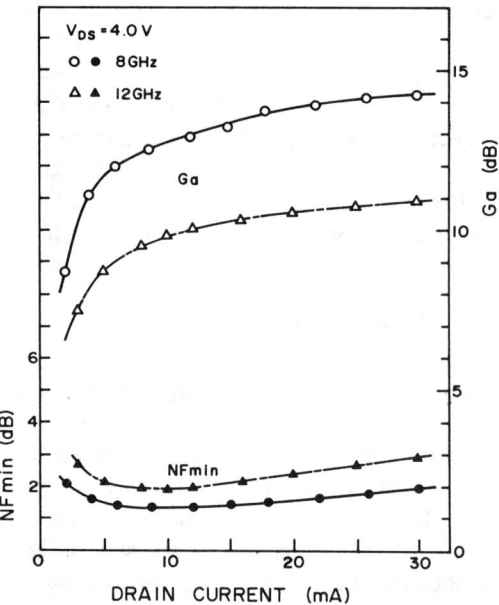

Fig. 7. Drain-current dependence of the minimum noise figure NF_{min} and the associated gain G_a at 8 and 12 GHz of the developed MESFET. The saturated drain current is 60 mA.

The developed FET's were applied to two-stage low-noise amplifiers of the 11.7-12.2-GHz band. Fig. 8 shows the performance of the developed amplifier at room temperature and at 0°C. The noise figures including the input and output isolators losses are 2.16 dB (the equivalent noise temperature T_e is 185 K) at room temperature and 1.94 dB (T_e: 163 K) at 0°C. Much lower noise figures can be achieved by further cooling of the amplifier.

The summary of RF performance of the FET itself and of two-stage low-noise amplifiers is shown in Table III. This performance is good enough to replace some of the parametric amplifiers.

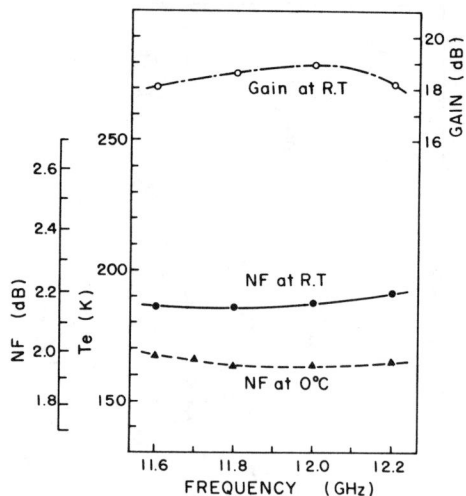

Fig. 8. Performance of the developed two-stage amplifier at room temperature and at 0°C.

TABLE III
SUMMARY OF THE PERFORMANCE OF THE DEVELOPED MESFET AND THE TWO-STAGE AMPLIFIER

	4 GHz		8 GHz		12 GHz	
	NF	Gain	NF	Gain	NF	Gain
Device	0.7	14.9	1.2	12.3	1.68	10.7
2-Stage Amplifier (0.5 GHz Band)	1.11 (84K)	29.0			2.16 (185K) 1.94* (163K)	18.9

*at 0°C (dB)

VI. SUMMARY

A new low-noise GaAs MESFET with a deep-recess structure has been developed and demonstrated to give less than a 2.0-dB noise figure at 12 GHz reproducibly. The source resistance was reduced to less than a half of that of the conventional flat-type MESFET by adopting a deep (≥ 0.3 μm) recess structure with a cylindrical shape of the recessed edge.

Using this new structure with a 0.5-μm self-aligned gate and a high-quality epitaxial layer, the noise figure was improved by more than 0.5 dB at 12 GHz compared with the conventional FET (NE 388). The best noise figures obtained were 0.7 dB at 4 GHz (14.9-dB gain) and 1.68 dB at 12 GHz (10.7-dB gain).

The developed low-noise FET's were applied to a two-stage amplifier of the 11.7-12.2-GHz band. The noise figure achieved was 2.16 dB (T_e: 185 K) at room temperature and 1.94 dB (T_e: 163 K) at 0°C. This performance was good enough to replace some of the parametric amplifiers, especially when the device was cooled further.

ACKNOWLEDGMENT

The authors wish to thank Dr. K. Suzuki and Dr. K. Mori for their helpful suggestions on microfabrication. They also wish to thank E. Nagata, S. Fukuda, and I. Haga for their valuable suggestions and discussions on the microwave measurement and characterizations as well as T. Tsuji, M. Ishikawa, and A. Higashisaka for their helpful suggestions on the fabrication process, T. Furutsuka for the noise-figure calculation, and Dr. K. Ayaki for the continuous support on this work and warm encouragement.

References

[1] H. F. Cooke, "Microwave FET's—A status report," in *1978 IEEE Int. Solid-State Circuits Conf., Dig. Tech. Papers*, pp. 116-117.

[2] M. Ogawa, K. Ohata, T. Furutsuka, and N. Kawamura, "Submicron single-gate and dual-gate GaAs MESFET's with improved low noise and high gain performance," *IEEE Trans. Microwave Theory Tech.*, vol. MTT-24, pp. 300-305, June 1976.

[3] K. Ohata, T. Nozaki, and N. Kawamura, "Improved noise performance of GaAs MESFET's with selectively ion-implanted n^+ source regions," *IEEE Trans. Electron Devices*, vol. ED-24, pp. 1129-1130, Aug. 1977.

[4] J. Wholey and M. Omori, "A low-noise microwave FET with self-aligned channels," in *Proc. 1978 Symp. GaAs and Related Compounds* (Inst. Phys., Conf. Series No. 45, Bristol and London, 1979), pp. 270-277.

[5] R. S. Butlin, A. J. Hughes, R. H. Bennett, D. Parker, and J. A. Turner, "J-band performance of 300 nm gate length GaAs FET's," in *1978 Int. Electron Devices Meet., Dig. Tech. Papers*, pp. 136-139.

[6] B. S. Hewitt, H. M. Cox, H. Fukui, J. V. DiLorenzo, W. O. Schlosser, and D. E. Iglesias, "Low-noise GaAs MESFET's: Fabrication and performance," in *1976 Symp. GaAs and Related Compounds* (Inst. Phys. Conf. Series No. 33a, Bristol and London, 1977), pp. 246-254.

[7] J. C. Vokes, W. P. Barr, J. R. Dawsey, B. T. Hughes, and S. J. W. Shrubb, "A low noise FET amplifier in coplanar waveguide," in *Proc. 1977 IEEE Int. Microwave Symp.*, pp. 185-186.

[8] C. Li, P. T. Chen, and P. H. Wang, "Sub-micrometer MESFET's fabricated on various GaAs substrates," in *1978 Symp. GaAs and Related Compounds* (Inst. Phys., Conf. Series No. 45, Bristol and London, 1979), pp. 353-360.

[9] H. Fukui, "Optimal noise figure of microwave GaAs MESFET's" *IEEE Trans. Electron Devices*, vol. ED-26, pp. 1032-1037, July 1979.

[10] H. H. Berger, "Contact resistance and contact resistivity," *J. Electrochem. Soc.*, vol. 119, pp. 507-514, Apr. 1972.

[11] T. Furutsuka, T. Tsuji, and F. Hasegawa, "Improvement of the drain breakdown voltage of GaAs power MESFET's by a simple recess structure," *IEEE Trans. Electron Devices*, vol. ED-25, pp. 563-567, June 1978.

[12] R. A. Pucel, D. J. Massé, and C. F. Krumm, "Noise performance of gallium arsenide field-effect transistors," *IEEE J. Solid-State Circuits*, vol. SC-11, pp. 243-255, Apr. 1976.

[13] R. W. H. Engelmann and C. A. Liechti, "Gunn domain formation in the saturated current region of GaAs MESFET's," in *1976 Int. Electron Devices Meet., Dig. Tech. Papers*, pp. 351-354.

Highly Reliable GaAs MESFET's with a Statistic Mean NF_{min} of 0.89 dB and a Standard Deviation of 0.07 dB at 4 GHz

T. SUZUKI, A. NARA, MASAAKI NAKATANI, AND TAKASHI ISHII

Abstract—High-performance and high-reliability low-noise GaAs MESFET's are studied from a practical point of view.

By optimizing the structure and the configuration of GaAs FET's and by developing techniques to form a reproducible thick submicrometer gate, GaAs FET's having improved characteristics have been made.

A mean minimum noise figure NF_{min} of 0.89 dB, a standard deviation of 0.07 dB at 4-GHz CW and a pulse input power capability of more than 0.4 and 2 W, respectively, and a failure rate, supported by field data, of less than 200 FIT have become practical.

Manuscript received April 28, 1979; revised October 10, 1979.
The authors are with the Semiconductor Laboratory, Mitsubishi Electric Corporation, 4-1 Mizuhara, Itami, 664 Hyogo, Japan.

I. INTRODUCTION

LOW-NOISE GaAs MESFET's are indispensable for use as microwave amplifiers and oscillators operating in the frequency region over 4 GHz [1]–[3].

Minimum noise figure NF_{min} of 0.6 dB at 4 GHz has been obtained in a laboratory. This value is nearly equal to the value calculated for GaAs FET's with a 0.5-μm gate length [4]. Nevertheless, GaAs FET's with noise figure less than 1 dB at 4 GHz, which are required for low-noise receivers in satellite communication systems, are at present difficult to obtain practically.

Concerning reliability, mean time to failure (MTTF) of 10^7–10^8 h at 70°C is deduced by thermal acceleration tests [5], [6]. However, as the failure of GaAs FET's in the field is not simply caused by thermal factor, reliability data supported by field tests are, in practice, more important.

In this paper, a study of practical low-noise GaAs FET's having superior performance and high reliability is described. Optimization of the structure and the configuration of device are studied. Devices are characterized with special emphasis of the statistical distribution of noise figure, dc surge pulse capability and RF input power capability. The reliability of improved devices has been examined by field tests.

II. IMPROVEMENTS OF GaAs MESFET'S

A. Optimization of Carrier Profile

Since low-noise GaAs MESFET's are operated at drain current as low as 10 to 15 mA under noise-matching conditions, the crystal quality and electrical characteristics of the active layer at the semi-insulating buffer-layer—active-layer interface are very important. The high net carrier density in the active layer and abrupt change of carrier profile are required [7], [8].

GaAs FET's fabricated from three kinds of epitaxial GaAs wafers with different carrier profiles were used. Each was produced by the controlled vapor phase GaAs epitaxial growth technique. Typical results are shown in Fig. 1. The relationship between minimum noise figure NF_{min}, associated gain G_a, and doping profile was also investigated. In the table inserted in Fig. 1, typical NF_{min} and G_a corresponding to the three carrier profiles are summarized. The best results for NF_{min} and G_a are both obtained from profile (B). The higher carrier density of the active region near the buffer layer provides superior performance and the decreased carrier density at the surface provides a higher breakdown voltage between source and gate.

B. Improvements of Structure and Configuration

To improve NF and G_a of GaAs FET's, the reduction of the source-to-gate capacitance C_{gs}, the lowering of the gate to source resistance R_s and the gate series resistance R_g, and the increase of the transconductance g_m [9]–[13] are very important. In Table I, the improvements of the structure and configuration of GaAs FET's for improving these physical factors are shown.

To improve reliability, the design of GaAs FET's should be based on failure analysis. Failure and degradation of GaAs FET's in the field are primarily caused by the accidental application of excess RF input power and dc surge pulse. The modes of failure, gate breakdown, and short circuit between source and drain near the gate pad, caused by thermally initiated process or electrical breakdown, have been observed [14]. Because of it, we therefore, decided to place the gate at position where the gate-to-drain spacing is larger than the gate-to-source spacing to prevent formation of short circuit and to form a gate-recess structure to suppress the concentration of

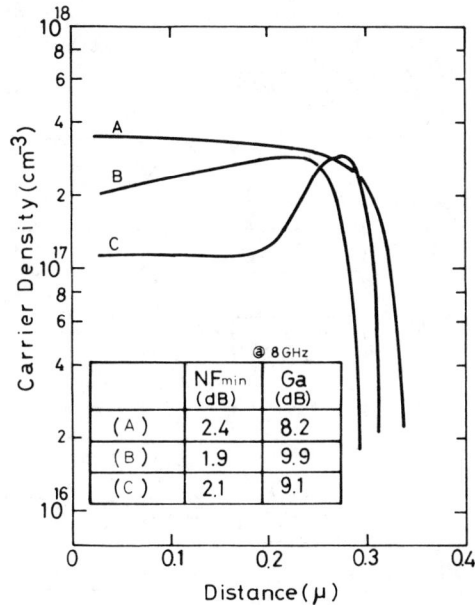

Fig. 1 Carrier profile of three kinds of epitaxial GaAs wafers and typical NF_{min} and G_a values corresponding to these three carrier profiles.

high electric field [15], [16]. R_s increases with increase of the width and depth of the recess, which, therefore, degrades the device performance. The width and depth of the recess should be optimized to balance the improvement of reliability and the degradation of performance.

Considering the above factors, the structure and the configuration of the GaAs FET's are optimized as follows for superior performance and high reliability:

1) narrow and deepen the recess gate portion to minimize the increase of R_s and suppress the concentration of high electric field;
2) use the asymmetric configuration of gate location to lower R_s and to prevent short circuits;
3) use a thick Al submicrometer gate to lower R_g and to obtain increased g_m;
4) use only one gate bonding pad in order to reduce parasitic capacitance and use two fingers to lower R_g and obtain increased g_m.

C. Fabrication

The top view and cross-sectional diagram of the resulting GaAs FET's founded on the improvements described are shown in Fig. 2.

The GaAs wafer is constructed from Cr-doped substrate and semi-insulating buffer and sulphur-doped N-type epitaxial layers. The ohmic metal of the source and drain electrodes, with specific sheet resistivity of less than 3×10^{-7} $\Omega \cdot cm^2$, was developed by a Au–Ge/Ni metallization system and the source to drain distance is 4 μm. The Al gate with two fingers of 200-μm width, 0.7-μm length, and 0.7-μm thickness was formed by using a practical 1-μm gate photomask.

Formation of the thick submicrometer gate using the 1-μm gate photomask was achieved by the accurate control of the photolithographic process, namely, the thickness of photo resist, the exposure time, the consistence of

TABLE I
IMPROVEMENTS OF GaAs FET'S FOR HIGH PERFORMANCE

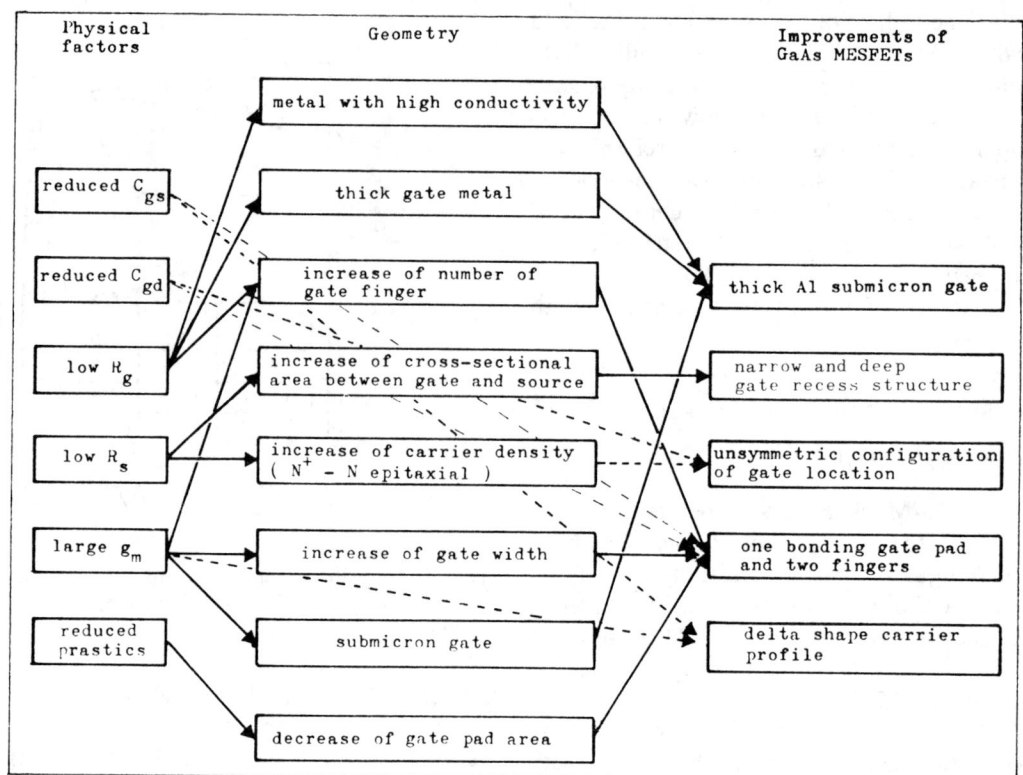

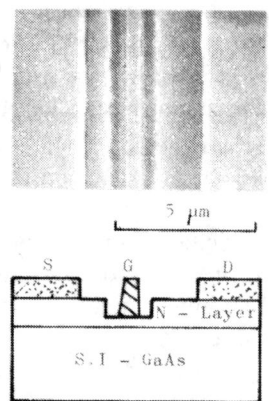

Fig. 2. Top view and cross-sectional diagram of the newly developed GaAs FET's.

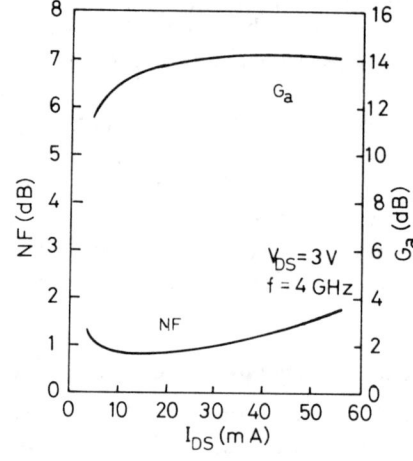

Fig. 3. Typical NF and G_a depending on I_{DS} of GaAs FET's at 4 GHz.

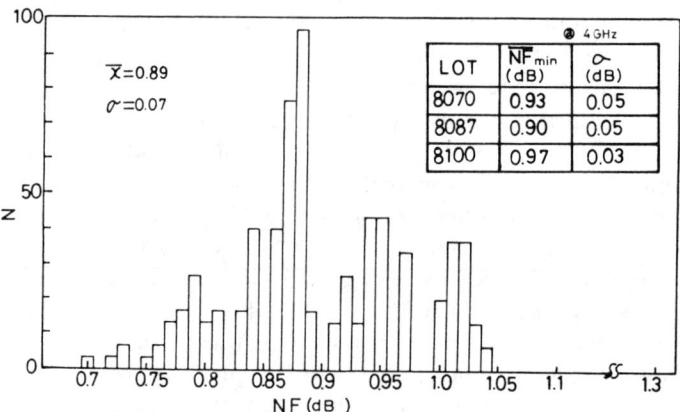

Fig. 4. Distribution of NF_{min} values for 700 devices randomly chosen from three lots.

a developer, and the temperature and the time of development. This newly developed technology made the process reproducible and the thick submicrometer gate was formed with high yield. The gate recess with 0.2–0.4-μm depth and 1.0–2.0-μm width is formed using an etchant of $5[CH(OH)COOH]_2 - 1H_2O_2$ of which etching rate is 100 Å/s, at 26.5 °C.

III. ELECTRICAL CHARACTERISTICS AND THEIR REPRODUCIBILITY

Fig. 3 shows typical NF and G_a as functions of drain current I_{DS} under the conditions of drain to source voltage V_{DS} of 3 V and at frequency of 4 GHz. NF_{min} of 0.9 dB is obtained with G_a of 13 dB at $I_{DS} = 10-15$ mA.

Fig. 4 shows the distribution of NF_{min} values for 700 pieces randomly chosen from three lots. The statistic

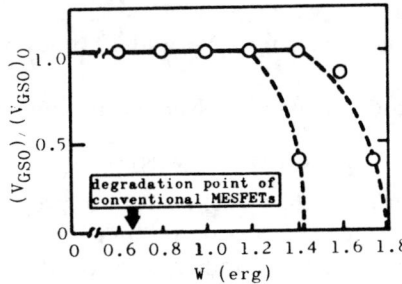

Fig. 5. The dc surge power capability.

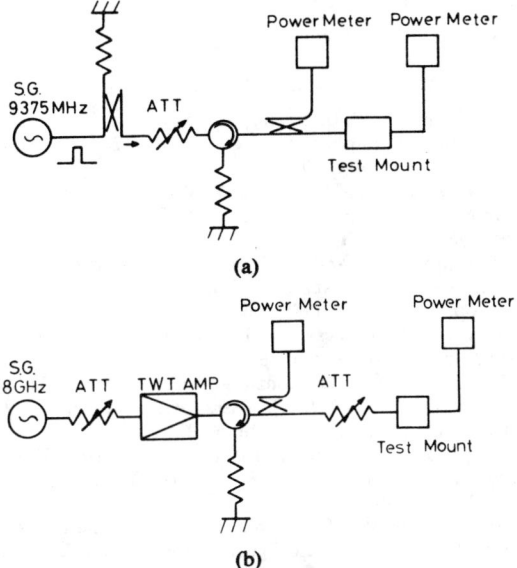

(a)

(b)

Fig. 6. Test circuit for examining RF input power capability. (a) Test circuit for CW input power capability. (b) Test circuit for pulse input power capability.

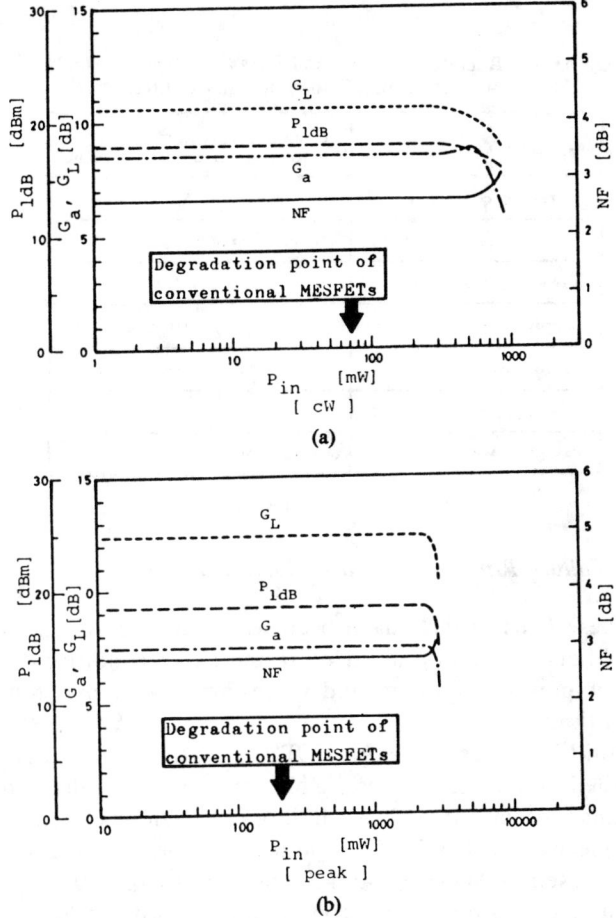

Fig. 7. Power capability of GaAs FET's against CW input power (a) and pulse input power (b).

mean NF_{min} of 0.89 dB is obtained with the standard deviation of 0.07 dB. As the best result, NF_{min} of 0.7 dB was obtained.

From the distribution, 99 percent of the devices provide NF_{min} of less than 1.1 dB at 4 GHz. As shown in the inserted table, the statistic mean NF_{min} and the standard deviation are from 0.9 to 0.97 dB and from 0.03 to 0.05 dB, respectively, for three lots. The newly developed technique to form thick submicrometer gate is confirmed by these results to be practical for use and the structure and configuration for high performance is also confirmed to be optimized.

IV. Reliability

A. Power Capability

Considering that the field failure of GaAs FET's is primarily caused by an accidental large RF power input and dc surge, a high input power capability is indispensable for high reliability. Therefore, the dc surge pulse capability was examined by a condenser discharge method. Fig. 5 shows the change of V_{GSO} for judging the degradation of GaAs FET's against the input energy W. The dc surge pulse capability was assured against energy levels as high as 1.2 erg, a level 2 times higher than those acceptable with conventional devices.

Fig. 6 (a) and (b) show the circuits for examining the pulse and CW power capabilities, respectively. The VSWR's of both test mounts are about 3. In the CW input power capability test, GaAs FET's were biased at $I_{DS}=40$ mA and $V_{DS}=6$ V (gain matching) at a frequency of 8 GHz. After operation for 2 h, the RF parameters of linear gain, G_L, G_a, and NF_{min}, used to judge degradation, were measured.

In the pulse input power capability test, the devices were biased at $I_{DS}=10$ mA and $V_{DS}=3$ V. Pulse power having 1-μs width and 1-kHz repetition rate was supplied at the center frequency of 9375 MHz. The RF parameters of G_L, G_a, and NF_{min} were measured after operation for 10 min.

Fig. 7 (a) and (b) show RF parameter dependency on CW input power and pulse input power, respectively. Degradation was found at about 0.4 W for CW devices as shown in Fig. 7 (a) and at about 2 W for pulsed devices as shown in Fig. 7 (b). These levels were more than 5 times as high as those of conventional GaAs FET's and 10 times as high for pulsed devices.

The power capability depends on the input power condition and on the bias condition. For example, pulse power capability was increased to 2.5 W by addition of a series resistance in the bias circuit and to 3.0 W by addition of a Zener diode in parallel with the bias circuit.

TABLE II
OPERATING RELIABILITY OF GaAs FET'S INSTALLED IN 4-GHz
MULTICHANNEL RADIO COMMUNICATION EQUIPMENTS

Representative equipment	4 GHz multichannel radio communication equipment
Operating condition	V_{CC} = 3 V, I_D = 15 - 25 mA
Enviromental condition	Ground, fixed
Number of devices	173
Operating time	11,260 - 23,310 hours
Component hours	2,947,980 hours
Number of failures	0
Failure rate	300 FIT, 60 % confidence level

Confirmed on Sept. 1 1979

B. Failure Rate Calculated by Field Data

These GaAs FET's have been used either in low-noise amplifiers under optimized noise matching conditions or in low-power amplifiers under optimized gain matching conditions.

For 173 pieces of GaAs FET's operating in the field, detailed data have been followed up. Results are summarized in Table II. No failure has been observed for 173 pieces of GaAs FET's. The total of component hours is at present 2 947 000 h and the failure rate of 300 FIT is calculated at a 60-percent confidence level. If the laboratory test results are taken into consideration, the failure rate will be less than 200 FIT.

V. CONCLUSION

Optimization of the structure and the configuration of GaAs FET's for high performance and high reliability was studied.

The optimized configuration contained a narrowly and deeply recessed structure, unsymmetric configuration of the gate location, thick submicrometer gate and delta shaped carrier profile.

By developing the technology to reproducibly form thick submicron gates, GaAs FET's with statistic mean NF_{min} of 0.89 dB and standard deviation of 0.07 dB at 4 GHz have become practical.

Concerning reliability, DC surge pulse capability of 1.2 erg was assured, a capability 2 times higher than that for conventional GaAs FET's. RF input power capability of 0.4 W for CW and 2.0 W for pulsed devices could also be guaranteed. These values were more than 5 times higher than conventional values for CW, and 10 times higher for pulsed.

The failure rate less than 200 FIT has become practical.

ACKNOWLEDGMENT

The authors wish to thank M. Ito for the preparation of GaAs epitaxial wafers, Y. Kadowaki and H. Hatakeyama for RF measurements. Reliability tests conducted by J. Watanabe, H. Matsumura and Y. Onodera are highly appreciated. The authors also thank Dr. K. Shirahata, Dr. S. Mitsui, and Dr. M. Otsubo for useful discussions and suggestions.

REFERENCES

[1] R. H. Knerr and C. B. Swan, "A low-noise gallium arsenide field effect transistor amplifier for 4 GHz radio," *Bell Syst. Tech. J.*, vol 57, pp. 497–490, Mar. 1978.
[2] C. L. Huang, S. L. Mason, R. W. Wong, L. J. Nevin, and J. S. Barrera, "Low noise X-band 0.5 μm GaAs FET amplifiers," in *Proc. ISSCC 79*, p. 112, Feb. 1979.
[3] W. Baechtold, "X- and Ku band amplifiers with GaAs Schottky barriers field effect transistors," *IEEE J. Solid State Circuits*, vol. SC-8, pp. 54–58, Feb. 1978.
[4] H. F. Cooke "Microwave field effect transistors in 1978," *Microwave J.*, pp. 43–48, Apr. 1978.
[5] T. Irie, I. Nagasako, H. Kohzu, and K. Sekido, "Reliability study of GaAs MESFET's," *IEEE Trans. Microwave Theory Tech.*, vol. MTT-24, no. 6, pp. 321–328, June 1976.
[6] D. Abbot and J. Turner, "Some aspects of GaAs MESFET reliability," *IEEE Trans. Microwave Theory Tech.*, vol. MTT-24, pp. 317–321, June 1976.
[7] R. E. Williams and D. W. Shaw, "Graded channel FET's: Improved linearity and noise figure," *IEEE Trans. Electron Devices*, vol. ED-25, pp. 600–605, June 1978.
[8] ——, "GaAs FET's with graded channel doping profiles," *Electron. Lett.*, vol. 13, no. 14, pp. 408–409, July 1977.
[9] K. Ohata, T. Nozaki, and N. Kawamura, "Improved noise performance of GaAs MESFET's with selectively ion-implanted n^+ source region," *IEEE Trans. Electron Devices*, vol ED-24, pp. 1129–1130, Aug. 1979.
[10] R. A. Pucel, D. J. Masse, and C. F. Krumm, "Noise performance of gallium arsenide field-effect transistors," *IEEE J. Solid-State Circuits*, vol. SC-11, pp. 243–255, Apr. 1976.
[11] H. Kodera, Y. Kaneda, and H. Sato, "A half-micron gate low noise GaAs MESFET and amplifiers," in *1977 IEEE MTT-S, Int. Microwave Symp. Dig.*, pp. 277–280, 1977.
[12] S. Takanashi, F. Murai, H. Kurono, M. Hirao and H. Kodera, "A half micron gate GaAs FET fabricated by chemical dry etching," in *Proc. 8th Conf. SSD (Tokyo, Japan)*, pp. 115–118, 1976.
[13] H. Statz, H. A. Haus, and R. A. Pucel, "Noise characteristics of gallium arsenide field effect transistors," *IEEE Trans. Electron Devices*, vol. ED-21, pp. 549–561, Sept. 1974.
[14] T. Suzuki, M. Otsubo, T. Ishii, and K. Shirahata, "Study of reliability of low noise GaAs MESFET's," in *Proc. Cont. 9th EUMC (Brighton, U.K.)*, pp. 331–335, Sept. 1979.
[15] T. Furutsuka, T. Tsuji, and F. Hasegawa, "Improvement of the drain breakdown voltage of GaAs power MESFETs by a simple recess structure," *IEEE Trans. Electron Devices*, vol. ED-24, pp. 563–567, June 1978.
[16] R. Yamamoto, A. Higashisaka, and F. Hasegawa, "Light emission and burn-out characteristics of GaAs power MESFETs," *IEEE Trans. Electron Devices*, vol. ED-25, pp. 563–573, June 1978.

Performance of GaAs MESFET's at Low Temperatures

CHARLES A. LIECHTI, SENIOR MEMBER, IEEE, AND
RODERIC B. LARRICK

Abstract—The noise- and *s*-parameters of a GaAs MESFET with 1-μm gate length are characterized versus temperature. At room temperature, the noise figure measured at 12 GHz is 3.5 dB. At 90 K, the noise figure decreases to 0.8 dB (T_e = 60 K). The associated gain is 8 dB. The design of a cooled amplifier for the 11.7–12.2-GHz communication band is discussed. At 60 K, the three-stage amplifier exhibits 1.6-dB noise figure (T_e = 130 K) and 31-dB gain.

INTRODUCTION

At room temperature, thermal noise sources dominate the noise performance of GaAs MESFET's in microwave amplifiers [1], [2]. By cooling the MESFET's to 77 K, a significant noise reduction has been observed in the 1–2-GHz range [3]–[5]. Also, a 2-dB noise-figure reduction has been reported for a 6-GHz MESFET amplifier when cooled to 190 K [6]. The purpose of this short paper is to present MESFET characteristics versus temperature that are relevant to amplifier applications. Based on these device data, the design of a cooled communication amplifier is outlined and the amplifier performance data versus temperature are discussed.

MESFET PERFORMANCE

The GaAs MESFET discussed in this short paper has been described in [7]. The gate is 1 μm long, 500 μm wide, and the channel is 0.2 μm thick. The active layer was grown by liquid-phase epitaxy directly on the (100) surface of a Cr-doped semi-insulating substrate. The criteria for the selection of the substrate are discussed in [8]. Tin was chosen as the donor impurity which has a negligible ionization energy (similar to sulfur [9]) at 1×10^{17} cm^{-3} doping density. Consequently, the free-carrier concentration in the channel is practically temperature independent.

Manuscript received October 1, 1975; revised January 26, 1976.
The authors are with the Solid State Laboratory, Hewlett-Packard Company, Palo Alto, CA 94304.

To characterize the MESFET performance at low temperature, the transistor chip is mounted in a test fixture in which miniature coaxial lines[1] are directly contacting the gate and drain pads on the chip. The source is grounded with 40-pH lead inductance. At 12 GHz, this fixture exhibits less than 0.1-dB insertion loss at the input and output ports. The backside of the MESFET substrate is mounted on a heat sink yielding a thermal resistance of 100 K/W. Under low-noise operating conditions, 100-mW dc power is dissipated in the MESFET, raising the channel temperature approximately 10 K above the ambient temperature. The temperature of the heat sink (ambient temperature) was monitored close to the chip with a copper–constantan thermocouple. The test fixture was suspended above a liquid-nitrogen bath, and the temperature was changed by varying the distance between the fixture and the liquid-nitrogen level.

The noise figure and associated gain of the GaAs MESFET were measured at 12-GHz versus ambient temperature. At each temperature, the gate bias and the RF tuning at the gate and drain[2] were adjusted for minimum noise figure. The solid curves in Fig. 1 show the result. The noise figure decreases from 3.5 dB at 300 K to 0.8 $\pm$ 0.5 dB at 90 K. This performance demonstrates the ultralow-noise capability of cooled GaAs MESFET's at frequencies as high as 12 GHz. The rate of noise-figure decrease is particularly large between 300 and 200 K, which makes thermoelectric cooling an attractive technique for improving noise performance. The gain associated with the minimum noise figure changes from 7.4 dB at 300 K to 8.3 dB at 90 K. With decreasing temperature, the reverse bias on the gate had to be increased to achieve lowest noise performance. At the optimum gate bias, the ratio of the actual drain current, I_{DS}, to the current at zero gate voltage, I_{DSS}, remained approximately independent of temperature for most FET's tested. Using this rule, the optimum gate bias can be determined from simple drain-current measurements on a cooled MESFET providing that the optimum bias at room temperature is known.

The dashed curves in Fig. 1 show the MESFET's noise figure and

[1] The inner conductor has a 100-μm diameter.
[2] The transistor is operated in common-source configuration.

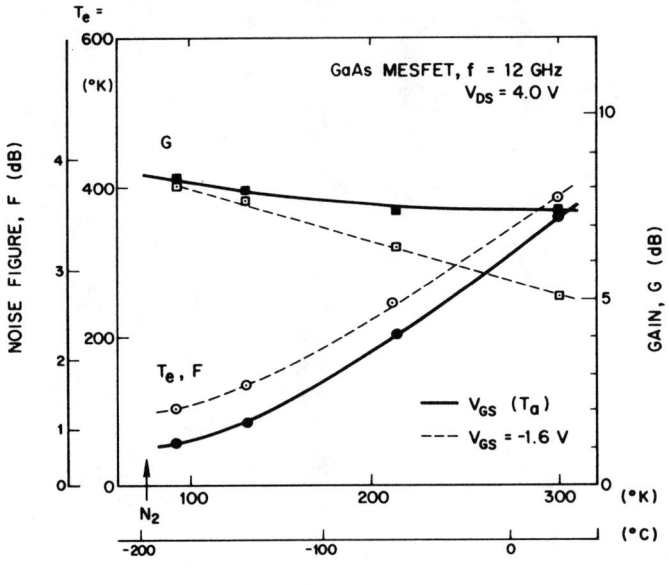

Fig. 1. Minimum noise figure, equivalent input noise temperature, and associated power gain of the GaAs MESFET versus ambient temperature. The source and load impedances and the gate bias, V_{GS}, were optimized at each temperature (solid line). Also, measurements with constant bias and fixed impedances were performed (broken line).

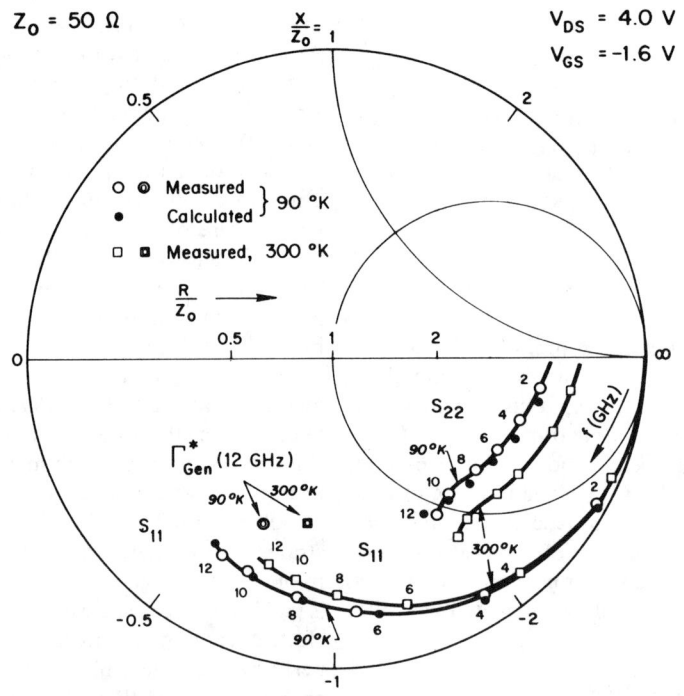

Fig. 2. Optimum generator reflection coefficient, Γ_{gen}, for minimum noise figure at 90 and 300 K ambient temperature. In this figure, the complex conjugate of Γ_{gen} is shown at 12 GHz. In addition, the measured s-parameters, s_{11} and s_{22}, are plotted versus frequency at the two temperatures. At 90 K, s-parameters calculated from the equivalent circuit (Fig. 4) are also shown.

gain versus temperature for constant gate bias and constant generator and load impedances. The gate bias ($V_{GS} = -1.6$ V), which yields minimum noise figure at 90 K, was chosen and the MESFET was tuned for minimum noise figure at room temperature. This could be a practical procedure in tuning the first stage of a cooled amplifier. However, this procedure does not lead to the minimum noise figure at 90 K, as illustrated in Fig. 1.

The variation in the optimum generator impedance versus temperature must also be considered. The generator impedances are shown in Fig. 2 for operation at 90 and 300 K. At the lower temperature, the optimum reactance is 20 percent smaller and the resistance 24 percent smaller. The MESFET's input impedance shows approximately the same relative change (Table I). The gate-to-source capacitance C_{gs}, shown in Fig. 4, increases by

TABLE I
ELEMENT VALUES IN THE EQUIVALENT CIRCUIT OF FIG. 4 THAT PROVIDE THE BEST AGREEMENT BETWEEN
CALCULATED AND MEASURED s-PARAMETERS AT 300 AND 90 K

Parameter P	P (300°K)	P (90°K)	$\frac{P(90) - P(300)}{P(300)}$	Parameter P	P (300°K)	P (90°K)	$\frac{P(90) - P(300)}{P(300)}$
C_{gs} (pF)	0.31	0.38	+0.23	g_m (mmho)	22	48	+1.2
R_i (Ω)	8.0	6.0	–	$\frac{g_m}{1 + g_m R_s}$ (mmho)	21	43	+1.1
R_s (Ω)	3.0	2.5	–	τ_0 (ps)	1	1	–
R_g (Ω)	3.0	1.5	–	C_{dg} (pF)	0.038	0.029	−0.24
$R_i + R_s + R_g$ (Ω)	14.0	10.0	−0.29	C_{dc} (pF)	0.003	0.01	+2.3
R_{ds} (Ω)	350	240	−0.31	L_g (nH)	0.05	0.05	–
R_d (Ω)	4.0	3.5	–	L_d (nH)	0.05	0.05	–
C_{ds} (pF)	0.11	0.10	(−0.09)	L_s (nH)	0.04	0.04	–

Note: The bias voltages are: $V_{DS} = 4.0$ V and $V_{GS} = -1.6$ V.

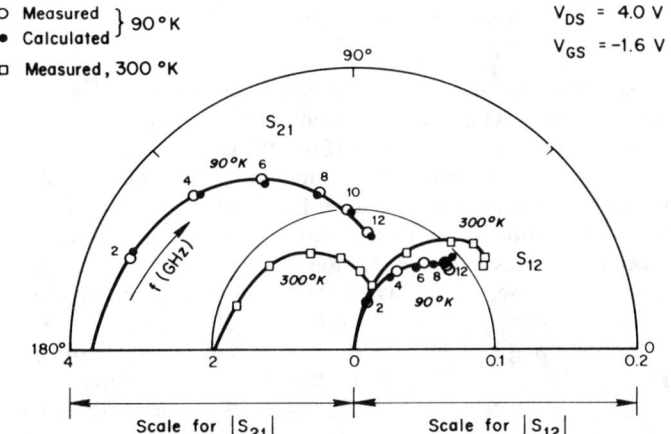

Fig. 3. Measured s-parameters, s_{12} and s_{21}, of the GaAs MESFET versus frequency at 90 and 300 K. At 90 K, s-parameters calculated from the equivalent circuit (Fig. 4) are also shown.

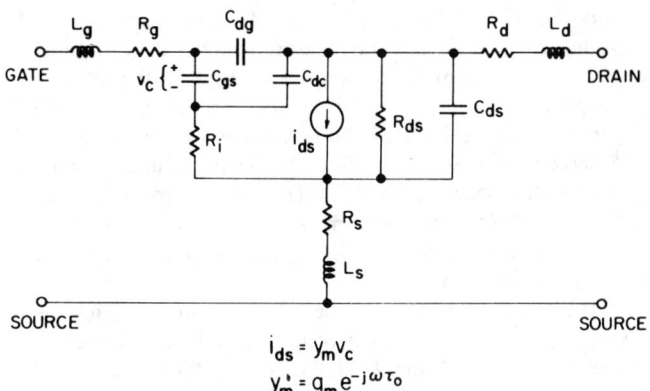

Fig. 4. Equivalent circuit for the GaAs MESFET. Element values for operation at 90 and 300 K are listed in Table I.

23 percent; and the sum of the intrinsic, source, and gate resistances, $R_i + R_s + R_g$, decreases by 29 percent. This latter change reflects the effect of an increasing electron drift mobility (+20 percent) and a larger conductive cross section in the active layer[3] at lower temperature.

The measured s-parameters of the MESFET are shown in Figs. 2 and 3 versus frequency for the two ambient temperatures, 90 and 300 K. They are described very accurately by the equivalent circuit illustrated in Fig. 4. Circuit-element values that yield best agreement between calculated and measured s-parameters are listed in Table I. The table also shows the change of each circuit component with temperature, which leads to the following discussion.

Between 300 and 90 K, the drain resistance, R_{ds} at $V_{GS} = -1.6$ V, decreases by 31 percent while the capacitance in parallel, $C_{ds} + C_{dc}$, remains constant. At $V_{GS} = 0$, R_{ds} decreases only by 20 percent. The temperature sensitivity of R_{ds} depends on the gate voltage. As the gate is reverse biased and V_{GS} approaches the cutoff voltage, $V_{GS(off)}$,[4] the temperature variation of $V_{GS(off)}$ (Fig. 5) causes R_{ds} to be more strongly temperature dependent; i.e., the widening of the conductive cross section in the active layer becomes more effective.

The parameter that varies most significantly with temperature is the intrinsic transconductance g_m. Between 300 and 90 K, g_m increases by 120 percent; a change also reflected in the variation of the forward transfer coefficient s_{21} (Fig. 3). g_m is related to the (extrinsic) low-frequency transfer admittance, $|y_{fs}|$, by

$$|y_{fs}| = \frac{g_m}{1 + g_m R_s}.$$

$|y_{fs}|$, as determined from the s-parameters, is listed in Table I, and values measured at 1 MHz are plotted in Fig. 5. The data agree within 10 percent. The temperature variation of g_m between

---

[3] The conductive cross section is larger because the measured cutoff gate voltage has increased (Fig. 5).

[4] $V_{GS(off)}$ is defined as the gate voltage that reduces the drain current to 1 mA. $V_{GS(off)}$ is temperature dependent because the depletion-layer width at the active layer to substrate interface apparently decreases with lower temperature for the particular MESFET investigated. The diffusion voltage of the Schottky barrier is practically temperature insensitive as reported in [11] and verified here with s_{11} measurements at $V_{GS} = V_{DS} = 0$.

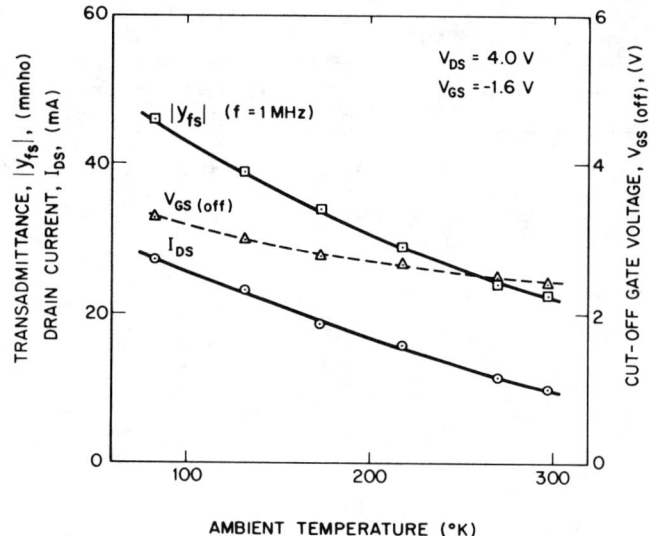

Fig. 5. Drain current, I_{DS}, transadmittance, $|y_{fs}|$, and cutoff gate voltage, $V_{GS(off)}$, of the GaAs MESFET versus ambient temperature.

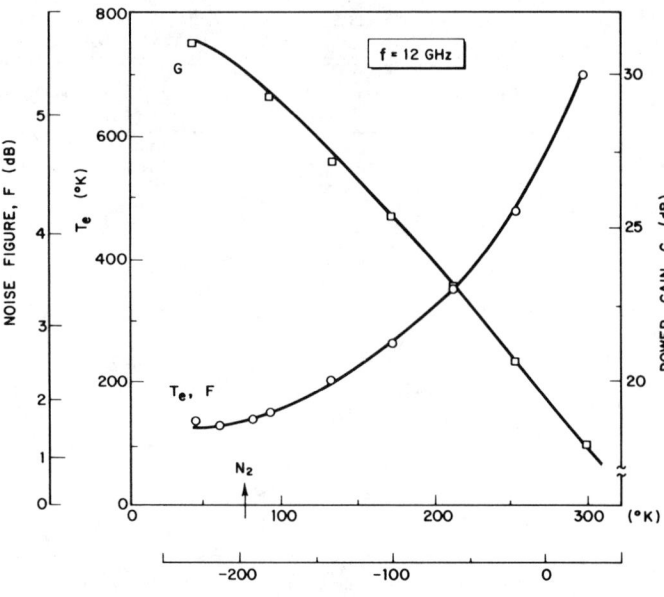

Fig. 6. Noise figure, equivalent input noise temperature, and power gain of the amplifier versus ambient temperature.

300 and 90 K can be traced to the following changes in the active-layer properties: a) the measured carrier drift mobility increases by 20 percent, b) the calculated peak drift velocity rises by 60 percent [10], and c) the cutoff gate voltage increases by 35 percent. On the basis of the parameter variations a) and b), a 45-percent increase in g_m is theoretically estimated with the simple model discussed in [12]. This calculated change agrees fairly well with measurements made at $V_{GS} = 0$. At this bias, the temperature dependence of $V_{GS(off)}$ can be neglected in a first-order approximation.

At the lower temperature, the drain-to-gate feedback capacitance, C_{dg}, is smaller (-24 percent) and the capacitance of the dipole layer in the channel [13], C_{dc}, is considerably larger. These changes are primarily due to the higher drain current ($+165$ percent, Fig. 5) at 90 K which results in a larger electron accumulation in the channel.[5] The negative space-charge, in turn, causes the depletion layer under the gate to widen toward the drain, thus reducing the drain-gate feedback capacitance.

Cooled Amplifier Design and Performance

Based on the measured temperature characteristics of the GaAs MESFET, an amplifier was designed for operation at 90 K ambient in the 11.7–12.2-GHz U.S. satellite communication band [14]. The amplifier consists of three low-noise stages with sapphire microstrip matching networks. The transistor chips are mounted directly on the ground plane of the amplifier package which is an effective heat sink. The gate bias of the first stage is adjusted to the optimum value for operation at 90 K, as discussed in the previous section. The noise match at the input port is accomplished with a low-loss structure of three elements: a cascaded transmission line next to the gate, a series interdigital capacitor, and an open-circuited shunt line.[6] The network is designed and

---

[5] The strong dependence of C_{dg} on I_{DS} is also experienced when the gate voltage is changed. At 90 K, C_{dg} is decreasing from 0.029 pF at 27 mA ($V_{GS} = -1.6$ V) to 0.007 pF at 110 mA ($V_{GS} = 0$). Notice that the smaller capacitance is obtained for the smaller drain-to-gate voltage, $V_{DS} - V_{GS}$.

[6] The series capacitor performs three functions: a) it is an element in the RF matching network, b) it acts as a dc block, and c), in conjunction with the bias network, it provides a high-pass filter which is required to protect the burn-out sensitive gate from surge pulses. The open-circuited stub can be easily adjusted in its location, length, and characteristic impedance and is, therefore, a convenient tuning element.

tuned for lowest noise figure at room temperature. Then the tuning is slightly changed to compensate for the shift in optimum generator impedance at 90 K (Fig. 2). The second and third stages are operated under bias and tuning conditions yielding higher gain than the first stage with an acceptable noise figure. No compensation is made for the variation in the MESFET's output impedance versus temperature (Fig. 2). The complete integrated circuit is enclosed in a waveguide below cutoff; the amplifier package is filled with helium and hermetically sealed.

The resulting amplifier performance versus temperature is documented in Figs. 6 to 9. At 300 K, the noise figure of the amplifier is 5.2 dB. The noise figure versus temperature curve has a steep slope at room temperature (Fig. 6). The slope decreases at lower temperatures, and at 60 K a lowest noise limit of 1.6 dB is reached[7] (equivalent input noise temperature $T_e = 130$ K). At lower temperature, e.g., 40 K, no further noise figure decrease is observed. Better noise performance can be obtained with an improved design. Based on 0.8-dB device noise figure (Fig. 2), 8.3-dB associated gain, and 0.2-dB loss in each of the matching circuits at the two ports, an amplifier noise figure of 1.2 dB is estimated at 90 K ambient temperature. This compares with 1.8 dB for the reported amplifier whose input and output matching circuits exhibited 0.5-dB loss. If the ultimate noise figure is not required, thermoelectric cooling of the amplifier is an economical and reliable approach for noise reduction. Cooling the first stage to 200 K and the following stages to 250 K is feasible and makes the 3.2-dB noise figure at 12 GHz possible (Fig. 6). In communication satellites, the use of radiation coolers for low-temperature (90 K) MESFET operation is of great interest.

Between 300 and 90 K, the gain of the amplifier increases by 11.5 dB from 17.8 to 29.3 dB. This gain variation is primarily due to the FET's increasing transconductance plotted in Fig. 7.

---

[7] The noise performance of this amplifier was measured independently at Hewlett-Packard and at the National Radio Astronomy Observatory in Charlottesville, VA. The noise figures as measured by the two laboratories agreed well with a maximum difference of 0.2 dB.

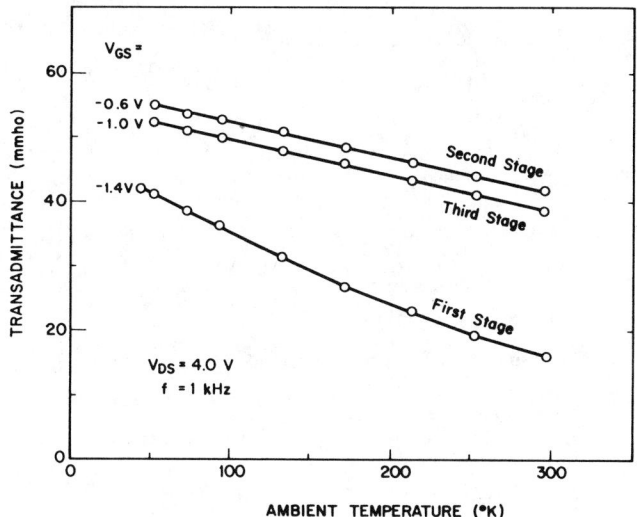

Fig. 7. Transadmittance versus ambient temperature for the MESFET's operated in the three amplifier stages.

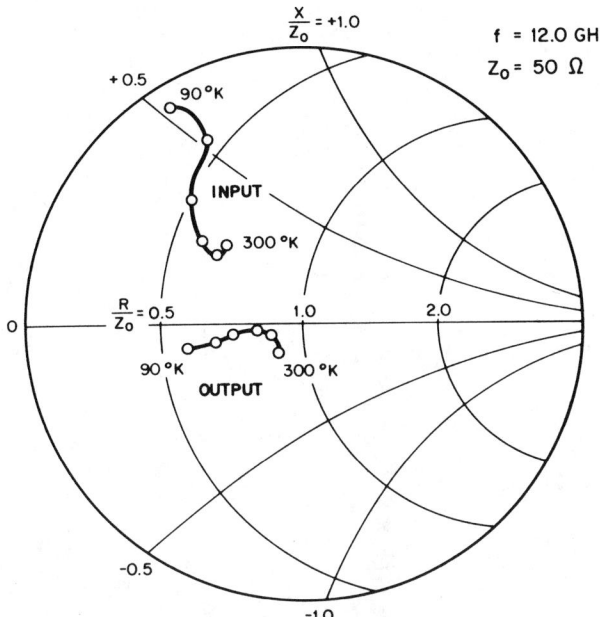

Fig. 8. Amplifier input and output impedances versus ambient temperature.

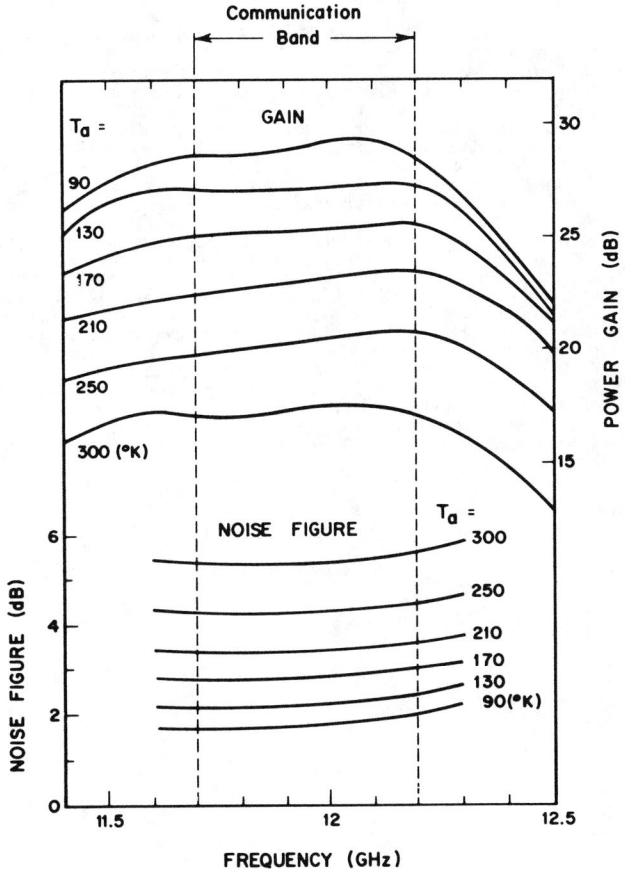

Fig. 9. Amplifier gain and noise figure versus frequency with ambient temperature as a parameter.

From data in Fig. 7, a gain increase of 12.2 dB is calculated. The reflection coefficients at the input and output of the amplifier are plotted in Fig. 8. The input VSWR rises from 1.7 at 300 K to 4.0 at 90 K. The amplifier output is approximately matched at 300 K. The output VSWR reaches 1.8 at 90 K. Fig. 9 illustrates the gain and noise figure versus frequency. The relatively small changes of the MESFET reactances prevent a frequency shift of the band center versus temperature.

CONCLUSION

It has been shown that GaAs MESFET's with 1-μm gate length are capable of very low-noise performance (NF = 0.8 dB at 12 GHz) when operated at liquid-nitrogen temperature. The MESFET's transconductance increases appreciably with decreasing temperature, raising the RF gain. The input and output impedances of the transistor are fairly insensitive to temperature variations. A three-stage amplifier was built, yielding 1.6-dB minimum noise figure at 12 GHz when operated at 60 K ambient. A 1.2-dB noise figure is estimated for an improved MESFET amplifier version operating at 90 K ambient temperature. This noise figure compares with 1.8 dB for uncooled and with 0.7 dB (T_e = 50 K) for cryogenic parametric amplifiers [15]. The MESFET's +10-dBm output power at 1-dB gain compression compares with -15 dBm [16] for the parametric amplifier. Bandwidths of 4 GHz have been demonstrated for low-noise MESFET amplifiers in C [17] and X band [7], [18] as compared to a 2-GHz bandwidth realized with parametric amplifiers at 10 GHz [16], [19]. In addition, a gain-versus-temperature slope of <0.1 dB/°C and gain stability of <0.1 dB/h are achieved without special care using FET's. Cooled GaAs MESFET amplifiers with their simple circuitry and relatively low cost are, therefore, expected to compete effectively with parametric amplifiers at frequencies up to 12 GHz.

ACKNOWLEDGMENT

The authors wish to thank Dr. D. Mellor for the design of the amplifier prototype and Dr. S. Weinreb and J. Davis of the National Radio Astronomy Observatory for noise measurements at low temperatures; especially for the amplifier performance characterization below 90 K.

REFERENCES

[1] W. Baechtold, "Noise behavior of GaAs field-effect transistors with short gate lengths," *IEEE Trans. Electron Devices*, vol. ED-19, pp. 674–680, May 1972.
[2] H. Statz, H. Haus, and R. Pucel, "Noise characteristics of GaAs

field-effect transistors," *IEEE Trans. Electron Devices*, vol. ED-21, pp. 549–562, Sept. 1974.
[3] B. Loriou, M. Bellec, and M. LeRouzic, "Performances à basse température d'un transistor hyperfréquences faible bruit à effet de champ," *Electron. Lett.*, vol. 6, pp. 819–820, Dec. 1970.
[4] J. Leost and B. Loriou, "Propriétés à basse température des transistors GaAs à effet de champ hyperfréquence," *Hyperfréquences*, vol. 54, pp. 514–522, Dec. 1974.
[5] J. Jimenez, J. Oliva, and A. Septier, "Very low noise cryogenic MESFET amplifier," in *Proc. 1973 European Microwave Conf.*, vol. 1, Paper A.5.2.
[6] P. Bura, "Operation of 6 GHz FET amplifier at reduced ambient temperature," *Electron. Lett.*, vol. 10, pp. 181–182, May 1974.
[7] C. Liechti and R. Tillman, "Design and performance of microwave amplifiers with GaAs Schottky-gate field-effect transistors," *IEEE Trans. Microwave Theory Tech.*, vol. MTT-22, pp. 510–517, May 1974.
[8] J. Barrera, "The importance of substrate properties on GaAs FET performance," in *Proc. 1975 Cornell Conf. Active Semiconductor Devices for Microwaves and Integrated Optics*, pp. 135–144.
[9] D. Eddolls, J. Knight, and B. Wilson, "The preparation and properties of epitaxial gallium arsenide," in *Proc. 1966 Int. Symp. GaAs* (Inst. Phys., Conf. Series No. 3, London), 1967, pp. 3–9.
[10] J. Ruch and W. Fawcett, "Temperature dependence of the transport properties of GaAs determined by a Monte Carlo method," *J. Appl. Phys.*, vol. 41, pp. 3843–3849, Aug. 1970.
[11] F. Padovani and G. Sumner, "Experimental study of gold-gallium arsenide Schottky-barriers," *J. Appl. Phys.*, vol. 36, pp. 3744–3747, Dec. 1965.
[12] P. Hower and G. Bechtel, "Current saturation and small-signal characteristics of GaAs field-effect transistors," *IEEE Trans. Electron Devices*, vol. ED-20, pp. 213–220, March 1973.
[13] R. Dawson, "Equivalent circuit of the Schottky-barrier field-effect transistor at microwave frequencies," *IEEE Trans. Microwave Theory Tech.*, vol. MTT-23, pp. 499–501, June 1975.
[14] C. Liechti, R. Larrick, and D. Mellor, "A cooled GaAs MESFET amplifier operating at 12 GHz with 1.6 dB noise figure," in *Proc. 1975 European Microwave Conf.*, pp. 306–309.
[15] L. Cuccia, "Status report: Modern low noise amplifiers in communication systems," *Microwave Systems News (MSN)*, vol. 4, pp. 120–132, Aug./Sept. 1974.
[16] P. Lombardo, LNR Communications, Inc., private communication.
[17] D. Ch'en and A. Woo, "A practical 4 to 8 GHz GaAs FET amplifier," *Microwave J.*, vol. 17, pp. 26 and 72, Feb. 1974.
[18] M. Walker, F. Mauch, and T. Williams, "Cover X-band with an FET amplifier," *Microwaves*, vol. 14, pp. 36–45, Oct. 1975.
[19] M. Lebenbaum, AIL Div., Cutler-Hammer Co., private communication.

GaAs Dual-Gate MESFET's

TAKASHI FURUTSUKA, MASAKI OGAWA, AND NOBUO KAWAMURA

Abstract—Performance of GaAs dual-gate MESFET, including high-frequency noise behavior, was analyzed on the basis of Statz's model. Under the design considerations developed from the analysis, fabrication and characterization of a prototype device were carried out. The present analysis was confirmed to reproduce satisfactorily the performance observed.

Minimum noise figure and associated gain observed in the device with two 1-μm gates were; 1.2 dB and 16.7 dB at 4 GHz, 2.2 dB and 16.3 dB at 8 GHz, and 3.2 dB and 12.6 dB at 12 GHz, respectively. More than 35-dB gain controllability was also obtained at 8 GHz.

I. INTRODUCTION

THE DEVELOPMENT of GaAs MESFET has stimulatedly been conducted since its invention in 1966 [1]. Many efforts have been concentrated on extending its high-frequency and high-power operating limits. As a result, GaAs MESFET's with single-gate structure have grown up to viable microwave transistors both for low-noise and high-power applications.

However, there is another attractive version, the dual-gate structure, which has a second gate between the gate and the drain. The advantages of the dual-gate, in contrast to the single-gate structure are 1) the increased functional capabilities due to the presence of two independent control gates such as gain control and signal mixing and 2) the reduced feedback or resulting improvement in power gain and stability.

The first report on GaAs dual-gate MESFET was made by Turner *et al.* in 1971 [2], followed by several pioneering papers describing its performance analysis, fabrication, and characterizations [3]-[7]. Performance analyses for optimum design of the device have not, however, been carried out thoroughly, especially for optimization of the noise behavior.

In the present paper, is described the high-frequency performance including noise behavior of the GaAs dual-gate MESFET analyzed on the basis of Statz's model [8], [9] which was shown to be applicable to a performance analysis of a device with 1-μm gate structure though its basis would become poor in case of a shorter gate [10]. Design considerations and fabrication of a dual-gate FET with two 1-μm gates are also described together with its excellent performance observed up to X-band.

II. PERFORMANCE ANALYSIS

A conceptual structure of a GaAs dual-gate MESFET and its symbolic representation are shown in Fig. 1(a) and (b), respectively. The first (G_1) and the second (G_2) Schottky gates are formed between two ohmic contacts, source and drain laid

Manuscript received September 26, 1977; revised February 28, 1978.
The authors are with Nippon Electric Company, Ltd., Central Research Laboratories, Miyazaki, Takatsu-ku, Kawasaki, Japan.

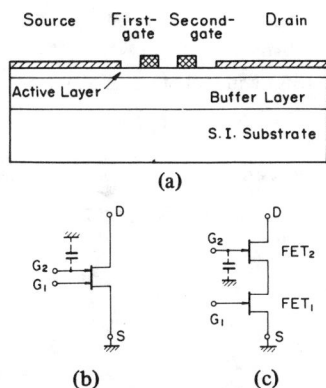

Fig. 1. GaAs dual-gate MESFET. (a) Cross section. (b) Symbolic representation. (c) Equivalent configuration composed of two single-gate FET's.

down on a thin n-type GaAs epitaxial layer grown on a semi-insulating substrate. The dual-gate device can equivalently be considered as cascode connection of two single-gate MESFET's, FET_1 and FET_2 having gates G_1 and G_2, respectively, as is shown in Fig. 1(c). The operation and characteristics of the composite (dual-gate) FET can, therefore, be realized by combining the analyzed characteristics of the two single-gate FET's operated under the same drain current. Performance analysis of the single-gate FET's which compose the dual-gate device was carried out by using Statz's model [8], [9] in which the effects of rised electron temperature and diffusion noise in the velocity saturated region were taken into account.

A. DC Characteristics

The conceptual configuration of the dual-gate FET used in the present analysis is shown in Fig. 2. It is composed of two single-gate FET's and of three parasitic resistors; FET_1, FET_2, source series resistor R_S, resistor giving resistance between two gates R_{IS}, and drain resistor R_D.

Using Statz's model, drain current I_{D1} and I_{D2} (subscript 1 or 2 in this paper specifies the quantity for FET_1 or FET_2, respectively) and channel lengths of carrier velocity unsaturated (L_{U1}, L_{U2}) and saturated (L_{S1}, L_{S2}) regions of both FET's are calculated as a function of gate bias (V_{G_1S}, V_{G_2S}) and of applied voltage across each single-gate FET ($V_{D1} - V_{S1}$, $V_{D2} - V_{S2}$). DC characteristics of the composite (dual-gate) FET for a given bias, V_{G_1S}, V_{G_2S}, and V_{DS} were obtained by adjusting $I_{D1} = I_{D2} = I_{DS}$ and voltage drops at the two FET's and at the three resistors sum up to V_{DS}.

In Fig. 3, are shown some of the calculated I_{DS}-V_{DS} characteristics of a prototype dual-gate FET with $L_{G1} = L_{G2} = 1$ μm, $L_{IS} = 1$ μm, $Z_1 = Z_2 = 300$ μm, $V_{P1} = V_{P2} = -2.5$ V, $N = 2.5 \times 10^{17}$ cm^{-3}, and $a = 0.14$ μm, where L_G, L_{IS}, Z, V_P, N, and a are

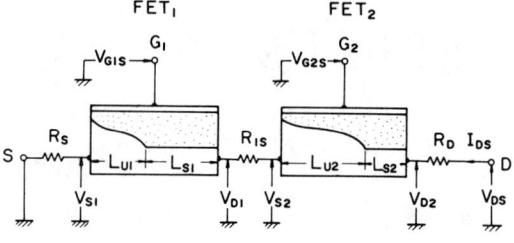

Fig. 2. Analytic model of GaAs dual-gate MESFET.

L_{u1}, L_{u2}: const. mobility region
L_{s1}, L_{s2}: const. velocity region

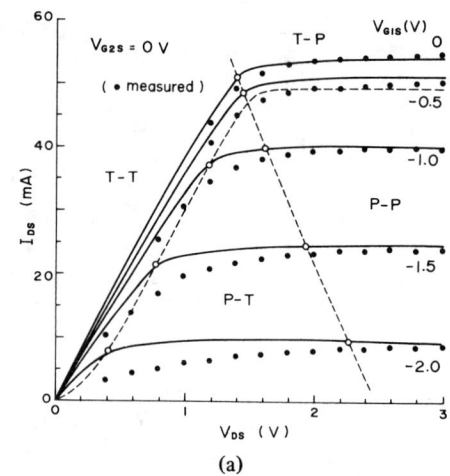

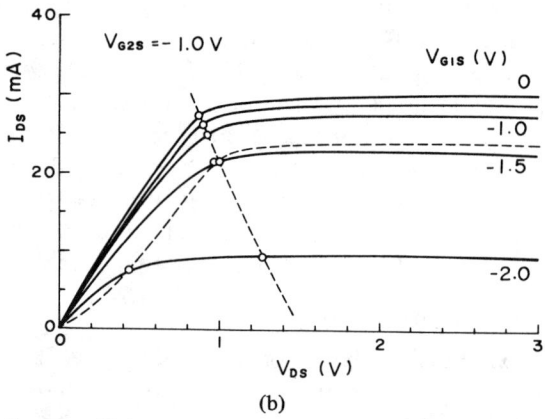

Fig. 3. Computed drain current I_{DS}–drain voltage V_{DS} characteristics as parameters of first gate bias V_{G1S} and second gate bias V_{G2S}. Measured values are also plotted. Device parameters are $L_{G1} = L_{G2} = 1$ μm, $L_{IS} = 1$ μm (L_G: gate length L_{IS}: spacing between two gates) and $V_{P1} = V_{P2} = -2.5$ V (V_P: pinchoff voltage). Dotted lines through open circles show boundaries between four operational modes: T-T, T-P, P-T, and P-P.

gate length, spacing between two gates, gate width, pinchoff voltage, impurity concentration in the channel, and active layer thickness, respectively. Values of material parameters used are: low-field mobility $\mu = 3800$ cm^2/V · s, saturation drift velocity $v_s = 1.5 \times 10^7$ cm/s, electric field at velocity saturation $E_s = 3.9$ kV/cm, and the same electron temperature versus electric field dependence as was used in [8] and [9]. Measured characteristics are also plotted in the figure and satisfactory agreement between them will be seen.

The dual-gate FET is known [7] to show four characteristic operational modes depending on bias conditions for FET$_1$ and

TABLE I
FOUR CHARACTERISTIC OPERATIONAL MODES OF A DUAL-GATE FET

	Operational Mode	
	FET$_1$	FET$_2$
T-T	triode	triode
T-P	triode	pentode
P-T	pentode	triode
P-P	pentode	pentode

triode : current-unsaturated operation
pentode : current-saturated operation

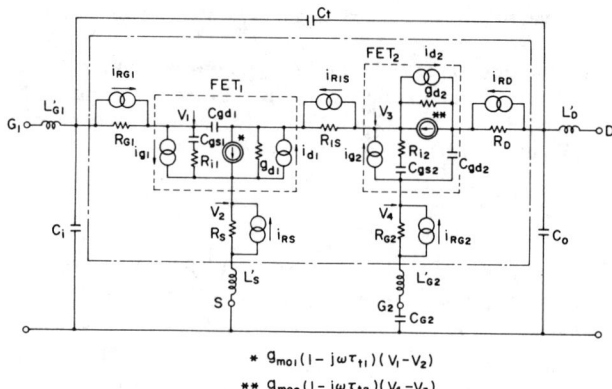

* $g_{mo1}(1 - j\omega\tau_{t1})(V_1 - V_2)$
** $g_{mo2}(1 - j\omega\tau_{t2})(V_4 - V_3)$

Fig. 4. Equivalent circuit of GaAs dual-gate MESFET. Noise sources are also shown. Circuit elements enclosed by the dot-dash line are taken into account for the analysis.

FET$_2$ as shown in Table I and in Fig. 3. The small transconductance in the triode-pentode (T-P) mode is due to small drain voltage for FET$_1$ caused by deep gate bias for FET$_2$, showing capability of gain control by V_{G2S}. The device is usually and usefully operated in the pentode-pentode (P-P) mode.

B. Equivalent Circuit Parameters and High-Frequency Characteristics

The equivalent circuit of the GaAs dual-gate MESFET is shown in Fig. 4 in which parasitic circuit elements and noise sources are also included. The intrinsic circuit elements for FET$_1$ and FET$_2$ which are enclosed by the dotted lines in Fig. 4 were analyzed on Statz's model and were given in terms of the bias dependent geometrical channel parameters (s and p in [8], [9], Appendix).

The gate resistance R_{Gi} ($i = 1$ or 2) was estimated by the following equation [11]:

$$R_{Gi} = \frac{1}{4} \cdot \frac{\rho Z_i}{3 t_G L_{Gi}} \quad (1)$$

in which a two-finger structure was assumed. Here $\rho(2.75 \times 10^{-6}$ Ω · cm), $t_G(0.5$ μm), L_G, and $Z(300$ μm) are specific resistivity of gate material (A1), thickness of the deposited gate film, gate length, and gate width, respectively.

The source series resistance R_S was also given by

$$R_S = (\sqrt{\rho_c R_{\square 1}} + L_{SG1} R_{\square 1})/Z_1 \quad (2)$$

where $R_{\square 1} = 1/(q\mu Na)$, L_{SG1} (1 μm), and ρ_c, being spacing

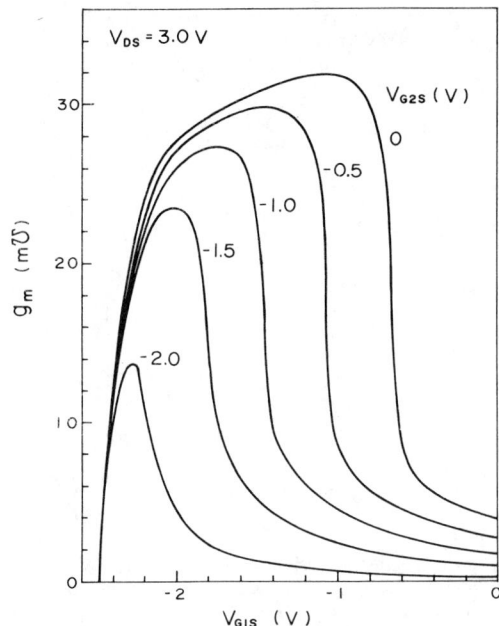

Fig. 5. Computed first gate bias V_{G1S} dependence of transconductance g_m as parameter of second gate bias V_{G2S}. Device parameters are same as those in Fig. 3.

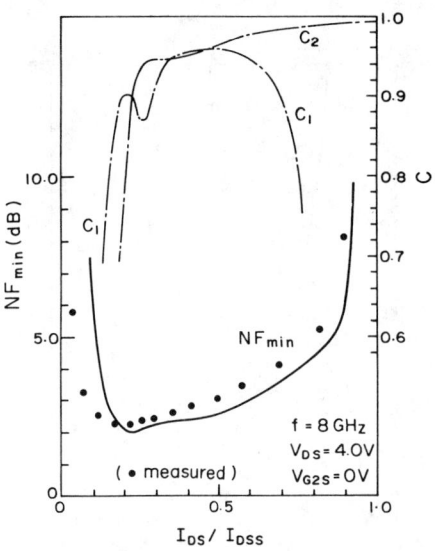

Fig. 6. Computed drain current (normalized to its saturation value) I_{DS}/I_{DSS} dependences of minimum noise figure NF_{min} and of correlation factors between noise sources C_i ($i = 1$ for FET_1, $i = 2$ for FET_2). Measured values of NF_{min} are also plotted. Bias conditions are $V_{DS} = 4.0$ V and $V_{G2S} = 0$ V. Device parameters are same as those in Fig. 3.

between source and first-gate, and specific ohmic contact resistance [12], respectively. The drain series resistance R_D was similarly estimated.

The electrical parameter of the composite dual-gate FET could be expressed in terms of the circuit elements thus obtained (see Appendix). As some of them, the transconductance to the first gate g_m and the drain conductance g_d of the dual-gate FET will be given below

$$g_m = g_{m01}/\{1 + g_{m01}R_S + g_{d1}(R_S + R_{IS})$$
$$+ g_{d1}(1 + g_{d2}R_D)/(g_{m02} + g_{d2})\}$$
$$\simeq g_{m01} \quad \text{(for P-P mode)} \tag{3}$$

$$g_d = g_{d2}/\{1 + g_{d2}R_D + (g_{m02} + g_{d2})(R_S + R_{IS})$$
$$+ (g_{m02} + g_{d2})(1 + g_{m01}R_S)/g_{d1}$$
$$\simeq g_{d1}/(g_{m02}/g_{d2}) \quad \text{(for P-P mode)}. \tag{4}$$

The variations of g_m with V_{G1S} are shown in Fig. 5 as a parameter of V_{G2S} for the same MESFET as shown in Fig. 3. The remarkably decreased g_m in the shallow bias region of V_{G1S} comes from the T-P mode operation of the device.

High-frequency performance of the dual-gate MESFET was calculated from the composite y-parameters which were obtained by cascade connection of the source grounded y-parameters of FET_1 and the gate grounded y-parameters of FET_2 together with incorporation of the parasitic circuit elements (see Appendix).

C. Noise Performance

Using the geometrical channel parameters analyzed in previous sections, the equivalent noise conductances and the correlation factors between noise sources in intrinsic FET's were calculated. All noise sources associating parasitic resistances shown in Fig. 4 were taken into account in evaluation

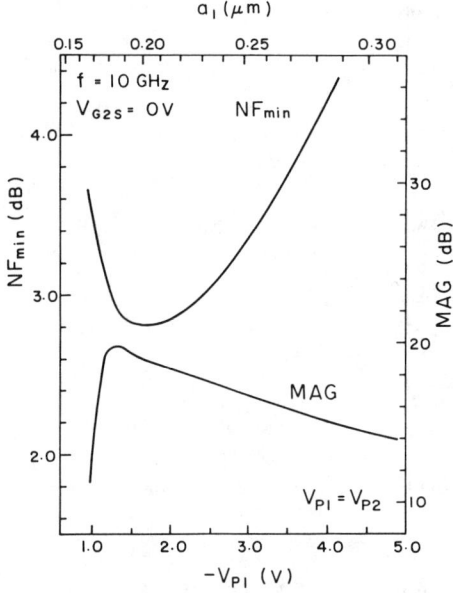

Fig. 7. Computed first gate pinchoff voltage V_{P1} dependences of minimum noise figure NF_{min} and of maximum available gain MAG at $V_{G2S} = 0$ V. $V_{P1} = V_{P2}$ also assumed.

of noise figure of the composite FET. Denoting each noise current which appears at drain by I_α (α specifies each noise source in Fig. 4), noise figure NF of the composite FET is defined by

$$NF = 1 + |(\bar{I}_{RS} + \bar{I}_{RG1} + \bar{I}_{RIS} + \bar{I}_{g1} + \bar{I}_{d1} + \bar{I}_{RG2} + \bar{I}_{RD}$$
$$+ \bar{I}_{g2} + \bar{I}_{d2})|^2/|\bar{I}_{in}|^2. \tag{5}$$

Here, $\bar{I}_{in}$ gives noise current at the drain, generated by the input signal source. These noise currents can be given in terms of the noise currents, i_α in Fig. 4, parasitic impedances, and y-parameters of intrinsic FET_1 and FET_2 [13].

The magnitude of NF depends on the signal source imped-

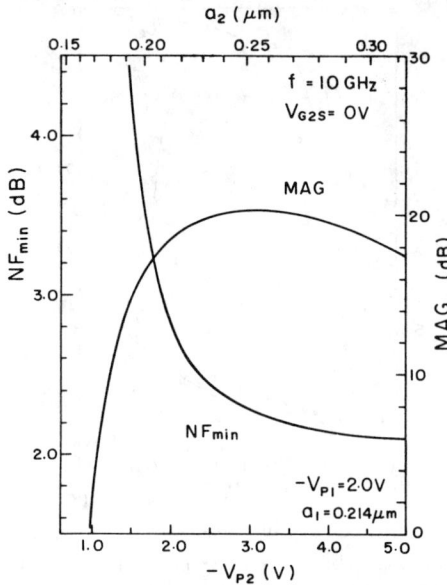

Fig. 8. Computed second gate pinchoff voltage V_{P2} dependences of NF_{min} and of MAG at $V_{G2S} = 0$ V.

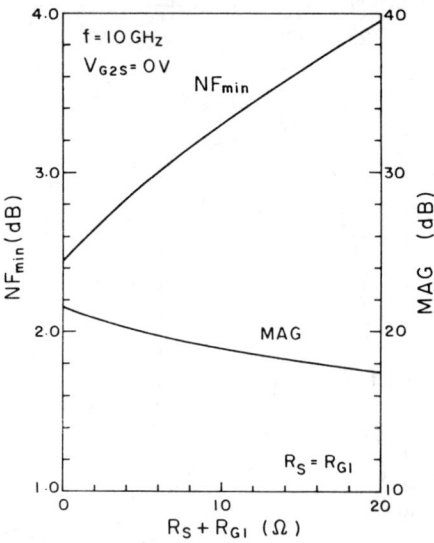

Fig. 10. Computed parasitic resistance, $R_S + R_{G1}$ (R_S: source series resistance, R_{G1}: first gate metallization resistance) dependences of NF_{min} and of MAG at $V_{G2S} = 0$ V.

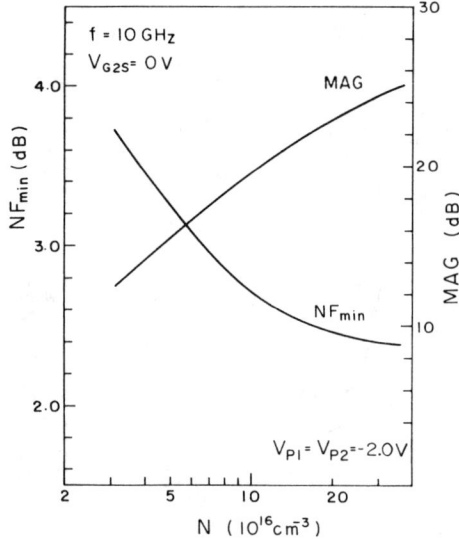

Fig. 9. Computed carrier concentration N dependences of NF_{min} and of MAG at $V_{G2S} = 0$ V.

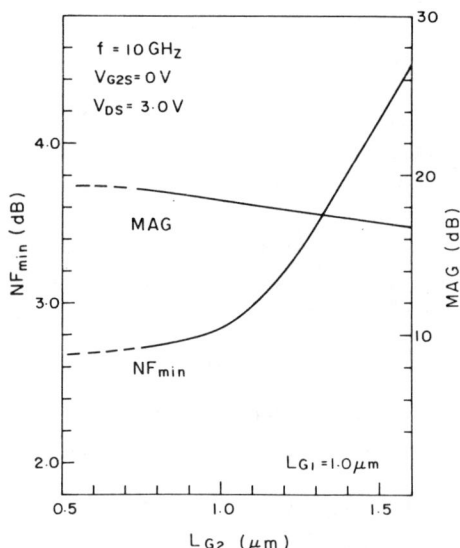

Fig. 11. Computed second gate length L_{G2} dependences of NF_{min} and of MAG at $V_{DS} = 3$ V and $V_{G2S} = 0$ V.

ance, so the impedance must be optimized in order to obtain the minimum noise figure NF_{min}. It was not easy to evaluate NF_{min} from (5) analytically. NF_{min} was, therefore, evaluated by iterative numerical calculations in which is chosen as initial value, a signal source impedance giving NF_{min} for FET_1 loaded with FET_2 operated under a given bias. The procedure used in the calculation of noise figure is described in the Appendix.

In Fig. 6, is shown the dependence of NF_{min} on drain current (correspondingly on V_{G1S}) together with the correlation factors C_i ($i = 1, 2$) calculated for the same device as shown in Fig. 3. As is shown in the figure, the calculated dependence of NF_{min} on V_{G1S} (I_{DS} in Fig. 6) reproduces well the experimental one and it will be found that V_{G1S} must be optimized for the lowest NF_{min}. The steep increase of NF_{min} in the range $I_{DS}/I_{DSS} > 0.9$ is due to the T-P mode operation of the device.

III. DUAL-GATE MESFET DESIGN

Based on the theoretical analysis, design considerations will be described here. Some of the results calculated are shown in Figs. 7-11, in which variations of NF_{min} and maximum available gain (MAG) with device parameters are presented. The parameters used in the calculation are listed in Table II. The features to be noted for optimum design of the FET are summarized below.

1) There is optimum pinchoff voltage (thickness of the active layer) for minimizing noise figure and maximizing gain. It is around -1.5 V in case of $V_{P1} = V_{P2}$ (Fig. 7).

2) By realizing $|V_{P2}| > |V_{P1}|$, NF_{min} and MAG behaviors are much improved [3], [7] (Fig. 8).

3) Increased impurity concentration N in the active layer, keeping V_{P1} and V_{P2} unchanged, leads to improvement of both NF_{min} and MAG. This is owing to the increased g_m.

TABLE II
THE DEVICE PARAMETERS USED FOR THE DEVICE DESIGN

parameters		values used	remarks
impurity concentration	N	8.0×10^{16} cm^{-3}	varied in Fig. 9
electron mobility	μ	4000 cm^2/Vsec	varied correspondingly to N in Fig. 9
electron saturation velocity	v_S	1.5×10^7 cm/sec	
built-in potential	ϕ_B	0.8 V	
pinch-off voltage	V_{Pi}	−2.0 V (i=1,2)	varied in Figs. 7,8 (through active layer thickness)
gate length	L_{Gi}	1.0 μm (i=1,2)	L_{G2}: varied in Fig. 11
gate width	Z_i	300 μm (i=1,2)	
S−G$_1$ spacing	L_{SG1}	1.0 μm	
G$_1$−G$_2$ spacing	L_{IS}	1.0 μm	
D−G$_2$ spacing	L_{DG2}	1.0 μm	

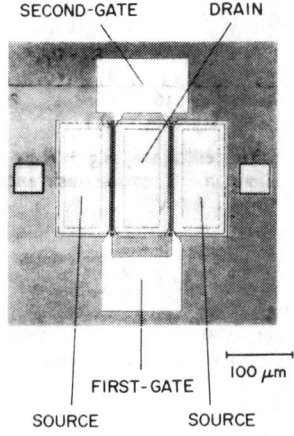

Fig. 12. Microphotograph of fabricated chip.

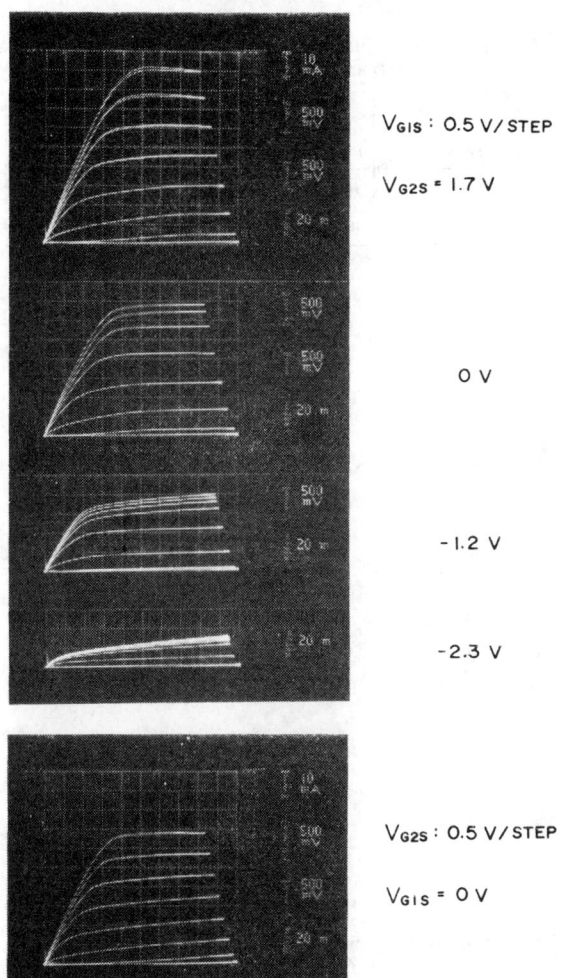

Fig. 13. Measured I_{DS}-V_{DS} characteristics showing current suppression by second gate bias V_{G2S}.

Upper limit of N will be set by a required breakdown voltage of the gate Schottky contact (Fig. 9).

4) Lower parasitic resistances R_S and R_G effectively improve NF_{min} and MAG (Fig. 10).

5) L_{G2} should also be reduced, as well as L_{G1}, to get the improved NF_{min} (Fig. 11).

IV. EXPERIMENTALS

Fabrication of a GaAs dual-gate MESFET was conducted under the design guide in the previous section. Structural parameters chosen are: $L_{G1} = L_{G2} = L_{IS} = 1$ μm, L_{SG1} (spacing between source and first gate) $= L_{DG2}$ (spacing between drain and second gate) = 1 μm, $Z_1 = Z_2 = 300$ μm, $V_{P1} = V_{P2} = -2 - -3$ V and a two-finger structure for each gate.

Vapor-grown GaAs epitaxial wafers with buffer layers [14] were used. The impurity concentrations in the active and buffer layers were 2-3 × 10^{17} cm^{-3} and ≲10^{14} cm^{-3}, respectively. Aluminum was used as a gate material. Ohmic contacts for source and drain were formed by alloying evaporated Ni-AuGe film onto the active layer. Microphotograph of the device fabricated is shown in Fig. 12.

A. DC Characteristics

Typical I_{DS}-V_{DS} characteristics are shown in Fig. 13, which clearly show the current control function of the second gate. In the P-P mode operation, g_d will be found remarkably reduced, that is one of the features of the dual-gate device.

B. High-Frequency Performance

The fabricated chip was mounted on a strip-line jig with characteristic impedance of 50 Ω and minimum noise figure NF_{min} and associated gain G_a were directly measured as a function of frequency. The results are shown in Fig. 14 together with unilateral gain U calculated from measured S-parameters. The high-frequency performance observed is summarized on Table III.

In Figs. 15 and 16, are shown variations of NF and G_a with V_{G1S} and V_{G2S} observed at 8 GHz. More than 35-dB gain control was found obtainable by changing V_{G2S} from 0 to −3 V.

V. DISCUSSIONS

DC and RF characteristics obtained in the present analysis were found to fall in quantitatively with experimental ones. However, some discrepancies between them were observed especially in marginal operating bias conditions. One of them is shown in Fig. 6 ($I_{DS}/I_{DSS} \lesssim 0.1$). The discrepancy is due to the graded impurity concentration near the interface between the active and buffer layers which is not taken into account in the present analysis.

215

TABLE III
MEASURED NOISE FIGURES NF AND POWER GAINS G AT TWO
DIFFERENT MATCHING CONDITIONS

Frequency / Matching	4 GHz G(dB)	4 GHz NF(dB)	8 GHz G(dB)	8 GHz NF(dB)	12 GHz G(dB)	12 GHz NF(dB)
Gain opt.	19.6	1.6	18.3	2.7	14.5	3.5
NF opt.	16.7	1.2	16.3	2.2	12.6	3.2

$V_{DS} = 4.0$ V, $V_{G2S} = 0$ V, $I_{DS} \simeq 10$ mA

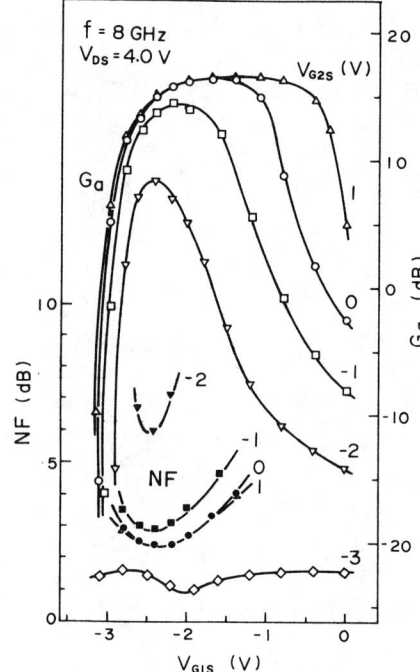

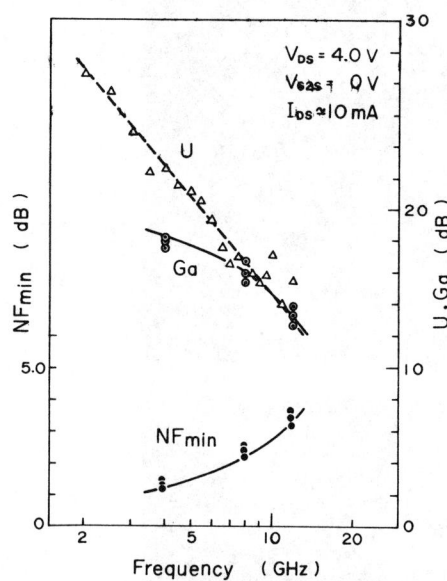

Fig. 14. Measured frequency dependences of minimum noise figure NF_{min}, associated gain G_a, and of unilateral gain U (calculated from measured S-parameters) at $V_{DS} = 4$ V, $V_{G2S} = 0$ V, and $I_{DS} = 10$ mA.

Fig. 15. Measured first gate bias V_{G1S} dependences of noise figure NF and of associated gain G_a as a parameter of second gate bias V_{G2S} at $V_{DS} = 4.0$ V and $f = 8$ GHz. Signal source impedance is not accurately optimized for noise minimum.

Because of the gradual channel approximation assumed in the constant mobility region, the validity of the present model is expected to become poor for a device with further reduced dimensions of $L_G \lesssim 5a$ [15].

VI. CONCLUSION

DC and high-frequency performances including noise behavior of a GaAs dual-gate MESFET were analyzed on the basis of Statz's model. Considerations for optimum design of the device were also developed.

With the aid of the design considerations, fabrication and characterization of a prototype device with two 1-μm gates were carried out.

The present analysis was confirmed to reproduce quantitatively the performances observed in the actual device.

The minimum noise figure and associated gain measured in the device are 1.2 dB and 16.7 dB at 4 GHz, 2.2 dB and 16.3 dB at 8 GHz, and 3.2 dB and 12.6 dB at 12 GHz, respectively. More than 35-dB gain control capability was also confirmed at 8 GHz together with only 4-dB noise figure degradation at 10-dB gain suppression from its maximum.

The improved gain in the dual-gate FET compared with a single-gate device will, in spite of the slightly degraded noise figure, suggest its usefulness for amplifier applications.

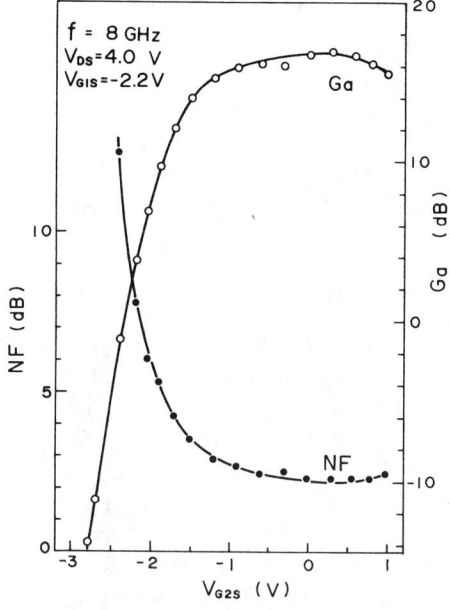

Fig. 16. Measured second gate bias V_{G2S} dependences of noise figure NF and of associated gain G_a at $V_{DS} = 4$ V, $V_{G1S} = -2.2$ V, and $f = 8$ GHz.

APPENDIX

The procedure of performance analysis described in Section II is shown in a flow-chart (Fig. 17). Equivalent circuit elements of FET_1 and FET_2 except C_{gd} were evaluated on Statz's model [8], [9]. Same expressions for parameters as were given in [8], [9] were used except for R_i and τ_t which were not given. They are also evaluated on the model as fol-

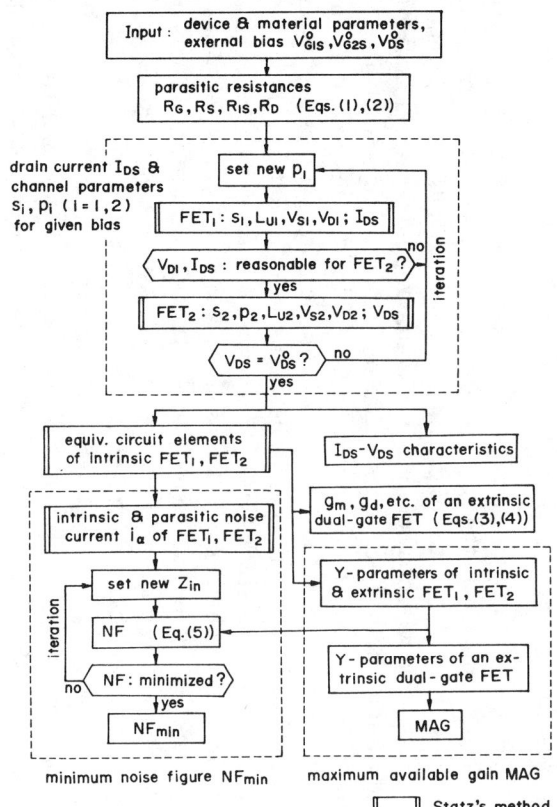

Fig. 17. Flow-chart of the analysis.

lows: R_i: mean value of channel resistance between source and gate depletion region boundary and τ_t: electron transit time under the gate region. Both R_i and τ_t were given as a function of channel parameters s and p. Gate-drain capacitance C_{gd} was approximated by $C_{gd} = \epsilon Z_i$ ($i = 1, 2$), where ϵ denotes the dielectric constant of GaAs.

To estimate the minimum noise figure of a dual-gate FET, noise current for each noise source in Fig. 4 was first calculated according to Statz's method. Noise currents at drain $\bar{I}_\alpha$ in (5) can then be obtained in terms of the noise currents i_α in Fig. 4, parasitic impedances (R_S, R_D, and R_G) and y-parameters of intrinsic FET$_1$ and FET$_2$ [13].

Noise figure NF was thus calculated and its minimum value NF_{min} was obtained by adjusting the signal source impedance Z_{in} for an optimum value by means of iterative numerical calculations.

Acknowledgment

The authors would like to express their appreciation to Dr. S. Asanabe and Dr. K. Ayaki for their continuous support and encouragment, to Y. Seki for supplying the GaAs epitaxial wafers, and to Dr. F. Hasegawa for his valuable discussions. They are also grateful to I. Nagasako for his advice in microwave measurements.

References

[1] C. A. Mead, "Schottky barrier gate field effect transistor," *Proc. IEEE*, vol. 54, pp. 307–308, Feb. 1966.
[2] J. A. Turner, A. J. Waller, E. Kelly, and D. Parker, "Dual-gate gallium-arsenide microwave field-effect transistor," *Electron. Lett.*, vol. 7, pp. 661–662, Nov. 1971.
[3] S. Asai, F. Murai, and H. Kodera, "The GaAs dual-gate FET with low noise and wide dynamic range," in *1973 IEEE Int. Electron Devices Conf., Dig. Tech. Papers*, pp. 64–67.
[4] C. A. Liechti, "Characteristics of dual-gate GaAs MESFET's," in *Proc. 1974 European Microwave Conf.*, pp. 87–91.
[5] —, "Performance of dual-gate GaAs MESFET's as gain-controlled amplifiers and high-speed modulators," *IEEE Trans. Microwave Theory Tech.*, vol. MTT-23, pp. 461–469, June 1975.
[6] R. Dean and R. Matarese, "Submicrometer self-aligned dual-gate GaAs FET," *IEEE Trans. Electron Devices*, vol. ED-22, pp. 358–360, June 1975.
[7] S. Asai, F. Murai, and H. Kodera, "GaAs dual-gate Schottky-barrier FET's for microwave frequencies," *IEEE Trans. Electron Devices*, vol. ED-22, pp. 897–904, Oct. 1975.
[8] H. Statz, H. A. Haus, and R. A. Pucel, "Noise characteristics of gallium arsenide field-effect transistors," *IEEE Trans. Electron Devices*, vol. ED-21, pp. 549–562, Sept. 1974.
[9] R. Pucel, H. Haus, and H. Statz, "Signal and noise properties of gallium arsenide microwave field-effect transistors," in *Advances in Electronics and Electron Physics*, vol. 38. New York: Academic Press, 1975, pp. 195–265.
[10] R. A. Pucel, D. J. Massé, and C. F. Krumm, "Noise performance of gallium arsenide field-effect transistors," *IEEE J. Solid-State Circuits*, vol. SC-11, pp. 243–255, Apr. 1976.
[11] P. Wolf, "Microwave properties of Schottky-barrier field-effect transistors," *IBM J. Res. Develop.*, vol. 14, pp. 125–141, Mar. 1970.
[12] M. Ogawa, K. Ohata, T. Furutsuka, and N. Kawamura, "Submicron single-gate and dual-gate MESFET's with improved low noise and high gain performance," *IEEE Trans. Microwave Theory Tech.*, vol. MTT-24, pp. 300–305, June 1976.
[13] S. Asai, S. Okazaki, and H. Kodera, "Optimized design of GaAs FET's for low-noise microwave amplifiers," *Solid-State Electron.*, vol. 19, pp. 463–470, June 1976.
[14] T. Nozaki, M. Ogawa, H. Terao, and H. Watanabe, "Multi-layer epitaxial technology for the Schottky barrier GaAs field-effect transistor," in *Proc. 1974 Int. Symp. Gallium Arsenide and Related Compounds*, pp. 46–54.
[15] D. P. Kennedy and R. R. O'Brien, "Computer aided two-dimensional analysis of the junction field-effect transistor," *IBM J. Res. Develop.*, vol. 14, pp. 95–116, Mar. 1970.

Failure Mechanisms and Reliability of Low-Noise GaAs FETs

By J. C. IRVIN and A. LOYA

(Manuscript received February 24, 1978)

The degradation and failure of low-noise GaAs FETs have been accelerated by various stress-aging techniques including storage at elevated temperatures with and without bias, exposure to humid atmospheres with and without bias, and temperature cycling. Several time-temperature-bias-induced catastrophic failure mechanisms have been observed, all involving the Al gate metallization. These mechanisms are Au-Al phase formation, Al electromigration, and electrolytic corrosion. Each of these processes results ultimately in an open gate. Accelerated aging also produces gradual, long-term degradation in both dc and RF characteristics, though the two are not always correlated. In fact, contrary to some expectations, contact resistance may increase almost two orders of magnitude without significant degradation in the noise figure or gain of a low-noise transistor. Besides contact resistance, other mechanisms such as traps in the channel are thought to play a role in the degradation of RF properties. It was found that all the important degradation mechanisms are bias-sensitive and that aging without bias gives erroneously long lifetime projections.

The cumulative failure distributions for the mechanisms observed approximate a log-normal relation with standard deviations between 0.6 and 1.4. The relevant degradation or failure processes have activation energies near 1.0 eV, which give rise to projected median lifetimes at 60°C (channel temperature) over 10^7 hours and corresponding failure rates (excepting infant mortality) under 40 FITs (40 per 10^9 device-hours) at 20 years of service.

I. INTRODUCTION

This paper describes the goals, experimental methods, and results of a study of the reliability of low-noise gallium arsenide field-effect transistors[1] involving about 1500 devices and 1.5 million device-hours of aging. The ultimate purpose of this work is twofold: (i) to calculate the probable failure rate as a function of time; (ii) to identify the failure and degradation mechanisms and propose corrective action where possible. To estimate the failure rate of the device for any given operational conditions, each of these mechanisms should be characterized in terms of the nature of its cumulative failure distribution, the median life and standard deviation of the distribution, and the activation energy of the mechanism.

Since the reliability of a GaAs FET depends intimately on its structural details, especially the choice of metallization, the structure of the devices studied is described in Section II of this report. The various acceleration methods, measurement techniques, and other aspects of the experimental program will be discussed in Section III. The failure modes observed may be categorized as either sudden or gradual. The former are marked by a complete collapse of dc and RF properties and are almost always associated with a failure of the gate metallization. They are the subject of Section IV. (Burn-out due to undesirable voltage pulses is considered a matter of handling technique or circuit design and is not investigated in the present work.) The gradual failures involve degradation of the important RF properties, especially the noise figure and the gain. There is an associated, though not well correlated, change in the observable dc characteristics. The gradual degradation of low-noise GaAs FETs is discussed in Section V. In Section VI, the pertinent failure statistics are summarized and some cumulative failure distributions are shown. Finally, estimated failure rates under typical operational conditions are presented in Section VII, together with some prognoses with regard to other operating environments.

II. THE STRUCTURE

Two slightly different versions of low-noise GaAs FETs were studied, differing primarily in the details of the gate bonding pad. In the earlier form, shown in Fig. 1, the Al gate metallization extends under the entire bonding area which is covered by a titanium-platinum-gold final metallization.[1] In the later version, the bonding area is separated laterally from the Al to which it is connected by the Ti-Pt-Au final metal. This is shown in Fig. 2. In both cases, the gate bonding pads, as well as the source and drain bonding pads, lie on the semi-insulating substrate. A schematic cross-sectional view of the source and drain contacts is given in Fig. 3. The ohmic contact consists of a layer of 88-percent Au/12-percent Ge, topped successively by a layer each of silver and gold and then alloyed. A final metallization of Ti-Pt-Au, as described above, is applied on top of the alloyed ohmic contact.

The active n-type layer is 3000 to 6000 Å thick with a donor density of approximately 1×10^{17} cm^{-3}. An n+ layer, about 3000 Å thick and with a donor density around 2×10^{18} cm^{-3} underlies the source and drain

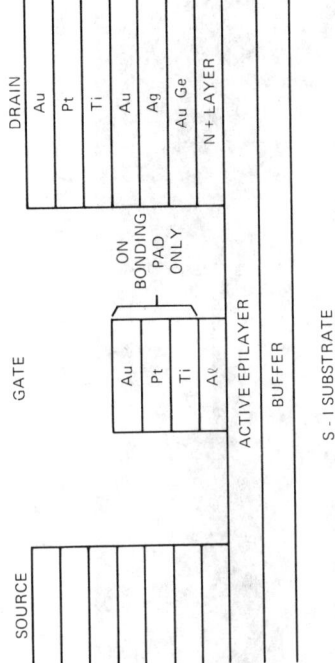

Fig. 3—Cross section of GaAs FET contact metallurgy (relative thicknesses not to scale).

contacts. A buffer layer with a donor density $< 10^{13}$ cm^{-3} and 2 to 5 μm thick separates the active layer from the semi-insulating (Cr-doped) substrate. Except in a few cases, no passivation layers were present on the finished chip. The chip size is 0.5 mm square and 50 μm thick. Each chip is bonded in a 2.5-mm square package which is hermetically sealable.

III. EXPERIMENTAL METHOD

To accelerate the degradation or failure of the GaAs FETs, a number of methods were used. The primary of these (about 1,000,000 device-hours) was aging at elevated temperatures, both with and without bias. The ambient was air at temperatures of 88°, 180°, 220°, 250°, and 275°C. The bias duplicated normal operating values consisting of 5V on the drain and a gate bias of -0.1 to -2 V, as necessary to produce 15 mA of drain current.[2] At this bias, the elevation of the channel temperature above ambient is estimated to be about 8°C. Due to the wide-band instability of GaAs FETs, RF oscillations will readily occur even at high temperatures. Such oscillations are in themselves sometimes destructive and they may also produce instantaneous, or by rectification, dc bias values of unknown and uncontrolled magnitudes. Thus considerable effort was devoted to the suppression of oscillations by various means. Dissipative media (Eccosorb), RC networks, and ferrite beads were employed, with various degrees of effectiveness. One principal difficulty was the incompatibility of some of the stabilizing components with the high temperatures involved and the fact that it is desirable to place such stabilization as near the FET as possible. Ferrite beads were the most effective and usually succeeded in quelling oscillation. Zener diodes were also employed in both the drain and gate supplies to protect the FET from destructive voltage transients.

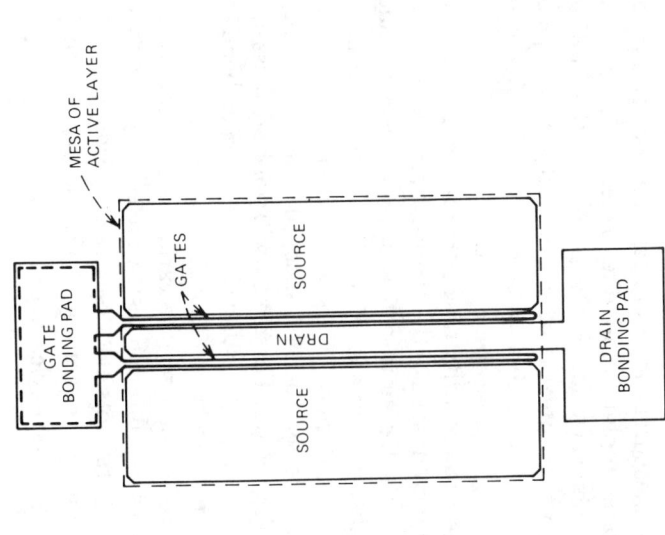

Fig. 1—Plan view of early low-noise GaAs FET.

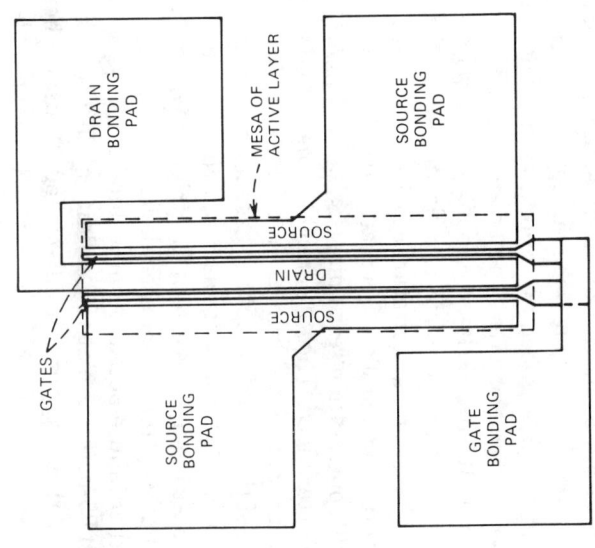

Fig. 2—Plan view of later model low-noise GaAs FET.

IV. CATASTROPHIC FAILURES

4.1 Au-Al phase formation

The most common cause of complete dc and RF failures encountered in this study was related to the formation of Au-Al compounds (a version of this on Si devices is known as purple plague). A dramatic example is shown in Fig. 4. In this case, the force of thermocompression bonding to the top Au layer has ruptured the integrity of the intervening Ti-Pt layer and caused contact between the Au top layer and wire and the bottom Al layer at the end of the gate structure. However, the loss of gate continuity is not due to embrittlement and subsequent parting of the bond nor to the high resistance of the Au-Al compound. Fed by the surplus of available Au, the Au-Al system (of which Au_5Al_2 is the favored end product) acts like a sink for the surrounding Al and has produced voids in the Al gate structure (the Kirkendall effect). The voids in the gate are visible in Fig. 4. The presence of Ga may catalyze this reaction

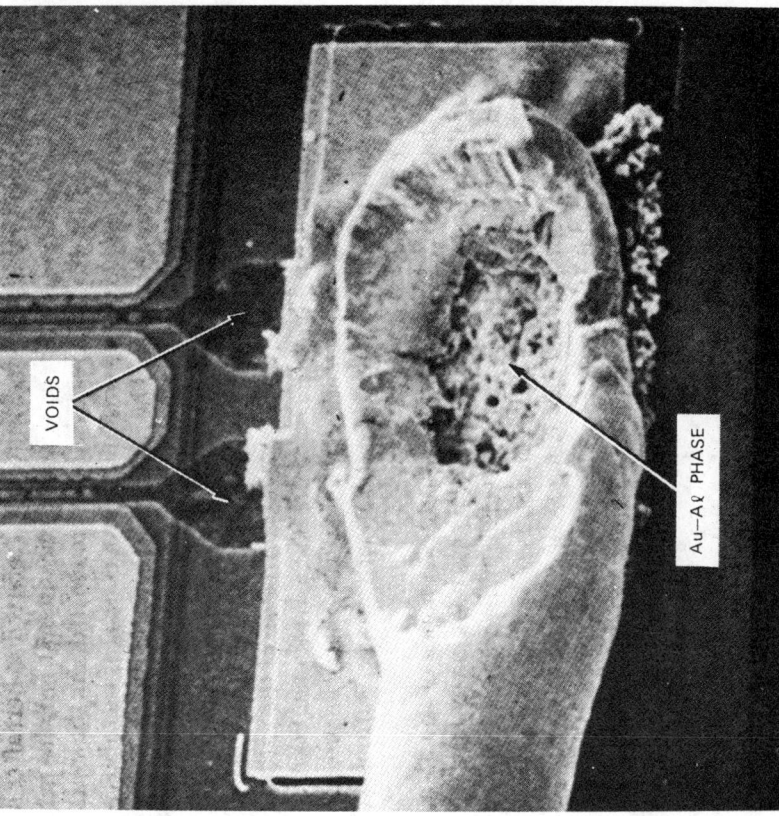

Fig. 4—SEM photo of Au-Al phase at gate-bonding pad of early model FET after aging at 250°C with bias.

While aging units under bias, the dc bias values could be monitored and were recorded daily. Catastrophic failures, that is, short or open circuits in the drain or gate, were thereby readily observed. RF properties could only be determined by periodic removal of the units and testing in either a tunable or a fixed-tuned amplifier. Thus, at intervals which ranged from 100 to 1000 hours, groups of FETs were temporarily removed from the aging environment and dc and RF measurements were performed. The dc characterization consisted of photographing the output characteristics from which the saturated drain current, I_{DSS}, and the low-field source-drain resistance, R_S, i.e., (dV_{DS}/dI_{DS}) at $V_G = V_{DS} = 0$, were determined. The RF parameters measured were the noise figure, NF, and the associated gain, G, at 4 GHz and 15 mA, 5V drain bias. In the case of the tunable amplifier used in the earlier stages of this study, the minimum NF was obtained; the NF obtained in the fixed-tuned amplifier (a modified Western Electric 652A2) was near-minimum, but not actually optimized for each device.

Another failure-acceleration technique used was storage under bias in air of 85-percent relative humidity at 85°C (referred to hereafter as 85/85). In these experiments (about 150,000 device-hours), only the gate was biased at $-4\frac{1}{2}$ or $-6V$ with respect to the grounded drain; the source floated. Some devices were aged without bias, of course, as was also the case at the higher temperatures. The reverse leakage and the continuity of the gate were checked hourly, then daily, and finally weekly in these experiments, which varied in duration from a few hours to a year. Periodic RF measurements were generally not performed on these devices since catastrophic failure due to electrolytic corrosion of the gate was the mechanism studied. The purpose of the 85/85 experiments was to determine the integrity of "hermetically sealed" packages, the presence of corrosive contaminants therein, and the effectiveness of various waterproofing or passivation coatings.

To test the security of the thermocompression bonds of the 25-μm diameter gold leads to the source, drain, and gate bonding pads as well as to test the hermetic seal, devices were cycled, under bias and without bias, between −40° and +125°C. The continuity of the bonds was tested before and after cycling as well as during cycling, in a few cases. Some devices were also thermal-shocked by alternate immersions in freezing and boiling water. These tests will not be discussed further, since in no case (out of 28,500 bond-cycles) was an open bond observed. In fact, no open bonds have been encountered among any of the over 1500 devices tested, before or after the various aging regimes described above. No centrifugal or vibration tests were employed in this program.

Lastly, a few lots of FETs have been aged without acceleration—that is, under normal operating dc bias (no RF) at room temperature (27°C), totaling 250,000 device-hours.

as does Si in the case of Si devices[3] (or other devices with SiO_2 layers). Figures 5 and 6 show other examples of Au-Al interaction leading to open gates. Note that the FETs of Figs. 5 and 6 employ the layout shown in Fig. 2, in which the gate bonding pad is separated laterally from the Al structure. However, Au and Al still were able to interdiffuse due to a slight mask misalignment. Both the devices shown in Figs. 4 and 5 were

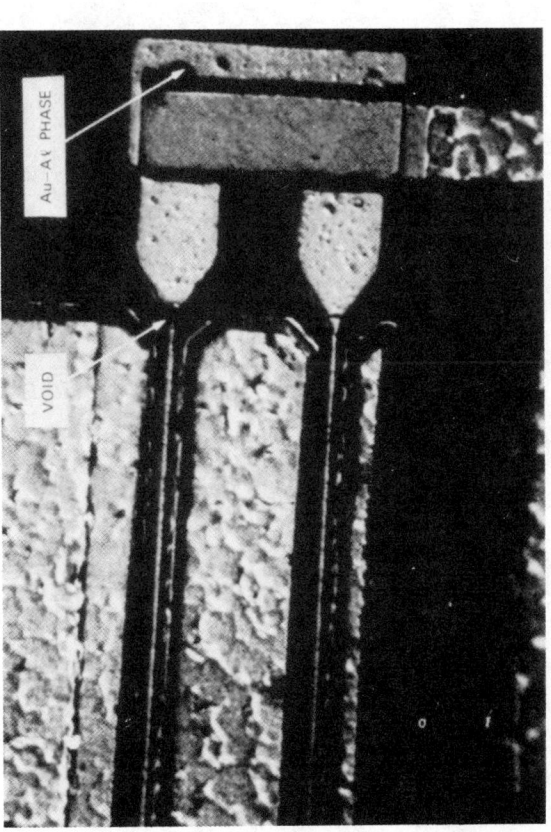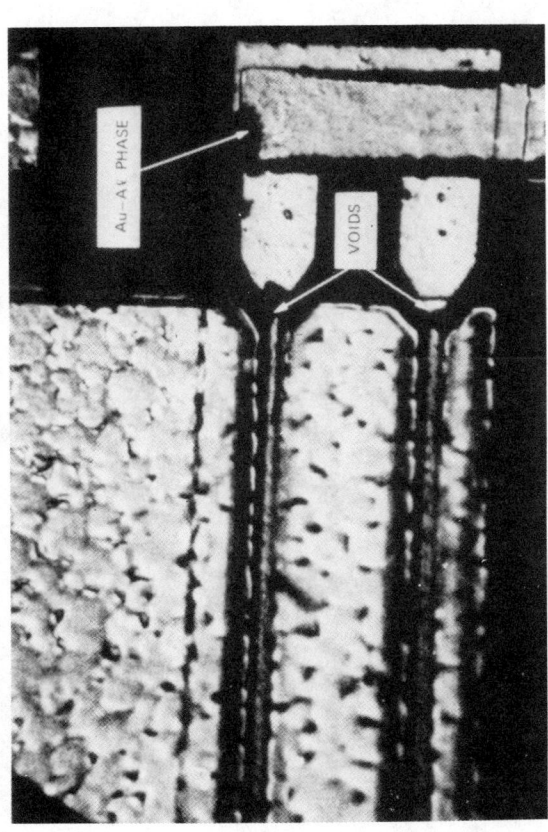

Fig. 5—Optical photos of Au-Al phase and consequent open gates in later model FETs after aging 144 hours at 250°C with bias.

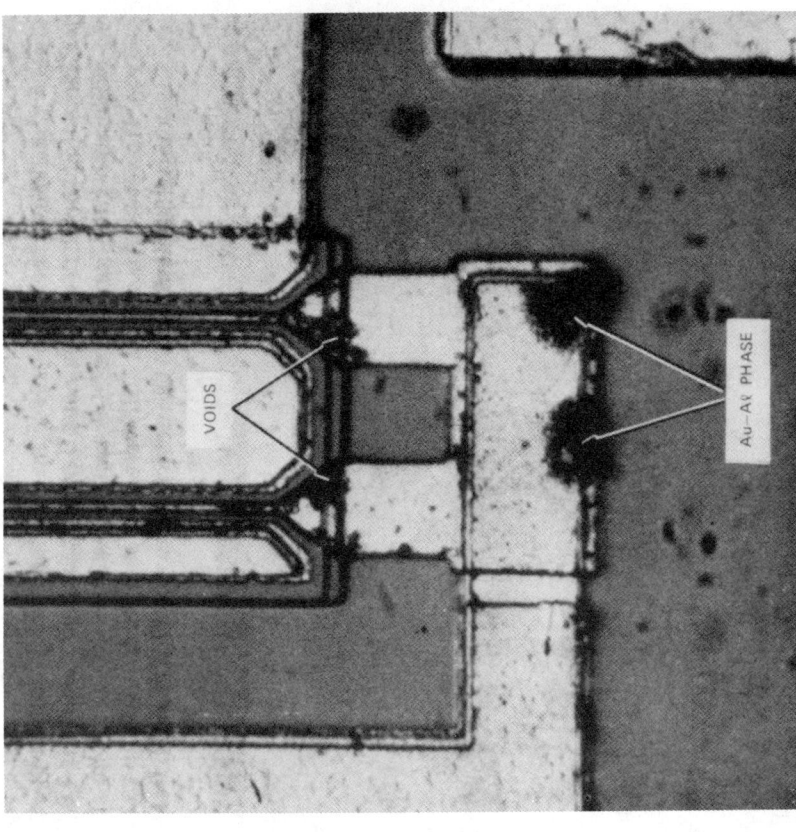

Fig. 6—Optical photo of Au-Al phase and consequent open gates in later model FET after aging at 144 hours at 250°C without bias. (This device is from a slice much more prone to Au-Al phase formation than the device of Fig. 5.)

aged under bias. The FET of Fig. 6 was aged at the same temperature (250°C) without bias.*

The detailed processes of this failure mechanism have not been fully unravelled, but a number of pertinent observations are summarized below.†

(*i*) Every FET that has failed due to an open gate (more than 100 have been examined) exhibits a Au-Al interaction site somewhere in the region where Au and Al overlap. This site may appear to be insignificantly small.

(*ii*) The median time to failure due to open gates is 3 to 10 times longer

---

* Even with perfect registration, Au-Al contact is expected to occur eventually due to diffusion through the Ti-Pt barrier, though no such cases have been observed so far in these experiments. A mask modification which eliminates the Au layer from the bridge between bonding pad and gate virtually prevents any Au-Al reaction.

† The authors are indebted to A. T. English for his assistance in analyzing the interdiffused gate structures.

among units aged without bias than among identical devices aged at the same temperature with bias.

(iii) In all bias-aged failures, a void appears in the gate stripe just where it broadens. This is the point of maximum current density in the gate metallization. Voids may also (but do not always) appear in the broad Al area or at the mesa step.

(iv) Failures among units aged without bias have voids scattered generally about the broad Al regions and frequently over the mesa step, but usually not at the aforementioned point of maximum current density.

It is apparent from these observations that there is a decided dependence on the presence of bias. It is important to note that, at 250°C, gate leakage currents at operating bias are one to two orders of magnitude larger than at room temperature, i.e., 20 to 200 μA instead of 1 to 5 μA. Even so, the maximum gate current densities during bias aging are calculated to be only 1×10^4 A/cm^2. This value is generally considered "safe" with regard to electromigration at 250°C. Furthermore, no correlation is found between failure and the gate currents of individual units during aging. Thermal dissipation in biased units in 250°C ambient raises the channel temperature to approximately 258°C. However, unbiased units in 275°C ambient have a much lower incidence of open gates than the 258°C bias-aged units. Thus, this aspect of self-heating cannot explain the bias dependence. Also, joule self-heating within the gate stripe itself is calculated to cause less than 1°C temperature rise, which, of course, also fails to justify much electromigration at these apparently modest current densities. However, current densities in the gate structure are not accurately calculable, since actual Al cross sections vary with the topography of the surface, especially at the mesa edge. It is known that electromigration is influenced by grain size and very little of the voluminous electromigration literature treats stripe widths as small as the 1-μm gates involved here. Thus, electromigration in conjunction with Au-Al phase formation is tentatively thought to be responsible for gate failures in bias aging. Electromigration would transport Al down the stripe away from the bonding pad while Au-Al phase formation causes diffusion in the opposite sense. Perhaps electromigration inhibits Al atoms near the gate throat from replacing the atoms just downstream in the wider portion of the structure (where the current density is less) which are being drawn by diffusion toward the Au-Al compound. The formation of voids may be accelerated by this tug-of-war situation.

If the above hypotheses regarding this gate failure mechanism are correct, the temperature dependence would be quite complex. The activation energy of Au-Al phase formation has been variously reported with values between 0.6 and 1.0 eV.[4] (In any case, the interaction with Ga may alter these values.) Al electromigration (at constant current density) is reported to have an activation energy of 0.5 to 0.7 eV.[5] However, as mentioned earlier, the gate leakage current itself is temperature-sensitive. The latter two effects together, it is calculated, should give the electromigration an effective activation energy of 1.2 eV.

The experimental situation, unfortunately, is not much clearer. On slices (of the type in Fig. 2) where Au-Al phase formation occurs, its occurrence is quite erratic, depending as it does on slight vagaries in alignment, lift-off, etching, and other details of pattern formation. Thus, among devices aged *without bias*, the median life (ML) varies greatly from slice to slice, though the activation energy observed is fairly consistent and near 1 eV. Electromigration varies as the second or third power of current density which, in turn, differs widely from one unit to another. However, no failures have been observed which appeared to be due to electromigration alone. When units are aged *under bias* at elevated temperature, both mechanisms are thought to be operative, though the degree of dominance by the one mechanism or the other probably varies both among devices and as a function of time during the course of void formation. Initially, phase formation is probably dominant, but when the cross-sectional area has been diminished enough and the local current density increases, electromigration becomes more important. In principle, it is inappropriate to use an activation energy to characterize this joint process consisting of two mechanisms. However, since both mechanisms are expected in this case to have an activation energy near 1 eV, as described above, it is a useful approximation to apply an "activation energy" to the combined effect. As expected, the experimental data are not entirely consistent, but are grouped about a value of 1.0 eV.

No typical ML can be cited for bias-aged devices, since many slices are entirely free of this mechanism. However, in the worst case, an ML of 94 hours at 250°C with bias has been observed. It is important to note that this mechanism has been observed in the present study in devices biasaged at temperatures as low as 180°C in times as short as 240 hours. Weaver and Brown detected Au-Al interdiffusion at 84°C in 3 hours.[6] Thus, Au-Al phase formation and subsequent destruction of GaAs FETs in which the choice of metallurgy and layout permits this combination cannot be dismissed as an exclusively high temperature phenomenon. However, an appropriate layout can completely eliminate the possibility of Au-Al phase formation.

4.2 Electrolytic corrosion

Figure 7 is an example of electrolytic gate corrosion. The corrosion shown was produced by a 2-hour exposure to an atmosphere of 85°C/85% RH with 6 V negative bias on the gate. The device was uncapped. This corrosion is clearly electrolytic, since in the absence of gate bias no sig-

intact. Water and chlorine are the chief offenders, and the trace amount of both which may be adsorbed on the package interior surface are apparently sufficient to produce corrosion. Patches of unremoved photoresist may also harbor enough impurities to cause corrosion. Figure 8 shows the corroded gate of a sealed device that failed after 240 hours under bias in 85/85 though no leak was detectable after removal from the chamber. A Krypton 85 radio-tracer technique was used for leak detection, which has a sensitivity in this case of 10^{-8} std cm^3/s. Though a leak below the detectable limit cannot be excluded and might have caused the corrosion, the Si experience[10] must be borne in mind and residual contamination suspected. This would be confirmed by the discovery of electrolytic corrosion in sealed devices aged at 80° to 90°C in dry air. Among the relatively few devices (30) aged in this manner in the present study, no corrosion has been observed. However, among a group of devices which had failed optical inspection due to unusually

Fig. 8—SEM photograph of corroded gate of FET after 1816 hours of aging, sealed and leak-tight, in humidity chamber. Analysis reveals traces of Cl at corrosion site.

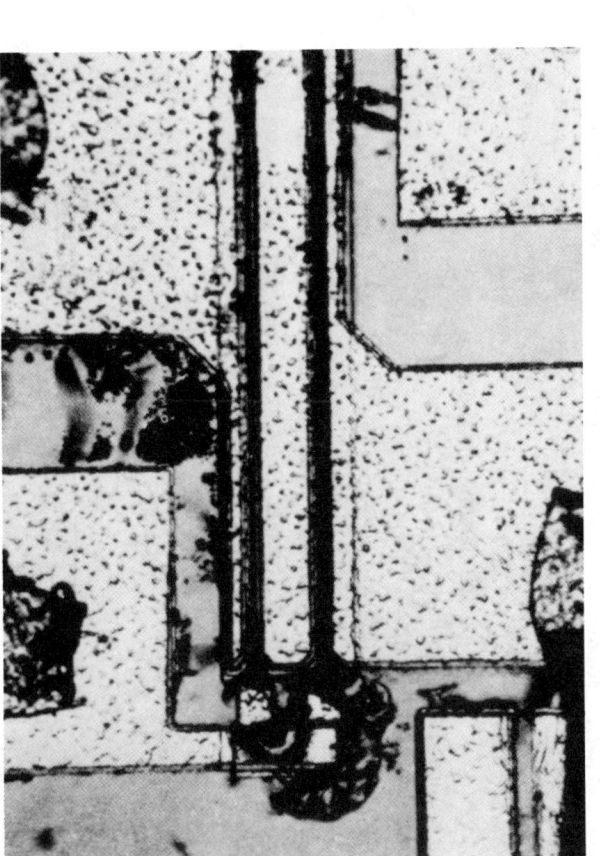

Fig. 7—Optical photograph of FET after 48 hours of aging, uncapped, in 85°C/85% RH humidity chamber, showing electrolytic corrosion of gate.

nificant corrosion is observed. Electrolytic corrosion of unprotected Al structures is well known, of course, from reliability studies of silicon devices.[7] The unusually close electrode spacing and consequent high fields in GaAs FETs as well as the minuteness of the Al gates make them prime candidates for this failure mechanism, in humid conditions. The acceleration factors relative to both varying humidity and temperature have already been reported in the Si device literature[8,9] and is summarized in Section VI.

Various passivation or protective coatings have been proposed for the prevention of electrolytic corrosion in GaAs FETs. Schemes that only coat the GaAs, such as grown oxides, would not be expected to be effective. The highly irregular topography of GaAs FETs complicates the task of achieving a continuous, impervious, pinhole-free, protective film. Equally important is the requirement that the film have small dielectric constant and low microwave loss; otherwise, the sacrifice in microwave performance is unacceptable. None of the films explored in this study fulfills all these specifications perfectly.

A hermetically sealed package can provide permanent protection against electrolytic corrosion from external humidity and without any sacrifice in RF performance, at least at frequencies where a package can be tolerated. It is suspected, however, as observed already among Si devices,[10] that residual impurities entrapped inside the package can produce destructive electrolytic corrosion, although the seal remains

large amounts of photoresist remaining on the chip and which were sealed and aged in 85/85, the incidence of corroded gates was 50 percent in 2000 hours. Altogether, the incidence of electrolytic gate corrosion among "clean" devices which passed optical inspection and which also passed the leak test before and after aging has been about 1 percent in 2000 hours.

V. GRADUAL DEGRADATION MECHANISMS

5.1 High-temperature effects

One of the earliest GaAs FET degradation mechanisms to be discussed was an increase in contact resistance.[11] According to well supported models,[12,13] Ga diffuses out of the crystal into the contact metallization at elevated temperatures. The resulting Ga vacancies probably act as acceptors, compensating the donors in the n-type lamina immediately adjacent to the contact, and thereby increase the contact resistance. The capacity of the metallization for absorbing Ga, or the effectiveness of an interposed barrier to the transport of the Ga, speed or inhibit the degradation process, respectively. In the ohmic contact structure described in Section II and illustrated in Fig. 3, the heavy final gold layer is the largest potential sink for migrating Ga, while the Ag and Ti-Pt layers act as an impeding barrier. As will be seen, however, other factors (such as the alloying cycle) must also play a role in the degradation of ohmic contacts.

By means of special test patterns and an appropriate computer program, the contact resistivity, ρ_c, and the channel resistance of a gateless device, $R(ch)$, were measured on certain FET slices before and after aging at 250°C, both with and without bias (0.3 A/cm, the same current per unit source width as in an operating device). A number of actual FETs from the same slice were also aged at the same temperature, with and without bias, and the usual parameters measured (R_S, I_{DSS}, NF, and G). Some slices (which will be designated Class I) showed virtually no change in any parameters after 500 hours, with or without bias, in either test patterns or actual FETs. Other slices (designated Class II), though nominally identical to Class I in design and fabrication, showed startling changes in certain dc parameters, as summarized for one slice in Table I. Typical values of R_S for unaged FETs were 15 to 30 ohms, of which the contact resistance contribution is only 0.2 to 0.5 ohms. The remainder of R_S is the resistance of the channel and of the semiconductor portions of the source and drain regions. Thus, it would appear plausible that a 5000-percent increase in contact resistance, as shown in Table I, would cause an approximate doubling of R_S. However, it is noted that the change in ρ_c was not affected by bias, whereas the change in R_S and I_{DSS} was very much bias dependent. (The latter bias dependence is also seen in Fig. 9.) The change in R_S (and corresponding change in I_{DSS}) is therefore not wholly attributable to contact deterioration.

The origin of the bias-dependent component of R_S has not been determined. One possibility is recombination enhanced defect formation,[14] though the hole production at the drain and under the gate seems too small for this effect. It is also suggested that the bias-dependent degradation may be related to the gate, since the test patterns, which were unaffected by bias, have no gates. It is also not understood what the essential difference is between Class I and Class II slices—and the continuous spectrum of behavior between these two extremes. It is thought that the least-controlled processing step may be the contact alloying, which is therefore tentatively blamed for at least a part of the slice-to-slice variation in aging behavior. Fortunately, perhaps because the degraded dc qualities of contacts are capacitively bypassed by RF signals,[15] these wide fluctuations in the degradation of dc parameters are not reflected in the RF performance.

Figure 9 shows the average values of R_S, I_{DSS}, NF, and G for two groups of FETs from the same slice (a Class II slice) aged at 250°C, one group with and the other without bias. Though both the dc and RF characteristics degrade faster with bias than without, it is seen that, while R_S doubles, the noise figure and gain only deteriorate by 0.2 to 0.5 dB. Another example is shown in Fig. 10, where NF and G degrade only 0.2 and 0.3 dB, respectively, while again R_S doubles. (The actual contact resistance increased 50-fold.) Two other interesting cases, both representative of many, are shown in Figs. 11 and 12. The devices of Fig. 11 suffered only negligible changes in R_S, I_{DSS}, and NF after 1300 hours of bias-aging at 250°C, though the gain declined about 0.6 dB. Figure 12 shows the data from a group of devices in which none of the measured dc or RF parameters changed significantly in 1500 hours of bias-aging. The results presented in Figs. 9 through 12 may be summarized as follows:

(i) There are significant differences among slices in the way the dc and RF characteristics change upon aging.

(ii) Radical deterioration of contact resistance (5000 percent) or of source-drain resistance (100 percent) are accompanied by only minor degradation of RF performance; conversely, NF and G may degrade slightly, even though R_S remains constant.

(iii) The median life at 250°C under bias, where failure is defined as

Table I — Changes in various parameters after 500 hours of aging at 250°C for a Class II slice

	Test Patterns			Actual FETs		
	ρ_c	$R(ch)$	R_S	I_{DSS}	NF	G
With bias	+5000%	0	+50%	−16%	0	0
Without bias	+5000%	0	+20%	−5%	0	0

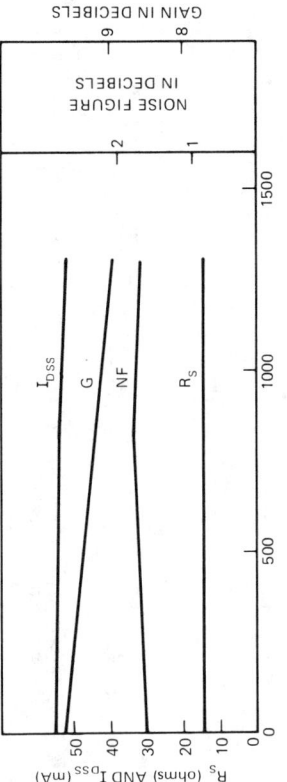

Fig. 11—Plot of median R_S, I_{DSS}, NF, and associated gain of a group of GaAs FETs from slice (3) as a function of aging time at 250°C with bias. (Note NF and G degrade slightly, though R_S in this case remains unchanged.)

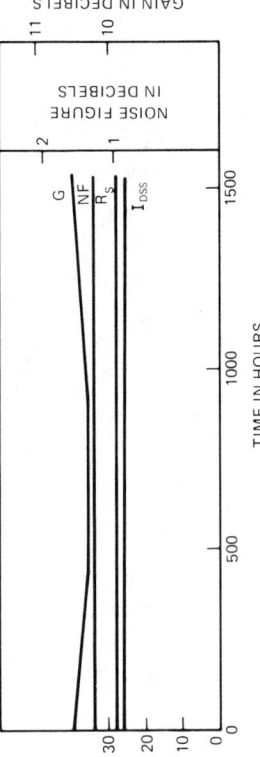

Fig. 12—Plot of median R_S, I_{DSS}, NF, and associated gain of a group of GaAs FETs from slice (4) as a function of aging time at 250°C with bias. (Note that all measured properties remain essentially unchanged.)

VI. FAILURE STATISTICS AND UNOBSERVABLES

6.1 Cumulative failure distributions

The statistics obtained from an aging study depend to some extent upon the definition of failure. A definition of failure can be tailored to a specific failure mechanism and thus be used to sort out data that are relevant exclusively to that mode. Two definitions have been used in various stages of the present investigation.

A. Catastrophic—collapse or radical change in dc output characteristics, usually due to a short or open circuit in one or more of the three electrodes. This definition is especially appropriate for study of the catastrophic mechanisms discussed in Section IV, but ignores any degradation of RF performance not associated with a large change in dc behavior.

B. RF degradation, exclusively—requires that any units that suffer catastrophic failure be subtracted from the population and not counted in the statistics. Figures 9, 10, 11, and 12 were based on such a population. The degree of permitted RF deterioration should be set with system requirements in mind. In this study, unless otherwise specified, a device

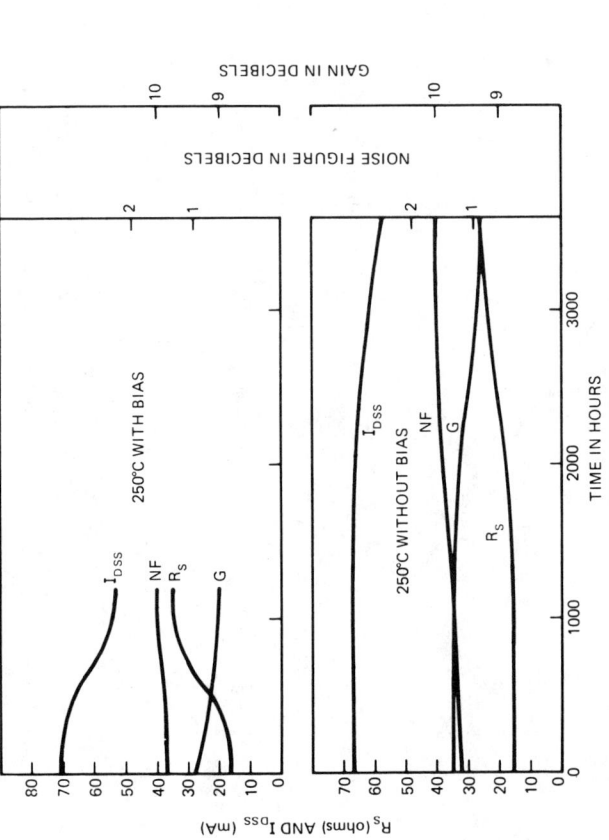

Fig. 9—Plot of median R_S, I_{DSS}, NF, and associated gain as a function of aging time at 250°C for two groups of GaAs FETs from slice (1): Group A aged with bias and Group B aged without bias. (Note acceleration due to bias.)

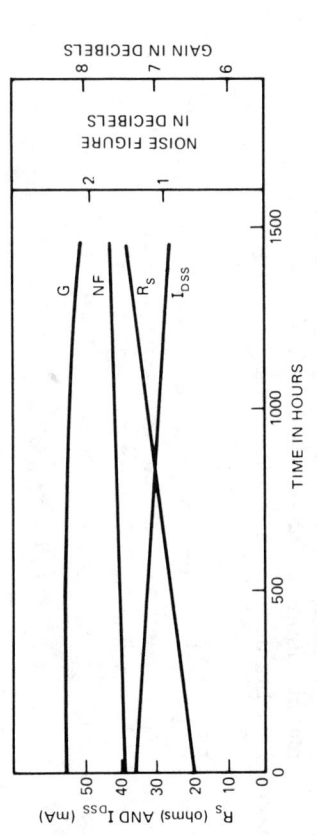

Fig. 10—Plot of median R_S, I_{DSS}, NF, and associated gain of a group of GaAs FETs from slice (2) as a function of aging time at 250°C with bias. (Note NF and G degrade only 0.2 dB, though R_S increases 90 percent.)

an NF or G degradation of equal to or more than 0.2 or 0.8 dB, respectively, is at least 1500 hours.

The observed activation energy of RF degradation is 0.8 to 1.0 eV. It may be noted that many diffusion phenomena within or on the surface of semiconductors, such as might produce traps or scattering centers, have activation energies near 1.0 eV.

was considered to have failed, RF-wise, if the noise figure increased 0.2 dB or more, or the associated gain changed (up or down) by 0.8 dB or more as measured at 4 GHz in a fixed-tuned amplifier.

Figure 13 is a log-normal plot of the cumulative percent failures of type B as a function of aging time for 31 devices representing four separate slices. The aging was performed with bias at 250°C air ambient. The channel temperature is estimated to be 8°C warmer. A few of the devices were sealed, but the majority were not. No difference has been observed in the aging behavior of sealed versus unsealed FETs at this temperature. The data are seen to fit reasonably a straight line, making allowance for the statistical vagaries of small samples, which means they approximate a log-normal distribution. The standard deviation estimate, s, of the line is about 1.3, obtained from the operational calculation

$$s = \ln[t(50)/t(16)],$$

where it is noted the natural logarithm is used and $t(50)$ and $t(16)$ are the times corresponding to 50 and 16 percent cumulative failure, respectively. The median life of this group is about 1700 hours. The results shown in Fig. 13 are typical of the RF degradation observed in this study.

Figure 14 shows the cumulative failure distributions at 250°C of two groups of units from one slice, one aged with bias and the other without bias. The MLs are about 100 hours and 350 hours, respectively, with nearly the same standard deviation of $s = 1$. The failures in this case are all type A and due to Au-Al phase formation, plus an apparent assist from electromigration in the biased group, as discussed in Section 4.1. This slice was unusually susceptible to the Au-Al phase problem and is chosen here to illustrate the failure statistics of that mechanism.

6.2 Humidity acceleration factors

Electrolytic corrosion is accelerated by increased humidity and temperature mainly as a result of and in proportion to the increased electrical conductivity of the surface. The problem has been most recently studied by Sbar and Kozakiewicz,[9] who give acceleration factors with respect to 85°C/85% RH for various encapsulations and temperature/humidity conditions. Though the absolute value of conductance on a GaAs surface may differ from that on a Si, Si_3N_4, or alumina surface, the temperature and humidity dependence are expected to be similar. For 60°C/5% RH (a choice which will be justified later), the Sbar-Kozakiewicz results indicate an acceleration factor of 2×10^5 with respect to 85/85. For a condition of 60°C/25% RH, the factor is about 10^4. Both values apply to an unencapsulated device. For a perfectly sealed device, of course, the external humidity has no effect. The only acceleration of electrolytic

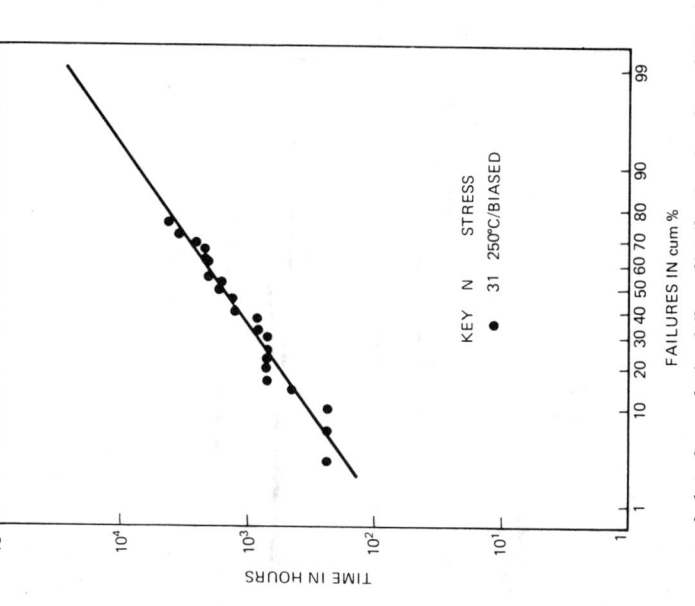

Fig. 13—Log-normal plot of cumulative failure distribution (RF degradation) of a group of GaAs FETs aged with bias at 250°C.

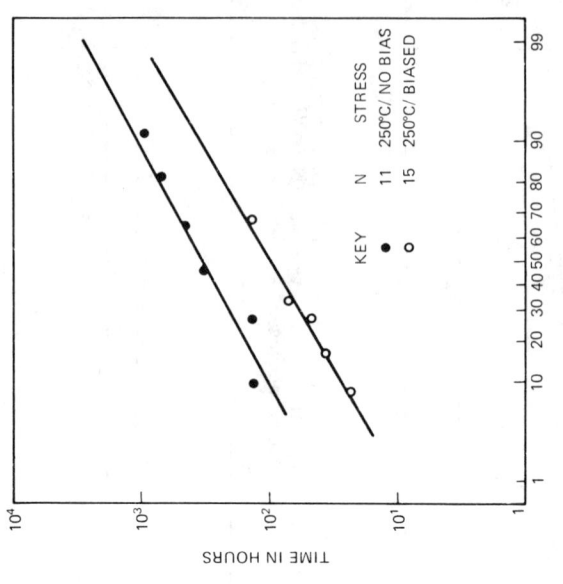

Fig. 14—Log-normal plot of cumulative failure distributions of two groups of GaAs FETs aged at 250°C with and without bias. (All failures were due to gate destruction by Au-Al phase formation plus, in one case, a bias-dependent factor.)

corrosion would be that due to the temperature elevation, assuming there are sufficient contaminants inside the package to produce an electrolyte, but that their release is not temperature-sensitive. The acceleration factor at 85°C relative to 60°C is approximately 3.

6.3 Statistical error

During the course of this study, numerous small changes have been made in the design or fabricational methods of the device under investigation. Since it was desirable to appraise the reliability aspects of each of these variations and both facilities and device are limited in supply, the strategy has been to age many small lots of FETs rather than fewer and larger lots. The relatively small sample size (10 to 20) raises questions about the validity of the statistics obtained. It should be pointed out, therefore, that with a sample size of 10, one can be 90 percent confident that the true ML (i.e., the ML of an infinitely large sample) would be somewhere between 45 and 225 percent of the observed value, assuming a log-normal distribution with a standard deviation of 1. Thus, in the next section where failure rates are calculated, error tolerances may be attached to the values given by noting that in the area where most of the data fall, a factor of 2 in ML produces a factor of approximately 2 to 3 in failure rate.

6.4 Possibility of unobserved failure mechanisms

A relatively small number of FETs have been aged under bias for periods of 0.5 to 1.5 years at moderate or room temperatures (more precisely, 30 units at 180°C for 6000 hours, 10 units at 88°C for 4600 hours, and 30 units at 27°C for 8000 to 14,000 hours, all in air ambient). Judging from the statistics obtained at higher temperatures, provided that only the same mechanisms prevail, no failures or degradation should be apparent in these modest times at lower temperature, except possibly for the Au-Al phase-migration problem. Indeed, this turns out to be the case: five units of one lot of 10 at 180°C have failed due to the Au-Al syndrome, but otherwise no degradation appears in any of the units.

The lack of any failures among the other units provides some lower bound to the activation energy of any failure mechanism which may be important at lower temperature but is obscured at higher temperatures by phenomena with higher activation energies. (This sort of insidious situation has been encountered in Pt-GaAs IMPATTs.[16]) It will be assumed that such a failure mechanism would have a log-normal failure distribution with a standard deviation of 1. It is noted furthermore that the absence of a single failure in a lot of 30 indicates with 90 percent confidence that the true percent failure cannot exceed 10 percent. Finally, it is noted that the ML of any such so-far-unobserved failure mechanism must be at least 1700 hours at 250°C, since that is the longest observed ML at that temperature due to recognized mechanisms. Combining these arguments leads to the conclusion that the minimum activation energy of a failure mechanism active at 180°C, but obscured at 250°C is 0.7 eV. However, the minimum activation energy for a failure mechanism dominant at 27° to 180°C but not yet observed in this program is only 0.03 eV. The latter figure is rather alarming. It means, in conjunction with the high temperature data, that if such a hypothetical mechanism exists, the projected ML at room temperature would be 30,000 hours and the ML would be only negligibly accelerated by elevated temperatures. The duration of the present reliability program is insufficient to rule out such a possibility. However, it is reassuring that other investigators have reported room temperature tests of low-noise GaAs FETs of similar metallurgy in excess of six years without any failures.[17]

6.5 Infant mortality

Few instances of infant mortality have been found in this study. There are two reasons: (i) rigorous optical inspection of all chips before mounting and discarding of any units which appear mechanically or electrically defective after mounting eliminate most devices that might otherwise be candidates for early failure; (ii) the small size of the samples used further diminishes the probability of encountering anomalous devices representing only a small proportion of the population. Thus, this work sheds no light on the nature of such early failures except that, with the present fabrication, inspection, and testing procedures, their occurrence is less than 1 percent. The failure and degradation mechanisms discussed here and the failure rates projected pertain to the main body of the population. It must be anticipated that some cases of infant mortality will accompany large-scale production and deployment of this device.

VII. ESTIMATION OF FAILURE RATES

Given the nature of the failure distributions at an elevated temperature and their respective activation energies, and making the all-important assumption that the mechanisms studied at elevated temperature are also the dominant ones at room temperature, and with the further assumption that the failure distribution is not temperature-dependent (i.e., it stays log-normal with the same s), it remains only to specify the operating conditions in order to calculate the probable failure rates in the field. The maximum ambient temperature in a Bell System radio relay application is 52°C (125°F). The corresponding maximum channel temperature would be 60°C. Though the annual average temperature would certainly be considerably lower, 60°C will be taken as the channel temperature for calculation of failure rates. Three separate cases will be considered.

7.1 Case I: No catastrophic mechanisms

In this case, it is assumed all fabricational steps have been faultless, assuring the absence of any contact between Au and Al, a contamination-free chip and package interior, and a leak-tight seal. The catastrophic failure mechanisms are therefore precluded, and only long-term degradation of RF properties is of concern. The median lifetime (type B) at 258°C was found to be about 1700 hours, and the associated activation energy will be taken as 0.8 eV. Thus, the projected ML at 60°C would be 4×10^7 hours. Taking a standard deviation of 1.0 and using Goldthwaite's curves,[18] the failure rate after 20 years of service is found to be less than 10^{-2} FIT (1 FIT = 1 failure in 10^9 device-hours). It should be noted that, with a standard deviation of 1.5, the projected failure rate would be 2 FITs and with $s = 2$, the failure rate is 30 FITs, i.e., the difference between $s = 1$ and $s = 2$ is more than 3 orders of magnitude in failure rate. Thus, an accurate knowledge of the standard deviation is vital to the accurate forecasting of failure rates. However, the low confidence levels of the statistics do not justify quibbling over the real value of s, and the predicted failure rates are small in any case (but do not include infant mortality).

7.2 Case II: Au-Al phase formation dominant

As mentioned in Section 4.1, the ML due to Au-Al phase formation at 250°C has been observed to be as short as 94 hours, though it exceeds observation times in many cases. Based on this shortest observed ML and the smallest observed activation energy of 0.5 eV, a worst-case prediction is obtained, indicating that for an unscreened product the failure rate could go over 10,000 FITs, i.e., 1 percent per 1000 device-hours. However, a reliability qualification test of each slice can be used to assure that the ML due to the Au-Al/electromigration syndrome is no less than 500 hours at 250°C. Assuming a relatively conservative value of 0.8 eV for the activation energy (from the wide range observed of 0.5 to 1.6 eV), an ML at 60°C of 1×10^7 hours is projected. Taking $s = 1$, as found in Fig. 14, gives an estimated failure rate of 0.6 FIT in a 20-year service period. Taking $s = 1.5$, as found in occasional slices also dominated by the Au-Al failure mechanism, gives a failure prediction of 40 FITs. The latter value is considered a realistic upper limit for devices of the type shown in Fig. 2 subjected to a reliability screening procedure (and is therefore the value quoted in the abstract).

7.3 Case III: Electrolytic corrosion dominant

If unsealed, unprotected low-noise GaAs FET chips were employed in an amplifier in which the housing was not hermetically sealed, electrolytic corrosion as described in Section 4.2 would be expected. Since in some radio relay applications, the waveguide is pressurized with 5-percent RH air, this value for the ambient will be considered for the first example. (It should be noted that 30°C/25% RH air becomes 5% RH air when heated, at constant water vapor content, to 60°C). For unprotected units in 85/85, an ML of about 3.5 hours due to electrolytic corrosion has been observed. Using the appropriate acceleration factor of 2×10^5, as described in Section 6.2, an ML of 7×10^5 hours at 60°C/5% RH is predicted. The corresponding failure rate is about 1000 FITs. For a second example, an atmosphere of 60°C/25% RH is chosen, which may be obtained by heating 32°C (90°F)/100% RH air up to 60°C. In this case, an ML of 3.5×10^4 hours and a failure rate of about 20,000 FITs after 5 years are predicted.

As a third example, it might be assumed that the GaAs FET chip is unprotected and the amplifier is hermetically sealed, but not adequately free of contaminants. Indeed, in view of the large amount of surface within an amplifier and the difficulty of giving it a high-temperature vacuum bakeout, it very likely would contain dangerous amounts of residual impurities. An ML of about 2000 hours has been observed in this study with contaminated packages at 85°C. The acceleration factor relative to 60°C in this case is only 3, as discussed in Section 6.2. Thus, an ML of 6000 hours might be anticipated for this amplifier with unsealed, unprotected FETs and a first-year failure rate of over 50,000 FITs.

It is emphasized that the above three examples of electrolytic corrosion assume unsealed, unprotected (unpassivated) devices. In the case of a clean, hermetically sealed device, electrolytic corrosion is effectively prevented, and no failures due to that mechanism are expected.

VIII. CONCLUSIONS

Two catastrophic failure mechanisms were found in this study of low-noise GaAs FETs, not including voltage transients which are considered primarily a problem of handling technique and circuit design. One of these mechanisms is Au-Al phase formation occurring at the junction of the Al gate and its Au bonding pad. This mechanism is enhanced by bias through what appears to be electromigration, though positive evidence of the latter is lacking. In a worst case, this mechanism could give rise to failure rates as high as 10,000 FITs, though with appropriate slice screening, values in the neighborhood of 1 to 50 FITs appear more likely. Proper design and fabrication methods can eliminate this mechanism entirely. The other catastrophic failure mechanism is electrolytic corrosion of the Al. In a humid environment or in a contaminated package, failure rates again in the order of 10,000 FITs might be anticipated. However, hermetic sealing in a contaminant-free package eliminates this problem. It is noted that both these failure mechanisms are related to the choice of an Al gate. They are not peculiar to GaAs FETs, but are well known as causes of failure in Si devices.

In the absence of catastrophic failure, a long-term, gradual degradation of noise figure and gain is observed. This effect is only weakly correlated with increase of contact resistance and is apparently more strongly influenced by other factors such as the formation of traps and scattering centers. The median lifetime due to this gradual RF degradation is estimated to be over 10^7 hours at a channel temperature of 60°C. The corresponding failure rate after 20 years of service is less than 2 FITs.

All the important failure modes were accelerated by the presence of drain and gate bias. Aging without bias would give erroneously optimistic predictions.

IX. ACKNOWLEDGMENTS

The authors are indebted to many colleagues whose contributions significantly aided this work. Special thanks are due to J. P. Beccone, W. L. Boughton, J. V. DiLorenzo, A. T. English, D. E. Iglesias, L. C. Luther, F. M. Magalhaes, W. C. Niehaus, Mrs. Y. C. Nielsen, R. H. Saul, and W. O. Schlosser.

REFERENCES

1. B. S. Hewitt, H. M. Cox, H. Fukui, J. V. DiLorenzo, W. O. Schlosser, and D. E. Iglesias, "Low Noise GaAs MESFETs: Fabrication and Performance," 1977 GaAs and Related Compounds (Edinburgh), 1976 (Inst. Phys. Conf. Ser. 33a), p. 246.
2. R. H. Knerr and C. B. Swan, "A Low-Noise GaAs FET Amplifier for 4 GHz Radio," B.S.T.J., 57, No. 3 (March 1978), p. 479.
3. B. Seliksonn, "Failure Mechanisms in Integrated Circuit Interconnect Systems," 6th Annual Proc. Rel. Phys. Symp. (IEEE) (1968), p. 201.
4. E. Philofsky, "Purple Plague Revisited," Solid-State Electronics, 13 (October 1970), p. 1391.
5. I. A. Blech and E. S. Meieran, "Electromigration in Thin Al Films," J. Appl. Phys. 40 (February 1969), p. 485.
6. C. Weaver and L. C. Brown, "Diffusion in Evaporated Films of Au-Al," The Phil. Mag., 7 (1961), p. 1.
7. B. Reich and E. B. Hamkim, "Environmental Factors Governing Field Reliability of Plastic Transistors and Integrated Circuits," 10th Annual Proc. Rel. Phys. Symp. (IEEE) (1972), p. 82.
8. D. S. Peck and C. H. Zierdt, Jr., "Temperature-Humidity Acceleration of Metal-Electrolysis Failure in Semiconductor Devices," 11th Annual Proc. Rel. Phys. Symp. (IEEE) (1973), p. 149.
9. N. L. Sbar and R. P. Kozakiewicz, "New Acceleration Factors for Temperature, Humidity, Bias Testing," to appear in 16th Annual Proc. Rel. Phys. Symp. (IEEE) (1978).
10. A. Shumka and R. R. Piety, "Migrated-Gold Resistive Shorts in Microcircuits," 13th Annual Proc. Rel. Phys. Symp. (IEEE) (1975), p. 93.
11. T. Irie, I. Nagasako, H. Kohza, and K. Sekido, "Reliability Study of GaAs MESFETs," IEEE Trans. on Microwave Th. and Tech., MTT-24 (June 1976), p. 321.
12. K. Ohata and M. Ogawa, "Degradation of Au-Ge Ohmic Contact to n-GaAs," 12th Annual Proc. Rel. Phys. Symp. (IEEE) (1974), p. 278.
13. A. Christou and K. Sleger, "Precipitation and Solid Phase Formation in Au(Ag)/Ge Based Ohmic Contacts for GaAs FETs," 6th Biennial Conf. on Active Microwave Semiconductor Devices and Circuits, Cornell, 1977.
14. L. C. Kimerling, "New Developments in Defect Studies in Semiconductors," IEEE Trans. on Nuclear Sci., NS-23 (1976), p. 1497.
15. J. C. Irvin and R. L. Pritchett, "Nonohmic Contacts for Microwave Devices," Proc. IEEE (Corres.), 58 (November 1970), p. 1845.
16. W. C. Ballamy and L. C. Kimerling, "Premature Failure in Pt-GaAs IMPATTS-Recombination Assisted Diffusion as a Failure Mechanism," Tech. Digest IEDM (IEEE) (1977), pp. 90–92.
17. D. A. Abbott and J. A. Turner, "Some Aspects of GaAs MESFET Reliability," IEEE Trans. on Microwave Th. and Tech. M-24 (June 1976), p. 317.
18. L. R. Goldthwaite, "Failure Rate Study for the Log-Normal Lifetime Model," Proc. 7th Nat'l. Symp. on Reliability and Quality Control, 208 (January 1961). [This curve was reprinted in the 9th Annual Proc. Rel. Phys. Symp. (IEEE) (1971), p. 78].

Determination of the Basic Device Parameters of a GaAs MESFET

By H. FUKUI

(Manuscript received July 28, 1978)

This paper describes a new technique to determine the basic properties of the active channel of a gallium arsenide (GaAs) metal-semiconductor field effect transistor (MESFET). The effective gate length, channel thickness, and carrier concentration are determined from dc parameters. A precise method of measuring the dc parameters is also given. The new techniques are demonstrated using a wide variety of sample devices. It is also shown that microwave performance parameters, such as the maximum output power and minimum noise figure, are well predicted by dc parameters. Calculated values of the intrinsic and extrinsic dc parameters, using simple analytical expressions developed in terms of the geometrical and material parameters of a device, are shown to be in excellent agreement with their measured values. These expressions can be used as a basis for device design.

I. INTRODUCTION

In a gallium arsenide (GaAs) metal-semiconductor field effect transistor (MESFET), the properties of the active channel are fundamental in describing its operation. The channel properties can be characterized by the four basic parameters: gate length, gate width, channel thickness, and channel doping.

In a recent paper,[1] the maximally obtainable value of channel current was defined as the maximum channel current, I_m. It was pointed out that I_m differs from either (fully open channel) saturation current, I_s, or zero-gate-bias drain current, which is often referred to as I_{dss}. Currents I_s and I_{dss} have conventionally been used to show upper limits of the drain current capability. However, neither I_s nor I_{dss} can represent the maximally obtainable value of channel current. It was emphasized that I_m plays an important role in determining the maximum capability of large-signal operation of the device. Simple expressions for I_m were then obtained in terms of the four basic channel parameters, as a result of an extended study of Shockley's gradual channel approximation[2] on Grebene-Ghandhi's two-section FET model[3] with Fukui's concept on the current limiting mechanism.[1]

Among the four basic channel parameters, the total gate width, Z, is usually a given factor or merely a scaling factor. Therefore, the other three parameters are noted to be the most crucial variables in the design work. For these three parameters, their effective values were adopted in Ref. 1. This was essential, especially for gate length. The effective gate length, L, may be either shorter or longer than the physical length of gate metallization, L_g, depending upon the gate junction topography. The channel thickness, a, and carrier concentration, N, represent their effective values in the active region of the channel.

In Ref. 1, practical expressions for the zero-gate-bias channel current, I_o, were also developed as functions of the basic channel parameters N, a, L, and Z, and an additional parasitic parameter of source series resistance, R_s. An approximate expression for the knee voltage, V_{kf}, corresponding to I_m in the drain I-V characteristic, was given by a combination of N, a, L, Z, R_s, and R_d on a semi-empirical basis, R_d being the drain series resistance. In addition, it has been known that the total pinch-off voltage, W_p, is determined by the Na^2 product and that W_p is equal to the sum of terminal pinch-off voltage, V_p, and Schottky-barrier built-in voltage, V_b.

It is now conceivable that N, a, and L may be determined from the measured values of I_m, I_o, V_{kf}, V_p, V_b, R_s, and R_d, provided that Z is known. The prime purpose of this paper is to present a new technique for carrying out this work. Throughout the paper, a transistor curve tracer is exclusively used as the tool necessary for measuring the dc parameters. However, test equipment of other types can, of course, be used as well.

There has been a common practice in which either a or N is determined from V_p, after knowing either N or a, respectively, and assuming an appropriate value of V_b. Also, Fair showed that an iterative analysis on I_o, R_s, and transconductance, g_m, makes it possible to determine N and a from known values of I_o, V_p, and terminal transconductance, g'_m.[4] However, as far as the author knows, there has been no published report on an evaluation technique for the effective gate length of a finished device. This paper presents such a technique as well as simultaneous determination of N and a, from known values of I_m, I_o, V_p, V_b, V_{kf}, R_s, and R_d.

The second purpose of this paper is to show prediction of the microwave performance parameters, such as the maximum output power and minimum noise figure, from the dc parameters and hence the basic channel parameters. To predict the minimum noise figure, the values of g_m and R_g, which is the gate series resistance, must be known. Therefore, the determination of g_m and R_g are also described in this paper. Once the detail of the structure outside the gate channel is given, the parasitic parameters, such as R_g, R_s, and R_d, can be analytically expressed in terms of the geometrical and material parameters of the corresponding sections of the device. The validity of such expressions is then examined with experimental results in this paper.

II. PRINCIPLE OF NEW TECHNIQUE

2.1 Analytical expressions for device dc parameters

To determine the basic channel parameters of a GaAs MESFET from the measured values of its dc parameters, expressions showing their relationships are essential. It has been well known that W_p and I_s are given by

$$W_p = \frac{qNa^2}{2\kappa\epsilon_0} \tag{1}$$

and

$$I_s = qv_s NaZ, \tag{2}$$

respectively,[5] in which q is the electronic charge, ϵ_0 is the permittivity of free space, κ is the specific dielectric constant, and v_s is the saturated velocity of electrons in n-GaAs. Substituting $q = 1.60 \times 10^{-19}$ C, $\epsilon_0 = 8.85 \times 10^{-14}$ F/cm, $\kappa = 12.5$ for GaAs, and a best fit value of $v_s = 1.4 \times 10^7$ cm/sec (Ref. 1) into (1) and (2) yields the following practical expressions:

$$W_p = 7.23 Na^2 = V_p + V_b \quad (V) \tag{3}$$

and

$$I_s = 0.224 ZNa \quad (A), \tag{4}$$

where N is in units of 10^{16} cm^{-3}, a is in μm, and Z is in mm.

As mentioned earlier, analytical expressions for I_m, I_o and V_{kf} have been derived in Ref. 1. First, an expression for I_m is given in the form

$$I_m = \beta I_s, \tag{5}$$

where β is the maximum channel opening factor. Parameter β is expressed approximately as

$$\beta \approx 1 - \frac{0.18}{a}\sqrt{\frac{L}{N}} \tag{6}$$

provided that a best-fit value of 0.29×10^4 V/cm is assumed for the critical electric field, E_c. Another approximate expression for I_m is shown as

$$\frac{I_m}{Z} \approx \frac{0.18 N^{1.3} a^{1.5}}{L^{0.28}} \quad \text{(A/mm)}. \tag{7}$$

Second, an expression for I_o is given in the form

$$I_o = \gamma I_s, \tag{8}$$

where

$$\gamma \approx 1 + \sigma - \sqrt{\delta + 2\sigma + \sigma^2} \tag{9}$$

$$\delta = \frac{V_b + 0.234 L}{W_p} \tag{10}$$

and

$$\sigma = \frac{0.0155 R_s Z}{a}. \tag{11}$$

Another approximate expression for γ is given by

$$\gamma \approx \left[1 - \frac{1}{\sqrt{W_p'}}\right]\left[1 - \frac{I_s R_s}{2\sqrt{W_p'}}\right], \tag{12}$$

where

$$W_p' = W_p + V_c. \tag{13}$$

In (13), V_c is a correction voltage that may vary from zero to a few tenths of a volt, depending upon the configuration of the channel structure. No analytical form is presently available for V_c.

Third, an expression for V_{kf} is given by

$$V_{kf} \approx (1 - \beta)^2 W_p + I_f(R_s + R_d) + V_c, \tag{14}$$

where I_f is the maximum value of total forward drain current, including the leakage current through the buffer layer and substrate.

2.2 Determination of basic channel parameters: Case I

If any one of the basic channel parameters N, a, and L is known, the other two can be determined from known values of V_p, V_b, and (I_m/Z) in a straightforward manner, using either a set of (3), (4), (5), and (6), or (3) and (7). For example, if the N value is known as a result of the epitaxial layer evaluation, a is readily determined by

$$a = \sqrt{\frac{V_p + V_b}{7.23 N}} \quad (\mu m), \tag{15}$$

as is well known. The maximum channel opening factor is evaluated as

$$\beta = \frac{12 I_m/Z}{\sqrt{N}(V_p + V_b)}. \tag{16}$$

Using this β value, L is determined by

$$L = 4.27(V_p + V_b)(1 - \beta)^2 \quad (\mu m). \tag{17}$$

In the case that L (or a) is known, similar evaluation for N and a (or N and L) can also be carried out as well.

Among dc parameters, I_m and W_p ($= V_p + V_b$) can be considered to be primary, because they are determined only by the active channel properties which are intrinsic. Other dc parameters, such as I_0 and V_{kf}, are secondary, since they are affected by extrinsic elements outside the gate channel region. The basic channel parameters can be determined only from the *primary* dc parameters, if any one of the three basic parameters for the device under evaluation is known. This is a characteristic of case I.

2.3 Determination of basic channel parameters: Case II

In this case, none of the three basic channel parameters is known. For the determination of these parameters, the *secondary* dc parameters play the major role and the primary dc parameters remain auxiliary. There are two ways to determine the basic channel parameters in this category.

The first method is based on parameter I_0. Rearranging (8) and (12) yields

$$I_s \approx \frac{\sqrt{V_p + V_b + V_c}}{R_s}\left[1 \mp \sqrt{1 - \frac{2I_0 R_s}{\sqrt{V_p + V_b + V_c} - 1}}\right]. \tag{18}$$

After knowing I_s, N and a can be determined from (3) and (4), respectively, as follows:

$$N = \frac{1}{V_p + V_b}\left[\frac{12 I_s}{Z}\right]^2 \quad (10^{16}\ \mathrm{cm}^{-3}) \tag{19}$$

and

$$a = 0.031\,(V_p + V_b)\frac{Z}{I_s} \quad (\mu m). \tag{20}$$

Now (17) can be used with (5) to obtain L as

$$L = 4.27\,(V_p + V_b)\left[1 - \frac{I_m}{I_s}\right]^2 \quad (\mu m). \tag{21}$$

The second method is to utilize parameter V_{kf}. Rewriting (14) yields

$$\beta \approx 1 - \sqrt{\frac{V_{kf} - I_f(R_s + R_d) - V_c}{V_p + V_b}}. \tag{22}$$

After knowing the β value, N and a can be determined by (19) and (20), respectively, provided that $I_s = I_m/\beta$. The L value can be directly determined from (14) and (17) as

$$L = 4.27\,[V_{kf} - I_f(R_s + R_d) - V_c] \quad (\mu m). \tag{23}$$

The first method demands the known values of Z, V_p, V_b, V_c, I_o, I_m, and R_s, whereas the second method requires Z, V_p, V_b, V_c, V_{kf}, I_f, R_s, and R_d already known. In the computation process of determining N, a, and L, the subtraction of two major terms is included in both cases. Therefore, chances of introducing an intolerable error are inevitable. Thus, taking only a single method is not advisable. The results obtained from one method have to be checked with the other method. The simplified relationship given in (7) could be conveniently used as a guide to examination and adjustment. Expressions (8) through (11) for I_o could also be applied for an additional checking of the determined values of N, a, and L, in comparison with the directly measured value of I_o. Some adjustments on temporarily determined parameters are often necessary to reach their most probable values.

III. MEASUREMENTS OF DC PARAMETERS

As previously mentioned, a transistor curve tracer is used as the test instrument in this paper. A good calibration of the measuring system is essential. Not only the curve tracer but also the test fixture must be taken into consideration. For example, lead resistances may introduce an intolerable measuring error in large-size devices. An excessive leakage current may mislead the determination of junction parameters. Instability and/or relatively low-frequency oscillation, often taking place in a high-performance device, are an annoying phenomena and require a special skill to suppress.

In the following sections, measuring methods for the dc parameters of a GaAs MESFET are described. Although some methods have been known or are easily derived from known methods, they are included with brief descriptions for completeness. The ideality parameter of a gate junction, n, and the open channel resistance, R_o, are needed neither for determination of the basic channel parameters nor for prediction of the microwave performance parameters. Nevertheless, they are secondarily obtained in the course of determination of the primary parameters. As they may be relevant to a further study of the device, their determination is also described.

3.1 Determination of gate barrier built-in voltage and ideality parameter

As is well known,[6] the forward current density, J, of a Schottky barrier junction for $V > 3kT/q$ is approximately written as

$$J = A^* T^2 \exp\left[-\frac{qV_b}{kT}\right]\exp\left[\frac{qV}{nkT}\right], \tag{24}$$

where A^* is the effective Richardson constant, T is the junction temperature in °K, k is the Boltzmann constant, n is the ideality parameter, and V is the forward bias voltage.

The extrapolated value of current density to zero bias gives the saturation current density, J_s. The barrier built-in voltage is then obtained from

$$V_b = \frac{kT}{q} \ln\left[\frac{A^* T^2}{J_s}\right]. \quad (25)$$

The ideality parameter is given by

$$n = \frac{q}{kT} \frac{\partial V}{\partial(\ln J)}. \quad (26)$$

Figure 1 is a multi-exposed photograph of the forward I-V characteristic of a gate junction at room temperature. As described in the caption, each curve corresponds to a current range in a decimal step for several orders of magnitude. In the highest current range, the gate was against three different connections, i.e., source and drain combined, source alone, and drain alone, in order to differentiate R_g, R_s, and R_d from each other later.

The I-V characteristic given in Fig. 1 is plotted as shown in Fig. 2. At high values of the gate bias, V_g, the gate current, I_g, tends to saturate due to the series resistance effect. At low values of V_g, I_g is often disturbed by a leakage current component around the gate periphery or with the package. In the middle range where the $\log I_g$ vs V_g characteristic is linear, two gate biases, $V_{g(m)}$ and $V_{g(m-1)}$ in V, can be chosen corresponding to $I_g = 10^m$ and $I_g = 10^{m-1}$ in A, respectively. Usually, m takes a negative value.

If the effective mass of electrons in n-GaAs were taken into account, the effective Richardson constant would be 8.7 A/cm^2/°K at room temperature.[7] Expression (25) for V_b can then be reduced to the following practical form at room temperature:

$$V_b = 0.768 - 0.06 \log J_s \quad (V), \quad (27)$$

in which

$$J_s = \frac{10^y}{L_g Z} \quad (10^{-7} \text{ A/cm}^2) \quad (28)$$

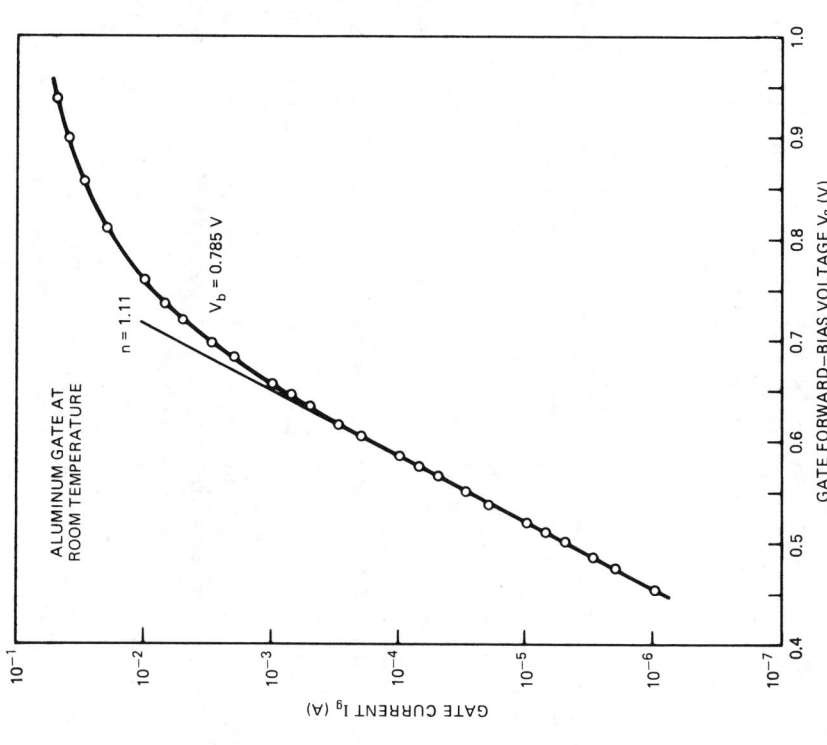

Fig. 2—The forward I-V characteristic of an aluminum gate diode at room temperature. The slope and the location of its linear portion on a semi-log paper give the ideality parameter, n, and the gate built-in voltage, V_b, respectively.

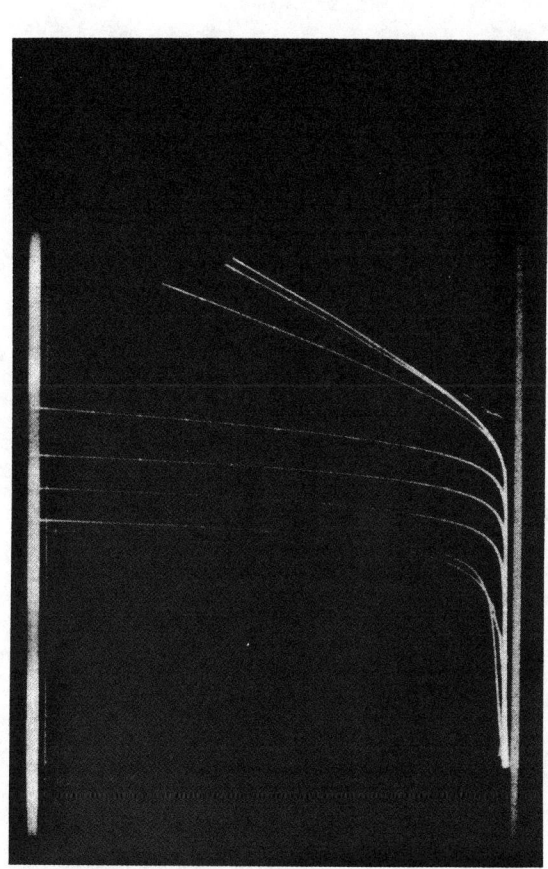

Fig. 1—A multi-exposed photograph of the forward I-V characteristic of a gate junction at room temperature. Each curve corresponds to a current range of decimal step. In the highest current range, three different ground connections are taken against the gate. They are the source and drain connected together, source alone, and drain alone. For all lower current ranges, the drain is connected to the source. The horizontal scale is in units of 0.1 V/div. The vertical scales are, left to right, 1 μA/div, 10 μA/div, 100 μA/div, 1 mA/div, 10 mA/div (source and drain together), 10 mA/div (source only), and 10 mA/div (drain only).

$$y = 12 + m - \frac{1}{1 - \frac{V_{g(m-1)}}{V_{g(m)}}} \quad (29)$$

and L_g is in units of μm and Z is in mm. A formula to be used for deriving n is deduced from (26) as

$$n = 16.8[V_{g(m)} - V_{g(m-1)}]. \quad (30)$$

3.2 Determination of pinch-off voltage, active channel resistance, and parasitic series resistances

Figure 3 shows the drain I-V characteristics in the so-called linear region. The characteristics were taken with the lowest scale of V_{ds} on the curve tracer, in which V_{ds} was the drain-source bias voltage. Each characteristic corresponds to a gate-source bias voltage, V_{gs}. The drain current, I_d, at $V_{ds} = 0.05$ V is then plotted as a function of V_{gs} as shown in Fig. 4. The terminal pinch-off voltage, V_p, can *temporarily* be determined by an extrapolation of the plot to the abscissa.

The current shown in Fig. 4 can now be converted into the resistance value as $R_{ds} = I_d/V_{ds}$. Such a resistance for $V_{ds} = 0.05$ V is shown in Fig. 5 as a function of V_p, V_b, and V_{gs}, in the same manner as used in Ref. 8, as compounded in parameter X defined as

$$X = \frac{1}{1 - \sqrt{\frac{V_b - V_{gs}}{V_b + V_p}}}. \quad (31)$$

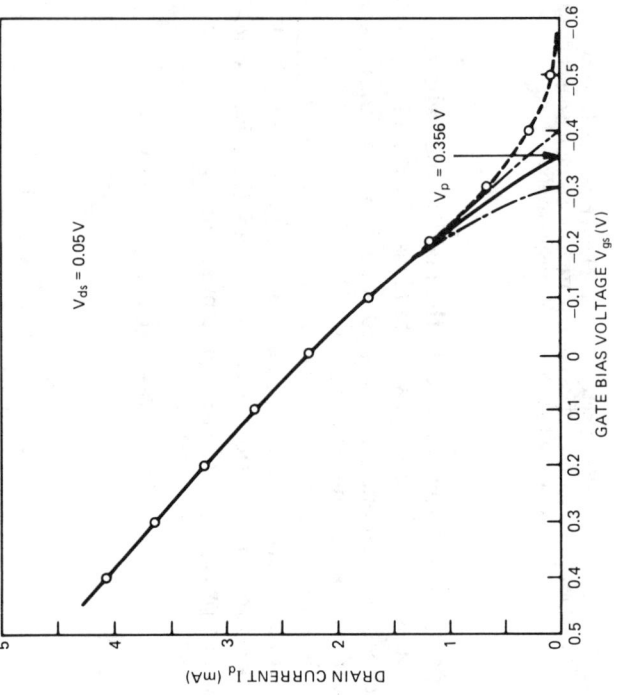

Fig. 3—An expanded view of the drain I-V characteristics in the so-called linear region near the origin on gate-bias offset mode (+0.4 V in this case).

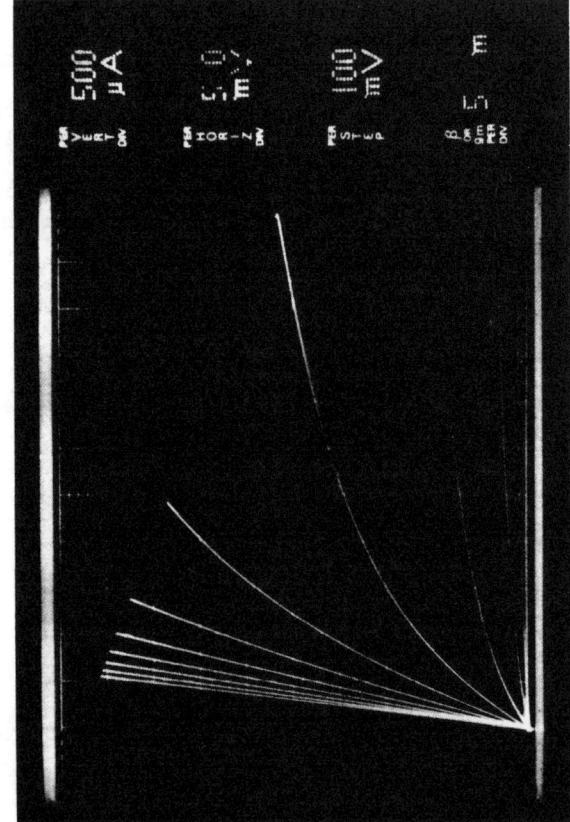

Fig. 4—Drain current as a function of gate bias voltage in both forward and reverse directions at a drain-source bias voltage, V_{ds}, of 0.05 V. An appropriate extrapolation of this curve to the abscissa gives the external (or terminal) pinch-off voltage, V_p. A misjudgment in the determination of V_p, as shown by the dash-dotted line, will cause a problem in the next step.

This plot should be a straight line. However, the plot may deviate from the line either upward or downward as X increases, as indicated by dashed lines in Fig. 5. Such a deviation depends upon the *temporarily* determined value of V_p. If the plot significantly departed from the straight line, the previously determined V_p had to be re-examined. Usually, a slight adjustment on the V_p value easily solves this problem and V_p is *finally* determined.

Such a way of determining V_p seems to be much more complicated and tedious than the conventional method. Usually, V_p is simply estimated from V_{gs} corresponding to the bottom line of the drain I-V characteristics. Indeed, the conventional method may not bring too much error in the determination of V_p in the case of thick active channels, i.e., devices with high pinch-off voltages. However, in devices with thin active channels, especially on buffer layers, the error may reach as high as 100 percent with the conventional method, as will be seen in Fig. 6. Therefore, the present method has been developed.

In Fig. 5, the linear extrapolation of the plot to the ordinate gives the value of $(R_s + R_d)$. The slope of the line is designated as R_o. The parameter R_oX represents the effective value of active channel resistance at a given V_{gs}.

As was mentioned previously, the forward gate current value is affected by a combination of series resistances at high current levels. Therefore, the slope of the I_g-V_g characteristic, measured at a current density of around 10^4 A/cm², gives an estimate of series resistance values. By measuring the gate current in three different ground connections, i.e, source and drain combined, source only, and drain only, three resistance values can be obtained. The differences between the last two values yields $(R_s - R_d)$. Since $(R_s + R_d)$ has been known, R_s and R_d are now readily separated. The gate series resistance R_g can be deduced from the first resistance value by subtracting the contribution of the paralleled R_s and R_d from the resultant.

3.3 Determination of specific voltages and currents

Figure 6 is a typical photograph of the drain I-V characteristics for negative gate potentials, taken with the nonoffset gate-bias mode of the curve tracer. In contrast with Fig. 6, Fig. 7 is an unconventional photograph of the drain characteristics of the same device when driven in the forward gate bias with the offset mode. As the positive value of gate offset is increased, the drain current increases. Beyond a certain value of the offset, V_f, however, the drain current no longer increases. Figure 7 shows such a state of offsetting.

The knee voltage of a drain I-V characteristic could be defined as the intercepting point between the extensions of two linear regions of the characteristic. The knee voltage of the I-V curve for V_f is denoted by V_{kf} as shown in Fig. 7. Also, the knee voltage for the zero-gate-bias curve is by V_{ko} as shown in Fig. 6. Total drain currents, I_f and I_{do}, for V_f and null gate bias, respectively, are measured at the corresponding knee voltages, V_{kf} and V_{ko}. Leakage current components, I_{pf} and I_{po}, are also measured at V_{kf} and V_{ko}, respectively, both for $V_{gs} = -V_p$. The maximum channel current, I_m, and zero-gate-bias channel current, I_o, are then evaluated as $I_f - I_{pf}$ and $I_{do} - I_{po}$, respectively.

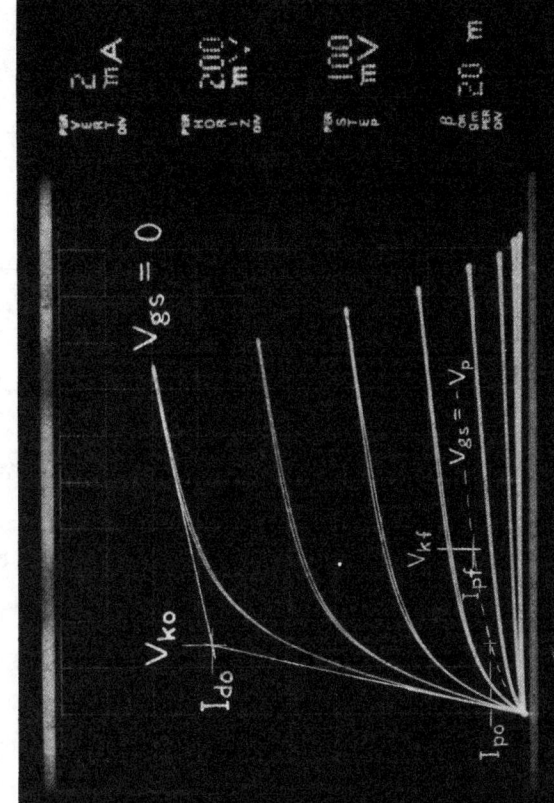

Fig. 6—A conventional drain I-V characteristics with nonoffset gate-bias mode as usual.

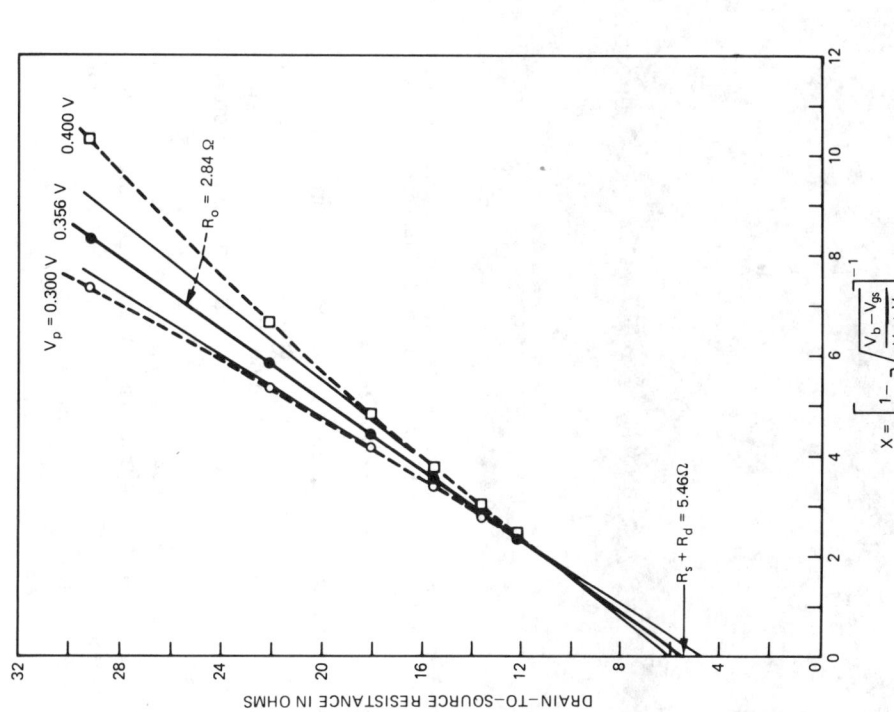

Fig. 5—Determination of the open channel resistance, R_o, and parasitic series resistances, R_s and R_d. The proper selection of V_p value in Fig. 4 (i.e., $V_p = 0.356$ V in this case) is essential for this determination, as illustrated in two wrong cases ($V_p = 0.300$ V and 0.400 V).

channel parameters, 11 GaAs MESFETs of various designs were chosen, as shown in Table I. The first five devices (A through E) were originally designed for high-power use[9] and the others (F through K) were for low-noise applications.[10]

All devices had a total device width of 0.5 mm, except for device E which had 1 mm. The distance between the source and drain electrodes was nominally 6 μm for devices A through E and 3 μm for devices F through K. The nominal gate length was 2.0 to 2.5 μm for devices A to D, 1.0 to 1.5 μm for device E, and 0.8 μm for devices F to K. The physical length of the gate electrode, however, was not necessarily equal to the effective gate length because the latter was subject to the shape of a gate junction.

All the sample devices had a multi-layer structure consisting of an undoped n-GaAs film 2 to 3 μm thick as the buffer, an n-GaAs channel, and n$^+$-GaAs layer, except for devices J and K. All the layers were grown sequentially on a semi-insulating GaAs substrate in an AsCl$_3$/Ga/H$_2$ CVD system.[11] After removing the n$^+$-GaAs layer and part of n-GaAs in the gate region, aluminum approximately 0.7 μm thick was deposited as the Schottky-gate metal. The ohmic contacts were formed with a 12 percent Ge/Au-Ag-Au system approximately 0.25 μm thick, alloyed at nearly 500°C. The final metallization was completed with a Ti-Pt-Au system 0.9 to 1.4 μm thick.

4.2 Results of measurements

First, dc parameters n, V_b, V_p, W_p, V_{kf}, V_f, I_f, I_{pf}, I_m, V_{ko}, I_{do}, I_{po}, I_o, R_o, R_s, R_d, R_g and g'_m were obtained in accordance with the measuring technique described in Section III. The measured values of these parameters are shown in Table I. Note that g'_m is an average value taken at approximately I_o.

Second, basic channel parameters L and N were deduced from each of the two methods described in Section 2.3. In the application of (18), (22), and (23), V_c was assumed to be zero for devices F to K and to be a single value of 0.17 V for devices F to K. The two deduced values for each of L and N were then averaged to obtain the most probable value. Using this mean value of N in (15), the most probable value of a was determined. All values mentioned here are shown in Table II.

4.3 Comparison of calculated and measured results

By inserting the determined values of N, a, and L into (4), (6), and (5), I_m was calculated. The calculated value of I_m was then compared with the measured value as shown in Table III. By adopting the measured value of V_b and R_s in (10) and (11), respectively, I_o was also calculated using (9), (4), and (8). The calculated I_o was compared with the measured value, again as shown in Table III. The comparison has shown excellent agreement between the calculated and measured

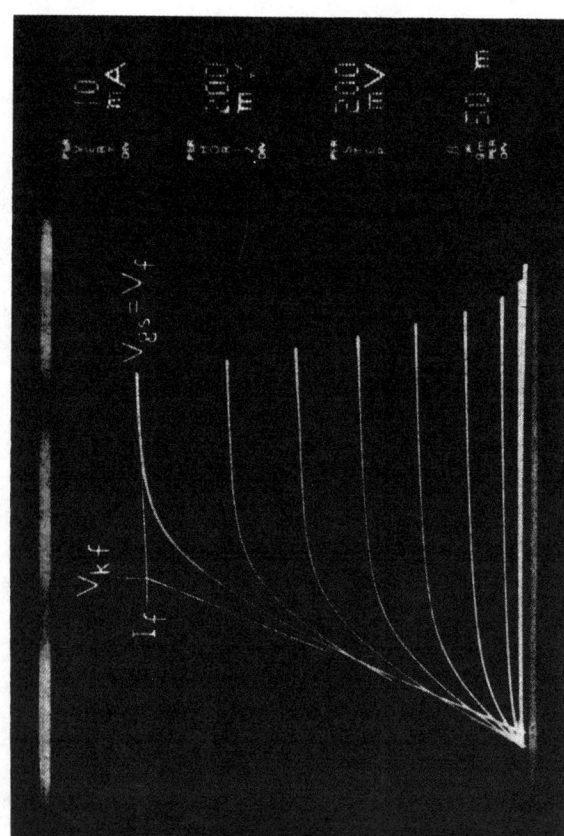

Fig. 7—A special drain I-V characteristics with a forward gate-bias offsetting of V_t. This is a critical value beyond which no increase in the drain current is observed.

3.4 Determination of transconductance

As is well known, the magnitude of the transconductance of a good device can be assumed to remain constant up to nearly the cutoff frequency. Therefore, the so-called dc transconductance can be considered a first-order approximation of the amplitude of microwave transconductance.

The following method of measuring g'_m and evaluating g_m is conventional. In the drain I-V characteristics, an increment of drain current, ΔI_d, between two adjacent curves for $V_{gs} = V_1$ and $V_{gs} = V_2$ at a given V_{ds} yields an average transconductance as

$$g'_m = \left| \frac{\Delta I_d}{V_2 - V_1} \right|. \qquad (32)$$

However, this value is a result of degradation due to R_s. The intrinsic value of the small-signal transconductance at the bias points, $V_{ds} = V_{ds}$ and $V_{gs} = (V_1 + V_2)/2$, can thus be derived as

$$g_m \approx \frac{g'_m}{1 - g'_m R_s}. \qquad (33)$$

IV. EXAMINATIONS OF NEW TECHNIQUE

4.1 Sample devices

To present the practical values of measured dc parameters and hence to demonstrate the new technique for determining the basic

Table II—Determination of the most probable values of active channel carrier concentration and thickness and effective gate length

Equations Used	Parameter		Device										
	Symbol	Units	A	B	C	D	E	F	G	H	I	J	K
	V_c	V	0	0	0	0	0	0.17	0.17	0.17	0.17	0.17	0.17
(18, 19)	N	10^{16} cm^{-3}	6.01	5.98	11.19	21.16	7.96	3.52	6.34	7.75	9.84	7.59	7.91
(18, 21)	L	μm	3.07	2.06	2.71	2.08	2.04	0.568	0.466	0.537	0.502	0.491	0.486
(21, 5, 19)	N	10^{16} cm^{-3}	5.86	6.15	10.58	20.13	7.78	3.51	6.78	7.76	9.83	7.23	8.31
(23)	L	μm	2.91	2.22	2.47	1.97	1.92	0.563	0.545	0.532	0.503	0.436	0.538
(averaged)	N	10^{16} cm^{-3}	5.94	6.07	10.89	20.65	7.87	3.52	6.56	7.76	9.84	7.41	8.11
(averaged)	L	μm	2.99	2.14	2.59	2.03	1.98	0.566	0.508	0.534	0.502	0.463	0.512
(15)	a	μm	0.391	0.394	0.233	0.114	0.314	0.273	0.165	0.140	0.134	0.155	0.138

Table III—Calculated values of the maximum channel current and zero-gate-bias channel current and their comparison with measured values

Equations used	Parameter		Device										
	Symbol	Units	A	B	C	D	E	F	G	H	I	J	K
(4, 5, 6)	I_m	mA	175	195	177	134	395	79.0	84.4	80.6	103	91.0	84.4
(4, 8, 9, 10, 11)	I_o	mA	118	133	99.0	36.2	272	31.0	18.5	11.0	23.1	18.4	12.1
I_m (cal)/I_m (meas)			1.000	0.999	1.000	1.000	1.000	1.000	0.999	1.001	1.000	1.000	0.999
I_o (cal)/I_o (meas)			0.965	1.002	0.971	0.953	1.069	1.058	0.973	0.916	1.076	1.054	0.966

Table I—Measured values of dc parameters

Parameter		Device										
Symbol	Units	A	B	C	D	E	F	G	H	I	J	K
Z	mm	0.5	0.5	0.5	0.5	1.0	0.5	0.5	0.5	0.5	0.5	0.5
z	mm	0.5	0.25	0.25	0.5	0.25	0.25	0.25	0.25	0.25	0.25	0.25
n		1.09	1.07	1.26	1.34	1.26	1.46	1.12	1.23	1.23	1.34	1.26
V_b	V	0.72	0.73	0.76	0.79	0.74	0.74	0.78	0.76	0.74	0.76	0.76
V_p	V	5.84	6.07	3.51	1.16	4.88	1.15	0.51	0.34	0.54	0.52	0.36
W_p	V	6.56	6.80	4.27	1.95	5.62	1.89	1.29	1.10	1.28	1.28	1.12
V_{kf}	V	1.95	1.95	2.10	1.55	1.85	0.75	0.59	0.68	0.77	1.08	0.83
V_f	V	1.49	1.34	1.80	1.49	1.62	1.39	0.97	0.96	1.12	1.30	1.14
I_f	mA	176	196	179	136	400	83.0	86.0	82.0	107	94.0	87.5
I_{pf}	mA	1	1	2	2	5	4.0	1.5	1.5	4	3.0	3.0
I_m	mA	175	195	177	134	395	79.0	84.4	80.5	103	91.0	84.5
V_{ko}	V	1.5	1.5	1.4	0.7	1.4	0.55	0.3	0.3	0.4	0.3	0.3
I_{do}	mA	122	134	104	39.5	257	32.0	20	13	24.0	19.5	14.5
I_{po}	mA	0	1	1	1.5	3	2.7	1	1	2.5	2	2
I_o	mA	122	133	103	38.0	254	29.3	19	12	21.5	17.5	12.5
R_o	Ω	4.1	3.0	2.9	3.4	1.2	4.0	2.9	2.9	2.5	2.7	3.0
R_s	Ω	4.8	3.7	4.5	5.2	1.8	2.7	1.5	2.3	2.3	3.8	2.9
R_d	Ω	2.4	3.6	4.0	2.8	1.7	2.7	1.9	2.4	2.3	4.8	3.4
R_g	Ω	3.8	1.7	1.7	4.5	4.4	13.7	3.8	3.7	4.0	4.5	3.8
g'_m	m℧	28	32	28	53	72	30	45	51	52	45	48

Table IV—Comparison of the most probable value of free carrier concentration in the active channel of an individual device with an uncorrected value of doping in the epitaxial layer of the corresponding slice

Parameter							Device					
Symbol	Units	A	B	C	D	E	F	G	H	I	J	K
N (slice) uncorrected	10^{16} cm^{-3}	6.7	7.4	14.2	24.6	8.8	4.1	7.1	9.0	10.4	8.1	11.2
N (slice)/N(device) uncor. most prob.		1.13	1.22	1.30	1.20	1.12	1.16	1.08	1.16	1.06	1.09	1.38

values in both cases. Note that the equations applied to calculate I_m and I_o had not been used to determine L, a, and N in Section 4.2.

The 11 devices used in this experiment were from 11 different slices. The free carrier concentration of each slice was evaluated by a doping profiler.[12] It has been recognized that this particular profiler gives an N-value 5 to 40 percent higher than the true value of N.[13] Also, a standard deviation of ± 3 percent in the doping across a wafer has been known for these slices.[11] Under such circumstances, the evaluated value of N for each device was compared with an uncorrected, representative N value obtained for the corresponding slice. The results shown in Table IV seem to be very reasonable. A consistent pattern of difference between the two N values is seen there, as expected from the above observation.

V. ANALYSES OF EXPERIMENTAL RESULTS

5.1 Schottky-barrier built-in voltage

The built-in voltage of an aluminum Schottky-barrier gate junction at room temperature has been expressed in an analytical form as a function of N in n-GaAs[1] as follows:

$$V_b = 0.706 + 0.06 \log N \quad (V), \quad (34)$$

where N is in units of 10^{16} cm^{-3}. This expression is compared with the measured value of V_b for all devices used, as shown in Fig. 8. This comparison indicates that (34) can be used for aluminum gates on n-GaAs at room temperature.

5.2 Transconductance

The small-signal transconductance of a GaAs MESFET has been described by the theoretical expression[5]

$$g_m = \frac{I_s}{2W_p}\left[1 - \frac{I_d}{I_s}\right], \quad (35)$$

where I_d is the dc drain bias current. Since g_m' was measured approximately at I_o, the theoretical value of g_m was also calculated for $I_d = I_o$ in (35). This calculated value was then compared with the measured value obtained using (33). This comparison led to a conclusion that the measured value of g_m would agree well with the predicted value if the latter were calculated by

$$g_m \simeq \frac{0.9 I_s}{2(V_p + V_b)}\left[1 - \frac{I_c}{I_s}\right], \quad (36)$$

In the above expressions, $a_{i=1,2,3}$, $\rho_{i=1,2,3}$ and $L_{i=2,3}$ are, respectively, the thickness, specific resistivity, and length of the GaAs epitaxial film at the corresponding place $i = 1$, 2 or 3, L_c is the length of the contacting metal electrode, and R_c is the specific contact resistance.

Parameters ρ and R_c are both functions of N. Using the experimental data taken by Matino on epitaxial n-GaAs films,[17] the doping dependence of ρ can be written in an analytical form

$$\rho \approx 0.11 \, N^{-0.82} \quad (\Omega\text{-cm}) \tag{41}$$

in an N range of 10^{-1} to $10^3 \times 10^{16}$ cm^{-3}. Based on the so-called Shockley method,[8] R_c was statistically investigated using monitor areas provided within the same slices as those fabricated for either high-power[9] or low-noise use.[10] An empirical expression for R_c was then found to be

$$R_c \approx 4 \, N^{-0.5} \quad (10^{-5} \, \Omega\text{-cm}^2) \tag{42}$$

for N values in the range of 3 to $10^3 \times 10^{16}$ cm^{-3}. This expression differs from that given by Heime et al.,[18] which is

$$R_c \approx 8 \, N^{-1} \quad (10^{-5} \, \Omega\text{-cm}^2). \tag{43}$$

However, both expressions give close values of R_c for N in the vicinity of the mid-10^{16} cm^{-3} range. As the N value increases, the difference in R_c between the two expressions becomes recognizable. This would give rise to the case of ohmic contacts formed on n$^+$-GaAs layers. An estimate by (43) would result in too-optimistic prediction of R_c.

Substitution of (41) and (42) into (37), (39), and (40) yields practical expressions for R_{co}, R_2, and R_3 as follows:

$$R_{co} \approx \frac{2.1}{Z a_1^{0.5} N_1^{0.66}} \quad (\Omega) \tag{44}$$

$$R_2 \approx \frac{1.1 \, L_2}{Z a_2 N_2^{0.82}} \quad (\Omega) \tag{45}$$

and

$$R_3 \approx \frac{1.1 \, L_3}{Z a_3 N_3^{0.82}} \quad (\Omega). \tag{46}$$

For the 11 sample devices with reasonable assumptions on N_1, a_1, L_2, N_2, a_2, L_3, N_3, and a_3, component resistances R_{co}, R_2, and R_3 were calculated using (44), (45), and (46), respectively. The predicted

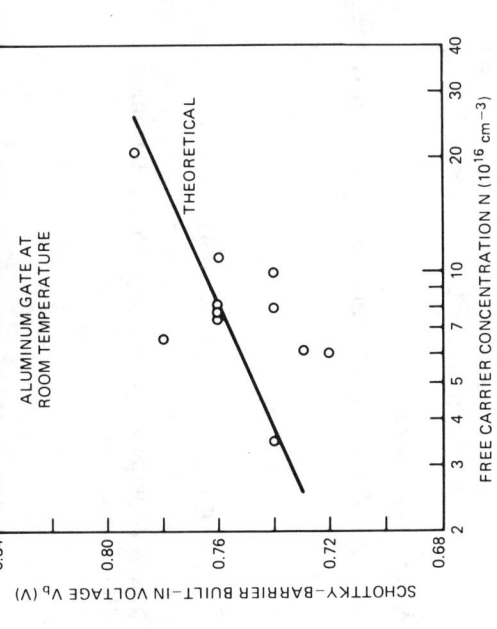

Fig. 8—Comparison of the measured values of gate built-in voltage for aluminum Schottky-barrier gates at room temperature, with a theoretical expression as a function of free carrier concentration in n-GaAs.

where I_c was the channel current, as shown in Table V.

5.3 Ohmic contacts and series channel resistance

The parasitic series resistance, R_s or R_d, consists of the ohmic contact resistance, R_{co}, and series channel resistance, R_{ch}, between the two concerned electrodes. These component resistances are now separately expressed. The expression for R_{co} given by Berger[14] and Murrmann and Widmann[15] can be simplified, as already shown by Macksey and Adams[16] as

$$R_{co} = \frac{1}{Z} \sqrt{\frac{R_c \rho_1}{a_1}} \coth \sqrt{\frac{\rho_1 L_c^2}{R_c a_1}} \approx \frac{1}{Z} \sqrt{\frac{R_c \rho_1}{a_1}}. \tag{37}$$

The series channel resistance can be further divided into two components which represent two individual parts of different structures in the space between the gate and one ohmic contacts. Thus,

$$R_{ch} \approx R_2 + R_3, \tag{38}$$

where

$$R_2 = \frac{\rho_2 L_2}{Z a_2} \tag{39}$$

and

$$R_3 = \frac{\rho_3 L_3}{Z a_3}. \tag{40}$$

239

Table VI—Predicted values of parasitic source (or drain) series resistance and its component resistances in comparison with measured value

Parameter		Device										
Symbol	Units	A	B	C	D	E	F	G	H	I	J	K
Z	mm	0.5	0.5	0.5	0.5	1.0	0.5	0.5	0.5	0.5	0.5	0.5
N_1	10^{16} cm^{-3}	100	100	50	100	100	100	100	100	100	7.4	8.1
a_1	μm	0.15	0.15	0.15	0.15	0.15	0.15	0.15	0.15	0.15	0.35	0.45
L_2	μm	2.1	2.1	2.1	2.1	2.5	0.85	0.75	0.85	0.85	0.75	0.75
N_2	10^{16} cm^{-3}	5.9	6.1	10.9	20.7	7.9	100	100	100	100	7.4	8.1
a_2	μm	0.35	0.35	0.2	0.1	0.35	0.1	0.1	0.1	0.1	0.25	0.35
L_3	μm	0	0	0	0	0	0.4	0.3	0.4	0.4	0.3	0.3
N_3	10^{16} cm^{-3}	—	—	—	—	—	3.5	6.6	7.8	9.8	7.4	8.1
a_3	μm	—	—	—	—	—	0.273	0.165	0.140	0.134	0.155	0.138
R_{co}	Ω	0.52	0.52	0.82	0.52	0.26	0.52	0.52	0.52	0.52	1.89	1.57
R_2	Ω	3.07	3.01	3.26	3.86	1.45	0.68	0.59	0.68	0.68	1.28	0.85
R_3	Ω	0	0	0	0	0	1.15	0.86	1.17	1.00	1.13	0.86
Predicted R_s, R_d	Ω	3.59	3.53	4.08	4.38	1.71	2.35	1.97	2.37	2.20	4.30	3.28
Measured $(R_s + R_d)/2$	Ω	3.6	3.65	4.25	4.5	1.75	2.7	1.7	2.35	2.3	4.3	3.15

Table VII—Determination of the effective resistivity for aluminum gate metallization

Parameter		Device										
Symbol	Units	A	B	C	D	E	F	G	H	I	J	K
Z	mm	0.5	0.5	0.5	0.5	1.0	0.5	0.5	0.5	0.5	0.5	0.5
z	mm	0.5	0.25	0.25	0.5	0.25	0.25	0.25	0.25	0.25	0.25	0.25
L_g	μm	2.5	2.0	2.0	2.0	1.2	0.8	0.8	0.8	0.8	0.8	0.8
h	μm	0.7	0.7	0.7	0.7	0.2	0.2	0.65	0.65	0.65	0.65	0.65
R_g	Ω	3.8	1.7	1.7	4.5	4.4	13.7	3.8	3.7	4.0	4.5	3.8
ρ_g	10^{-6} Ω-cm	4.0	5.7	5.7	3.8	5.1	5.3	4.9	4.8	5.0	5.6	4.9

value of R_s (or R_d) was thus obtained as the sum of these resistances. The average of the measured values of R_s and R_d for each device was then compared with the predicted value, as shown in Table VI. They were in good agreement.

5.4 Gate series resistance

Since the input signal applied to the feeding end of the gate travels along the gate metallization to the other end, the gate must be considered as a distributed network. The effective value of the gate metallization resistance in a lumped equivalent circuit is, therefore, different from the dc value measured from one end to the other. As theoretically analyzed by Wolf,[19] this effective value, R_g, is one-third of the end-to-end dc resistance as a first-order approximation. Thus,

$$R_g \approx \frac{\rho_g z^2}{3 L_g h Z}, \quad (47)$$

where ρ_g is the specific resistivity, L_g is the mean length, h is the mean height, and z is the unit width of the gate metallization.

By substituting the nominal values of Z, z, L_g and h into (47) in conjunction with the measured value of R_g, ρ_g was evaluated as shown in Table VII. The range from 3.8 to 5.7 × 10^{-6} Ω-cm gives a mean value of 5.0 × 10^{-6} Ω-cm. Although this value is higher than a bulk aluminum resistivity of 2.8 × 10^{-6} Ω-cm, it seems to be very reasonable for such a fine, thin, and scaly structure of gate metallization. Based on this finding, a practical expression for R_g can be given by

$$R_g \approx \frac{17 z^2}{L_g h Z} \quad (\Omega), \quad (48)$$

where Z and z are in units of mm and L_g and h are in μm.

It may be noted that annealing of a device sometimes results in an improvement in the effective value of ρ_g. The above measured values were obtained without additional heat treatment.

5.5 Open channel resistance

The open channel resistance mentioned in Section III can be expressed as

$$R_o = \frac{L}{q \mu_0 N a Z}, \quad (49)$$

where μ_0 is the low-field mobility of electrons.[8] Thus, μ_0 of an individual device can be obtained from the measured value of R_o in conjunction with Z, L, a, and N, which have already been determined. On the other hand, μ_0 is related to ρ in the form

$$\mu_0 = \frac{1}{q N \rho}. \quad (50)$$

Substituting (41) into (50) yields μ_0 as a function of N

$$\mu_0 \approx 5.7 \, N^{-0.18} \quad (10^3 \text{cm}^2/\text{V-sec}). \quad (51)$$

The μ_0 value deduced from the measured value of R_o and that predicted by (51) are shown in Table VIII for all the sample devices. The evaluation of the latter value was based on the average value of N shown in Table II. In the high-power devices, these two values of μ_0 were close enough to confirm Matino's results[17] and to support the use of (41) and (51). In the low-noise devices, however, the μ_0 value deduced from R_o appeared to be approximately one-half the predicted value by (51). Such a discrepancy might be caused by a special two-dimensional gate-recess structure of these devices, and/or be due to an increased influence of the transition layer as the active layer was thinned, this transition layer being between the active and buffer layers. This is subject to further investigation.

As shown in Table VIII, the μ_0 value deduced from the measured value of R_o using (49) differs from one device to the other in some degree. In accordance with the two-piece approximation of the v-E characteristic, however, μ_0 is assumed to be constant by definition and is equal to v_s/E_c. Substituting the aforementioned best-fit values of v_s and E_c into this yields a fixed value of 4.8×10^3 cm^2/V-sec for μ_0, which is much greater than that deduced from R_o by (49) and that calculated by (51) as a function of N in the normal range. On the contrary, the fixed values of v_s and E_c have been satisfactory to evaluate I_m for a wide range of N, as seen in the previous sections. Such a conflicting situation must be reconciled. This is also subject to further study. Nevertheless, the following can be applied in practice, for the time being. The fixed values of v_s and E_c are appropriate for the evaluation of I_m for all devices. The μ_0 value provided by (51) is adequate in (49) to calculate R_o for devices with plane gates. However, a suitable correction factor is necessary for μ_0 given by (51) to make μ_0 effective for nonplane gate devices. For example, this factor was 0.5 for the low-noise devices used as samples.

VI. PREDICTION OF MICROWAVE PERFORMANCE

6.1 Maximum output power

As was previously mentioned, devices A through E were originally designed for high-power use. The maximum available output power, $P_{\max}$, of these devices were measured at 4 GHz in a coaxial system. A double-slug tuner was provided in each of the input and output circuits to obtain the conjugate match. The data taken at a drain bias of 12V are shown in Table IX.

The maximum output power delivered from a GaAs MESFET operating at a drain-source voltage of V_{ds} is approximately given by

$$P_{\max} \approx \frac{I_m}{4} (V_{ds} - V_{kf}) \quad (52)$$

if V_{ds} is sufficiently lower than the drain-source breakdown voltage. By substituting the measured values of I_m and V_{kf} into (52), $P_{\max}$ was calculated for $V_{ds} = 12$V. Furthermore, $P_{\max}$ was predicted using the geometrical and material parameters, shown in Tables II and VI, to calculate I_m and V_{kf}. Both predicted values of $P_{\max}$ are shown in Table IX.

As seen in Table IX, there is excellent agreement between the measured and predicted values of $P_{\max}$ in devices C, D, and E at this drain bias voltage. However, the measured values of devices A and B were substantially smaller than the predicted values. This discrepancy could be caused by the saturation effect in output power as V_{ds} increased. This problem of power saturation will be discussed in a separate paper. Without power saturation mechanisms, *the maximum output power capability at microwave frequencies can be predicted by dc parameters as well as by device geometrical and material parameters*.

6.2 Minimum noise figure

Devices F through K were originally designed for low-noise amplifiers. The minimum noise figure, $F_{\min}$, of these devices was measured at 5.92 GHz in a coaxial system with double-slug tuners. The measured values are shown in Table X.

Table VIII—Comparison between the low-field electron mobility value deduced from the measured value of open channel resistance by (49) and that predicted from the most probable value of free-carrier concentration in the active channel by (51)

Parameter	Device										
μ_0 in 10^3 cm^2/V-sec	A	B	C	D	E	F	G	H	I	J	K
Calculated by (49)	3.9	3.7	4.4	3.2	4.2	1.8	2.0	2.1	1.9	1.9	1.9
Calculated by (51)	4.1	4.1	3.7	3.3	3.9	4.5	4.1	3.9	3.8	4.0	3.9

Table IX—Predicted and directly measured values of the maximum output power at 4 GHz and 12V drain bias

Parameter	Device				
$P_{\max}$ @ 12 volts	A	B	C	D	E
Measured directly (W)	0.40	0.40	0.44	0.36	1.00
Predicted from measured dc para's (W)	0.439	0.491	0.437	0.350	1.001
Predicted from geo. and mat. para's (W)	0.439	0.493	0.440	0.353	1.006

Table X—Predicted and directly measured values of the minimum noise figure at 5.92 GHz

Parameter	Device						
F_{min} @ 5.92 GHz	F	G	H	I	J	K
Measured directly (dB)	2.22	1.51	1.84	1.74	1.75	1.76
Predicted from measured dc para's (dB)	2.21	1.50	1.80	1.73	1.75	1.76
Predicted from geo. and mat. para's (dB)	2.12	1.56	1.79	1.70	1.72	1.80

A simple expression for F_{min} has been derived as

$$F_{min} \approx 10 \log [1 + KfL \sqrt{g_m(R_g + R_s)}] \quad (dB), \quad (53)$$

where f is the frequency of interest in GHz and K is the fitting factor.[20]

This fitting factor, which represents the channel material properties, ranges from 0.25 to 0.3 in most cases. Substituting (33) into (53) with a typical K-value of 0.27 yields

$$F_{min} \approx 10 \log \left[1 + 0.27 \, fL \sqrt{\frac{g'_m(R_g + R_s)}{1 - g'_m R_s}} \right] \quad (dB). \quad (54)$$

Using the measured values of g'_m, R_g, and R_s, and deduced value of L in (54), F_{min} was calculated for devices F to K. The minimum noise figure was also predicted from the geometrical and material parameters, using the values shown in Tables II, VI, and VII to calculate g_m, R_s, and R_g.

These predicted values are compared with the directly measured value in Table X. The agreement is excellent between them. This supports the idea that *the dc characterization of a low-noise GaAs MESFET makes it possible to predict F_{min} at microwave frequencies with a remarkably high accuracy. Also, once the geometrical and material parameters are given for a device, its F_{min} can be calculated as well.* It would be worthwhile to note that, if the operating frequency approaches the cutoff frequency of a device or frequencies where the skin effect on the gate metallization becomes significant, an additional term is required in (54) for an improved accuracy.[20]

VII. SUMMARY OF RELATIONSHIPS

Table XI is a summary of the relationships between the dc and rf performance parameters and the geometrical and material parameters of a GaAs MESFET.

VIII. CONCLUSIONS

This paper complements the recent study of a new model of the GaAs MESFET.[1] A new technique has been introduced in which the basic channel parameters, such as the effective gate length, channel doping, and channel thickness, are determined from the so-called dc

Table XI—Summary of relationships

dc Parameter	Equations	Participating Geometrical and Material Parameters
V_b	(34)	N
W_p'	(3)	N, a
V_p'	(3, 34)	N, a, Z
I_s	(4)	N, a, L
β	(6)	N, a, L, Z
I_m	(5, 6, 4) or (7)	$N, N^+, a_1, a_2, a_3, L_2, L_3, Z$
R_s, R_d	(44, 45, 46)	Z, z, h, L_{gs}, ρ_g
R_g'	(48)	N, a, L, Z, R_s
I_o	(8, 9, 10, 11, 4)	N, a, Z, R_s, V_c
V_{hf}	or (8, 12, 13, 3, 4) (14, 3, 4, 5, 6, 44, 45, 46)	$N, a, L, Z, R_s, R_d, V_c$

Performance Parameter	Equations	Above Parameters plus Bias Parameter
g_m'	(36, 3, 4)	N, a, L, Z, I_c
g_m	(36, 33, 3, 4, 44, 45, 46)	N, a, L, Z, R_s, I_c
P_{max}	(52, 3, 4, 5, 6, 14, 44, 45, 46)	$N, a, L, Z, R_s, R_d, V_{ds}$
F_{min}	(54, 36, 3, 4, 44, 45, 46, 48)	N, a, L, R_s, R_g, f

parameters. Also, a precise technique developed for measuring the dc parameters was shown. Using 11 sample devices chosen from a wide variety of designs, usefulness of the new techniques was demonstrated.

The determined values of the basic channel parameters for the sample devices were used to calculate their dc parameters, such as the maximum channel current, zero-gate-bias channel current, and transconductance in the simple, analytical expressions recently obtained.[1] Their predicted values were then compared with the measured values in excellent agreement for all devices used. Practical expressions, in terms of device geometrical and material parameters, developed for parasitic resistances were verified in good agreement with measured values on all sample devices.

Using the sample devices, it was demonstrated that the maximum output power and minimum noise figure at microwave frequencies can be predicted by dc parameters as well as by device geometrical and material parameters through simple, analytical expressions. In other words, proper dc characterization of a GaAs MESFET makes it possible to predict the microwave power handling capability and minimum noise property.

Finally, the relationships between the dc and rf performance parameters of a GaAs MESFET and its geometrical and material parameters were summarized with the relevant equations. This summary would be very useful as a handy reference for the design and optimization processes of a GaAs MESFET.

IX. ACKNOWLEDGMENTS

The author is grateful to J. V. DiLorenzo, H. M. Cox, L. A. D'Asaro,

B. S. Hewitt, L. C. Luther, W. C. Niehaus, W. Robertson, P. F. Sciortino, J. A. Seman, and J. R. Velebir for fabricating the sample devices used in this paper. He is also thankful to J. P. Beccone, D. E. Iglesias, F. M. Magalhaes, and W. O. Schlosser for preparing the microwave test equipment used for the output power and noise measurements, and to J. E. Kunzler and L. J. Varnerin for their support of the GaAs MESFET project.

REFERENCES

1. H. Fukui, "Channel Current Limitations in GaAs MESFETs," unpublished work.
2. W. Shockley, "A unipolar 'field-effect' transistor," Proc. IRE, *40* (November 1952), pp. 1365–1376.
3. A. B. Grebene and S. K. Ghandhi, "General Theory for Pinched Operation of the Junction Gate FET," Solid-State Elec., *12* (July 1969), pp. 573–585.
4. R. B. Fair, "Graphical Design and Iterative Analysis of the dc Parameters of GaAs FET's," IEEE Trans. Electron Devices, *ED-21* (June 1974), pp. 357–362.
5. R. A. Pucel, H. A. Haus, and H. Statz, "Signal and Noise Properties of Gallium Arsenide Microwave Field-Effect Transistors," *Advances in Electronics and Electron Physics*, vol. 38, New York: Academic Press, 1975, pp. 195–265.
6. S. M. Sze, *Physics of Semiconductor Devices*, New York: Wiley-Interscience, 1969, p. 393.
7. C. R. Crowell, J. C. Sarace, and S. M. Sze, "Tungsten-Semiconductor Schottky-Barrier Diodes," Trans. Metallurgical Soc. AIME, *233*, (March 1965), pp. 478–481.
8. P. L. Hower, W. W. Hooper, B. R. Cairns, R. D. Fairman, and D. A. Tremere, "The GaAs Field-Effect Transistor," *Semiconductors and Semimetals*, vol. 7, New York: Academic Press, 1971, pp. 147–200.
9. W. C. Niehaus, H. M. Cox, B. S. Hewitt, S. H. Wemple, J. V. DiLorenzo, W. O. Schlosser, and F. M. Magalhaes, "GaAs Power MESFETs," *Gallium Arsenide and Related Compounds (St Louis), 1976*, Conf. Series No. 33b, Bristol and London: The Institute of Physics, 1977, pp. 271–280.
10. B. S. Hewitt, H. M. Cox, H. Fukui, J. V. DiLorenzo, W. O. Schlosser, and D. E. Iglesias, "Low-Noise GaAs MESFETs: Fabrication and Performance," *Gallium Arsenide and Related Compounds (Edinburgh), 1976*, Conf. Series No. 33a, Bristol and London: The Institute of Physics, 1977, pp. 246–254.
11. H. M. Cox and J. V. DiLorenzo, "Characteristics of an $AsCl_3/Ga/H_2$ Two-Bubbler GaAs CVD System for MESFET Applications," *Gallium Arsenide and Related Compounds (St Louis), 1976*, Conf. Series No. 33b, Bristol and London: The Institute of Physics, 1977, pp. 11–22.
12. G. L. Miller, "A Feedback Method for Investigating Carrier Distributions in Semiconductors," IEEE Trans. Electron Devices, *ED-19* (October 1972), pp. 1103–1108.
13. H. M. Cox, private communication.
14. H. H. Berger, "Contact Resistance on Diffused Resistors," 1969 IEEE ISSCC Digest of Technical Papers, February 1969, pp. 160–161.
15. H. Murrmann and D. Widmann, "Current Crowding on Metal Contacts to Planar Devices," 1969 IEEE ISSCC Digest of Technical Papers, February 1969, pp. 162–163.
16. H. Macksey and R. Adams, "Fabrication Processes for GaAs Power FET's," Proc. Fifth Cornell Conf. on Active Semiconductor Devices for Microwave and Integrated Optics, 1975, pp. 255–264.
17. H. Matino, "A Study of GaAs Microwave Semiconductor Devices," Doctoral Dissertation (Japanese), 1972.
18. K. Heime, U. König, E. Kohn, and A. Wortmann, "Very low resistance Ni-AuGe-Ni contacts to n-GaAs," Solid-State Elec. *17* (1974), pp. 835–837.
19. P. Wolf, "Microwave Properties of Schottky-Barrier Field-Effect Transistors," IBM J. Res. Develop, *14* (March 1970), pp. 125–141.
20. H. Fukui, "Optimal Noise Figure of Microwave GaAs MESFETs," unpublished work.

Experimental Measurement of Microstrip Transistor-Package Parasitic Reactances

ROBERT J. AKELLO, MEMBER, IEEE, BRIAN EASTER, MEMBER, IEEE, AND I. M. STEPHENSON

Abstract—A resonance method of measurement is described for the determination of the parasitic reactances of a microwave transistor package mounted in microstrip. Results for two types of package obtained from normal-sized and from scaled-up models are presented. The influence of the parasitics on the characteristics of a typical microwave FET chip is briefly discussed.

I. Introduction

THE parasitic reactances associated with the package or mounting can seriously limit the performance of a microwave semiconductor device and need to be accurately known for good circuit and device modeling. Some diode packages in coaxial mounts, such as the S4 have been examined thoroughly [1], [2], but relatively sparse data are available on packages for microstrip application [3], [12]. The rapid advance in the performance of gallium arsenide FET's highlights the need for the accurate characterization and improvement of packages and mountings suitable for hybrid MIC's. In this paper two types of transistor packages will be considered, the leadless inverted device (LID) and the S2 package proposed by James *et al.* [4].

Previous papers [5]–[7] have described the measurement of the small reactances and susceptances of microstrip junctions and discontinuities, using a resonance technique and a close approach to a substitution procedure. In this procedure the change in resonant frequency is observed when the unknown element is introduced into a microstrip resonant circuit, only light coupling through noncritical connections being required for accurate determination of resonance. This method has the advantage of largely avoiding the problems entailed in the measurement of microstrip circuits through a coaxial-to-microstrip transition. While this approach often cannot be directly used for active devices, due to the low-circuit Q-factor that would result, it can be usefully applied to the study of the package parasitics. In addition to examining normal-sized packages with the active element appropriately disabled or disconnected, the method has been used to study scaled-up models of the package and circuit, Stycast material of the correct permittivity being employed in place of the ceramics of the package and circuit substrate. This is a quick and accurate method

Manuscript received June 2, 1976; revised October 4, 1976.
The authors are with the School of Electronic Engineering Science, University College of North Wales, Dean Street, Bangor, Caerns, Wales, Great Britain.

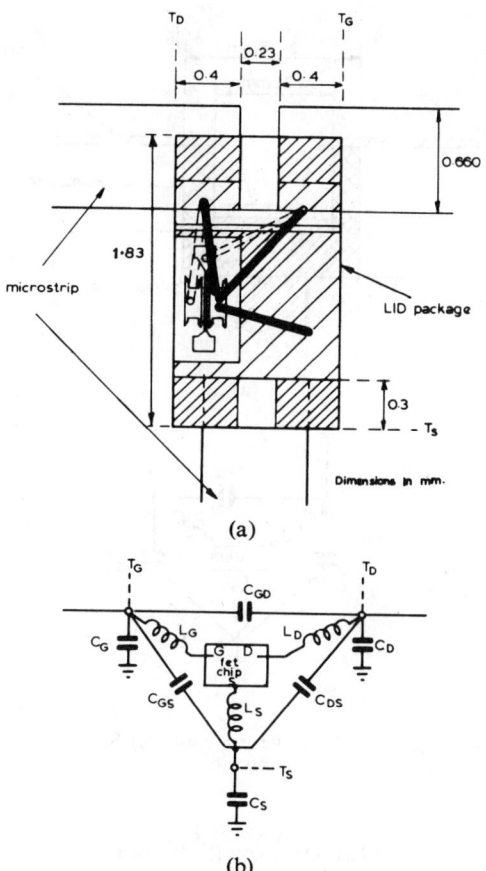

Fig. 1. (a) Bottom view of specially bonded normal size LID mounted on a 0.660-mm microstrip. T_G, T_D, and T_S are taken for the reference planes. Note: in a normal device, and the X8 model, the drain and gate wires were in the positions shown by the broken lines. (b) Equivalent circuit of LID package mounted as in Fig. 1(a) Note: C_G, C_D, and C_S include the microstrip capacitances under the contact pads of LID.

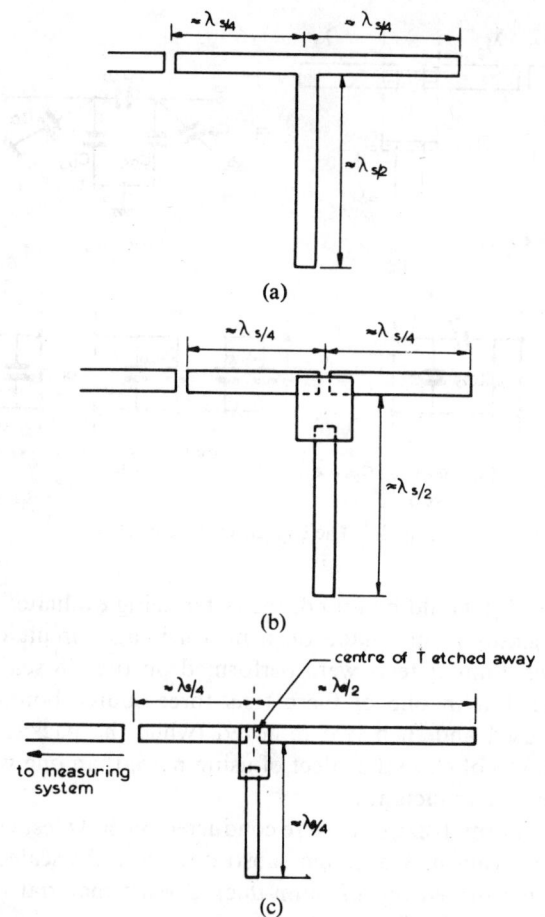

Fig. 2. (a) Reference T-junction resonator. (b) Configuration for determining $(L_G + L_D)$. (c) Test section for measuring $(L_G + L_S)$; for measuring $(L_D + L_S)$ the mirror image of the above is used.

for examining the effect of rearranging bonds and connections, as changes in both can be made without removing the scaled-up package from the microstrip circuit. Quite small effects can be directly observed without the variabilities inherent if the package must be removed and replaced. An important general advantage of the technique described is that the package may be evaluated in a mounting configuration close to that in which the device will be employed.

II. THE LID PACKAGE

The general features of the LID package are shown in Fig. 1(a) which shows the arrangement of a typical FET chip and the bond wires. The package is normally mounted on the upper surface of a microstrip circuit and is connected to three 50-Ω microstrip lines on 0.660-mm alumina. Also indicated are the reference planes for the assumed equivalent circuit of Fig. 1(b).

A. LID Inductance

A modified device is employed with the three bonds connected to the same point as indicated in Fig. 1(a). When mounted, the device forms a modified microstrip T junction which may be characterized by the method of [7]. $(L_G + L_D)$ is first determined employing the configuration of Fig. 2(b), where the "top" of the T forms a half-wave resonator with L_G and L_D at the position of the current maximum. The stem of the T is approximately half-wave in length and presents a high impedance which is connected at a low-impedance point; thus L_S, etc., have very little effect on the resonance. In order to gain the advantages of the substitution method, a reference T-junction resonator was first measured [Fig. 2(a)]. Using the equivalent-circuit values of [7] for the end and gap effect and also the T junction in conjunction with the physical length of the resonator, the microstrip velocity was determined for the measured resonant frequency. The junction of the T was then etched away and the LID package mounted, and the resonant frequency redetermined, enabling $(L_G + L_D)$ to be found. The effect of the capacitances C_G, C_D, and C_S is small, as they are virtually at a voltage zero. If necessary, their presence can be allowed for. If the stem of the T does not present an ideal open circuit to the top section, the method relies on the symmetry of the T configuration. However, it can be shown that moderate departures (e.g., 5°–10° of electrical length) from this condition have only a slight effect and can be taken into account if necessary.

By using an L-shaped resonator [Fig. 2(c)], again with a λ/2 arm to terminate the third port in an effective open circuit, the values of $(L_G + L_S)$, and in a similar way

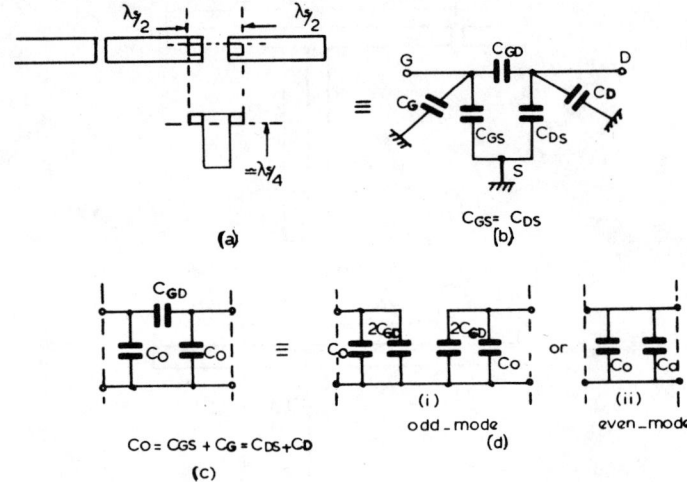

Fig. 3. The capacitive π-network.

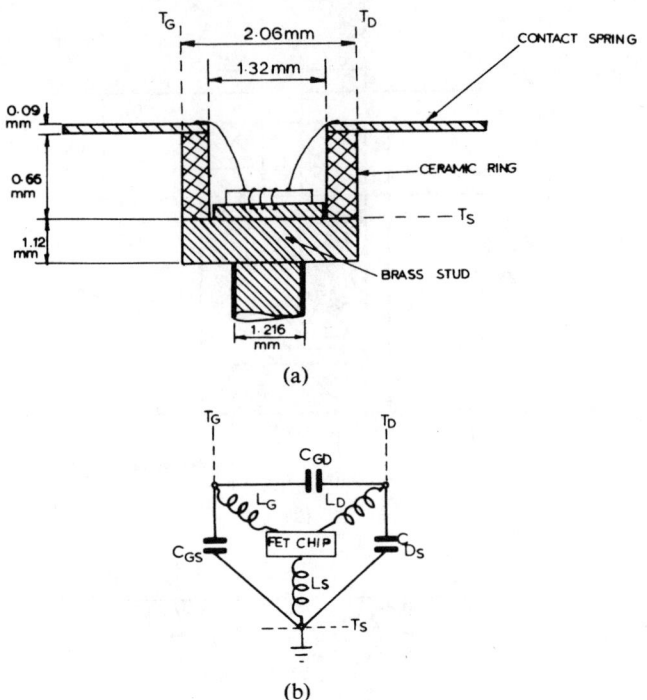

Fig. 4. (a) S2 package bond-wire layout. (b) S2 package equivalent circuit.

($L_D + L_S$), could be found, the latter being evaluated from the measurements made on a mirror-image circuit of the former. Similar tests were performed on the X8 scaled-up model, but in one of these tests three source bond-wires were used and then two removed (while the package was *in situ*) to observe the effect of using more than one wire to reduce the inductance.

The normal-size tests were conducted on 50-Ω test circuits on 0.66-mm-thick alumina substrates while the scaled tests were performed on 5.29-mm-thick Stycast material having relative permittivity of 10.

B. Capacitance Measurements

The capacitances of an empty package were initially measured using the General Radio type 1616—a three-terminal low-frequency capacitance bridge. They were then measured at microwave frequencies. As the device was not mounted in microstrip for the low-frequency measurements, the capacitance value can differ from that obtained under normal mounting conditions. The resonance technique used for the microwave measurements is based on the capacitive π-network model for a gap in microstrip line [8]–[10].

The LID in the microstrip configuration of Fig. 3(a) has an equivalent circuit of Fig. 3(b) where the ground connection of the source pad is provided by the $\lambda/4$-line in Fig. 3(a). On resonating the circuit, the odd (lower) and even (higher) resonance frequencies (typically, 6.86 and 7.24 GHz) that correspond to the equivalent circuits of Fig. 3(d)(i) and (ii), respectively, were observed and used to evaluate the capacitances C_0 and $2C_{GD}$. As C_G or C_D cannot be physically separated from either C_{GS} or C_{DS}, C_0 and C_{GD} are the effective parasitic capacitances that appear at the package terminals when the package is mounted in microstrip. It can be shown that the length of the $\lambda/4$ line to the source contact-pads has only small influence on the capacitance values. For example, an error of (say) 10° in the electrical length of this line at 7 GHz would cause an error of 4 percent in the value of the capacitance C_0.

In Table I, results are shown for both the real-size and scaled-up LID. As the reference plane is taken at the edge of the package, the inductance values include the inductances of the lead wires and those of the package metallization.

TABLE I
LID PACKAGE REACTANCE

Parameter	Normal size device		X8 Scaled model, values scaled to normal size	
	Measured value	Form of device	Measured value	Form of device
(1)* L_G, nH	0.66+	Standard GAT2 device with all leads bonded to source pad. One source bond wire.	0.68	X8 LID packaged with dummy chip. All leads bonded to source pad. One source bond wire.
(2)* L_D, nH	0.61+		0.54	
(3)* L_S, nH	0.760		0.76	
(4)* L_S, nH	–	–	0.65+	As above but 3 source bond wires.
(5)** C_{GD}, pF	0.026	Empty LID package	0.023	X8 Empty LID package
(6)* C_{GD}, pF	0.038+		–	
(7)** C_0, pF	0.038		0.037	
(8)* C_0, pF	0.104+		–	

* Measured at microwave frequencies.
** Measured at low frequencies.
+ Values used in Section 4.

III. THE S2 (CRC) PACKAGE

A. Inductance Measurements

The S2 package, shown in Fig. 4, has a threaded stud that connects the source bond-wires to the ground plane of the microstrip line; therefore, unlike the LID, the source must always be grounded. Hence, the T- and L-shaped

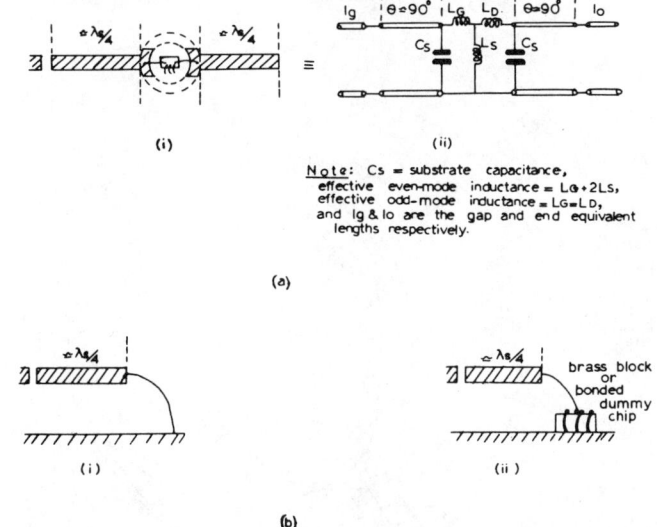

Fig. 5. Microstrip configuration for measuring S2 package inductances.

TABLE II
THE MEASURED S2 PACKAGE REACTANCE

TYPE	PARAMETER DEFINITION	MEASURED VALUE Inductance in nH capacitance in pF
L_1	Inductance of gate bond-wire (L_G) from Fig. 5a; $L_G = L_D$.	0.525+
L_2	Inductance of gate bond wire and three source wire. ($L_G + L_S$)	0.768
L_3	Inductance of three source bond wires.	0.12+
L_{4A}	Inductance of short-circuiting wire in Fig. 5b (i)	0.595
L_{4B}		0.516*
L_5	Inductance of short-circuiting wire on brass-block (Fig. 5b (ii).	0.507
L_{6A}	Inductance with three source bond-wires bonded on top of the dummy chip (Fig. 5b (iii). ($L_G + L_S$)	0.63
L_{6B}		0.842*
L_7	Apparent inductance of three source bond wires $L_7 = (L_{6A} - L_5)$	0.123
L_8	Inductance of gate bond-wire and 2 source wires.	0.707
L_9	Apparent inductance of two source bond wires $L_9 = (L_8 - L_5)$	0.2
L_{10}	Inductance of gate bond-wire and one source bond wire.	0.761
L_{11}	Apparent inductance of one source wire $L_{11} = (L_{10} - L_5)$	0.256
C_{1A}	The gate to drain capacitance (C_{GD}) measured on the l.f. bridge @ 1kHz.	0.004
C_{1B}		0.003*+
C_2	The capacitance $C_0 = C_{GS} + C_G$ $= C_{DS} + C_D$ measured at microwave frequencies (Fig.3)	0.097*+

* Values found from measurements on real-size models. Other values (without asterisks) are from scale-up models.

+ Values used in Section 4.

TABLE III
CHANGE IN INDUCTANCE WITH WIRE SPACING AND WITH USE OF RIBBON IN S2 PACKAGE

L, nH	COMMENTS	WIRE SPACING 1.0	1.5	2.5	3.5
L_1	Inductance of bond wire on a brass dummy chip.	0.476	-	-	-
L_2	Inductance of three source bond-wires	0.178	0.166	0.141	0.108
L_3	Inductance of two source bond wires.	0.195	0.194	0.156	0.115
L_4	Inductance of one source bond wire.	0.274	-	-	-
L_5	Inductance of copper ribbon of width = 2.5 mm.	0.094	-	-	-
L_6	Inductance of copper ribbon of width 5mm	0.036	-	-	-

Note: All given figures are scaled-down values.

resonators previously used are not appropriate. The values of the three inductances of the bond-wires in Fig. 4(b) cannot be directly found separately, but may be deduced with reasonable accuracy by a series of experiments based on Fig. 5, where both or either of the normal-size and/or scaled-up models are used whenever convenient. Fig. 5(a)(i) shows the microstrip configuration used to determine L_G, L_D, and L_S. Fig. 5(a)(ii) is its electrical equivalent circuit. This circuit resonates in odd and even modes at frequencies corresponding to the effective value of inductance that terminates the $\lambda/4$ microstrip line. Using the known data for the end and gap effect [7], the values of L_G and L_S were calculated.

The method is usable only when the circuit is symmetrical. This condition was achieved by positioning of the dummy chip at the center of the package, thus making $L_G = L_D$, and then adjusting the lengths of the $\lambda/4$ lines in the scaled-up circuit. It can be shown that since the inductances being evaluated are at voltage nodes, slight asymmetry in the circuit causes very small error in the inductance values. The method has the advantage that it gives the overall inductance values including those due to the mutual inductance. However, it is not possible to separate the mutuals from the self-inductances. The results are given in Table II (L_1–L_3).

Further experiments to check the measured inductances of this package were carried out based on Fig. 5(b), where in Fig. 5(b)(i), a bond wire was connected from the gate terminal to the package stud and resonated in the microstrip system as shown. This wire is slightly longer than the normal gate bond-wire but of the same order of magnitude in inductance, the value of which is L_4. The inductance L_5 of the normal gate bond-wire was deduced from Fig. 5(b)(ii) in a scaled-up model, where a brass block of the same height as the semiconductor chip was used to shortcircuit the $\lambda/4$ line. The total inductance in this system is that of the bond wire plus that of the block which is of negligible value as compared to the former; so, we assume $L_G \simeq L_5$. A scaled-up dummy chip was also used to simulate the FET with the

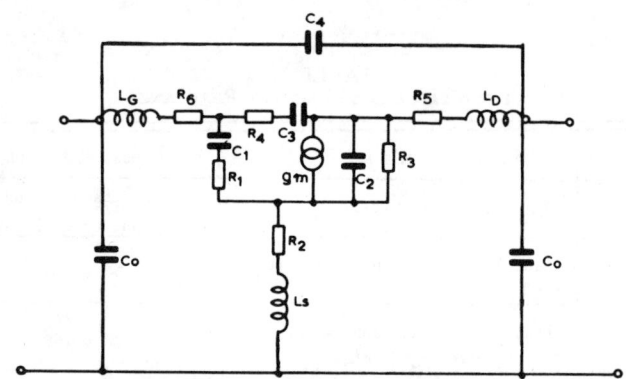

Fig. 6. Equivalent circuit of the packaged FET. $R_1 = 6.5$; $R_2 = 3.7$; $R_3 = 192$; $R_4 = 658$; $C_1 = 0.5$; $C_2 = 0.2$; $C_3 = 0.014$; $g_{m0} = 43$; $g_m = g_{m0}e^{-j\omega\tau_0}$; $\tau_0 = 5$ ps; $R_5 = R_6 = 0$. Resistance in ohms, capacitances in picofarads. C_0, C_4, L_G, L_D, and L_S taken from the package measurements (Tables I and II). Other values from [11].

gate and source bond-wires bonded on the same pad [as in Fig. 4(a)], the total inductance being $L_{6A} = (L_G + L_S)$. A standard FET (bonded at Plessey Co. Ltd., Towcester) was used in a real-size model for the same purpose, but with the drain bond-wires terminated on a microstrip line of length $\simeq \lambda/2$ to present a near open-circuit at the drain end on the

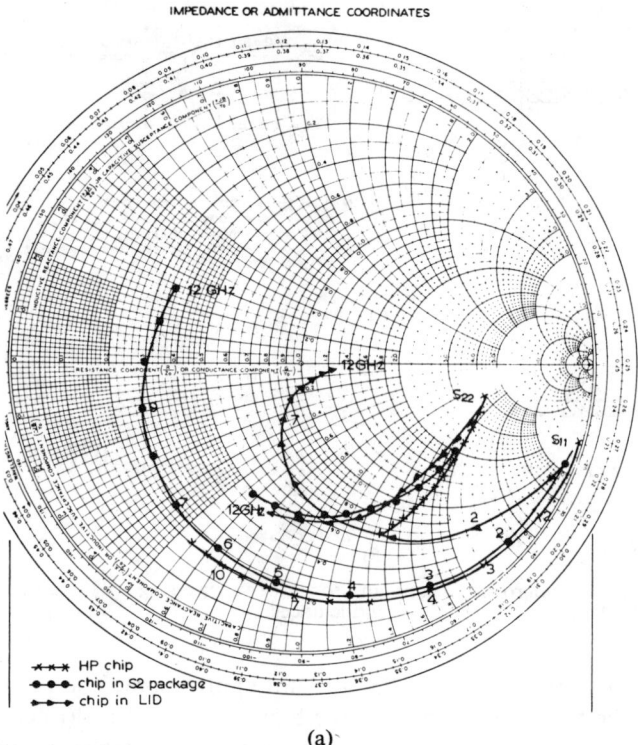

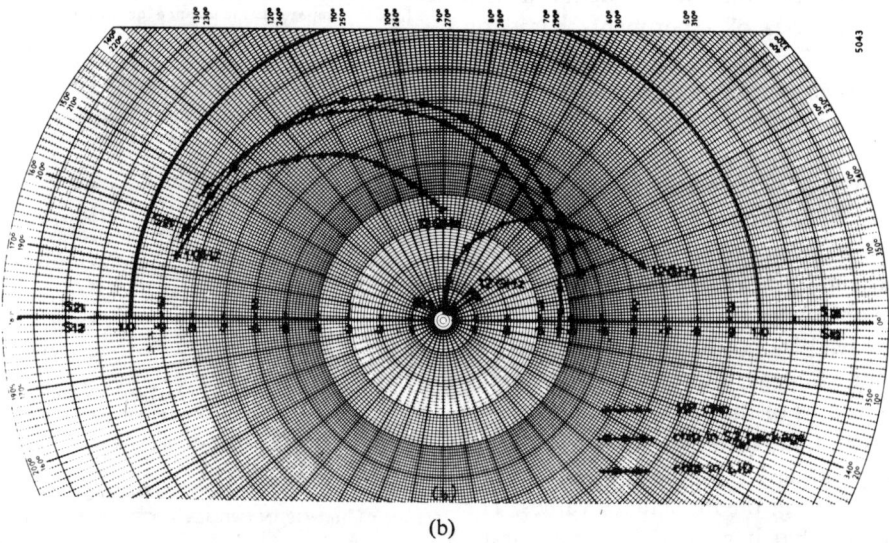

Fig. 7. (a) Effect of package parasitics on chip S-parameters (S_{11} and S_{22}). (b) Effect of package parasitics on chip S-parameters (S_{12} and S_{21}).

chip. The resulting inductance value $L_{6B} = (L_G + L_S)$ is seen to agree within an acceptable limit with L_2 from Fig. 5(a).

To show the effect of using more than one source bond-wire, after obtaining L_{6A} [from Fig. 5(b)(ii)], the number of wires was reduced by plucking them off one by one while the package was in position. The results (L_8–L_{11}) given in Table II show a reduction in the source inductance with an increase in the number of wires. Experiments were also conducted on the same circuit to show the effect of various spacings of the source bond-wires and also of the effect of conducting ribbon bonds on the source inductance. The results are shown in Table III.

It can be inferred that the changes seen in measuring the inductance L_{6A}, L_8, and L_{10} are predominantly due to the change in L_S, but it should be noted that the measurement is in terms of $(L_G + L_S)$.

B. Capacitance Measurement

It was proposed to measure the S2 capacitances C_{GD}, C_{GS}, and C_{DS} using the same microwave method as used for the LID capacitances. However, due to the low value of C_{GD}, combined with low Q-factor of microstrip, only a single resonance was observed as would be expected from the value in Table II (C_{1B}). The low-frequency capacitance value of C_{GD} was thus used, whereas C_{GS} and C_{DS} were calculated from the observed resonance. The obtained values for these capacitances are in Table II.

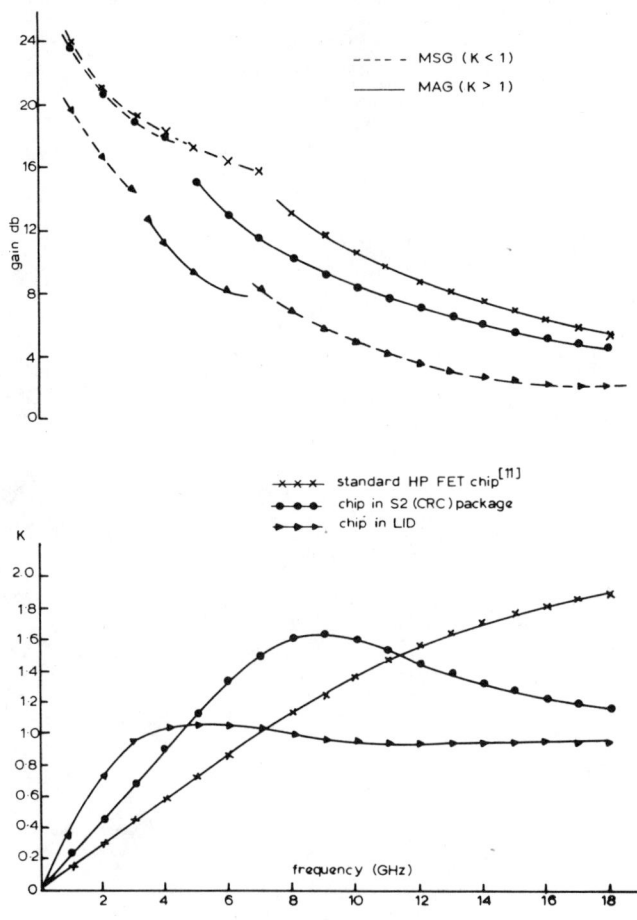

Fig. 8. Gain and stability factor (K).

IV. EFFECTS OF THE MEASURED PACKAGE PARASITIC REACTANCES ON FET CHIP CHARACTERISTICS

To demonstrate the effect of the measured package reactances on chip performance, the Hewlett-Packard 1-μm MESFET chip [11], with the parameters shown in Fig. 6 was taken and the values for the parasitics (indicated in Tables I and II) added using the Z-, Y-, and S-matrices. The overall S-parameters, gain (MSG and MAG) and stability factor (K) were computed for the chip in the two types of package. The results are given in Figs. 7 and 8. Further analysis using this computation showed that most of the effects on the chip characteristics are due to L_S and C_{GD}.

V. CONCLUSIONS

a) The resonance method has been shown capable of accurately measuring very small package inductances and capacitances at microwave frequencies with the package mounted in a configuration similar to that of the circuit normally used.

b) From Tables I–III, it is seen that there is a reasonable agreement between the results obtained from the real-size and scaled-up models. This gives confidence in using scaled-up models, especially in cases where real-size devices are too small for easy handling and where quick changes in circuit configuration are necessary.

c) The results have been used to show the improvement in the performance of a typical chip embedded in the S2-type package (as compared to the LID package) in terms of broad-band matching (S_{11} and S_{22} of Fig. 7). The improvements stem from reduced values of both feedback capacitance C_{GD} and the source lead inductance L_S. Thus, apart from the higher gain obtained in the S2-type package, the device remains potentially stable over a broader band of frequency.

ACKNOWLEDGMENT

The authors wish to thank their colleagues Dr. A. Gopinath for his valuable discussions and help in CAD, P. D. Cooper (now of The Plessey Co. Ltd., Roke Manor) for his work on the LID scaled-up model, J. A. Turner of The Plessey Co. Ltd., for providing the packaged devices, and E. W. Parry and J. Tame for the preparation and fabrication of the circuits.

REFERENCES

[1] W. J. Getsinger, "The packaged and mounted diode as a microwave circuit," *IEEE Trans. Microwave Theory Tech.*, vol. MTT-14, pp. 58–69, Feb. 1966.
[2] R. P. Owens, "Mount independent equivalent circuit of the S4 diode package," *Electron. Lett.*, vol. 7, pp. 580–582, 23 Sept. 1971.
[3] T. Yahara, Y. Kadowaki, H. Hoshika, and K. Shirahata, "Broad-band 180° phase shift section in X-band," *IEEE Trans. Microwave Theory Tech.*, vol. MTT-23, pp. 307–309, Mar. 1975.
[4] D. S. James, R. J. P. Douville, R. W. Breithaupt, and A. L. Van Koughnett, "A 12 GHz FET amplifier for communications satellite applications," in *Proc. of 1974 European Microwave Conf.*, Montreux, Switzerland, pp. 97–101.
[5] B. Easter, J. G. Richings, and I. M. Stephenson, "Resonant techniques for accurate measurements of microstrip properties and equivalent circuits," Paper B.7.5. *Proceedings of the 1973 European Microwave Conference* (Brussels).
[6] B. Easter and I. M. Stephenson, "Resonant techniques for establishing the equivalent circuits of small discontinuities in microstrip," *Electron. Lett.*, vol. 7, pp. 582–584, 23 Sept. 1971.
[7] B. Easter, "The equivalent circuit of some microstrip discontinuities," *IEEE Trans. Microwave Theory Tech.*, vol. MTT-23, pp. 655–660, Aug. 1975.
[8] A. A. Oliner, "Equivalent circuits for discontinuities in balanced strip transmission lines," *IRE Trans. Microwave Theory Tech.*, vol. MTT-3, pp. 134–143.
[9] S. B. Cohn, "Direct-coupled-resonator filters," *IRE Proc.*, pp. 187–196, Feb. 1957.
[10] P. Benedek and P. Sylvester, "The equivalent capacitances for microstrip gaps and steps," *IEEE Trans. Microwave Theory Tech.*, vol. MTT-20, pp. 729–733, Nov. 1972.
[11] G. D. Vendelin and M. Omori, "Try CAD for accurate GaAs MESFET models," *Microwaves*, vol. 14, pp. 58–70, June 1975.
[12] J. Cohen and M. Gilden, "A mathematical model of a varactor package in microstrip line," *IEEE Trans. Microwave Theory Tech.*, vol. MTT-21, pp. 412–413, June 1973.

Part IV
Low-Noise Amplifier Design

Description of the Noise Performance of Amplifiers and Receiving Systems

Sponsored by
IRE Subcommittee 7.9 on Noise
H. A. Haus, *Chairman*

R. Adler
Zenith Radio Corporation
Chicago, Ill.

R. S. Engelbrecht
Bell Telephone Laboratories
Murray Hill, N. J.

S. W. Harrison
General Telephone and
Electronics Laboratories, Inc.
Bayside, N. Y.

H. A. Haus
Massachusetts Institute of Technology
Cambridge, Mass.

M. T. Lebenbaum
Airborne Instruments Laboratory
A Division of Cutler-Hammer, Inc.
Deer Park, N. Y.

W. W. Mumford
Bell Telephone Laboratories
Whippany, N. J.

I. Introduction

IN GENERAL the output noise of a receiving system contains components contributed not only by the termination at the input of the receiving system but also by the receiving system itself. Furthermore, the output signal-to-noise ratio[1] of the system will depend not only on the output noise but also on the nature of the signal that is impressed upon the input of the receiver. Hence, any meaningful evaluation of the noise performance of a receiver when used in a particular system must include considerations of the sources that contribute to the output noise, the bandwidth and gain of the receiving system in all of its responses, the nature of the signal and the efficacy of the output utilization circuit. *It is evident that no single number can describe completely how well a given receiver will perform in all kinds of systems.*

What, then, are the pertinent attributes of a receiver, and how are they measured and quoted? From the viewpoint of the designer of the receiver, the attributes must be readily measurable. From the viewpoint of the designer of the system, the numbers quoted by the receiver designer must be such that the output signal-to-noise-ratio (SNR) under operating conditions can be calculated.

It is the responsibility of the designer of the system to match his signal to the bandwidth of the receiver and to know what penalty is paid when he doesn't. In general matching presents no hardship, since appropriate matching filters, which tend to optimize the output SNR, can always be introduced into the system. It is also the responsibility of the system designer to employ the best possible utilization circuit at the output of the receiver. The nature of the utilization circuit will depend upon the nature of the signal, whether it is AM, FM, single sideband, double sideband, sky noise, etc.

Let us assume that the system designer has considered carefully these important problems, that he has matched the noise bandwidth to his signal bandwidth and is using the most efficient detection for his signal. What else must he know about a given receiver to predict its noise performance? He must know the gain-frequency characteristics of the receiver. Does it have only one response or multiple responses? He must know how much of the output noise is attributable to the receiver itself. Thus the pertinent attributes that should be measured and quoted by the designer of the receiver are:

1) The gain at each of the responses,
2) The bandwidth,
3) The *effective input noise temperature*.[2]

The effective input noise temperature of the receiving system can be used with an input termination noise temperature such as antenna temperature to compute an *operating noise temperature*,[2] T_{op}, which is a system characteristic and is a function of the signal and the environment, etc. The operating noise temperature is a simple number for the designer of the system to use in

---

[1] We use the signal-to-noise power ratio as a measure of quality of the output response. This is reasonable in amplifiers or systems that we define as having a single output response, namely systems in which any frequency component entering the system at any one of the input frequencies produces a response at one single corresponding output frequency. An example of a case we do not intend to cover is the double-sideband degenerate parametric amplifier.

[2] IRE standard definitions of these italicized terms may be found in the preceding Standard.

evaluating his system noise performance, since kT_{op} is the power per unit signal output bandwidth required of an input signal to make the output SNR unity.

The terms operating noise temperature and effective input noise temperature have been defined but the concepts require further discussion. The relationship between operating noise temperature, *noise temperature*[3] of the input termination, effective input noise temperature and *noise factor*[3] will be brought out during the discussion.

The discussion here will deal with systems having single as well as multiple input responses, but only a single output response. A system with multiple input responses and a single output response is one in which several different input frequencies in the different input bands produce an output at one single output frequency in the output band.[4]

II. Operating Noise Temperature

The noise performance of a particular system may be evaluated in terms of its output SNR *under operating conditions*. Now the output signal power per unit bandwidth S_o can always be expressed as a signal power per unit output bandwidth, S_i, available at the input terminals multiplied by the signal gain G_s. The signal gain is defined (see definition of operating noise temperature) as the ratio of

1) the signal power delivered at the specified output frequency into the output circuit (under operating conditions) to
2) the signal power available at the corresponding input frequency or frequencies to the system (under operating conditions) at its accessible input terminations.

Stated in another way, S_i may be thought of as the output signal power per unit bandwidth referred to the input. In a similar manner, the output noise power per unit output signal bandwidth, N_o, can be referred to the input by dividing by G_s and can then be related to the *operating noise temperature*,[5] T_{op}, as follows:

$$\frac{N_o}{G_s} = kT_{op} \quad (1)$$

---

[3] IRE standard definitions of these italicized terms may be found in "IRE Standards on Electron Tubes: Definitions of Terms, 1957, 57 IRE 7.S2," Proc. IRE, vol. 45, pp. 983–1010; July, 1957. For convenience, these standard definitions are included as the Appendix to this paper.

[4] A parametric amplifier is an example where inputs at the so-called signal and idler frequency produce an output at the signal frequency.

[5] The use of temperature to express noise powers is particularly convenient when the noise powers are small and when the frequencies are in the normal frequency region, say up to 10^3 kMc. At these frequencies, the available output noise power per unit bandwidth from a resistor may be expressed as $P_n = kT$, where k is Boltzmann's constant, and T is the absolute temperature of the resistor. The use of temperature as a measure of receiving system noise does not infer that the sources of noise are necessarily thermal; only that the output noise power can be accounted for by a noiseless receiver with an input termination at a specified temperature.

Hence, the output SNR is

$$\frac{S_o}{N_o} = \frac{S_o/G_s}{N_o/G_s} = \frac{S_i}{kT_{op}}.$$

From this, it is clear that two receiving systems will exhibit the same output SNR if they have the same S_i/T_{op} ratio; also, that two receiving systems having the same available input signal power per unit output bandwidth must have the same operating noise temperature to produce the same SNR at the output.

III. Derivation of General Equation for Output Noise Power

We shall now derive the general expression for the noise power per unit bandwidth flowing into the output termination of a linear multiport transducer (see Fig. 1). This output noise power, at a specified output frequency, arises from two sources:

1) The contributions to the output noise power due to noise power available from the impedance terminations that are connected to all accessible ports except the output port under operating conditions. These contributions can be described by assigning appropriate noise temperatures to the input terminations at all of the various responses.
2) All other contributions to the output noise power. These include noise generated within the receiver components as well as noise resulting from any frequency conversions internal to the receiving system. Noise generated in the load and reflected at the receiver output may also contribute to this output noise power.

If we denote the portion of the output noise power per unit bandwidth described under 1) by N_{io} and that described under 2) by N_N, we can write the total output noise power per unit bandwidth N_o as

$$N_o = N_{io} + N_N. \quad (2)$$

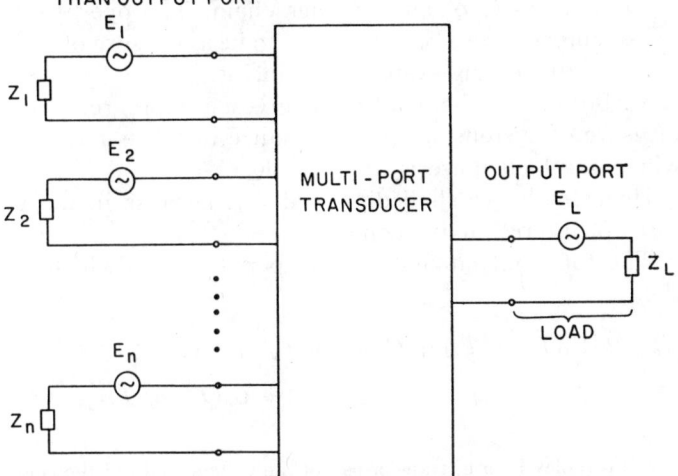

Fig. 1—Equivalent circuit of multiport transducer. The E_i's are the open-circuit signal and/or noise voltages of the input terminations. E_L is the open-circuit noise voltage of the load.

If T_{in} is the noise temperature at the nth response, and there is no correlation between the noises at different responses, then

$$N_{io} = k(T_{i1}G_{12} + T_{i2}G + \cdots + T_{in}G_n) \quad (3)$$

where G_n is the transducer gain of the nth response and is the ratio of 1) the output power delivered to the utilization circuit at the specified output frequency to 2) the corresponding input power available at the nth input response.

In a similar manner, we can express the noise attributable to the receiver in terms of the effective input noise temperature, T_e, which *by definition* is common to all responses. Thus

$$N_N = kT_e(G_1 + G_2 + \cdots + G_n) + V_L. \quad (4)$$

Here, N_L is a term that takes into account the contribution of the noise generated in the load and reflected at the output of the receiver, since the effective input noise temperature definition does not include such a contribution. In order to evaluate N_L one must know the impedance and the equivalent noise temperature of the load, and the output impedance of the receiver. In most cases, N_L will be negligible compared to the other terms in this expression. The contribution N_L may be of importance in systems with insufficient gain, or in systems with an output impedance with a negative real part. In such cases, N_o of (2) is to be interpreted as the net noise power passing the cross section of the output port.[6]

A few words of explanation may be in order for the reason why T_e excludes the load noise contribution and T_{op} includes it. In excluding the load noise from the definition of T_e the effective input noise temperature of a single response two-port is brought into direct correspondence with its noise figure F according to the formula $T_e = 290(F-1)$. The noise figure cascading formula of single response two-ports may thus be adapted to T_e. The cascading formula facilitates the evaluation of T_e of an amplifier chain. The operating noise temperature T_{op} is intended to be a measure of the noise of a receiving system. The inclusion in T_{op} of the contribution of the load-noise gives a more realistic measure of system noise performance in those cases in which the load noise is appreciable.

Henceforth, we shall disregard N_L, because in most cases of interest it is negligible.

The total output noise power per unit bandwidth is then

$$N_o = k[G_1(T_{i1} + T_e) + G_2(T_{i2} + T_e) + \cdots + G_n(T_{in} + T_e)]. \quad (5)$$

[6] If complex Fourier transforms v of the voltage and i of the current can be defined, the net output power is

$$\frac{1}{2} \operatorname{Re} \int_{\omega_1}^{\omega_2} v^* i \, d\omega.$$

We can now characterize the noise performance of the receiving system in terms of the operating noise temperature T_{op} which, from (1), is given by

$$T_{op} = \frac{N_o}{kG_s}. \quad (6)$$

T_{op} may be seen to be a number which characterizes the noise performance of a receiving system under operating conditions. From a knowledge of T_{op}, one may compute the required input signal power to give a desired SNR at the output.

We must now show that the quantities in (6) can be readily measured and then apply these equations to various practical systems.

IV. Measurement of Parameters

In the discussion so far, the terms effective input noise temperature and operating noise temperature have been used, and these are defined for noise power per unit bandwidth at a specified output frequency. In practice, however, finite bandwidths are employed, and the average noise power per unit bandwidth in a specified output band is a more meaningful quantity. The latter term can be referred to the input and related to a temperature in a similar manner as was done for noise power per unit bandwidth at a specified output frequency. The equivalent temperature is then defined as the *average effective input noise temperature* $\overline{T}_e$.[2]

The measurement of both bandwidth and gain are covered in most standard references. One must remember that both are in terms of power, not voltage. For the multiple-response receiver, the input bandwidth must be referred to.

It is of fundamental importance to establish clearly how the temperature, $\overline{T}_e$, which characterizes the noise performance of the receiver, can be determined by measurement. Since most modern noise generators used in noise measurements (noise diodes, gas discharge lamps, hot and cold loads) generate broad-band noise with constant amplitude across the band the direct measurement method is one in which noise is injected *equally* into *all* responses. In other words, the measurement conditions are usually such that the noise temperatures of all the input terminations are equal, such that

$$T_i = T_{i1} = T_{i2} \cdots = T_{in}. \quad (7)$$

Hence, if such a broad-band noise generator is connected to the input terminals of the receiver, the expression for the total noise power is

$$N_{To} = k(T_i + \overline{T}_e)(B_1G_{01} + B_2G_{02} + \cdots + B_nG_{0n}) \quad (8)$$

where B_n is the *noise bandwidth* and G_{0n} is the transducer gain at the reference frequency f_{0n} of the nth response.[7]

[7] For a discussion of these quantities see "IRE Standard on Methods of Measuring Noise in Linear Two Ports, 1959, 59 IRE 20 S1," Proc. IRE, vol. 48, pp. 61–68; January, 1960.

For a limiting bandwidth B_N common to all responses, (8) becomes

$$N_{T_o} = kB_N(T_i + \overline{T}_e)(G_{01} + G_{02} + \cdots + G_{0n}). \quad (9)$$

To measure $\overline{T}_e$, the output noise powers for two different temperatures of the input terminations T_i (hot) and T_i (cold) are observed. One defines a quantity Y by

$$Y = \frac{N_{T_o}(\text{hot})}{N_{T_o}(\text{cold})}. \quad (10)$$

From (9) one finds

$$Y = \frac{T_i(\text{hot}) + \overline{T}_e}{T_i(\text{cold}) + \overline{T}_e}. \quad (11)$$

Solving for $\overline{T}_e$ one has

$$\overline{T}_e = \frac{T_i(\text{hot}) - YT_i(\text{cold})}{Y - 1} \quad (12)$$

and, hence, one can compute the *average effective input noise temperature* of the receiver from the calculated value of Y and the known values of the two different input termination noise temperatures.

It is clear that if the noise generator terminates all receiver input responses and if the gain at each of the responses remains the same between the two measurements of N_{T_o} (hot) and N_{T_o} (cold), $\overline{T}_e$ may be computed directly from the two measurements without concern for the response characteristic of the receiver. It should also be pointed out that T_i (hot) and T_i (cold) may be actual (physical) temperatures if hot and cold bodies are used for the measurement, or may be the noise temperatures of gas discharge lamps or noise diodes. (The noise temperature is the temperature of a passive system having an available noise power per unit bandwidth equal to that of the actual generator employed.) The *effective input noise temperature*, T_e, which is defined at a specific output frequency, can be determined by including a filter between the receiver output and the power measuring device. The bandwidth of the filter must be sufficiently narrow so that the receiver characteristics, such as gain, noise, are constant over the band. If T_e of the receiver depends on the impedance and the noise temperature of the output termination at frequencies other than the specified output frequency, care must be taken that the insertion of the filter does not change T_e. When the spot frequency is near the center of the passband of the receiver, the temperature so obtained may not differ greatly from the average.

V. Calculation of T_{op} for Specific Systems

The expression for the operating noise temperature T_{op} was given in (6). Again, in practice, the noise performance of a receiving system will be characterized by an *average operating noise temperature*,[1] $\overline{T}_{\text{op}}$, which for the simple case of a square response with uniform gain G_s can be written as

$$\overline{T}_{\text{op}} = \frac{N_{T_o}}{kB_oG_s} \quad (13)$$

where B_o is the output signal bandwidth.[8] If B_N is common to all responses, the total output noise power is given in general by

$$N_{T_o} = kB_N[G_1(\overline{T}_{i1} + \overline{T}_e) + G_2(\overline{T}_{i2} + \overline{T}_e) + \cdots + G_n(\overline{T}_{in} + \overline{T}_e)] \quad (14)$$

where $\overline{T}_{in}$ is the averaged noise temperature and G_n is the transducer gain for the nth input response.

A. Single Response Receiver[9]

A simple case to consider is that in which the receiver has only a single response. In this case, all G's in (14) except the first are zero, and $G_1 = G_s$. Then

$$N_{T_o} = kB_NG_s(\overline{T}_{i1} + \overline{T}_e) \quad (15)$$

and so

$$\overline{T}_{\text{op}} = \frac{B_N}{B_o}(\overline{T}_{i1} + \overline{T}_e). \quad (16)$$

The bandwidth ratio, B_N/B_o, is equal to or greater than unity. The lowest operating noise temperature occurs when the noise bandwidth, B_N, matches the signal bandwidth, B_o. It is convenient to assume that the system designer will take care of this in his system design and so (15) becomes

$$\overline{T}_{\text{op}} = \overline{T}_{i1} + \overline{T}_e.$$

B. Multiple Response Receivers; Signal Input at One Response Only

For the case where the input signal occupies only response number 1, $G_s = G_1$, and, from (13) and (14), the expression for the average operating noise temperature becomes

$$\overline{T}_{\text{op}} = (\overline{T}_{i1} + \overline{T}_e) + \frac{G_2}{G_1}(\overline{T}_{i2} + \overline{T}_e)$$

$$+ \frac{G_3}{G_1}(\overline{T}_{i3} + \overline{T}_e) + \cdots + \frac{G_n}{G_1}(\overline{T}_{in} + \overline{T}_e) \quad (17)$$

where it has been assumed that the noise bandwidth is matched to the signal bandwidth.

---

[8] B_o is the bandwidth of the signal delivered to the output utilization circuit. (In the case of several coherent output signal responses —appearing in different frequency bands—B_o denotes the bandwidth of the signal in any one response. In the case of a superheterodyne receiving system, B_o denotes the signal bandwidth appearing in the intermediate frequency amplifier.)

[9] By single response receiver is meant any receiver in which only one frequency at the accessible input terminals corresponds to a single output frequency, regardless of the complexity of the gain-frequency characteristics.

For the special case in which, under operating conditions, the noise temperatures of the input terminations at all input responses are equal, (17) reduces to

$$\overline{T}_{op} = (\overline{T}_i + \overline{T}_e)\left(1 + \frac{G_2}{G_1} + \cdots + \frac{G_n}{G_1}\right). \quad (18)$$

C. Multiple Response Receivers, Signal Input at More Than One Response

If we now wish to evaluate $\overline{T}_{op}$ for the case in which the received input signal is distributed over more than one input response, we note that only G_s can be affected in the equation

$$\overline{T}_{op} = \frac{N_{To}}{kB_o G_s}.$$

N_{To} is, of course, unaffected in a linear system and we assume that B_o, the signal output bandwidth, remains unaffected also.

When the portions of the input signal that are received by the various responses are totally *uncorrelated*, with their powers denoted by $S_{i1}, S_{i2}, \cdots, S_{in}$,

$$S_o = S_i G_s = S_{i1} G_1 + S_{i2} G_2 + \cdots + S_{in} G_n \quad (19)$$

or

$$G_s = \frac{S_{i1} G_1 + S_{i2} G_2 + \cdots + S_{in} G_n}{S_{i1} + S_{i2} + \cdots + S_{in}} = \frac{S_o}{S_i}.$$

We again obtain $\overline{T}_{op}$ from (13) and (14) by substituting G_s:

$$\overline{T}_{op} = \frac{N_{To}}{kB_o G_s} = \frac{B_N[G_1(\overline{T}_{i1} + \overline{T}_e) + G_2(\overline{T}_{i2} + \overline{T}_e) + \cdots + G_n(\overline{T}_{in} + \overline{T}_e)]}{B_v \left[\dfrac{S_{i1} G_1 + S_{i2} G_2 + \cdots + S_{in} G_n}{S_{i1} + S_{i2} + \cdots + S_{in}}\right]}. \quad (20)$$

For the simple case where $S_{i1} = S_{i2} = \cdots = S_{in}$, we obtain $G_s = (G_1 + G_2 + \cdots G_n)/n$, and with $B_N = B_o$,

$$\overline{T}_{op} = \frac{G_1(\overline{T}_{i1} + \overline{T}_e) + G_2(\overline{T}_{i2} + \overline{T}_e) + \cdots + G_n(\overline{T}_{in} + \overline{T}_e)}{\dfrac{1}{n}(G_1 + G_2 + \cdots + G_n)}. \quad (21)$$

For the equally simple case in which the gains are equal, at all responses $G_1 = G_2 = \cdots = G_n$, and the S_i's are arbitrary but uncorrelated, we have $G_s = G_1$ and

$$\overline{T}_{op} = (\overline{T}_{i1} + \overline{T}_e) + (\overline{T}_{i2} + \overline{T}_e) + \cdots + (\overline{T}_{in} + \overline{T}_e). \quad (22)$$

$\overline{T}_e$, of course, is obtained as before from (12).

From the above discussion it can be seen that when the received input signal is distributed over several responses *incoherently*, G_s is never larger than it would be if the received signal were entirely in the response exhibiting the largest gain. Hence, for the case in which all response gains are equal, G_s and thereby $\overline{T}_{op}$ [(22)] are independent of how the uncorrelated input signal is distributed over the various responses. With $\overline{T}_{op}$ (and B_o) constant, the output signal-to-noise power ratio

$$\frac{S_o}{N_o} = \frac{S_i}{k\overline{T}_{op} B_o} \quad (23)$$

is seen to depend only on the *total input signal power*.

In receiving man-made signals, this total input signal power is fixed by the transmitter's capability. If we choose to spread it incoherently over several frequency bands—corresponding to the several responses of our receiving system—the power input to each response simply drops and nothing is gained.

However, if no limitation of transmitter power exists (as, for instance, in the broad-band radiometry case) the total input signal to our receiving system is proportional to the number of its input responses. This results in a corresponding improvement of S_o/N_o over some other receiving system having the same $\overline{T}_{op}$ but only one input response. For instance, assuming equal gain and equal signal densities in all input responses, S_o/N_o of an n input response receiving system is equal to that obtained with a single response system having an operating noise temperature equal to $1/n$ times that of the n response system.

Let us now briefly consider the case in which the portions of the input signal that are received by the various responses are partially or totally *correlated*. For this case, the gain G_s will be a more complex function of the various response gains $G_1, G_2, \cdots, G_n$, and depend also on the degree of correlation as well as the combining process at the output of the receiving system. However, with all contributing factors specified, one can always determine G_s and $\overline{T}_{op}$ from their basic definitions.

D. Relation Between Various Noise Temperatures and Noise Factor

In this paper, the concepts $\overline{T}_{op}$, T_{op}, $\overline{T}_e$, and T_e have been discussed. These terms have simple relationships to F and $\overline{F}$, the noise figure (factor) and average noise figure (factor).[3] In this section these relationships will be discussed.

For the case of a *single response receiver*,

$$\overline{T}_e = (\overline{F} - 1)290,$$
$$T_e = (F - 1)290. \tag{24}$$

The expression for $\overline{F}$ for measurement purposes can be found from (12) and (24):

$$\overline{F} = \frac{\left[\dfrac{T_i(\text{hot})}{290} - 1\right] - Y\left[\dfrac{T_i(\text{cold})}{290} - 1\right]}{Y - 1}. \tag{25}$$

If the "cold" temperature is a 290°K load, we may write (25) as

$$\overline{F} = \frac{\left[\dfrac{T_i(\text{hot})}{290} - 1\right]}{Y - 1}. \tag{26}$$

For single response receivers, then, concepts of average effective input noise temperature, $\overline{T}_e$, and average noise factor, $\overline{F}$, are equally acceptable provided the IRE definitions are adhered to. For low noise receivers $\overline{T}_e$ is probably the more convenient.

If the contribution of the load-noise can be neglected, the operating noise temperature and the average operating noise temperature for the single response receiver can be written as

$$T_{\text{op}} = T_i + T_e$$
$$\overline{T}_{\text{op}} = \overline{T}_i + \overline{T}_e \tag{27}$$

so that

$$T_{\text{op}} = T_i + (F - 1)290$$
$$\overline{T}_{\text{op}} = \overline{T}_i + (\overline{F} - 1)290. \tag{28}$$

For a multiple response receiver, the noise factor can be simply and directly related to operating noise temperature only for the special case when the generator is 290°K and the signal is in one response only. For this case

$$T_{\text{op}} = 290F.$$

E. Conclusion

The foregoing indicates that one very important characteristic of a receiving system is the signal-to-noise ratio, S_o/N_o, at the output of the system. The system designer can evaluate S_o/N_o from the knowledge of the average operating noise temperature $\overline{T}_{\text{op}}$, the output signal bandwidth B_o, the total input signal power S_i, and the signal gain G_s.

The component designer must supply data of sufficient generality so that the values of these quantities can be computed. For this purpose he must supply information on the average effective input noise temperature, $\overline{T}_e$, which he can measure by (12). He must specify the gains of the various responses and their signal and noise bandwidths. The system designer can use these values to calculate his particular system's average operating noise temperature, $\overline{T}_{\text{op}}$, by inserting them together with his signal output bandwidth, B_o, and his particular input termination noise temperatures, T_i's, into the general equations (13) and (14), or any appropriate simpler form. (For instance: (16) for the single response receiver; (18) for the multiple input response receiver with signal in only one response; (22) for multiple input responses with equal gains and uncorrelated input signals which are arbitrarily distributed, etc.) From this value of $\overline{T}_{\text{op}}$, the output SNR may be calculated from (23).

APPENDIX[10]

IRE DEFINITION OF NOISE FIGURE
(Taken from 57 IRE 7.S2, July PROC. IRE)

Noise Factor (Noise Figure) (of a Two-Port Transducer). At a specified input frequency the ratio of 1) the total noise power per unit bandwidth at a corresponding output frequency available at the output *Port* when the Noise Temperature of its input termination is standard (290° K) at all frequencies (Reference: Definition for Average Noise Factor) to 2) that portion of 1) engendered at the input frequency by the input termination at the *Standard Noise Temperature* 290° K).

Note 1: For heterodyne systems there will be, in principle, more than one output frequency corresponding to a single input frequency, and vice versa; for each pair of corresponding frequencies a *Noise Factor* is defined. 2) includes only that noise from the input termination which appears in the output via the principal-frequency transformation of the system, *i.e.*, via the signal-frequency transformation(s), and does not include spurious contributions such as those from an unused image-frequency or an unused idler-frequency transformation.

Note 2: The phrase "available at the output *Port*" may be replaced by "delivered by system into an output termination."

Note 3: To characterize a system by a *Noise Factor* is meaningful only when the admittance (or impedance) of the input termination is specified.

Noise Factor (Noise Figure), Average (of a Two-Port Transducer). The ratio of 1) the total noise power delivered by the transducer into its output termination when the *Noise Temperature* of its input termination is standard (290° K) at all frequencies, to 2) that portion of 1) engendered by the input termination.

---

[10] For clarity, the underlined words have been added to the IRE Standard Definitions.

Note 1: For heterodyne systems, 2) includes only that noise from the input termination which appears in the output via the principal-frequency transformation of the system, *i.e.*, via the signal-frequency transformation(s), and does not include spurious contributions such as those from an unused image-frequency or an unused idler-frequency transformation.

Note 2: A quantitative relation between the *Average Noise Factor* $\overline{F}$ and the *Spot Noise Factor* $F(f)$ is

$$\overline{F} = \frac{\int_0^\infty F(f)G(f)df}{\int_0^\infty G(f)df}$$

where f is the input frequency, and $G(f)$ is the transducer gain, *i.e.*, the ratio of 1) the signal power delivered by the transducer into its output termination, to 2) the corresponding signal power available from the input termination at the input frequency. For heterodyne systems, 1) comprises only power appearing in the output via the principal-frequency transformation *i.e.*, via the signal-frequency transformation(s) of the system; for example, power via unused image-frequency or unused idler-frequency transformation is excluded.

Note 3: To characterize a system by an *Average Noise Factor* is meaningful only when the admittance (or impedance) of the input termination is specified.

Noise Factor (Noise Figure), Spot. See:

Noise Factor (Noise Figure) (of a Two-Port Transducer).

Note: This term is used where it is desired to emphasize that the *Noise Factor* is a point function of input frequency.

Noise Temperature (at a Port). The temperature of a passive system having an available noise power per unit bandwidth equal to that of the actual *Port*, at a specified frequency.

Note: See *Thermal Noise*.

Noise Temperature, Standard. The standard reference temperature T_o for noise measurements is 290° K.

Note: $kT_o/e = 0.0250$ volt, where e is the magnitude of the electronic charge and k is Boltzmann's constant.

Stability and Power Gain of Tuned Transistor Amplifiers*

ARTHUR P. STERN†, SENIOR MEMBER, IRE

Summary—The transistor is a nonunilateral device which, if appropriately terminated, can become unstable at frequencies where its "internal feedback" is sufficiently large. At such frequencies, the maximum power gain is infinite and the transistor may oscillate. This paper discusses the maximum power gain realizable as a function of a required degree of stability.

A "stability factor" is defined in terms of the transistor parameters and terminations (the admittance matrix is used as an example, but the approach is analogous using other representations). The maximum stable power gain of an isolated amplifier stage and the terminating admittances required for the realization of this maximum power gain are then computed as functions of the stability factor. The computations are extended to include bandwidth requirements and limitations. (It is found that, although bandwidth requirements *may* impose limitations on the power gain, there is no simple relationship tying together bandwidth and power gain.) The treatment of multistage amplifiers is outlined with the conclusion that the gain realizable in an n-stage amplifier is smaller than n times the gain of a one-stage amplifier having the same stability factor as the stages of the n-stage amplifier. The respective advantages of different representations for different circuit configurations are discussed.

In an appendix, the theoretical considerations are applied to tuned transistor amplifiers in common-emitter and common-base configurations. The stability factor is related to the tolerances in transistor parameters and terminating impedances. Examples are given for the maximum realizable stable gain as function of parameter tolerances.

Introduction

TRANSISTORS are *nonunilateral* devices; a signal applied to the output port of a transistor amplifier results in a response at the input port. The existence of *internal feedback* is expressed by the fact that, if the transistor is described by the "z," "y," "h," or "g" matrices, the matrix element having the subscript 12 (*i.e.*, z_{12}, y_{12}, h_{12}, or g_{12}) is different from zero.

If properly terminated at its ports, a device having sufficient *internal* feedback may become unstable (*i.e.*, may oscillate) even in the absence of *external* feedback. For example, in the "tuned plate-tuned grid" vacuum tube oscillator, feedback is provided entirely by the grid-to-plate capacitance of the tube.

It has been shown[1,2] that transistors exhibit *potential instability* within certain frequency ranges. The frequency range of potential instability is different for the three transistor configurations (common-emitter, -base and -collector).[3]

At frequencies where the transistor configuration is unconditionally stable (*i.e.*, where the transistor cannot become unstable, no matter what passive terminations are used), the maximum available power gain can be calculated.[1] However, at frequencies, where potential instability exists, the transistor may oscillate and the maximum power gain is infinite. Consequently, at such frequencies, the unqualified "maximum available power gain" is not a very useful concept. On the other hand, engineers engaged in the development of transistor circuitry do know very well that a practical measure for the maximum power gain realizable in a stable amplifier is indeed useful and very much needed. Furthermore, experience shows that such a maximum stable gain does actually exist.

Various attempts have been made to define a practical measure indicative of the maximum realizable power gain. For example, it has been shown that the transistor has a finite maximum gain if conjugately matched at its output, the termination at the input being either a pure resistance[4] or a resistance in combination with a reactance tuning out the reactive component of h_{11}.[5] It also has been thought that the "neutralized gain" may be a useful measure of stable amplifying capability and, besides other expressions, Mason's U function[6] has been suggested as a practical indication of attainable stable power gain.

These and other approximations for the maximum realizable stable power gain have the common advantage of presenting the practicing engineer with a measure useful as a reference. Their common disadvantage lies in the fact that the physical conditions under which these gain expressions can actually be realized are chosen essentially arbitrarily and are quite different from the situation most frequently occurring (and most desirable) in practice: the case of an unneutralized amplifier using impedance transforming ("matching") tuned circuits at both input and output ports.

---

* Original manuscript received by the IRE, August 15, 1956; revised manuscript received, November 13, 1956.
Presented at WESCON Convention, Los Angeles, Calif., August 21–24, 1956.
† Electronics Lab., General Electric Co., Syracuse, N. Y.
[1] J. G. Linvill, "The Relationship of Transistor Parameters to Amplifier Performance," presented at the IRE-AIEE Conference on Transistor Circuits, Univ. of Pennsylvania, Philadelphia, Pa.; February 17, 1955.
J. G. Linvill and L. G. Schimpf, "Design of tetrode transistor amplifiers," *Bell Syst. Tech. J.*, vol. 35, p. 813; July, 1956.
[2] A. P. Stern, C. A. Aldridge, and W. F. Chow, "Internal feedback and neutralization of transistor amplifiers," Proc. IRE, vol. 43, pp. 838–847; July, 1955.

[3] A. P. Stern, "Considerations on the Stability of Active Elements and Applications," 1956 IRE Convention Record, Part 2, pp. 46–52.
[4] R. L. Pritchard, "High frequency power gain of junction transistors," Proc. IRE, vol. 43, pp. 1075–1085; September, 1955.
[5] W. N. Coffey and R. L. Pritchard, private communication.
[6] S. J. Mason, "Power gain in feedback amplifiers," IRE Trans., vol. CT-1, pp. 20–25; June, 1954.

In order to obtain an expression for the maximum realizable stable power gain which is clean and acceptable from both practical and conceptual points of view, it seems desirable to tie in the maximum realizable power gain with a measure indicating the stability of the amplifier. If this is done, the maximum power gain is finite for stable amplifiers and can appear as a *function of the degree of stability* of the amplifier.

The purpose of this paper is to discuss the maximum power gain of transistor amplifiers using a measure of stability which appears to be both simple and practical. The following considerations can be applied directly to other active nonunilateral two-port elements.[7]

The Stability Factor "k"

The following calculations make use of the "y" matrix elements (the admittance parameters) of the transistor and external circuit elements are represented as admittances. This procedure is arbitrary; analogous calculations can be carried out using the "z" matrix elements and external impedances or the "h" (or "g") matrix elements and suitable combinations of admittances and impedances. It will be shown that these various representations have their respective advantages depending on the types of resonant circuits used in the tuned amplifier.

If y_{ij} are the admittance parameters of the transistor, one can write, separating real and imaginary components:

$$y_{11} = g_{11} + jb_{11} \quad (1)$$
$$y_{22} = g_{22} + jb_{22} \quad (2)$$
$$y_{12}y_{21} = M + jN. \quad (3)$$

The absolute value of the "internal loop gain" ($y_{12}y_{21}$) is

$$L = |y_{12}y_{21}| = \sqrt{M^2 + N^2}. \quad (4)$$

It has been shown[1-3] that if

$$L + M \geq 2g_{11}g_{22} \quad (5)$$

the transistor *may* oscillate without external feedback, if properly terminated.

If the transistor is terminated by a generator admittance Y_G and a load admittance Y_L (Fig. 1), both Y_G

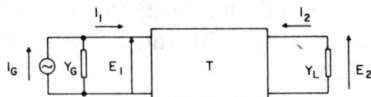

Fig. 1—Transistor with terminating admittances.

and Y_L may be complex:

---

[7] An approach similar to the one presented in this paper has been used independently by G. Bahrs in his doctoral thesis at Stanford University, Stanford, Calif.

$$Y_G = G_G + jB_G \quad (6)$$
$$Y_L = G_L + jB_L. \quad (7)$$

The transistor and the conductive components of the terminating admittances can be considered as constituting a new two terminal pair element C (Fig. 2). Applying

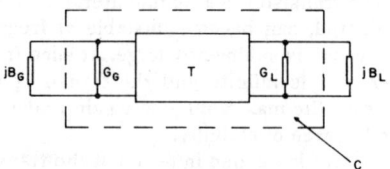

Fig. 2—Transistor and conductive terminating components forming composite network C.

inequality (5) to C, one finds the condition of potential instability:

$$L + M \geq 2(g_{11} + G_G)(g_{22} + G_L). \quad (8)$$

Note that (8) does not imply actual instability: in order that the arrangement be actually unstable, C must be terminated by suitable susceptances B_G and B_L. However, in most tuned amplifiers, the terminating (or coupling) susceptances are variable and are actually varied when "tuning up" the amplifier. In arrangements having reasonably narrow band, the terminating susceptances will assume a wide range of values. Consequently, it can be assumed that, in most cases, during the tuneup procedure, B_G and B_L will assume values resulting in oscillation, provided that (8) is satisfied. Therefore, for many practical purposes, it can be stated that, if (8) is satisfied, the amplifier is not only *potentially unstable*, but is very likely to become *actually unstable.*

Unconditional stability can be achieved if

$$(g_{11} + G_G)(g_{22} + G_L) > (L + M)/2. \quad (9)$$

In other words, for stability, one requires

$$(g_{11} + G_G)(g_{22} + G_L) = k(L + M)/2 \quad (10)$$

where $k > 1$. The larger k, the more remote is the likelihood of instability. Consequently, k can be considered as a measure of stability and will be called *stability factor*.

For any individual transistor, whose parameters are known, stability can be achieved by chosing G_G and G_L such as to make k slightly larger than 1. However, considering that transistor parameters (*i.e.*, the y_{ij}) vary with transistor age, dc bias, temperature, etc., that transistors belonging to the same type have slightly different parameters and, finally, that certain variations in G_G and G_L must also be tolerated (especially if the terminations are other transistor amplifier stages), it is usually desirable to design the amplifier with k well in excess of unity in order to insure interchangeability of components and stability under all reasonable conditions of operation. (Values of k up to 10 may be desirable, depending on the various tolerances involved.)

An infinite number of combinations of G_G and G_L correspond to the same value of k. Although the degree of stability of all these arrangements is the same, they may lead to different values of power gain. In the following section, the power gain will be maximized for a given value of k.

Analysis of a One-Stage Amplifier

The Power Gain

The transducer gain G_T is defined as the ratio of the power delivered to the load admittance Y_L to the available power of the generator having an admittance Y_G. The transducer gain is:

$$G_T = \frac{4|y_{21}|^2 G_G G_L}{|(y_{11} + Y_G)(y_{22} + Y_L) - y_{12}y_{21}|^2}. \quad (11)$$

Defining:

$$g_{11} + G_G = G_1, \quad (12)$$
$$g_{22} + G_L = G_2, \quad (13)$$
$$b_{11} + B_G = B_1, \quad (14)$$
$$b_{22} + B_L = B_2, \quad (15)$$

one finds for a given stability factor k:

$$G_1 = k(L + M)/2G_2. \quad (16)$$

Then, the transducer gain is:

$$GT = \frac{N}{D} = \frac{4|y_{21}|^2[k(L+M)/2G_2 - g_{11}](G_2 - g_{22})}{[B_1 B_2 - k(L+M)/2 + M]^2 + [B_1 G_2 + B_2 k(L+M)/2G_2 - N]^2} \quad (17)$$

It is desired to find the value of G_2 which leads to maximum G_T for a given k. In order to find G_T as a function of G_2 alone, it is convenient to eliminate the two other variables B_1 and B_2. Since B_1 and B_2 occur only in the denominator D of (17), this can be done by determining the values of B_1 and B_2 which minimize the denominator. It should be remembered that, B_1 and B_2 representing susceptances, must be positive or negative *real* numbers.

The process of minimizing D in (17) can be simplified by a change of variables. One defines

$$y_1 = B_1 G_2 + B_2 k(L + M)/2G_2 \quad (18)$$
$$y_2 = B_1 B_2. \quad (19)$$

D can then be written:

$$D = (y_2 - A)^2 + (y_1 - N)^2 \quad (20)$$

where

$$A = k(L + M)/2 - M. \quad (21)$$

Note that for $k > 1$, A is always positive. B_1 and B_2 can be expressed in terms of the new variables y_1 and y_2:

$$B_1 = (y_1 + \sqrt{\Delta})/2G_2 \quad (22)$$
$$B_2 = (y_1 - \sqrt{\Delta})G_2/k(L + M) \quad (23)$$

where

$$\Delta = y_1^2 - 2k(L + M)y_2. \quad (24)$$

Since G_2 and $k(L+M)$ are always real numbers, B_1 and B_2 will be real, whenever

$$\Delta \geq 0. \quad (25)$$

In the $y_1 - y_2$ plane, the curve $\Delta = 0$ separates the region of real values of B_1 and B_2 from the region in which these quantities are complex. The equation

$$\Delta = y_1^2 - 2k(L + M)y_2 = 0 \quad (26)$$

is that of a parabola in the $y_1 - y_2$ plane (Fig. 3), the

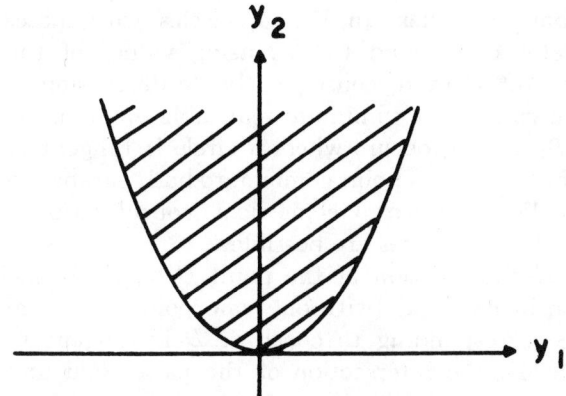

Fig. 3—Parabola separating region of complex B_1 and B_2 (shaded) from region in which B_1 and B_2 are real.

shaded area "inside" the parabola representing the region of complex values of B_1 and B_2 and the area "external" to the parabola being the region of real values of B_1 and B_2.

In a three-dimensional coordinate system with axes D, y_1 and y_2, (20) represents a paraboloid of revolution, whose axis is perpendicular to the $y_1 - y_2$ plane and intersects this plane at the point $Q(N, A)$. In the same space, (26) is that of a parabolic cylinder. [Fig. 4 shows the parabola (26) and the projection of "constant D circles" on the $y_1 - y_2$ plane.] The "interior" of this cylinder is the space within which D cannot lie if B_1 and B_2 are real and one is interested to determine the minimum value of D "outside" the parabolic "wall." [The vertex Q of the paraboloid having coordinates (N, A), where A is always positive for $k > 1$, will be "inside" the parabolic cylinder, since it corresponds to infinite gain, which of course, is excluded, in view of the assumptions regarding stability.]

An analysis of the geometrical situation shows that the minimum of D (i.e., the maximum of the gain) will

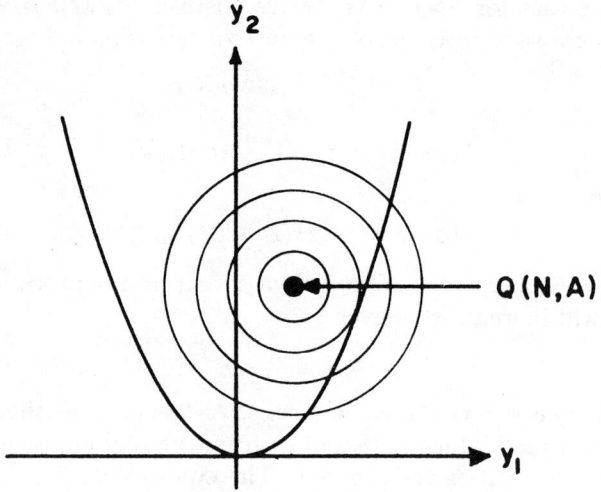

Fig. 4—Parabola (26) and projections of constant D circles on the y_1-y_2 plane.

occur along the intersection of the paraboloid with the parabolic cylinder. In Fig. 4, circles with increasing diameter correspond to increasing values of the denominator D (and, consequently, to decreasing values of the gain). The minimum realizable value of D with real B_1 and B_2 occurs where a circle is tangent to the parabola, i.e., at a point common to both paraboloid and parabolic cylinder. Geometrical considerations also show that two cases are possible:

1) If the location of the point $Q(N, A)$ is such as shown in Fig. 5(a) (with only one point P_0 at which a circle corresponding to constant D is tangent to the parabola), the intersection of the paraboloid and the parabolic cylinder has **one** minimum at point P_0 corresponding to $D = D_0$.

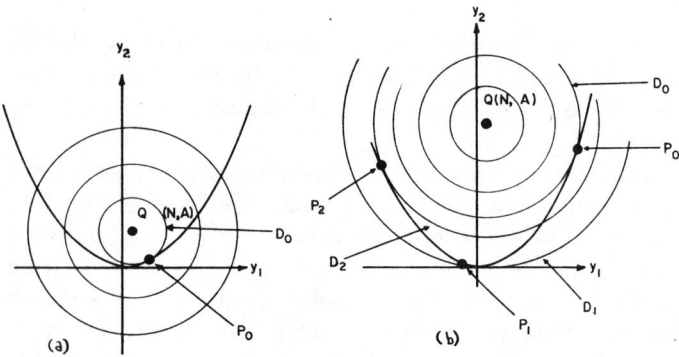

Fig. 5—Situations leading to one minimum (a) and three extrema (b) respectively.

2) If there are three points at which "constant D circles" are tangent to the parabola, there is a minimum at P_0 corresponding to a low value D_0 of D. There is a second minimum at P_2 corresponding to the higher level D_2 of D and there is a maximum at P_1 corresponding to an even higher level D_1 of D. In this case the maximum gain is given by the smaller minimum occurring at P_0.

These qualitative considerations can be put into quantitative form. Since D_0 occurs along the intersection of the paraboloid with the parabolic cylinder, one must determine the minimum of (20) with condition (26).

Defining the variable z as

$$z = y_1/\sqrt{k(L+M)} \tag{27}$$

D can be written:

$$D = z^4/4 + [k(L+M) + 2M]z^2/2 \\ - 2N\sqrt{k(L+M)}z + A^2 + N^2. \tag{28}$$

The minimum of D can be found by setting

$$0 = dD/dz = z^3 + [k(L+M) + 2M]z \\ - 2N\sqrt{k(L+M)}. \tag{29}$$

As expected from the foregoing qualitative analysis, (29) leads to two possibilities:

1) (29) has one real root z_0, the other two being conjugate complex. In this case z_0 corresponds to D_0.

2) (29) has three real roots. Two roots (z_0 and z_2) represent minima of D, z_0 corresponding to the smaller minimum, while the third root (z_1) is a maximum.

In either case, z_0 leads to the maximum realizable gain. Substituting z_0 into (28), the minimum value D_0 of D is found. Note that, since both z_0 and D are functions of k and ($y_{12}y_{21}$) only, D_0 depends only on k and ($y_{12}y_{21}$) and is independent of other transistor parameters and of conductive terminations.

Assuming now that D_0 is known, the gain (maximized with regard to B_1 and B_2) can be written:

$$G_T = (4|y_{21}|^2/D_0)[k(L+M)/2G_2 - g_{11}](G_2 - g_{22}). \tag{30}$$

Generator and load conductances are assumed to be positive and one must have $G_1 > g_{11}$ and $G_2 > g_{22}$. In other words:

$$k(L+M)/2g_{11} > G_2 > g_{22}. \tag{31}$$

Setting $dG_T/dG_2 = 0$, one finds that for maximum realizable power gain:

$$G_2 = \sqrt{k(L+M)/2}\sqrt{g_{22}/g_{11}} \tag{32}$$

and, using (16):

$$G_1 = \sqrt{k(L+M)/2}\sqrt{g_{11}/g_{22}}. \tag{33}$$

Substituting these expressions into (30), the maximum realizable gain, corresponding to the stability factor k, is found:

$$G_{\max k} = (4|y_{21}|^2/D_0)[\sqrt{k(L+M)/2} - \sqrt{g_{11}g_{22}}]^2. \tag{34}$$

Eq. (34) expresses the maximum power gain realizable with a given stability factor k. As k increases, $G_{\max,k}$ decreases. Consequently, one can say that increased stability can be achieved by sacrificing gain. It is, however, important to know, that moderate gain does not necessarily imply improved stability: k is determined by the product $(G_1 G_2)$ and this product can be chosen such that while k is small (close to instability), the gain is also small due to a disadvantageous ratio (G_1/G_2). In this case gain is sacrificed without improvement in stability.

From (32) and (33) the required values of the terminating conductances are found:

$$G_G = G_1 - g_{11}$$
$$= g_{11}[\sqrt{k(L+M)/2}/\sqrt{g_{11}g_{22}} - 1] \quad (35)$$
$$G_L = G_2 - g_{22}$$
$$= g_{22}[\sqrt{k(L+M)/2}/\sqrt{g_{11}g_{22}} - 1].^8 \quad (36)$$

The required values of the susceptances can be determined, using (22), (23), and (27):

$$B_1 = B_G + b_{11} = G_1 z_0/\sqrt{k(L+M)} \quad (37)$$
$$= z_0\sqrt{k(L+M)}/2G_2$$
$$B_2 = B_L + b_{22} = G_2 z_0/\sqrt{k(L+M)} \quad (38)$$

where G_1 and G_2 are given by (32) and (33). Eqs. (37) and (38) show that B_1 and B_2 have identical sign (that of z_0). This, of course, is not necessarily true for B_G and B_L.

Bandwidth Limitations

The preceding computations were carried out without regard to possible selectivity requirements. In reality, however, when designing a tuned amplifier, the bandwidth requirements are very important and, usually, the following problem must be solved: "design a stable amplifier with given center frequency and bandwidth and obtain as high gain as possible."

Simple considerations show that realizing the maximum gain imposes certain limitations on the bandwidth and, on the other hand, requiring a certain bandwidth imposes certain restrictions on the gain. These limitations belong to two categories:

1) Limitations due to the transistor (device limitations) and
2) Limitations due to external circuit elements.

Some aspects of the gain vs bandwidth relationship can be explained by considering the transistor terminated by synchronous single tuned circuits. Since the y parameters were used in the preceding calculations, it is convenient to use parallel tuned circuits, as shown in Fig. 6. It is assumed that the transistor parameters

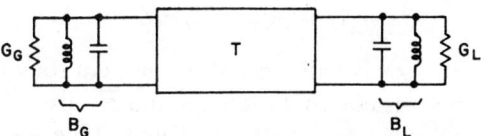

Fig. 6—Transistor with parallel tuned terminations

are frequency independent throughout the frequency band of interest.

The device limitations can be understood *qualitatively* in the following manner. The values of G_G and G_L leading to maximum gain are given by (35) and (36). Maximum gain will be realized only if the values of B_1 and B_2 are those given by (37) and (38). Considering now, say, the output selectivity one may assume, for example, that the required value of $B_L(=B_2-b_{22})$ is positive. This implies that the load susceptance must be capacitive and since the load consists of L_L and C_L in parallel, the required value of B_L can be realized only if $\omega_0 C_L > (B_2 - b_{22})$, ω_0 being the center frequency of the amplifier. Consequently, for maximum gain the tuned circuit must contain a capacitance C_L exceeding the value of $(B_2 - b_{22})/\omega_0$. However, this and the possible capacitive component of the transistor driving point admittance impose an upper limit on L_L. This results in an upper limit of bandwidth realizable with the maximum gain given by (34). If larger bandwidth is required, a smaller capacitance C_L will be used, implying that the denominator of the gain expression in (17) will not be D_0 but larger and, therefore, the gain will be reduced. Similarly, if B_L is required to be negative, meaning an inductive load susceptance, there is a maximum admissible value of L_L. This again imposes an upper limit on the bandwidth realizable with the calculated maximum gain. Analogous considerations do apply, of course, to the input selectivity.

The bandwidth limitations resulting from device properties are not important in the case of most amplifiers having reasonably narrow bandwidth. Their analytical treatment is complicated and will not be dealt with here.

The limitations imposed by the required bandwidth upon the realizable gain due to external circuit elements are caused by the finite "Q" of practical inductances and can be calculated.

In the circuit of Fig. 7, g_{x1} and g_{x2} represent the

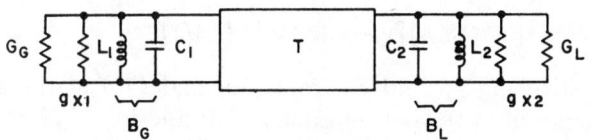

Fig. 7—Transistor with tuned circuits having finite "Q."

equivalent parallel loss conductances of inductances L_1 and L_2. Q_0 being the "unloaded Q" of the inductances and assuming that $Q_{01} = Q_{02} = Q_0$, one has:

---

[8] Eqs. (35) and (36) show that G_G and G_L will be positive if the bracketed expressions on the right hand side of these equations are positive, *i.e.*, if the stability factor k of the terminated amplifier exceeds the stability factor k_0 of the unloaded amplifier. If the unloaded amplifier is stable in itself, for positive terminations, one must have $k > k_0$.

$$Q_0 = 1/\omega_0 L_1 g_{x1} = 1/\omega_0 L_2 g_{x2}. \qquad (39)$$

The transistor is still treated in terms of its y parameters but the loss conductances g_{x1} and g_{x2} are considered to be parts of g_{11} and g_{22} respectively. One has:

$$g_{11} = g_{11}' + g_{x1} \qquad (40)$$
$$g_{22} = g_{22}' + g_{x2} \qquad (41)$$

where g_{11}' and g_{22}' are the short circuit input and output conductances of the transistor itself. The transducer gain is:

$$G_T = (4|y_{21}|^2/D_0)[k(L+M)/2G_2 - g_{11}' - g_{x1}](G_2 - g_{22}' - g_{x2}). \qquad (42)$$

Assuming that the tuned circuits at the input and the output are synchronously tuned but may have different fractional bandwidths $B^{(1)}$ and $B^{(2)}$ ($B = \Delta\omega/\omega_0$, where $\Delta\omega$ is the 3-db bandwidth):

$$B^{(1)} = (G_G + G_{\text{in}})/Q_0 g_{x1} \qquad (43)$$
$$B^{(2)} = (G_L + G_{\text{out}})/Q_0 g_{x2}. \qquad (44)$$

G_{in} and G_{out} are the conductive components of the transistor driving point admittances when the transistor is terminated.

$$G_{\text{out}} = g_{22} - (MG_1 + NB_1)/(G_1^2 + B_1^2). \qquad (45)$$

Using (37) and defining:

$$W = \frac{M + z_0 N/\sqrt{k(L+M)}}{1 + z_0^2/[k(L+M)]} \qquad (46)$$

one finds:

$$G_{\text{out}} = g_{22} - [2W/k(L+M)]G_2. \qquad (47)$$

Similarly:

$$G_{\text{in}} = g_{11} - (MG_2 + NB_2)/(G_2^2 + B_2^2). \qquad (48)$$

Using (38), one finds:

$$G_{\text{in}} = g_{11} - W/G_2. \qquad (49)$$

Eqs. (48) and (49) can be substituted into (43) and (44), yielding:

$$B^{(1)} = [k(L+M)/2G_2 Q_0 g_{x1}][1 - 2W/k(L+M)] \qquad (50)$$
$$B^{(2)} = [G_2/Q_0 g_{x2}][1 - 2W/k(L+M)]. \qquad (51)$$

Substituting g_{x1} and g_{x2} from (50) and (51) in the gain expression (42), G_T is obtained as a function of G_2 alone. Equating (dG_T/dG_2) to zero, the value of G_2 leading to maximum gain is found:

$$G_2 = \sqrt{k(L+M)/2}\sqrt{(1 - P/Q_0 B^{(1)})/(1 - P/Q_0 B^{(2)})}$$
$$\cdot \sqrt{g_{22}'/g_{11}'}. \qquad (52)$$

The corresponding value of G_1 is:

$$G_1 = \sqrt{k(L+M)/2}\sqrt{(1 - P/Q_0 B^{(2)})/(1 - P/Q_0 B^{(1)})}$$
$$\cdot \sqrt{g_{11}'/g_{22}'}. \qquad (53)$$

where:

$$P = 1 - 2W/k(L+M). \qquad (54)$$

Substituting G_2 in expression (42), the maximum realizable gain is:

$$G_{\text{max},k} = (4|y_{21}|^2/D_0)[\sqrt{(k/2)(L+M)} \cdot \sqrt{(1-P/Q_0 B^{(1)})(1-P/Q_0 B^{(2)})} - \sqrt{g_{11}' g_{22}'}]^2. \qquad (55)$$

If $B^{(1)} = B^{(2)} = B$, (55) becomes:

$$G_{\text{max},k} = (4|y_{21}|^2/D_0)[(1 - P/Q_0 B)\sqrt{k(L+M)/2} - \sqrt{g_{11}' g_{22}'}]^2. \qquad (56)$$

Eqs. (55) and (56) show that the finite Q of external circuit elements reduces the maximum realizable gain at *small bandwidths*, whereas the device limitations have been shown to reduce the maximum realizable gain at large bandwidths.

Multistage Amplifiers

The method used in the previous sections to determine the maximum realizable transducer gain maintaining a given degree of stability can also be applied to multistage amplifiers. In such amplifiers, interstage coupling is usually performed by "impedance matching"[9] tuned transformers. The problem is to determine the turns ratio m of the transformers and the conductances terminating the amplifier at its input (G_G of the first stage) and its output (G_L of the last stage) terminal pairs. When carrying out such calculations for multistage amplifiers, the following points should be observed:

1) The expression to be maximized is the transducer gain of the amplifier. The transducer gain of the multistage amplifier is, however, *not* the product of the transducer gains of the individual stages, but is the transducer gain of the first stage multiplied by the product of the actual power gains of the remaining stages.

2) The interstage transformer having turns ratio m transforms impedances in both directions. Therefore, since individual stages are terminated by the transformed driving point admittances of adjoining stages, load and source admittances of individual stages cannot be chosen independently. The condition for maximum over-all gain of the amplifier, all stages being operated with a certain de-

---

[9] The word "matching" is inappropriately used in an amplifier with prescribed stability factor k, since, in order to insure stability, the stages are deliberately mismatched.

sired stability factor k, does not imply that the individual stages are operated for maximum individual gain. Consequently, the maximum over-all gain of an n-stage amplifier, all of whose stages have the stability factor k, is less than n times the maximum gain realizable with a single stage having the same degree of stability.

It should be mentioned that, already in the case of two-stage amplifiers (the analysis of which is outlined in Appendix II) the algebraic and numerical evaluation of the equations becomes exceedingly involved. The complete evaluation of such or more complicated cases (for example, that of iterative stages) should be handled with the aid of appropriate computing machinery.

The Use of Different Matrix Representations

In the preceding sections, the discussion was based on the use of the admittance parameters. Since condition (5) can be stated in analogous form in terms of the "z," "h," and "g" parameters, all calculations could be carried out in terms of these other matrix representations. In fact, the calculations could have been carried out in terms of generalized "k" parameters, provided that the generator immittance is given the dimension of k_{11} and load immittance the dimension of k_{22}.

However, the different matrix representations have their respective advantages, depending on the nature of the terminating (or interstage coupling) tuned circuits. The following sets of parameters should be used with different types of tuned circuits:

1) y parameters with parallel tuned circuits at input and output (or parallel-parallel interstage networks)—see Fig. 8(a).
2) z parameters with series tuned circuits at input and output (or series-series tuned interstage networks)—see Fig. 8(b).
3) h parameters with series tuned circuit at the input and parallel tuned circuit at the output (or parallel-series tuned interstage networks)—see Fig. 8(c).
4) g parameters with parallel tuned circuit at the input and series tuned circuit at the output (or series-parallel tuned interstage networks)—see Fig. 8(d).

With transistors, cases 1) and 3) have practical importance. Cases 2) and 4) are almost never used because the transistor driving point immittances would lead to inconvenient values of inductances and capacitances.

Some numerical calculations are carried out in Appendix I, using the h parameters.

Conclusion

It has been shown that using an appropriately defined "stability factor" k, the "maximum gain" of a transistor

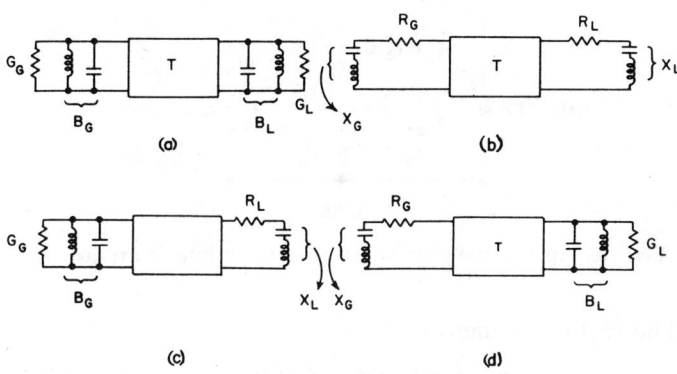

Fig. 8—Various tuned amplifier configurations.

amplifier is a meaningful concept even at frequencies where the transistor exhibits potential instability. The maximum gain corresponding to a given stability factor k has been computed for a one-stage amplifier having tuned generator and load admittances. The gain limitations due to bandwidth requirements have also been discussed for this case.

Appendix I

Some Numerical Examples

We consider a transistor having the following nominal device parameters:

Low-frequency common-base short circuit current amplification: $a_0 = 0.99$.
Cutoff frequency of the common-base short circuit current amplification: $\omega_a/2\pi = 10$ mc.
Collector capacitance: $C_c = 10$ uuf.
Base spreading resistance: $r_b' = 100$ ohms.
Emitter diffusion resistance (at 1-ma emitter current): $r_e = 25$ ohms.

We are interested in the maximum power gain realizable with this transistor in *common emitter* configuration with various given values of the stability factor k. The operating frequency is assumed to be $\omega/2\pi = 0.5$ mc and, consequently: $\rho = \omega/\omega_a = 0.05$.

The circuit used is shown in Fig. 9 and calls for the

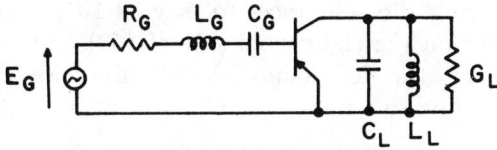

Fig. 9—Common emitter circuit with series tuned input and parallel tuned output.

use of the h parameters. Using the approximate equivalent circuit of Fig. 10 the parameters of interest can be calculated.

The real component of h_{11} is:

$$r_{11} \cong r_b' + r_e(1 - a_0)/\rho^2 \cong 200 \text{ ohms.} \tag{57}$$

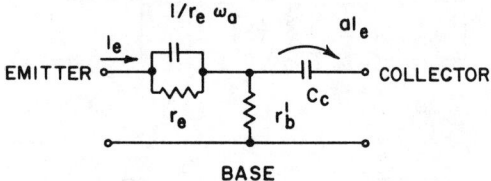

Fig. 10—Approximate equivalent circuit of a junction transistor.

The real component of h_{22} is:

$$g_{22} \cong \omega_a C_c \cong 6 \times 10^{-4} \text{ mhos.} \quad (58)$$

The "internal loop amplification" ($h_{12}h_{21}$) is:

$$h_{12}h_{21} \cong [\omega_a C_c r_e + j\omega_a C_c r_e (1 - a_0)/\rho](1/j\rho) \quad (59)$$
$$\cong 0.06 - j\,0.3.$$

The device is potentially unstable, since

$$[|h_{12}h_{21}| + \text{Re}\,(h_{12}h_{21})]/2 \cong 0.19 > r_{11}g_{22} \cong 0.12. \quad (60)$$

The choice of k depends on the admissible tolerances in device parameters and in the load. For the *nominal* parameters and corresponding terminations we want to have

$$(R_g + r_{11})(G_L + g_{22}) = k[|h_{12}h_{21}| + \text{Re}\,(h_{12}h_{21})]/2. \quad (61)$$

Since Im $(h_{12}h_{21}) \gg$ Re $(h_{12}h_{21})$, the right-hand side of (61) will vary like [see (59)]

$$\text{Im}\,(h_{12}h_{21}) \cong \omega_a^2 C_c r_e/\omega. \quad (62)$$

We may, for example, assume that ω_a, C_c, r_e will each have a tolerance (in the upper direction) of 3 per cent. Furthermore, we may assume that the two factors in the left-hand side of (61) will vary by no more than 3 per cent each in the lower direction (these tolerances are too small to be realistic and they are also inexact, since both r_{11} and g_{22} are functions of the device parameters). We would then require $k \cong (1.03)^6 \cong 1.2$. Computing the corresponding maximum gain, we find

$$G_{\max,1.2} \cong 36 \text{ db.}$$

By increasing the tolerances to, say, +10 per cent in each factor on the right-hand side of (62) and to −10 per cent on the left-hand side of (61), we require $k \cong 1.7$, leading to

$$G_{\max,1.7} \cong 33 \text{ db.}$$

By increasing the tolerances even further, to +30 per cent in ω_a and r_e, +50 per cent in C_c, −10 per cent in $(G_G + g_{11})$ and $(G_L + g_{22})$, we require $k \cong 4$. This leads to

$$G_{\max,4} \cong 30 \text{ db.}$$

A further increase in tolerances may require $k = 8$ and leads to

$$G_{\max,8} \cong 28 \text{ db.}$$

The example shows that the tolerances in device parameters (ω_a, C_c), emitter current (r_e is principally a function of the emitter current) and terminations (G_G and G_L) have considerable influence on the maximum realizable stable gain.

Similar calculations can be carried out for the *common base* stage. Here we have:

$$r_{11} \cong r_e \cong 25 \text{ ohms} \quad (63)$$
$$g_{22} \cong \omega^2 C_c^2 r_b' \cong 10^{-7} \text{ mhos} \quad (64)$$
$$h_{12}h_{21} \cong -j\omega C_c r_b' a_0 \cong 3 \cdot 10^{-3}. \quad (65)$$

The device is potentially unstable, since:

$$[|h_{12}h_{21}| + \text{Re}\,(h_{12}h_{21})]/2 \cong 15 \times 10^{-4}$$
$$r_{11}g_{22} \cong 25 \times 10^{-7}.$$

Calculating the maximum gain for the same values of k as in the common emitter case, we find:

$$k = 1.2 \quad G_{\max,1.2} \cong 44 \text{ db}$$
$$k = 1.7 \quad G_{\max,1.7} \cong 36 \text{ db}$$
$$k = 4.0 \quad G_{\max,4} \cong 28 \text{ db}$$
$$k = 8.0 \quad G_{\max,8} \cong 25 \text{ db.}$$

Note that in the common base case, these values of k correspond to *different tolerances* in the device parameters than in the common emitter case. With the particular set of device parameters chosen, the common base gain is higher for low (impractical) values of k, whereas at higher values of k, the common emitter gain is superior.

Note furthermore, that for a different circuit (for example, if parallel tuned circuits are used at both input and output, where the y parameters should be employed instead of the h parameters used in the above example), different values of k will correspond to the same permissible tolerances in the device parameters and different values of maximum gain may be realized.

APPENDIX II

OUTLINE OF THE ANALYSIS OF A TWO-STAGE AMPLIFIER

The schematic circuit of an amplifier consisting of two stages T_1 and T_2 is shown in Fig. 11.

We designate by:

P_G = the power available from source Y_G.
P_L = the power delivered to load Y_L.
P_{i2} = the power delivered by T_1 to T_2.
$G_{T,\text{total}}$ = the transducer gain of the two-stage amplifier.
G_{T1} = the transducer gain of T_1.
G_{P2} = the actual power gain of T_2 (=power delivered to Y_L divided by input power of T_2).

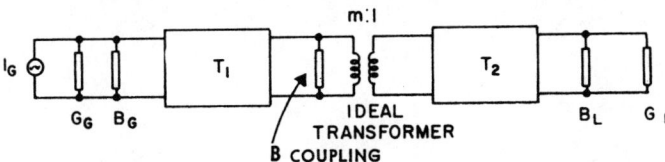

Fig. 11—Two stage amplifier with interstage impedance transformation (in practice the ideal transformer would, of course, be part of the coupling susceptance).

Then:

$$G_{T,\text{total}} = \frac{P_L}{P_G} = \frac{P_{i2}}{P_G} \cdot \frac{P_L}{P_{i2}} = G_{T1} \cdot G_{P2}. \quad (66)$$

Defining G_1 and G_2 in the sense used throughout the body of this paper and indicating by a second subscript whether the quantity considered belongs to T_1 or T_2, one has according to (10) and (30):

$$G_{T1} = \frac{4|y_{21}|^2}{D_0} (G_{11} - g_{11})$$

$$\left[\frac{k(L+M)}{2G_{11}} - g_{22} \right]. \quad (67)$$

The power gain of T_2 is:

$$G_{P2} = \frac{|y_{21}|^2 (G_{22} - g_{22})}{g_{11}[G_{22}^2 + B_{22}^2] - MG_{22} - NB_{22}}. \quad (68)$$

For any given value of G_{22}, G_{P2} will be maximum if

$$B_{22} = N/2g_{11}. \quad (69)$$

Substituting (69) into (68), G_{P2} can then be written:

$$G_{P2} = \frac{4g_{11}|y_{21}|^2 (G_{22} - g_{22})}{(2g_{11}G_{22} - M)^2 - L^2}. \quad (70)$$

The transducer gain of the complete amplifier is:

$$G_{T,\text{total}} = \frac{16g_{11}|y_{21}|^4}{D_0} (G_{11} - g_{11}) \left[\frac{k(L+M)}{2G_{11}} - g_{22} \right]$$

$$\cdot \frac{G_{22} - g_{22}}{(2g_{11}G_{22} - M)^2 - L^2}. \quad (71)$$

The load admittance of T_1 is $(1/m^2)$ times the input admittance of T_2 and the generator admittance of T_2 is m^2 times the output admittance of T_1. Furthermore, the same stability factor is required for both stages and (10) applies to both stages. In view of (69) and (46) one can write:

$$G_{21} = \frac{k(L+M)}{2G_{11}} = g_{22} + \frac{1}{m^2}\left(g_{11} - \frac{MG_{22} + N^2/2g_{11}}{G_{22}^2 + N^2/4g_{11}^2}\right) (72)$$

and

$$G_{12} = \frac{k(L+M)}{2G_{22}} = g_{11} + m^2\left(g_{22} - \frac{W}{G_{11}}\right). \quad (73)$$

These two equations establish relationships between $G_L(=G_{22}-g_{22})$, $G_G(=G_{11}-g_{11})$ and m. m can be eliminated by multiplying (72) by (73). One finds:

$$\left[\frac{k(L+M)}{2G_{22}} - g_{11}\right]\left[\frac{k(L+M)}{2G_{11}} - g_{22}\right]$$

$$= \left(g_{22} - \frac{W}{G_{11}}\right)\left(g_{11} - \frac{MG_{22} + N^2/2g_{11}}{G_{22}^2 + N^2/4g_{11}^2}\right). \quad (74)$$

Eq. (74) relates G_{11} to G_{22} (or G_L to G_G). One of these quantities can be expressed as a function of the other and substituted into the gain expression (71). Then, in principle, the maximum over-all transducer gain can be determined. It is, however, obvious that this procedure involves computational difficulties due to the complicated nature of the resulting gain expression.

Acknowledgment

The author is pleased to acknowledge the valuable advice and interest of Paul Weiss from the General Electric Company's Electronics Laboratory.

Stability and Power-Gain Invariants of Linear Twoports*

J. M. ROLLETT[†], MEMBER, IRE

Summary—It is shown that the stability of a linear twoport is invariant under arbitrary lossless terminations, under interchange of input and output, and under "immittance substitution," a transformation group involving the arbitrary interchanging of impedance and admittance formulations at both ports. The quantity

$$k = \frac{2 \, Re\,(\gamma_{11}) \, Re\,(\gamma_{22}) - Re\,(\gamma_{12}\gamma_{21})}{|\gamma_{12}\gamma_{21}|}$$

(where the γ may be any of the conventional immittance z, y, or hybrid h, g matrix parameters) is the simplest invariant under these transformations, and describes uniquely the degree of stability, provided $Re(\gamma_{11})$, $Re(\gamma_{22}) \geq 0$; the larger k is, the greater the stability, and in particular $k = 1$ defines the boundary between unconditional and conditional stability. The quantity k is thus the basic invariant stability factor. Its definition is also extended to include the effect of terminating immittances, which may be padding resistances or source and load immittances, or both.

Certain power-gain functions, including the maximum available power gain, are shown to be invariant under immittance substitution, and k is identified as a function of ratios between them, where they exist. This provides a fundamental way of determining k, apart from calculating it from matrix parameters, and indicates that it is a measure of an inherent physical property.

Introduction

TWO centers of interest in modern circuit theory are the passivity and stability of linear networks.[1] The concept of the passivity (or activity) of linear twoports has been greatly illuminated by Mason's invariant U function,[2,3] but no similar invariant has so far been proposed for stability. The present work (suggested by the analogy between the two concepts) shows that there is an invariant stability factor which uniquely characterizes the degree of conditional or unconditional stability of a linear twoport, much as Mason's invariant U function uniquely characterizes its passivity or activity.

The stability of linear twoports has been discussed by many authors.[1,4] A measure of stability, including the effect of source and load immittances, was first proposed by Stern[5,6] and Bahrs,[7,8] and has been discussed by Venkateswaran and Boothroyd.[9] The invariant stability factor introduced here is, in its basic form, identical with

---

* Received by the PGCT, May 13, 1961; revised manuscript received, October 15, 1961.
† British Telecommunications Research Ltd., Taplow Court, Taplow, near Maidenhead, Berks., England.
[1] An introduction and bibliography are given by E. F. Bolinder, "Survey of some properties of linear networks," IRE TRANS. ON CIRCUIT THEORY, vol. CT-4, pp. 70–78; September, 1957.
[2] S. J. Mason, "Power Gain in Feedback Amplifiers," Res. Lab. of Electronics, Mass. Inst. Tech., Cambridge, Tech. Rept. 257, August 25, 1953; IRE TRANS. ON CIRCUIT THEORY, vol. CT-1, no. 2, pp. 20–25, June, 1954.
[3] S. J. Mason, "Some properties of three-terminal devices," IRE TRANS. ON CIRCUIT THEORY, vol. CT-4, pp. 330–332; December, 1957.
[4] F. B. Llewellyn, "Some fundamental properties of transmission systems," PROC. IRE, vol. 40, pp. 271–283; March, 1952.
[5] A. P. Stern, "Considerations on the stability of active elements and applications to transistors," 1956 IRE NATIONAL CONVENTION RECORD, pt. 2, pp. 46–52.
[6] A. P. Stern, "Stability and power gain of tuned transistor amplifiers," PROC. IRE, vol. 45, pp. 335–343; March, 1957.
[7] G. S. Bahrs, "Amplifiers Emplying Potentially Unstable Elements," Electronics Res. Lab., Stanford University, Stanford, Calif., Tech. Rept. 105; May 7, 1956.
[8] G. S. Bahrs, "Stable amplifiers employing potentially unstable transistors," 1957 IRE WESCON CONVENTION RECORD, pt. 2, pp. 185–189.
[9] S. Venkateswaran and A. R. Boothroyd, "Power gain and bandwidth of tuned transistor amplifier stages," Proc. IEE, vol. 106B, Suppl. no. 15, pp. 518–529; 1959.

a quantity first defined by Aurell,[10] without any indication of its possible significance, and is the inverse of a "criticalness factor" defined by Linvill and Schimpf.[11] It differs from Stern's in that it has only one unique value (a property half recognised by Aurell but not mentioned by Linvill and Schimpf), whereas Stern's takes on four different values, depending on the formulation used.[9]

The Stability Transformation Group

A linear twoport is unconditionally stable if, with arbitrary passive terminations, its characteristic frequencies remain in the left half of the complex frequency plane. An equivalent statement is that the real part of the immittance looking in at only one of the two ports remains positive with arbitrary passive terminations at the other, provided also that the characteristic frequencies of the twoport with ideal terminations (infinite immittances, *i.e.*, open or short circuits, as appropriate) lie in the left half-plane. This last condition will be assumed to hold in what follows.

The stability of a twoport increases with lossy terminations, but remains constant with lossless terminations. Thus any quantity which indicates conditional or unconditional stability must (if it exists) be invariant under arbitrary lossless terminations. Furthermore, since (with the proviso above) stability depends on the positive realness of the immittance looking in at either port, it must be invariant under interchange of input and output; and since the immittance may be expressed either as impedance or admittance, it must be invariant under the interchange of impedance and admittance formulations at either port. The arbitrary interchanging of impedance and admittance formulations is carried out most easily by substituting any one set of the conventional impedance, admittance or hybrid matrix[12] parameters (z, y, h, g) for any other, and it is easy to show that the substitution operators form a group (isomorphic with the "vierergruppe"). This transformation group will be called *immittance substitution*.

The quantity characteristic of stability we are looking for is therefore invariant under arbitrary lossless terminations, under interchange of input and output, and under immittance substitution. These transformations may be combined to form a single infinite group, which is the "direct product" of the three separate groups, and contains all the transformations associated with stability.

THE INVARIANT STABILITY FACTOR

The well-known criterion for a linear twoport to be unconditionally stable[1,4] is, in admittance parameters,

$$2g_{11}g_{22} \geq |y_{12}y_{21}| + Re(y_{12}y_{21}), \qquad (1)$$

[10] C. G. Aurell, "Representation of the general linear four-terminal network and some of its properties," *Ericsson Tech.*, vol. 11, pp. 155–179; 1955.
[11] J. G. Linvill and L. G. Schimpf, "The design of tetrode transistor amplifiers," *Bell Sys. Tech. J.*, vol. 35, pp. 813–840; July, 1956.
[12] R. F. Shea, Ed., "Principles of Transistor Circuits," John Wiley and Sons, Inc., New York, N. Y.; 1953.

provided

$$g_{11}, g_{22} \geq 0$$

where $g_{11} = Re(y_{11})$, etc. Any quantity indicative of conditional or unconditional stability must include all the information contained in (1), which is clearly invariant under arbitrary lossless terminations and under interchange of input and output. Now it can be shown that the quantity k, defined by

$$k \triangleq \frac{2g_{11}g_{22} - Re(y_{12}y_{21})}{|y_{12}y_{21}|}, \qquad (2)$$

is invariant also under immittance substitution. It is therefore invariant under the complete group of transformations associated with stability. As it is the simplest such invariant, to which all others are trivially related (in the mathematical sense), it will be called the *invariant stability factor*. Since the value of k remains unchanged when z, h or g parameters are substituted for y parameters, it is convenient to generalize the notation, and so γ will be used for any of the z, y, h, g twoport parameters. In this notation

$$k \triangleq \frac{2\rho_{11}\rho_{22} - Re(\gamma_{12}\gamma_{21})}{|\gamma_{12}\gamma_{21}|}, \qquad (3)$$

where $\rho_{11} = Re(\gamma_{11})$, etc. The value of k lies between $+\infty$ and -1, if $\rho_{11}, \rho_{22} \geq 0$.

The criterion for unconditional stability may now be written as

$$k \geq 1 \qquad (4)$$

provided

$$\rho_{11}, \rho_{22} \geq 0,$$

and this is the fundamental property of the invariant stability factor. Furthermore, when k is positive and large compared with unity, the degree of unconditional stability is high; when k is only just greater than unity, the twoport is near the boundary between unconditional and conditional stability defined by $k = 1$. When $1 > k \geq -1$, the twoport is in the region of conditional stability, and it is always possible to choose terminations which will lead to negative real input or output immittances or which will result in oscillations; that is, the characteristic frequencies can be located in the right half-plane or on the real frequency axis.

If external immittances Γ_1, Γ_2 are added to the twoport so that they may be regarded as part of it from the viewpoint of stability, the definition of the stability factor may be extended, and an over-all stability factor K can be defined by

$$K \triangleq \frac{2(P_1 + \rho_{11})(P_2 + \rho_{22}) - Re(\gamma_{12}\gamma_{21})}{|\gamma_{12}\gamma_{21}|}, \qquad (5)$$

where $P_1 = Re\,(\Gamma_1)$, etc. The external immittances may be padding resistances or source and load immittances or both. The over-all stability factor K is useful in characterizing the stability of practical tuned amplifiers, working between source and load immittances whose real parts are known.

It is important to relate the invariant stability factor k, which so far has been introduced as a pure number, to other basic properties of the twoport. The next section shows how it is related to various invariant power gain functions, and indicates how it can in principle be determined, when it is positive, without knowing the matrix parameters. When it is negative, it can be determined indirectly by adding known padding resistances to make the overall stability factor positive,[13] or by calculation from matrix parameters.

Invariant Power-Gain Functions

There are three power-gain functions of interest which are invariant under immittance substitution and are directly related to the invariant stability factor. These are the maximum available power gain,[14] the maximum stable power gain,[15] and a function defined below as the minimum conjugate-termination transducer gain; each is introduced in turn.

Maximum Available Power Gain

The maximum available power gain[14] G_{MA} of a twoport is obtained when the input and output ports are simultaneously matched to their conjugate immittances. This is only possible if the device obeys the unconditional stability criterion (1). The expression for G_{MA} is

$$G_{MA} = \frac{|\gamma_{21}|^2}{2\rho_{11}\rho_{22} - Re\,(\gamma_{12}\gamma_{21}) + \sqrt{\{[2\rho_{11}\rho_{22} - Re\,(\gamma_{12}\gamma_{21})]^2 - |\gamma_{12}\gamma_{21}|^2\}}}, \qquad (6)$$

or, in terms of the stability factor,

$$G_{MA} = \left|\frac{\gamma_{21}}{\gamma_{12}}\right| \frac{1}{k + \sqrt{(k^2 - 1)}}$$

$$= \left|\frac{\gamma_{21}}{\gamma_{12}}\right| [k - \sqrt{(k^2 - 1)}]. \qquad (7)$$

As expected on general physical grounds, G_{MA} is invariant under immittance substitution. Since the factor involving k is invariant, the other factor $|\gamma_{21}/\gamma_{12}|$ is also invariant. It is discussed and identified in the next section.

The optimum load immittances, which provide the simultaneous conjugate match, may be conveniently expressed using k. If they are denoted by Γ_{1opt} at the input and Γ_{2opt} at the output, then

$$\Gamma_{1opt} = \frac{|\gamma_{12}\gamma_{21}|\sqrt{(k^2 - 1)}}{2\rho_{22}} + j\left[\frac{Im\,(\gamma_{12}\gamma_{21})}{2\rho_{22}} - \sigma_{11}\right], \qquad (8)$$

(where $\sigma_{11} = Im\,(\gamma_{11})$, etc.) and similarly for Γ_{2opt}. The total self immittances are particularly concise:

$$\Gamma_{1opt} + \gamma_{11} = \frac{\gamma_{12}\gamma_{21} + |\gamma_{12}\gamma_{21}|\,[k + \sqrt{(k^2 - 1)}]}{2\rho_{22}} \qquad (9)$$

and similarly for $(\Gamma_{2opt} + \gamma_{22})$.

The optimum load immittances (8) are invariant under immittance substitution, as would be expected on general physical grounds. However, the total self immittances (9) are *not* invariant; that is, although the expressions for them are similar, using different matrix parameters, their physical values are different.

The expression for maximum available power gain (7) is made up of two factors, the one involving k being reciprocal (i.e., invariant with respect to interchange of forward and reverse transfer parameters), while the other (which is discussed below) is nonreciprocal. Thus the factor $[k - \sqrt{(k^2 - 1)}]$ is the maximum efficiency[10,16] of the reciprocal part of the twoport, while its inverse $[k + \sqrt{(k^2 - 1)}]$ may be called the *minimum reciprocal attenuation*, provided the unconditional stability criterion holds. (It must be remembered that a device can be both reciprocal and active,[17] and that it then necessarily violates the unconditional stability criterion.) The minimum reciprocal attenuation is purely a function of k, and this gives an insight into the nature of the physical property of which k is a measure.

If the maximum available power gain in the reverse direction, found by interchanging the two ports, is denoted by G^r_{MA}, then we have

$$\frac{G_{MA}}{G^r_{MA}} = \left|\frac{\gamma_{21}}{\gamma_{12}}\right|^2 \qquad (10)$$

and

$$\sqrt{(G_{MA} G^r_{MA})} = k - \sqrt{(k^2 - 1)}, \qquad (11)$$

which enable $|\gamma_{21}/\gamma_{12}|$ and k to be determined, provided $k > 1$ and $\rho_{11}, \rho_{22} > 0$.

Maximum Stable Power Gain

The maximum stable power gain[15] is defined as follows. If $k < 1$ or if ρ_{11} or $\rho_{22} < 0$, then lossy padding immittances

---

[13] Thus if P_a, P_b are placed successively at one port, and the corresponding (positive) over-all stability factors K_a, K_b are measured, then the basic stability factor is given by $k = (P_a K_b - P_b K_a)/(P_a - P_b)$.
[14] "IRE Standards on Electron Tubes," Proc. IRE, vol. 45, pp. 983–1010; July, 1957.
[15] M. A. Karp, "Power gain and stability," IRE Trans. on Circuit Theory, vol. CT-4, pp. 339–340; December, 1957.

[16] This quantity is discussed in a paper which has just been brought to the attention of the author: S. Venkateswaran, "An invariant stability factor and its physical significance," IEE Mono. No. 468E, September, 1961, to be republished in Proc. IEE, pt. C.
[17] J. Shekel, "Reciprocity relations in active 3-terminal elements," Proc. IRE, vol. 42, pp. 1268–1270; August, 1954.

Γ_1 and Γ_2 can always be placed at the two ports so as to make the real parts of the self parameters $(P_1 + \rho_{11})$, $(P_2 + \rho_{22}) > 0$, and the over-all stability factor $K \to 1$, i.e., so that the over-all twoport approaches the boundary between unconditional and conditional stability. The maximum available power gain then tends towards its maximum stable value,

$$G_{MA} \to \left| \frac{\gamma_{21}}{\gamma_{12}} \right| \triangleq G_{MS}, \tag{12}$$

and G_{MS} is called the *maximum stable power gain*.[15] It is also invariant under immittance substitution.[9]

The use of lossy elements means that K for the padded twoport must be greater than k for the basic twoport. This implies that G_{MS} is only defined for devices where $k \leq 1$. However, it is useful to extend the definition of G_{MS} to include devices with $k > 1$, and this is done by observing that, in principle, negative resistances could be chosen for the terminations Γ_1, Γ_2, so as to allow K to be less than k. The stability boundary $K = 1$ can then again be approached, so that G_{MS} can be defined as the maximum stable power gain for all devices.

Having identified the quantity $|\gamma_{21}/\gamma_{12}|$, we can write, in general, for a device obeying the unconditional stability criterion,

$$G_{MA} = G_{MS}/[k + \sqrt{(k^2 - 1)}]; \tag{13}$$

i.e., the maximum available power gain is given by dividing the maximum stable power gain by the minimum reciprocal attenuation. The value of G_{MS} may be found as indicated in the previous section.

Minimum Conjugate-Termination Power Gain

Here we introduce a power-gain function which suggests a method for determining k over a wider range of values than the previous method (11) allows.

Consider the following sequence of operations. Conjugately match at one port, with the second arbitrarily terminated in Γ. Then remove the arbitrary termination Γ and conjugately match at the second port, thereby destroying the conjugate match at the first port (provided the device is nonunilateral). The transducer gain[14] is now a function of the termination Γ and has a minimum which occurs when Γ is lossless. This may be called the *minimum conjugate-termination transducer gain G_{CT}*, where

$$G_{CT} \triangleq \frac{|\gamma_{21}|^2}{4\rho_{11}\rho_{22} - 2\,Re\,(\gamma_{12}\gamma_{21})}, \tag{14}$$

or, in terms of invariants already discussed,

$$G_{CT} = \left| \frac{\gamma_{21}}{\gamma_{12}} \right| \frac{1}{2k} = G_{MS}/2k. \tag{15}$$

This quantity has been introduced by Linvill and Schimpf,[11] who call it "P_{o0}/P_{i0}", but make no mention of its extremum properties. Its importance in the present context lies in the fact that it enables us to determine k, except when $k \leq 0$. For, although the device is not unconditionally stable for $0 \leq k < 1$, it can easily be shown that when $\Gamma \to \infty$, the sequence of operations outlined above always leads to the total port immittances at both ports having positive real parts.

Thus denoting by G_{CT}^r the reverse quantity, found by interchanging input and output, we have

$$k = 1/2\sqrt{(G_{CT}G_{CT}^r)} \tag{16}$$

for $k > 0$, and provided $\rho_{11}, \rho_{22} > 0$.

DISCUSSION

An invariant stability factor has been introduced, together with its associated transformation group, and its essential properties described. Its relationships with certain invariant power-gain functions have been investigated, and shown to lead to basic methods of determining it, without calculating it from matrix parameters.

This suggests that k is a measure of a fundamental physical property, despite the fact that it can only be determined indirectly when it is negative.[18] Although k is the simplest stability invariant, to which all others are trivially related, in the mathematical sense, it may turn out that some function of k can be more closely identified with a physical property than k itself. However, its invariant properties ensure that no other quantity can convey more information about stability, and it is in this sense that k is unique.

The extension of the definition to include the effects of terminating immittances in the over-all stability factor K should prove useful in the design of a common class of amplifier, i.e., those which are unneutralized, resistively mismatched and reactively tuned. In this connection it is worth pointing out that the transducer gain G_T

$$G_T = \frac{4P_1P_2 |\gamma_{21}|^2}{|(\Gamma_1 + \gamma_{11})(\Gamma_2 + \gamma_{22}) - \gamma_{12}\gamma_{21}|^2}, \tag{17}$$

(where Γ_1, Γ_2 are source and load immittances and $P_1 = Re\,(\Gamma_1)$, etc.) is also invariant under immittance substitution.[19] If a manageable relation between the over-all K and G_T could be found, it would enable gain to be exchanged with stability on a quantitative basis; this has so far eluded the present author.

ACKNOWLEDGMENT

This work was started while the author was at Marconi's Wireless Telegraph Company Ltd., and was first issued there as "Invariants in Linear Two-Ports," Interim Technical Memorandum No. 755, July, 1960. The author is indebted to the Chief of Research, M. W. T. Co., and to the Director of Research, British Telecommunications Research Ltd., for permission to publish it.

---

[18] An exactly similar situation exists in the case of Mason's invariant U function.
[19] The ratio of forward to reverse transducer gain is the square of the maximum stable power gain, cf. (10). The derivation of this result in reference [17] appears to be fallacious.

STABILITY CONSIDERATIONS OF LOW-NOISE TRANSISTOR AMPLIFIERS WITH SIMULTANEOUS NOISE AND POWER MATCH

Les Besser
Farinon Electric
San Carlos, California 94070

Abstract

Microwave transistor amplifiers may be simultaneously matched for optimum noise and input/output VSWR. This paper demonstrates a combination of mapping techniques, computer optimization and stability considerations through two amplifier designs (70 MHz and 4000 MHz) to achieve these goals.

Discussion

One of the basic problems existing in low-noise transistor amplifier design is the significant difference between the desired source impedances for optimum noise and input match. Although recently device manufacturers have made progress in transforming the impedances closer to each other by either modifying the device geometries or diffusion process, at 4 GHz typically a 2 - 3 dB noise figure degradation is suffered when the input and output are simultaneously matched for minimum reflection in a 50Ω system (see Fig. 1). In such case, some form of an isolator is needed at the input to maintain a reasonable input match when the device is operating at its minimum noise figure. The cost and the associated losses of the isolator may make this approach undesirable.

Using mapping techniques, the effect of feedback can be plotted for the circuit and noise parameters. For example, Figures 2 and 3 show the variation of optimum noise source impedance, minimum noise figure, and the conjugate of the input reflection coefficient for various lossless shunt and series feedback elements at 4 GHz. The plots show that capacitive elements will typically reduce the minimum noise figure and move the optimum noise source away from the optimum power match. Conversely, inductive feedback usually has the opposite effect. Combining the two feedbacks in some cases can move the optimum noise source impedance toward the desired direction, albeit with a small change in the optimum noise figure (this is particularly true at VHF and UHF frequencies where the transistors have higher gain).

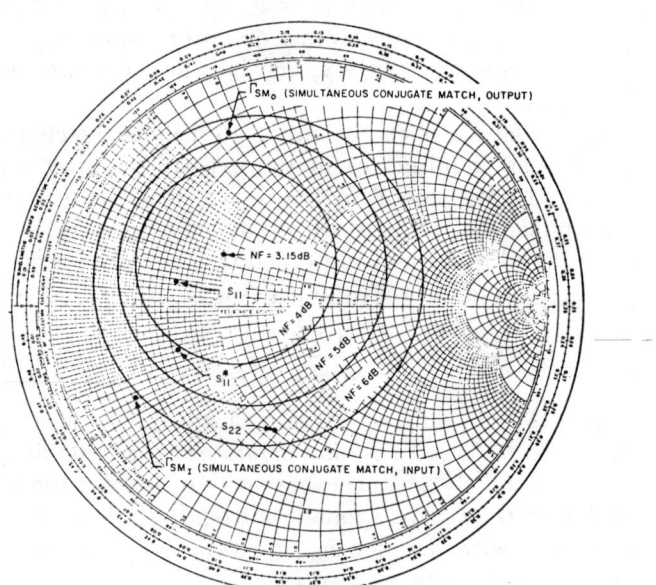

Figure 1
Constant Noise Circles and Simultaneous Conjugate Match of the NEC V-222 at 4 GHz

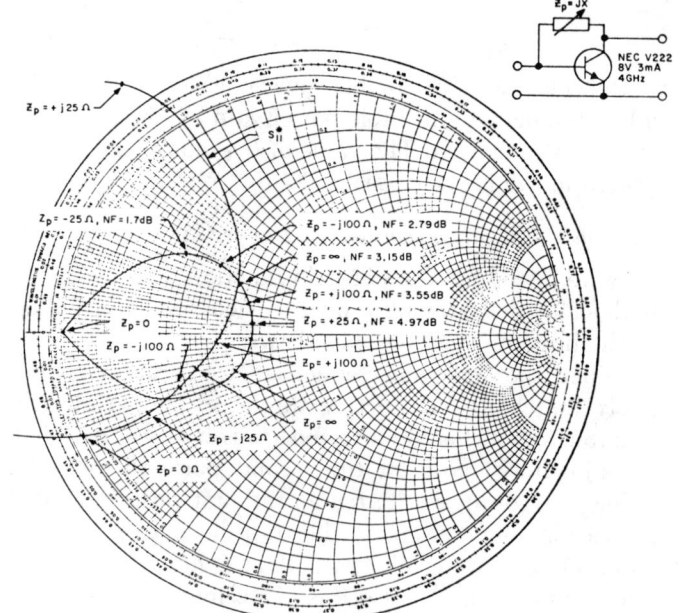

Figure 2
Minimum Noise Figure,
Optimum Noise Source Impedance,
and S_{11}^* vs. Parallel Feedback

Lossless feedback elements can transform the input and optimum noise source impedances until simultaneous match or an acceptable compromise is reached.[1] However, the reactive feedback will change the minimum noise figure as well as the optimum noise source impedance. In addition the gain and the stability factor of the circuit will also change. The latter change in particular will affect the design procedure, as described below. If the circuit reaches a potentially unstable state, both the load and source terminations must be selected with care to prevent oscillation.

The plots may mislead the casual observer. For example in the extreme case, if the shunt feedback is reduced to zero ohm the computed minimum noise figure is reduced to zero. But obviously, the device has no gain and cannot be used for practical purposes. This suggests that the gain of the device should also be considered and the noise measure[2] may be a better parameter for the evaluation.

Reprinted from *IEEE MTT-S Int. Microwave Symp.*, 1975, pp. 327-329.

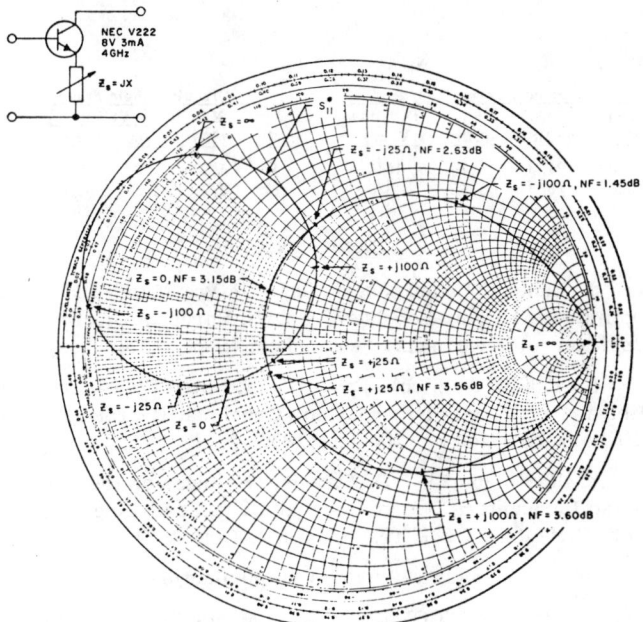

Figure 3
Minimum Noise Figure, Optimum Noise Source Impedance,
and S_{11}^* vs. Series Feedback

The transformation of the noise parameters[3,4] is computed through two-port additions as follows:

$$Y_{opt} = \frac{G_n' + R_n' (Re (Y_c'))^2}{R_n'} - j\, Imag(Y_c')$$

$$F_{min}' = 1 + 2R_n'(Real(Y_c') + Real(Y_{opt}'))$$

R_n', G_n' and Y_c' are the functions of R_n, G_n and Y_c of the active device and the noise transformation parameters, which are:

$$N_s = \begin{bmatrix} 1 & \vline & z_{11}^a - \dfrac{z_{21}^a}{z_{21}^b} z_{11}^b \\ \hline 0 & \vline & 1 + \dfrac{z_{21}^a}{z_{21}^b} \end{bmatrix} \quad \text{for series connections}$$

$$N_p = \begin{bmatrix} 1 + \dfrac{y_{21}^a}{y_{21}^b} & \vline & 0 \\ \hline y_{11}^a - \dfrac{y_{21}^a}{y_{21}^b} y_{11}^b & \vline & 1 \end{bmatrix} \quad \text{for parallel connections}$$

where the "a" superscripts refer to active two port and the "b" superscripts refer to the passive two-port.

Since there is a strong interaction between the feedback elements and the load and source terminations, each of the plots (e.g. Fig. 2 and 3) is only valid for a specific combination of element values. Therefore, the plots should only be used to determine the initial values for the subsequent computer optimization.

Although the reactive feedback will typically reduce the transducer power gain and the stability factor, the Maximum Stable Gain (MSG) of the circuit may still be very close to GMAX of the transistor.

The computer aided design procedure is the following:

1. Map the optimum noise source impedance, the optimum noise figure, and the conjugate of s_{11} as functions of the series and shunt feedback.

2. Select a combination of feedback elements that transforms the optimum input match and optimum noise source impedance toward the same region without causing a significant increase of the optimum noise figure. Select the appropriate matching circuit configuration and determine the initial component values from the Smith Chart.

3. Optimize the above circuit for simultaneous input match and noise figure.

4. Run a stability analysis on the resulting circuit. If the circuit is potentially unstable, plot the stability circle for the output plane and select the load impedance in the stable area. Compute GMAX or MSG, whichever is applicable.

5. Perform a final optimization to achieve simultaneously low noise and low input VSWR. The value of GMAX (or MSG) of Step 4 should be the gain specified during optimization.

6. If the results do not satisfy the desired specifications, change the relative values of the weighting factors assigned to GMAX, VSWR, and noise figure, and repeat step 5.

This design approach was tested at two extreme frequencies: at 70 MHz with an NEC VO21 transistor, and at 4000 MHz using an NEC V222 device. A general purpose microwave circuit optimization program, COMPACT,[5] was used through commercial timesharing of United Computing Systems. The final optimization of the 4 GHz amplifier stage is shown in Table 2. The target of this run is to achieve 7.5 dB gain, low noise figure as well as low input/output VSWR. The optimized circuits are shown in Figure 4 and 5. The circuit performances are summarized in Table 1. Note that in both cases the actual noise figure is within a few tenths of a dB of the optimum while the input VSWR is kept at low values.

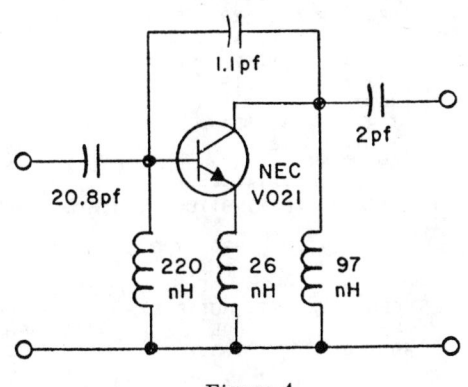

Figure 4
70 MHz Low Noise Amplifier

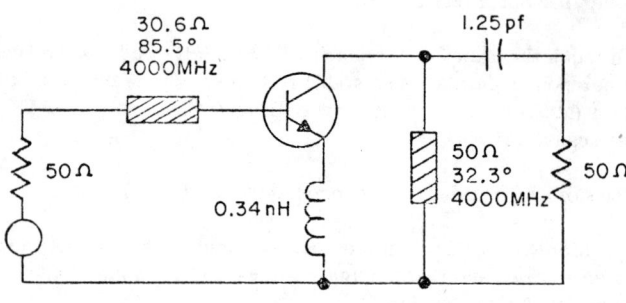

Figure 5
4000 MHz Low Noise Amplifier

TABLE I

	Device Parameters			Circuit (Single Stage)			
Freq. MHz	GMAX or MSG	NF$_{opt}$ (dB)	DC Bias V/mA	Shown in Fig.	NF (dB)	VSWR Input	VSWR Output	Gain (dB)
70	21.8	1.1	10/5	4	1.23	1.6	1.6	20
4000	8.3	3.15	8/3	5	3.45	1.8	1.6	7.5

Conclusion

Simultaneous match for minimum noise and VSWR of bipolar transistor amplifiers may be obtained by introducing appropriate lossless feedback to the active device. However, the gain and the stability of the circuit must be monitored and considered during the computer optimization. A similar effort is being undertaken elsewhere for GaAs FET amplifiers.

References

1. J. Enberg, "Simultaneous Input Power Match and Noise Optimization Using Feedback, Fourth European Microwave Conference, Montreaux, Switzerland, Sept. 1974.

2. H. Fukui, "Available Power Gain, Noise Figure and Noise Measure of Two-Ports and Their Graphical Representations", IEEE Transaction on CT, June 1966.

3. K. Hartmann, J.J.D. Strautt, "Changes on the Four Noise Parameters Due to General Changes of Linear Two-Port Circuits", IEEE Trans. on ED, Oct. 1974.

4. George Vendelin, Private Communications.

5. Les Besser, "COMPACT User Manual", Version 3.3, April 1975.

6. George Vendelin, "Feedback Effects on the Noise Performance of GaAs MESFET's", IEEE International Microwave Conference, May 1975.

```
        CIRCUIT OPTIMIZATION WITH  5 VARIABLES
   INITIAL CIRCUIT ANALYSIS:


       POLAR S-PARAMETERS IN   50.0 OHM SYSTEM

    F        S11         S21         S12         S22       S21     K
   MHZ   (MAGN)ANGL) ( MAGN)ANGL) ( MAGN)ANGL) (MAGN)ANGL)  DB    FACT.

  4000.0 ( .37< -77) ( 1.93< -61) ( .269< -11) ( .48< -39)  5.72   .78      Initial Analysis
                       NOISE FIGURE DATA
   FREQ.    OPT. NOISE FIG.   OPT. NOISE SOURCE    ACTUAL NF   NORMALIZED
    MHZ         DB             MAGN.   ANGLE          DB          RN
   4000.0      3.42             .33     179           3.51       .228
   OPTIMIZATION BEGINS WITH FOLLOWING VARIABLES AND GRADIENTS

       VARIABLES:              GRADIENTS:

       ( 1):    40.000         ( 1):    5.867                             Initial Variables
       ( 2):   110.000         ( 2):    5.069                                    and
       ( 3):      .600         ( 3):    -.276                             Partial Gradients
       ( 4):    50.000         ( 4):   12.958
       ( 5):     2.000         ( 5):   -1.966
       ERR. F.=   3.528  CUM. CPU TIME (INCL. PREMIUM)=  .45 SECONDS
              ----◆◆◆◆----
       ( 1):    30.679         ( 1):     .005
       ( 2):    85.519         ( 2):    -.044                             Final Variables
       ( 3):      .343         ( 3):    -.015                                    and
       ( 4):    32.354         ( 4):     .043                             Partial Gradients
       ( 5):     1.251         ( 5):     .060
       ERR. F.=    .142  CUM. CPU TIME (INCL. PREMIUM)= 5.91 SECONDS
              ----◆◆◆◆----
   FRACTIONAL TERMINATION WITH ABOVE VALUES. FINAL ANALYSIS   FOLLOWS
                                                                          Final Analysis
  4000.0 ( .27< -30) ( 2.36< -3) ( .225< 27) ( .21< 55)  7.47   1.04

                       NOISE FIGURE DATA
   FREQ.    OPT. NOISE FIG.   OPT. NOISE SOURCE    ACTUAL NF   NORMALIZED
    MHZ         DB             MAGN.   ANGLE          DB          RN
   4000.0      3.31             .35     164           3.47       .233
```

Table 2 - Final Optimization Printout of COMPACT for the 4 GHz Amplifier Stage

Power Waves and the Scattering Matrix

K. KUROKAWA, MEMBER, IEEE

Abstract—This paper discusses the physical meaning and properties of the waves defined by

$$a_i = \frac{V_i + Z_i I_i}{2\sqrt{|\text{Re } Z_i|}}, \quad b_i = \frac{V_i - Z_i^* I_i}{2\sqrt{|\text{Re } Z_i|}}$$

where V_i and I_i are the voltage at and the current flowing into the ith port of a junction and Z_i is the impedance of the circuit connected to the ith port. The square of the magnitude of these waves is directly related to the exchangeable power of a source and the reflected power. For this reason, in this paper, they are called the power waves. For certain applications where the power relations are of main concern, the power waves are more suitable quantities than the conventional traveling waves. The lossless and reciprocal conditions as well as the frequency characteristics of the scattering matrix are presented.

Then, the formula is given for a new scattering matrix when the Z_i's are changed. As an application, the condition under which an amplifier can be matched simultaneously at both input and output ports as well as the condition for the network to be unconditionally stable are given in terms of the scattering matrix components. Also a brief comparison is made between the traveling waves and the power waves.

I. INTRODUCTION

THE CONCEPT of traveling waves along a transmission line and the scattering matrix of a junction of transmission lines are well known and they play important roles in the theory of microwave circuits. However, the traveling wave concept is more closely related to the voltage or current along the line than to the power in a stationary state. If a circuit which terminates a line at the far end is not matched to the characteristic impedance of the line, even if the circuit has no source at all, we have to consider two waves traveling in opposite directions along the line. This makes the calculation of power twice as complicated. For this reason, when the main interest is in the power relation between various circuits in which the sources are uncorrelated, the traveling waves are not considered as the best independent variables to use for the analysis. A different concept of waves is introduced. The incident and reflected power waves a_i and b_i are defined by

$$a_i = \frac{V_i + Z_i I_i}{2\sqrt{|\text{Re } Z_i|}}, \quad b_i = \frac{V_i - Z_i^* I_i}{2\sqrt{|\text{Re } Z_i|}} \quad (1)$$

where V_i and I_i are the voltage and the current flowing into the ith port of a junction and Z_i is the impedance looking out from the ith port. The positive real value is chosen for the square root in the denominators. These

Manuscript received September 8, 1964.
The author is with the Bell Telephone Labs., Inc., Murray Hill, N. J.

power waves were first introduced by Penfield [1][1] for the discussion of noise performance of negative resistance amplifiers and later they were used for the discussion of actual noise measure of linear amplifiers by Kurokawa [2]. However, since it was not their main objective, the meaning of these waves and the properties of the corresponding scattering matrix were only briefly discussed. At about the same time, Youla [3] studied the same waves; however, his Z_i's were limited to have positive real part only. More recently, Youla and Paterno used these waves to study the attenuation error in mismatched systems [4].

The purpose of this paper is to present the physical meaning of the waves defined by (1) as well as the properties of the scattering matrix based on this new wave concept. Some of the properties such as the lossless condition for the matrix have been discussed in the previous papers. However, for the sake of completeness, they are included in this paper also.

II. PHYSICAL MEANING

Since the waves defined by (1) are closely related with the exchangeable power [5] of a generator, we have to discuss briefly what it is. For this purpose, let us consider the equivalent circuit of a linear generator, as shown in Fig. 1, in which Z_i is the internal impedance and E_o is the open circuit voltage of the generator. The power P_L into a load Z_L is given by $\text{Re } Z_L |I_i|^2$, where I_i is the current into the load. Since the magnitude of the current is equal to $|E_o/(Z_L+Z_i)|$, P_L is given by

$$P_L = \text{Re } Z_L \left|\frac{E_o}{Z_L + Z_i}\right|^2 = \frac{R_L |E_o|^2}{(R_L + R_i)^2 + (X_L + X_i)^2} \quad (2)$$

$$= \frac{|E_o|^2}{4R_i + \frac{(R_L - R_i)^2}{R_L} + \frac{(X_L + X_i)^2}{R_L}} \quad (3)$$

where R_L and R_i are the real parts of Z_L and Z_i, respectively, and X_L and X_i are the imaginary parts. With $R_i > 0$, we can easily see from (3), that the denominator becomes minimum when

$$R_L = R_i, \quad X_L = -X_i \quad (4)$$

The corresponding maximum power P_L is

$$P_a = \frac{|E_o|^2}{4R_i}, \quad (R_i > 0) \quad (5)$$

[1] In the original definition, Re Z_i is taken instead of $|\text{Re } Z_i|$ in the square root of the denominator of (1) (cf Section VIII).

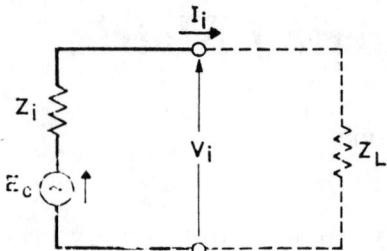

Fig. 1. Equivalent circuit of a linear generator.

This maximum power is called the available power of the generator. When the real part of Z_i is negative, P_L becomes infinite as R_L and X_L approach $-R_i$ and $-X_i$, respectively as we can see from (2). In this case, (5) no longer represents the maximum power that can be drawn from the generator. However, the expression given in (5) remains finite and the power represented by it is called the exchangeable power P_e of the generator, for any nonzero R_i. That is,

$$P_e = \frac{|E_o|^2}{4R_i}, \qquad (R_i \lessgtr 0) \tag{6}$$

Thus, for $R_i > 0$, the exchangeable power is the maximum power that the generator can supply. With $R_i < 0$, the exchangeable power is no longer equal to the maximum possible power flow into the load, which is infinite. However, regardless of the sign of R_i it can be considered as the stationary value of the expression P_L with respect to a small change of the load impedance Z_L. This can be easily seen from (3), in which R_L and X_L appear only in the second-order terms of the difference between R_L and R_i and of the difference between X_L and $-X_i$.

Now, we are in a position to discuss the waves defined by (1). In the discussion of electric circuits, the voltage and current at the terminals are generally chosen as the independent variables. However, one may equally well choose any linear transformation of them as long as the transformation is not singular, i.e., as long as the inverse transformation exists. The waves defined by (1) are the result of just one of an infinite number of such linear transformations.

With a fixed Z_i, if V_i and I_i are given, a_i and b_i are readily calculated from (1). On the other hand, if a_i and b_i are given, V_i and I_i are obtained from the inverse transformation

$$V_i = \frac{p_i}{\sqrt{|\operatorname{Re} Z_i|}} (Z_i^* a_i + Z_i b_i),$$

$$I_i = \frac{p_i}{\sqrt{|\operatorname{Re} Z_i|}} (a_i - b_i) \tag{7}$$

where p_i is defined by

$$p_i = \begin{cases} 1 & \text{when } \operatorname{Re} Z_i > 0 \\ -1 & \text{when } \operatorname{Re} Z_i < 0 \end{cases} \tag{8}$$

Thus, any result in terms of one set of variables can easily be converted to that in terms of the other set of variables. This justifies the use of the waves a_i and b_i defined by (1) in place of the terminal voltage and current for any analysis. Referring to Fig. 1, the voltage at the generator terminal is given by

$$V_i = E_o - Z_i I_i$$

Inserting this into the first expression in (1), and taking the square of the magnitude, we have

$$|a_i|^2 = \frac{|E_o|^2}{4|R_i|}$$

which is equivalent to

$$P_e = p_i |a_i|^2 \tag{9}$$

It is worth noting that, when E_0 is equal to zero, a_i becomes zero also.

Next, let us consider $|a_i|^2 - |b_i|^2$. Direct substitution of (1) into this expression gives

$$|a_i|^2 - |b_i|^2$$
$$= \frac{(V_i + Z_i I_i)(V_i^* + Z_i^* I_i^*) - (V_i - Z_i^* I_i)(V_i^* - Z_i I_i^*)}{4|R_i|}$$
$$= \frac{(Z_i + Z_i^*)(V_i I_i^* + V_i^* I_i)}{4|R_i|} = p_i \operatorname{Re}\{V_i I_i^*\}$$

from which we have

$$\operatorname{Re}\{V_i I_i^*\} = p_i(|a_i|^2 - |b_i|^2) \tag{10}$$

The left-hand side of (10) expresses the power which is actually transferred from the generator to the load. Therefore, this is called the actual power from the generator (or to the load). Equation (10) shows that the actual power is equal to $p_i(|a_i|^2 - |b_i|^2)$. Since $-|b_i|^2$ is always negative whether the load contains some source or not, the magnitude of the exchangeable power of a generator $|a_i|^2$ can be identified as the maximum power that the generator can supply when $R_i > 0$, and as the maximum power that the generator can absorb when $R_i < 0$.

For a moment, let us confine ourselves to the case where the real part of the internal impedance of the generator is positive, i.e., p_i is equal to 1. Then, (9) and (10) can be interpreted as follows. The generator is sending the power $|a_i|^2$ toward a load, regardless of the load impedance. However, when the load is not matched, i.e., if (4) is not satisfied, a part of the incident power is reflected back to the generator. This reflected power is given by $|b_i|^2$ so that the net power absorbed in the load is equal to $|a_i|^2 - |b_i|^2$. Associated with these incident and reflected powers, there are waves a_i and b_i, respectively.

To help understand the meaning of the incident and reflected powers, let us consider a new equivalent circuit of the generator in which we see these powers separately. Suppose that a new generator and load are connected to two arms of a three-port circulator and they are matched to the circulator impedance and that a lossless circuit which transforms the circulator impedance into Z_i is connected, as shown in Fig. 2. The maximum power we can obtain from the third arm of the circulator is equal to the power the new generator supplies toward the circulator. Because the lossless circuit does not consume any power, this maximum power must be equal to the maximum power which the outside load Z_L can absorb. Since the change of the load impedance Z_L does not affect the load condition of the generator at arm 1, the available power $|a_i|^2$ must be equal to the power which the generator is sending to the circulator. Further, since the net power to the load Z_L is equal to $|a_i|^2 - |b_i|^2$ and as no power comes back to the generator at arm 1, the balance $|b_i|^2$ must be absorbed in the load connected to arm 2 of the circulator. Thus we see that the incident power from the original generator is the power which the internal generator in this equivalent circuit is producing and the reflected power is the power which the internal load is absorbing.

Since one may well argue that, using an arbitrary constant C, $|a_i|^2 + C$ is the incident power from a generator while $|b_i|^2 + C$ is the reflected power, the above interpretation of incident and reflected powers is somewhat arbitrary. However, we set C equal to zero so that the maximum power a load can absorb is equal to the incident power which the generator sends to the load. This situation is very similar to that of the Poynting vector $E \times H$. Using an arbitrary vector function X, $E \times H + \nabla \times X$ can be considered as the transmission power density; whenever it is integrated over a closed surface the contribution from the last term $\nabla \times X$ disappears. Nevertheless, we generally consider that the power density is expressed by $E \times H$, so that there is no energy flow where there is no electric or magnetic field.

Extending our discussion to the case where the real part of the internal impedance of the generator may be negative, we say that the generator is sending the power $p_i |a_i|^2$ toward the load regardless of the load impedance and, when the load is not matched, $p_i |b_i|^2$ is reflected back so that the net power absorbed in the load is given by $p_i(|a_i|^2 - |b_i|^2)$. Associated with these incident and reflected powers, there are the incident and reflected waves a_i and b_i. Since the incident power to a load is equal to the exchangeable power of the generator connected to the load, $p_i |a_i|^2$ may also be called the exchangeable power to the load. The reason why, for the discussion of powers, we do not consider the incident and reflected powers directly but through the waves a_i and b_i lies in the fact that there is a linear relation between a_i's and b_i's and this can be used advantageously as we shall see in the following sections. There is no such relation between powers.

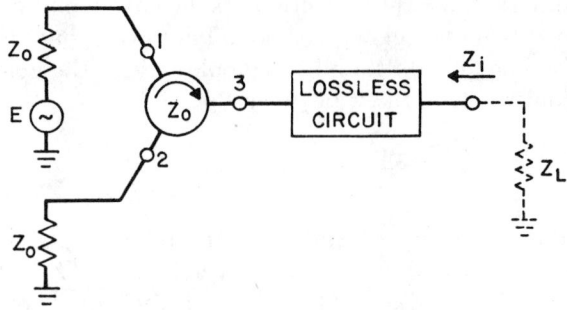

Fig. 2. New equivalent circuit of a generator.

III. Reflection Coefficients and Scattering Matrix

When we consider two quantities such as voltage and current, we take the ratio, an impedance. Similarly, since we have two quantities a_i and b_i, let us define the ratio s

$$s = \frac{b_i}{a_i} \quad (11)$$

and call it the power wave reflection coefficient.[2] Further, let us call the square of its magnitude, i.e., $|s|^2$, the power reflection coefficient. Using (1) and the relation $V_i = Z_L I_i$, s can be expressed in terms of impedances.

$$s = \frac{Z_L - Z_i^*}{Z_L + Z_i} \quad (12)$$

Substituting $Z_i = R_i + jX_i$, $Z_L = R_L + jX_L$ into (12), s can be rewritten in the form

$$s = \frac{R_L + j(X_L + X_i) - R_i}{R_L + j(X_L + X_i) + R_i} \quad (13)$$

Comparing this expression with that of the conventional voltage reflection coefficient, we see that s corresponds to the vector drawn from the center of the Smith chart to the point where the normalized impedance is given by $[R_L + j(X_L + X_i)]/R_i$. In other words, if the reactance part of Z_i is added to Z_L and normalized with respect to the real part of Z_i, then the corresponding point on the Smith chart gives the magnitude and the phase of the power wave reflection coefficient. From this, the following important property of s is derived: When R_i and R_L have the same sign, $|s| < 1$ and when they have opposite signs, $|s| > 1$.

The power reflection coefficient is given by

$$|s|^2 = \left| \frac{Z_L - Z_i^*}{Z_L + Z_i} \right|^2 \quad (14)$$

When the matching condition (4) is satisfied, the power reflection coefficient becomes zero, as is expected.

---

[2] When Z_i is real and positive, this is a voltage reflection coefficient.

s and $|s|^2$ are the reflection coefficients looking into the load from the generator side. The corresponding reflection coefficients s' and $|s'|^2$ looking into the generator from the load must be given by

$$s' = \frac{Z_i - Z_L^*}{Z_i + Z_L}, \qquad |s'|^2 = \left|\frac{Z_i - Z_L^*}{Z_i + Z_L}\right|^2$$

where the subscripts i and L are interchanged. s' is not necessarily equal to s. However, since $|Z_i - Z_L^*| = |Z_i^* - Z_L| = |Z_L - Z_i^*|$, $|s'|^2$ is always equal to $|s|^2$. Thus the power reflection coefficient remains the same when roles of generator and load are interchanged.[3] $1 - |s|^2$ is called the power transmission coefficient and this also remains constant when we interchange the role of generator and load. It is worth noting that the power transmission coefficient times the exchangeable power is equal to the actual power, or that the actual power divided by the power transmission coefficient is the exchangeable power.

Next, to define the scattering matrix, let us consider a linear n-port network and let a, b, v and i be vectors whose ith components are a_i, b_i, V_i, and I_i at the ith port of the network respectively. Then, a and b can be written in terms of v and i as follows:

$$a = F(v + Gi), \qquad b = F(v - G^+i) \qquad (15)$$

where F and G are the diagonal matrices whose ith diagonal components are given by $1/2\sqrt{|\text{Re} Z_i|}$ and Z_i, respectively, and $+$ indicates, in general, the complex conjugate transposed matrix. Since there is a linear relation between v and i given by

$$v = Zi \qquad (16)$$

where Z is the impedance matrix, and since a and b are the result of a linear transformation of v and i, there must be a linear relation between a and b. Let us write it in the form

$$b = Sa \qquad (17)$$

and call this S the power wave scattering matrix. Elimination of a, b, and v from (15), (16), and (17) gives

$$F(Z - G^+)i = SF(Z + G)i$$

from which the following expression of S can be obtained.

$$S = F(Z - G^+)(Z + G)^{-1}F^{-1} \qquad (18)$$

Similarly, Z can be expressed in terms of S

$$Z = F^{-1}(I - S)^{-1}(SG + G^+)F \qquad (19)$$

where I is a unit matrix. In Sections IV and V, we shall consider the conditions which S has to satisfy in order to represent a reciprocal network and a lossless network, respectively.

---

[3] In this interchange, only the place of the (zero-impedance) voltage source and the (zero-impedance) load current meter are reversed, leaving the generator and load impedance stationary.

IV. Reciprocal Condition

It is a well-known fact that the impedance matrix Z representing a reciprocal network has to satisfy the relation

$$Z = Z_t \qquad (20)$$

where the subscript t indicates the transposed matrix. The corresponding relation for S is given by

$$S_t = PSP \qquad (21)$$

where P is a diagonal matrix with its ith diagonal component being p_i. The proof of (21) will be given in Appendix I. Equation (21) is equivalent to

$$S_{ji} = p_i p_j S_{ij} \qquad (22)$$

which implies that, if the signs of $\text{Re } Z_i$ and $\text{Re } Z_j$ are the same, S_{ij} is equal to S_{ji} and, if they are opposite, S_{ij} is equal to $-S_{ji}$. For either case

$$|S_{ij}|^2 = |S_{ji}|^2 \qquad (23)$$

Now, suppose that all the circuits except the one connected to the ith port of the network have no source. Since the power from the jth circuit to the network is generally given by $p_j(|a_j|^2 - |b_j|^2)$ and $a_j(j \neq i)$ is equal to zero, the power to the jth circuit from the network is given by $p_j|b_j|^2$. Further, in this case, b_j is equal to $S_{ji}a_i$ and hence, the ratio of the actual power $p_j|b_j|^2$ into the load j to the exchangeable power $p_i|a_i|^2$ from the source i is equal to $p_i p_j |S_{ji}|^2$. However, because of (23), the value of this ratio does not change when the subscripts i and j are interchanged. Thus, we conclude that the relation between the actual power into a load and the exchangeable power from the source stays constant when the roles of source and load are interchanged in a reciprocal network. This is a power reciprocal relation.

It is interesting to note that there is no such reciprocal theorem in general between the exchangeable power to a load and the exchangeable power from the source, nor between the actual power to a load and the actual power from the source. The actual power into the jth circuit is given by $p_j|b_j|^2$ and the power transmission coefficient at this point by $1 - |S_{jj}|^2$. Therefore, the exchangeable power to the jth circuit is equal to $p_j|b_j|^2/(1 - |S_{jj}|^2)$. The ratio of the exchangeable power into the jth circuit to that from the ith circuit is given by $p_j p_i |S_{ji}|^2/(1 - |S_{jj}|^2)$. However, since $|S_{jj}|$ is not necessarily equal to $|S_{ii}|$, the value of this ratio does not necessarily stay constant when the roles of source and load are interchanged. Similarly, since the actual power from the ith circuit is given by $p_i|a_i|^2(1 - |S_{ii}|^2)$, the ratio of the actual power into the jth circuit to that from the ith circuit is equal to $p_i p_j |S_{ji}|^2/(1 - |S_{ii}|^2)$. This again does not remain constant when the subscripts i and j are interchanged. However, the ratio of the exchangeable power to the load j to the actual power from the source i is given by

$p_i p_j |S_{ji}|^2 / (1 - |S_{jj}|^2)(1 - |S_{ii}|^2)$ and remains constant when the subscripts are interchanged.

The foregoing discussion is readily applicable to a pair of antennas. It is a well-known fact that there exists a power reciprocal relation between the transmitting and receiving antennas. However, it seems to be less well understood that it is between the exchangeable power from the source and the actual power to the load that the reciprocal theorem generally holds. Unless the matching conditions for both antennas are satisfied, the reciprocal theorem does not necessarily hold between the actual powers nor between the exchangeable powers.

V. Lossless Condition

In this section, let us consider the condition which a scattering matrix has to satisfy in order to represent a lossless network. The actual power into the network from the ith circuit is given by $p_i(|a_i|^2 - |b_i|^2)$. Therefore, the total power into the network is

$$\sum_i p_i(|a_i|^2 - |b_i|^2).$$

When the network is lossless, this total power must be zero, hence we have

$$\sum_i p_i(|a_i|^2 - |b_i|^2) = 0$$

which we can rewrite in a matrix form as follows

$$a^+ P a - b^+ P b = 0$$

Substitution of (17) gives

$$a^+(P - S^+ P S)a = 0$$

Since a is arbitrary, this means

$$S^+ P S = P. \tag{24}$$

Equation (24) is the condition that the scattering matrix representing a lossless network has to satisfy.

For a simple example, let us consider a two-port junction. Equation (24) gives three independent conditions

$$p_1 |S_{11}|^2 + p_2 |S_{21}|^2 = p_1$$
$$p_1 S_{11} S_{12}^* + p_2 S_{21} S_{22}^* = 0$$
$$p_1 |S_{12}|^2 + p_2 |S_{22}|^2 = p_2 \tag{25}$$

From the second condition, we have

$$|S_{11}|^2 |S_{12}|^2 = |S_{21}|^2 |S_{22}|^2$$

Combining the first and last conditions in (25) with this equation, we obtain

$$\frac{p_2}{p_1}(1 - |S_{22}|^2)|S_{11}|^2 = \frac{p_1}{p_2}(1 - |S_{11}|^2)|S_{22}|^2$$

which is equivalent to

$$|S_{11}|^2 = |S_{22}|^2 \tag{26}$$

Equation (26) shows that, for a lossless two-port junction, the power reflection coefficient at one port is equal to that at the other port. From this conclusion, we see that the power reflection coefficient as well as the power transmission coefficient remain constant regardless of the position of the reference plane we take along a lossless transmission system. This means that we can choose any convenient plane as the reference plane for power discussion in a lossless transmission system.

The fact that the exchangeable power is preserved during a nonsingular lossless transformation can also be easily shown using the above result. Let a_2 be zero for a moment. The exchangeable power from the output port 2 is given by

$$\frac{p_2 |b_2|^2}{1 - |S_{22}|^2} = \frac{p_2 |S_{21}|^2 |a_1|^2}{1 - |S_{11}|^2} = p_1 |a_1|^2$$

where we have used the first condition in (25). The right-hand side of this equation is just the exchangeable power to the lossless junction. Thus, we have shown that the exchangeable powers are the same at the input and output of a lossless two-port junction provided that $|S_{11}|^2 = |S_{22}|^2 \neq 1$. If $|S_{11}|^2 = |S_{22}|^2 = 1$, the input and output ports are effectively disconnected inside the junction, which is of no practical interest.

Inserting (26) back into the first and last expressions in (25), and comparing the result, we have

$$|S_{12}|^2 = |S_{21}|^2 \tag{27}$$

This is a kind of power reciprocal theorem. However, it is only the lossless condition that we have used for the derivation. Therefore, even a nonreciprocal twoport junction has to satisfy the power reciprocal theorem if it is lossless.

Coming back to (24), and multiplying both sides by $(PS)^{-1} = S^{-1} P^{-1}$ from the right and SP^{-1} from the left, we have

$$SPS^+ = P \tag{28}$$

Equations (24) and (28) are equivalent to each other. However, sometimes one may find (28) being more convenient than (24). The example is found in the discussion of the actual noise measure of linear amplifiers.

When the network is lossy, the total power into the network must be positive and hence

$$a^+(P - S^+ P S)a \geq 0$$

Thus, for a passive network, $(P - S^+ P S)$ must be positive definite or positive semidefinite.

VI. Frequency Characteristic of Lossless Reciprocal Network

When a junction under consideration is lossless as well as reciprocal, (21) and (24) must be satisfied simultaneously. Further in this case, corresponding to the well-known relation for $\partial Z / \partial \omega$,

$$i^+ \frac{\partial Z}{\partial \omega} i = j \int (\mu H^* \cdot H + \epsilon E^* \cdot E) dv \qquad (29)$$

we can derive a relation for $\partial S/\partial \omega$ as we shall do in Appendix II. It is given by

$$a^+ j \left(K^+ + S^+ K^+ S + 2 S^+ P \frac{\partial S}{\partial \omega} \right) a$$
$$= \int (\mu H^* \cdot H + \epsilon E^* \cdot E) dv \qquad (30)$$

where K is a diagonal matrix with the ith diagonal component being

$$\left\{ \frac{\partial}{\partial \omega} (Z_i^* - Z_i) \right\} \bigg/ |Z_i^* + Z_i|,$$

and E and H are the electric and magnetic field respectively. The integral in the right-hand side of (30) extends all over the junction region and represents twice the stored energy in the network, hence it is positive. Thus, we see that

$$j \left(K^+ + S^+ K^+ S + 2 S^+ P \frac{\partial S}{\partial \omega} \right)$$

has to be positive definite. The first and second terms represent the effect of the possible change of the terminal impedance Z_i's. It is interesting to note that both terms disappear when all the imaginary parts of Z_i's remain constant. When this is the case, (30) reduces to

$$a^+ \left(j 2 S^+ P \frac{\partial S}{\partial \omega} \right) a = \int (\mu H^* \cdot H + \epsilon E^* \cdot E) dv \qquad (31)$$

For a lossless oneport network, $|S_{11}| = 1$ and S_{11} can be written in the form $e^{-j\phi}$. Therefore, in this case the above relation reduces to

$$\frac{\partial \phi}{\partial \omega} = \frac{1}{2 p_1 |a_1|^2} \int (\mu H^* \cdot H + \epsilon E^* \cdot E) dv$$

However, since $b_1 = e^{-j\phi} a_1$ and $\partial \phi / \partial \omega$ give the time delay between b_1 and a_1 of the wave envelope, which can be interpreted as the energy delay, the interpretation of (31), when applied to a one-port network, is as follows. The time required for an incident energy to enter the network and leave again is the total stored energy divided by the exchangeable power of the source. When the real part of the source impedance is negative, the required time becomes negative. This is what we expect if no oscillation takes place. In most cases where we connect a negative resistance to a oneport network of which the losses are negligible, oscillations occur and therefore it is impossible to observe the above phenomenon directly. For multiport networks, even if some of the impedances have negative real parts, a stable operation becomes possible and (30) with the corresponding p_i's being negative gives a more realistic condition for $\partial S/\partial \omega$.

One might think that the imaginary parts of the circuit impedances Z_i could be considered to be part of the junction and that K could therefore always be set equal to zero without loss of generality. This is not necessarily the case, for the imaginary parts of the circuit impedances Z_i may not have the frequency dependence of ordinary passive networks. Examples are $-L$ and $-C$.

VII. Change of Circuit Impedance

Suppose that the impedances of the circuits connected to the junction under consideration are changed from Z_i to Z_i' ($i = 1, 2, \cdots, n$). Then the incident and reflected waves have to be redefined accordingly. The scattering matrix S' connecting these new power wave vectors is, of course, different from the original one. However, it is expressible in terms of the original S and the power wave reflection coefficient r_i of Z_i' with respect to Z_i^*, i.e.,

$$S' = A^{-1} (S - \Gamma^+)(I - \Gamma S)^{-1} A^+ \qquad (32)$$

where Γ and A are the diagonal matrices with their ith diagonal components being r_i and $(1 - r_i^*)\sqrt{|1 - r_i r_i^*|}/|1 - r_i|$, respectively. An outline of the derivation is given in Appendix III. Essentially the same formula (for Re $Z_i > 0$) is also derived by Youla and Paterna [4] using a different approach.

There are a number of applications of this formula. Consider, for example, a twoport amplifier whose source and load impedances, Z_1 and Z_2, respectively, have positive real parts and let us obtain the condition under which both input and output ports can be matched simultaneously without changing the signs of the real parts of the source and/or load impedances. The matching conditions for input and output ports are given by $S_{11}' = 0$ and $S_{22}' = 0$, respectively. Using (32), the condition $S_{11}' = 0$ provides

$$r_1^* = \frac{S_{11} + r_2 (S_{12} S_{21} - S_{11} S_{22})}{1 - r_2 S_{22}} \qquad (33)$$

Similarly, $S_{22}' = 0$ provides

$$r_2^* = \frac{S_{22} + r_1 (S_{12} S_{21} - S_{11} S_{22})}{1 - r_1 S_{11}} \qquad (34)$$

For simultaneous matching, (33) and (34) have to be satisfied at the same time. Thus, the problem is reduced to that of finding the solutions of the simultaneous equations and checking whether or not they satisfy the appropriate conditions which ensure that the real parts of the source and load impedances remain positive. As explained in connection with (13), the latter conditions are given by $|r_1| < 1$ and $|r_2| < 1$, respectively. For the check of these conditions, a straightforward but lengthy calculation is necessary. From it, we see that when $|S_{12} S_{21}| \neq 0$ the necessary and sufficient condition for simultaneous matching to be possible is given by

$$2|S_{12}S_{21}| < 1 + |S_{12}S_{21} - S_{11}S_{22}|^2 - |S_{11}|^2 - |S_{22}|^2 \qquad (35)$$

When $|S_{12}S_{21}| = 0$, the same condition is given by

$$|S_{11}| < 1 \quad \text{and} \quad |S_{22}| < 1 \qquad (36)$$

The transducer gain under the simultaneously matched condition is

$$|S_{21}'|^2 = \frac{|S_{21}|}{|S_{12}|}(k \pm \sqrt{k^2 - 1}) \quad \text{for} \quad |S_{12}S_{21}| \neq 0 \qquad (37)$$

where

$$k = \frac{1 + |S_{12}S_{21} - S_{11}S_{22}|^2 - |S_{11}|^2 - |S_{22}|^2}{2|S_{12}S_{21}|} \qquad (38)$$

The upper sign applies when

$$B = |S_{22}|^2 - |S_{11}|^2 - 1 + |S_{12}S_{21} - S_{11}S_{22}|^2 \qquad (39)$$

is positive, and the lower sign when $B < 0$. When $|S_{12}S_{21}| = 0$, the same gain is

$$|S_{21}'|^2 = \frac{|S_{21}|^2}{(1 - |S_{11}|^2)(1 - |S_{22}|^2)} \qquad (40)$$

An amplifier is said to be unconditionally stable if the real parts of its input and output impedances remain positive when the load and source impedances, respectively, are changed arbitrarily, but keeping their real parts positive. Let us next consider the condition for an amplifier to be unconditionally stable. For the input impedance, we require $|S_{11}'|$ to be less than 1 when r_2 is changed arbitrarily, but keeping $|r_2| < 1$. Similarly, for the output impedance, $|S_{22}'| < 1$ is required when $|r_1| < 1$. Using (32), a little manipulation shows that the necessary and sufficient conditions are given by

$$|S_{12}S_{21}| < 1 - |S_{11}|^2$$
$$|S_{12}S_{21}| < 1 - |S_{22}|^2$$
$$2|S_{12}S_{21}| < 1 + |S_{12}S_{21} - S_{11}S_{22}|^2 - |S_{11}|^2 - |S_{22}|^2 \qquad (41)$$

The last condition is identical with that under which simultaneous matching is possible when $|S_{12}S_{21}| \neq 0$. It is interesting to note that simultaneous matching is possible for any amplifier which is unconditionally stable, but the reverse is not necessarily true.

From the first two conditions in (41), it can be shown that B, as given by (39), is negative when the amplifier is unconditionally stable. Therefore, the lower sign applies in this case on the right-hand side of (37). Furthermore, since $|S_{12}|/|S_{21}|$ is invariant to changes in source and load impedances, as can be shown from (32), and $|S_{21}'|^2$ in (37) is similarly invariant (from physical reasoning), k is also invariant to changes in source and load impedances.

VIII. Choice of Phase

The phase of the incident wave a_i is equal to that of the open circuit voltage E of the ith circuit and the phase of the reflected wave b_i is that of $E - 2\{\text{Re } Z_i\}I_i$. When the ith circuit has no source, b_i has the phase of the voltage across the resistance in the series representation of the circuit. However, since it is only the square of the magnitude of the waves that we have used for the power discussion in Section II, an arbitrary phase could be assigned to each wave without changing the power relation. Thus, in place of (1), we could define the waves a_i and b_i by

$$a_i = \frac{V_i + Z_i I_i}{2\sqrt{|\text{Re } Z_i|}} e^{j\phi_i}, \quad b_i = \frac{V_i - Z_i^* I_i}{2\sqrt{|\text{Re } Z_i|}} e^{j\psi_i} \qquad (42)$$

respectively, where ϕ_i and ψ_i are arbitrary angles. The scattering matrix S in this case is defined through the relation

$$b = Sa$$

where a and b are the vectors with their ith components being a_i and b_i given by (42). The reciprocal condition (21) is replaced by

$$S_t = (MN)^{-1} P S P M N$$

where M and N are the diagonal matrices whose ith diagonal components are $e^{j\phi_i}$ and $e^{j\psi_i}$, respectively. The lossless condition (24) remains the same. However, the equation corresponding to (30) has two additional terms

$$2S^+ PS \frac{\partial M}{\partial \omega} M^{-1} + 2S^+ PN \frac{\partial N^{-1}}{\partial \omega} S,$$

in the bracket of the left-hand side of (30). The original form used by Penfield [3] is just a special case of the above definition. The phases were chosen so that, for $\text{Re } Z_i > 0$, $e^{j\phi_i} = e^{j\psi_i} = 1$ and for $\text{Re } Z_i < 0$, $e^{j\phi_i} = e^{j\psi_i} = -j$. In this case MN is equal to P and the reciprocal relation takes a simple form: $S_t = S$.

Another interesting choice of the phases is given by

$$e^{j\phi_i} = \sqrt{\frac{|Z_i|}{Z_i}}, \quad e^{j\psi_i} = \sqrt{\frac{|Z_i|}{-Z_i^*}}$$

where

$$-\frac{\pi}{2} \leq \phi_i \leq \frac{\pi}{2}, \quad -\frac{\pi}{2} \leq \psi_i \leq \frac{\pi}{2}$$

The significance of this choice lies in the following fact. When we replace every quantity appearing in the definition of the waves (42) by the corresponding dual quantity, the waves stay the same. Thus

$$a_i = \frac{V_i + Z_i I_i}{2\sqrt{|\text{Re } Z_i|}} \sqrt{\frac{|Z_i|}{Z_i}} = \frac{I_i + Y_i V_i}{2\sqrt{|\text{Re } Y_i|}} \sqrt{\frac{|Y_i|}{Y_i}}$$

$$b_i = \frac{V_i - Z_i^* I_i}{2\sqrt{|\text{Re } Z_i|}} \sqrt{\frac{|Z_i|}{-Z_i^*}} = \frac{I_i - Y_i^* V_i}{2\sqrt{|\text{Re } Y_i|}} \sqrt{\frac{|Y_i|}{-Y_i^*}}$$

The traveling waves defined by (43) in Section IX have this property. However, the power waves defined by (1) have not.

IX. Comparison with Traveling Waves

The traveling waves along a transmission line can be defined by

$$a(z) = \frac{V(z) + Z_0 I(z)}{2\sqrt{Z_0}}, \qquad b(z) = \frac{V(z) - Z_0 I(z)}{2\sqrt{Z_0}} \qquad (43)$$

where $V(z)$ and $I(z)$ are the voltage and current at a point z along the line and Z_0 is the characteristic impedance. If we consider Z_0 with a positive real value, the expression for the power waves becomes identical with that for the traveling waves. Therefore, all the conditions which the scattering matrix for traveling waves must satisfy in order to represent certain networks can be obtained if we set P equal to a unit matrix I in the corresponding conditions for the power wave scattering matrix. Thus, the lossless condition becomes $S^+ S = I$, the reciprocal condition $S_t = S$ and (35), (36), and (41) stay the same. However, the interpretation of these results is different. For example, let us consider the condition $|S_{ij}|^2 = |S_{ji}|^2$ for a reciprocal network. Assuming that all the characteristic impedances of the lines are real and positive, the direct interpretation of this condition is as follows. The power coming out from the jth port when the incoming power into the ith port is unity is equal to the power coming out from the ith port when the incoming power into the jth port is unity, provided that all the circuits connected to the far ends of the lines are matched to the line characteristic impedances. However, the last restriction is generally too stringent for practical applications. And it is only after a little manipulation that we discover the power reciprocal relation given in Section IV. Thus, according to the particular problem we have, a choice must be made between the traveling wave and power wave representation. For instance, if we want to discuss the properties of a junction irrespective of the impedances connected to the terminals, the traveling waves may be more convenient. On the other hand, for the power relation between circuits connected through a junction, the power wave representation is more suitable. One may ask then what is the relation between the traveling waves and the power waves. When Z_0 is real and positive, there is no difference in the expressions of the power waves and the traveling waves. Therefore, in this case, the net power in the z direction is given by $|a(z)|^2 - |b(z)|^2$. However, when Z_0 is complex, the situation is different. $|a(z)|^2 - |b(z)|^2$ is calculated to be $\operatorname{Re}\{Z_0 V^* I\}/|Z_0|$, which is not equal to the power $\operatorname{Re}\{VI^*\}$. Thus, each traveling wave cannot be considered to bring the power expressed by the square of the magnitude. Further, since the traveling wave reflection coefficient is given by $(Z_L - Z_0)/(Z_L + Z_0)$ and the maximum power transfer takes place when $Z_L = Z_0^*$, where Z_L is the load impedance, it is only when there is a certain reflection in terms of traveling waves that the maximum power is transferred from the line to the load. Thus, we have seen that, in general, the traveling wave concept is not so closely related with the power.

X. Conclusion

The physical meaning of power waves and the properties of the scattering matrix are presented. Although the power waves are the result of just one of an infinite number of possible linear transformations of voltage and current, it has been shown that, for certain applications, they give a clearer and more straightforward understanding of the power relations between circuit elements connected through a multiport network.

Appendix I

Let us prove the reciprocal condition (21). Using (18), (20), and the obvious relations $F_t = F$, $G_t = G$, the left-hand side of (21) can be rewritten in the form

$$\begin{aligned} S_t &= F_t^{-1}(Z+G)_t^{-1}(Z-G^+)_t F_t \\ &= F_t^{-1}(Z_t + G_t)^{-1}(Z_t - G_t^+) F_t \\ &= F^{-1}(Z+G)^{-1}(Z-G^+) F \end{aligned}$$

We wish to prove that this last expression is equal to the right-hand side of (21), which is given by

$$PSP = PF(Z - G^+)(Z + G)^{-1} F^{-1} P$$

To do so, since $P = P^{-1}$, we have only to prove the following equation.

$$(Z - G^+) F P F (Z + G) = (Z + G) F P F (Z - G^+).$$

Performing the matrix product, the above equation becomes

$$ZFPFZ + ZFPFG - G^+FPFZ - G^+FPFG$$
$$= ZFPFZ - ZFPFG^+ + GFPFZ - GFPFG^+,$$

of which the first terms in both sides are the same and the last terms are equal to each other. Thus, all that we have to prove is

$$ZFPFG - G^+FPFZ = -ZFPFG^+ + GFPFZ$$

or

$$ZFPF(G + G^+) = (G^+ + G)FPFZ \qquad (44)$$

Since

$$FPF(G + G^+) = \tfrac{1}{2} I$$
$$(G^+ + G)FPF = \tfrac{1}{2} I,$$

the validity of (44) is obvious. This completes the proof of (21).

Appendix II

The derivation of (30) will be given briefly. From Maxwell's equations and an appropriate definition of voltage and current at the reference planes, after a little manipulation, we have

$$i^+ \frac{\partial v}{\partial \omega} + v^+ \frac{\partial i}{\partial \omega} = j \int (\mu H^* \cdot H + \epsilon E^* \cdot E) dv \quad (45)$$

from which (29) is derived. The expressions for v and i in terms of a and b are

$$v = 2PF(G^+ a + Gb), \quad i = 2FP(a - b)$$

Substituting these expressions, the left-hand side of (45) becomes

$$i^+ \frac{\partial v}{\partial \omega} + v^+ \frac{\partial i}{\partial \omega} = 4 \left\{ a^+ \left(F^+ \frac{\partial FG^+}{\partial \omega} + GF^+ \frac{\partial F}{\partial \omega} \right) a \right.$$
$$+ a^+ (F^+ FG^+ + GF^+ F) \frac{\partial a}{\partial \omega}$$
$$- b^+ \left(F^+ \frac{\partial FG}{\partial \omega} + G^+ F^+ \frac{\partial F}{\partial \omega} \right) b$$
$$\left. - b^+ (F^+ FG + G^+ F^+ F) \frac{\partial b}{\partial \omega} \right\} \quad (46)$$

Since

$$F^+ FG^+ + GF^+ F = \tfrac{1}{2} P$$
$$F^+ FG + G^+ F^+ F = \tfrac{1}{2} P$$
$$F^+ \frac{\partial FG^+}{\partial \omega} + GF^+ \frac{\partial F}{\partial \omega} = \tfrac{1}{4} K$$
$$F^+ \frac{\partial FG}{\partial \omega} + G^+ F^+ \frac{\partial F}{\partial \omega} = -\tfrac{1}{4} K$$

the right-hand side of (46) reduces to

$$a^+ Ka + b^+ Kb + 2 \left(a^+ P \frac{\partial a}{\partial \omega} - b^+ P \frac{\partial b}{\partial \omega} \right)$$

Using $b = Sa$, this can be rewritten in the form

$$a^+ Ka + a^+ S^+ KSa + 2 \left(a^+ P \frac{\partial a}{\partial \omega} - a^+ S^+ P \frac{\partial S}{\partial \omega} a \right.$$
$$\left. - a^+ S^+ PS \frac{\partial a}{\partial \omega} \right)$$

Because of (24), the first and last terms in the bracket cancel each other. Therefore, (45) becomes

$$a^+ \left(K + S^+ KS - 2S^+ P \frac{\partial S}{\partial \omega} \right) a$$
$$= j \int (\mu H^* \cdot H + \epsilon E^* \cdot E) dv$$

which is equivalent to (30).

Appendix II

An outline of the derivation of (32) will be given in this appendix. From (18),

$$S' = F'(Z - G'^+)(Z + G')^{-1} F'^{-1} \quad (47)$$

where F' and G' represent F and G, respectively, when Z_i is replaced by Z_i' everywhere. Substituting (19) into (47) and using Γ, defined by

$$\Gamma = (G' - G)(G' + G^+)^{-1}, \quad (48)$$

S' can be rewritten in the form

$$F' F^{-1} (I - S)^{-1} (S - \Gamma^+)(I - \Gamma^+)(I - \Gamma)$$
$$\cdot (I - S\Gamma)^{-1}(I - S) F F'^{-1}$$

Since

$$(I - S)^{-1}(S - \Gamma^+)(I - \Gamma^+)^{-1}$$
$$= (I - \Gamma^+)^{-1}(S - \Gamma^+)(I - S)^{-1}$$
$$(I - \Gamma)(I - S\Gamma)^{-1}(I - S)$$
$$= (I - S)(I - \Gamma S)^{-1}(I - \Gamma)$$

S' becomes

$$S' = A^{-1}(S - \Gamma^+)(I - \Gamma S)^{-1} A^+$$

where A is a diagonal matrix defined by

$$A = F'^{-1} F (I - \Gamma^+).$$

Calculation of the ith diagonal component A_i shows that

$$A_i = \frac{1 - r_i^*}{|1 - r_i|} \sqrt{|1 - r_i r_i^*|}$$

where r_i is the ith diagonal component of Γ and (referring (48)) is given by

$$r_i = \frac{Z_i' - Z_i}{Z_i' + Z_i^*}.$$

From this, r_i is interpreted as the power wave reflection coefficient of Z_i' with respect to Z_i^*.

References

[1] Penfield, P., Jr., Noise in negative resistance amplifiers, *IRE Trans. on Circuit Theory*, vol. CT-7, Jun 1960, pp 166–170.
[2] Kurokawa, K., Actual noise measure of linear amplifiers, *Proc. IRE*, vol. 49, Sep 1961, pp 1391–1397.
[3] Youla, D. C., On scattering matrices normalized to complex port numbers, *Proc. IRE*, vol. 49, July 1961, p 1221.
[4] Youla, D. C., and P. M. Paterno, Realizable limits of error for dissipationless attenuators in mismatched systems, *IEEE Trans. on Microwave Theory and Techniques*, vol. MTT-12, May 1964, pp 289–299.
[5] Haus, H. A., and R. B. Adler, *Circuit Theory of Linear Noisy Networks*. New York: Wiley, 1959.

Scattering Parameter Approach to the Design of Narrow-Band Amplifiers Employing Conditionally Stable Active Elements

C. S. GLEDHILL AND M. F. ABULELA

Abstract—In terms of scattering parameters, the equation of transducer power gain is shown to be capable of representation as a family of circles of constant gain from which the design of load and source terminations to achieve a restricted bandwidth can be obtained. This is an extension of an earlier approach which only allowed either load reflection coefficient or source reflection coefficient to be considered in a given design. Through the use of a specification statement of VSWR, it is shown how a marginal stability factor can be derived. From the study of the interaction between the input and output reflection coefficients, a detuning factor is analytically derived to correlate the interaction between the input and output reflection coefficients. Either of these factors can be chosen and used to select optimum input and output reflection coefficients which provide stable operation for an amplifying stage that is to employ a conditionally stable active element. An example using these factors is given.

I. Introduction

Over the past few years, there has been an increasing interest in the transistor two-port scattering parameters [1]–[4] further extended to linear integrated circuits [5]. Their use in amplifier design has been formalized [6], [7], and in particular Bodway [7] has carried out a full investigation from a power-gain point of view into the unconditionally stable case of the active two-port. Under normal circumstances, each stage of an amplifier is to be operated under the dc bias conditions which will provide the highest value of maximum unconditionally stable transducer power gain [7]. For an active element that shows a conditionally stable (or unstable) quiescent operating point, the logical solution is to look for another dc one which shows unconditional stability.

However, situations may arise when one must use the active element under predetermined dc quiescent values imposed by a higher priority criterion, e.g., noise figure in a front-end stage. Such a case may yield ac parameters showing conditional stability, and while the unconditionally stable case provides unique values of source and load reflection coefficients at the maximum transducer power gain, the conditionally stable case does not. This is why it is

Manuscript received February 9, 1972; revised June 22, 1973.
The authors are with the University of Manchester Institute of Science and Technology, P. O. Box 88, Manchester, England.

necessary to look for a new criterion on which to base the termination selection. Such a criterion will therefore be required: 1) to provide complete stabilization for a conditionally stable case; 2) to be analytically related to the characterization of the active element; 3) to be related to the specification requirements.

In this short paper, the case of the conditionally stable two-port is considered. Starting with a reexamination of the transducer power-gain equation, followed by a discussion of the stabilization approaches, the design formulas incorporating the stabilization factors and bandwidth requirements are given in terms of the scattering parameters.

Relationship Between Gain and Stability in Terms of Scattering Parameters

The definitions of gain have been formulated as follows [7]:

$$\text{power gain} = G = \frac{\text{power delivered to load}}{\text{power into two-port}}$$

$$= \frac{|S_{21}|^2 (1 - |r_2|^2)}{|(1 - |S_{11}|^2) + |r_2|^2 (|S_{22}|^2 - |\Delta|^2) - 2 \operatorname{Re}(r_2 C_2)|} \quad (1)$$

$$\text{available power gain} = G_A = \frac{\text{power available at the output}}{\text{power available from the generator}}$$

$$= \frac{|S_{21}|^2 (1 - |r_1|^2)}{|(1 - |S_{22}|^2) + |r_1|^2 (|S_{11}|^2 - |\Delta|^2) - \operatorname{Re}(r_1 C_1)|} \quad (2)$$

$$\text{transducer power gain} = G_T = \frac{\text{power delivered to load}}{\text{power available from generator}}$$

$$= \frac{|S_{21}|^2 (1 - |r_1|^2)(1 - |r_2|^2)}{|1 - r_1 S_{11} - r_2 S_{22} + r_1 r_2 \Delta|^2} \quad (3)$$

in which

$$r_i = \frac{Z_i - Z_{i0}}{Z_i + Z_{i0}^*} \quad (4)$$

$$\Delta = S_{11} S_{22} - S_{12} S_{21} \quad (5)$$

$$C_1 = S_{11} - S_{22}^* \Delta \quad (6)$$

$$C_2 = S_{22} - S_{11}^* \Delta. \quad (7)$$

Re indicates real part of and * indicates conjugate. Further, the stable and unstable regions of operation have been shown to be defined in the r_i plane circles of the general form:

$$\left| r_i - \frac{C_i^*}{|S_{ii}|^2 - |\Delta|^2} \right|^2 = \left| \frac{S_{12} S_{21}}{|S_{ii}|^2 - |\Delta|^2} \right|^2 \quad (8)$$

or

$$|r_i - r_{si}|^2 = |\rho_{si}|^2. \quad (9)$$

Stable operation occurs outside the ρ_{si} circle, shown in Fig. 1, and since for physically realizable two-port source and load reflection coefficients, $|r_i|$ must be less than unity, it follows that unconditionally stable operation is possible if the ρ_{si} circle lies completely outside the unity circle in the r_i plane, and the origins are stable, i.e.,

$$||\rho_{si}| - |r_{si}|| > 1 \quad \text{and} \quad |s_{ii}| < 1.$$

Under these conditions simultaneous conjugate matching of both ports is possible, and the maximum transducer gain is achieved [7]. Alternatively, the situation can occur where the ρ_{si} circle is fully or partially included in the unity circle, thus yielding a case of conditional stability.

II. CONSTANT POWER GAIN AND STABILITY CIRCLES

From (1)–(3) it can be seen that for the design of the load and source impedances, first r_2 can be optimized from consideration of the equation of power gain (1) and then r_1 can be presented as families of circles of constant power gains. From (1)

$$\frac{1}{g_2} = -D_2 + \left[\frac{B_2 - 2 \operatorname{Re}(r_2 C_2)}{1 - |r_2|^2} \right] \quad (10)$$

in which

$$g_2 = \frac{G}{|S_{21}|^2} \quad (11)$$

$$B_2 = 1 + |S_{22}|^2 - |S_{11}|^2 - |\Delta|^2 \quad (12)$$

$$D_2 = |S_{22}|^2 - |\Delta|^2. \quad (13)$$

Rearranging (10),

$$\left| r_2 - \frac{g_2}{1 + D_2 g_2} C_2^* \right|^2 = \frac{|(1 - 2K|S_{12} S_{21}|g_2 + |S_{12} S_{21}|^2 g_2^2)|}{|1 + D_2 g_2|^2} \quad (14)$$

in which

$$K = \frac{(1 + |\Delta|^2 - |S_{11}|^2 - |S_{22}|^2)}{2 |S_{12} S_{21}|} \quad (15)$$

(which has been defined as the invariant stability factor [9]). Equation (14) represents a family of circles defining load reflection coefficients giving the same power gain.

For physically realizable loads, $|r_2| \leq 1$ or $(1 - |r_2|^2) \geq 0$, and g_2 becomes infinite if

$$D_2(1 - |r_2|^2) = B_2 - 2 \operatorname{Re}(r_2 C_2). \quad (16)$$

This condition can be rewritten as

$$\left| r_2 - \frac{C_2^*}{|S_{22}|^2 - |\Delta|^2} \right|^2 = \left| \frac{S_{12} S_{21}}{|S_{22}|^2 - |\Delta|^2} \right|^2 \quad (17)$$

i.e., the same equation as (8), which defined the general stability condition:

Alternatively, families of circles defining source reflection coefficients giving the same available power gain could have been obtained, starting from (2). However, an equation relating both input and output reflection coefficients to the efficiency of power exchange can only be obtained from the transducer power-gain equation.

Starting from (3) for the transducer power gain

$$|r_1 - R_{gT1}|^2 = |\rho_{gT1}|^2 \quad (18)$$

in which

$$R_{gT1} = \frac{(1 - r_2 S_{22})(S_{11}^* - r_2^* \Delta^*)}{\{|S_{11} - r_2 \Delta|^2 + (1 - |r_2|^2)/g_t\}} \quad (19)$$

$$\rho_{gT1} = \left| \frac{\{[|S_{11} - r_2 \Delta|^2 - |1 - r_2 S_{22}|^2][(1 - |r_2|^2)/g_t] + [(1 - |r_2|^2)/g_t]^2\}^{1/2}}{|S_{11} - r_2|^2 + (1 - |r_2|^2)/g_t} \right| \quad (20)$$

$$g_T = \frac{G_T}{|S_{21}|^2}. \quad (21)$$

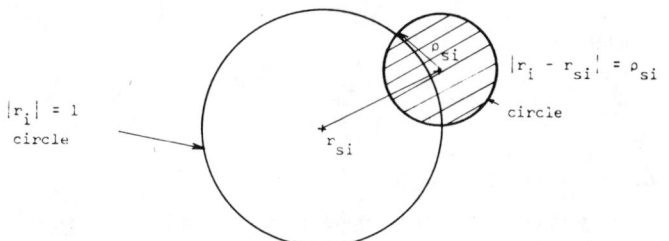

Fig. 1. Stability circles in r_i plane. Unstable region is shaded.

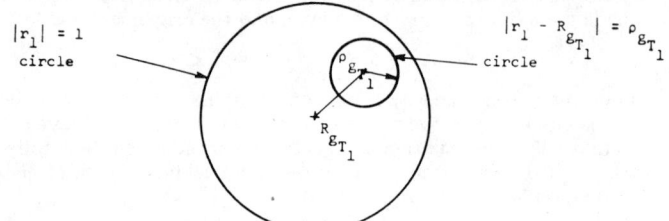

Fig. 2. Constant transducer power-gain circle in r_1 plane.

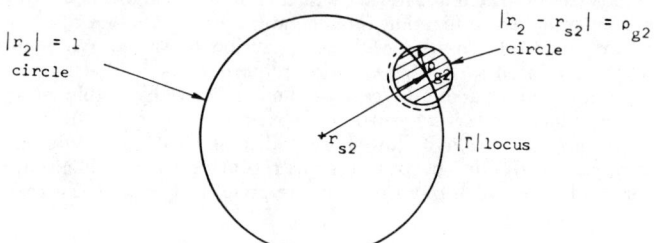

Fig. 3. Stabilization in r_2 plane.

From this equation, the circles shown in Fig. 2 can be drawn. Given an output termination and a specified transducer power gain, the input reflection coefficient, and hence the source impedance, can be calculated or can be obtained graphically. On the other hand, (14) only relates the output reflection coefficient to the power gain which, by definition, does not include the source conditions.

III. STABILIZATION APPROACH

Since it is always true that satisfactory stable operation will only be possible outside the stability circle (g_i equals ∞ circle) and inside the unit circle, there are two possible ways to achieve stable operation, namely:

a) to alter the network so that the stability circle lies completely outside the unit circle;
b) to limit the stage gain to a certain value, determined as follows.

A. Load Reflection Coefficient

Solution a), preceding, implies that one deliberately makes

$$\left| \frac{|C_2| - |S_{12}S_{21}|}{D_2} \right| > 1 \tag{22}$$

or

$$\{[(1 - |S_{11}|^2)(|S_{22}|^2 - |\Delta|^2) + |S_{12}S_{21}|]^{1/2} - |S_{12}S_{21}|\}$$
$$> ||S_{22}^2| - |\Delta|^2|. \tag{23}$$

This is only possible by feedback, and the design equations are better derived from a three-port parameter approach [10].

For the second proposed solution, b), preceding, two possibilities exist.

i)

$$|r_2| \leq |\Gamma| \tag{24}$$

in which

$$|\Gamma| = M ||r_{S2}| - \rho_{S2}| \tag{25}$$

$$M = \frac{\nu(1 + ||r_{S2}| - \rho_{S2}|)^{-1} - 0.5}{1 - \nu(1 + ||r_{S2}| - \rho_{S2}|)} \tag{26}$$

and $\nu = 1 - (\text{min VSWR})/(\text{max VSWR})$. M is called the marginal stability factor, determined by the maximum tolerable deviation (ν) of VSWR. This particular value of $|r_2| = |\Gamma|$ will ensure that the two-port is stable irrespective of the sign of D_2 [7]. In fact, this is equivalent to the approach to the loading of unstable amplifiers by Stern [8].

ii)

$$|r_2| \leq |\Gamma| \leq |P \pm (P^2 - Q)^{1/2}| \tag{27}$$

(this is derived in Appendix II), in which

$$P = \frac{\Delta + S_{11}S_{22} + \gamma S_{12}S_{21}}{2 \cdot S_{22} \cdot \Delta} \tag{28}$$

$$Q = \frac{S_{11}}{S_{22}\Delta} \tag{29}$$

and

$$\frac{1}{\gamma} = \delta_{r2} \quad (\text{reflection detuning factor}). \tag{30}$$

In this, the reflection detuning factor is defined as the ratio of the fractional change in the two-port input reflection coefficient to the fractional change in the output load reflection coefficient, and the value of $|r_2|$ given by (27) is dependent on the maximum tolerable value of this that the design sensitivity requirement may impose. From (1) and (11) it is evident that G_2 will be a maximum when $2 \operatorname{Re}(r_2 C_2)$ is a maximum, i.e., when the phase angle of r_2 is equal to the phase angle of C_2^*. Therefore, the optimum load reflection coefficient is $r_2 = |\Gamma| \arg C_2^*$ which is identical to

$$r_2 = |\Gamma| \arg \frac{C_2^*}{D_2} \tag{31}$$

which, from (11) shows that r_2 will have the same phase angle as r_{S2} (see Fig. 3). Note [7] that if D_2 is negative, the region of stable operation is inside the gain equals ∞ circle which also includes the origin, and therefore the conditions of (26), (27), and (31) still apply.

It will now be shown from consideration of the load quality factor Q that the optimum value of phase angle implicit in (31) may not always be obtainable.

The relationship between Q and r_2 is given by [11]

$$Q = \left| \frac{r_2(\omega_0) - r_2^*(\omega_0)}{1 - r_2(\omega_0)^2} \right| \tag{32}$$

giving

$$Q = \frac{1}{2} \frac{|\Gamma| \sin \phi_r}{1 - |\Gamma|^2}. \tag{33}$$

Hence

$$\phi_r = \sin^{-1}\left[2Q \frac{1}{|\Gamma|} - |\Gamma|\right]. \tag{34}$$

If $\Delta\omega_s$ and $\Delta\omega_p$ are defined, respectively, to be the desired stage bandwidth and the obtainable bandwidth, since $Q = (\omega_0)/(\Delta\omega_p)$, if $\Delta\omega_s \leq \Delta\omega_p$,

$$\phi_r = \arg\left(\frac{C_2^*}{D_2}\right). \tag{35}$$

But if $\Delta\omega_s > \Delta\omega_p$, the required phase angle cannot be obtained, the nearest approach to it being given from (34) by

$$\phi_r = \sin^{-1}\left[\frac{2\omega_0}{\Delta\omega_p}\left(\frac{1}{|\Gamma|} - |\Gamma|\right)\right] \tag{36}$$

in which, from (32),

$$\Delta\omega_p = \frac{\omega_0(1 - |\Gamma|^2)}{||\Gamma| \arg (C_2^*/D_2) - |\Gamma| \arg (C_2/D_2)|}. \tag{37}$$

B. Source Reflection Coefficient

The normalized transducer power gain g_T can be rewritten as

$$g_T = \frac{G_T}{|S_{21}|^2} = \frac{a(1 - |r_1|^2)}{|b + r_1 d|^2} \tag{38}$$

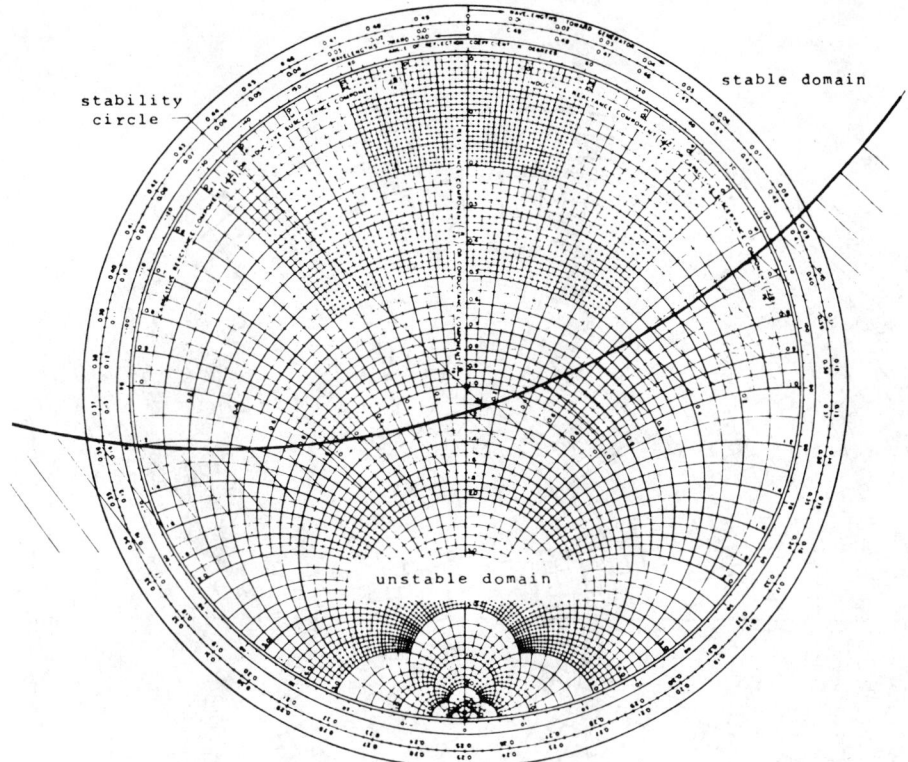

Fig. 4. r_1 plane.

in which

$$a = 1 - |r_2|^2 \quad (39)$$
$$b = 1 - r_2 S_{22} \quad (40)$$
$$d = r_2 \Delta - S_{11}. \quad (41)$$

If (38) was subject to restraints viz.

$$|r_1| \leq 1 \quad (42)$$

and

$$g_T \leq m \quad (43)$$

in which m is some value determined from consideration of the r_2 plane, then this value can be used to define the gain circle in (18) and the locus of permissible reflection coefficients found. Also, using the simultaneous condition imposed by the stage bandwidth $\Delta\omega_s$, there is obtained

$$\frac{\omega_0}{\Delta\omega_s} = \frac{|r_1(\omega_0) - r_1^*(\omega_0)|}{1 - |r_1(\omega_0)|^2} \quad (44)$$

and the particular circle

$$|r_1 - R_{g_T=m}| = \rho_{g_T=m}. \quad (45)$$

The solution of these last two equations fully defines and determines the magnitude and phase of r_1. In general, there will be two values, but realizability conditions allow only the value $|r_1| \leq 1$.

IV. Example

Defining a narrow-band amplifier as one having a bandwidth <10 percent of the midband frequency, and also one for which the linear approximations of the behavior of the physical components with respect to frequency hold, a typical example was designed and built [10]. A transistor, type BF 180, operated at $I_C = 1.0$ mA and $V_{CE} = 11$ V was designed to give the maximum gain over a $2\frac{1}{2}$-percent bandwidth at 800 MHz. The measured scattering parameters, with reference impedance of 50, were

$$S_{11} = 0.233 \angle 106.04° \quad S_{12} = 0.163 \angle 86.17°$$
$$S_{21} = 1.032 \angle -135.23° \quad S_{22} = 0.989 \angle -32.94°$$

from which it can be seen that the device is not potentially unstable.

Using (8) and (9) stability circles were drawn, as shown in Figs. 4 and 5, from which it can be seen that the device is conditionally stable at both input and output.

Taking the inherent minimum VSWR of the amplifier including connectors to be 1.1:1, and specifying the maximum VSWR to be 1.13:1, we obtained, from (26), the marginal stability factor to be zero, which in turn determined the input reflection coefficient r_1 to be at the origin of the r_1 plane.

From (27) r_2 was derived and shown in Fig. 5 as two curves for various values of r_2 between 0.01 and 1.0. From these r_2 was chosen to be 0.2, giving $r_2 = 0.824 \angle 37.53°$. From (3) the transducer power gain, G_T, was found to be 9.39 dB.

Using the methods of [11], the design was finalized as shown in Fig. 6 with the measured response shown in Fig. 7. It should be noted that the expression for Q, [11, eq. (1)], was derived in the Appendix to that paper by a method which was not generally true, and the following derivation is preferred.

Assume that the driving-point impedance at a given port is given by $R + jX$, which when normalized to Z_0 is $r + jx$. Then, $\Gamma = (r + jx - 1)/(r + jx + 1)$ and $\Gamma^* = (r - jx - 1)/(r - jx + 1)$. Hence by simple manipulation

$$\frac{|\Gamma - \Gamma^*|}{1 - |\Gamma|^2} = \frac{x}{r} = Q.$$

Of course, considered at the particular frequency at which $x = 0$, Q apparently equals zero, which is precisely why the alternative form $Q(\omega) = \omega/2\Delta\omega$ has been derived as an approximate formula for engineering design. The derivation, however, employs the formula $Q = x/r$.

Appendix I

Marginal stability factor based on the tolerable deviation of VSWR. If $|\Gamma'|$ is the radius of the circle, centered at the origin, which is tangential to the g equals ∞ circle, then the corresponding

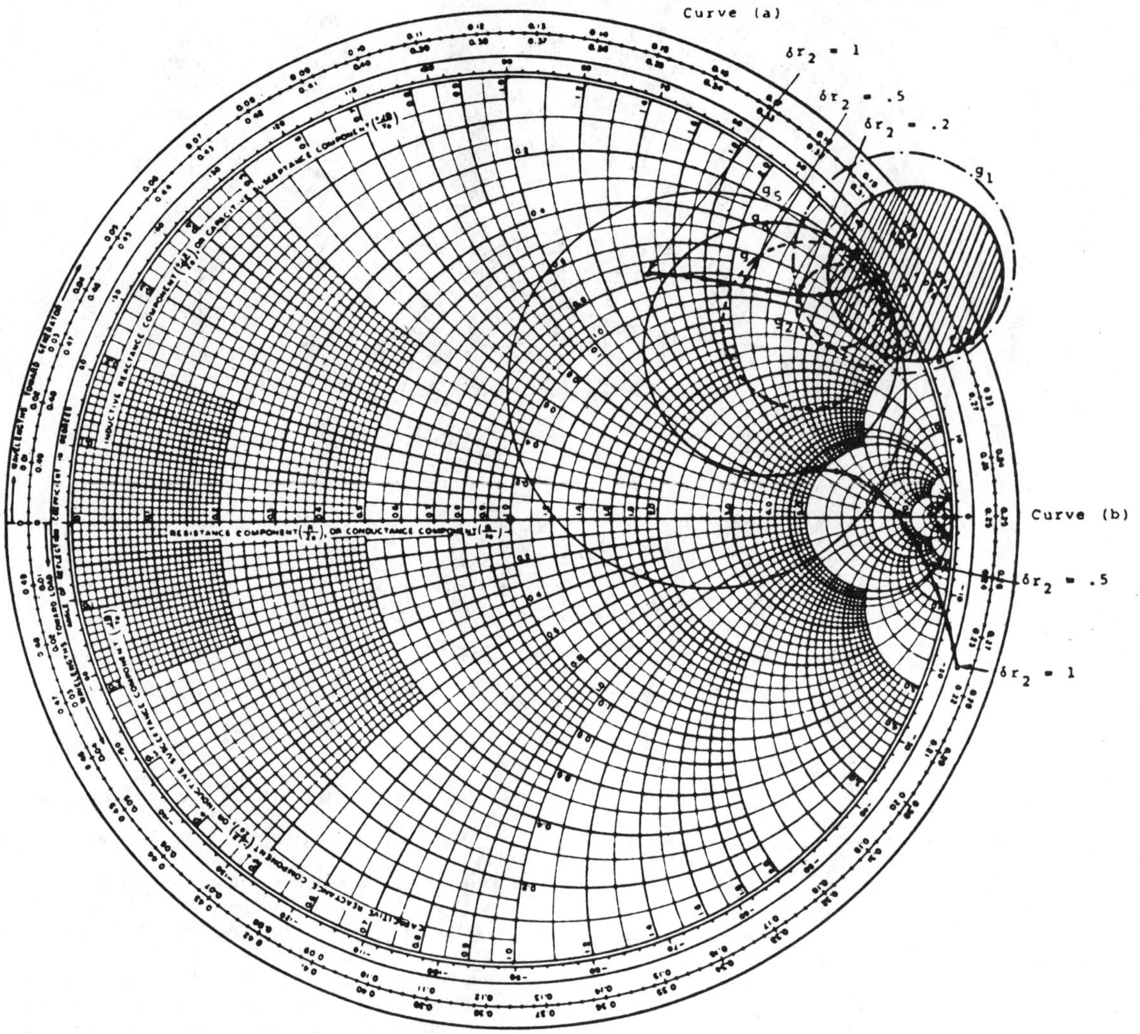

Fig. 5. Stability and gain circles in r_2 plane. Constant transducer power-gain circles q. $q_2 = 9.4$ dB; $q_3 = 6.4$ dB; $q_4 = 4.1$ dB; $q_5 = 1.1$ dB. All for $r_1 = 0$.

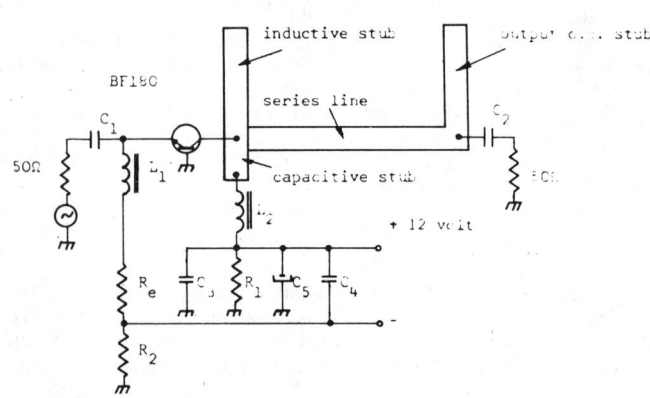

Fig. 6. Circuit diagram of stabilized 800-MHz narrow-band single-stage amplifier designed using two-port S-parameters. $v_{CC} = +12$ V; $R_1 = 18$ kΩ, $R_2 = 3.9$ kΩ, $R_e = 1$ kΩ; $L_1 = L_2 = 0.15$ μH (RF choke); $C_1 = 0.1$ μF disk ceramic; C_2 200 pF ceramic; $C_3 = C_4 = 300$ pF; $C_5 = 10$ μF/16 V (electrolytic); electrical length of capacitive branch equals 8.186 cm; electrical length of inductive branch equals 10.49 cm; series line electrical length equals 11.26 cm; and the open-circuit stub length equals 10.49 cm.

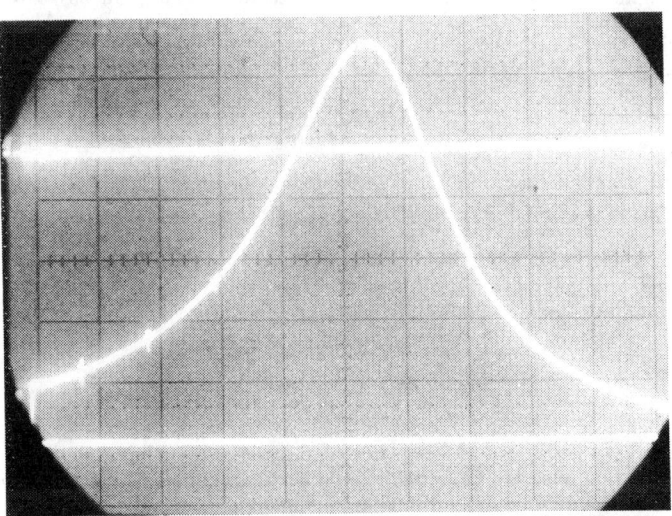

Fig. 7. Display of the single-stage 800-MHz narrow-band amplifier designed using two-port S-parameters. Vertical center line is at 800 MHz. Upper horizontal trace is the 3-dB marker. Frequency markers are 10 MHz apart.

VSWR is given by

$$\text{VSWR}\,|_{\rho=\infty} = \frac{1+|\Gamma'|}{1-|\Gamma'|} = \frac{1+||r_s|-\rho_s|}{1-||r_s|-\rho_s|} = S.$$

Let $b = ||r_s| - \rho_s|$, and if ν equals maximum tolerable deviation of the VSWR of the system.

$$\text{VSWR}\,|_{\text{stable}} = (1-2\nu)S = S'$$

giving

$$|\Gamma| = \frac{2\nu(1+b)-b}{2-2\nu(1+b)}.$$

By definition, $|\Gamma| = M\,||r_s| - \rho_s| = M \cdot b$

$$M = \frac{2\nu(1+1/b)-1}{2-2\nu(1+b)}$$

$$= \frac{\nu(1+(||r_s|-\rho_s|)^{-1})-0.5}{1-\nu(1+||r_s|-\rho_s|)}.$$

Appendix II

Reflection Detuning Factor

The detuning factor δ_r is defined as the fractional change in the two-port input reflection coefficient with respect to the fractional change in the output load reflection coefficient, i.e.,

$$\delta_{r2} = \left(\frac{\partial \Gamma_1}{\Gamma_1}\right) \bigg/ \left(\frac{\partial r_2}{r_2}\right).$$

But

$$\Gamma_1 = S_{11} - \frac{S_{12}S_{21}}{S_{22}-(1/r_2)} = \frac{S_{11}-r_2\Delta}{1-r_2 S_{22}}$$

$$\frac{\partial \Gamma_1}{\partial r_2} = \frac{S_{12}S_{21}}{[S_{22}-(1/r_2)]^2} \cdot \frac{1}{r_2^2}$$

$$\delta_{r2} = \frac{S_{12}S_{21}}{(1-r_2 S_{22})^2} \cdot \frac{r_2/(S_{11}-r_2\Delta)}{1-r_2 S_{22}}$$

$$\delta_{r2} = \frac{r_2 S_{12} S_{21}}{(1-r_2 S_{22})(S_{11}-r_2\Delta)}.$$

Hence for a specified value of δ_{r2}

$$S_{11} - r_2\Delta - r_2 S_{11} S_{22} + r_2^2 \Delta S_{22} - \frac{r_2}{\delta_{r2}} S_{12}S_{21} = 0$$

$$|r_2| = \left| \left(\frac{\Delta + S_{11}S_{22} + (S_{12}S_{21})/\delta_{r2}}{2 \cdot S_{22} \cdot \Delta}\right) \right.$$
$$\left. \pm \left| \left(\frac{\Delta + S_{11}S_{22} + (S_{12}S_{21})/\delta_{r2}}{2 \cdot S_{22} \cdot \Delta}\right)^2 - \left(\frac{S_{11}}{S_{22}\Delta}\right) \right|^{1/2} \right|.$$

This is the limiting value of the load reflection coefficient.

References

[1] D. Leed and O. Kummer, "A loss and phase set for measuring transistor parameters and two-port networks between 5 and 250 MHz," *Bell Syst. Tech. J.*, pp. 841–884, May 1961.
[2] D. Leed, "An insertion loss, phase and delay measuring set for characterizing transistors and two-port networks between 0.25 and 4.2 GHz," *Bell Syst. Tech. J.*, pp. 194–202, Mar. 1966.
[3] F. Weinert, "Scattering parameters speed design of high-frequency transistor circuits," *Electronics*, vol. 40, pp. 78–88, Sept. 5, 1966.
[4] W. H. Froehner, "Quick amplifier design with scattering parameters," *Electronics*, vol. 41, pp. 100–109, Oct. 16, 1967.
[5] M. F. Abulela, "A study of linear integrated circuits," M.S. thesis, Manchester Univ., Manchester, England, ch. 4, pp. 74–86, Oct. 1970.
[6] K. Kurokawa, "Power waves and the scattering matrix," *IEEE Trans. Microwave Theory Tech.*, vol. MTT-13, pp. 194–202, Mar. 1965.
[7] G. Bodway, "Two-port power flow analysis using generalized scattering parameters," *Microwave J.*, vol. 10, pp. 61–69, May 1967.
[8] A. P. Stern, "Stability and power gain of tuned transistor amplifiers," *Proc. IRE*, vol. 45, pp. 335–343, Mar. 1957.
[9] J. M. Rollett, "Stability and power gain invariants of linear two-ports," *IRE Trans. Circuit Theory*, vol. CT-9, pp. 29–32, Mar. 1962.
[10] M. F. Abulela, "Studies of some aspects of linear amplifier design in terms of measurable two-port and three-port scattering parameters," Ph.D. dissertation, Manchester Univ., Manchester, England, 1972.
[11] C. S. Gledhill and M. F. Abulela, "Notes on the conjugate matched two-port as a UHF amplifier," *IEEE Trans. Microwave Theory Tech.*, vol. MTT-20, pp. 289–292, Apr. 1972.
[12] J. G. Linvill and J. F. Gibbons, *Transistors and Active Circuits*. New York: McGraw-Hill, 1961.

Low-Noise Design of Microwave Transistor Amplifiers

R. S. TUCKER, STUDENT MEMBER, IEEE

Abstract—Parameters are derived for circles of constant overall noise figure on the source admittance plane of a preamplifier cascaded with a noisy main amplifier. It is shown that the noise figure and noise measure of an amplifier can be expressed in terms of the scattering parameters of a lossless two-port network connected at the input of the amplifier. Examples are given which demonstrate how this network can be synthesized to meet amplifier noise specifications.

I. INTRODUCTION

The theory of noise in linear two-port networks has been developed extensively in recent decades. It is known, for example, that the noise figure [1] and noise measure [2] of a two-port network can be minimized by appropriate selection of the source admittance and that loci of constant noise figure and constant noise measure are circles on the source admittance plane. Graphical constructions of this type are useful in the design of coupling networks for low-noise microwave transistor amplifiers [3] where the source admittance must be carefully controlled for optimum noise performance. The overall noise performance of a preamplifier cascaded with a noisy main amplifier has been investigated by Baechtold and Strutt [4] who have found the optimum source admittance for the preamplifier. Hartmann and Strutt [5] have presented general formulas for noise parameters of two-ports under a number of network transformations.

This short paper analyzes the amplifier considered by Baechtold and Strutt [4] and presents loci of constant overall noise figure on the source admittance plane. Furthermore, it is shown that a lossless two-port coupling network, connected at the input of an amplifier, can be designed to meet specifications on amplifier noise figure or noise measure by relating the noise figure of the amplifier to the transducer power gain of the network. Consequently, the graphical construction of noise figure or noise measure loci may be avoided.

II. THEORY

The amplifier considered here is shown in Fig. 1. A preamplifier is driven from the source admittance Y_s which is provided by the lossless two-port coupling network N. The preamplifier has noise figure F and available gain G_a and is cascaded with a main amplifier of noise figure F'.

Constant Noise Figure Loci

The overall noise figure of the amplifier in Fig. 1 is

$$F_t = F + \frac{E}{G_a} \quad (1)$$

where $E = F' - 1$ is the excess noise figure of the main amplifier. The noise figure of the preamplifier is given by [1]

$$F = F_{\min} + \frac{R_{ef}}{G_s}[(G_s - G_{of})^2 + (B_s - B_{of})^2] \quad (2)$$

Manuscript received October 21, 1974; revised February 10, 1975. This work was supported by the Australian Radio Research Board.
The author was with the Department of Electrical Engineering, University of Melbourne, Parkville, Vic., Australia. He is now with the Department of Electrical Engineering and Computer Sciences, University of California, Berkeley, Calif. 94720.

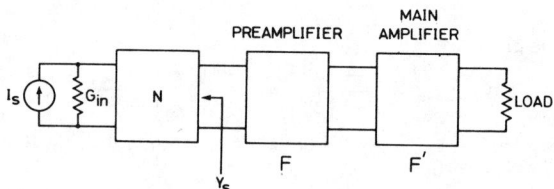

Fig. 1. The amplifier.

where $F_{\min}$ is the minimum noise figure of the preamplifier, R_{ef} is a parameter, and $Y_{of} = G_{of} + jB_{of}$ is the source admittance which produces $F_{\min}$. Similarly [2], the available gain of the preamplifier is given by

$$\frac{1}{G_a} = \frac{1}{G_{a_{\max}}} + \frac{R_{eg}}{G_s}[(G_s - G_{og})^2 + (B_s - B_{og})^2] \quad (3)$$

where $G_{a_{\max}}$ is the maximum available gain, R_{eg} is a parameter, and $Y_{og} = G_{og} + jB_{og}$ is the source admittance which produces $G_{a_{\max}}$. Loci of constant noise figure can be obtained by substituting (2) and (3) in (1) and rearranging. Hence

$$(G_s - G_{op})^2 + (B_s - B_{op})^2 = G_{RP}^2 \quad (4)$$

where

$$G_{op} = \frac{2G_{of}R_{ef} + 2G_{og}ER_{eg} + F_t - F_{\min} - \dfrac{E}{G_{a_{\max}}}}{2(R_{ef} + ER_{eg})} \quad (5)$$

$$B_{op} = \frac{B_{of}R_{ef} + B_{og}ER_{eg}}{R_{ef} + ER_{eg}} \quad (6)$$

and

$$G_{RP} = \frac{1}{2(R_{ef} + ER_{eg})}\left[\left(F_t - F_{\min} - \frac{E}{G_{a_{\max}}}\right)^2 \right.$$
$$+ 4(G_{of}R_{ef} + G_{og}ER_{eg})\left(F_t - F_{\min} - \frac{E}{G_{a_{\max}}}\right)$$
$$\left. - 4R_{ef}R_{eg}E\{(G_{of} - G_{og})^2 + (B_{of} - B_{og})^2\}\right]^{1/2}. \quad (7)$$

In practice, E can be assumed independent of Y_s if the preamplifier is unilateral or if the main amplifier is adjusted for a fixed E by appropriate choice of a lossless coupling network connected between the preamplifier and main amplifier. Under these assumptions, (4) describes circles of constant overall noise figure on the source admittance plane with centers at $Y_{op} = G_{op} + jB_{op}$ and radii G_{RP} as shown in Fig. 2. As an illustration, in Fig. 3 circles are plotted on the Smith chart [2] for an 11-dB overall amplifier noise figure. The preamplifier uses an AT-561 transistor [6] with a collector current of 3 mA, a collector to emitter voltage of 10 V, and a frequency of 6 GHz. The circles are drawn with F' as a parameter, varying from 11 to 13.3 dB. For F' greater than 13.3 dB, an overall noise figure of 11 dB cannot be obtained.

In addition to describing the noise performance of a preamplifier cascaded with a noisy main amplifier, the foregoing results can be used to find the noise measure of a single amplifier stage. If the main amplifier of Fig. 1 comprises a large number of stages, each identical to the preamplifier, then the overall noise figure of the amplifier is

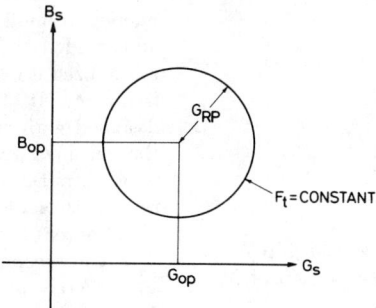

Fig. 2. Locus of constant overall noise figure on the source admittance plane.

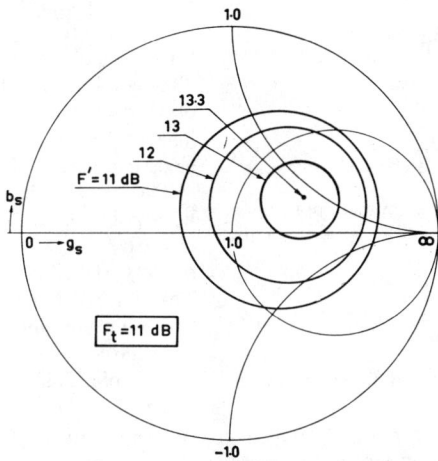

Fig. 3. Loci of constant overall noise figure on the Smith chart as a function of main amplifier noise figure. The preamplifier uses an AT-561 transistor and the chart is normalized to 20 mmho.

$$F_t = 1 + \frac{F - 1}{1 - 1/G_a}$$
$$= 1 + M \quad (8)$$

where M is the noise measure of each stage. For a large number of stages, the noise figure of the main amplifier is also equal to F_t. Thus

$$E = M. \quad (9)$$

Substituting (8) and (9) in (5) and (6), one obtains the centers of circles of constant noise measure as found by Fukui [2]. The radii of these circles[1] are obtained from (8) and (9) in (7).

Input Network Design

As was mentioned in the introduction, loci of constant noise figure and noise measure find application in the design of amplifier coupling networks. In the narrow-band case, a few noise figure loci on the source admittance plane are usually adequate to enable an appropriate design to be selected. With broad-band amplifiers, however, noise calculations are necessary at more than one frequency and a large number of circles may be required. The resulting design procedure can be quite cumbersome and tedious. An alternative approach is to directly synthesize the networks from noise figure specifications.

As the first step in a method of coupling network synthesis (4) is written in the form

$$F_t = F_{tm} + \frac{R_{ef} + ER_{eg}}{G_s}[(G_s - G_o)^2 + (B_s - B_o)^2] \quad (10)$$

where

[1] There are errors in Fukui's formula [2] for this parameter.

$$F_{tm} = F_{\min} + \frac{E}{G_{a_{\max}}} + 2(R_{ef} + ER_{eg})G_o - 2(G_{of}R_{ef} + G_{og}ER_{eg}) \quad (11)$$

$$G_o = \frac{1}{R_{ef} + ER_{eg}}[(R_{ef} + ER_{eg})(G_{of}^2 R_{ef} + G_{og}^2 ER_{eg})$$
$$+ R_{ef}ER_{eg}(B_{og} - B_{of})^2]^{1/2} \quad (12)$$

and

$$B_o = B_{op}. \quad (13)$$

The minimum overall noise figure of the amplifier is F_{tm} and is obtained when $Y_s = G_o + jB_o$, the optimum preamplifier source admittance. The minimum noise measure of a single-stage amplifier[1] $M_{\min}$ is obtained from (11) with $F_{tm} = 1 + M_{\min}$ and $E = M_{\min}$

$$M_{\min} = [M_2 + (M_2^2 - M_1 M_3)^{1/2}]/M_1 \quad (14)$$

where

$$M_2 = \left(1 - \frac{1}{G_{a_{\max}}} + 2R_{eg}G_{og}\right)(F_{\min} - 1 - 2R_{ef}G_{of})$$
$$+ 2R_{eg}R_{ef}(|Y_{og}|^2 + |Y_{of}|^2 - 2B_{og}B_{of}) \quad (15)$$

and M_1 and M_3 are as defined in [2].

The source admittance for minimum noise measure $Y_{om} = G_{om} + jB_{om}$ [2] can be obtained by substituting $E = M_{\min}$ in (12) and (13).

In Fig. 4, a lossless two-port input coupling network N with scattering matrix S is connected between the source conductance G_{in} and the admittance $Y_o^* = G_o - jB_o$ where the asterisk represents the complex conjugate and S is normalized to G_{in} at the input port and to Y_o^* at the output port. Generally Y_o^* is different from the input admittance of the preamplifier, but the connection of Fig. 4 enables the overall noise figure of the amplifier of Fig. 1 to

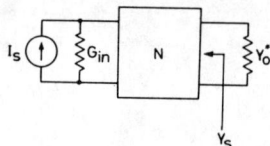

Fig. 4. Lossless coupling network.

be related to scattering parameters of N. This relationship is obtained by writing the inverse of the transducer power gain of N in the form

$$\frac{1}{|s_{21}(j\omega)|^2} = 1 + \frac{1}{4G_sG_o}[(G_s - G_o)^2 + (B_s - B_o)^2] \quad (16)$$

where

$$|s_{21}(j\omega)|^2 + |s_{11}(j\omega)|^2 = 1 \quad (17)$$

and by substituting (16) in (10)

$$|s_{21}(j\omega)|^2 = \frac{4G_o(R_{ef} + ER_{eg})}{(F_t - F_{tm}) + 4G_o(R_{ef} + ER_{eg})}. \quad (18)$$

The corresponding normalizing admittance at the output port of N is given by (12) and (13). For a single-stage amplifier $E = 0$ is substituted into (18) to obtain

$$|s_{21}(j\omega)|^2 = \frac{4\alpha}{F - F_{\min} + 4\alpha} \quad (19)$$

where $\alpha = G_o R_{ef}$ is the invariant noise parameter defined by Lange [7].

Substituting (8) and (9) in (18) and putting $F_{tm} = 1 + M_{\min}$, one obtains the transducer power gain as a function of the noise measure M

$$|s_{21}(j\omega)|^2 = \frac{4G_{opm}(R_{ef} + MR_{eg})}{(M - M_{\min}) + 4G_{opm}(R_{ef} + MR_{eg})}. \quad (20)$$

The corresponding normalizing admittance $Y_{opm}^* = G_{opm} - jB_{opm}$ is obtained by substituting (8) and (9) in (12) and (13). It should be noted that Y_{opm} is a function of M and is different from Y_{om} except when $M = M_{\min}$. In the following section, however, it will be seen that in a practical low-noise design problem where $M \simeq M_{\min}$ at all frequencies in the amplifier passband, Y_{opm} and Y_{om} are nearly identical.

III. DESIGN EXAMPLES

The two examples below use a GaAs Schottky-barrier-gate field-effect transistor which was described recently [3]. The design method is simpler and gives greater insight into the broad-band low-noise design problem than the alternative graphical technique which requires the construction of noise figure or noise measure loci.

First, consider an amplifier comprising identical cascaded stages with each stage designed on a near-minimum noise measure basis. It is assumed that the output of each transistor is coupled to the following transistor by a lossless two-port network and that transistors are the only source of amplifier noise. Using (20), a specification of $M = M_{\min}$ for each stage of the amplifier gives $|s_{21}|^2 = 1$ across the amplifier passband. This requirement of perfect match between the network N and the admittance Y_{opm}^* cannot be met, in general, due to the gain–bandwidth limitations imposed by Y_{opm}^* [9]. In order to achieve a realistic gain–bandwidth requirement of N, a 0.5-dB maximum increase from optimum overall amplifier noise figure is specified for the present problem.

If the number of amplifier stages is large, then from (8) the maximum allowable noise measure for each stage is

$$M = 1.122 M_{\min} + 0.122. \quad (21)$$

Using (21) and (17) in (20), the input coupling network reflection coefficient magnitude $|s_{11}|$ has been computed as a function of frequency from 8 to 12 GHz. The corresponding locus of Y_{opm}^* is almost identical with Y_{om}^* which is given in [3]. In Fig. 5, $|s_{11}|$, normalized to G_{in}, and the equivalent VSWR is plotted against frequency. If the input network for each stage of the amplifier is designed with reflection coefficient less than or equal to $|s_{11}|$, then the specification on noise performance will be met. A convenient network in this case would be of the Chebyshev impedance-matching type with a maximum VSWR of 1.6 and designed to match into Y_{opm}^* across the amplifier passband [3].

In the second example, a single-stage amplifier is designed with an objective of amplifier noise figure equal to the minimum noise figure of the transistor at 12 GHz. The amplifier noise figure and the amplifier gain are both to be constant across the passband of 8–12 GHz. With $F = 4.5$ dB in (19), $|s_{21}|^2$ has been computed and is plotted against frequency in Fig. 6. The normalizing admittance for the input coupling network is Y_{of}^* which can be obtained from the plot of Y_{of} in [3]. To allow for gate bonding wire inductance, the series reactance of a 0.15-nH inductance is added to $Z_{of}^* = 1/Y_{of}^*$. This combination is modeled with an equivalent circuit consisting of the series connection of a 20-Ω resistor and an open-circuited length of transmission line of characteristic impedance 6.21 Ω and electrical length of 0.063 λ at 10 GHz.

The transducer power gain of the input network is approximated by a polynomial expression [8] which is prescribed to have a slope of 3 dB/octave over the range 6–12 GHz, rising to a value of 0 dB at 12 GHz. The approximated gain function is plotted in Fig. 6. The input network has been synthesized in distributed form using this function and, subject to gain-bandwidth limitations [9], is terminated with the equivalent circuit described in the previous paragraph. This equivalent circuit is omitted in the final amplifier realization. The complete amplifier is shown in Fig. 7. In Fig. 8 the transistor minimum noise figure $F_{\min}$ and the computed amplifier

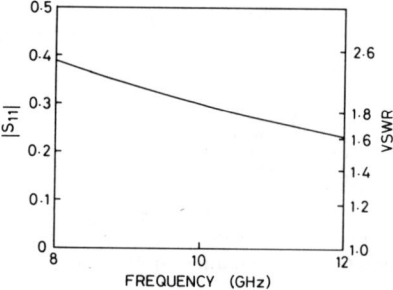

Fig. 5. Input coupling network reflection coefficient and VSWR for near-minimum noise measure design.

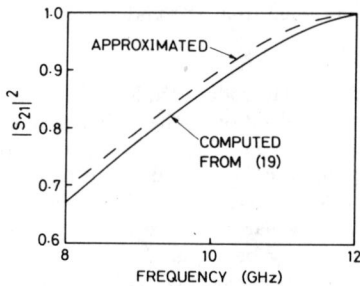

Fig. 6. Input coupling network transducer power gain for single-stage amplifier.

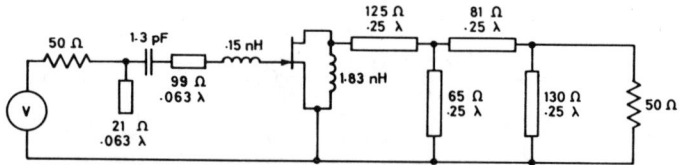

Fig. 7. Circuit of single-stage amplifier. Transmission lines are specified by characteristic impedance and electrical length at 10 GHz.

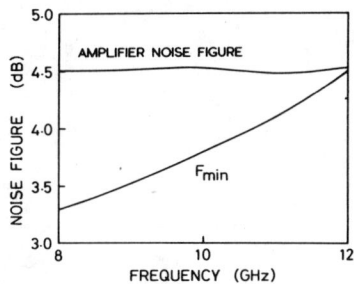

Fig. 8. Minimum noise figure of the transistor and noise figure of the single-stage amplifier.

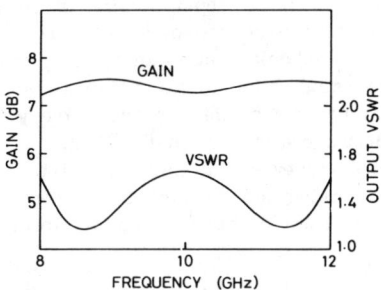

Fig. 9. Gain and output VSWR of single-stage amplifier.

noise figure are plotted against frequency. The amplifier noise figure is close to 4.5 dB across the passband as expected. It is worth noting that the approximated function of Fig. 6 represents a good impedance match between the input coupling network and $Y_{of}{}^*$ at 12 GHz, but at other frequencies the function falls off in magnitude and the match is degraded. The gain–bandwidth limitations associated with the function are therefore considerably less severe than for a Chebyshev function which approximates a good match across the entire passband. This property of the approximated function permits a lower amplifier noise figure at 12 GHz than is generally available with a Chebyshev coupling network, but the improvement at 12 GHz is achieved at a cost of increased noise figure at lower frequencies.

As well as producing the prescribed noise figure characteristic, the input network affects the available gain of the transistor. For this example, the available gain of the transistor is almost constant across the amplifier passband. The lossless output matching network in Fig. 7 has been designed by a direct synthesis technique [8] using a Chebyshev gain characteristic with 0.25 dB of ripple. The computed amplifier gain and output VSWR is shown in Fig. 9.

IV. CONCLUSION

The loci of constant overall noise figure derived in this short paper can be used to graphically investigate the noise performance of an amplifier for various preamplifier source admittances and main amplifier noise figures. This has been illustrated with an example. An input coupling network synthesis method has been described for noise figure and noise measure specifications. The method enables a wide variety of broad-band design problems to be handled since it reduces the low-noise design problem to an impedance-matching problem which can be solved by well-known analytical techniques. Advantages of the method are related to its simplicity compared with existing low-noise design methods and to the information it can provide regarding noise figure limitations for a given transistor and bandwidth.

REFERENCES

[1] "IRE standards on methods of measuring noise in linear two ports, 1959," *Proc. IRE*, vol. 48, pp. 60–68, Jan. 1960.
[2] H. Fukui, "Available power gain, noise figure, and noise measure of two-ports and their graphical representations," *IEEE Trans. Circuit Theory*, vol. CT-13, pp. 137–142, June 1966.
[3] C. A. Liechti and R. L. Tillman, "Design and performance of microwave amplifiers with GaAs Schottky-gate field-effect transistors," *IEEE Trans. Microwave Theory Tech.*, vol. MTT-22, pp. 510–517, May 1974.
[4] W. Baechtold and M. J. O. Strutt, "Noise in microwave transistors," *IEEE Trans. Microwave Theory Tech. (Special Issue on Noise)*, vol. MTT-16, pp. 578–585, Sept. 1968.
[5] K. Hartmann and M. J. O. Strutt, "Changes of the four noise parameters due to general changes of linear two-port circuits," *IEEE Trans. Electron Devices*, vol. ED-20, pp. 874–877, Oct. 1973.
[6] K. Hartmann and M. J. O. Strutt, "Scattering and noise parameters of four recent microwave bipolar transistors up to 12 GHz," *Proc. IEEE* (Lett.), vol. 61, pp. 133–135, Jan. 1973.
[7] J. Lange, "Noise characterization of linear two ports in terms of invariant parameters," *IEEE J. Solid-State Circuits*, vol. SC-2 pp. 37–40, June 1967.
[8] R. S. Tucker, "Synthesis of broadband microwave transistor amplifiers," *Electron. Lett.*, vol. 7, pp. 455–456, Aug. 12, 1971.
[9] D. C. Youla, "A new theory of broad-band matching," *IEEE Trans. Circuit Theory*, vol. CT-11, pp. 30–50, Mar. 1964.

FEEDBACK EFFECTS ON THE NOISE PERFORMANCE OF GaAs MESFETS

George D. Vendelin
Varian Associates
611 Hansen Way
Palo Alto, California 94303

Abstract

A technique is presented for evaluating the effect of lossless feedback on the four noise parameters of a transistor. The feedback effects on noise parameters are presented for a FMT940 GaAs MESFET at 4 and 8 GHz.

Introduction

The gain and stability factor of GaAs MESFETs are often varied by feedback effects which are introduced via the package or by the circuit design requirements [1]. Common-lead inductance (L_s) and gate-drain feedback capacitance (C_f) are usually the effects of packaging the device. In addition to the gain and stability factor, the noise parameters of the transistor are also changed. A technique for predicting the resulting changes in the four noise parameters will be presented for the two general cases of lossless feedback, series feedback X_s and parallel feedback X_f (see Fig. 1). From applying this technique to the measured noise parameters of a Fairchild FMT940 GaAs MESFET, the conditions required for varying the noise performance are presented. The result of this analysis will indicate the reasons for the higher noise figure and higher available gain which is measured for the FMT940 (stripline package) compared to the FMT980 (coaxial package) transistor.

Noise Parameters

The noise parameters for any two-port are given by either of the following relations, where the first set of parameters are usually given by the device data sheet.

$$F = F_{min} + \frac{R_n}{G_s} |Y_s - Y_{on}|^2 \quad (1)$$

$$F = 1 + \frac{G_n}{G_s} + \frac{R_{no}}{G_s} |Y_s + Y_{cor}|^2 \quad (2)$$

The equivalence between these two parameters sets are given by:

$$\left.\begin{array}{l} F_{min} = 1 + 2R_{no}(G_{cor} + G_{on}) \\ R_n = R_{no} \\ G_{on} = \left[\frac{G_n + R_{no}G_{cor}^2}{R_{no}}\right]^{1/2} \\ B_{on} = -B_{cor} \end{array}\right\} \quad (3)$$

$$\left.\begin{array}{l} G_n = R_n(G_{on}^2 - G_{cor}^2) \\ R_{no} = R_n \\ G_{cor} = \frac{F_{min}-1}{2R_n} - G_{on} \\ B_{cor} = -B_{on} \end{array}\right\} \quad (4)$$

Hartmann and Strutt [2] have given the formulas for series and parallel feedback effects in terms of the second noise parameter set. A computer program has been written which uses the input noise parameters of Eq. (3), converts to the second parameter set Eq. (4), adds the lossless feedback transformations given by [2], and converts the result back to the noise parameters of Eq. (3). The results of this analysis are given below for the FMT940 transistor at 4 and 8 GHz. A similar analysis has been presented by Engberg [3] for a UHF bipolar transistor using the second parameter set Eq. (4) and the noise equivalent circuits given by Rothe and Dahlke [4].

Computed Noise Effects for FMT940

The following noise parameters and s-parameters are typical for the FMT940 at 4 and 8 GHz.

Noise Parameters and S-Parameters
for FMT 940 at V_{DS} = 5 volts, I_{DS} = 10 mA

f GHz	F_{min} dB	R_n ohm	G_{on} mmho	B_{on} mmho
4	3.3	100	5	-13
8	5.0	35	34	-18

f GHz	S_{11}	S_{21}	S_{12}	S_{22}
4	.86 ∠ -76	.82 ∠ 98	.031 ∠ 42	.87 ∠ -50
8	.71 ∠ -143	.84 ∠ 25	.035 ∠ 23	.86 ∠ -101

The effect of adding lossless feedback has been plotted in Figures 2-6. For series feedback, the noise figure will be raised by a small inductance and lowered by capacitance or by reducing the common lead inductance. For parallel feedback, the noise figure can be reduced by increasing the input-output capacitance. Naturally the gain is correspondingly reduced and the minimum noise measure remains invariant to lossless feedback effects, where M_{min} is given by

$$M_{min} = \frac{F_{min} - 1}{1 - 1/g_{av}} \quad (5)$$

and the resulting noise figure of an infinite chain of amplifiers is given by

$$F_{tot} = M_{min} + 1 \quad (6)$$

The variations of Z_{on} given in Figures 5 and 6 are particularly significant for low-noise amplifier designs. By proper adjustment of the feedback elements, the condition for simultaneous gain and noise match can be found. This allows the amplifier designer to simultaneously achieve minimum input VSWR and optimum noise performance. Particular attention to stability is required for a useful amplifier design.

Computed Noise Effects for FMT 980 at 8 GHz

The reduced noise figure of the FMT980 (coaxial package) at 8 GHz is partially a result of the lower common-lead inductance of this package. From measuring short circuited packages, the common lead inductance is estimated at 0.16 nH and 0.08 nH for the stripline and coaxial package respectively. From Figure 2, this would correspond to a change in noise figure of about 0.2 dB, which also results in a change in available gain of about 1 dB. When the calculation for optimum noise performance F_{tot} is performed with the above changes in F_{min} and g_{av}, both transistors have a predicted noise figure of $F_{tot} = 6.0$ dB. Since the data sheet noise figure is 4.0 dB for the FMT980, the additional discrepancy is due to the package and fixture losses. When these losses are subtracted, the corrected value for F_{tot} is about 5.1 dB for both packages.

Conclusions

A technique has been described for analyzing the feedback effects on the noise performance of low-noise bipolar and field-effect transistor amplifiers. Computer programs which simultaneously predict gain and noise performance will enable low-noise feedback designs to follow. The technique may also be applied to the design of low-noise amplifiers with a minimum input VSWR.

References

(1) Les Besser, "Design Considerations of a 3.1-3.5 GHz GaAs FET Feedback Amplifier", Digest of Technical Papers, 1972 IEEE-GMTT International Microwave Symposium, pp. 230-232.

(2) K. Hartmann and M.J.O. Strutt, "Changes of the Four Noise Parameters Due to General Changes of Linear Two-Port Circuits", IEEE Trans. on Electron Devices, October 1973, pp. 874-877.

(3) Jakob Engberg, "Simultaneous Input Power Match and Noise Optimization Using Feedback", Digest of Technical Papers, European Microwave Conference, September 1974, pp. 385-389.

(4) H. Rothe and W. Dahlke, "Theory of Noisy Fourpoles", Proc. IRE, Vol. 44, June 1956, pp. 811-818.

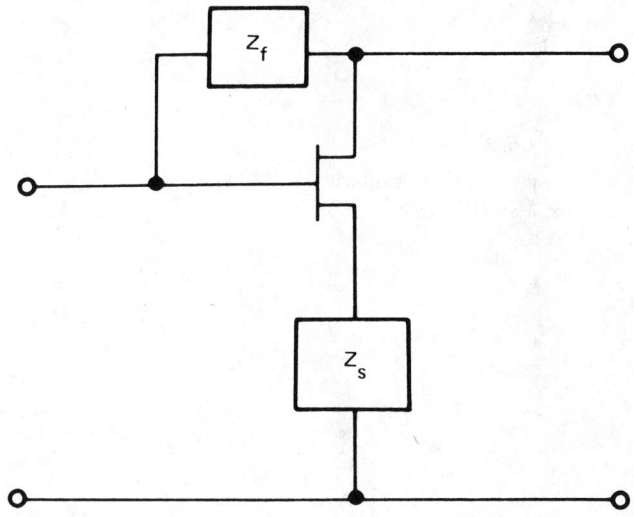

Figure 1. Parallel and series feedback for FET.

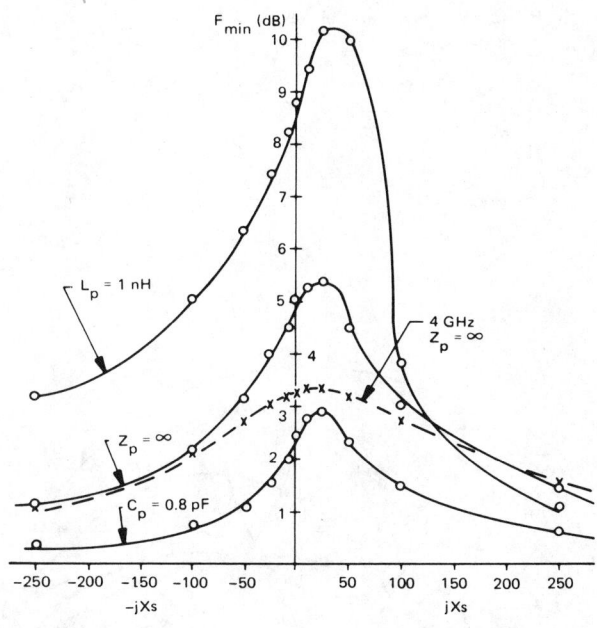

Figure 2. F_{min} vs series feedback for FMT 940 at 4 and 8 GHz.

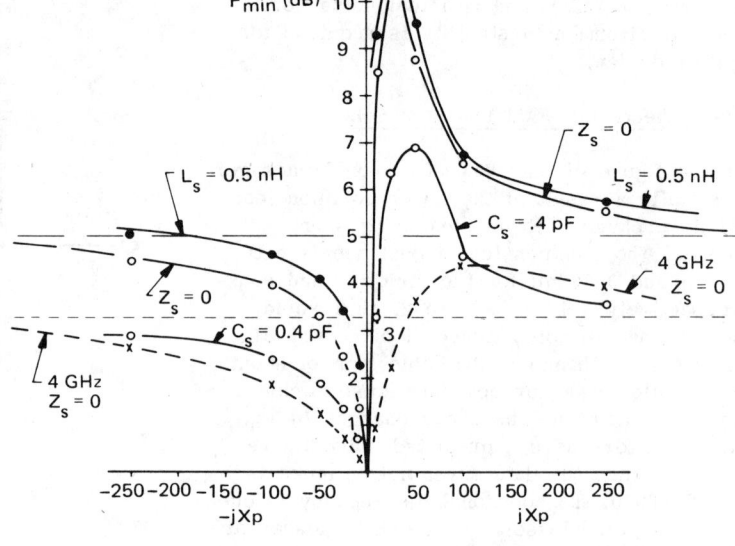

Figure 3.

F_{min} vs parallel feedback for FMT 940 at 4 and 8 GHz.

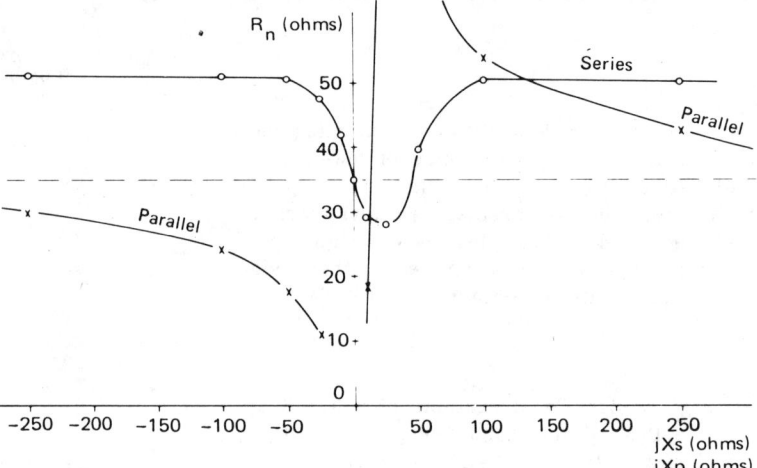

Figure 4.

R_n vs series and parallel feedback for FMT 940 at 8 GHz.

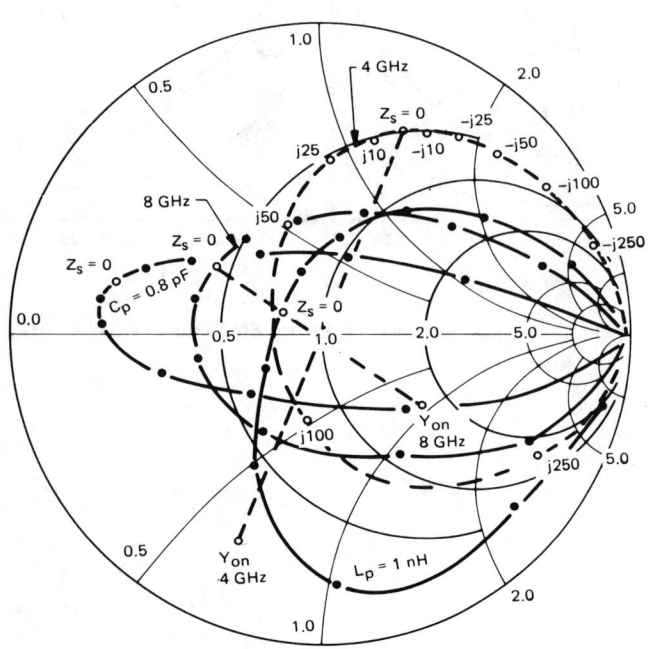

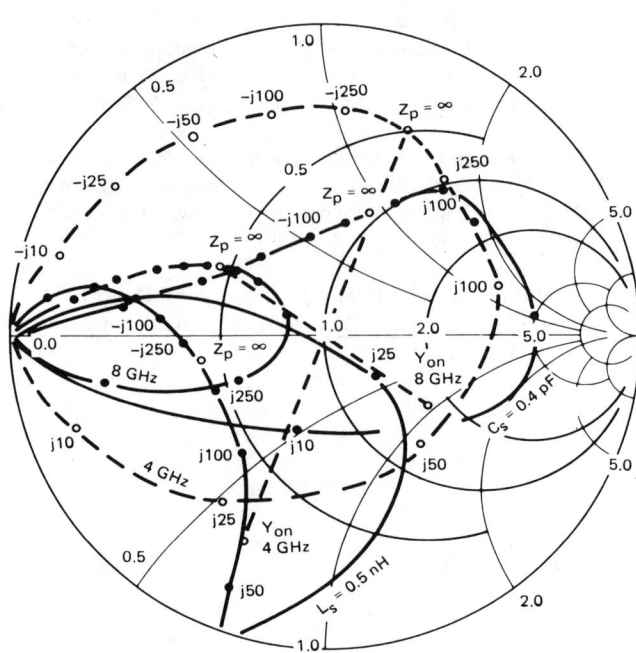

Figure 5. Optimum source impedance vs series feedback for FMT 940 at 4 and 8 GHz.

Figure 6. Optimum source impedance vs parallel feedback for FMT 940 at 4 and 8 GHz.

AN INTRODUCTION TO SIMULATION AND OPTIMIZATION

J. W. BANDLER
Group on Simulation, Optimization and Control
and Department of Electrical Engineering
McMaster University, Hamilton, Canada

Invited Paper

Abstract

A review of recent work in simulation and optimization is made with the aim of introducing the designer to the benefits of automating optimal design procedures and to indicate limitations imposed by the current state of the art.

Introduction

This paper is directed to the engineer interested in using computer aids to modeling and design, and considering the application of optimization techniques. Limitations on the size and scope of problems which can be approached from the optimization point of view as imposed by the current state of the art are also indicated.

It seems that the gap between theoretical developments and their practical implementation is in danger of widening. With the plethora of literature in optimization methods and computer-aided circuit design, particularly articles laying claim to superiority of technique, a confused impression is created.

With these thoughts in mind, the author will attempt to direct the microwave reader to work which appears relevant, useful, instructive or stimulating within the domain of activity of the respective authors.

Review

It is felt that Calahan's book on computer-aided network design[1] is a good indicator of current trends and possible future developments in computer based circuit and system design techniques and philosophies. The collection of articles[2] considered by Director to be benchmarks in simulation and optimization is also recommended, again not so much for the details as for its point of view. Szentirmai's reprint volume[3] deals with various aspects of filter design, and appears representative of numerical advances in that area.

Complementary survey articles on optimization techniques are those by Bandler[4] and Charalambous[5]. Also appearing in the IEEE Transactions on Microwave Theory and Techniques and somewhat complementary in the areas of simulation and sensitivity analysis are papers by Bandler and Seviora[6] and Monaco and Tiberio[7]. A pragmatic article of particular interest to microwave designers is one by Perlman and Gelnovatch[8].

Analysis

Effort is being directed at solving larger systems more efficiently. See, for example, Wexler et al[9] and others[10,11]. As far as engineering design is concerned, it is important to stress that it is generally inefficient to put a conventional simulation program into an optimization loop without taking certain things into account. Assuming, for example, that the program exploits sparsity in the computations the question of efficient computation of the effects of parameter changes (indispensable to design) arises. In general, for economical and physical reasons, not all possible design variables or degrees of freedom are always utilized. Setting up the necessary equations and recomputing the entire response every time a relatively small number of parameters is changed will result in a much larger computing bill than is necessary. In considering the value of an analysis routine for design purposes, then, the manner in which the effects of component variation are handled is crucial.

Sensitivity

A much debated topic in the circuit literature, particularly in time-domain analysis, efficient sensitivity evaluation is a cornerstone to automatic design[2]. Branin[12] has dispelled some of the mystique shrouding the adjoint network method[2,6] by a compact, abstract presentation. The adjoint network approach whereby, for example, the first-order sensitivities of the output of a circuit may be efficiently evaluated with respect to all designable components using the results of only two circuit analyses has, however, been a powerful motivating force.

Extensions and applications of the adjoint network concept abound in the literature[7,13]. In the frequency domain for linear circuits, at least, it appears that, by suitable mathematical manipulations, higher-order sensitivities[14], large-change sensitivities[15], sensitivities with respect to frequency[16] etc., are available relatively efficiently by suitable programming. The most widely acclaimed optimization methods[5], however, require only first derivatives. Furthermore, the value of second- and higher-order sensitivities at points possibly far from the optimum has not been established.

Formulation

There appear to be two principal approaches to the formulation of design objectives. On the one hand, some designers attempt to approximate ideal performance specifications which, by definition, are unattainable. This approach requires the least preparation of the problem, but the results tend to be somewhat ambiguous in the context of meeting specifications and subsequent assignment of tolerances. On the other hand, more insight can be brought to bear if design problems are cast in the form of meeting or exceeding realistic performance specifications[4]. One can go a step further, exploiting more fully one's prior knowledge or insight into the problem at hand, by devising artificial specifications[17] in an attempt to anticipate more closely the actual optimum performance realizable by the configuration and thereby permit its more rapid evaluation. Optimal assignment of manufacturing tolerances appears to be more well-defined in the context of realistic specifications.

Objectives

The ubiquitous least squares objective[2,18], usually employed in conjunction with error-prone data or ideal

---

This work was supported by the National Research Council of Canada under Grant A7239.

Reprinted from *IEEE MTT-S Int. Microwave Symp.*, 1976, pp. 204-206.

specifications in the context, for example, of modeling or design, respectively, is probably the simplest to implement. Particularly in filter design, however, non-Euclidean measures of error have been widely applied historically. See, for example, Szentirmai[3]. Numerical approximation methods for minimax (Chebyshev) or near minimax solutions, contrary to prevailing assumptions, can, for all practical purposes be realized almost as easily as least squares solutions[19]. On paper, at least, they produce more impressive-looking responses. One reason is that one or more trial runs are usually required in practice to verify a solution. Once a run, for example, using a least squares objective has been performed, sufficient information about the properties of the problem is often available to allow one to subsequently force at least a near minimax solution with relatively little additional effort[19].

Algorithms

It is known that a well-conditioned problem in terms of selection of a well-behaved objective function and nonredundant variables which have been properly scaled allows the conventional steepest descent method to perform adequately. The Newton method, which may be viewed as steepest descent with respect to a different norm[20] is generally less sensitive to scaling but, unlike steepest descent, is affected by the properties of the second derivatives and convexity.

Modern gradient methods[21,22] attempt to overcome the limitations of the basic steepest descent and Newton methods, as do analogous methods in the minimax optimization of a set of functions[19,23,24]. In minimax problems, in particular, classical assumptions about the number or character of the equal (or active) extrema vis a vis the number of independent variables need not and, in general, do not hold.

Current efforts in optimization[20] are directed at developing robust algorithms, however, anticipation and alleviation of ill-conditioning, where possible, is desirable.

Centering

Centering a design usually implies the process of finding a nominal design somehow influenced by manufacturing tolerances and, possibly, post production tuning[25,26]. The procedure may involve optimal assignment of component tolerances, maximization of production yield, design subject to a specified yield, etc. The problem could be a worst-case one with design variables assumed independent; it might involve correlated elements, statistical distributions, and so on. A numbe of relevant works will provide the interested reader with further details[27]. It should be emphasized that, in general, all design parameters: nominal values, tolerances, tuning ranges and so on will interact in defining an optimal design[25,26]. A solution obtained from a least squares or minimax approximation in the usual sense does not necessarily provide the best nominal values. The centering problem is generally significantly more expensive to solve, requiring careful preparation.

Software

An excellent survey of both available and proprietary general purpose software for circuit designers has been made by Kaplan[18]. The article, however, appears limited to developments in the U.S. and probably places undue emphasis on least squares objectives. A number of optimization programs with documentation is available from the present author[28]. Two collections[29,30] of reprints, reports, notes and programs should also be mentioned. Documented listings of very useful optimization programs are also available from the U.K. Atomic Energy Research Establishment[31], and the Numerical Optimization Centre[32]. See also pp.242-243 of Gill and Murray[20].

Should one use a commercially available analysis and design package, for example, through a time-sharing facility? It is felt that current optimization features in these packages are generally weak, so that their use will probably be expensive in the long run.

New algorithms or packages should be tested on suitable examples and compared with respect to features, flexibility, ease of use, convergence to known solutions, memory required and running times. This is particularly appropriate in optimal design where, over an extended period of use, enormous numbers of simulations might be required.

Techniques which appear different may sometimes be alternative implementations of the same basic algorithm[5]. This is, understandably, often not realized at the time by the proponents of the techniques. As the state of the art advances, unification takes place and the techniques can be put into better perspective. See also Branin[12] and Bonfatti et al[11].

Conclusions

Having assimilated the essential past achievements (regrettably inadequately referenced because of limited space) where might one find indicators of possible new developments? Three additional recent works may be singled out: an advance in minimax algorithms where derivatives are not required[33], an advance in efficient design in the time domain employing sensitivity information[34], and an advance in centering which takes account of many uncertainties relevant to the microwave area[35]. A number of sessions at this year's IEEE International Symposium on Circuits and Systems (Munich, Germany, Apr. 1976) promise further achievements in simulation and optimization in all areas covered by this paper[36].

References

[1] D.A. Calahan, Computer-Aided Network Design (Revised Edition). New York: McGraw-Hill, 1972.

[2] S.W. Director, Ed., Computer-Aided Circuit Design: Simulation and Optimization. Stroudsburg, Penn.: Dowden, Hutchinson and Ross, 1973.

[3] G. Szentirmai, Ed., Computer-Aided Filter Design. New York: IEEE Press, 1973.

[4] J.W. Bandler, "Optimization methods for computer-aided design," IEEE Trans. Microwave Theory Tech., vol.MTT-17, Aug. 1969, pp.533-552.

[5] C. Charalambous, "A unified review of optimization," IEEE Trans. Microwave Theory Tech., vol. MTT-22, Mar. 1974, pp. 289-300.

[6] J.W. Bandler and R.E. Seviora, "Current trends in network optimization," IEEE Trans Microwave Theory Tech., vol. MTT-18, Dec. 1970, pp. 1159-1170.

[7] V.A. Monaco and P. Tiberio,"Computer-aided analysis of microwave circuits," IEEE Trans. Microwave Theory Tech., vol.MTT-22, Mar. 1974, pp. 249-263.

[8] B.S. Perlman and V.G. Gelnovatch, "Computer aided design, simulation and optimization," in Advances in

Microwaves vol.8, L. Young and H. Sobol, Eds. New York: Academic Press, 1974.

[9] A. Wexler et al, "Solution of large, sparse systems in design and analysis", IEEE Int. Microwave Symp. Digest (Palo Alto, Calif., May 1975).

[10] G.D. Hachtel, R.K. Brayton and F.G. Gustavson, "The sparse tableau approach to network analysis and design", IEEE Trans. Circuit Theory, vol. CT-18, Jan. 1971, pp.101-113.

[11] F. Bonfatti, V.A. Monaco and P. Tiberio, "Microwave circuit analysis by sparse-matrix techniques", IEEE Trans. Microwave Theory Tech., vol.MTT-22, Mar.1974, pp.264-269.

[12] F.H. Branin, Jr., "Network sensitivity and noise analysis simplified", IEEE Trans. Circuit Theory, vol. CT-20, May 1973, pp.285-288.

[13] G.C. Temes, R.M. Ebers and R.N. Gadenz, "Some applications of the adjoint network concept in frequency domain analysis and optimization", Computer Aided Design, vol. 4, Apr. 1972, pp. 129-134.

[14] A.K. Seth and P.H. Roe, "Hybrid formulation of explicit formulae for higher order network sensitivities", IEEE Trans. Circuits Syst., vol. CAS-22, May 1975, pp.475-478.

[15] R.N. Gadenz, M.G. Rezai-Fakhr and G.C. Temes, "A method for the computation of large tolerance effects", IEEE Trans. Circuit Theory, vol. CT-20, Nov.1973, pp.704-708.

[16] J.W. Bandler, M.R.M. Rizk and H. Tromp, "Efficient calculation of exact group delay sensitivities", IEEE Trans. Microwave Theory Tech., vol. MTT-24, April 1976.

[17] C. Charalambous and J.W. Bandler, "New algorithms for network optimization", IEEE Trans. Microwave Theory Tech., vol. MTT-21, Dec. 1973, pp.815-818.

[18] G. Kaplan, "Computer-aided design", IEEE Spectrum, vol.12, Oct.1975, pp.40-47.

[19] J.W. Bandler, C. Charalambous, J.H.K. Chen and W.Y. Chu, "New results in the least pth approach to minimax design", IEEE Trans. Microwave Theory Tech., vol. MTT-24, Feb.1976.

[20] P.E. Gill and W. Murray, Eds., Numerical Methods for Constrained Optimization. New York: Academic Press, 1974.

[21] R. Fletcher, "Fortran subroutines for minimization by quasi-Newton methods", Atomic Energy Research Establishment, Harwell, England, Report AERE-R7125,1972.

[22] P.E. Gill and W. Murray, "Quasi-Newton methods for unconstrained optimization", J.Inst.Math.Applics., vol.9, 1972, pp.91-108.

[23] K. Madsen, H. Schjaer-Jacobsen and J. Voldby, "Automated minimax design of networks", IEEE Trans. Circuits Syst., vol. CAS-22, Oct.1975, pp.791-796.

[24] C. Charalambous and A.R. Conn, "Optimization of microwave networks", IEEE Trans. Microwave Theory Tech., vol. MTT-23, Oct. 1975, pp.834-838.

[25] J.F. Pinel, K.A. Roberts and K. Singhal, "Tolerance assignment in network design", Proc. IEEE Int. Symp. Circuits Syst. (Newton, Mass., Apr. 1975),pp.317-320.

[26] J.W. Bandler, P.C. Liu and H. Tromp, "A nonlinear programming approach to optimal design centering, tolerancing and tuning", IEEE Trans. Circuits Syst., vol. CAS-23, Mar. 1976.

[27] "Tolerance assignment in network design", Special Session, Proc. IEEE Int. Symp. Circuits Syst. (Newton, Mass., Apr. 1975), pp.317-336.

[28] Group on Simulation, Optimization and Control, Faculty of Engineering, McMaster University, Hamilton, Canada L8S 4L7.

[29] W. Kinsner, Ed., "Notes on numerical methods in engineering analysis and design of fields, circuits and systems", Faculty of Engineering, McMaster Univ., Hamilton, Canada, Report SOC-93, June 1975.

[30] J.W. Bandler, Ed., "Notes on numerical methods of optimization with applications in optimal design", Faculty of Engineering, McMaster Univ., Hamilton, Canada, Report SOC-113, Nov.1975.

[31] Computer Sciences and Systems Division, Atomic Energy Research Establishment, Harwell, England OX11 ORA.

[32] Numerical Optimization Centre, Hatfield Polytechnic, Hatfield, Hertfordshire, England.

[33] K. Madsen, O. Nielsen, H. Schjaer-Jacobsen and L. Thrane, "Efficient minimax design of networks without using derivatives", IEEE Trans. Microwave Theory Tech., vol. MTT-23, Oct. 1975, pp.803-809.

[34] R.K. Brayton and S.W. Director, "Computation of delay time sensitivities for use in time domain optimization", IEEE Trans. Circuits Syst., vol. CAS-22, Dec. 1975, pp. 910-920.

[35] J.W. Bandler, P.C. Liu and H. Tromp, "Integrated approach to microwave design", IEEE Trans. Microwave Theory Tech., to appear.

[36] Regular and Special Sessions, IEEE International Symposium on Circuits and Systems, Munich, Germany, Apr. 1976.

Optimum Gain-Bandwidth Limitations of Transistor Amplifiers as Reactively Constrained Active Two-Port Networks

WALTER H. KU, MEMBER, IEEE, AND WENDELL C. PETERSEN, SENIOR MEMBER, IEEE

Abstract—The design of broad-band high-frequency transistor amplifiers is a difficult and challenging outstanding problem in active network theory of great practical importance. New and explicit optimum gain-bandwidth limitations of high-frequency transistor amplifiers for arbitrary prescribed transistor gain rolloff characteristics are presented. The transistor is modeled as a reactively constrained active two-port network. The limitations derived and presented are applicable to the design of broad-band small-signal as well as high-power transistor amplifiers. Realization of a class of practical broad-band matching networks are also presented. The explicit gain-bandwidth limitations and the new realization results presented form the basis for a new broad-banding theory for high-frequency transistor amplifiers and represent a significant advancement in the design theory of active networks.

I. INTRODUCTION

THE DESIGN of broad-band high-frequency transistor amplifiers is a difficult and challenging theoretical problem in active network theory of great practical importance. The difficulties are largely due to the *dual* requirements for broad-band matching the prescribed reactive constraints as well as for broad-band impedance transformation with large impedance ratios. In addition, the broad-band matching networks must be designed with prescribed tapered magnitude characteristics to compensate for the inherent transistor gain rolloffs with frequency which are intrinsic with the high-frequency bipolar and field-effect transistors (FET's).

At present, a rigorous design theory for broad-band transistor amplifiers is not available and designers must rely on computer-aided and experimental cut-and-try techniques which are often inadequate. This paper presents new and explicit results on the optimum gain-bandwidth limitations of broad-band transistor amplifiers for *arbitrary* prescribed transistor gain rolloff characteristics. These important optimum limitations delineate *explicitly* the amplifier gain-bandwidth tradeoffs for a given reactive constraint and a prescribed transistor gain taper. Thus, by applying these limitations, it is possible to know *a priori* the broad-banding capability of a given bipolar or FET. Optimum limitations derived and presented in this paper are applicable to the design of both bipolar and FET amplifiers used for linear small-signal and high-power large-signal applications. In addition, this paper presents practical realizations of a class of broad-band matching networks for prescribed gain rolloff characteristics and impedance transformation ratios.

The problem of designing broad-band high-frequency transistor amplifiers has been of interest for many years and is a basic problem in the theory of active networks. Recently, high-power bipolar transistors in the UHF and low microwave frequencies and both bipolar and FET's in the microwave frequencies extending to *C* and *S* band have become available. With the introduction of these new transistors, the conventional design techniques become inadequate and a number of new problems are presented to the circuit designer. First of all, at UHF and microwave frequencies, the conventional transistor equivalent circuits are not adequate to characterize the device behavior over a broad band of frequencies unless a very complicated equivalent circuit is used. It becomes apparent that a practical design theory must be based on the measured transistor characteristics for both small-signal (receiver) and large-signal (transmitter) applications. The prescribed transistor parameters are measured under the specified operating conditions and thus represent the actual characteristics of the transistor under the actual design conditions. This point needs to be emphasized since meaningful analytical design must take into account the transistor behavior under actual operating conditions specified for a given amplifier design.

For linear small-signal transistor amplifiers, present design procedures are based on the measured scattering coefficients of the transistor [1]–[6]. However, at present a rigorous and complete broad-banding theory is not as yet available although some results have recently been presented [7], [8]. For broad-band large-signal high-power transistor amplifier design, presently available design theory is even more inadequate. Based on large-signal impedance or scattering characterizations of the transistor, current design techniques must rely heavily on experimental cut-and-try procedures on a breadboard both for transistor characterizations as well as final amplifier design. Some computer optimization programs are available. However, the analytical broad-band design theory and gain-bandwidth limitations are still not complete.

The gain-bandwidth limitations and the new broad-banding theory presented in the following sections are based on various equivalent circuits of the bipolar and FET which are derived from measured or experimentally derived RF transistor characteristics under actual operating conditions. In addition to these equivalent circuits, which exhibit explicitly the inherent *reactive constraints* as well as *impedance level* of the transistor, we incorporate also the intrinsic *gain rolloff characteristics* of the transistor. The latter characteristics are also derived from measured gain versus frequency characteristics of the transistor under actual operating conditions. It should be emphasized at the outset that the equivalent circuits used in this paper are based on the measured small-signal or large-signal impedance or scattering parameters of the transistor. Computer-aided optimization techniques are used to finalize the broad-band amplifier design if the measured results should significantly deviate from the assumed equivalent circuits and gain taper characteristics; i.e., if the actual transistor resistive or reactive characteristics are not constant and/or the gain taper characteristics should deviate significantly from a constant slope across a very wide frequency band of interest.

II. Design of High-Frequency Transistor Amplifiers

In this section, the basic problem of designing high-frequency transistor amplifiers is presented to serve as an introduction to the broad-band matching theory for transistor amplifiers. At high frequencies, it is more convenient and desirable to characterize the transistors in terms of their measured scattering parameters, and this characterization has become a standard starting point for the design of high-frequency active and passive devices and circuits.

Consider a general linear active two-port described in terms of its scattering parameters. Let a_1 and a_2 denote the incident waves and b_1 and b_2 the reflected waves at the respective ports, and let the scattering matrix of the active two-port be defined with respect to 50-Ω normalization at both ports 1 and 2. Then

$$\begin{bmatrix} b_1 \\ b_2 \end{bmatrix} = \begin{bmatrix} s_{11}(j\omega) & s_{12}(j\omega) \\ s_{21}(j\omega) & s_{22}(j\omega) \end{bmatrix} \begin{bmatrix} a_1 \\ a_2 \end{bmatrix}, \quad b = \underline{S}a. \quad (1)$$

The ports are terminated in arbitrary passive impedances and the reflection factors of these terminations are represented by s_g and s_L, which are again normalized to 50 Ω.

The derivation of the amplifier transducer power gain expression can be simplified by using the concept of complex normalization since the scattering formulation will now embody the properties of the external networks as well as the intrinsic characteristics of the active two-port. Invoking the matrix transformation relationship for the scattering matrix under a change of normalization, it has been shown that the transducer power gain for the general transistor amplifier is given by [2]

$$G_t(\omega) = \frac{|s_{21}(j\omega)|^2(1 - |s_g(j\omega)|^2)(1 - |s_L(j\omega)|^2)}{|1 - s_g(j\omega)s_{11}(j\omega) - s_L(j\omega)s_{22}(j\omega) + s_g(j\omega)s_L(j\omega)\Delta_s(j\omega)|^2} \quad (2)$$

where Δ_s denotes the determinant of the S-matrix

$$\Delta_s = \det [\underline{S}] = s_{11}s_{22} - s_{12}s_{21}. \quad (3)$$

The design problem involved in a high-frequency transistor amplifier is now clearly exhibited by (2). The $s_{ij}(j\omega)$, $i,j = 1,2$, are the *measured* scattering parameters of the transistor over the frequency band of interest and the reflection coefficients $s_g(j\omega)$ and $s_L(j\omega)$ represent the passive source and load matching networks, respectively. The design problem is that of designing the matching networks represented by $s_g(j\omega)$ (or $Z_g(j\omega)$) and $s_L(j\omega)$ (or $Z_L(j\omega)$) with a prescribed set of measured $s_{ij}(j\omega)$'s of the transistor such that the transducer power gain given by (2) has some prescribed realizable response and gain-bandwidth characteristics. This general broad-band matching problem for the bipolar and FET transistor amplifier is shown schematically in Fig. 1 in which s_g and s_L are represented by the Darlington equivalent of the one-port input and output matching networks.

It is noted that the general broad-band matching problem defined in Fig. 1 is considerably more complicated than that treated by the Bode–Fano–Youla optimum broad-band matching theory [9]–[11]. The broad-band matching problem under consideration is that of a constrained active two-port whereas the classical optimum broad-band matching theory is applicable only to a constrained one-port. The complexity introduced by the constrained active two-port (transistor) is due to the coupling of the input and output of the prescribed transistor in that the match at port 1 is dependent on the matching network at port 2 and vice versa. An additional complication of the present transistor amplifier matching problem is that the inherent transistor gain rolloffs with respect to frequency must be compensated by using matching networks with prescribed tapered-magnitude characteristics.

It should be noted that Bode–Fano-type of integral restrictions due to reactive constraints is valid for arbitrary passive and lossless matching networks. In particular, *ideal transformers* are admissible as part of the matching networks since they are reciprocal, passive, and lossless two-ports. Thus, in the derivations of the optimum gain-bandwidth limitations, the actual impedance transformation ratio does not appear explicitly since ideal transformers (assumed to be available) can always be used in theory.

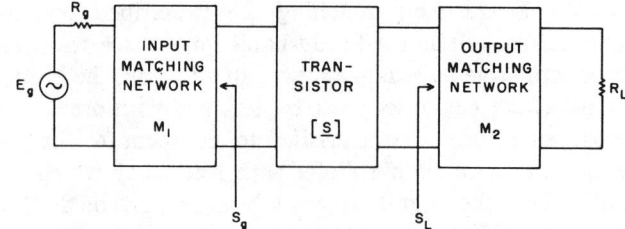

Fig. 1. Broad-band matching problem for bipolar and FET amplifiers.

When we impose the additional practical restriction that no ideal transformer is allowed, the design problem becomes much more demanding even though the theoretical gain-bandwidth limitations are not affected. The reason why the optimum limitations are not affected by actual impedance transformations is that in theory the ideal transformer can be approximated arbitrarily closely over a finite frequency band (excluding dc) by increasing the complexity of the approximating transformer network. In practice, however, this may not be acceptable and the amplifier designers are then faced with the practical realizations which can satisfy the dual requirements of matching reactive constraints and impedance transformation.

Referring to Fig. 1 for the general broad-banding problem under consideration, the added complexity of a constrained active two-port can be handled approximately by assuming that the active two-port is *unilateral* but we will take the gain rolloff characteristics, which are directly due to the intrinsic coupling in transistor, into account separately in the design of the broad-band matching networks. This is a satisfactory solution based on practical considerations. It has been found that this assumption is valid for many practical design examples for both microwave bipolar and FET amplifiers. If the transistor is assumed to be unilateral, then $s_{12} = 0$ and the transducer power gain expression contained in (2) reduces to

$$G_t(\omega^2) = \frac{|s_{21}|^2(1 - |s_g|^2)(1 - |s_L|^2)}{|1 - s_g s_{11} - s_L s_{22} + s_g s_L s_{11} s_{22}|^2} \quad (4)$$

or

$$G_t(\omega^2) = |s_{21}|^2 \cdot \frac{(1 - |s_g|^2)}{|1 - s_g s_{11}|^2} \cdot \frac{(1 - |s_L|^2)}{|1 - s_L s_{22}|^2}. \quad (5)$$

As shown in (5), the gain expression becomes greatly simplified in the sense that the constrained two-port matching problem is in effect "uncoupled." As stated previously, even though we are using a simplified uncoupled model, the coupling will be taken into account by using the specified *tapered* magnitude characteristics. The gain tapering is a direct consequence of the fact that the transistor maximum available gain decreases with increasing frequency. Thus, for a flat overall gain response, matching networks with tapered magnitude characteristics must be used. The exact frequency dependency of the taper or gain slope varies with specific types of transistors (bipolar or FET) as well as with power

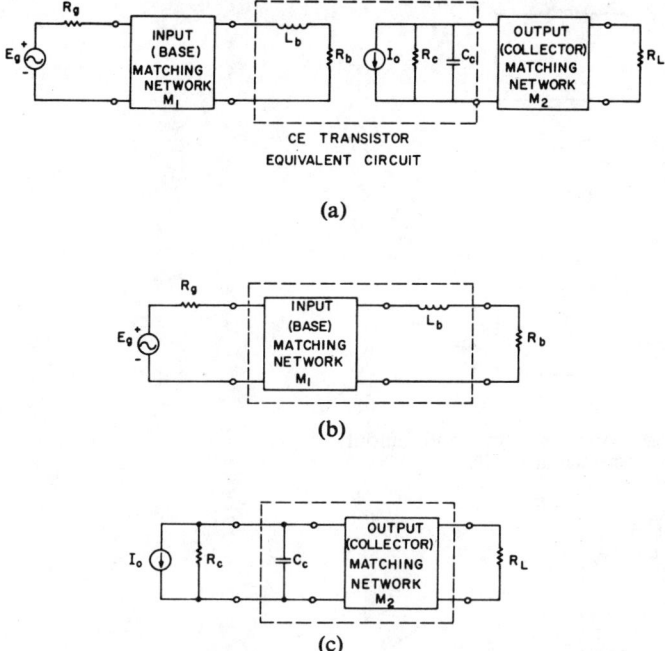

Fig. 2. Broad-band matching of high-frequency bipolar transistor amplifiers.

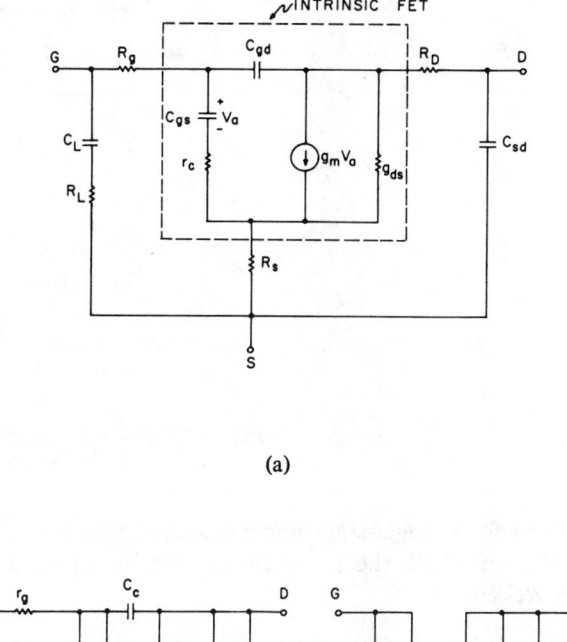

Fig. 3. Equivalent circuits of microwave FET's.

levels (linear small-signal amplifiers or large-signal power amplifiers).

Fig. 2(a) shows the broad-band matching problem of a typical high-frequency bipolar transistor amplifier. For convenience, a common-emitter transistor is shown but the circuit configuration is equally valid for the common-base configuration as well. The input matching (network M_1 shown in Fig. 2(b)) must be designed to provide simultaneously 1) broad-band matching to equalize the series reactive constraint represented by the series $R_b L_b$ combination, and 2) broad-band impedance transformation of source and load impedance levels ($R_g = 50\ \Omega$, and R_b is of the order of 1 Ω for a high-power class-C transistor with the exact value depending on the power capability of the transistor).

Fig. 2(c) shows the output matching network M_2 which is at the collector. The reactive constraint is now a shunt capacitance C_c. The impedance transformation in this case is from R_c to $R_L = 50\ \Omega$. If this amplifier is part of a cascaded power amplifier, then R_g and R_L must be modified. In fact if the power amplifier shown in Fig. 2 is followed by even higher power amplifier stages, then R_L must be modified to an equivalent impedance represented by a series $R_b L_b$ circuit.

Fig. 3(a) shows the equivalent circuit of microwave FET's including the parasitic elements of the package. Fig. 3(b) and (c) represent simplified circuit models. In particular, the one shown in Fig. 3(c) is the unilateral model but now the R's and C's are strictly equivalent circuit elements derived from the measured scattering parameters rather than device equivalent circuits. For some devices such as the MESFET's the input can be modeled more closely by a series RC circuit. Figs. 4 and 5 show schematically the broad-band matching

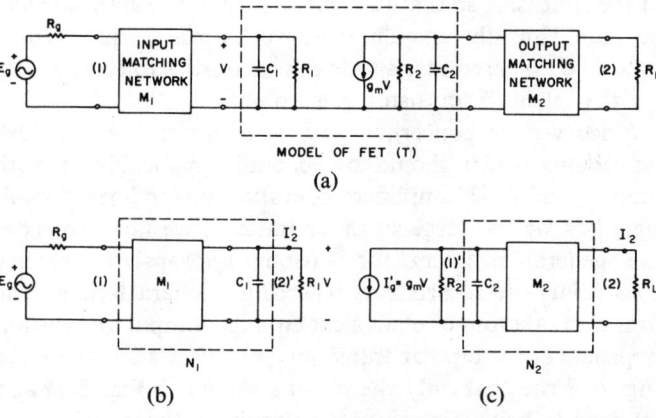

Fig. 4. Broad-band matching of single-stage microwave FET amplifier.

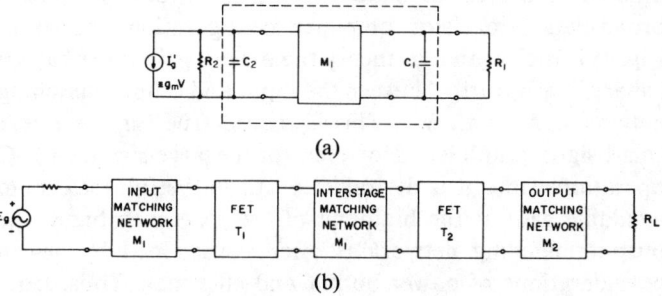

Fig. 5. Broad-band matching of cascaded FET amplifier.

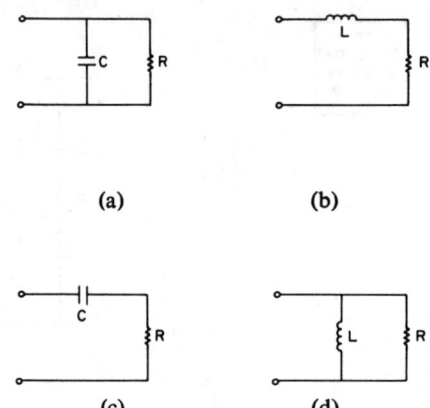

(a)　(b)

(c)　(d)

Fig. 6. Equivalent circuit models of transistor input and output impedances for bipolar transistors and FET's.

problems for a single-stage and a cascaded microwave FET amplifier in which the input circuit can be replaced by a series RC circuit.

III. Optimum Gain-Bandwidth Limitations for Broad-Band Transistor Amplifiers

In this section, the theoretical gain-bandwidth limitations for broad-band transistor amplifiers are derived. It is well known that if the transistor amplifier is conjugate matched over a broad frequency band, the maximum available gain will have a tapered characteristic as a function of frequency. As a result, it is necessary to selectively *mismatch* the input of the transistor so that the overall gain of the amplifier will be flat. Thus the matching networks must exhibit some prescribed tapered magnitude characteristics to compensate for the intrinsic transistor gain rolloff.

Since we are concerned with optimum gain-bandwidth limitations which should be generally applicable to both bipolar and FET amplifiers operating under linear small-signal as well as large-signal conditions, we must consider the general problem for various appropriate reactive constraints and an arbitrary tapered gain characteristic. The four general types of equivalent circuits of input and output impedances for bipolar transistors and FET's are shown in Fig. 6. Note that only the circuits shown in Fig. 6(a) and (c) need to be studied since the circuits in Fig. 6(b) and (d) are the respective dual networks of those shown in Fig. 6(a) and (c). The circuits in Fig. 6(a) and (c) are fundamentally different and must be treated separately. From the general broad-band transistor amplifier configuration shown in Fig. 1, it is clear that in theory the overall gain taper can be "shared" arbitrarily between the input and output matching networks, M_1 and M_2. This is indeed true for the linear small-signal amplifiers. However, for the large-signal class-C operations since it is desirable to utilize the full maximum available gain at the high end of the frequency band, the output matching network is often constrained by design considerations of power output and efficiency. Thus, from these practical considerations, only the input matching network is at our disposal to compensate for the inherent gain rolloff of the high-power transistors.

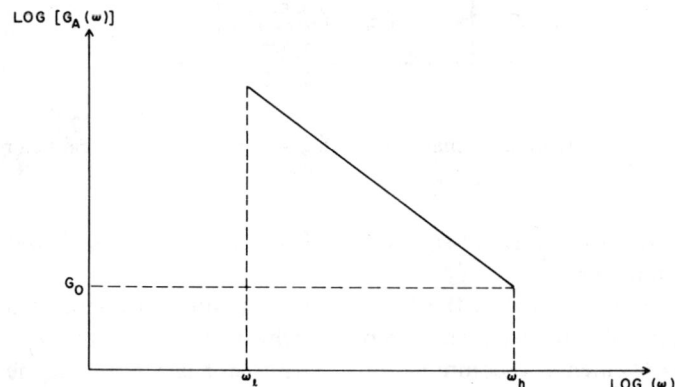

Fig. 7. Intrinsic gain taper characteristic of uncompensated transistor amplifier.

The intrinsic gain taper characteristic of an uncompensated transistor amplifier is shown in Fig. 7. Over the frequency band of interest from ω_l to ω_h, this gain characteristic $G_A(\omega)$ is represented by a straight line on a log-log scale. The gain function $G_A(\omega)$ is given by

$$G_A(\omega) = G_0 \left(\frac{\omega}{\omega_h}\right)^{-k}, \qquad \omega_l \leq \omega \leq \omega_h \qquad (6)$$

where G_0 is the gain at $\omega = \omega_h$, the coefficient k is given by

$$k = \frac{x}{10 \log 2} \cong \frac{x}{3} \qquad (7)$$

and the parameter x denotes the *slope* of the line in decibels per octave. Since the desired gain of the overall amplifier G_t is to be constant from ω_l to ω_h, the gain of the input matching network actually used $G_M(\omega)$ must be

$$G_M(\omega) = \begin{pmatrix} K\left(\frac{\omega}{\omega_h}\right)^k, & \omega_l \leq \omega \leq \omega_h \\ 0, & \text{otherwise} \end{pmatrix} \qquad (8)$$

where K is the gain-scale factor which must be less than or equal to unity because of the passivity constraint. Using (6) and (8), the overall transistor amplifier gain becomes

$$G_T = G_M(\omega) G_A(\omega), \qquad \text{for } \omega_l \leq \omega \leq \omega_h. \qquad (9)$$

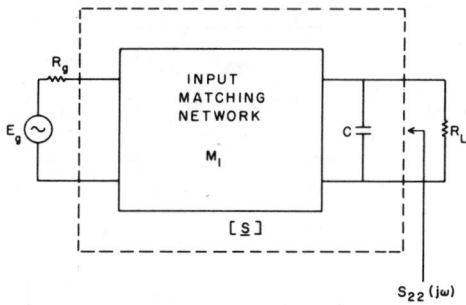

Fig. 8. Broad-band input matching network for equivalent "low-pass" reactive constraint.

It is necessary to treat the cases shown in Fig. 6(a) and (c) separately. Consider first the "low-pass" reactive constraint case shown in Fig. 6(a). The input matching network is shown in Fig. 8, in which the scattering matrix $\underline{S}^M$ is normalized to R_g at port 1 and R_L at port 2. It is assumed that R_g and R_L are both positive real constants. Note that $\underline{S}^M$ includes the equivalent input capacitance of the transistor. The fundamental integral restriction for this case was originally derived by Bode [9] and is given by

$$\int_0^\infty \ln \frac{1}{|s_{22}^M(j\omega)|^2} d\omega \leq \frac{2\pi}{\tau} \quad (10)$$

where

$$\tau = R_L C. \quad (11)$$

Since the two-port matching network including C is reciprocal and lossless, the $\underline{S}^M$-matrix must be unitary. Thus

$$|s_{22}^M(j\omega)|^2 = 1 - |s_{12}^M(j\omega)|^2 = 1 - G_M(\omega). \quad (12)$$

Inserting (14) into the integral in (12) and using the tapered characteristic of (10) we obtain

$$\int_0^\infty \ln \left(\frac{1}{1 - G_M(\omega)} \right) d\omega$$

$$= \int_{\omega_l}^{\omega_h} \ln \left(\frac{1}{1 - K \left(\frac{\omega}{\omega_h} \right)^k} \right) d\omega \leq \frac{2\pi}{\tau}. \quad (13)$$

Without loss of generality, τ can be normalized so that $\omega_h = 1.0$ rad/s. By defining the normalized τ, $\tau_N = \omega_h \tau$, the restriction in (15) becomes

$$\int_{\omega_l}^1 \ln \left(\frac{1}{1 - K\omega^k} \right) d\omega \leq \frac{2\pi}{\tau_N}. \quad (14)$$

The integral restriction (14) is the fundamental limitation for a prescribed gain characteristic with arbitrary gain taper. The parameters contained in (14) are the normalized bandwidth $(1 - \omega_l)$, the gain-scale factor K, the time constant of the reactive constraint τ_N, and the slope parameter k.

For $k = 0$, corresponding to the ideal flat-gain case with no taper, the optimum gain-bandwidth limitation is given by (using the equality in (14))

$$\left(\ln \frac{1}{1 - K} \right) \cdot (1 - \omega_l) = \frac{2\pi}{\tau_N}, \quad \text{with } K < 1 \quad (15)$$

For the flat-gain case, the gain and bandwidth tradeoff as a function of τ is clearly shown in (15). To express the gain K and the normalized bandwidth $(1 - \omega_l)$, explicitly, (15) is given equivalently by

$$K = 1 - \exp \left(-\frac{2\pi}{(1 - \omega_l)\tau_N} \right), \quad \text{with } K < 1 \quad (16)$$

and

$$(1 - \omega_l) = \frac{2\pi}{\tau_N} \frac{1}{\ln(1/(1 - K))}, \quad \text{with } K < 1. \quad (17)$$

Note that the limiting value of K equal to 1 cannot be attained.

For $k = 1$, corresponding to a prescribed gain slope of 3 dB/oct, (14) can be solved analytically to give the following explicit gain-bandwidth limitation

$$\left(\frac{1 - K}{K} \ln(1 - K) - \frac{1 - K\omega_l}{K} \ln(1 - K\omega_l) \right.$$

$$\left. + (1 - \omega_l) \right) = \frac{2\pi}{\tau_N} \quad (18)$$

or, equivalently,

$$\frac{(1 - K)^{1-K}}{(1 - K\omega_l)^{1-K\omega_l}} = \exp \left(K \left\{ \frac{2\pi}{\tau_N} - (1 - \omega_l) \right\} \right). \quad (19)$$

For $k = 2$, corresponding to a prescribed gain slope of 6 dB/oct, (14) can also be solved analytically to yield the following explicit gain-bandwidth limitation

$$\left(\frac{1 - \sqrt{K}}{\sqrt{K}} \ln(1 - \sqrt{K}) - \frac{1 + \sqrt{K}}{\sqrt{K}} \ln(1 + \sqrt{K}) \right.$$

$$- \frac{1 - \omega_l\sqrt{K}}{\sqrt{K}} \ln(1 - \omega_l\sqrt{K})$$

$$\left. + \frac{1 + \omega_l\sqrt{K}}{\sqrt{K}} \ln(1 + \omega_l\sqrt{K}) + 2(1 - \omega_l) \right) = \frac{2\pi}{\tau_N} \quad (20)$$

or, equivalently,

$$\frac{(1 - \sqrt{K})^{1-\sqrt{K}}(1 + \omega_l\sqrt{K})^{1+\omega_l\sqrt{K}}}{(1 + \sqrt{K})^{1+\sqrt{K}}(1 - \omega_l\sqrt{K})^{1-\omega_l\sqrt{K}}}$$

$$= \exp \left(\sqrt{K} \left\{ \frac{2\pi}{\tau_N} - 2(1 - \omega_l) \right\} \right). \quad (21)$$

It is noted that in contrast to the flat-gain result, the gain-bandwidth limitations for the tapered gain cases presented in (18) and (21) are well defined for the limiting case of $K = 1$ since $\lim_{x \to 0} x \ln x = 0$. The general tradeoffs of gain and bandwidth are of course still present in the sense that by decreasing the gain from unity, the bandwidth will be increased.

The explicit gain-bandwidth limitations derived in (18), (19) and (20), (21) are plotted in Fig. 9 with the gain scale factor K as a parameter. The two sets of curves correspond

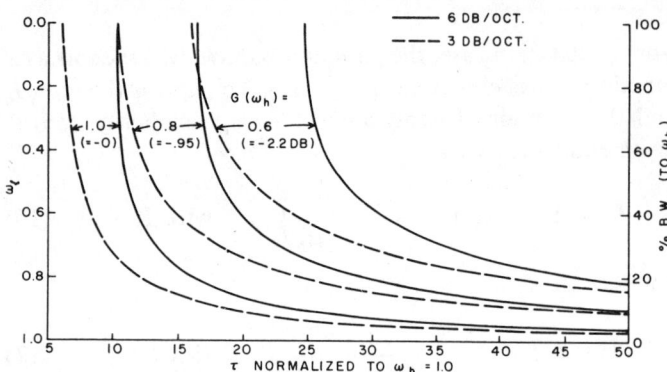

Fig. 9. Optimum gain-bandwidth limitations for tapered magnitude gain characteristics for equivalent "low-pass" reactive constraint.

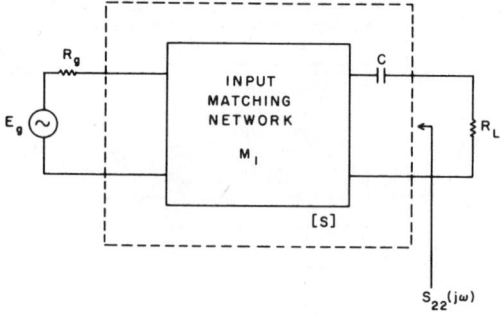

Fig. 10. Broad-band input matching network for equivalent "high-pass" reactive constraint.

to the cases of $k = 1$ (3 dB/oct) and $k = 2$ (6 dB/oct). As expected, these curves show that increasing K, increasing τ, or decreasing k all tend to decrease the bandwidth. With the aid of the curves plotted in Fig. 9, it is also simple to check what values of τ are acceptable once k, K, and bandwidth are prescribed. For example, if an amplifier with a 50-percent bandwidth is desired for $K = 1$ and $k = 2$, then the normalized τ_N must be less than or equal to 11.8. In practice, it must be somewhat less than this value since the result represents the optimum case, which can only be approximated by an actual network. As another example, let $k = 2$, $K = 1$, and suppose that the desired bandwidth is 200 percent with a denormalized ω_h of $2\pi \times 100$ MHz. Then the maximum normalized τ_N is 10.25. Thus we can determine *a priori* which transistors are acceptable for a specific amplifier application.

In applying these results to practical amplifier design, the time constant τ and the gain slope k are prescribed, then these results present explicit relationships between the gain and the bandwidth.

For other values of gain slope between 3 dB/oct and 6 dB/oct (k between 1 and 2), interpolation of the curves in Fig. 9 can be used to given an estimate of the gain-bandwidth relationship for a given τ. If more accuracy is needed, then the exact integral restriction for the general taper case contained in (14) can be solved numerically. A computer program has been written to solve the integral for arbitrary slope parameter k.

The results presented in (13)–(21) are applicable to the dual case shown in Fig. 6(b). It is only necessary to set $\tau = L/R_L$ in place of $\tau = R_L C$ in (10)–(21) and the result is presented in Fig. 9. The series R_L equivalent circuit shown in Fig. 6(b) is used to represent the input impedances of high-power bipolar transistors.

A similar derivation will now be presented for the circuit shown in Fig. 6(c), which represents a basically "high-pass" reactive constraint. This equivalent circuit of a transistor input impedance is valid for some types of FET's, such as the MESFET [6]. The matching network is shown in Fig. 10, and as before the scattering matrix $\underline{S}^M$ is normalized to R_g and R_L at the respective ports and the series capacitance C is included in the two-port network $\underline{S}^M$. In this case, the integral restriction is given by

$$\int_0^\infty \frac{1}{\omega^2} \ln |s_{22}{}^M(j\omega)|^2 \, d\omega \geq -2\pi\tau \qquad (22)$$

where

$$\tau = R_L C. \qquad (23)$$

Since $G_A(\omega)$ is still given by (6), the definition of $G_M(\omega)$ in (8) and the unitary condition (12) are still valid. Substitution of (6) and (12) into (22) yields

$$\int_{\omega_l}^{\omega_h} \frac{1}{\omega^2} \ln \left[\frac{1}{1 - K\left(\frac{\omega}{\omega_h}\right)^k} \right] d\omega \leq 2\pi\tau. \qquad (24)$$

As in the previous case, τ can be normalized without loss in generality such that $\omega_h = 1.0$ rad/s. Then (24) becomes

$$\int_{\omega_l}^{1} \frac{1}{\omega^2} \ln \left[\frac{1}{1 - K\omega^k} \right] d\omega \leq 2\pi\tau_N. \qquad (25)$$

Choosing equality so that ω_l is as small as possible, the solution to (25) for $k = 0$ (flat-gain case with no taper) is given by

$$\omega_l = \frac{\ln(1 - K)}{\ln(1 - K) - 2\pi\tau_N}, \qquad \text{with } K < 1 \qquad (26)$$

or, equivalently,

$$K = 1 - \exp\left[\frac{2\pi\tau_N \omega_2}{\omega_l - 1}\right], \qquad \text{with } K < 1. \qquad (27)$$

Similar to the flat-gain limitation derived in (15)–(17), the limiting value of $K = 1$ cannot be attained.

For $k = 1$, corresponding to a 3-dB/oct average gain rolloff, (25) can be integrated to give

$$(1 - K) \ln(1 - K) - \frac{(1 - K\omega_l)}{\omega_l} \ln(1 - K\omega_l)$$

$$- K \ln \omega_l = 2\pi\tau_N \qquad (28)$$

or, equivalently,

$$\frac{(1 - K)^{1-K}}{\omega_l{}^k (1 - K\omega_l)^{(1-K\omega_l)/\omega_l}} = \exp[2\pi\tau_N]. \qquad (29)$$

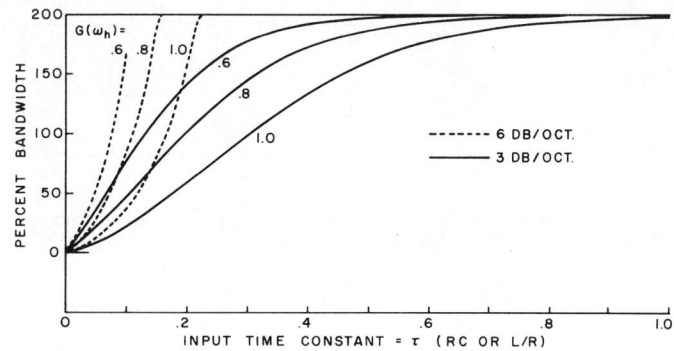

Fig. 11. Optimum gain-bandwidth limitations for tapered magnitude gain characteristics for equivalent "high-pass" reactive constraint.

For $k = 2$, corresponding to a 6-dB/oct gain rolloff, (25) can be integrated to yield

$$(1 - \sqrt{K}) \ln(1 - \sqrt{K}) + (1 + \sqrt{K}) \ln(1 + \sqrt{K})$$

$$- \frac{(1 - \omega_l\sqrt{K})}{\omega_l} \ln(1 - \omega_l\sqrt{K})$$

$$- \frac{(1 + \omega_l\sqrt{K})}{\omega_l} \ln(1 + \omega_l\sqrt{K}) = 2\pi\tau_N \quad (30)$$

or, equivalently,

$$\frac{(1 - \sqrt{K})^{1-\sqrt{K}}(1 + \sqrt{K})^{1+\sqrt{K}}}{(1 - \omega_l\sqrt{K})^{(1-\omega_l\sqrt{K})/\omega_l}(1 + \omega_l\sqrt{K})^{(1+\omega_l\sqrt{K})/\omega_l}}$$
$$= \exp[2\pi\tau_N]. \quad (31)$$

For the series RC or equivalent "high-pass" reactive constraint case, the explicit gain-bandwidth limitations presented in (30)–(33) are plotted in Fig. 11 with K as a parameter. These curves show that decreasing τ or k, or increasing K all tend to make the bandwidth smaller. This corresponds exactly with the equivalent "low-pass" shunt RC case, except that the effect of τ is now reversed. Such a reversal should be expected since C is now a series capacitor instead of a shunt capacitor. As an example, for $K = 1$, $k = 1$, and a bandwidth of 100 percent, the minimum acceptable value for the normalized time constant is 3.05. As another example, suppose $k = 2$, $N = 1$, and bandwidth of 200 percent is desired with a denormalized ω_h of $2\pi \times 100$ MHz. The normalized τ, τ_N, should be 0.221. The results in (24)–(31) and Fig. 11 are applicable to the equivalent circuit of Fig. 6(d) by setting $\tau = L/R$.

IV. Practical Design of Broad-Band Matching Networks with Prescribed Tapered Magnitude Characteristics

In this section, the practical design of broad-band matching networks with various tapered magnitude characteristics prescribed by the intrinsic transistor gain slopes is considered. A systematic design method is presented for a class of practical and broad-band ladder-type matching networks with arbitrary gain slopes and impedance transformation ratio. As discussed in detail in the previous sections, these ladder networks must meet the dual requirements of prescribed gain tapering and impedance transformation since for most practical amplifier design the use of broad-band transformer is not feasible and/or desirable. Computer-aided optimized realizations of various low-pass ladder matching networks for both the input and the output matching networks of the transistor amplifiers are also presented.

Fig. 12 shows the general low-pass ladder matching network configuration designed to approximate a prescribed tapered magnitude gain characteristic. This has been used by Pitzalis and Gilson [12], [13]. This ladder network must provide not only the prescribed gain slope but also provide a broad-band impedance transformation of r [see Fig. 12]. If there is no taper (flat-gain response), the dual matching problem reduces to that of a relatively simple design problem of broad-band impedance transformer. This problem has been studied extensively by Matthaei [14], Cristal [15], and Levy [16].

Broad-band impedance transformers with no gain taper can be used for transistor output matching networks with flat-gain characteristics and also for some input matching networks in which the transistor gain characteristics are relatively constant with frequencies over the frequency band of interest. For the case of flat response, a simple frequency transformation can be used [14]. However, there is no known transformation to design a low-pass ladder with an arbitrary taper and some approximation to the prescribed tapered characteristic is needed. Pitzalis and Gilson [12] have presented a set of tables of low-pass element values for $n = 4$, using computer-aided optimizations. In the following, a systematic method is presented which can be generalized to arbitrary n. Let the gain function for an n-element low-pass ladder network be given by

$$|s_{21}(\omega)|^2 = \frac{1}{1 + [Q(\omega^2)]^2} \quad (32)$$

where

$$Q(\omega^2) = a_0 + a_1\omega^2 + \cdots + a_{n/2}\omega^n. \quad (33)$$

Note that n must be an even number, and that $((n/2) + 1)$ constants are available to approximate the desired taper. This is less than the $(n + 1)$ that would be available if an nth-order polynomial was used in the denominator, but the advantage gained by using (32) is that the resulting $|s_{21}(\omega)|^2$

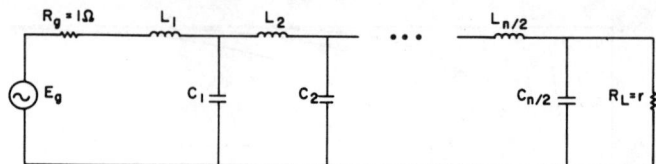

Fig. 12. General low-pass ladder matching network configuration to approximate prescribed tapered magnitude gain characteristic.

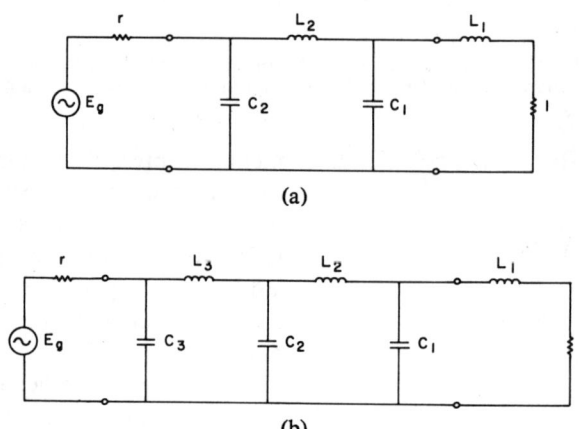

Fig. 13. Broad-band low-pass tapered magnitude matching network configurations with (a) $n = 4$ and (b) $n = 6$.

is realizable by a low-pass ladder network. Once an approximation is found from (32) an optimization procedure can be used to adjust the actual element values. Since the initial approximation will be reasonably close, the optimization process will converge rapidly.

To choose the polynomial $Q(\omega^2)$ in (32), let $G(\omega)$ be the desired realizable gain function with a prescribed taper. Then, set

$$G(\omega_k) = \frac{1}{1 + \left(\sum_{i=0}^{n/2} a_i(\omega_k^{2i}) \right)^2}, \quad k = 0, 1, \cdots, \frac{n}{2}. \quad (34)$$

The conditions given in (34) can be rewritten as

$$\sum_{i=0}^{n/2} a_i \omega_k^{2i} = C_k, \quad \text{for } k = 0, 1, \cdots, \frac{n}{2} \quad (35)$$

where

$$C_k = \sqrt{\frac{1}{G(\omega_k)} - 1}. \quad (36)$$

The C_k are all real numbers since $G(\omega)$ is bounded by one for all ω. Writing (35) in matrix notation gives

$$\underline{\Omega} a = C \quad \text{or} \quad a = \underline{\Omega}^{-1} C \quad (37)$$

where

$$\underline{\Omega} = \begin{bmatrix} 1 & \omega_0^2 & \omega_0^4 & \cdots & \omega_0^n \\ 1 & \omega_1^2 & \omega_1^4 & \cdots & \omega_1^n \\ \vdots & \vdots & \vdots & \cdots & \vdots \\ 1 & \omega_{n/2}^2 & \omega_{n/2}^4 & \cdots & \omega_{n/2}^n \end{bmatrix} \quad (38)$$

$$a = [a_0 \quad a_1 \quad \cdots \quad a_{n/2}]^T$$

and

$$C = [C_0 \quad C_1 \quad \cdots \quad C_{n/2}]^T.$$

At this point, the problem reduces to choosing the values for ω_k. Since the impedance ratio of the transformer is fixed, the gain at dc is specified and $\omega_0 = 0$ rad/s should be used. A convenient way to choose the remaining ω_k is to space them evenly over the band of interest, using either a linear or a logarithmic frequency scale. From experience, it is found that the approximation using a log scale is preferred. Once the vector a is known, (32) gives the transfer function of the transformer. Proceeding in the usual manner, $s_{21}(p)s_{21}(-p)$ is obtained by making the substitution $\omega^2 = -p$ in (32) and then since the transformer is lossless and reciprocal

$$s_{11}(p)s_{11}(-p) = 1 - s_{21}(p)s_{21}(-p). \quad (39)$$

Next, $s_{11}(p)s_{11}(-p)$ is factored such that all of the poles of $s_{11}(p)$ are in the left half-plane, and the input impedance to the transformer is found by

$$z(p) = \frac{1 + s_{11}(p)}{1 - s_{11}(p)}. \quad (40)$$

Equation (40) can then be used to synthesize a low-pass ladder network using Cauer or long divison techniques.

A computer program has been written which does this entire synthesis procedure automatically. For the low-pass $n = 4$ and $n = 6$ matching network configurations shown in Fig. 13, typical response curves generated by this synthesis

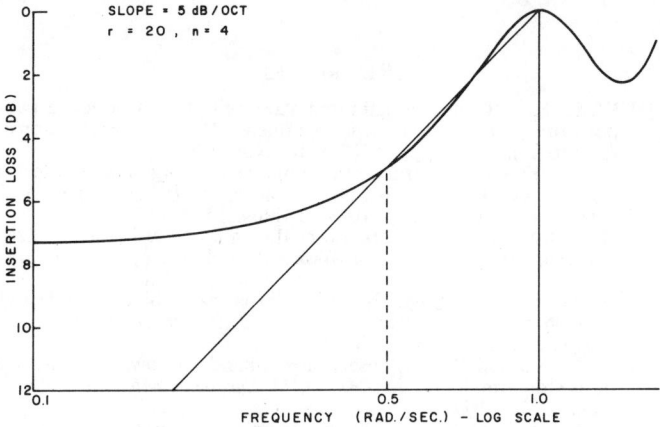

Fig. 14. Gain characteristic of typical low-pass tapered magnitude matching network *without* optimization ($n = 4$; $r = 20$; gain slope = 5 dB/OCT).

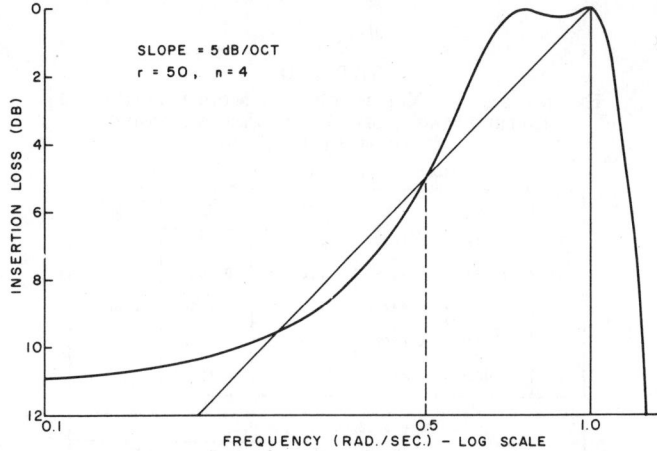

Fig. 16. Gain characteristic of typical low-pass tapered magnitude matching network *without* optimization ($n = 4$; $r = 50$; gain slope = 5 dB/OCT).

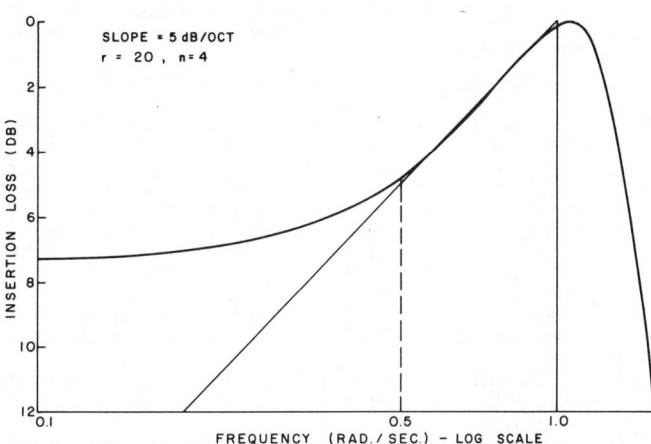

Fig. 15. Optimized gain characteristic of low-pass tapered magnitude matching network ($n = 4$; $r = 20$; gain slope = 5 dB/OCT).

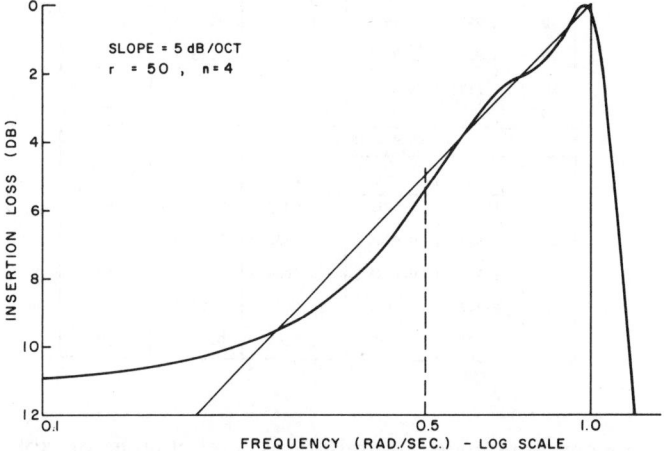

Fig. 17. Optimized gain characteristic of low-pass tapered magnitude matching network ($n = 4$; $r = 50$; gain slope = 5 dB/OCT).

technique are shown in Figs. 14 and 16 for a selected set of design parameters. The corresponding results after computer aided optimization are presented in Figs. 15 and 17. Explicit element values for $n = 4$ and $n = 6$ low-pass tapered-magnitude broad-band matching networks are presented in Tables I and II, respectively. These low-pass results extend the results published by Pitzalis and Gilson for the $n = 4$ case [13]. From the results presented in Tables I and II, it is clear that the added complexity of the matching networks yields better match as well as higher amplifier gain-bandwidth products.

V. Conclusions

In this paper, we have derived the optimum gain-bandwidth limitations for high-frequency transistor amplifiers with arbitrary prescribed transistor gain rolloff characteristics. These limitations are applicable to both bipolar and FET's in small-signal linear as well as large-signal high-power broad-band amplifier applications. A synthesis technique has also been presented for practical realizations of a class of low-pass lumped broad-band matching networks with tapered magnitude characteristics which provide broad-band reactive matching and impedance transformation simultaneously. Using this synthesis technique together

TABLE I
Explicit Element Values for $n = 4$ Modified Low-Pass Tapered Magnitude Broad-Band Matching Networks (Fig. 17(a))

SLOPE = 6 dB/oct.							
r	L_1	C_1	L_2	C_2	L_3	C_3	MAX DEV. (dB)
20	5.979	0.2110	33.266	0.0745	20.333	0.0002	0.17
30	6.917	0.1836	37.024	0.0557	28.275	0.0073	0.09
40	5.416	0.2085	35.003	0.0480	35.145	0.0123	0.05
50	6.088	0.1948	42.345	0.0480	56.911	0.0140	0.04
60	6.450	0.1865	46.559	0.0452	69.781	0.0146	0.03

SLOPE = 5 dB/oct.							
r	L_1	C_1	L_2	C_2	L_3	C_3	MAX DEV. (dB)
20	5.022	0.2445	27.049	0.0812	20.001	0.0113	0.08
30	5.145	0.2307	29.524	0.0671	32.392	0.0196	0.04
40	5.720	0.2158	34.830	0.0619	47.182	0.0205	0.03
50	6.116	0.2062	38.678	0.0567	58.768	0.0191	0.03
60	6.697	0.1939	42.275	0.0521	69.333	0.0175	0.06

SLOPE = 4 dB/oct.							
r	L_1	C_1	L_2	C_2	L_3	C_3	MAX DEV. (dB)
20	4.519	0.2699	21.735	0.0943	21.868	0.0279	0.03
30	5.157	0.2482	27.084	0.0826	35.890	0.0285	0.02
40	5.780	0.2301	31.124	0.0725	46.984	0.0251	0.05
50	6.042	0.2227	44.189	0.0559	57.158	0.0224	0.07
60	6.115	0.2200	36.556	0.0613	66.624	0.0203	0.10

TABLE II
Explicit Element Values for $n = 6$ Modified Low-Pass Tapered Magnitude Broad-Band Matching Networks (Fig. 17(b))

SLOPE = 6 dB/oct.; $\omega_\ell = 0.5$

r	L_1	C_1	L_2	C_2	MAX. DEV. (dB)
20	5.0594	0.1490	1.6077	0.0390	0.56
30	1.0933	0.8515	6.8788	0.1522	0.20
40	1.4321	0.8977	8.5999	0.1336	0.06
50	1.6925	0.8669	10.0387	0.1202	0.18
60	1.8808	0.8185	10.6713	0.1181	0.35

SLOPE = 5 dB/oct.; $\omega_\ell = 0.5$

r	L_1	C_1	L_2	C_2	MAX. DEV. (dB)
20	4.449	0.2467	19.1281	0.0442	0.25
30	4.8121	0.2439	28.425	0.0460	0.07
40	1.6970	0.8907	8.7657	0.1426	0.25
50	1.9251	0.8206	9.7081	0.1347	0.45
60	2.1578	0.7564	11.1335	0.1190	0.66

SLOPE = 4 dB/oct.; $\omega_\ell = 0.5$

r	L_1	C_1	L_2	C_2	MAX. DEV. (dB)
20	1.230	1.0016	5.6849	0.2074	0.05
30	1.662	0.9206	7.4451	0.1739	0.26
40	1.964	0.8214	8.8904	0.1520	0.50
50	2.247	0.7404	10.545	0.1299	0.75
60	2.504	0.6755	12.1364	0.11388	0.99

with computer-aided optimizations, a set of tables of explicit element values is also presented. Both the optimum limitations and the practical realizations presented will provide the basis for a rigorous design theory for broad-band bipolar and FET amplifiers.

References

[1] W. H. Ku, "Unilateral gain and stability criterion of active two-ports in terms of scattering parameters," *Proc. IEEE* (Correspondence), vol. 54, pp. 1617–1618, Nov. 1966.

[2] ——, "Design of transistor amplifiers using the scattering parameters of active two-ports," in *Proc. 1st Asilomar Conf. Circuits and Systems*, pp. 101–111, Nov. 1967.

[3] G. E. Bodway, "Two-port power flow analysis using generalized scattering parameters," *Microwave J.*, vol. 10, pp. 61–69, May 1967.

[4] W. H. Ku, "A design theory for broadband bipolar and FET transistor amplifiers," in *Proc. 1971 IEEE Int. Symp. Electrical Network Theory*, Sept. 1971.

[5] W. H. Ku and P. L. Clouser, "Broadband microwave field-effect transistor amplifier," in *Proc. 1971 European Microwave Conf.*, Aug. 23–28, 1971.

[6] C. A. Liechti and R. L. Tillman, "Applications of GaAs schottky-gate FET's in microwave amplifiers," in *Digest Int. Solid-State Circuits Conf.*, vol. VII, 1973.

[7] R. S. Tucker, "Gain-bandwidth limitations of microwave transistor amplifiers," *IEEE Trans. Microwave Theory Tech.*, vol. MTT-21, pp. 322–327, May 1973.

[8] W. H. Ku and W. C. Petersen, "Optimum gain-bandwidth limitations and a new broadbanding theory for high-frequency transistor amplifiers," in *Proc. IEEE Int. Symp. Circuits and Systems*, pp. 539–543, Apr. 1974.

[9] H. W. Bode, *Network Analysis and Feedback Amplifier Design*. New York: Van Norstrand, 1945.

[10] R. M. Fano, "Theoretical limitations on the broadband matching of arbitrary impedances, *J. Franklin Inst.*, vol. 249, pp. 57–83, Jan. 1960; pp. 139–155, Feb. 1960.

[11] D. C. Youla, "A new theory of broadband matching," *IEEE Trans. Circuit Theory*, vol. CT-11, pp. 30–50, Mar. 1964.

[12] O. Pitzalis, Jr. and R. A. Gilson, "Tables of impedance matching networks which approximate prescribed attenuation versus frequency slopes," *IEEE Trans. Microwave Theory Tech.*, vol. MTT-19, pp. 381–386, Apr. 1971.

[13] O. Pitzalis, Jr. and R. A. Gilson, "Broadband microwave class-C transistor amplifiers," *IEEE Trans. Microwave Theory Tech.*, Nov. 1973.

[14] G. L. Matthaei, "Tables of Chebyshev impedance-transforming networks of low-pass filter form," *Proc. IEEE*, vol. 52, pp. 939–963, Aug. 1964.

[15] E. G. Cristal, "Tables of maximally flat impedance-transforming networks of low-pass-filter form," *IEEE Trans. Microwave Theory Tech.* (Correspondence), vol. MTT-13, pp. 693–695, Sept. 1965.

[16] R. Levy, "Explicit formulas for Chebyshev impedance-matching networks, filters, and interstages," *Proc. IEEE*, vol. 53, pp. 939–963, Aug. 1964.

MICROWAVE OCTAVE-BAND GaAs FET AMPLIFIERS[*]

Walter H. Ku
M.E. Mokari-Bolhassan[**]
Wendell C. Petersen
School of Electrical Engineering
Cornell University
Ithaca, New York 14853

Allen F. Podell
Bruce R. Kendall
HPA Division
Hewlett-Packard Company
Page Mill Road
Palo Alto, California 94304

Abstract

The design of microwave broadband amplifiers using the 1μ-gate GaAs field-effect transistors covering the 4-8 GHz and 7-14 GHz octave bands is presented. The braodband matching networks of these amplifiers consist of lumped and/or distributed circuit elements. Using analytical and computer-aided optimization techniques, a typical octave-band amplifier has been designed with a nominal power gain of 8 dB with a maximum deviation of ±0.07dB covering the 7-14 GHz band based on the measured scattering parameters of a 1μ GaAs FET chip. For a packaged 1μ GaAs FET, a 4-8 GHz band amplifier has been designed with a gain of 7.2 dB ± 0.2 dB.

Introduction

This paper presents some results on the design of microwave broadband amplifiers using the 1μ-gate GaAs field-effect transistors covering the 4-8 GHz and 7-14 GHz octave bands. The GaAs FET chips used in the broadband amplifier design have been described by Liechti and Tillman[1]. The packaged GaAs FETs have been recently developed by HPA and use HP style 60 packages. Based on the basic design theory and gain-bandwidth limitations described previously[2,3], the results on the octave-band GaAs FET amplifiers are summarized in the following sections.

Broadband Amplifier Design for GaAs FET Chips

The measured scattering parameters of the HP 1μ-gate GaAs FET chip are presented in Table I. The lumped unilaterial equivalent circuit of the chip is shown in Figure 1. The input equivalent circuit is a series RC circuit and the output equivalent circuit can be approximated by a shunt RC circuit. The design of lumped matching networks with n=4 is illustrated in Figures 2, 3, and 4. For the circuit shown in Figure 2, the initial matching networks are designed to give broadband matching of the reactive constraints as well as broadband impedance transformation. In addition, the input matching network is designed to provide the required gain taper. The initial and optimized element values are presented as follows (Circuit in Figure 2):

Initial	Optimized
$C_1 = 0.2776$ pF	$C_1 = 0.2489$ pF
$L_1 = 2.239$ nH	$L_1 \simeq \infty$
$L_2 = 0.8975$ nH	$L_2 = 0.5230$ nH
$L_3 = 1.746$ nH	$L_3 = 0.8080$ nH
$C_2 = 0.6343$ pF	$C_2 \simeq \infty$
$L_4 = 2.006$ nH	$L_4 = 1.387$ nH
$L_5 = 1.547$ nH	$L_5 = 2.187$ nH
$C_3 = 0.2959$ pF	$C_3 = 0.9397$ pF

It is noted that the initial responses of the matching networks have equiripple characteristics (Figure 3). The predicted initial response is calculated by assuming $s_{12}=0$ and is nearly minimax as shown in Figure 4.

Two lumped/distributed broadband amplifiers for the GaAs FET chips have been designed. The amplifier circuits are shown in Figures 5 and 7 and the amplifier responses are presented in Figures 6 and 8. The optimized circuit elements for Figure 5 are given by

$Z_{01} = 57.41\Omega$ $\ell_1 = 0.361$ cm
$Z_{02} = 42.18\Omega$ $\ell_2 = 0.356$ cm
$Z_{03} = 54.7\Omega$ $\ell_3 = 0.367$ cm
$L_2 = 1.5624$ nH
$Z_{04} = 40.98\Omega$ $\ell_4 = 0.359$ cm
$Z_{05} = 56.76\Omega$ $\ell_5 = 0.499$ cm

For the GaAs FET chips, all the amplifiers are designed for the 7-14 GHz octave band. For the simplified matching networks shown in Figure 7, the amplifier gain response has a nominal power gain of 8 dB with a maximum deviation of ±0.07 dB (Figure 8).

Broadband Amplifier Design for Packaged GaAs FETs

For the packaged GaAs FETs, the measured scattering parameters are presented in Table II and the lumped equivalent circuit is shown in Figure 9. For both the input and output matching networks, two simultaneous reactive constraints are imposed.

For the lumped amplifier design shown in Figure 10, the element values are given by

Initial	Optimized
$C_1 = 0.5490$ pF	$C_1 = 0.5350$ pF
$L_1 = 1.0017$ nH	$L_1 = 1.0546$ pF
$C_2 = 1.8098$ pF	$C_2 = 1.7152$ pF
$L_2 = 0.6730$ nH	$L_2 = 0.7273$ nH
$C_3 = 0.6696$ pF	$C_3 = 1.0802$ pF
$L_3 = 0.6139$ nH	$L_3 = 0.4276$ nH
$L_4 = 1.7228$ nH	$L_4 = 0.9007$ nH
$C_4 = 0.4237$ pF	$C_4 = 0.4182$ pF
$L_5 = 1.2341$ nH	$L_5 = 0.2369$ nH

The results are presented in Figures 11 and 12.

For the lumped/distributed amplifier design shown in Figures 13 and 14, the optimized element values are given by

$Z_{01} = 59.24\Omega$ $\ell_1 = 0.595$ cm
$Z_{02} = 66.94\Omega$ $\ell_2 = 0.459$ cm
$Z_{03} = 20.21\Omega$ $\ell_3 = 0.970$ cm
$Z_{04} = 50.86\Omega$ $\ell_4 = 0.568$ cm
$C_1 = 1.671$ pF

---

[*] Portion of this work was supported by the NSF Grant GK-31012X and Air Force RADC Post-Doctoral Program under Contract No. F30602-72-C-0497.

[**] On leave from the faculty of the Department of Electrical Engineering, Pahlavi Univ., Shiraz, IRAN.

$Z_{05} = 49.09\Omega$ $\ell_5 = 0.521$ cm
$Z_{06} = 39.74\Omega$ $\ell_6 = 1.08$ cm
$Z_{07} = 17.50\Omega$ $\ell_7 = 0.1875$ cm

For the packaged GaAs FETs, all the amplifiers are designed for the 4-8 GHz octave band. For the simplified matching networks shown in Figure 15, the amplifier gain response has a nominal power gain of 7.2 dB and gain flatness is ±0.2 dB across the 4-8 GHz band (Figure 16).

Analytical and Computer-Aided Design Techniques

The design of the broadband GaAs FET amplifiers presented in this paper is based on both analytical and computer-aided optimization techniques. Based on the unilateral FET models, the broadband matching networks are designed analytically to provide for 1) reactance(s) absorption to satisfy the inherent reactive constraints; 2) broadband impedance transformation; and 3) tapering of the gain characteristics to account for the intrinsic gain roll-off of the FETs. The computer-aided optimization techniques are used only to account for the inaccuracies of the FET models (lumped equivalent circuits and the unilateral assumption of $s_{12} = 0$.)

As shown in Table III, the intrinsic gain slope of the FET chip used in our design is approximately 6.75 dB/octave. We have chosen to compensate for this taper in the input matching network. The maximun available gain of the chip is approximately 7.98 dB at 14 GHz.

Refering to Figure 1 for the lumped model of the FET chip, the initial analytical design involves the design of the input matching network to absorb the capacitance of 0.46 pF, to transform 9.66 Ω to 50Ω, and to provide the required 6.75 dB/octave slope. The output matching network is designed to be flat and it provides the impedance transformation from 503.4Ω to 50Ω and the absorption of the shunt capacitance of 0.072 pF.

The maximum available gain of the packaged GaAs FET is presented in Table IV. The gain roll-off is approximately 5.78 dB/octave. The maximum available gain is approximately 7.515 dB at 8 GHz.

Results presented in this paper show that the lumped initial designs are very close to the optimized gain responses for the octave-band FET amplifiers. For the lumped/distributed designs presented, approximations are needed in transforming lumped and distributed elements. Several new classes of distributed networks have recently been developed. These new structures can be used for the design of broadband transistor amplifiers directly in the distributed domain.

References

1. C. A. Liechti and R. L. Tillman, "Applications of GaAs Schottky-Gate FET's in Microwave Amplifiers," *Digest of International Solid-State Circuits Conference*, Vol.VII, Philadelphia, 1973.
2. W.H. Ku, W.C. Petersen, and A.F. Podell, *Proc.IEEE/GMTT International Microwave Symposium*, pp.357-359, 1974.
3. W.H. Ku and W.C. Petersen, *Proc. of IEEE International Symposium on Circuits and Systems*, pp. 539-543, 1974.

TABLE I

MEASURED SCATTERING PARAMETERS OF A HP 1µ-GATE GaAs FET CHIP

FREQ (GHz)	S_{11} MAG	S_{11} ANG	S_{12} MAG	S_{12} ANG	S_{21} MAG	S_{21} ANG	S_{22} MAG	S_{22} ANG
7	.826	-88.9	.0358	37.4	1.69	100.5	.834	-19.3
8	.805	-96.5	.0375	33.2	1.56	92.6	.830	-21.4
9	.787	-103.1	.0389	29.7	1.44	85.2	.828	-23.6
10	.771	-108.9	.0399	26.6	1.34	78.3	.826	-25.5
11	.759	-114.1	.0407	23.9	1.25	71.8	.826	-27.8
12	.748	-118.6	.0412	21.5	1.16	65.6	.826	-29.9
13	.739	-122.7	.0416	19.4	1.09	59.7	.827	-31.9
14	.731	-126.3	.0419	17.5	1.03	54.1	.829	-34.0

TABLE II

MEASURED SCATTERING PARAMETERS OF A PACKAGED HP 1µ-GATE GaAs FET

FET# = -3   UD = 4   ID = .035   UG = 0

FREQ (MHz)	11 MAG	11 ANG	12 MAG	12 ANG	21 MAG	21 ANG	22 MAG	22 ANG
1000	.976	-33.4	.014	73.6	3.371	151.7	.685	-11.7
1500	.982	-46.9	.021	62.6	2.979	138.5	.677	-17.3
2000	.898	-61.4	.025	53.9	3.140	124.9	.659	-22.3
2500	.825	-76.9	.029	44.6	2.924	111.6	.636	-27.5
3000	.812	-97.2	.032	36.5	2.773	99.9	.617	-33.0
3500	.741	-112.5	.034	33.8	2.698	87.4	.585	-37.6
4000	.712	-129.9	.036	25.9	2.589	74.7	.563	-42.4
4500	.741	-148.6	.038	28.3	2.484	63.1	.520	-47.9
5000	.682	-161.7	.038	16.7	2.376	51.6	.485	-55.9
5500	.695	-177.2	.040	12.6	2.229	38.9	.457	-64.0
6000	.668	168.1	.043	6.9	2.111	26.9	.412	-74.7
6500	.712	152.4	.045	.3	1.987	15.3	.370	-83.3
7000	.659	145.2	.047	-5.0	1.810	5.3	.320	-95.9
7500	.695	135.7	.048	-11.0	1.698	-6.4	.266	-109.9
8000	.674	126.7	.050	-16.8	1.649	-16.1	.244	-132.2
8500	.653	117.5	.051	-21.9	1.529	-26.3	.249	-148.8
9000	.669	108.3	.054	-26.9	1.478	-36.8	.246	-162.0
9500	.657	99.8	.058	-30.7	1.376	-46.5	.224	-170.2
10000	.665	91.6	.065	-35.3	1.351	-57.2	.179	-165.7
10500	.647	80.3	.076	-43.0	1.332	-68.9	.191	197.3
11000	.641	66.3	.084	-53.5	1.269	-94.7	.221	106.0
11500	.619	52.4	.094	-60.3	1.162	-99.7	.252	92.8
12000	.595	44.0	.118	-70.4	.998	-112.0	.255	70.9

TABLE III

MAXIMUM AVAILABLE GAIN (MAG) AND OPTIMIZED GAIN (FIG. 7) GaAs FET CHIP

FREQ (GHz)	MAG (dB)	OPT. GAIN (dB)
7.0	14.7327	8.0343
7.5	13.8394	7.9830
8.0	13.1167	7.9847
8.5	12.4965	8.0084
9.0	11.8915	8.0103
9.5	11.3853	8.0350
10.0	10.8747	8.0162
10.5	10.4311	7.9955
11.0	10.0424	8.0238
11.5	9.5893	7.9737
12.0	9.2085	7.9873
12.5	8.8310	7.9889
13.0	8.5491	8.0535
13.5	8.2715	8.0505
14.0	7.9848	7.9247

TABLE IV

MAXIMUM AVAILABLE GAIN (MAG) AND OPTIMIZED GAIN (FIG. 15) PACKAGED GaAs FET

FREQ (GHz)	MAG (dB)	OPT. GAIN (dB)
4.0	13.293	7.1429
4.5	13.038	7.2443
5.0	11.725	7.3146
5.5	11.235	7.2172
6.0	10.224	7.2483
6.5	10.112	7.3805
7.0	8.375	6.9729
7.5	8.064	7.2465
8.0	7.515	7.0052

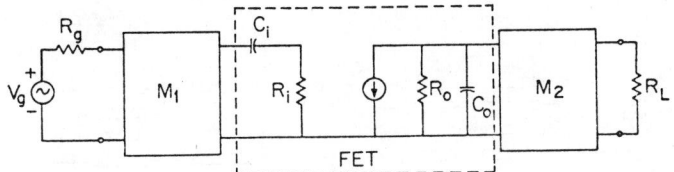

$C_i = 0.46\ \text{PF}$ $C_o = 0.072\ \text{PF}$
$R_i = 9.66\ \Omega$ $R_o = 503.4\ \Omega$
$R_g = 50.0\ \Omega$ $R_L = 50.0\ \Omega$

Figure 1 Unilateral FET Equivalent Circuit

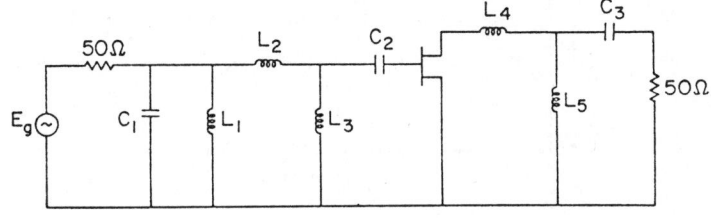

Figure 2 Lumped Circuit Amplifier

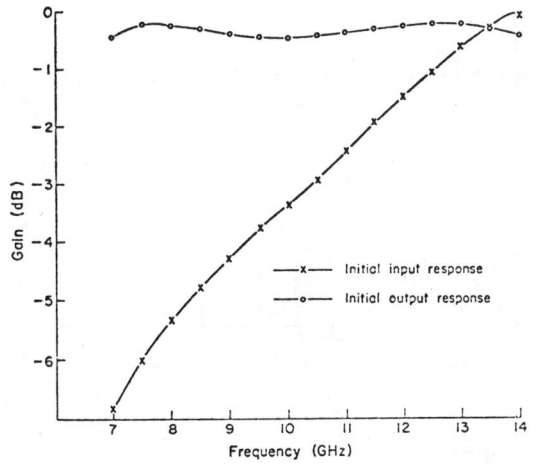

Figure 3 Initial Response-Matching Networks

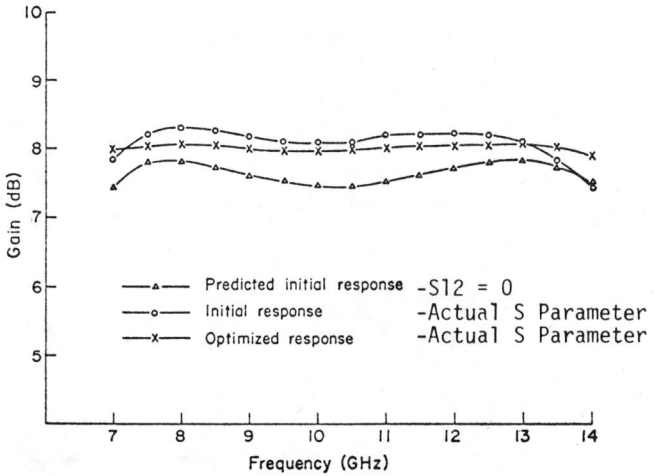

Figure 4 Frequency Response-Lumped Amplifier

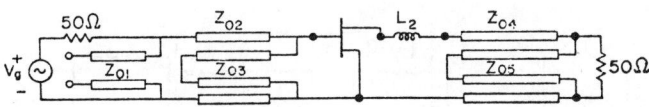

Figure 5 Distributed-Lumped Amplifier 1

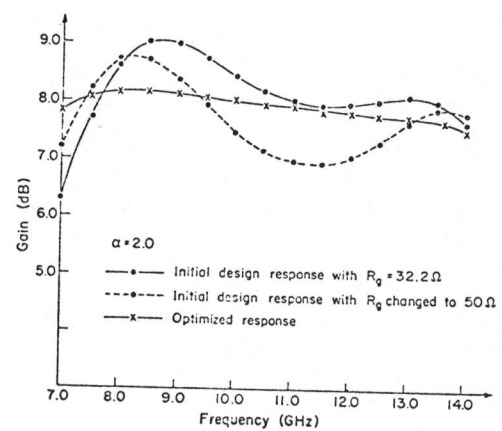

Figure 6 Frequency Response - D.L.A. #1

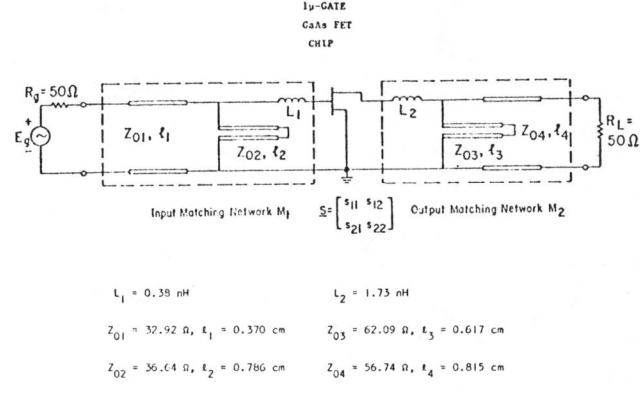

1µ-GATE
GaAs FET
CHIP

$\ell_1 = 0.38\ \text{nH}$ $\ell_2 = 1.73\ \text{nH}$
$Z_{01} = 32.92\ \Omega,\ \ell_1 = 0.370\ \text{cm}$ $Z_{03} = 62.09\ \Omega,\ \ell_3 = 0.617\ \text{cm}$
$Z_{02} = 36.04\ \Omega,\ \ell_2 = 0.786\ \text{cm}$ $Z_{04} = 56.74\ \Omega,\ \ell_4 = 0.815\ \text{cm}$

Figure 7 Distributed-Lumped Amplifier 2

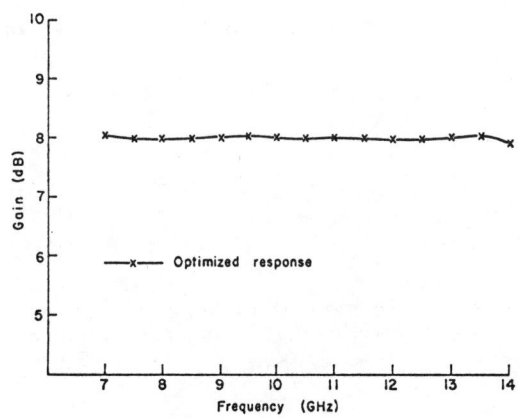

Figure 8 Frequency Response - D.L.A. #2

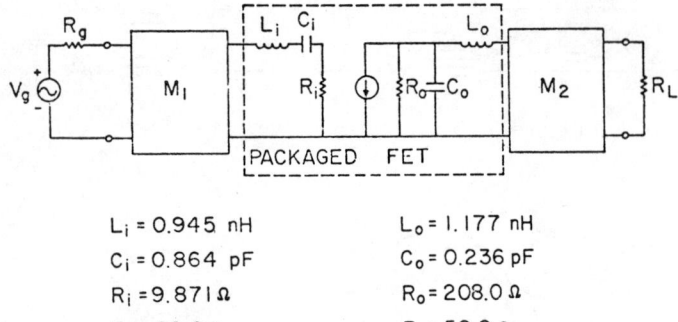

$L_i = 0.945$ nH
$C_i = 0.864$ pF
$R_i = 9.871\ \Omega$
$R_g = 50.0\ \Omega$

$L_o = 1.177$ nH
$C_o = 0.236$ pF
$R_o = 208.0\ \Omega$
$R_L = 50.0\ \Omega$

Figure 9 Unilateral Packaged FET Equivalent Circuit

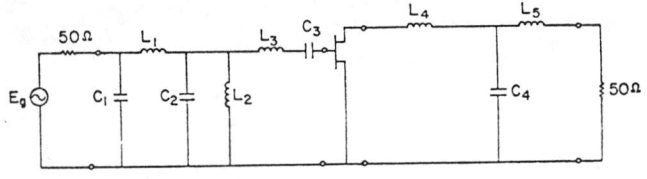

Figure 10 Lumped Circuit Amplifier

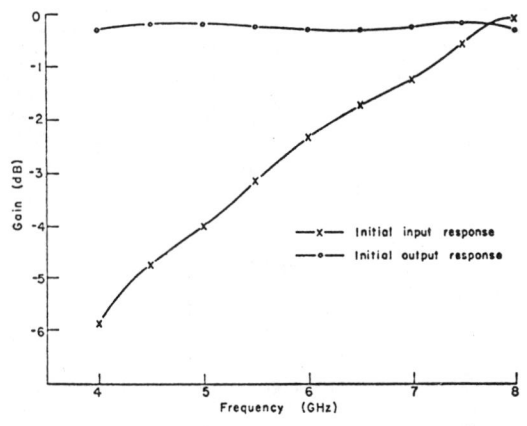

Figure 11 Initial Response-Matching Networks

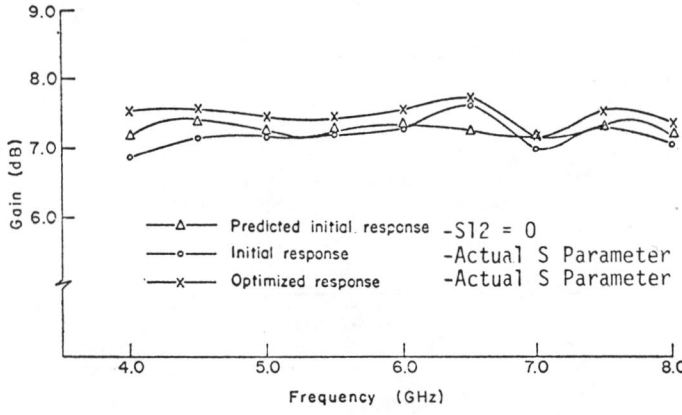

Figure 12 Frequency Response-Lumped Amplifier

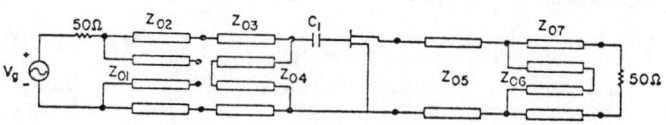

Figure 13 Distributed-Lumped Amplifier 3

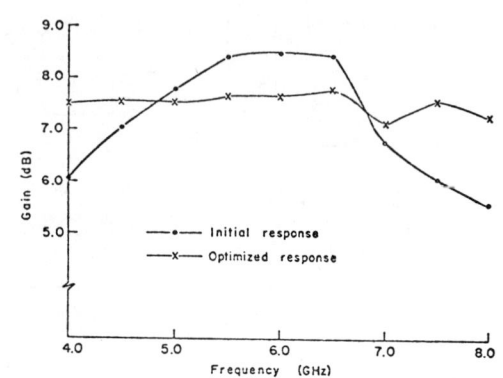

Figure 14 Frequency Response - D.L.A. #3

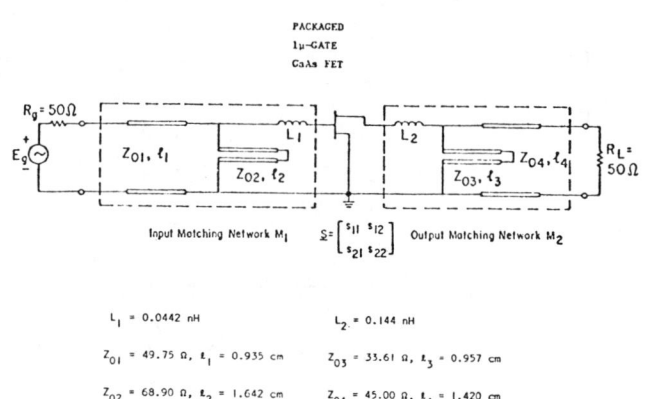

PACKAGED
1μ-GATE
GaAs FET

$L_1 = 0.0442$ nH
$Z_{01} = 49.75\ \Omega,\ \ell_1 = 0.935$ cm
$Z_{02} = 68.90\ \Omega,\ \ell_2 = 1.642$ cm

$L_2 = 0.144$ nH
$Z_{03} = 33.61\ \Omega,\ \ell_3 = 0.957$ cm
$Z_{04} = 45.00\ \Omega,\ \ell_4 = 1.420$ cm

Figure 15 Distributed-Lumped Amplifier 4

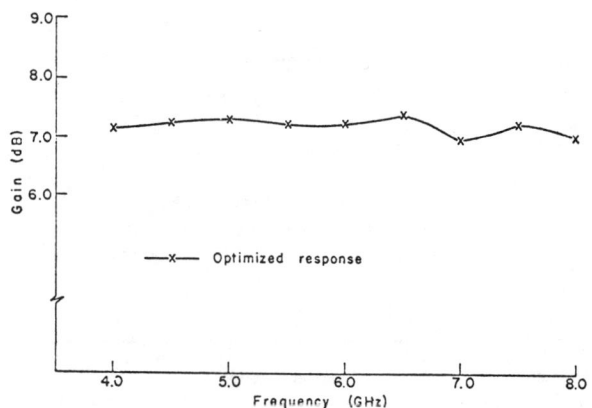

Figure 16 Frequency Response - D.L.A. #4

Part V
Practical Amplifier Techniques

A Wide-Band Low Noise *L*-Band Balanced Transistor Amplifier

R. S. ENGELBRECHT, MEMBER, IEEE, AND K. KUROKAWA, MEMBER, IEEE

Abstract—The design principles, construction details, and experimental results of a balanced transistor amplifier suitable for precise wide-band applications in the low microwave frequency range are described. Each amplifier stage consists of a pair of electrically similar transistors whose input and output signals are combined by 3-dB directional couplers. This eliminates the need for tuning adjustments.

For a wide frequency range. the main advantages of the balanced design over more conventional multistage amplifiers are 1) improved input and output impedance matching, gain flatness, phase linearity, gain compression, and intermodulation characteristics, 2) possible designing of the amplifier simultaneously for minimum noise figure and good input match, 3) relatively little effect on overall amplifier gain and matching by changes in the distribution of transistor impedance characteristics, provided that transistors can be selected in similar pairs. The amplifier gain is easily controlled over a wide range by the dc bias with little degradation of the gain flatness and impedance matches.

To obtain these advantages, a second transistor is required for each amplifier stage.

A four-stage balanced amplifier was designed and constructed in printed circuit form for *L*-band operation, using Western Electric GF-40037 transistors in the common emitter configuration. The transistors were paired in a test circuit so that their input or output impedances were within about ten per cent. (With the current GF-40037 distribution, this takes less than five random tries on the average.) Without any tuning adjustments. the following characteristics were obtained over a 20 per cent (and 60 per cent) frequency band in the 0.8 to 1.6-Gc/s range:

- Gain = 20 dB ± 0.2 dB (± 0.5 dB)
- Reverse Loss > 50 dB (> 40 dB)
- Phase < ± 1° from linear (± 6°)
- VSWR In < 1.10 (< 1.2)
- VSWR Out < 1.10 (< 1.2)
- Noise Figure ≅ 6 dB
- Gain Compression ≤ 0.25 dB at 0-dBm output signal level
- Third-Order Intermodulation: With −3-dBm output power from two signals at frequencies f_A and f_B, power out at $2f_A - f_B$ and $2f_B - f_A \simeq -50$ dBm.

The 3-dB bandwidth points are at 650 Mc and 1700 Mc, and the voltage standing-wave ratios (VSWR's) over this range are under 2.

The transistors are operated at $I_e = 2$ to 4 mA and $V_{CB} = 5$ to 6 volts, depending on the stage. The total dc power consumption for the four-stage amplifier is about 50 mA at 7.5 volts.

I. INTRODUCTION

TRANSISTOR amplifiers for use in the low microwave range are currently receiving widespread interest because of their greater simplicity and higher power handling capability, as compared with other solid-state microwave amplifiers. Available microwave transistors, such as the Western Electric GF-40037 (L-2254), have been used in the past two years for designing amplifiers with approximately 5-dB gain per stage and 7-dB noise figures in the 1- to 1.5-Gc/s frequency range.[1] These amplifiers are similar to lower-frequency tuned RF amplifiers in design and require numerous tuning adjustments for obtaining wide-band flat gain characteristics and good input and output impedance matching. In addition to the circuit complexity and construction expense introduced by the tuning mechanisms, it has been found that a considerable amount of time is required to adjust amplifiers for flat gain within a few tenths of a dB and for input and output VSWR's of <1.3 over 20 per cent frequency bands. Such performance characteristics, which are typical of precise microwave receiving systems, must often be maintained by the amplifier over a period of years. Although this problem of circuit stability with time is greatly alleviated by the use of distributed tuning networks, it has become apparent that a drastically new approach to microwave transistor amplifier design is desirable.

In this paper the design principles and experimental results obtained with the first model of a balanced transistor amplifier designed for operation in the 1- to 1.5-Gc/s range will be discussed. In this amplifier design, transistors are selected in pairs for operation in each amplifier stage. The input and output signals of each transistor pair are combined by means of 3-dB directional couplers. As will be shown, the need for tuning adjustments can thereby be eliminated. In spite of this simplification, the input and output impedance matches and the gain flatness of the balanced amplifier are equal or superior to those obtained with the more conventional amplifiers after extensive tuning. Additional advantages are considerably improved intermodulation characteristics and phase linearity and the possibility of designing the amplifier for minimum noise figure operation while maintaining an impedance match at the amplifier input. Furthermore, the amplifier gain can be controlled over wide ranges by the dc bias with little effect on gain flatness or impedance matching. Finally, for those applications in which the insertion gain or phase of the amplifier are not critical, superior reliability is obtained; the failure of any one transistor drops the over-all gain by approximately 6 dB.

Manuscript received November 6, 1964.
The authors are with Bell Telephone Laboratories, Inc., Murray Hill, N. J.

[1] Hamasaki, J., A wideband, high gain, transistor amplifier at l-band, *Digest of Technical Papers*, 1963 Internat'l Solid State Circuits Conf. pp 46–47.

The above advantages are obtained at the expense of an increased number of transistors, since approximately twice as many are needed for a given amplifier gain. This is offset to some extent by the fact that the balanced amplifier design does not depend critically on a given transistor distribution as long as transistors can be readily obtained in pairs.

II. Principles of Operation

A. Design Concept

As shown in Fig. 1, each stage of the balanced amplifier consists of a pair of transistors T_a and T_b whose inputs are connected to two ports forming a conjugate pair of a 3-dB directional coupler and whose outputs are similarly connected to conjugate ports of another 3-dB coupler. As is well known, 3-dB couplers of excellent electrical characteristics are obtained by means of a coupled pair of parallel transmission lines of properly chosen geometry. Their electrical characteristics are symmetrical about the frequency f_0, at which the coupling region is $\lambda/4$ long. Over a broad band around f_0, the properties of the 3-dB coupler are such that a signal entering the input port 1 will be split into two signals of nearly equal amplitude and 90° relative phase emerging from ports 2 and 4, each being 3 dB down in power from the input signal.

Assuming for the moment that the input and output impedances of transistors T_a and T_b are matched to the 3-dB couplers and that they have identical insertion gain and phase characteristics, the two signals from ports 2 and 4 will be amplified and recombined in the output coupler where they emerge as a single output signal from port 1′. The insertion gain between ports 1 and 1′ is that of either transistor. No power will appear at port 3 of the input coupler nor port 3′ of the output coupler. The input and output impedance matches at ports 1 and 1′, respectively, depend only on the input and output coupler characteristics. In principle, the couplers can be constructed so that these impedance matches are frequency independent.[2,3]

In the discussion which follows, we shall assume that ideal 3-dB couplers are used in the circuit of Fig. 1 and that any deviation from the idealized behavior described above results from imbalances between transistors T_a and T_b. This assumption is justified in Appendix I, where it is shown that actual 3-dB couplers exhibit nearly ideal characteristics over a frequency band of ±30 per cent around f_0, as well as by experiment.

B. Match

With the 3-dB couplers of Fig. 1 assumed to be ideal, the only contributions to an impedance mismatch seen

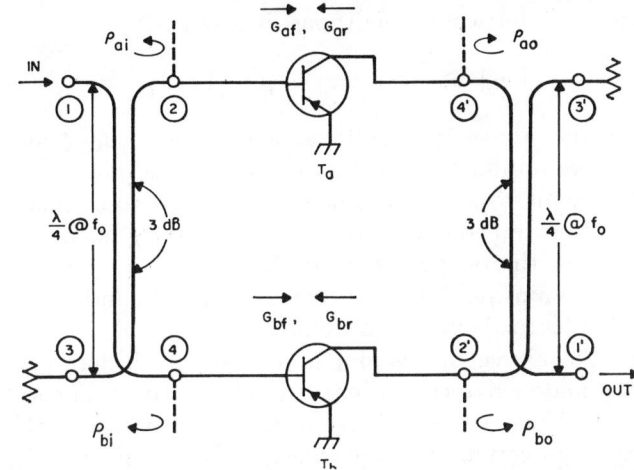

Fig. 1. Schematic representation of single-stage balanced amplifier.

at port 1 of the input coupler are due to mismatches of the transistor inputs and outputs and any mismatch in the load impedance connected to port 1′ of the output coupler, i.e., the input to the next stage in a cascaded amplifier. We shall now consider these three contributions separately, as if they existed one at a time.

If we denote the (complex) voltage reflection coefficient between port 2 of the input coupler and the input impedance of transistor T_a by ρ_{ai} and the corresponding one of transistor T_b by ρ_{bi}, then the magnitude of the reflection coefficient seen looking into port 1 is given by

$$|\rho_I| = \tfrac{1}{2}|\rho_{ai} - \rho_{bi}| \quad (1)$$

Similarly, if we denote the reflection coefficients between the transistor outputs and the output coupler by ρ_{a0} and ρ_{b0}, respectively, we obtain a reflection component at port 1 of the input coupler with magnitude

$$|\rho_{II}| = \tfrac{1}{2}|G_{af}G_{ar}\rho_{a0} - G_{bf}G_{br}\rho_{b0}| \quad (2)$$

where G_{af}, G_{bf}, G_{ar}, and G_{br} denote the complex (amplitude and phase) forward and reverse voltage gains, respectively, of transistors T_a and T_b, when operated between identical source and load impedance levels.

Finally, if we let ρ_c represent the voltage reflection coefficient seen by port 1′ of the output coupler, a reflection at port 1 of the input coupler will be obtained of magnitude

$$|\rho_{III}| = \tfrac{1}{4}|(G_{af} + G_{bf})(G_{ar} + G_{br})\rho_c| \quad (3)$$

From the above, it is clear that $|\rho_I|$ can be minimized if transistors T_a and T_b are selected to have nearly identical input impedances. $|\rho_{II}|$ and $|\rho_{III}|$ are small whenever the products $(G_{af}G_{ar})$ and $(G_{bf}G_{br})$ are much less than unity (implying transistors with high reverse losses). Typically, with paired GF-40037 transistors, $|\rho_I| < 0.05$ is readily obtainable. With $(G_{af}G_{ar})$ and $(G_{bf}G_{br})$ typically 0.4, $|\rho_{II}|$ and $|\rho_{III}|$ are less than 0.02.

Assuming that the above three contributions are reasonably small, interactions between them may be neg-

---

[2] Jones, E. M. T., and J. T. Bolljahn, Coupled-strip-transmission-line filters and directional couplers, *IRE Trans. on Microwave Theory and Techniques*, vol 4, Apr 1956, pp 75–81.

[3] Cohn, S. B., Characteristic impedance of broadside-coupled strip transmission lines, *IRE Trans. on Microwave Theory and Techniques*, vol 8, Nov 1960, pp 633–637.

lected and the total reflection at port 1 of the input coupler, under worst conditions, is given by

$$|\rho_1| \cong |\rho_\text{I}| + |\rho_\text{II}| + |\rho_\text{III}| \qquad (4)$$

With the above assumed values of $|\rho_\text{I}|$, $|\rho_\text{II}|$, and $|\rho_\text{III}|$, we obtain $|\rho_1| < 0.1$ under worst conditions.

The same type of reasoning can be used to evaluate the output impedance mismatch seen looking back into port $1'$ of the output coupler. Reflections at this port can be minimized by selecting transistors T_a and T_b for nearly identical output impedances.

In the above discussion, it was assumed that perfectly matched terminations are connected to ports 3 and $3'$. These terminations are required to absorb the average reflection from the transistor input mismatches and transistor output mismatches, respectively. Thus, for instance, with unity voltage applied to port 1 of the input coupler, the voltage appearing at port 3 of the input coupler consists of three components of magnitude,

$$|V_\text{I}| = \tfrac{1}{2}|\rho_{ai} + \rho_{bi}|$$

$$|V_\text{II}| = \tfrac{1}{2}|G_{af}G_{ar}\rho_{a0} + G_{bf}G_{br}\rho_{b0}|$$

$$|V_\text{III}| = \tfrac{1}{4}|(G_{af} + G_{bf})(G_{ar} - G_{br})\rho_c|$$

C. Gain and Phase

Assuming again ideal 3-dB couplers and transistors with reasonably well-matched input and output impedances, the insertion power gain between ports 1 and $1'$ of Fig. 1 is given by

$$G = \tfrac{1}{4}(G_{af} + G_{bf})^2 \qquad (5)$$

where G_{af} and G_{bf} are, as noted before, the forward voltage gains of transistors T_a and T_b. If these two voltage gains differ from each other in either their amplitude or phase, a power component of magnitude $\tfrac{1}{4}(G_{af}-G_{bf})^2$ will appear at $3'$ (assuming unity power incident on port 1 of the input coupler).

The transfer phase between ports 1 and $1'$ is given by the phase angle of $G_{af}+G_{bf}$ plus the transmission phase characteristics of the two 3-dB couplers. As shown in Appendix I, the latter contribution to the transfer phase is almost linear over a ± 30 per cent frequency band around f_0. Over this frequency range, the phase slope of the two 3-dB couplers is almost perfectly represented by a section of uniform transmission line, approximately $130°$ long at f_0.

The insertion gain and phase between ports 1 and $1'$ are also affected by mismatches in the transistor input and output impedances. To a first approximation, the insertion gain is reduced by the factor

$$[1 - \tfrac{1}{2}(|\rho_{ai}|^2 + |\rho_{bi}|^2)][1 - \tfrac{1}{2}(|\rho_{a0}|^2 + |\rho_{b0}|^2)]$$

As the high frequency limits of transistors are approached, their gain decreases at approximately 6 dB per octave. In order to obtain flat gain vs. frequency characteristics, a relatively simple tuned filter can be used which has no effect at the highest frequency of interest but which absorbs power at the lower frequencies. This filter, which is described in more detail later, is obtained by a pair of identical networks, one coupled to port $2'$ and the other to port $4'$ of the output coupler in Fig. 1.

D. Cascading

The input and output impedances of typical microwave transistors vary rapidly with frequency. When cascading several stages it is, therefore, very difficult to adjust conventional transistor amplifiers so as to have negligible gain ripples over a wide band. The circuit of Fig. 1, on the other hand, can be cascaded as often as necessary to obtain any desired over-all amplifier gain without encountering this problem. Since the impedance match between stages will be good (provided that nearly identical transistors are used as pairs) the gain and phase ripples which normally result from multiple interstage reflections can be kept at a minimum.

If the transistor impedances are approximately matched to the circuit at the highest frequency of interest, any reflections at the lower frequencies (again assumed to be identical from both transistors) are dissipated in the terminations connected to ports 3 or $3'$ in Fig. 1. They will thus aid in reducing the 6-dB-per-octave gain slope characteristic.

A simple approximation to the input and output impedance match of the over-all amplifier can be obtained by assuming that all stages have similar mismatch characteristics. Using the notation developed in Section A, one can readily show that the worst possible input reflection coefficient ρ_in of an amplifier with N identical stages is given by

$$|\rho_\text{in}| = (|\rho_\text{I}| + |\rho_\text{II}|) \cdot \sum_{n=0}^{N-1} \left|\frac{(G_{af} + G_{bf})(G_{ar} + G_{br})}{4}\right|^n \qquad (6)$$

(where the output of the over-all amplifier is assumed matched). More realistically, one would expect that the contributions from the N stages add rms-wise. For this case, one expects more nearly the following input reflection coefficient:

$$|\rho_\text{in}| = \left\{(|\rho_\text{I}|^2 + |\rho_\text{II}|^2) \cdot \sum_{n=0}^{N-1} \left|\frac{(G_{af} + G_{bf})(G_{ar} + G_{br})}{4}\right|^{2n}\right\}^{1/2} \qquad (7)$$

Experimentally, the product $\tfrac{1}{4}(G_{af}+G_{bf})(G_{ar}+G_{br})$ is approximately 0.4 for the L-band transistor amplifiers under discussion. For this case, the summation in (7) converges rapidly, with the first stage alone contributing about 90 per cent of $|\rho_\text{in}|$.

Similarly, an estimate of the gain ripples vs. frequency can be made by assuming a cascade of identical stages. Considering only the interactions between adjacent stages, it is easily shown that at each interstage connection a ripple component is introduced on the power gain of magnitude

$$\frac{\Delta G}{G} = \pm \frac{1}{2} \left| (\rho_{a0} - \rho_{b0})(\rho_{ai} - \rho_{bi}) \right| \qquad (8)$$

Assuming rms addition for the contributions from N interstage connections, we obtain a total power gain ripple of $\sqrt{N}(\Delta G/G)$. With appropriate selection of GF-40037 transistors in pairs, the factors $(\rho_{a0}-\rho_{b0})$ and $(\rho_{ai}-\rho_{bi})$ are 0.1 or better. With $N=4$, we thus obtain total power gain ripples of approximately ± 0.01 (or ± 0.04 dB).

E. Intermodulation and Compression

The upper limit to the range of signal levels which can be handled by an amplifier is usually determined by intermodulation and/or gain compression in the last amplifier stage. The type of intermodulation of most concern in broadband amplifiers with multiple frequency channels is usually one in which two strong signals of frequency f_A and f_B produce third-order intermodulation signals at frequencies $2f_A - f_B$ and $2f_B - f_A$, also within the over-all pass band of the amplifier, where they might interfere with wanted weak signals. The gain compression, on the other hand, is a decrease in gain as the signal output power level approaches the power from the dc supply.

Since the signal power is shared equally between two transistors in the balanced amplifier design, one would expect that a given amount of compression or intermodulation is reached at a 3-dB-higher signal level (as compared to measurements obtained with the same type of transistors in a conventional "single-string" circuit). The actual improvement in intermodulation measured with the balanced amplifier is slightly better than expected.

F. Minimum Noise Figure Operation

As is well known, the noise figure of a linear two-port transducer as a function of its source admittance Y_s is given by the expression[4]

$$F = F_{\min} + \frac{R_n}{G_s} \left| Y_s - Y_{\min} \right|^2 \qquad (9)$$

where $F_{\min}$ is the minimum noise figure obtained as a function of source admittance, $Y_{\min}$ is the particular complex source admittance at which $F_{\min}$ is obtained, and R_n is a constant of the two-port with the units of a resistance. In general, the admittance $Y_{\min}$ is not equal to, and may be substantially different from, the source admittance necessary for producing maximum available gain, i.e., conjugate match. For this reason, a simple two-port transducer usually cannot be tuned simultaneously for input impedance match and minimum noise figure. However, this can be accomplished in the balanced amplifier, provided that the two transistors selected as a pair have nearly identical noise characteristics and, therefore, require nearly identical $Y_{\min}$'s. (For the case of ideal couplers and transistors with identical noise characteristics, the noise figure of the circuit of Fig. 1 is that of either transistor.) In this case, it is only necessary to provide suitable mismatching (of an identical nature) between ports 2 and 4 of the input coupler and the inputs of transistors T_a and T_b, respectively, such that they both see a source admittance $Y_{\min}$. Since the two mismatches are identical, the input impedance seen at port 1 remains matched, with the reflection being absorbed by the termination on port 3. It is, therefore, possible to operate the circuit of Fig. 1 with minimum noise figure and matched input impedance simultaneously. In practice it has been found that with available L-band transistors, an improvement in noise figure of 0.5 to 1 dB can be obtained when tuning the transistor inputs for minimum noise figure rather than for maximum available gain.

The transducer gain of active two-ports adjusted for minimum noise figure is usually lower than its maximum available gain, due to the reflection introduced at the input. As a result, a compromise between minimum noise figure and maximum available gain is needed for the first stage in a cascaded amplifier in order to obtain the minimum over-all noise figure. If all stages (including their transistors) are identical in a cascade forming a high gain amplifier, this optimum is obtained when the individual stages are adjusted for minimum noise measure,

$$M = \frac{F - 1}{1 - \dfrac{1}{G}} \qquad (10)$$

where, as noted above, both G and F (and hence M) are functions of source admittance Y_S. The source admittance which is presented to each transistor should, therefore, be somewhere "between" the conjugate match and the minimum noise figure cases. Obviously, just as in the case of minimum noise figure, the input impedance of each stage (port 1 of Fig. 1) remains matched.

G. Gain Control

The gain of the balanced transistor amplifier can be controlled over a wide range by controlling either the dc emitter current or collector voltage to the transistors. It has been found in practice that a pair of microwave transistors (GF-40037) which was selected for identical characteristics at one dc operating point will remain very similar over an appreciable range of operating conditions. The gain of each stage in the balanced amplifier

---

[4] IRE Subcommittee 7.9 on Noise, Representation of noise in linear twoports, *Proc. IRE*, vol 48 Jan 1960, pp 69–74.

can, therefore, be controlled by changing the operating point on both transistors simultaneously without appreciably affecting the input and output impedance matches. A slight tilt in the gain vs. frequency characteristic and a change in input to output phase is usually observed as the gain is varied over a wide range, but virtually no gain or phase ripples are introduced.

H. Reliability

The use of balanced transistor pairs in every amplifier stage introduces a certain amount of redundancy. As shown above, this redundancy can be readily used for obtaining superior performance. On the other hand, for those applications where precise gain and phase performance are not essential, the same redundancy provides an extra degree of reliability. Thus, if one transistor (say T_a in Fig. 1) fails, the insertion gain between ports 1 and 1' drops by approximately a factor of 4 (6 dB) in accordance with (5). Here it was assumed that the two transistors were initially identical and that after failure the gain G_{af} dropped to zero (infinite loss). If the failure occurs in the first or last amplifier stage, the amplifier input or output impedance match is necessarily degraded, following the reasoning of Section A. However, if the failure occurs in an internal amplifier stage, the effect on input and output impedance match of the over-all amplifier is greatly reduced due to the isolation of the adjacent stages [as seen from (3)].

III. Design Details

Since the performance characteristics of the balanced amplifier depend strongly on the 3-dB directional couplers, the design details and experimental results of individual 3-dB couplers will be discussed first. Next, the characteristics of Western Electric GF-40037 transistors are briefly presented and their socket design and output impedance transformer described. Following this, it is shown how the gain slope is leveled without deterioration of the other characteristics. Finally, the dc bias circuit is described.

A. 3-dB Coupler

The design principles of 3-dB couplers are well established.[2,3] To obtain symmetrical coupling lines with proper odd- and even-mode impedances, the line configuration shown in Fig. 2 is employed. The dielectric material is cross-linked polyethylene ($\epsilon = 2.32$) clad with 2-ounce copper (2.6 mils thickness).

The thickness of the center dielectric board is held closely at 0.027 ± 0.0005 inch, in order to ensure proper coupling. A change of this thickness by 1 mil produces approximately 0.1-dB change in coupling. On the other hand, the conductor width in the coupling region is not as critical, i.e., a 10-mil difference in width produces about the same 0.1-dB change. To compensate for end effects of the coupling region, a small capacitive pad is added to each corner of the coupler as shown in Fig. 3.

The results obtained with this 3-dB coupler are quite satisfactory. The measured coupling is shown in Fig. 4. (The dotted lines show the calculated values, neglecting end effects of the coupling region.) The standing-wave ratio (SWR) looking into one port (using type N female panel mount connectors as stripline-coax transducers) when all the other ports are terminated by 50 ohms is given in Fig. 5. Figure 6 shows the same SWR with two of the ports open circuited as shown in the inset. For any identical passive impedances connected to ports 2 and 4, the SWR seen at port 1 is thus between those of Figs. 5 and 6. The input and output SWR's of a balanced amplifier should, therefore, be less than that in Fig. 6, provided that transistors with identical characteristics are used in pairs. From these figures, it is seen that, over a wide frequency range, the input and output reflection of a balanced amplifier can be considered to result only from differences in the transistor characteristics.

The loss and phase characteristics of the circuit consisting of two 3-dB couplers, as shown in Fig. 7, were also measured and are plotted in Figs. 8 and 9, respectively. One stage of the balanced amplifier is essentially this circuit with a pair of transistors inserted at the junction points 2-4' and 4-2' of the two 3-dB couplers.

B. Transistor Characteristics and Compensation Circuits

As shown in Fig. 10, the input and output impedances of GF-40037 transistors are generally scattered on the Smith chart in the shaded and cross-hatched areas, respectively. For use in the balanced-type configuration, the input impedance seems to be satisfactory. However, in order to reduce the possible mismatch losses at the output, the output impedance must be modified. This is presently done by mounting the transistor on a disk with a thin folded line connected to the collector pin as shown in Fig. 11. The thin line acts approximately as a lumped inductance connected in series with the transistor output impedance. Therefore, the resultant output impedance approaches the center of the Smith chart (50 ohms). The best inductance value can be selected for each transistor from disks prepared with several different inductances. After this improvement, the output impedance generally shows VSWR's of less than 2 over a 200-Mc bandwidth at L band. This SWR is small enough provided the reflection is nearly the same for the two transistors in a pair.

One additional compensation circuit is necessary. This is due to the fact that for available transistors the gain at L band tends to decrease with increasing frequency. In order to get a flat gain response, the excess signal power at the lower frequencies has to be absorbed in some circuit element without deteriorating the impedance balance discussed above. As shown in Fig. 12, a simple circuit is employed for this purpose. It consists of a resistance connected in series to a parallel-resonant circuit. The resonant frequency is adjusted to be at or slightly above the highest frequency of interest. At this frequency, essentially no power is absorbed. At lower

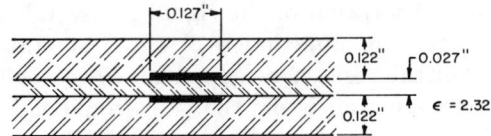

Fig. 2. Cross section of 3-dB coupler configuration.

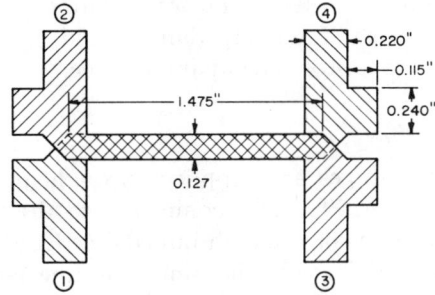

Fig. 3. Top view of 3-dB coupler center conductors.

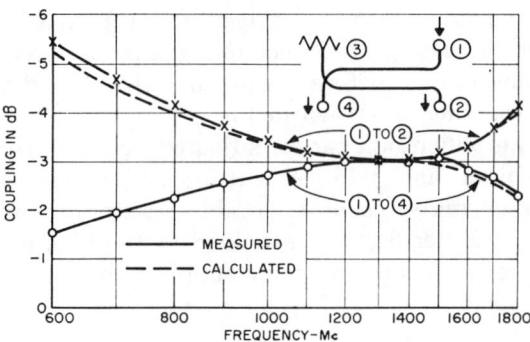

Fig. 4. Coupling of 3-dB coupler vs. frequency.

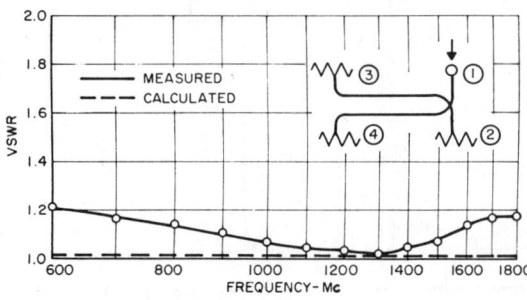

Fig. 5. SWR looking into one port when other ports are terminated.

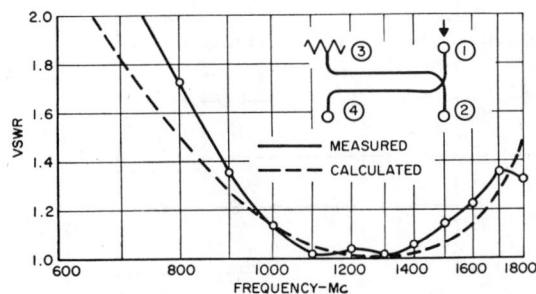

Fig. 6. SWR looking into one port when two opposite ports are open circuited.

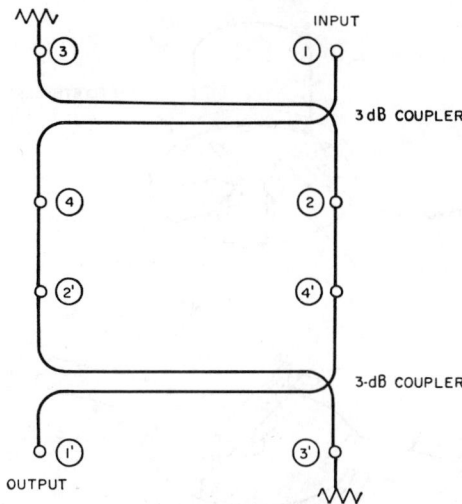

Fig. 7. Connection of two 3-dB couplers.

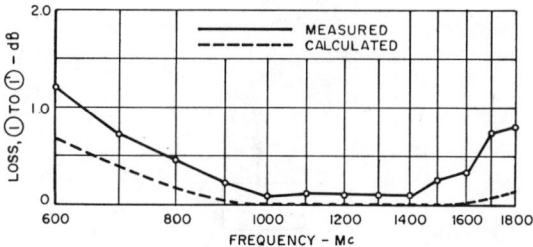

Fig. 8. Insertion loss of the circuit of Fig. 7.

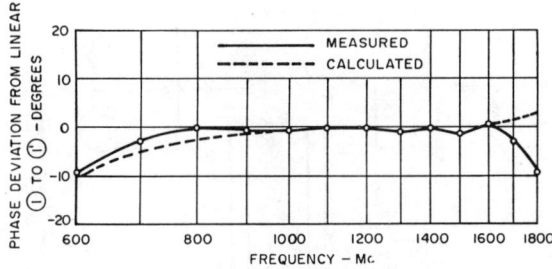

Fig. 9. Phase deviation from linearity of the circuit of Fig. 7.

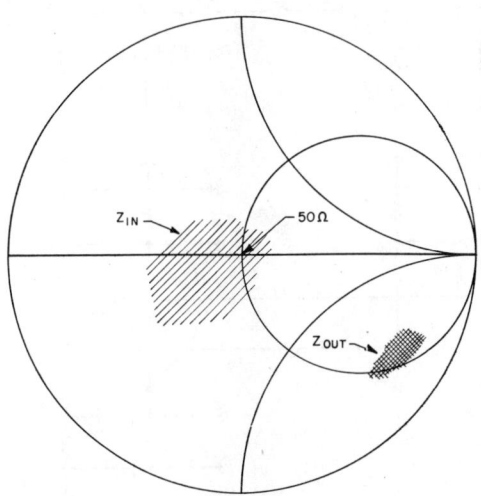

Fig. 10. Typical input- and output-impedance distribution of GF-40037 transistors.

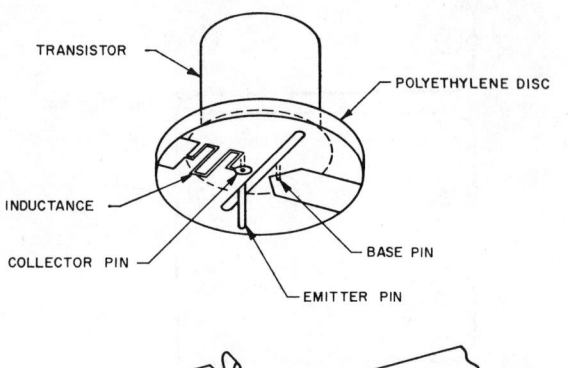

Fig. 11. Transistor socket.

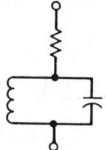

Fig. 12. Absorption filter for gain leveling.

frequencies, absorption of the signal power takes place. The characteristic impedance of the resonant circuit and value of the resistor are determined in such a way that for average transistors the gain-frequency characteristic becomes level in the amplifier. The circuit is connected to the main RF circuit at the output side of each transistor and as close to the socket as possible. This position was selected in order that the frequency dependence of the filter impedance could partly cancel that of the transistor output impedance.

C. DC Circuitry

The bias voltages are supplied through high impedance lines ($Z_0 \simeq 150$ ohms) connected to the RF lines. Since the far ends of the high impedance lines are effectively grounded for RF and since the line lengths are approximately a quarter wavelength, the presence of these bias circuits does not appreciably affect the RF circuits. To provide dc isolation between stages, the RF lines have cuts which are bridged by mica capacitors. A two- or three-stage RC decoupling filter is inserted in each bias circuit before connecting those of different stages together as shown in Fig. 13.

The amplifier gain can be adjusted over a fairly wide range (more than ± 25 per cent in dB) by changing the emitter currents, using a variable resistor in the base circuit. A Zener diode is provided at the bias input to eliminate variations in the dc supply voltage.

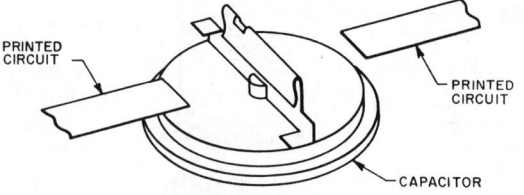

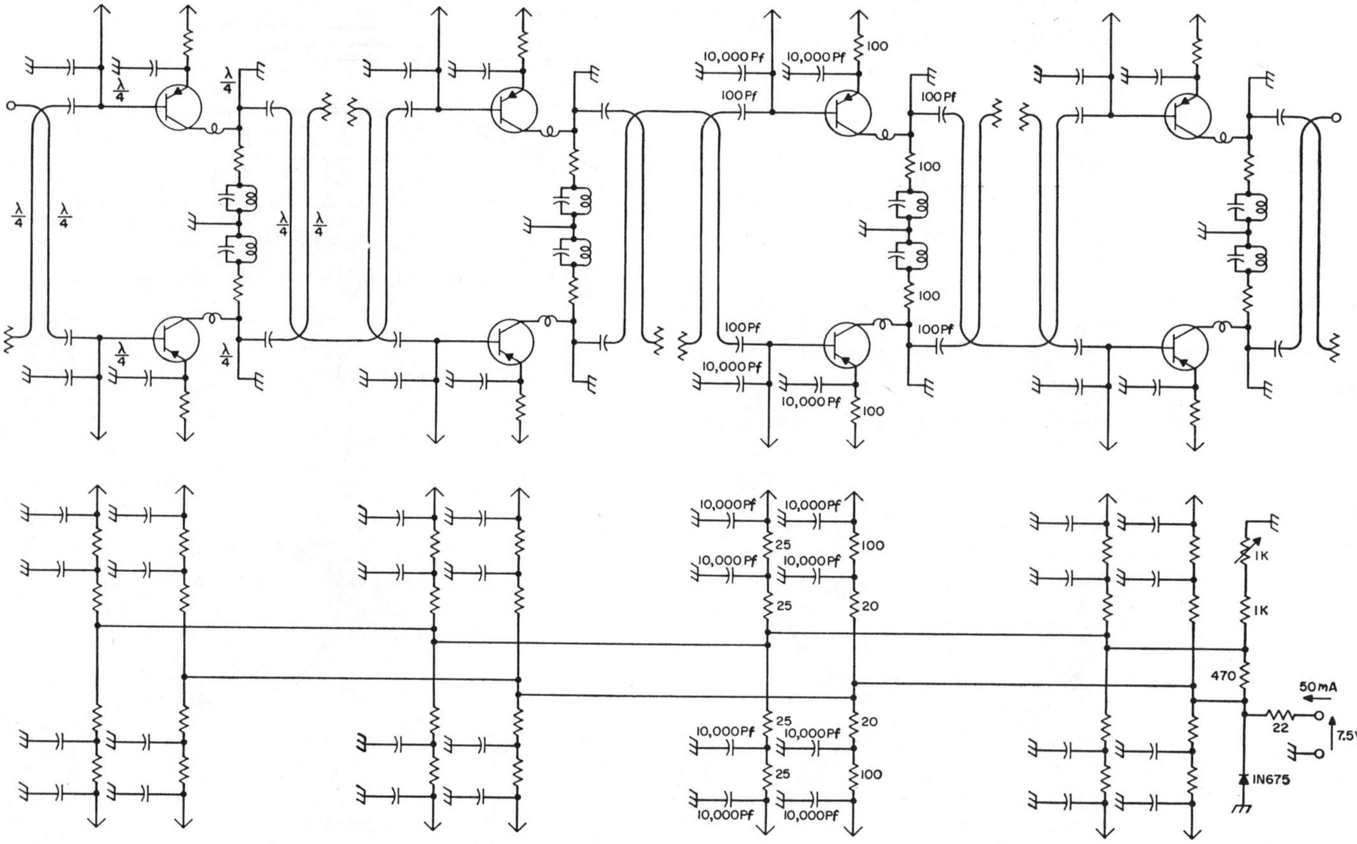

Fig. 13. Schematic diagram of four-stage balanced amplifier.

IV. Experimental Results

A. Single-Stage Amplifier

As a building block of multistage amplifiers, a single-stage balanced amplifier was built first and the paired characteristics of transistors were checked. In Fig. 14, the gain vs. frequency curves over a 200-Mc bandwidth at L band are shown for three different transistor pairs. Over the same bandwidth, the corresponding phases are linear within $\pm 1°$, which is about the limit of measurement accuracy.

The input and output SWR's were less than 1.2 for all pairs. Since the selection of matched pairs was made from a limited number of transistors, it is expected to be relatively easy to obtain pairs which show better SWR's, e.g., less than 1.1, from a larger lot of transistors such as encountered in mass production.[5]

B. Four-Stage Amplifier

Using some of the pairs selected and checked as described above, a four-stage balanced amplifier was assembled and tested. The gain and phase over a 200-Mc bandwidth at L band for one set of transistors are shown in Figs. 15 and 16, respectively. The gain is flat within ± 0.15 dB and the phase nonlinearity less than $\pm 1.0°$. Nearly the same results were obtained with several different sets of transistors. The noise figure and the reverse loss were measured from 1000 Mc to 1800 Mc and are plotted in Figs. 17 and 18. Figures 19 and 20 show the measured gain and phase over an extended frequency range and Fig. 21 the corresponding input and output SWR's.

One-dB gain compression takes place when the output signal power reaches $+5$ dBm as shown in Fig. 22. At 0-dBm output the gain compression is about 0.2 dB. The third-order intermodulation output at $2f_A - f_B$ was measured at three different frequencies. As shown in Fig. 23, when the total output power in signals f_A and f_B is -10 dBm, the output power at $2f_A - f_B$ is less than -77 dBm. Correspondingly, for 0-dBm total power at f_A and f_B, the output at $2f_A - f_B$ is about -50 dBm. The intermodulation and compression characteristics can be further improved by increasing the emitter current on the last amplifier stages (3 mA in the above case). A photograph of the assembled amplifier is shown in Fig. 24. A disassembled amplifier can be seen in Fig. 25.

V. Conclusions

The balanced transistor amplifier design described in this memorandum has been demonstrated to be capable of providing excellent electrical characteristics over a wide frequency band in the low microwave range. Without any tuning adjustments, the impedance matches, gain flatness, phase linearity, and intermodulation characteristics are equal or superior to those obtainable with conventional tuned amplifiers. The required transistor

[5] Cf. Appendix II.

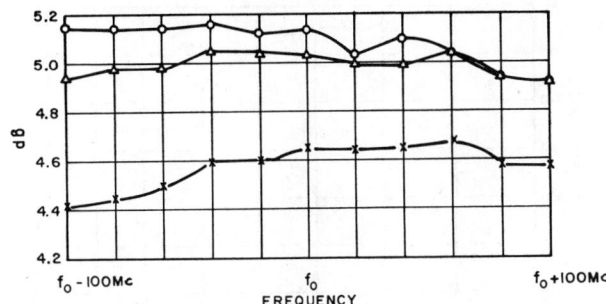

Fig. 14. Gain vs. frequency curves for three different transistor pairs in single stage.

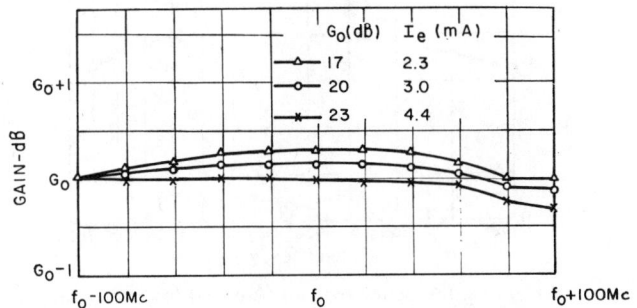

Fig. 15. Gain characteristic of four-stage amplifier.

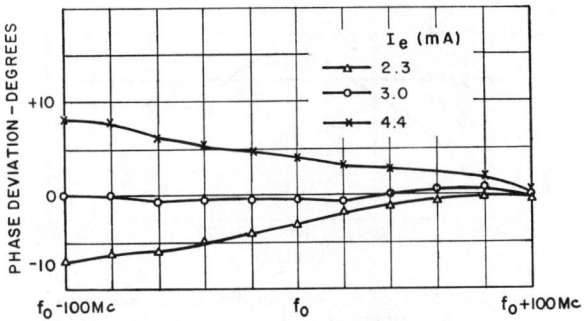

Fig. 16. Phase characteristic of four-stage amplifier.

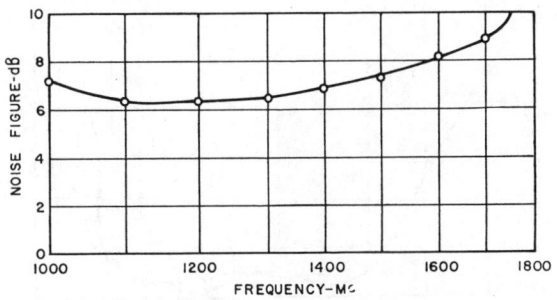

Fig. 17. Noise figure of four-stage amplifier.

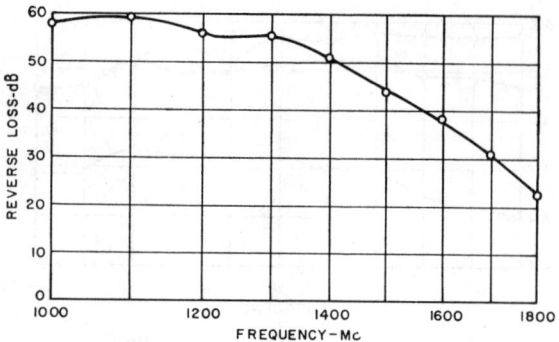

Fig. 18. Reverse loss of four-stage amplifier.

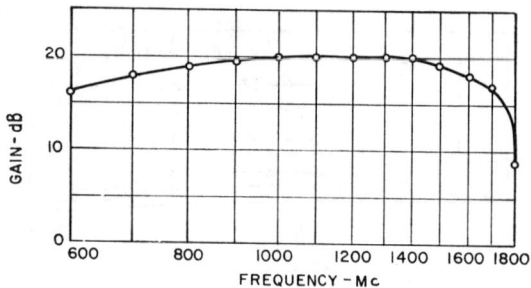

Fig. 19. Gain vs. frequency over an extended frequency range.

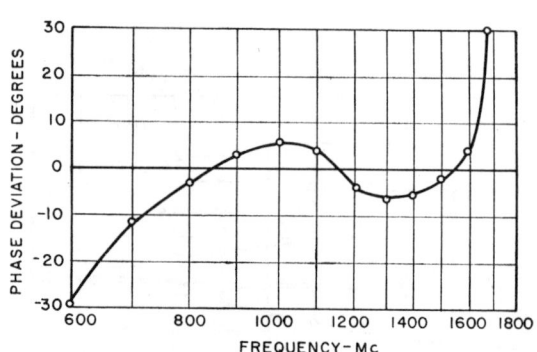

Fig. 20. Phase deviation from linear vs. frequency over an extended frequency range.

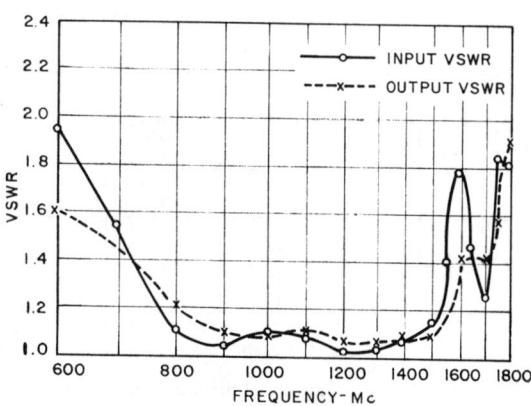

Fig. 21. Input and output SWR of four-stage amplifier.

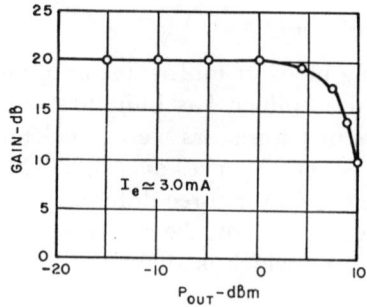

Fig. 22. Gain compression characteristic of four-stage amplifier.

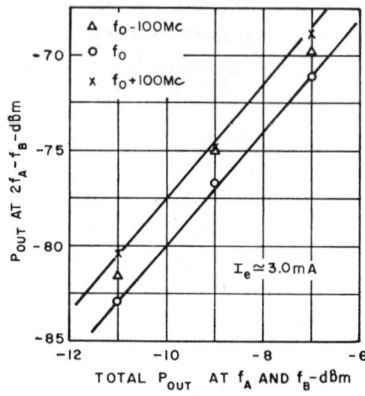

Fig. 23. Intermodulation output of four-stage amplifier.

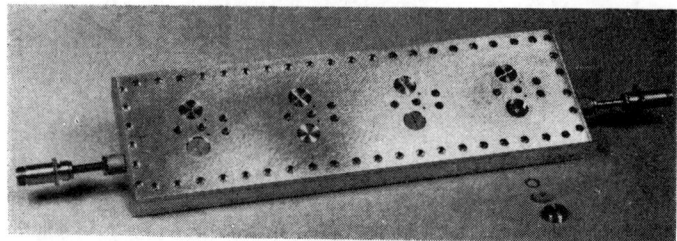

Fig. 24. Four-stage amplifier.

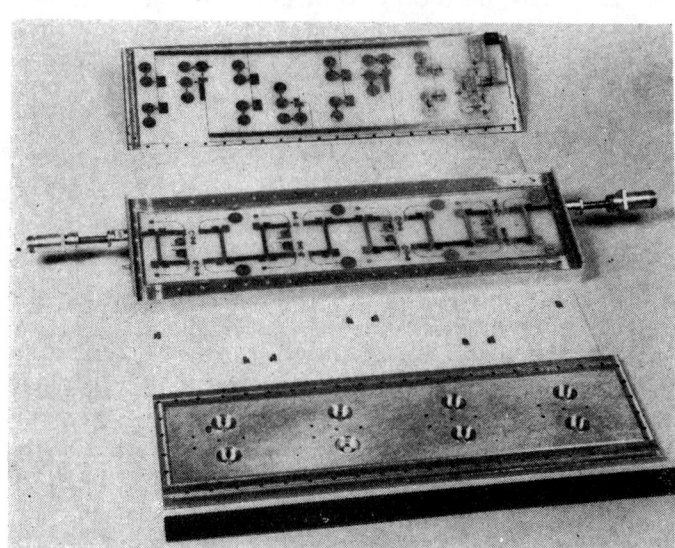

Fig. 25. Disassembled four-stage amplifier.—Top: Bias Circuit. Middle: RF Circuit.

selection into pairs has been shown to require only a few random tries with available mass-produced L-band transistors.

The circuit involves no critical components and is readily produced by conventional microwave printed circuit techniques. The gain per stage is approximately 6 dB maximum and can be controlled to any lower value by an external dc bias control. As has been shown, the circuit of one stage can be cascaded to obtain any desired over-all gain, provided that feedback through the bias circuit is kept sufficiently low (the present design is adequate up to about 50- to 60-dB over-all gain).

Although the same circuit techniques can be applied with little difficulty to other microwave ranges, presently available transistors provide useful gain only up to about 2 to 3 Gc/s. For this reason, the main effort on microwave transistor amplifiers remains at L band for the time being. It is currently directed toward the development of an integrated balanced amplifier using the same electrical principles and providing electrical performance characteristics similar to those described. It appears that existing thin-film techniques can be used for the dc biasing circuit, whereas new techniques are currently being investigated for obtaining the necessary distributed and lumped RF components.

Note Added in Proof: Since the submission of this manuscript, successful operation of both an L-band integrated amplifier and a four-stage S-band (2–3 Gc/s) amplifier has been obtained, using the design principles described in this paper. The results are summarized in Eisele, Engelbrecht and Kurokawa's paper.[6]

Appendix I
Three-dB Coupler Properties

The following is a summary of the more important electrical properties of 3-dB couplers which are of interest in the balanced amplifier design. It is assumed that the couplers are constructed by means of a symmetrical pair of coupled transmission lines imbedded in a uniform and homogeneous dielectric and supporting only TEM modes. As has been shown in the paper, this type of coupler is easily constructed in printed-circuit form by means of two coupled strip-transmission-line conductors surrounded by flat ground planes.

The characteristic impedance Z_0 of this coupler is identical at all four ports.[2] This characteristic impedance is frequency independent and is solely determined by the geometry of the coupling region. Thus, with Z_0 connected to ports 2, 3, and 4 of the input coupler in Fig. 1, the impedance as seen into port 1 is also equal to Z_0 at all frequencies. We shall now evaluate the mis-

[6] Eisele, K. M., R. S. Engelbrecht, and K. Kurokawa, Balanced transistor amplifiers for precise wideband microwave applications, *Conference Digest, 1965 Internat'l Solid State Circuits Conf.*, Feb 17–19, 1965.

match seen looking into port 1 (relative to Z_0) when identical mismatches are placed on ports 2 and 4 of a 3-dB coupler. With voltage V_1 incident at port 1, and voltage appearing at port 2 is given by

$$\frac{V_2}{V_1} = \frac{j \sin \theta}{\cos \theta + j\sqrt{2} \sin \theta} \quad (11)$$

where θ is the electrical length of the coupling region in radians ($\theta = \pi/2$ at the frequency f_0 at which the coupling length is $\lambda/4$) and the voltage appearing at port 4 is given by

$$\frac{V_4}{V_1} = \frac{1}{\cos \theta + j\sqrt{2} \sin \theta} \quad (12)$$

If we denote the reflection coefficient of the (identical) mismatches attached to ports 2 and 4 by ρ_0, then a component $\rho_0 V_2$ is reflected at port 2 and a component $\rho_0 V_4$ is reflected at port 4. The transmission coefficient of these two components back to port 1 is given again by (11) and (12). Hence, the total reflection seen looking into port 1 is obtained as

$$\frac{V_1(\text{refl.})}{V_1(\text{inc.})} = \frac{(1 - \sin^2 \theta)\rho_0}{(\cos \theta + j\sqrt{2} \sin \theta)^2} \quad (13)$$

or

$$\left|\frac{V_1(\text{refl.})}{V_1(\text{inc.})}\right| = \frac{|\rho_0|}{1 + 2 \tan^2 \theta} \quad (14)$$

The VSWR corresponding to (14) is shown in Fig. 6, calculated for the case of $\rho_0 = 1$. Thus, if we exclude the possibility of connecting negative impedances to ports 2 and 4, the VSWR at port 1 is seen to be excellent over a wide range of frequencies at f_0 for arbitrary but identical mismatches on ports 2 and 4.

Next, we will evaluate the contribution of the two 3-dB couplers in Fig. 1 to the insertion gain and phase between ports 1 and 1'. For this purpose, we shall assume that the transistors T_a and T_b are absent and that port 2 is connected directly to port 4' and port 4 directly to port 2'. Assuming the two 3-dB couplers of Fig. 1 to be identical, transmission between ports 2' and 1' is given by (11) and the transmission between ports 4' and 1' is given by (12). The total voltage appearing at port 1' is, therefore,

$$\frac{V_1'}{V_1} = \frac{2j \sin \theta}{(\cos \theta + j\sqrt{2} \sin \theta)^2} = \left|\frac{V_1'}{V_1}\right| e^{-j\beta} \quad (15)$$

The amplitude and phase of (15) are tabulated in Table I.

Again, we find that over a ±30 per cent frequency band about f_0, the transmission is nearly 100 per cent in amplitude and almost perfectly linear in phase. The

TABLE I

f/f_0	$|V_1'|/|V_1|$	$20 \log_{10} |V_1'|/|V_1|$	β
0.5	0.943	−0.51 dB	19.5°
0.6	0.977	−0.20 dB	35.6°
0.7	0.995	−0.05 dB	50.4°
0.8	0.999	−0.01 dB	64.1°
0.9	1.00	0	77.2°
1.0	1.00	0	90.0°
1.1	1.00	0	102.8°
1.2	0.999	−0.01 dB	115.9°
1.3	0.995	−0.05 dB	129.6°
1.4	0.977	−0.20 dB	144.4°
1.5	0.943	−0.51 dB	160.5°

phase slope with frequency is approximately that of a uniform piece of transmission line, approximately 130° long.

From the above, it is reasonable to make the following approximations when calculating the balanced amplifier performance:

1) Over a wide frequency range near f_0, about ±20 to ±30 per cent, the input match seen into port 1 is nearly perfect if identical mismatches are connected to ports 2 and 4 with their reflection coefficients not exceeding 0.5 (VSWR less than 3). The actual impedance mismatches at the amplifier input and output are thus primarily determined by an imbalance in the transistor input or output impedances.

2) For determining the insertion gain and phase between ports 1 and 1' the 3-dB couplers can, over a frequency range of about ±20 to ±30 per cent, be considered to be unity amplitude and linear phase networks. The insertion gain and phase between 1 and 1' are, therefore, primarily determined by those of transistors T_a and T_b.

Appendix II

Starting with 29 GF-40037 transistors coming from the same batch in the Western Electric Co., Laureldale, Pa., production line and meeting the present RF specifications, the number of random trials required to get a

TABLE II

Transistor Pair	Input Return Loss	Output Return Loss	Gain	No. of Tries Necessary for Obtaining Pair
Run I				
D6494–D6444	25 dB	21 dB	7.3 dB	4
D6765–D6787	25	23	6.4	8
D6436–D6439	20	25	7.1	2
D6210–D6472	22	25	7.7	2
D6767–D6462	25	22	6.9	2
			Average, Run I	3.6
Run II				
D6773–D6372	22	25	6.2	1
D6767–D6494	25	22	6.6	4
D6402–D6462	20	25	6.4	1
D6472–D6769	24	25	6.2	2
D6439–D6731	22	25	6.1	9
			Average, Run II	3.4

matched pair with input and output return losses both larger than 20 dB (VSWR ≤ 1.22) are tabulated in Table II. The single-stage balanced circuit described in Section IV A is used for this purpose. When the number of remaining transistors becomes too small, the selection is expected to be difficult. Therefore, the experiment was stopped after 5 pairs were selected, and a new run with 29 transistors started. From the table, we see that three to four random tries are necessary to obtain a matched pair such that either the input or the output VSWR is less than 1.12 and both are less than 1.22.

Acknowledgment

The authors wish to thank L. D. Gardner, T. E. Saunders, and J. A. MacPhee for their help in obtaining the amplifier models and R. Meyer for performing the experiments.

An Integrated 4-GHz Balanced Transistor Amplifier

THOMAS E. SAUNDERS AND PAUL D. STARK, MEMBER, IEEE

Abstract—This paper presents performance data and design information on a broadband 4-GHz balanced transistor amplifier being developed for possible use in microwave radio relay systems. The balanced stripline circuitry and passive components are integrally fabricated on a 1.5-inch-square alumina substrate using thin-film technology. A comprehensive description is presented of the circuit design, mechanical fabrication techniques, and long-term stability tests.

Three-stage amplifiers give 15 dB of gain at 4 GHz with a 3-dB bandwidth of 1000 MHz. Input and output VSWR's were below 1.05 with a noise figure of 7 dB. A mean time to failure of more than 10^6 hours has been indicated for a complete three-stage device by data obtained on accelerated component aging tests.

INTRODUCTION

TRANSISTOR AMPLIFIERS for use in the low microwave range have become very competitive with other semiconductor devices and with traveling-wave tubes in low-level applications. Features such as high gain, wide bandwidth, low noise, and wide dynamic range can be realized simultaneously. Recent developments in microwave printed circuit techniques and improvements in microwave transistor technology have resulted in L-band amplifiers that give about 9 dB of gain per stage, essentially flat over a 600-MHz band. These amplifiers can provide noise figures of less than 3 dB and output power levels in excess of 10 mW.

Transistor amplifiers have been reported that are useful at frequencies up to 2.7 GHz.[1] The amplifier to be described here extends the frequency range to above 4 GHz. An earlier design has been reported upon previously.[2] In addition, it is the first application of thin-film integrated circuitry for use above L band.

This paper presents a comprehensive description of the electrical design and mechanical fabrication and performance data of a 4-GHz balanced transistor amplifier, suitable for use in the receiver section of broadband microwave radio repeater stations. The discussion will be in two major sections. In the first section the electrical design of the amplifier, its physical description as it relates to performance, and performance data will be presented. The second section will discuss the thin-film technology and the mechanical aspects of the design with some discussion on reliability.

Manuscript received November 18, 1966.
The authors are with Bell Telephone Labs., Inc., Murray Hill, N. J.
[1] K. M. Eisele, R. S. Engelbrecht, and K. Kurokawa, "Balanced transistor amplifiers for precise wideband microwave application," *Digest of Technical Papers, 1965 Internat'l Solid-State Circuits Conf.*, vol. 8, pp. 18–19.
[2] T. E. Saunders and P. D. Stark, "An integrated 4-GHz balanced transistor amplifier," *Digest of Technical Papers, 1966 Internat'l Solid-State Circuits Conf.*, vol. 9, pp. 18–19.

THE BALANCED TRANSISTOR AMPLIFIER

Design Philosophy

Early microwave transistor amplifiers were of an unbalanced design.[3] Each stage required tuning adjustments to optimize performance, and since several stages were required to obtain the desired gain, the tuning adjustments interacted. It was thus extremely difficult to tune such an amplifier for flat wideband gain and good match. In addition, small changes in transistor characteristics caused wide variations in amplifier performance.

These difficulties were largely overcome by the use of balanced circuitry in the design of the amplifier.[4] The balanced amplifier consists essentially of two electrically similar transistors joined at input and output by 3-dB hybrid directional couplers, as shown in Fig. 1. The couplers have the property that if the two output ports are terminated in equal impedances, all of the reflected energy appears at the fourth port. Then if the transistors are identical in the balanced amplifier but not necessarily well matched to the circuit impedance, the reflected energy can be made to appear in the terminating resistor at the fourth port of the coupler. Thus no reflected energy will be observed at the input and the amplifier will have a good match. Even if the transistors are not identical, the input reflection coefficient will be only one-half the vector difference between the reflection coefficients of the individual transistors. The good input and output match permit many such units to be cascaded for high gain with little interaction between stages.[5]

Circuit Discription

In practice, the balanced amplifier assumes a more complex form. Figure 2 is the schematic of a 4-GHz single-stage amplifier that has been realized in integrated form using tantalum thin-film technology on a ceramic substrate.

The transistors are operated in the common emitter configuration. Each emitter is heavily bypassed with about 500 pF of capacitance and dc power is supplied by way of a 200-ohm, 500-pF distributed RC component that effectively decouples the amplifier from the power supply.

[3] J. Hamasaki, "A wideband high-gain transistor amplifier at L-Band," *Digest of Technical Papers, 1963 Internat'l Solid-State Circuits Conf.*, vol. 6, pp. 46–47.
[4] R. S. Engelbrecht and K. Kurokawa, "A wide-band low noise L-Band balanced transistor amplifier," *Proc. IEEE*, vol. 53, pp. 237–247, March 1965.
[5] K. Kurokawa, "Design theory of balanced transistor amplifiers," *Bell Sys. Tech. J.*, vol. 44, pp. 1675–1698, October 1965.

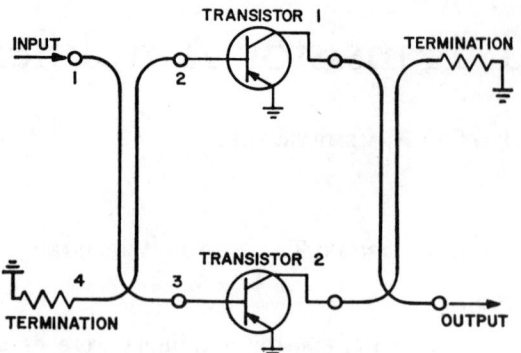

Fig. 1. Simplified schematic of a single-stage balanced amplifier.

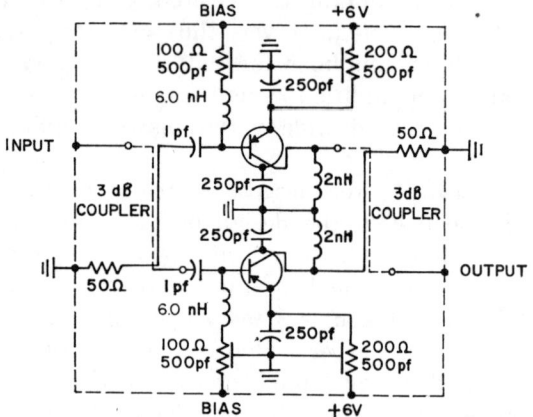

Fig. 2. Schematic diagram of a single-staged 4-GHz amplifier.

Base bias is supplied through a similar distributed RC component.

The series capacitor and shunt inductance in the base circuit and the shunt inductor in the collector circuit are chosen to match the average transistor to the 50-ohm circuit impedance and no tuning adjustments are required. The transistors are selected in pairs for good input and output match.

Physical Aspects of the Amplifier

With the exception of the transistors and one conductor of each coupler, all of the circuit components are integrally fabricated on a ceramic substrate. The actual physical layout is very similar to the schematic, and many of the components can be identified by their respective positions on the substrate, as shown in Fig. 3.

The ceramic substrate is 1.5 inches square, 0.024 inch thick, and glazed on one side to provide a microscopically smooth surface for good thin-film components. The substrate material is high-density alumina and has advantages of high strength, low loss, excellent stability, and, since it can be formed easily in the green state, low cost. The latter point is important because fifteen holes through the substrate are required, which would be very expensive to obtain in glass or finished ceramic.

One conductor of each coupler is on the amplifier substrate and one is on each of the two smaller substrates that can be seen in Fig. 3. The 50-ohm terminating resistors appear as mesh-like devices at the edge of the substrate and are connected to ground by the springs that clip over the edge of the board. These terminations have a VSWR less than 1.1 to above 6 GHz.

The series capacitor in the base circuit consists of a narrow gap in the 50-ohm transmission line. The gap is approximately 0.003 inch wide and 0.4 inch long, but patterned in a meandering fashion to fit in the 0.120 inch linewidth. The nominal value of this capacitor is 1 pF. The base shunt inductor is 0.005 inch wide and 0.375 inch long and has an inductance of about 6 nH. The "cold" end of this inductor is bypassed to ground and connected to the external bias supply through a 100-ohm, 500-pF distributed RC component. The collector is compensated with a shunt inductor of about 2-nH nominal value, which takes the form of a strap between collector and ground. The emitter bypass capacitors are wide and short for effective bypassing and are situated under the grounding springs. The emitter is also connected to the external supply through a distributed RC component and the power supply connections are made to the four small pads at the edges of the substrate.

Two large holes position the transistors, which are connected to the circuit by a welding or soldering operation. The transistors and small coupler substrates are bonded to the amplifier substrate by a small amount of epoxy.

The back side of the substrate has three ground pads, which are connected to the front-side circuitry by means of plated through holes. The circuit is grounded at appropriate points with six multifingered springs, eyeletted to the board. In addition, the springs serve to hold the substrate in the proper position between the 1/8-inch spaced ground planes.

Figure 4 shows the 50-ohm stripline configuration and a view of the coupler construction. In order to have symmetry in the vertical direction in the coupler region, the base substrate is offset from center by 0.017 inch so that the midplane of the coupler is midway between the ground planes. The asymmetry of the transmission line causes no difficulty as far as performance is concerned, though it does complicate the mechanical support of the substrate.

The dielectric constant of the substrate is about 8.5. The transmission line is not completely surrounded by the ceramic, so the effective dielectric constant is considerably less, about 1.69. Thus the propagation velocity is about 70 percent of that in a standard air line. The ceramic serves two functions. One is to support the stripline with little effect on its characteristics and the other is to provide a surface on which passive thin-film components can be deposited.

All conductors are copper plated to about 0.4 mil thickness for low loss. At 4 GHz, the skin depth is about 0.05 mil so no significant improvement in circuit loss would be expected if thicker plating were used.

Fig. 3. Completed 4-GHz transistor amplifier card.

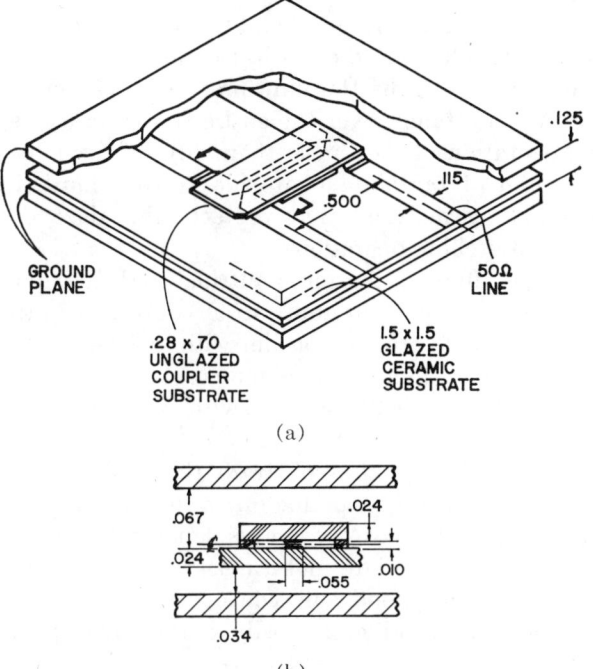

Fig. 4. Ceramic substrate geometry, 3-dB directional coupler.

Performance

The performance of the balanced amplifier depends very strongly on the performance of the 3-dB coupler. The ceramic substrate coupler used here has a coupling value of 3 ±0.1 dB over a 1000-MHz band centered at 4.0 GHz with about 30 dB of isolation between conjugate ports. The insertion loss of the coupler is quite low; two couplers in tandem have less than 0.25-dB loss. In the band of interest, the coupler has a match of about 35 dB return loss.

Single-stage amplifiers typically provide a gain of 5 dB at 4.0 GHz with a variation of less than 0.5 dB over the 3.7- to 4.2-GHz band of interest, as shown in Fig. 5. The frequency dependence of the gain is due to the tuning required to obtain a match between the transistors and the 50-ohm circuit impedance. The input and output VSWR's are typically less than 1.05 (greater than 30-dB return loss).

Several three-stage amplifiers have been built for evaluation for use in the receiver section of radio relay stations. This application tentatively requires an average gain between 11 and 15 dB with a 7-dB noise figure. In addition, the amplifier is required to have input and output return losses greater than 30 dB in WR-229 waveguide.

Figure 6 shows the typical gain and match characteristics of such an amplifier. Note the two gain curves. The upper curve is the gain obtained from a simple three-stage unit. The midband gain was about 15 dB with a variation of 1.2 dB over the band. In order to improve the gain flatness, a simple absorptive filter was used, which reduced the gain by 1 dB at midband but had little effect at band edge. This expedient reduced the midband gain to 14 dB with a variation of less than 0.6 dB over the 3.7- to 4.2-GHz band, as shown by the lower curve. The input and output return loss was in excess of 30 dB over the band.

The gain-leveling filter takes the form of a resistively loaded stripline resonator coupled to a strip transmission line. A balanced circuit was used; i.e., two filters were employed between 3-dB hybrids to obtain a good match. It should be noted that a more complex matching network or improved packaging of the transistor could result in a broader gain curve, which would eliminate the necessity of a gain leveler.

A high degree of isolation is also provided by the amplifier. The reverse loss is greater than 36 dB and is a smooth, monotonic function of frequency. There is no condition of input and output load impedances that will cause the amplifier to oscillate; thus the amplifier is unconditionally stable.

The gain of the amplifier can be decreased by reducing the total emitter current through control of the base-to-emitter bias. In practice, the first two stages are biased for low noise performance, and the output stage for low intermodulation.

Multistage noise figures of 7 dB are typical. The best transistors have given 5.7 dB of gain with a 5.6-dB noise figure in a single stage. This implies that multistage noise figures of 6.6 dB are now possible.

The three-stage amplifiers show less than 0.1 dB of gain compression at an output level of 0 dBm. The 0.5-dB compression point is typically 7 dBm, though output levels as high as 10 dBm have been obtained at the 0.5-dB compression point.

Intermodulation measurements were made with three test signals each at −15-dBm input level. The third-order product of the $A + B - C$ type at the output was at a level of −44 dBm, or more than 48 dB down from the total output level. AM to PM conversion was less than 0.1 degree/dB at 0-dBm output level.

Gain versus temperature characteristics are of interest since units in actual service may have to operate over a relatively large temperature excursion. For a temperature range of 40 to 140°F, the gain changes by 1.4 dB, decreasing with increasing temperature.

The multistage amplifiers include zener diodes to

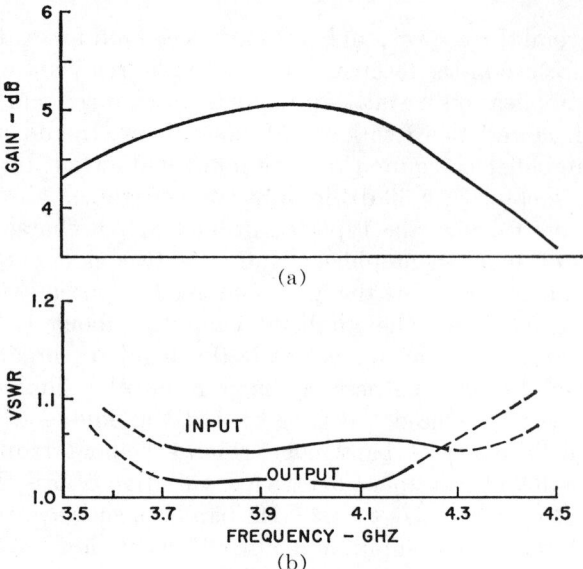

Fig. 5. Gain and match of a single-stage balanced transistor amplifier.

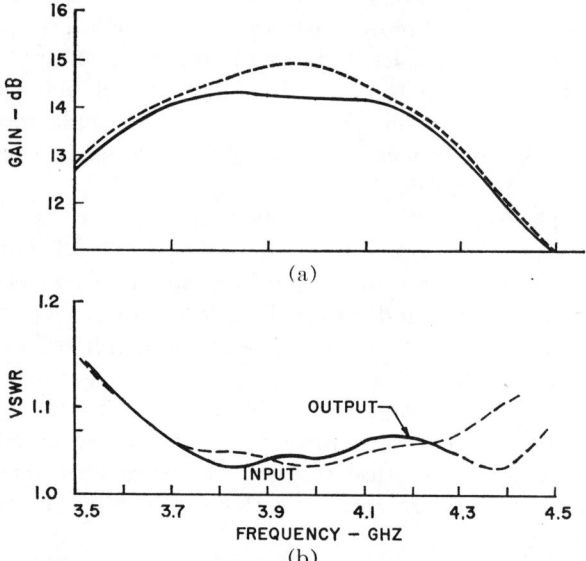

Fig. 6. Gain and match of a three-stage amplifier.

regulate the operating voltage of the transistors and to permit the units to operate from a 12-volt supply. With a ±1.5-volt variation about the nominal 12-volt supply, the gain of the multistage changes less than 0.1 dB.

MECHANICAL DESIGN ASPECTS

Tantalum Integrated Circuitry

The fabrication of the single-stage amplifier is based on tantalum thin-film technology, which permits resistors and capacitors and distributed RC components to be integrally fashioned on the surface of a ceramic substrate. Special techniques have been developed that make it possible to copperplate the thin films to obtain low-loss microwave conductors. The overall process is complex, requiring approximately thirty-five operations to complete the passive thin-film components and microwave printed circuitry. Later assembly operations, such as transistor and ground spring attachment, are required to make the amplifier operational.

As stated previously, the amplifier is fabricated on a high alumina ceramic substrate (typically 96 percent), glazed on one side and provided with the necessary preformed holes to accommodate the transistors and through connections to the far side. The substrate measures $1\frac{1}{2}$ inches by $1\frac{1}{2}$ inches by 0.024 inch thick and is typically commercially reproducible to ±1 percent in linear dimensions and to ±0.002 inch in thickness at reasonable cost. The alumina base and glaze are extremely stable and provide high strength and relative high thermal conductivity with low electrical-loss characteristics.

Every substrate, prior to the first film deposition, undergoes a series of tests to insure that it is free of cracks and flaws and meets established tolerances on size and thickness, flatness, and surface finish. Thus those substrates that have low probability of becoming good amplifiers are eliminated early in the process.

The first step in the thin-film process is the deposition of 400 Å of tantalum over the entire substrate in a sputtering operation. This film is thermally oxidized to produce a film of tantalum pentoxide, a very hard, inert, nonconducting material that protects the glaze during subsequent etching operations.

A 2700-Å film of tantalum nitride is then deposited by sputtering the tantalum in a controlled nitrogen atmosphere. This produces a film with a sheet resistivity of 15 ohms per square and a temperature coefficient of resistivity of about −100 ppm/°C with excellent long-term stability.

In those areas of the substrate that will ultimately contain the microwave conductors, a dual layer of 200 Å of chromium and 3000 Å of gold is deposited by evaporation through appropriate metal masks. The chromium is required to obtain good adhesion of the gold layer. The gold is a good conductor, easily plated, and prevents oxidation of the deposited film.

The circuit pattern in the conductive and resistive layers is generated using photoetching techniques. Immediately following the evaporation processes the substrates are coated with KTFR photoresist material and baked. The pattern is exposed through high-resolution glass negatives, developed, and again baked before etching. The negatives are keyed to the same registration points used in all the processing steps. The coated substrates are then etched using selective solutions that define the tantalum resistor and capacitor base-electrode areas and the conductor geometry.

The resistor and capacitor areas can now be anodized to form a protective oxide layer on the resistors and the lower dielectric layer of the capacitors. The tantalum nitride acts as an anode when suspended in a weak citric acid electrolyte. An anodizing current density of approximately 1 mA/cm² of exposed film is typical. Specific voltage limits maintain desired thicknesses of the oxide. The pentoxide layers are stabilized by baking them at 250°C for five hours in air.

The evaporated gold layer on the conductor regions is now electroplated with copper and then gold-flashed. The result is a very low resistivity film, 0.0004 inch thick, for the low-loss microwave conductors. The gold-flash layer prevents oxidation of the heavy copper as well as providing a good base material to make attachments.

Circuits of this type typically require that resistive elements be held to within ±5 percent of the specified value. A trim anodizing operation is utilized to adjust the resistors; the equipment is designed to automatically monitor the process as it proceeds, and then interrupt it when the desired value is attained.

The duplex capacitor structure is completed by vapor-depositing silicon monoxide dielectric of about 5000-Å thickness and then layers of chromium (200-Å) and gold (3000-Å) to form the counterelectrodes. The result is a dual-dielectric capacitor consisting of tantalum pentoxide and silicon monoxide. The dielectric properties of the capacitor are determined principally by the SiO layer. The result is a design factor of about 0.06 pF/(mil)^2 that allows capacitors of the order of 100 pF to be fabricated with minimum registration difficulties. In addition, the dual-dielectric construction reduces the probability of shorts due to film defects; the chance of an SiO defect coinciding with a Ta_2O_5 defect is extremely small statistically. Figure 7 is a cutaway view of the thin-film buildup.

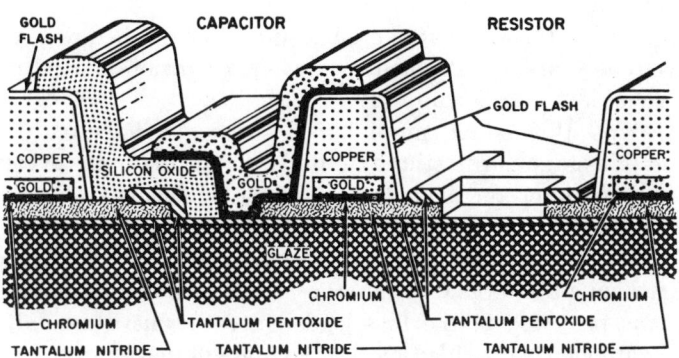

Fig. 7. Sectional view of thin-film circuitry.

Post-Description Fabrication Techniques

The completed substrate must undergo several additional operations following the film processing to make the amplifier operational. Two small coupler substrates that provide the top coupler conductors, and that have been previously processed, are fitted with two precision-ground ceramic spacers of the correct height and appliqued to the main substrate. This is accomplished using precision assembly fixtures that utilize the processing datum points to insure accurate registration of the three substrates. Noncorrosive epoxy is employed in these operations. Solder-plated copper tabs attached with a parallel gap welder are used to connect the top coupler lines to the main substrate, as shown in Fig. 8 (where an earlier circuit design is illustrated). This method provides high-strength bonds that will endure thermal cycling without damaging the plated film or the glazed surface.

The three ground-pad areas require connection to the housing in such a way that the path length to ground is short in order to minimize the series inductance, and such that the contact is distributed over a large area. Preformed multiple-finger beryllium copper springs that are loosely eyeletted to the main substrate perform this function and also serve to position the substrate between the ground planes established by the enclosure. Three of these springs are required on each side of the substrate.

The 50-ohm mesh termination resistors also require a connection to ground with similar requirements to those just mentioned. This is provided by a spring, which slips over the edge of the circuit board and snaps into place and provides grounding to both ground planes.

Mounting in the enclosure also requires the use of two support tabs, one being located at each side of the substrate. The beryllium copper details support the board and also serve to reference the circuit board to the enclosure when epoxied in place.

Finally the transistors are welded in place using the parallel gap welder mentioned previously. The emitter tabs are 0.001-inch nickel flashed with gold and solder-plated 0.0005 inch thick. The base and collector leads are also solder-plated but formed from 0.0015-inch gold-flashed Kovar.

Enclosures—Hardware Details

Special fixtures are required to evaluate the circuit board and characterize transistor performance. Figure 9 shows the amplifier in its test enclosure. The amplifier is supported in the frame by the two support tabs and the two RF connectors. The frame is $\frac{1}{8}$ inch thick to set the proper spacing between the ground planes that is formed by the lids. The grounding springs serve to position the substrate between the ground planes.

On the areas comprising the transmission lines, couplers, and transistors, the ground-plane spacing is maintained at 0.125 inch. There is a 0.034-inch gap beneath and a 0.067-inch gap above the substrate. In the regions of the grounding springs, the spacing is reduced to 0.034 to the upper ground plane in order to utilize the same spring on top and bottom of the substrate. Bias is supplied by means of small coaxial connectors, and contact to the circuit conductors is by means of miniature bellows.

General Radio Type "900-BT" precision connectors are used with appropriate tapered transition sections machined within the connector flange and set into the box to provide a smooth transition from the connector (14 mm) to the smaller dimension stripline configuration. In testers, two of the modified connectors and a section of ceramic with plated 50-ohm line give VSWR's less than 1.1 up to 7 GHz.

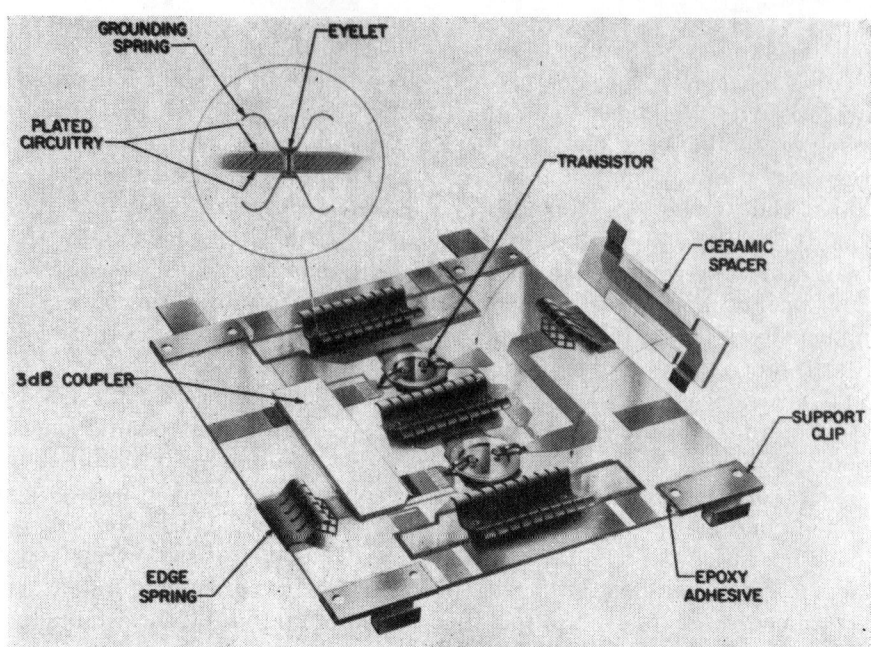

Fig. 8. 4-GHz integrated transistor amplifier circuit.

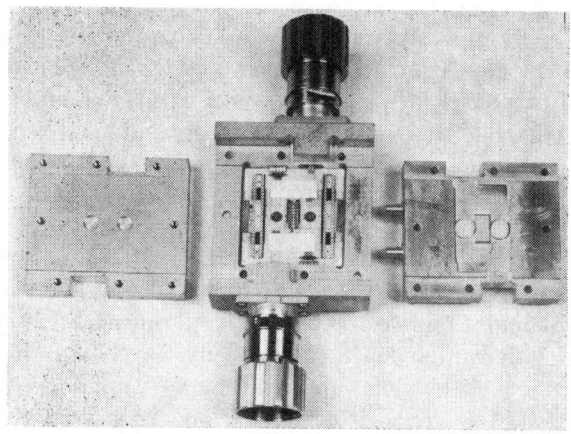

Fig. 9. Single-stage amplifier and test enclosure.

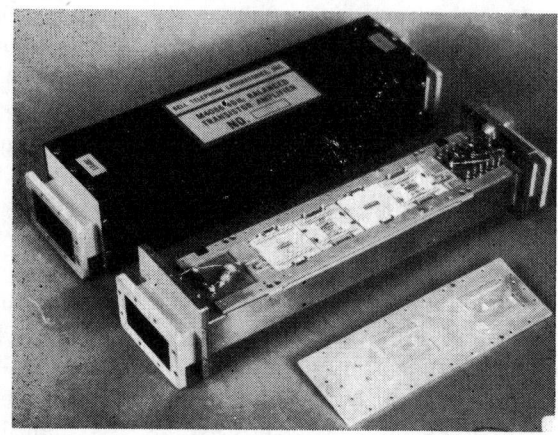

Fig. 10. Three-stage 4-GHz integrated transistor amplifier.

Two access plugs are provided in the lid over the transistor locations to permit selection of transistors. These plugs have plastic foam pads that press the transistors into place on the circuit for tests. After a pair has been selected, they may then be permanently attached to the circuit.

All metal parts of the enclosure, springs, bellows, and circuits, are gold-flashed to prevent corrosion and permit reliable pressure contacts.

The three-stage amplifier is shown in Fig. 10. The single-stage cards (three amplifiers and one gain leveler) are mounted on the side rack of the enclosure frame and interconnections are made with simple straps welded to the circuitry. The cards are powered and biased in parallel utilizing a flexible wiring harness; no additional decoupling was provided in the bias leads since the distributed RC components provided more than enough decoupling.

The housing allows the boards to be inverted one next to the other so that inputs and outputs can be aligned and can be connected as mentioned. Both covers are removable and uniform loading is accomplished by tapping one cover and utilizing clearance holes in the other cover and the frame, and then systematically torquing the screws. Radio-frequency leakage is minimized by means of multiple-fingered strip shielding. The bias circuitry, namely the resistors and zener diode, can be seen at each end of the amplifier in Fig. 10. Standard waveguide to coaxial transducers that are connected to the stripline by short lengths of airline are employed. Three tuning screws are provided in each transducer for optimizing input and output match. The complete three-stage amplifier including the metal cover measures 13.5 inches long, 3.25 inches wide, and 2.75 inches thick.

Stability and Reliability

Accelerated aging tests on the passive circuitry components have indicated extremely low failure rates and corresponding long lifetimes. Resistors tested for 10 000 hours show that changes of ±0.1 percent can be expected after over one billion hours at normal operating conditions. Capacitors and RC distributed networks similarly tested also indicate satisfactory operation in the order of a billion hours before failure.

These results indicate that a single-stage amplifier card comprising a total of ten passive components will have a mean time to failure of 10^8 hours. Based on the low failure rate of the germanium transistors, it is reasonable to predict a mean lifetime of more than 100 years for a complete three-stage amplifier.

Other amplifiers identical in mechanical construction but which operate in a different frequency range have been exposed to rather severe thermal and mechanical shock tests. Temperature cycling between −65°F and +165°F for over 100 cycles and shocking by removing from 212°F air-oven to liquid nitrogen did not affect the electrical performance, and equally as important, no effects were detectable on the film components, couplers, circuitry, and associated hardware. In addition, a typical three-stage device has been drop-tested numerous times at a distance of approximately six inches with no measurable change in performance.

The thin-film techniques outlined here are adaptable to mass production methods at a potentially low cost. The circuitry is electrically and mechanically stable and all indications are that long life can be expected that would be comparable with that of antennas, waveguide runs, and other mechanical hardware.

Applications of Integrated Circuit Technology to Microwave Frequencies

HAROLD SOBOL, SENIOR MEMBER, IEEE

Abstract—**Integration techniques suitable for microwave circuits have been developed. Various aspects of the technology of integration of microwave circuits are reviewed and the reasons for choosing the hybrid approach instead of the monolithic approach and thin-film metallization instead of thick-film are discussed. Design data relating circuit performance to substrate roughness and thickness of thin-film metal adhesion layers are presented. Propagation and radiation characteristics of microstrip lines are discussed. Design equations for thin-film lumped-element passive components are given. Examples of various microwave integrated circuits are shown.**

Introduction

THE PAST several years have witnessed rapid progress in solid-state devices for microwave and millimeter wave applications. Avalanche and transferred electron diodes have produced watts of CW power and kilowatts of pulsed power. Microwave transistors are available as power amplifiers at S band and as small signal amplifiers at C band. Significant advances also have been made in Schottky barrier, varactor, and PIN diodes.

The solid-state device is usually fabricated as a semiconducting chip or die with a volume of the order of 500–5000 cubic milli-inches. This very small size leads to electrical, thermal, and mechanical problems. The task of applying signals to and extracting output power from the chips is by no means trivial. The parasitic reactance encountered in connecting the active device to a circuit can seriously limit the performance. Adequate heat sinking must be provided to prevent the dissipated power from excessively heating the devices and consequently degrading performance and reliability The semiconductor die requires specialized handling techniques and protection to prevent mechanical and environmental damage.

The conventional method of treating the electrical, thermal, and mechanical problems is to mount and bond the active die in a suitably designed coaxial, pill, stripline, or microstrip package. The packaged device is then incorporated in an appropriate circuit. The use of discrete devices in conventional waveguide, coaxial or stripline circuits, suffices for many applications but leaves much to be

Manuscript received January 7, 1971.
The author is with the RCA Corporation, David Sarnoff Research Center, Princeton, N. J. 08540.

desired in others. The transverse dimensions of most conventional circuitry is much greater than the packaged device dimensions and discontinuities are introduced that limit performance. In addition, the large size and expensive machining or electroforming of conventional circuitry are undesirable in many applications. The reliability and mechanical integrity of conventional microwave circuitry is a matter of concern in certain space and military applications that require meeting severe vibration, shock, and temperature specifications. Large subassemblies that are bolted together cannot easily meet these requirements.

Many of the circuit-device interface problems and the size and cost problems of the discrete device and conventional circuit approach can be solved by utilizing microelectronic techniques for microwave circuits. Integrated circuits have revolutionized the low frequency domain because of their ability to solve these very problems. During the past several years, the microwave industry has been very active in introducing integration into the microwave bands. At this time, nearly every low and medium power microwave function has been demonstrated utilizing microwave integrated circuits (MICs). This paper will discuss the integration techniques, circuit forms and technology used for MICs, and will present examples of various devices and subsystems.

INTEGRATION TECHNIQUES

MICs like lower frequency integrated circuits (ICs) can be made in monolithic or hybrid form. In the monolithic circuit, active devices are grown *in situ* on or in a semiconducting substrate, and passive circuitry is either deposited on the substrate or grown in it. In the hybrid circuit, active devices are attached to a ceramic, glass, or ferrite substrate, containing the passive circuitry. Monolithic ICs have been successful for digital and linear applications where all required circuit components can be simultaneously fabricated. In many cases, the same device (bipolar or MOS transistors) can be utilized for amplifiers, diodes, resistors, and capacitors with no loss in performance. Many digital circuits require large arrays of identical devices. Thus it is possible in conventional ICs to obtain very high packing densities and associated low cost.

Microwave circuits are quite different from lower frequency circuits. There are very few applications that require densely packed arrays of identical devices and there is little opportunity to utilize active devices for passive components. Digital and video circuits require the storage of only potential energy and consequently no inductors are needed. Most microwave applications demand band definition and this necessitates storage of kinetic as well as potential energy; thus the packing density of typical microwave circuits (measured in terms of active real-estate to total real-estate) is far less than in conventional digital ICs. As an example, a typical microwave transistor amplifier utilizes only 5 percent of the area for the active device. One hundred of these amplifiers can be simultaneously fabricated on a 1-in silicon wafer. The same wafer on the other hand could accommodate more than 2000 transistors. If an overall yield of 10 percent is achieved, a wafer could yield 200 transistors or ten single-stage monolithic amplifiers. It is not apparent that the monolithic approach will offer any cost advantage in microwave circuits. Further, since devices are grown in or on the substrate, it is necessary to use top contact devices which can be a disadvantage in certain applications.

Subsystems for low frequency or digital applications can usually be constructed with similar devices for many of the functions. On the other hand, a microwave subsystem usually requires a variety of devices. One approach to monolithic subsystems would be to use identical processing (epitaxy, diffusions) for all devices, but to use various surface geometries. As an example, a family of PIN diodes

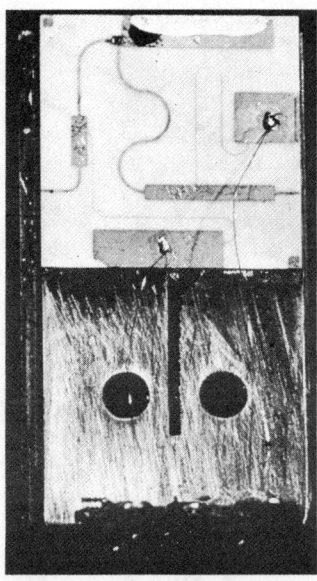

Fig. 1.   *L*-band microstrip transistor amplifier, 1 by 1 by 0.010 in alumina substrate, Cr–Au lines and ground plane.

could be used for IMPATT, switching, and tuning varactor applications. Most likely such an approach would not lead to optimum performance as too many comprises would be involved. Furthermore, while it is theoretically possible to grow GaAs on Si and vice versa, it is not practical at this time; therefore, optimum combinations of materials could not be utilized in the monolithic subsystem.

Thus it appears that conceptually, the monolithic approach is not well suited for microwave subsystems and circuits. In addition to the problems discussed previously, process difficulties, low yields, and poor performance have seriously limited the application of the monolithic technology to microwave circuits.

The hybrid technology permits the use of many varieties of devices and overcomes many of the difficulties of the monolithic technology. To date, the hybrid form of technology is used almost exclusively in the frequency range from 1 to 15 GHz. However, the monolithic technology has potential advantages that warrant consideration for future applications. A most important advantage of the monolithic approach is the reduction and control of the parasitic inductance encountered in joining the circuit and the device. This advantage is extremely important at high frequencies and makes the monolithic circuit appear attractive for millimeter-wave circuits. The main emphasis of the present paper is on the hybrid approach.

CIRCUIT TECHNIQUES

Considerable cost reduction of passive circuitry can be achieved by the use of microelectronic techniques in place of the conventional circuit fabrication by machining and electroforming. Photolithography and screening are the most popular methods of defining circuitry for hybrid integrated circuits. In order to realize the full potential of these techniques, it is necessary to use circuit forms in which the propagation properties are determined by the definition of conductors and circuit elements in a single plane.

The microstrip transmission line shown in the transistor amplifier of Fig. 1 is the most popular form of circuit for MICs. The line consists of a strip conductor separated from the ground plane by a dielectric layer. All circuit definition is performed in the plane of the strip conductor. The impedance and length of the lines determine the circuit properties. The major part of the propagating field is confined to the region of the dielectric below the strip conductor and the propagation approximates a TEM mode.

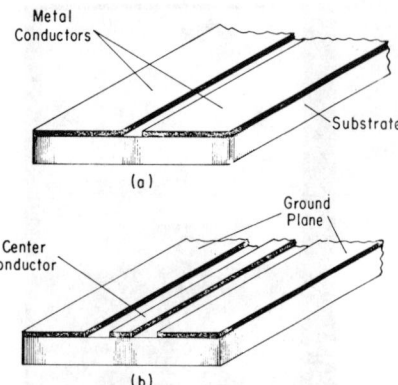

Fig. 2. Non-TEM lines for microwave integrated circuits. (a) Slotline. (b) Coplanar waveguide.

Fig. 3. Array of 3-part UHF lumped element hybrids on 1 by 0.75-in sapphire substrate, Cr–Au lines and inductors and SiO_2 capacitors.

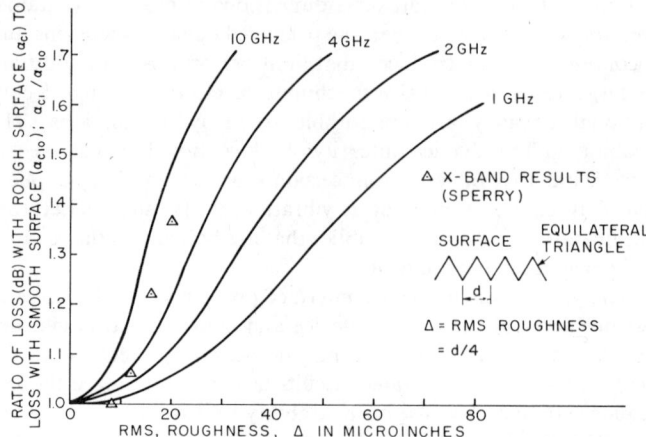

Fig. 4. Ratio of loss of microstrip on rough surface substrate (α_{cr}(dB)) to loss on smooth surface substrate (α_{c10}(dB)) as a function of rms roughness.

Two other forms of distributed circuits that have recently been introduced are illustrated in Fig. 2. Fig. 2(a) illustrates the slot line [1], and Fig. 2(b) the coplanar waveguide line [2]. Because both these lines have conductors on only one side of the substrate, they generally permit shunt mounting of devices without requiring holes through the substrate, as in the case of microstrip lines. They also have longitudinal as well as transverse RF magnetic fields, and have polarization properties that are useful for nonreciprocal ferrite devices. At this time, these lines have not yet been used to the same extent as microstrip. Some interesting circuits made by combining the circuits of Fig. 2(a) and (b) with microstrip lines have been described [3].

Conventional stripline circuits are usually defined by photolithography and can be considered as a class of hybrid integrated circuits. Stripline systems that qualify as MICs can utilize either standard circuit boards [4] or alumina substrates [5].

An alternate circuit form utilizes lumped-element components that are a small fraction of a wavelength in size and behave as capacitors, inductors, or resistors. The values of the components are independent of frequency over the range of interest. This type of circuit was usually avoided at microwave frequencies in the past because conventional fabrication techniques could not provide coils and capacitors that were sufficiently small to behave as true lumped elements. However, it is now possible to fabricate such elements by use of the technology of line definition developed in the semiconductor industry.

The lumped-element approach always results in smaller circuits than the distributed approach. Although the smaller size may be an advantage in some cases, microstrip circuits are often sufficiently small for most applications. However, there are two key advantages to be gained from the use of lumped-element circuits, both of which are a consequence of small size. These advantages are reduced cost and the adaptability of the lumped-circuit component in a hybrid subsystem.

Because of the small size of lumped-element circuits, many devices can be fabricated simultaneously on a ceramic wafer. As an example, Fig. 3 shows an array of 3-port lumped-element hybrid circuits formed on a 1 by 3/4-in sapphire wafer. The devices are batch-fabricated in the same manner as transistors. Consequently, because optimum use is made of batch-fabrication, the cost of lumped-element circuits tends to be less than that of microstrip lines, which are usually made one at a time.

The passive circuit loss in all forms of MICs will tend to be an order of magnitude higher than the loss in conventional circuits. This increased loss is primarily due to the great reduction in conductor surface area.

Technology of Hybrid MICs

Two types of hybrid technologies have evolved for low-frequency circuits, the thick-film and thin-film approaches. The general names thick and thin films cannot be applied in microwave applications since, regardless of technology used, conductor films must be of the order of three to five skin depths thick. However, the processes used for the two technologies can be considered for microwave circuits. The low-frequency thick-film conductor approach involves the application of a paste containing metal particles to a substrate and the subsequent firing of the film. The conductor pattern may be defined by printing the paste through a screen or by etching after the film is fired. The low-frequency thin-film approach involves the vacuum deposition of a thin film by evaporation or sputtering. The conductor patterns are usually defined by etching. Plating may be used in both techniques to increase the conductor thickness to be required three to five skin depths.

The fired films require a glass phase in the ceramic substrate for adhesion to the substrate. This phase is usually not present in high-quality microwave substrates such as the 99.5-percent body alumina; however, it is present in the lower purity substrates. Therefore, in order to achieve good adherence of fired films to substrates, it may be necessary to use a mechanical bond which can be obtained by using a rough surface on the ceramic substrate.

The vacuum deposited films adhere to substrates by a chemical bond that depends on the formation of an oxide between the con-

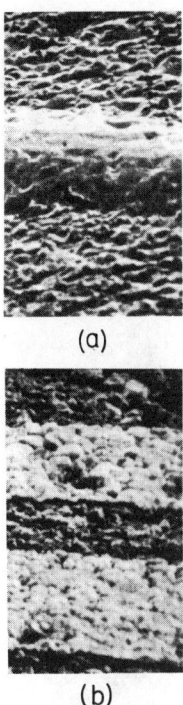

(a)

(b)

Fig. 5. Scanning electron microscope photographs. (a) Fired Au line on alumina. (b) Deposited Au line on alumina. ×200 magnification.

ductor and the substrate. The oxide is usually formed in a thin layer of a relatively high-resistivity metal as chromium, titanium, or tantalum. The chemical bond can be formed on a very smooth substrate surface.

There are several factors that make the deposited metal approach considerably more desirable for microwaves than the fired metal technique. These include substrate surface finish, metal film conductivity, and film porosity. These factors result in higher loss and poorer definition in a fired film than in a deposited film.

The effect of surface roughness on microwave loss was studied twenty years ago by Morgan [6]. Theoretical eddy current losses for equilateral triangular grooves, transverse to current flow, were derived from Morgan's work and are shown in Fig. 4. Also shown in Fig. 4 are the experimental data presented by Sperry Rand, Inc., [7]. The correlation between theory and experiment is good, in spite of the neglect of grooves parallel to current flow. It is apparent from Fig. 4 that rms surface finishes of the order of 5 micro-inches or less are required at X band if the percentage increase in loss (decibels) due to rough surfaces is to be kept below 5 percent. An rms surface finish of 10–20 micro-inches is suitable at L band. Since rougher surfaces are often used in order to achieve a fired-film mechanical bond, the loss of the fired-film line for X-band applications will tend to be considerably higher (40–65 percent) than the deposited film line on a smooth substrate. Fired films have been used on lower purity alumina substrates with 5–10-microinch finish, but these substrates tend to introduce more loss than the 99.5 percent body alumina.

The dc conductivity of the fired film (thick film) lines tends to be significantly less than the bulk metal. As an example, fired gold lines may have conductivities as low as 10 percent of the conductivity of bulk gold. Fig. 5(a) shows a photomicrograph of a fired film gold line. The rough texture and the structure due to suspended metal particles in the glass phase is evident. The relatively smooth texture of the deposited film is illustrated in Fig. 5(b). The conductivity of an evaporated and plated gold film is usually greater than 50 percent of the bulk conductivity of gold. Accordingly, the loss of the fired film may be up to 2.5 times higher than the loss of the deposited film.

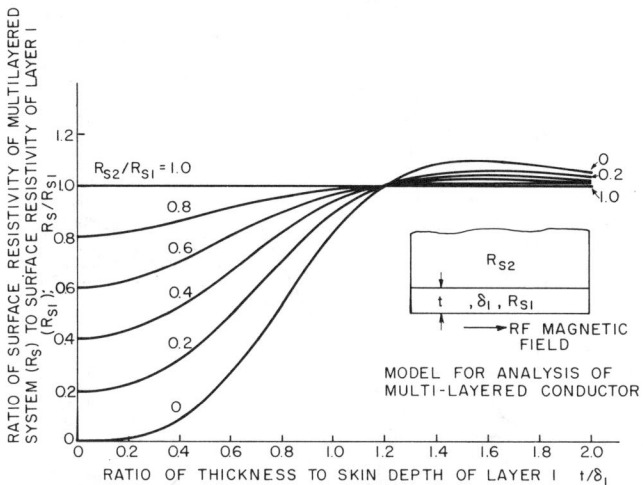

Fig. 6. Ratio of surface resistivity of multilayered system (R_s) to surface resistivity of adhesion layer (R_{s1}) as function thickness of adhesion layer (t/δ_1).

The structure of the fired film is relatively "porous" as illustrated in Fig. 5(a). It is difficult to achieve straight edges with this type of film and the ragged edges and hemispherical shape of the conductors contribute to loss. Etching of fired films is rather difficult since the film is a two-phase system and both a glass etch and a metal etch must be used. The fired film system cannot be used for lumped elements or high impedance lines where extremely fine resolution is required.

Experimental results [8], [9] on fired film microstrip lines have tended to confirm the high loss and, in general, the loss measured is two times the loss of a similar deposited film line. For this reason, the bulk of the microstrip work has been done with deposited metal systems on smooth substrates.

The thin film of high resistivity metal used for adhesion is a source of additional loss in the deposited metal system. The loss of the multilayered metal system shown later in Fig. 7 may be calculated by an extension of the analysis presented by Ramo and Whinnery [10]. The real part of the surface impedance R_s of the multilayered metal system is given by

$$R_s = R_{s1} \frac{\left[1 + \dfrac{R_{s2}}{R_{s1}}\right]^2 \exp(4t/\delta_1) + 2\left[\left(\dfrac{R_{s2}}{R_{s1}}\right)^2 - 1\right] \sin \dfrac{2t}{\delta_1} - \left[\dfrac{R_{s2}}{R_{s1}} - 1\right]^2}{\left[1 + \dfrac{R_{s2}}{R_{s1}}\right]^2 \exp(4t/\delta_1) - 2\left[\left(\dfrac{R_{s2}}{R_{s1}}\right)^2 - 1\right] \cos \dfrac{2t}{\delta_1} + \left[\dfrac{R_{s2}}{R_{s1}} - 1\right]^2} \quad (1)$$

where

R_{s1} surface resistivity of film 1 (adhesion layer)
R_{s2} surface resistivity of film 2 (conductor)
t thickness of film 1
δ_1 skin depth of film 1.

Equation (1) is plotted in Fig. 6 for a range of resistance ratios. The increased loss in a chromium-gold microstrip system over the pure

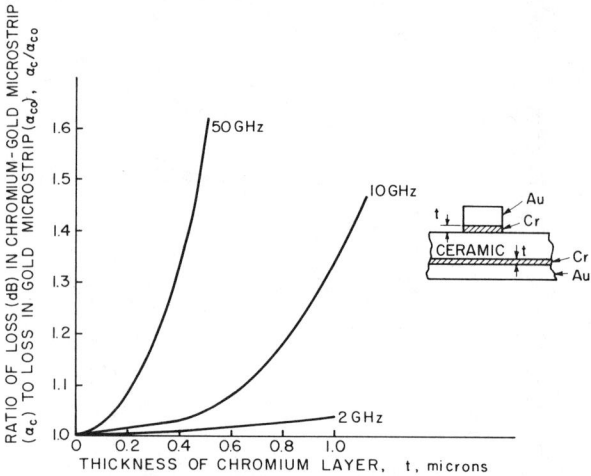

Fig. 7. Ratio of loss (α_c) of Cr–Au microstrip to loss of Au microstrip (α_{c0}) as function of Cr thickness.

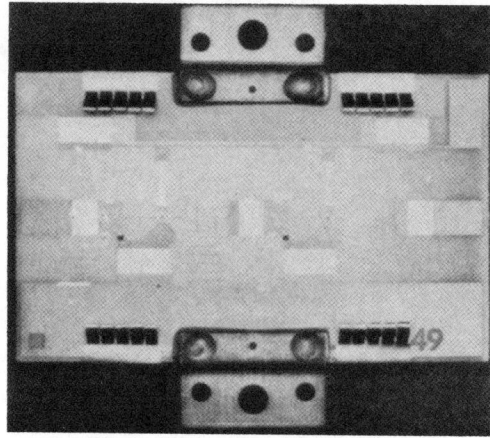

Fig. 8. Two-stage L-band transistor amplifier with tantalum nitride resistors and interdigitated capacitors. (Courtesy of Bell Laboratories, Inc.)

gold microstrip system [11] is shown in Fig. 7. The calculations account for the change in resistivity of the thin chromium films as a function of thickness of the film [12]. Chemical bonds are achieved with adhesive layers from 100–500 Å thick. Hence the typical adhesion layer will produce only a negligible increase in loss at frequencies well into the millimeter bands.

Deposited Film Systems

There are a variety of metal systems that may be used for deposited film circuits. The choice of a system is determined by the particular passive elements required in a circuit, the temperature range that the circuit will be subjected to during processing and operation, the compatability with the active devices and in some cases the availability of processing equipment. The tantalum system [13] offers the possibility of producing capacitors, resistors, and rudimentary interconnections (subsequently plated up with a high conductivity conductor) with a single material. Other refractory materials, notably hafnium and titanium, may be similarly used; however, none have achieved the success of the tantalum system to date. Tantalum and many of its compounds are very stable and can be anodized to form dielectric films and can be reactively sputtered to form resistive films. The tantalum films are usually sputtered onto a substrate. Tantalum pentoxide (Ta_2O_5) capacitors and tantalum nitride (Ta_2N) resistors have been made with extremely fine accuracies. The L-band transistor amplifier fabricated by Bell Laboratories, shown in Fig. 8, is an example of a tantalum system MIC. Interdigitated capacitors [14] with a Ta_2O_5 flash are utilized. The conductors are gold plated over a layer of tantalum.

Alternative metal systems based on evaporated chromium or titanium as the adhesion layer are very common. The following systems are metal layer combinations that have been successfully used.

Cr–Au

Cr–Cu–Au

Cr–Cu

Cr–Cu–Ni–Au

Ti–Pd–Au

Ti–Pt–Au.

The gold top surface is excellent for ultrasonic bonding, while the copper top surface usually requires soldering for bonding. The Ti Pd Au system is metallurgically stable under high temperature

Fig. 9. Four-stage transistor low noise broad-band amplifier (70–700 MHz) with NiCr resistors and SiO_2 capacitors.

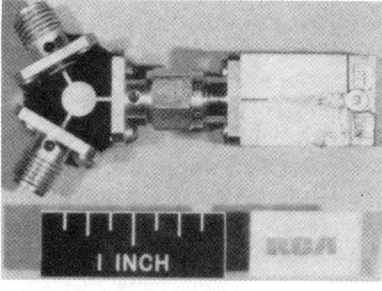

Fig. 10. X-band transferred electron amplifier with GaAs diode and circulator; 2-GHz bandwidth, 12-dB gain.

conditions while the Cr–Au system will result in a high resistivity conductor if exposed to elevated temperatures for long periods.

Evaporated nichrome resistors are usually utilized with this metal system. Chromium and titanium resistors do not have the long-term stability of the nichrome resistors. Deposited SiO_2 [15] produced by the pyrolitic decomposition of silane (SiH_4) is a popular dielectric material for MICs based on the above metal systems.

Fig. 9 is a broad-band (50–700 MHz) amplifier fabricated with SiO_2 capacitors, NiCr resistors, and Cr–Au interconnections. The four transistors are mounted on gold pads with an Si–Sb–Au eutectic and ultrasonically wire bonded with Al wires. Fig. 10 is an X-band transferred electron amplifier utilizing Cr–Cu–Au microstrip lines. A circulator is utilized to separate input and output signals.

TABLE I
Conductor Characteristics

Material	Dc Relative to Cu	Skin Depth δ at 2 GHz (μ)	Surface Resistivity (ohms/square $\times 10^{-7}\sqrt{f}$)	Coefficient of Thermal Expansion ($\alpha_T/°C \times 10^{-6}$)	Adherence to Dielectric Film or Substrate	Method of Deposition[a]
Ag	0.95	1.4	2.5	21	poor	E, Sc
Cu	1.0	1.5	2.6	18	very poor	E, P
Au	1.36	1.7	3.0	15	very poor	E, P
Al	1.60	1.9	3.3	26	very poor	E
W	3.2	2.6	4.7	4.6	good	Sp, EB, V
Mo	3.3	2.7	4.7	6.0	good	EB, Sp
Cr	7.6	2.7	4.7	9.0	good	E
Ta	9.1	4.0	7.2	6.6	very good	EB, Sp
Mo–Mn	$5 \times$ Mo				very good	Sc

[a] E = evaporation; EB = electron-beam evaporation; Sc = screening; P = plating; V = vapor-phase; Sp = sputtering.

TABLE II
Properties of Dielectric Films

Material	ε_r	Dielectric Strength (V/cm)	Microwave Q
SiO (evaporated)	6–8	4×10^5	30
SiO$_2$ (deposited)	4	10^7	1000
AL$_2$O$_3$ (anodized or evaporated)	7–10	4×10^6	
Ta$_2$O$_5$ (anodized or sputtered)	22–25	6×10^6	100
Si$_3$N$_4$ (vapor-phase; sputtered)	7.6	10^7	
	6.5	10^7	

TABLE III
Properties of Resistive Films

Material	Resistivity (ohms/square)	T.C.R. (%/°C)	Stability
Cr (evaporated)	10–1000	−0.1 to +0.1	poor
NiCr (evaporated)	40–400	+0.002 to +0.1	good
Ta (sputtered in A–N)	4–100 +	−0.01 to +0.01	excellent
Cr–SiO (evaporated or cermet)	up to 600	−0.005 to −0.02	fair
Ti (evaporated)	5 to 2000	−0.1 to +0.1	fair

Tables I, II and III summarize the properties of typical metal, dielectric, and resistor materials used for MICs.

SUBSTRATES

Substrates used for MICs must have the following general properties:

1) low loss at microwave frequencies;
2) good adherence for conductors;
3) polished surfaces (2–5-microinch finish) and relative freedom from voids;
4) no deformation during processing of circuit.

The selection of a substrate material also depends on the expected circuits dissipation, the circuit function, and the type of circuit to be used.

Substrates for microstrip circuits should have relatively high dielectric constants (9 or higher) for size reduction. The dielectric constant must be uniform from batch to batch of substrate material and must not vary significantly with temperature.

Substrates with practically no surface voids are used for thin-film lumped-element circuits. In addition, the substrate should be amenable to sawing or scribing for separation of the individual circuits after batch-fabrication.

Compound substrates are occasionally used in MICs for ferrite

Fig. 11. 16-W 225–400-MHz transistor amplifier. Transistor mounted on BeO substrate. Lumped-element circuits on sapphire substrates.

TABLE IV
Properties of Substrates

Material	Tan at 10 GHz	ε_r	k (W/cm · °C)	Applications
Alumina	2×10^{-4}	9.6–9.9	0.2	microstrip, suspended substrate
Sapphire	10^{-4}	9.3–11.7	0.4	microstrip, lumped element
Glass	4×10^{-4}	5	0.01	lumped element
Beryllium oxide	10^{-4}	6	2.5	compound substrates
Rutile	4×10^{-4}	100	0.02	microstrip
Ferrite/garnet	2×10^{-4}	1316	0.03	microstrip, slot line, coplanar compound substrates

components or for high thermal-dissipation circuits. Although it is possible to make entire microstrip systems on ferrite substrates, in some applications a ferrite or garnet disk can be inserted in an aluminum substrate. This technique is employed typically when circulators are used in systems. High-power devices require good heat sinks; when the device cannot be mounted directly on a metal heat sink, it may be mounted on beryllium oxide and then electrically connected to a passive circuit fabricated on a lower thermal-conductivity substrate. Fig. 11 is a photograph of a 16-W CW UHF transistor amplifier with the device mounted on a BeO substrate and the lumped-element circuitry on sapphire substrates.

The properties and applications of some popular substrates are given in Table IV.

The use of high dielectric constant substrates seems appealing as a step towards achieving reduced size circuitry. However, the modal problems discussed in the next section places constraints on substrate thickness with the result that extremely narrow conductors are required for reasonable impedance levels. These narrow conductors tend to be lossy and also introduce severe definition problems. In addition, the dielectric constant of materials such as rutile is quite temperature dependent and this can lead to problems in many applications. For these reasons, materials such as rutile have not been used to any great extent in MICs.

CIRCUIT DESIGN

The designer of MICs does not have the luxury or the ease of trimming or tuning of his circuits once they are fabricated. A cavity or conventional lumped-element circuit can be adjusted and peaked to desired performance with adjustable stubs or capacitors. Since these variable elements are not compatible with MIC technology, the circuit design must be carefully carried out in order to eliminate the need for extensive post-fabrication adjustments. It is therefore important that the properties of the passive and active devices used are well understood.

Theoretical data [16]–[23], suitable for accurate design of microstrip circuits, have been presented only recently even though microstrip lines have been used and studied for nearly twenty years [24].

MICROSTRIP TRANSMISSION LINE

The fundamental propagation mode of a microstrip line such as that shown in Fig. 1 is TEM. At low microwave frequencies, most of the energy propagates in the dielectric below the strip conductor. The remainder of the energy propagates in a fringe field, some of which extends above the dielectric in the air space. The properties of TEM propagation in microstrip can be determined by conformal mapping. The proper mapping to use for microstrip was proposed by Wheeler [1], [17] who derived relationships describing the phase velocity (or effective dielectric constant) and impedance of TEM microstrip propagation. This analysis was carried out for wide strips (width of strip conductor w to thickness of dielectric layer h ratio greater than unity) and for narrow strips (w/h less than unity). The phase velocity V_{ph} in terms of the effective dielectric constant ε_{reff} is given as

$$V_{ph} = \frac{3 \times 10^{10}}{\sqrt{\varepsilon_{reff}}} \quad (\text{cm/s}). \quad (2)$$

The effective dielectric constant is expressed in terms of a dielectric filling factor q as

$$q = \frac{\varepsilon_{reff} - 1}{\varepsilon_r - 1} \quad (3)$$

where ε_r is the dielectric constant of the insulating layer.

The dielectric filling factor for wide strips is given by

$$qd = d - \ln\frac{d+c}{d-c}$$
$$+ \frac{0.732}{\varepsilon_r}\left[\ln\frac{d+c}{d-c} - \cosh^{-1}(0.358d + 0.595)\right]$$
$$+ \frac{\varepsilon_r - 1}{\varepsilon_r}\left(0.386 - \frac{1}{2(d-1)}\right) \quad (4)$$

where $d = 1 + \sqrt{1+c^2}$ and c is found implicitly from

$$\frac{\pi w}{2h} = c - \sinh^{-1} c$$

where w is the width of the strip conductor and h is the thickness of the dielectric. The impedance of the wide strip line is given as

$$Z_0 = \frac{377 h}{\sqrt{\varepsilon_r}\, w}\left[1 + \frac{h}{\pi w}\left\{2\ln 4 + (\varepsilon_r + \frac{(\varepsilon_r+1)}{\varepsilon_r}\ln\frac{\pi e}{2}\left(\frac{w}{2h} + 0.94\right)\right.\right.$$
$$\left.\left. + \frac{\varepsilon_r - 1}{\varepsilon_r^2}\ln\frac{e\pi^2}{16}\right\}\right]^{-1}. \quad (5)$$

For the narrow strip case, Wheeler shows that the effective dielectric constant is

$$\varepsilon_{reff} = \frac{\varepsilon_r + 1}{2}\left(1 + \frac{\varepsilon_r - 1}{\varepsilon_r + 1}\frac{\ln\frac{\pi}{2} + \frac{1}{\varepsilon_r}\ln\frac{4}{\pi}}{\ln\frac{8h}{w}}\right) \quad (6)$$

and the impedance is

$$Z_0 = \frac{377}{2\pi\sqrt{\frac{\varepsilon_r + 1}{2}}}\left[\ln\frac{8h}{w} + \frac{1}{8}\left(\frac{w}{2h}\right)^2\right.$$
$$\left. - \frac{1}{2}\frac{\varepsilon_r - 1}{\varepsilon_r + 1}\left(\ln\frac{\pi}{2} + \frac{1}{\varepsilon_r}\ln\frac{\pi}{4}\right)\right]. \quad (7)$$

Wheeler also presents implicit equation for deriving the h/w values in terms of the desired impedances. This type of data can be very useful for computer aided design. For the wide line,

$$\frac{w}{h}\frac{\pi}{2} = \frac{377\pi}{2\sqrt{\varepsilon_r}\,Z_0} - 1 - \ln\left(\frac{377\pi}{\sqrt{\varepsilon_r}\,Z_0} - 1\right)$$
$$+ \frac{\varepsilon_r - 1}{2\varepsilon_r}\left[\ln\left(\frac{377\pi}{2\sqrt{\varepsilon_r}\,Z_0} - 1\right) + 0.293 - \frac{0.517}{\varepsilon_r}\right] \quad (8)$$

and for the narrow line,

$$2\frac{h}{w} = \frac{1}{4}e^{h'} - \frac{1}{2}e^{-h'}$$

where

$$h' = \sqrt{\frac{\varepsilon_r + 1}{2}}\frac{Z_0}{60} + \frac{\varepsilon_r - 1}{\varepsilon_r + 1}\left(0.226 + \frac{0.120}{\varepsilon_r}\right). \quad (9)$$

Curves [18] of impedance and wavelength are shown in Figs. 12 and 13.

Some recent work has been presented on a wave analysis [21], [22] of microstrip propagation. A microstrip line was enclosed in a metal container and a mixed TE and TM solution was used. The use of the metal walls permits a straightforward specification of boundaries, certainly simpler to handle than the open boundaries of a microstrip in free space. The enclosed line is practical since most circuits will be used in a package. Microstrip propagation is analyzed by moving the metal walls far from the line. This analysis has shown that the wavelength based on a dispersion-free TEM solution may be as much as 5 percent too high at 10 GHz, but will be correct in the 2-to-3-GHz range and below.

The loss per unit length of a microstrip line [18] can be shown to be inversely proportional to the substrate thickness h; consequently, the Q_0 (unloaded Q) of a microstrip resonator is directly proportional to h. Furthermore, a TEM analysis shows the Q_0 attainable for microstrip resonators to be proportional to the square root of

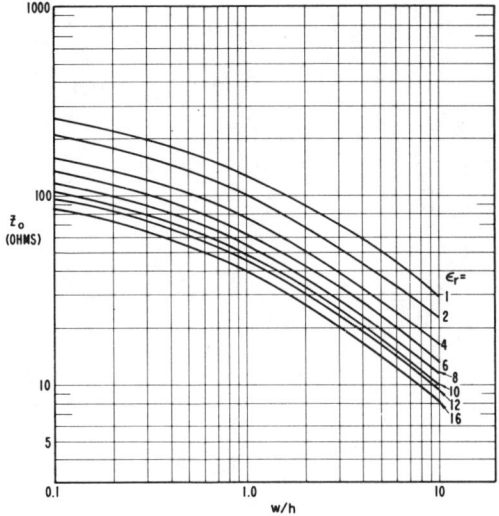

Fig. 12. Impedance of microstrip as function of width of conductor to thickness of substrate.

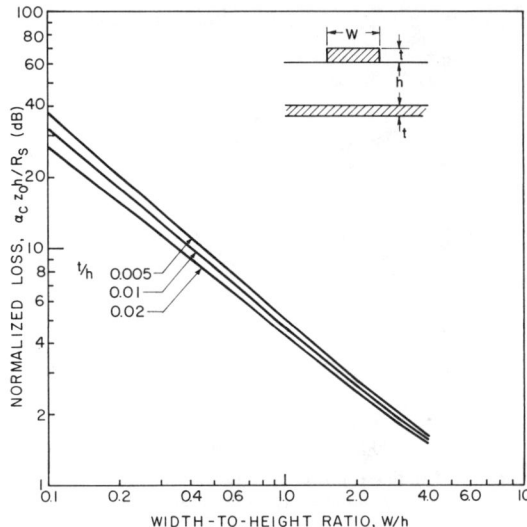

Fig. 14. Normalized loss of microstrip as function of width of conductor to thickness of substrate (after Pucel, Massé, and Hartwig [23]).

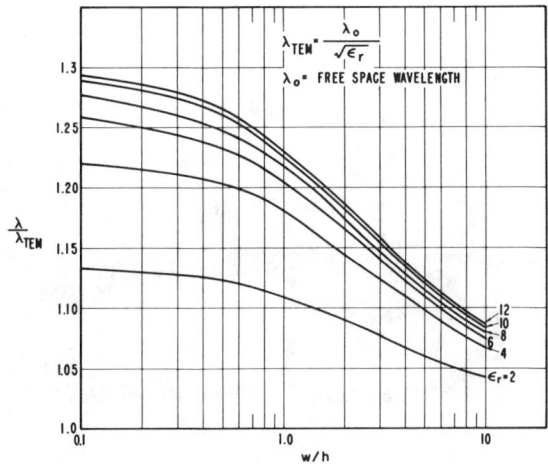

Fig. 13. Normalized guide wavelength of microstrip as function of width of conductor to thickness of substrate.

Fig. 15. Times 4 multiplier (2.125 GHz to 8.5 GHz) with low-pass filter, coupled line bandpass filter and open circuit idler stub.

frequency. Quarter-wave resonators on 25-mil-thick alumina substrates have Q_0 values of about 200 at 2 GHz, while Q_0 of about 400 is achievable at 10 GHz using 50-Ω resonators. Lower impedance lines will yield higher Q_0 values.

Theoretical curves of loss of microstrip line have been presented and verified by Pucel et al. [23], and are reproduced in Fig. 14. The loss per wavelength of the TEM line has been shown by Caulton et al. [18], to be independent of dielectric constant to a first approximation.

There are a number of higher order modes that can exist in microstrip. The presence of the metal enclosure used in practice and in the analysis will alter many of these modes. A particular family, surface modes can be easily excited when using high dielectric constant substrates unless precautions are taken. The cutoff frequency of the lowest order TE_1 surface mode is [25]

$$f_{TE_1} = \frac{3 \times 10^{10}}{4h\sqrt{\varepsilon_r - 1}}, \quad h \text{ in cm} \quad (10)$$

which occurs when the substrate thickness is approximately $\lambda_0/(4\sqrt{\varepsilon_r})$.

In order to avoid working near this cutoff, relatively thin substrates must be used with high dielectric constant substrates. Consequently, in order to achieve reasonable high impedance levels, extremely narrow lines will be required and this, in turn, leads to definition problems.

Open-circuit microstrip stubs are frequently used in filters and matching networks as shown in the idler circuit of the times four multiplier illustrated in Fig. 15. An ideal lossless stub with a true open-circuit loading presents a pure susceptance at the junction of the stub and the main line.[1] However, the actual boundary condition at the open end of the line is not an open circuit but instead is a complex load that can be represented by a conductance G due to radiation of the principal mode and a susceptance B due to energy stored in the higher order modes. Because of the complex load and distributed losses in the stub, the actual stub admittance presented at the junction of the stub and the main line, is complex.

The susceptance at the open end B can usually be represented by a hypothetical extension Δl of the stub, where Δl is approximately 0.3 to 0.5 substrate thickness [26]. The simple theory for infinitely wide plates results in an extension of $0.44 h$.

The radiation conductance can be calculated by treating the problem as radiation from an aperture [27]–[30]. Sobol [27] has shown that the conductance G can be approximated as

$$G \approx \frac{\sqrt{\varepsilon_{\text{reff}}}}{180}\left(\frac{w}{\lambda}\right)^2 = \frac{(\varepsilon_{\text{reff}})^{3/2}}{180}\left(\frac{w}{\lambda_0}\right)^2. \quad (11)$$

---

[1] The exact location of the junction will not coincide with the geometric junction but may be displaced by a distance of the order of the substrate thickness.

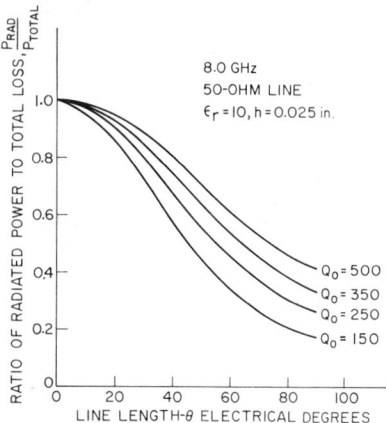

Fig. 16. Ratio of radiated power to total loss of open microstrip stub as function of stub length.

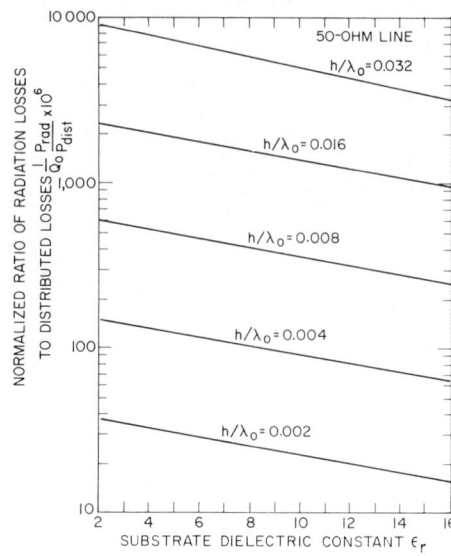

Fig. 17. Ratio of radiated power to distributed loss of open 90° microstrip stub as function of dielectric constant.

The ratio of the losses due to radiation to the total stub losses at 8 GHz is plotted in Fig. 16 for a 50-Ω line on a 0.025-in thick alumina substrate as a function of stub length. This plot illustrates the importance of considering radiation losses. For a line with $Q_0 = 250$, 60 percent of the total losses of a 45° line are due to radiation while 30 percent of the total losses of a 90° line are due to radiation. The value of B used is taken from the results of Napoli and Hughes [26], where $\Delta l \approx 0.4 h$.

The ratio of the radiation losses to the distributed losses of a quarter-wave resonator is

$$\frac{P_{\text{rad}}}{P_{\text{dist}}} = \frac{Q_0 Z_0 (\varepsilon_{\text{reff}})^{3/2} \left(\frac{w}{\lambda_0}\right)^2}{45\pi \left(1 + 4 \frac{\Delta l}{h} \frac{h}{\lambda_0} \sqrt{\varepsilon_{\text{reff}}}\right)}. \quad (12)$$

Fig. 17 shows $P_{\text{rad}}/P_{\text{dist}}$ for a quarter-wave resonator as a function of the substrate dielectric constant ε_r and frequency. The results shown are for a 50-Ω line and are in agreement with the data presented by Denlinger [31].

Lewin [29] considers further cases of radiation from microstrip bends and short circuited stubs.

Microwave circuit elements such as filters, couplers, and hybrids can be made by using coupled microstrip lines. Fig. 15 is an example of an X-band filter fabricated with coupled lines. Several papers [32]-[34] have been published on the coupled microstrip lines. In dealing with these structures, it is necessary to consider the even and odd mode velocities and impedances. It is difficult to achieve high directivity in coupled microstrip lines because of differences between even and odd mode velocities.

LUMPED-ELEMENT MICROWAVE COMPONENTS

Lumped-element microwave networks are used where small size is desired and in particular for those devices that are to be batch-fabricated. The lumped-element networks can be made in single layer circuits by using interdigitated capacitors and meander line inductors or in three-layer circuits using sandwich capacitors and spiral conductors. The single layer circuits are larger than the three-layer and tend to introduce distributed effects at lower frequencies than the three-layer circuits. Lumped-elements can be combined with microstrip circuits.

In order to exhibit true lumped behavior, a circuit element must be small enough that there is no appreciable phase shift across its entire length. If this requirement is satisfied the low-frequency design equations can be utilized with good success [11], [35]. Fig. 19 is an S-band 1-W transistor amplifier that utilizes rectangular

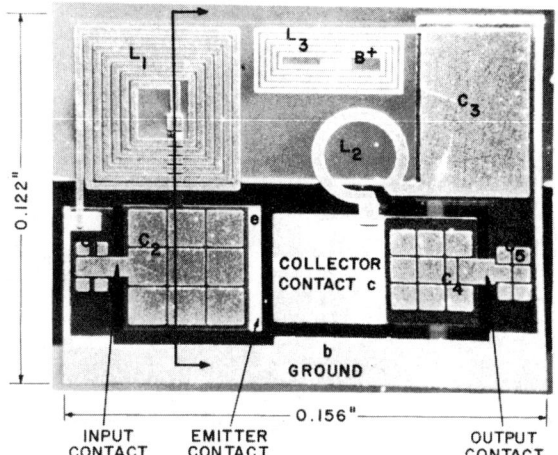

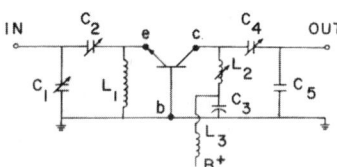

Fig. 18. 1-W S-band lumped-element transistor amplifier.

spiral coils as RF chokes, a single turn inductor for tuning and sandwich capacitors. The capacitors are made variable by segmenting the upper plates. In this way the circuit can be tuned over wide frequency ranges by bonding the appropriate capacitors into the circuit at the same time that the transistor is bonded. The amplifier of Fig. 18 is made entirely on a sapphire substrate. A higher power amplifier shown earlier in Fig. 11 is made by using separate input and output circuits on individual sapphire substrates and mounting these substrates and the transistor on a BeO substrate.

When designing a lumped-element network, it is vitally important to account for all parasitic reactance encountered in the circuit. The simplest inductor is a straight strip. The inductance L per unit length of such a structure (in nanohenries per centimeter) is given by [35]

$$L\left(\frac{\text{nH}}{\text{cm}}\right) = 2\left[\ln\frac{l}{w} + 1.193 + 0.224\frac{w}{l}\right] \quad (13)$$

where l is the length and w is the width of the strip in centimeters.

The resistance r per unit length of the strip inductor (in ohms per centimeter can be determined by assuming that the RF current flow is contained within a skin depth at the upper and lower surfaces of the strip, as follows:

$$r = \frac{\pi K}{w}\sqrt{f \cdot \rho} \quad (14)$$

where f is the frequency in gigahertz, ρ is the resistivity in ohm-centimeters, and w is the width of the strip in centimeters. K is a factor between 1 and 2 which accounts for current crowding at the corners of the strip [35]. The ratio of the Q of the inductor to the total inductance in nanohenries is given by:

$$\frac{Q}{L \cdot l} = 2\frac{w}{l}\frac{1}{K}\sqrt{\frac{f}{\rho}}. \quad (15)$$

A strip inductor is typically used in applications requiring inductances of 0.5 to 4 nH. The strip-conductor equation can be used for a single-turn coil if the width of the strip is much less than the diameter of the turn.

Larger inductance values may be achieved by use of flat spiral inductors. The total inductance L_t of the flat spiral (in nanohenries) is given by:

$$L_t(\text{nH}) = 393\frac{a^2 n^2}{8a + 11c} \quad (16)$$

where a is the average radius of the coil in cm.

$$a = \frac{d_0 + d_i}{4}$$

d_0 is the outer diameter in centimeters, d_i is the inner diameter in centimeters, $c = (d_0 - d_i)/2$ in centimeters, and n is the number of turns.

It can be shown [35] that the maximum Q of a circular spiral is obtained when the ratio of inner diameter to outer diameter d_i/d_0 is 0.2. For this ratio, $Q_{\max}$ of a spiral is expressed as follows:

$$\frac{Q_{\max}}{\sqrt{L_t}} = \frac{2.4w}{K}\sqrt{\frac{f}{\rho d_0}} \quad (17)$$

where d_0 and w are in cm, L_t is in nanohenries, f is in GHz, and ρ in ohm-centimeters.

Circular spiral inductors with Q's of 100 at 2 GHz have been fabricated. Square spirals have lower Q values than round spirals, and can be used to obtain the largest inductance in a given area. These inductors are useful for RF chokes. Square spirals 0.080 in on a side have been made with inductances up to 100 nH.

Connection to the center of a spiral coil is usually made by running a section of conductor under the coil. A dielectric film is used to insulate the feedthrough from the main coil winding. The self-resonance of a coil can be estimated as the frequency at which the total unwound electrical length of the spiral (modified by dielectric loading) is one-quarter wavelength.

The inductance equations presented thus far apply to an unshielded inductor. The inductance per unit length is reduced as a ground plane is brought in the vicinity of inductor. The reduction in inductance can be estimated by considering the inductor and the ground as a transmission line. When the characteristic impedance Z_0 of this line is less than $300/\sqrt{\varepsilon_{\text{reff}}}$, the effective inductance L_{eff} (in nanohenries) is given by:

$$L_{\text{eff}} \approx \frac{Z_0\sqrt{\varepsilon_{\text{reff}}} \cdot l}{30} \quad (18)$$

where l is in centimeters.

To obtain the unshielded inductance value, it is necessary that the distance to a ground plane beneath an inductor on alumina or sapphire be greater than 20 times the width of the inductor conductor. Similarly, a ground in the same plane as the inductor should be five widths away.

A capacitor may be considered as a short length of the transmission line with an open-circuit load. The driving point impedance of this structure is given by:

$$Z_m = \frac{2R_0}{3} + \frac{1}{Q_d \omega C_0} - j\frac{1}{\omega C_0} \quad (19)$$

where

R_0 = RF resistance of the capacitor plates,
C_0 = parallel plate capacitance,
Q_d = dielectric Q.

Thus the equivalent circuit for a capacitor is two resistors (one representing dielectric loss and the other conductor loss) in series with a capacitor. The total Q of the capacitor is given by:

$$Q_{\text{total}} = \frac{Q_d Q_s}{Q_d + Q_s} \quad (20)$$

where

$$Q_s = \frac{3}{2\omega C_0 R_0}.$$

Generally the dielectric Q or Q_d of most materials varies little with frequency in the microwave range, while Q_s varies inversely as the product of frequency to the three halves and capacitance. Typical Q_0 values for microstrip [36] and lumped-element components are shown in Fig. 19.

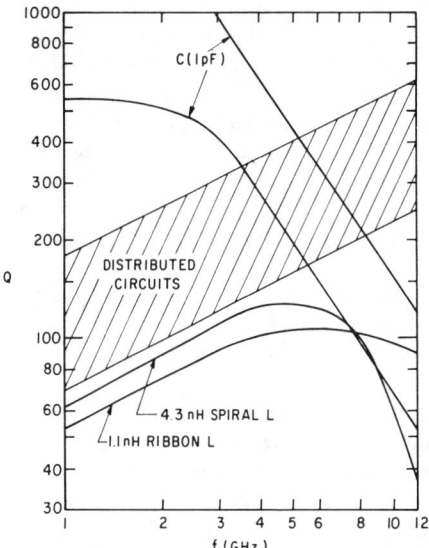

Fig. 19. Q_0 of microstrip resonators and lumped elements as function of frequency.

Active Devices

Active devices can be mounted in MICs in a variety of ways. The most common method is to mount the semiconductor chip directly in the circuit, using the same technique as in mounting a device in a package. Examples of this mounting are shown in Figs. 8, 9, 11, and 18. The transistors are eutectic mounted on gold pads and wire bonds are used to connect the circuit to the device contacts. Usually several bond wires are used in parallel to reduce the parasitic inductance. For devices with large contact area, it is possible to use thermo-compression bonding and foil in place of the wire bonds. Shunt-mounted devices can be mounted directly on ground (thermal and electrical) or on posts to achieve good thermal mounts. Beam leads provide an excellent mounting scheme for low dissipation microwave devices and both diodes and transistors have been mounted using this technique. However, this technique has inherent thermal limitations and cannot be used for power devices.

An alternate for incorporating devices into circuits is to use a chip carrier [37] as illustrated in Fig. 20. The transistor carrier of Fig. 20 is used for L-band applications and has excellent thermal properties. The carrier has specific lengths of transmission line on the emitter and collector leads to provide impedance transformation to a specific level. The carrier is mounted in a microstrip circuit and foils are used to connect the carrier to the circuit.

In many cases the entire circuit will be enclosed in a hermetic package so that it may not be necessary to use hermetic chip packages for the devices. In other applications, hermetic chip packages are required. A hermetic package suitable for microstrip circuits is shown in Fig. 21.

As mentioned previously, tuning of MICs after fabrication is difficult and it is important to obtain accurate characterization of devices. One of the more difficult problems in device characterization is the definition of reference planes. It is important to characterize the device in a fixture that closely approximates the final circuit to avoid shift in reference planes in going from the text fixture to a circuit. The test fixtures must be completely characterized to account for all parasitics.

Scattering parameters are convenient to use for the design of linear circuits. Because these parameters are determined with ports terminated in 50 Ω rather than an open or a short circuit, many of the stability problems encountered in y and z characterizations are eliminated.

In the design of nonlinear circuits, such as power amplifiers, the linear circuit analysis cannot be used and it is necessary to measure the device parameters under dynamic conditions at the desired frequency and power level. A system such as that shown in Fig. 22 can be used for this measurement. The stubs are used to tune the device to the desired operating points. The device jig is removed, and the impedance presented at the device plane is then determined. A circuit is synthesized to approximate the required impedances over the desired frequency band. The realization of this circuit is then fabricated in either distributed or lumped-element form.

System and Subsystem Applications

The application of MICs for systems and subsystems is illustrated in Figs. 23 and 24. Both figures illustrate modules for phased-array systems. Fig. 23 is the Texas Instruments, Inc., MERA [38] module, which includes a transmitter, receiver, and two phase shifters, including logic. The transmitter input is at S band and the output is an 0.7-W pulsed signal at X band. Fig. 24 is an RCA S-band CW power module [39] that has a 1.5-GHz input and a 12-W 10-percent bandwidth output at 3 GHz.

These are but two subsystem applications of the technology. There are many other modules that have been fabricated incorporat-

Fig. 20. L-band transistor carrier for microstrip applications.

Fig. 21. Hermetic transistor package for microstrip applications.

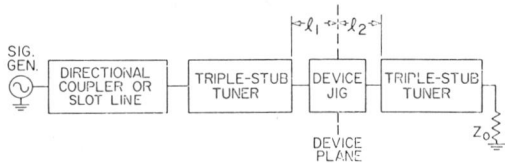

Fig. 22. System for measurement of dynamic characteristics of devices.

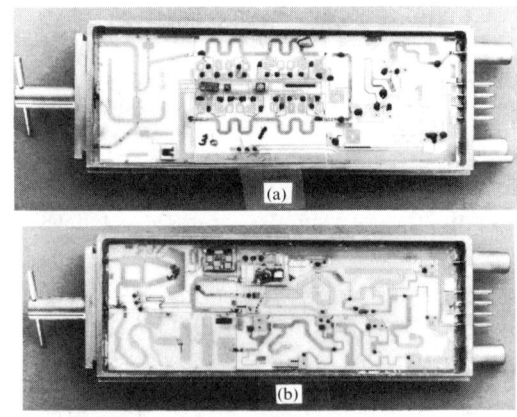

Fig. 23. MERA module. (a) S-band preamplifier phase shifters and receiver multiplier. (b) TR switch, mixer, IF amplifier, power amplifier, and multiplier. (Courtesy Texas Instruments, Inc.)

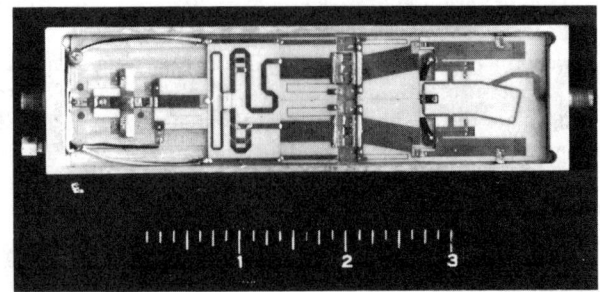

Fig. 24. S-band power module including L-band preamplifiers, power amplifier, power splitter, multipliers, and combiner.

ing practically every microwave semiconductor device and performing nearly every medium and low-power microwave function.

The designer of modules faces the problem of partitioning the module. He must decide how many functions to include in each block of the module and also what impedance levels are required at the block terminals. There are no general rules at this time for partitioning, each module must be treated individually in terms of economics and performance requirements.

Conclusions

This paper has reviewed various aspects of MICs although there is much that was not covered. The technology has advanced significantly in the past several years and has been met with enthusiasm in the industry. There are problems that remain; however, the prospects are for a bright future. In addition, the technology developed has application in other areas of the microwave industry. As an example, the techniques for depositing microstrip lines on ceramic can be used for planar circuits for vacuum traveling-wave amplifiers. In addition, deposited circuits are currently being used for surface wave acoustic devices.

Acknowledgment

The author would like to express his gratitude to his colleagues at RCA Solid State Division, Somerville, N. J.; RCA Microwave Department, Harrison, N.J.; RCA Laboratories and RCA Electronic Components, Princeton, N. J., for many useful discussions and for permission to use their photographs in this paper. The author wishes to thank Dr. M. Caulton for his suggestions and careful reading of the manuscript. Thanks are also due Bell Telephone Laboratories and Texas Instruments, Inc., for permission to exhibit photographs of their microwave integrated circuits.

References

[1] S. B. Cohn, "Slot-line on alternative transmission medium for integrated circuits," *IEEE G-MTT Int. Microwave Symp. Dig.*, May 1968, pp. 104–109.

[2] C. P. Wen, "Coplanar waveguide, a surface strip transmission line suitable for nonreciprocal gyromagnetic device applications," *IEEE G-MTT Int. Microwave Symp. Dig.*, May 1969, pp. 110–114.

[3] F. C. deRonde, "A new class of microstrip directional couplers," *IEEE G-MTT Int. Microwave Symp. Dig.*, May 1970, pp. 184–189.

[4] B. R. Halford, "Low noise microstrip mixer on a plastic substrate," *IEEE G-MTT Int. Microwave Symp. Dig.*, May 1970, pp. 206–211.

[5] R. S. Englebrecht and K. Kurokawa, "A wide-band low noise *L*-band balanced transistor amplifier," *Proc. IEEE*, vol. 53, Mar. 1965, pp. 237–247.

[6] S. P. Morgan, "Effect of surface roughness on eddy current losses at microwave frequencies," *J. Appl. Phys.*, vol. 20, Apr. 1949, pp. 352–362.

[7] Nonpublished results obtained by Sperry Rand, Inc.

[8] R. N. Patel, "Microwave conductivity of thick-film conductors," *Electron. Lett.*, vol. 6, July 1970, pp. 455–456.

[9] F. Z. Keister, "An evaluation of materials and processes for integrated microwave circuits," *IEEE Trans. Microwave Theory Tech.*, vol. MTT-16, July 1968, pp. 469–475.

[10] S. Ramo and J. R. Whinnery, *Fields and Waves in Modern Radio*. New York: Wiley, 1953, pp. 249–250.

[11] H. Sobol, "Technology and design of hybrid microwave integrated circuits," *Solid-State Technol.*, Feb. 1970.

[12] A. A. Milgram and C. S. Lu, "Preparation and properties of chromium films," *J. Appl. Phys.*, vol. 39, May 1968, pp. 2851–2856.

[13] D. A. McLean, N. Schwartz, and E. D. Tidd, "Tantalum-film technology," *Proc. IEEE*, vol. 52, Dec. 1964, pp. 1450–1462.

[14] G. D. Alley, "Interdigitated capacitors for use in lumped element microwave integrated circuits," *IEEE G-MTT Int. Microwave Symp. Dig.*, May 1970.

[15] N. Goldsmith and W. Kern, "The deposition of vitreous silicon dioxide films from silane," *RCA Rev.*, vol. XXVIII, Mar. 1967, pp. 153–165.

[16] H. A. Wheeler, "Transmission-line properties of parallel wide strips by a conformed-mapping approximation," *IEEE Trans. Microwave Theory Tech.*, vol. MTT-12, May 1964, pp. 280–289.

[17] H. A. Wheeler, "Transmission-line properties of parallel strips separated by a dielectric sheet," *IEEE Trans. Microwave Theory Tech.*, vol. MTT-13, Mar. 1965, pp. 172–185.

[18] M. Caulton, J. J. Hughes, and H. Sobol, "Measurements on the properties of microstrip transmission lines for microwave integrated circuits," *RCA Rev.*, vol. 27, Sept. 1966, pp. 377–391.

[19] C. P. Hartwig, D. Massé, and R. A. Purel, "Frequency dependent behavior of microstrip," *IEEE G-MTT Int. Microwave Symp. Dig.*, May 1968.

[20] H. E. Stinehelfer, "An accurate calculation of uniform microstrip transmission lines," *IEEE J. Solid-State Circuits*, vol. SC-3, June 1968, pp. 101–106.

[21] G. I. Zysman and D. Varon, "Wave propagation in microstrip transmission lines," *IEEE G-MTT Int. Microwave Symp. Dig.*, May 1969, pp. 3–7.

[22] R. Mittra and T. Itoh, "A new technique for the analysis of the dispersion characteristics of microstrip lines," *IEEE Trans. Microwave Theory Tech.*, vol. MTT-19, Jan. 1971, pp. 47–56.

[23] R. A. Pucel, D. J. Massé, and C. P. Hartwig, "Losses in microstrip," *IEEE Trans. Microwave Theory Tech.*, vol. MTT-16, June 1968, pp. 342–350.

[24] F. Assadourian and E. Rimeir, "Simplified theory of microstrip transmission systems," *Proc. IRE*, vol. 40, Dec. 1952, pp. 1651–1657.

[25] R. E. Collins, *Field Theory of Guided Waves*. New York: McGraw-Hill, 1960, pp. 470–474.

[26] L. S. Napoli and J. J. Hughes, "Open-end fringe effect of microstrip lines on alumina," to be published in *RCA Rev.*

[27] H. Sobol, "Radiation conductance of open-circuit microstrip," to be published.

[28] S. Ramo and J. R. Whinnery, *Fields and Waves in Modern Radio*. New York: Wiley, 1953, pp. 526–536.

[29] L. Lèwin, "Radiation from discontinuities in stripline," *Proc. Inst. Elec. Eng.* (London), vol. 107, Feb. 1960, pp. 163–170.

[30] N. Marcuvitz, *Waveguide Handbook*, vol. 10 (Radiation Laboratory Series). New York: McGraw-Hill, 1951, p. 179.

[31] E. J. Denlinger, "Radiation from microstrip resonators," *IEEE Trans. Microwave Theory Tech.* (Corresp.), vol. MTT-17, Apr. 1969, pp. 235–236.

[32] T. G. Bryant and J. A. Weiss, "Parameters of microstrip transmission lines and of coupled pairs of microstrip lines," *IEEE Trans. Microwave Theory Tech.*, vol. MTT-16, Dec. 1968, pp. 1021–1027.

[33] M. K. Krage and G. I. Haddad, "Characteristics of coupled microstrip transmission lines—I: Coupled-mode formulation of inhomogeneous lines," *IEEE Trans. Microwave Theory Tech.*, vol. MTT-18, Apr. 1970, pp. 217–222.

——, "Characteristics of coupled microstrip transmission lines—II: Evaluation of coupled-line parameters," *IEEE Trans. Microwave Theory Tech.*, vol. MTT-18, Apr. 1970, pp. 222–228.

[34] A. Schwarzmann, "Microstrip plus equations adds up to fast design," *Electronics*, vol. 40, Oct. 1967, pp. 109–114.

[35] M. Caulton, S. P. Knight, and D. A. Daly, "Hybrid integrated lumped-element microwave amplifier," *IEEE J. Solid-State Circuits*, vol. SC-3, June 1968, pp. 59–66.

[36] M. Caulton and H. Sobol, "Microwave integrated circuit technology," *IEEE J. Solid-State Circuits*, vol. SC-5, Dec. 1970, pp. 292–303.

[37] E. Belohoubek, D. Stevenson, and A. Rosen, "Hybrid integrated 10-W CW broadband source at *S*-band," *IEEE Int. Solid-State Circuits Conf. Dig. Tech. Papers*, Feb. 1969, pp. 124–125.

[38] T. M. Hyltin, "Microwave integrated electronics for radar and communication systems," *Microwave J.*, vol. 11, Feb. 1968, pp. 51–55.

[39] E. Belohoubek *et al.*, "*S*-band CW power module for phased arrays," *Microwave J.*, vol. 13, July 1970, p. 29.

Design and Performance of Microwave Amplifiers with GaAs Schottky-Gate Field-Effect Transistors

CHARLES A. LIECHTI, MEMBER, IEEE, AND ROBERT L. TILLMAN

Abstract—The design and performance of an X-band amplifier with GaAs Schottky-gate field-effect transistors are described. The amplifier achieves 20 ± 1.3-dB gain with a 5.5-dB typical noise figure (6.9 dB maximum) over the frequency range of 8.0–12.0 GHz. The VSWR at the input and output ports does not exceed 2.5:1. The minimum output power for 1-dB gain compression is +13 dBm, and the intercept point for third-order intermodulation products is +26 dBm. The design of practical wide-band coupling networks is discussed. These networks minimize the overall amplifier noise figure and maintain a constant gain in the band.

I. INTRODUCTION

IT HAS BEEN demonstrated that GaAs field-effect transistors with Schottky-barrier gates (MESFET's) combine low-noise properties with high gain and large dynamic range in a frequency range that has not been invaded by bipolar transistors [1]–[3]. Recent advances in GaAs materials and device technology have made possible the fabrication of reliable MESFET's with 0.2-μm channels and 1-μm gate structures with reasonable yield and in satisfactory quantities.

Several successful microwave FET amplifier designs have been reported [3]–[10]. Some have proven that very wide band amplification is feasible with these devices in X and Ku bands [9], [10]. These designs are based on empirical network topologies (except [10]), and the circuit elements have been computer optimized for maximum gain. This paper describes networks that are especially suitable for coupling power in and out of microwave FET's. The choice of the particular circuit topology is justified and performance limits are outlined. A step-by-step procedure for designing an amplifier with minimum noise figure and constant gain will be described. This design is based on established matching network theory and requires the use of a computer only for very broad bandwidths. The experimental results prove that high-gain and low-noise figures can be simultaneously achieved over a band as wide as 8–12 GHz.

The topics are treated in the following sequence. First, the transistor properties relevant to the circuit design are described. Second, the principles and procedures for the network synthesis are presented, and the basic amplifier construction is outlined. Third, the experimental results from an X-band amplifier are discussed leading to the final conclusions.

II. MESFET CHARACTERISTICS

The MESFET structure is illustrated in the scanning electron micrograph of Fig. 1. The dc current flows from drain to source in an 0.2-μm-thin epitaxial layer, doped with 1×10^{17} donors/cm^3, and grown on a semi-insulating GaAs substrate. The current is controlled by the depth of a depletion layer that extends beneath the Schottky-barrier gate. The gate is a 1-μm-wide and 500-μm-long metal stripe. Source and drain are alloyed Au–Ge ohmic contacts. They are separated from the gate by 1 and 2 μm, respectively. Outside the active device area, the epitaxial film is removed by mesa etching. This permits a large gate contact pad to be located on the semiinsulating substrate where it contributes only a small fraction to the overall gate to source capacitance. The drain and gate are contacted with an overlay metallization which runs over the mesa step. The transistor chip has a size of 0.010×0.025 in. The chips are tested from dc through 13 GHz in a fixture that makes pressure contact to the overlay metallization with negligible parasitic reactances [2].

The noise figure F_a of an amplifier which consists of a large number of cascaded identical stages with gain G and noise figure F is

$$F_a = 1 + \frac{F - 1}{1 - 1/G} = 1 + M \tag{1}$$

where M is the noise measure. In order to minimize the amplifier noise figure, it is necessary to minimize the noise measure rather than the noise figure of each stage. This distinction is essential in finding the optimum dc bias of the transistor. With 4-V drain voltage, the MESFET in this example exhibits maximum gain at 80 mA, minimum noise measure at 30 mA, and minimum noise figure at 17-mA drain current. At $I_{DS} = 30$ mA, the source admittance has been optimized, first, for minimum noise figure, second, for minimum noise measure, and third, for maximum gain [11]. The associated noise figures and gains are plotted in Fig. 2. Since the MESFET exhibits high gain in this frequency range, the source admittances for minimum noise measure and noise figure are nearly identical. The noise figure increases from 3.3 dB at 8 GHz

---

Manuscript received August 21, 1973; revised December 6, 1973. This work was supported jointly by Hewlett-Packard Company and the U. S. Army Electronics Command, Fort Monmouth, N. J., under Contract DAAB07-72-C-0322.

C. A. Liechti is with the Solid-State Laboratory, Hewlett-Packard Company, Palo Alto, Calif. 94304.

R. L. Tillman was with the Solid-State Laboratory, Hewlett-Packard Company, Palo Alto, Calif. 94304. He is now with the Microwave Technology Center, Hewlett-Packard Company, Palo Alto, Calif. 94304.

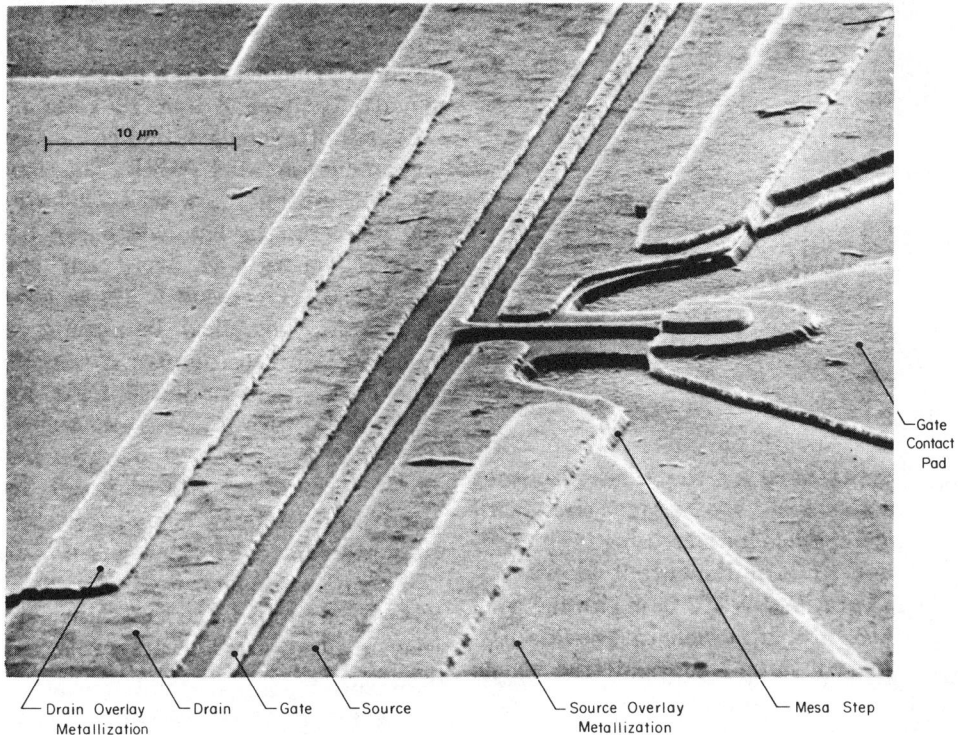

Fig. 1. Scanning electron micrograph of the MESFET's center section.

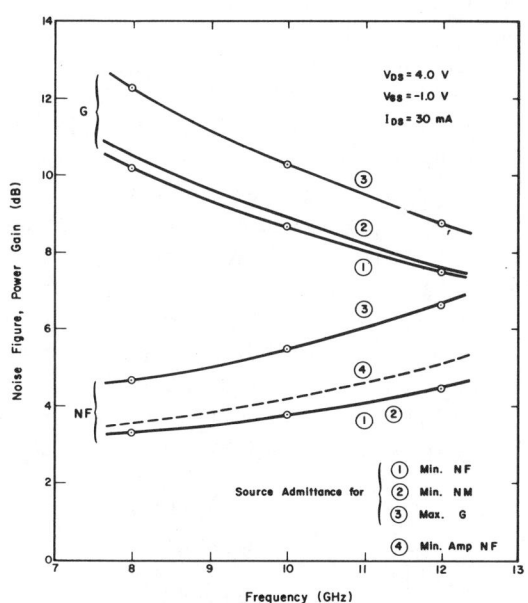

Fig. 2. MESFET power gain and noise figure versus frequency for source admittances that yield minimum noise figure ①, minimum noise measure ②, and maximum gain ③. Curve ④ shows the minimum noise figure for a large number of identical cascaded stages.

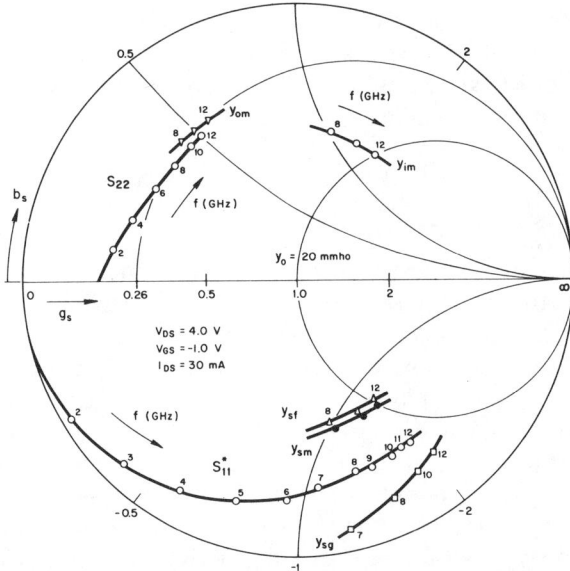

Fig. 3. Optimum source admittance that yields minimum noise figure Y_{sf}, minimum noise measure Y_{sm}, and maximum gain Y_{sg}. Also plotted are the MESFET scattering parameters s_{11}^*, s_{22}, the MESFET's output admittance Y_{om}, for an input noise measure match, and the equivalent noise input admittance $Y_{im} = Y_{sm}^*$.

to 4.5 dB at 12 GHz, and the associated gain (curve ②) drops with a 4.7-dB/octave slope from 10.4 to 7.6 dB. With these data, the lowest possible total amplifier noise figure can be predicted from (1). It is shown in curve ④.

The optimum source admittances referenced to the input plane of the MESFET chip are shown in the Smith chart of Fig. 3 with frequency as a parameter. Y_{sf} applies to the case of minimum noise figure, Y_{sm} to minimum noise measure, and Y_{sg} to maximum gain. For comparison, the complex conjugate of the scattering parameter s_{11} is also plotted. Y_{sg} and Y_{sm} and the associated noise figures are substantially different. Consequently, the conventional circuit optimization on the basis of MESFET s parameters is not adequate for a low-noise amplifier design.

The output scattering parameter s_{22} is shown in the upper half of the Smith chart. For a noise measure match

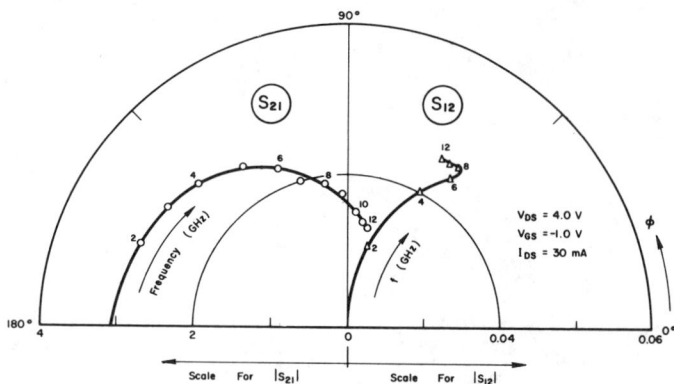

Fig. 4. MESFET scattering parameters s_{12} and s_{21} versus frequency.

at the input (source admittance Y_{sm}), the MESFET's output admittance Y_{om} lies on a constant conductance circle with increasing capacitive susceptance for increasing frequency. In a limited band around 10 GHz, the output admittance can be modeled as a resistor, $R_{om} = 192\ \Omega$, in parallel with a capacitor, $C_{om} = 0.16$ pF. In order to provide the complete set of s parameters, s_{12} and s_{21} are plotted versus frequency in Fig. 4.

For frequencies sufficiently above the instability limit (6.5 GHz), the MESFET can approximately be treated as a unilateral device. This is demonstrated in Fig. 3 with the proximity of Y_{sg} to $s_{11}{}^*$ and Y_{om} to s_{22}. The unilateral assumption breaks the MESFET equivalent circuit into two decoupled networks, a series RC at the input, and a parallel RC at the output. This approximation is utilized in the design techniques of Section III.

III. AMPLIFIER DESIGN

A. Input Matching Network

For receiver preamplifiers, the design objective is minimum amplifier noise figure rather than maximum gain. The main task is to synthesize the optimum source admittance Y_{sm} for minimum transistor noise measure at the FET input plane. An approximate solution can be found in the following way. An equivalent noise input admittance Y_{im} is defined as the complex conjugate of Y_{sm}. The two admittances are plotted in Fig. 3 versus frequency. Y_{im} can be approximated with a capacitance C_{im} in series with a resistance R_{im} with good accuracy over an octave bandwidth. The values, $C_{im} = 0.57$ pF and $R_{im} = 21\ \Omega$, are determined in Fig. 3.[1] The task consists of designing a Chebyshev impedance-matching network to match the generator admittance Y_0 to the noise input admittance Y_{im}. Chebyshev prototypes are selected to maximize the bandwidth and provide a good match at the band edges.

The first step is to establish the number of resonators required to synthesize Y_{sm} with a specified accuracy between the lower and upper band edges, f_l and f_u. The accuracy limits are determined by the tolerated excess noise figure. The minimum source admittance deviation from Y_{sm} which causes a maximum noise figure increase of 0.5 dB between 8 and 12 GHz has been determined.[2] This admittance deviation can be expressed in terms of a maximum tolerated VSWR. The computation yields 1.3:1. In synthesizing Y_{sm} with a Chebyshev impedance-matching network, the bandwidth can be computed using the VSWR versus load decrement curves plotted in [12, fig. 4.09-3]. The result is illustrated in Fig. 5 showing the maximum theoretical bandwidth w versus center frequency f_0 with the number of resonators n as a parameter. The curves are computed for a load with $C_{im} = 0.57$ pF in series with $R_{im} = 21\ \Omega$. A single resonator structure ($n = 1$) consisting of a series inductance and an impedance inverter (transformer) has an 8.9–11.1-GHz bandwidth. A more complex circuit with two additional resonators ($n = 3$) extends the band from 6 to 14 GHz.

The network shown in Fig. 6 was found to provide wide-band impedance matching, amplifier stability, and easy realization. The MESFET's equivalent noise input admittance is resonated with the series inductance L_1 and coupled to a network consisting of two $\lambda/4$ resonators. The resistance R_{im} can be considered the load resistance and $L_1 - C_{im}$ the last resonator of a three-stage bandpass filter coupling R_{im} to the source resistance Z_0. All resonators are resonant at the band center frequency f_0 and are coupled alternately by impedance inverters and admittance inverters. The inverters perform two functions. They convert the circuit connected to their output into its dual form at their input. This property allows the use of only one kind of resonator (e.g., series circuits). In addition, the inverters act as impedance transformers. This enables the designer to choose the characteristic impedance of the $\lambda/4$ resonators and adjust the inverter constants, K_{12}, J_{23}, and K_{34}, to ensure that the resonator reactance slopes at f_0 and the source resistance have the relationship prescribed by the Chebyshev impedance-matching prototype. If the line impedances are chosen to be equal to the source impedance Z_0 then the inverter constants are [12]

$$K_{12} = Z_0 \left(\frac{\pi w}{4 g_1 g_2} \frac{R_{im}}{Z_0 \delta} \right)^{1/2} \qquad (2)$$

$$J_{23} = \frac{1}{Z_0} \frac{\pi w}{4} \left(\frac{1}{g_2 g_3} \right)^{1/2} \qquad (3)$$

$$K_{34} = Z_0 \left(\frac{\pi w}{4 g_3 g_4} \right)^{1/2} \qquad (4)$$

where w is the relative bandwidth:

$$w = (f_u - f_l)/f_0 \qquad (5)$$

---

[1] The actual MESFET input admittance, $Y_{ig} = Y_{sg}{}^*$, with $C_{ig} = 0.58$ pF in series with $R_{ig} = 7\ \Omega$ has a considerably smaller series resistance than Y_{im}.

[2] The direction and magnitude of the worst admittance deviation are found graphically by plotting the constant noise figure contours in the source admittance plane at various frequencies [11]. The four noise parameters, required for this plot, are determined from data given in Figs. 2 and 3.

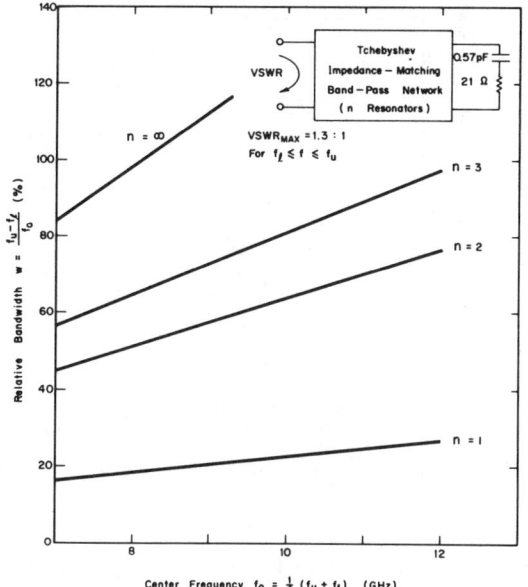

Fig. 5. Maximum bandwidth w versus center frequency f_0 for Chebyshev impedance-matching networks with n resonators.

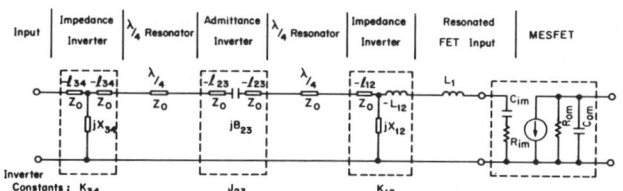

Fig. 6. MESFET with input matching network.

f_0 the center frequency:

$$f_0 = \tfrac{1}{2}(f_u + f_l) \qquad (6)$$

δ the decrement of the load impedance at f_0:

$$\delta = (2\pi f_0 R_{im} C_{im})/w \qquad (7)$$

and g_1, g_2, g_3, and g_4 the normalized element values of the Chebyshev impedance-matching prototype for the decrement δ [12].

The elements shown in the circuit diagram of Fig. 6 can now be computed using the following formulas [12].

1) *Series Inductance:*

$$L_1 = \frac{1}{(2\pi f_0)^2 C_{im}}. \qquad (8)$$

2) *Impedance Inverters:*

$$X_{ik} = \frac{K_{ik}}{1 - (K_{ik}/Z_0)^2} \qquad (9)$$

$$l_{ik} = \frac{\lambda}{4\pi} \tan^{-1}\left(\frac{2X_{ik}}{Z_0}\right) \qquad (10)$$

$$L_{12} \approx \frac{K_{12}}{2\pi f_0}, \qquad (\text{for } K_{12} \ll Z_0). \qquad (11)$$

3) *Admittance Inverter:*

$$B_{23} = \frac{J_{23}}{1 - (J_{23}Z_0)^2} \qquad (12)$$

$$l_{23} = \frac{\lambda}{4\pi} \tan^{-1}(2B_{23}Z_0) \qquad (13)$$

where λ is the wavelength of the transmission line at f_0.

The synthesis for an X-band amplifier with

$$f = 8.0\text{--}12.0 \text{ GHz}$$
$$R_{im} = 21 \text{ }\Omega$$
$$C_{im} = 0.57 \text{ pF}$$
$$Z_0 = 50 \text{ }\Omega$$

yields the network that is shown in Fig. 7. The resulting computed source admittance Y_s at the MESFET input deviates most from the target Y_{sm} at 12 GHz. At this frequency, it raises the noise figure 0.2 dB[3] above the design objective shown in Fig. 2. Computer optimizing this circuit results in the element values shown in parentheses in Fig. 7. Only small changes are necessary to achieve optimum performance. The microstrip realization illustrated in Fig. 8 shows that all network elements are easily realized. The two inductances are approximated with short bonding wires and the series capacitance by a narrow gap between the ends of adjacent transmission lines. The shunt inductance facilitates gate bias insertion, and the series capacitance acts as a dc block. Below the passband, the shorted shunt line and the series capacitance form a high-pass filter at the amplifier input. This structure reflects low frequency signals and protects the Schottky-barrier gate from burnout.

B. Output Coupling Network

In general, an amplifier is required to have low output VSWR and constant power gain in the design band. However, the available power gain of the MESFET, when matched for minimum noise measure, exhibits a slope of approximately 4.7 dB/octave (Fig. 2). Therefore, an ideal output network has to couple all available power to the load at the upper band edge f_u and provide a frequency dependent attenuation that compensates the FET's gain slope for $f < f_u$.

A simple network suitable for this purpose is shown in Fig. 9. The output admittance of the transistor, R_{om} in parallel with C_{om}, is resonated with the series inductance L_o at f_u. The transformer matches the resulting resistance to the load Z_0. At f_u, no power is dissipated in the resistor R_1 since it is connected in series with a short-circuited shunt stub that is a quarter-wave long at this frequency. Below f_u, the power transferred to the load decreases because the impedance of the resonated FET output

---

[3] This excess noise figure was computed with formula 1 of [11]. The four noise parameters were determined from data presented in Figs. 2 and 3.

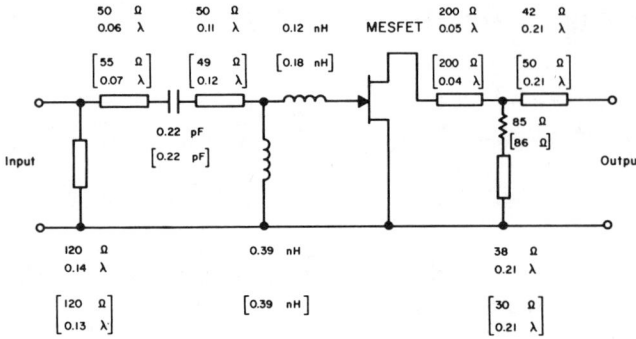

Fig. 7. Single-stage X-band amplifier circuit. Element values calculated with the formulas and curves of Section III are compared with computer optimized values (in parentheses). The transmission lines are defined by their characteristic impedances and lengths which are expressed as fractions of a wavelength at 10 GHz.

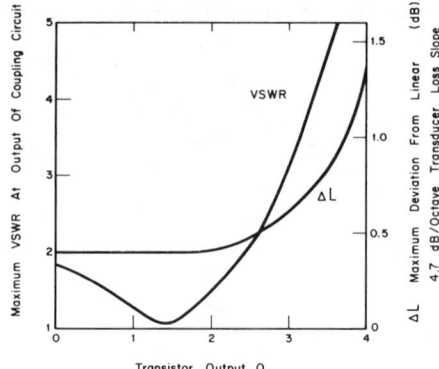

Fig. 10. Transducer loss and VSWR (in plane T) of output coupling circuit versus transistor output Q. For each Q value, the circuit has been optimized to yield a perfect power match at the upper band edge and exhibit a 4.7-dB/octave slope in transducer loss within the $f_u/f_l = 1.5:1$ frequency range.

loss target is less than 0.5 dB and the output VSWR does not exceed 2:1. For higher Q values, a more complex circuit is required to shape the coupling loss versus frequency curve and keep the output match within acceptable limits. The optimum element values are plotted in Fig. 11. Using these curves, the X-band circuit, as shown in Fig. 7, has been designed for

$$R_{om} = 192 \ \Omega$$

$$C_{om} = 0.16 \text{ pF}$$

$$f_u = 12 \text{ GHz}$$

which yields a Q of 2.3. The inductance is too large (0.9 nH) to be realized in lumped form and is modeled as a short high-impedance transmission line. It consists of a bonding wire which runs parallel to and above the ground plane. At the output, the ideal transformer is approximated with a 42-Ω $\lambda/4$ transmission line, which is sufficiently broad band for this case. The accuracy of this design for the exact MESFET model with bilateral characteristics has been evaluated. For this purpose, the output circuit has been computer optimized with the MESFET represented by its s parameters and with the input circuit of Fig. 7 (values

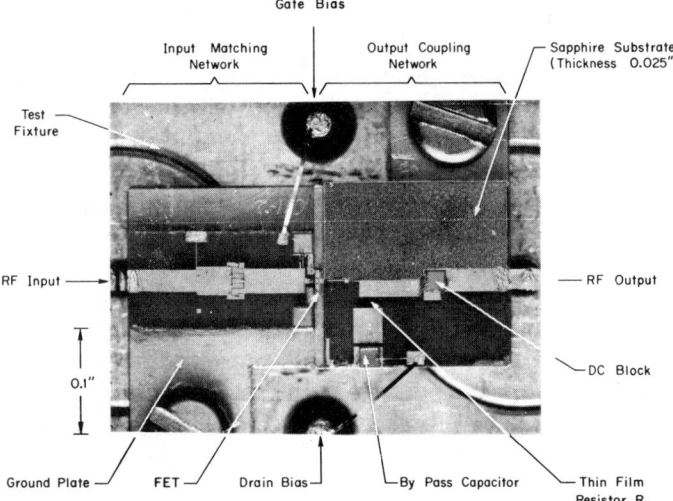

Fig. 8. Realization of the single-stage X-band amplifier.

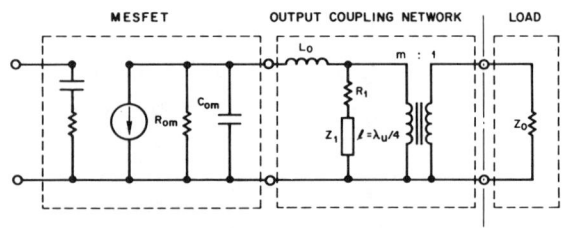

Fig. 9. MESFET with output coupling network.

changes rapidly and the shunt branch with the resistor becomes increasingly lossy.

The resistance R_1 and the characteristic impedance of the transmission line Z_1 were varied until the best approximation to a 4.7-dB/octave slope for the output circuit transducer loss was obtained. Fig. 10 shows the results optimized for a frequency range of $f_u/f_l = 1.5:1$ with the MESFET output Q

$$Q = 2\pi f_u R_{om} C_{om} \qquad (14)$$

as a variable. It is seen that this simple network is useful as long as Q does not exceed 2.5. Under this condition, the maximum deviation from the 4.7-dB/octave transducer

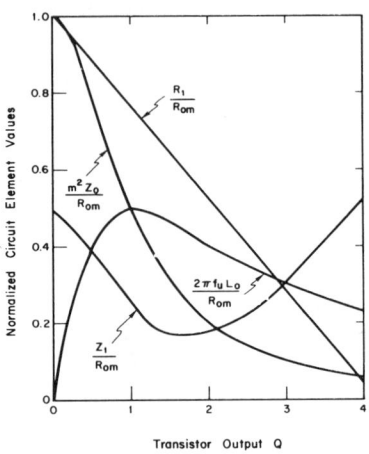

Fig. 11. Optimum values for circuit elements of the output coupling network versus transistor output Q.

stated in parentheses) connected. The resulting element values are shown in parentheses in Fig. 7. They are very close to the values obtained from Fig. 11. The microstrip realization is illustrated in Fig. 8.

C. Amplifier Construction

The basic building block is a single-stage amplifier which is approximately matched to a 50-Ω source and load and uses the circuits described in the previous sections. Such an amplifier unit is shown in Fig. 8. The circuits with resistors, interdigitated capacitors, and transmission lines are fabricated on 0.025-in sapphire substrates. The substrates are soldered onto a ground plate and are separated by a ridge on which the transistor chip is mounted. In this configuration, the transistor source can be grounded with less than 50-pH lead inductance, and RF power leakage from the FET output to input is minimized. RF bypass and dc blocking capacitors are soldered onto the substrates, and the devices are interconnected with thermocompression bonded gold wires. After assembly, the single-stage units are mounted in a test fixture where they are tuned and characterized.

For the X-band amplifier, three identical single-stage units were cascaded. This modular approach does not achieve the widest possible bandwidth, but it has the following practical advantages.

1) The design and fabrication are limited to two circuits.
2) The tuning of a few variables on a single-stage amplifier is efficient and convergent.
3) In final assembly, stages can be selected according to their tuned performance (e.g., lowest noise unit selected for first stage).
4) In the case of device failure, each stage can be easily exchanged with only minor retuning of the amplifier.

The assembled stages are built into a metallic enclosure that acts as a waveguide below cutoff, thus suppressing modes that cause resonances and RF leakage from output to input.

IV. X-BAND AMPLIFIER PERFORMANCE

The gain and noise figures of the three-stage amplifier are plotted versus frequency in Fig. 12. From 8.0 to 12.0 GHz, the power gain is 20.7 dB with less than ±1.3-dB variation. A single stage yields a constant gain of 7 dB up to 12 GHz. This is only 0.6 dB less than the maximum possible gain predicted on the basis of the measured device gain (Fig. 2) under the operating conditions for minimum noise measure. The gain variation is close to the minimum value of 1.2 dB that can be achieved with the simple output circuit (Fig. 10). The reverse insertion loss $|s_{12}|^{-2}$ does not drop below 50 dB (Fig. 13) which indicates that for high-gain requirements more than three stages can be cascaded without loss of stability. The maximum local gain slope is 0.005 dB/MHz, and the gain dependence on ambient temperature under constant bias conditions is −0.07 dB/°C. Improved stability can be obtained with a temperature-controlled gate bias circuit. The transmission

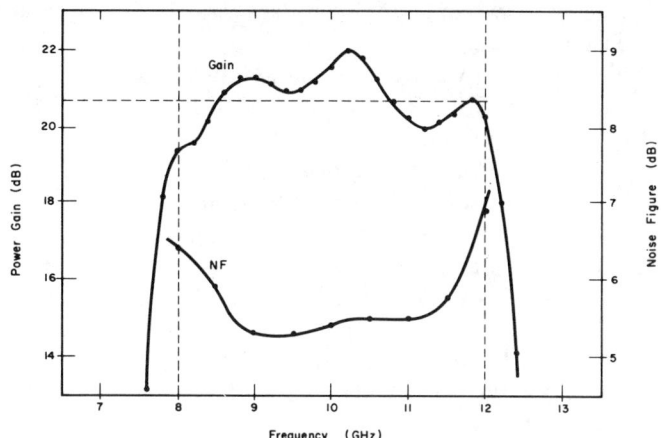

Fig. 12. Power gain and noise figure versus frequency for the three-stage X-band amplifier.

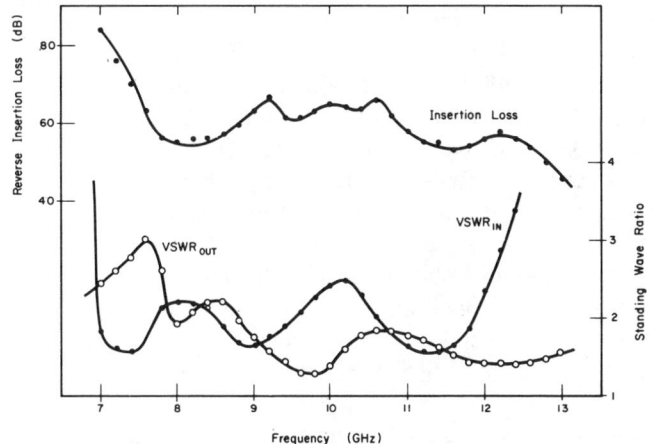

Fig. 13. Input and output VSWR and reverse insertion loss versus frequency for the three-stage X-band amplifier.

phase deviation from linearity is less than ±14° between 8 and 12 GHz.

The most noteworthy characteristic of this amplifier is the low noise figure which does not exceed 5.9 dB over the wide range of 8.5–11.5 GHz. At 10 GHz, for example, the first stage has a noise figure of 4.6 dB with 7.0-dB gain. The transistor used in this stage, which is biased for minimum noise measure, exhibits a noise figure of 3.9 dB with 9.0-dB associated (device) gain. The 0.7-dB difference in noise figure is caused by attenuation in the coaxial to microstrip transition, attenuation in the input network (approximately 0.3 dB), and by a mismatch with respect to the optimum admittance for lowest noise measure. The contributions from the following stages raise the overall amplifier noise figure to 5.4 dB. At the lower and upper band edge, the noise match of the input stage degrades, which causes the noise figure to increase. At 12 GHz, the maximum value of 6.9 dB is reached. In comparison, a narrow-band design that was optimized at 8 GHz yielded a minimum noise figure of 3.9 dB with 25-dB gain. The lowest possible amplifier noise figure, as determined from Fig. 2 (curve ④), is 3.6 dB at this frequency. At the edges of an 800-MHz band, the noise figure degrades to 4.4 dB.

These narrow-band amplifier data are similar to results reported in [9].

The standing wave ratios at the input and output ports are shown in Fig. 13. The input VSWR does not exceed 2.4:1 even though the circuit is designed for a noise match as opposed to an impedance match. This performance clearly demonstrates the broad-band matching capability of this input network. The output VSWR is 1.4:1 at 12 GHz and 2:1 at 8 GHz. The simple output coupling circuit provides a good impedance match at 12 GHz and is capable of attenuating the excess power at lower frequencies rather than reflecting it. This characteristic is essential for the success of the modular approach in which pretuned stages are cascaded.

The large signal properties of the three-stage amplifier are also noteworthy. In order to increase the dynamic range, the drain current of the MESFET in the output stage was raised to 50 mA. The drain voltage was left unchanged at 4 V. The minimum output power for 1-dB gain compression is +13 dBm reached at 10 GHz. At this frequency, the third-order intermodulation intercept is +26 dBm. In a different circuit specially designed for a power output stage, +20-dBm output power at 5-dB gain[4] was obtained with a single transistor. The MESFET, biased at 6-V drain voltage and 43-mA drain current, exhibited a drain-efficiency ($P_{\text{RF-out}}/P_{\text{dc-in}}$) of 39 percent.

To determine the nuclear radiation hardness of the amplifier, packaged MESFET's have been subjected to the fast neutron radiation of an unmoderated pulsed uranium reactor.[5] A flux of 5×10^{14} neutrons/cm^2 yields a 10-percent decrease in transconductance. A ten times higher neutron fluence causes extensive lattice damage (traps) such that most free carriers are removed, and the drain current drops to zero. Results published for GaAs JFET's with the same channel doping and of similar structure [13] show the same parameter degradation at an order of magnitude higher neutron fluence. The results, however, are difficult to compare because in [13] a moderated reactor with a spectrum rich in low energetic neutrons was used. The MESFET's were also subjected to the 1.25-MeV gamma radiation of a cobalt-60 source. Absorbed doses up to 10^8 rad (GaAs) caused no detectable change in transconductance.

V. CONCLUSIONS

Optimum low-noise amplifier performance is obtained with MESFET's biased for minimum noise measure and with an input coupling network designed to synthesize the optimum source admittance for lowest noise measure. The input circuit consists of a series inductance and quarter-wave resonators alternately coupled by impedance and admittance inverters. The network design is based on Chebyshev impedance-matching prototypes. A series inductance and a transformer match the MESFET's output admittance to the load at the upper band edge. A lossy shunt stub provides frequency dependent attenuation compensating the MESFET's gain slope. Optimized design curves are presented for this output coupling network. The circuits yield amplifier stages that are stable, approximately matched to 50 Ω, and easy to realize in microstrip.

An 8.0–12.0-GHz amplifier was built with three identical cascaded stages. The amplifier exhibits a maximum noise figure of 6.9 dB (5.5 dB typical) with 20.7 ± 1.3-dB gain. The VSWR at the input and output ports does not exceed 2.5:1. The minimum output power for 1-dB gain compression is +13 dBm, and the intercept point for third-order intermodulation products is +26 dBm.

These experimental results demonstrate that amplifiers with GaAs MESFET's can be designed to yield high-gain and low-noise properties over wide bandwidths. In this frequency range (>8 GHz), bipolar transistors have insufficient gain for low-noise amplifiers ($G \leq 4$ dB with 4-dB noise figure at 8 GHz), and tunnel diodes exhibit very limited dynamic range (IM intercept point < -10 dBm). For ultra-low-noise receiver requirements ($NF \leq 3$ dB), parametric amplifiers still hold a unique position. However, if the parametric amplifier is excluded because of its complexity and cost, then the MESFET is presently the only solid-state device suitable for low-noise amplification in X and Ku bands with acceptable dynamic range. The data reported in this paper confirm that MESFET amplifiers meet specifications that are equivalent to those imposed on low-noise traveling wave tube amplifiers. The MESFET amplifier qualifies, therefore, as the solid-state replacement in this frequency range.

ACKNOWLEDGMENT

The authors wish to thank R. Larrick for his measurement assistance and valuable experimental contributions, Dr. J. Barrera and R. Drabin for supplying the GaAs epitaxial layers, E. Gowen for the MESFET fabrication, D. Hollars for the microcircuit assembly, and Dr. E. Graham of Sandia Laboratories for the neutron and gamma irradiation experiments. They also wish to thank Dr. R. Archer, A. Podell, R. Van Tuyl, J. Dupre, and J. Kesperis for helpful suggestions and comments.

REFERENCES

[1] W. Baechtold et al., "Si and GaAs 0.5μm-gate Schottky-barrier field-effect transistors," *Electron. Lett.*, vol. 9, pp. 232–234, May 1973.
[2] C. A. Liechti, E. Gowen, and J. Cohen, "GaAs microwave Schottky-gate field-effect transistor," in *ISSCC Dig. Tech. Papers*, 1972, pp. 158–159.
[3] N. G. Bechtel, W. W. Hooper, and D. Mock, "X-band GaAs FET," *Microwave J.*, vol. 15, pp. 15–19, Nov. 1972.
[4] P. L. Clouser, and V. V. Risser, "C-band FET amplifiers," in *ISSCC Dig. Tech. Papers*, 1970, pp. 52–53.
[5] W. Baechtold, "Ku-band GaAs FET amplifier and oscillator," *Electron. Lett.*, vol. 7, pp. 275–276, May 1971.
[6] S. Arnold, "Single and dual gate FET integrated amplifiers in C-band," in *IEEE Int. Microwave Symp. Dig. Tech. Papers*, 1972, pp. 233–234.
[7] L. Besser, "Design considerations of a 3.1–3.5 GHz GaAs FET feedback amplifier," in *IEEE Int. Microwave Symp. Dig. Tech. Papers*, 1972, pp. 230–232.
[8] R. Zuleeg, E. W. Bledl, and A. F. Behle, "Broadband GaAs field effect transistor amplifier," Air Force Avionics Lab.,

---

[4] The small-signal gain was 7.5 dB at 10 GHz.
[5] Reactor SPR II of Sandia Laboratories, Albuquerque, New Mex.

Wright-Patterson Air Force Base, Ohio, Tech. Rep. AFAL-TR-73-109, Mar. 1973.

[9] W. Baechtold, "X- and Ku-band amplifiers with GaAs Schottky-barrier field-effect transistors," *IEEE J. Solid-State Circuits* (*Special Issue on Microwave Integrated Circuits*), vol. SC-8, pp. 54–58, Feb. 1973.

[10] C. A. Liechti and R. L. Tillman, "Application of GaAs Schottky-barrier FETs in microwave amplifiers," in *ISSCC Dig. Tech. Papers*, 1973, pp. 74–75.

[11] H. Fukui, "Available power gain, noise figure, and noise measure of two-ports and their graphical representations," *IEEE Trans. Circuit Theory*, vol. CT-3, pp. 137–142, June 1966.

[12] G. L. Matthaei, L. Young, and E. M. T. Jones, *Microwave Filters, Impedance-Matching Networks, and Coupling Structures*. New York: McGraw-Hill, 1964.

[13] A. F. Behle and R. Zuleeg, "Fast neutron tolerance of GaAs JFET's operating in the hot electron range," *IEEE Trans. Electron Devices* (Corresp.), vol. ED-19, pp. 993–995, Aug. 1972.

Performance of Dual-Gate GaAs MESFET's as Gain-Controlled Low-Noise Amplifiers and High-Speed Modulators

CHARLES A. LIECHTI, SENIOR MEMBER, IEEE

Abstract—This paper describes the microwave performance of GaAs FET's with two 1-μm Schottky-barrier gates (dual-gate MESFET). At 10 GHz the MESFET, with an inductive second-gate termination, exhibits an 18-dB gain with −26-dB reverse isolation. Variation of the second-gate potential yields a 44-dB gain-modulation range. The minimum noise figure is 4.0 dB with 12-dB associated gain at 10 GHz. Pulse modulation of an RF carrier with a 65-ps fall and a 100-ps rise time is demonstrated. The dual-gate MESFET with high gain and low noise figure is especially suited for receiver amplifiers with automatic gain control (AGC) as an option. The MESFET is equally attractive for subnanosecond pulsed-amplitude modulation (PAM), phase-shift-keyed (PSK), and frequency-shift-keyed (FSK) carrier modulation.

I. INTRODUCTION

THE gallium arsenide field-effect transistor with Schottky-barrier gate (GaAs MESFET) is the next generation of small-signal low-noise transistors. Owing to the high electron mobility and high peak-drift-velocity in GaAs, the MESFET exhibits shorter transit times than silicon transistors. The useful frequency range is extended more than a factor of 2 over present bipolar transistors. The gain and noise performance of single-gate GaAs MESFET's have been extensively characterized [1]–[6], and the MESFET potential in low-noise amplifiers has been clearly demonstrated [7]–[16].

Another member in the MESFET family with very attractive gain and noise performance, stability, and modulation capabilities is the dual-gate MESFET. This transistor has two parallel gate electrodes between source and drain. The dual-gate MESFET is normally operated in common-source configuration. The RF input signal is applied between the first gate and the source. The output signal between drain and source is coupled to the load. The second gate is usually RF grounded. The superior stability of the dual-gate MESFET has been demonstrated [17], [18]. In addition, it has been shown that dc bias applied to the second gate varies the RF gain [17]–[22]. A negative bias beyond the pinch-off voltage yields a large insertion loss, and positive bias yields a gain maximum that is considerably higher than the gain of a single-gate device with equal dimensions. Many characteristics are qualitatively analogous to those of dual-gate Si MOSFET's used in UHF receivers [23]–[26]. Both MESFET's and MOSFET's have traditionally been characterized as two-port devices with the second gate RF grounded. A more general approach deals with the dual-gate MESFET as a three-port device [27]. If one connects an RF impedance between the second gate and source, then the properties of the resulting two-port can easily be expressed as functions of this impedance. It will be shown that this impedance has a strong influence on many MESFET parameters. With a proper choice, the gain or stability can be optimized or the noise measure minimized.

The purpose of this paper is to discuss the characteristics of the dual-gate MESFET as a high-gain low-noise amplifier with voltage-controlled gain and as a high-speed RF modulator. First, measured three-terminal s-parameters will be described, and the forward gain and reverse isolation will be computed for various impedances connected between the second gate and source. In addition, the gain versus frequency will be characterized. Second, the bias at the second gate will be varied, and the resulting gain, input impedance, and transmission-phase change will be discussed. Third, the noise performance and power capability will be compared with single-gate MESFET's. Fourth, the high-speed modulation capability will be demonstrated. A major objective is to show clearly the differences between the single-gate and the dual-gate MESFET performance. Realizing each FET version's unique features will lead, at the end of this paper, to obvious conclusions for their application.

II. DEVICE DESCRIPTION AND SIGNAL FLOW GRAPH

The MESFET is fabricated on a semi-insulating Cr-doped GaAs substrate that is covered with a thin n-type epitaxial layer. The conducting layer is grown from Sn-doped Ga solution on the ⟨100⟩ substrate face. The doping is 7×10^{16} cm^{-3}, and the layer thickness is 0.2 μm in the channel region. A microphotograph of the chip is shown in Fig. 1, and a magnified view of the center section is illustrated in Fig. 2. Alloyed ohmic contacts form the source and drain, and rectifying Schottky-contacts form the two gates. The gates are metal stripes, 1 μm wide and 400 μm long, running in parallel between the ohmic contacts. The first gate is separated by 1 μm from the source and by 2 μm from the second gate. Outside the active area,

---

Manuscript received August 30, 1974; revised December 10, 1974.
The author is with the Solid State Laboratory, Hewlett-Packard Laboratories, Palo Alto, Calif. 94304.

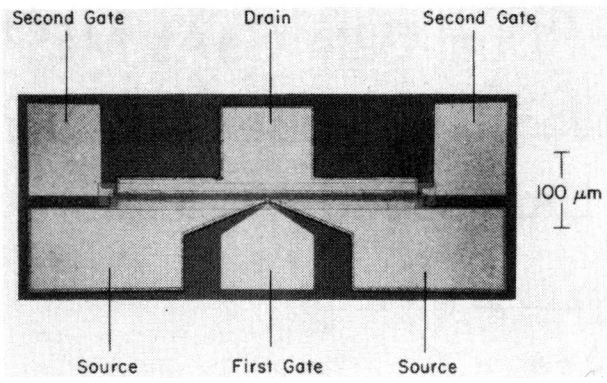

Fig. 1. Microphotograph of the dual-gate MESFET chip.

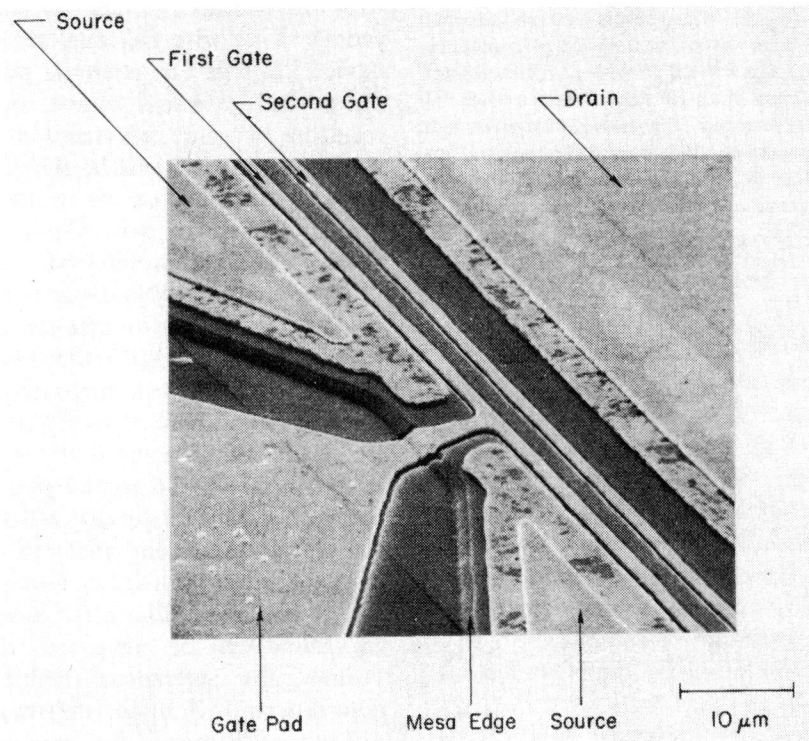

Fig. 2. Scanning-electron micrograph of the dual-gate MESFET's center section.

the epitaxial film is removed by mesa etching. This permits large gate pads to be located on the semi-insulating substrate where they contribute only a small fraction to the interelectrode capacitance.

An artist's concept of the dual-gate MESFET is shown in Fig. 3(a). It is helpful to visualize the device as two separate single-gate MESFET's connected in cascade as shown in Fig. 3(b) [18], [23]. The output current of the first MESFET flows directly into the channel of the second MESFET. If V_{D_1S}, the potential between the two gates, is larger than the threshold voltage for current saturation $V_{D(\text{sat})}$ [Fig. 3(c)], then the first transistor acts essentially as an ideal current source. The bias applied at the second gate V_{G_2S} controls the effective drain voltage V_{D_1S} of the first transistor. V_{D_1S} adjusts to establish the proper gate-to-source bias $V_{G_2S} - V_{D_1S}$ that allows the second transistor to carry the dc current from the first transistor. With positive V_{G_2S} no dc current is flowing into the second gate as long as this voltage stays about 0.5 V below the drain voltage and the first-gate bias is zero or negative ($V_{G_1S} \leq 0$). The model of Fig. 3(b) will frequently be referenced to interpret measured results.

The dual-gate MESFET is characterized as a three-port device with the source as a common terminal. The transistor is measured in a test fixture with drain and gate pads pressed against the center-conductor endings of miniature coaxial lines. The source pads are grounded in this fixture with negligible lead inductance [2]. s-parameters measured at 10 GHz in a 50-Ω system are illustrated in the signal flow graph of Fig. 4. The reference plane for port 1 and port 2 is located at the first gate, and the reference plane for port 3 runs through the center of the

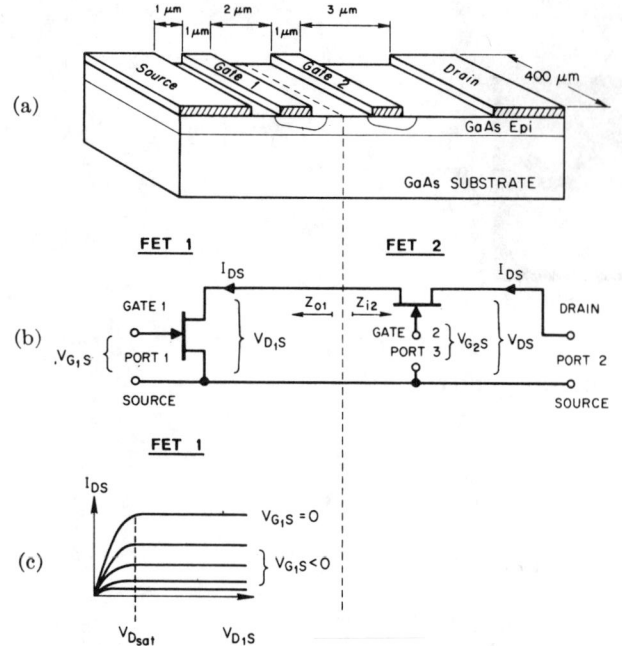

Fig. 3. The dual-gate MESFET is modeled as two single-gate MESFET's connected in cascade. The first MESFET is operated with a common source. Its drain current feeds the source of the second MESFET. The drain current versus drain voltage of the first MESFET is shown in (c).

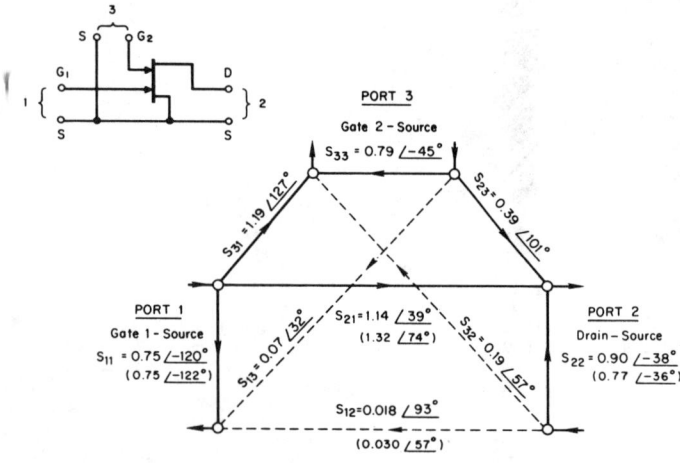

Fig. 4. Three-port signal flow graph of the dual-gate MESFET at 10 GHz. Operating conditions: $V_{DS} = 4.5$ V; $V_{G_1S} = 0$ V; $V_{G_2S} = 2.0$ V; $I_{DSS} = 46$ mA; $Z_0 = 50\ \Omega$; $f = 10$ GHz. The s-parameters of a comparable single-gate MESFET are listed in parentheses.

chip perpendicular to the gates. The dc bias has been chosen to yield high gain and allow a simultaneous image match at port 1 and port 2 for a 50-Ω termination at port 3. As expected, strong signal coupling is experienced from gate 1 to drain. But, in addition, nearly the same amount of power is fed from the first gate to the second gate. The reverse couplings are weak, especially s_{12}. The forward coupling from port 3 to port 2 is less pronounced because the output impedance Z_{o1} connected in series with the second-gate capacitance [Fig. 3(b)], keeps the extrinsic transconductance of the second gate low. Z_{o1}, acting also as a series feedback impedance, enhances the reverse signal flow from port 2 to port 3. In comparison, the parameters s_{11}, s_{12}, s_{21}, and s_{22} of a single-gate MESFET with the same geometry are listed in parentheses in Fig. 4. The three-port s-parameters fully describe the dual-gate MESFET's small-signal behavior. They are the basis for computations carried out in the following section.

III. GAIN VERSUS SECOND-GATE TERMINATION

The MESFET characterized as a three-port device is converted to a two-port device by terminating the second gate with an impedance Z_3. The first gate is now considered the input port (port 1) and the drain is considered the output port (port 2). The s-parameters s_{ik}' of this two-port are related to the parameters s_{ik} of the original three-port network by [28]

$$s_{ik}' = s_{ik} + \{(s_{i3} \cdot s_{3k})/[(1/\Gamma_3) - s_{33}]\} \quad (1)$$

where Γ_3 is the reflection coefficient of the load Z_3 with respect to the reference impedance Z_0

$$\Gamma_3 = (Z_3 - Z_0)/(Z_3 + Z_0). \quad (2)$$

Knowing s_{ik}', the maximum available forward gain G_f, and associated reverse isolation, G_r, between the image-matched input and output port can be computed [29]

$$G_f = \left|\frac{s_{21}'}{s_{12}'}\right| [k \mp (k^2 - 1)^{1/2}] \quad (3)$$

and

$$G_r = \left|\frac{s_{12}'}{s_{21}'}\right| [k \mp (k^2 - 1)^{1/2}] \quad (4)$$

with the stability factor k

$$k = \frac{1 + |s_{11}' \cdot s_{22}' - s_{12}' \cdot s_{21}'|^2 - |s_{11}'|^2 - |s_{22}'|^2}{2 \cdot |s_{12}'| \cdot |s_{21}'|}. \quad (5)$$

The negative sign in front of the square root applies if

$$1 - |s_{11}' \cdot s_{22}' - s_{12}' \cdot s_{21}'|^2 + |s_{11}'|^2 - |s_{22}'|^2 > 0. \quad (6)$$

For an RF-shorted second gate, the dual-gate MESFET represents a cascode circuit with a common-source input stage driving a common-gate output stage [Fig. 3(b)]. In Table I, the s-parameters of the dual-gate MESFET with grounded second gate are compared with those of its single-gate counterpart; s_{11}' and $|s_{21}'|$ are practically equal, i.e., the first transistor in Fig. 3(b) essentially determines the input impedance and the magnitude of the forward transfer coefficient s_{21}'. In the dual-gate MESFET, s_{21}' has a smaller phase angle due to the increased electrical length of this device. The magnitudes of s_{12}' and s_{22}' are very different in the two transistors. Signal feedback is considerably reduced, and the output impedance is increased in the dual-gate MESFET. A simple low-frequency ($f < 1$ GHz) analysis of the cascode circuit shows why this is so. A voltage v_{ds}, applied to port 2 in Fig. 3(b), is fed back through the drain–source re-

TABLE I
s-Parameters, Forward Gain, Reverse Isolation, and Stability Factor for a Single-Gate MESFET and for a Dual-Gate MESFET with RF-Grounded Second Gate

	s'_{11}	s'_{12}	s'_{21}	s'_{22}	G_f	G_r	k
Dual-Gate MESFET	0.73/−117°	0.011/82°	1.39/44°	0.94/−36°	16 dB	−26 dB	1.7
Single-Gate MESFET	0.75/−122°	0.030/57°	1.32/74°	0.77/−36°	11 dB	−22 dB	2.1

Operating Conditions: Dual-gate MESFET: $V_{DS} = 4.5$ V; $V_{G_1S} = 0$; $V_{G_2S} = 2.0$ V; $I_{DSS} = 46$ mA; $\Gamma_3 = -1$; $f = 10$ GHz. Single-gate MESFET: $V_{DS} = 4.0$ V; $V_{G_1S} = 0$; $I_{DSS} = 56$ mA; $f = 10$ GHz.

sistance of the second MESFET. The resulting drain voltage at the first transistor v_{d_1s} is smaller by the factor μ

$$\mu = v_{ds}/v_{d_1s} = 2 + g_m R_{ds} \approx 12. \quad (7)$$

R_{ds} is the single-gate output resistance and g_m is the transconductance. Consequently, the voltage fed back to port 1 over the first transistor's drain–gate capacitance is reduced by this factor. The capacitance between drain and gate of the second MESFET does not contribute to feedback because the second gate is RF grounded. The enhancement of the dual-gate MESFET's output impedance is also related to the fact that $v_{d_1s} = v_{ds}/\mu$. At low frequencies, v_{d_1s} causes a drain current at port 2 which is equal to $v_{ds}/(\mu R_{ds})$. The output resistance of the dual-gate MESFET is, therefore, μ times larger than it is in the single-gate version. At high frequencies, the capacitances in the equivalent circuit have also to be considered. As a result, the output resistance starts to fall off at microwave frequencies to a final value equal to the output resistance of a single-gate MESFET.

The contours of constant forward gain G_f and associated reverse isolation G_r have been computed using the s-parameters listed in Fig. 4. They are plotted in the Γ_3 plane in Fig. 5. With the second gate grounded ($\Gamma_3 = -1$), the maximum available forward gain is 16 dB with −26-dB reverse isolation. In comparison, the single-gate MESFET exhibits 11-dB gain with −22-dB isolation (Table I). Up to 21-dB gain can be obtained from the dual-gate MESFET with an inductive termination Z_3. An inductance reduces the reactive mismatch at the input to the channel of the second gate. The reverse isolation decreases, however, and the limits are set by the boundary $k = 1$ beyond which the transistor is potentially unstable. On the other hand, a termination Z_3 can be found that yields forward loss. This happens when the signal flowing via gate 2 to the drain interferes destructively with the signal directly fed from gate 1 to drain.

The frequency dependence of the gain for an RF-grounded second gate is plotted in Fig. 6. Mason's unilateral gain is calculated from the measured s-parameters. The computation yields a 21-dB gain at 10 GHz and a slope of 6 dB per octave.[1] The maximum available gain has been determined with tuned measurements.[2] The gain is 16 dB at 10 GHz and the slope, approximately 5.5

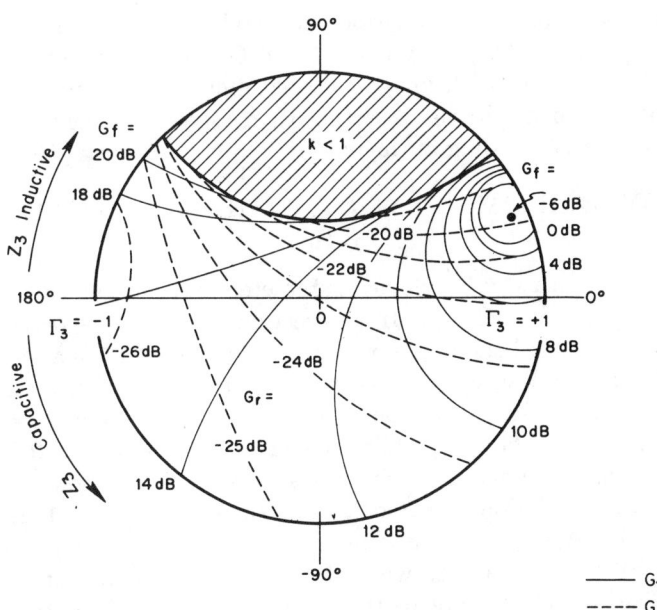

Fig. 5. Tuned forward gain G_f and associated reverse isolation G_r with port 1 as input and port 2 as output port are plotted versus Γ_3 at 10 GHz. Γ_3 is the reflection coefficient of the load impedance Z_3 connected to port 3. In the crosshatched area the transistor is potentially unstable. $V_{DS} = 4.5$ V; $V_{G_1S} = 0$; $V_{G_2S} = 2.0$ V; $Z_0 = 50$ Ω; $f = 10$ GHz.

dB/octave. A single-gate MESFET exhibits 11-dB maximum available gain and a slope of 5.5 dB/octave. Below 6 GHz, both MESFET versions are potentially unstable ($k < 1$).[3]

In a practical amplifier application, the second gate does not have to be terminated with a frequency-independent impedance. Moving along the $|\Gamma_3| = 1$ circle in Fig. 5 from an inductive Z_3 to a capacitive Z_3 decreases gain

---

[1] The slope is expected to increase at frequencies above X band.
[2] The measured results agree well with G_f calculated from measured s-parameters according to (3).

[3] That is, both ports cannot be simultaneously image matched. However, stable operation is possible at any frequency, providing that a drain load with a sufficiently large conductance is chosen.

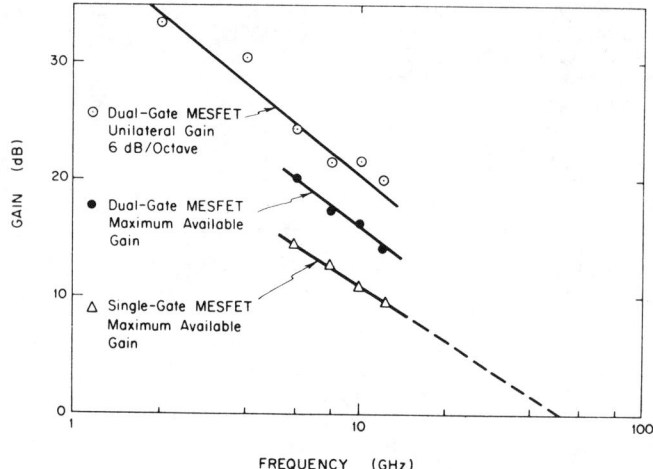

Fig. 6. Calculated unilateral gain and directly measured maximum available gain are plotted versus frequency for a dual-gate MESFET with RF-grounded second gate. $V_{DS} = 4.5$ V; $V_{G_1S} = 0$; $V_{G_2S} = 2$ V; $\Gamma_3 = -1$. The maximum available gain of a single-gate MESFET is shown for comparison.

and increases stability. Z_3 as a series-resonant circuit with an appropriate reactance slope could compensate the MESFET's intrinsic gain slope and stabilize the transistor at the lower frequency end ($f < 6$ GHz). In this way, a constant gain with image-matched input and output can be obtained without resorting to resistive feedback [9] or resistive loading of the output coupling network [16].

IV. GAIN MODULATION WITH SECOND-GATE BIAS

Up to this point, device parameters were discussed for fixed dc-bias voltages. In this section, the influence of the second-gate bias V_{G_2S} on forward gain, reverse isolation, input impedance, and transmission phase will be investigated. At 10 GHz, the second-gate termination $\Gamma_3 = 1 \angle 153°$ is chosen, yielding 18-dB gain at $V_{G_2S} = 2.0$ V under image-matched conditions (Fig. 5). The associated reverse isolation is -26 dB. Next, the second-gate voltage is decreased; and the gain, reverse isolation, and input VSWR are measured without readjustment of the tuners. The parameters are plotted in Fig. 7 with solid curves. The forward gain drops from $+18$ dB to -26 dB. This large gain variation can be explained by referring to the model shown in Fig. 3(b). V_{G_2S} controls the effective drain voltage V_{D_1S} of the first transistor. Decreasing V_{D_1S} increases the output conductance, real $(1/Z_{o1})$, of the first transistor. When this conductance starts to be comparable in magnitude to the input conductance of the second transistor, the transferred RF power begins to decrease. As V_{D_1S} drops below $V_{D(sat)}$, the first transistor acts as an RF voltage-controlled series resistance [Fig. 3(c)]. The RF voltage across this resistance modulates the depletion-layer width under the second gate. By decreasing V_{G_2S}, the dc drain current I_{DS} decreases, and in turn the modulating voltage drop across the series resistance has to decrease. This decrease of the modulating voltage and the decreasing transconductance of the second gate cause the continued gain drop. At -2.5 V, the

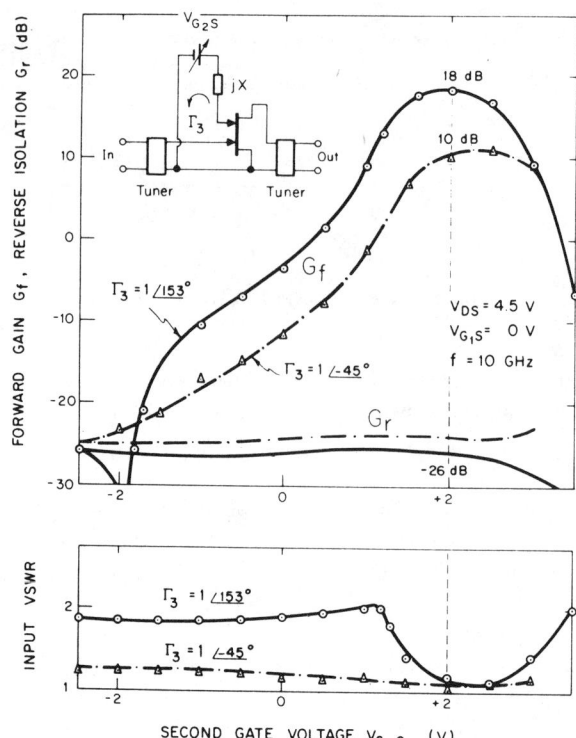

Fig. 7. Forward gain G_f, reverse isolation G_r, and input VSWR versus second gate voltage V_{G_2S} for two reactive terminations at the third port. The first gate and the drain are image matched at $V_{G_2S} = 2.0$ V.

channel under the second gate is completely depleted and the drain current is cut off. At this voltage, the dual-gate MESFET is a passive, reciprocal device ($G_f = G_r$) in which the drain to first-gate capacitance C_{dg_1} dominates the coupling between the ports. At $V_{G_2S} = -2.0$ V, shortly before the cutoff voltage is reached, the destructive interference of two signals at the output causes a very high forward insertion loss. The two signals result from power coupling over C_{dg_1} in one case and from voltage modulation of the channel current in the other case.

The reverse isolation is -26 dB and remains approximately constant over the entire gate-bias range. The effective drain to first-gate capacitance C_{dg_1} is essentially independent of V_{G_2S}. The same observation was made by Maeda [22]. Also, the input impedance versus V_{G_2S} stays within close limits because the reverse coupling coefficients s_{12} and s_{13} in Fig. 4 remain small in the bias range. The input mismatch plotted in Fig. 7 never rises above 2.1:1.

The forward transmission phase of the dual-gate MESFET is plotted in Fig. 8 versus second-gate voltage. The phase change for a 20-dB gain variation is 75°. If the RF driver is mismatched at $V_{G_2S} = 2$ V, an additional phase change would have to be considered due to the variation of the MESFET's input impedance. In some applications, a phase change of this magnitude is not acceptable. Fortunately, an optimum second-gate termination exists ($\Gamma_3 = 1 \angle -45°$) that minimizes the phase and input-impedance variation. With this Γ_3, a gain variation from its maximum of $+10$ dB to -10 dB (Fig. 7) changes the phase by only 5° (Fig. 8). The input VSWR reaches a maximum of 1.3:1,

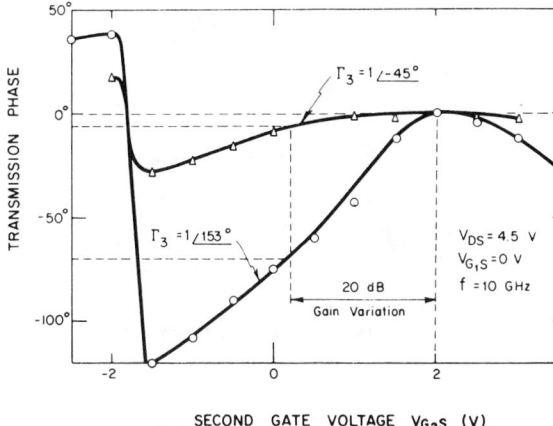

Fig. 8. Transmission phase versus second-gate voltage for two reactive terminations at the third port. The measurement conditions are identical with those in Fig. 7.

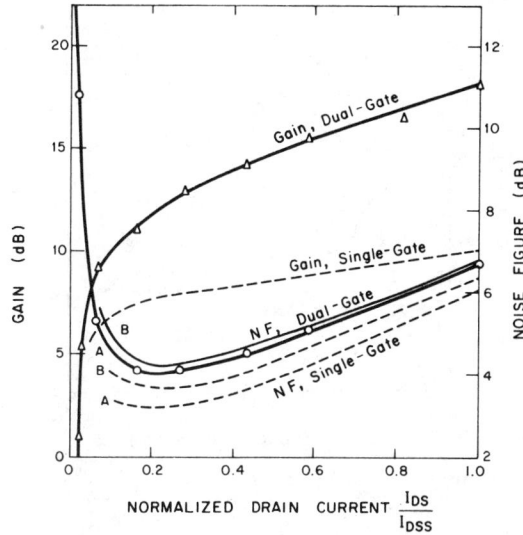

Fig. 9. Minimum noise figure and associated gain versus normalized drain current for single-gate and dual-gate MESFET'S. The current is varied by changing the first-gate bias voltage. Curves A show the spot noise figure measured on individual transistors. Curves B represent calculated noise figures for infinite chains of identical stages. $V_{DS} = 4.5$ V; $V_{G_2S} = 2.0$ V; $\Gamma_3 = 1\angle 141°$; $f = 10$ GHz.

virtually eliminating phase changes due to multiple reflections between the modulator and RF driver. The dual-gate MESFET is, therefore, a modulator with large gain-control range and minimum associated transmission phase change, high unilateral gain, and practically gain-independent input impedance.

V. NOISE PERFORMANCE AND POWER CAPABILITY

The noise figure of a dual-gate MESFET at a fixed frequency is a strong function of the drain current. Therefore, this characteristic is discussed first and compared to the performance of a single-gate MESFET with similar geometry. A second-gate termination is chosen ($\Gamma_3 = 1\angle 141°$) that yields a low noise figure and simultaneously high gain. Fig. 9 illustrates the noise figure and gain performance versus drain current I_{DS}, at 10 GHz. The current is changed by varying the first-gate voltage V_{G_1S}. At $V_{G_1S} = 0$ the current is I_{DSS}, and the maximum available gain would be 20 dB (Fig. 5). In adjusting the tuner at the MESFET input for minimum noise figure, the gain drops to 18.2 dB yielding a 6.6-dB noise figure. In decreasing the drain current,[4] the noise figure decreases nearly linearly to a minimum of 4.0 dB with 12-dB associated gain. The single-gate MESFET exhibits the same noise figure versus drain-current characteristic (Fig. 9). Its noise-figure minimum is 3.2 dB with 8.0-dB gain.[5] Despite the considerably lower gain of the single-gate MESFET, an infinite chain of cascaded single-gate MESFET'S still has a 0.6-dB lower noise figure than a similar chain of dual-gate MESFET'S. The difference in noise figure decreases, however, if the loss in the input matching and interstage coupling is considered.

In a dual-gate MESFET, the noise-figure minimum can be approached only if the output of the first transistor is noise-matched to the input of the second transistor [Fig. 3(b)]. The input impedance Z_{i2} can be adjusted by varying the external impedance Z_3. On the other side, the output impedance of the first transistor Z_{o1} can be varied with the potential V_{D_1S}. This potential is controlled by the second-gate bias. In Fig. 10, the noise figure and gain are plotted versus V_{G_2S}. The noise figure reaches its minimum at 2 V. By decreasing V_{G_2S}, it rises, first slowly, and then, below 0.5 V, very rapidly. It is known that the noise figure of single-gate MESFET's is very insensitive to the drain voltage as long as V_{D_1S} is above $V_{D(\text{sat})}$ [Fig. 3(c)], [5], [6]. Therefore, it is assumed that the noise figure of the first transistor in Fig. 3(b) stays constant for $V_{G_2S} > 0.5$ V, and the gain and overall noise-figure variation is solely due to the change in the output conductance of the first stage. This conductance in turn determines the impedance mismatch between the stages and the noise power generated by the second transistor.

Van der Ziel proposed to improve the noise figure of dual-gate FET's by neutralizing the drain-to-gate capacitance of the first transistor in Fig. 3(b) and by providing a tuned matching network between the two transistors [30]. To make this improvement possible, an ohmic contact between the two gates, a neutralization circuit, a tuning reactance to ground, and dc-blocking capacitors are proposed in addition to the RF-grounding and dc-biasing circuit at the second gate. The realization of this is a rather demanding task at X band. Asai's technique to improve the dual-gate noise figure requires a longer second gate and thicker second channel [17], [18]. This approach still needs theoretical justification. For device fabrication, a critical second alignment step is required

---

[4] The tuning was optimized at each drain current.
[5] The compared single-gate MESFET has not been fabricated on the same chip. However, its gate length, gate-to-source spacing, channel thickness, and epitaxial material characteristics closely match the dual-gate MESFET's parameters.

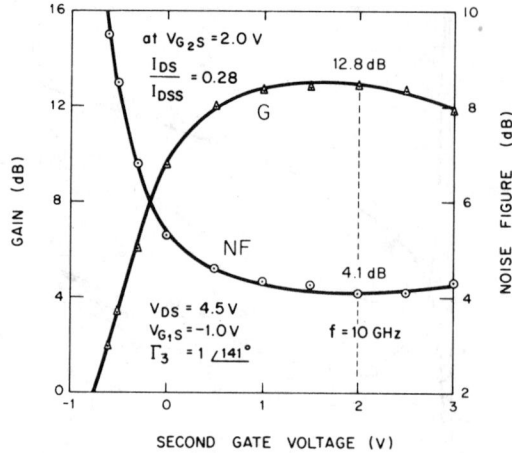

Fig. 10. Minimum noise figure and associated gain versus second-gate voltage.

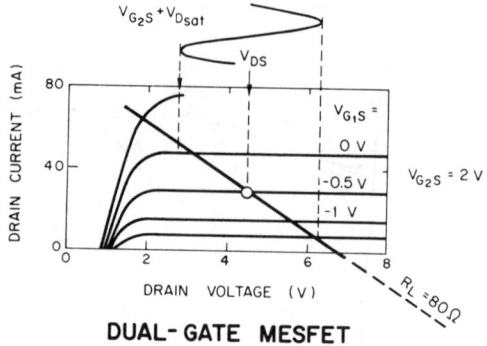

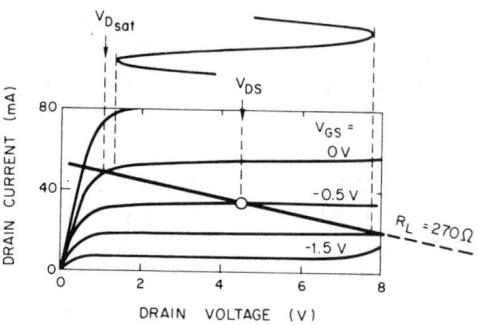

Fig. 11. Static drain-current versus drain-voltage characteristics, dc bias points, and load lines for the two MESFET versions. The MESFET's are operated with identical dc bias, power consumption, and small-signal gain. The maximum drain-voltage swing for "linear" operation is illustrated.

which raises the cost of the dual-gate over the single-gate MESFET. In contrast, it is demonstrated here that an optimized second-gate termination and second-gate dc bias yield a dual-gate noise figure of 4.0 dB at 10 GHz, only 0.8 dB above the minimum noise figure of the compared single-gate MESFET.

After comparing the noise characteristics, the power capability of the two MESFET versions are discussed. For this purpose, a single-gate and a dual-gate MESFET with approximately equal gate width and drain current I_{DSS} are chosen. Both transistors are operated at the same drain and gate bias ($V_{DS} = 4.5$ V, $V_{G_1S} = V_{GS} = -0.5$ V) with about equal dc power dissipation (130 mW). In addition, the two MESFET's are operated at the same small-signal gain, i.e., they both yield 9-dB gain at 10 GHz. The single-gate MESFET is image matched at both ports. For the dual-gate MESFET, the input is image matched, the output capacitance is parallel resonated, and the load resistance is adjusted for equal gain ($R_L = 80$ Ω). The second gate is terminated with $\Gamma_3 = 1 \angle 153°$ and biased to $V_{G_2S} = 2.0$ V. The resulting output power for 1-dB gain compression is +9 dBm for the single-gate and +10 dBm for the dual-gate MESFET.

Under the stated conditions, both MESFET's have about the same power capability. The drive conditions are illustrated in Fig. 11. The static drain-current versus drain-voltage characteristics, the dc bias points, and the load lines of the two MESFET's are shown. For the dual-gate MESFET, the maximum drain-voltage swing is limited by $V_{DS} - (V_{G_2S} + V_{D(\text{sat})})$. If V_{DS} drops below $V_{G_2S} + V_{D(\text{sat})}$, the second transistor in the model of Fig. 3(b) is driven into the resistive region which raises the drain to second-gate capacitance more than an order of magnitude. The output is now shunted by the second-gate termination which drastically decreases the RF output impedance. If V_{DS} is reduced below V_{G_2S}, the second gate is sufficiently forward biased to draw dc current. In the single-gate version, the maximum drain-voltage swing is larger since it is limited only by $V_{DS} - V_{D(\text{sat})}$. For equal gain performance, however, the load resistance R_L must be considerably higher.[6] The larger R_L offsets the advantage of the larger voltage swing $\hat{v}_{ds}$ because the delivered RF power is proportional to $\hat{v}_{ds}^2/R_L$.

VI. HIGH-SPEED AMPLITUDE MODULATION OF AN RF CARRIER

The dual-gate MESFET as a modulator offers a very fast switching response. This capability is demonstrated here by pulse-amplitude modulating a microwave carrier. The basic test circuit is illustrated in Fig. 12(a). The output of an 8-GHz signal generator is fed directly into the first gate without impedance transformation. A pulse generator switches the second gate between -2.5 V (off) and $+1.5$ V (on). This pulse is synchronized with the 8-GHz signal, so the resulting RF burst can be observed on a sampling scope. The drain and the two gates are connected to 50-Ω output lines in order to compare the three waveforms simultaneously on the sampling scope. The physical arrangement of the test circuit in the vicinity of the MESFET chip is shown in Fig. 12(b). The transmission

---

[6] The load resistance is 270 Ω. The ratio of the two load resistances would be four if the dual-gate MESFET had the same equivalent circuit (with no feedback elements) as its single-gate counterpart with the exception of a negligibly small output conductance.

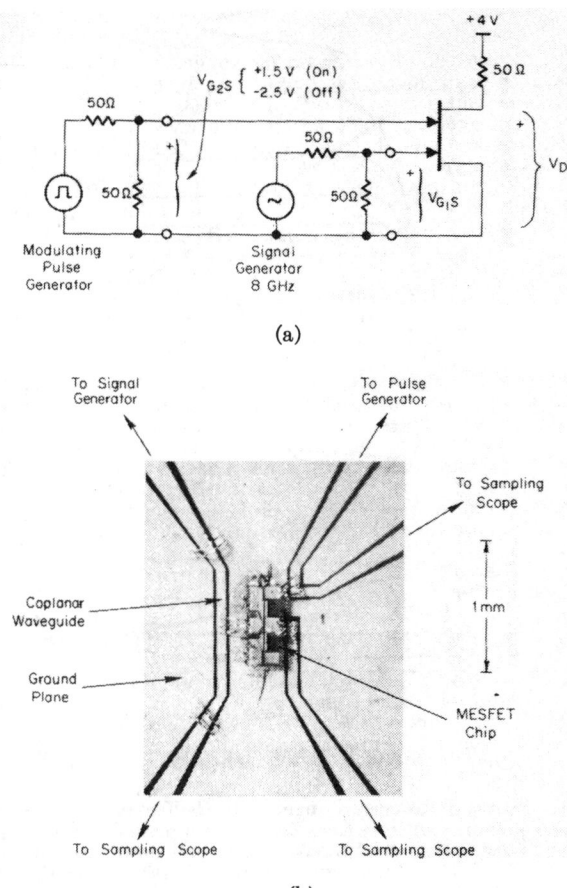

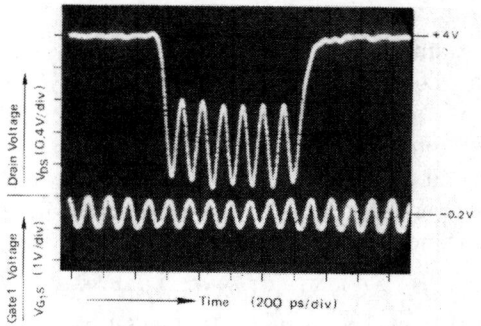

Fig. 13. Pulsed-amplitude modulation of an 8-GHz carrier. The RF input voltage at the first gate is shown in the lower trace and the output-voltage waveform at the drain is shown in the upper trace.

Fig. 12. Pulsed-amplitude modulation of an 8-GHz carrier with the dual-gate MESFET. (a) Illustrates the basic test circuit. (b) Shows the arrangement of the 50-Ω coplanar transmission lines on a sapphire substrate in the vicinity of the MESFET chip.

lines are 50-Ω coplanar waveguides [31] on a sapphire substrate. The coplanar lines are tapered from a size convenient for connector mating close to the substrate edge to a size suitable for short wire bonds to the chip [32]. The coplanar waveguide also has the advantage of a continuous ground plane on the substrate surface, so that low-inductance grounding of the MESFET source can be achieved.

The voltage waveforms are shown in the oscilloscope display of Fig. 13. On the horizontal time axis, the scale is 200 ps/div. The sinusoidal waveform at the bottom is the 8-GHz signal voltage that is applied to the first gate. The drain voltage is shown in the upper trace. As long as the second-gate voltage is low, the MESFET is turned off; the output potential is at the supply voltage, and no RF signal is coupled to the output. When the second-gate voltage is pulsed to a positive potential, the MESFET is switched on; drain current flows causing a voltage drop in the 50-Ω load resistor. The MESFET transmits the RF input signal. The RF burst is about 7 cycles long and full switching is accomplished in less than one RF cycle. With the RF signal turned off, a drain-voltage fall time of 65 ps and rise time of 100 ps is measured. This compares with a 100-ps rise time and 125-ps fall time of the modulating pulse. Only a fraction of the applied voltage swing drives the MESFET through its active region which results in the slightly faster output-switching response. The switching speed in this experiment is limited by the pulse generator and not by the MESFET. It demonstrates the MESFET's capability to switch from gain to high insertion loss within approximately one RF cycle at X-band frequencies.

VII. CONCLUSIONS

The GaAs MESFET with two 1-μm Schottky-barrier gates and an inductive second-gate termination exhibits 18-dB stable gain at 10 GHz. This compares with 11-dB maximum available gain for the single-gate counterpart. The minimum noise figure is 4.0 dB with 12-dB associated gain as opposed to a 3.2-dB noise figure and 8.0-dB gain in the single-gate version. The output impedance of the dual-gate MESFET is, in general, higher and dependent on the second-gate termination. Both MESFET versions deliver approximately 10-mW output power at 1-dB gain compression if they are operated with identical dc bias, power consumption, and small-signal gain. The dual-gate version offers the added feature of gain modulation with 44-dB gain control and less than 70-ps modulating response time.

The application potential of the single-gate versus dual-gate MESFET in small-signal amplifiers can now be clearly outlined. The signal-gate MESFET has a slightly lower noise measure. If the lowest possible amplifier noise figure is required, then the single-gate version has to be used in the first stage. However, for all the other stages, the dual-gate version has definite advantages. First, it yields higher gain, thereby reducing the number of required stages and the cost of an amplifier. Second, it enables automatic gain control over a large dynamic range with minimum transmission phase shift. Third, it allows gain-slope compensation and low-frequency stabilization with a simple series-resonant circuit between second gate and source.

The dual-gate MESFET is also a modulator with a high on-to-off gain ratio, practically gain-independent input impedance, and ultrafast switching speed. These features make the device very attractive for subnanosecond pulsed-amplitude modulation (PAM). They are also very useful in applications requiring phase- or frequency-shift keyed carrier modulation (PSK) and (FSK). Biphase modulation, for example, can be easily realized by operating two dual-gate MESFET's in parallel with a common drain output. The 0°- and 180°- shifted carriers are fed to the first gates and complimentary modulating pulses are applied to the second gates. The capability of performing multiple functions opens a wide basis for innovations in the application of the dual-gate MESFET in microwave systems.

ACKNOWLEDGMENT

The author wishes to thank R. Larrick for his measurement assistance and valuable experimental contributions, R. Van Tuyl and J. Lai for making the high-speed switching test, Dr. J. Barrera and R. Drabin for supplying the GaAs epitaxial layers, E. Gowen for the MESFET fabrication, E. Talbert for the fabrication of the test fixture. and D. Hollars for the microwave integrated-circuit assembly. A special acknowledgment is also due Dr R. Archer, Dr. R. Engelmann, and A. Podell for helpful suggestions and comments, and to E. Miller and S. Ybarra for their typing and drafting assistance in preparing the manuscript.

REFERENCES

[1] W. Baechtold, W. Walter, and P. Wolf, "X and Ku band GaAs MESFET," *Electron. Lett.*, vol. 8, pp. 35–37, Jan. 1972.
[2] C. A. Liechti, E. Gowen, and J. Cohen, "GaAs microwave Schottky-gate field-effect transistor," in *IEEE Int. Solid-State Circuits Conf. Digest Tech. Papers* (1972), pp. 158–159.
[3] N. G. Bechtel, W. W. Hooper, and D. Mock, "X-band GaAs FET," *Microwave J.*, vol. 15, pp. 15–19, Nov. 1972.
[4] W. Baechtold *et al.*, "Si and GaAs 0.5μm-gate Schottky-barrier field-effect transistors," *Electron. Lett.*, vol. 9, pp. 232–234, May 1973.
[5] W. Baechtold, "Noise behavior of GaAs field-effect transistors with short gate lengths," *IEEE Trans. Electron Devices*, vol. ED-19, pp. 674–680, May 1972.
[6] G. Brehm and G. Vendelin, "Biasing FETs for optimum performance," *Microwaves*, vol. 13, pp. 38–44, Feb. 1974.
[7] P. L. Clouser and V. V. Risser, "C-band FET amplifiers," in *IEEE Int. Solid-State Circuits Conf. Digest Tech. Papers* (1970), pp. 52–53.
[8] W. Baechtold, "Ku-band GaAs FET amplifier and oscillator," *Electron. Lett.*, vol. 7, pp. 275–276, May 1971.
[9] L. Besser, "Design considerations of 3.1–3.5 GHz GaAs FET feedback amplifier," in *IEEE Int. Microwave Symp. Digest Tech. Papers* (1972), pp. 230–232.
[10] R. Zuleeg, E. W. Bledl, and A. F. Behle, "Broadband GaAs field effect transistor amplifier," Air Force Avionics Lab., Wright-Patterson AFB, Ohio, Tech. Rep. AFAL-TR-73-109, Mar. 1973.
[11] W. Baechtold, "X- and Ku-band amplifiers with GaAs Schottky-barriers field-effect transistors," *IEEE J. Solid-State Circuits (Special Issue on Microwave Integrated Circuits)*, vol. SC-8, pp. 54–58, Feb. 1973.
[12] C. A. Liechti and R. L. Tillman, "Application of GaAs Schottky-gate FETs in microwave amplifiers," in *IEEE Int. Solid-State Circuits Conf. Digest Tech. Papers* (1973), pp. 74–75.
[13] N. Slaymaker and J. Turner, "Microwave FET amplifiers with center frequencies between 1 and 11 GHz," in *3rd European Microwave Conf. Digest Tech. Papers* (1973).
[14] G. Vendelin, J. Archer, and G. Bechtel, "A low-noise integrated S-band amplifier," in *IEEE Int. Solid-State Circuits Conf. Digest Tech. Papers* (1974), pp. 176–177.
[15] D. Ch'en and A. Woo, "A practical 4 to 8 GHz GaAs FET amplifier," *Microwave J.*, vol. 17, pp. 26 and 72, Feb. 1974.
[16] C. Liechti and R. Tillman, "Design and performance of microwave amplifiers with GaAs Schottky-gate field-effect transistors," *IEEE Trans. Microwave Theory Tech.*, vol. MTT-22 pp. 510–517, May 1974.
[17] S. Asai *et al.*, "Single- and dual-gate GaAs Schottky-barrier FETs for microwave frequencies," presented at the 4th Conf. Solid-State Devices, Tokyo, Japan, Aug. 1973.
[18] S. Asai, F. Murai, and H. Kodera, "The GaAs dual-gate FET with low noise and wide dynamic range," in *IEEE Int. Electron Devices Conf. Digest Tech. Papers* (1973), pp. 64–67.
[19] J. Turner, A. Waller, E. Kelly, and D. Parker, "Dual-gate GaAs microwave FET," *Electron. Lett.*, vol. 7, pp. 661–662, Nov. 1971.
[20] J. Turner and S. Arnold, "Schottky-barrier FET's···next low-noise designs," *Microwaves*, vol. 11, pp. 44–49, Apr. 1972.
[21] S. Arnold, "Single and dual-gate GaAs FET integrated amplifiers in C-band," in *IEEE Int. Microwave Symp. Digest Tech. Papers* (1972), pp. 233–234.
[22] M. Maeda and Y. Minai, "Application of dual-gate GaAs FET to microwave variable-gain amplifier," in *IEEE Int. Microwave Symp. Digest Tech. Papers* (1974), pp. 351–353.
[23] H. Kleinman, "Application of dual-gate MOS field-effect transistors in practical radio receivers," *IEEE Trans. Broadcast Telev. Receivers*, vol. BTR-13, pp. 72–81, July 1967.
[24] E. Hesse, "Untersuchungen an P-Kanal-MOS-Feldeffekt-Tetroden," *Nachrichtentech. Z.*, vol. 7, pp. 491–494, Oct. 1970.
[25] R. Ronen and L. Strauss, "The silicon-on-sapphire MOS tetrode—Some small-signal features, LF to UHF," *IEEE Trans. Electron Devices*, vol. ED-21, pp. 100–109, Jan. 1974.
[26] H. Sigg, D. Pitzer, and T. Cauge, "D-MOS for UHF linear and nanosecond switching applications," in *IEEE Int. Conv. Digest Tech. Papers* (1974), Session 32/4, pp. 1–7.
[27] C. Liechti, "Characteristics of dual-gate GaAs MESFET's," in *4th European Microwave Conf. Digest Tech. Papers* (1974), pp. 87–91.
[28] G. Bodway, "Circuit design and characterization of transistors by means of three-port scattering parameters," *Microwave J.*, vol. 11, pp. 55–63, May 1968.
[29] ——, "Two port power flow analysis using generalized scattering parameters," *Microwave J.*, vol. 10, pp. 61–69, May 1967.
[30] A. Van der Ziel and K. Takagi, "Improvement in the tetrode FET noise figure by neutralization and tuning," *IEEE J. Solid State Circuits* (Corresp.), vol. SC-4, pp. 170–172, June 1969.
[31] C. Wen, "Coplanar waveguide: A surface strip transmission line suitable for nonreciprocal gyromagnetic device applications," *IEEE Trans. Microwave Theory Tech. (1969 Symposium Issue)*, vol. MTT-17, pp. 1087–1090, Dec. 1969.
[32] R. Van Tuyl and C. Liechti, "High-speed integrated logic with GaAs MESFET's," *IEEE J. Solid-State Circuits (Special Issue on Semiconductor Memory and Logic)*, vol. SC-9, pp. 269–276, Oct. 1974.

Low-Noise Receiver Design Trends Using State-of-the-Art Building Blocks

HERMAN C. OKEAN, FELLOW, IEEE, AND ALEXANDER J. KELLY, SENIOR MEMBER, IEEE

Invited Paper

Abstract—The current state-of-the-art and the various design tradeoffs encompassing the variety of low-noise microwave and millimeter-wave receiver "building blocks" which have evolved during the past two decades are described. Key examples of these are the high-idler noncryogenic parametric amplifier, the gallium arsenide field-effect transistor (GaAs FET) amplifier, and the image-enhanced Schottky-diode mixer.

It is then shown how this inventory of building blocks can best be integrated into optimum receiver configurations for application in a multiplicity of future and present microwave and millimeter-wave communications, radar, and radiometer systems.

I. Introduction

THE low-noise receiver has evolved as the essential element in defining the performance level of a variety of microwave and millimeter-wave systems applications. Generically, the low-noise receiver "front end" consists of one or more stages of low-noise microwave or millimeter-wave amplification followed by a heterodyne frequency converter, the latter translating the received signals to the appropriate frequency range for signal processing. In the last two decades, a variety of "building blocks" for low-noise front ends have evolved in the direction of lower noise performance, solid-state implementation, smaller size, lighter weight, and longer life maintenance-free operation. Some of these, such as the maser, the low-noise traveling-wave tube, and the tunnel diode amplifier are diminishing in importance and currently experience at best limited usage, whereas others such as the high-idler noncryogenic parametric amplifier, the gallium arsenide field-effect transistor (GaAs FET) amplifier, and the image-enhanced Schottky-diode mixer are the primary constituents of current and future front ends.

Improvements in the noise performance of the receiver front end have resulted in corresponding improvements in the performance of microwave and millimeter-wave communications, radar, and radiometer systems, by virtue of greater communications predetection signal-to-noise ratio, increased radar range or reduced radar transmitter power, and decreased radiometer minimum detectable temperature, respectively. It is the purpose of this paper to assess the

Manuscript received September 15, 1976; revised September 30, 1976. Some of the equipment described in this paper was developed under sponsorship of the NASA Goddard Space Flight Center, Greenbelt, MD; Air Force Avionics Laboratory, Dayton, OH; Naval Electronics Laboratory Center, San Diego, CA; Naval Air Systems Command, Washington, DC; and Rome Air Development Center, Rome, NY.

The authors are with the Department of Research and Development, LNR Communications, Inc., Hauppauge, NY 11787.

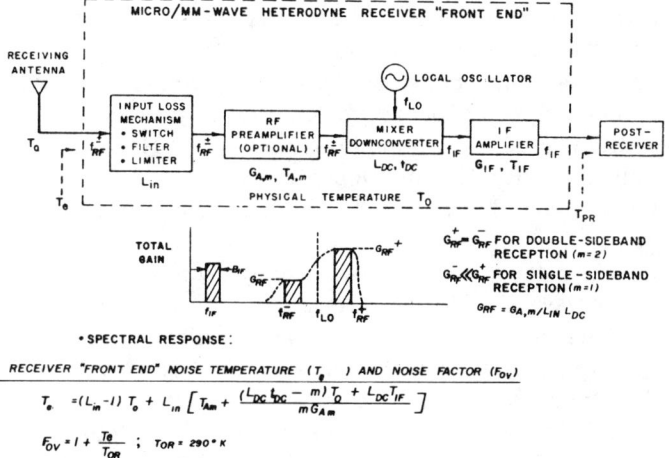

Fig. 1. Block diagram of generic single-channel heterodyne front end.

state-of-the-art in the various low-noise front-end building blocks and to show how this inventory of building blocks can be best integrated into optimum receiver configurations tailored to each of the foregoing systems applications.

II. Description of General Receiver Front End

A. Functional Description

A generic single-channel low-noise receiver front end, pertinent to each of the microwave/millimeter-wave systems applications under consideration, is of the heterodyne type in which one or both RF input sidebands are translated via a single downconversion, into a single IF band. Depicted in block diagram form in Fig. 1, this generic front end consists of the following:

—Input coupling network forming interface between receiving antenna feed and subsequent front-end components, and providing filtering, switching, limiting, and/or test signal injection functions, depending upon particular receiver application.

—Single or multistage RF amplifier centered at microwave or millimeter-wave carrier frequency and, in the latter case, often representable as the cascade of an ultra-low-noise RF preamplifier and a moderately low-noise high dynamic range RF postamplifier.

—Frequency downconverter, including input power divider or demultiplexer loss, resistive mixer with self-contained or externally provided local oscillator (LO) and

IF amplifier, centered at a convenient IF range for demodulation, processing, further downconversion, etc., depending upon the system application.

All succeeding circuits providing the foregoing further downconversion, processing, and/or demodulation functions as well as possible further demultiplexing and/or power division, and in aggregate comprising the postreceiver, are not considered part of the front end under consideration and are not within the province of this discussion (assuming sufficient total front-end RF/IF gain to offset any significant postreceiver contributions to overall receiver noise performance). Furthermore, the most general low-noise front end can consist of a multiplicity of RF amplifier/downconverter channels demultiplexed off a common receiving antenna feed or of many RF downconverter channels demultiplexed off a common RF amplifier. However, each individual RF/IF input/output path or "channel" is representable by the generic single-channel block diagram of Fig. 1.

Within the context of Fig. 1, most microwave and millimeter-wave front ends utilize either *single-* or *double-*sideband heterodyne reception, in which input frequencies $f_{RF}^\pm = f_{LO} \pm f_{IF}$, situated in *one* or *both* sidebands of the downconverter LO frequency f_{LO} are translated down to the IF (f_{IF}), with overall conversion gains $G_{0v}^\pm$, respectively. In the majority of receiving systems applications, including communications and telemetry links, coherent radar and electronic warfare, discrete information-bearing signals in one input sideband must be received and processed unambiguously with respect to extraneous signals in the other "image" or unwanted sideband.

Therefore, in single-sideband (SSB) front ends, a high degree of rejection is generally presented to the unwanted image sideband by the selectivity of the RF preamplifier by the presence of a "preselector" bandpass filter preceding the mixer and/or by the use of a mixer configuration with inherent image rejection properties ($G_{0v}^- \ll G_{0v}^+$).

In contrast to the aforementioned, there are other receiving system applications including radio astronomy, radiometry, and noncoherent radar, in which the received signal is broad-band noise appearing in both input sidebands. In these cases, sensitivity is maximized by the use of a double-sideband (DSB) receiver ($G_{0v}^+ \approx G_{0v}^-$).

B. General Formulation of Front-End Noise Performance

The noise performance of the generic heterodyne front end (Figs. 1 and 2) is formulated [1], [2] in terms of that of each of its constituent building blocks in Fig. 1, as characterized by the familiar parameters, broad-band input noise temperature (T_e), and noise factor (F). These formulations make use of the foregoing assumptions on $G_{0v}^\pm$ for SSB and DSB reception.

A more meaningful measure of overall antenna/receiver system sensitivity relevant to the aforementioned microwave/millimeter-wave applications is *system operating noise temperature* T_{op}, which in turn is expressible [1], [2]

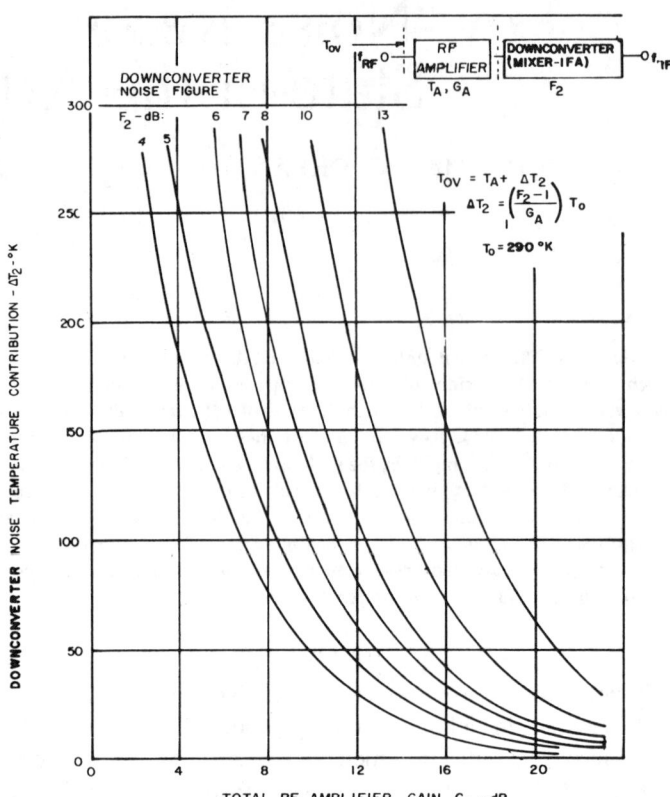

Fig. 2. Downconverter contribution to overall front-end noise temperature.

in terms of T_e and antenna temperature T_a as indicated in Fig. 1.

It is immediately clear from Fig. 1 that for antenna temperature T_a equal to $T_{OR} = 290$ K (as is often the case), system operating noise temperature and receiver noise figure are very simply related by

$$F = T_{op}/T_{OR}. \quad (1)$$

The equivalent antenna temperature T_a is a strong function of antenna orientation relative to the earth, antenna elevation, and received signal frequency [3] with the following particular cases applicable to the majority of microwave and millimeter-wave system applications:

—Ground-based antenna, vertical elevation pointing at "cold" sky (satellite communications, space telemetry radar, radio astronomy): $T_a < 10$ K and $T_a = 15$–75 K for frequency between 1 and 20 GHz and 20 and 100 GHz, respectively (except for absorption region between 50 and 60 GHz, for which $T_a = 290$ K).

—Ground-based antenna, horizontal elevation (radar, point-to-point communications): $T_a = 125$–250 K over 1–100 GHz, and about 290 K at or above 20 GHz.

—Airborne, ground-looking antenna (radar, radiometric mapping, communications): $T_a = 300$ K over 1–100 GHz.

In any given microwave and millimeter-wave system application, the optimum tradeoff between sensitivity and system cost and complexity occurs at a level of receiver noise performance (T_e) compatible with antenna temperature, e.g., $T_e \sim T_a$. Therefore, in light of the preceding, it is seen that certain systems applications (satellite communications links, space research, radio astronomy) require the ultimate in receiver noise performance ($T_e \lesssim 50$ K) whereas others can effectively utilize moderately low-noise performance ($T_e \approx 300$–1000 K). This, in turn, impacts the choice of receiver front-end configuration for a given application, as will be discussed in a subsequent section of this paper.

The foregoing formulations of overall front-end noise performance underscore the importance of minimizing total functional (duplexer, limiter, switch, filter, etc.) input circuit losses (L_{in}). Hence $1.05 \leq L_{in} \leq 1.25$ (0.25–1 dB) in most representative system applications. Even more importantly, they emphasize the role of and frequent necessity for low-noise RF preamplification, with sufficient RF preamplifier gain (G_A) provided to minimize contributions of the downconverter (and possibly of the RF postamplifier immediately preceding it) to overall receiver noise figure, so that the noise figure of the RF amplifier defines that of the front end.

This introduces an important tradeoff in overall front-end design, that is, downconverter noise performance versus required RF amplifier gain for a given degree of downconverter noise contribution suppression. This tradeoff is exemplified in Fig. 2, which depicts the second-stage noise temperature contribution ΔT_2 as a function of second-stage noise figure and RF amplifier gain. Fig. 2 provides the rationale for some of the specific front-end alternatives such as dedicated low-noise downconverter and single-stage low-noise RF amplifier versus demultiplexed multichannel downconverter and multistage low-noise RF amplifier or single-stage low-noise RF preamplifier plus moderate noise RF postamplifiers, as will be described in subsequent sections of this paper.

C. Other Aspects of Low-Noise Receiver Performance

In addition to noise performance, other important receiver front-end parameters include RF/IF bandwidth and dynamic range.

The RF input bandwidth (B_{RF}) of the generic heterodyne receiver front end is related to the output IF bandwidth (B_{IF}), as a function of the particular mode of mixer operation under consideration (Fig. 1), with the majority of cases encompassing the following:

—Tunable LO, broad-band image mixer with overlapping signal, LO, and image bands. Suitable for either DSB or SSB reception, the latter requiring the use of properly phased dual mixers for image rejection and possible image enhancement ($B_{RF} \gg B_{IF}$, $B_{IF} \approx 30$–200 MHz).

—Fixed LO, broad-band image mixer with separate signal and image sidebands. Suitable for either DSB or SSB reception, the latter requiring a fixed preselector isolator at the input for image rejection under matched-image termination ($B_{IF} = B_{RF}$ SSB and B_{RF} DSB $= 2f_{IF}^+ > 2B_{IF}$).

—Fixed LO, reactive image-enhanced SSB mixer with properly situated preselector at RF input to provide optimum reactive termination to mixer diodes over image band for minimum SSB conversion loss, while simultaneously providing high image rejection ($B_{RF} = B_{IF}$).

Typically, microwave and millimeter-wave heterodyne front ends utilizing tunable LO operation can be configured with RF bandwidths of an octave or more, but those operating as SSB or folded DSB translators exhibit maximum RF bandwidths of 5–20 percent. In each case, the front-end RF bandwidth capability is determined by the specific type of building blocks employed therein, as described in a subsequent section.

The dynamic range of the generic microwave/millimeter-wave front end is characterized at the low end by noise performance and upper end by the onset of nonlinearity, as defined by one or more of the following generally interrelated [4] parameters, depending upon the specific system application:

—Deviation from input/output linearity exemplified by input or output level at which 1-dB gain compression occurs.

—Two-tone third-order intermodulation (IM) intercept point, referenced to input or output (P_{inIM}, P_{oIM}) representing an extrapolation of the levels of the two IM products (at frequencies $2f_1 - f_2$ and $2f_2 - f_1$) generated by two equal in-band tones at f_1 and f_2 to the point where tones and IM products become equal.

—AM-to-PM conversion coefficient defined by the degree of output phase distortion encountered over a 1-dB variation in input amplitude about some nominal level.

For the type of nonlinearities associated with most microwave and millimeter-wave amplification and conversion devices, the preceding parameters are precisely related to one another [4]. Therefore, the third-order output IM intercept point is customarily chosen as the nonlinearity parameter characterizing most microwave and millimeter-wave receivers and their constituent components. In fact, the output intercept point of the generic front ends (Fig. 1) can be expressed in terms of that of its key constituents, the RF amplifier, downconverter mixer, and IF amplifier, as given by [5]

$$P_{oIM_{ov}} = P_{oIM_{IF}} \left(1 + \frac{P_{oIM_{dc}} L_{dc}}{P_{oIM_A}} + \frac{P_{oIM_{IF}} L_{dc}}{G_{IF} P_{oIM_A}}\right)^{-1} \quad (2)$$

where

$P_{oIM_A}, P_{oIM_{dc}}, P_{oIM_{IF}}$ third-order output intercept points of RF amplifier, mixer downconverter, and IF amplifier, respectively;

L_{dc}, G_{IF} downconverter mixer conversion loss and IF amplifier gain.

The implication of the preceding relationship is that in many microwave and millimeter-wave front ends, particularly those employing high-gain IF amplifiers, it is the output IF amplifier dynamic range capability that defines that of the entire front end. Therefore, in many cases the RF amplifier for the generic front end can be of relatively modest large-signal capability. The impact of the

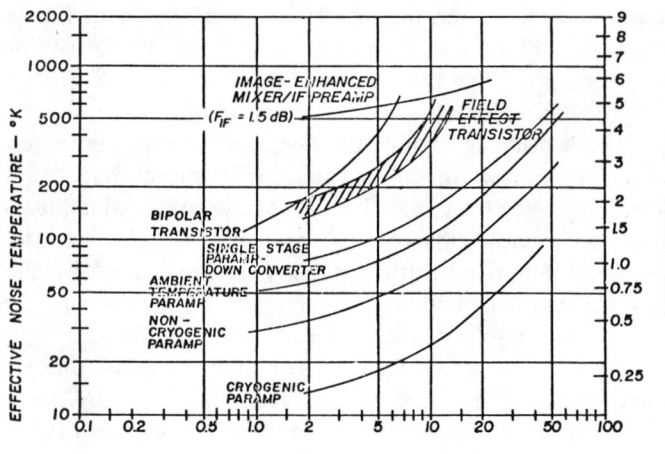

Fig. 3. Noise performance of state-of-the-art front-end building blocks.

preceding on the choice of front-end components will be described in a subsequent section.

III. State-of-the-Art Receiver Building Blocks

A. General Comparative State-of-the-Art

A variety of building blocks have evolved during the past two decades for use in low-noise microwave and millimeter-wave front ends. Some of these such as the traveling-wave maser and the low-noise linear traveling-wave tube amplifier see only limited usage and are being almost completely supplanted by solid-state amplifiers, such as the parametric amplifier, the bipolar and FET amplifier, and to a lesser extent, the tunnel diode amplifier. In addition, the original point-contact diode mixers have given way to sophisticated broad-band image and image-enhanced low-loss Schottky-diode (and more recently FET) mixers.

Fig. 3 summarizes the current state-of-the-art of receiver front-end performance [6]–[45] as a function of input signal frequency.

In the past few years significant progress has been made in translating laboratory results into reliable operational hardware. Much of this progress has been as a result of advances in semiconductor technology (particularly applied to varactor, mixer, Gunn diodes, and MESFET transistors; lower loss ferrite circulators and isolators; and new and maturing microwave circuit structures and design techniques).

From among the aforementioned building blocks, the most recent emphasis has been upon the following:

—Continually decreasing noise temperatures in noncryogenic parametric amplifiers with noise temperatures as low as 30–90 K for frequencies from 2 to 15 GHz.

—Ultralow-noise parametric amplifiers in the 10–40-GHz region.

—High-reliability spacecraft-qualified miniature parametric amplifiers, operated at ambient or elevated temperatures.

—Low-noise FET's as front ends and as postamplifiers for paramps.

—Image-recovery mixers as low-noise front ends and as second stages for parametric amplifiers.

The aforementioned key building blocks are described in more detail in the following paragraphs.

B. Description of Individual Building Blocks

1) Parametric Amplifiers and Converters: The parametric amplification mechanism arising from the multiple-frequency interactions within a variable-capacitance varactor, pumped at f_p, has led to two extremely low-noise amplifier/converter configurations, depicted in block diagram form in Fig. 4. These configurations are as follows:

—The positive-resistance upper sideband upconverter (USUC), in which inputs at f_s are converted to sum-frequencies $f_u = f_p + f_s$, with gain $\leq (f_u/f_s)$.

—The circulator-coupled negative-resistance parametric amplifier in which negative resistance, generated at f_s under proper nominal short-circuited termination of the idler at $f_i = f_p - f_s$, is capable of unlimited reflection, but at the expense of bandwidth and operational gain stability. Two paramp configurations are possible: the nondegenerate (SSB) with $f_p = 2f_s$ and the degenerate (DSB) with $f_{i0} = f_{s0} = f_p/2$.

The paramp noise temperature cited in Fig. 3 are for SSB nondegenerate configuration, the corresponding DSB degenerate paramp noise temperature would be about 70 percent of that given.

The USUC is useful primarily over the UHF/L-band input frequency range (0.1–1.0 GHz) where sufficiently wide-band low-loss circulators are not always available. Circulator-coupled paramps, on the other hand, have been utilized from 1 to about 50 GHz and are feasible up to about 100 GHz. Therefore, for microwave and millimeter-wave applications, the remainder of this discussion will focus upon the circulator coupled negative resistance parametric amplifier.

The key design alternatives underlying the development of most current paramps [6]–[20] relate primarily to their impact on paramp noise performance and are enumerated as follows:

—Degenerate versus nondegenerate, depending on application as to whether DSB or SSB reception is required.

—Pump frequency (nondegenerate): The noise performance (and bandwidth capability) of a circulator-coupled paramp improve with increasing pump frequency up to a broad optimum value, beyond which degradation is encountered. The optimum pump frequency selection is further impacted by the realizeability of the required pump source in solid-state form.

—Varactor quality: The lower the varactor loss content (the higher the cutoff frequency) the lower the paramp noise temperature. However, this dependence is weak, particularly at low signal frequencies, where circuit rather than semiconductor losses predominate.

—Circulator implementation: Waveguide (lowest loss) versus stripline (more compact and wider band).

—Number of stages (gain per stage) (nondegenerate): Both gain bandwidth and operational stability considera-

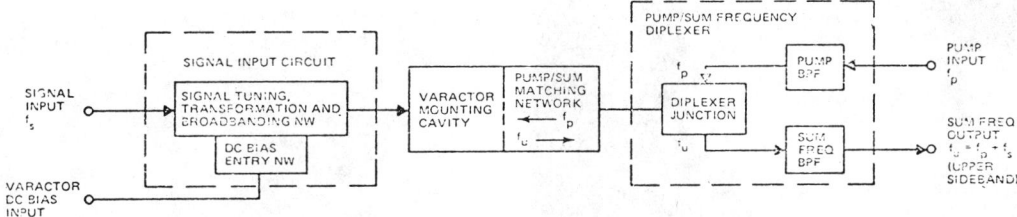

A- VARACTOR UPPER SIDEBAND UPCONVERTER

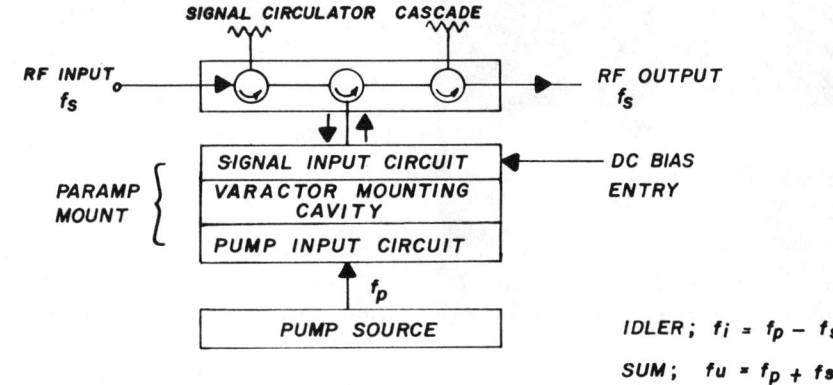

B - NEGATIVE RESISTANCE PARAMETRIC AMPLIFIER

Fig. 4. Block diagram of parametric amplifier and upconverter.

tions dictate that paramp gain per stage be limited to 14 dB maximum for moderate to wide-band applications (greater than 10 percent fractional bandwidth) whereas somewhat higher passband gain levels (18-dB maximum) per stage be considered only for extremely narrow-band (less than 5 percent fractional bandwidth) applications.

—Physical temperature: Due to the absence of shot noise, the noise temperature of the parametric amplifier is directly related to its absolute operating temperature. Therefore, based upon the use of GaAs varactor technology, paramps have been readily operated at cryogenic (20 K), room ambient, or elevated (320 K) temperatures.

Within the context of the foregoing design alternatives, single and multistage parametric amplifiers have exhibited noise temperatures as low as 10–100 K over the entire microwave range and well into the millimeter-wave region, concurrently with bandwidth capability from 5 percent to almost half-octave and with output IM intercept points from −20 to 0 dBm. Particular examples of the current paramp art (Fig. 3) include the following.

a) Cryogenic Paramps: Refrigerated to 20 K and capable of virtually noise-free amplication ($T_e = 10$–50 K) over the microwave and much of the millimeter-wave [11] frequency range, these have essentially supplanted the maser for use in large satellite communications and radio astronomy ground stations, but are costly and require periodic maintenance of the closed cycle mechanical refrigerators.

b) Advanced Noncryogenic Paramps: Utilization of ultrahigh-quality low-parasitic-content GaAs varactors, solid-state millimeter-wave pump sources ($f_p = 50$–100 GHz), extremely low-loss stripline and waveguide ferrite circulators, and efficient thermoelectric (Peltier) cooling modules has resulted in noncryogenic ultralow-noise amplification with near cryogenic performance ($T_e = 30$–90 K) at frequencies from S to K_u band, with 5–15 percent bandwidths and in completely packaged RF amplifier assemblies consisting of dual-stage paramps and possibly transistor postamps, along with self-contained solid-state pump sources, Peltier modules, and dc power and control circuits. Examples of the preceding include the following:

i) Deployment of hundreds of RF amplifier assemblies consisting of dual-stage paramp/transistor postamplifiers and providing typical noise temperatures of 45 K in the 3.7–4.2-GHz common carrier band in communications satellite (3.7–4.2 GHz) earth stations throughout the world.

ii) Fully packaged and operational [12] K_u-band two-stage 26-dB-gain paramp (Fig. 5) demonstrating noise temperatures of 76–84 K over the 14.7–15.2-GHz NASA Downlink Band by virtue of the use of a solid-state 30-mW 96-GHz pump source comprising a 48-GHz Gunn oscillator driving a highly efficient varactor doubler, a single-pass high isolation waveguide circulator with a forward loss of 0.07 dB and a high-quality GaAs chip varactor in a single-ended raised idler configuration, forming the highest idler resonance (81 GHz) ever reported.

c) Single-Stage Paramp Assemblies: For less demanding low-noise requirements, coupled with stringent size, cost, and/or environmental constraints, a preferable composite RF amplifier configuration consists of a single-stage paramp followed by a moderately low-noise postamplifier, and incorporating self-contained pump sources, dc power regulation and thermal stabilization circuits, and operated at a slightly elevated temperature. As an example, a weatherproof assembly incorporating a single-stage 3.7–4.2-GHz paramp plus transistor postamplifier provided typical noise

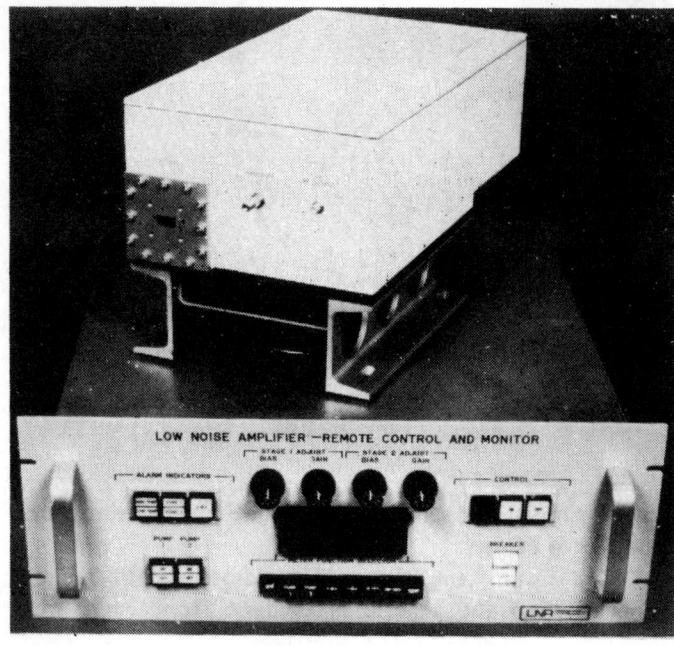

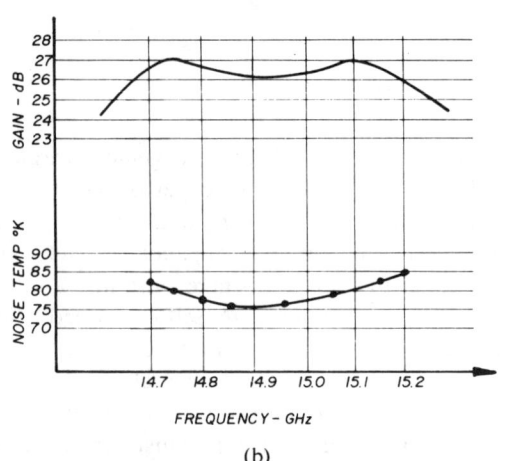

Fig. 5. Ultralow-noise noncryogenic parametric amplifier assembly. (a) Photograph. (b) Measured performance.

temperatures of 75–85 K when operated in small shipboard, oil rig, and unattended Alaska Bush earth terminals.

d) Spaceborne Paramps: Single-stage paramps addressing the stringent demands of space operation, with the attendant requirements on small size and weight, low power drain, immunity to severe shock and vibration, precise thermal stabilization, and ultralong life, are exemplified by the development for NASA of a fully operational spacecraft-qualified S-band low-noise parametric amplifier which is completely self-contained and requires less than 8 W of dc prime power. A gain of 18 dB, noise temperature of less than 45 K over the 2.2–2.3-GHz band (Fig. 6), were achieved in a 21-oz package which withstood severe vibration and shock and demonstrated operation in a vacuum environment. Other miniaturized single-stage paramps for space applications [13], [14] at frequencies as high as K_u band, exhibited noise temperatures as low as 150 K.

e) Millimeter-Wave Paramps: A K_a-band paramp assembly, developed [15] for satellite communications in the

Fig. 6. S-band space qualified amplifier (NASA space shuttle prototype).

35–40-GHz low atmospheric attenuation window, and incorporating [Fig. 7(a)] a directly integrated 96-GHz Gunn oscillator/varactor tripler pump source and self-contained dc power regulation, control and elevated temperature thermal stabilization circuits for 0–60°C operation, is electronically tunable, covering 36.5–38.5 GHz, and has a noise temperature of 350 K. Under its intended bimodal operation, the paramp is switch-tuned upon external voltage command to either of two specific frequency slots separated by over 1 GHz, as shown in Fig. 7(b), each providing 15 ± 0.5-dB gain over an instantaneous bandwidth of 150 MHz. However, under broad-band alignment this K_a-band paramp exhibited [Fig. 7(b)] 14-dB minimum gain with 1000-MHz instantaneous bandwidth. In addition, a solid-state pumped single stage operating in the 55–65-GHz region has been demonstrated [16] with less than a 6-dB noise figure. Finally, the varactor, circulator, and circuit technology for a 94-GHz paramp has been developed [17], but realization of useful performance awaits development of a sufficiently high-power 170–200-GHz CW pump source.

An additional application of parametric amplifiers in millimeter-wave low-noise front ends is as microwave IF preamplifiers used in conjunction with low-loss millimeter-wave mixer downconverters. Of particular advantage in radio astronomy receivers [21], the use of an IF amplifier consisting of a moderate gain, 20 percent or greater bandwidth paramp stage followed by a transistor postamplifier can provide a 1–3-dB improvement in downconverter noise figure as compared with the use of an all-transistor IF amplifier. This approach is especially advantageous at cryogenic temperatures, utilizing cryogenic mixer and paramp technology [22].

2) Paramp/Downconverter Assemblies: A logical extension of the completely packaged self-contained parametric RF amplifier assemblies described previously is the integration of an uncooled single-stage paramp and a low-loss (possibly image-enhanced) mixer/downconverter/transistor IF amplifier (and associated power, control, and thermal stabilization circuits) in a single self-contained enclosure (with or without a self-contained LO for the mixer) which is compact and light enough for antenna

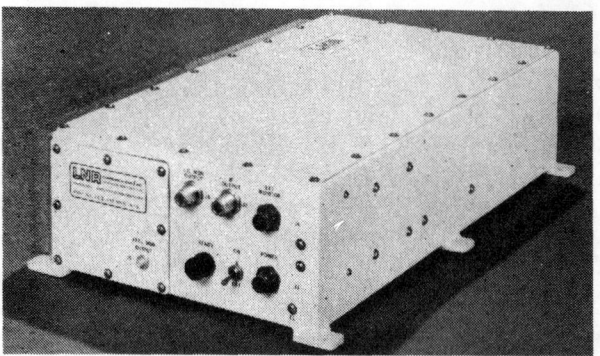

Fig. 8. Low-noise 11.7–12.2-GHz integrated parametric amplifier downconverter.

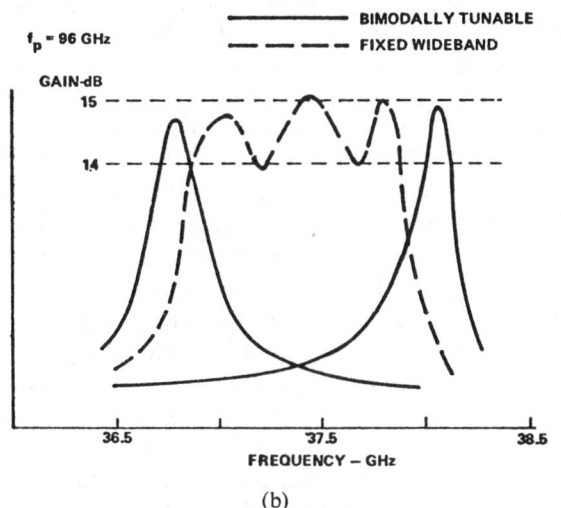

Fig. 7. Miniature K_a-band paramp assembly. (a) Photograph. (b) Measured gain response.

mounting. Such units, of particular use at high microwave and millimeter-wave input frequencies (7–42 GHz) at which low-noise transistor RF postamplifiers are either not always cost-effective (7–12 GHz), or do not exist (12–42 GHz), are available as operational hardware in the 7.25–7.75-GHz military communications and 11.7–12.2-GHz communications technology satellite (CTS) bands (Fig. 8), exhibiting overall noise temperatures (with 1-GHz IF) of 115 and 220 K, respectively, and usable with IF's from 70 MHz to 2 GHz.

A similar single-stage paramp/downconverter configuration has been implemented for shipboard satellite communications receiver usage in the 35–40-GHz range [15]. Utilizing the previously described "bimodally" tunable K_a-band paramp design, and incorporating a balanced GaAs Schottky-diode mixer and a low-noise UHF transistor amplifier [34], this low-noise front end exhibits an overall SSB noise factor of less than 4 dB, an overall conversion gain of 34 ± 1 dB, and an instantaneous RF/IF bandwidth of 150 MHz, which is bimodally switch-tunable to either sideband of the externally provided 37.5-GHz LO. The overall paramp/downconverter enclosure, including dc power supply and thermal stabilization, measures only 6 × 5 × 4 in.

3) Transistor Amplifiers: Transistor amplifiers find application in the generic heterodyne front ends under consideration (Fig. 1) at frequencies from VHF to K_u band either as RF or IF amplifiers.

Significant recent advances [6]–[10], [23]–[26] in the realization of low-noise GaAs FET and Si bipolar transistor amplifiers has yielded, in the state-of-the-art limit, superior noise performance (T_e = 75–600 K, F = 1–5 dB) to all but the paramp and maser, at frequencies up to about 15 GHz, with Si bipolars providing lower noise performance through lower S band and GaAs FET's having the edge above 3 GHz. These advances have been achieved primarily on the strength of refinements in Si and GaAs device processing, resulting in smaller electrode geometries, reduced parasitics, and consequently higher maximum frequencies of operation. In GaAs FET devices, unlike bipolars, shot noise does not completely dominate, and the thermal noise contribution is more significant, so that potential noise figure improvement at reduced physical temperatures is a reality.

Current state-of-the-art in bipolar and GaAs FET amplifier production hardware is summarized as follows [6]–[10], [23]–[26]:

Frequency	Bandwidth	Device	Geometry	Maximum Noise Figure
VHF/UHF	octave	bipolar	—	1.5 dB
2 GHz	<20 percent	bipolar	—	2.5 dB
4 GHz	<20 percent	bipolar	—	3.5 dB
1–2 GHz	octave	bipolar	—	3 dB
2–4 GHz	octave	bipolar	—	4 dB
3.95 GHz	500 MHz	GaAs FET	1-μm gate	2.5 dB (uncooled)
3.95 GHz	500 MHz	GaAs FET	1-μm gate	1.8 dB (cooled to −40°C)
4 GHz	2 GHz	GaAs FET	1-μm gate	4.5 dB (uncooled)
8 GHz	1.2 GHz	GaAs FET	1-μm gate	5–6 dB (uncooled)
12 GHz	1.2 GHz	GaAs FET	1-μm gate	5.5 dB
12 GHz	—	GaAs FET	1-μm gate	1.2 dB (cooled to 40 K)
14 GHz	500 MHz	GaAs FET	0.5-μm gate	5 dB (uncooled)
15 GHz	6 GHz	GaAs FET	0.5-μm gate	5–6.5 dB

The preceding results are obtained in multistage configurations with typical gain per stage decreasing from 12 to 6 dB from VHF to K_u band and with output IM intercept points above +10 dBm. Construction is typically in stripline or coax at UHF and the lower microwave frequencies where various encapsulated device form factors predominate, but is almost exclusively microstrip above C band, for maximum compatibility with the unencapsulated device embedding necessary for optimum performance. Input and output matching is accomplished by a combination of reciprocal lossless network techniques, use of ferrite isolators at input and/or output, and implementation of the balanced configuration [26] in which a pair of identical amplifiers are coupled through the in phase and quadrature parts of 0–90° 3-dB hybrids to the input and output interfaces.

It is expected that the further refinement of 0.5-μm Schottky-barrier MESFET gate geometry will result in even further reduction in GaAs FET amplifier noise figure and in extension of the maximum usable frequency range to beyond 20 GHz. Concurrent improvements in GaAs FET processing technology should eliminate any remaining questions on device operational stability and reliability.

4) Tunnel Diode Amplifiers: The tunnel diode amplifier (TDA) is a negative-resistance amplifier (by virtue of the negative-slope region in its dc current–voltage characteristic) usually operated in the circulator-coupled-reflection mode [2], [6], [27]. Its primary noise mechanism is that of shot noise due to dc current flow through the degenerate p-n tunnel diode junction, so that there is negligible advantage in below-room-temperature operation.

Shot-noise constant N and, hence, overall TDA noise temperature $T_A \doteq 300N$, are primarily functions of the semiconductor material comprising the tunnel diode. For the three commonly used materials (GaSb, Ge, GaAs), the shot-noise constants N are approximately 0.8, 1.2, and 2.0, respectively. Hence GaSb tunnel diodes yield lowest noise operation, followed by Ge and GaAs diodes. (The reverse is true for RF saturation capability; therefore, Ge tunnel diodes are generally chosen as the best compromise between low noise and high saturation.) GaSb diodes are also difficult to manufacture and hence find very limited usage.

Since the noise performance of TDA's is essentially limited by the constants of the diode semiconductor material, little future improvement therein is anticipated. Despite the extreme simplicity, low power drain, and half to full octave bandwidth capability of the TDA, its current role [2], [6] as a moderately low-noise ($F \doteq 4.0$–7.0 dB from 2 to 25 GHz) RF amplifier is being steadily supplanted by the low-noise bipolar transistor and FET RF amplifier below K_u band, and, in the millimeter region, by the low-conversion-loss image-enhanced mixer without RF preamplification, by virtue of both the significant improvements in noise performance of the latter two components and the limited dynamic range capability of the TDA (IM intercept point: −30 to −10 dBm). In addition, TDA's are being replaced by reduced-cost miniaturized paramps in certain applications (e.g., radar receivers) in which the resulting improvement in sensitivity is of significant benefit. Therefore, the usefulness of the TDA in a limited number of receiver applications at X and K_u band as an RF pre- or postamplifier is expected to ultimately become negligible.

5) Mixer IF Amplifiers: Advances in Schottky-barrier mixer diode technology, first in Si and with even greater success in GaAs [6]–[10], has led to the evolution of the low-noise Schottky-diode mixer IF amplifier as the basic heterodyne downconverter in microwave and millimeter-wave front ends. Furthermore, coupling state-of-the-art mixer diodes with advanced circuit techniques, it is possible to configure simple mixer front ends which are competitive with more complex front ends employing RF preamplification, for both SSB and DSB reception, particularly in the millimeter-wave region up to 100 GHz.

The noise figure of this basic downconverter is essentially the product of the IF amplifier noise figure and the mixer conversion loss, be it SSB or DSB. The mixer conversion loss, in turn, is strongly dependent [28]–[34] upon the terminating impedances presented to the mixer diode(s) at the various higher order conversion products, e.g., idlers at frequencies $mf_{LO} \pm f_{IF}$, associated with a given IF (f_{IF}) and the mth order ($m = 1,2,3,\cdots$) harmonics of the LO (f_{LO}), particularly upon that at the image $f_i = 2f_{LO} - f_s$ corresponding to a given RF input signal f_s. In specific, the conventional matched-image (resistively terminated input over $f_{LO} \pm f_{IF}$) mixer in its various degrees of complexity (single ended, balanced, double balanced) is inherently a DSB converter providing equal conversion loss to RF inputs at f_s and f_i, e.g., $f_{LO} \pm f_{IF}$. It is therefore capable of exhibiting, in the state-of-the-art, impressively low DSB noise performance, particularly in the millimeter region ($F_{DSB} = 3$–7 dB over 1–100 GHz, assuming 1.5 dB F_{IF}), including the effects of circuit and diode losses on DSB conversion loss. The preceding results are dependent upon providing properly phased reactive terminations at said higher order idlers.

The implementation of the matched-image mixer for SSB reception (via a resistively isolated input preselector or a properly phased image-reject dual mixer configuration) degrades the preceding values by at least 3 dB. The matched-image mixer has the advantage, however, of half to full octave (or greater) bandwidth capability in both SSB and DSB applications. Practical cost-effective implementations of such Schottky-diode mixers, either in balanced or single-ended configurations, when integrated with a low-noise (1.5-dB) IF preamplifier, achieve SSB noise factors ranging from approximately 6 to 10 dB over the 1–100-GHz range [6], [35]–[37].

It has long been known [29]–[34] that proper reactive termination of the image frequency ($2f_{LO} - f_{RF}$) along with that of the higher order idlers can further reduce the conversion loss and noise factor of SSB downconverters by typically 1–2 dB. The advent of high-quality low-parasitic-content GaAs Schottky diodes has led to the practical realization of this potential in the image-enhanced and image-recovery mixer. The reactive image termination causes energy converted to the image frequency (only

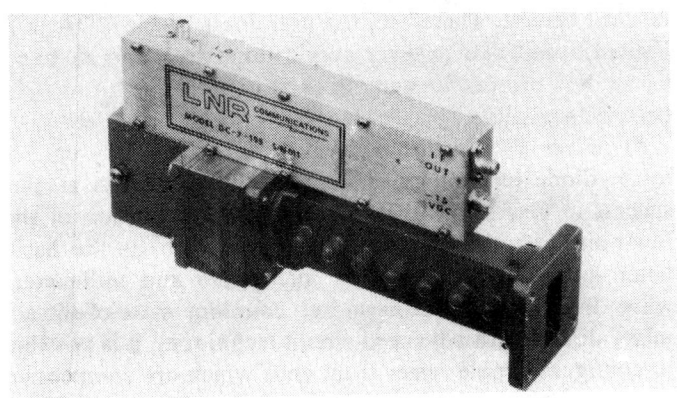

Fig. 9. X-band image-enhanced downconverter.

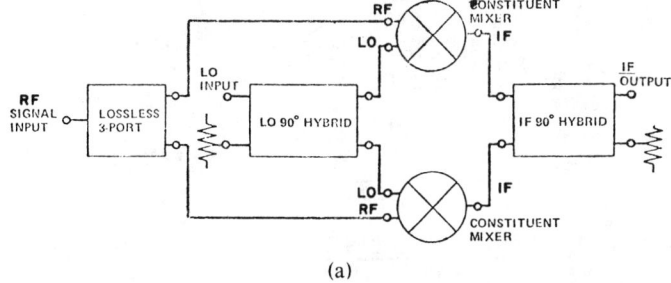

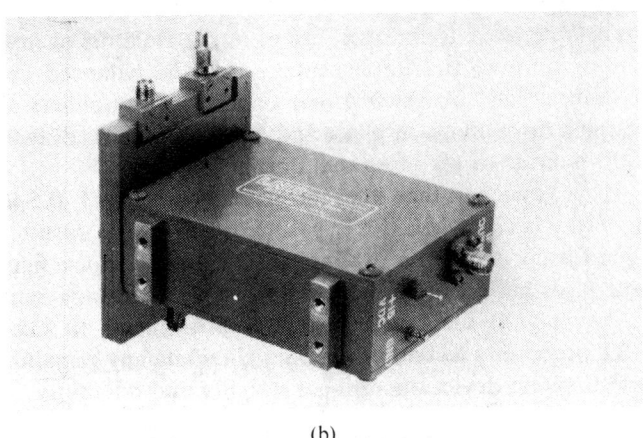

Fig. 10. Image-recovery mixer configuration. (a) Mixer block diagram. (b) Integrated mixer IF amplifier.

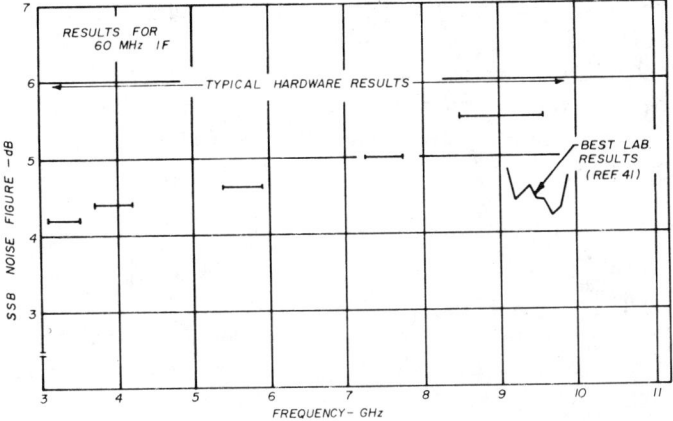

Fig. 11. Measured noise performance of image-recovery mixers.

possible upon simultaneous properly phased reactive termination of the higher order idlers) to be reflected back into the mixer and reconverted to IF.

Reactive image termination is accomplished either by proper location of a signal frequency bandpass filter [31]–[33], [38], [39] at the signal input to a single-ended or balanced mixer, or by use of a unique image-recovery [40]–[43] configuration, incorporating a pair of identical constituent mixers with properly phased signal and LO inputs and IF output. These two image-enhancement techniques are associated with fixed LO, one-to-one RF to IF frequency translator and tunable LO, narrow-band IF modes of mixer operation.

The filter-type reactive image mixer, requires a fixed finite separation between the RF signal and image sidebands, and therefore generally utilizes wide-band IF's ranging from UHF to S band in order to minimize the signal frequency insertion loss of the input filter. Therefore, the practical implementation of reactive image termination mixers imposes a limitation on maximum achievable fractional signal circuit bandwidth of the order of 10 percent or less due to the narrow bandwidth restriction of a properly located filter in the signal input circuit to provide the required stopband reactive termination over the image band. However, in many communications applications, this bandwidth limitation can readily be accommodated so that many filter-type image-enhanced mixers integrated with low-noise transistor IF amplifiers have been implemented as practical hardware. A typical X-band mixer of this type is depicted in Fig. 9, wherein the waveguide input filter provides the reactive image termination, and the diode mount is designed to suppress the higher order idlers. The X-band unit shown exhibits a typical overall noise figure of 5 dB, while providing 40-dB image rejection.

The image-recovery mixer [40]–[43] depicted in Fig. 10 consists of a pair of balanced mixers coupled to a common input signal, LO and IF output ports through a lossless in-phase signal input network, a quadrature LO hybrid, and a quadrature IF output hybrid, respectively. This configuration provides a reactive image termination by virtue of its symmetry with the common signal junction appearing as a short circuit at the image. Typical measured performance of off-the-shelf mixers of this type are depicted along with best X-band laboratory results [34], [41] in Fig. 11.

More generally, the image-recovery technique, is applicable to low-frequency IF (30–70 MHz) and offers input RF bandwidths of 10 percent or greater and direct conversion front-end noise figures of 4–7 dB from C to K band. In both matched and reactive image mixer configurations, IM intercept points of 0–+20 dBm or better can be readily realized, simultaneously with low-noise performance. One can project further improvement in *both* SSB and DSB mixer IF amplifier noise performance and the consequent realization of competitive low-noise front ends without RF preamplification, particularly in the millimeter range, ($F = 3$–6 dB over 1–100 GHz) based upon the use of still higher quality, lower parasitic content Schottky diodes, the realization of computer-aided circuit design techniques for image enhancement over wider RF bandwidths, and the use of ultralow-noise *parametric* IF amplifiers ($F_{IF} \leq 1$ dB).

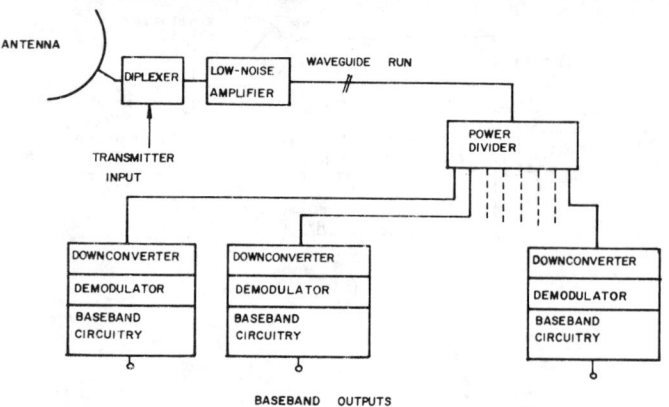

Fig. 12. Block diagram of satellite earth station receiving system.

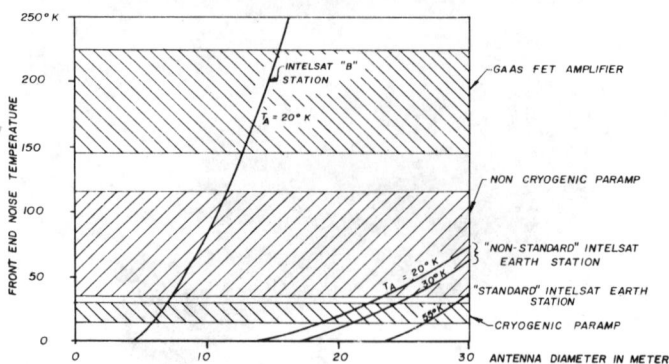

Fig. 13. Tradeoff between front-end noise temperature and antenna diameter for a 4-GHz INTELSAT earth station.

Still further improvement is achievable for operation of the mixer parametric IF amplifier at cryogenic temperatures [22] or in unique dual antiparallel diode subharmonically pumped [44]–[45] configurations.

IV. Importance of Receiver Noise Performance in Various System Applications

The need for state-of-the-art noise figure in microwave and millimeter-wave receiver applications arises for one of two reasons. In systems where front-end noise figure may be traded off against transmitter power and/or antenna size, the most economical system design often entails utilizing the lowest noise receiver available. In many applications system performance goals may only be achieved by utilizing the best available front end, antenna, and transmitter. These points are best illustrated with the following specific examples.

A. Satellite Communications

The demands of the satellite communications industry [6]–[8] have provided much of the impetus for the advancement in the low-noise art over the last decade, particularly in view of practical limitations on transmitter power and antenna size.

Other practical design considerations are as follows.

—SSB reception with image rejection is necessary to suppress interfering signals in the image passband.

—An input filter preceding the RF preamplifier or mixer is required to suppress out-of-band leakage from the system transmitter.

—High linear dynamic range is required to suppress intermodulation and cross modulation generated by time-coincident input signals and AM-to-PM conversion on strong signals.

—A high degree of amplitude flatness and phase linearity (delay flatness) in the receiver passband response is specified to prevent distortion of wide-band signals.

—Sufficiently high net RF/IF conversion gain to offset losses incurred in back-end multiplexing is necessary.

—RF bandwidth must be sufficiently large to permit multiple access.

—The IF bandwidth is chosen to accommodate wide-band FM signal modulation.

Fig. 12 shows a typical INTELSAT receiving system. The low-noise RF amplifier (LNA) comprises two paramp stages at the input, followed by three transistor stages, providing an overall gain of 55 dB. The transistor stages are biased for increasingly high dynamic range to minimize IM distortion and AM-to-PM conversion. The overall LNA gain is sufficient to reduce the noise temperature contribution of the following conventional power divider/downconverter array to less than 3 K. Use of more advanced mixers in the downconverter array to further reduce this contribution is less economical than increasing the gain of the LNA with relatively inexpensive bipolar transistor stages.

The figure of merit for a satellite receiving system [6]–[8] is G/T, which is the antenna gain divided by the system noise temperature T:

$$T = T_A + T_r = \frac{T_a}{L_F} + \left(1 + \frac{1}{L_F}\right) T_0 + T_r$$

where T_A is the equivalent antenna noise temperature, L_F is the feed loss, and T_r is the receiver noise temperature.

For a standard INTELSAT earth station,

$$G/T \text{ (dB)} \geq 40.7 + 20 \log (f_{\text{GHz}}/4).$$

Fig. 13 shows the required receiver noise temperature, to provide the required G/T as a function of antenna diameter, with antenna noise temperature as a parameter, assuming antenna efficiency of 67 percent. Also shown is the requirement for an INTELSAT B earth station wherein the G/T may be 10 dB lower.

A standard INTELSAT station must provide the required G/T at elevation angles down to 5° above the horizon. Because the antenna temperature increases significantly at low angles, the G/T can only be met at this time with a cryogenic paramp. For a large number of INTELSAT stations, however, which are nonstandard and have a favorable elevation angle, noncryogenic paramps are used on the basis of their lower purchase and maintenance costs.

Due to the variations in predetection bandwidth and signal-to-noise requirements for acceptable link performance, the high G/T of a standard INTELSAT station is not always required. This opens up a wider range of tradeoffs in implementing small earth stations for specialized traffic. As described in the preceding section, there are three types of low-noise amplifiers in use with various 4-GHz earth station receiving systems. These are cryogenic paramps, noncryogenic paramps (including units that are thermoelectrically cooled for the lowest noise, and lower cost designs that are temperature stabilized at the highest specified ambient), and GaAs FET amplifiers.

For the higher microwave satellite bands, there is a more diverse application of low-noise technology. For the 11.7–12.2-GHz CTS applications, the previously described (Fig. 8) Paraconverter® low-noise front end combines a single-stage advanced noncryogenic parametric amplifier, image-enhanced mixers, and low-noise microwave transistor IF amplifiers to provide a 200 K overall noise temperature in a housing that is directly mountable at the antenna feed. In this approach, the entire satellite band is translated to a lower microwave band, and conventional postreceivers are utilized for further downconversion and demodulation. The rapid strides in GaAs FET technology, the availability of low-noise mixer preamplifiers, and the lack of legal restriction on satellite radiated power will open a significant range of tradeoffs to system designers as the 11.7–12.2-GHz band is made available for commercial use. The choice will be made purely on the basis of economics. That is to say, the relative costs of satellite RF power, a tracking system for a highly directional earth station antenna, and a state-of-the-art low-noise parametric amplifier ($T_r < 100$ K) will have to be weighed.

As millimeter frequencies, through K_a band, the parametric amplifier is the sole practical choice, since limitations on satellite power, combined with high atmospheric attenuation, dictate the use of the lowest noise front end available.

B. Radar

Although state-of-the-art low-noise components find application in radar, they do so only in specialized applications.

Front-end noise figure enters into radar performance in two ways. The radar range equation [46] is

$$R_{max} = k \cdot \left[\frac{P_t G^2}{S_{min}}\right]^{1/4}$$

where R_{max} is the maximum range at which a target of given cross section can be detected, P_A is the peak transmitter power, G is the antenna gain expressed as a power ratio, and S_{min} is the minimum detectable signal (defined as the effective system input noise temperature).

The second way in which noise figure enters into radar performance is in parameter estimation such as angular location.

The theoretical angular accuracy of a radar is [46]

$$\delta\theta = \frac{0.628\theta_B}{(2S/N)^{1/2}}$$

where θ_B is the antenna beamwidth and S/N is the signal-to-noise ratio. Finally, improved receiving noise performance increases the probability of detection and decreases the false alarm rate. With all other parameters fixed, the signal-

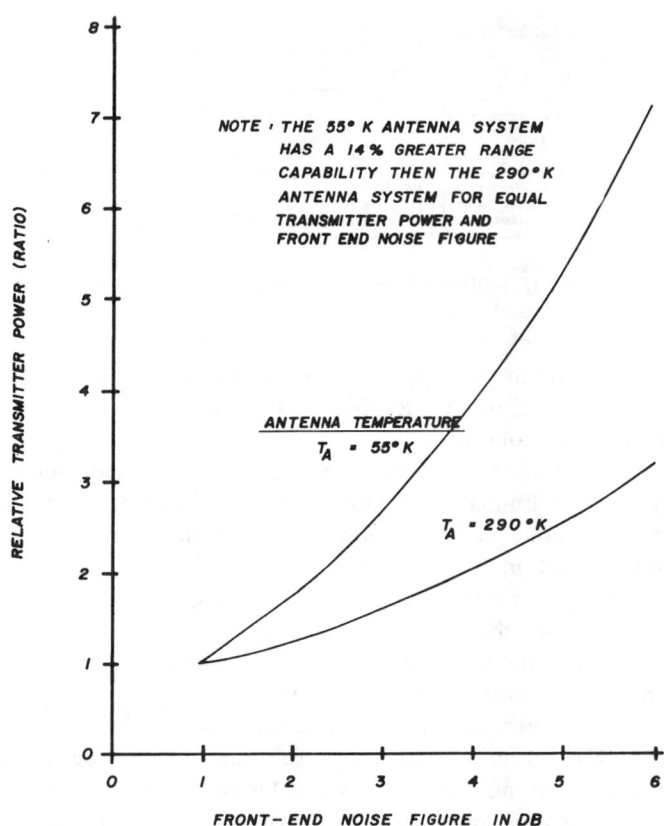

Fig. 14. Tradeoff between transmitter power and noise figure to achieve a given radar system range capability.

to-noise ratio varies inversely with the system noise temperature.

From these two equations it is seen that range increases with the fourth root of signal-to-noise, while angular accuracy varies as its square root. Moreover, it is seen that peak transmitter power can be directly traded off against system noise temperature. Fig. 14 shows this tradeoff for two antenna noise temperatures, 290 and 55 K. This plot shows relative transmitter power, to achieve a given range, versus front-end noise figure. A 290 K antenna noise temperature is typical of an air-to-ground radar, where the main beam illuminates the ground. In this case, it is generally more cost-effective to concentrate on maximizing transmitter power while utilizing a relatively low-cost conventional mixer preamp as the front-end noise-determining element. With the lower antenna temperature, as with a ground-to-air or air-to-air radar where the antenna "sees" the cold sky, a low-noise receiver can significantly enhance system performance.

Although the tradeoff between system noise temperature and transmitter power is important in a radar, the system performance can be more easily improved by increasing the antenna diameter. Therefore, state-of-the-art noise figure becomes important only when there is a constraint on antenna size, the latter due to location (for example, in an aircraft), to economics (as for a phased array), or to a requirement on a minimum width beam to achieve a required coverage (such as a cosecant-squared elevation pattern).

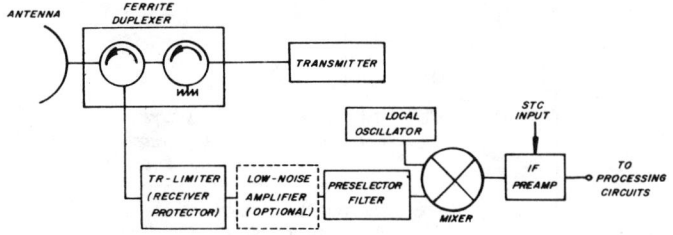

Fig. 15. Block diagram of a typical radar front-end.

Based upon the representative block diagram of Fig. 15, the practical design considerations for radar front ends include the following:
—Duplexer to separate transmitted and received signals.
—Input limiter preceding RF preamplifier or mixer to protect same from burnout during transmitter leakage pulse duration.
—Severe environmental requirements, including temperature, shock, vibration, etc.
—Phase and gain matching of front ends in sets of two or three for monopulse tracking radars.
—Size constraints for airborne systems.
—SSB reception with image rejection, often a desirable capability to minimize effects of interference in image band.
—High dynamic range of linear operation to inhibit large-signal distortion.
—High degree of amplitude flatness and phase linearity in the receiver passband response to minimize distortion in those systems using wide-band "chirped" FM pulses.
—Sufficiently wide RF bandwidth to accommodate system-derived requirement on transmitter carrier frequency diversity.
—Receiver blanking and/or sensitivity time control desirable to prevent overload during transmission, and from short-range returns.

Based upon the preceding section, the three low-noise front-end components that find application in radar front ends are parametric amplifiers, transistor amplifiers (bipolar and GaAs FET), and image-recovery mixers.

At X band and above, the image-recovery mixer (Figs. 10 and 11) has a number of advantages that make it particularly suitable for radar receivers, namely, its low-noise performance and its high dynamic range. The former eliminates the need for an RF amplifier with its additional size and weight, whereas the latter (1-dB-gain compression level at inputs greater than -15 dBm) eliminates the need for receiver desensitization circuitry at RF.

Below X band, improved system sensitivity can be obtained without excessive added size, weight, or cost by using a GaAs FET or bipolar transistor RF amplifier followed by a SSB mixer to avoid the contribution of image-band noise generated by the generally wide-band RF amplifier. Finally, in radar deployments where maximum receiver sensitivity is a necessity, a parametric RF amplifier must be used.

An example of the effects of X-band front-end noise figure on radar range capability, presented graphically in

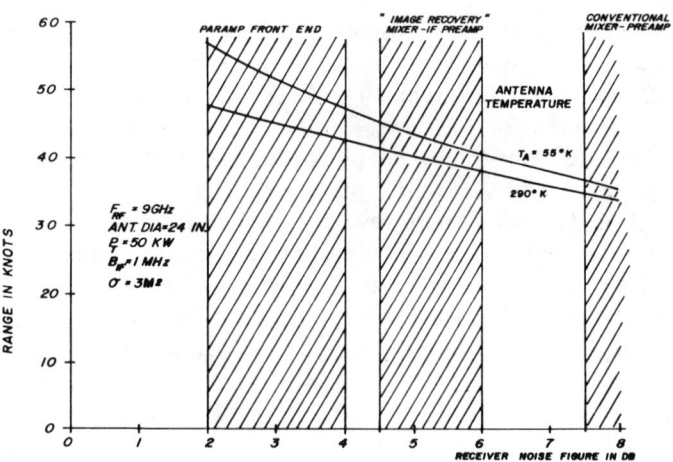

Fig. 16. Range capability versus noise figure for typical X-band radar parameters.

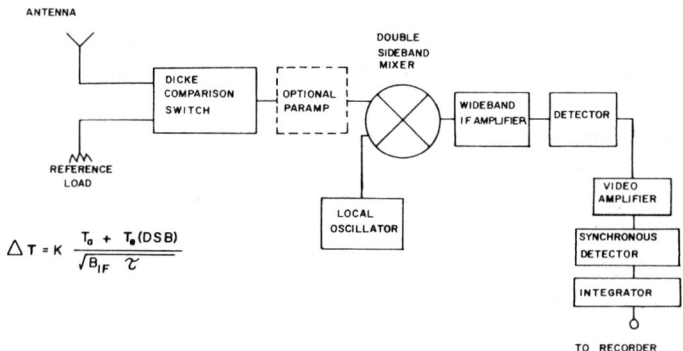

Fig. 17. Block diagram of a radiometric receiver.

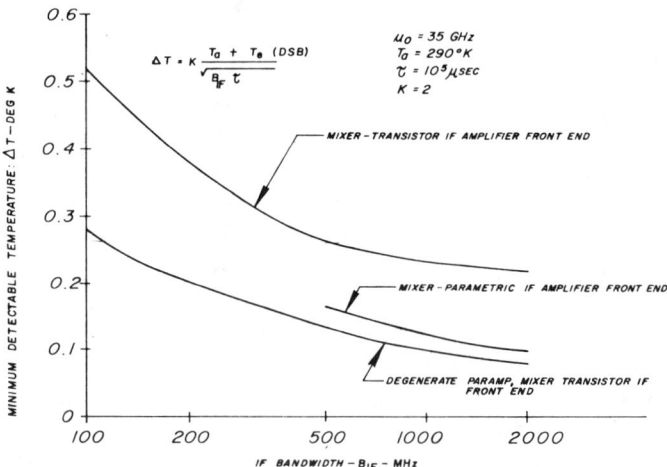

Fig. 18. Sensitivity of alternative K_a-band radiometric receivers.

Fig. 16, summarizes the impact of low-noise receiver technology on radar performance.

C. Radiometric Receivers

Radiometric receivers find use both in radio astronomy and in certain mapping and guidance systems. Fig. 17 depicts a typical radiometric receiver, and Fig. 18 formulates its sensitivity in terms of minimum detectable temperature, that is, the change in receiver-input broad-band noise temperature level that produces a change in detector output equal in amplitude to the internally generated rms fluctuation. Shown in Fig. 18 are the key radiometer parameters and their relation to sensitivity.

Radiometric receivers may be located in a relatively controlled stable environment for radio astronomy, or may be required to meet full military environmental specifications for airborne mapping or guidance applications. This can significantly affect the choice of front-end components. Other key constraints [21] are the following:

—DSB reception to maximize ability to detect low-level broad-band noise spectra.

—Utilization of Dicke configuration including square-wave-driven switch and synchronous detector, the former alternately connecting the receiver input to the receiving antenna and a reference load and thereby reducing the effects of short-term variation in receiver gain and noise temperature stability.

—Sufficiently large IF (predetection) bandwidth to enhance the radiometer detection sensitivity.

—Sufficiently large RF bandwidth to include both sidebands of width B_{IF}.

To illustrate the tradeoffs available in radiometric receiver design [21] based upon the previously presented (Fig. 3) building block state-of-the-art, the sensitivity of the following three alternative 35-GHz front ends is presented as a function of IF bandwidth in Fig. 18:

—A degenerate paramp followed by a mixer and transistor IF amplifier.

—A mixer and transistor IF amplifier.

—A mixer and a parametric IF amplifier.

Fig. 18 illustrates a key point relative to the millimeter-wave front-end state-of-the-art. Above 40 GHz, the judicious use of advanced mixer technology, including cryogenic cooling, can result in the realization of a comparable or lower noise front end than with a millimeter paramp, due to practical limitations on pump frequency and varactor cutoff which limit the noise figure achievable with millimeter paramps.

V. Future Trends in Low-Noise Receivers

Based upon the preceding assessment of the current state-of-the-art in microwave and millimeter front ends, one may project the following trends.

—Evolutionary improvements in GaAs technology toward highly reliable, extremely small metal semiconductor (MES) configurations will extend the frequency range of all devices.

—The noncryogenic parametric amplifier will continue to provide the lowest noise performance in a cost-effective reliable front end because of its basically noise-free amplification mechanism. Cryogenic paramps will only be used where the ultimate in noise-free reception is required. The predominant paramp applications will be in satellite communications ground station receivers.

—MESFET devices will play an increasingly important role, through X band, as RF preamplifiers and postamplifiers for parametric RF preamplifiers, as well as microwave IF amplifiers. MESFET RF preamplifiers will play an increasingly dominant role in small earth station point-to-point communications receivers as well as radar receivers.

—For many moderately low-noise receiver applications, particularly in the millimeter-wave radar and radiometric region, simple front ends consisting of ultralow-loss Schottky-barrier mixers followed by ultralow-noise transistor or parametric IF amplifiers will eliminate the need for RF preamplification. A particularly useful application for this configuration is in millimeter radio astronomy installations.

In each case, advances in the receiver art will be ultimately linked to further concurrent advances not only in semiconductor device technology but in related areas such as ferrite circulator technology, advanced planar, and mixed media transmission line techniques and circuit and structural design concepts.

Acknowledgment

The authors wish to thank Dr. F. Arams and S. Okwit, P. Lombardo, W. Hollis, and J. DeGruyl for their contributions in the preparation of this paper.

References

[1] R. Adler et al., "Description of the noise performance of amplifiers and receiving systems," Proc. IEEE, vol. 51, pp. 436–442, Mar. 1963.

[2] H. C. Okean and P. P. Lombardo, "Noise performance of M/W and MM-wave receivers," Microwave J., pp. 41–50, Jan. 1973.

[3] D. C. Hogg and W. W. Mumford, "The effective noise temperature of the sky," Microwave J., vol. III, pp. 80–85, Mar. 1960.

[4] G. L. Heiter, "Characterization of nonlinearities in microwave devices and systems," IEEE Trans. Microwave Theory Tech., vol. MTT-21, pp. 797–805, Dec. 1973.

[5] D. E. Norton, "The cascading of high dynamic range systems," Microwave J., pp. 57–71, June 1973.

[6] C. L. Cuccia, "Status report: Modern low noise amplifiers in communications systems," Microwave Syst. News, pp. 120–132, Aug./Sept. 1974, and pp. 79–89, Oct./Nov. 1974.

[7] C. L. Cuccia and C. Hellman, "The low-cost low-capacity earth terminal," Microwave Syst. News, pp. 19–42 June/July 1975.

[8] C. L. Cuccia and C. Hellman, "Status report: Microwave communications update—1976," Microwave Syst. News, pp. 20–53, Feb./Mar. 1976.

[9] F. R. Arams et al., "State-of-art low noise receivers for microwave and millimeter communications," to be presented at National Telecommunications Conference, Dallas, TX, Nov. 1976.

[10] W. Hollis, "Practical state-of-the-art of microwave and millimeter wave receiver front ends," Commun. News, Oct. 1976.

[11] J. Edrich, "20°K-cooled parametric amplifier for 46 GHz with less than 60°K noise temperature," 1973 IEEE—G-MTT Int. Symp. Digest, pp. 72–74, June 1973.

[12] H. C. Okean, J. A. DeGruyl, and E. Ng, "Ultra low noise K_u-band parametric amplifier assembly," 1976 IEEE S-MTT Int. Microwave Symp. Digest, pp. 82–84, June 1976.

[13] E. Kraemer, J. Leeper, and J. Whelehan, "K_u-band spacecraft parametric amplifier," 1974 IEEE S-MTT Int. Symp. Digest, pp. 222–224, June 1974.

[14] A. D'Ambrosio, "A SHF parametric amplifier for space applications: Design implementation," Alta Frequenza, vol. XLIII, pp. 876–883, 1974.

[15] H. C. Okean et al., "Electronically-tunable, low noise K_a-band paramp—downconverter satellite communications receiver," 1975 IEEE S-MTT Int. Microwave Symp. Digest, pp. 43–45, May 1975.

[16] J. Whelehan et al., "A nondegenerate millimeter wave parametric amplifier with a solid-state pump source," 1973 IEEE G-MTT Int. Symp. Digest, pp. 75–77, June 1973.

[17] H. C. Okean, J. R. Asmus, and L. J. Steffek, "Low noise 94 GHz parametric amplifier development," 1973 IEEE G-MTT Int. Microwave Symp. Digest, pp. 78–79, June 1973.

[18] H. C. Okean and L. J. Steffek, "Low loss, 3MM junction circulator," 1973 IEEE G-MTT Int. Microwave Symp. Digest, pp. 80–81, June 1973.

[19] S. D. Lacey, B. T. Hughes, and J. C. Vokes, "A low noise room

temperature 12 GHz parametric amplifier," *1974 IEEE S-MTT Int. Symp. Digest*, pp. 220–221, June 1974.

[20] T. Okajima *et al.*, "18 GHz paramps with triple-tuned gain characteristics," *IEEE Trans. Microwave Theory Tech.*, vol. MTT-20, pp. 812–819, Dec. 1972.

[21] LNR Staff, "Millimeter wave radiometric sensitivity," *Microwave J.*, pp. 56–57, Jan. 1973.

[22] S. Weinreb and A. R. Kerr, "Cryogenic cooling of mixers for millimeter and centimeter wavelengths," *IEEE J. Solid-State Circuits*, pp. 58–63, Feb. 1973.

[23] C. A. Liechti and R. L. Tillman, "Design and performance of microwave amplifiers with GaAs Schottky gate field effect transistors," *IEEE Trans. Microwave Theory Tech.*, vol. MTT-22, pp. 510–517, May 1974.

[24] N. K. Osbrink *et al.*, "Transistor technology at microwave frequencies," *Microwave J.*, pp. 27–29, Nov. 1975.

[25] J. T. Lindauer and N. K. Osbrink, "GaAs FET amplifiers are closing fast on the low-noise narrow band leaders," *Microwave Syst. News*, pp. 63–67, May 1976.

[26] M. G. Walker *et al.*, "MESFET amplifiers go to 18 GHz," *Microwave Syst. News*, pp. 39–48, May 1976.

[27] H. C. Okean, "Tunnel diodes," ch. 8 in *Semiconductors and Semimetals*, vol. 7B, Willardson and Beer, Eds. New York: Academic Press, 1971.

[28] H. C. Torrey and C. A. Whitmer, "Crystal rectifiers," *Rad. Lab. Series*, vol. 15. New York: McGraw-Hill, 1948.

[29] R. J. Mohr and S. Okwit, "A note on the optimum source conductance of crystal mixers," *IRE Trans. Microwave Theory Tech.*, vol. MTT-8, pp. 662–627, Nov. 1960.

[30] A. A. M. Saleh, *Theory of Resistive Mixers*, The M.I.T. Press, 1971.

[31] S. Egami, "Nonlinear, linear analysis and computer-aided design of resistive mixers," *IEEE Trans. Microwave Theory Tech.*, vol. MTT-21, pp. 270–275, Mar. 1973.

[32] M. R. Barber, "Noise figure and conversion loss of the Schottky barrier mixer diode," *IEEE Trans. Microwave Theory Tech.*, vol. MTT-15, pp. 629–635, Nov. 1967.

[33] G. B. Stracca, F. Aspesi, and T. D'Angelo, "Low noise microwave down-converter with optimum matching at idle frequencies," *IEEE Trans. Microwave Theory Tech.*, vol. MTT-21, Aug. 1973.

[34] A. J. Kelly, H. C. Okean, and S. J. Foti, "Low noise microwave and millimeter wave integrated circuit mixers," *IEEE S-MTT Int. Microwave Symp. Digest*, pp. 146–148, May 1975.

[35] L. T. Yuan, "A low noise broadband K_a-band waveguide mixer," *IEEE S-MTT Int. Microwave Symp. Digest*, pp. 272–273, May 1975.

[36] T. G. Phillips and K. B. Jefferts, "Millimeter wave receivers and their application to radio astronomy," *1974 IEEE S-MTT Int. Microwave Symp. Digest*, pp. 116–117, June 1974.

[37] M. V. Schneider and G. T. Wrixon, "Development and testing of a receiver at 230 GHz," *1974 IEEE S-MTT Int. Microwave Symp. Digest*, pp. 120–122, June 1974.

[38] L. E. Dickens and D. W. Maki, "An integrated circuit balanced mixer, image and sum enhanced," *IEEE Trans. Microwave Theory Tech.*, vol. MTT-23, pp. 276–281, Mar. 1975.

[39] Y. Konishi, K. Uenakada, N. Yazawa, N. Hoshino, and T. Takahashi, "Simplified 12-GHz low-noise converter with mounted planar circuit in waveguide," *IEEE Trans. Microwave Theory Tech.*, vol. MTT-22, pp. 451–455, Apr. 1974.

[40] T. H. Oxley, J. F. Lord, K. J. Ming, and J. Clarke, "Image recovery mixers," *Proceedings 1971 European Microwave Conf.*

[41] A. J. Kelly and H. C. Okean, "Wideband image recovery mixers," *1974 GOMAC Digest of Papers*, pp. 240–241, June 1974.

[42] D. Neuf, "A quiet mixer," *Microwave J.*, vol. 16, pp. 29–32, May 1973.

[43] L. E. Dickens and D. W. Maki, "A new 'phased-type' image enhancement mixer," *1975 IEEE S-MTT Int. Microwave Symp. Digest*, pp. 149–151.

[44] M. V. Schneider and W. W. Snell, Jr., "Harmonically pumped strip line downconverter," *IEEE Trans. Microwave Theory Tech.*, vol. MTT-23, pp. 271–275, Mar. 1975.

[45] M. Cohn *et al.*, "Harmonic mixing with an antiparallel diode pair," *IEEE Trans. Microwave Theory Tech.*, vol. MTT-23, pp. 667–673, Aug. 1975.

[46] M. I. Skolnik, *Introduction to Radar Systems*. New York: McGraw-Hill, 1962.

Low-Noise and Linear FET Amplifiers for Satellite Communications

DAVID A. COWAN, PAUL MERCER, AND ABE B. BELL, MEMBER, IEEE

Abstract—FET amplifiers with ambient noise figures as low as 4.8 dB at 12 GHz, 35-dB gain, and intercept points as high as +28 dBm have been developed for use in communications satellites. Predicted mean time to failure is in excess of 10^6 h.

Introduction

ALTHOUGH microwave FET's have been available for many years they have not become generally accepted for use in systems where ultra-high reliability is required (for example, in communications satellites). Until recently, little information on GaAs FET reliability was available; however, accelerated life testing performed by manufacturers [1], [2], coupled with a reasonable accumulation of long-term device testing, has demonstrated that GaAs FET's can be highly reliable devices with predicted mean time to failure (MTTF) in excess of 10^8 h.

Two 12-GHz FET amplifier types have been developed by Spar Technology Limited, for use in communications satellites. These amplifiers utilize packaged 1-µm gate-length GaAs FET devices which have undergone stringent qualification testing to ensure their long-term reliability. A low-noise and a linear amplifier design have been developed. The low-noise amplifier has an ambient temperature noise figure as low as 4.8 dB with a gain of 35 dB, while for the linear amplifiers, the gain is 35 dB with a third-order intermodulation intercept point as high as +28 dBm. Both amplifiers have good gain flatness, high-temperature stability, and low-group delay. Small size and low weight are of prime importance, and each amplifier operates from a +15-V dc supply, utilizing an integral switching type power converter for minimum power consumption.

FET Specification and Qualification

The transistor selected for the FET amplifiers was the Nippon NE 24406, a 1-µm gate-length transistor in a hermetic ceramic package. No significant performance penalties are encountered from using a packaged device for at least up to 5-percent operating bandwidths at this frequency. The microwave parameters of the transistors are specified at a frequency of 12 GHz and are based on typical results of characterization in a microstrip test mount. Devices are

---

Manuscript received May 13, 1977; revised July 27, 1977. This work was supported in part by the Canadian Department of Industry Trade and Commerce.

D. A. Cowan was with Spar Technology Limited, Saint Anne de Bellevue, Quebec, Canada H9X 3R2. He is now with Continental Microwave Limited, Dustable, England.

P. Mercer was with Spar Technology Limited, Saint Anne de Bellevue, Quebec, Canada H9X 3R2. He is now with MA Electronics Canada Limited, Mississauga, Canada.

A. B. Bell is with Spar Technology Limited, Saint Anne de Bellevue, Quebec, Canada H9X 3R2.

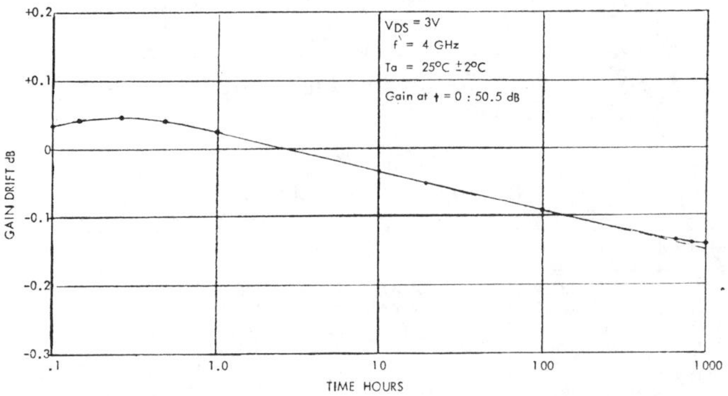

Fig. 1. Gain drift for five-stage amplifier.

specifically selected for either low noise figure or high intercept point.

Stringent qualification testing of flight devices included power burn-in of all devices for 336 h at elevated temperatures (100 C° ambient) with both dc and RF parameters monitored and checked for drift during this period; ±5-percent ΔI_{DSS} was permitted, and ±0.1-dB ΔMAG (maximum available gain). Short- and medium-term stability were evaluated by selecting five sample devices from each wafer and building them into a five-stage amplifier. The amplifiers were tested over a period of 1000 h and the gain drift with time was monitored; a maximum amplifier gain drift of 0.1 dB per decade of time was permitted. Typical measured value was 0.08 dB/decade. For long-term stability, measurements (Fig. 1) have shown that gain tends to stabilize during the first 1000 h [3]; extrapolating these figures to a spacecraft lifetime of seven years results in a maximum FET gain change of 0.25 dB for a six-stage amplifier (following a 600-h burn-in). Results of accelerated life testing [1], [2] tend to support this small change, together with such life test data as is currently available.

Other qualification tests included a wafer reliability assessment based on accelerated life testing at 295°C on sample devices, Group A testing on all devices and Group B testing on sample devices, using MIL-STD-750 methods. The Group A tests covered measurement of dc parameters such as drain current and gate-reverse current and RF measurement of MAG. The more extensive Group B tests for lot qualification covered mechanical, thermal, and operational life testing for 1000 h at 100°C followed by dc parameter drift measurements. SEM (scanning electron microscope) inspection was carried out on sample devices from each wafer, together with 100-percent X-ray inspection of devices after packaging. All flight devices were checked for short-term ($t < 100$ min) drift characteristics (dc and RF) following the power burn-in.

TABLE I
S-Parameter Measurements—12-GHz FETA

F (GHz)	S11		S12		S21		S22	
11.0	.621	133.2	.033	-54.6	1.337	-38.7	.810	-138.5
11.5	.612	120.7	.032	-67.7	1.28	-52.1	.828	-150.8
11.6	.613	118.7	.032	-70.6	1.27	-54.6	.832	-153.4
11.7	.614	116.2	.031	-73.4	1.26	-57.2	.839	-159.7
11.9	.615	110.9	.036	-79.7	1.241	-63.0	.846	-160.6
12.1	.614	106.9	.029	-86.6	1.218	-69.0	.85	-165.8
12.3	.607	102.5	.029	-98.3	1.197	-75.1	.875	-170.9
12.4	.605	100.1	.027	-104.5	1.178	-78.6	.880	-173.9

Note: NE 24406 transistor—computer corrected transistor S-parameters (50-Ω mount).

TABLE II
S-Parameter Measurements—12-GHz FETA

F (GHz)	S11		S12		S21		S22	
11.0	.380	37.9	.062	-161.8	2.13	-145.7	.783	117.6
11.5	.418	-11.2	.081	164.4	2.55	-179.9	.560	85.5
11.6	.431	-20.0	.083	156.1	2.62	172.3	.500	74.8
11.7	.446	-30.3	.084	147.9	2.66	163.4	.438	61.7
11.9	.466	-47.8	.089	130.9	2.69	145.8	.344	23.7
12.1	.463	-63.2	.090	111.0	2.60	126.7	.354	-27.7
12.3	.446	-74.8	.088	88.8	2.39	107.0	.475	-69.9
12.4	.435	-78.6	.084	77.6	2.25	97.9	.540	-85.0

Note: NE 24406 transistors with single-stage input and output matching transformers—computer corrected S-parameters (partially matched mount).

TABLE III
Low-Noise Amplifier—Gain Budget

COMPONENT	GAIN (dB)	LOSS (dB)
Input & Output Connections including Stress Relief		-0.7
Isolators Qty 4		-1.6
PIN Attenuator		-4.0
Low-Noise Two-Stage Amplifier Module	+13.0	
High-Gain Two-Stage Amplifier Modules Qty 2	+30.0	
Net Gain	+36.7 dB	

Device Characterization

If computer-aided design techniques are to be used for amplifier modules it is essential to perform accurate device S-parameter measurements. At 12 GHz the parameters depend to a large extent on the test mount used for characterization (for example, the S_{12} value for the transistor may be considerably modified by feedback within the test mount); the test-mount configuration should, therefore, resemble closely the mounting configuration to be used in amplifier modules. The test mount used comprised microstrip transmission lines at input and output to the transistor with printed bias networks and beam-lead capacitors for dc isolation. The width of the test-mount cavity was less than $\lambda/2$ at the maximum frequency of operation to eliminate waveguide feedback modes. The mount was characterized on an automatic network analyzer before mounting the transistors, and an equivalent circuit was derived; this was used in the correction of transistor S-parameters to a convenient design reference plane, in conjunction with a network analysis program used on a time sharing computer.

Initial characterization was carried out using 50-Ω characteristic impedance input and output lines; for more accurate measurements a partially matched mount, comprising single stage input and output matching transformers, was used. Typical corrected S-parameter measurements are given in Table I and Table II.

Optimum source impedance for low-noise performance was determined empirically and found to be close to a conjugate input match to the transistor. The optimum load impedance for intermodulation products was also determined empirically by tuning for maximum saturated power output. Typical bias conditions were $V_{DS} = 4.5$ V, $I_{DS} = 10$ mA, for low-noise devices, and $V_{DS} = 5.0$ V, $I_{DS} = 25$ mA, for low intermodulation product devices. Typical MAG was found to be in excess of 8 dB for low-noise devices and 9 dB for other devices.

Amplifier Design

A single-ended design approach, with isolator coupling, was used. Advantages of the single-ended approach include lower cost (a significant figure in the case of flight-qualified devices), higher reliability due to a lower component count, and lower power consumption.

A modular design approach has been developed for MIC (microwave integrated circuit) FET amplifiers and offers many advantages when compared with a single integrated unit. Individual modules are tuned in test mounts prior to "drop-in" integration into the amplifier assembly, enabling optimum performance to be achieved from each module and reducing the time required for tuning. Repairs can be quickly and conveniently carried out by simply replacing modules, and problem areas which arise on integration can be readily identified. Module types used in the amplifiers include low-noise, high-gain, and linear two-stage amplifiers, isolators and p-i-n attenuators.

Each of the two amplifier types contains three two-stage amplifier modules (six FET's) with isolator modules between them and at the input and output ports to the amplifiers. A p-i-n attenuator module, located either before or after the second amplifier module (depending on the amplifier type), is used to achieve compensation of the amplifier gain change with temperature. The p-i-n attenuator can provide a dynamic range of several decibels with a very small gain slope variation. The position of the attenuator modules was chosen to have minimum effect on noise figure of the low-noise amplifier and on intermodulation products in the linear amplifier.

The schematic for the low-noise amplifier is shown in Fig. 2. Biasing and matching transistors for low-noise or low-intermodulation products rather than maximum gain, together with circuit losses (including isolators), and p-i-n attenuator losses, results in the need for six stages to achieve 35-dB overall gain. The design gain margin is then 2 dB, based on the gain budget shown in Table III.

The three two-stage amplifier module types were designed using typical measured S-parameter data, with the aid of a computer optimization program. (There is a brief descrip-

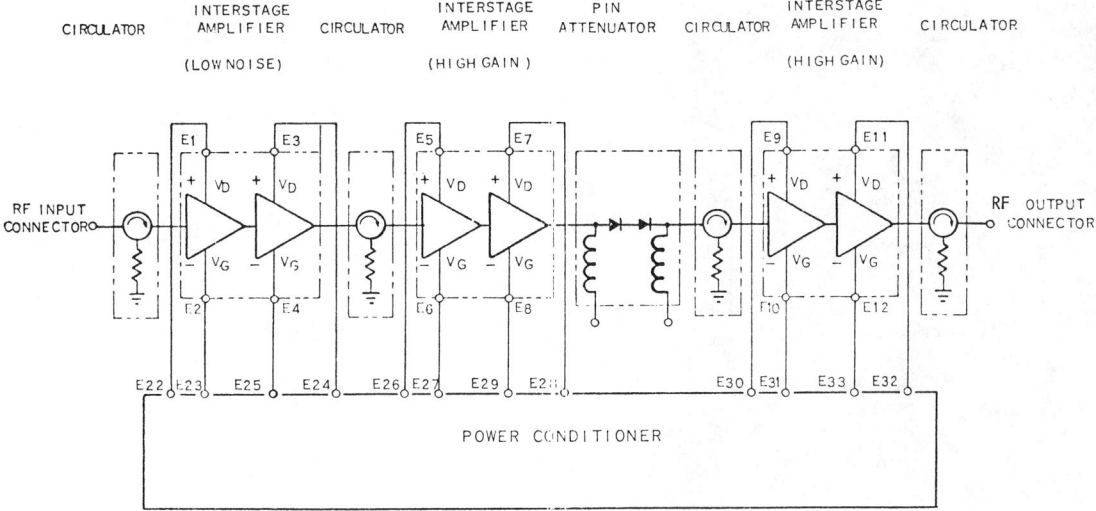

Fig. 2. Low-noise amplifier schematic.

tion of the program given in the Appendix.) Matching is accomplished using distributed elements only, except that ferrite beads are required on the dc bias filtercons to provide resistive loading at low frequencies and thus ensure stability; a printed low-pass filter structure ensures that the bias network is purely reactive at 12 GHz. MOS beam lead capacitors, which have low-reflection coefficients when mounted in series in a 50-Ω transmission line, are used to achieve dc isolation.

The MIC isolator modules are three-port circulators printed on yttrium garnet substrates, with a stubmatched 50-Ω chip resistor to terminate the third port. The design achieves low output to input coupling by suppressing radiation from the circulator disc. A minimum isolation of 26 dB with less than a 0.3-dB insertion loss is achieved.

Each p-i-n attenuator module comprises two series-mounted beam lead diodes (a "fail safe" configuration, since p-i-n diodes normally fail short circuit) spaced approximately one-quarter wavelength apart to achieve low VSWR. The drive circuit for the diodes consists of a thermistor-controlled current source whose resistor elements are computer optimized to achieve the best temperature compensation curve fit.

Power Converter Design

The common-source FET mounting configuration used in the amplifier requires both positive and negative voltage bias rails for each transistor. High stability in both the short and long term is important for good amplifier gain stability. High efficiency is a prime requirement due to the maximum available power of 1.5 W for each amplifier.

The power converter design employs a pulsewidth modulated buck-type switching regulator, resulting in efficient conversion of the available +15-V supply voltage to a +5.7-V drain supply and −5.0-V gate supply. A slow start-up circuit is employed to prevent voltage overshoots; this is desirable for reliable FET operation since voltage spikes can cause damage. High regulation is obtained by using high-stability components and a feedback loop with high gain. Filter networks at input and output to the converter result in low ripple on the bias rails and prevent current overshoots.

Biasing of individual transistors is achieved through series resistors in the positive-drain supply, and through two resistors comprising a voltage divider network to each gate in the negative supply. The p-i-n attenuator drive circuit derives its power from the +5.7-V rail. The power converter efficiency is in excess of 75 percent over the full operating temperature range.

Mechanical Design and Layout

Amplifier and p-i-n attenuator modules are constructed with 0.025-in thick 99.5-percent alumina substrates, soldered to nickel-plated Kovar carriers; and the isolator modules are constructed with 0.025-in garnet substrates soldered to a nickel-plated nickel–iron alloy carrier. In both cases the thermal-expansion coefficient of substrate and carrier are closely matched to reduce stresses during temperature cycling. Stress-relieved bonds are used to interconnect modules, both for the RF line and the ground plane. The amplifier housing is machined from aluminum and is nickel plated.

The amplifier layout is determined primarily by RF considerations. All modules are designed to be in cut-off waveguide sections when the amplifier cover is on, resulting in low reverse feedback and high stability. The power converter is completely isolated from the RF circuits, with the bias supplies to transistors being fed through filtercons mounted on the Kovar carrier. A photograph of the linear amplifier is shown in Fig. 3.

Overall size of the amplifier is approximately three and one-half inches square by one inch deep, and the weight is 0.84 lb. Vibration testing showed that all stress margins were adequate.

Amplifier Reliability

All components and materials used in the amplifiers are subjected to the usual screening requirements for use in high

Fig. 3. Linear FET amplifier.

TABLE IV
WORST-CASE GAIN CHANGE OVER SEVEN-YEAR LIFE

T °C	5	25	45
ΔL_{PIN} dB	+.76 −.35	+.35 −.20	+.14 −.07
ΔG_{FET} dB	+.00 −.24	+.00 −.24	+.00 −.24
ΔG_{BIAS} dB	+.26 −.09	+.26 −.09	+.26 −.09
ΔG_{TOT} dB	+.61 −1.09	+.46 −.68	+.33 −.47

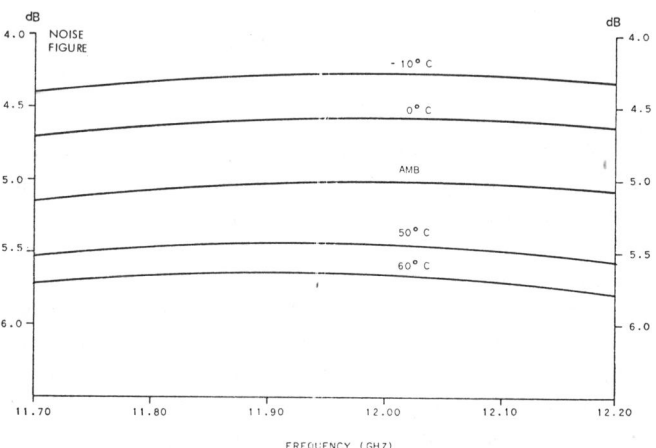

Fig. 5. Noise figure—low-noise amplifier.

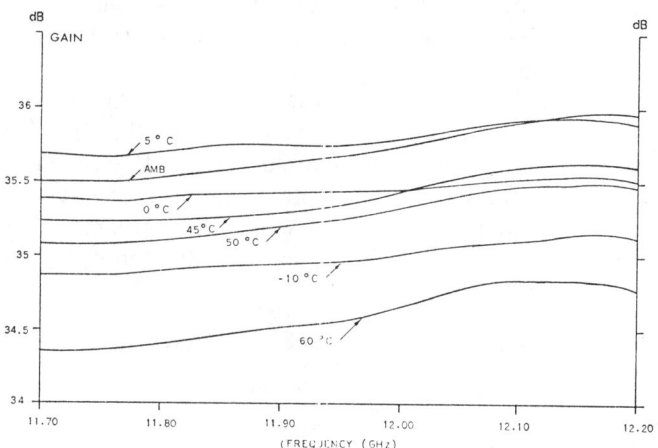

Fig. 4. Gain—linear FET amplifier.

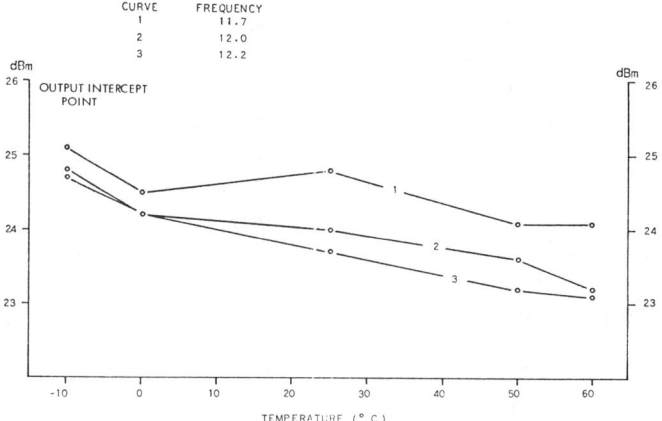

Fig. 6. Intercept point—linear amplifier (10-MHz frequency separation).

reliability applications, and rigorous inspection is carried out by quality control personnel at all stages of manufacture and test.

A computation of MTTF and a worst-case analysis were carried out. For the complete amplifier, including power converter, a failure rate of 876 FITS (failures in 10^9 h) is predicted; this includes a contribution of 362 FITS for the power converter, and assumes a failure rate of 64 FITS for each transistor (maximum channel temperature 82°C). Worst-case analysis for a seven-year operation shows the most significant performance change to be due to ageing of components in the p-i-n attenuator drive circuit. Summing the effects of power converter-bias network, FET, and p-i-n attenuator ageing, results in a worst-gain stability of +0.61 dB, −1.09 dB, $T_{amb} = 5°C$ as shown in Table IV. The worst-case change of the noise figure is 0.4 dB, and of the third-order intercept point is 0.21 dB. These figures are sufficiently small to ensure that specification will be met over the full spacecraft lifetime.

Such data as have so far been published on GaAs FET sensitivity to radiation and particle bombardment [4], [5], suggest that these devices are relatively immune. It is estimated that the ionizing radiation dose absorbed by the devices will be limited to 10^5 rad by the spacecraft structure and amplifier and receiver enclosures, and that particle fluence will not exceed $2 \times 10^{13}/cm^2$. These figures are well below the thresholds at which damage to FET's has been recorded.

MEASURED PERFORMANCE

All modules were tuned and tested individually prior to integration and found to meet or exceed their design goals.

TABLE V
SPECIFICATION AND PERFORMANCE SUMMARY

Parameter	Specification		Measured Performance (Worst Case Over Spec. Temperature Range)	
Operating Frequency Range 11.7 to 12.2 GHz Unless stated otherwise	Parameter Specification	Applicable Temperature Range (°C)	Low Noise Amplifier	Linear Amplifier
Midband Gain	35 dB + 2, -0 dB	25	38.65 dB	35.75 dB
Gain Variation over Frequency Band	1.0 dB peak-to-peak maximum	0-50	0.40 dB	0.50 dB
Gain Stability	1.25 dB peak-to-peak maximum	5-45	0.80 dB	0.50 dB
Gain Slope	.004 dB/MHz maximum	0-50	0.003 dB/MHz	0.0032 dB/MHz
Noise Figure	7 dB maximum (low noise) 10 dB maximum (linear)	0-50	5.6 dB	9.5 dB
Port VSWR	1.25:1 maximum	0-50	1.11:1 Input Port 1.10:1 Output Port	1.20:1 Both Ports
Group Delay Variation	0.5 ns maximum	0-50	0.4 ns	0.5 ns
Linearity	Third-Order Intercept Point +17 dBm minimum (low noise) +23 dBm minimum (linear)	0-50	+20.3 dBm	+23.2 dBm
Spurious Outputs	-75 dBm maximum	0-50	None detectable	None detectable
Power Consumption	1.5W maximum	0-50	1.09W	1.27W
Voltage Variation	15V ± 5%	0-50	No measureable effect on Gain Frequency Response, Noise Figure, Linearity.	
Weight	0.75 lb maximum	-	0.84 lb	0.84 lb

Following integration into the amplifier housing, a little additional tuning was required to achieve the overall amplifier specifications. This is due primarily to the small mismatches introduced by the flexible module interconnections and due to evanescent radiation modes which tend to degrade performance by mutual coupling between modules.

The significant electrical parameters were recorded over several temperature cycles with the extremes ±10°C in excess of the 0–50°C design temperature range. A summary of the test results for the first development models of both amplifier types is given in Table V. Using better FET's subsequently obtained from the supplier, a 0.4-dB noise-figure improvement and a 4-dB intercept-point improvement, were achieved. Design specification values are given in the table, and it can be seen that all important requirements are met.

Referring to Table V, the midband gains for the low-noise and linear amplifiers are 38.65 and 35.78 dB, respectively, with 0.4- and 0.5-dB gain flatness over the 500-MHz operating frequency band. The higher gain in the low-noise unit is due to higher module gains than the design target. The worst-case gain slope is 0.003 dB/MHz. The high stability of the gain slope with temperature is demonstrated in the Fig. 4 plot of the linear amplifier gain and is a result of the design strategy of employing a constant bias on the FET's and a well-matched p-i-n diode attenuator for temperature compensation. Within the optimized 5–45°C temperature range, gain variations of 0.8 and 0.5 dB for the low-noise and linear amplifiers were measured.

The noise figure characteristic of the low-noise amplifier is given in Fig. 5. A worst-case value of 5.6 dB was measured at 50°C. This performance was achieved by the selection of front-end FET's which have a 3.9-dB noise figure at 25°C, and by minimizing the loss of the input isolator.

The group-delay variation was very small. At the temperature extremes, the variation was 0.4 and 0.5 ns over the 500-MHz band, for the low-noise and linear amplifiers, respectively.

The third-order intercept point of the linear amplifier is shown for several operating frequencies in Fig. 6. The worst-case intercept point is 23.2 dBm. No linearity variation was observed with changing carrier separation (0.5 to 50 MHz Δf was tested) indicating effective decoupling of the converter supply from the RF circuitry.

The amplifier power consumptions are well below the 1.5-W maximum, with a power consumption variation of less than 5 percent over the operating temperature range. This reflects the low-current requirement of the p-i-n attenuator, the use of constant bias on the FET's, and the relatively constant converter efficiency over the temperature range.

Radiated susceptibility testing was carried out on the low-noise amplifier and showed that the unit was essentially connector limited. With a 1-V/m field strength, an output power of −67-dBm maximum was observed with incidence around the input port.

Conclusion

Design and measured data have been given for a low-noise and a highly linear 12-GHz FET amplifier. It is felt that these amplifiers represent the state-of-the-art for high-reliability amplifiers in this frequency range, particularly for the gain slope and gain stability, and noise figure and group delay characteristics. Reliability predictions indicate that the probability of survival for such amplifiers over communications satellite lifetime, is at least as high as alternative types of amplifiers that are currently available, and it is likely that they will find increasing application in high-reliability systems.

Appendix

The computer optimization program used is an in-house program called COSMIC-K, which stands for computer optimization of simple microwave integrated circuits, kronos time-sharing. Although written for microwave circuits, it accommodates all lumped or distributed circuit elements that can be represented by a complex 2 × 2 matrix, including RLC circuits, microstrip lines, and S-parameters.

The program consists of a network analysis section and a minimum-seeking algorithm, plus the necessary connective software. The minimum-seeking algorithm follows the literature [6]. In order to try to guarantee finding a global minimum and not a local minimum, a limited grid search is used.

To apply the optimization capability, the user indicates which of the circuit elements are to be varied and over what permissible limits. He also inputs weights and limits on up to four circuit characteristics to be optimized. The program then finds the element values at which the circuit performance comes closest to the desired characteristics.

Acknowledgment

The precise machine work done under the supervision of A. Thivierge, as well as the module fabrication, processing, and assembly work developed for our requirements by R. E. Cardinal and J. Bignet, are gratefully acknowledged. I. Edward designed the power conditioner and J. Prevost did an excellent job tuning and testing the amplifiers.

References

[1] H. Kohzu, I. Nagasako, M. Ogawa, and N. Kawamura, "Reliability studies of one-micron Schottky-gate GaAs FET," in *1975 Int. Electron Devices Meeting, Dig. Tech. Papers*, pp. 247–250.

[2] D. Abbot and J. Turner, "Some aspects of GaAs FET reliability," in *1975 Int. Electron Devices Meeting, Dig. Tech. Papers*, pp. 253–246.

[3] Nippon Electric Company, "Test Report on NE 24406 for RCA Ltd.," Sept. 1976.

[4] C. Liechti and R. Tillman, "Design and performance of microwave amplifiers with GaAs Schottky-gate field-effect transistors," *IEEE Trans. Microwave Theory and Tech.*, vol. MTT-22, pp. 510–517, May 1974.

[5] A. Behle and R. Zuleeg, "Fast neutron, tolerance of GaAs JFET's operating in the hot electron range," *IEEE Trans. Electron Devices*, vol. ED-10, pp. 993–995, Aug. 1972.

[6] V. G. Gelnovatch and I. L. Chase, "DEMON—An optimal seeking computer program for the design of microwave circuits," *IEEE J. Solid-State Circuits*, vol. SC-5, pp. 303–309, Dec. 1970.

A Low-Noise Gallium Arsenide Field Effect Transistor Amplifier for 4 GHz Radio

By R. H. KNERR and C. B. SWAN

(Manuscript received June 14, 1977)

A low-noise amplifier for 4 GHz radio has been designed and is in manufacture. The noise figure is ≤2 dB and the gain is typically 10 dB. Input and output return losses are ≥25 dB. The insertion loss with failure of either the power supply or the low-noise transistor is typically 5 to 8 dB. The amplifier uses a single gallium arsenide field effect transistor in conjunction with a passive failsafe by-pass network utilizing circulators. This approach permits the noise figure and the gain flatness to be optimized for each amplifier without compromising the input and output matches. It is concluded that this single-transistor amplifier design has significant advantages both in performance and in simplicity over the balanced amplifier design.

I. INTRODUCTION

Gallium arsenide Field Effect Transistors (GaAs FETs) are effecting a revolution in both the design philosophy and the performance capability of new microwave systems. In addition, these devices can often provide an economical means for significantly upgrading the performance of existing systems. Such is the case with the 4 GHz radio system, where an RF preamplifier with a maximum noise figure of 2 dB is achieved with GaAs FETs. In this application, each common multichannel amplifier permits the output power of typically five transmitters to be dropped 4 dB, from 5 watts to 2 watts, while still maintaining the system thermal noise objective for 1500 channels. This significantly increases the life of the transmitter amplifier triodes, thus improving the overall system reliability.

II. GENERAL DESIGN CONSIDERATIONS

The use of the GaAs FET amplifier as an RF preamplifier for FM systems requires low intermodulation as well as a low noise figure. In addition, since the amplifier is common to several channels (including the protection channel), reliability is of utmost importance. The two most serious failure mechanisms envisioned are: (i) transistor failure and (ii) power supply failure. With either type of failure, the GaAs FET amplifier inherently exhibits an unacceptable transmission loss (>20 dB) for radio applications. Use of a balanced amplifier with two transistors coupled with input and output 3 dB hybrid couplers would reduce the gain by only 6 dB for failure of a single transistor. But this redundancy and extra cost gives no relief for loss of the dc supply voltage for the transistors. Schemes, without active devices, for reducing the loss to <10 dB for either type of failure and which apply to the balanced as well as the single-ended amplifier are shown schematically in Fig. 1 and 2. The signal reflected from the unpowered FETs is fed to the output by interconnecting the normally terminated arms of the coupler (Fig. 2) or isolator (Fig. 1). We have designed, constructed, and evaluated both balanced and single-ended amplifiers.

The requirements for this application are shown in Table I. The choice of the design approach to meet these requirements was based on a de-

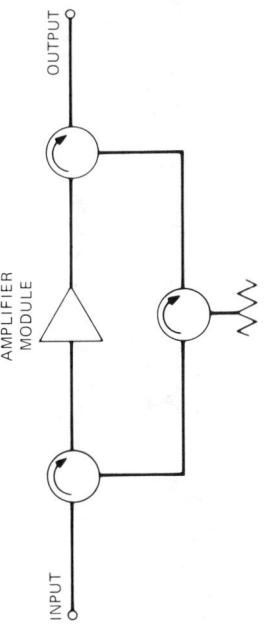

Fig. 1—Single-ended amplifier with provision for unpowered transmission.

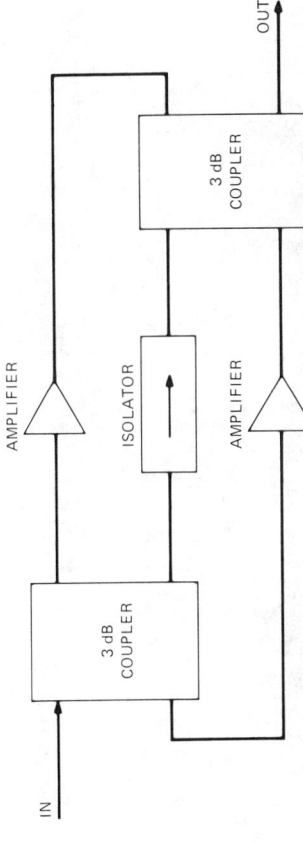

Fig. 2—Balanced amplifier with provision for unpowered transmission.

Reprinted with permission from *Bell Syst. Tech. J.*, vol. 57, pp. 479–490, Mar. 1978. Copyright © 1978 American Telephone and Telegraph Company.

Table I — Electrical requirements for 3.7 GHz to 4.2 GHz amplifier

	Max.	Min.	Units
Input return loss	—	25	dB
Output return loss	—	25	dB
Noise figure	2.0	—	dB
Gain	11.0	8.0	dB
Gain flatness	±0.5	—	dB
Intermodulation (2A-B intercept)	—	23	dBm
Unpowered insertion loss	10	—	dB

Table II — Single-ended versus balanced amplifier

	Single	Balanced
Gain	Same (8–11 dB)	Same
Unpowered loss	≥25 dB	≥20 dB
Output return loss	≥25 dB	≥17 dB
Input return loss	6–8 dB loss	2–5 dB gain
Transistor failure		3 dB advantage
Intermodulation (2A-B intercept)	0.3 dB advantage	
Noise figure	1	2
Transistors required	0	2
Couplers required	3	1
Circulators required		

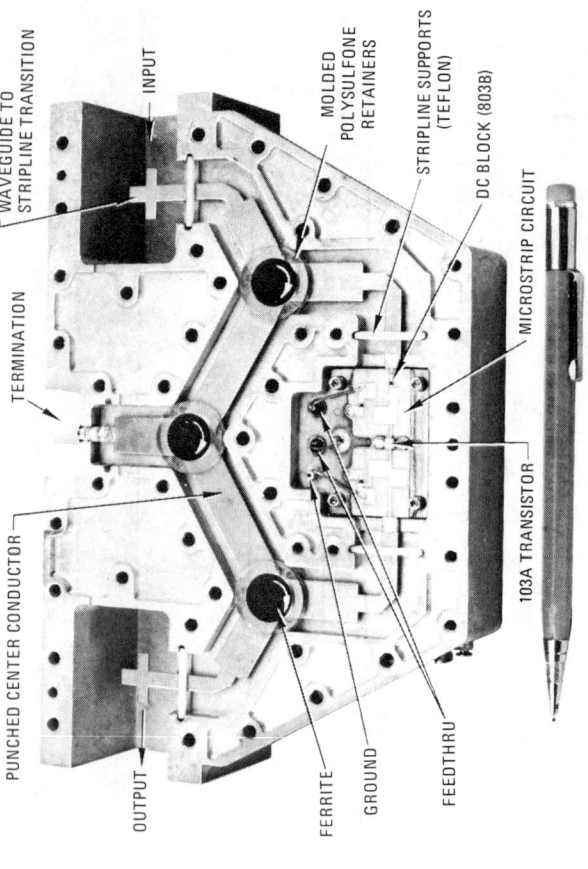

Fig. 3—4 GHz MIC amplifier (open).

tailed comparison of the capabilities of the two amplifiers. Based on our laboratory experience, Table II compares the performances that we consider practical in manufacture. We realized that meeting the intermodulation and failsafe requirements with a single transistor would allow significant cost savings. The single-ended GaAs FET amplifier reported here not only meets these requirements but also has match and noise figure advantages. This results from the low loss input circulator which allows us to independently optimize the input circuit match and the transistor source impedance for minimum noise.

III. AMPLIFIER MODULE

The GaAs FET is mounted in a microstrip circuit (Fig. 3). This transmission line permits easy mounting of the transistor and MOS dc-blocking capacitors. The amplifier module per se has no adjustments. Tuning screws near the input and output of the module and in the circulator arms are used to adjust the amplifier for optimum noise figure and gain flatness. This feature compensates for variations in transistor parameters as well as for manufacturing tolerances of the piece parts.

3.1 The GaAs FET output circuit design

In a first order approximation the output circuit elements were determined using BAMP.* Supplying the S-parameters and the input re-

---
* Basic Analysis and Mapping Program.

flection coefficient (Γ_{MNF}) that results in a minimum noise figure, circles of constant gain are drawn (Fig. 4). If the output circuit reflection coefficient equals Γ_{ML}, optimum gain is obtained. Any deviation from Γ_{ML} results in a loss of gain corresponding to the values indicated on the circles of Fig. 4. Strictly speaking, the set of circles is only valid for one frequency (in our specific case 4 GHz), and a corresponding set would have to be drawn for each frequency under consideration. Since the S-parameter variation over the 12.5 percent frequency band of interest is smooth and relatively small, one set of circles suffices to demonstrate that the output impedance, shown in a dashed line, is reasonably close to match. The actual circuit which produced the impedance was trimmed empirically for bandwidth and flatness of gain.

3.2 The input circuit

The theory of noisy four poles has been treated extensively in the literature.[1-5] It essentially says that the noise figure of the four pole depends solely on the impedance of the input circuit. The noisy four pole is completely characterized by the S-, Y-, or Z-parameters, the source reflection coefficient (Γ_{MIN}) at which the noise figure is minimum (NF_{MIN}), and the equivalent noise resistance (R_n). The measurement of R_n is somewhat cumbersome and is described in Ref. 1. Once the parameters are known, circles of constant noise figure[4,5] can be drawn (Fig.

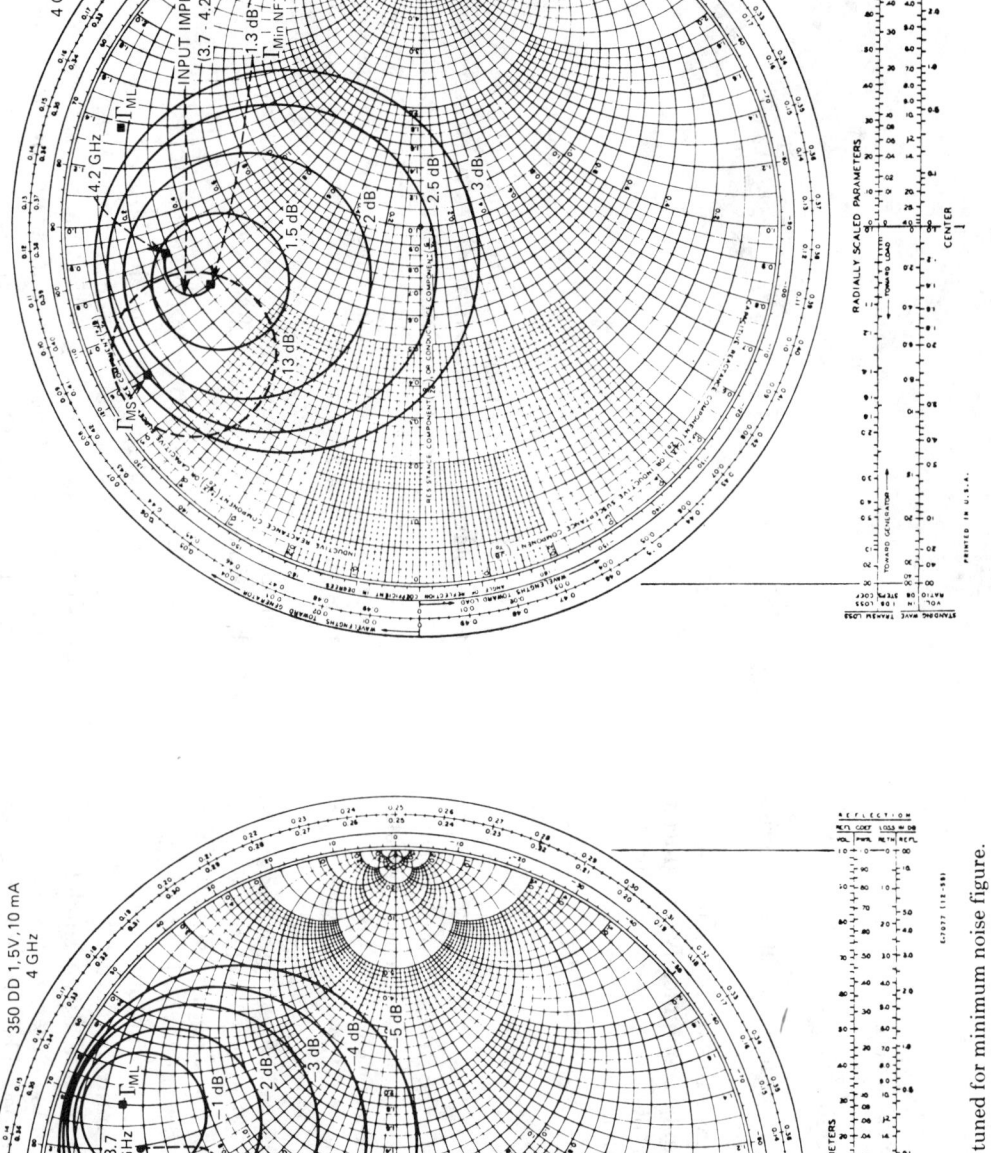

Fig. 4—Circles of constant gain with input tuned for minimum noise figure.

Fig. 5—Circles of constant noise figure.

5, solid circles). This set of circles is very insensitive to frequency and independent of the load impedance. The spread of the circles increases with increasing R_n. In our specific case $R_n = 14\ \Omega$. Γ_{ML}, the maximum gain load impedance, has been explained in the output circuit design. The reflection coefficient, Γ_{MS}, in Fig. 5 represents the reflection coefficient of the source that would yield maximum gain, which in our case is about 15.5 dB. It is quite obvious that the points for optimum noise figure and optimum gain are significantly apart. A set of circles similar to the ones in Fig. 4 can be constructed around Γ_{MS}, assuming that the load reflection coefficient is Γ_{ML}. To keep Fig. 5 from becoming overcrowded, only one circle is shown. It is seen that a gain of about 13 dB for optimum noise figure versus 15.5 dB for optimum match can be obtained. This figure, of course, is further reduced by broadbanding and gain flattening, as can be deduced from the source impedance trace in Fig. 5. The performance of the single-ended amplifier module is shown in Fig. 6. The gain of 11.6 dB and corresponding noise figure of 1.5 dB are in good agreement with the values that can be extrapolated from Figs. 4 and 5.

3.3 Final circuit

For the amplifier to be manufacturable, some adjustability is required to compensate for variations in transistor parameters as well as mechanical tolerances on all components. This adjustability is not readily provided in the microstrip circuit, but can be economically introduced in the air dielectric stripline circuit. Pairs of tuning screws are thus located in the air line just in front and just after the microstrip module (Fig. 7). These permit tuning of the amplifier for optimum noise figure and gain flatness.

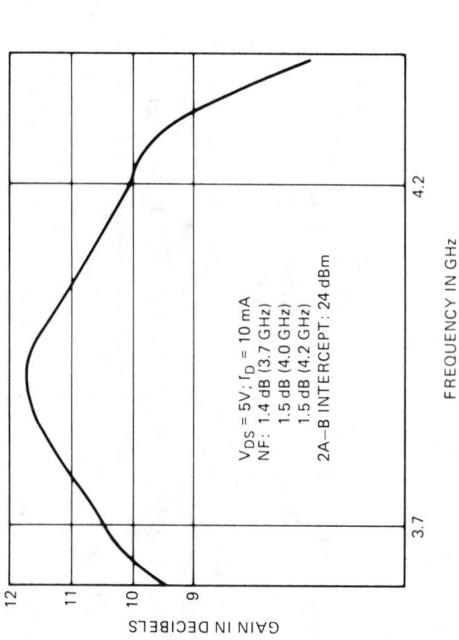

Fig. 6—Performance of amplifier module.

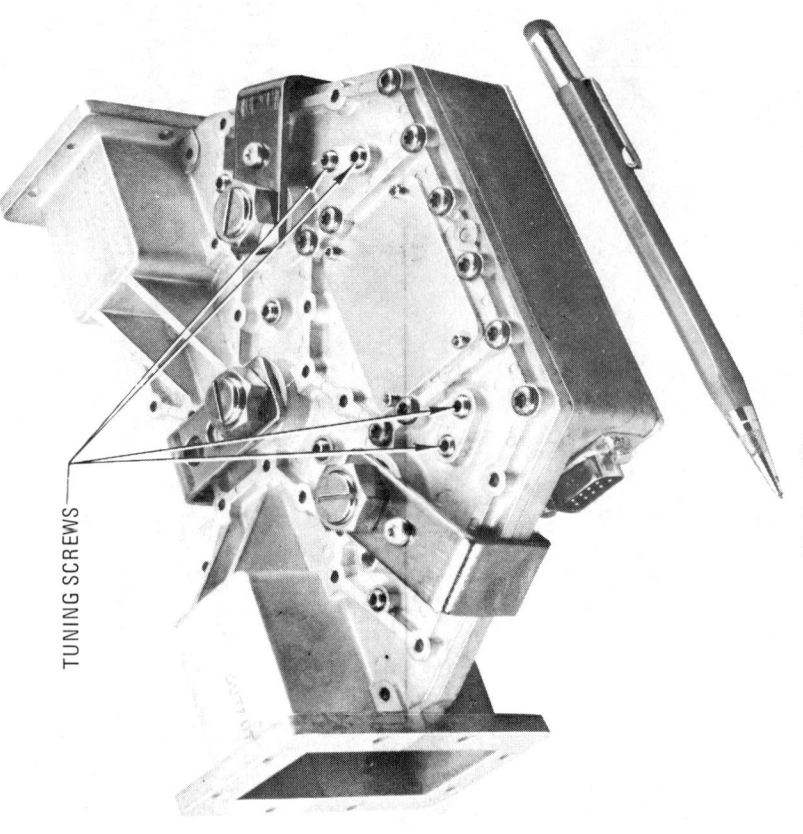

Fig. 7—4 GHz MIC amplifier.

IV. FAILSAFE BYPASS CIRCUIT

When the GaAs FET is unpowered, both the gate and drain circuits appear approximately as open circuits. The transmission loss typically exceeds 20 dB. If the transistor fails, we expect a short circuit. In either case, the input and output return losses at 4 GHz are typically 2 to 4 decibels.

The provision of three circulators, as shown in Fig. 1, provides an effective passive by-pass circuit. In the normal state, the relatively small reflected input signal is recombined with the amplifier signal at the output of the transistor. This appears as a small ripple on the gain characteristic which can be compensated by output tuning. In the unpowered or failed state, both the gate and drain circuits are "switched" to open or short circuits. The input signal, with relatively small loss, is then directed to the drain circuit of the GaAs FET where it is reflected to the output circulator and directed to the load. The total insertion loss is typically 5 to 8 dB.

The circulators for the bypass circuit and the waveguide-to-stripline transition[6] were developed in air dielectric stripline (Fig. 8). This simple technology assures minimum circuit losses, low cost parts and assembly, and very high yields. The intermediate circulator is terminated with 50 ohms to provide >25 dB isolation. Since this isolation is only maintained over the 3.7–4.2 GHz band, positive feedback can cause the amplifier to oscillate at lower frequencies. The "low-pass filter" on the output substrate (Fig. 9) eliminated this oscillation which, for our particular by-pass loop, occurred at about 800 MHz.

V. POWER REGULATOR AND ALARM CIRCUIT

The dc operating point for the GaAs FET is a compromise between minimum noise and acceptable linearity. A regulator automatically sets the gate voltage so that $I_D = 15$ mA and $V_{DS} = 4.8$ volts. All GaAs FETs are thus powered identically and require no bias adjustment in manufacture. The amplifier (Fig. 10) operates from a -24 volt supply at 60 milliamperes.

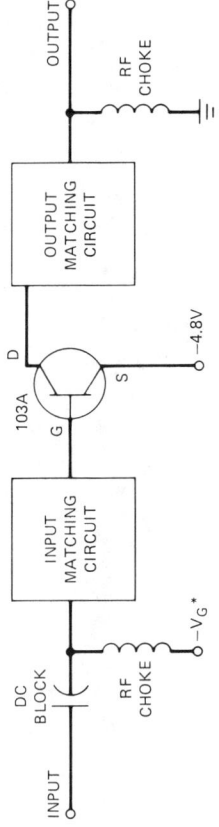

*V_G AUTOMATICALLY ADJUSTED FOR I_{DS} = 15 mA

Fig. 10—Amplifier module (schematic).

VII. PHYSICAL DESIGN

The completed amplifier is shown in Figs. 3 and 7. The aluminum housing is die cast in two parts. The stripline center conductor is stamped in a single piece from sheet brass. Interlocking molded plastic locating rings are used to locate both the circulator ferrites and the center conductor in the lower housing channel. The printed circuit board with power regulator and alarm circuits (Fig. 11) is mounted on the bottom side of the lower housing.

VIII. AMPLIFIER PERFORMANCE AND TESTS

8.1 Tests

In order to meet the requirements in Table I, the amplifier was subjected to several tests, most of which used straightforward test procedures. Special test sets were constructed for noise figure and inter-

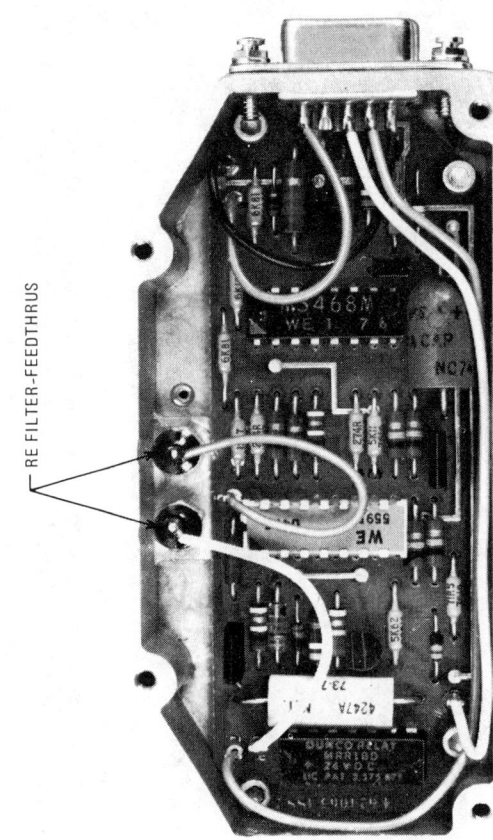

Fig. 11—Power regulator and alarm circuit of 4 GHz MIC amplifier.

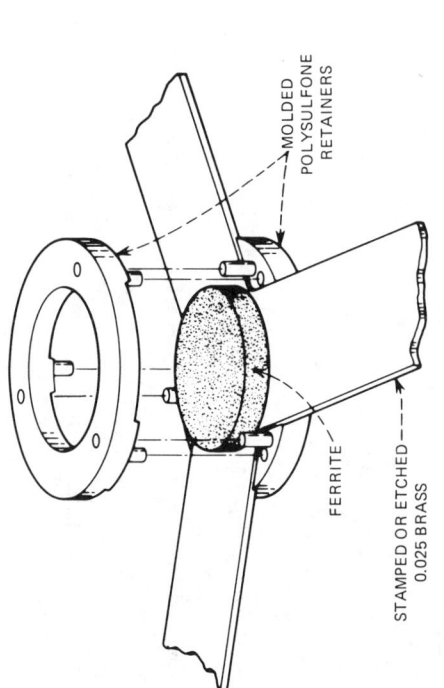

Fig. 8—4 GHz stripline circulator.

In case of transistor or power supply failure (I_D < 5 mA or I_D > 25 mA), a contact to ground is provided which energizes a remote alarm.

VI. THE LOW-NOISE TRANSISTOR

The GaAs FET was developed at the Murray Hill, New Jersey Laboratory.[7] The gate length and width are 0.8 μm and 2 × 250 μm. The typical noise figure is about 1.2 to 1.4 dB at 4 GHz.

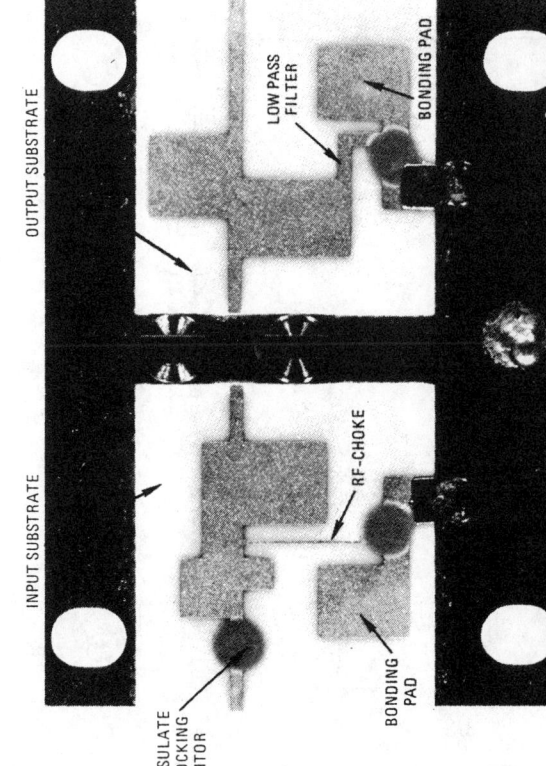

Fig. 9—Amplifier module without transistor.

IX. CONCLUSION

We have demonstrated a simple 4 GHz microwave amplifier design which achieves a noise figure of 2 dB in manufacture. This has been achieved with a single low noise GaAs field effect transistor in conjunction with a passive failsafe by-pass circuit. It is concluded that the single-ended amplifier with input and output isolator has significant advantages both in performance and in simplicity over the balanced amplifier design for this application. The housing and major piece parts are die replicated so as to fit together with minimal assembly effort. Tuning screws are provided to accommodate variations in transistor characteristics and to allow relaxed piece part tolerances.

X. ACKNOWLEDGMENTS

We gratefully acknowledge the team efforts of Bell Laboratories and Western Electric engineers in the development of this amplifier. We mention especially the efforts of J. J. Kostelnick, G. M. Keltz, G. M. Palmer, and J. L. Brown. The GaAs FET was developed by J. V. DiLorenzo and W. O. Schlosser. L. F. Moose and L. J. Varnerin, Jr, provided essential coordination and technical direction for the program.

REFERENCES

1. H. Rothe and W. Dahlke, "Theorie rauschender Vierpole," Archiv der Elektrischen Übertragung, 9 (March, 1955), pp. 117-121.
2. H. Rothe, "Die Theorie rauschender Vierpole und ihre Anwendung", Nachrichten-technische Fortschritte, Issue 2 (1955), pp. 24-26.
3. A. G. Th. Becking, H. Groendijk, and K. S. Knol, "The Noise Factor of Four-Terminal Networks," Philips Res. Rep. 10, 1955, pp. 349-357.
4. H. Rothe and W. Dahlke, "Theory of Noisy Four Poles," Proc. IRE, 44, June 1956, pp. 811-818.
5. H. Fukui, "Available Power Gain, Noise Figure, and Noise Measure of Two-Ports and Their Graphical Representations," IEEE Trans. Circ. Theory, CT-13, No. 2 (June 1966), pp. 137-142.
6. R. H. Knerr, "A New Type of Waveguide-to-Stripline Transition," IEEE Trans. Microw. Theory Tech. MTT-16, No. 3 (March 1968), pp. 192-194.
7. B. S. Hewitt, H. M. Cox, H. Fukui, J. V. DiLorenzo, W. O. Schlosser, and D. E. Iglesias, "Low Noise GaAs MESFET's—Fabrication and Performance," 6th International Symposium on GaAs and Related Compounds, Edinburgh, September 19-22, 1976.

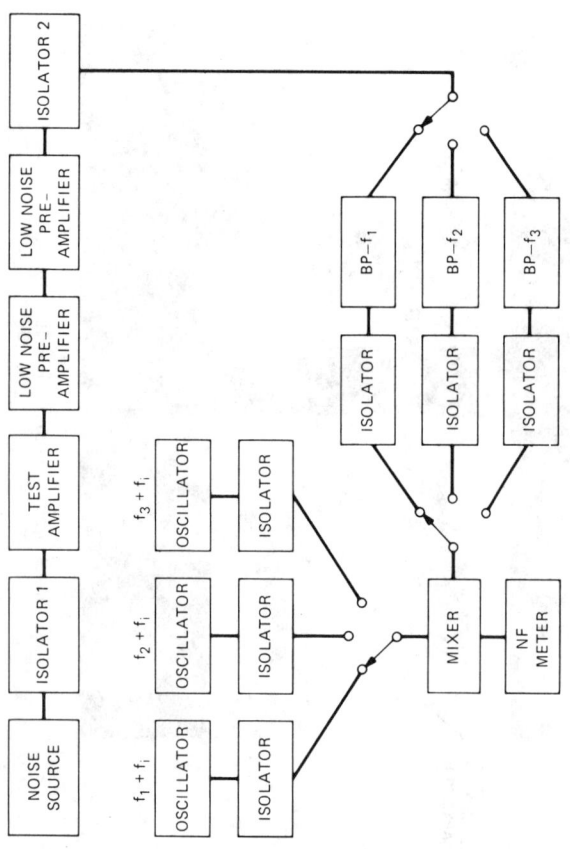

Fig. 12—Complete three-frequency NF test set.

modulation measurements. The noise figure can be accurately and rapidly measured at three frequencies in the test set shown in Fig. 12. Intermodulation (IM) tests were done using a three-tone measurement set. Since not all stations are air-conditioned, a humidity test was started. The amplifier was placed in an 85°C (185°F)—85% humidity environment for two months with DC bias applied. No change in performance was detected. Six field trial models were cycled over the temperature range of 4°C (40°F) to 60°C (140°F) with no significant change in performance.

8.2 Performance

The amplifier is in manufacture at Western Electric Company and meets the requirements summarized in Table I. Typical performance values obtained are:

NF: 1.6–1.8 dB
Input and output return loss: 28 dB
2A-B intercept: 26 dBm
Unpowered transmission loss: 5–8 dB
Gain: 10 dB.

We find that the amplifier tuning arrangement permits the present spread in transistor parameters to be accommodated easily.

The Matched Feedback Amplifier: Ultrawide-Band Microwave Amplification with GaAs MESFET's

KARL B. NICLAS, MEMBER, IEEE, WALTER T. WILSER, MEMBER, IEEE, RICHARD B. GOLD, MEMBER, IEEE, AND WILLIAM R. HITCHENS

Abstract—An ultrawide-band amplifier module has been developed that covers the frequency range from 350 MHz to 14 GHz. A minimum gain of 4 dB was obtained across this 40:1 bandwidth at an output power of 13 dBm. The amplifier makes use of negative and positive feedback and incorporates a GaAs MESFET that was developed with special emphasis on low parasitics. The transistor has the gate dimensions 800 by 1 μm. The technology and RF performance of the GaAs MESFET are discussed, as are the design considerations and performance of the single-ended feedback amplifier module.

I. INTRODUCTION

TWO-THIRDS of a century have gone by since, in 1913, Alexander Meissner was granted a patent on a feedback oscillator circuit by the German patent office. In the same year Edwin H. Armstrong presented a paper on regenerative circuits and in 1914 Lee DeForest, without whose earlier invention of the triode the feedback circuit would have been meaningless, filed a patent application on the regenerative circuit.

Ever since the invention of the original regenerative circuit, a great number of new feedback circuits with a multitude of applications have emerged. One particularly significant application is the use of negative feedback to control the gain and the input and output impedance of an amplifier [1]–[3]. It is, therefore, not surprising that the principle of negative feedback with its wide bandwidth potential, low input and output reflection coefficients, and good gain flatness has made its entry into the field of microwave amplifiers and, quite recently, into that of GaAs MESFET amplifiers [4].

This paper describes the use of both negative and positive feedback to extend the bandwidth of microwave amplifiers far beyond that reported to date. In order to accomplish this goal, a GaAs MESFET was developed with special emphasis on reduction of parasitics. Experimental amplifier modules exhibit a bandwidth of more than five octaves covering a frequency band from 350 MHz to 14 GHz. Minimum gain over this band is 4 dB at 13 dBm of output power. The amplifier makes use of "frequency controlled" feedback and simple matching techniques. The theory behind the basic feedback ampli-

Manuscript received September 4, 1979; revised November 19, 1979.
The authors are with Watkins-Johnson Company, Palo Alto, CA 94304.

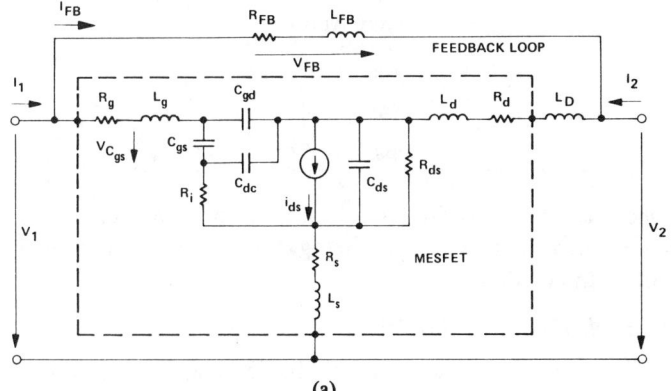

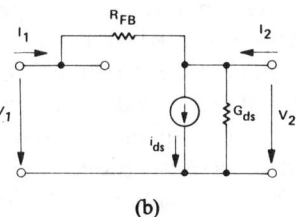

Fig. 1. Circuit diagram of the basic feedback amplifier. (a) High-frequency model. (b) Low-frequency model.

fier circuit is discussed in detail, as are the fabrication and performance of the amplifier modules. The technology, performance, and model of the GaAs MESFET are also described. Finally, gain, reflection coefficients, and reverse isolation of the computer model are compared to the measured data of two amplifier modules.

II. BASIC FEEDBACK AMPLIFIER CIRCUIT

The parasitic elements of a GaAs MESFET, as shown in the schematic of Fig. 1 (a), restrict the amplifier bandwidth capability. Minimization of these parasitics was a major goal of the device development, but there are obvious practical limitations. For this reason, we looked for supporting techniques to extend the bandwidth of the "conventional" negative feedback amplifier to higher frequencies. A practical answer was found in the introduction of the drain inductance L_D and the feedback induc-

TABLE I
Elements of the Transistor Model

INTRINSIC ELEMENTS		EXTRINSIC ELEMENTS	
g_m =	54 m mhos	R_g =	1.5 ohm
τ_0 =	3.5 psec	L_g =	.19 nH
C_{gs} =	.67 pF	R_s =	.9 ohm
C_{gd} =	.017 pF	L_s =	.151 nH
C_{dc} =	.032 pF	C_{ds} =	.082 pF
R_i =	4.4 ohm	R_d =	2 ohm
R_{ds} =	200 ohm	L_d =	.143 nH

DC BIAS CONDITIONS

V_{DS} = 4V
V_{GS} = -1V
I_{DS} = 100 mA

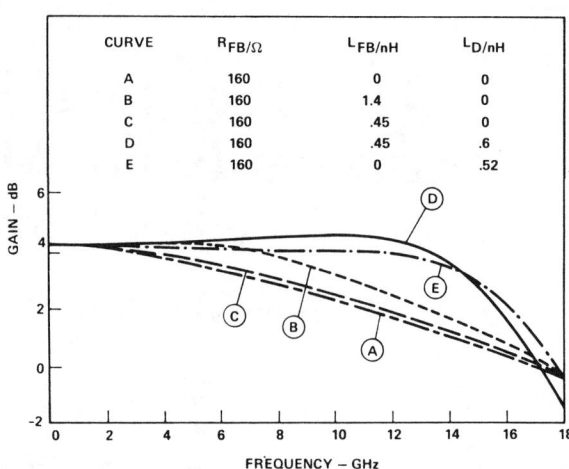

Fig. 2. Computed small-signal gain for various combinations of R_{FB}, L_{FB}, and L_D.

tance L_{FB}. These inductances and the parasitic elements of the GaAs MESFET are being employed to frequency control the feedback.

A. High-Frequency Model

The circuit diagram of the basic feedback amplifier making use of frequency controlled feedback is shown in Fig. 1(a). The amplifier's active device, the GaAs MESFET, is presented in form of its equivalent circuit, whose element values are listed in Table I [5]. The values of the reactive elements are small enough that at frequencies below 1.5 GHz all reactive elements can be neglected for determination of such quantities as gain, input and output VSWR and reverse isolation, i.e., up to 1.5 GHz the transistor can be represented by its dc model. However, the element values in Table I are such that aside from neglecting R_g, R_d, R_s, and maybe R_i, any further simplification of the high-frequency model of Fig. 1(a) leads to erroneous results at frequencies above 1.5 GHz. But even with $R_g = R_d = R_s = R_i = 0$, the admittance matrix of the basic feedback amplifier is so complicated that the computer provides the only efficient tool for obtaining solutions.

The basic feedback amplifier (Fig. 1 (a)) contains two series inductors, L_D in the drain line and L_{FB} in the feedback loop, that have been inserted to extend the amplifier's bandwidth. L_D was chosen to compensate for the capacitive portion of the output impedance at the upper band edge. It improves the output VSWR and eliminates the reactive component of the output impedance at this frequency. L_{FB} reduces the effectiveness of the negative feedback with increasing frequency.

The purpose of these two series inductors are best demonstrated by comparing their influence on insertion gain and bandwidth of the amplifier of Fig. 1. This comparison is presented in Fig. 2 for five selected combinations of R_{FB}, L_{FB}, and L_D. The curves clearly demonstrate the influence of L_D and L_{FB} on the bandwidth and gain response of the basic feedback amplifier. The inductor L_D is mainly responsible for the extended band coverage while the feedback loop provides the flat gain response.

For better understanding of the feedback amplifier's behavior, the vector diagrams of several important voltages as they appear across certain terminals of the amplifier have been drawn at selected frequencies. They are shown in Fig. 3(a) for the case $R_{FB} = 160$ Ω, $L_{FB} = 0.45$ nH, $L_D = 0.6$ nH, and in Fig. 3(b) for the case of $R_{FB} = 160$ Ω, $L_{FB} = L_D = 0$. All voltages are normalized to $V_S/2$, which is that portion of the source voltage V_S that appears across a 50-Ω load when the signal source is terminated with such a load. Comparing the vector diagrams of Fig. 3 (a) one notices that at 13.75 GHz V_1 and V_2 are in phase, while at very low frequencies they are 180° out of phase. The feedback current I_{FB} has advanced 180° with respect to the signal source current I_S, as is shown in Fig. 4. The signal source current is the current that flows from a 50-Ω source into a 50-Ω load. At 13.75 GHz the ratio V_2/V_1 reaches its maximum and this frequency marks the point of optimum positive feedback.

In order to obtain a more detailed comparison between the behavior of the "conventional negative feedback amplifier" and the amplifier that makes use of controlled feedback, we have plotted the input voltage V_1 and the output voltage V_2 in Fig. 5 as a function of frequency. It can be seen that the influence of L_D and L_{FB} on the magnitude of the input voltage is not very pronounced. Noticeable phase differences exist above 9 GHz, however, with a crossover point at 13.75 GHz. The output voltage of both the conventional negative feedback amplifier and that of the frequency controlled feedback amplifier are almost identical up to 4 GHz. The reactive elements L_D and L_{FB} maintain a nearly constant gain response to almost 14 GHz (curve C) however, which is the reason for the use of reactive control. Also shown in Fig. 5 (curve A) are the input (V_1) and output (V_2) voltages of the amplifier without feedback ($R_{FB} = \infty$). This curve in particular demonstrates the bandwidth potential of the transistor

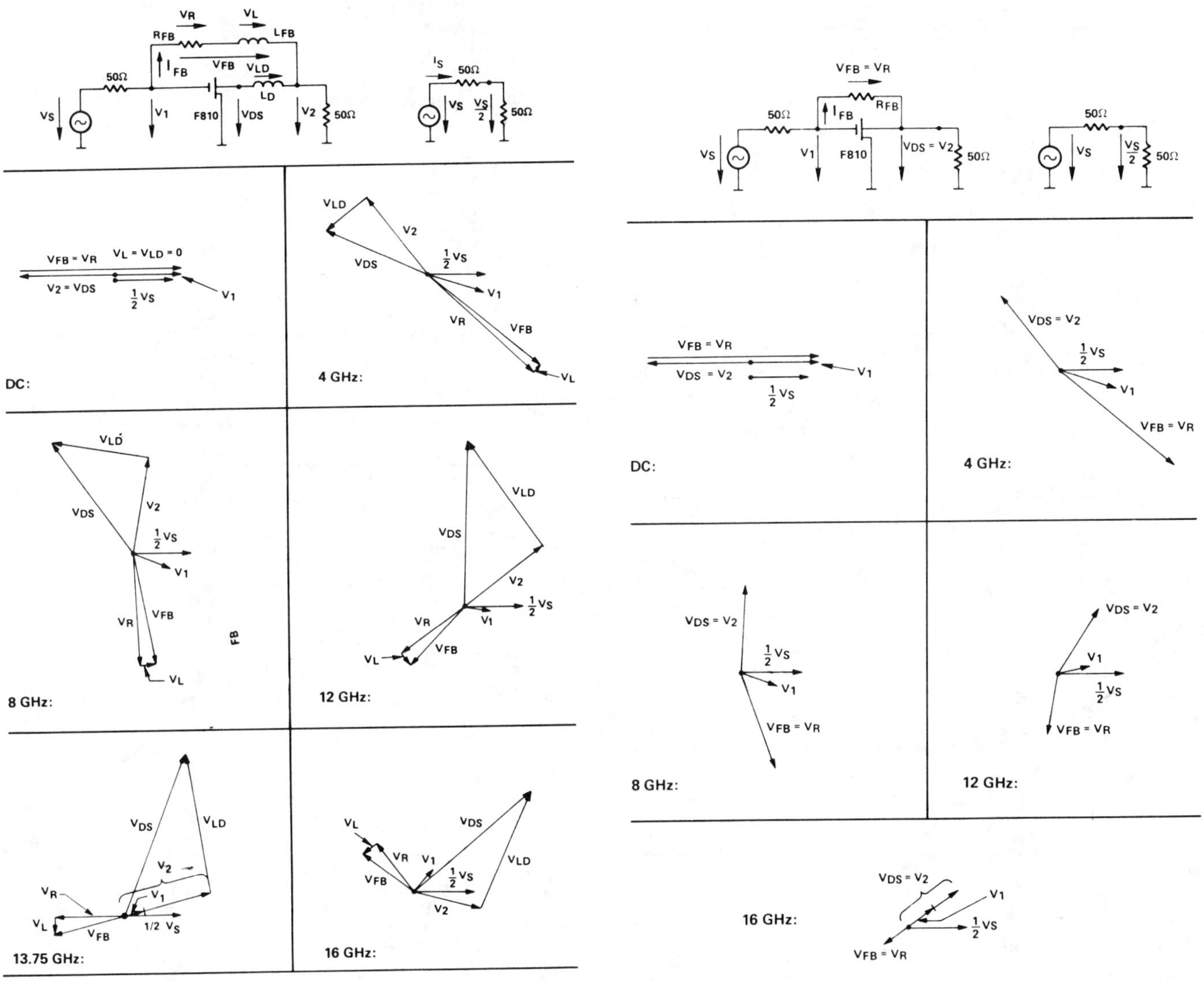

Fig. 3. Voltage vector diagrams of (a) the amplifier with frequency-controlled feedback, and (b) the conventional negative feedback amplifier.

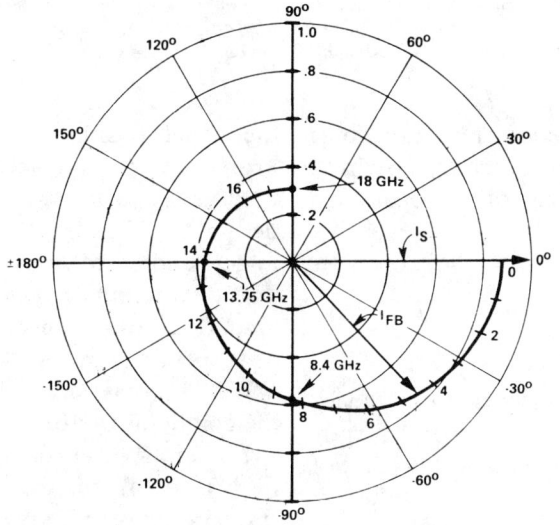

Fig. 4. Magnitude and phase of the normalized feedback current.

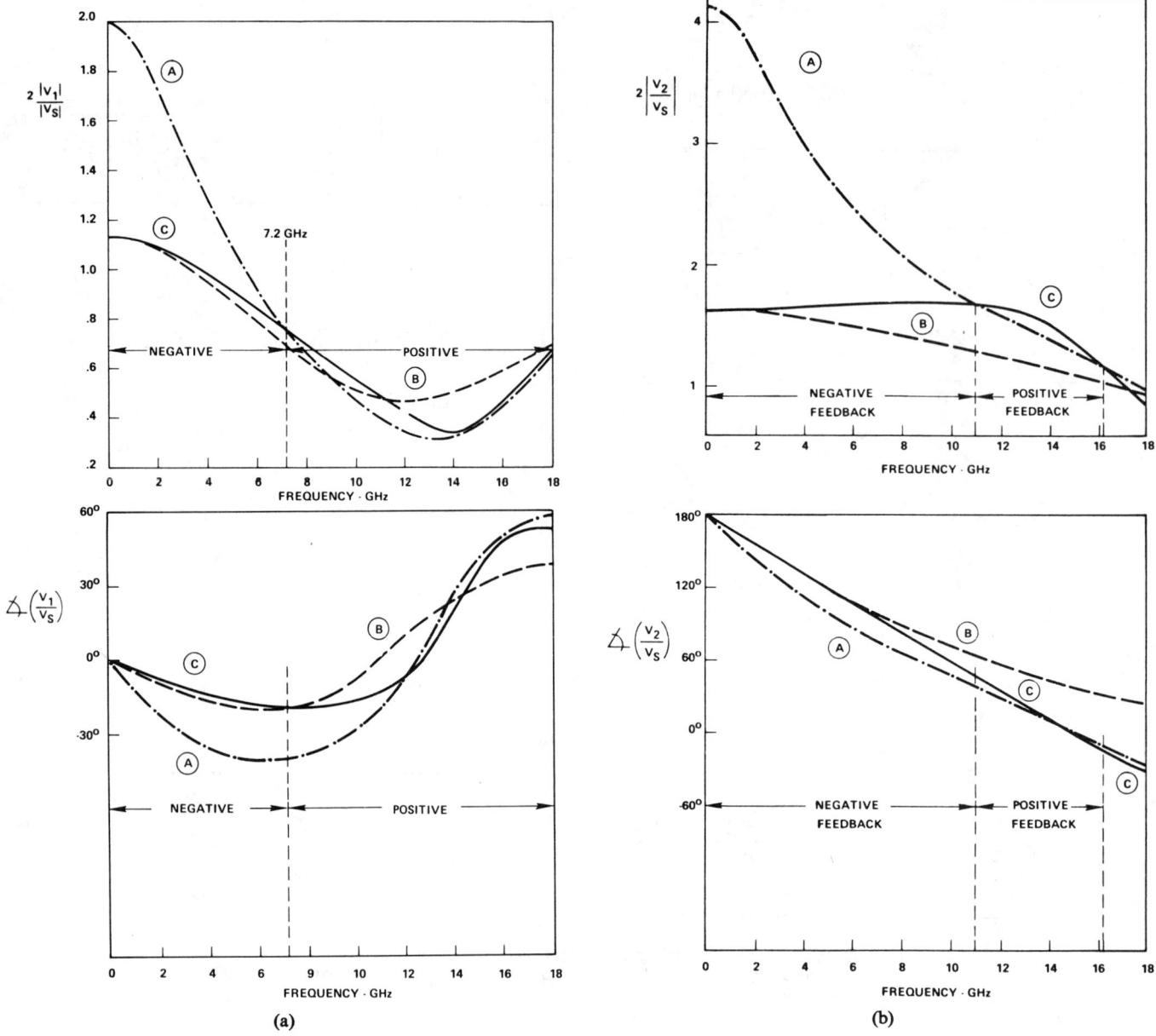

Fig. 5. Normalized input (a) and output (b) voltages. Curve A: the transistor only, $R_{FB} = \infty$; curve B: the conventional negative feedback amplifier, $R_{FB} = 160$ Ω, $L_{FB} = L_D = 0$; curve C: the amplifier with frequency controlled feedback, $R_{FB} = 160$ Ω, $L_{FB} = 0.45$ nH, $L_D = 0.6$ nH.

used in our experiments. It also shows the frequency range of positive feedback, i.e., the band in which the feedback elevates the insertion gain above that of the open-loop amplifier (11–16.2 GHz). The feedback loop causes such a "positive range" for the input voltage for all frequencies above 7.2 GHz.

A. Low-Frequency Model

As discussed earlier, the reactive elements of the transistor model under study are relatively low and consequently the circuit diagram of Fig. 1 (a) can be reduced to that of Fig. 1 (b) for frequencies below 1.5 GHz. The usefulness of the low-frequency model is obviously restricted to the very low end of the frequency band. However, the use of the model is justified on the basis that it yields two simple expressions for the feedback resistor R_{FB} ((A.10), (3a)) that we found to be extremely valuable for our amplifier design (see Section IV-A). The model also provides an understanding of the tradeoffs between match and gain. The scattering parameter matrix of the feedback amplifier's low-frequency model (A.4)–(A.9) is derived in its general form in the Appendix. Input and output VSWR become identical if the feedback resistor satisfies the condition (A.10)

For this special case the S parameters are presented in (A.11)–(A.14) of the Appendix. To demonstrate the tradeoff between VSWR and gain, we assume

$$G_{ds}Z_0 \ll 1; \quad G_{ds} \ll g_m. \tag{1}$$

Using the general set of equations (A.5)–(A.9) of the Appendix, we find the S parameters

$$S_{11} = S_{22} = \frac{1}{\Sigma} \left[\frac{R_{FB}}{Z_0} - g_m Z_0 \right] \quad (2a)$$

$$S_{12} = \frac{2}{\Sigma} \quad (2b)$$

$$S_{21} = \frac{-2}{\Sigma} \left[g_m R_{FB} - 1 \right] \quad (2c)$$

with

$$\Sigma = 2 + g_m Z_0 + \frac{R_{FB}}{Z_0}. \quad (2d)$$

They are plotted in Fig. 6, which shows the relationship between gain, VSWR, transconductance g_m, and feedback resistance R_{FB}. Only the lower section of the right half of the diagram is of practical interest. In this region the amplifier yields the highest-gain–lowest-VSWR combinations. The curves demonstrate that gain at low frequencies can be significantly increased due to an increase of the reflection coefficients. Ideal matching yields the lowest gain. The curves of Fig. 6 further reveal that by varying the feedback resistor the gain can be changed between total attenuation and maximum gain.

The condition for ideal match ($S_{11} = S_{22} = 0$) requires

$$R_{FB} = g_m Z_0^2. \quad (3a)$$

The associated gain is

$$G = 20 \log(g_m Z_0 - 1). \quad (3b)$$

C. Circuit Objectives and Requirements

The circuit objective was to extend the bandwidth of the negative feedback amplifier from its upper frequency limit of about 6–14 GHz. The choice of the two series inductances L_D and L_{FB} made this possible. These values were determined as follows.

1) L_D was chosen to compensate for the capacitive component of the GaAs MESFET's output impedance so that resonance occurs at the upper band edge. This measure simultaneously results in a marked improvement in the output match.

2) L_{FB} was chosen, in cooperation with L_D, to adjust the S parameters of the feedback amplifier so that optimum positive feedback exists at the upper band edge. This condition coincides with the input (V_1) and output (V_2) voltage being in phase and the feedback current I_{FB} advanced by 180° with respect to its phase at very low frequencies. The degree of feedback is mainly controlled by the feedback resistor R_{FB}.

Once steps 1) and 2) are accomplished and the broadband potential of the feedback amplifier shown in Fig. 1(a) is nearly exhausted, a third important step is added; i.e., a simple input and output matching network to further improve the input and output VSWR. In this case one follows published design techniques. Since we chose to use distributed rather than lumped elements, we were

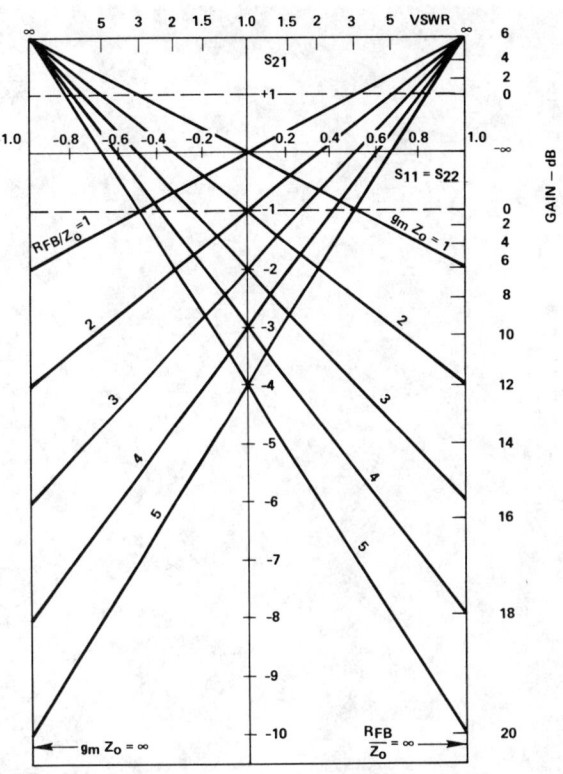

Fig. 6. Gain and VSWR of the low-frequency model for $G_{ds} = 0$ and $R_{FB} = g_m Z_0^2$.

confined to short series transmission lines and short open-ended shunt stubs due to the enormous bandwidth we set out to cover. More details on the matching networks will be found in Section IV.

III. Device Technology and Performance

A. Technology

The GaAs MESFET used in this study, the WJ-F810, is the 800-μm gate width device shown in Fig. 7. The chip size is 320 by 370 μm. The 1-μm long gate is centered in a 4-μm source–drain channel.

The GaAs MESFET's were fabricated using a self-aligned etched aluminum gate process described earlier [6]. The n-type active layer was grown by liquid phase epitaxy on Cr-doped substrates The epitaxial layer was doped with Sn to a concentration of 1.0×10^{17} cm^{-3}.

Special emphasis was paid to the reduction of parasitic elements, particularly capacitances which would limit broad-band performance of the transistor. The gate and drain pads were made as small as practical to minimize the input, output, and feedback capacitances. By using 0.5-mil wire for all bond connections we were able to easily bond to these small pads. The minimal size of the source pad is important for other applications which use the common-gate configuration. In this common source application, however, the source pad size is of little concern because both the source pad and the back of the chip are at RF ground. To further minimize the capacitances, a gate–drain spacing of 1.5 μm was chosen, in contrast to 0.9 μm for an earlier device [6]. This spacing is a com-

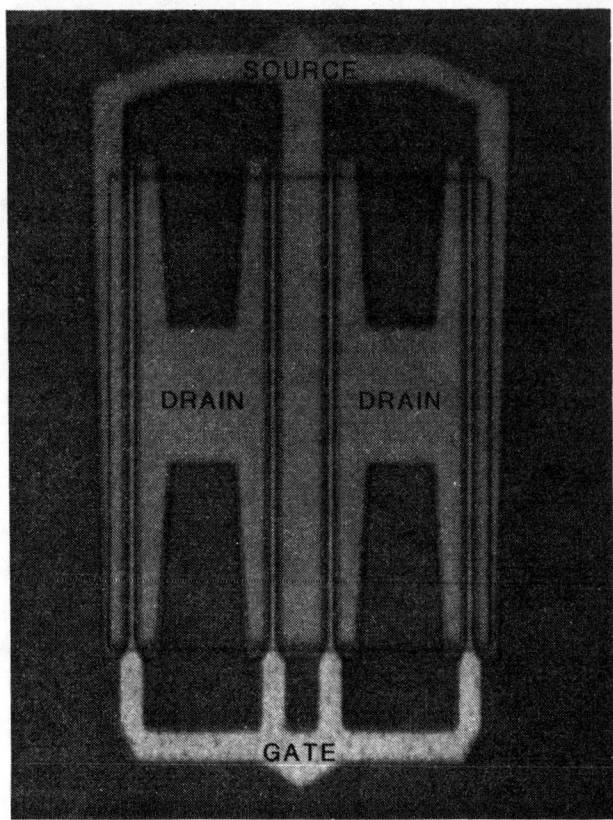

Fig. 7. GaAs MESFET chip (0.32 × 0.36 mm).

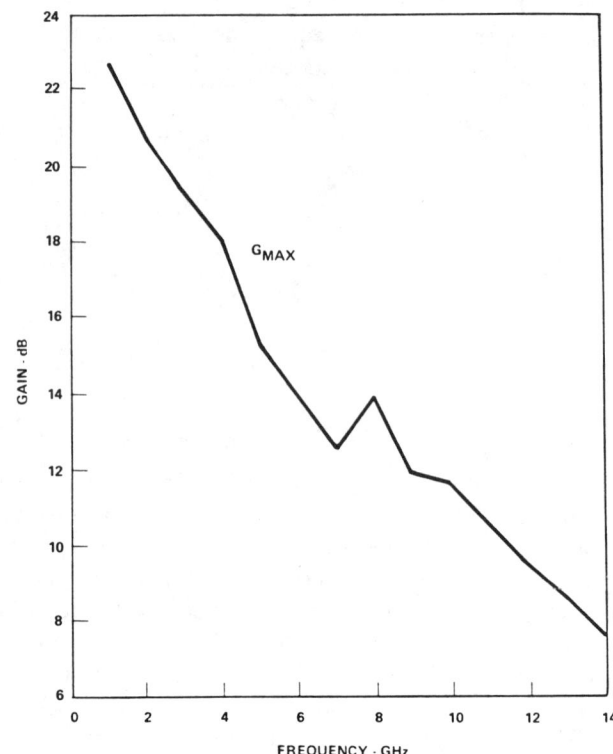

Fig. 8. Measured maximum available gain of the GaAs MESFET.

promise between increased resistance and decreased capacitance.

Most low-noise GaAs FET's have a gate width of about 300 μm. Larger gate widths lead to a linear increase in the transconductance g_m as well as intrinsic capacitances; the cutoff frequency f_t is only slightly affected. The 800-μm gate width was chosen to give substantially higher g_m than available in standard low-noise devices. This high g_m has proven to be an essential element in circuit designs for feedback amplifiers.

B. Transistor Performance and Device Model

The GaAs MESFET has a saturated drain-source current I_{DSS} of 170–240 mA, gate-source pinchoff voltage V_p of 5–7 V, and dc transconductance g_m at 1/2 I_{DSS} of 50–60 mmhos. The range of the drain-source bias voltage was between 4 and 6 V, depending on output power requirements. Fig. 8 presents the maximum available gain of the device between 1 and 14 GHz.

The device model is shown in Fig. 1 (a) and the quantities of the model elements are given in Table I. The agreement between the measured S parameters and those computed using the model was excellent. Table II compares the parasitic elements of the intrinsic transistor model of the device discussed in this paper with two of our GaAs MESFET's which have been described elsewhere ([6], [7]). The element values are normalized to the gate width for reasons of comparison. It can be seen that the topology of the WJ-F810 has lead to a significant

TABLE II
ELEMENTS OF TRANSISTOR MODELS NORMALIZED TO GATEWIDTH

PARAMETER	UNIT	WJ-F110	WJ-F1010	WJ-F810
W_{GATE}	μm	300	1000	800
L_{GATE}	μm	1	1	1
g_m/W	mmhos/mm	73	68	68
C_{gs}/W	pF/mm	1.37	.94	.84
C_{gd}/W	pF/mm	.033	.055	.021
C_{dc}/W	pF/mm	.097	.047	.040
C_{ds}/W	pF/mm	.41	.24	.10

reduction in the magnitudes of the normalized capacitive elements making this device highly suitable for broadband applications.

IV. AMPLIFIER DESIGN AND PERFORMANCE

A. Small-Signal Design

The first step in designing an ultrabroad-band feedback amplifier is the selection of the feedback resistor R_{FB}. According to Table I the transconductance g_m and the drain-source conductance G_{ds} of our device are 0.054 mhos and 0.005 mho, respectively. If it is desired that input and output VSWR be identical, we find from (A10)–(A14) that, at very low frequencies, a gain of 4.2 dB and an input and output VSWR of 1.2:1 will be obtained for $R_{FB} = (g_m + G_{ds})Z_0^2 = 147.5$ Ω. Raising the

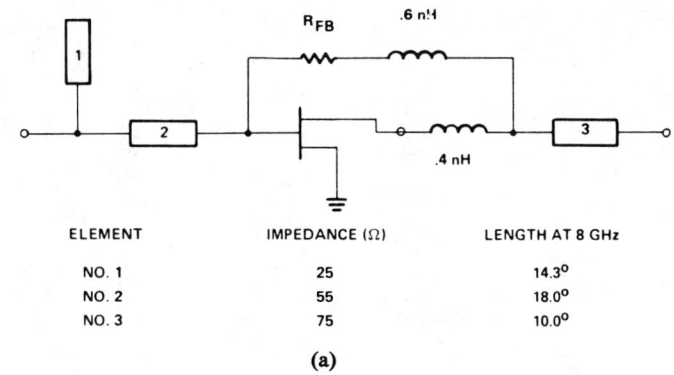

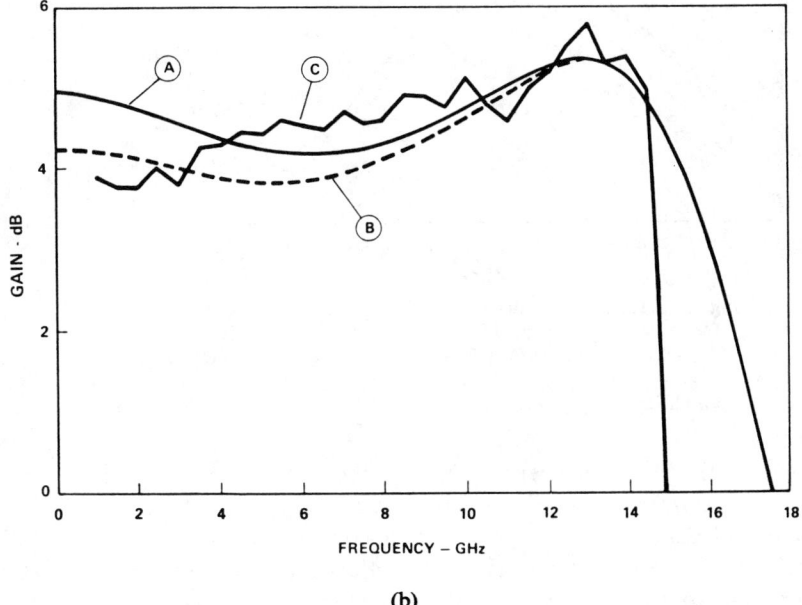

Fig. 9. Schematic (a) and gain curves (b) of the matched feedback amplifier. Curve A: computed for $R_{FB} = 180\ \Omega$; curve B: computed for $R_{FB} = 160\ \Omega$; curve C: measured for $R_{FB} = 160\ \Omega$ ($V_{DS} = 4$ V, $I_{DS} = 53$ mA).

feedback resistor to $R_{FB} = 160\ \Omega$ increases the gain to 4.6 dB as calculated with (A.7). Using (A.5) and (A.8) the input VSWR for this case computes to 1.27:1 and the output VSWR to 1.13:1.

The selection of L_D and L_{FB} as well as the influence of these two inductors on gain has been described in detail in Section II. The gain response of the basic feedback amplifier shown in Fig. 1 (a) with $R_{FB} = 160\ \Omega$ is nearly flat to almost 14 GHz (Fig. 2, curve D). However, the corresponding input reflection coefficient is rather poor above 8 GHz. In order to improve the input match over the upper portion of the band, we employed an open-circuit shunt stub and a series transmission line. Due to the feedback, these two components had a negative influence on the output match, so we inserted a series transmission line connected to the output terminal of the basic amplifier to counteract the degradation of the output reflection coefficient. The introduction of the input and output matching networks led to the matched feedback amplifier.

The schematic of the matched feedback amplifier is shown in Fig. 9 (a). The selection of $L_{FB} = 0.6$ nH and $L_D = 0.4$ nH deviates from the optimum values of $L_{FB} = 0.45$ nH and $L_D = 0.6$ nH discussed in Section II. This change was made to attain a more practical circuit layout. Since L_{FB} bridges most of the physical distance between the transistor's gate terminal and the node between L_D and the series transmission line, it becomes somewhat impractical to make L_{FB} smaller than L_D. In addition, a higher feedback resistor of $R_{FB} = 180\ \Omega$ was inserted to obtain optimum gain flatness for the new set of inductors. These measures constitute a necessary tradeoff between practicality and optimum performance. The computed small signal gain of the amplifier between 50-Ω impedances is plotted as curve A of Fig. 9 (b), while curve B represents the computed small signal gain for $R_{FB} = 160\ \Omega$. The computed reflection coefficients of the practical version of the matched amplifier (Fig. 9 (a)), the negative feedback amplifier (Fig. 3 (b)), and the basic feedback amplifier (Fig. 3 (a)) are plotted in Fig. 10 as curves A, B,

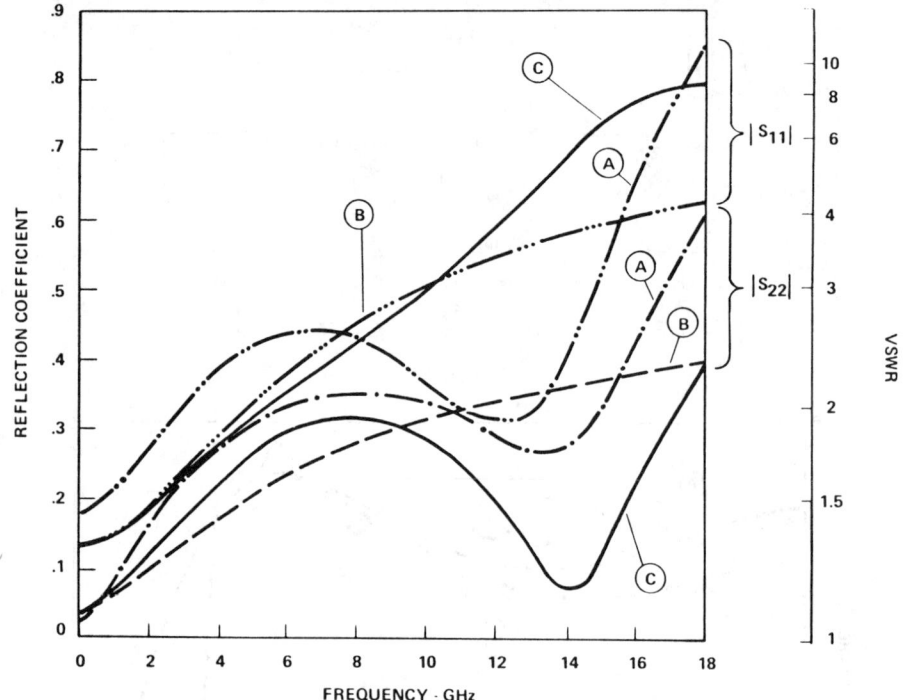

Fig. 10. Input and output reflection coefficients. Curve A: matched feedback amplifier of Fig. 9 (a) with $R_{FB}=180$ Ω, $L_{FB}=0.6$ nH, $L_D=0.4$ nH; curve B: conventional negative feedback amplifier of Fig. 3 (b) with $R_{FB}=160$ Ω, $L_{FB}=L_D=0$; curve C: basic feedback amplifier of Fig. 3 (a) with $R_{FB}=160$ Ω, $L_{FB}=0.45$ nH, $L_D=0.6$ nH.

and C, respectively. The input reflection coefficient of the matched amplifier shows a significant improvement brought about by the matching circuits.

However, the output reflection coefficient has experienced a slight degradation due to the influence of the input matching circuit on the output impedance brought about by the feedback loop. This is particularly pronounced in the area of relatively strong feedback, i.e., above 11 GHz. The amplifier is unconditionally stable up to 13.8 GHz. Reverse isolation computed between dc and 18 GHz has a minimum value of 9.7 dB at 17 GHz for $R_{FB}=160$ Ω and 10 dB at 16 GHz for $R_{FB}=180$ Ω.

B. Amplifier Fabrication and Performance

Fused silica, 0.015 in in thickness, was used as substrate material for the input and output circuits. The circuit pattern was etched into a thin gold film, while the feedback resistor was subsequently etched into a tantalum nitride film which was deposited below the gold.

The measured small-signal gain of the amplifier is plotted in Fig. 9 (b) (curve C). The feedback resistor of this unit measured $R_{FB}=160$ Ω instead of the desired 180 Ω. A comparison of the measured small-signal gain and the gain computed for $R_{FB}=160$ Ω (curve B) shows excellent agreement at frequencies up to 14.5 GHz. Beyond this frequency the actual measured gain dropped rather abruptly. Below 1 GHz the drop in gain was due to the influence of the internal dc biasing network (not shown in the schematic of Fig. 9 (a)). The measured reverse isolation had its minimum value of $|S_{12}|^2_{min}=14$ dB at 1 GHz. This isolation is somewhat better than the 11.2 dB computed with (A.6). Maximum measured reverse isolation was $|S_{12}|^2_{max}=21$ dB at 10.5 GHz which compares to the maximum computed value of 22.4 dB at 9 GHz. The reflection coefficients of the actual amplifier did not exceed $|S_{11}|_{max}=0.38$ for the input and $|S_{22}|_{max}=0.37$ for the output terminal between 2 GHz and 14.4 GHz. The computed values were $|S_{11}|_{max}=0.42$ and $|S_{22}|_{max}=0.37$, respectively.

The measured small-signal gain of an identical amplifier module except for $R_{FB}=225$ Ω is plotted as curve B in Fig. 11. Biasing of this amplifier was accomplished by means of external bias networks. The drop in the measured gain below 1 GHz was caused by the 50-pF dc blocking capacitor in the feedback loop and by the bias networks. The capacitor serves the purpose of separating the drain bias from the gate bias potential. The computed gain is plotted as curve A, while curve C shows the gain at 13 dBm of output power. This module covers a 40:1 bandwidth ranging from 350 MHz to 14 GHz, or almost 5 1/3 octaves. Maximum input and output reflection coefficients between 350 MHz and 14 GHz were $|S_{11}|_{max}=0.45$ and $|S_{22}|_{max}=0.54$, respectively. A minimum reverse isolation of $|S_{12}|^2_{min}=11.5$ dB was measured at 14 GHz. Computed and measured reflection coefficients and minimum reverse isolations are in good agreement.

V. Conclusion

The design of an ultrawide-band amplifier has been described which makes use of both negative and positive feedback. The frequency dependence of the feedback is controlled by two inductors, one in series with the feed-

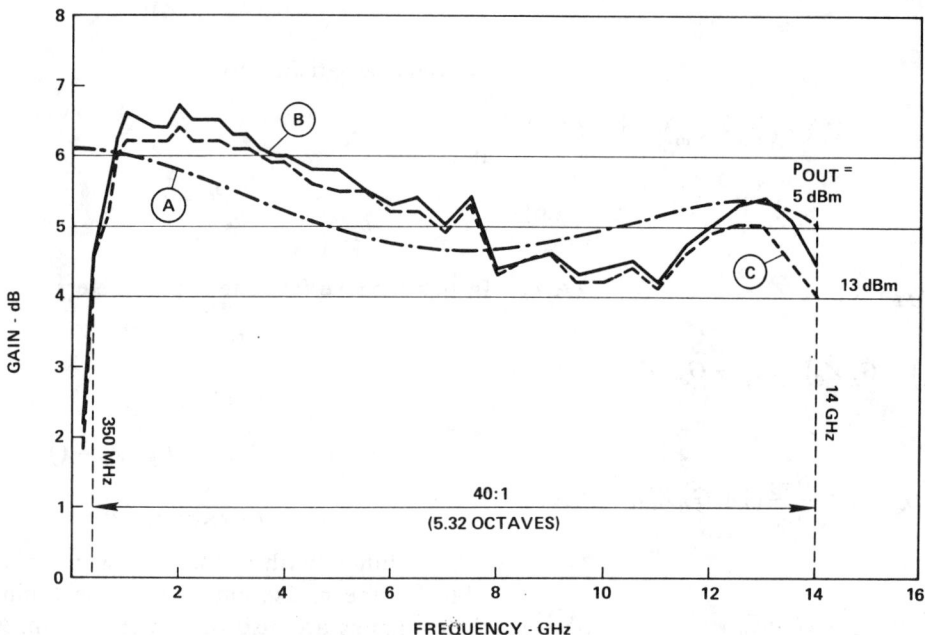

Fig. 11. Gain of the matched feedback amplifier for $R_{FB}=225$ Ω. Curve A: computed small signal gain; curve B: measured gain at 5 dBm of output power ($V_{DS}=5$ V, $I_{DS}=72$ mA); curve C: measured gain at 13 dBm of output power ($V_{DS}=5$ V, $I_{DS}=72$ mA).

back resistor and the other in series with the output of the GaAs MESFET. A detailed study of the effect of the inductors on the characteristics of the feedback amplifier has been made and guidelines have been presented on how to determine their magnitudes. In addition to the feedback circuitry, simple matching networks have been employed to improve the input and output reflection coefficients. The GaAs MESFET used in the experiments was designed with major emphasis on reducing its parasitics. These efforts resulted in a device with ultrawide bandwidth capability.

Two experimental amplifier modules have been described that cover the frequency band 1.0–14.5 GHz and 350 MHz–14.0 GHz with 3.8 dB and 4.2 dB of minimum gain, respectively. Maximum reflection coefficients of these single-ended units were $|S_{11}|_{max}=0.38$ and $|S_{11}|_{max}=0.45$ for the respective input terminals and $|S_{22}|_{max}=0.37$ and $|S_{22}|_{max}=0.54$ for the respective output terminals. Minimum reverse isolations across these bands were $|S_{12}|^2_{min}=14$ dB and $|S_{12}|^2_{min}=11.5$ dB, respectively. The agreement between the computed and the measured data of small-signal gain, reflection coefficients, and reverse isolation was very good.

The use of the matched feedback amplifier with GaAs MESFET's provides several advantages over usual techniques for broad-band microwave amplification. It has an exceptionally wide bandwidth and much lower reflection coefficients than regular single-ended amplifiers of comparable bandwidths. It can be constructed with a very simple and small circuit and is relatively easy to cascade. These advantages make feedback amplifiers prime candidates for monolithic applications.

Appendix

The equivalent circuit of a GaAs MESFET feedback amplifier as shown in Fig. 1 (a) can be reduced to the model of Fig. 1 (b) for frequencies at which the reactive elements of the transistor and the matching networks may be neglected. A discussion of the limitations of the low-frequency model is found in Section II.2. Under these conditions and the assumption that R_d and R_s of Fig. 1 (a) are very small compared to the feedback resistor R_{FB} and the load resistor $R_L=Z_0$ voltages and currents are described by the simple conductance matrix

$$\begin{bmatrix} I_1 \\ I_2 \end{bmatrix} = \begin{bmatrix} G_{FB} & -G_{FB} \\ (g_m - G_{FB}) & (G_{FB} + G_{ds}) \end{bmatrix} \begin{bmatrix} V_1 \\ V_2 \end{bmatrix} \quad \text{(A.1)}$$

where

$$G_{FB} = R_{FB}^{-1} \quad \text{(A.2a)}$$

$$G_{ds} = R_{ds}^{-1} \quad \text{(A.2b)}$$

and

$$i_{gs} = g_m V_{gs}. \quad \text{(A.3)}$$

Using elementary algebra, the matrix (A.1) converts into the scattering parameter matrix

$$S_{ij} = \begin{bmatrix} S_{11} & S_{12} \\ S_{21} & S_{22} \end{bmatrix}. \quad \text{(A.4)}$$

Its elements are

$$S_{11} = \frac{1}{\Sigma}\left[\frac{R_{FB}}{Z_0}(1+G_{ds}Z_0)-(g_m+G_{ds})Z_0\right] \quad (A.5)$$

$$S_{12} = \frac{2}{\Sigma} \quad (A.6)$$

$$S_{21} = \frac{-2}{\Sigma}[g_m R_{FB}-1] \quad (A.7)$$

$$S_{22} = \frac{1}{\Sigma}\left[\frac{R_{FB}}{Z_0}(1-G_{ds}Z_0)-(g_m+G_{ds})Z_0\right] \quad (A.8)$$

with

$$\Sigma = 2+(g_m+G_{ds})Z_0+\frac{R_{FB}}{Z_0}(1+G_{ds}Z_0). \quad (A.9)$$

For the condition

$$\frac{R_{FB}}{Z_0} = (g_m+G_{ds})Z_0 \quad (A.10)$$

we find

$$S_{11} = -S_{22} = \frac{G_{ds}Z_0^2}{\Sigma}(g_m+G_{ds}) \quad (A.11)$$

$$S_{12} = \frac{2}{\Sigma} \quad (A.12)$$

$$S_{21} = -\frac{2}{\Sigma}[g_m(g_m+G_{ds})Z_0^2-1] \quad (A.13)$$

$$\Sigma = 2+(g_m+G_{ds})(2+G_{ds}Z_0)Z_0. \quad (A.14)$$

The reflection coefficients S_{11} and S_{22} improve with decreasing drain-source conductance G_{ds}. The general (A.5)–(A.9) and the special (A.10)–(A.14) equations for the S parameters demonstrate that gain, input and output VSWR, and reverse isolation are all fixed quantities once the value for the feedback resistor R_{FB} is chosen.

The ideal matching condition

$$S_{11} = S_{22} = 0 \quad (A.15)$$

can only be satisfied for

$$G_{ds} = 0$$

and

$$\frac{R_{FB}}{Z_0} = g_m Z_0. \quad (A.16)$$

In this case we find the S parameters

$$S_{11} = S_{22} = 0 \quad (A.17)$$

$$S_{12} = \frac{1}{g_m Z_0+1} \quad (A.18)$$

$$S_{21} = -(g_m Z_0-1). \quad (A.19)$$

Acknowledgment

The authors wish to thank R. Pereira, who was responsible for the measurement and the tuning of the amplifiers. Thanks are also due to A. Hallin, K. Lutz, and K. Lindstedt, who fabricated the GaAs MESFET devices and to J. Martin, who assembled the circuits. The authors are indebted to S. Rose, who was responsible for device testing, and to M. Walker for many helpful discussions.

References

[1] H. S. Black, "Stabilized feedback amplifiers," *Elec. Eng.*, vol. 53, pp. 114–120, Jan. 1934.
[2] F. E. Terman, *Radio Engineer's Handbook*. New York: McGraw-Hill, 1943, pp. 395–406.
[3] M. S. Ghausi, *Principles and Designs of Linear Active Circuits*. New York: McGraw-Hill, 1965, pp. 363–370.
[4] E. Ulrich, "Use of negative feedback to slash wideband VSWR," *Microwaves*, pp. 66–70, Oct. 1978.
[5] R. Dawson, "Equivalent Circuit of the Schottky-barrier field-effect transistor at microwave frequencies," *IEEE Trans. Microwave Theory Tech.*, vol. MTT-23, pp. 499–501, June 1975.
[6] K. B. Niclas, R. B. Gold, W. T. Wilser, and W. R. Hitchens, "A 12-18 GHz medium power GaAs MESFET amplifier," *IEEE J. Solid State Circuits*, vol. SC-13, pp. 520–527, Aug. 1978.
[7] K. B. Niclas, W. T. Wilser, R. B. Gold, and W. R. Hitchens, "Application of the two-way balanced amplifier concept to wideband power amplification using GaAs MESFET's," *IEEE Trans. Microwave Theory Tech.*, to be published.

Cryogenically Cooled GaAs FET Amplifier with a Noise Temperature Under 70 K at 5.0 GHz

JOHN PIERRO

Abstract—A 4.5–5.0-GHz gallium arsenide field-effect transistor (GaAs FET) amplifier cryogenically cooled to approximately 70 K is described. A noise temperature of under 70 K is achieved over the band. Power gain for the two-stage amplifier is 20 dB. A noise analysis is performed to predict noise-temperature dependence on the temperature of the amplifier.

INTRODUCTION

The GaAs FET is presently one of the most attractive devices available to the designer of low-noise microwave amplifiers. A rapid succession of improvements in fabrication technology and device design has resulted in reliable devices that can be produced in quantity with noise and gain performance that would have been viewed as incredible several years ago. Uses for FET's are not restricted to low-noise amplifiers. They have shown great potential as high-power amplifiers, oscillators, mixers, and switches [1].

Since most of the noise generated by the FET is thermal, a substantial noise figure reduction may be realized if one simply cools the device. Unlike the bipolar transistor, whose gain drops with decreasing temperature, the FET gain increases, producing the desirable effect of further enhancing its output signal-to-noise ratio (SNR) for a given fixed input SNR. Here it is manifesting an ideal combination of properties.

This short paper describes a two-stage single-ended GaAs FET amplifier cryogenically cooled to 70 K designed to cover the 4.5–5.0-GHz range. Its purpose is to serve as a low-noise second stage to a parametric amplifier in a radio astronomy receiver. The amplifier has a nominal gain of 20 dB at the cold temperature and a noise temperature under 70 K over the full band.

In an effort to acquire an ability to predict (albeit roughly) the noise temperature at the cold temperature and to gain some insight into the mechanisms which are responsible for the noise-temperature improvement, an analysis of a rather simplified GaAs FET noise model was performed. Key results of this work are presented. The model's validity is restricted but is useful when the device is biased at the point of minimum noise figure.

To corroborate the analysis, single-stage amplifier noise-temperature measurements were made and presented here.

AMPLIFIER DESCRIPTION

The amplifier utilizes microstrip circuitry etched on 25-mil-thick alumina. The GaAs FET is the NE 244. S-parameter and two-port noise parameter measurements were made at 300 K and provided the basis for the design. The S parameters and optimum generator reflection coefficient are presented in Table I. As shown in [2], S_{11}, S_{22}, and S_{12} are practically invariant with

Manuscript received May 17, 1976; revised July 26, 1976. This work was supported in part by the National Radio Astronomy Observatory, Socorro, NM. The work at AIL was performed under the direction of M. Lebenbaum, Division Director, and A. Anselmo, Department Head.
The author is with AIL, A Division of Cutler-Hammer, Melville, NY 11746.

TABLE I
MEASURED S PARAMETERS AND OPTIMUM GENERATOR REFLECTION COEFFICIENT

Freq. (GHz)	S_{11} /θ	S_{21} /θ	S_{12} /θ	S_{22} /θ	Γ_{opt}
4.0	.94 /-57	1.58 /116	.009 /52	.92 /-31	-
4.5	.92 /-63.5	1.44 /109	.006 /180	.90 /-32.5	.7 /+90
5.0	.90 /-71	1.33 /97	.005 /161	.87 /-41.5	.7 /+100°
5.5	.86 /-79	1.30 /83.5	.006 /142	.83 /-48	-

Note: $V_{DS} = 3.0$ V; $I_D = 10$ mA; $T_{AMB} = 300$ K.

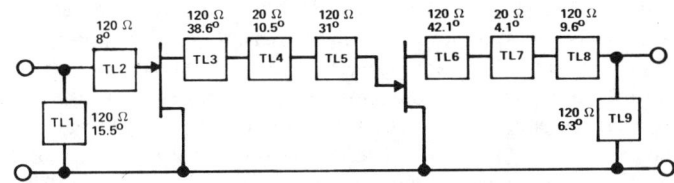

Fig. 1. Circuit configuration. All degrees referred to 4.7 GHz.

temperature in this frequency range. The forward transducer gain, $|S_{21}|^2$, is strongly dependent on temperature, however. Insight into the S_{21} dependence may be acquired by investigating the low-frequency transconductance g_m variation with temperature since $|S_{21}|$ is proportional to g_m. We show in the next section that the transconductance is approximately proportional to $T^{-1/2}$ from 70 to 300 K.

Input, interstage, and output matching networks were designed using the room-temperature transistor data and Smith chart matching techniques. The final values of the network elements were arrived at with the aid of computer optimization. The final circuit appears in Fig. 1.

The input and output impedances are of necessity different from 50 Ω. Voltage standing wave ratios (VSWR's) of 4:1 to 7:1 are generally measured with GaAs FET amplifiers in this frequency range. The high VSWR's are a result of noise match and stability requirements. High output VSWR's are also a consequence of the shortcomings of realizable practical matching networks. To obviate these drawbacks, two cryogenic isolators were integrated with the amplifier so that the input/output VSWR's would be under 1.3:1.

A severe problem is destructive stresses that develop at electrical connections at the cold temperature due to the dissimilar coefficients of contraction of the various materials and components which comprise the complete amplifier. Since the matching networks are of the distributed type, realized in microstrip, they presented no problem. Attention was focused on the bias resistors and bypass capacitors. We chose chip resistors that utilized a resistance element deposited on an alumina substrate. The capacitors were made of a ceramic material—not alumina—whose coefficient of contraction seems to be very similar to alumina. Strain reliefs at the input and output connections were incorporated to alleviate the stress at these points. It is important to allow adequate clearance between the amplifier housing, which is typically aluminum, and the substrate. Aluminum contracts four times as much as alumina. Inadequate

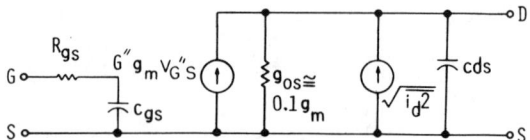

Fig. 2. GaAs MESFET noise model.

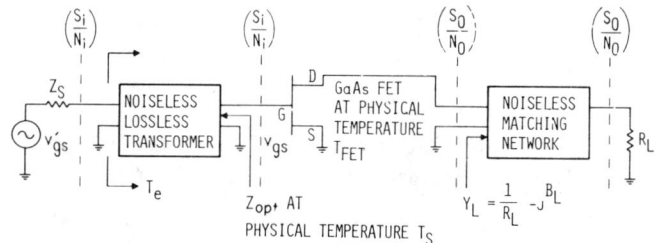

Fig. 3. Single-stage amplifier with definition of symbols for noise calculation.

clearance at room temperature could result in a fractured substrate at the cold temperature.

Proper heat sinking of the transistors was necessary to ensure that the transistor would be cooled to nearly the same temperature as the amplifier housing. Of concern are the transistor-substrate thermal path and substrate-housing thermal path. By soldering the NE 24406 package carrier to the substrate's ground plane, we reduced the thermal resistance of this path to an absolute minimum. The thermal resistance of the NE 24406 with this mounting arrangement is conservatively estimated to be 50°C/W. At minimum noise figure bias, the device is dissipating about 30 mW. Therefore, the channel temperature is at most 2° higher than the substrate temperature. To minimize the thermal resistance of the substrate-housing interface, two measures were taken. We selected a substrate for flatness and machined the shelf on which it is placed extremely flat. The surface finish of the shelf was on the order of 0.1 mil. To further guarantee uniform contact, a 2-mil sheet of indium was placed between the substrate and the housing.

Noise Analysis Model and Key Results

Far more rigorous and thorough noise analyses of the GaAs FET appear in the literature, but they tend to yield equations that cannot easily be applied unless the user first determines a host of constants and parameters for his particular device. Secondly, none has, to the author's knowledge, given sufficient emphasis to the temperature dependence of GaAs FET noise figure to the point of generating user-oriented practical results.

The simplified model (Fig. 2) analyzed here emphasizes the important role the low-frequency transconductance g_m plays in the noise temperature of the device and the effects of impedance matching and physical temperature on its "noisiness." For this simple model, therefore, the gate noise generator of van der Ziel [3], the thermal noise generators corresponding to the source and gate contact resistances, the interelectrode capacitances, and source inductances have been neglected.

After van der Ziel [3] we assumed the channel noise to be represented by a noise generator of the form

$$\overline{i_d^2} = K_c K T_{\text{FET}} g_m \Delta f, \qquad K_c = 2.4$$

with the usual meanings attached to the various quantities. As pointed out [4] this model is valid only when the mobility of the majority carriers in the channel is constant. At low bias currents ($I_D \approx 0.2 I_{DSS}$) where minimum noise figure typically occurs, the assumption of constant mobility is a valid one.

We calculated the noise figure of the amplifier illustrated in Fig. 3 using the basic definition of noise figure F, i.e.,

$$F = \frac{\left(\dfrac{S_i}{N_i}\right)}{\left(\dfrac{S_o}{N_o}\right)}.$$

The symbols are defined in Fig. 3.

The transconductance is a function of transistor operating point and its physical temperature through its dependence on mobility, and the mobility of the majority carriers in the channel is a function of temperature. For physical temperatures greater than approximately 60 K, the mobility may be described by

$$a(T_{\text{FET}})^x$$

where a is a constant and the exponent x assumes a range of values depending on the material and doping level. Curves for the mobility of charge carriers in a doped semiconductor appear in the literature [5]. A frequently used value for the exponent is $-3/2$. From measurements made on actual transistors at AIL over the temperature range of 300–70 K and from [6], it appears that the exponent is closer to $-1/2$. The lower exponent is due to velocity saturation of the carriers in the channel. The mobility for a typical semiconductor is maximum at about 60 K and decreases below 60 K because impurity scattering comes into play.

By postulating a function for the transconductance g_m of the form $K(T_{\text{FET}})^{-1/2}$ where T_{FET} is the physical temperature of the GaAs FET, we get the following result for the noise temperature T_e of the GaAs FET:

$$T_e = 290 \left(\frac{0.6}{g_m R'}\right) \times \left(\frac{T_{\text{FET}}}{T_s}\right)^{3/2} \qquad (1)$$

where g_m is the low-frequency transconductance at temperature T_s (300 K, for convenience), and R' is given by

$$\frac{R_s \left|1 - \left(\dfrac{R_s}{R_s + j\omega L_1}\right)\right|^2}{\omega^2 C_{gs}^2 |Q|^2}$$

where

ω radian frequency, radians per second;
R_s 50 Ω;
C_{gs} gate-source capacitance of FET;
$|Q|^2 = |Z_{\text{opt}} + Z_{\text{in}}|^2$;
Z_{in} input impedance of FET;
Z_{opt} optimum source impedance for minimum noise temperature.

At 4.7 GHz, R' is calculated to be approximately 15 Ω. The transconductance g_m is 26.5 mmho and T_s is 300 K. The reactance $j\omega L_1$ is the lumped-element equivalent reactance of the shorted stub $TL1$ in the input matching network (Fig. 1). The equation for T_e is plotted in Fig. 4 as a function of T_{FET}.

Noise-Temperature Measurements

Noise-temperature measurements on the single-stage 10-dB FET amplifier were performed using an AIL 20 K parametric amplifier having 15-dB gain as the second stage. With this con-

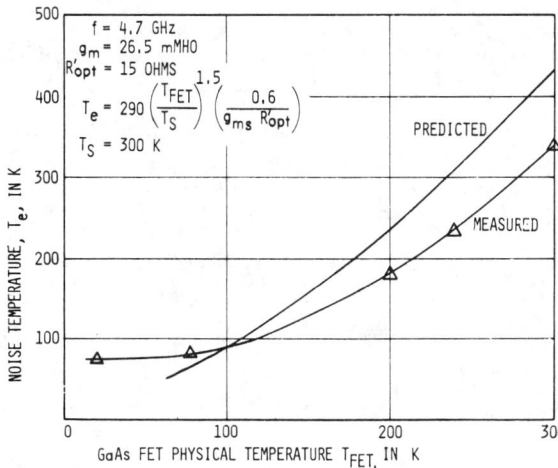

Fig. 4. GaAs FET amplifier noise temperature.

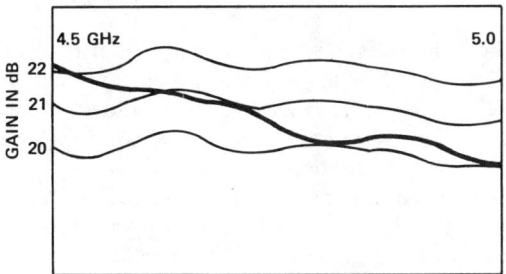

Fig. 5. Gain response of two-stage GaAs FET amplifier.

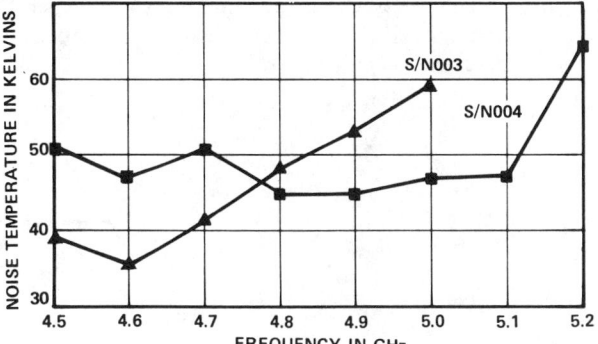

Fig. 6. Noise temperature of two-stage GaAs FET amplifier.

figuration, the second-stage contribution in the noise-temperature measurements was no more than 10 K.

Noise-temperature measurements on the 20-dB FET amplifier were made with an AIL precision receiver and mixer–preamplifier as the second stage. The total correction to the noise-temperature measurement was on the order of 30 K.

The noise source for the preceding measurements was the AIL 7009 hot/cold noise generator.

DISCUSSION OF RESULTS

The single-stage and two-stage amplifiers required little alignment at room temperature to get the desired performance. Adjustment of biases was all that was required. The single-stage amplifier had 8.5 ± 0.2-dB gain and a noise figure of 3.5 dB maximum over the band. The two-stage unit had 16 dB of gain at midband with a gain tilt of about -3 dB/500 MHz. Its noise figure was 2.6 dB maximum. The lower noise figure was achieved through transistor selection.

The gain tilt associated with the two-stage unit at first seemed undesirable but later proved to be an advantage. The intended application for this unit is to serve as a second stage to a cooled paramp. By introducing an opposite gain tilt in the paramp through a slight bias adjustment, we achieve a flat overall gain response. Since the paramp is called on to provide 3 dB more gain at the upper band edge, the noise contribution of the FET to the overall paramp FET noise temperature is cut in half. This measure provides a nearly flat noise temperature for the combination.

The units were installed in a liquid helium refrigerator equipped with 70 and 20 K stations. Cool-down time was approximately 3–4 h for the units.

The passband of the single-stage unit tended to skew as the temperature was lowered. The single-stage unit was required to have a gain flatness of 0.5 dB peak to peak from 4.5 to 5.0 GHz at 20 K. To achieve this flatness, the output matching network was retuned to introduce an opposite tilt at room temperature. At each temperature of interest the gate bias was adjusted for minimum noise figure, which coincidentally was also the point of maximum gain. We observed that the gate bias had to be steadily increased as the temperature was lowered. The explanation for this is that since the saturation current I_{DSS} and the pinchoff voltage V_p are increasing with decreasing temperature, larger negative gate bias is required so that the drain current I_D could be held to 20 percent of the saturation drain current. This is the point of minimum noise figure in the GaAs MESFET. It is the point at which diffusion noise is insignificant.

We performed similar bias adjustment on the two-stage unit to achieve minimum noise figure. The tilt we observed at room temperature was still present at 70 K but smaller by 1 dB.

In Fig. 4 the noise temperature predicted by analysis is plotted along with the measured amplifier noise-temperature data. Although the variation seems to follow the actual variation of noise temperature, it would seem plausible to expect the predicted noise temperature to be perhaps 30 to 50 percent lower than the measured data since we have neglected the gate current and contact resistance noise generators. It is not and this may be so for several reasons. Foremost among them may be that the actual value of K_c, assumed to be 2.4, may be considerably lower than that value. Secondly, as explained in [4] there exists a strong correlation (almost unity) between the drain noise and induced gate noise, which leads to a high degree of cancellation in the noise output of the GaAs FET. Since we have omitted the gate noise generator, it seems plausible that the noise temperature of the FET as predicted from the analysis would be pessimistic in spite of the fact that the contact resistance noise generators were omitted.

Fig. 5 depicts the gain response of the 20-dB GaAs FET amplifier cooled to approximately 70 K, and Fig. 6 is a plot of noise temperature over the 4.5–5.0-GHz band. The amplifier noise temperature is well under 70 K. Preliminary results from a production run indicate that this performance is rather typical for the units and is in no way unusual.

With a paramp cooled to 20 K preceding the two-stage FET amplifier, we have reliably and repeatedly achieved noise temperatures under 16 K over the band. The overall gain of the combination was 36 dB.

The output power at the 1-dB compression point was well in excess of $+5$ dBm for the FET amplifiers. Input/output VSWR's were well below 1.3:1.

CONCLUSION

Data show that it is presently feasible to construct low-noise GaAs FET amplifiers with performance comparable to uncooled parametric amplifiers. They exhibit the qualities of easy alignment, require little testing time, and demonstrate a high degree of gain stability. Saturation powers well in excess of those associated with parametric amplifiers are easily attained. The simplicity of the approach makes the cooled GaAs FET amplifier an attractive candidate as a second stage to a parametric amplifier. Using $1/2$-μm GaAs FET's, it should be possible to realize noise temperatures well under 40 K in this frequency range.

The first-order analysis presented in this short paper for a simplified noise model of the FET yields results that correlate well with measurements.

ACKNOWLEDGMENT

The author wishes to thank L. Hernandez who did his usual outstanding job of assembling the unit.

REFERENCES

[1] Special Issue on Microwave Field Effect Transistors, *IEEE Trans. Microwave Theory Tech.*, vol. MTT-24, June 1976.
[2] C. A. Liechti and R. B. Larrick, "Performance of GaAs MESFET's at low temperatures," *IEEE Trans. Microwave Theory Tech.*, vol. MTT-24, June 1976.
[3] A. van der Ziel, "Thermal noise in field-effect transistors," *Proc. IRE*, vol. 50, pp. 1808–1812, August 1962.
[4] Pucel *et al.*, "Noise and gallium arsenide FET's," *IEEE J. Solid-State Circuits*, April 1976.
[5] R. M. Rose, L. A. Shepard, and John Wulff, *The Structure and Properties of Materials*, vol. IV. New York: Wiley. pp. 106–107.
[6] Technical Staffs, C.E.L. Inc. and NEC Ltd., "The design, performance and application of the NEC V244 and V388 gallium arsenide field effect transistors," Feb. 1976.

Low-Noise Cooled GASFET Amplifiers

SANDER WEINREB, FELLOW, IEEE

Abstract—Measurements of the noise characteristics of a variety of gallium–arsenide field-effect transistors at a frequency of 5 GHz and temperatures of 300 K to 20 K are presented. For one transistor type detailed measurements of dc parameters, small-signal parameters, and all noise parameters (T_{min}, R_{opt}, X_{opt}, g_n) are made over this temperature range. The results are compared with the theory of Pucel, Haus, and Statz modified to include the temperature variation. Several low-noise amplifiers are described including one with a noise temperature of 20 K over a 500-MHz bandwidth. A theoretical analysis of the thermal conduction at cryogenic temperatures in a typical packaged transistor is included.

I. INTRODUCTION

THE PRESENT state of the art for microwave low-noise amplifiers is shown in Fig. 1. The gallium–arsenide field-effect transistor (GASFET) amplifier does not yet achieve the noise temperature of the very best parametric amplifiers but is equal to or better than many paramps manufactured 10 years ago. In addition, the GASFET has higher stability and lower cost because of two inherent advantages: 1) it is much less critical to circuit impedance than a negative resistance amplifier such as a paramp; 2) it is powered by dc whereas the paramp requires a power oscillator and tuned circuits at several times the frequency of operation.

There are systems, particularly those requiring large-area antennas such as radio astronomy or space communications, where no present device operating at room temperature has sufficiently low noise. This is shown clearly in Fig. 1 where, in the 0.5–20-GHz range, the 300 K paramp performance is typically an order of magnitude greater than the natural noise limits of galactic, cosmic, and atmospheric noise. The lowest noise, highest cost solution is a maser or parametric up-converter-into-maser system operating at 4 K. Intermediate in cost and performance are paramps and GASFET's cooled to 20 K by closed-cycle helium refrigerators which are now available [10] at a cost of under $5000 and a weight less than 45 kg.

Several reports of the noise temperature of cryogenically cooled GASFET amplifiers have been made [4]–[8] but, for the most part, these are for one specific device, do not determine the four noise parameters which characterize the noise of a linear two-port ([4] is an exception to this), do not report the device dc and small-signal parameters as a function of temperature, and do not attempt to correlate the results with theory. An attempt will be made to do the above in this paper and to present some information regarding the following questions.

Manuscript received April 3, 1980; revised June 24, 1980. The National Radio Astronomy Observatory is operated by Associated Universities, Inc., under contract to the National Science Foundation.
The author is with National Radio Astronomy Observatory, Charlottesville, VA 22903.

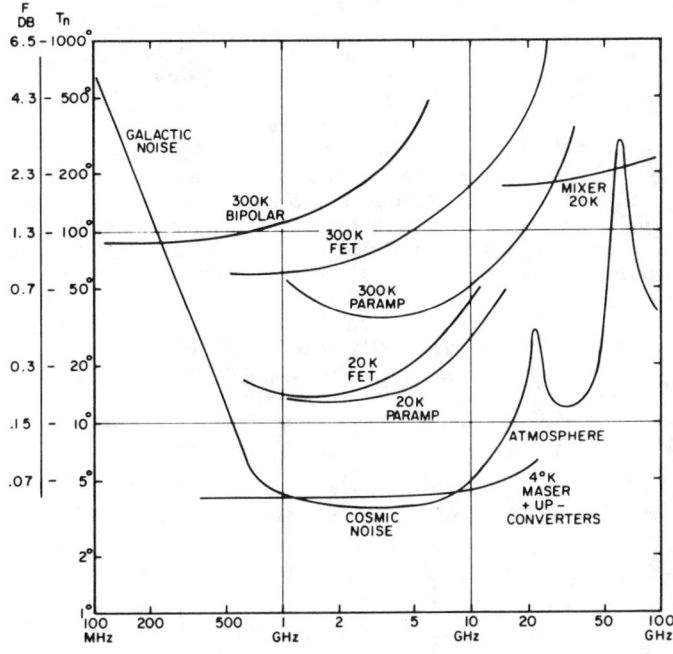

Fig. 1. Noise figure 10 log F, and noise temperature $T_n = 290°(F-1)$, versus frequency for various 1980 state-of-the-art low-noise devices. The 300 K bipolar transistor, FET, and paramp values are taken from manufacturers data sheets [1]–[3], the 20 K FET curve is from the data of this paper plus data of others [4]–[6] at 0.6, 1.4, and 12 GHz, respectively. The 20 K paramp, 4 K maser (including parametric up-converter at lower frequencies), and 20 K mixer results (which are SSB and include IF noise) are from systems in use at National Radio Astronomy Observatory (NRAO). The natural noise limitations due to galactic noise, the cosmic background radiation, and atmospheric noise are for optimum conditions and are taken from [9] plus points at 22 GHz and 100 GHz measured at NRAO.

1) Considering the dc power dissipated and thermal resistance problems, what is the actual physical temperature of the FET channel? What is the lowest physical temperature which can be achieved in the channel?
2) What is the noise improvement factor for cooling of presently available low-noise GASFET's? Are one manufacturer's devices superior for some fortuitous reason?
3) Can the present room-temperature GASFET noise theory be applied at cryogenic temperatures?
4) Can a GASFET be specifically designed for the best performance at cryogenic temperatures?
5) Is there any difference in the circuit design for a cryogenic amplifier?

II. THERMAL RESISTANCE

The room-temperature thermal resistance of a typical low-noise GASFET is specified on manufacturers' data sheets and is of the order of 100 K/W for a chip and 200

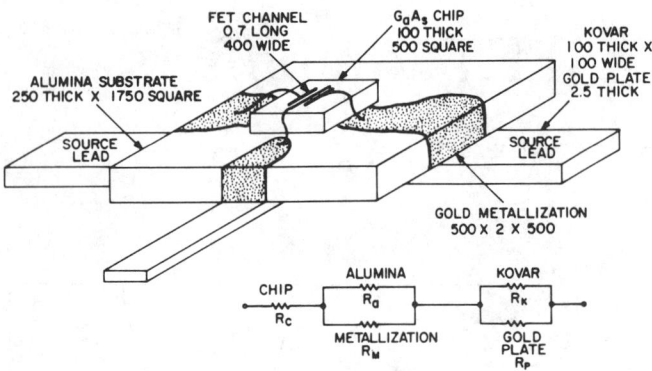

Fig. 2. Typical GASFET chip in 1.75-mm² package with top cover removed; all dimensions are in micrometers. Heat flow is from the FET channel spreading through the GaAs chip into the gold-metallized alumina substrate and out the source leads. An equivalent electrical circuit is also shown with thermal resistance values in Table I. Some manufacturers do not connect the source metallization to the chip.

TABLE I
THERMAL RESISTANCE OF PACKAGED GASFET IN K/W AND IN PARENTHESIS MATERIAL CONDUCTIVITIES IN W/K·cm

Component (Dimensions in μm)	R	Temperature			
		300°	77°	20°	4°
FET Channel (0.7 x 400)	R_c	120 (0.44)	12 (4.4)	13 (4.1)	840 (.06)
Alumina Substrate 250 x (1.750)²	R_a	55 (0.35)	13 (1.5)	85 (0.23)	3900 (.005)
Gold Metalization 500 x (500 x5)	R_m	335 (3)	285 (3.5)	62 (16)	45 (22)
Total $R_c + R_a // R_m$	R_t	169	24	49	885
Kovar in Source Leads 250 x (100 x 1000)	R_K	76 (.165)	156 (.08)	625 (.02)	4170 (.003)
Gold Plate on Source Leads 250 x 2.5 x 2200	R_P	76 (3)	65 (3.5)	14 (16)	10 (22)
Total Including Source Leads $R_t + R_K // R_P$	R_T	207	89	63	895
Add for Epoxy Bond of Chip 25 x (500 x 500)	—	50 (.02)	100 (.01) EST	330 (.003) EST	1000 (.001)

K/W for a packaged device. These values produce a heating of 5 or 10 K for a typical low-noise dc power dissipation of 50 mW and do not significantly effect the room-temperature performance. However, at cryogenic temperatures the situation may be drastically different because the thermal conductivity of most materials changes by orders of magnitude; pure metals and crystalline substances become better thermal conductors while alloys and disordered dielectrics become worse.

An analysis of the heat flow in a typical 1.75 mm² packaged GASFET sketched in Fig. 2 has been performed using the thermal resistance equations of Cooke [11] with material thermal conductivities published in various references (GaAs [12], alumina and gold [13], Kovar and iron alloys [14]). Results are summarized in Table I which also gives material conductivities used in the calculations.

At temperatures down to 20 K the total thermal resistance decreases substantially for the configuration of Fig. 2. The heat flow medium shifts in the substrate from alumina to gold metallization (assumed 5 μm thick). It is thus important that the gold metallization and plating be thick, pure, and free of voids. A case designed for cryogenic operation should have pure silver or copper source leads and a sapphire or crystalline quartz substrate.

It is important that the chip be solder-bonded to a metallized substrate with the metallization continuing to the source leads. This is not the case in all commercially available devices. A calculation of a solder-joint thermal resistance shows it to be negligible even at 4 K. However, a silver-loaded epoxy joint of 25-μm thickness would add 300 K/W at 20 K [16].

It is also important that the total heat path from chip, thru package to amplifier case, and on to cooling station be carefully considered; this often conflicts with the desired microwave design. A chip GASFET soldered to a high-purity copper amplifier case is an excellent solution to thermal problems above 20 K but is not a necessity; a packaged device can be used.

At 4 K the thermal problem within the GaAs chip is quite severe due to boundary scattering of phonons [17, p. 149] which produces a thermal resistance increasing as T^{-3} for temperatures below 20 K. A channel with 20–50 mW of power dissipation will stabilize at a temperature of ~15 K even if the chip boundaries are at 4 K; hence little is gained compared to 20 K cooling. The value of the chip thermal resistance at 4 K given in Table I is only a rough approximation as the problem becomes complex. The thermal conductivity is no longer a point property of the material; the heat conduction is by acoustical waves and wave transmission and reflection at boundaries must be considered. However, an effective thermal conductivity dependent upon the object size can be defined (see Callaway [15]) and has been used in Table I with the size parameter set equal to a gate length of 0.7 μm. It should be noted that the chip thermal resistance would not be significantly reduced by immersion in normal liquid helium which has insufficient thermal conductivity for the area and heat flux involved (Though, super-fluid helium at a temperature below 2.2 K would be effective).

There is a possibility that a detailed study of the heat conduction mechanism from the channel at 20 K would show increased heating due to the small size effects discussed above which are certainly present at 4 K. Experimental evidence against this, however, is the fact that for most devices evaluated the amplifier noise temperature variation with dc bias power dissipation is small at cryogenic temperatures and similar to the variation at room temperature.

A different conclusion regarding self-heating at cryogenic temperatures was reached by Sesnic and Craig [56] who predict large self-heating for the 4 K–77 K temperature range due to poor conductivity of the Kovar source leads. Their paper did not consider the strong effects of plating on the source leads or the boundary-scattering decrease in chip thermal conductivity. These erroneous results were applied by Brunet–Brunol [57] who then attributed the lack of change of GASFET electrical characteristics below 77 K to self-heating.

III. DC Characteristics Versus Temperature

The dc characteristics of a GASFET can be analyzed to determine parameters such as transconductance and input resistance which enter directly into the noise temperature equation, and also device fabrication parameters such as channel thickness a and carrier density N, which affect the noise temperature in a more complex manner. In addition, by measuring the variation of dc parameters with temperature, the variation of material parameters such as mobility μ and saturation velocity v_s, can be determined; these also enter into the noise theory.

The curves of drain current versus drain voltage at steps of gate voltage for three different manufactures of GASFET's at 300 K and 23 K are shown in Fig. 3. In general, there is only a mild change in the characteristics of all devices tested with most changes occurring between 300 K and 80 K. The dominant effects are an increase in transconductance, saturation current, and drain conductance.

A more detailed analysis of a sample device, the Mitsubishi MGF 1412, will be performed using, with some modification, the methods of Fukui [18]. Microwave noise measurements of the identical device are described in the next section. The total channel width Z was measured with a microscope and found to be 400 μm in two 200-μm stripes and the gate length L has a published [34] value of 0.7 μm (the gate length has only a minor effect on the quantities evaluated in this section, but is important for the noise analysis). All dc data were measured at five temperatures utilizing 0.1-percent accuracy digital meters. The results are given in Table II and discussed below.

A. Forward-Biased Gate Characteristics

The forward-biased gate junction was first evaluated to find the barrier potential V_B, Schottky ideality factor n, and gate-plus-source resistance $R_g + R_s$. This was performed by measuring the gate-to-source voltage V_{gs} for forward gate currents of 0.1 μA–10 mA in decade steps, all with drain current $I_d = 0$. The results were fitted to a normal Schottky-barrier current–voltage characteristic with V_B replaced by V_B/n as suggested by Hackam and Harrop [19]. The resulting values for V_B show little variation with temperature, in agreement with the results of others [19], [20]. The n factor increases by a large amount as temperature decreases; this is due to tunneling through the narrow, forward-biased depletion layer [21], [22] and does not effect the reverse diode characteristics and normal FET operation. When tunneling is present the n factor will vary slightly with current (i.e., the I–V characteristic is not exponential) if the doping density is not uniform. This limits the accuracy of the determination of n and hence, of V_B and $R_g + R_s$. The n values given in Table II are for the 1–10-μA range.

The value of $R_g + R_s$ is determined from the I_g-V_{gs} measurements in the 0.1–20-mA range. At these currents the voltage drop across $R_g + R_s$ becomes significant compared to the exponential ideal diode characteristic, and hence $R_g + R_s$ can be determined. However, R_g at these

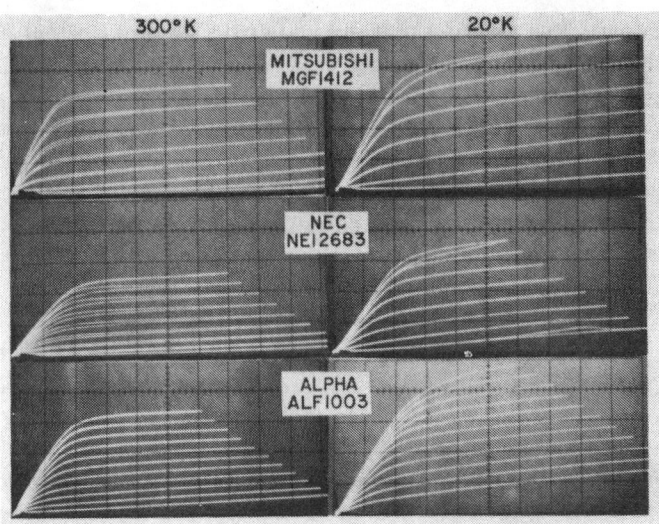

Fig. 3. Drain current, 20 mA per vertical division, versus drain voltage, 0.5 V per horizontal division, for 0.2 V steps of gate voltage for 3 manufacturers of GASFET's at 300 K and 20 K. The top curve in each photograph is at 0 V gate voltage.

TABLE II
DC Parameters Versus Temperature for the Mitsubishi MGF 1412 GASFET (Units Are in Volts, Ohms, Milliamperes, and Millimhos

Quant	300°K	228°K	151°K	81°K	21°K
V_B	.796	.811	.834	.810	.782
n	1.125	1.216	1.47	2.0	7.2
R_g	2.66	1.7	1.5	1.4	1.2
R_s	2.3	2.3	2.1	2.1	2.2
R_d	2.1	2.4	2.5	2.4	2.4
R_t	8.8	8.8	9.3	11	10–14
V_p	1.226	1.151	1.156	1.172	1.159
I_o	67.8	74.3	79.1	84.7	86.2
I_s	255	279	297	318	324
g_m'	39	44	47	50	55
g_m	43	49	52	56	62
v_s/v_{so}	1.0	1.13	1.18	1.27	1.32
g_m/g_{mo}	1.0	1.14	1.21	1.30	1.44

currents is nonlinear due to the distributed gate metallization resistance. A distributed ladder network of resistors and diodes must be considered. At high currents the potential drop across the metallization resistance produces a reverse bias which results in less current through the diodes at the end of the ladder away from the gate connection point; the effective value of R_g decreases with current. This situation is described by a nonlinear differential equation which has been solved and the results have been used to find the constant, low-current value of $R_g + R_s$. The solution will be described in a separate publication [54]. The small-signal value of $R_g + R_s$ at 10-mA bias current was also measured at 100 kHz and 10 MHz. Values equal to the dc slope resistance were obtained, thus assuring that thermal errors are not present in the dc measurement.

To separate R_g and R_s and also determine the drain series resistance R_d, the gate-to-source voltage V_{gsd} with drain connected to source was measured along with the gate-to-drain voltage V_{gd} with the source open; both of these values are at a gate current of 10 mA. By differencing these quantities with V_{gs}, also measured at $I_g = 10$ mA, R_s and R_d are obtained as

$$R_s = \Delta R_1 + \sqrt{(\Delta R_1)^2 + \Delta R_1 \cdot \Delta R_2} \quad (1)$$

and

$$R_d = \Delta R_2 + R_s \quad (2)$$

where $\Delta R_1 = (V_{gs} - V_{gsd})/0.01$ and $\Delta R_2 = (V_{gd} - V_{gs})/0.01$. Note that since only voltage differences are measured the values can be highly accurate and do not depend on removing the exponential portion of the I_g-V_g characteristic, as is the case with the determination of $R_g + R_s$, and therefore R_g.

The values of R_s and R_d show little variation with temperature. This is to be expected, since for a highly doped semiconductor, both the carrier density and mobility show little variation with temperature even at temperatures as low as 2 K [23], [24]. The mobility is limited by impurity scattering and carriers do not "freeze out" because the impurity band overlaps the conduction band. The gate resistance R_g decreases from 2.7 Ω at 300 K to 1.2 Ω at 21 K since it is primarily due to the resistivity of aluminum which decreases by a large amount dependent upon its purity. The theoretical resistance of the gate metallization, using the formula of Wolf [25] is 1.7 Ω at 300 K and <0.001 Ω at 21 K indicating that ∼1 Ω is due to impurities or semiconductor resistance.

B. Drain Voltage–Current Characteristic

A second check on mobility variation is obtained by measurements in the linear region (i.e., no velocity saturation) of the I_d-V_d characteristic. The drain current I_d was measured for $V_{gs} = 0$ and $V_d = 50$ or 100 mV. $R_t = V_d/I_d$ is reported in Table II and shows somewhat more variation with temperature than R_s or R_d (which are contained in R_t). In particular the value at 21 K is dependent upon past history; it is lower by ∼30 percent after forward biasing the gate junction. This has not been explained. No such "memory" effect is observed in the device at normal bias levels.

The saturated drain current I_0 at $V_{gs} = 0$ and $V_d = 3$ V was measured along with the gate voltage $-V_p$ to bring the drain current down to 2 mA; V_p is a good approximation to the pinch-off voltage of the device. These quantities, measured at 300 K, can be used to determine the doping density N, and channel thickness a of the device through the well-established relations

$$V_p + V_B = \frac{qNa^2}{2\kappa\varepsilon_0} \quad (3)$$

$$I_s = qv_s NaZ \quad (4)$$

where $q = 1.6 \times 10^{-19}$ C, $\varepsilon_0 = 8.85 \times 10^{-14}$ F/cm, $\kappa = 12.5$ for GaAs, $Z = 0.4$ mm, and the saturated velocity v_s is assumed to be 1.4×10^7 cm/s. The quantity I_s is the open-channel saturation current and is related to I_0 [18] by

$$I_s = I_0/\gamma \quad (5)$$

where

$$\gamma = 1 + \sigma - \sqrt{\delta + 2\sigma + \sigma^2} \quad (6)$$

$$\delta = (V_B + 0.234 L)/(V_B + V_p) \quad (7)$$

$$\sigma = 0.0155 R_s Z/a \quad (8)$$

and a and L are in micrometers and Z is in millimeters. The channel thickness a is not initially known for use in σ but since it is a moderately small correction factor, an initial estimate can be used and later iterated. Solving (3) and (4) for a and N gives $a = 0.10$ μm and $N = 2.9 \times 10^{17}$/cm^3, in good agreement with information supplied by the manufacturer [26].

The measured insensitivity to temperature of $V_p + V_B$ verifies through (3) that N is not a function of temperature. The temperature dependence of I_s must then be due to a change in saturation velocity v_s; its value relative to the 300 K value is given in Table II. The results are in general agreement with the increase in saturation velocity measured by Ruch and Kino [27] in the 340 K–140 K range.

Finally, the transconductance g'_m was measured by taking 50-mV increments in V_g above and below the low-noise bias point of $V_d = 5$ V, $I_d = 10$ mA. This must be corrected for effects of source resistance to give the true transconductance $g_m = g'_m/(1 - g'_m R_s)$; both g'_m and g_m are given in Table II in units of millimhos. The values are approximately 35 percent lower than those given by the approximate theoretical expression

$$g_m = \frac{I_s}{2(V_p + V_B)} \cdot \frac{1}{(1 - I_d/I_s)} \quad (9)$$

and the exact theoretical curve given in [28, fig. 12(c)] (which is a little closer). Fukui had a similar problem in the analysis of his data [18, p. 787]. The increase in measured transconductance with decreasing temperature follows the saturation velocity increase determined from measurements of I_0, V_p, and V_B.

IV. MICROWAVE PERFORMANCE VERSUS TEMPERATURE

A. Gain and Noise Measurement Procedure

A block diagram of the test configuration used for measurements of gain and noise temperature of cooled amplifiers is shown in Fig. 4. Auxiliary equipment such as a network analyzer, reflectometer, and spectrum analyzer were used for impedance, return loss, and spurious oscillation measurements. At a later stage of the work an HP 9845 calculator and HP 346 avalanche noise source were incorporated to allow a swept frequency plot of 50 noise temperature and gain measurements to be performed in

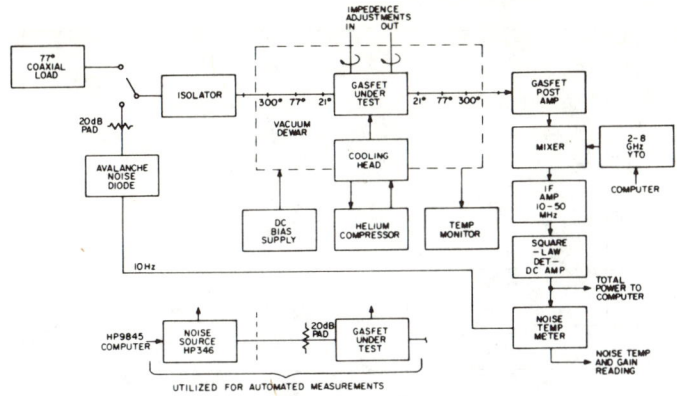

Fig. 4. Test setup for noise and gain measurements of cooled amplifiers. The automated measurement procedure gives a swept-frequency output of noise temperature (corrected for second-stage contribution) and gain; this replaced the hot load-cold load manual method after corroboration of results was established.

Fig. 6. Test amplifier with adjustable source and load resistance. The spacing between the $\lambda/4 = 15.9$-mm long rectangular slab transmission line and ground plane is adjusted by turning the circular threaded slug. DC blocking capacitors are realized by teflon coating the SMA connector center pin where it enters the slab. The slab transmission line is supported by the coaxial connector, a chip capacitor, and a teflon bushing held by a nylon screw. The transistor is mounted by its source leads on a 6.3-mm diameter threaded copper slug with center 5.7 mm from the end of each $\lambda/4$ line.

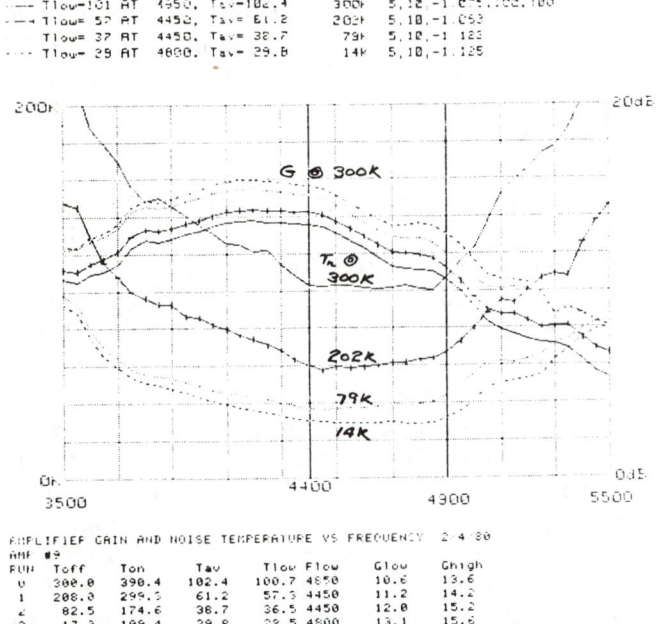

Fig. 5. Noise temperature and gain versus frequency for the Mitsubishi MGF 1412 transistor at several physical temperatures. This is a direct copy of the on-line output of an automated measuring system utilizing an Hewlett-Packard 9845 computer. The results are first plotted on a CRT and then can be copied on a thermal printer. Tabulated below the plot are noise-source-off temperature, noise-source-on temperature, average amplifier noise temperature between markers at 4.4 and 4.9 GHz, lowest noise and its frequency, and lowest and highest gain between marker frequencies. The lowest and average noise temperatures are also given above the plot area along with a space for user comments which in this case are physical temperature, drain voltage and current, and gate voltage.

20 s; a typical output from this automated system is shown in Fig. 5. The amplifiers under test were cooled by a commercially available [29] closed-cycle refrigerator with a capacity of 3 W at 20 K and a cool-down time of 5 h.

A large error in the noise temperature measurement of a mismatched amplifier can result if the source impedance changes when switching from hot load to cold load. For this reason an isolator is used between the loads and amplifier input. The total loss from 77 K liquid-nitrogen cooled load to the amplifier was accurately measured and corrected for. This included 0.62 dB in the coaxial switch, isolator, and SMA hermetic vacuum feed thru; 0.11 dB in a stainless-steel outer-conductor beryllium–copper inner-conductor coaxial line transition from 300 K to 20 K; and 0.07 dB in 10 cm of internal coaxial line at 20 K.

For the automated measurements an HP 346 calibrated avalanche diode was used in cascade with a modified HP 8493A coaxial attenuator. No isolator is necessary. The attenuator is modified for cryogenic use [30] by replacing the conductive-rubber internal contacts by bellows contacts [31], drilling a vacuum vent hole in the body, and tapping the body for mounting a temperature sensor [32]. The change in attenuation of the modified 20-dB attenuator upon cooling from 300 K to 77 K is a decrease of 0.05 ± 0.03 dB; it is assumed that negligible change will occur upon further cooling to 20 K as this is typical for most resistance alloys. The attenuator may be purchased calibrated from Hewlett Packard (Option 890); however, the contact modification necessitates a recalibration (attenuation at 300 K decreased by 0.05 dB) and was performed with a digital power meter. For an amplifier with 30 K noise temperature the cooling of the attenuator from 300 K to 20 K provides a factor of $(300+30)/(20+30) \sim 6$ decrease in error of the amplifier noise temperature measurement due to noise source or attenuator errors. For this case a noise temperature error of $\pm 2.3°$ results from an uncertainty of noise source excess noise plus attenuator error of ± 0.2 dB. The agreement between amplifier noise measurements made with the hot/cold loads and the HP 346 noise source was within ± 2 K for cooled amplifiers and ± 5 K for uncooled amplifiers.

In order to determine the optimum source resistance and load resistance as a function of temperature, a test amplifier having variable impedance quarter-wave transformers at input and output was constructed and is shown in Fig. 6. A schematic of this amplifier and of all other

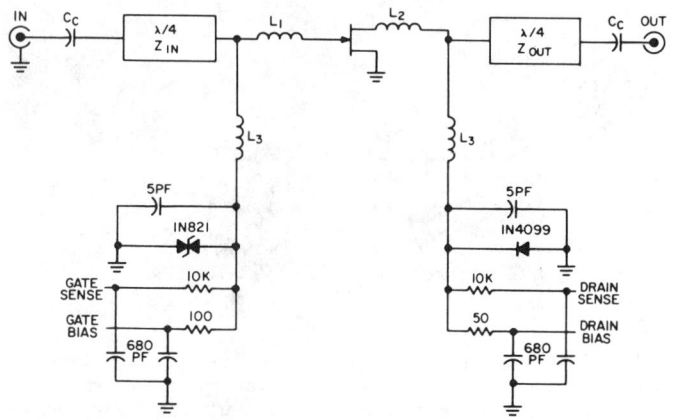

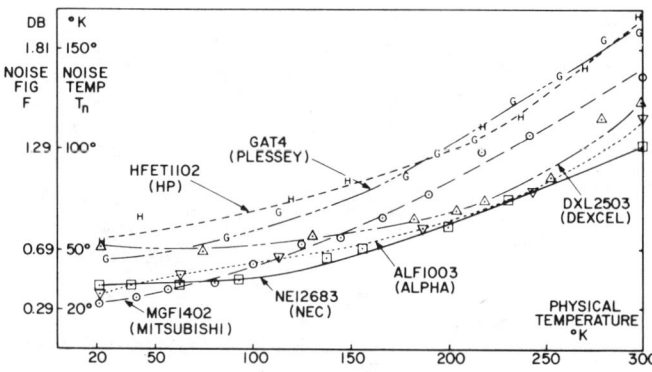

Fig. 7. Schematic of all amplifiers discussed in this paper. The bias decoupling inductance L_3 is realized by $\lambda/4$, $Z_0 = 100$ Ω line in the test amplifier; in other amplifiers it is a small coil described in the text.

Fig. 9. Noise temperature versus physical temperature for several manufacturers of GASFET's. Only one or two samples of the HP, Dexcel, Alpha, and Plessey devices were evaluated, and units from another batch may give better cryogenic performance.

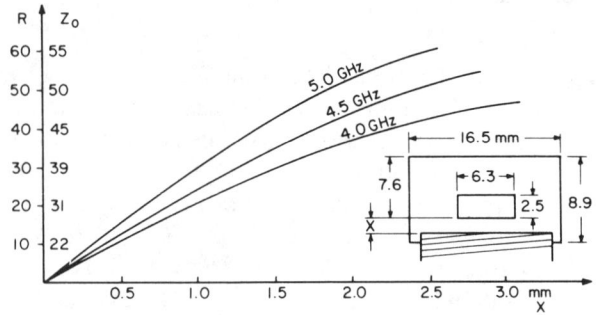

Fig. 8. Characteristic impedance Z_0 and source resistance, $R = Z_0^2/50$, as a function of spacing X from threaded-slug ground plane to $\lambda/4$ slab transmission line having cross section shown in figure.

amplifiers described in this paper is shown in Fig. 7. The variable impedance line consists of a slab transmission line having cross section as shown in Fig. 8 and a movable ground plane realized as a threaded slug with a diameter of $\lambda/4$. The slug is slotted and has a thread-tightening screw to assure ground contact and to lock the position. Calibration of the characteristic impedance versus slug position was performed by utilization of a miniature coaxial probe inserted in place of the FET and connected to a network analyzer; results are given in Fig. 8. The loss of the $\lambda/4$ line was calculated to be less than 0.025 dB at $Z_0 = 22$ Ω using the microstrip loss formulas of Schneider [33].

B. Results

As a first step, the noise temperature of several commercially available GASFET's was measured at a frequency of ~4.5 GHz from 300 K to 20 K; results are shown in Fig. 9. At each temperature bias and source resistance were optimized for minimum noise; the changes were not large. For all units the peak gain was ~12 dB at 300 K and increased by 1 or 2 dB at 20 K. The measurement frequency was chosen for minimum noise at 300 K and no change in the optimum frequency was noted as temperature decreased.

On the basis of these tests and because of its high burnout level [34], the Mitsubishi device was selected for detailed study and for use in the construction of amplifiers needed for use in radio astronomy. A selected version of the MGF 1402, the MGF 1412, became available and was used for further tests; it has lower noise temperature (~100 K) at 5 GHz and 300 K, but has a noise temperature at 20 K close to that of the MGF 1402. Of approximately 15 samples of the MGF 1412 tested at 20 K, the noise temperature was between 15 K and 27 K.

The gain and noise temperature of a MGF 1412 at four temperatures between 300 K and 14 K is shown in Fig. 5; dc parameters for the same device are given in Table II. As the amplifier is cooled, the gain increases by 2.0 dB. This is less than the 3.0-dB increase of g_m because of internal negative feedback (which stabilizes the gain against g_m changes) and possibly because of a reduction in parallel output resistance. As the noise temperature decreases, the bandwidth for a given increase in noise temperature increases, but there is very little other change in frequency response of either gain or noise temperature.

A complete comparison of noise theory and experiment requires the measurement of four noise parameters of the device. There are many sets of four parameters which can be compared. The set which is most directly measured is the minimum noise temperature $T_{\min}$, the optimum source impedance $R_{opt} + jX_{opt}$, and the noise conductance g_n. These relate to the measured data in Fig. 10 of noise temperature T_n as a function of source impedance $R + jX$ by

$$T_n = T_{\min} + T_0 \cdot g_n \cdot \frac{\left[(R - R_{opt})^2 + (X - X_{opt})^2\right]}{R} \quad (10)$$

where $T_0 = 290$ K.[1]

The values of R, R_{opt}, X, X_{opt}, and g_n are dependent upon the choice of reference plane between device and

---
[1] In this paper $T_0 = 290$ K independent of ambient temperature which will be denoted as $t \cdot T_0$. Several papers in this field normalize quantities which are not functions of temperature to ambient temperature (using the symbol T_0), producing a normalized quantity which is a function of temperature and confusing the reader.

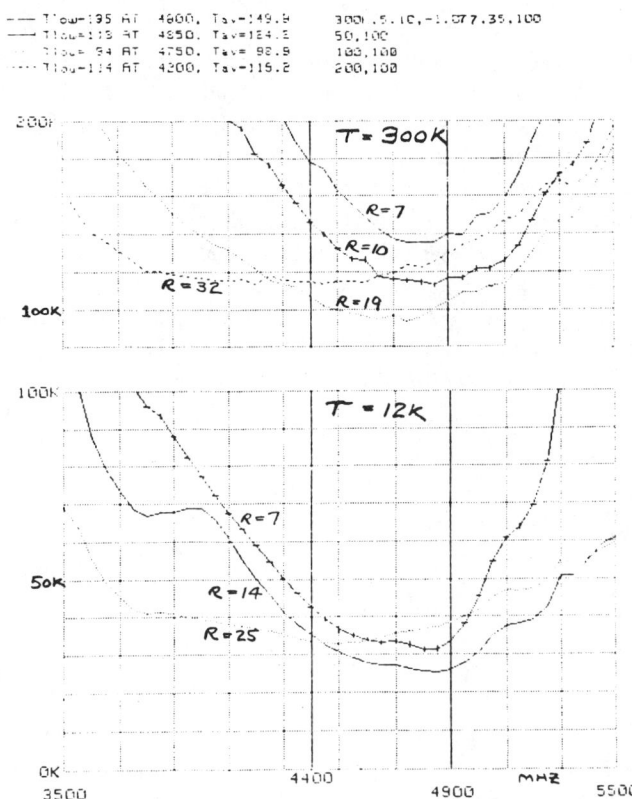

Fig. 10. Noise temperature versus frequency for various source resistances and Mitsubishi MGF 1412 GASFET at 300 K (top) and 12 K (bottom, note change in temperature scale.) Device bias was $V_{ds}=5$, $I_d=10$ mA at both temperatures.

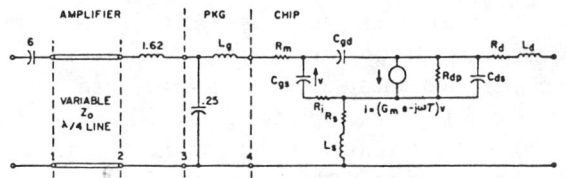

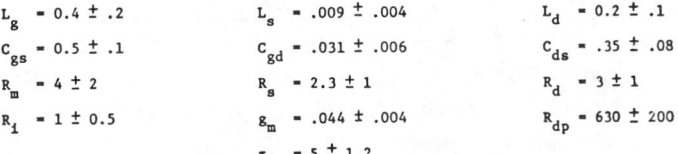

ELEMENT VALUES IN nH, pF, OHMS, MHOS, AND ps.

L_g = 0.4 ± .2	L_s = .009 ± .004	L_d = 0.2 ± .1
C_{gs} = 0.5 ± .1	C_{gd} = .031 ± .006	C_{ds} = .35 ± .08
R_m = 4 ± 2	R_s = 2.3 ± 1	R_d = 3 ± 1
R_i = 1 ± 0.5	g_m = .044 ± .004	R_{dp} = 630 ± 200
	τ = 5 ± 1.2	

Figure 11 - Equivalent circuit of amplifier input circuit and Mitsubishi MGF 1412 transistor. The element values are based upon DC resistance measurements, 1-MHz capacitance measurements, and three sets of S-parameters measured in different test fixtures. The resistance, R_i, cannot be separated from R_m by these measurements since the reactance of C_{gd} is so high; the value of R_i is an estimate. The error values are conservative estimates based upon analysis of all the data.

TABLE III
MEASURED AND THEORETICAL 4.9-GHz NOISE PARAMETERS OF MITSUBISHI MGF 1412 AT 300 K AND 20 K

Symbol	Expt Pkg 300°K	Expt Chip 300°K	Pucel Theory 300°K	Fukui Formula 300°K	Expt Pkg 20°K	Expt Chip 20°K	Pucel Theory 20°K
T_{min}	91±3	91±3	72	151	25±2	25±2	16
R_{opt}	20±2	63±6	74	22	15±1	47±3	41
$X_{opt} = -X_c$	55±5	101±9	119	42	58±5	107±10	104
$1/g_n$	100±20	205±40	668	727	370±30	800±100	1800
R_n	34±8	69±18	29	3.1	10±2	17±3	6.9
r_n	3.8±.6	15±3	8.2	-38	0.6±.2	2.6±1	0.9
R_c	-4±5	-31±15	8.1	167	0.9±4	-13±10	7.9

source but T_n and T_{min} are not for a lossless coupling circuit. Values for two reference planes defined in Fig. 11 are presented in Table III. The first and fifth columns are for a reference plane at the gate-lead case interface of the packaged device; the second and sixth columns are for the GASFET chip and thus the effects of case capacitance and gate bonding wire inductance have been removed. These "chip" columns should be compared with the theoretical columns which will be discussed in Section V.

C. Impedance Measurements

The noise theory to be discussed in the next section expresses the four noise parameters in terms of material parameters, dc bias values, device dimensions, and the equivalent circuit elements R_m, R_i, R_s, g_m, and C_{gs}, shown in Fig. 11. The values of $R_m + R_i = R_g$, R_s, and g_m, all measured at dc, are given in Table II. The capacitance, $C_{gs} + C_{gd}$, was measured at 1 MHz as a function of V_{gs} with a precision capacitance bridge [35]; a value of 0.75 pF was obtained at the low-noise bias of $V_{gs} = -1.1$ V. This includes case capacitance, which was then removed by measuring a defective device with the gate bonding wire lifted off the chip; a value of 0.25 pF was obtained for the case capacitance.

The above dc resistances and 1-MHz capacitances, together with a skin effect correction to R_m and an estimate of R_i, will be used in the comparison of theoretical and measured noise parameters. Attempts were made to determine $R_m + R_i$ and C_{gs} by microwave S-parameter measurements, to be described next, but the results were variable and more confidence is placed in the low-frequency measurements. The skin-depth of aluminum at 5 GHz is 1.2 μm which should be compared with the gate metallization thickness of 0.7 μm. The dc value of R_g is 2.7 Ω and its value at 5 GHz is estimated to be 5 Ω; this estimate is partially influenced by the S-parameter results which gave 4.3 to 12.7 Ω. The gate-charging resistance R_i, is estimated to be 1 Ω by noting that R_s and R_d have values of 2.3 and 2.1 Ω and R_i occurs over a length somewhat shorter than R_s and R_d. It should be noted that R_i is of second order importance to the noise results but the fact that R_m is determined by subtracting R_i from the measured R_g is of somewhat greater importance. The final value of R_m is 4±2 Ω and this error value limits the accuracy of comparison of noise theory and experiment.

The S parameters of a packaged device were measured from 2 to 18 GHz on computer-corrected automatic network analyzers at two organizations [36], [37] with different test fixtures and calibration procedures; a third set of S parameters is reported on the MGF 1402 data sheet. The input capacitance at 2 GHz determined from each of the three S-parameter sets was within 0.06 pF of the 1-MHz capacitance measurement. However, above 8 GHz the measured S parameters diverge, probably due to reference plane definition problems. The COMPACT program [59] was used to optimize circuit elements to give minimum weighted mean-square error between circuit and measured S parameters in the 2–10-GHz range. Some of the elements determined in this way have large variability dependent upon the data set used and details of the fitting procedure. As a relevant example, the value of R_g changed from 12.7 Ω to 7.8 Ω dependent upon whether R_s was fixed at 2.3 Ω or allowed to vary in the optimization procedure (which gave $R_s = 1.0$ Ω). Using S parameters measured with a different analyzer gave $R_g = 6.7$ and 5.4 for R_s fixed and variable, respectively.

The reasons for the variability of the R_g value are its small resistance relative to the reactance of C_{gs} and also the strong dependence of the input resistance upon the small feedback elements C_{gd}, L_s, and R_s. Our conclusion is that it is very difficult to determine R_g from S-parameter measurements at normal bias values; a careful measurement of S_{11} with $V_{ds}=0$ (and hence $g_m=0$ and no feedback effects) may be more successful.

V. Noise Theory

A. Summary of Present GASFET Noise Theory

A comprehensive paper covering the history, dc, and small-signal properties, and an exhaustive but not conclusive treatment of noise in microwave GASFET's was written in 1975 by Pucel et al. [28]. The noise treatment includes the induced gate noise mechanism of Van der Ziel [38], the hot-electron effects introduced by Bachtold [39], and presents as new, the mechanism of high-field diffusion noise due to dipole layers drifting through the saturated-velocity portion of the channel. This latter mechanism along with the thermal noise of parasitic resistances R_m and R_s in the input circuit was found to be the dominant source of noise in modern, short-gate-length GASFET's at room temperature.

The above paper predicts the correct dependence of noise upon drain current and, by adjustment of a material parameter D, which describes the diffusion noise but is not accurately know from other theory or experiments, a good fit to measured 4-GHz data is presented. However, the theory does not agree with experimental data at the low end of the microwave spectrum (<3 GHz) where a linear dependence of noise temperature upon frequency is predicted but not observed. It was suggested that this noise may be due to traps at the channel–substrate interface but devices with a buffer layer such as the NEC 244 also show the excess low-frequency noise [40]. The temperature dependence of the low microwave frequency noise also does not agree with the trap assumption [41]. A recent work by Graffeuil [42] attributes the low-frequency noise to frequency dependence of hot-electron noise and also matches experimental data without invoking the high field diffusion noise which is dominant to the Pucel et al. theory.

The four noise parameters $T_{\min}$, R_{opt}, X_{opt}, and g_n define the noise properties of any linear two-port and all other noise parameters can be derived from them. Other noise parameters which appear in GASFET noise theory are the noise resistances R_n and r_n, and correlation impedance $R_c + jX_c$. Here, R_n is proportional to the total mean square noise voltage in series with the device input, and r_n is proportional to the portion of this noise voltage which is uncorrelated with the current noise represented by g_n. (It should be remarked that the symbol R_n is used to represent a noise voltage source that is not in series with the input but is called the "input noise resistance" in some papers [38], [41].) The following relations between these quantities can be easily derived using the definitions given by Rothe and Dahlke [43]:

$$R_n = g_n |Z_{\text{opt}}|^2 \tag{11}$$

$$R_c = \frac{T_{\min}}{T_0} \frac{1}{2g_n} - R_{\text{opt}} \tag{12}$$

$$X_c = -X_{\text{opt}} \tag{13}$$

$$r_n = g_n(R_{\text{opt}}^2 - R_c^2) \tag{14}$$

$$r_n = \frac{T_{\min}}{T_0}\left(R_{\text{opt}} - \frac{T_{\min}}{T_0}\frac{1}{4g_n}\right). \tag{15}$$

The experimental results of the previous section will be compared with the Pucel et al. theory and also with the empirical equations of Fukui which are based upon fitting noise data to Bell Telephone Laboratory GASFET's at 1.8 GHz [44]. The theoretical results for the four noise parameters $T_{\min}$, R_{opt}, X_{opt}, and g_n, which are most directly measureable are expressed in terms of intermediate parameters r_n and R_c as follows [28]:

$$g_n = K_g / g_m X_{gs}^2 \tag{16}$$

$$r_n = t(R_m + R_s) + K_r / g_m \tag{17}$$

$$R_c = R_m + R_s + K_c R_i \tag{18}$$

$$R_{\text{opt}} = (R_c^2 + r_n / g_n)^{1/2} \tag{19}$$

$$X_{\text{opt}} = K_c |X_{gs}| \tag{20}$$

$$T_{\min} = 2T_0 g_n (R_c + R_{\text{opt}}) \tag{21}$$

where t is the ratio of device physical temperature to 290 K, $X_{gs} < 0$ is the reactance of the gate-to-source capacitance, and the three K's are noise coefficients which are given in [28] as functions of bias, device dimensions, and material parameters. K_g is proportional to the squared magnitude of the equivalent current source in the input circuit and represents the induced gate circuit noise current as well as the correlated[2] current needed to represent noise in the drain circuit. The uncorrelated noise resistance r_n contains a first term due to thermal noise in $R_m + R_s$ (and hence proportional to t as suggested in [28]) and a second term which, through K_r, is a measure of the noise voltage necessary to represent uncorrelated drain current noise.

---

[2] The words correlated and uncorrelated can refer to correlation between voltage and current sources in a Rothe–Dahlke noise representation or to correlation between gate current noise and drain current noise; an asterisk (*) will be used when the latter is meant.

TABLE IV
EXPERIMENTAL AND THEORETICAL NOISE COEFFICIENTS OF
MGF 1412 AT 300 K AND 20 K

Qty	Expt 300°K	Theory 300°K	Expt 20°K	Theory 20°K
K_g	0.89	0.27	0.33	0.15
K_c	1.55	1.83	1.65	1.60
K_r	0.36	.072	0.13	.03
P	2.50	.98	1.02	0.403
R	0.64	.26	0.27	0.082
C	0.893	0.961	0.917	0.935

It is important to note that the K noise coefficients are independent of frequency except for the possible frequency dependence of material parameters (high field diffusion coefficient D, saturated velocity v_s, hot-electron noise coefficient δ, and mobility μ). The noise coefficients are convenient for describing the measured noise performance of a device but three other coefficients P, R, and C are more directly related to the noise generation mechanisms. The mean-square drain current noise is proportional to P and the mean-square gate current noise is proportional to R; the correlation* coefficient between these two is C. The relations between the coefficients are given in [28] and are repeated below in an algebraically simplified form

$$K_g = R + P - 2C\sqrt{RP} \tag{23}$$

$$K_g K_c = P - C\sqrt{RP} \tag{24}$$

$$K_g K_r = RP(1 - C^2). \tag{25}$$

B. Comparison of Experimental Results with Theory

The experimental results and theory are compared in Tables III and IV. The theoretical values are obtained using the measured values of g_m and I_s from Table II and R_i, R_m, R_s, and C_{gs} from Fig. 11 (rather than values which could have been computed from device dimensions, and material parameters). The device channel thickness, $a = 0.10$ μm, and doping density $N = 2.9 \times 10^{17}$, were determined from the dc measurements. The gate length, $L = 0.7$ μm, was measured; this value is also reported in [34]. The material parameter values are those used in [28] ($D = 35$ cm²/s, $\delta = 1.2$, $E_s = 2.9$ kV/cm, $v_s = 1.3 \times 10^7$ cm/s, and $\mu_0 = 4500$ cm²/V·s). The device was operated with $V_{ds} = 5$ V, $I_d = 10$ mA, and $V_{gs} = -1.081$ at 300 K and -1.123 at 20 K; these values gave minimum noise at both temperatures. It is, of course, the experimental values corrected to the chip reference plane which should be compared with theoretical results, which do not include package parasitics.

The last column of Table III gives the theoretical results of Pucel et al. modified in the following way for application at 20 K.

1) The thermal noise in the parasitic resistances $R_m + R_s$, has been reduced by the factor $t = 20/290$ in (17). This reduces T_{min} from 72 K to 40 K.

2) The transconductance g_m has been replaced by the 20 K measured value (an increase from 0.043 to 0.062 mhos) and the saturation velocity v_s, and saturation field E_s, have been increased by a factor of 1.32 as deduced from the measured dc parameter changes. This further reduces T_{min} from 40 K to 26.5 K.

3) Thermal noise within the channel has been reduced by multiplying the Pucel factors P_0, R_0, and S_0 used to calculate the K noise coefficients by t; this brings T_{min} to 15.7 K.

4) As points of additional theoretical interest, if t is made equal to zero both for the external and internal thermal noise, T_{min} reduces from 15.7 K to 11.2 K; i.e., the theoretical coefficient of noise temperature versus physical temperature is 0.22 K per K at cryogenic temperatures. Also, if $D = 0$ (no high-field diffusion noise) $T_{min} = 5.0$ K. Thus at 20 K and at the experimental bias value which gives lowest noise temperature, approximately 1/3 of the total noise is contributed by each of the mechanisms of thermal, high-field diffusion, and hot-electron noise. This is conceptually only a rough approximation since the noise contributions to T_{min} are not additive; at each step above (i.e., $t = 0$ or $D = 0$) the source impedance has a new optimum which changes the contributions from remaining noise mechanisms. Perhaps a more meaningful comparison would be for T_n at a fixed source impedance.

Several authors [41], [42], [53] find the hot-electron noise coefficient δ to be a strong function of temperature but this is primarily because the nonthermal portion of the hot electron noise has been normalized to ambient temperature (see footnote 1). Our results suggest that a convenient form for the electron temperature T_e is

$$T_e = T_0[t + f(E)] \tag{26}$$

where $T_0 \equiv 290$ K (and cancels out in the noise figure expression) and $f(E)$ is the nonthermal noise dependent upon electric field but only weakly dependent upon temperature. For the last column of Table III, $f(E) = \delta(E/E_s)^3$ has been used with $\delta = 1.2$ and $E_s = 3800$ V/cm (compared to $E_s = 2900$ V/cm at 300 K). A better fit to experimental data would be achieved if either δ were higher or E_s lower at 20 K. Experimental evidence for the increase of $f(E)$ by a factor of ~2.5 at cryogenic temperatures is contained in our own unpublished measurements on millimeter-wave GaAs mixer diodes and also in the measurements of Keen [55].

Further insight into the noise temperature limitation at cryogenic temperatures can be gained by examining an approximate form of the minimum noise temperature equation

$$T_{min} = 2T_0 \cdot \sqrt{K_g} \cdot \omega C_{gs} \cdot \sqrt{\frac{t(R_m + R_s)}{g_m} + \frac{K_r}{g_m^2}} \tag{27}$$

which is valid for $K_g R_c^2 / X_{gs}^2 \ll t(R_m + R_s)g_m + K_r$; this approximation produces ~10-percent error for the MGF 1412 data. At room temperature the first term under the large radical is dominant and K_r is not important; the noise voltage generator in the input circuit is dominated by the thermal noise of $R_m + R_s$. At cryogenic temperatures where $t \to 0$ the second term K_r/g_m^2 becomes dominant (even though g_m has increased) and the noise temperature is limited by the amount of nonthermal noise coupled into the gate circuit and uncorrelated* with the drain current noise. The coefficient K_r is proportional to $1 - C^2$ where C, the correlation* coefficient, is near 1 and thus small changes in C are likely to have large effects on the cryogenic noise temperature; this will not be true at room temperature. The larger variability of the cryogenic noise temperature from one device to another may be due to this effect.

The agreement between experimental results and the theory of Pucel et al. for values of $T_{\min}$, R_{opt}, and X_{opt} is good at both 300 K and 20 K—especially if the error range for R_m is considered. Other samples of the MGF 1412 gave a lower noise temperature at 20 K (as low as 15 K) than the particular device which was evaluated in detail; this would improve the agreement with theory. The agreement with the theory is marred by a factor ~ 3 disagreement in the value of g_n. It is not known at present whether this is an experimental artifact or a failure of the theory. The value of g_n was determined from the noise temperature versus source resistance characteristic but it was also checked, with good agreement, by measuring the noise-temperature bandwidth of the data of Fig. 10. The discrepancy in g_n leads to an even larger discrepancy in the correlation resistance R_c, which depends, through (12), on the difference between $1/g_n$ and R_{opt}. The negative value of R_c which results from the measurements is physically possible (it only means the input voltage and current noise sources have a negative real part in their correlation coefficient) but is certainly in disagreement with the theory.

The experimental values of the K coefficients in Table IV were obtained by solving (16)–(21) in terms of the measured noise parameters. The coefficients P, R, and C were then obtained by solving (23)–(25) by an iteration method. The discrepancy between experimental and theoretical R_c and g_n further propagate into discrepancies in the noise coefficients. In addition, the experimental value of K_r at 300 K is subject to large error because the measured data is insensitive to K_r due to the dominance of thermal noise.

The agreement between the experimental results and the empirical equations of Fukui [44] is poor. This may be due to the fact that Fukui's equations are derived from measurements at 1.8 GHz where low-frequency noise generation mechanism has a large effect; thus the formula predicts a higher than observed noise temperature at 4.9 GHz. It may also be true that some other variables in the transistor fabrication (such as the doping profile near the substrate interface or gate metallization thickness) effect the equations. It should be noted that the Fukui noise parameter equations require the values of g_m and C_{gs} at zero gate bias; thus the measured zero bias values of 0.098 mhos and 0.75 pF have been used in Table III.

VI. Examples of Cryogenic GASFET Amplifiers

Several models of GASFET amplifiers for the 5-GHz frequency range and for use at 20 K have been designed. All use 1.75 mm² packaged GASFET's (usually the Mitsubishi MGF 1412), gold-plated copper (for thermal conductivity and solderability) metal parts (except for some brass contact tabs), and microstrip transmission lines with teflon-coated fiber-glass dielectric [45]. In order to avoid large thermal stresses and mechanical failures, tight and rigid connections are avoided. Metallized ceramic substrates are also avoided because of possible cracking of the ceramic or metal-film solder-joints after repeated temperature cycling.

All amplifiers utilize an external dc power regulator which automatically adjusts the gate voltage to maintain a set drain current. This requires one TL075BCM quad-operational amplifier chip per GASFET stage and provides buffered monitoring of V_{ds}, I_d, and V_{gs}. As shown in Fig. 7 bias protection circuitry is included in the microwave chassis. The 1N821 voltage-reference zener diode utilized for gate protection contains a diode which prevents forward conduction of the zener diode and thus allows the GASFET gate to be forward biased for testing; the zener diode limits negative gate bias to approximately -6 V. Both the 1N821 and 1N4099 diodes have sharper zener characteristics at 20 K than at 300 K.

A. Single-Stage Basic Amplifier

A single-stage amplifier, for use as a second stage following a cooled-paramp first stage, was required for the front ends of the very large array radio telescope [46]. The unit need not be optimized for minimum noise as the first stage gain is 15 dB. However, it must have a gain which is flat within ± 0.5 dB over the 4.5–5-GHz range, an output return loss $\geqslant 10$ dB, and must be highly reliable since 54 units are required in the 27-element, dual-polarization array. An input match is not required since the input will be connected through a short cable to the 5-port circulator of the paramp.

A photograph of the amplifier and some of the key parts is shown in Fig. 12. Utilizing a thermostatically-controlled hot plate, joints are soldered as follows: 1) GASFET source leads to mounting stud with pure indium solder [47] (for good thermal conductivity) and a flux [48]; 2) chip capacitors to chassis with silver-alloy flux-core solder [49]; and 3) zener-diodes and connector ground with low-temperature solder [50]. Other components are then soldered to the chip-capacitors with silver-alloy solder and a small soldering iron.

The input and output $\lambda/4$ transformers are 4.1 mm wide $\times$ 11.2 mm long $\times$ 0.75 mm thick microwave circuit board [45] and are held in place through slotted holes with 2-56 nylon screws. The slotted holes allow the transformer position relative to the GASFET to be adjusted to tune the center frequency of operation. Brass tabs under each

Fig. 12. Single-stage basic amplifier; scale on photograph is in units of mm. See Fig. 7 for schematic, Fig. 14 for close-up of transistor mounting stud, and text for construction details.

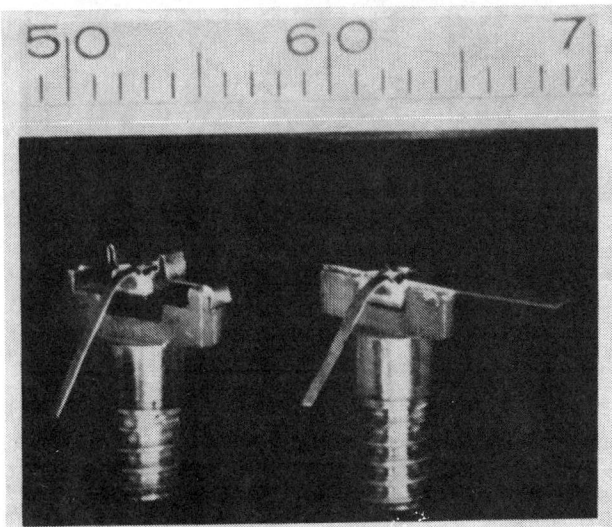

Fig. 14. Transistor mounting studs for feedback (left) and basic (right) amplifiers. Scale is in millimeters.

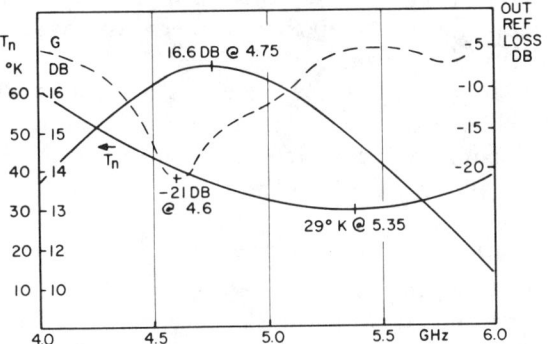

Fig. 13. Noise temperature, gain, and output return loss for single-stage basic amplifier at a temperature of 20 K.

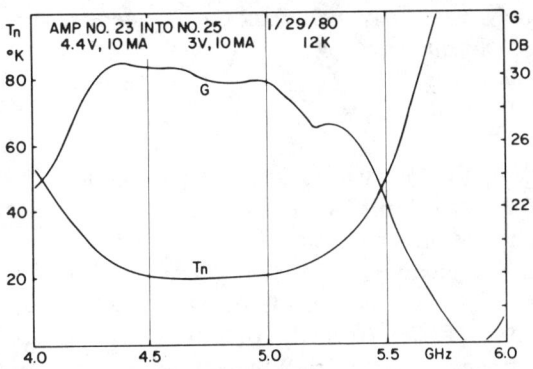

Fig. 15. Gain and noise temperature of cascade of isolator, feedback amplifier, isolator, and basic amplifier, all at 12 K. Noise temperature is 20 K over a 500-MHz bandwidth and <25 K over a 800-MHz frequency range. Output return loss is >10 dB from 4.5 to 5.1 GHz.

nylon screw make connections to the SMA input and output connectors and also to the GASFET gate and source leads. Two layers of 0.02-mm thick polyester tape [51] are placed between each connector tab and the transmission line to form a dc blocking capacitor. Bias voltage for both gate and drain is fed through coils formed of 3 turns of 0.2-mm diameter phosphor–bronze wire wound with 1.25-mm inner diameter; these have high impedance relative to the circuit and are not critical.

The completed amplifier is tuned by sliding the transformers and slightly bending gate and drain leads. The drain is tuned for maximum output return loss (>20 dB) at band center and the gate is tuned for flat-gain. This will result in a minimum noise frequency f_{min}, 600 MHz above the gain center frequency f_0, as shown in Fig. 13. This offset is predicted by the noise theory. For a source impedance consisting of an inductor in series with the source resistance, $f_{min}/f_0 = \sqrt{K_c}$, assuming the drain circuit is broad band compared to the gate circuit. (Neither of these assumptions is quite true for this amplifier.) This offset may be removed with source-inductance feedback as is described next.

B. Single-Stage Feedback Amplifier

A small modification of the previously described amplifier results in a unit with little frequency offset between peak gain and minimum noise temperature. The modification, described in Fig. 14, is to increase the source lead inductance by widening the GASFET mounting stud and also bending small loops in the source leads. This change has little direct effect upon noise temperature but the input circuit can now be tuned to a lower resonant frequency (4.25 GHz with $V_d=0$, 4.70 GHz with $V_d=4.4$) giving both optimum noise and maximum gain at ∼4.75 GHz. The gain and noise temperature of this amplifier, cascaded with the amplifier of the previous section, is shown in Fig. 15; cooled isolators are included at the input and between the two amplifiers. Output return loss is >10 dB from 4.5 to 5 GHz without an output isolator.

The use of source lead inductance to improve input match for a ∼1.5-GHz amplifier has been previously described [5], [52]. At 5 GHz the situation is more complex because of effects of gate to drain capacitance and

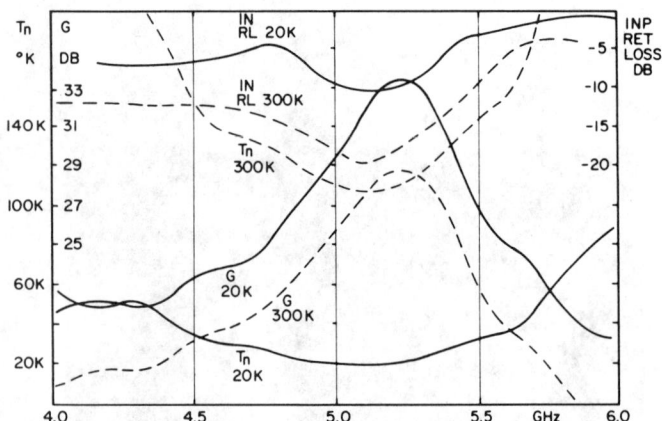

Fig. 16. Noise temperature, gain, and input return loss for two-stage feedback narrow-band amplifier described in Section IV-C at 20 K and 300 K. Output return loss at 5.0 GHz was 17 dB at 20 K and 25 dB at 300 K.

increased sensitivity of noise temperature to the reactance of the driving source. The FET source lead inductance L_s produces an effective impedance z_{FB} added in series with the input to the gate

$$z_{FB} = \frac{g_m L_S}{C_{gs}} \cdot \frac{1}{1 + y_d/y_1} \quad (28)$$

where y_d is the transistor output admittance for $g_m = 0$ and y_1 is the load admittance presented to the transistor. For the amplifier under consideration y_1 is adjusted so z_{FB} is primarily capacitive, the input resonant frequency is increased, and the noise-gain frequency offset is reduced to zero. At this value of y_1 the output is also matched (but $y_1 \neq y_d^*$ since y_d is for $g_m = 0$) but the input is not matched. In the next section a two-stage amplifier, matched by feedback, will be described.

C. Two-Stage Feedback Narrow-Band Amplifier

An amplifier was constructed by combining in one case the single-stage feedback amplifier and the single-stage basic amplifier previously described. (With the exception that input and output $\lambda/4$ transformer width was 6 mm; interstage $\lambda/2$ line was 4.1 mm wide.) The amplifier was then tuned at the desired frequency of operation, 5.0 GHz, for minimum noise by adjusting input inductance and for maximum input return loss by adjustment of first-stage source lead inductance and drain inductance. In this case, the feedback is used to achieve input match, z_{FB} is primarily resistive, and little attention was paid to the gain versus frequency response as the application was narrow band. The resulting gain, noise temperature, and input return loss at 300 K and 20 K are shown in Fig. 16. The results are flawed by the decrease in input return loss at 20 K; this is probably due to the change in g_m effecting z_{FB}. Another version of this amplifier was used with an input cooled isolator to achieve input match and a noise temperature of 17 K at 5.0 GHz

It is particularly desirable for cryogenic GASFET amplifiers to achieve match by feedback (or perhaps by a balanced configuration) since ferrite devices usually do not function well at both room and cryogenic temperatures. Our goal has been to construct an amplifier which is wide band, matched, performs well at both room and cryogenic temperatures, and hence does not utilize ferrite isolators. All of the above characteristics have not yet been achieved in a single unit; work will continue in this direction.

VII. Conclusions

All of the questions posed in the Introduction have not been answered and in some cases the answers are "hints" based on insufficient data; more work, both experimental and theoretical, is needed. Some conclusions that may be ventured are as follows.

A. Thermal Considerations

1) At temperatures above 15 K there is no fundamental problem in the cooling of a FET but attention must be paid to the fact that materials have vastly different thermal conductivity at cryogenic temperatures. In particular alumina, Kovar, and epoxy (whether silver loaded or not) become near insulators.

At temperatures below 15 K the thermal conductivity between the FET channel and the chip is greatly reduced due to boundary scattering of phonons and little can be done to alleviate this problem.

B. DC Characteristics and Material Properties

1) Changes in pinch-off voltage, barrier potential, and linear resistances within FET have been found, experimentally, to be small (see Fig. 3 and Table II) in the range 300 K–20 K. This implies that the changes in carrier density and mobility are small.

2) As the temperature was reduced from 300 K to 20 K, saturation current increased by 30–50 percent for the FET's of three manufacturers (see Fig. 3). This implies that the saturation velocity increases as the device is cooled; the differences between devices are due to either different doping density or some other constituent of the GaAs.

3) For the same 3 devices the output resistance decreased from 30–200 percent upon cooling; this has not been explained.

C. Small-Signal Characteristics

Transconductance increased by the same factor as saturation current and this causes an increase in RF gain that is somewhat smaller due to negative feedback and a concurrent decrease in output resistance. There is little other change in the gain versus frequency characteristic (see Fig. 5).

D. Noise

1) The noise temperature improvement factor for cooling from 300 K to 20 K varies from 3 to 5 for the six types of transistors tested at 5 GHz (see Fig. 9). It is not known

why the improvement factor varies; more detailed investigation is needed of devices with small improvement factors. The variation may be due to the stronger dependence of noise temperature upon the gate–drain noise correlation coefficient C at cryogenic temperatures.

2) The results for one device, the Mitsubishi MGF 1412, correlate fairly well with the noise theory of Pucel et al. [28] both at 300 K and 20 K. The noise improvement is due to the reduction in thermal noise of the channel and parasitic resistances and the increase in transconductance. The nonthermal noise power (hot-electron and high-field diffusion noise) may remain constant or increase at cryogenic temperature; the experimental data is of insufficient accuracy and content to make this judgement. A study of noise variation with bias has recently been performed [58] and forms a further test of the theory.

E. Amplifiers

1) An amplifier has been constructed with performance close to that of the best cooled paramps at 5 GHz (see Fig. 15).

2) It appears feasible, at least for narrow-bandwidths ($\sim$2 percent) at 5 GHz, to design unbalanced amplifiers which are matched by feedback rather than ferrite devices, and operate well at all temperatures from 300 K to 20 K (see Fig. 16).

3) The reliability of amplifiers constructed with the techniques described in Section VI has been excellent. At the time of this writing, 15 of the basic amplifiers have been constructed and subjected to a total of over forty 300 K to 20 K temperature cycles with no failures.

ACKNOWLEDGMENT

The author wishes to thank C. R. Pace for assistance with the construction and measurements, J. Granlund for the solution of the nonlinear differential equation for the forward-biased gate and for finding the transistor equivalent circuit elements from S-parameter measurements, and T. Brookes for several helpful discussions and the program for calculating results of the Pucel et al. noise theory. A. R. Kerr suggested the cooled-attenuator noise measurement procedure and the modification of the attenuator contact. The author appreciates the S-parameter measurements performed by R. Hamilton of Avantek, Inc. and R. Lane of California Eastern Laboratories.

REFERENCES

[1] *Diode and Transistor Designer's Catalog*, Hewlett Packard Co., Palo Alto, CA, 1980.
[2] *NEC Microwave Transistor Designer's Guide*, California Eastern Laboratories, Santa Clara, CA.
[3] Data Sheets, LNR Communications, Hauppauge, NY.
[4] D. M. Burns, "The 600 MHz noise performance of GaAs Mesfet's at room temperature and below," M. S. Thesis, Dep. Elec. Eng., Univ. of California, Berkeley, CA, Dec. 1978.
[5] D. Williams, S. Weinreb, and W. Lum, "L-band cryogenic GaAs FET amplifier," *Microwave J.*, vol. 23, no. 10, October 1980.
[6] C. A. Liechti and R. A. Larrick, "Performance of GaAs MESFET's at low temperatures," *IEEE Trans. Microwave Theory Tech.*, vol. MTT-24, pp. 376–381, 1976.
[7] J. Pierro and K. Louie, "Low temperature performance of GaAs MESFETs at L-band," 1979 *Int. Microwave Symp. Dig.* (Orlando, FL), IEEE Cat. No. 79CH1439-9, pp. 28–30.
[8] R. E. Miller, T. G. Phillips, D. E. Iglesias, and R. H. Knerr, "Noise performance of microwave GaAs F.E.T. amplifiers at low temperatures," *Electron. Lett.*, vol. 13, no. 1, pp. 10–11, Jan. 6, 1977.
[9] J. D. Kraus, *Radio Astronomy*. New York: McGraw Hill, 1966, ch. 7, p. 237.
[10] Model 21 Cryodyne, CTI-Cryogenics Inc., Waltham, MA, 02154.
[11] H. F. Cooke, "Fets and bipolars differ when the going gets hot," *Microwaves*, pp. 55–60, Feb. 1978. Also printed as *High-Frequency Transistor Primer, Part III*, Thermal Properties, Avantak Corp., Santa Clara, CA.
[12] M. G. Holland, "Phonon scattering in semiconductors from thermal conductivity studies," *Phys. Rev.*, vol. 134, pp. A471–480, Apr. 20, 1964.
[13] *American Institute of Physics Handbook*, 3rd ed. New York: McGraw Hill, 1972.
[14] *Thermal Conductivity of Solids at Room Temperature and Below*, Monograph 131, National Bureau of Stds, Boulder, CO, Sept. 1973.
[15] J. Callaway, "Model for lattice conductivity at low temperatures," *Phys. Rev.*, vol. 113, pp. 1046–1051, Feb. 15, 1959.
[16] C. L. Reynolds and A. C. Anderson, "Thermal conductivity of an electrically conducting epoxy below 3K," *Rev. Sci. Instru.*, vol. 48, no. 12, p. 1715, Dec. 1977.
[17] C. Kittel, *Introduction to Solid State Physics*, 5th ed. New York: Wiley, 1976.
[18] H. Fukui, "Determination of the basic device parameters of a GaAs Mesfet," *BSTJ*, vol. 58, no. 3, pp. 771–797, March 1979.
[19] R. Hackam and P. Harrop, "Electrical properties of nickel-low doped n-type gallium arsenide Schottky barrier diodes," *IEEE Trans. Electron Devices*, vol. ED-19, no. 12, pp. 1231–1238, Dec. 1972.
[20] D. Vizard, "Cryogenic DC characteristics of millimeter-wavelength Schottky barrier diodes," Appelton Laboratory, Slough, U.K., unpublished.
[21] F. A. Padovani and R. Stratton, "Field and thermionic-field emission in Schottky barriers," *Solid-State Electron.*, vol. 9, pp. 695–707, 1966.
[22] T. Viola and R. Mattauch, "High-frequency noise in Schottky barrier diodes," Res. Labs Eng. Sci. Univ. of Virginia, Charlottesville, VA, Rep. EE-4734-101-73J, Mar. 1973.
[23] M. Giterman, L. Krol, V. Medvedev, M. Orlova, and G. Pado, "Impurity-band conduction in n-GaAs," *Soviet Phys. Solid State*, vol. 4, no. 5, pp. 1017–1018, Nov. 1962.
[24] O. Emel'yanenko, T. Tagunova, and D. Naselov, "Impurity zones in P- and N-type gallium arsenide crystals," *Soviet Phys. Solid State*, vol. 3, no. 1, pp. 144–147, July 1961.
[25] P. Wolf, "Microwave properties of Schottky-barrier field-effect transistors," *IBM J. Res. Develop.*, vol. 14, pp. 125–141, Mar. 1970.
[26] A. Nara, Mitsubishi Semiconductor Laboratory, Hyogo, Japan, private communication.
[27] J. Ruch and G. Kino, "Transport properties of GaAs," *Phy. Rev.*, vol. 174, no. 3, pp. 921–927, Oct. 15, 1968.
[28] R. Pucel, H. Haus, and H. Statz, "Signal and noise properties of gallium arsenide microwave field-effect transistors," *Adv. in Electronics and Electron Physics*, vol. 38, L. Morton, Ed. New York: Academic, 1975.
[29] Model 350 Cryodyne, CTI-Cryogenics, Inc., Waltham, MA.
[30] A. R. Kerr, Goddard Inst. for Space Studies, New York, NY, private communication.
[31] Type 2156 Bellows, Servometer Corp., Cedar Grove, NJ.
[32] Type DT-500-CV-DRC, Lake Shore Cryotronics, Westerville, OH.
[33] M. Schneider, "Microstrip lines for microwave integrated circuits," *BSTJ* vol. 48, pp. 1421–1444, May 1969.
[34] T. Suzuki, A. Nara, M Nakatoni, and T. Ishii, "Highly reliable GaAs MESFET's with a static mean NF_{min} of 0.89 dB and a standard deviation of .07dB at 4 GHz," *IEEE Trans. Microwave Theory Tech.*, vol MTT-27 no. 12, pp. 1070–1074, Dec. 1979.
[35] Model 75D, Boonton Electronics, Parsippany, NJ.
[36] R. Lane, California Eastern Laboratories, Santa Clara, CA, private communication.
[37] R. Hamilton, Avantek Corp., Santa Clara, CA, private communication.

[38] A. van der Ziel, "Thermal noise in field-effect transistors," *Proc. IRE*, vol. 50, pp. 1808–1812, 1962.
[39] W. Baechtold, "Noise behavior of GaAs field-effect transistors with short gate lengths," *IEEE Trans. Electron Devices*, vol. ED-19, pp. 674–680, May 1972.
[40] NE244 Data Sheet, California Eastern Labs, Santa Clara, CA.
[41] K. Takagi and A. van der Ziel, "High frequency excess noise and flicker noise in GaAs FET's," *Solid-State Electron.*, vol. 22, pp. 285–287, 1979.
[42] J. Graffeuil, "Static, dynamic, and noise properties of GaAs Mesfets," Ph.D. thesis, Univ. Paul Sabatier, Toulouse, France.
[43] H. Rothe and W. Dahlke, "Theory of noisy fourpoles," *Proc. IRE*, vol. 44, no. 6, pp. 811–818, June 1956.
[44] H. Fukui, "Design of microwave GaAs MESFET's for broad-band low-noise amplifiers," *IEEE Trans. Microwave Theory Tech.*, vol. MTT-27, no. 7, pp. 643–650, July 1979.
[45] Type D-5880 RT/Duroid, .031" dielectric, 1 oz. 2 side copper, Rogers Corp., Chandler, AZ.
[46] S. Weinreb, M. Balister, S. Maas, and P. J. Napier, "Multiband low-noise receivers for a very large array," *IEEE Trans. Microwave Theory Tech.*, vol. MTT-25, no. 4, pp. 243–248, Apr. 1977.
[47] Indalloy No. 4 Solder, 100% Indium, 157°C, Indium Corp. of America, Utica, NY.
[48] #30 Supersafe Flux (water soluable), Superior Flux and Mfg. Co., Cleveland, OH.
[49] SN62 Solder, 62% Tin, 36% Lead, 2% Silver, 179°C, Multicore Solders, Westbury, NY.
[50] 20E2 Solder, 100°C, Alpha Metals, Jersey City, NJ.
[51] #74 Polyester Electrical Tape, 3M Co., Minneapolis, MN.
[52] L. Nevin and R. Wong, "*L*-band GaAs FET amplifier," *Microwave J.* vol. 22, no. 4, p. 82, Apr. 1979.
[53] J. Frey, "Effects of intervalley scattering on noise in GaAs and InP field effect transistors," *IEEE Trans. Electron Devices*, vol. ED-23, no. 12, pp. 1298–1303, Dec. 1976.
[54] J. Granlund, "Resistance associated with FET gate metallization," *IEEE Trans. Electron Devices*, to be published.
[55] N. J. Keen, "The role of the undepleted epitaxial layer in low noise Schottky barrier diodes for millimeter wave mixers," Max-Planck-Institute for Radio Astronomy, Bonn, W. Germany, to be published.
[56] S. Sesnic and G. Craig, "Thermal effects in JFET and MOSFET devices at cryogenic temperatures," *IEEE Trans. Electron Devices*; vol. ED-19, no. 8, pp. 933–942, Aug. 1972.
[57] D. Brunet-Brunol, "Etude et réalisation d'amplificateur a transistor a effet de champ a l'GaAs refroidi a trés basse température," *Rev. Phys. Appl.*, vol. 13, no. 4, pp. 180–187, Apr. 1978.
[58] S. Weinreb and T. M. Brookes, "Characteristics of low-noise GaAs MESFET's from 300 K to 20 K," in *Proc. European Microwave Conf.*, 1980.
[59] *COMPACT* Network Analysis Program, Compact Engineering Inc., Palo Alto, CA.

GaAs IC DIRECT-COUPLED AMPLIFIERS

Derry Hornbuckle
Hewlett Packard Company
1400 Fountain Grove Parkway
Santa Rosa, California 95404

ABSTRACT

Performance of six different direct-coupled GaAs IC amplifiers, with bandwidths up to 4.5 GHz, is described, along with that of one similar ac-coupled amplifier. Statistical data on gain, bandwidth, distortion, and noise is presented. The effectiveness of inductor peaking, resistive loads, and variable feedback is discussed.

Introduction

Previously reported GaAs MESFET monolithic integrated-circuit amplifiers have been narrow-band designs with passive tuning elements in C- and X-band.[1,2,3] A single exception is the report by Van Tuyl[4] of dc-coupled amplifiers with 4-GHz upper corner (-1 dB) frequencies. For this design approach, the absence of passive tuning restricts the frequency range to about 4 GHz with present geometries, but makes possible multi-octave applications and reduces circuit area by one to two orders of magnitude, with a similar effect on cost.

This paper presents experimental results on amplifiers which extend the direct-coupled design approach in several ways, including gain, bandwidth, and output power. All are capable of driving a 50-ohm load. (The voltage amplifiers of Ref. 4 use small monitor FET's at input and output, avoiding the bandwidth limitation associated with driving a large output FET for gain into 50 ohms.)

Amplifier Descriptions

Each amplifier consists of one or more of either the open-loop or the feedback stages of Figure 1, or variations to be described, driving a relatively large output FET. Table 1 summarizes the amplifier configurations.

All designs were based on computer simulations with either ASTAP[5] or OPNODE[6]; additional details are given elsewhere in this volume.[8]

Figure 2 shows typical measured gain and bandwidth for the seven amplifier types, which are further described in the following sections.

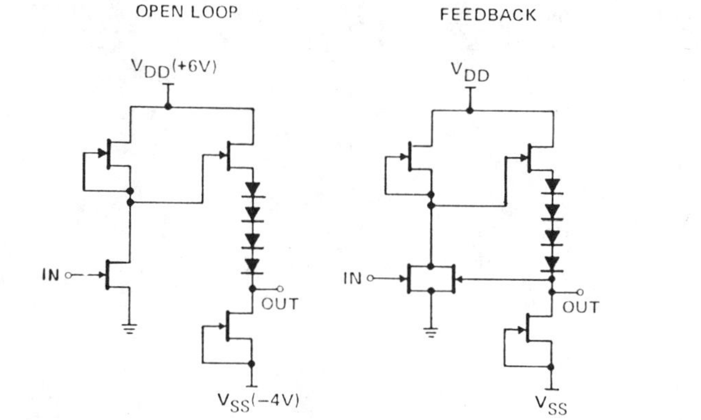

Fig. 1 Direct-Coupled GaAs IC Amplifier Stages.

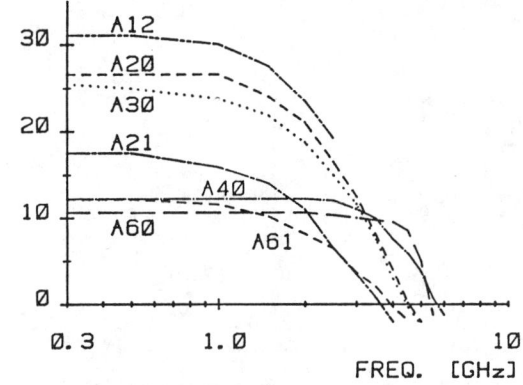

Fig. 2 Frequency Response of GaAs IC Amplifiers; R_L=50 ohms, V_{DD}=+6V, V_{SS}=-4V.

TABLE 1

CIRCUIT TYPE	STAGE 1	STAGE 2	STAGE 3	OUTPUT FET WIDTH (μ)	NOTES
A12	OL-30*	OL-30	FB-40	200	AC-Coupled
A20	OL-40	FB-40	FB-80	400	
A21	FB-60	FB-120	-----	1000	
A30	OL-60	OL-120	-----	400	Variable Feedback Around Stage 2
A40	FB-60	FB-120	-----	400	
A60	FB-120	-----	-----	500	Inductor Peaking
A61	OL-120	-----	-----	500	Bulk-Resistor Load, Inductor Peaking

* "OL" indicates open-loop stage (see Figure 1); "FB" inducates feedback stage. Number indicates device width in microns (width of inverter FET for open-loop stage, width of inverter FET plus feedback FET for feedback stage). Source-follower and current-source FET's are same width, active-load FET is half this width.

Reprinted from *IEEE MTT-S Int. Microwave Symp. Digest*, 1980, pp. 387-389.

Amplifier Performance

Wide-Bandwidth Amplifiers

The highest bandwidth, 4.5 GHz, was obtained with the aid of a single peaking inductor between the inverter and source-follower portions of a feedback-amplifier stage (A60-Fig. 3 (a)). Theoretical and experimental data were used to design the 13 nH square-spiral inductor, which takes up as much space as all the active circuitry on the chip, despite utilizing 3-micron lines and spaces. A chip photograph is shown in Figure 4 (a). This circuit achieved better bandwidth at lower power dissipation than the best totally active design (A40-Figures 3 (b) and 4 (b)).

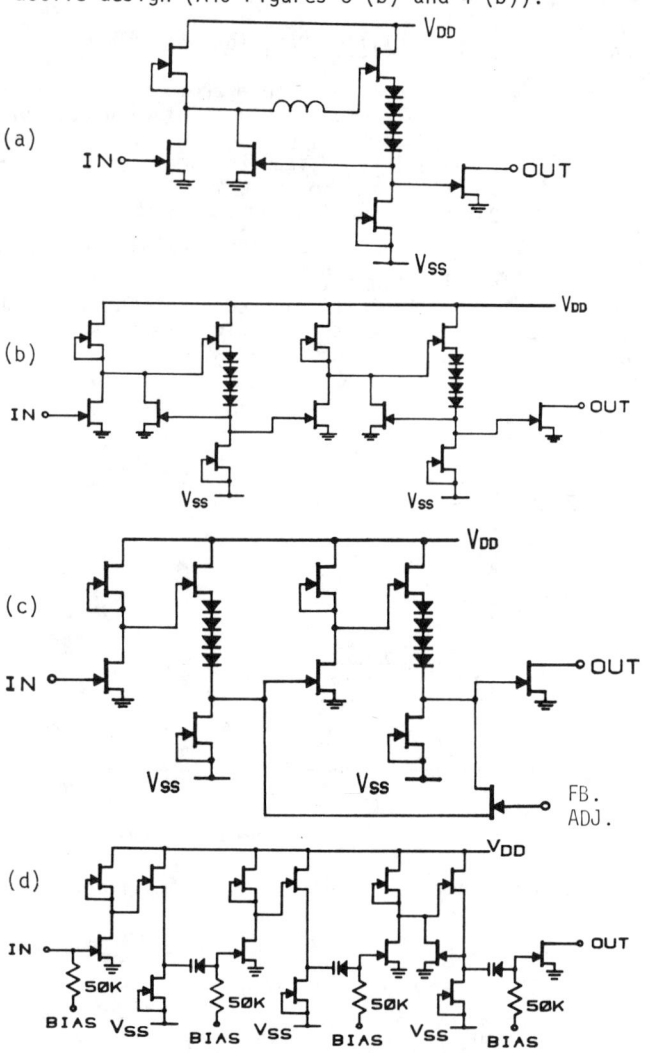

Fig. 3 Circuit Schematics For (a) A60, (b) A40 and A21, (c) A30, and (d) A12 Amplifiers.

Low-Distortion Amplifiers

Low-distortion applications were addressed with three different approaches: a) a large output FET, b) feedback, and c) a resistor in place of the active load. The large-output-FET amplifier, designated A21, was identical schematically to Figure 3 (b), but used a 1000-micron output FET in place of a 400-micron FET. It provided 10 dBm output power into 50 ohms with low harmonic distortion (Figure 5 (a)). A second low-distortion design (A30) used variable feedback around a full stage, the stage which contributes the most to distortion (Figure 3 (c)). The third low-distortion design (A61) made use of a bulk GaAs 300 ohm resistor in place of the less linear active-load device of the A60 version, with the same peaking inductor. Both of the latter approaches were less successful than the 1000-micron output design for minimizing distortion (Figure 5 (a) and (b)). Other amplifiers achieved harmonic performance similar to the 1000-micron-output version only at 5 dBm output power (Figure 5 (b)), except for the inductor-peaked (A60) design. Energy storage in the inductor helped to overcome the slew-rate effect limiting high-frequency output power, so that this circuit performed better than the 1000-micron-output design at high frequencies (Figure 5 (a)).

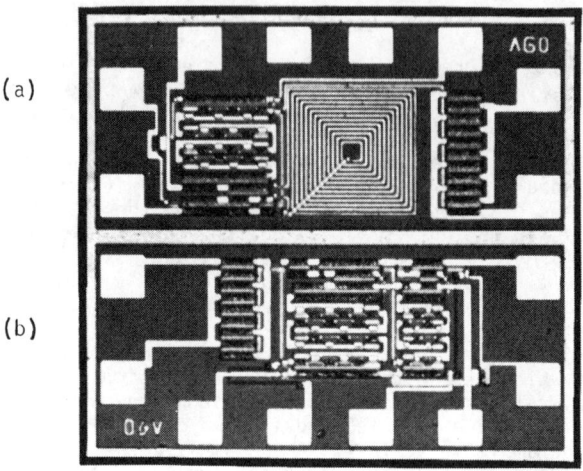

Fig. 4 Photograph of (a) A60 and (b) A40 Amplifiers; Chip Size is 300 x 650 Microns

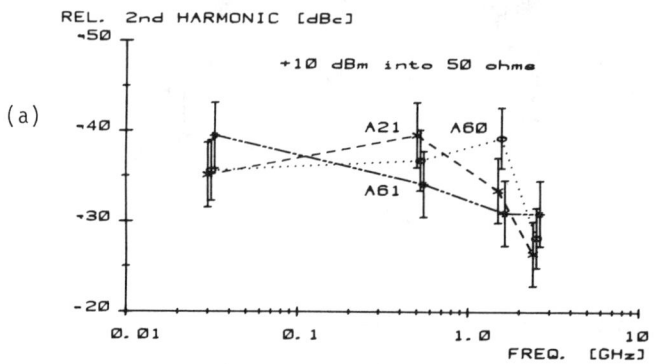

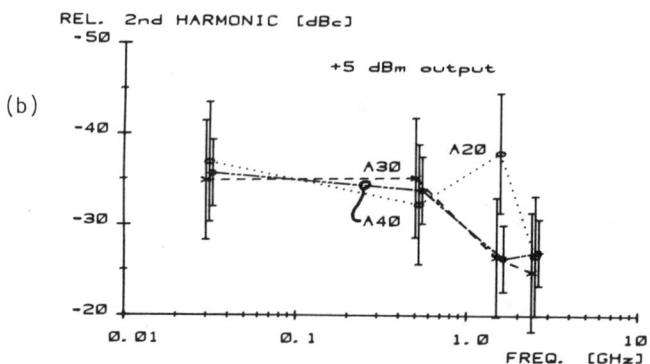

Fig. 5 Harmonic performance of various amplifiers at (a) +10 dBm and (b) +5 dBm output. Data is average for typically 20 chips; bars indicate one standard deviation above and below average.

AC-Coupled Amplifier

A low power-dissipation, ac-coupled design (A12-Fig. 3(d)) took advantage of varactor-diode coupling capacitors to eliminate the level-shift diodes and V_{SS} power supply; the three-stage amplifier had 30 dB gain and 2 MHz to 1.5 GHz bandwidth at the -3 dB points. Average dissipation was 420 mW, 30% lower than for any of the other designs. Dissipated power could be further reduced to about half this value by eliminating the buffer stages (source-follower and current-source MESFET's); however, bandwidth would be cut roughly in half as well.

Gain Variations

Chip-to-chip variations of gain were greatest for the amplifiers with the most stages, as expected. For example, the three-stage dc-coupled amplifier (A20) had a standard deviation of gain equal to 2.42 dB at 500 MHz (for 119 chips from 7 wafers), and 1.98 dB at 2.4 GHz -- which is one measure of bandwidth variation, since 2.4 GHz is above the amplifier's corner frequency. By comparison, the standard deviation for the 500 MHz gain of the single-stage inductor-peaked amplifier (A60) was 0.93 dB (for 81 chips from 5 of the above 7 wafers).

Noise Figure

Noise figure was close to computer simulations for white noise relative to a 50 ohm input termination; some representative values are: 16.8 dB average (0.84 dB standard deviation) for the dc-coupled three-stage design (A20) at 1.5 GHz, and 16.4 dB average (0.68 dB standard deviation) for the inductor-peaked active-load version (A60) at 1.5 GHz. This moderately high noise figure is due both to the contributions of the active-load and source-follower FET's, and to the lack of impedance matching to the amplifier's high optimum-noise input impedance. The 1/f-noise corner frequency was similar to that reported for discrete FET's (7), approximately 100 MHz, regardless of amplifier type.

Microwave Wafer Testing

Microwave characterization of over five hundred circuits, from a number of wafers, has formed the statistical base for the data presented above. The key elements of the test system which made this volume of testing possible were a thin-film probe card, which maintained 50-ohm lines to within 1 mm of the IC wafer; and a Hewlett Packard 8566A Spectrum Analyzer, which performed all gain, power, harmonic, and noise measurements under calculator control.

Conclusion

Production of the type of circuits described, along with companion linear circuits under investigation, should make possible compact, low-cost versions of systems which heretofore could only be realized in hybrid form.

Acknowledgement

The author is grateful to Rory Van Tuyl for ideas, knowledge, and encouragement; to Alejandro Chu for process development and initial wafer production; to Carol Coxen and Annie Fowler for fabrication; and to Carl Hart for testing.

References

[1] R.S. Pengelly and J.A. Turner, "Monolithic Broadband GaAs FET Amplifiers," Electronics Letters, 13 May 1976, Vol. 12 No. 10, pp. 251-252.

[2] J.G. Oakes, et al., "Directly Implanted GaAs Monolithic X-band RF Amplifier Utilizing Lumped Element Technology," 1979 IEEE Gallium Arsenide Integrated Circuit Symposium Research Abstracts, paper 22.

[3] J.L. Vorhaus, R.A. Pucel, and P. Ng, "GaAs FET Monolithic Integrated Circuit Power Amplifiers," ibid., paper 23.

[4] R. Van Tuyl, "A Monolithic Integrated 4-GHz Amplifier," 1978 IEEE International Solid-State Circuits Conference, pp. 72-73.

[5] Advanced Statistical Analysis Program, IBM Corporation, White Plains, New York 10604.

[6] OPNODE Program, Hewlett Packard Company, Sunnyvale, California 94088.

[7] E. Ulrich, "Use Negative Feedback to Slash Wideband VSWR," Microwaves, Vol. 17, No. 10, Oct. 1978, pp. 66-70.

[8] R.L. Van Tuyl, D. Hornbuckle, and D.B. Estreich, "Computer Modeling of Monolithic GaAs IC's", 1980 IEEE/MTT Symposium Digest (this volume).

20-GHz Band Monolithic GaAs FET Low-Noise Amplifier

ASAMITSU HIGASHISAKA AND TAKAYUKI MIZUTA

Abstract—A 20-GHz band monolithic GaAs FET low-noise amplifier has been developed. Design and fabrication were performed by obtaining the transmission properties of the microstrip lines on a semi-insulating GaAs substrate. The developed monolithic amplifier consists of a submicron gate GaAs MESFET and the input and output distributed matching circuits on a semi-insulating GaAs substrate measuring 2.75 mm × 1.45 mm. A noise figure of 6.2 dB and an associated gain of 7.5 dB were obtained at 21 GHz without any additional tuning adjustments.

I. INTRODUCTION

MUCH PROGRESS has been made in C- and X-band GaAs MESFET's in the past several years [1]–[4], and some of the parametric amplifiers and traveling wave tubes are being replaced by GaAs MESFET's [1], [4]. On the other hand, demands for GaAs FET amplifiers operating at K-band or higher are increasing for satellite communication systems applications. One problem is how the input and output matching circuits should be realized to exhibit the inherent superiority of GaAs MESFET at such a high frequency range. A decrease in the wavelength inevitably requires good accuracy of the pattern size for the matching circuits; for example, a small error in the length of the bonding wires brings about a serious deterioration in the microwave performance. As long as distributed or lumped element circuits constructed on alumina substrates are used in the conventional manner, it is difficult to obtain a good performance with a reasonable reproducibility.

In order to overcome these problems, monolithic GaAs FET amplifiers have been proposed and demonstrated [5]–[7], where in all active and passive elements are integrated on a semi-insulating substrate. In this approach, all bonding wires are eliminated and a fine matching circuit pattern is obtainable because the refined techniques developed for GaAs MESFET's are used for the fabrication of the matching circuits.

The merit of the monolithic approach is enhanced at higher frequency ranges than K-band, because the matching circuit on the GaAs substrate becomes smaller for higher frequencies. Most of the previous reports are, however, concerned with X-band amplifiers [5], [7].

The purpose of this paper is to demonstrate a K-band (20-GHz) monolithic GaAs FET low-noise amplifier, whose matching circuits are formed in a distributed form on a semi-insulating substrate. In order to obtain the fundamental data which are necessary in designing the distributed circuits on a semi-insulating GaAs substrate, the dielectric constant ϵ_r of GaAs crystal and the transmission properties of the microstrip line such as the characteristic impedance (Z_0) the slow-wave factor (λ_g/λ_0) and the attenuation constant (α) were also investigated. These fundamental data are shown in the next section. Section III shows the design and the fabrication process of a 20-GHz band GaAs FET monolithic low-noise amplifier. The microwave performance—power gain (G_a), noise figure (NF_{min}) and their frequency response—is shown in Section IV.

II. MEASUREMENTS OF TRANSMISSION PROPERTIES OF MICROSTRIP LINES ON AN SI GaAs SUBSTRATE

The semi-insulating (SI) GaAs substrate was confirmed to be a low-loss high-permittivity medium suitable as a substrate for microstrip lines. In order to obtain the fundamental data for the design of microstrip lines on a SI GaAs substrate, the dielectric constant (ϵ_r) of SI GaAs substrate and the transmission properties of the microstrip lines such as the characteristic impedance, the wavelength and the attenuation constant are measured. In the experiments, Cr-doped SI GaAs substrate having resistivity of more than 10^8 $\Omega \cdot$cm are used. Strip conductors were prepared by vacuum evaporation of Cr (600 A)/Au (2000 A) film followed by Au-plating of about 3 μm in thickness. Measurements were carried out for the frequency range of 2–12 GHz.

A. Relative Dielectric Constant (ϵ_r) of SI GaAs

The relative dielectric constant (ϵ_r) is an important parameter for the substrate medium of a microstrip line. Several measurements on ϵ_r of GaAs have been reported so far [8], [9] giving ϵ_r measurements of 10 to 13. In this work, in order to ascertain the previous data and to get a basic datum for the design of the microstrip lines, the relative dielectric constant ϵ_r of up to data GaAs substrate was obtained by measuring the resonant frequency of rectangular blocks of SI GaAs [10]. The basic resonator structure is shown in the inset of Fig. 1, where the two broad faces are metalized with Cr–Au film, but the side walls are left uncoated. The resonant frequencies were measured over the frequency range of 2 to 12 GHz by

---

Manuscript received June 3, 1980; revised August 28, 1980.
A. Higashisaka is with Basic Technology Research Laboratories, Nippon Electric Company, Ltd., Takatsuku, Kawasaki 213, Japan.
T. Mizuta is with Microwave and Satellite Communications Division, Nippon Electric Company, Ltd., Midoriku, Yokohama 226, Japan.

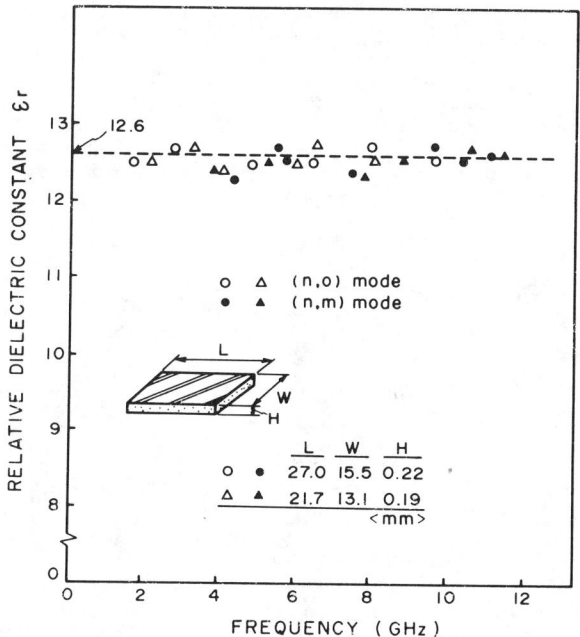

Fig. 1. Measured relative dielectric constant ϵ_r of semi-insulating GaAs.

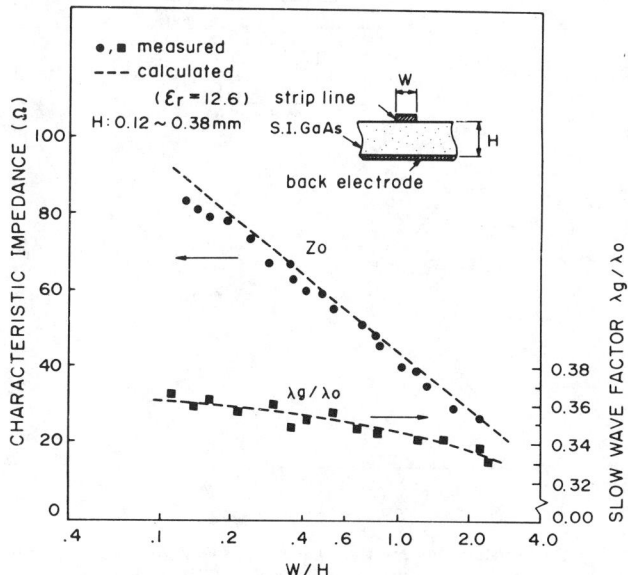

Fig. 2. Characteristic impedance Z_0, slow-wave factor λ_g/λ_0 versus dimensions of the microstrip lines constructed on SI GaAs substrate.

observing sharp minima on the transmission spectra with a Hewlett-Packard network analyzer.

The ϵ_r obtained is illustrated in Fig. 1 as a function of the resonant frequency. There are some scatter in the measured points, which is probably due to a coupling error [10], but the relative dielectric constant of SI GaAs is found to be about 12.6. This value agrees well with that recently measured by Neidert [9].

B. Characteristic Impedance (Z_0) and Slow-Wave Factor (λ_g/λ_0)

The relationship between the microstrip line size and the characteristic impedance is important in designing the actual microstrip circuits. The characteristic impedance was obtained by treating a sample as a $\lambda_g/4$ transformer. From the real part (R_I) of the input impedance of the sample terminated with 50-Ω load at the resonant frequency, the characteristic impedance (Z_0) was calculated by the following equation:

$$Z_0 = (R_I \, 50)^{1/2}. \qquad (1)$$

In our experiments, R_I was measured for a lot of samples with different linewidths (W), and with different substrate thicknesses (H) for the frequency range of 1.5 to 4 GHz.

Solid circles in Fig. 2 show the measured characteristic impedance as a function of the linewidth to the substrate thickness ratio (W/H). The broken line represents the theoretical one [11] given calculated by assuming ϵ_r to be 12.6. Experimental and the theoretical results show a good agreement, indicating that the dielectric constant of 12.6 can be adopted in designing the microstrip lines on a SI GaAs substrate.

Another important factor which is indispensable in designing the actual microstrip circuits is the wavelength in the microstrip line. The wavelength was measured by observing the resonant frequency of a half-wavelength microstrip line for the frequency range of 3.8 to 8.2 GHz. The measured wavelength (λ_g) normalized by the free space wavelength (λ_0)—slow-wave factor—is shown in Fig. 2 by solid rectangles. Measured λ_g/λ_0 agrees well with the theoretical one (broken line).

By using the experimental or theoretical results predicted in Fig. 2 the actual microstrip lines are designed. For example, the W/H value which gives Z_0 of 50 Ω is about 0.75, that is, when the substrate thickness is 200 μm, the linewidth is 150 μm. The slow-wave factor is, in this case, about 0.35. A wavelength at 20 GHz is calculated from the slow-wave factor to be about 5.25 mm.

C. Attenuation Constant (α)

Attenuation constant of the microstrip lines constructed on a SI GaAs substrate was measured and compared with that of microstrip lines on an alumina substrate. In this work, the attenuation constant was derived from the unloaded Q-value of a half-wavelength microstrip resonator.

Fig. 3 shows the attenuation per wavelength of samples with different linewidths and substrate thicknesses as a function of the measured frequency. Since the matching circuit becomes smaller for a higher frequency, the attenuation per wavelength ($\alpha\lambda_g$) is adopted in the figure. The thickness of the strip conductor (Au) is about 3 μm.

The attenuation ($\alpha\lambda_g$) of the microstrip line with 50-Ω characteristic impedance constructed on 0.38-mm thick SI GaAs substrate is about 0.185 dB at 4 GHz and 0.13 dB at 15 GHz, these values are almost the same as those of the microstrip line constructed on the alumina substrate. It is confirmed from this experimental result that the dielectric loss of SI GaAs is as small as that of the alumina substrate and the transmission loss of the microstrip line is almost that of the conductor.

It should be noted that the conductor loss is greater for

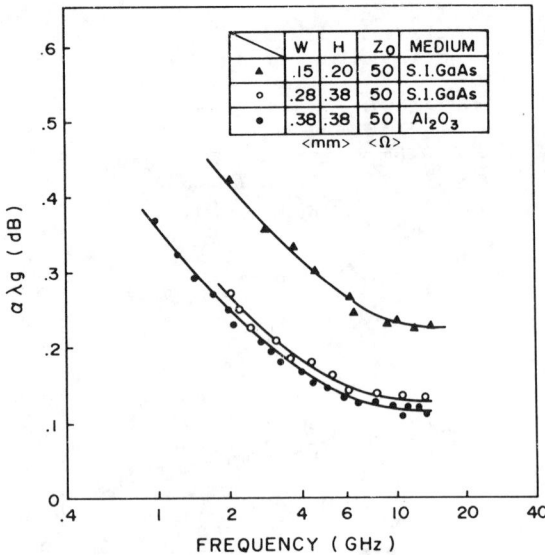

Fig. 3. Relationship between the transmission loss and frequency measured on microstrip lines constructed on a SI GaAs substrate and on a Al_2O_3 substrate.

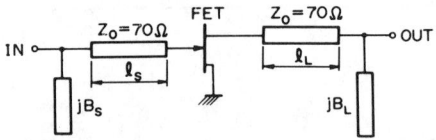

Fig. 4. Equivalent circuit of a 20-GHz band monolithic low-noise amplifier.

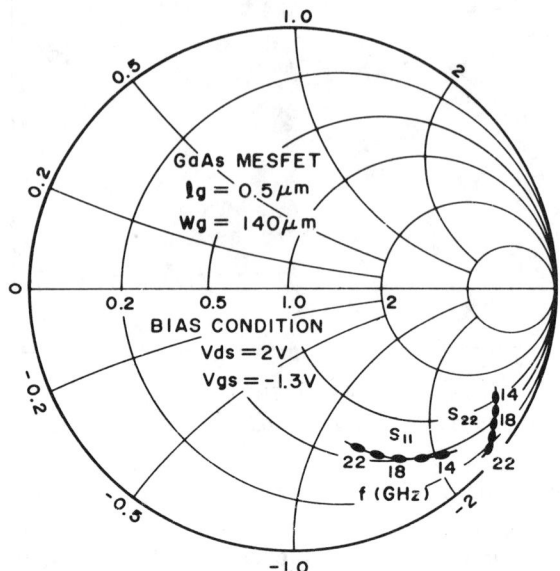

Fig. 5. Small-signal S-parameters (S_{11}, S_{22}) of a 0.5-μm gate low-noise GaAs MESFET with a gate width of 140 μm.

microstrip lines with a narrower linewidth on a thinner substrate. However, even for the practical SI GaAs microstrip line with the substrate thickness of 0.2 mm ($Z_0 = 50$ Ω), the attenuation per wavelength is about 0.22 dB at 15 GHz and higher frequencies. This is small enough from a practical point of view.

III. Design and Fabrication of Monolithic GaAs FET Amplifier

A. Design of Monolithic Amplifier

The merit of a monolithic GaAs FET amplifier is enhanced at higher frequencies such as K-band, because 1) the matching circuit can be made smaller for higher frequencies [7], and 2) it is not easy to construct such a high-frequency amplifier in a conventional manner. From these points of view, a 20-GHz band monolithic GaAs FET low-noise amplifier was adopted in this work to demonstrate its feasibility.

In order to obtain an adequate gain at 20-GHz band, a low-noise high-gain GaAs MESFET with a submicron gate was required as an active device. We have already developed submicron gate low-noise FET (NE388) [10]. The GaAs MESFET used in the monolithic amplifier was designed in the same manner as the previous discrete device having a carrier density of 2.0–2.5×10^{17} cm^{-3} and a pinch-off voltage of 1.5–2.5 V. The gate width is 140 μm (one-half of NE388). Equivalent circuit of the developed monolithic amplifier is shown in Fig. 4. The matching circuits are constructed with microstrip lines and parallel tuning stubs.

The matching circuits were determined using a Smith Chart based on the small-signal S-parameters (S_{11}, S_{22}) of the active device. Fig. 5 shows the frequency loci of the input and output impedance of the active device, which were derived from the measured S-parameters of the discrete device (NE388) by subtracting the bonding wire inductance and by doubling the impedance values.

The input and output matching circuits were designed not to achieve the wide-band operation, but to get the maximum gain at 20 GHz, since the main objective of this work is to test the feasibility of the monolithic approach of a low-noise GaAs FET amplifier at 20 GHz.

The characteristic impedance of the microstrip lines between the active device (GaAs FET) and the parallel tuning stubs were chosen to be 70 Ω by considering both the line length and the transmission loss—high characteristic impedance reduces the line length needed for the same impedance transformation but gives a larger loss. The tuning stubs are capacitive stubs with the length of $\lambda g/8$.

B. Device Fabrication

The basic structure and the fabrication process of the active device used in the monolithic amplifier are almost the same as those of the discrete device [12]. A vapor phase epitaxial layer grown successively on a high-purity epitaxial buffer layer is used. The gate electrode was formed by an aluminum film strip having a length of 0.5 μm and a width of 140 μm. The source and drain ohmic contacts were prepared by alloying a 0.15-μm thick gold–germanium film covered by a thin nickel film.

The strip line conductor was made of a chromium–gold film, which as thickened to 3 μm by gold-plating in order to reduce the conductor loss. At the same time, the source and drain regions were also metalized.

Fig. 6. SEM photograph of the 20-GHz band monolithic amplifier constructed on (a) SI GaAs substrate measuring 2.75 mm × 1.45 mm, and (b) magnified view of GaAs MESFET portion involved in the monolithic amplifier.

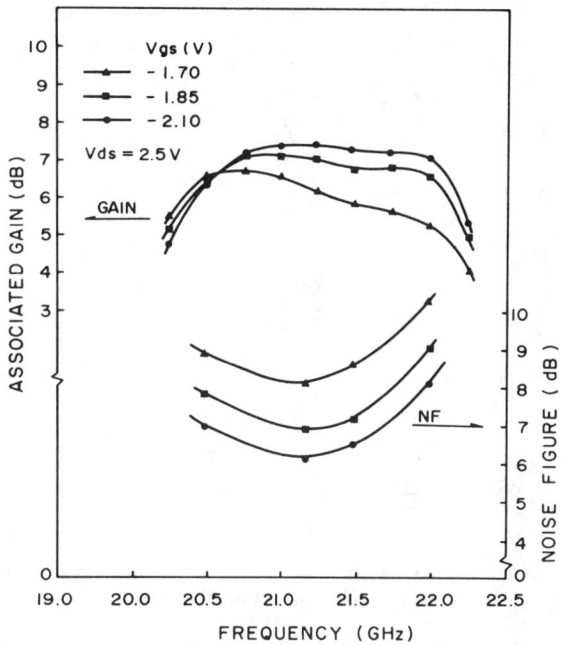

Fig. 7. Power gain and noise figure versus frequency of a monolithic GaAs FET amplifier (gate bias dependence).

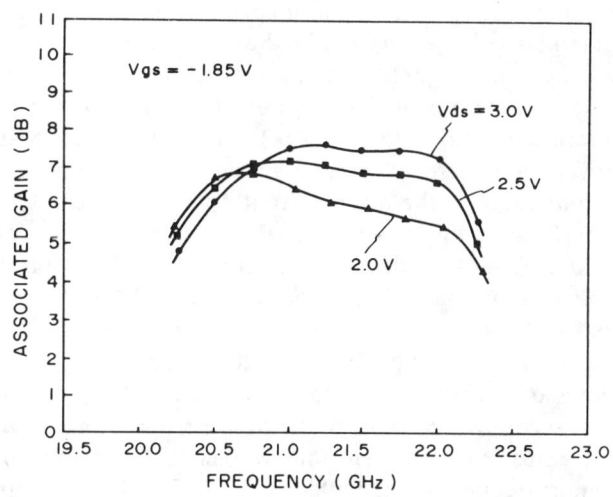

Fig. 8. Power gain response versus frequency as a function of drain bias voltage.

Fig. 6(a) shows the SEM photograph of the 20-GHz band monolithic GaAs FET low-noise amplifier formed on a 2.75-mm × 1.45-mm × 0.2-mm chip. The FET is in the center of the chip. The magnified photograph of the active FET portion is shown in Fig. 6(b). The 70-μm microstrip lines, are bent rectangularly to reduce the chip length.

The source regions are spreading gradually from the FET portion toward both sides, and grounded via a gold film plated on chip side. This technique, which can reduce the source lead inductance, is very important in obtaining a good performance at K-band and higher frequencies. According to our measurements, the source-ground reactance was about 4 Ω at 20 GHz, which corresponds to the inductance of about 32 pH.

IV. MICROWAVE PERFORMANCE OF THE DEVELOPED AMPLIFIER

A. Measurements

In the measurements, the input and the output terminals of the amplifier were connected to the external circuits with 50-Ω characteristic impedance with the mesh ribbons. The drain and the gate bias were applied through the dummy ports of the isolators attached at the input and output connectors of the test jig. DC block was achieved by coxial to waveguide transformers.

B. Microwave Performance

Power gain and noise figure response versus frequency of the monolithic GaAs FET low-noise amplifier are shown in Fig. 7 (gate bias dependence) and in Fig. 8 (drain bias dependence). The device is as made and no additional tuning adjustments were made inside or outside the chip. Although the performance degraded at lower gate bias or lower drain bias the following performance was obtained at the bias condition of $V_{ds} = 2.5$ V, $V_{gs} = -2.1$ V:

center frequency $\quad f_c = 21.3$ GHz
bandwidth $\quad$ BW = 1.7 GHz (20.5–22.2 GHz)
minimum noise figure $\quad$ NF$_{min}$ = 6.2 dB
associated power gain $\quad G_a = 7.5$ dB.

The center frequency is somewhat higher than the designed value, which is probably due to the fact that the matching circuits were designed by using the S-parameters of the discrete device. The bandwidth was not satisfactory

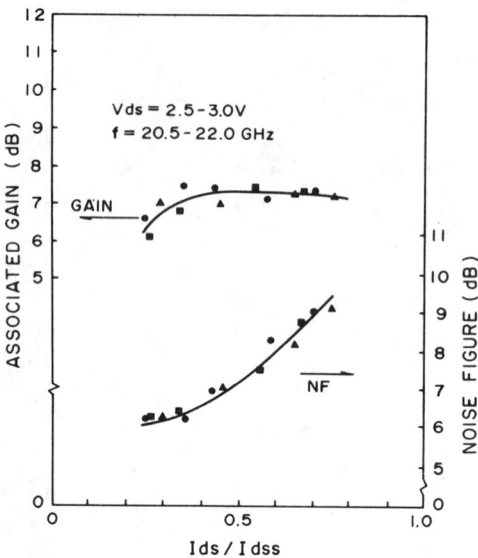

Fig. 9. Drain current dependence of the power gain and the noise figure of a 20-GHz band monolithic amplifier.

enough; this is simply because no attempt was made in the matching circuit design to achieve a broad-band response.

Power gain and noise figure depend more strongly on the gate bias voltage than on the drain bias voltage. Fig. 9 shows the noise figures and associated gains of several chips, as a function of drain current (I_{ds}) normalized to the drain saturation current (I_{dss}). The noise figure was improved with decrease of the drain current and showed a minimum value at the drain current I_{ds} of 0.2 to 0.3 I_{dss}. The best noise figure was 6.2 dB at 21 GHz with an associated gain of 7.5 dB. This performance is comparable or better than that of the conventional 20-GHz band GaAs FET low-noise amplifiers with discrete device [13].

One of the features of the monolithic approach is that it can give an uniform performance with a good reproducibility. According to our measurements on more than twenty chips from different four wafers, the scattering of the center frequency was within ±0.2 GHz, which is only 2 percent deviation. On the other hand, power gain and noise figure showed a little larger scatter of ±0.3 dB as shown in Fig. 9. These differences are probably attributable to the slight difference in the active device's performance, such as the gate length or the carrier concentration.

It is confirmed from these experimental results that, if only the active device (FET) is fabricated uniformly from chip to chip, the monolithic amplifier with a uniform performance can be obtained with a sufficient reproducibility.

V. Summary

It was demonstrated that a 20-GHz band monolithic GaAs FET low-noise amplifier can deliver a comparable or better performance than that of a conventional amplifier with a discrete GaAs FET. It was designed well, and the uniformity of the performance was good (center frequency: ±0.2 GHz; noise figure: ±0.3 dB).

The design and fabrication were performed after measuring transmission properties of the microstrip lines on the semi-insulating GaAs substrate. It was found that the dielectric constant of 12.6 can be used satisfactorily for the design of the microstrip lines on GaAs substrates. The dielectric loss of semi-insulating GaAs was negligibly small. The transmission line loss, most of which is due to conductor loss, was about 0.22 dB per wavelength at 15 GHz or higher frequencies for a 50-Ω stripline (3-μm thick gold) on a 0.2-mm thick GaAs substrate. This is small enough from a practical point of view.

The developed monolithic amplifier on a 2.75-mm× 1.45-mm chip exhibited the minimum noise figure of 6.2 dB with an associated gain of 7.5 dB at 21-GHz band without any external matching. The bandwidth at 1-dB gain compression was 1.7 GHz. From these results, monolithic amplifiers seem very promising in obtaining a superior performance with a sufficient reproducibility at K-band or higher frequencies.

Acknowledgment

The authors wish to thank Dr. F. Hasegawa for his valuable advices and discussions throughout this work. They would like to thank S. Fukuda and I. Haga for their valuable contributions in the design of the matching circuits, and Dr. H. Kato and S. Sugiura for their useful suggestions in measuring the transmission constants of the microstrip line. They would also like to thank Dr. K. Ayaki and A. Masuda for their constant support and encouragement on this work.

References

[1] K. Ohata, H. Ito, F. Hasegawa, and Y. Fujiki, "Super low noise GaAs MESFETs with a deep recess structure," in *1979 Int. Electron Device Meeting Dig. Tech. Papers*, Dec. 1979, pp. 277–280.
[2] J. V. DiLorenzo and W. R. Wisseman, "GaAs power MESFET's: design, fabrication and performance," *IEEE Trans. Microwave Theory Tech.*, vol. MTT-27, pp. 367–378, May 1979.
[3] A. Higashisaka, T. Furutsuka, Y. Aono, Y. Takayama, and F. Hasegawa, "Power GaAs MESFET's with a graded recess structure," *Japan. J. Appl. Phys.*, Supplement 19-1, pp. 339–343, 1980.
[4] P. T. Ho, C. M. Pham, and R. L. Mencik, "A 10 Watt, C-band FET amplifier for TWTA Replacement," in *1979 IEEE MTT-S Int. Microwave Symp. Dig.*, 1979, pp. 128–130.
[5] J. A. Higgins, A. Cupta, G. Robinson, and D. R. Chen, "Microwave GaAs FET monolithic circuits," in *ISSCC Dig. Tech. Papers*, 1979, pp. 120–121.
[6] V. Sokolov, R. E. Williams and D. W. Shaw, "X-band monolithic GaAs push-pull amplifiers," in *ISSCC Dig. Tech. Papers*, 1979, pp. 118–119.
[7] R. A. Pucel, P. Ng, and J. Vorhaus, "An X-band GaAs FET monolithic power amplifier", in *1979 IEEE MTT-S Int. Microwave Symp. Dig.*, 1979, pp. 387–389.
[8] K. S. Champlin, R. J. Erlandson, G. H. Glover, P. S. Hange, and T. Lu, "Search for resonance behavior in the microwave dielectric constant of GaAs", *Appl. Phys. Lett.*, vol. 11, pp. 348–349, Dec. 1967.
[9] R. E. Neidert, "Dielectric constant of semi-insulating gallium arsenide," *Electron. Lett.*, vol. 16, pp. 245–246, Mar. 1980.
[10] P. H. Ladbrooke, M. H. N. Potok, and E. H. England, "Coupling errors in cavity resonance measurements on MIC dielectrics," *IEEE Trans. Microwave Theory Tech.*, vol. MTT-21, pp. 360–362, Aug. 1973.
[11] H. A. Wheeler, "Transmission line properties of parallel strips

separated by a dielectric sheet", *IEEE Trans. Microwave Theory Tech.*, vol. MTT-13, pp. 172–185, Mar. 1965.

[12] M. Ogawa, K. Ohata, T. Furutsuka, and N. Kawamura, "Submicron single-gate and dual-gate MESFET's with improved low noise and high gain performance," *IEEE Trans. Microwave Theory Tech.*, vol. MTT-24, pp. 300–305, June 1976.

[13] M. M. Nowak, P. A. Terzian, and R. D. Fairman, "K-band F.E.T. amplifiers," *Electron. Lett.*, vol. 13, pp. 159–160, Mar. 1977.

Advanced microwave circuits

Computer-aided design and innovative physical configurations help increase performance and reduce costs

New developments in microwave circuitry are expected to meet future systems requirements for increased RF power, higher frequencies, wider bandwidth, lower noise levels, and smaller high-density packaging at lower cost. Advanced microwave circuitry is slated for use in diverse areas such as satellite communications, navigational equipment for ships, radar systems, and phased-array modules for information transfer systems. (See "High-frequency components play catch-up," *Spectrum*, November 1976, pp. 31–35.)

In many cases, microstrip circuitry and discrete active devices used today cannot do the job. The following developments in microwave technology promise to fill the bill:
- GaAs FETs are emerging as major circuit elements in applications related to power, low noise, gain and phase control, high efficiency, fast-tuning filters, and mixers.
- Monolithic circuits hold promise for high-volume, low power, ultrawide-band applications at higher microwave frequencies.
- Hybrid circuits are indicated for cost-effective designs at moderate quantities especially at frequencies below X band and in high power applications.

Most present microwave components and subsystems use standard microstrip circuits that are defined on many substrates, interfaced with separately packaged active devices, and housed in some form of metal container. Earlier attempts to integrate several functions on a common substrate were in most cases not successful, since such circuits required hand-trimming. It was more convenient for a manufacturer to divide a module into small subcomponents that could be tested easily.

One example of a complex multielement microwave module is RCA's high-gain power amplifier for satellite communications. A common metal housing encloses a series of GaAs FET amplifiers, isolators, and combining and bias networks. The unit has a total gain of 55 dB and a power output of 8.5 W, with 30 percent overall efficiency from 3.7 to 4.2 GHz. Individual amplifiers and subcomponents can be tested separately and trimmed for optimum performance. Final trimming can also be performed on the complete amplifier to reduce gain and phase ripples and to increase the power output and efficiency of the module. Time-consuming hand-trimming is justified in components that operate very close to the limits of the state of the art and where the expense of such a design can be afforded.

There are, however, increasing microwave applications in which cost and size are dominant factors.

Two main trends to overcome the limitations of present microwave components can be identified: the monolithic approach that promises cost-effective mass production of very large numbers of modules (10^5 to 10^6), and the hybrid approach that can also provide small sized modules, but is more versatile and can be used for smaller quantities. In both cases, GaAs devices, as well as advanced computer-aided measurement, design, and optimization, will be important factors in bringing these technologies to fruition.

GaAs FETs replacing Si diodes, bipolar transistors

For many applications above a few gigahertz, silicon diodes and bipolar transistors have been replaced with GaAs devices. There are still some special cases—high-power PIN diodes, millimeter-wave IMPATTS, and bipolar transistors for class-C power amplifiers in the lower microwave frequency range, for example—where the earlier devices maintain a strong hold. However, many of today's systems are being designed entirely around GaAs devices. Their great popularity can be attributed to their frequency response and versatility.

Present GaAs single-chip power FETs can handle up to 8 W at 6 GHz and as much as 1 W at 15 GHz. Even higher power outputs have been achieved when several chips have been combined on the same carrier. Using internal lumped-element matching networks and four parallel chips, Mitsubishi of Japan has obtained 10 W output at 10 GHz.

Power FETs provide not only low intermodulation and linear performance, but also very high efficiencies. Recent work at RCA has shown that GaAs FETs can operate at a power-added efficiency of 72 percent with 1.2 W output at 2.45 GHz. The high efficiency results from proper shaping of voltage and current waveforms and most likely can be improved still further. The FET also offers low noise performance. Figures as low as 1.5 dB have been

[1] In addition to good power-handling capabilities, FETs offer low-noise performance. Noise figures as low as 1.5 dB have been reported at X-band frequencies.

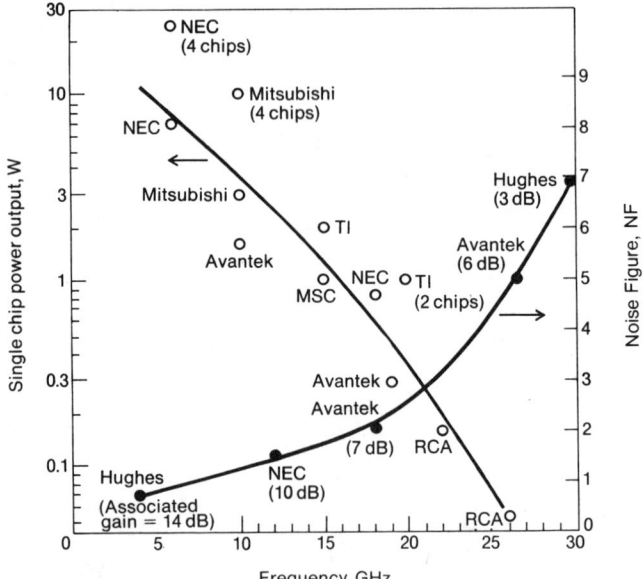

E.F. Belohoubek   RCA Laboratories

reported at X-band frequencies (Fig. 1).

Aside from functioning as amplifiers, GaAs FETs also make efficient microwave oscillators. Efficiencies of 19 percent have been reported at 16 GHz. The noise behavior of these oscillators, if properly designed with high-Q resonators, can approach that of the much lower-efficiency transferred-electron oscillators.

By adding an additional gate, one can form a new device, the so-called dual-gate FET. This has become a key component in a wide variety of sophisticated system applications. Dual-gate FETs are used as mixers, switches, amplitude-controlled amplifiers, limiters, and discriminators, to name some important examples. Another interesting application lies in the use of three to four amplitude-controlled dual-gate FET amplifiers to form a wide-band phase shifter that also has amplitude control. Compared with the usual diode-switched phase shifters, these units offer true analog control (infinite bit number) and simultaneously permit continuous adjustment of the output amplitude. These features will be important in phased-array modules, where the presently used hand-trimming to obtain given gain and phase shift values will soon be replaced by computer-controlled adjustments of dual-gate FET phase-shifter/amplitude-control modules.

Other GaAs FET applications are in high-speed modulators, where the FETs act as fast switches to replace diodes and also to provide extra gain and, more recently, ultrafast-tuning high-Q filters. In the latter, FETs are used as negative-resistance elements to compensate for losses of varactor-tuned lumped-element circuits, and the result is compact, rapidly tunable filter elements. In addition, complex switching networks using FETs in integrated form are under study for satellite communications. The list of possible uses is by no means exhausted.

The reliability of GaAs devices is not as well understood and subject to control as that of some of the more mature Si devices. However, in recent life tests, GaAs FETs for satellite communications have shown high reliability. Their choice for both the low-noise and power-amplifier sections of the RCA Satcom satellite indicates their great potential. The solid-state GaAs FET amplifier weighs less and has simpler power-supply requirements than does the TWT, and yet it maintains comparable efficiency.

Monolithic technology advancing

The versatility of the GaAs FET is a major reason why the development of monolithic GaAs multifunction chips is being spurred. Silicon technology is much more mature and, in the form of silicon-on-sapphire, it may find some application in the lower microwave frequency range. However, the vast majority of potential monolithic circuit applications are based on GaAs.

A monolithic module is a chip in which devices and circuit components are integrated to form a complete functional unit. It may be as simple as a single amplifier stage with input and output matching or as complex as a complete receiver front end, consisting of an oscillator, mixer, filter, IF amplifier, and all the bias networks. The major advantage of the monolithic configuration is its very small size, the possibility for mass fabrication and, as a result of that, low cost.

However, to obtain low cost there must be a real need for very large numbers of modules. Only then can the large capital and development outlays be justified to bring this technology to fruition. Furthermore, only if the overall chip is relatively small can cost savings be achieved. This would allow for the fabrication of many chips at the same time by large wafer processing techniques, as pioneered by the Si LSI process. This means that the circuit has to be compressed into a very small area by the use of either lumped-element technology or circuits in which capacitors and inductors are replaced by active devices. These monolithic circuits will be best suited for frequencies at X band and above. Microwave circuits at lower frequencies, even in lumped element form, require a rather large substrate compared with that used by active devices.

Another important consideration in monolithic circuit design is power handling. For low-noise and medium-power applications, the GaAs substrate (which is a much poorer heat conductor than Si) can satisfactorily conduct the heat generated by the active devices to the outside heat sink. In power applications, the requirement for a very thin substrate for heat transfer directly interferes with the need for a certain minimum substrate thickness to keep the circuit Q high and power losses at an acceptable level. These considerations indicate that the best candidates for monolithic integration will be low-power circuits.

An example of a state-of-the-art chip development is Hewlett-Packard's monolithic wide-baseband amplifier design. It consists of a two-stage amplifier with only one peaking inductor between stages to increase the bandwidth. The amplifier has a gain of approximately 10 dB from 300 MHz to 4.5 GHz. Multioctave low-power amplification can be obtained from extremely small chips (300 × 650 micrometers) at the expense of a relatively high noise figure, and limited upper-frequency response.

The more conventional approach to monolithic amplifiers is exemplified by Texas Instruments' monolithic amplifier, in which lumped inductors and capacitors are used as matching elements. Both the three-stage and the four-stage versions of the monolithic amplifier have the same chip size, 1 × 4 mm. The four stage amplifier uses a FET structure with 2400-micrometer gate width to generate approximately 1 W from 8.6 to 9.2 GHz. The overall gain of the chip is 27 dB. This design probably approaches the upper limit in power handling for a monolithic chip of this size.

Another example of a high level of integration is Rockwell International's 8-GHz receiver front end. It includes a preamplifier, mixer, and IF amplifier. The single-ended mixer uses a dual-gate FET, followed by a common-drain output stage. The components are still made and tested individually, but since they use the same process technology they all could be combined eventually on a common GaAs substrate.

Plessey is working on a complete monolithic receiver chip that uses an image-rejection mixer for the 2.7- to 3.5-GHz band. The image-rejection feature requires a 90° phase shift between the RF inputs to two balanced mixers. Because the size of a 3-dB quadrature coupler becomes prohibitive for monolithic integration at lower frequencies, Plessey developed a special active phase-splitter/amplifier (Fig. 2). This chip contains the low-noise preamplifier and active quadrature splitter, using a silicon nitride overlay as well as interdigital capacitors and spiral inductors. Here also, the final goal is full integration of all receiver front-end components on one chip.

To achieve full-phase and amplitude control, a monolithic GaAs module has been designed at the RCA David Sarnoff Research Laboratory with four dual-gate FET amplifiers. The amplifiers are used with in-line 90° and 180° splitters and combiners. A dual-gate amplifier covering the frequency range from 4.0 to 8.0 GHz has already been realized in monolithic form (Fig. 3), and the feasibility of its operation has been demonstrated with separate discrete couplers. Eventually, active phase splitters, similar to those used by Plessey in its image-reject mixer and by TI in its paraphase amplifier, will permit the integration of phase and amplitude control on a common GaAs chip. Since the phase control achieved in such a module would be fully continuous (there is no flyback phenomenon at 360°), frequency translation without sidebands could be obtained.

A Role for Hybrid Technology

The term hybrid, as used for lower-frequency components, encompasses single-substrate passive circuit boards to which one of

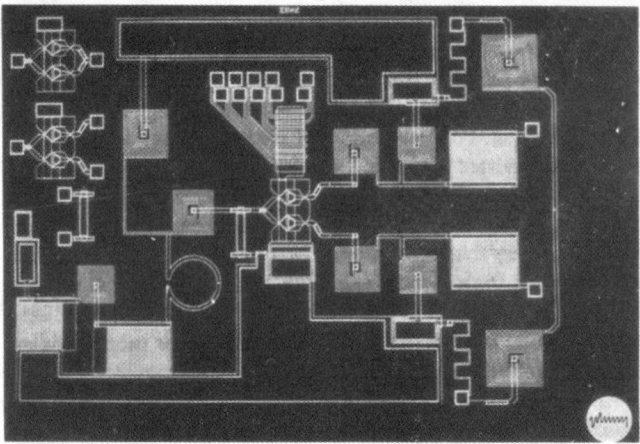

[2] Plessey developed this special active phase-splitter/amplifier. The chip contains a low-noise preamplifier and active quadrature splitter and uses a silicon nitride overlay, as well as interdigital capacitors and spiral inductors.

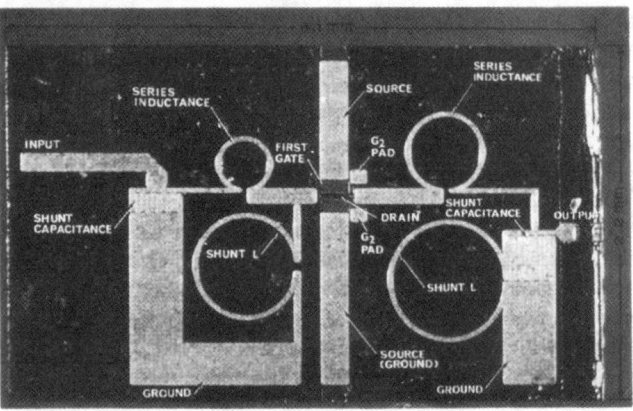

[3] Four dual-gate FET amplifiers with in-line 90° and 180° splitters combiners can be used in the design of a small wide-band module that provides continuous phase and amplitude control. A single dual-gate amplifier (shown above) covering the frequency range from 4.0 to 8.0 GHz has been realized in monolithic form.

several active chips are attached. In microwave technology, as discussed here, a similar definition can be used: Hybrid circuits differentiate smaller, more integrated modules from standard microwave components and true monolithic approaches. The choice between microwave hybrid and monolithic modules may not always be clear-cut, but there are definite conditions that favor one or the other.

Since hybrids use different materials and processing steps for the circuits and the devices, each circuit and each device can be separately optimized. The circuit substrates can cost less and have better heat dissipation and lower loss properties than substrates used for device fabrication.

To achieve very small size, light weight, high reliability, and low cost, the ideal hybrid can be a single substrate (alumina, quartz, or BeO, as the application demands) that has all the circuit components—lumped inductors, capacitors, resistors, etc.—in integrated form. Because lumped elements would be used wherever possible, the substrate would be very small and therefore suited for low-cost batch processing.

Active devices in such a design can be flip-chip mounted to this substrate to form the complete operating module. The GaAs FETs used here, unlike Si bipolars, have the advantage that all contact points are accessible on the top surface. These contacts when equipped with plated bumps, have the triple function of providing the mechanical support, the electrical connections, and the heat sink. Flip-chip bump mounting eliminates the commonly used wire bonds and reduces parasitics and the manual labor content. The circuit substrates could have either wrap-around grounds or be equipped with via holes or special metal septums to provide the necessary isolation between stages and the low-inductance ground returns for the devices.

In the examples of state-of-the-art hybrids that follow, not all of the above features are necessarily incorporated, but with the features illustrated, the advantages of hybrid-module technology become quite clear.

An example of a single-stage hybrid amplifier is a design from Watkins-Johnson of Palo Alto, Calif. By reducing parasitics and using positive and negative feedback, the manufacturer has obtained extremely wide-band performance from 350 MHz to 14 GHz. The single-stage amplifier has a minimum gain of 4 dB over this frequency range with a power output of 20 mW.

A very wide-band hybrid circuit at the upper frequency limit of today's devices is Avantek's experimental six-stage hybrid amplifier covering 13 to 28 GHz. It has a gain ranging between 23 and 32 dB, power output of 20 mW, and a maximum noise figure of 11.7 dB. To ease the problems associated with cascading many amplifier stages, this design uses balanced amplifiers with two active devices per stage throughout. Each stage—including two quadrature couplers, input- and output-matching networks, and two GaAs FETs—is assembled on a common ceramic substrate. This construction is very cost-effective and is being used in a wide variety of commercial amplifiers.

Similarly, a small amplifier module with very wide bandwidth, but higher output power, has been developed by TI. The single-stage amplifier covers a frequency range from 6 to 18 GHz, with power output of 200 mW and a small-signal gain of 6 dB. The average power-added efficiency is 18 percent over the band. This amplifier is a building block for a planned multistage wide-band power amplifier at the 1-W level. The unit is still fabricated more along the lines of a standard microwave module, although the packaged device has been replaced by a low-parasitic chip carrier. The separate mounting of the active device on a metal ridge between the input- and output-matching circuits is necessary for heat dissipation.

A hybrid 16-GHz medium-power amplifier from RCA Laboratories achieves proper heat sinking, together with a very low inductance ground return, while providing high-Q matching circuits. In the experimental implementation, discrete capacitors and bond wires are still used as matching elements close to the device pellet. Later configurations will use thin-film lumped-element circuits directly on the BeO carrier. BeO is used to provide good heat sinking for the active device. A copper septum divides the input and output sides of the carrier and ensures a very low inductance ground return. This is extremely important for the efficient operation of relatively large power FET pellets, especially at higher frequencies.

The GaAs FET pellet is bump-mounted upside down onto the carrier. This configuration eliminates the commonly used bond wires and their associated parasitics, provides excellent heat sinking for the pellet, and also reduces the highly skilled labor required in the assembly. The advantages of bump mounting active devices are slowly becoming more widely recognized and accepted. This approach is exceptionally well suited for hybrid fabrication, especially when reproducibility and reliability are of key importance.

It should be pointed out that a counterpart to the BeO carrier with integral septums exists in the monolithic circuit version. Here, Raytheon, among others, has pioneered the plated via hole approach in GaAs. The holes, approximately 50.8 micrometers in diameter, are at the source contacts of power FETs to provide good RF ground returns. The inherent difficulty of this approach is that it leads to very thin GaAs wafers, which then cannot simultaneously carry the required high-Q circuit elements for matching of the devices. This is a major reason why high-power applications will be much more easily handled with hybrid approaches.

Another TI hybrid integration is a wide-band tunable FET oscillator that uses a single alumina substrate containing the FET, two varactor chips, and associated circuitry. The use of two separate varactors, one in the gate and another in the source lead of the FET, achieves a bandwidth of nearly an octave at X band, with a power output of more than 2 dBm. This circuit is much more easily realized in hybrid than monolithic form because of the incompatibility of the process requirements for FETs and varactors.

As an unconventional use of the FET, consider the RCA symmetric resonant element that is tunable in frequency and has a Q that can be adjusted independently (Fig. 4). The element consists of a single alumina substrate, 5 × 5 mm, that contains two FETs and two varactors. In principle, the negative resistance of the FET is used to compensate for the circuit and varactor losses and to permit the achievement of arbitrary Q values. Several such elements can be cascaded to form complex filter arrangements (Fig. 5). Such a filter is not only extremely small in size and lightweight, but also offers tuning speeds that are orders of magnitude faster than comparable YIG filters.

The present tuning range extends from 9.0 to 9.6 GHz, but work toward multigigahertz bandwidth coverage is under way. Although independent tuning voltages for each varactor and FET pair are required, the currents are extremely low (on the order of microamperes), and thus digital circuitry, together with D/A converters, can easily provide the necessary tracking voltages. Such filters can also fulfill more complex functions, such as changing not only the frequency, but also the bandwidth or skirt selectivity on command—features that could become very important as microwave systems become more sophisticated.

Integrated circuits, monolithic as well as hybrid, will offer not only better performance, but also much smaller volume and lower weight than present microwave IC components and TWTs. In the next few years, the useful frequency range for GaAs FETs can be expected to extend to 30 GHz.

Hybrid circuits, in many cases, will be made as small and light in weight as monolithic circuits, while offering better power-handling capability and lower cost for small or moderately large quantities.

For further reading

An analysis of a GaAs FET microwave oscillator designed for 16 GHz with an efficiency of 19 percent is dealt with in "Ku-Band FET Oscillator," F.N. Sechi and J.E. Brown, *IEEE ISSCC Digest*, February 1980, pp. 124–125.

Various monolithic implementations of wide-baseband monolithic amplifiers are treated in "GaAs IC Direct-Coupled Amplifiers," D. Hornbuckle, *IEEE MTT-S Digest*, May 1979, pp. 416–418.

A single-stage hybrid amplifier that uses both positive and negative feedback and operates on a 350 MHz to 14 GHz band is discussed in "The Matched Feedback Amplifier: Ultra-Band Microwave Amplitication with GaAs MESFETs," K.B. Niclas *et al*, *IEEE Transactions on Microwave Theory Technology*, Vol. MTT-28, no. 4, April 1980, pp. 285–294.

An analysis of the use of GaAs FETs in ultra-fast-tuning high-Q filter application is the subject of "High-Speed, Varactor-Tunable Microwave Filter Element," A. Presser, *IEEE MTT-S Digest*, May 1979, pp. 416–418.

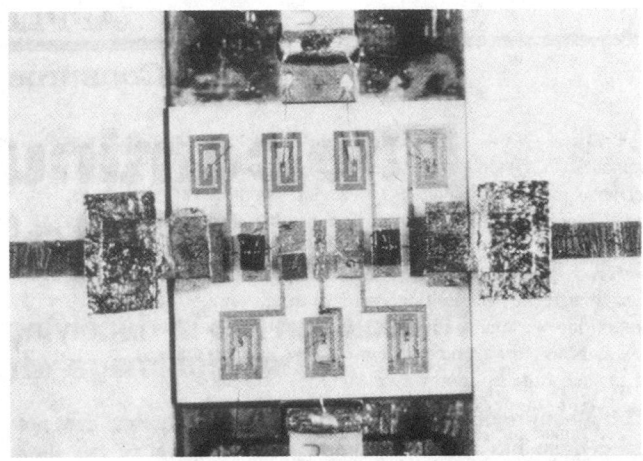

[4] This single alumina substrate, 5 × 5mm, contains two FETs and two varactors to form a symmetric resonant element that is tunable in frequency and that has an independently adjustable Q. In principle, the negative resistance of the FET is used to compensate for the circuit and varactor losses and to permit the achievement of arbitrary Q values.

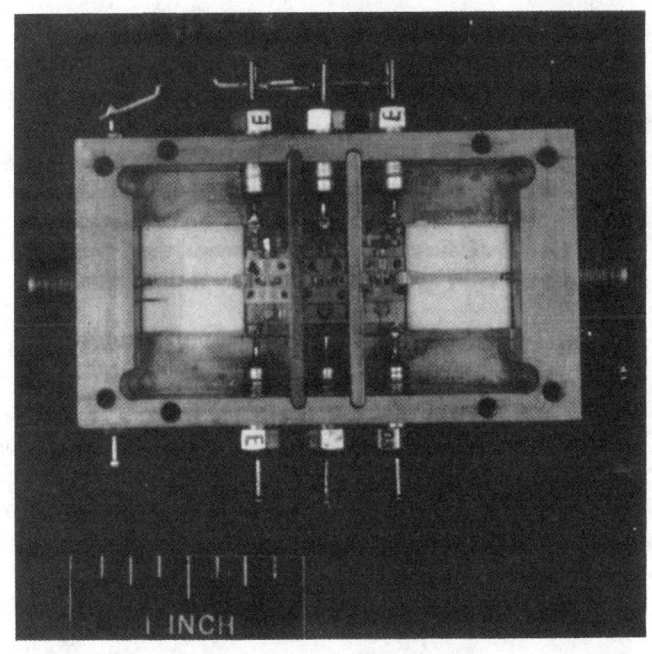

[5] Several symmetric resonant elements can be cascaded to form complex filter arrangements, as illustrated by this three-element band-pass filter. Such a filter is small in size, light in weight, and offers faster tuning speeds than do comparable YIG filters.

Design Considerations for Monolithic Microwave Circuits

ROBERT A. PUCEL, FELLOW, IEEE

MTT National Lecture Invited Paper

Abstract—Monolithic microwave integrated circuits based on silicon-on-sapphire (SOS) and gallium arsenide technologies are being considered seriously as viable candidates for satellite communication systems, airborne radar, and other applications. The low-loss properties of sapphire and semi-insulating GaAs substrates, combined with the excellent microwave performance of metal-semiconductor FET's (MESFET's), allows, for the first time, a truly monolithic approach to microwave integrated circuits. By monolithic we mean an approach wherein all passive and active circuit elements and interconnections are formed into the bulk, or onto the surface of the substrate by some deposition scheme, such as epitaxy, ion implantation, sputtering, evaporation, and other methods.

The importance of this development is that microwave applications such as airborne phased-array systems based on a large number of identical circuits and requiring small physical volume and/or light weight, may, finally, become cost effective.

The paper covers in some detail the design considerations that must be applied to monolithic microwave circuits in general, and to gallium arsenide circuits in particular. The important role being played by computer-aided design techniques is stressed. Numerous examples of monolithic circuits and components which illustrate the design principles are described. These provide a cross section of the world-wide effort in this field. A glimpse into the future prospects of monolithic microwave circuits is made.

I. INTRODUCTION

THE LAST two to three years have witnessed an intensive revival in the field of analog monolithic microwave integrated circuits (MMIC's), that is, microwave circuits deposited on a semiconductor substrate, or an insulating substrate with a semiconductor layer over it. In this paper, we shall address the design and technology considerations of monolithic microwave integrated circuits as well as the potential applications of these circuits to microwave systems, such as satellite communications and phased-array radar, as well as instrumentation.

It is important that the reader understand what we mean by the term "monolithic" circuit. By monolithic, we mean an approach wherein all active and passive circuit elements or components and interconnections are formed into the bulk, or onto the surface, of a semi-insulating substrate by some deposition scheme such as epitaxy, ion implantation, sputtering, evaporation, diffusion, or a combination of these processes and others.

It is essential that the full implication of this definition

Manuscript received January 16, 1981. This work summarizes the lecture given by the National Lecturer throughout the United States, Canada, and Europe, during 1980–1981.

The author is with the Research Division, Raytheon Company, Waltham, MA 02254.

be understood, since it strikes at the very core of why one would want to design and fabricate a microwave monolithic circuit. The reasons are embedded in the following promising attributes of the monolithic approach:

1) low cost;
2) improved reliability and reproducibility;
3) small size and weight;
4) multioctave (broad-band) performance; and
5) circuit design flexibility and multifunction performance on a chip.

The importance of this development is that systems applications based on a large number of identical components, for instance, space-borne phase-array radars planned for the future which require lightweight and reliable, low-cost transmit–receive modules, may finally become cost effective. One might consider this type of application as the microwave system analog of the computer (which spurred the growth of the silicon digital monolithic circuit market), since both require a large number of identical circuits.

Maximum cost effectiveness, as well as improved reliability, derives in part from the fact that wire bonding is eliminated in MMIC's, at least within the chip itself, and is relegated to less critical and fewer locations at the periphery of the chip. Wire bonds have always been a serious factor in reliability and reproducibility. Furthermore, wire bonding, being labor intensive, is not an insignificant factor in the cost of a circuit.

Small size and volume, and their corollary, light weight, are intrinsic properties of the monolithic approach. Small size allows batch processing of hundreds of circuits per wafer of substrate. Since the essence of batch processing is that the cost of fabrication is determined by the cost of processing the entire wafer, it follows that the processing cost per chip is proportional to the area of the chip. Thus, the higher the circuit count per wafer, the lower the circuit cost.

The elimination of wire bonding and the embedding of active components within a printed circuit eliminate many of the undesired parasitics which limit the broad-band performance of circuits employing packaged discrete devices. The monolithic approach will certainly ease the difficulty of attaining multioctave performance. Furthermore, such broad-banding approaches as distributed amplifier stages, heretofore shunned as too wasteful of active elements, will now become feasible, because a cost penalty will not accrue from the prolific use of low-gain stages, and because the unavoidable parasitics associated with the active devices will be incorporated in the propagating circuit and rendered less harmful.

The small circuit size intrinsic to the monolithic approach will enable circuit integration on a chip level, ranging from the lowest degree of complexity such as an oscillator, mixer, or amplifier, to a next higher "functional block" level, for example a receiver front end or a phase shifter. A still higher level of circuit complexity, for example, a transmit–receive module, will be integrated, most likely in multichip form.

So far we have discussed only the virtues of the monolithic approach. Now let us consider some of its disadvantages and problem areas. These are principally the following:

1) unfavorable device/chip area ratio;
2) circuit tuning (tweaking) impractical;
3) trouble-shooting (debugging) difficult;
4) suppression of undesired RF coupling (crosstalk), a possible problem; and
5) difficulty of integrating high power sources (IMPATT's)

The first item refers to the fact that only a small fraction of the chip area is occupied by devices, hence the high processing cost and lower yield associated with active device fabrication is unavoidably applied to the larger area occupied by the circuitry. A corollary of this is that the lower yield processes of device fabrication dominate the overall chip yield. Although these problems diminish as the chip size becomes smaller, that is, for higher frequencies, they are absent in the hybrid approach where the circuit and device technologies are separated.

The second and third items are related and can be considered together. The small chip sizes characteristic of the monolithic approach make it virtually impossible to tune ("tweak") and troubleshoot circuits. Indeed, to want to do so would violate one of the precepts of this approach, namely, to reduce costs by minimizing all labor-intensive steps. What then can be done about these very real problems?

First, it is necessary to minimize the need for tweaking. This can be done by adopting a design philosophy which leads to circuits that are insensitive to manufacturing tolerances in the active devices and physical dimensions of the circuit components. This will be a difficult compromise to accept on the part of the circuit designer, who expects the ultimate in performance from each active device by circuit tuning. However, here computer-aided design (CAD) techniques come to the rescue. Not only will CAD techniques play a major, if not mandatory, role in monolithic circuit design, they will also be used to assess the effect of tolerances on circuit performance during the design phase—and rather easily. CAD program for doing this

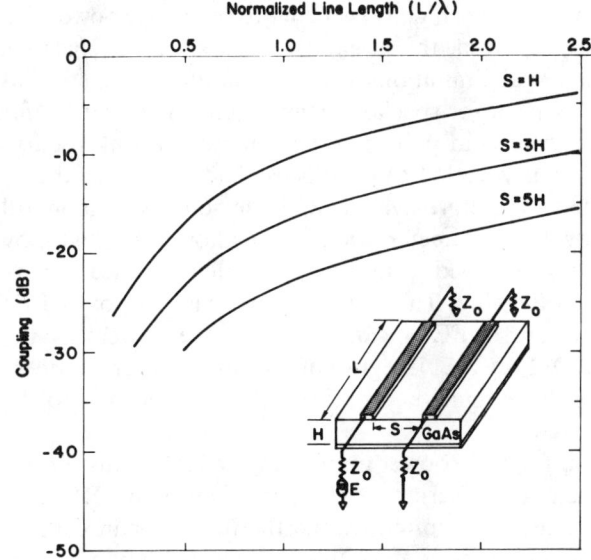

Fig. 1. Calculated coupling between adjacent parallel microstrip lines as a function of spacing and frequency.

reside on many internal computer systems and are also available commercially [4].

The use of CAD also helps alleviate the problem of troubleshooting a working circuit. Until microwave probes suitable for monolithic circuits become practical [19], troubleshooting must be based on terminal RF measurements of the circuit, usually the input and output ports. If a certain component is suspected of being faulty, it is a simple matter of building this defect into the CAD data file and comparing the resultant calculated circuit response with that measured. This can be done for a series of suspected faults, and convergence to the true fault can be achieved rather expeditiously.

The potential problem of undesirable RF coupling within the circuit is real because of the small chip sizes involved. To illustrate this point, Fig. 1 is a theoretical calculation of the coupling between two parallel microstrip lines on a GaAs substrate, one of which is excited by a generator. Both lines are matched at either end. Shown is the fraction of power coupled from the excited line to the adjacent line as a function of line length and line spacing. It is obvious that the coupling can become unacceptably high for long line lengths approaching a wavelength or more. Even for short lines, of the order of a quarter-wavelength or less, a feedback problem may exist if, say, a high-gain amplifier exists in one of the lines. In practice, line spacings of the order of three substrate thicknesses or more ($S > 3H$) have been found adequate in most cases. This proximity "rule" plays a major role in determining the chip area and, hence, the chip cost. This restriction on circuit packing density, somewhat unique to MMIC's, can be alleviated measurably if direct-coupled circuitry is used, that is, if no distributed or lumped componentry is involved. We shall see examples of this approach later.

Turning to the fifth item, though both low-noise and power FET circuitry can easily be integrated on the same chip, where very high powers, more precisely, power densities are involved, the monolithic approach may face some fundamental limitations. These limitations are associated with the need for special means of removing heat from the device. A case in point is the diamond heatsink used with millimeter-wave IMPATT diodes. Though it would be desirable to integrate avalanche diode sources in monolithic circuits for millimeter wave applications, the high-power densities involved cannot be handled by heat transfer through the chip. This is not a problem with power FET's, but of course, FET's cannot deliver the powers available from IMPATT's. Integration of high power sources in monolithic circuits is a problem that, as yet, has not been addressed.

Even for FET power amplifiers, tradeoffs must be made between good thermal performance and good RF design. For example, to minimize the thermal resistance through the substrate, it is desirable to use as thin a wafer as practical. However, a thin wafer increases the circuit skin effect losses, and hence the attenuation. Furthermore, since heat-sinking requires metallization of the chip bottomside, additional parasitic capacitance to ground is introduced and corrections must be made to planar inductors to account for "image" currents in the ground plane.

Despite these limitations on power, it is possible that with on-chip power combining techniques applied to FET's which are thermally isolated from each other [17], power outputs of the order of 10-W CW or so may be realizable from a single chip at the lower microwave frequencies, that is, at X-band.

II. MMIC's—A Brief History

The concept of MMIC's is not new. Its origin goes back to 1964 to a government-funded program based on silicon technology, which had as its objective a transmit–receive module for an aircraft phased-array antenna. Unfortunately, the results were disappointing because of the inability of semi-insulating silicon to maintain its semi-insulating properties through the high-temperature diffusion processes. Thus, very lossy substrates resulted, which were unacceptable for microwave circuitry [12]. Because of these and other difficulties the attempt to form a monolithic circuit based on a semiconductor substrate lay dormant till 1968 when Mehal and Wacker [15] revived the approach by using semi-insulating gallium arsenide (GaAs) as the base material and Schottky barrier diodes and Gunn devices to fabricate a 94-GHz receiver front end. However, it was not until Plessey applied this approach to an X-band amplifier, based on the Schottky-gate field-effect transistor, or MESFET (MEtal-Semiconductor Field-Effect Transistor), as the key active element that the present intense activity began [16].

What brought on this revival? First, the rapid development of GaAs material technology, namely, epitaxy and ion implantation, and the speedy evolution of the GaAs FET based on the metal Schottky gate during the last decade led to high-frequency semiconductor device performance previously unattained. A few examples are high-efficiency and high-power amplifier performance through Ku-band, low-noise amplifiers, variable-gain dual-gate amplifiers, and FET mixers with gain. The dual-gate FET will play a major role in MMIC's because of its versatility as a linear amplifier whose gain can be controlled either digitally or in analog fashion. With dual-gate FET's, multiport electronically switchable RF gain channels are feasible. Second, resolution of many troublesome device reliability problems made FET's more attractive for systems applications. Third, recognition of the excellent microwave properties of semi-insulting GaAs (approaching that of alumina), removed the major objection of silicon. Fourth, hybrid circuits were becoming very complex and labor intensive because of the prolific use of wire bonds, and hence too costly. Fifth, the emergence of clearly defined and discernible systems applications for MMIC's became more apparent. Thus it was the confluence of all of these factors, and others, which stimulated the development of GaAs MMIC's within the last five years.

III. Silicon or Gallium Arsenide?

It is ironic that this revival based on GaAs technology has, in turn, restimulated the interest in silicon MMIC's— but now based on the silicon-on-sapphire (SOS) approach [13]. There are understandable reasons for this. First, the use of sapphire as a substrate eliminates the losses associated with semi-insulating silicon mentioned earlier. Second, silicon technology is an extremely well developed technology—much more so than GaAs. Third, the availability of the simpler MESFET technology, developed in GaAs, could now be used in place of the more complex bipolar technology, which, however, was still available should it be needed. Nevertheless, gallium arsenide has the "edge" for reasons to be discussed next.

Table I lists some of the pertinent physical and electrical properties of GaAs and silicon (n-type) in their insulating and semiconducting states, as well as that of sapphire and alumina. As is evident from this table, as a high-resistivity substrate, semi-insulating GaAs, sapphire, and alumina are, for all practical purposes, comparable. Also evident is that the carrier mobility of gallium arsenide is over six

TABLE I
Some Properties of Semiconductors and Insulators

Property	GaAs	Silicon	Semi-insulating GaAs	Semi-insulating Silicon	Sapphire	Alumina
Dielectric Constant	12.9	11.7	12.9	11.7	11.6 (C-axis)	9.7
Density (gm/cc)	5.32	2.33	5.32	2.33	3.98	3.89
Thermal Cond. (watts/cm-°K)	0.46	1.45	0.46	1.45	0.46	0.37
Resistivity (ohm-cm)	---	---	$10^7 - 10^9$	$10^3 - 10^5$	$>10^{14}$	$10^{11} - 10^{14}$
Elec. Mobility (cm^2/v-sec.)	4300*	700*	---	---	---	---
Sat. Elec. Vel. (cm/sec.)	1.3×10^7	9×10^6	---	---	---	---

* At $10^{17}/cm^3$ doping

times that of silicon. For this reason and others, GaAs MESFET's are operable at higher frequencies and powers than silicon MESFET's of equivalent dimensions. For example, silicon MESFET's based on 1-μm gate technology will be limited in operation to upper S-band at best, whereas GaAs MESFET's operate well at X-band and higher. Therefore, it is highly likely that the performance of 1-μm gate silicon MESFET's will be matched, and perhaps exceeded by that of 2-μm gate GaAs MESFET's at S-band. The near-future availability of much larger GaAs wafers, approaching 3.5 in in diameter [20], obtained by the Czochralski method, will overcome the size limitations imposed by the present 1-in wafers grown by the Bridgman method. The early success of direct-coupled FET analog circuitry [11], [21], which leads to high component density at S-band, will also help overcome wafer size limitations in GaAs. Finally, the proven success of gigahertz high-speed GaAs logic circuitry will allow, for the first time, complete integration of logic and analog microwave circuitry. This opens up the feasibility of high-speed signal processing on a chip.

We do not wish to imply that MMIC work based on SOS technology should be diminished; however, we believe its major role will be found in the range below 2 GHz, for example, in IF circuitry and other applications. In light of this conclusion, we shall direct the following discussion to the GaAs approach. However, much of what we shall say, as will be obvious to the reader, will also apply, with minor changes, to the SOS approach or to other approaches which may emerge in the future. Nevertheless, we maintain that before this decade is over, it will be GaAs monolithic integrated circuits that will exert the greatest influence on the way solid-state device circuitry is used in microwave systems.

IV. THE GALLIUM ARSENIDE APPROACH

A cornerstone of the monolithic approach will be the availability of a highly reproducible device technology. This in turn is related, in part, to the control of the starting material, especially the active (semiconducting) layer.

Two general techniques are available for forming this layer on GaAs substrates, namely, epitaxy and ion implantation. Of the two approaches, the former at present is more widely used and developed. In this approach a doped single crystal semiconducting layer is deposited on a semi-insulating GaAs substrate, usually with an intervening high-resistivity epitaxial "buffer" layer to screen out diffusion of impurities from the substrate during the active layer growth. With ion implantation, dopant atoms are implanted directly into the surface of a semi-insulating GaAs substrate. This procedure requires a higher state of purity of the substrate—a problem at present.

Expitaxy does not have the control or flexibility associated with implantation. With implantation, more uniform conducting layers are possible over a large area—more uniform in doping level as well as in thickness. Furthermore, with implantation, selective doping is easy, that is, formation of different doping profiles in different parts of the wafer is easy to achieve, whereas with epitaxy it is difficult. The potential device reproducibility achievable with implantation is a definite advantage for it.

It should be added that implantation can also be used in conjunction with epitaxy. One such application is the isolation implant, wherein oxygen is implanted in the unused portions of the epitaxial layer to produce a high-resistivity region within the epitaxial layer onto which microwave circuitry may be situated. Thus a truly planar surface is maintained, since no mesa etching is required to remove the undesired regions of epitaxial layer. This also eliminates yield problems associated with metallization patterns extending over mesa steps.

It is likely that, once substrate purity reaches the necessary level for ion implantation (as it is approaching with unintentionally doped Czochralski-pulled crystals), ion implantation will supplant epitaxy as the preferred method for monolithic circuits.

The processing technology used for FET fabrication is also applicable to the monolithic circuit elements. The high degree of dimensional definition associated with FET photolithography is more than adequate for the circuit elements.

V. GENERAL DESIGN CONSIDERATIONS

We turn now to a discussion of the design considerations for MMIC's.

A. Constraints on Chip Size

Present substrate sizes corresponding to that of GaAs boules are approximately 1 in in diameter, though larger boules approaching 3 in in diameter are now being grown by the Czochralski method. Given the expected limits on substrate size, it is instructive to estimate the circuit count/wafer achievable as a function of frequency, since the processing cost per circuit is inversely proportional to this density.

We assume that the maximum linear dimension per circuit will fall between $\lambda_g/10$ and $\lambda_g/4$, where λ_g is the wavelength of the propagation mode (microstrip-coplanar, etc.) in GaAs. The lower limit takes into account the approximate maximum size of lumped elements; the upper limit, the typical maximum size of distributed elements. It seems reasonable to assume that in the vicinity of 10 to 20 GHz some distributed elements of the order of $\lambda_g/4$ (for example, hybrid and branch line couplers) will be used. Therefore, above this frequency range, linear circuit dimensions of the order of $\lambda_g/4$ will be the rule. Let us choose 16 GHz as the demarcation frequency. We then postulate a "linear" admixture of lumped- and distributed-element weighting so that we obtain $\lambda_g/10$ at 1 GHz and $\lambda_g/4$ at 16 GHz as the probable linear dimension of a circuit function "chip."

Fig. 2 is a plot of the approximate density of these circuits as a function of frequency for two sizes of wafer. (The 2-in square wafer corresponds to a 3-in diameter

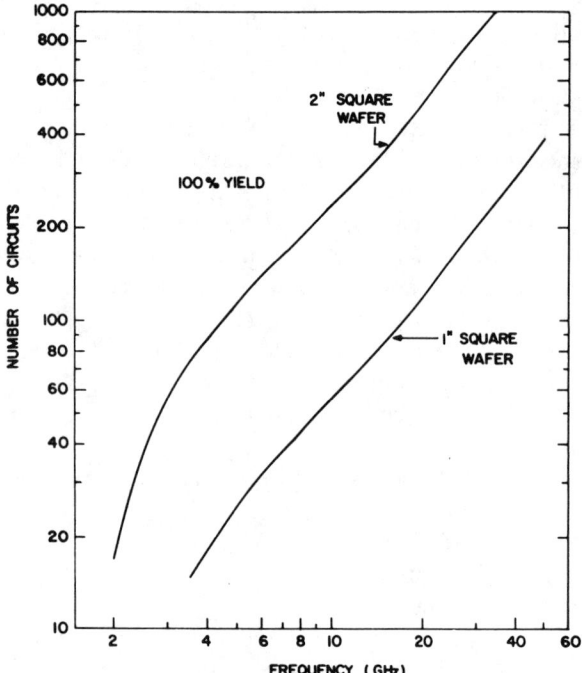

Fig. 2. Estimated number of circuits per wafer taking dicing and edge waste into account.

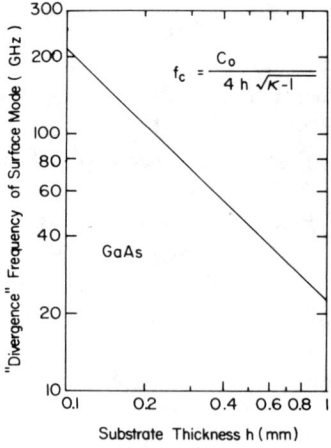

Fig. 3. Frequency of onset of lowest order TE surface wave on a GaAs substrate as a function of substrate thickness.

wafer.) The circuit density estimates take into account edge and cutting waste, but not "proximity effects," wafer yield, and other factors which will reduce these numbers.

A yield factor is associated with each fabrication step, the overall yield being the product of the individual yield factors. Thus, since active devices generally require the most processing steps (about 40 for an FET), the overall yield is determined by the device processing technology. The "proximity effect," that is, the RF coupling problem mentioned earlier, will put stringent limitations on how closely packed the signal lines may be, and hence how much circuitry can be compressed into the chip area, which is fixed by wavelength or lumped-element dimensions as just described.

The circuit count estimate must be modified for very low microwave frequencies (below C-band) if active components such as FET's are used to simulate resistors and capacitors and if inductors are dispensed with because tuning is not necessary. In this so-called direct-coupled design, packing densities approaching those normally associated with digital circuitry is possible [21], that is, much higher than that indicated in Fig. 2. However, it must be cautioned that this circuit approach is not suitable for all monolithic applications, for example, high-efficiency power amplifiers or low-noise circuits. The reason is that the use of active (FET) devices as resistive elements in the gate and drain circuits introduces high dc power dissipation and mismatch, as well as additional noise [11].

It is appropriate at this time to point out that the size advantages of GaAs MMIC's will be lost if proper packaging techniques are not used. Perhaps the efficient techniques adopted for low-frequency and digital circuitry can be suitably modified for microwave applications. Much thought must be devoted to this very important problem.

B. Constraints on Wafer Thickness

So far we have discussed requirements on the substrate area. There also are constraints imposed on the substrate thickness. Some of these are:

1) volume of material used;
2) fragility of wafer;
3) thermal resistance;
4) propagation losses;
5) higher order mode propagation;
6) impedance-linewidth considerations; and
7) thickness tolerance versus impedance tolerance.

Obviously, to keep material costs down one wishes to use as thin a substrate as can be handled without compromising the fragility. Thermal considerations also require the thinnest wafer possible. On the other hand, a thin wafer emphasizes the effect of the ground plane. For example, propagation losses increase inversely with substrate thickness in the case of microstrip. Furthermore, the Q-factor and inductance of thin-film inductors decrease with decreasing substrate thickness. In contrast, undesired higher-order surface mode excitation is inhibited for thinner substrates.

Fig. 3 is a graph of the frequency denoting the onset of the lowest order (TE) surface mode as a function of substrate thickness. For example, for a substrate thickness of 0.1 mm (4 mils) the "safe" operating frequency range is below 200 GHz. It appears that, for presently contemplated circuit applications, surface mode propagation is not a limiting factor in the choice of substrate thickness. The linewidth dimensions for a given impedance level of some propagation modes, such as microstrip, are proportional to substrate thickness. Therefore, thicker substrates alleviate the effect of thickness and linewidth tolerances.

The point being made here is that the choice of substrate thickness is a tradeoff of the factors listed above, being strongly dependent on the frequency of operation and the

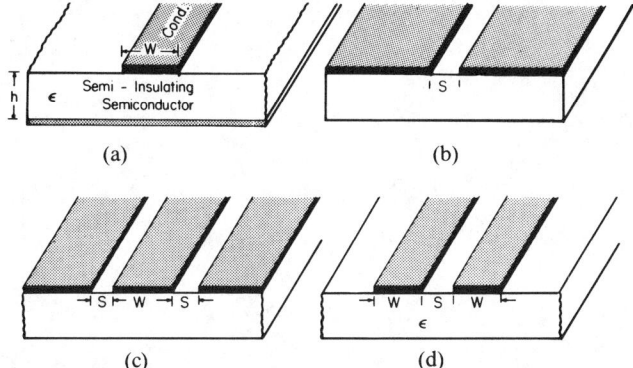

Fig. 4. Four candidate propagation modes for monolithic circuits. (a) Microstrip (MS). (b) Slot line (SL). (c) Coplanar waveguide. (d) Coplanar strips (CS).

power dissipation of the circuit. It is true that perhaps the most important of the factors is the thermal consideration. We believe that in the frequency range up to 30 GHz a substrate thickness of the order of 0.1 mm to 0.15 mm is appropriate for power amplifier circuits, with thicknesses up to 0.6 mm tolerable for low-noise amplifiers and similar circuits, provided a satisfactory means of dicing the thicker wafers can be found.

C. Propagation Modes

At microwave frequencies, the interconnections between elements on a high dielectric constant substrate such as GaAs, where considerable wavelength reduction occurs, must be treated as waveguiding structures. On a planar substrate, four basic modes of propagation are available, as illustrated in Fig. 4. The first mode (Fig. 4(a)) is microstrip (MS), which requires a bottomside ground metallizaton. Its "inverse," slot line (SL), is shown in Fig. 4(b). The third mode is the coplanar waveguide (CPW) shown in Fig. 4(c); it consists of a central "hot" conductor separated by a slot from two adjacent ground planes. Its "inverse," the coplanar stripline (CS), is illustrated in Fig. 4(d); here, one of the two conductors is a ground plane. Both the coplanar waveguide and coplanar strips are generally considered to be on infinitely thick substrates. Of course, this condition cannot be met. We shall see the implication of this later.

Of the four modes, only the slot line is not TEM-like. For this reason, and because it uses valuable "topside" area, we do not expect slot line to be a viable candidate for monolithic circuits, except possibly in special cases.

The principal losses of microstrip and the coplanar modes consist of ohmic losses. Since the coplanar structures are, in essence, "edge-coupled" devices, with high concentration of charge and current near the strip edges, the losses tend to be somewhat higher than for microstrip, as verified by experiment [5].

The lack of a ground plane on the topside surface of the microstrip structure is a considerable disadvantage when shunt element connections to the hot conductor are required. However, as we shall see later, there are ways to overcome this disadvantage. Table II summarizes, in a qualitative way, the features of the four modes of propagation illustrated in Fig. 4.

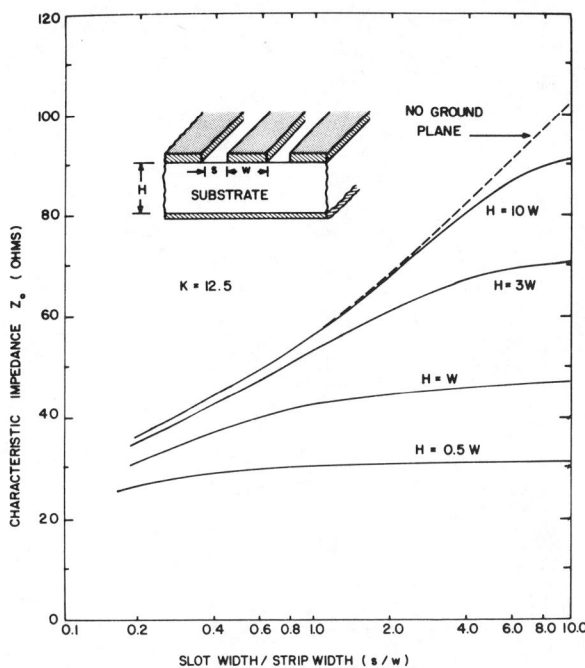

Fig. 5. Effect of ground plane on characteristic impedance of a coplanar waveguide.

TABLE II
QUALITATIVE COMPARISON OF PROPAGATION MODES

	MICROSTRIP	COPLANAR WAVEGUIDE	COPLANAR STRIPS	SLOT LINE
Attenuation Loss	low	medium	medium	high
Dispersion	low	medium	medium	high
Impedance Range (ohms)	10-100	25-125*	40-250*	high
Connect Shunt Elements	diff.	easy	easy	easy
Connect Series Elements	easy	easy	easy	diff.

*Infinitely thick substrate

The impedance range achievable with CPW and CS is somewhat greater than for MS, particularly at the higher end of the impedance scale, provided an infinitely thick substrate is assumed for CPW and CS. This range is reduced considerably when practical substrate thicknesses are used and the bottomside of the chip is metallized. Fig. 5 shows how the high impedance end of the scale is lowered when substrates of the order of 0.1–0.25 mm thick are mounted on a metal base (for heat-sinking purposes). The considerable reduction in Z_0 makes the design of monolithic circuitry with CPW nearly as dependent on substrate thickness as with MS, at least at the high end of the impedance scale.

Microstrip has its own unique restriction on the realizable impedance range. This is dictated by technology considerations. The limitation stems from the fact that for MS

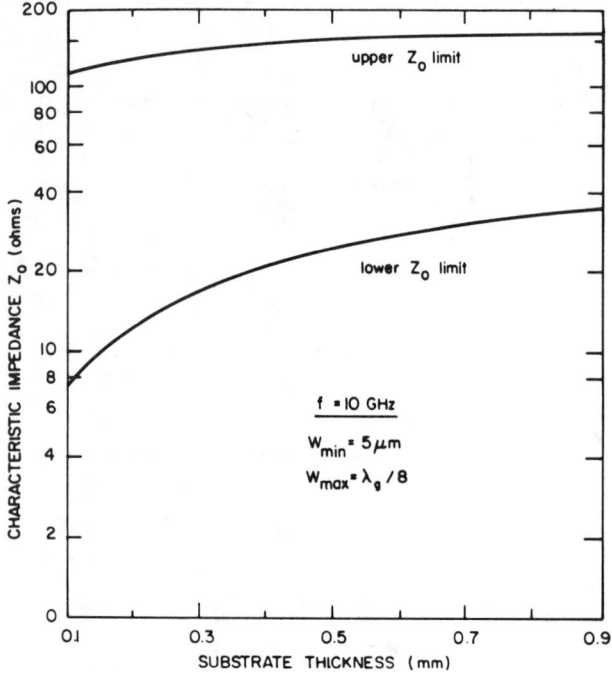

Fig. 6. Range of characteristic impedance of microstrip on GaAs substrate as a function of substrate thickness.

the characteristic impedance Z_0 is a function of the ratio W/H (see Fig. 4). The highest achievable impedance is determined by the smallest linewidth W that can be realized with acceptable integrity over a long length, say, a quarter of a wavelength. Our experience is that a minimum linewidth of 5 µm is reasonable. With this restriction, and an additional limit imposed on the maximum linewidth to be well below a quarter-wavelength, say, one-eighth-wavelength, the realizable range of characteristic impedance as a function of substrate thickness and frequency is constrained within the range indicated by Fig. 6

It is evident that the usable impedance range for a 0.1-mm thick substrate is approximately 10–100 Ω, and somewhat less for thicker substrates and higher frequencies. For higher frequencies, the lower curve moves "up." This limited impedance range is a severe restriction in the design of matching networks, a problem not faced in the hybrid approach.

Weighing all of these factors, we believe that of the four candidate modes, MS and CPW are the most suitable for GaAs monolithic circuits, with preference toward MS. Indeed, there will be instances where both modes may be used on the same chip to achieve some special advantage. The transition from one mode to the other is trivial. Most of the examples of MMIC's to be described later are based on MS.

Fig. 6(a) is a graph of the wavelength of a 50-Ω MS line on GaAS as a function of frequency, with dispersion neglected. The wavelength of CPW is similar. Fig. 7(b) illustrates the attenuation of MS as a function of characteristic impedance and substrate thickness at 10 GHz. The loss in decibels per centimeter increases as the square root

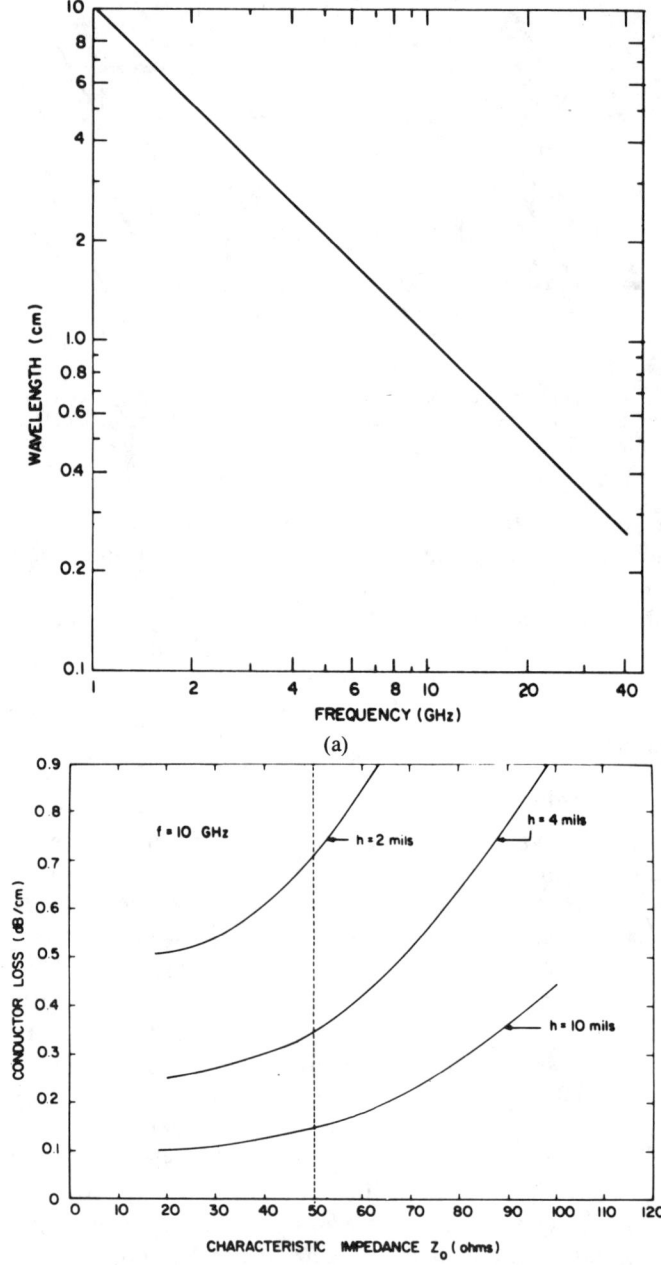

Fig. 7. (a) Wavelength as a function of frequency for microstrip on a GaAs substrate for $h = 0.1$ mm. (b) Conductor loss of microstrip on a GaAs substrate as a function of substrate thickness and characteristic impedance for $f = 10$ GHz.

of frequency. The loss per wavelength, on the other hand, decreases as the square root of frequency. Note the inverse dependence of loss on substrate thickness.

D. *Low Inductance Grounds and Crossovers*

Microstrip and coplanar waveguide are adequate for interconnections that do not require conductor crossovers or that are not to contact the bottomside ground metallization. Often, however, such connections are needed. In particular, with MS, which does not have any topside ground planes, some means of achieving a low-inductance ground is essential.

Fig. 8. 50-μm diameter "via" hole etched in a GaAs wafer.

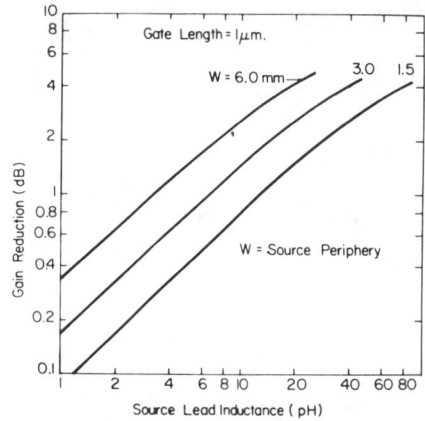

Fig. 9. Calculated gain reduction of a GaAs power FET as a function of source lead inductance.

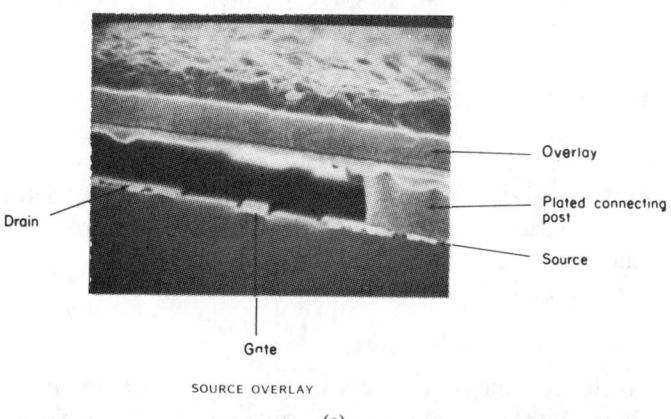

Fig. 10. (a) SEM microphotograph of a segment of source overlay (airbridge) of a power FET showing gate and drain contacts. (b) Top view of a GaAs power FET showing an air-bridge overlay connecting all source pads.

Two general methods of grounding are available: 1) the "wrap-around" ground; and 2) the "via" hole ground. The former requires a topside metallization pattern near the periphery of the chip which can be connected to the chip ground. The "via" hole technique, on the other hand, allows placement of grounds through the substrate where desired. Holes are chemically milled through the substrate until the top metallization pattern is reached. These holes are subsequently metallized at the same time as the ground plane to provide continuity between this plane and the desired topside pad. A microphotograph of such a "via" hole etched through a test wafer of 50-μm thickness is shown in Fig. 8. The hole diameter, in this case, is only 50 μm, much smaller than those normally used in monolithic circuits. The estimated inductance of a via hole is approximately 40–60 pH/mm of substrate thickness. Examples of circuits using both grounding techniques will be described later.

Low inductance grounds are especially important in source leads of power FET's. An inductance in the source lead manifests itself as resistive loss in the gate circuit, and hence a reduction in power gain. To illustrate this, Fig. 9 is a graph of the calculated gain reduction as a function of source lead inductance for an unconditionally stable power FET, corresponding to a power output of 1, 2, and 4 W ($W = 1.5$, 3.0, and 6.0 mm).

The second interconnect problem arises when it is necessary to connect the individual cells of a power FET without resorting to wire bonds. A requirement is that these interconnects also have a low inductance. Here the so-called "air-bridge" crossover is useful. This crossover consists of a deposited strap which crosses over one or more conductors with an air gap in between for low capacitive coupling.

Fig. 10(a) is a cross-sectional view of a source crossover which interconnects two adjacent source pads of a power FET. The air gap is approximately 4 μm. Clearly shown is the 1-μm gate and the larger drain pad underneath the crossover. Fig. 10(b) is a closeup, angular view of a power FET which employs an airbridge (overlay) interconnect bus.

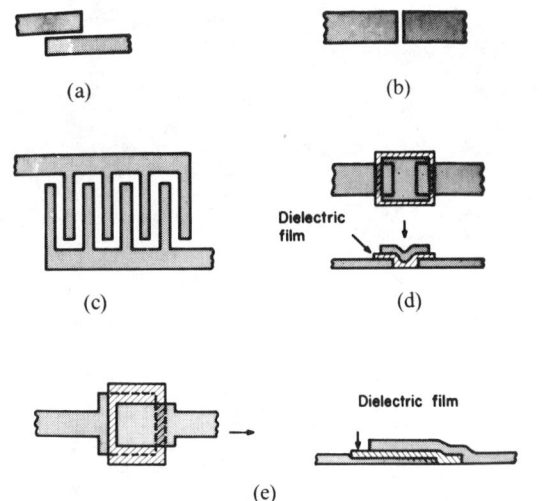

Fig. 11. Some planar capacitor designs. (a) Broadside coupled. (b) End coupled. (c) Interdigitated. (d) End-coupled overlay. (e) Overlay.

It is evident that the airbridge technology allows one to interconnect all cells without recourse to wire-bonding and therefore nicely satisfies the criterion for a monolithic circuit. Airbridge interconnects, of course, are also useful for microstrip and other crossovers. A good example is a planar spiral inductor, which requires a contact to the inner terminal.

E. Thin-Film Components

A flexible monolithic circuit design philosophy must include both lumped elements (dimensions <0.1 wavelength) and distributed elements, that is, elements composed of sections of transmission line. Lumped elements, R's, C's, and L's, are also useful for the RF circuitry, and in some cases mandatory, as for example, in thin-film resistive terminations for couplers. Lumped thin-film capacitors are absolutely essential for bias bypass applications, because of the large capacitance values required. Planar inductors can be extremely useful for matching purposes, especially at the lower end of the microwave band where stub inductors are very large, physically.

The choice of lumped or distributed elements depends on the frequency of operation. Lumped elements are suitable through X-band up to, perhaps, 20 GHz. It is likely, however, that beyond this frequency range, distributed elements will be preferred. It is difficult to realize a truly lumped element, even at the lower frequencies, because of the parasitics to ground associated with thin substrates. In this section we shall review the design principles of planar lumped elements.

1) Planar Capacitors: There are a variety of planar capacitors suitable for monolithic circuits—those achieved with a single metallization scheme, and those using a two-level metallization technology in conjunction with dielectric films. Some possible geometries for planar capacitors are shown in Fig. 11. The first three, which use no dielectric film and depend on electrostatic coupling via the substrate, generally are suitable for applications where low values of capacitance are required (less than 1.0 pF) for instance, high-impedance matching circuits. The last three geometries, the so-called overlay structures which use dielectric films, are suitable for low-impedance (power) circuitry and bypass and blocking applications. Capacitance values as high as 10 to 30 pF are achievable in small areas.

Two sources of loss are prevalent in planar capacitors, conductor losses in the metallization, and dielectric losses of the films, if used. Since the first three schemes illustrated in Fig. 11 are edge-coupled capacitors, high charge and current concentrations near the edges tend to limit the Q-factors. At X-band, typical Q-factors measured to date have been in the range of 50, despite the fact that no dielectric losses are present. The last three geometries distribute the current more uniformly throughout the metal plates because of the intervening film. However, even here, Q-factors only in the range of 50–100 are typical at X-band (10 GHz) because of dielectric film losses. Let us turn now to a more detailed analysis of the overlay structures depicted in Fig. 11, in particular the structure in Fig. 11(e).

First, we review briefly some general requirements of dielectric films for the overlay geometry. Some properties of dielectric films of importance are 1) dielectric constant, 2) capacitance/area, 3) microwave losses, 4) breakdown field, 5) temperature coefficient, 6) film integrity (pinhole density, stability over time), and 7) method and temperature of deposition. This last requirement is obviously important, because the technology used for film deposition must be compatible with the technology used for the active device (FET). Dielectric films which easily satisfy this criterion are SiO_x and Si_3N_4.

Some useful figures of merit for dielectric films are the capacitance-breakdown voltage product

$$F_{cV} = \left(\frac{C}{A}\right)V_b \tag{1a}$$

$$= \kappa \epsilon_0 E_b \tag{1b}$$

$$\cong (8-30) \times 10^3 \text{ pF} \cdot \text{V/mm}^2$$

and the capacitance-dielectric Q-factor product

$$F_{cq} = \left(\frac{C}{A}\right)Q_d \tag{1c}$$

$$= \frac{(C/A)}{\tan \delta_d} \tag{1d}$$

where C/A is the capacitance per unit area, V_b is the breakdown voltage, E_b is the corresponding breakdown field, κ is the dielectric constant, and $\tan \delta_d$ is the dielectric loss tangent. Breakdown fields of the order of 1–2 MV/cm are typical of good dielectric films. Dielectric constants are in the order of 4–20. Loss tangents can range from 10^{-1} to 10^{-3}. It is desirable to have as high figures of merit as possible. Table III is a list of candidate films and their properties.

We return, now, to the overlay structure of Fig. 11(e). A closeup perspective view is shown in Fig. 12. Taking into

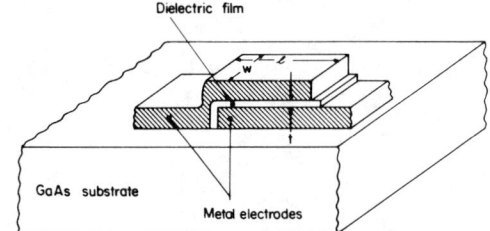

Fig. 12. Perspective of an overlay thin-film capacitor.

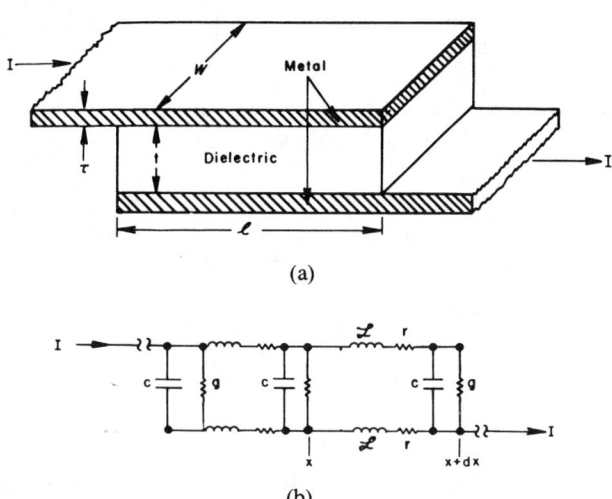

Fig. 13. Diagrams relevant to analysis of impedance of a thin-film capacitor. (a) Thin-film capacitor. (b) Circuit model.

TABLE III
PROPERTIES OF SOME CANDIDATE DIELECTRIC FILMS

DIELECTRIC	κ	TCC (ppm/°C)	C/A* (pF/mm²)	(C/A)·Q_d	(C/A)·V_b	COMMENTS
SiO	4.5–6.8	100–500	300	low	medium	Evaporated
SiO$_2$	4–5		50	200	medium	medium	Evaporated, CVD, or Sputtered
Si$_3$N$_4$	6–7	25–35	300	high	high	Sputtered or CVD
Ta$_2$O$_5$	20–25	0–200	1100	medium	high	Sputtered and Anodized
Al$_2$O$_3$	6–9	300–500	400	high	high	CVD, anodic oxidation, sputtering
Schottky-Barrier Junction	12.9	--	550	very low	high	Evaporated Metal on GaAs
Polyimide	3–4.5	–500	35	high	--	Spun and Cured Organic Film

*Film thickness assumed = 2000 Å, except for polyimide, 10,000 Å.

account the longitudinal current paths in the metal contacts, one may analyze this device as a lossy transmission line as indicated in Fig. 13. In Fig. 13(b), ℓ and r represent the inductance and resistance per unit length of the metal plates, and c and g denote the capacitance and conductance per unit length of the dielectric film. The relation between g and c is determined by the loss tangent, $g = \omega c \tan \delta_d$. The series resistance in the plates is determined by the skin resistance if the metal thickness exceeds the skin depth, or the bulk metal resistance if the reverse is true. Usually, the bottom metal layer is evaporated only, and hence is about 0.5 μm thick, which may be less than the skin depth. The top metal is normally built up to a thickness of several micrometers or more by plating.

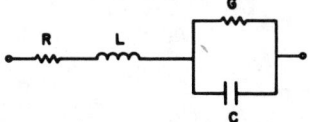

Fig. 14. Equivalent circuit of a thin-film capacitor. $R = 2/3\, rl$. $C = cl$. $G = \omega c \tan \delta$. $L = \ell l$. r = resistance/length (electrodes). c = capacitance/length. ℓ = inductance/length (electrodes). $\tan \delta$ = loss tangent of dielectric film.

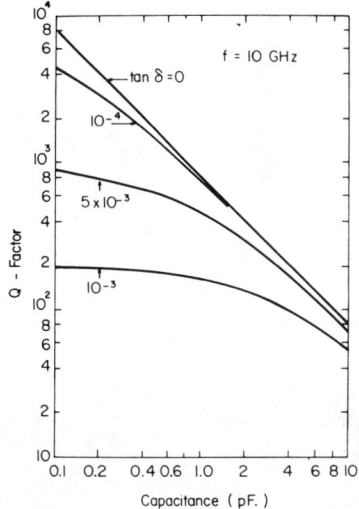

Fig. 15. Quality factor of a square thin-film capacitor as a function of capacitance and dielectric loss tangent for $f = 10$ GHz.

For a well-designed capacitor, the longitudinal and transverse dimensions are small compared with a wavelength in the dielectric film. In this case, a good approximation to the capacitor is the equivalent circuit shown in Fig. 14. When the skin loss condition prevails, the Q-factor corresponding to these losses is given by the expression

$$Q_c = \frac{3}{2\omega R_s (C/A) l^2} \qquad (2)$$

where R_s is the surface skin resistivity and l is the electrode length (see Fig. 12). Note the strong dependence on electrode length. This arises because of the longitudinal current path in each electrode. Note that if one electrode, say the bottom plate, is very thin, Q_c is decreased.

The dielectric Q-factor is $Q_d = 1/\tan \delta$, and the total Q-factor is given by the relation $Q^{-1} = Q_d^{-1} + Q_c^{-1}$. Fig. 15 is a graph of the calculated Q-factor as a function of capacitance for various loss trangents. Note that for a 1-pF capacitor, and no dielectric losses, the predicted Q-factor is approximately 800! Yet, experimentally, values more like one-tenth of this are obtained, suggesting that dielectric films are extremely lossy—much more so than their bulk counterparts. No satisfactory explanation for this observation has yet been advanced.

2) Planar Inductors: Planar inductors for monolithic circuits can be realized in a number of configurations, all achieved with a single-layer metallization scheme. Fig. 16 illustrates various geometries that can be used for thin-film inductors. Aside from the high-impedance line section, all

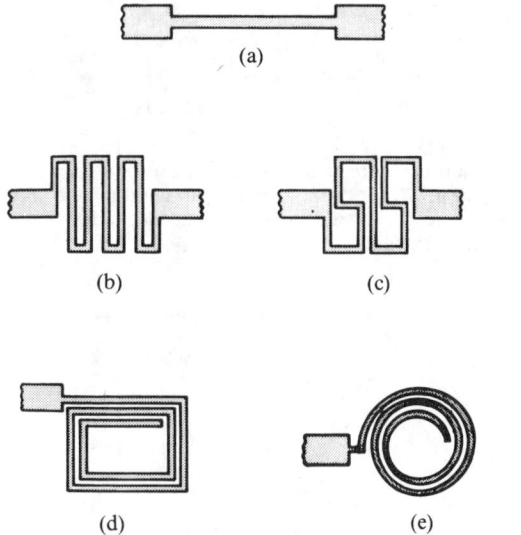

Fig. 16. Some planar inductor configurations. (a) High-impedance line section. (b) Meander line. (c) S-line. (d) Square spiral. (e) Circular spiral.

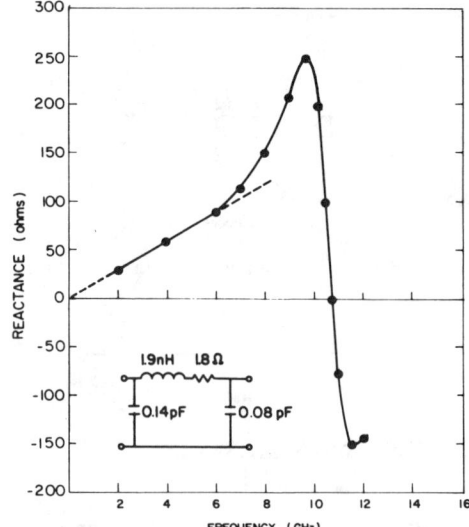

MEASURED REACTANCE OF A TEN-SEGMENT SQUARE-SPIRAL GROUNDED INDUCTOR ON A 0.1 MM THICK SI-GaAs SUBSTRATE

Fig. 18. Measured reactance of a ten-segment square spiral inductor on a 0.1-mm thick GaAs substrate (equivalent circuit shown in inset).

of the structures depend on mutual coupling between the various line segments to achieve a high inductance in a small area. In any multisegment design, one must insure that the total line length is a small fraction of a wavelength, otherwise the conductor cannot be treated as "lumped." Unfortunately, this latter condition is not often satisfied. Fig. 17 is a SEM photograph of a multisegment square-spiral inductor. Note the crossover connections.

When thin substrates are used, corrections must be made to the calculated inductance to account for the ground plane. These corrections are always in the direction to reduce the inductance, and are typically in the range of 15 percent, though for large-area inductors, the reduction can be as high as 30 percent.

Typical inductance values for monolithic circuits fall in the range from 0.5 to 10 nH. The higher values are difficult to achieve in strictly lumped form because of intersegment fringing capacitance. A more serious problem is that of shunt capacitance to ground, especially in the case of the microstrip format. This capacitance to ground can become important enough to require its inclusion in determining the performance of the inductor.

An illustration of the serious effect of capacitance to ground is demonstrated by the data of Fig. 18. This is a graph of the measured reactance of a ten-segment square spiral inductor as a function of frequency. The inductor is approximately 0.4 mm square, consisting of segments 1 mil wide, separated by 1 mil (see Fig. 17). The inductance, as designed, was nominally 1.9 nH. Note that above 10 GHz the reactance becomes capacitive! The equivalent circuit, as deduced from two-port S-parameters, is shown in the inset. The substrate thickness was 0.1 mm.

Of course, the inductor is usable, provided all of the parasitics indicated in Fig. 18 are taken into account.

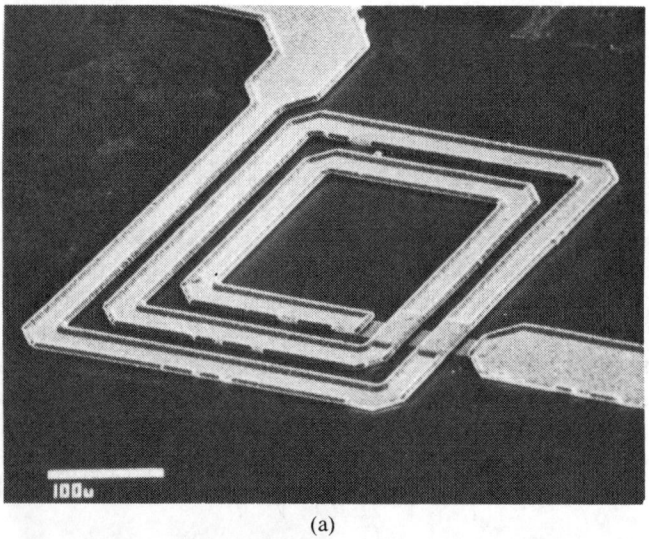

(a)

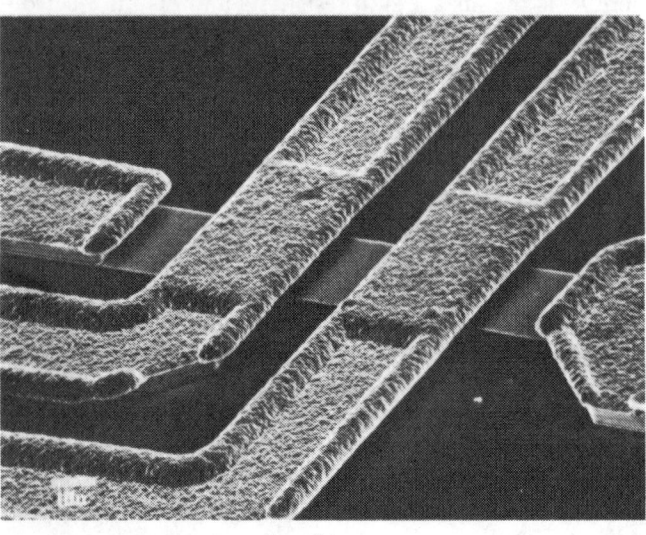

(b)

Fig. 17. SEM photographs of a thin-film square spiral inductor showing air-bridge crossovers.

Unfortunately, these parasitics are not known in advance. Thus, in a computer-aided approach, corrections to the circuit in which the inductor appears must be made in later iterations. This can become a costly procedure. It is often more sensible to use an inductive transmission line segment whose electrical behavior is known in advance.

Some of the skin losses in the inductor reside in the ground plane (assuming a metallized bottom side) and increase as the ground plane approaches the film inductor (not unlike shielding losses). However, the dependence on substrate thickness is mild, since most of the losses reside in the film turns, because of their small cross section.

In practice, inductor Q-factors of the order of 50 are observed at X-band, with higher values at higher frequencies. There appears to be no way to improve the Q-factor significantly, because of the highly unfavorable ratio of metal surface area to dielectric volume.

Somewhat higher Q-factors are achievable with microstrip resonant stub sections. These are more properly considered as distributed inductors, or more correctly, as distributed resonant elements. Three sources of loss are important here, skin losses, dielectric losses, and radiation losses. For microstrip stubs, the skin losses are those associated with microstrip, as are the dielectric losses. Skin losses vary inversely with the substrate thickness, and increase as the line impedance increases. Assuming negligible dielectric losses, one may show that the conductor Q-factor for a quarter-wave open circuit stub is given by

$$Q_c = \frac{27.3}{(\alpha \lambda g)} \quad (3)$$

where $(\alpha \lambda g)$ is the loss in the line section in decibels per wavelength. Since $(\alpha \lambda g)$ decreases as $f^{-1/2}$, Q_c increases as the square root of frequency, as for thin-film inductors. On the other hand, radiation losses from the open circuit end vary as [8]

$$Q_r = \frac{R}{(fh)^2} \quad (4)$$

where h is the substrate thickness and R is a function of w/h and the dielectric constant of the substrate. (The radiation factor R is considerably larger for a quarter-wave stub grounded at its far end.) Note that the radiation Q decreases as the square of the frequency and the substrate thickness h. Thus any attempt to increase the conductor Q-factor by increasing the frequency and substrate thickness is eventually overcompensated by the decrease in radiation Q. Fig. 19 illustrates this fact for practical substrate thicknesses. Thus, above X-band, open-circuit stub resonators are dominated by radiation losses, unless the substrate is less than 0.25 mm thick. This radiation also can cause coupling to adjacent circuits. A way to overcome both problems is to use a ring resonator.

The choice then as to whether reactive lumped elements or distributed elements should be used must be considered for each individual application. If high-Q narrow-band

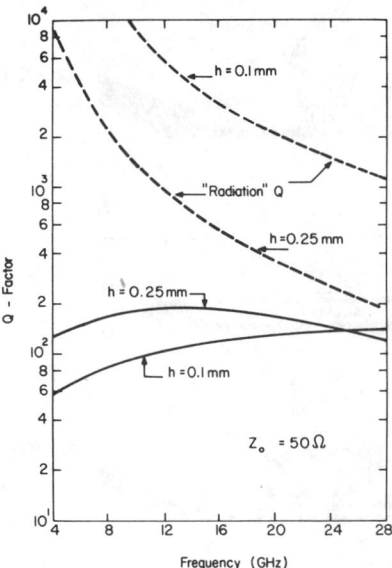

Fig. 19. Quality factor of a quarter-wave microstrip resonator on a GaAs substrate.

circuits are to be realized, distributed elements are recommended, provided space is available. On the other hand, broad-band circuits are probably easier to design with lumped elements, though even here synthesis techniques based on transmission line stubs are now available. Some circuits are more readily designed with distributed elements. Examples are four-port couplers and power combiners/dividers.

3) Planar Loads: Planar loads are essential for terminating such components as hybrid couplers, power combiners and splitters, and the like. Some factors to be considered in the design of such loads are: 1) the sheet resistivity available; 2) thermal stability or temperature coefficient of the resistive material; 3) the thermal resistance of the load; and 4) the frequency bandwidth. Other applications of planar resistors are bias voltage dividers and dropping resistors. However, such applications should be avoided in monolithic circuits, where power conservation is usually an objective.

Planar resistors can be realized in a variety of forms but fall into three categories: 1) semiconductor films; 2) deposited metal films; and 3) cermets. Resistors based on semiconductors can be fabricated by forming an isolated land of conducting epitaxial film on the substrate, for example, by mesa etching or by isolation implant of the surrounding conducting film. Another way is by implanting a high-resistivity region within the semi-insulating substrate. Metal film resistors are formed by evaporating a metal layer over the substrate and forming the desired pattern by photolithography. These techniques are illustrated in Fig. 20. Cermet resistors are formed from films consisting of a mixture of metal and a dielectric. However, because of the dielectric, they are expected to exhibit an RC frequency dependence similar to that of carbon resistors, which may be a problem in the microwave band.

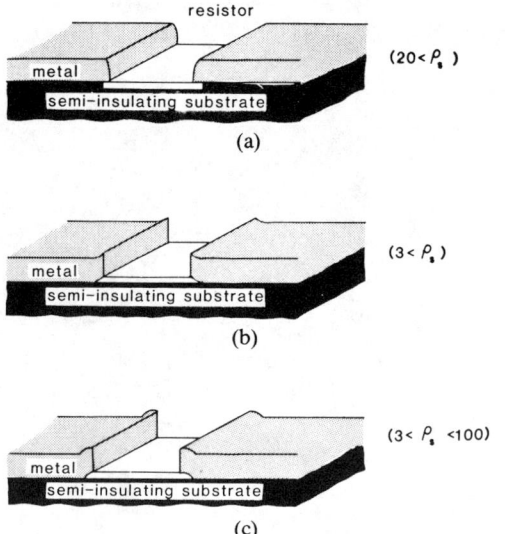

Fig. 20. Examples of planar resistor designs. (a) Implanted resistor. (b) Mesa resistor. (c) Deposited resistor.

TABLE IV
PROPERTIES OF SOME RESISTIVE FILMS

MATERIAL	RESISTIVITY ($\mu\Omega$-cm)	TCR (ppm/°C)	METHOD OF DEPOSITION	STABILITY	COMMENTS
Cr	13 (BULK)	+3000 (BULK)	EVAPORATED SPUTTERED	G-E	EXCELLENT ADHERENCE TO GaAs
Ti	55-135	+2500	EVAPORATED SPUTTERED	G-E	EXCELLENT ADHERENCE TO GaAs
Ta	180-220	-100 TO +500	SPUTTERED	E	CAN BE ANODIZED
Ni Cr	60-600	200	EVAP. (300°C) SPUTTERED	G-E	STABILIZED BY SLOW ANNEAL AT 300°C
TaN	280	-180 TO -300	REACTIVELY SPUTTERED	G	CANNOT BE ANODIZED
Ta$_2$N	300	-50 TO -110	REACTIVELY SPUTTERED	E	CAN BE ANODIZED
BULK GaAs	3-100 ohms/sq.	+3000	EPITAXY OR IMPLANTATION	E	NONLINEAR AT HIGH CURRENT DENSITIES

Metal films are preferred over semiconducting films because the latter exhibit a nonlinear behavior at high dc current densities and a rather strong temperature dependence—as some metal films do. Not all metal films are suitable for monolithic circuits, since their technology must be compatible with that of GaAs. Table IV lists some candidate metal films along with GaAs.

A problem common to all planar resistors used as microwave loads is the parasitic capacitance attributable to the underlying dielectric region and the distributed inductance of the film, which makes such resistors exhibit a frequency dependence at high frequencies. If the substrate bottomside is metallized, one may determine the frequency dependence by treating the load as a very lossy microstrip line.

For low thermal resistance, one should keep the area of the film as large as possible. To minimize discontinuity effects in width, the width of the resistive film load should not differ markedly from the width of the line feeding it. This means that the resistive element should be as long as possible to minimize thermal resistance. This length is specified by the sheet resistivity of the film and is given by

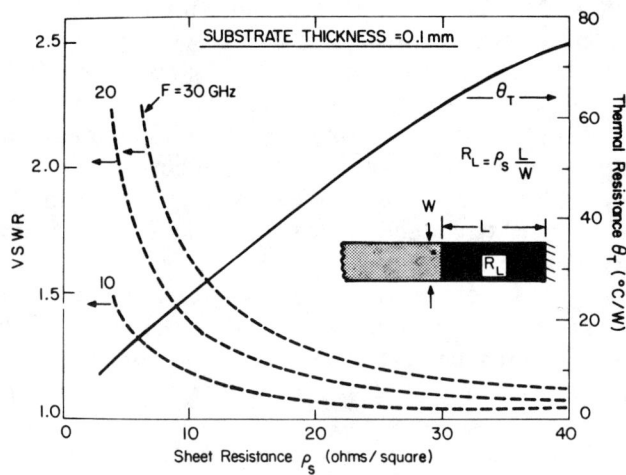

Fig. 21. Thermal resistance and VSWR of a planar resistor as a function of sheet resistance and frequency.

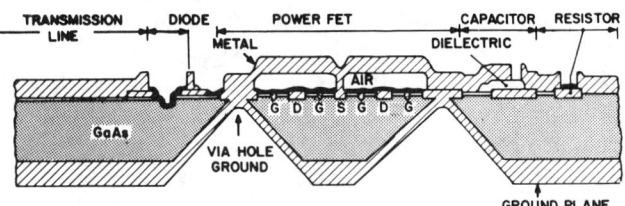

Fig. 22. Composite sketch illustrating technologies used in monolithic circuits.

the formula

$$l = \frac{wR}{\rho_s} \quad (5)$$

where w is the width of the film, R the desired load resistance, and ρ_s the sheet resistance of the film.

If one increases the length of the load (by decreasing the sheet resistivity) to achieve a low thermal resistance, one may get into trouble because the load may begin to exhibit the behavior of a transmission line (albeit a very lossy one) rather than a lumped resistor. Fig. 21 shows how the VSWR increases dramatically at low values of ρ_s because the length of the load becomes too large. Also shown is the thermal resistance. Clearly, a tradeoff is necessary between VSWR and thermal resistance.

All of the technologies we have discussed above are conveniently summarized in the cross-sectional view of a hypothetical monolithic circuit shown in Fig. 22.

4) Transmission Line Junction Effects: The many junctions and bends required of transmission lines in monolithic circuits to achieve close packing introduce unwanted parasitic inductances and capacitances. Fig. 23 illustrates some of the circuit representations of these junctions. Since such discontinuities cannot be avoided, but only minimized, the frequency dependencies must be taken into account, especially when the frequency is above *X*-band. It is particularly important to include junction effects in any broad-band design, that is, octave bandwidths. Unfortunately, though much work has been done on this topic, the results are not generally in a form useful for the circuit

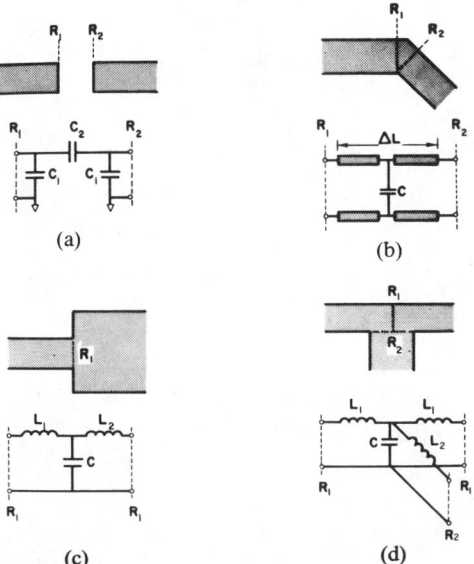

Fig. 23. Some microstrip discontinuities and their equivalent circuits. (a) Gap. (b) Bend. (c) Width discontinuity. (d) Tee junction.

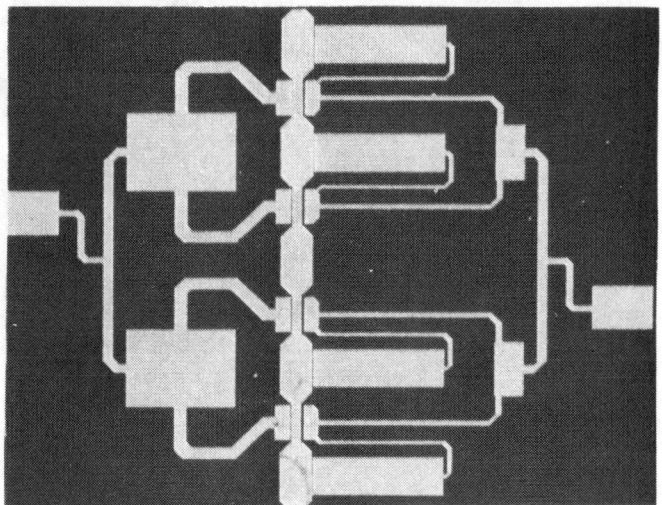

Fig. 24. Monolithic GaAs four-FET X-band power combiner. Chip size is 4.8×6.3×0.1 mm. (Raytheon Company.)

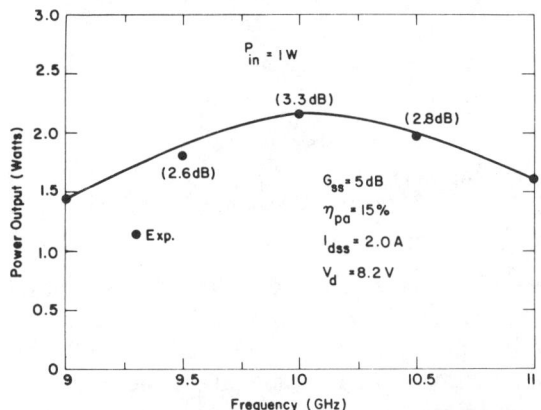

Fig. 25. Power output–frequency response for monolithic GaAs four-FET power combiner.

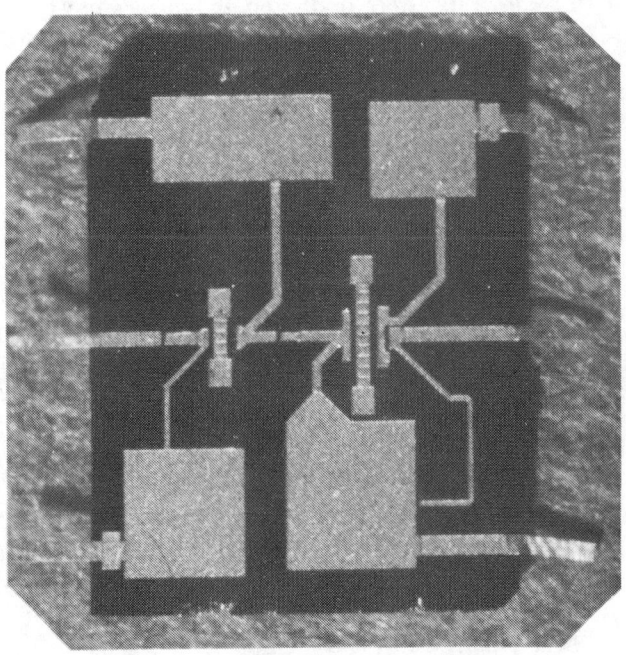

Fig. 26. Two-stage GaAs monolithic X-band amplifier. Chip size is 2.5×3.2×0.1 mm. (Raytheon Company.)

designer. As a consequence, computer-aided design programs do not incorporate corrections for junctions at present.

VI. Examples of Monolithic Circuits

We shall present examples of some practical monolithic circuits which demonstrate the design principles discussed above. These circuits are representative of the research being conducted at laboratories around the world.

Fig. 24 is a photograph of a GaAs chip containing a single-stage four-FET power combiner designed at Raytheon (Research Division) [17]. This amplifier, an X-band microwave circuit, was the first to dispense with wire bonds on the chip by use of "via" holes for grounding the source pads. Built on a chip 4.8×6.3×0.1 mm in size, and using a microstrip format with on-chip matching to a 50-Ω system, the circuit exhibited a 5-dB small-signal gain at 9.5 GHz and a saturated CW power output of 2.1 W at 3.3-dB gain (see Fig. 25). Bias was supplied through bias tees via the RF terminals. Although large by present standards, the chip area could be reduced by 30 percent if the capacitive stubs were replaced by thin-film capacitors, which were not available at the time.

An extension of this technology to a two-stage X-band power amplifier also designed at this laboratory [22] is shown in Fig. 26. In this circuit, thin-film capacitors, based on SiO or Si_3N_4 technology, were incorporated on the chip for RF blocking and bias applications. Another innovation, clearly evident in the future, is the use of extended integral (grown) beam leads, an offshoot of the airbridge technology. The beam leads allow off-chip bonding of the RF and dc supply connections to the chip, thus avoiding damage to the chip. The amplifier, built on a 2.5×3.2×0.1-mm chip, exhibited a saturated CW power output of 550 mW and 8.5-dB gain at 9.5 GHz and a small-signal gain of 10 dB.

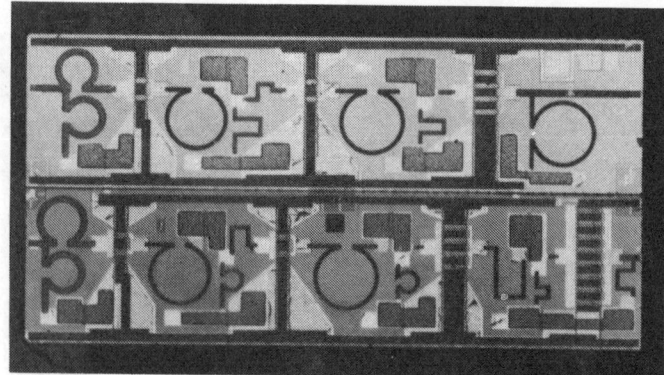

Fig. 27. Three- and four-stage GaAs monolithic X-band power amplifiers. Circuit sizes are 1.0×4.0×0.1 mm (Courtesy, W. Wisseman, Texas Instruments, Inc.)

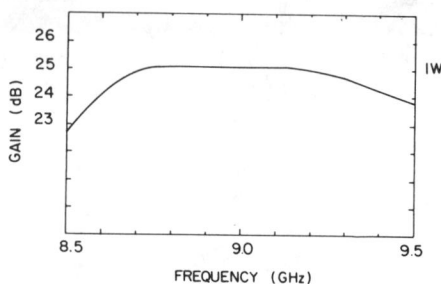

Fig. 28. Measured power gain–frequency response of four-stage amplifier of Fig. 27. (Courtesy, W. Wisseman, Texas Instruments, Inc.)

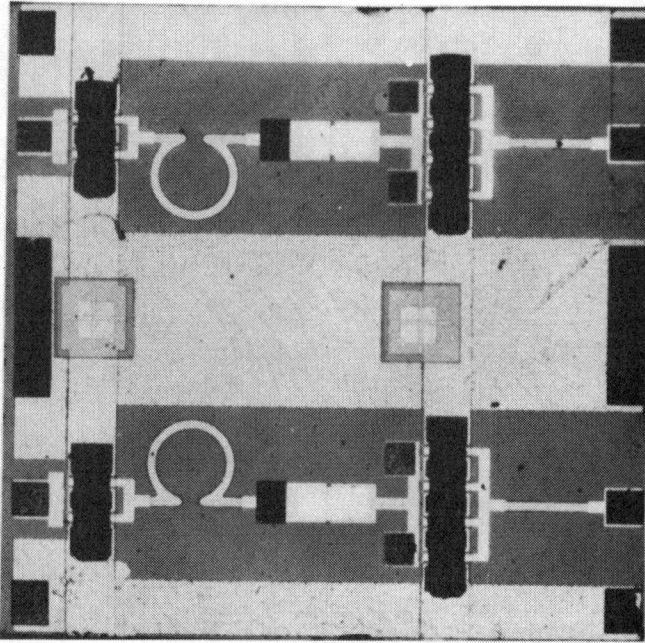

Fig. 29. Two-stage GaAs monolithic X-band push–pull amplifier. Chip size is 2.0×2.0×0.1 mm. (Courtesy, W. Wisseman, Texas Instruments, Inc.)

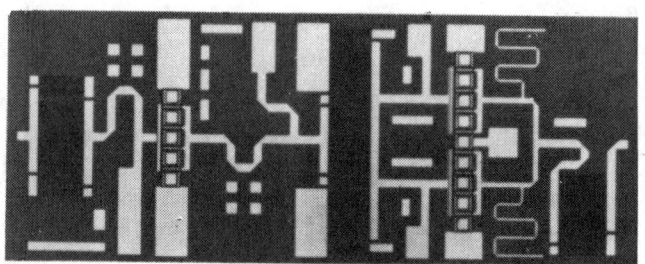

Fig. 30. Two-stage 5.7–11-GHz GaAs monolithic power amplifier. Chip size is 2.0×4.75×0.1 mm. (Courtesy, J. Oakes, Westinghouse.)

Turning to results obtained at other laboratories, Fig. 27 represents some of the research at Texas Instruments [18]. Shown is a chip containing side-by-side X-band amplifiers: the top, a three-stage FET amplifier; the bottom, a four-stage amplifier. Each chip is 1×4×0.1 mm in size. Both designs are based on a lumped-element approach which uses spiral inductors, clearly evident in the photographs, and thin-film capacitors of the end-coupled variety (Fig. 11(d)). Grounding is achieved by means of a metallized peripheral strip, and bias connections are made by wirebonds to pads on the chip. The three-stage amplifier delivers 400 mW at 23-dB gain and the four-stage delivers 1 W at 27-dB gain and 15–17-percent power-added efficiency in the 8.8 to 9.2-GHz range (see Fig. 28).

Another circuit reported by this laboratory [18] is the push–pull amplifier shown in Fig. 29. Each channel is a two-stage power amplifier, again based on the lumped-element approach, situated on a 2.0×2.0×0.1-mm chip. Although not monolithic in the strict sense of the word because inductive wire bonds interconnect the two channels, the design is unique in that a "virtual" ground is achieved by connection of the corresponding source pads of the adjacent channels; thus the need for a low inductance ground for the sources is avoided. Over 12-dB gain was obtained at 9.0 GHz with a combined CW power output of 1.4 W. All three amplifiers interface with a 50-Ω system.

An octave bandwidth GaAs amplifier designed at Westinghouse (R. and D. Center) is shown in Fig. 30. This circuit, similar to the one reported by Degenford et al. [7], consists of 1200-μm and 2400-μm periphery power FET's in cascade formed by selective ion implantation into a semi-insulating substrate. Built on a 2.0×4.75×0.1-mm chip, the circuit is based on a microstrip format with via holes, and makes liberal use of interdigitated capacitors. Source pads are grounded individually with vias. The amplifier produces a power output of 28±0.7 dBm at a gain of 6±0.7 dB across the 5.7 to 11-GHz band.

Another monolithic wideband amplifier is the 4–8-GHz eight-stage GaAs circuit reported by TRW [3] shown in Fig. 31. The design, based on the lumped-element approach (spiral inductors and SiO$_2$ thin-film capacitors) uses a coplanar feed at the input and output 50-Ω ports, with coplanar ground planes extending the full length of the 2.5×5.0-mm chip.

A departure from the GaAs approach is the SOS three-stage L-band amplifier built at Raytheon (Equipment Division) [13] (Fig. 32). This circuit, occupying a chip 7.5×7.5 ×0.46 mm in size, delivers 200-mW CW output at 20-dB gain at 1.3 GHz. The circuit uses spiral inductors. Dielec-

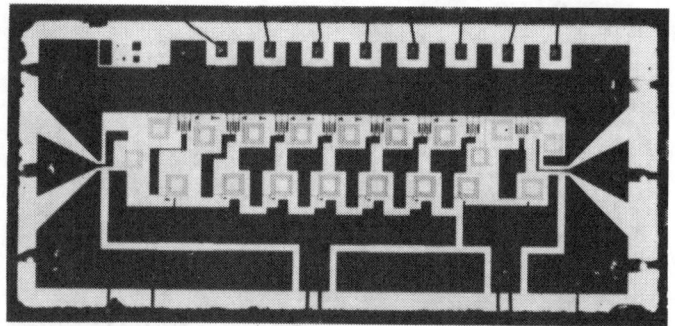

Fig. 31. Eight-stage 4–8-GHz GaAs monolithic amplifier. Chip size is 2.5×5.0 mm. (Courtesy, A. Benavides, T.R.W., Inc.)

Fig. 32. Three-stage L-band silicon-on-sapphire amplifier. Chip size is 7.5×7.5×0.46 mm. (Courtesy, D. Laighton, Raytheon Company.)

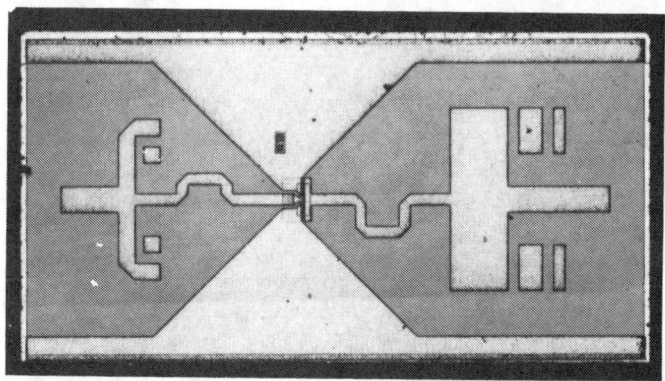

Fig. 33. Single-stage 20-GHz GaAs monolithic low-noise amplifier. Chip size is 2.75×1.95×0.15 mm. (Courtesy, A. Higashisaka, Nippon Electric Company.)

tric (SiO_2) films are used for capacitors and conductor crossovers.

So far we have described power amplifiers only. The first monolithic low-noise amplifier was reported by NEC (Central Research Laboratories) [10] (Fig. 33). This is a one-stage circuit on a 2.75×1.95×0.15-mm GaAs chip. The matching circuits use microstrip lines and stubs to interface with a 50-Ω system through bias tees. Large topside pads are used for the source RF grounds. The circuit, using a 0.5-μm gate, exhibited a noise figure of 6.2 dB and an associated gain of 7.5 dB in the 20.5–22.2-GHz band.

Most of the circuits we have described so far are based on the lumped-element or the microstrip approach or on a

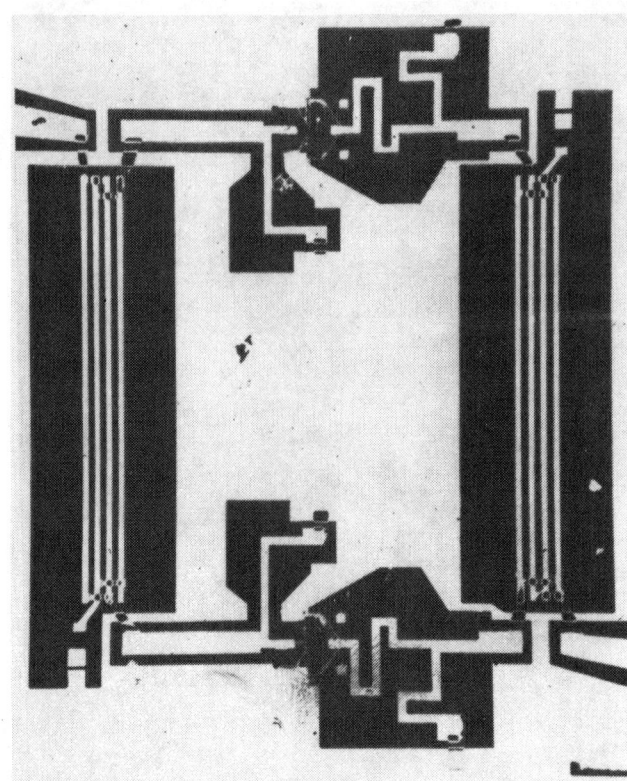

Fig. 34. X-band GaAs monolithic balanced amplifier using coplanar coupler. Chip size is 4.0×4.0 mm. (Courtesy, E. M. Bastida, CISE SpA.)

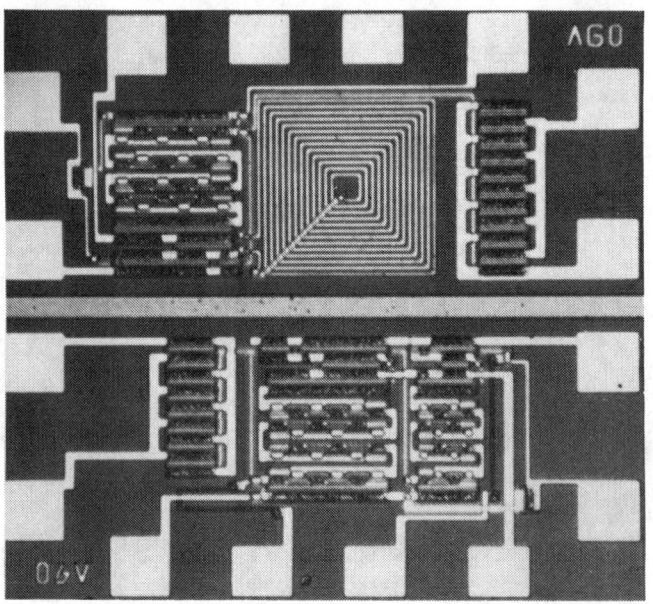

Fig. 35. Multistage direct-coupled GaAs monolithic amplifiers. Circuit sizes are 300×650 μm. (Courtesy, D. Hornbuckle, Hewlett Packard.)

combination of the two. Fig. 34 is a photograph of an X-band circuit using coplanar waveguides. This is a balanced amplifier reported by CISE SpA [2] built on a 4.0×4.0-mm GaAs chip, which uses two 90°, 3-dB broadband couplers. The couplers employ CPW rather than MS to obviate the need for micron-line spacings which are necessary with MS couplers. Lumped inductors and thin film (SiO_2) capacitors are used for RF matching and bypass. The circuit utilizes 0.8-μm gate MESFET's and has demonstrated a gain slightly below 10 dB between 8.5 and 11 GHz.

The next circuits, Fig. 35 represent a complete departure from the design philosophy considered so far. Shown is a photograph of two wide-band (0–4.5 GHz) amplifiers de-

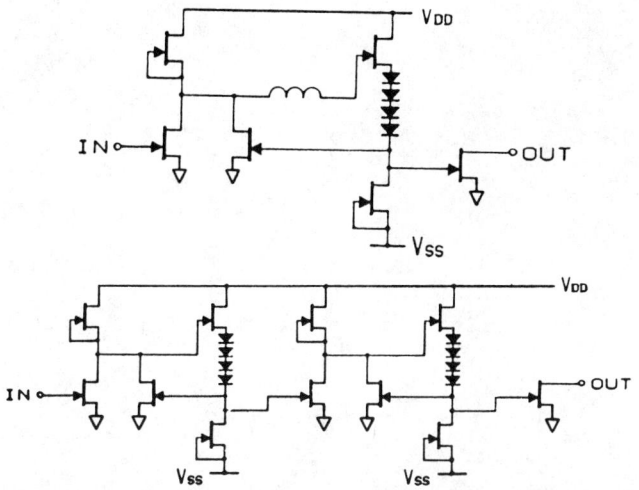

Fig. 36. Circuit schematics for direct-coupled amplifiers shown in Fig. 35. (Courtesy, D. Hornbuckle, Hewlett Packard.)

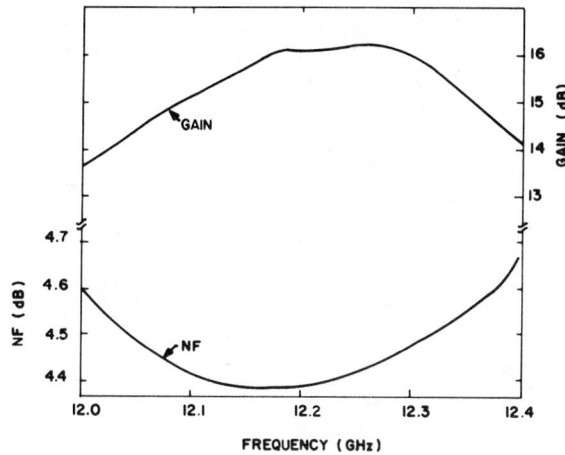

Fig. 38. Performance curves for receiver front end shown in Fig. 37. (Courtesy, P. Harrop, LEP.)

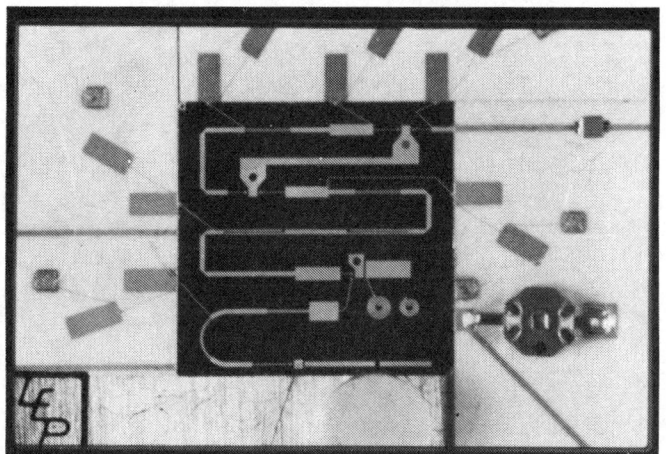

Fig. 37. 12-GHz GaAs monolithic receiver front end. Chip size is 1.0×1.0 cm. (Courtesy, P. Harrop, LEP.)

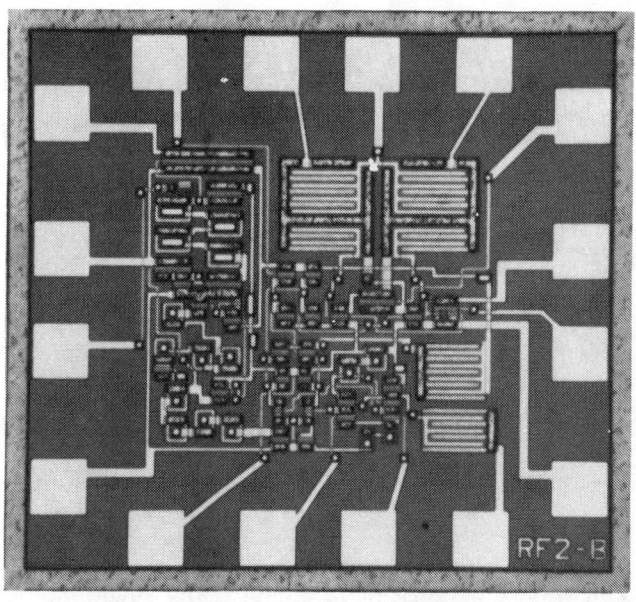

Fig. 39. Direct-coupled GaAs monolithic FET RF signal generation chip. Chip size is 600×650 μm. (Courtesy, R. Van Tuyl, Hewlett Packard.)

signed at Hewlett Packard [11]. What is unique about these circuits is the fact that, except for the spiral inductor, MESFET's are used throughout as active devices and as replacements for resistors and capacitors. The elimination of lumped elements, in conjunction with a direct-coupled circuit approach, allows a very high circuit packing density. Fig. 36 illustrates the circuit complexity achieved in each 0.3×0.65-mm area. Both amplifiers exhibited a gain in excess of 10 dB over the band.

Up until now we have described circuits which earlier we referred to as the lowest level of complexity. The next series of circuits represent integration on a functional block level. The first circuit (Fig. 37) is an integrated receiver front end on a GaAs chip intended for 12-GHz operation. This was reported by LEP [9]. The circuit, deposited on a large 1-cm square chip of GaAs, consists of a two-stage low-noise 12-GHz MESFET amplifier, an 11-GHz FET oscillator, and a dual-gate FET mixer. The matching circuits use microstrip lines and quarter-wave dc blocks. The oscillator is stabilized by an off-chip dielectric resonator. Bias circuits are included on the surrounding alumina substrate. Preliminary results are summarized in Fig. 38. The circuit is intended for a potential consumer market for domestic satellite-to-home TV reception planned for Europe.

Another example of the functional block approach is the monolithic GaAs FET RF signal generation chip (Fig. 39) designed at Hewlett Packard [21]. An extremely high degree of integration was achieved by use of the direct-coupled approach described earlier. Contained within the 0.65× 0.65-mm chip is the circuit shown in the schematic of Fig. 40. The local oscillator is resonated by an off-chip inductor which is tuned over the 2.1–2.5-GHz range by an on-chip Schottky barrier junction capacitor. The circuit is intended for an instrument application.

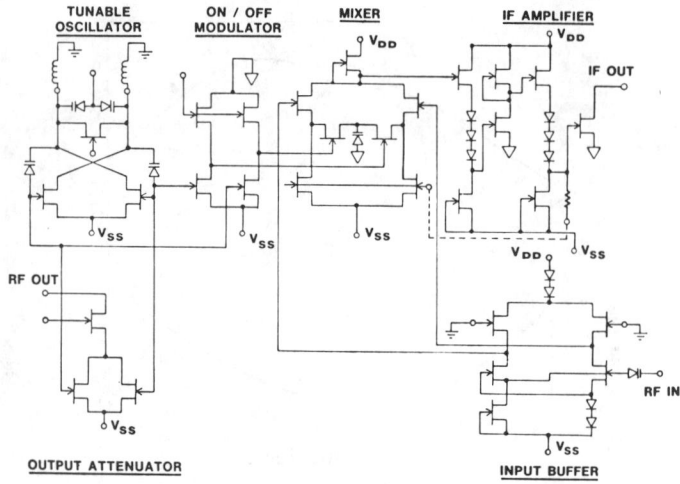

Fig. 40. Schematic for direct-coupled signal generation chip shown in Fig. 39. (Courtesy, R. Van Tuyl, Hewlett Packard.)

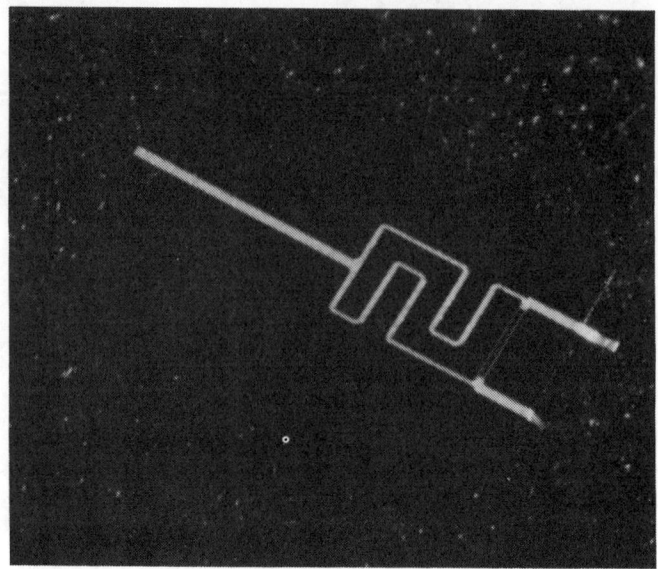

Fig. 42. GaAs monolithic X-band Wilkinson combiner/divider. Chip size is 1.5×2.5×0.1 mm. (Raytheon Company.)

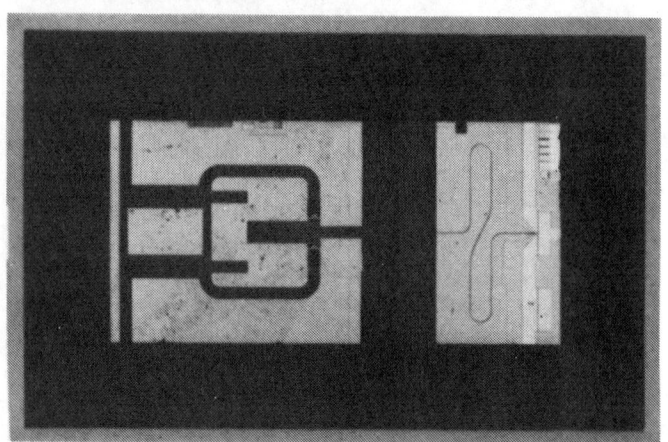

Fig. 41. GaAs monolithic mixer/IF circuit for millimeter-wave receiver applications. Chip size is 2.7×5.3×0.18 mm. (Courtesy, R. Sudbury, Lincoln Laboratories.)

Our final functional block circuit is the monolithic balanced Schottky-barrier diode mixer/IF FET preamplifier chip illustrated in Fig. 41. This MS circuit, reported by Lincoln Laboratories [6], is built on a 2.7×5.3×0.18-mm GaAs chip in MS format. The circuit operates between a 31-GHz signal source and a 2-GHz IF output. An external oscillator signal is injected through one of the coupler ports. The circuit exhibits an overall gain of 4 dB and a single-sideband noise figure of 11.5 dB.

We now turn to some special passive components fabricated in monolithic form. The first is a Wilkinson combiner/divider reported by Raytheon [23] shown in Fig. 42. Built on a 1.5×2.5×0.1-mm chip, the circuit uses a thin-film titanium balancing resistor and was designed to operate at a center frequency of 9.5 GHz. Note the extended beam leads. As an illustration of the extremely good electrical balance that one can achieve with the high-resolution photolithography intrinsic to the monolithic approach, we show in Fig. 43 a graph of the power division and phase balance measured for the two 3-dB ports.

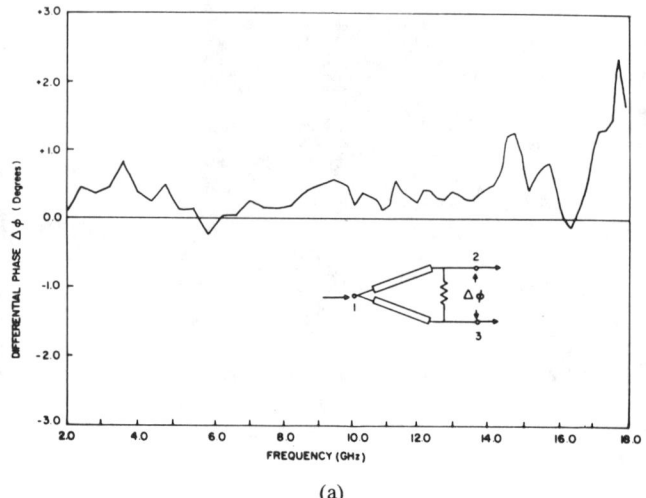

(a)

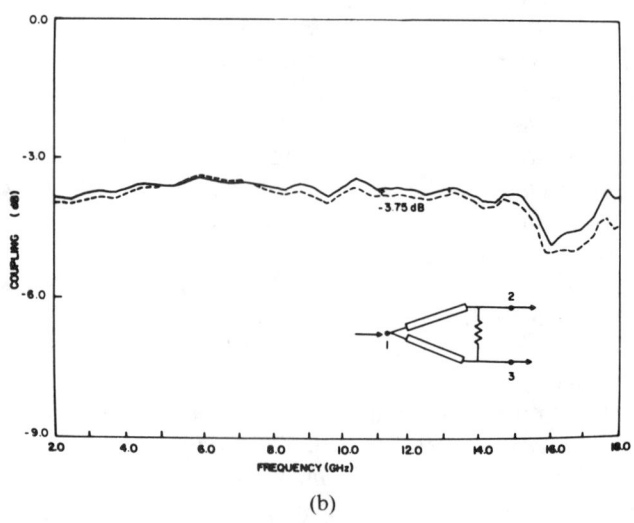

(b)

Fig. 43. Measured phase and power balance of Wilkinson divider shown in Fig. 42.

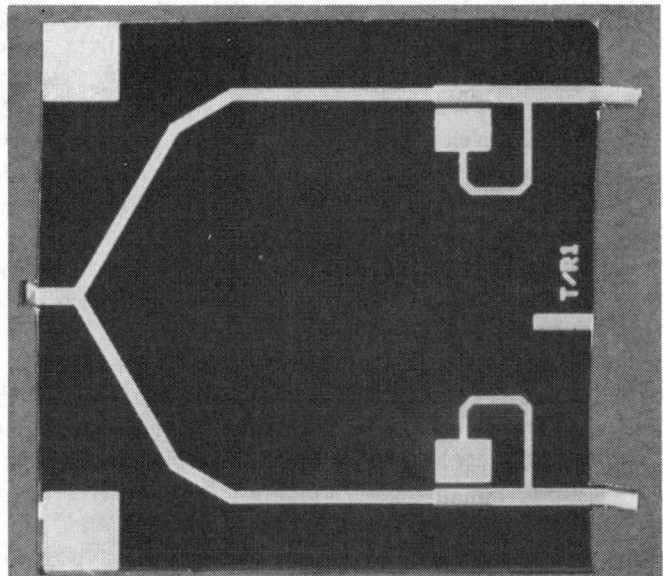

Fig. 44. GaAs monolithic X-band transmit/receive switch. Chip size is 3.0×3.0×0.1 mm. (Raytheon Company.)

Another component designed at this laboratory is the all-FET T/R switch shown in Fig. 44 [1]. This switch, intended for phased-array applications at X-band, requires no dc hold power in either state. Built on a 3.0×3.0×0.1-mm chip, the switch exhibits an isolation in excess of 33 dB between the transmitter and receiver ports in the 7–13-GHz range, and an insertion loss as low as 0.5 dB within this band. An alternative approach, also using FET's, was reported by McLevige et al. [14]. Both approaches utilize the change in source–drain resistance with gate bias.

The examples we have shown, though not exhaustive, are representative of the work reported so far (December 1980) and are intended to give the reader a good perspective of the advances made in the field during the last two years. Needless to say, the next several years will see the emergence of a still higher level of circuit integration in this rapidly developing field.

VII. Future Developments

We have so far concerned ourselves primarily with the technical aspects of monolithic circuits—their technology, design considerations, and microwave performance. Problems have been described and their solutions demonstrated. This is as it should be in the early stage of development of a new technical venture. No major unsolvable technical problems are evident; therefore, on the basis of technical considerations alone, there is no reason why the steady rate of progress already established cannot be maintained, indeed, accelerated.

What then will determine the future course of progress? The answer is simple—cost! Because the development of MMIC's requires a large capital investment and involves time-consuming and expensive processing technology plus a sophisticated testing procedure, the future development of this field will rest squarely on the as yet unproven expectations of reductions in cost and, to a lesser extent, improvements in reliability and reproducibility accruing from the monolithic approach.

The matter of cost reductions, in turn, rests on the answers to two questions.

1) Will the many complex technology steps required of MMIC's lend themselves to a high-yield production process?

2) Will a mass market develop in the microwave system area—a mass market necessary to capitalize on the high-volume low-cost attributes of batch processing?

Both of these requirements were eventually satisfied for silicon technology. Will this happen for gallium arsenide microwave technology? Time will tell. Since the silicon development was helped along by a vast domestic market (radios, TV's, etc., and more important, the commercial computer) and military markets, what are the expected large-volume markets for MMIC's?

Two potentially large markets appear to be developing, one military, the other consumer. In the military area, one such market includes electronically scanned radar systems, especially airborne and space-borne systems being planned for the future. For it is in the phased-array antenna, which may require modules as high as 10^5 in number, that we find a microwave system analog of the computer, which gave impetus to the growth of the silicon IC market. The antenna module requirements have already spurred developments of such module subsystems as transmitters, low-noise receivers, phase shifters, and transmit–receive (T/R) switches, some examples of which were described earlier. Here, along with cost, important design performance criteria will be reliability and small weight and size.

Another military application is in ECM systems, which require low-cost high-gain broad-band amplifiers. The difficult technical problems and projected high manufacturing cost associated with the hybrid integrated approach to this task have in essence mandated the use of monolithic circuits. Finally, the possibility of merging high-speed GaAs digital and microwave circuitry on the same chip may encourage use of such circuitry in signal processing at the RF level.

Turning to the nonmilitary markets, one potentially large outlet may be receiver front ends for the direct satellite-to-home-TV consumer market. Numerous such systems are being planned, for example, in Europe. We have described earlier one circuit intended for this market.

A third potential market, though much smaller in size, is instrumentation. Here cost and possibly circuit packing density are most important. Several examples of circuits earmarked for this application have been described.

We have not said much of the millimeter-wave spectrum. It is perhaps premature to do so, as this field itself is in its early stage of development. Here monolithic applications might develop, more for technical reasons than for economic reasons, because of the important role played by undesired packaging parasitics associated with discrete devices at these high frequencies. It is not unlikely that here too, as at lower frequencies, military applications may spur initial development. Now we turn to the question of costs.

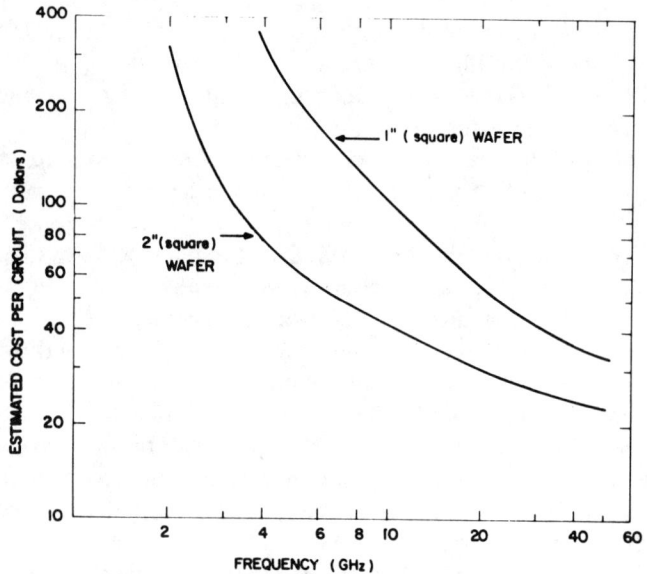

Fig. 45. Zero-order estimate of circuit cost as a function of frequency and wafer size.

Cost being the crucial item that it is, what are the factors contributing to it? They are the following:

1) materials;
2) materials processing;
3) circuit/device technology;
4) circuit assembly and packaging; and
5) testing (dc and RF).

These items, as is evident, do not include the important but nonrecurring costs such as capital investment, engineering, and mask design. In the materials category we include cost of substrate qualification, epitaxial growth and/or ion implantation, and profile evaluation, among others. Testing includes both dc and RF testing at the wafer probe level as well as circuit performance evaluation at the jig or package level.

Can a dollar figure be attached to these costs? At this stage, no! It is hazardous, at best, to attempt an accurate cost analysis based on laboratory experience, for large volume production, because ultimate module costs will be directly dependent on circuit yield in a manufacturing environment.

It is helpful, nevertheless, to attempt at least a "zeroth" order estimate of potential circuit costs, not so much to obtain an absolute level of cost, but to pinpoint the high cost items in the list above. To do this we have estimated the number of available circuits per wafer as a function of circuit operating frequency. This estimate was shown in Fig. 2. In the context of our present discussion, a circuit is equivalent to one submodule, for example, a transmitter stage or two, a phase shifter, etc. Using this estimate, we have determined the cost per circuit as a function of operating frequency. This data is shown in Fig. 45.

The cost estimates were derived by assumption of a 50-percent processing yield, independent of frequency. The base cost includes material cost, fully loaded labor cost, and circuit qualification at the dc and RF level. We feel that the cost estimates shown are useful guidelines but they should not be considered accurate in any absolute sense. For instance, depending on circuit complexity, current laboratory yields at X-band range from near zero to 20 percent. The development of a 50-percent yield fabrication process technology, deemed necessary, requires much additional experience and substantial simplification of monolithic circuit fabrication techniques.

Adjustments may be necessary at either end of the frequency scale. For example, in the range below 3–4 GHz, a drastic cost reduction may ensue, at least for some circuit applications, if the direct-coupled approach can be used. At the other end of the scale, above, say, 10–12 GHz, the cost figures should be elevated. The reason is that, because of the necessity of submicron gate technology, the lower throughput of the ultrahigh resolution electron beam (EB) lithography will increase costs substantially. Here what will be needed is optimization of the processing technology by appropriate merging of the EB lithography for the active devices and the higher throughput photolithography which is more than adequate for the circuit elements. This problem has not yet been addressed.

On the basis of our cost analysis, certain definite conclusions can be reached about the expected relative cost of the several items listed above. First, the two materials factors, under large production lots (>100 K parts) will contribute a negligible amount to the total cost—of the order of 5 percent or less. Second, next to wafer processing, the cost of packaging and microwave testing will be the largest cost factor. Indeed, because these latter costs will be fairly independent of the frequency band, and because of the decreasing processing cost per circuit with increasing frequency and wafer size, it is expected that packaging and testing will be the dominant cost factor at the higher frequencies (perhaps above 10 GHz).

It seems evident that, in light of this conclusion, the reduction of assembly and testing costs will be of paramount importance and must be addressed rather soon. Not only must as many functions as possible be integrated on one chip, consistent with high yield, but RF testing of chips and monolithic circuits and modules must be automated, just as dc tests have been. This will be very difficult because RF probes small enough for chip use are still in the laboratory stage, and their extension to performance tests on an entire circuit are nonexistent.

VIII. Conclusions and Summary

Monolithic microwave circuits based on gallium-arsenide technology have finally become a practical reality. Owing its origin to early experiments based on silicon bipolar technology, the gallium-arsenide approach, except for some scattered results in the sixties, emerged as a serious development only within the last three years.

The factors most responsible for this rapid growth can be traced to: 1) the development of the Schottky-barrier field-effect transistor; 2) the excellent microwave properties of semi-insulating GaAs as a low-loss substrate; 3) the perfection of GaAs epitaxy and ion implantation; 4) the establishment of GaAs crystal pulling facilities capable of large-diameter crystal growth; and 5) the emergence of potential systems applications for monolithic microwave circuits.

We have attempted to demonstrate in this paper some of the many design considerations and tradeoffs that must be made to optimize the performance of GaAs monolithic microwave circuits. Attention has been focussed, primarily, on the nondevice aspects of monolithic circuit design.

Despite the small physical size of the circuitry, interconnections between components often must be treated as wave-propagating structures because of the high dielectric constant of GaAs, which reduces the wavelength within the substrate. Both coplanar waveguide and microstrip lines, as well as combinations of both, are appropriate for monolithic circuits.

A typical circuit design may use both distributed and lumped-element components. Lumped elements, it was shown, are not truly lumped, because of built-in parasitics arising from the dielectric substrate. These must be taken into account at X-band and higher frequencies. A major drawback of thin-film inductors and capacitors is the limited Q-factor achieved to date. Much has yet to be learned about loss reduction in thin dielectric films.

We have shown that MMIC's are realized rather easily. Via hole grounding and source airbridge interconnections are eminently suited for them. Computer-aided design techniques are a "must" to reduce the number of iterations necessary.

Many examples of monolithic circuits have been shown which demonstrate the design principles described. These circuits, representing a world-wide cross section of the efforts in this field, have emerged within the last two to three years, and demonstrate the variety of circuit applications amenable to the monolithic approach. The promising attributes of the monolithic technology to cut fabrication costs, improve reliability and reproducibility, and reduce size and weight will overcome many of the shortcomings of the hybrid approach.

We have argued that, based on the cost considerations, the potential markets for MMIC's will be for the most part systems requiring large quantities of circuits of the same type. Because of this, and because of the large capital expenditures required of an organization to become a viable contender for these markets, it is most likely that the major efforts in MMIC's will eventually reside in the systems houses themselves.

ACKNOWLEDGMENT

The progress reported in this paper represents the cumulative effort of many people, too numerous to mention individually. However, the author wishes to express his deep appreciation to his associates at Raytheon, whose work is described here, and to his colleagues at many other laboratories who so graciously gave him permission to use their photographs and latest results to help describe their research.

REFERENCES

[1] Y. Ayasli, R. A. Pucel, J. L. Vorhaus, and W. Fabian, "A monolithic X-band single-pole, double-throw bidirectional GaAs FET Switch," in *GaAs IC Symp. Res. Abstracts*, 1980, paper no. 21.

[2] E. M. Bastida, G. P. Donzelli, and N. Fanelli, "An X-band monolithic GaAs balanced amplifier," in *GaAs IC Symp. Res. Abstracts*, 1980, paper no. 25.

[3] A. Benavides, D. E. Romeo, T. S. Lin, and K. P. Waller, "GaAs monolithic microwave multistage preamplifier," in *GaAs IC Symp. Res. Abstracts*, 1980, paper no. 26.

[4] L. Besser, "Synthesize amplifiers exactly," *Microwave Syst. News*, pp. 28–40, Oct. 1979.

[5] D. Ch'en and D. R. Decker, "MMIC's the next generation of microwave components," *Microwave J.*, pp. 67–78, May 1980.

[6] A. Chu, W. E. Courtney, L. J. Mahoney, G. A. Lincoln, W. Macropoulos, R. W. Sudbury, and W. T. Lindley, "GaAs monolithic circuit for millimeter-wave receiver application," in *ISSCC Dig. Tech. Pap.*, pp. 144–145, 1980.

[7] J. E. Degenford, R. G. Freitas, D. C. Boire, and M. Cohn, "Design considerations for wideband monolithic power amplifiers," in *GaAs IC Symp. Res. Abstracts*, 1980, paper no. 22.

[8] E. Denlinger, "Losses of microstrip lines," *IEEE Trans. Microwave Theory Tech.*, vol. MTT-28, pp. 513–522, June 1980.

[9] P. Harrop, P. Lesarte, and A. Collet, "GaAs integrated all-front-end at 12 GHz," in *GaAs IC Symp. Res. Abstracts*, 1980, paper no. 28.

[10] A. Higashisaka, in *1980 IEEE MTT-S Workshop Monolithic Microwave Analog IC's.*

[11] D. Hornbuckle, "GaAs IC direct-coupled amplifiers," in *1980 IEEE MTT-S Int. Microwave Symp. Dig.*, pp. 387–388.

[12] T. M. Hyltin, "Microstrip transmission on semiconductor substrates," *IEEE Trans. Microwave Theory Tech.*, vol. MTT-13, pp. 777–781, Nov. 1965.

[13] D. Laighton, J. Sasonoff, and J. Selin, "Silicon-on-sapphire (SOS) monolithic transceiver module components for L- and S-band," in *Government Microcircuit Applications Conf. Dig. Pap.*, vol. 8, 1980, pp. 299–302.

[14] W. V. McLevige and V. Sokolov, "A monolithic microwave switch using parallel-resonated GaAs FET's," in *GaAs IC Symp. Res. Abstracts*, 1980, paper no. 20.

[15] E. Mehal and R. W. Wacker, "GaAs integrated microwave circuits," *IEEE Trans. Microwave Theory Tech.* vol. MTT-16, pp. 451–454, July 1968.

[16] R. S. Pengelly and J. A. Turner, "Monolithic broadband GaAs FET amplifiers," *Electron. Lett.* vol. 12, pp. 251–252, May 13, 1976.

[17] R. A. Pucel, J. L. Vorhaus, P. Ng, and W. Fabian, "A monolithic GaAs X-band power amplifier," in *IEDM Tech. Dig.* 1979, pp. 266–268.

[18] V. Sokolov and R. E. Williams, "Development of GaAs monolithic power amplifiers," *IEEE Trans. Electron Devices*, vol. ED-27, pp. 1164–1171, June 1980.

[19] E. Strid and K. Reed, "A microstrip probe for microwave measurements on GaAs FET and IC wafers," in *GaAs IC Symp. Res. Abstracts*, 1980, paper no. 31.

[20] R. N. Thomas, "Advances in bulk silicon and gallium arsenide materials technology," in *IEDM Tech. Dig.*, 1980, pp. 13–17.

[21] R. Van Tuyl, "A monolithic GaAs FET RF signal generation chip," in *ISSCC Dig. Tech. Pap.* 1980, pp. 118–119.

[22] J. L. Vorhaus, R. A. Pucel, Y. Tajima, and W. Fabian, "A two-stage all-monolithic X-band power amplifier," in *ISSCC Dig. Tech. Pap.*, pp. 74–75, 1980.

[23] R. C. Waterman, W. Fabian, R. A. Pucel, Y. Tajima, and J. L. Vorhaus, "GaAs monolithic Lange and Wilkinson couplers," *GaAs IC Symp. Res. Abstracts*, 1980, paper no. 30.

Author Index

A
Abulela, M. F., 284
Akello, R. J., 244
Anonymous, 32

B
Bandler, J. W., 297
Bell, A. B., 377
Belohoubek, E. F., 426
Besser, L., 272

C
Cooke, H. F., 68
Cowan, D. A., 377
Cox, H. M., 190
Cruz-Emeric, J. A., 110

D
Dahlke, W., 10
DiLorenzo, J. V., 190

E
Easter, B., 244
Engelbrecht, R. S., 316

F
Friis, H. T., 6
Fujiki, Y., 194
Fukui, H., 18, 87, 100, 176, 184, 190, 230
Furutsuka, T., 211

G
Gledhill, C. S., 284
Gold, R. B., 389

H
Hasegawa, F., 194
Haus, H. A., 150
Hawkins, R. J., 113
Hewitt, B. S., 190
Higashisaka, A., 420
Hitchens, W. R., 389
Hornbuckle, D., 417
Hsu, T-H., 119

I
IRE Subcommittee 7.9 on Noise, 41, 252
Irvin, J. C., 218

Ishii, T., 200
Itoh, H., 194
Iversen, S., 24

K
Katoh, H., 54
Kawamura, N., 211
Kelly, A. J., 363
Kendall, B. R., 311
Knerr, R. H., 383
Krumm, C. F., 164
Ku, W. H., 300, 311
Kuhn, N., 58
Kurokawa, K., 275, 316

L
Lane, R. Q., 47, 49
Larrick, R. B., 205
Liechti, C. A., 128, 205, 346, 354
Livingstone, R. D., 110
Loya, A., 218
Luther, L. C., 190

M
Malaviya, S. D., 102
Malpass, J. C., 110
Massé, D. J., 164
McNamara, D. A., 110
Mercer, P., 377
Meys, R. P., 28
Mitama, M., 54
Mizuta, T., 420
Mokari-Bolhassan, M. E., 311

N
Nakatani, M., 200
Nara, A., 200
Niclas, K. B., 389

O
Ogawa, M., 211
Ohata, K., 194
Okean, H. C., 363

P
Penfield, P., Jr., 26
Petersen, W. C., 300, 311
Pierro, J., 399
Podell, A. F., 311
Pucel, R. A., 150, 164, 430

R

Rollett, J. M., 268
Rothe, H., 10

S

Sannino, M., 52
Saunders, T. E., 327
Seman, J. A., 190
Snapp, C. P., 119
Sobol, H., 334
Stark, P. D., 327
Statz, H., 150
Stephenson, I. M., 244
Stern, A. P., 259
Strid, E. W., 61

Suzuki, T., 200
Swan, C. B., 383

T

Tillman, R. L., 346
Tucker, R. S., 290

V

van der Ziel, A., 102, 110
Velebir, J. R., Jr., 190
Vendellin, G. D., 294

W

Weinreb, S., 403
Wilser, W. T., 389

Subject Index

A

Advanced microwave circuits, 426
Airborne radar, 430
Amplifiers
  balanced, 327, 383
  broadband, 176, 300, 311, 316, 389, 417
  cascaded, 164
  cooled, 205, 399, 403
  cryogenically cooled, 399
  direct-coupled, 417
  FET, 164, 300, 377, 383, 399, 403
  for microwave radio, 383
  gain-bandwidth limitations, 300
  high frequency, 300
  integrated circuits, 417, 420
  K-band, 420
  L-band, 316, 327, 334
  linear, 377
  low-noise, viii, 1, 176, 205, 272, 290, 294, 354, 377, 383, 403, 420
  lumped-circuit, 311, 334
  matched feedback, 389
  MESFET, 128, 389
  microwave, 1, 128, 290, 311, 316, 327, 346, 383, 389, 399
  multistage, 259
  narrow-band, 284
  noise of input-matching network, 61
  noise performance, 252
  octave-band, 311
  parametric, 1, 194, 363, 399
  power gain, 259
  reflection type, 1
  stability, 259, 272
  transistor, viii, 259, 272, 290, 300, 311, 316, 327
  tuned, 259
  wave representation of noise, 26
  wideband *see* broadband
  X-band, 346
Astronomy
  radio, 1, 399
Automatic design, 297

B

Balanced transistor amplifiers, 383
  integrated, 327
  L-band, 316
Bandpass filters, 426
Bibliography
  FET, 128
Biographies
  H. Fukui (author), 461
Bipolar transistors, 1, 184
  microwave, 113, 119
  replacement by GaAs FET's, 426
Broadband amplifiers
  high frequency, 300
  low-noise, 176, 316
  microwave, 311, 389, 417
Broadband receivers
  noise measurement, 58
Brownian motion, 1

C

Circuits *see* Networks
Circulators, 1
Communication satellites, 1, 430
  FET amplifiers, 377
Computer aided design, 297, 377
  GaAs FET amplifiers, 311
Computer programs
  noise measurement, 47
Cryogenically cooled GaAs FET amplifiers, 399

D

Detuning factor, 284
Diffusion noise, 1
Diodes
  Gunn, IMPATT and tunnel, 1
Direct-coupled amplifiers
  GaAs IC, 417
Directional couplers, 316, 327
Distributed systems
  measurement, 28
Dual-gate MESFET, 1
  GaAs, 211, 354

E

Electron tubes, 41
  grid controlled, 10

F

Failure mechanisms
  of low-noise GaAs FET, 218
Feedback
  effect on noise figure, 24, 294
  of GaAs MESFET, 294
Feedback amplifiers, 403
  matched, 389
FET amplifiers
  dual-gate, 426
  linear, 377

Field effect transistors, 1
  feedback effects, 294
  GaAs, 150, 164, 218
  microwave, 128
  package mounted in microstrip, 244
  Schottky-barrier, 164
Flicker noise, 1
Fourpoles
  noisy, 1, 10
  see also Twoports
Four-terminal networks
  noise figure, 6
Frequency-shift-keyed carrier modulation, 354
Fukui, H. (author)
  biography, 461

G

GaAs FET, 426
  dual-gate, 354
  failure mechanisms, 218
  low-noise, 218, 383

H

High-frequency transistor amplifiers
  broadband, 300
High-power amplification, 1
High-speed modulators, 354
Hybrid circuits
  MIC, 334

I

IC see Integrated circuits
IGFET, 128
IMPATT diodes, 1
Input-matching network
  to low-noise amplifier, 61
Integrated circuits
  application to microwave frequencies, 334, 417, 420
  direct-coupled amplifiers, 417
  microwave, 430
  monolithic, 1, 420, 426, 430
  4-GHz balanced transistor amplifiers, 327
Invariant power-gain functions, 268
IRE Subcommittee on Noise, 32, 41, 252

J

JFET, 128

K

K-band amplifiers, 420

L

L-band balanced transistor amplifiers, 327
  low-noise, 316
  microstrip, 334
Linear FET amplifiers
  for satellite communications, 377
Linear generator
  equivalent circuit, 275
Linear microwave devices
  noise, 28
Linear twoports, viii
  IRE standards, 32, 41
  noise parameter measurement, 54
  representation of noise, 26, 28
Long-distance communication, viii
Losses in noise-matching networks, 61
Low-noise amplification
  at microwave frequencies, 1
Low-noise amplifiers, viii
  broadband, 176
  feedback effects, 316
  for satellite communications, 377
  gain-controlled, 354
  microwave, 272, 290, 354, 403, 420
  stability, 272
Low-noise receivers, 363
"Low-noise window", viii
Low-pass ladder networks, 300

M

Matched feedback amplifiers, 389
Matrices see Scattering matrix properties
Measurements
  device noise parameters, 47, 54
  losses in noise-matching networks, 61
  noise figure, 58
  parasitic reactances of microwave transistors, 244
MESFET, 346, 430
  basic parameters, 230
  dual-gate, 1, 211, 354
  feedback effects, 294
  GaAs, 1, 54, 128, 176, 184, 190, 194, 211, 230, 420
  low-temperature performance, 205
  microwave amplification, 389
  reliability, 200, 218
  Si, 128
MIC (microwave integrated circuits), 430
  amplifiers, 383
  hybrid, 334
Microstrip lines, 334, 420
Microstrip transistor-package parasitic reactances
  measurement, 244
Microstrip tuners, 54
Microwave amplifiers, 1, 128, 311
  integrated, 327
  low-noise, 290, 346
  X-band, 346
Microwave circuits
  advanced, 426
  application of IC, 334
  monolithic, 430
  receivers, 363
Microwave communication, 363
Microwave devices
  linear, 28

noise, 28
Microwave FET
  1976 status, 128
Microwave radio
  use of GaAs FET amplifier, 383
Microwave transistors, viii
  amplifiers, 272, 290
  bipolar, 113, 119, 184, 426
  equivalent circuit in HF region, 100
  field effect, 128, 164, 176, 244, 311, 383, 403
  Ge, 87
  header parasitics, 102
  HF measurement, 68
  low noise, 383
  noise, 1, 87, 102
  noise measure, 100
  noise measurement, 47, 52, 102
  n-p-n, 18, 119
  parasitic reactances, 244
  planar, 18, 68,
  Si, 18, 68, 87
  theory and design, 68
Millimeter communications, 363
Millimeter receivers, 363
Minimum noise figure
  of transistors, 110
Mismatch relations
  for two networks, 6
Mixer circuits
  noise, 10
Modulation noise, 1
Modulators
  high-speed, 354
Monolithic GaAs FET amplifiers
  low-noise, 420
Monolithic microwave circuits
  design considerations, 430
Multipliers, 334

N

Narrow-band amplifiers
  design using scattering parameters, 284
Networks see Advanced microwave circuits, Amplifiers, Balanced transistor amplifiers, Bandpass filters, Broadband amplifiers, Cryogenically cooled GaAs FET amplifiers, Direct-coupled amplifiers, Feedback amplifiers, FET amplifiers, Fourpoles, Four-terminal networks, High-frequency transistor amplifiers, High-speed modulators, Hybrid circuits, Input-matching network, Integrated circuits, K-band amplifiers, L-band balanced transistor amplifiers, Linear FET amplifiers, Linear twoports, Low-noise amplifiers, Low-pass ladder networks, Matched feedback amplifiers, MIC, Microstrip tuners, Microwave amplifiers, Microwave circuits, Mixer circuits, Monolithic GaAs FET amplifiers, Monolithic microwave circuits, Narrow-band amplifiers, Noise-matching networks, Noncryogenic parametric amplifiers, Octave-band GaAs FET amplifiers, Parametric amplifiers, Printed circuits, Reflection-type amplifiers, Thin-film circuits, Three-pole networks, Tuned transistor amplifiers, Twoports, X-band amplifiers
Nielsen's and related noise equations
  applied to microwave bipolar transistors, 113
Noise
  amplifiers, 26, 252
  fourpoles, 10
  GaAs FET, 150, 164
  GaAs MESFET, 176
  IRE Subcommittees, 32, 41
  linear microwave devices, 28
  linear twoports, 26, 32, 41, 54
  measurement, 47, 54, 58
  microwave, 49, 87
  receiving systems, 252
  see also Diffusion noise, Flicker noise, Generation-recombination noise, Minimum noise figure, Modulation noise, Noise figure or Noise factor, Noisy fourpoles, Shot noise, Stationary noise sources, Thermal noise
Noise figure or Noise factor, 1, 10, 41
  effect of feedback, 24
  four-terminal networks, 6
  GaAs FET, 150
  GaAs MESFET, 184
  IRE definition, 252
  measurement, 6, 32, 47, 49, 52, 58, 102
  microwave devices, 28
  microwave transistors, 68, 87, 102, 119, 184, 290
  minimum, 47, 52, 110, 113, 184, 190, 200
  Neilsen's expression, 113
  of radio receivers, 6
  two-ports, 18, 24, 252
Noise-matching networks
  measurement of losses, 61
Noise measure
  microwave transistors, 100
  twoports, 18
Noise measurement, 32, 49
  computer programs, 47
  of UHF MOS structure, 47
Noise measurement systems
  use of standard equipment, 58
Noise parameter measurement
  improved computational method, 54
Noise power, 6
Noise sources, 1
Noise temperature, 252
  cryogenic cooled GaAs FET amplifiers, 399
  microwave devices, 28
Noise-wave source measurement, 28
Noisy fourpoles, 26
  linear, 1
  theory, 10
Noncryogenic parametric amplifiers
  low-noise, 363

O

Octave-band GaAs FET amplifiers, 311
Optimization
    broadband HF transistor amplifiers, 300
    using computers, 294

P

PAM *see* Pulse-amplitude modulation
Parametric amplifiers, 1, 194, 399
    noncryogenic, 363
Parasitic reactances
    of microwave transistors, 244
Phased-array systems
    airborne, 430
Phase-shift-keyed carrier modulation, 354
Planar transistors
    microwave, 18, 68
Power gain
    twoports, 18
Power transistors, 68
Power waves, 275
Printed circuits
    L-band amplifiers, 316
PSK *see* Phase-shift-keyed carrier modulation
Pulse-amplitude modulation, 354

R

Radar systems, viii, 1, 363
    airborne, 430
Radio astronomy, 1, 399
Radio astronomy receivers
    parametric amplifiers, 399
Radiometer receivers, 363
Radio microwave systems, 383
Radio receivers
    noise figures, 6
Radio relay systems, 327
Receivers
    low-noise, 363
    radio, 6
Receiving systems
    noise performance, 252
Reflection coefficients, 275
    optimization, 284
Reflection-type amplifiers, 1
Reliability
    of GaAs MESFET, 200
    of low-noise GaAs FET, 218
RF carrier
    high-speed AM, 354
Rothke-Dahlke noise model
    for noisy twoport, 26

S

Satellite communication, 1, 430
    low-noise and linear FET amplifiers, 377
Scattering matrix properties, 275
Scattering parameters
    narrow-band amplifiers, 284
Schottky-barrier GaAs FET, 290
    for microwave amplifiers, 346
    low-noise performance, 164
Schottky's theorem, 1
Semiconductor devices *see* Bipolar transistors, Diodes, Dual-gage MESFET, Field-effect transistors, GaAs FET, GaAs MESFET, Germanium transistors, Gunn diodes, IGFET, JFET, MESFET, Microwave FET, Microwave transistors, Planar transistors, Power transistors, Schottky-barrier GaAs FET, Silicon diodes, Silicon transistors, Transistors, Unencapsulated transistors, Unipolar transistors, Varactors
Shot noise, 1
Silicon diodes
    replacement by GaAs FET's, 426
Silicon-on-sapphire (SOS) technology, 430
Silicon transistors
    MESFET, 128
    microwave, 18, 68
Simulation using computer techniques, 297
Spot-noise parameters
    measurement, 32
Stability factor
    of linear twoports, 268
    of transistors, 259, 272
Standards of IRE
    on Methods of Measuring Noise, 32, 41
Stationary noise sources
    representations, 41

T

Terrestrial communication, 1
Thermal noise, 1
Thin-film circuits, 334
Three-pole networks
    with feedback, 24
Transducers, 32
    noise factor, 252
    power gain, 284
Transistors
    accurate expression for noise figure, 110
    noise parameters, 294
    stability and power gain, 259
    static induction, 1
    unencapsulated, 100
    unipolar, 1
Traveling-wave tubes
Tuned transistor amplifiers
    stability and power gain, 259
Tuners
    available gain, 61
    microstrip, 54
    varactor, 49
Tunnel diodes, 1
Twoports, 290
    active, viii, 300

available power gain, 18
linear, 1, 26, 41, 54, 268
noise figure, 6, 10, 24
noise representation, 41
noisy, 1
stability and power gain, 268
stabilization, 284
stationary sources, 41

U

Ultrawide-band microwave amplification
  with GaAs MESFET's, 389
Unencapsulated transistors
  equivalent circuit in HF region, 100
Unipolar transistors, 1

V

Varactors, 1
Varactor tuners
  microwave, 49

W

Wave representation
  of amplifier noise, 26
  of noise properties of linear microwave devices, 28
Wideband amplifiers *see* Broadband amplifiers

X

X-band amplifiers
  with GaAs Schottky-gate FET, 346

Editor's Biography

Hatsuaki Fukui was born in Yokohama, Japan. He was graduated summa cum laude in electrical engineering from Miyakojima Technical College, Osaka, Japan, and received the Doctor of Engineering degree in electrical engineering from Osaka University, Osaka.

From 1949 to 1954 he did research on microwave electron tubes at Osaka City University, Osaka, and went to Shimada Physical and Chemical Industrial Company, Ltd., Tokyo, as a design engineer of microwave test equipment for a year. In 1955 he joined Tokyo Tsushin Kogyo, Ltd. (former name of Sony Corporation), Tokyo, to work as a pioneer in the semiconductor industry. In the Semiconductor Division he headed a group developing new transistors for use in the world's first commercial solid-state radio and television receivers. In 1960 he was in charge of the entire Esaki tunnel-diode operation. A year later he was transferred to the Engineering Division as Manager of the Advanced Technology Department where future generations of consumer electronics products were developed. In 1962 he joined Bell Telephone Laboratories, Murray Hill, NJ. He first worked on microwave semiconductor devices, such as Ge and Si bipolar transistors, GaAs bulk-effect devices, and Si avalanche diodes, and their circuit applications. From 1966 to 1973 he was engaged in the research and development of electro-optical devices and subsystems for future PICTUREPHONE ® use, which included storage tubes, cathode-ray tubes, phosphors, plasma display devices, Si diode-array camera tubes, and charge-coupled imaging devices. He also supervised the development of new techniques for the vacuum deposition of III-V compounds. From 1973 to 1981 he was involved in the GaAs MESFET development project, working on device modeling, design, fabrication, characterization, and reliability study. Most recently he has been active in the field of long-wavelength lasers as Supervisor of the Lightwave Transmitter and Source Coordination Group. He is the author or co-author of three technical books and some 60 articles in Japanese, and of approximately 40 technical papers published in English.

Dr. Fukui is a Senior Member of the IEEE, a member of the Microwave Theory and Techniques Society serving on the Editorial Board, and a member of the Electron Devices Society serving for the IEEE Standards Committee (P642) on Microwave Transistors Characterization. He was a member of the Steering Committee of the Institute of Television Engineers of Japan from 1973 to 1974. He was the recipient of the 1980 Microwave Prize from the IEEE Microwave Theory and Techniques Society and a recipient of the Inada Award from the Institute of Electrical Communication Engineers of Japan in 1960. His name has been listed in Marquis' "Who's Who in the World" and other biographical references.